## Critical Values of t

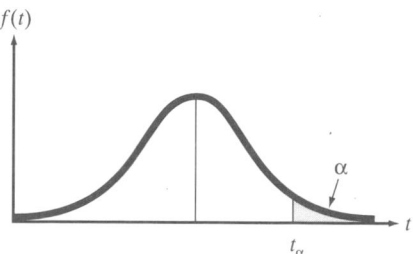

| $v$ | $t_{.100}$ | $t_{.050}$ | $t_{.025}$ | $t_{.010}$ | $t_{.005}$ | $t_{.001}$ | $t_{.0005}$ |
|---|---|---|---|---|---|---|---|
| 1 | 3.078 | 6.314 | 12.706 | 31.821 | 63.657 | 318.31 | 636.62 |
| 2 | 1.886 | 2.920 | 4.303 | 6.965 | 9.925 | 22.326 | 31.598 |
| 3 | 1.638 | 2.353 | 3.182 | 4.541 | 5.841 | 10.213 | 12.924 |
| 4 | 1.533 | 2.132 | 2.776 | 3.747 | 4.604 | 7.173 | 8.610 |
| 5 | 1.476 | 2.015 | 2.571 | 3.365 | 4.032 | 5.893 | 6.869 |
| 6 | 1.440 | 1.943 | 2.447 | 3.143 | 3.707 | 5.208 | 5.959 |
| 7 | 1.415 | 1.895 | 2.365 | 2.998 | 3.499 | 4.785 | 5.408 |
| 8 | 1.397 | 1.860 | 2.306 | 2.896 | 3.355 | 4.501 | 5.041 |
| 9 | 1.383 | 1.833 | 2.262 | 2.821 | 3.250 | 4.297 | 4.781 |
| 10 | 1.372 | 1.812 | 2.228 | 2.764 | 3.169 | 4.144 | 4.587 |
| 11 | 1.363 | 1.796 | 2.201 | 2.718 | 3.106 | 4.025 | 4.437 |
| 12 | 1.356 | 1.782 | 2.179 | 2.681 | 3.055 | 3.930 | 4.318 |
| 13 | 1.350 | 1.771 | 2.160 | 2.650 | 3.012 | 3.852 | 4.221 |
| 14 | 1.345 | 1.761 | 2.145 | 2.624 | 2.977 | 3.787 | 4.140 |
| 15 | 1.341 | 1.753 | 2.131 | 2.602 | 2.947 | 3.733 | 4.073 |
| 16 | 1.337 | 1.746 | 2.120 | 2.583 | 2.921 | 3.686 | 4.015 |
| 17 | 1.333 | 1.740 | 2.110 | 2.567 | 2.898 | 3.646 | 3.965 |
| 18 | 1.330 | 1.734 | 2.101 | 2.552 | 2.878 | 3.610 | 3.922 |
| 19 | 1.328 | 1.729 | 2.093 | 2.539 | 2.861 | 3.579 | 3.883 |
| 20 | 1.325 | 1.725 | 2.086 | 2.528 | 2.845 | 3.552 | 3.850 |
| 21 | 1.323 | 1.721 | 2.080 | 2.518 | 2.831 | 3.527 | 3.819 |
| 22 | 1.321 | 1.717 | 2.074 | 2.508 | 2.819 | 3.505 | 3.792 |
| 23 | 1.319 | 1.714 | 2.069 | 2.500 | 2.807 | 3.485 | 3.767 |
| 24 | 1.318 | 1.711 | 2.064 | 2.492 | 2.797 | 3.467 | 3.745 |
| 25 | 1.316 | 1.708 | 2.060 | 2.485 | 2.787 | 3.450 | 3.725 |
| 26 | 1.315 | 1.706 | 2.056 | 2.479 | 2.779 | 3.435 | 3.707 |
| 27 | 1.314 | 1.703 | 2.052 | 2.473 | 2.771 | 3.421 | 3.690 |
| 28 | 1.313 | 1.701 | 2.048 | 2.467 | 2.763 | 3.408 | 3.674 |
| 29 | 1.311 | 1.699 | 2.045 | 2.462 | 2.756 | 3.396 | 3.659 |
| 30 | 1.310 | 1.697 | 2.042 | 2.457 | 2.750 | 3.385 | 3.646 |
| 40 | 1.303 | 1.684 | 2.021 | 2.423 | 2.704 | 3.307 | 3.551 |
| 60 | 1.296 | 1.671 | 2.000 | 2.390 | 2.660 | 3.232 | 3.460 |
| 120 | 1.289 | 1.658 | 1.980 | 2.358 | 2.617 | 3.160 | 3.373 |
| $\infty$ | 1.282 | 1.645 | 1.960 | 2.326 | 2.576 | 3.090 | 3.291 |

*Source:* This table is reproduced with the kind permission of the Trustees of Biometrika from E. S. Pearson and H.O. Hartley (eds.), *The Biometrika Tables for Statisticians*, Vol. 1, 3d ed., Biometrika, 1966.

# STATISTICS

# STATISTICS

## Ninth Edition

**James T. McClave**
*Info Tech, Inc.*
*University of Florida*

**Terry Sincich**
*University of South Florida*

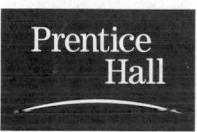

**Prentice Hall**   Upper Saddle River, New Jersey 07458

**Library of Congress Cataloging-in-Publication Data**
McClave, James T.
  Statistics/James T. McClave, Terry Sincich.—9th ed.
    p.   cm.
  Includes bibliographical references and index.
  ISBN 0-13-065598-8
  ISBN 0-13-066060-4 (Annotated Instructor's Edition)
  1. Statistics. I. Sincich, Terry. II. Title.

QA276.12.M4 2003
519.5—dc21                              2001055495


Acquisition Editor: Quincy McDonald
Editor in Chief: Sally Yagan
Vice President/Director of Production and Manufacturing: David W. Riccardi
Executive Managing Editor: Kathleen Schiaparelli
Senior Managing Editor: Linda Mihatov Behrens
Project Management: Elm Street Publishing Services, Inc.
Manufacturing Buyer: Alan Fischer
Manufacturing Manager: Trudy Pisciotti
Marketing Manager: Angela Battle
Marketing Assistant: Rachel Beckman
Assistant Editor of Media: Vince Jansen
Assistant Managing Editor, Math Media Production: John Matthews
Media Production Editor: Donna Crilly
Editorial Assistant/Supplements Editor: Joanne Wendelken
Art Director: John Christiana
Interior and Cover Designer: Jon Boylan
Art Editor: Thomas Benfatti
Managing Editor, Audio/Video Assets: Grace Hazeldine
Creative Director: Carole Anson
Director of Creative Services: Paul Belfanti
Photo Editor: Beth Boyd
Interior Photos: Photodisc
    Page 1: Photographer: Karl Weatherly; page 19: Scenics of America/Photolink; page 107: Photographer: Karl Weatherly;
    page 173: F. Schussler/Photolink; page 217: J. Luke/Photolink; page 261: Photographer: Karl Weatherly;
    page 287: Scenics of America/Photolink; page 327: S. Alden/Photolink; page 379: Photographer: Karl Weatherly;
    page 441: D. Falconer/Photolink; page 509: Photographer: Karl Weatherly; page 575: Photographer: Gary Irving;
    page 705: Photographer: Gary Irving; page 739: Photographer: Gary Cole
Cover Photo: © Lawrence Manning/Corbis
Art Studio: Elm Street Publishing Services, Inc.

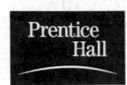

© 2003 by Prentice-Hall, Inc.
Upper Saddle River, NJ 07458

Printed in the United States of America
10  9  8  7  6  5  4  3  2  1

ISBN 0-13-065598-8

Pearson Education LTD.
Pearson Education Australia PTY, Limited
Pearson Education Singapore, Pte. Ltd
Pearson Education North Asia Ltd
Pearson Education Canada, Ltd.
Pearson Educación de Mexico, S.A. de C.V.
Pearson Education—Japan
Pearson Education Malaysia, Pte. Ltd

# CONTENTS

## 8  Inferences Based on a Single Sample: Tests of Hypothesis    327

## 9  Inferences Based on Two Samples: Confidence Intervals and Tests of Hypotheses    379

## 10  Analysis of Variance: Comparing More Than Two Means    441

## 11  Simple Linear Regression    509

## 12  Multiple Regression and Model Building    575

## 13  Categorical Data Analysis    705

## 14    Nonparametric Statistics    739

### Appendix A:  Tables    793

### Appendix B:  Data Sets on CD-ROM That Accompanies Text    823

### Appendix C:  Calculation Formulas for Analysis of Variance    829

### Answers to Selected Exercises    835

### Index    847

# PREFACE

## STATISTICS IN THE NEW MILLENIUM

This ninth edition of *Statistics* is an introductory text emphasizing inference, with extensive coverage of data collection and analysis as needed to evaluate the reported results of statistical studies and make good decisions. As in earlier editions, the text stresses the development of statistical thinking and the assessment of credibility and value of the inferences made from data, by both those who consume and those who produce them. It assumes a mathematical background of basic algebra.

A briefer version of the book, *A First Course in Statistics*, is available for single semester courses that include minimal coverage of regression analysis, analysis of variance, and categorical data analysis.

## NEW IN THE NINTH EDITION

- Exercise sets have been revised to provide a greater variety in level of difficulty. In addition to "Learning the Mechanics" exercises, "Applied Exercises" are categorized into "Basic," "Intermediate," and "Advanced" at the end of each section.
- All printouts from statistical software (SAS, SPSS, MINITAB, and STATISTIX) have been revised to reflect the latest versions of the software.
- Chapter 12 (Multiple Regression and Model Building) now includes a more detailed discussion of multicollinearity, with a real-data example and guidelines on detecting and dealing with the problem (Section 12.14).
- The sections in Chapter 13 (Categorical Data Analysis) have been reorganized, with emphasis on whether one or two qualitative variables are analyzed.
- An introductory section on distribution-free tests (Section 14.1) is now included in Chapter 14 (Nonparametric Statistics).

## TRADITIONAL STRENGTHS

We have maintained the pedagogical features of *Statistics* that we believe make it unique among introductory statistics texts. These features, which assist the student in achieving an overview of statistics and an understanding of its relevance in the social and life sciences, business, and everyday life, are as follows:

- **Use of Examples as a Teaching Device**—Almost all new ideas are introduced and illustrated by data-based applications and examples. We believe that students better understand definitions, generalizations, and abstractions *after* seeing an application.
- **Exploring Data with Statistical Computer Software**—Each statistical analysis method presented is demonstrated using output from four leading statistical software packages: SAS, SPSS, MINITAB, or STATISTIX. In addition, output and key stroke instructions for the TI-83 Graphing Calculator are covered in optional boxes.

- **Statistics in Action**—Each chapter concludes with a case study on a contemporary, controversial, or high-profile issue. Data from the study are presented for analysis and questions prompt the students to evaluate the findings and to think through the statistical issues involved. Additional cases appear on the Web site.

- **Real Data Exercises**—The text includes more than 1,200 exercises illustrated by applications in almost all areas of research. All the applied exercises employ the use of current real data extracted from a wide variety of publications (e.g., newspapers, magazines, journals, and the Internet). Many students have trouble learning the mechanics of statistical techniques when all problems are couched in terms of realistic applications. For this reason, all exercise sections are divided into four parts:

  *Learning the Mechanics* Designed as straightforward applications of new concepts, these exercises allow students to test their ability to comprehend a mathematical concept or a definition.

  *Applying the Concepts—Basic* Based on applications taken from a wide variety of journals, newspapers, and other sources, these short exercises help students begin to develop the skills necessary to diagnose and analyze real-world problems.

  *Applying the Concepts—Intermediate* Based on more detailed real-world applications, these exercises require students to apply their knowledge of the technique presented in the section.

  *Applying the Concepts—Advanced* These more challenging real-data exercises require students to utilize their critical thinking skills.

- **Quick Review**—Each chapter ends with a list of key terms and formulas, with reference to the page number where they first appear.

- **Language Lab**—Following the Quick Review is a pronunciation guide to Greek letters and other special terms. Usage notes are also provided.

- **Student Projects**—Presented at the end of each chapter, these projects emphasize gathering data, analyzing data, and/or report writing.

## FLEXIBILITY IN COVERAGE

The text is written to allow the instructor flexibility in the coverage of topics through "optional" sections in relevant chapters and occasional footnotes. Suggestions for covering two topics, probability and regression, follow.

- **Probability and Counting Rules**—One of the most troublesome aspects of an introductory statistics course is the study of probability. Probability poses a challenge for instructors because they must decide on the level of presentation, and students find it a difficult subject to comprehend. We believe that one cause for these problems is the mixture of probability and counting rules that occurs in most introductory texts. We have included the counting rules in a separate optional section at the end of Chapter 3 (Probability). In addition, all exercises that require the use of counting rules are marked with an asterisk (*). Thus, the instructor can control the level of coverage of probability.

- **Multiple Regression and Model Building**—This topic represents one of the most useful statistical tools for the solution of applied problems. Although an entire text could be devoted to regression modeling, we feel that we have presented coverage that is understandable, usable, and much more comprehensive than the presentations in other introductory statistics texts. We devote two full chapters to discussing the major types of inferences that can be derived from a regression analysis, showing how these results appear in the output from statistical software and, most important, how to select multiple regression models to be used in an analysis. Thus, the instructor has the choice of a one-chapter coverage of simple linear regression (Chapter 11), a two-chapter treatment of simple and multiple regression (excluding the optional sections in Chapter 12 on model building), or complete coverage of regression analysis, including model building and regression diagnostics. This extensive coverage of such useful statistical tools will provide added evidence to the student of the relevance of statistics to real-world problems.

- **Footnotes**—Although the text is designed for students with a non-calculus background, footnotes explain the role of calculus in various derivations. Footnotes are also used to inform the student about some of the theory underlying certain methods of analysis. These footnotes allow additional flexibility in the mathematical and theoretical level at which the material is presented.

## ACKNOWLEDGMENTS

This book reflects the efforts of a great many people over a number of years. First, we would like to thank the following professors, whose reviews and comments on this and prior editions have contributed to the ninth edition:

### Reviewers of Previous Editions of *Statistics*

Bill Adamson
*South Dakota State University*

Ibrahim Ahmad
*Northern Illinois University*

David Atkinson
*Olivet Nazarene University*

Mary Sue Beersman
*Northeast Missouri State University*

William H. Beyer
*University of Akron*

Patricia M. Buchanan
*Pennsylvania State University*

Dean S. Burbank
*Gulf Coast Community College*

Kathryn Chaloner
*University of Minnesota*

Hanfeng Chen
*Bowling Green State University*

Gerardo Chin-Leo
*The Evergreen State College*

Linda Brant Collins
*Iowa State University*

Brant Deppa
*Winona State University*

John Dirkse
*California State University—Bakersfield*

N. B. Ebrahimi
*Northern Illinois University*

John Egenolf
*University of Alaska—Anchorage*

Dale Everson
*University of Idaho*

Christine Franklin
*University of Georgia*

Rudy Gideon
*University of Montana*

Victoria Marie Gribshaw
*Seton Hill College*

Larry Griffey
*Florida Community College*

David Groggel
*Miami University at Oxford*

John E. Groves
*California Polytechnic State University—San Luis Obispo*

Dale K. Hathaway
*Olivet Nazarene University*

Shu-ping Hodgson
*Central Michigan University*

Jean L. Holton
*Virginia Commonwealth University*

Soon Hong
*Grand Valley State University*

Ina Parks S. Howell
*Florida International University*

Gary Itzkowitz
*Rowan College of New Jersey*

John H. Kellermeier
*State University College at Plattsburgh*

Timothy J. Killeen
*University of Connecticut*

William G. Koellner
*Montclair State University*

James R. Lackritz
*San Diego State University*

Diane Lambert
*AT&T/Bell Laboratories*

Edwin G. Landauer
*Clackamas Community College*

James Lang
*Valencia Junior College*

Glenn Larson
*University of Regina*

John J. Lefante, Jr.
*University of South Alabama*

Pi-Erh Lin
*Florida State University*

R. Bruce Lind
*University of Puget Sound*

Rhonda Magel
*North Dakota State University*

Linda C. Malone
*University of Central Florida*

Allen E. Martin
*California State University—Los Angeles*

Rick Martinez
*Foothill College*

Brenda Masters
*Oklahoma State University*

Leslie Matekaitis
*Cal Genetics*

E. Donice McCune
*Stephen F. Austin State University*

Mark M. Meerschaert
*University of Nevada—Reno*

Greg Miller
*Steven F. Austin State University*

Satya Narayan Mishra
*University of South Alabama*

Christopher Morrell
*Loyola College in Maryland*

A. Mukherjea
*University of South Florida*

Thomas O'Gorman
*Northern Illinois University*

Bernard Ostle
*University of Central Florida*

William B. Owen
*Central Washington University*

Won J. Park
*Wright State University*

John J. Peterson
*Smith Kline & French Laboratories*

Ronald Pierce
*Eastern Kentucky University*

Betty Rehfuss
*North Dakota State University—Bottineau*

Andrew Rosalsky
*University of Florida*

C. Bradley Russell
*Clemson University*

Rita Schillaber
*University of Alberta*

James R. Schott
*University of Central Florida*

Susan C. Schott
*University of Central Florida*

George Schultz
*St. Petersburg Junior College*

Carl James Schwarz
*University of Manitoba*

Mike Seyfried
*Shippensburg University*

Arvind K. Shah
*University of South Alabama*

Lewis Shoemaker
*Millersville University*

Charles W. Sinclair
*Portland State University*

Robert K. Smidt
*California Polytechnic State University—San Luis Obispo*

Vasanth B. Solomon
*Drake University*

W. Robert Stephenson
*Iowa State University*

Dan Voss
*Wright State University*

Thaddeus Tarpey
*Wright State University*

Augustin Vukov
*University of Toronto*

Kathy Taylor
*Clackamas Community College*

Dennis D. Wackerly
*University of Florida*

Barbara Treadwell
*Western Michigan University*

Matthew Wood
*University of Missouri—Columbia*

### Reviewers Involved with the Ninth Edition of *Statistics*

Ann Cascarelle
*St. Petersburg College*

Mir Mortazavi
*Eastern New Mexico University*

Dale K. Hathaway
*Olivet Nazarene University*

Steve Nimmo
*Morningside College (Iowa)*

Gary S. Itzkowitz
*Rowan College of New Jersey*

James R. Schott
*University of Central Florida*

Special thanks are due to our ancillary authors, Nancy Shafer Boudreau, Mark Dummeldinger, Engin Sungur, Anne Drougas, Augustin Vukov, Singh Kelly, and to typist Kelly Barber, several of whom have worked with us for many years. Sudhir Goel has done an excellent job of accuracy checking the ninth edition and has helped us to ensure a highly accurate, clean text. Finally, the Prentice Hall staff of Quincy McDonald, Joanne Wendelken, Angela Battle, Amy Lysik, Vince Jansen, Linda Behrens, and Alan Fischer and Elm Street Publishing Services' Martha Beyerlein, Melissa Morgan, and Irene Sotiroff helped greatly with all phases of the text development, production, and marketing effort.

## SUPPLEMENTS FOR THE INSTRUCTOR

### Annotated Instructor's Edition (AIE) (ISBN 0-13-066060-4)

Marginal notes placed next to discussions of essential teaching concepts include:

- Teaching Tips—suggest alternative presentations or point out common student errors
- Exercises—reference specific section and chapter exercises that reinforce the concept
- Short Answers—section and chapter exercise answers are provided next to the selected exercises

### Instructor's Solutions Manual by Nancy S. Boudreau (ISBN 0-13-066071-X)

Solutions to all of the even-numbered exercises are given in this manual. Careful attention has been paid to ensure that all methods of solution and notation are consistent with those used in the core text. Solutions to the odd-numbered exercises are found in the *Student's Solutions Manual*.

### *Test Bank* by Mark Dummeldinger (ISBN 0-13-066072-8)

The *Test Bank* includes more than 1,000 problems that correlate to problems presented in the text.

### *Instructor's Notes* by Mark Dummeldinger (ISBN 0-13-066074-4)

This printed resource contains suggestions for using the questions at the end of the *Statistics in Action* cases as the basis for class discussion and a complete short answer book with letter of permission to duplicate for student use.

### NEW *TestGen-EQ* (ISBN 0-13-066078-7)

*TestGen-EQ* is a computerized test bank that allows you to view and edit test bank questions, transfer them to exams, and print in a variety of formats. The program offers many options for organizing and displaying test banks and tests, and a built-in random number and test generator allows instructors to create multiple versions of exams.

### NEW *McClaveTutor 5.0* (ISBN 0-13-066075-2)

Based on the highly successful *MathPro 5.0* software, *McClaveTutor* is a Web-based statistics tutorial and testing package. Matched specifically to the text, students are invited to practice an unlimited amount of problems correlated to examples in the text. Students having difficulty are given step-by-step guidance, along with video of a professor explaining the relevant topic.

Instructors may also create their own tests and quizzes for students, or randomly generate tests and quizzes using *McClaveTutor*'s algorithmically based question generator. After the students take free-response quizzes and tests assigned by the instructor, the results are graded by the program and can be reported to *McClaveTutor*'s course management system for easy reference by the instructor.

### NEW *WebCT/BlackBoard* by Engin Sungur

The *BlackBoard* and *WebCT* sites offer course-compatible content including the *Student's Solutions Manual*, technology help, quizzes, and lecture content. These sites also include Web-based statistical applets (STATLETS™) designed specifically for the text by StatPoint, LLC, which encourage students to develop their conceptual understanding of statistics by providing accessible computational and graphing capabilities matched with specific questions to reinforce understanding of key concepts from the text.

### NEW *CourseCompass* by Engin Sungur

*CourseCompass* is the perfect course management solution that combines quality Prentice Hall content with state-of-the-art *BlackBoard* technology! It is a dynamic, interactive online course management tool powered by *BlackBoard*. This exciting product allows you to teach with market-leading Prentice Hall content in an easy-to-use customizable format. This site offers course-compatible content including the *Student's Solutions Manual*, technology help, quizzes, and lecture content. This site also includes Web-based statistical applets (STATLETS™) designed specifically for the text by StatPoint, LLC, which encourage students to develop their conceptual understanding of statistics by providing accessible computational and graphing capabilities matched with specific questions to reinforce understanding of key concepts from the text.

### Companion Website (0-13-066073-6)

Located at *http://www.prenhall.com/mcclave*, the *Statistics* (ninth edition) *Companion Website* is an online, interactive study guide, matched to each chapter of the text, with an online syllabus builder, technology help, quizzes, objectives, Internet destinations, PowerPoint downloads, and the text's data files available for download.

## SUPPLEMENTS FOR THE STUDENT

### Data CD (ISBN 0-13-066076-0)

The text is accompanied by a CD that contains files for all of the data sets marked with a CD icon in the text. These include data sets for text examples, exercises, and *Statistics in Action* cases. All data files are saved in ASCII format for easy importing into statistical software (e.g., SAS, SPSS, MINITAB, or STATISTIX). A list of the data sets, with file names and variables, is provided in Appendix B.

### Student's Solutions Manual by Nancy S. Boudreau (ISBN 0-13-066068-X)

Fully worked-out solutions to all of the odd-numbered exercises are provided in this manual. Careful attention has been paid to ensure that all methods of solution and notation are consistent with those used in the core text.

### NEW *TI-83 Manual* by Singh Kelly (ISBN 0-13-066919-9)

### NEW *MINITAB Manual* by Augustin Vukov (ISBN 0-13-066922-9)

### NEW *Microsoft Excel Manual* by Anne Drougas (ISBN 0-13-066921-0)

Each spiral-bound companion manual works hand-in-glove with the text. Step-by-step keystroke-level instructions, with screen captures, provide detailed help for using the technology to work pertinent exercises. A cross-reference chart indicates which text examples are included and the exact page reference in both the text and technology manual. Output with brief instructions is provided for selected odd-numbered exercises to reinforce the examples.

### NEW *McClaveTutor 5.0* (0-13-066075-2)

Based on the highly successful *MathPro 5.0* software, *McClaveTutor* is a Web-based statistics tutorial and testing package. Matched specifically to the text, students are invited to practice an unlimited amount of problems correlated to examples in the text. Students having difficulty are given step-by-step guidance, along with video of a professor explaining the relevant topic.

### NEW *WebCT/BlackBoard* by Engin Sungur

The *BlackBoard* and *WebCT* sites offer course-compatible content including the *Student's Solutions Manual*, technology help, quizzes, and lecture content. These sites also include Web-based statistical applets (STATLETS™) designed specifically for the text by StatPoint, LLC, which encourage students to develop their conceptual understanding of statistics by providing accessible computational and graphing capabilities matched with specific questions to reinforce understanding of key concepts from the text.

### NEW *CourseCompass* by Engin Sungur

*CourseCompass* is the perfect course management solution that combines quality Prentice Hall content with state-of-the-art *BlackBoard* technology! It is a dynamic, interactive online course management tool powered by *BlackBoard*. This exciting product allows you to learn with market-leading Prentice Hall content in an easy-to-use customizable format. This site offers course-compatible content, including the *Student's Solutions Manual*, technology help, quizzes, and lecture content. This site also includes Web-based statistical applets (STATLETS™) designed specifically for the text by StatPoint, LLC, which encourage students to develop their conceptual understanding of statistics by providing accessible computational and graphing capabilities matched with specific questions to reinforce understanding of key concepts from the text.

### Companion Website

Located at *http://www.prenhall.com/mcclave*, the *Statistics* (ninth edition) *Companion Website* is an online, interactive study guide, matched to each chapter of the text, with technology help, quizzes, objectives, Internet destinations, PowerPoint downloads, and the text's data files available for download.

# LIST OF DATA SOURCES

*ABC News*
*Academy of Management Journal*
*Adapted Physical Activity Quarterly*
Alachua County (Florida) Property
  Appraiser's Office
American Association for Marriage and
  Family Therapy
*American Association of Nurse
  Anesthetists Journal*
American Automobile Association
*American Cancer Society*
*American Education Research Journal*
*American Journal of Audiology*
*American Journal of Dance Therapy*
*American Journal of Orthopsychiatry*
*American Journal of Political Science*
*American Journal of Psychiatry*
*American Journal of Psychology*
*American Journal of Public Health*
*American Journal of Speech-Language
  Pathology*
*American Journal on Mental Retardation*
*American Psychologist*
American Psychosomatic Society
*American Rifleman*
*American Sociological Review*
American Scientist
American Statistical Association
*American Statistician*
*AmStat News*
*Animal Behavior*
*Annals of the Association of American
  Geographers*
*Annals of Internal Medicine*
*Annals of Tourism Research*
*Applied Animal Behaviour Science*
*Applied Psycholinguistics*
*Applied Spectroscopy*
*Aquatic Botany*
*Arctic and Alpine Research*
*Archives of Disease in Childhood*
*Archives of Internal Medicine*
Association for the Advancement of
  Applied Sport Psychology
*Astronomical Journal*
*Astronomy*
*Australian Journal of Zoology*
Bayonet Point Hospital (St. Petersburg,
  Florida)
*Beanie World Magazine*
*Benefits Quarterly*
Biogen, Inc.
*Brain and Behaviour Evolution*
*Brain and Language*
*British Journal of Sports Medicine*
*Business Week*
CACI Marketing Systems
*Card Fax*

*Car and Driver*
*Cell Biology*
Centers for Disease Control and
  Prevention
*Chance*
*Chemosphere*
*Chest*
*Children and Youth Services Review*
*Christianity Today*
*Chromatographia*
*Clinical Kinesiology*
*Clinical Psychology Review*
*Cognitive Aging Conference*
College Entrance Examination Board
*College Student Journal*
*Collegium Antropologicum*
*Communication Research*
*Community Mental Health Journal*
*Consumer Reports*
*Consumer's Research*
*Dermatology*
Dini Miller, Virginia Tech University
*Discover*
*Ear & Hearing*
*Ecological Applications*
*Economic Botany*
*Educational Technology*
Elizabeth Schreiber, University of Kansas
*Entrepreneur*
*Environmental Entomology*
Environmental Protection Agency
*Environmental Science & Engineering*
*Environmental Science & Technology*
*Ethology*
*European Journal of Operational
  Research*
*Exceptional Children*
*Explore* (University of Florida)
Federal Trade Commission
*Fisheries Science*
Florida Attoney General's Office
Florida Department of Health and
  Administrative Services
Florida Employer Opinion Survey
*Florida Scientist*
Florida Shredder's Association
*Florida Trend*
Food and Drug Administration
*Forbes*
*Fortune*
Frances VanVoorhis
*Genetical Research*
*Geoderma*
*Geography*
*Geological Magazine*
*Geriatric Psychiatry*
*Health Education Journal*
*Health Psychology*

Herman Kelting
Hillsborough County (Florida) Property
  Appraiser's Office
Hillsborough County (Florida) Water
  Department
*Howard Journal of Criminal Justice*
*Human Factors*
*IBM Journal of Research and
  Development*
*IEEE Engineering in Medicine and
  Biology Magazine*
*IEEE Transactions*
*IEEE Transactions on Semiconductor
  Manufacturing*
*IEEE Transactions on Speech and Audio
  Processing*
*Industrial Engineering*
Info Tech, Inc.
*Interfaces*
*International Journal of Circuit Theory
  and Applications*
*International Journal of Radiation
  Oncology and Biological Physics*
International Rhino Federation
Johnson & Johnson, Inc.
*Journal of the Academy of Rehabilitative
  Audiology*
*Journal of Adlerian Theory, Research and
  Practice*
*Journal of Advanced Nursing*
*Journal of Agricultural, Biological, and
  Environmental Statistics*
*Journal of Allergy and Clinical
  Immunology*
*Journal of the American Medical
  Association*
*Journal of the American Mosquito
  Control Association*
*Journal of the American Statistical
  Association*
*Journal of Applied Biology*
*Journal of Applied Biomechanics*
*Journal of Applied Business Research*
*Journal of Applied Ecology*
*Journal of Applied Phycology*
*Journal of Applied Physics*
*Journal of Applied Psychology*
*Journal of Behavioral Medicine*
*Journal of Business Communications*
*Journal of Business & Economic Statistics*
*Journal of Business Ethics*
*Journal of Business Finance and
  Accounting*
*Journal of Child Psychology and
  Psychiatry*
*Journal of Colloid and Interface Science*
*Journal of Communication Disorders*
*Journal of Computer Information Systems*

*Journal of Consulting and Clinical Psychology*
*Journal of Drug Issues*
*Journal of Economics*
*Journal of Educational Statistics*
*Journal of Ethnopharmacology*
*Journal of Experimental Psychology*
*Journal of Experimental Zoology*
*Journal of Family Violence*
*Journal of Fish Biology*
*Journal of Gang Research*
*Journal of Genetic Psychology*
*Journal of Geography*
*Journal of Hazardous Materials*
*Journal of Head Trauma Rehabilitation*
*Journal of Human Stress*
*Journal of Information Science*
*Journal of Infrastructure Systems*
*Journal of Intellectual Disability Research*
*Journal of Leisure Research*
*Journal of Marketing Research*
*Journal of Marriage and the Family*
*Journal of the Mosquito Control Association*
*Journal of the National Cancer Institute*
*Journal of Negro Education*
*Journal of Nonverbal Behavior*
*Journal of Nursing Education*
*Journal of Nutrition*
*Journal of Occupational Psychology*
*Journal of Organic Chemistry*
*Journal of Orthopsychiatry*
*Journal of Parapsychology*
*Journal of Peace Research*
*Journal of Political Economy*
*Journal of Portfolio Management*
*Journal of Psychology*
*Journal of Psychology and Aging*
*Journal of Quality Technology*
*Journal of Quantitative Criminology*
*Journal of Real Estate Research*
*Journal of Rehabilitation*
*Journal of Research in Music Education*
*Journal of Retailing*
*Journal of Shellfish Research*
*Journal of Socio-Economics*
*Journal of Soil and Water Conservation*
*Journal of Speech, Language, and Hearing Research*
*Journal of Sport and Social Issues*
*Journal of Sport Behavior*
*Journal of Sports Medicine and Physical Fitness*
*Journal of STAR Research*
*Journal of Statistics Education*
*Journal of Studies in Technical Careers*
*Journal of Testing and Evaluation*
*Journal of Visual Impairment & Blindness*
*Journal of Wildlife Management*
*Journal of Zoology*

Kelly Uscategui, University of Connecticut
*Ladies Home Journal*
Lei Lei, Rutgers University
*Library Acquisitions: Practice & Theory*
*Liquid Assets: The International Guide to Fine Wines*
Major League Baseball
*Management Science*
*Marine Technology*
*Mathematical Geology*
*Memory & Cognition*
Merck Research Labs
*Meteoritics*
*Minneapolis Star and Tribune*
Minnesota Department of Transportation
Moffitt Cancer Research Center (University of South Florida)
*Motivation and Emotion*
National Aeronautics and Space Administration
National Center for Education Statistics
National Center for Environmental Health
National Direct Student Loan Program
National Earthquake Information Center
National Highway Traffic Safety Administration
National Institute of Environmental Health Services
National Oceanic and Atmospheric Administration
*National Wildlife*
*Nature*
*Naval Research Logistics*
*Networks*
*Newark Star-Ledger*
*New England Journal of Medicine*
*New Jersey Business*
*New Scientist*
*New York Times*
*Newsweek*
*Nilson Report*
*Optometry and Vision Science*
*Organizational Behaviour and Human Decision Processes*
*Organizational Science*
*Parade Magazine*
*Pediatric Exercise Science*
*Perception & Psychophysics*
*Physical Therapy*
*Phytopathology*
*Planta Medica*
*Political Science & Politics*
*Principles of Operation Management*
*Proceedings of the National Academy of Sciences*
*Production and Inventory Management Journal*
*Professional Geographer*
*Progress in Natural Science*

*Psychological Assessment*
*Psychological Reports*
*Psychology and Aging*
*Psychosomatic Medicine*
*Quality Congress Transactions*
*Quality Engineering*
*Quality Management Journal*
*Quality Progress*
*Queuing Models*
*Rehabilitative Psychology*
*Research Review*
Robinson Appraisal Co., Inc.
*Science*
*Science Education*
*Science News*
*Scientific American*
*Sleep*
*Social Psychology Quarterly*
*Sociological Methods & Research*
*Sociology of Sport Journal*
*Soil Science*
*Sporting News*
*Sports Illustrated*
*Statistical Abstract of the United States*
*Statistical Bulletin*
*Statistics in Sport*
*St. Petersburg Times*
Susan Farber, *Identical Twins Reared Apart*
*Tampa Tribune*
*Teaching Psychology*
*Teaching Sociology*
*Technometrics*
*The Bell Curve: Intelligence and Class Structure in American Life*
*The Condor*
*The Movie Times*
*Time*
*UCLA Journal of Environmental Law & Policy*
University of Florida
University of South Florida Office of Research
U.S. Army Corps of Engineers
U.S. Bureau of Labor Statistics
U.S. Chess Federation
U.S. Department of Energy
U.S. Golf Association
*U.S. News & World Report*
*USA Today*
*USF Magazine*
*Wall Street Journal*
*Washington Post*
*Wetlands*
*Wine Spectator*
*Work and Occupations*
*World Archaeology*
*World Development*
*World Economy*
*Zoological Science*

# HOW TO USE THIS BOOK

## To the Student

The following pages will demonstrate how to use this text in the most effective way to make studying easier and to understand the connection between statistics and your world.

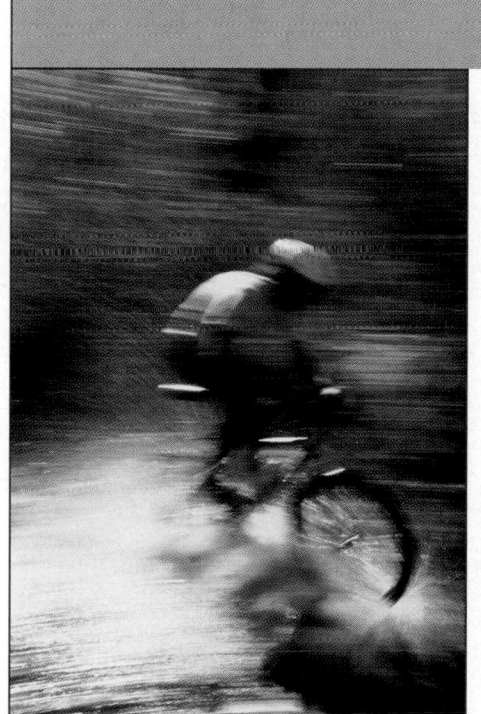

### CHAPTER 3

## Probability

### Contents

### Statistics in Action

Game Show Strategy: to Switch or not to Switch?

### Chapter Openers Show Links Between Old and New Topics

- **Where We've Been** quickly reviews how information learned previously applies to the chapter at hand.

- **Where We're Going** highlights how the chapter topics fit into your growing understanding of statistics.

### Where We've Been

We've identified inference, from a sample to a population, as the goal of statistics. And we've seen that to reach this goal, we must be able to describe a set of measurements. Thus, we explored the use of graphical and numerical methods for describing both quantitative and qualitative data sets and for phrasing inferences.

### Where We're Going

Now that we know how to phrase an inference about a population, we turn to the problem of making the inference. What is it that permits us to make the inferential jump from sample to population and then to give a measure of reliability for the inference? As you'll see, the answer is *probability*. This chapter is devoted to a study of probability—what it is and some of the basic concepts of the theory behind it.

Page 107

## Easy to Locate Boxes Highlight Important Information

- Definitions, Strategies, Key Formulas, and other important information is highlighted for quick reference.

- Helps students to prepare for quizzes and tests by reviewing the highlighted information.

---

### 4.4  THE BINOMIAL RANDOM VARIABLE

Many experiments result in *dichotomous* responses—i.e., responses for which there exist two possible alternatives, such as Yes–No, Pass–Fail, Defective–Nondefective, or Male–Female. A simple example of such an experiment is the coin-toss experiment. A coin is tossed a number of times, say 10. Each toss results in one of two outcomes, Head or Tail, and the probability of observing each of these two outcomes remains the same for each of the 10 tosses. Ultimately, we are interested in the probability distribution of $x$, the number of heads observed. Many other experiments are equivalent to tossing a coin (either balanced or unbalanced) a fixed number $n$ of times and observing the number $x$ of times that one of the two possible outcomes occurs. Random variables that possess these characteristics are called **binomial random variables**.

Public opinion and consumer preference polls (e.g., the CNN, Gallup, and Harris polls) frequently yield observations on binomial random variables. For example, suppose a sample of 100 students is selected from a large student body and each person is asked whether he or she favors (a Head) or opposes (a Tail) a certain campus issue. Suppose we are interested in $x$, the number of students in the sample who favor the issue. Sampling 100 students is analogous to tossing the coin 100 times. Thus, you can see that opinion polls that record the number of people who favor a certain issue are real-life equivalents of coin-toss experiments. We have been describing a **binomial experiment**; it is identified by the following characteristics.

---

#### Characteristics of a Binomial Random Variable

1. The experiment consists of $n$ identical trials.
2. There are only two possible outcomes on each trial. We will denote one outcome by $S$ (for Success) and the other by $F$ (for Failure).
3. The probability of $S$ remains the same from trial to trial. This probability is denoted by $p$, and the probability of $F$ is denoted by $q$. Note that $q = 1 - p$.
4. The trials are independent.
5. The binomial random variable $x$ is the number of $S$'s in $n$ trials.

---

**Page 189**

---

**EXAMPLE 5.8**  Find the value of $z$, call it $z_0$, in the standard normal distribution that will be exceeded only 10% of the time. That is, find $z_0$ such that $P(z \geq z_0) = .10$.

**Solution**

In this case we are given a probability, or an area, and asked to find the value of the standard normal random variable that corresponds to the area. Specifically, we want to find the value $z_0$ such that only 10% of the standard normal distribution exceeds $z_0$ (see Figure 5.14).

**FIGURE 5.14**
*Areas Under the Standard Normal Curve for Example 5.8*

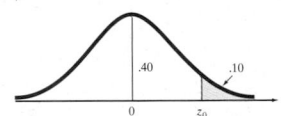

We know that the total area to the right of the mean $z = 0$ is .5, which implies that $z_0$ must lie to the right of (above) 0. To pinpoint the value, we use the fact that the area to the right of $z_0$ is .10, which implies that the area between $z = 0$ and $z_0$ is $.5 - .1 = .4$. But areas between $z = 0$ and some other $z$ value are exactly the types given in Table IV. Therefore, we look up the area .4000 in the body of Table IV and find that the corresponding $z$ value is (to the closest approximation) $z_0 = 1.28$. The implication is that the point 1.28 standard deviations above the mean is the 90th percentile of a normal distribution. ∎

**EXAMPLE 5.9**  Find the value of $z_0$ such that 95% of the standard normal $z$ values lie between $-z_0$ and $+z_0$—that is, $P(-z_0 \leq z \leq z_0) = .95$.

**Solution**

Here we wish to move an equal distance $z_0$ in the positive and negative directions from the mean $z = 0$ until 95% of the standard normal distribution is enclosed. This means that the area on each side of the mean will be equal to $\frac{1}{2}(.95) = .475$, as shown in Figure 5.15. Since the area between $z = 0$ and $z_0$ is .475, we look up

**FIGURE 5.15**
*Areas Under the Standard Normal Curve for Example 5.9*

---

## Learn by Example Approach

- New ideas are introduced and illustrated with data-based applications and examples. Complete solutions and explanations are also provided to help students prepare for end-of-section exercises.

- Tangible applications and examples help students to digest theory because they are always provided *before* definitions, generalizations, or abstractions.

- All examples are numbered for easy reference.

- The end of the solution is marked with a ∎ symbol.

---

**Page 232**

**Using the TI-83 Graphing Calculator**

**How to Graph a Normal Probability Plot**

Start from the home screen.

Step 1 Enter your data into **List 1** (*Recall the Lists are accessed by pressing* **STAT ENTER**. *Always begin from a "clear" List*).

Step 2 Access the "Stat Plot" menu.
*Press* **2ⁿᵈ ENTER**
*Press* **ENTER**
*Press* **ENTER**
Arrow down and right to last graph.
*Press* **ENTER**
*Set* Data List: **L1**
Data Axis: **X**

Step 3 *Press* **Zoom 9**
Your data will be displayed against the residuals. If you see a "generally" linear relationship your data are near normal.

Example Using a Normal Probability Plot test whether or not the data are normally distributed.

9.7  93.1  33.0  21.2  81.4  51.1
43.5  10.6  12.8   7.8  18.1  12.7

Step 1 Enter Data into List1 (see screen below).

**TI-83 Graphing Calculator Key Stroke Level Instruction Contained in Boxes**

- Gives students step-by-step instruction on how to use the TI-83.

- Helps students to see how technology can be used as a substitute for hand computation.

- Teaches students valuable skills they may use in later careers.

**Page 240**

## Emphasis on Real Data

- Most of the 1200+ exercises contain data or information taken from current newspaper articles, magazines, journals, and the internet. Statistics are all around you.

- Real data is also integrated into examples, chapter content, graphs, tables, etc.

## Computer Output Integrated Throughout

Exploring Data with Statistical Computer Software

- Statistical software packages, such as SPSS, MINITAB, SAS, or STATISTIX crunch data quickly so students can spend time analyzing the results. Learning how to interpret statistical output will prove helpful in future classes or on the job.

- When computer output appears in examples, the solution explains how to read and interpret the output.

- Output and instructions for the TI-83 graphing calculator are also provided.

**EXAMPLE 5.11**

The EPA mileage ratings on 100 cars, first presented in Chapter 2 (p. 000), are reproduced in Table 5.2. Numerical and graphical descriptive measures for the data are shown on the MINITAB and SPSS printouts, Figure 5.17a–c. Determine whether the EPA mileage ratings are from an approximate normal distribution.

EPAGAS

**Table 5.2   EPA Gas Mileage Ratings for 100 Cars (miles per gallon)**

| | | | | | | | | | |
|---|---|---|---|---|---|---|---|---|---|
| 36.3 | 41.0 | 36.9 | 37.1 | 44.9 | 36.8 | 30.0 | 37.2 | 42.1 | 36.7 |
| 32.7 | 37.3 | 41.2 | 36.6 | 32.9 | 36.5 | 33.2 | 37.4 | 37.5 | 33.6 |
| 40.5 | 36.5 | 37.6 | 33.9 | 40.2 | 36.4 | 37.7 | 37.7 | 40.0 | 34.2 |
| 36.2 | 37.9 | 36.0 | 37.9 | 35.9 | 38.2 | 38.3 | 35.7 | 35.6 | 35.1 |
| 38.5 | 39.0 | 35.5 | 34.8 | 38.6 | 39.4 | 35.3 | 34.4 | 38.8 | 39.7 |
| 36.3 | 36.8 | 32.5 | 36.4 | 40.5 | 36.6 | 36.1 | 38.2 | 38.4 | 39.3 |
| 41.0 | 31.8 | 37.3 | 33.1 | 37.0 | 37.6 | 37.0 | 38.7 | 39.0 | 35.8 |
| 37.0 | 37.2 | 40.7 | 37.4 | 37.1 | 37.8 | 35.9 | 35.6 | 36.7 | 34.5 |
| 37.1 | 40.3 | 36.7 | 37.0 | 33.9 | 40.1 | 38.0 | 35.2 | 34.8 | 39.5 |
| 39.9 | 36.9 | 32.9 | 33.8 | 39.8 | 34.0 | 36.8 | 35.0 | 38.1 | 36.9 |

**FIGURE 5.17a**
*MINITAB Histogram for Mileage Data*

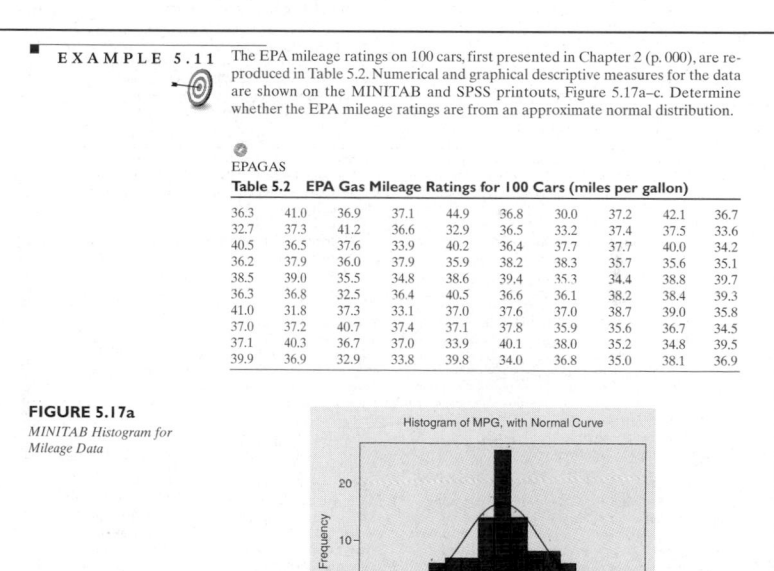

Histogram of MPG, with Normal Curve

**FIGURE 5.17b**
*MINITAB Descriptive Statistics for Mileage Data*

Descriptive Statistics: MPG

| Variable | N | Mean | Median | TrMean | StDev | SE Mean |
|---|---|---|---|---|---|---|
| MPG | 100 | 36.994 | 37.000 | 36.992 | 2.418 | 0.242 |

| Variable | Minimum | Maximum | Q1 | Q3 | | |
|---|---|---|---|---|---|---|
| MPG | 30.000 | 44.900 | 35.625 | 38.375 | | |

## Lots of Exercises for Practice

Every section in the book is followed by an Exercise Set divided into four parts:

• **Learning the Mechanics** Designed as straight foward applications of new concepts, these exercises allow students to test their ability to comprehend a mathematical concept or presentation.

• **Applying the Concepts—Basic** Based on applications taken from a wide variety of journals, newspapers and other sources, these short exercises help the student to begin developing the skills necessary to diagnose and analyze real-world problems.

---

### ✔ EXERCISES 6.15–6.33

**Learning the Mechanics**

**6.15** Suppose a random sample of $n$ measurements is selected from a population with mean $\mu = 100$ and variance $\sigma^2 = 100$. For each of the following values of $n$, give the mean and standard deviation of the sampling distribution of the sample mean $\bar{x}$.
  **a.** $n = 4$
  **b.** $n = 25$
  **c.** $n = 100$
  **d.** $n = 50$
  **e.** $n = 500$
  **f.** $n = 1,000$

**6.16** Suppose a random sample of $n = 25$ measurements is selected from a population with mean $\mu$ and standard deviation $\sigma$. For each of the following values of $\mu$ and $\sigma$, give the values of $\mu_{\bar{x}}$ and $\sigma_{\bar{x}}$.
  **a.** $\mu = 10, \sigma = 3$
  **b.** $\mu = 100, \sigma = 25$
  **c.** $\mu = 20, \sigma = 40$
  **d.** $\mu = 10, \sigma = 100$

**6.17** Consider the probability distribution shown here.

| $x$ | 1 | 2 | 3 | 8 |
|------|----|----|----|----|
| $p(x)$ | .1 | .4 | .4 | .1 |

  **a.** Find $\mu, \sigma^2$, and $\sigma$.
  **b.** Find the sampling distribution of $\bar{x}$ for random samples of $n = 2$ measurements from this distribution by listing all possible values of $\bar{x}$, and find the probability associated with each.
  **c.** Use the results of part **b** to calculate $\mu_{\bar{x}}$ and $\sigma_{\bar{x}}$. Confirm that $\mu_{\bar{x}} = \mu$ and that $\sigma_{\bar{x}} = \sigma/\sqrt{n} = \sigma/\sqrt{2}$.

**6.18** Will the sampling distribution of $\bar{x}$ always be approximately normally distributed? Explain.

**6.19** A random sample of $n = 64$ observations is drawn from a population with a mean equal to 20 and standard deviation equal to 16.
  **a.** Give the mean and standard deviation of the (repeated) sampling distribution of $\bar{x}$.
  **b.** Describe the shape of the sampling distribution of $\bar{x}$. Does your answer depend on the sample size?
  **c.** Calculate the standard normal $z$-score corresponding to a value of $\bar{x} = 15.5$.
  **d.** Calculate the standard normal $z$-score corresponding to $\bar{x} = 23$.

**6.20** Refer to Exercise 6.19. Find the probability that
  **a.** $\bar{x}$ is less than 16
  **b.** $\bar{x}$ is greater than 23
  **c.** $\bar{x}$ is greater than 25
  **d.** $\bar{x}$ falls between 16 and 22
  **e.** $\bar{x}$ is less than 14

**6.21** A random sample of $n = 100$ observations is selected from a population with $\mu = 30$ and $\sigma = 16$. Approximate the following probabilities:
  **a.** $P(\bar{x} \geq 28)$
  **b.** $P(22.1 \leq \bar{x} \leq 26.8)$
  **c.** $P(\bar{x} \leq 28.2)$
  **d.** $P(\bar{x} \geq 27.0)$

**6.22** A random sample of $n = 900$ observations is selected from a population with $\mu = 100$ and $\sigma = 10$.
  **a.** What are the largest and smallest values of $\bar{x}$ that you would expect to see?
  **b.** How far, at the most, would you expect $\bar{x}$ to deviate from $\mu$?
  **c.** Did you have to know $\mu$ to answer part **b**? Explain.

---

**6.23** Consider a population that contains values of $x$ equal to $0, 1, 2, \ldots, 97, 98, 99$. Assume that the values of $x$ are equally likely. For each of the following values of $n$, use the computer to generate 500 random samples and calculate $\bar{x}$ for each sample. For each sample size, construct a relative frequency histogram of the 500 values of $\bar{x}$. What changes occur in the histograms as the value of $n$ increases? What similarities exist? Use $n = 2, n = 5, n = 10, n = 30$, and $n = 50$.

**Applying the Concepts—Basic**

**6.24** According to the American Automobile Association (AAA), the average daily meal and lodging costs for a family of four is $213 (*Travel News*, May 11, 1999). Assume that the standard deviation of such costs is $15. Consider a random sample of 30 families of four and their travel expenses.
  **a.** Describe the sampling distribution of $\bar{x}$, the average daily meal and lodging costs for the sample of families. In particular, how is $\bar{x}$ distributed and what are the mean and variance of $\bar{x}$?
  **b.** Find the probability that $\bar{x}$ is greater than $200.

**6.25** The ocean quahog is a type of clam found in the coastal waters of the mid-Atlantic states. A federal survey of offshore ocean quahog harvesting in New Jersey revealed an average catch per unit effort (CPUE) of 89.34 clams. The CPUE standard deviation was 7.74 (*Journal of Shellfish Research*, June 1995). Let $\bar{x}$ represent the mean CPUE for a sample of 35 attempts to catch ocean quahogs off the New Jersey shore.
  **a.** Compute $\mu_{\bar{x}}$ and $\sigma_{\bar{x}}$. Interpret their values.
  **b.** Sketch the sampling distribution of $\bar{x}$.
  **c.** Find $P(\bar{x} > 88)$.
  **d.** Find $P(\bar{x} < 87)$.

**6.26** Refer to *The American Statistician* (May 2001) study of female students who suffer from bulimia, Exercise 2.27 (p. 39). Recall that each student completed a questionnaire from which a "fear of negative evaluation" (FNE) score was produced. (The higher the score, the greater the fear of negative evaluation.) Suppose the FNE scores of bulimic students have a distribution with mean $\mu = 18$ and standard deviation $\sigma = 5$. Now consider a random sample of 45 female students with bulimia.
  **a.** What is the probability that the sample mean FNE score is greater than 17.5?
  **b.** What is the probability that the sample mean FNE score is between 18 and 18.5?
  **c.** What is the probability that the sample mean FNE score is less than 18.5?

**6.27** Refer to the *Chance* (Winter 2001) examination of Standardized Admission Test (SAT) scores of students who pay a private tutor to help them improve their results, Exercise 2.81 (p. 70). On the SAT-Mathematics test, these students had a mean score change of +19 points, with a standard deviation of 65 points. In a random sample of 100 students who pay a private tutor to help them improve their results, what is the likelihood that the sample mean score change is less than 10 points?

**Applying the Concepts—Intermediate**

**6.28** The *College Student Journal* (Dec. 1992) investigated differences in traditional and nontraditional students, where nontraditional students are generally defined as those 25 years old or older. Based on the study results, we can assume that the population mean and standard deviation for the GPA of all nontraditional students is $\mu = 3.5$ and $\sigma = .5$. Suppose that a random sample of $n = 100$ nontraditional students is selected from the population of all nontraditional students, and the GPA of each student is determined. Then $\bar{x}$, the sample mean, will be approximately normally distributed (because of the Central Limit Theorem).
  **a.** Calculate $\mu_{\bar{x}}$ and $\sigma_{\bar{x}}$.
  **b.** What is the approximate probability that the nontraditional student sample has a mean GPA between 3.40 and 3.60?
  **c.** What is the approximate probability that the sample of 100 nontraditional students has a mean GPA that exceeds 3.62?
  **d.** How would the sampling distribution of $\bar{x}$ change if the sample size $n$ were doubled from 100 to 200? How do your answers to parts **b** and **c** change when the sample size is doubled?

**6.29** Interpersonal violence (e.g., rape) generally leads to psychological stress for the victim. *Clinical Psychology Review* (Vol. 15, 1995) reported on the results of all recently published studies of the relationship between interpersonal violence and psychological stress. The distribution of the time elapsed between the violent incident and the initial sign of stress has a mean of 5.1 years and a standard deviation of 6.1 years. Consider a random sample of $n = 150$ victims of interpersonal violence. Let $\bar{x}$ represent the mean time elapsed between the violent act and the first sign of stress for the sampled victims.
  **a.** Give the mean and standard deviation of the sampling distribution of $\bar{x}$.
  **b.** Will the sampling distribution of $\bar{x}$ be approximately normal? Explain.
  **c.** Find $P(\bar{x} > 5.5)$.
  **d.** Find $P(4 < \bar{x} < 5)$.

**6.30** The Computer-Assisted Hypnosis Scale (CAHS) is designed to measure a person's susceptibility to hypnosis. In computer-assisted hypnosis, the computer serves as a facilitator of hypnosis by using digitized speech processing coupled with interactive involvement with the hypnotic subject. CAHS scores range from 0 (no susceptibility) to 12 (extremely high susceptibility). A study in *Psychological Assessment* (Mar. 1995) reported a mean CAHS score of 4.59 and a standard deviation of 2.95 for University of Tennessee undergraduates. Assume that $\mu = 4.29$ and $\sigma = 2.95$ for this population.

• **Applying the Concepts—Intermediate** Based on more detailed real-world applications, these exercises require the student to apply their knowledge of the technique presented in the section.

• **Applying the Concepts—Advanced** These more challenging real-data exercises require the student to utilize their critical thinking skills.

## Statistics In Action Case Studies Explore High Interest Issues

- Each chapter concludes with a data-based case study that showcases controversial, contemporary, or high-profile issues.

- **Focus** questions are provided to help you evaluate the findings.

- Additional cases are available on the companion website located at **www.prenhall.com/mcclave**

---

**STATISTICS IN ACTION**
*IQ, Economic Mobility, and the Bell Curve*

In their controversial book *The Bell Curve* (Free Press, 1994), Professors Richard J. Herrnstein (a Harvard psychologist who died while the book was in production) and Charles Murray (a political scientist at MIT) explore, as the subtitle states, "intelligence and class structure in American life." *The Bell Curve* heavily employs statistical analyses in an attempt to support the authors' positions. Since the book's publication, many expert statisticians have raised doubts about the authors' statistical methods and the inferences drawn from them. (See, for example, "Wringing *The Bell Curve*: A cautionary tale about the relationships among race, genes, and IQ," *Chance*, Summer 1995.) In the Statistics in Action of Chapter 12, we explore a few of these problems.

One of the many controversies sparked by the book is the authors' tenet that level of intelligence (or lack thereof) is a cause of a wide range of intractable social problems, including constrained economic mobility. "America has taken great pride in the mobility of generations," state Herrnstein and Murray, "but this mobility has its limits. ... The son of a father whose earnings are in the bottom five percent of the [income] distribution has something like one chance in twenty (or less) of rising to the top fifth of the income distribution and almost a fifty-fifty chance of staying in the bottom fifth. He has less than one chance in four of rising above even the median income. ... Most people at present are stuck near where their parents were on the income distribution in part because [intelligence], which has become a major predictor of income, passes on sufficiently from one generation to the next to constrain economic mobility."

The measure of intelligence chosen by the authors is the well known Intelligent Quotient (IQ). Numerous tests have been developed to measure IQ; Herrnstein and Murray use the Armed Forces Qualification Test (AFQT), originally designed to measure the cognitive ability of military recruits.

Psychologists traditionally treat IQ as a random variable having a normal distribution with mean $\mu = 100$ and standard deviation $\sigma = 15$. This distribution, or **bell curve**, is shown in Figure 5.28.

In their book, Herrnstein and Murray refer to five cognitive classes of people defined by percentiles of the normal distribution. Class I ("very bright") consists of those with IQs above the 95th percentile; Class II ("bright") are those with IQs between the 75th and 95th percentiles; Class III ("normal") includes IQs between the 25th and 75th percentiles; Class IV ("dull") are those with IQs between the 5th and 25th percentiles; and Class V ("very dull") are IQs below the 5th percentile. These classes are also illustrated in Figure 5.28.

### Focus

a. Assuming that the distribution of IQ is accurately represented by the bell curve in Figure 5.28, determine the proportion of people with IQs in each of the five cognitive classes defined by Herrnstein and Murray.

b. Although the cognitive classes above are defined in terms of percentiles, the authors stress that IQ scores should be compared with $z$-scores, not percentiles. In other words, it is more informative to give the difference in $z$-scores for two IQ scores than it is to give the difference in percentiles. To demonstrate this point, calculate the difference in $z$-scores for IQs at the 50th and 55th percentiles. Do the same for IQs at the 94th and 99th percentiles. What do you observe?

c. Researchers have found that scores on many intelligence tests are decidedly nonnormal. Some distributions are skewed toward higher scores, others toward lower scores. How would the proportions in the five cognitive classes differ for an IQ distribution that is skewed right? Skewed left?

**FIGURE 5.28**
*The Distribution of IQ*

## Robust End of Chapter Review Material

- Each chapter ends with information designed to help students check their understanding of the material, study for tests, and expand their knowledge of statistics.
- **Quick Review** provides a list of key terms and formulas with page number references.
- **Language Lab** helps students learn the language of statistics through pronunciation guides, descriptions of symbols, names, etc.
- **Supplementary Exercises** review all of the important topics covered in the chapter.

Exercises marked with  require a computer for solution.

Data sets for use with  the problems are available on free data CD packaged free with every new text.

---

### QUICK REVIEW

**Key Terms**
*Note: Starred (*) terms are from the optional sections to this chapter*

| | | |
|---|---|---|
| Bar graph   22 | Measures of central | Quartiles*   76 |
| Bivariate relationship*   86 | tendency   44 | Range   56 |
| Box plots*   76 | Measures of relative | Rare-event approach*   81 |
| Central tendency   44 | standing   70 | Sample standard deviation   57 |
| Chebyshev's Rule   61 | Measures of variation or | Sample variance   57 |
| Class   20 | spread   55 | Scattergram*   86 |
| Class frequency   20 | Median   46 | Scatterplot*   86 |
| Class relative frequency   20 | Middle quartile*   76 | Skewness   48 |
| Dot plot   30 | Modal class   48 | Standard deviation   57 |
| Empirical Rule   62 | Mode   48 | Stem-and-leaf display   30 |
| Hinges*   76 | Mound-shaped distribution   61 | Symmetric distribution   48 |
| Histogram   31 | Numerical descriptive | Upper quartile*   76 |
| Inner fences*   77 | measures   44 | Variability   44 |
| Interquartile range*   76 | Outer fences*   77 | Whiskers*   77 |
| Lower quartile*   76 | Outlier*   75 | $z$-score   72 |
| Mean   44 | Percentile   71 | |
| Measurement classes   32 | Pie chart   23 | |

**Language Lab**

| Symbol | Pronunciation | Description |
|---|---|---|
| $\sum$ | sum of | Summation notation; $\sum_{i=1}^{n} x_i$ represents the sum of the measurements $x_1, x_2, \ldots, x_n$, |
| $\mu$ | mu | Population mean |
| $\bar{x}$ | $x$-bar | Sample mean |
| $\sigma^2$ | sigma-squared | Population variance |
| $\sigma$ | sigma | Population standard deviation |
| $s^2$ | | Sample variance |
| $s$ | | Sample standard deviation |

---

### ✓ SUPPLEMENTARY EXERCISES 2.116–2.140

*Note: Starred (*) exercises refer to the optional sections in this chapter.*

**Learning the Mechanics**

**2.116** Construct a relative frequency histogram for the data summarized in the table below.

| Measurement Class | Relative Frequency | Measurement Class | Relative Frequency |
|---|---|---|---|
| .00–.75 | .02 | 5.25–6.00 | .15 |
| .75–1.50 | .01 | 6.00–6.75 | .12 |
| 1.50–2.25 | .03 | 6.75–7.50 | .09 |
| 2.25–3.00 | .05 | 7.50–8.25 | .05 |
| 3.00–3.75 | .10 | 8.25–9.00 | .04 |
| 3.75–4.50 | .14 | 9.00–9.75 | .01 |
| 4.50–5.25 | .19 | | |

**2.117** Consider the following three measurements: 50, 70, 80. Find the $z$-score for each measurement if they are from a population with a mean and standard deviation equal to
- **a.** $\mu = 60, \sigma = 10$
- **b.** $\mu = 60, \sigma = 5$
- **c.** $\mu = 40, \sigma = 10$
- **d.** $\mu = 40, \sigma = 100$

**2.122** For each of the following data sets, compute $\bar{x}$, $s^2$, and $s$:
- **a.** 13, 1, 10, 3, 3
- **b.** 13, 6, 6, 0
- **c.** 1, 0, 1, 10, 11, 11, 15
- **d.** 3, 3, 3, 3
- **e.** For each data set in parts **a–d**, form the interval $\bar{x} \pm 2s$ and calculate the percentage of the measurements that fall in the interval.

**\*2.123** Construct a scatterplot for the data listed here. Do you detect any trends?

| Variable #1: | 174 | 268 | 345 | 119 | 400 | 520 | 190 | 448 | 307 | 252 |
|---|---|---|---|---|---|---|---|---|---|---|
| Variable #2: | 8 | 10 | 15 | 7 | 22 | 31 | 15 | 20 | 11 | 9 |

**2.124** The following data sets have been invented to demonstrate that the lower bounds given by Chebyshev's Rule are appropriate. Notice that the data are contrived and would not be encountered in a real-life problem.
- **a.** Consider a data set that contains ten 0s, two 1s, and ten 2s. Calculate $\bar{x}$, $s^2$, and $s$. What percentage of the measurements are in the interval $\bar{x} \pm s$? Compare this result to Chebyshev's Rule.
- **b.** Consider a data set that contains five 0s, thirty-two

---

### STUDENT PROJECTS

Many dependent variables in all areas of research serve as the subjects of regression modeling efforts. We list five such variables here:

1. Crime rate in various communities
2. Daily maximum temperature in your town
3. Grade point average of students who have completed one academic year at your college
4. Gross Domestic Product of the United States
5. Points scored by your favorite football team in a single game

Choose one of these dependent variables or choose some other dependent variable for which you want to construct a prediction model. There may be a large number of independent variables that should be included in a prediction equation for the dependent variable you choose. List three potentially important independent variables, $x_1$, $x_2$, and $x_3$, that you think might be (individually) strongly related to your dependent variable. Next, obtain 10 data values, each of which

consists of a measure of your dependent variable $y$ and the corresponding values of $x_1$, $x_2$, and $x_3$.

a. Use the least squares formulas given in this chapter to fit three straight-line models—one for each independent variable—for predicting $y$.

b. Interpret the sign of the estimated slope coefficient $\hat{\beta}_1$ in each case, and test the utility of each model by testing $H_0$: $\beta_1 = 0$ against $H_a$: $\beta_1 \neq 0$. What assumptions must be satisfied to ensure the validity of these tests?

c. Calculate the coefficient of determination $r^2$ for each model. Which of the independent variables predicts $y$ best for the 10 sampled sets of data? Is this variable necessarily best in general (that is, for the entire population)? Explain.

Be sure to keep the data and the results of your calculations, since you will need them for the Student Projects section in Chapter 12.

## Additional End of Chapter Resources

**Student Projects** provide challenging projects for further exploration individually or with a group of students. These projects provide good practice in gathering and analyzing data and report writing—skills that will be important in future classes and in the workplace.

# CHAPTER 1

# Statistics, Data, and Statistical Thinking

## Contents

## Statistics in Action

A "20/20" View of Survey Results: Fact or Fiction?

### ☞ Where We're Going

Statistics? Is it a field of study, or is it a group of numbers that summarizes the state of our national economy, the performance of a football team, or the social conditions in a particular locale? Or, as one popular book (Tanur *et al.*, 1989) suggests, is it "a guide to the unknown"? We'll see in Chapter 1 that each of these descriptions is applicable in understanding what statistics is. We'll also see that there are two areas of statistics: *descriptive statistics*, which focuses on developing graphical and numerical summaries that describe some phenomenon, and *inferential statistics*, which uses these numerical summaries to assist in making decisions. The primary theme of this text is inferential statistics. Thus, we'll concentrate on showing how you can use statistics to interpret data and use them to make decisions. Many jobs in industry, government, medicine, and other fields require you to make data-driven decisions so understanding these methods offers you important practical benefits.

## 1.1 THE SCIENCE OF STATISTICS

What does statistics mean to you? Does it bring to mind batting averages, Gallup polls, unemployment figures, numerical distortions of facts (lying with statistics!)? Or is it simply a college requirement you have to complete? We hope to persuade you that statistics is a meaningful, useful science whose broad scope of applications to business, government, and the physical and social sciences is almost limitless. We also want to show that statistics can lie only when they are misapplied. Finally, we wish to demonstrate the key role statistics plays in critical thinking—whether in the classroom, on the job, or in everyday life. Our objective is to leave you with the impression that the time you spend studying this subject will repay you in many ways.

The *Random House College Dictionary* defines *statistics* as "the science that deals with the collection, classification, analysis, and interpretation of information or data." Thus, a statistician isn't just someone who calculates batting averages at baseball games or tabulates the results of a Gallup Poll. Professional statisticians are trained in *statistical science*. That is, they are trained in collecting numerical information in the form of **data**, evaluating it, and drawing conclusions from it. Furthermore, statisticians determine what information is relevant in a given problem and whether the conclusions drawn from a study are to be trusted.

> **DEFINITION 1.1**
> **Statistics** is the science of data. This involves collecting, classifying, summarizing, organizing, analyzing, and interpreting numerical information.

In the next section, you'll see several real-life examples of statistical applications that involve making decisions and drawing conclusions.

## 1.2 TYPES OF STATISTICAL APPLICATIONS

Statistics means "numerical descriptions" to most people. Monthly housing starts, the failure rate of liver transplants, and the proportion of African-Americans who feel brutalized by local police all represent statistical descriptions of large sets of data collected on some phenomenon. Often the data are selected from some larger set of data whose characteristics we wish to estimate. We call this selection process *sampling*. For example, you might collect the ages of a sample of customers at a video store to estimate the average age of *all* customers of the store. Then you could use your estimate to target the store's advertisements to the appropriate age group. Notice that statistics involves two different processes: (1) describing sets of data and (2) drawing conclusions (making estimates, decisions, predictions, etc.) about the sets of data based on sampling. So, the applications of statistics can be divided into two broad areas: *descriptive statistics* and *inferential statistics*.

> **DEFINITION 1.2**
> **Descriptive statistics** utilizes numerical and graphical methods to look for patterns in a data set, to summarize the information revealed in a data set, and to present that information in a convenient form.

> ### DEFINITION 1.3
> **Inferential statistics** utilizes sample data to make estimates, decisions, predictions, or other generalizations about a larger set of data.

Although we'll discuss both descriptive and inferential statistics in the following chapters, the primary theme of the text is **inference**.

Let's begin by examining some studies that illustrate applications of statistics.

**Study 1**   "1999–2000 Financial Report, H. Lee Moffitt Cancer Center & Research Institute" (*Today's Tomorrow*, Winter 2000/2001)

The Moffitt Cancer Center at the University of South Florida treated a total of 28,196 patients during the 2000 fiscal year. The center classified the type of cancer treated for each patient. The results, extracted from the center's annual financial report, are summarized in the graphic, Figure 1.1. From the graph, you can clearly see that 19% of the patients were treated for breast cancer and 13% for prostate cancer. Since Figure 1.1 describes the type of cancer treated in all Moffitt patients in fiscal year 2000, the graphic is an example of descriptive statistics.

**FIGURE 1.1**

*Distribution of Moffitt Cancer Patients*

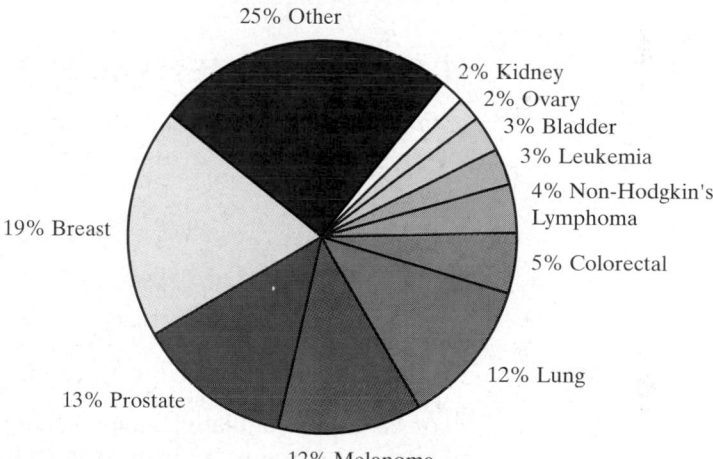

**Study 2**   "Does Height Influence Progression through Primary School Grades?" (*The Archives of Disease in Childhood*, April 2000)

Researchers from the University of Melbourne (Australia) analyzed the heights of over 2,800 students in primary school (grades 1 through 6). After dividing the students into three equal groups based on age (youngest third, middle third, and oldest third) within each grade level, the researchers found that the oldest group of students were the shortest in height, on average—and that this phenomenon was mostly due to the 133 children (mostly boys) who had been held back a grade level. From the analysis, the researchers inferred "that older boys within grades were relatively shorter than their younger peers" and that "when boys experienced school difficulties, height [is] one factor influencing the final decision to [hold back the student a grade level]." Thus, inferential statistics was applied to arrive at this conclusion.

**Study 3**   "The Significance of Belief and Expectancy Within the Spiritual Healing Encounter" (*Social Science & Medicine*, July 1995)

"Spiritual healing techniques have been a fundamental component of the healing rituals of virtually all societies since the advent of man," says researcher Daniel P. Wirth. What effect, if any, do spiritual healing encounters have on the mental and physical health of patients in today's society? And do the effects depend on the patient's expectations? To answer these questions, Wirth conducted a scientific study of 48 individuals residing in a northern California suburb. The mental and physical health of each subject was determined via a battery of standardized tests. Each individual then underwent a spiritual healing encounter that involved a "magnetic laying-on-of-hands approach." One week following the healing session, subjects were retested.

An analysis of the data revealed a significant difference between the pretest and posttest health scores for the subjects. Further analysis led Wirth to conclude that "high healer and patient expectancy [are] important ... predictors as well as facilitators of the healing process."

Like Study 2, this study is an example of the use of inferential statistics. Wirth used data from 48 individuals to make inferences about the beliefs and health improvements of all patients who experience a spiritual healing encounter.

These studies provide three real-life examples of the uses of statistics. Notice that each involves an analysis of data, either for the purpose of describing the data set (Study 1) or for making inferences about a data set (Studies 2 and 3).

## 1.3    FUNDAMENTAL ELEMENTS OF STATISTICS

Statistical methods are particularly useful for studying, analyzing, and learning about *populations*.

---

**DEFINITION 1.4**

A **population** is a set of units (usually people, objects, transactions, or events) that we are interested in studying.

---

For example, populations may include (1) *all* employed workers in the United States, (2) *all* registered voters in California, (3) *everyone* who is afflicted with AIDS, (4) *all* the cars produced last year by a particular assembly line, (5) the entire stock of spare parts available at United Airlines' maintenance facility, (6) *all* sales made at the drive-in window of a McDonald's restaurant during a given year, and (7) the set of *all* accidents occurring on a particular stretch of interstate highway during a holiday period. Notice that the first three population examples (1-3) are sets (groups) of people, the next two (4-5) are sets of objects, the next (6) is a set of transactions, and the last (7) is a set of events. Also notice that each set includes all the units in the population.

In studying a population, we focus on one or more characteristics or properties of the units in the population. We call such characteristics *variables*. For example, we may be interested in the variables age, gender, and/or the number of years of education of the people currently unemployed in the United States.

---

**DEFINITION 1.5**

A **variable** is a characteristic or property of an individual population unit.

The name "variable" is derived from the fact that any particular characteristic may vary among the units in a population.

In studying a particular variable it is helpful to be able to obtain a numerical representation for it. Often, however, numerical representations are not readily available, so the process of measurement plays an important supporting role in statistical studies. **Measurement** is the process we use to assign numbers to variables of individual population units. We might, for instance, measure the performance of the President by asking a registered voter to rate it on a scale from 1 to 10. Or we might measure workforce age simply by asking each worker, "How old are you?" In other cases, measurement involves the use of instruments such as stop watches, scales, and calipers.

If the population you wish to study is small, it is possible to measure a variable for every unit in the population. For example, if you are measuring the GPA for all incoming first-year students at your university, it is at least feasible to obtain every GPA. When we measure a variable for every unit of a population, it is called a **census** of the population. Typically, however, the populations of interest in most applications are much larger, involving perhaps many thousands or even an infinite number of units. Examples of large populations are those following Definition 1.4, as well as all graduates of your university or college, all potential buyers of a new fax machine, and all pieces of first-class mail handled by the U.S. Post Office. For such populations conducting a census would be prohibitively time-consuming and/or costly. A reasonable alternative would be to select and study a *subset* (or portion) of the units in the population.

> **DEFINITION 1.6**
>
> A **sample** is a subset of the units of a population.

For example, instead of polling all 120 million registered voters in the United States during a presidential election year, a pollster might select and question a sample of just 1,500 voters (see Figure 1.2). If he is interested in the variable "presidential preference," he would record (measure) the preference of each sampled vote.

After the variable(s) of interest for every unit in the sample (or population) are measured, the data are analyzed, either by descriptive or inferential statistical methods. The pollster, for example, may be interested only in *describing* the voting patterns of the sample of 1,500 voters. More likely, however, he will want to use the information in the sample to make inferences about the population of all 120 million voters.

> **DEFINITION 1.7**
>
> A **statistical inference** is an estimate, prediction, or some other generalization about a population based on information contained in a sample.

That is, *we use the information contained in the smaller sample to learn about the larger population.*\* Thus, from the sample of 1,500 voters, the pollster may estimate

---

\*The terms *population* and *sample* are often used to refer to the sets of measurements themselves, as well as to the units on which the measurements are made. When a single variable of interest is being measured, this usage causes little confusion. But when the terminology is ambiguous, we'll refer to the measurements as *population data sets* and sample data sets, respectively.

**FIGURE 1.2**

*A Sample of Voter Registration Cards for All Registered Voters*

Population                                    Sample

1,500 voter ID cards

120 million voter ID cards

the percentage of all the voters who would vote for each presidential candidate if the election were held on the day the poll was conducted, or he might use the results to predict the outcome on election day.

**EXAMPLE 1.1**

According to *USA Today* (Dec. 30, 1999), the average age of viewers of MSNBC cable television news programming is 50 years. Suppose a rival network executive hypothesizes that the average age of MSNBC news viewers is less than 50. To test her hypothesis, she samples 500 MSNBC news viewers and determines the age of each.

a. Describe the population.
b. Describe the variable of interest.
c. Describe the sample.
d. Describe the inference.

**Solution**

a. The population is the set of units of interest to the cable TV executive, which is the set of all MSNBC news viewers.
b. The age (in years) of each viewer is the variable of interest.
c. The sample must be a subset of the population. In this case, it is the 500 MSNBC news viewers selected by the executive.

d. The inference of interest involves the *generalization* of the information contained in the sample of 500 viewers to the population of all MSNBC viewers. In particular, the executive wants to estimate the average age of the viewers in order to determine whether it is less than 50 years. She might accomplish this by calculating the average age in the sample and using the sample average to estimate the population average. ∎

**EXAMPLE 1.2**

"Cola wars" is the popular term for the intense competition between Coca-Cola and Pepsi displayed in their marketing campaigns. Their campaigns have featured movie and television stars, rock videos, athletic endorsements, and claims of consumer preference based on taste tests. Suppose, as part of a Pepsi marketing campaign, 1,000 cola consumers are given a blind taste test (i.e., a taste test in which the two brand names are disguised). Each consumer is asked to state a preference for brand A or brand B.

a. Describe the population.
b. Describe the variable of interest.
c. Describe the sample.
d. Describe the inference.

**Solution**

a. The population of interest is the collection or set of all consumers.
b. The characteristic that Pepsi wants to measure is the consumer's cola preference as revealed under the conditions of a blind taste test, so *cola preference* is the variable of interest.
c. The sample is the 1,000 cola consumers selected from the population of all cola consumers.
d. The inference of interest is the *generalization* of the cola preferences of the 1,000 sampled consumers to the population of all cola consumers. In particular, the preferences of the consumers in the sample can be used to *estimate* the percentage of all cola consumers who prefer each brand. ∎

The preceding definitions and examples identify four of the five elements of an inferential statistical problem: a population, one or more variables of interest, a sample, and an inference. But making the inference is only part of the story. We also need to know its **reliability**—that is, how good the inference is. The only way we can be certain that an inference about a population is correct is to include the entire population in our sample. However, because of *resource constraints* (i.e., insufficient time and/or money), we usually can't work with whole populations, so we base our inferences on just a portion of the population (a sample). Thus, we introduce an element of *uncertainty* into our inferences. Consequently, whenever possible, it is important to determine and report the reliability of each inference made. Reliability, then, is the fifth element of inferential statistical problems.

The measure of reliability that accompanies an inference separates the science of statistics from the art of fortune-telling. A palm reader, like a statistician, may examine a sample (your hand) and make inferences about the population (your life). However, unlike statistical inferences, the palm reader's inferences include no measure of reliability.

Suppose, like the cable TV executive in Example 1.1, we are interested in estimating the average age of a population of cable TV viewers from the average age of a sample of viewers. Using statistical methods, we can determine a *bound on the estimation error.* This bound is simply a number that our estimation error (the difference between the average age of the sample and the average age of the population of viewers) is not likely to exceed. We'll see in later chapters that this bound is a measure of the uncertainty of our inference. The reliability of statistical inferences is discussed throughout this text. For now, we simply want you to realize that an inference is incomplete without a measure of its reliability.

> **DEFINITION 1.8**
> A **measure of reliability** is a statement (usually quantified) about the degree of uncertainty associated with a statistical inference.

Let's conclude this section with a summary of the elements of both descriptive and inferential statistical problems and an example to illustrate a measure of reliability.

---

### Four Elements of Descriptive Statistical Problems

1. The population or sample of interest
2. One or more variables (characteristics of the population or sample units) that are to be investigated
3. Tables, graphs, or numerical summary tools
4. Identification of patterns in the data

### Five Elements of Inferential Statistical Problems

1. The population of interest
2. One or more variables that are to be investigated
3. The sample of population units
4. The inference about the population based on information contained in the sample
5. A measure of reliability for the inference

---

**EXAMPLE 1.3**

Refer to Example 1.2, in which the cola preferences of 1,000 consumers were indicated in a taste test. Describe how the reliability of an inference concerning the preferences of all cola consumers in the Pepsi bottler's marketing region could be measured.

### Solution

When the preferences of 1,000 consumers are used to estimate those of all consumers in the region, the estimate will not exactly mirror the preferences of the population. For example, if the taste test shows that 56% of the 1,000 consumers preferred Pepsi, it does not follow (nor is it likely) that exactly 56% of all cola drinkers in the region prefer Pepsi. Nevertheless, we can use sound statistical reasoning (which we'll explore later in the text) to ensure that the sampling procedure will generate estimates that are almost certainly within a specified limit of the true percentage of all consumers who prefer Pepsi. For example, such reasoning might assure us that the estimate of the preference for Pepsi is almost certainly within 5% of the actual population preference. The implication is that the actual preference for Pepsi is between 51% [i.e., $(56 - 5)$%] and 61% [i.e., $(56 + 5)$%]—that is, $(56 \pm 5)$%. This interval represents a measure of reliability for the inference.  ∎

## 1.4   TYPES OF DATA

You have learned that statistics is the science of data and that data are obtained by measuring the values of one or more variables on the units in the sample (or population). All data (and hence the variables we measure) can be classified as one of two general types: *quantitative data* and *qualitative data*.

Quantitative data are data that are measured on a naturally occurring numerical scale.* The following are examples of quantitative data:

1. The temperature (in degrees Celsius) at which each in a sample of 20 pieces of heat-resistant plastic begins to melt
2. The current unemployment rate (measured as a percentage) for each of the 50 states
3. The scores of a sample of 150 law school applicants on the LSAT, a standardized law school entrance exam administered nationwide
4. The number of convicted murderers who receive the death penalty each year over a 10-year period

> **DEFINITION 1.9**
> **Quantitative data** are measurements that are recorded on a naturally occurring numerical scale.

In contrast, qualitative data cannot be measured on a natural numerical scale; they can only be classified into categories.† Examples of qualitative data include:

1. The political party affiliation (Democrat, Republican, or Independent) in a sample of 50 voters
2. The defective status (defective or not) of each of 100 computer chips manufactured by Intel
3. The size of a car (subcompact, compact, mid-size, or full-size) rented by each of a sample of 30 business travelers
4. A taste-tester's ranking (best, worst, etc.) of four brands of barbecue sauce for a panel of 10 testers

Often, we assign arbitrary numerical values to qualitative data for ease of computer entry and analysis. But these assigned numerical values are simply codes: They cannot be meaningfully added, subtracted, multiplied, or divided. For example, we might code Democrat = 1, Republican = 2, and Independent = 3. Similarly, a taste tester might rank the barbecue sauces from 1 (best) to 4 (worst). These are simply arbitrarily selected numerical codes for the categories and have no utility beyond that.

---

*Quantitative data can be subclassified as either *interval data* or *ratio data*. For ratio data, the origin (i.e., the value 0) is a meaningful number. But the origin has no meaning with interval data. Consequently, we can add and subtract interval data, but we can't multiply and divide them. Of the four quantitative data sets listed above, (1) and (3) are interval data, while (2) and (4) are ratio data.

†Qualitative data can be subclassified as either *nominal data* or *ordinal data*. The categories of an ordinal data set can be ranked or meaningfully ordered, but the categories of a nominal data set can't be ordered. Of the four qualitative data sets listed above, (1) and (2) are nominal and (3) and (4) are ordinal.

> **DEFINITION 1.10**
> **Qualitative data** are measurements that cannot be measured on a natural numerical scale; they can only be classified into one of a group of categories.

**EXAMPLE 1.4**

Chemical and manufacturing plants often discharge toxic-waste materials such as DDT into nearby rivers and streams. These toxins can adversely affect the plants and animals inhabiting the river and the river bank. The U.S. Army Corps of Engineers conducted a study of fish in the Tennessee River (in Alabama) and its three tributary creeks: Flint Creek, Limestone Creek, and Spring Creek. A total of 144 fish were captured, and the following variables were measured for each:

1. River/creek where each fish was captured
2. Species (channel catfish, largemouth bass, or smallmouth buffalofish)
3. Length (centimeters)
4. Weight (grams)
5. DDT concentration (parts per million)

Classify each of the five variables measured as quantitative or qualitative.

**Solution**

The variables length, weight, and DDT concentration are quantitative because each is measured on a numerical scale: length in centimeters, weight in grams, and DDT in parts per million. In contrast, river/creek and species cannot be measured quantitatively: They can only be classified into categories (e.g., channel catfish, largemouth bass, and smallmouth buffalofish for species). Consequently, data on river/creek and species are qualitative.

     As you would expect, the statistical methods for describing, reporting, and analyzing data depend on the type (quantitative or qualitative) of data measured. We demonstrate many useful methods in the remaining chapters of the text. But first we discuss some important ideas on data collection.

## 1.5   COLLECTING DATA

Once you decide on the type of data—quantitative or qualitative—appropriate for the problem at hand, you'll need to collect the data. Generally, you can obtain data in four different ways:

1. Data from a *published source*
2. Data from a *designed experiment*
3. Data from a *survey*
4. Data from an *observational study*

     Sometimes, the data set of interest has already been collected for you and is available in a **published source**, such as a book, journal, or newspaper. For example, you may want to examine and summarize the divorce rates (i.e., number of divorces per 1,000 population) in the 50 states of the United States. You can find this data set (as well as numerous other data sets) at your library in the *Statistical Abstract of the United States*, published annually by the U.S. government. Similarly, someone who is interested in monthly mortgage applications for new home construction would find this data set in the *Survey of Current Business*, another government publication. Other examples of published data sources include *The Wall Street Journal* (financial data) and *The Sporting News* (sports information). The Internet (World Wide Web) now provides a medium by which data from published sources are readily obtained.*

A second method of collecting data involves conducting a **designed experiment**, in which the researcher exerts strict control over the units (people, objects, or things) in the study. For example, a recent medical study investigated the potential of aspirin in preventing heart attacks. Volunteer physicians were divided into two groups—the *treatment* group and the *control* group. In the treatment group, each physician took one aspirin tablet a day for one year, while each physician in the control group took an aspirin-free placebo made to look like an aspirin tablet. The researchers, not the physicians under study, controlled who received the aspirin (the treatment) and who received the placebo. As you'll learn in Chapter 10, a properly designed experiment allows you to extract more information from the data than is possible with an uncontrolled study.

Surveys are a third source of data. With a **survey**, the researcher samples a group of people, asks one or more questions, and records the responses. Probably the most familiar type of survey is the political poll conducted by any one of a number of organizations (e.g., Harris, Gallup, Roper, and CNN) and designed to predict the outcome of a political election. Another familiar survey is the Nielsen survey, which provides the major networks with information on the most watched programs on television. Surveys can be conducted through the mail, with telephone interviews, or with in-person interviews. Although in-person surveys are more expensive than mail or telephone surveys, they may be necessary when complex information is to be collected.

Finally, observational studies can be employed to collect data. In an **observational study**, the researcher observes the experimental units in their natural setting and records the variable(s) of interest. For example, a child psychologist might observe and record the level of aggressive behavior of a sample of fifth graders playing on a school playground. Similarly, a zoologist may observe and measure the weights of newborn elephants born in captivity. Unlike a designed experiment, an observational study is one in which the researcher makes no attempt to control any aspect of the experimental units.

Regardless of which data collection method is employed, it is likely that the data will be a sample from some population. And if we wish to apply inferential statistics, we must obtain a *representative sample*.

---

**DEFINITION 1.11**

A **representative sample** exhibits characteristics typical of those possessed by the target population.

---

For example, consider a political poll conducted during a presidential election year. Assume the pollster wants to estimate the percentage of all 120 million registered voters in the United States who favor the incumbent President. The pollster would be unwise to base the estimate on survey data collected for a sample

---

*With published data, we often make a distinction between the *primary source* and a *secondary source*. If the publisher is the original collector of the data, the source is primary. Otherwise, the data is secondary source data.

of voters from the incumbent's own state. Such an estimate would almost certainly be *biased* high; consequently, it would not be very reliable.

The most common way to satisfy the representative sample requirement is to select a random sample. A **random sample** ensures that every subset of fixed size in the population has the same chance of being included in the sample. If the pollster samples 1,500 of the 120 million voters in the population so that every subset of 1,500 voters has an equal chance of being selected, he has devised a random sample. The procedure for selecting a random sample is discussed in Chapter 3. Here, however, let's look at two examples involving actual sampling studies.

---

**EXAMPLE 1.5**

"Have you ever seen anything that you believe was a spacecraft from another planet?" This was the question put to 1,500 American adults in a national poll conducted by ABC News and *The Washington Post*. The pollsters used random-digit telephone dialing to contact adult Americans, until 1,500 responded. Ten percent (i.e., 150) of the respondents answered that they had, in fact, seen an alien spacecraft (*Chance*, Summer 1997). No information was provided on how many adults were called and, for one reason or another, did not answer the question.

a. Identify the data collection method.
b. Identify the target population.
c. Are the sample data representative of the population?

### Solution

a. The data collection method is a telephone survey: 1,500 adults were contacted via random-digit dialing of telephone numbers.
b. Since the pollsters conducted a *national* survey, presumably the target population is all American adults.
c. Because the 1,500 respondents clearly make up a subset of the target population, they do form a sample. Whether or not the sample is representative is unclear, since we are given no information on the 1,500 respondents. However, telephone surveys (and surveys in general) often suffer from **nonresponse bias**. It is possible that many adults who chose not to respond would have answered the question "no," leading to a percentage of affirmative answers less than 10%. ∎

---

**EXAMPLE 1.6**

Psychologists at the University of Tennessee carried out a study of the susceptibility of people to hypnosis (*Psychological Assessment*, Mar. 1995). In a random sample of 130 undergraduate psychology students at the university, each experienced both traditional hypnosis and computer-assisted hypnosis. Approximately half were randomly assigned to undergo the traditional procedure first, followed by the computer-assisted procedure. The other half were randomly assigned to experience computer-assisted hypnosis first, then traditional hypnosis. Following the hypnosis episodes, all students filled out questionnaires designed to measure a student's susceptibility to hypnosis. The susceptibility scores of the two groups of students were compared.

a. Identify the data collection method.
b. Is the sample data representative of the target population?

### Solution

a. The researchers controlled which type of hypnosis—traditional or computer-assisted—the students experienced first. Consequently, a designed experiment was used to collect the data for the two groups.

b. The sample of 130 psychology students was randomly selected from all psychology students at the University of Tennessee. If the target population is *all University of Tennessee psychology students*, it is likely that the sample is representative. However, the researchers warn that the sample data should not be used to make inferences about other, more general, populations. ▬ ■

## 1.6 THE ROLE OF STATISTICS IN CRITICAL THINKING

According to H. G. Wells, author of such science-fiction classics as *The War of the Worlds* and *The Time Machine,* "*Statistical thinking* will one day be as necessary for efficient citizenship as the ability to read and write." Written more than a hundred years ago, Wells' prediction is proving true today.

The growth in data collection associated with scientific phenomena, business operations, and government activities (quality control, statistical auditing, forecasting, etc.) has been remarkable in the past several decades. Every day the media present us with published results of political, economic, and social surveys. In increasing government emphasis on drug and product testing, for example, we see vivid evidence of the need to be able to evaluate data sets intelligently. Consequently, each of us has to develop a discerning sense—an ability to use rational thought to interpret the meaning of data. This ability can help you make intelligent decisions, inferences, and generalizations; that is, it helps you *think critically* using statistics.

> **DEFINITION 1.12**
> **Statistical thinking** involves applying rational thought to assess data and the inferences made from them critically.

To gain some insight into the role statistics plays in critical thinking, let's look at a *New York Times* (June 17, 1995) article titled "The Case for No Helmets." The article, written by the editor of a magazine for Harley-Davidson bikers, lists several reasons why motorcyclists should not be required by law to wear helmets. First, he argued that helmets may actually kill, since in collisions at speeds greater than 15 miles an hour, the heavy helmet may protect the head but snap the spine. Second, the editor cited a "study" that claimed "nine states without helmet laws had a lower fatality rate (3.05 deaths per 10,000 motorcycles) than those that mandated helmets (3.38)." Finally, he reported that "in a survey of 2,500 [at a rally], 98% of the respondents opposed such laws."

You can use "statistical thinking" to help you critically evaluate the arguments presented in this article. For example, before you accept the 98% estimate, you would want to know how the data were collected for the study cited by the editor of the biker magazine. If a survey was conducted, it's possible that the 2,500 bikers in the sample were not selected at random from the target population of all bikers, but rather were "self-selected." (Remember, they were all attending a rally—maybe even a rally for bikers who oppose the law.) If the respondents were likely to have strong opinions regarding the helmet law (e.g., strongly oppose the law), the resulting estimate is probably biased high.

You'd also want more information about the study comparing the motorcycle fatality rate of the nine states without a helmet law to those states that mandate helmets: Were the data obtained from a published source? Were all 50 states included in the study? That is, are you seeing sample data or population data? Furthermore, do the helmet laws vary among states? If so, can you really compare the fatality rates?

These questions actually led to the discovery of two scientific and statistically sound studies on helmets. The first, a UCLA study of nonfatal injuries, disputed the charge that helmets shift injuries to the spine. The second study reported a dramatic *decline* in motorcycle crash deaths after California passed its helmet law.

In the remaining chapters of the text, you'll become familiar with the tools essential for building a firm foundation in statistics and statistical thinking.

## QUICK REVIEW

### Key Terms

| | | |
|---|---|---|
| Census  5 | Nonresponse bias  12 | Representative sample  11 |
| Data  2 | Observational study  11 | Sample  5 |
| Descriptive statistics  2 | Population  4 | Statistical inference  5 |
| Designed experiment  11 | Published source  10 | Statistical thinking  13 |
| Inference  3 | Qualitative data  10 | Statistics  2 |
| Inferential statistics  3 | Quantitative data  9 | Survey  11 |
| Measure of reliability  8 | Random sample  12 | Variable  4 |
| Measurement  5 | Reliability  7 | |

## EXERCISES 1.1–1.27

### Learning the Mechanics

**1.1**  What is statistics?

**1.2**  Explain the difference between descriptive and inferential statistics.

**1.3**  List and define the five elements of an inferential statistical analysis.

**1.4**  List the four major methods of collecting data and explain their differences.

**1.5**  Explain the difference between quantitative and qualitative data.

**1.6**  Explain how populations and variables differ.

**1.7**  Explain how populations and samples differ.

**1.8**  What is a representative sample? What is its value?

**1.9**  Why would a statistician consider an inference incomplete without an accompanying measure of its reliability?

**1.10**  Define statistical thinking.

**1.11**  Suppose you're given a data set that classifies each sample unit into one of four categories: A, B, C, or D. You plan to create a computer database consisting of these data, and you decide to code the data as A = 1, B = 2, C = 3, and D = 4. Are the data consisting of the classifications A, B, C, and D qualitative or quantitative? After the data are input as 1, 2, 3, or 4, are they qualitative or quantitative? Explain your answers.

### Applying the Concepts—Basic

**1.12**  Colleges and universities are requiring an increasing amount of information about applicants before making acceptance and financial aid decisions. Classify each of the following types of data required on a college application as quantitative or qualitative.

  **a.** High school GPA

  **b.** High school class rank

  **c.** Applicant's score on the SAT or ACT

  **d.** Gender of applicant

  **e.** Parents' income

  **f.** Age of applicant

**1.13**  Classify the following examples of data as either qualitative or quantitative:

  **a.** The bacteria count in the water at each of 30 city swimming pools

  **b.** The occupation of each of 200 shoppers at a supermarket

  **c.** The marital status of each person living on a city block

  **d.** The time (in months) between auto maintenance for each of 100 used cars

Did you ever notice that, no matter where you stand on popular issues of the day, you can always find statistics or surveys to back up your point of view—whether to take vitamins, whether day care harms kids, or what foods can hurt you or save you? There is an endless flow of information to help you make decisions, but is this information accurate, unbiased? John Stossel decided to check that out, and you may be surprised to learn if the picture you're getting doesn't seem quite right, maybe it isn't.

Barbara Walters gave this introduction to a March 31, 1995, segment of the popular prime-time ABC television program, "20/20." The story is titled "Facts or Fiction?—Exposés of So-Called Surveys." One of the surveys investigated by ABC correspondent John Stossel compared the discipline problems experienced by teachers in the 1940s and those experienced today. The results: In the 1940s, teachers worried most about students talking in class, chewing gum, and running in the halls. Today, they worry most about being assaulted! This information was highly publicized in the print media—daily newspapers, weekly magazines, Ann Landers' column, *The Congressional Quarterly*, and *The Wall Street Journal*, among others—and referenced in speeches by a variety of public figures, including former first lady Barbara Bush and former Education Secretary William Bennett.

"Hearing this made me yearn for the old days when life was so much simpler and gentler, but was life that simple then?", asks Stossel. "Wasn't there juvenile delinquency [in the 1940s]? Is the survey true?" With the help of a Yale School of Management professor, Stossel found the original source of the teacher survey—Texas oilman T. Colin Davis—and discovered it wasn't a survey at all. Davis had simply identified certain disciplinary problems encountered by teachers and published his list in a conservative newsletter—a list he admitted was not obtained from a statistical survey, but from Davis' personal knowledge of the problems in the 1940s ("I was in school then") and his understanding of the problems today ("I read the papers").

Stossel's commonsense questions of the teacher "survey" is a clear example of the critical thinking process in action. Several more misleading surveys were presented on the ABC program. Listed here, most of these surveys were conducted by businesses or special interest groups with specific objectives in mind.

The "20/20" segment ended with an interview of Cynthia Crossen, author of *Tainted Truth*, an exposé of misleading and biased surveys. Crossen warns:

> If everybody is misusing numbers and scaring us with numbers to get us to do something, however good [that something] is, we've lost the power of numbers. Now, we know certain things from research. For example, we know that smoking cigarettes is hard on your lungs and heart, and because we know that, many people's lives have been extended or saved. We don't want to lose the power of information to help us make decisions, and that's what I worry about.

### Focus

a. Consider the false March of Dimes report on domestic violence and birth defects. Discuss the type of data required to investigate the impact of domestic violence on birth defects. What data collection method would you recommend?

b. Refer to the American Association of University Women (AAUW) study of self-esteem of high school girls. Explain why the results of the AAUW study are likely to be misleading. What data might be appropriate for assessing the self-esteem of high school girls?

c. Refer to the Food Research and Action Center study of hunger in America. Explain why the results of the study are likely to be misleading. What data would provide insight into the proportion of hungry American children?

| Reported Information (Source) | Actual Study Information |
| --- | --- |
| • Eating oat bran is a cheap and easy way to reduce your cholesterol count (Quaker Oats) | • Diet must consist of nothing but oat bran to achieve a slightly lower cholesterol count. |
| • 150,000 women a year die from anorexia. (Feminist group) | • Approximately 1,000 women a year die from problems that were likely caused by anorexia. |
| • Domestic violence causes more birth defects than all medical issues combined. (March of Dimes) | • No study—false report. |
| • Only 29% of high school girls are happy with themselves, compared to 66% of elementary school girls. (American Association of University Women) | • Of 3,000 high school girls, 29% responded to the statement, "I am happy the way I am." Most answered "Sort of true" and "Sometimes true." |
| • One in four American children under age 12 is hungry or at risk of hunger. (Food Research and Action Center) | • Based on responses to questions: "Do you ever cut the size of meals?", "Do you ever eat less than you feel you should?", "Did you ever rely on limited numbers of foods to feed your children because you were running out of money to buy food for a meal?" |

**1.14** A food-products company is considering marketing a new snack food. To see how consumers react to the product, the company conducted a taste test using a sample of 100 randomly selected shoppers at a suburban shopping mall. The shoppers were asked to taste the snack food and then fill out a short questionnaire that requested the following information:

1. What is your age?
2. Are you the person who typically does the food shopping for your household?
3. How many people are in your family?
4. How would you rate the taste of the snack food on a scale of 1 to 10, where 1 is least tasty?
5. Would you purchase this snack food if it were available on the market?
6. If you answered yes to part **5**, how often would you purchase the product?

a. Identify the data collection method.
b. Classify the data generated for each question as quantitative or qualitative. Justify your classifications.

**1.15** A poll is to be conducted in which 2,000 individuals are asked whether the 1999 presidential impeachment trial of Bill Clinton was conducted fairly. The 2,000 individuals are selected by random-digit telephone dialing and asked the question over the phone.

a. What is the relevant population?
b. What is the variable of interest? Is it quantitative or qualitative?
c. What is the sample?
d. What is the inference of interest to the pollster?
e. What method of data collection is employed?
f. How likely is the sample to be representative?

**1.16** Consider the set of all students enrolled in your statistics course this term. Suppose you're interested in learning about the current grade point averages (GPAs) of this group.

a. Define the population and variable of interest.
b. Is the variable qualitative or quantitative?
c. Suppose you determine the GPA of every member of the class. Would this represent a census or a sample?
d. Suppose you determine the GPA of 10 members of the class. Would this represent a census or a sample?
e. If you determine the GPA of every member of the class and then calculate the average, how much reliability does this have as an "estimate" of the class average GPA?
f. If you determine the GPA of 10 members of the class and then calculate the average, will the number you get necessarily be the same as the average GPA for the whole class? On what factors would you expect the reliability of the estimate to depend?
g. What must be true in order for the sample of 10 students you select from your class to be considered a random sample?

**1.17** The Cutter Consortium surveyed 154 U.S. companies to determine the extent of their involvement in electronic commerce (called *e-commerce*). Four of the questions they asked follow. (*Internet Week*, Sept. 6, 1999.)

1. Do you have an overall e-commerce strategy?
2. If you don't already have an e-commerce plan, when will you implement one?
3. Are you delivering products over the Internet?
4. What was your company's total revenue in the last fiscal year?

a. For each question, determine the variable of interest and classify it as quantitative or qualitative.
b. Do the data collected for the 154 companies represent a sample or a population? Explain.

## Applying the Concepts—Intermediate

**1.18** *The American Association of Nurse Anesthetists Journal* (Feb. 2000) published the results of a study on the use of herbal medicines before surgery. Of 500 surgical patients that were randomly selected for the study, 51% used herbal or alternative medicines (e.g., garlic, ginkgo, kava, fish oil) against their doctor's advice prior to surgery.

a. Do the 500 surgical patients represent a population or a sample? Explain.
b. If your answer was sample in part **a**, is the sample representative of the population? If you answered population in part **a**, explain how to obtain a representative sample from the population.
c. For each patient, what variable is measured? Is the data collected quantitative or qualitative?

**1.19** All highway bridges in the United States are inspected periodically for structural deficiency by the Federal Highway Administration (FHWA). Data from the FHWA inspections are compiled into the National Bridge Inventory (NBI). Several of the nearly 100 variables maintained by the NBI are listed below. Classify each variable as quantitative or qualitative.

a. Length of maximum span (feet)
b. Number of vehicle lanes
c. Toll bridge (yes or no)
d. Average daily traffic
e. Condition of deck (good, fair, or poor)
f. Bypass or detour length (miles)
g. Route type (interstate, U.S., state, county, or city)

**1.20** Refer to Exercise 1.19. The most recent NBI data were analyzed, and the results published in the *Journal of Infrastructure Systems* (June 1995). Using the FHWA inspection ratings, each of the 470,515 highway bridges in the United States was categorized as structurally deficient, functionally obsolete, or safe. About 26% of the bridges were found to be structurally deficient, while 19% were functionally obsolete.

a. What is the variable of interest to the researchers?
b. Is the variable of part **a** quantitative or qualitative?
c. Is the data set analyzed a population or a sample? Explain.
d. How did the researchers obtain the data for their study?

**1.21** Does a massage enable the muscles of tired athletes to recover from exertion faster than usual? To answer this question, researchers recruited eight amateur boxers to participate in an experiment (*British Journal of Sports Medicine*, April 2000). After a 10-minute workout in which each boxer threw 400 punches, half the boxers were given a 20-minute massage and half just rested for 20 minutes. Before returning to the ring for a second workout, the heart rate (beats per minute) and blood lactate level (micromoles) were recorded for each boxer. The researchers found no difference in the means of the two groups of boxers for either variable.
   **a.** Identify the data collection method used by the researchers.
   **b.** Identify the experimental units of the study.
   **c.** Identify the variables measured and their type (quantitative or qualitative).
   **d.** What is the inference drawn from the analysis?
   **e.** Comment on whether this inference can be made about all athletes.

**1.22** A Gallup Youth Poll was conducted to determine the topics that teenagers most want to discuss with their parents. The findings show that 46% would like more discussion about the family's financial situation, 37% would like to talk about school, and 30% would like to talk about religion. The survey was based on a national sampling of 505 teenagers, selected at random from all US. teenagers.
   **a.** Describe the sample.
   **b.** Describe the population from which the sample was selected.
   **c.** Is the sample representative of the population?
   **d.** What is the variable of interest?
   **e.** How is the inference expressed?
   **f.** Newspaper accounts of most polls usually give a *margin of error* (i.e., plus or minus 3%) for the survey result. What is the purpose of the margin of error and what is its interpretation?

**1.23** A group of University of South Florida surgery researchers believe that the drug aprotinin is effective in reducing the blood loss of burn patients who undergo skin replacement surgery (*USF Office of Research Annual Report, 1997–1998*). In a clinical trial of 14 burn patients, half were randomly assigned to receive the drug and half a placebo (no drug). A preliminary analysis of the patients' blood loss revealed that the group with the drug had a significant reduction of bleeding.
   **a.** Identify the data collection method used for this study.
   **b.** Does the study involve descriptive or inferential statistics? Explain.
   **c.** What is the population (or sample) of interest to the researchers?

**1.24** A first-year chemistry student conducts an experiment to determine the amount of hydrochloric acid necessary to neutralize 2 milliliters (ml) of a basic solution. The student prepares five 2-ml portions of the solution and adds a known concentration of hydrochloric acid to each. The amount of acid necessary to achieve neutrality of the solution is recorded for each of the five portions.
   **a.** Describe the population of interest to the student.
   **b.** What is the variable of interest? Is it quantitative or qualitative?
   **c.** Describe the sample.
   **d.** Describe the data collection method.

## Applying the Concepts—Advanced

**1.25** *USA Today* (Aug.14,1995) reported on a study that suggests "frequently 'heading' the ball in soccer lowers players' IQs." A psychologist tested 60 male soccer players, ages 14–29, who played up to five times a week. Players who averaged 10 or more headers a game had an average IQ of 103, while players who headed one or fewer times per game had an average IQ of 112.
   **a.** Describe the population of interest to the psychologist.
   **b.** Identify the variables of interest.
   **c.** Identify the type (qualitative or quantitative) of the variables, part **b**.
   **d.** Describe the sample.
   **e.** What is the inference made by the psychologist?
   **f.** Discuss possible reasons why the inference, part **e**, may be misleading.

**1.26** Are men or women more adept at remembering where they leave misplaced items (like car keys)? According to University of Florida psychology professor Robin West, women show greater competence in actually finding these objects (*Explore*, Fall 1998). Approximately 300 men and women from Gainesville, Florida, participated in a study in which each person placed 20 common objects in a 12-room "virtual" house represented on a computer screen. Thirty minutes later, the subjects were asked to recall where they put each of the objects. For each object, a recall variable was measured as "yes" or "no."
   **a.** Identify the population of interest to the psychology professor.
   **b.** Identify the sample.
   **c.** Does the study involve descriptive or inferential statistics? Explain.
   **d.** Are the variables measured in the study quantitative or qualitative?

**1.27** The Holocaust—the Nazi Germany extermination of the Jews during World War II—has been well documented through film, books, and interviews with the concentration camp survivors. But a recent Roper poll found that one in five U.S. residents doubted the Holocaust really occurred. A national sample of more than 1,000 adults and high school students were asked, "Does it seem possible or does it seem impossible to you that the Nazi extermination of the Jews never happened?" Twenty-two percent of the adults and 20% of the high school students responded "Yes." (*New York Times*, May 16, 1994).

**a.** Describe the population of interest to the pollsters.

**b.** What is the variable of interest? Is it quantitative or qualitative?

**c.** Describe the sample.

**d.** Identify the data collection method employed.

**e.** What inference was made by the pollsters?

**f.** Comment on the reliability of the inference. [*Hint:* Examine, closely, the survey question.]

 **STUDENT PROJECTS**

Scan your daily newspaper or weekly news magazine or search the Internet for articles that contain numerical data. The data might be a summary of the results of a public opinion poll, the results of a vote by the U.S. Senate, or a list of crime rates, birth or death rates, etc. For each article you find, answer the following questions:

**a.** Do the data constitute a sample or an entire population? If a sample has been taken, clearly identify both the sample and the population; otherwise, identify the population.

**b.** What type of data (quantitative or qualitative) has been collected?

**c.** What is the data source?

**d.** If a sample has been observed, is it likely to be representative of the population?

**e.** If a sample has been observed, does the article present an explicit (or implied) inference about the population of interest? If so, state the inference made in the article.

**f.** If an inference has been made, has a measure of reliability been included? What is it?

**g.** Use your answers to questions d–f to critically evaluate the article.

**REFERENCES**

*Careers in Statistics.* American Statistical Association, Biometric Society, Institute of Mathematical Statistics and Statistical Society of Canada, 1995.

Cochran, W.G. *Sampling Techniques*, 3rd ed. New York: Wiley, 1977.

Deming, W. E. *Sample Design in Business Research.* New York: Wiley, 1963.

*Ethical Guidelines for Statistical Practice.* American Statistical Association, 1995.

Hansen, M. H., Hurwitz, W. N., and Madow, W. G. *Sample Survey Methods and Theory*, Vol. 1. New York: Wiley & Sons, Inc., 1953.

Kirk, R. E. ed. *Statistical Issues: A Reader for the Behavioral Sciences.* Monterey, Ca.: Brooks/Cole, 1972.

Kish, L. *Survey Sampling.* New York: Wiley & Sons, Inc., 1965.

Scheaffer, R., Mendenhall, W., and Ott, R. L. *Elementary Survey Sampling*, 2nd ed. Boston: Duxbury, 1979.

Tanur, J. M., Mosteller, F, Kruskal, W H., Link, R. E, Pieters, R. S., and Rising, G. R. *Statistics: A Guide to the Unknown* (E. L. Lehmann, special editor). San Francisco: Holden-Day, 1989.

*What Is a Survey?* Section on Survey Research Methods, American Statistical Association, 1995.

Yamane, T. *Elementary Sampling Theory,* 3rd ed. Englewood Cliffs, N.J.: Prentice-Hall, 1967.

# CHAPTER 2

# Methods for Describing Sets of Data

## Contents

## Statistics in Action

The "Eye Cue" Test: Does Experience Improve Performance?

### ✊ Where We've Been

In Chapter 1 we looked at some typical examples of the use of statistics and we discussed the role that statistical thinking plays in supporting decision making. We examined the difference between descriptive and inferential statistics and described the five elements of inferential statistics: a population, one or more variables, a sample, an inference, and a measure of reliability for the inference. We also learned that data can be one of two types—quantitative or qualitative.

### ☞ Where We're Going

Before we make an inference, we must be able to describe a data set. We can do this by using graphical and/or numerical methods, which we discuss in this chapter. As we'll see in Chapter 7, we use some sample numerical descriptive measures to estimate the values of corresponding population descriptive measures. Therefore, our efforts in this chapter will ultimately lead us to statistical inference.

Suppose you wish to evaluate the mathematical capabilities of a class of 1,000 first-year college students based on their quantitative Scholastic Aptitude Test (SAT) scores. How would you describe these 1,000 measurements? Characteristics of interest include the typical or most frequent SAT score, the variability in the scores, the highest and lowest scores, the "shape" of the data, and whether or not the data set contains any unusual scores. Extracting this information isn't easy. The 1,000 scores provide too many bits of information for our minds to comprehend. Clearly we need some method for summarizing and characterizing the information in such a data set. Methods for describing data sets are also essential for statistical inference. Most populations are large data sets. Consequently, we need methods for describing a data set that let us make descriptive statements (inferences) about a population based on information contained in a sample.

Two methods for describing data are presented in this chapter, one *graphical* and the other *numerical*. Both play an important role in statistics. Section 2.1 presents both graphical and numerical methods for describing qualitative data. Graphical methods for describing quantitative data are illustrated in Sections 2.2, 2.8, and 2.9; numerical descriptive methods for quantitative data are presented in Sections 2.3–2.7. We end this chapter with a section on the *misuse* of descriptive techniques.

## 2.1    DESCRIBING QUALITATIVE DATA

Consider a study of aphasia published in the *Journal of Communication Disorders* (Mar. 1995). Aphasia is the "impairment or loss of the faculty of using or understanding spoken or written language." Three types of aphasia have been identified by researchers: Broca's, conduction, and anomic. They wanted to determine whether one type of aphasia occurs more often than any other, and, if so, how often. Consequently, they measured aphasia type for a sample of 22 adult aphasiacs. Table 2.1 (p. 21) gives the type of aphasia diagnosed for each aphasiac in the sample.

For this study, the variable of interest, aphasia type, is qualitative in nature. Qualitative data are nonnumerical in nature; thus, the value of a qualitative variable can only be classified into categories called *classes*. The possible aphasia types—Broca's, conduction, and anomic—represent the classes for this qualitative variable. We can summarize such data numerically in two ways: (1) by computing the *class frequency*—the number of observations in the data set that fall into each class; or (2) by computing the *class relative frequency*—the proportion of the total number of observations falling into each class.

**DEFINITION 2.1**
A **class** is one of the categories into which qualitative data can be classified.

**DEFINITION 2.2**
The **class frequency** is the number of observations in the data set falling in a particular class.

**DEFINITION 2.3**
The **class relative frequency** is the class frequency divided by the total number of observations in the data set, i.e.,

$$\text{class relative frequency} = \frac{\text{class frequency}}{n}$$

⊘ APHASIA
**TABLE 2.1  Data on 22 Adult Aphasiacs**

| Subject | Type of Aphasia |
|---------|-----------------|
| 1 | Broca's |
| 2 | Anomic |
| 3 | Anomic |
| 4 | Conduction |
| 5 | Broca's |
| 6 | Conduction |
| 7 | Conduction |
| 8 | Anomic |
| 9 | Conduction |
| 10 | Anomic |
| 11 | Conduction |
| 12 | Broca's |
| 13 | Anomic |
| 14 | Broca's |
| 15 | Anomic |
| 16 | Anomic |
| 17 | Anomic |
| 18 | Conduction |
| 19 | Broca's |
| 20 | Anomic |
| 21 | Conduction |
| 22 | Anomic |

*Source:* Li, E. C., Williams, S. E., and Volpe, R. D., "The effects of topic and listener familiarity of discourse variables in procedural and narrative discourse tasks." *Journal of Communication Disorders*, Vol. 28, No. 1, Mar. 1995, p. 44 (Table 1).

Examining Table 2.1, we observe that 5 aphasiacs in the study were diagnosed as suffering from Broca's aphasia, 7 from conduction aphasia, and 10 from anomic aphasia. These numbers—5, 7, and 10—represent the class frequencies for the three classes and are shown in the summary table, Table 2.2.

Table 2.2 also gives the relative frequency of each of the three aphasia classes. From Definition 2.3, we know that we calculate the relative frequency by dividing the class frequency by the total number of observations in the data set. Thus, the relative frequencies for the three types of aphasia are

$$\text{Broca's:} \quad \frac{5}{22} = .227$$

$$\text{Conduction:} \quad \frac{7}{22} = .318$$

$$\text{Anomic:} \quad \frac{10}{22} = .455$$

**TABLE 2.2  Summary Table for Data on 22 Adult Aphasiacs**

| Class | Frequency | Relative Frequency |
|-------|-----------|--------------------|
| **Type of Aphasia** | **Number of Subjects** | **Proportion** |
| Broca's | 5 | .227 |
| Conduction | 7 | .318 |
| Anomic | 10 | .455 |
| **Totals** | **22** | **1.000** |

From these relative frequencies we observe that nearly half (45.5%) of the 22 subjects in the study are suffering from anomic aphasia.

Although the summary table of Table 2.2 adequately describes the data of Table 2.1, we often want a graphical presentation as well. Figures 2.1 and 2.2 show two of the most widely used graphical methods for describing qualitative data—bar graphs and pie charts. Figure 2.1 shows the frequencies of aphasia types in a **bar graph** produced with SAS. Note that the height of the rectangle, or "bar," over each class is equal to the class frequency. (Optionally, the bar heights can be proportional to class relative frequencies.)

**FIGURE 2.1**

*SAS Bar Graph for Data on 22 Aphasiacs*

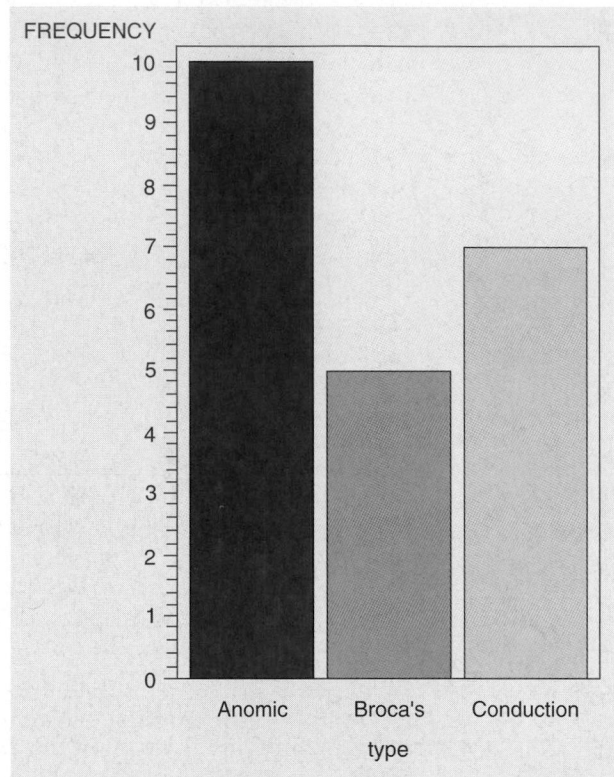

**FIGURE 2.2**

*SPSS Pie Chart for Data on 22 Aphasiacs*

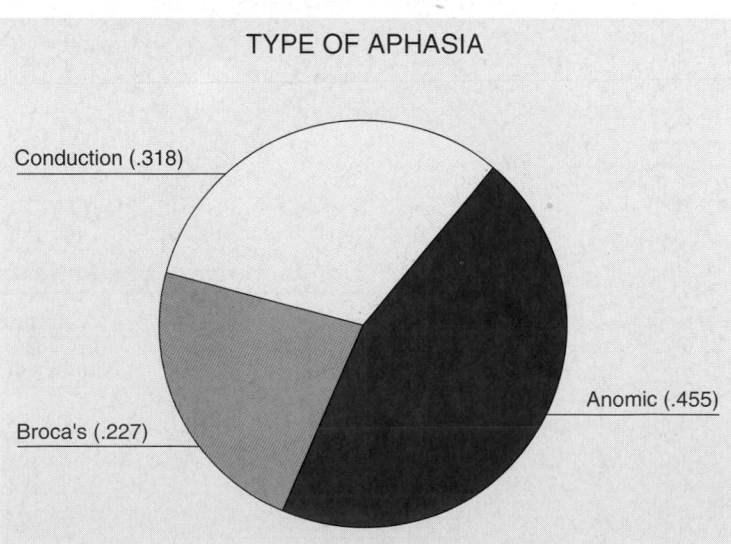

In contrast, Figure 2.2 (p. 22) shows the relative frequencies of the three types of aphasia in a **pie chart** generated with SPSS. Note that the pie is a circle (spanning 360°) and the size (angle) of the "pie slice" assigned to each class is proportional to the class relative frequency. For example, the slice assigned to anomic aphasia is 45.5% of 360°, or ( 455)(360°) = 163.8°.

Let's look at a practical example that requires interpretation of the graphical results.

**EXAMPLE 2.1**

BLOODLOSS

A group of cardiac physicians in southwest Florida have been studying a new drug designed to reduce blood loss in coronary bypass operations. Blood loss data for 114 coronary bypass patients (some who received a dosage of the drug and others who did not) are saved in the **BLOODLOSS** file. Although the drug shows promise in reducing blood loss, the physicians are concerned about possible side effects and complications. So their data set includes not only the qualitative variable DRUG, which indicates whether or not the patient received the drug, but also the qualitative variable COMP, which specifies the type (if any) of complication experienced by the patient. The four values of COMP recorded in Appendix B are (1) redo surgery, (2) post-op infection, (3) both, or (4) none.

a. Figure 2.3, generated using SAS, shows summary tables for the two qualitative variables, DRUG and COMP. Interpret the results.
b. Interpret the SAS output shown in Figures 2.4 and 2.5.

**Solution**

a. The top table in Figure 2.3 is a summary frequency table for DRUG. Note that exactly half (57) of the 114 coronary bypass patients received the drug and half did not. The bottom table in Figure 2.3 is a summary frequency table for COMP. The class relative frequencies are given in the Percent column. We see that almost 75% of the 114 patients had no complications, leaving about 25% who experienced either a redo surgery, a post-op infection, or both.
b. Figure 2.4 displays bar graphs for the data. The four bars in the top graph represent the frequencies of COMP for the 57 patients who did not receive the drug; the four bars in the bottom graph represent the frequencies of COMP for the 57 patients who did receive a dosage of the drug. The graph clearly shows that patients who did not receive the drug suffered fewer complications. The exact percentages are displayed in the SAS summary tables of Figure 2.5 (p. 25). Over 82% of the patients who got the drug had no complications, compared to about 68% for the patients who got no drug.

**FIGURE 2.3**

*SAS Summary Tables for DRUG and COMP*

The FREQ Procedure

| DRUG | Frequency | Percent | Cumulative Frequency | Cumulative Percent |
|------|-----------|---------|----------------------|--------------------|
| NO | 57 | 50.00 | 57 | 50.00 |
| YES | 57 | 50.00 | 114 | 100.00 |

| COMP | Frequency | Percent | Cumulative Frequency | Cumulative Percent |
|------|-----------|---------|----------------------|--------------------|
| BOTH | 4 | 3.51 | 4 | 3.51 |
| INFECT | 12 | 10.53 | 16 | 14.04 |
| NONE | 86 | 75.44 | 102 | 89.47 |
| REDO | 12 | 10.53 | 114 | 100.00 |

**FIGURE 2.4**

*SAS Bar Graphs for
COMP by Value of DRUG*

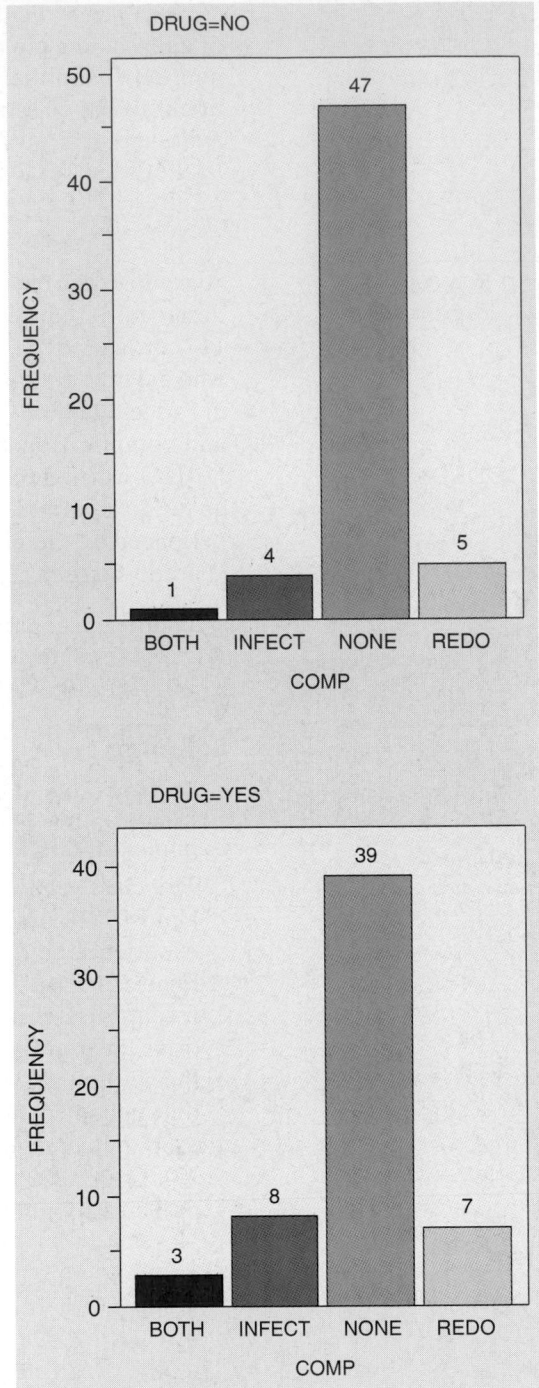

Although these results show that the drug may be effective in reducing blood loss, they also imply that patients on the drug may have a higher risk of complications. But before using this information to make a decision about the drug, the physicians will need to provide a measure of reliability for the inference. That is, the physicians will want to know whether the difference between the percentages of patients with complications observed in this sample of 114 patients is generalizable to the population of all coronary bypass patients. ■

**FIGURE 2.5**

*SAS Summary Tables for COMP by Value of DRUG*

-------------------------------------------------- DRUG=NO --------------------------------------------------

| COMP | Frequency | Percent | Cumulative Frequency | Cumulative Percent |
|------|-----------|---------|----------------------|--------------------|
| BOTH | 1 | 1.75 | 1 | 1.75 |
| INFECT | 4 | 7.02 | 5 | 8.77 |
| NONE | 47 | 82.46 | 52 | 91.23 |
| REDO | 5 | 8.77 | 57 | 100.00 |

-------------------------------------------------- DRUG=YES --------------------------------------------------

| COMP | Frequency | Percent | Cumulative Frequency | Cumulative Percent |
|------|-----------|---------|----------------------|--------------------|
| BOTH | 3 | 5.26 | 3 | 5.26 |
| INFECT | 8 | 14.04 | 11 | 19.30 |
| NONE | 39 | 68.42 | 50 | 87.72 |
| REDO | 7 | 12.28 | 57 | 100.00 |

 EXERCISES 2.1–2.15

## Learning the Mechanics

**2.1** Complete the following table.

| Grade on Statistics Exam | Frequency | Relative Frequency |
|--------------------------|-----------|--------------------|
| A: 90–100 | | .08 |
| B: 80–89 | 36 | |
| C: 65–79 | 90 | |
| D: 50–64 | 30 | |
| F: Below 50 | 28 | |
| **Total** | 200 | 1.00 |

**2.2** A qualitative variable with three classes (X, Y, and Z) is measured for each of 20 units randomly sampled from a target population. The data (observed class for each unit) are listed below.

Y  X  X  Z  X  Y  Y  Y  Y  X  X  Z  X
Y  Y  X  Z  Y  Y  Y  X

**a.** Compute the frequency for each of the three classes.
**b.** Compute the relative frequency for each of the three classes.
**c.** Display the results, part **a**, in a frequency bar graph.
**d.** Display the results, part **b**, in a pie chart.

## Applying the Concepts—Basic

**2.3** The International Rhino Federation estimates that there are 13,585 rhinoceroses living in the wild in Africa and Asia. A breakdown of the number of rhinos of each species is reported in the accompanying table.

| Rhino Species | Population Estimate |
|---------------|---------------------|
| African Black | 2,600 |
| African White | 8,465 |
| (Asian) Sumatran | 400 |
| (Asian) Javan | 70 |
| (Asian) Indian | 2,050 |
| **Total** | 13,585 |

*Source:* International Rhino Federation, July 1998.

**a.** Construct a relative frequency table for the data.
**b.** Display the relative frequencies in a bar graph.
**c.** What proportion of the 13,585 rhinos are African rhinos? Asian?

**2.4** *USA Today* (Dec. 31, 1999) reported on a Gallup Survey which asked: "Do you think cigarette smoking is one of the causes of lung cancer?" The accompanying table shows the results of the 1999 survey compared to the results of a similar survey conducted 45 years earlier.

| Is Smoking a Cause of Lung Cancer? | 1954 Survey | 1999 Survey |
|-----------------------------------|-------------|-------------|
| Yes | 41% | 92% |
| No | 31% | 6% |
| No opinion | 28% | 2% |

**a.** Construct a pie chart for the 1954 survey results.
**b.** Construct a pie chart for the 1999 survey results.

**c.** Compare the two graphs. Do the graphs show a dramatic shift in people's opinion on whether smoking causes lung cancer?

**2.5** Driver-side and passenger-side air bags are installed in all new cars to prevent serious or fatal injury in an automobile crash. However, air bags have been found to cause deaths in children and small people or people with handicaps in low-speed crashes. Consequently, in 1998 the federal government began allowing vehicle owners to request installation of an on-off switch for air bags. The table describes the reasons for requesting the installation of passenger-side on-off switches given by car owners in 1998 and 1999.

| Reason | Number of Requests |
|---|---|
| Infant | 1,852 |
| Child | 17,148 |
| Medical | 8,377 |
| Infant & Medical | 44 |
| Child & Medical | 903 |
| Infant & Child | 1,878 |
| Infant & Child & Medical | 135 |
| **Total** | **30,337** |

*Source:* National Highway Transportation Safety Administration, Sept. 2000.

**a.** What type of variable, quantitative or qualitative, is summarized in the table? Give the values that the variable could assume.

**b.** Calculate the relative frequencies for each reason.

**c.** Display the information in the table in an appropriate graph.

**d.** What proportion of the car owners who requested on-off air bag switches gave Medical as one of the reasons?

**2.6** In *Psychology and Aging* (Dec. 2000), University of Michigan School of Public Health researchers studied the roles that elderly people feel are the most important to them in late life. The accompanying table summarizes the most salient roles identified by each in a national sample of 1,102 adults, 65 years or older.

| Most Salient Role | Number |
|---|---|
| Spouse | 424 |
| Parent | 269 |
| Grandparent | 148 |
| Other relative | 59 |
| Friend | 73 |
| Homemaker | 59 |
| Provider | 34 |
| Volunteer, club, church member | 36 |
| **Total** | **1,102** |

*Source:* Krause, N., and Shaw, B. A. "Role-specific feelings of control and mortality," *Psychology and Aging*, Vol. 15, No. 4, Dec. 2000 (Table 2).

**a.** Describe the qualitative variable summarized in the table. Give the categories associated with the variable.

**b.** Are the numbers in the table frequencies or relative frequencies?

**c.** Display the information in the table in a bar graph.

**d.** Which role is identified by the highest percentage of elderly adults? Interpret the relative frequency associated with this role.

**2.7** The Moffitt Cancer Center at the University of South Florida treats over 25,000 patients a year. The graphic (reproduced from Figure 1.1) describes the types of cancer treated in Moffitt's patients during fiscal year 2000.

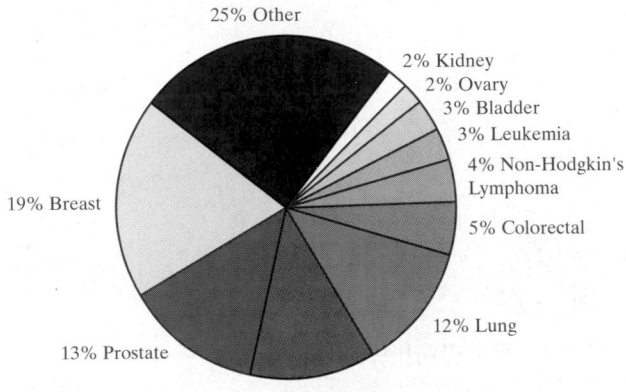

**a.** What type of graph is portrayed?

**b.** Which type of cancer is treated most often at Moffitt?

**c.** What percentage of Moffitt's patients are treated for melanoma, lymphoma, or leukemia?

## Applying the Concepts—Intermediate

**2.8** Archaeologists excavating the ancient Greek settlement at Phylakopi classified the pottery found in trenches (*Chance*, Fall 2000). The accompanying table describes the collection of 837 pottery pieces uncovered in a particular layer at the excavation site. Construct and interpret a graph that will aid the archaeologists in understanding the distribution of the pottery types found at the site.

| Pot Category | Number Found |
|---|---|
| Burnished | 133 |
| Monochrome | 460 |
| Slipped | 55 |
| Painted in curvilinear decoration | 14 |
| Painted in geometric decoration | 165 |
| Painted in naturalistic decoration | 4 |
| Cycladic white clay | 4 |
| Conical cup clay | 2 |
| **Total** | **837** |

*Source:* Berg, I., and Bliedon, S. "The Pots of Phylakopi: Applying Statistical Techniques to Archaeology," Vol. 13, No. 4, Fall 2000.

**2.9** *Forbes* magazine periodically conducts a salary survey of chief executive officers. In addition to salary information, *Forbes* collects and reports personal data on the CEOs, including level of education. Do most CEOs have advanced degrees, such as masters degrees or doctorates? The data in the table represent the highest degree obtained for each of the top 25 best-paid CEOs of 1999. Use a graphical method to summarize the highest degree obtained for these CEOs. What is your opinion about whether most CEOs have advanced degrees?

⊘ CEODEGREES

| CEO | Company | Degree |
|---|---|---|
| 1. Michael D. Eisner | Walt Disney | Bachelors |
| 2. Mel Karmazin | CBS | Bachelors |
| 3. Stephen M. Case | America Online | Bachelors |
| 4. Stephen C. Hilbert | Conseco | none |
| 5. Craig R. Barrett | Intel | Doctorate |
| 6. Millard Drexler | GAP | Masters |
| 7. John F. Welch, Jr. | General Electric | Doctorate |
| 8. Thomas G. Stemberg | Staples | Masters |
| 9. Henry R. Silverman | Cendant | JD |
| 10. Reuben Mark | Colgate-Palmolive | Masters |
| 11. Philip J. Purcell | Morgan Stanley Dean Witter | Masters |
| 12. Scott G. McNealy | Sun Microsystems | Masters |
| 13. Margaret C. Whitman | eBay | Masters |
| 14. Louis V. Gerstner, Jr. | IBM | Masters |
| 15. John F. Gifford | Maxim Integrated Products | Bachelors |
| 16. Robert L. Waltrip | Service Corp. International | Bachelors |
| 17. M. Douglas Ivester | Coca-Cola | Bachelors |
| 18. Gordon M. Binder | Amgen | Masters |
| 19. Charles R. Schwab | Charles Schwab | Masters |
| 20. William R. Steere, Jr. | Pfizer | Bachelors |
| 21. Nolan D. Archibald | Black & Decker | Masters |
| 22. Charles A. Heimbold, Jr. | Bristol-Myers Squibb | LLB (law) |
| 23. William L. Larson | Network Associates | JD |
| 24. Maurice R. Greenberg | American International Group | LLB (law) |
| 25. Richard Jay Kogan | Schering-Plough | Masters |

*Source: Forbes*, May 17, 1999.

**2.10** Audiologists have recently developed a rehabilitation program for hearing-impaired patients in a Canadian home for senior citizens (*Journal of the Academy of Rehabilitative Audiology*, 1994). Each of the 30 residents of the home were diagnosed for degree and type of sensorineural hearing loss, coded as follows: 1 = hear within normal limits, 2 = high-frequency hearing loss, 3 = mild loss, 4 = mild-to-moderate loss, 5 = moderate loss, 6 = moderate-to-severe loss, and 7 = severe-to-profound loss. The data are listed in the accompanying table. Use a graph to portray the results. Which type of hearing loss appears to be the most prevalent among nursing home residents?

⊘ HEARLOSS

| | | | | | | | | | | | | |
|---|---|---|---|---|---|---|---|---|---|---|---|---|
| 6 | 7 | 1 | 1 | 2 | 6 | 4 | 6 | 4 | 2 | 5 | 2 | 5 |
| 1 | 5 | 4 | 6 | 6 | 5 | 5 | 5 | 2 | 5 | 3 | 6 | 4 |
| 6 | 6 | 4 | 2 | | | | | | | | | |

*Source:* Jennings, M. B., and Head, B. G. "Development of an ecological audiologic rehabilitation program in a home-for-the-aged." *Journal of the Academy of Rehabilitative Audiology*, Vol. 27, 1994, p. 77 (Table 1).

**2.11** *Choice* magazine provides new-book reviews each issue. A random sample of 375 *Choice* book reviews in American history, geography, and area studies was selected and the "overall opinion" of the book stated in each review was ascertained (*Library Acquisitions: Practice and Theory*, Vol. 19, 1995). Overall opinion was coded as follows: 1 = would not recommend, 2 = cautious or very little recommendation, 3 = little or no preference, 4 = favorable/recommended, 5 = outstanding/significant contribution. A summary of the data is provided in the accompanying bar graph.

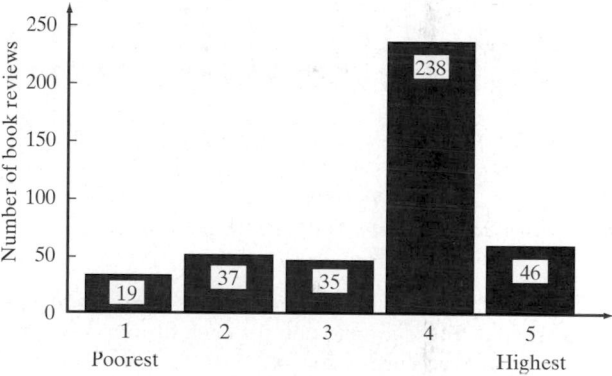

*Source:* Carlo, P.W., and Natowitz, A. "*Choice* book reviews in American history, geography, and area studies: An analysis for 1988–1993." *Library Acquisitions: Practice & Theory*, Vol. 19, No. 2, 1995, p. 159 (Figure 1).

**a.** Find the opinion that occurred most often. What proportion of the books reviewed had this opinion?
**b.** Do you agree with the following statement extracted from the study: "A majority (more than 75%) of books reviewed are evaluated favorably and recommended for purchase."?

## Applying the Concepts—Advanced

**2.12** Drowsy driving leads to thousands of automobile crashes each year. A report on driver fatigue and drowsiness, sponsored by the National Center on Sleep Disorders Research (NCSDR) and the National Highway Traffic Safety Administration (NHTSA), found that the time of day when the accident occurs depends on the age of the drowsy driver. Based on data collected from police reports for accidents attributed to drowsy driving or the driver falling asleep, bar graphs showing the distribution

a) 25 and under

b) 26–45 years

c) 46–65 years

d) 65 and over

*Source*: National Highway Traffic Safety Administration, 1996.

of the time of day in which the crash occurred were constructed. These graphs, one for each of four driver age groups (25 and under, 26 to 45, 46 to 65, and over 65), are displayed above. Compare the time-of-day distributions. What patterns do the graphs reveal?

**2.13** Marine scientists who study dolphin communication have discovered that bottlenose dolphins exhibit an individualized whistle contour known as the *signature whistle*. A study was conducted to categorize the signature whistles of ten captive adult bottlenose dolphins in socially interactive contexts (*Ethology*, July 1995). A total of 185 whistles were recorded during the study period; each whistle contour was analyzed and assigned to a category using a contour similarity (CS) technique. The results are reported in the accompanying table. Use a graphical method to summarize the results. Do you detect any patterns in the data that might be helpful to marine scientists?

 DOLPHIN

| Whistle Category | Number of Whistles |
| --- | --- |
| Type a | 97 |
| Type b | 15 |
| Type c | 9 |
| Type d | 7 |
| Type e | 7 |
| Type f | 2 |
| Type g | 2 |
| Type h | 2 |
| Type i | 2 |
| Type j | 4 |
| Type k | 13 |
| Other types | 25 |

*Source:* McCowan, B., and Reiss, D. "Quantitative comparison of whistle repertoires from captive adult bottlenose dolphins (Delphiniae, *Tursiops truncatus*): A re-evaluation of the signature whistle hypothesis." *Ethology*, Vol. 100, No. 3, July 1995, p. 200 (Table 2).

**2.14** According to *Benford's Law*, certain digits (1, 2, 3, . . . , 9) are more likely to occur as the first significant digit in a randomly selected number than other digits. For example, the law predicts that the number "1" is the most likely to occur (30% of the time) as the first digit. In a study reported in the *American Scientist* (July–Aug. 1998) to test Benford's Law, 743 first-year college students were asked to write down a six-digit number at random. The first significant digit of each number was recorded and its distribution summarized in the table on p. 29.

⊘ DIGITS

| First Digit | Number of Occurrences |
|---|---|
| 1 | 109 |
| 2 | 75 |
| 3 | 77 |
| 4 | 99 |
| 5 | 72 |
| 6 | 117 |
| 7 | 89 |
| 8 | 62 |
| 9 | 43 |
| **Total** | 743 |

*Source:* Hill, T. P "The First Digit Phenomenon." *American Scientist,* Vol. 86, No. 4, July–Aug. 1998, p. 363 (Figure 5).

**a.** Describe the first digit of the "random guess" data with an appropriate graph.

**b.** Does the graph support Benford's Law? Explain.

**2.15** Owing to several major ocean oil spills by tank vessels, Congress passed the 1990 Oil Pollution Act, which requires all tankers to be designed with thicker hulls. Further improvements in the structural design of a tank vessel have been proposed since then, each with the objective of reducing the likelihood of an oil spill and decreasing the amount of outflow in the event of a hull puncture. To aid in this development, *Marine Technology* (Jan. 1995) reported on the spillage amount and cause of puncture for 50 recent major oil spills from tankers and carriers. The data are reproduced in the table below.

**a.** Use a graphical method to describe the cause of oil spillage for the 50 tankers.

**b.** Does the graph, part **a**, suggest that any one cause is more likely to occur than any other? How is this information of value to the design engineers?

⊘ OILSPILL

| Tanker | Spillage (metric tons, thousands) | Cause |
|---|---|---|
| *Atlantic Empress* | 257 | Collision (C) |
| *Castillo De Bellver* | 239 | Fire/Explosion (FE) |
| *Amoco Cadiz* | 221 | Hull Failure (HF) |
| *Odyssey* | 132 | FE |
| *Torrey Canyon* | 124 | Grounding (G) |

| Tanker | Spillage | Cause |
|---|---|---|
| *Sea Star* | 123 | C |
| *Hawaiian Patriot* | 101 | HF |
| *Independenta* | 95 | C |
| *Urquiola* | 91 | G |
| *Irenes Serenade* | 82 | FE |
| *Khark 5* | 76 | FE |
| *Nova* | 68 | C |
| *Wafra* | 62 | G |
| *Epic Colocotronis* | 58 | G |
| *Sinclair Petrolore* | 57 | FE |
| *Yuyo Maru No 10* | 42 | C |
| *Assimi* | 50 | FE |
| *Andros Patria* | 48 | FE |
| *World Glory* | 46 | HF |
| *British Ambassador* | 46 | HF |
| *Metula* | 45 | G |
| *Pericles G. C.* | 44 | FE |
| *Mandoil II* | 41 | C |
| *Jakob Maersk* | 41 | G |
| *Burmah Agate* | 41 | C |
| *J. Antonio Lavalleja* | 38 | G |
| *Napier* | 37 | G |
| *Exxon Valdez* | 36 | G |
| *Corinthos* | 36 | C |
| *Trader* | 36 | HF |
| *St. Peter* | 33 | FE |
| *Gino* | 32 | C |
| *Golden Drake* | 32 | FE |
| *Lonnis Angelicoussis* | 32 | FE |
| *Chryssi* | 32 | HF |
| *Ionnis Challenge* | 31 | HF |
| *Argo Merchant* | 28 | G |
| *Heimvard* | 31 | C |
| *Pegasus* | 25 | Unknown (U) |
| *Pacocean* | 31 | HF |
| *Texaco Oklahoma* | 29 | HF |
| *Scorpio* | 31 | G |
| *Ellen Conway* | 31 | G |
| *Caribbean Sea* | 30 | HF |
| *Cretan Star* | 26 | U |
| *Grand Zenith* | 26 | HF |
| *Athenian Venture* | 26 | FE |
| *Venoil* | 26 | C |
| *Aragon* | 24 | HF |
| *Ocean Eagle* | 21 | G |

*Source:* Daidola, J. C. "Tanker structure behavior during collision and grounding." *Marine Technology*, Vol. 32, No. 1, Jan. 1995, p. 22 (Table 1).

## 2.2 GRAPHICAL METHODS FOR DESCRIBING QUANTITATIVE DATA

Recall from Section 1.4 that quantitative data sets consist of data that are recorded on a meaningful numerical scale. For describing, summarizing, and detecting patterns in such data, we can use three graphical methods: *dot plots, stem-and-leaf displays,* and *histograms*. Since most statistical software packages can be used to construct these displays, we'll focus here on their interpretation rather than their construction.

⊘ EPAGAS

**TABLE 2.3    EPA Mileage Ratings on 100 Cars**

| | | | | | | | | | |
|------|------|------|------|------|------|------|------|------|------|
| 36.3 | 41.0 | 36.9 | 37.1 | 44.9 | 36.8 | 30.0 | 37.2 | 42.1 | 36.7 |
| 32.7 | 37.3 | 41.2 | 36.6 | 32.9 | 36.5 | 33.2 | 37.4 | 37.5 | 33.6 |
| 40.5 | 36.5 | 37.6 | 33.9 | 40.2 | 36.4 | 37.7 | 37.7 | 40.0 | 34.2 |
| 36.2 | 37.9 | 36.0 | 37.9 | 35.9 | 38.2 | 38.3 | 35.7 | 35.6 | 35.1 |
| 38.5 | 39.0 | 35.5 | 34.8 | 38.6 | 39.4 | 35.3 | 34.4 | 38.8 | 39.7 |
| 36.3 | 36.8 | 32.5 | 36.4 | 40.5 | 36.6 | 36.1 | 38.2 | 38.4 | 39.3 |
| 41.0 | 31.8 | 37.3 | 33.1 | 37.0 | 37.6 | 37.0 | 38.7 | 39.0 | 35.8 |
| 37.0 | 37.2 | 40.7 | 37.4 | 37.1 | 37.8 | 35.9 | 35.6 | 36.7 | 34.5 |
| 37.1 | 40.3 | 36.7 | 37.0 | 33.9 | 40.1 | 38.0 | 35.2 | 34.8 | 39.5 |
| 39.9 | 36.9 | 32.9 | 33.8 | 39.8 | 34.0 | 36.8 | 35.0 | 38.1 | 36.9 |

For example, the Environmental Protection Agency (EPA) performs extensive tests on all new car models to determine their mileage ratings. Suppose that the 100 measurements in Table 2.3 represent the results of such tests on a certain new car model. How can we summarize the information in this rather large sample?

A visual inspection of the data indicates some obvious facts. For example, most of the mileages are in the 30s, with a smaller fraction in the 40s. But it is difficult to provide much additional information on the 100 mileage ratings without resorting to some method of summarizing the data. One such method is a dot plot.

## Dot Plots

A MINITAB **dot plot** for the 100 EPA mileage ratings is shown in Figure 2.6. The horizontal axis of Figure 2.6 is a scale for the quantitative variable in miles per gallon (mpg). The numerical value of each measurement in the data set is located on the horizontal scale by a dot. When data values repeat, the dots are placed above one another, forming a pile at that particular numerical location. As you can see, this dot plot verifies that almost all of the mileage ratings are in the 30s, with most falling between 35 and 40 miles per gallon.

**FIGURE 2.6**

*A MINITAB Dot Plot for the EPA Mileage Ratings on 100 Cars*

## Stem-and-Leaf Display

Another graphical representation of these same data, a MINITAB **stem-and-leaf display**, is shown in Figure 2.7. In this display the *stem* is the portion of the measurement (mpg) to the left of the decimal point, while the remaining portion to the right of the decimal point is the *leaf*.

In Figure 2.7, the stems for the data set are listed in the second column from the smallest (30) to the largest (44). Then the leaf for each observation is listed to the right in the row of the display corresponding to the observation's stem. For example, the leaf 3 of the first observation (36.3) in Table 2.3 appears in the row corresponding to the stem 36. Similarly, the leaf 7 for the second observation (32.7) in Table 2.3 appears in the row corresponding to the stem 32, while the leaf 5 for the third observation (40.5) appears in the row corresponding to the stem 40. (The stems and leaves for these first three observations are highlighted in Figure 2.7.)

**FIGURE 2.7**

*MINITAB Stem-and-Leaf Display for the EPA Mileage Ratings on 100 Cars*

```
Stem-and-leaf of MPG          N = 100
Leaf Unit = 0.10

    1    30   0
    2    31   8
    6    32   5799
   12    33   126899
   18    34   024588
   29    35   01235667899
   49    36   01233445566777888999
  (21)   37   000011122334456677899
   30    38   0122345678
   20    39   00345789
   12    40   0123557
    5    41   002
    2    42   1
    1    43
    1    44   9
```

Typically, the leaves in each row are ordered as shown in the MINITAB stem-and-leaf display.

The stem-and-leaf display presents another compact picture of the data set. You can see at a glance that the 100 mileage readings were distributed between 30.0 and 44.9, with most of them falling in stem rows 35 to 39. The six leaves in stem row 34 indicate that six of the 100 readings were at least 34.0 but less than 35.0. Similarly, the eleven leaves in stem row 35 indicate that eleven of the 100 readings were at least 35.0 but less than 36.0. Only five cars had readings equal to 41 or larger, and only one was as low as 30.

The definitions of the stem and leaf for a data set can be modified to alter the graphical description. For example, suppose we had defined the stem as the tens digit for the gas mileage data, rather than the ones and tens digits. With this definition, the stems and leaves corresponding to the measurements 36.3 and 32.7 would be as follows:

| Stem | Leaf | | Stem | Leaf |
|:----:|:----:|---|:----:|:----:|
| 3 | 6 | | 3 | 2 |

Note that the decimal portion of the numbers has been dropped. Generally, only one digit is displayed in the leaf.

If you look at the data, you'll see why we didn't define the stem this way. All the mileage measurements fall in the 30s and 40s, so all the leaves would fall into just two stem rows in this display. The resulting picture would not be nearly as informative as Figure 2.7.

## Histograms

A STATISTIX **histogram** for these 100 EPA mileage readings is shown in Figure 2.8. The horizontal axis of Figure 2.8, which gives the miles per gallon for a given automobile, is divided into *intervals* commencing with the interval from 30.0–31.5 and proceeding in intervals of equal size to 43.5–45.0 mpg. The vertical axis gives the number (or *frequency*) of the 100 readings that fall in each interval.

**FIGURE 2.8**

*STATISTIX Histogram for EPA Mileage Data*

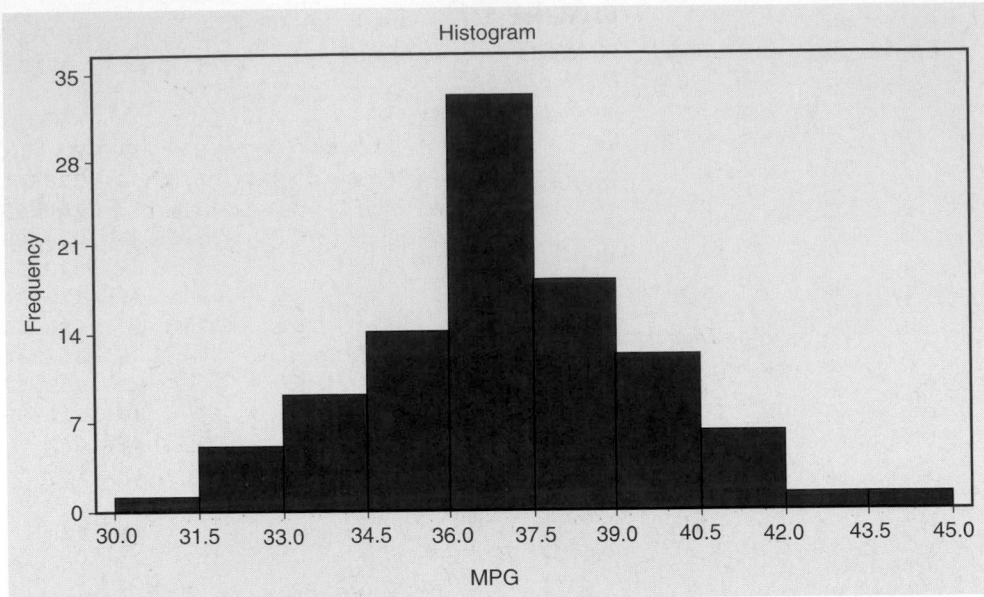

It appears that about 33 of the 100 cars, or 33%, obtained a mileage between 36.0 and 37.5. This interval contains the highest frequency, and the intervals tend to contain a smaller number of the measurements as the mileages get smaller or larger.

Histograms can be used to display either the frequency or relative frequency of the measurements falling into specified intervals known as **measurement classes**. The measurement classes, frequencies, and relative frequencies for the EPA car mileage data are shown in the summary table, Table 2.4.

**TABLE 2.4   Measurement Classes, Frequencies, and Relative Frequencies for the Car Mileage Data**

| Measurement Class | Frequency | Relative Frequency |
|---|---|---|
| 30.0–31.5 | 1 | .01 |
| 31.5–33.0 | 5 | .05 |
| 33.0–34.5 | 9 | .09 |
| 34.5–36.0 | 14 | .14 |
| 36.0–37.5 | 33 | .33 |
| 37.5–39.0 | 18 | .18 |
| 39.0–40.5 | 12 | .12 |
| 40.5–42.0 | 6 | .06 |
| 42.0–43.5 | 1 | .01 |
| 43.5–45.0 | 1 | .01 |
| **Totals** | 100 | 1.00 |

By summing the relative frequencies in the intervals 34.5–36.0, 36.0–37.5, and 37.5–39.0, you can see that 65% of the mileages are between 34.5 and 39.0. Similarly, only 2% of the cars obtained a mileage rating over 42.0. Many other summary statements can be made by further study of the histogram and accompanying summary table.

When interpreting a histogram (say, the histogram in Figure 2.8), consider two important facts. First, the proportion of the total area under the histogram that falls above a particular interval on the *x*-axis is equal to the relative frequency of measurements falling in the interval. For example, the relative frequency for the class interval 36.0–37.5 is .33. Consequently, the rectangle above the interval contains .33 of the total area under the histogram.

Second, imagine the appearance of the relative frequency histogram for a very large set of data (say, a population). As the number of measurements in a data set is increased, you can obtain a better description of the data by decreasing the width of the class intervals. When the class intervals become small enough, a relative frequency histogram will (for all practical purposes) appear as a smooth curve (see Figure 2.9).

**FIGURE 2.9**

*The Effect of the Size of a Data Set on the Outline of a Histogram*

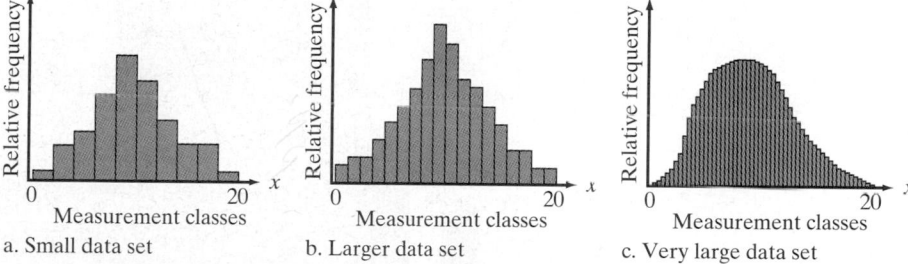

a. Small data set    b. Larger data set    c. Very large data set

While histograms provide good visual descriptions of data sets—particularly very large ones—they do not let us identify individual measurements. In contrast, each of the original measurements is visible to some extent in a dot plot and clearly visible in a stem-and-leaf display. The stem-and-leaf display arranges the data in ascending order, so it's easy to locate the individual measurements. For example, in Figure 2.7 we can easily see that two of the gas mileage measurements are equal to 36.3, but can't see that fact by inspecting the histogram in Figure 2.8. However, stem-and-leaf displays can become unwieldy for very large data sets. A very large number of stems and leaves causes the vertical and horizontal dimensions of the display to become cumbersome, diminishing the usefulness of the visual display.

**EXAMPLE 2.2**

The data in Table 2.5 give, by state, the percentages of the total number of college or university student loans that are in default.
a. Use a statistical software package to create a histogram for these data.
b. Use a statistical software package to create a stem-and-leaf display for these data.
c. Compare and interpret the two graphic displays of these data.

⦿ LOANDEFAULT
**TABLE 2.5    Percentage of Student Loans (per State) in Default**

| State | % | State | % | State | % | State | % |
|-------|-----|-------|-----|--------|-----|---------|------|
| Ala. | 12.0 | Ill. | 9.3 | Mont. | 6.4 | R.I. | 8.8 |
| Alaska | 19.7 | Ind. | 6.7 | Nebr. | 4.9 | S.C. | 14.1 |
| Ariz. | 12.1 | Iowa | 6.2 | Nev. | 10.1 | S. Dak. | 5.5 |
| Ark. | 12.9 | Kans. | 5.7 | N.H. | 7.9 | Tenn. | 12.3 |
| Calif. | 11.4 | Ky. | 10.3 | N.J. | 12.0 | Tex. | 15.2 |
| Colo. | 9.5 | La. | 13.5 | N. Mex. | 7.5 | Utah | 6.0 |
| Conn. | 8.8 | Maine | 9.7 | N.Y. | 11.3 | Vt. | 8.3 |
| Del. | 10.9 | Md. | 16.6 | N.C. | 15.5 | Va. | 14.4 |
| D.C. | 14.7 | Mass. | 8.3 | N. Dak. | 4.8 | Wash. | 8.4 |
| Fla. | 11.8 | Mich. | 11.4 | Ohio | 10.4 | W Va. | 9.5 |
| Ga. | 14.8 | Minn. | 6.6 | Okla. | 11.2 | Wis. | 9.0 |
| Hawaii | 12.8 | Miss. | 15.6 | Oreg. | 7.9 | Wyo. | 2.7 |
| Idaho | 7.1 | Mo. | 8.8 | Pa. | 8.7 | | |

*Source:* National Direct Student Loan Program.

**FIGURE 2.10**

*SPSS Frequency Histogram for Student Loan Default Rate*

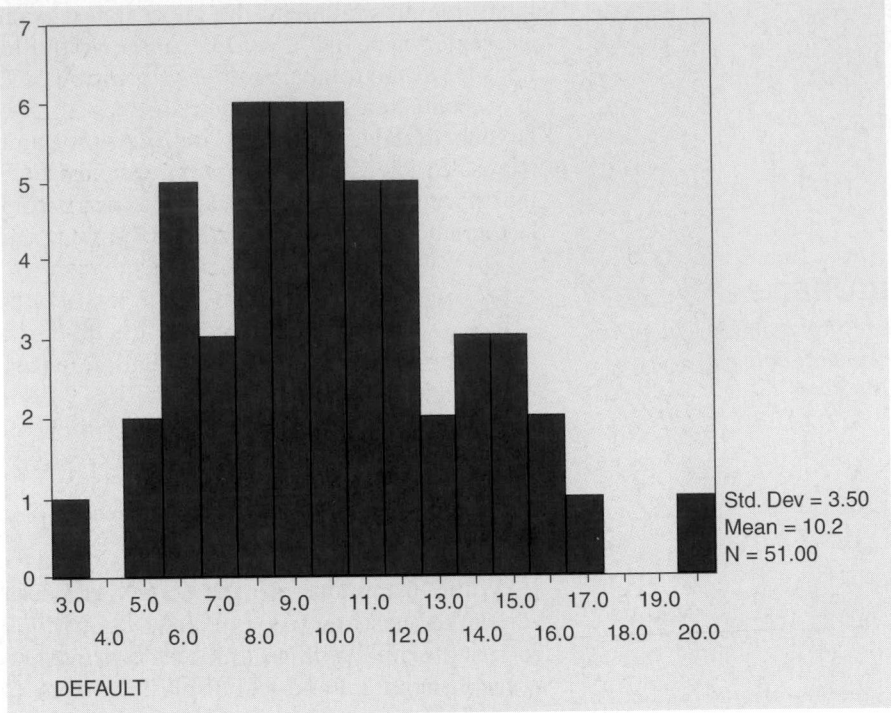

### Solution

a. We used SPSS to generate the frequency histogram in Figure 2.10. Note that 18 classes were formed by the SPSS program. The classes are identified by their *midpoints* rather than their endpoints. Thus, the first interval has a midpoint of 3, the second of 4, and so on. The corresponding measurement classes based on these midpoints are therefore 2.5–3.5, 3.5–4.5, etc. Note that the measurement classes with midpoints of 8, 9, and 10 (ranging from 7.5 to 10.5) contain 18 of the 51 default measurements.

b. We also used SPSS to generate the stem-and-leaf display in Figure 2.11. Note that the stem, which is the *second* column in the printout, has been defined as

**FIGURE 2.11**

*SPSS Stem-and-Leaf Display for Defaulted Student Loan Data*

```
DEFAULT Stem-and-Leaf Plot

 Frequency      Stem  &  Leaf

     1.00         0  .  2
     4.00         0  .  4455
     9.00         0  .  666667777
    12.00         0  .  888888899999
     9.00         1  .  000011111
     7.00         1  .  2222223
     7.00         1  .  4444555
     1.00         1  .  6
     1.00    Extremes    (>=20)

 Stem width:     10.0
 Each leaf:         1 case(s)
```

the number *two* places to the left of the decimal. The leaf is the third column in the printout, and is the number one place to the left of the decimal. (The digit to the right of the decimal is not shown.) SPSS also indicates the frequency of measurements (first column of the printout) in each stem.

c. As is usually the case for data sets that are not too large (say, fewer than 100 measurements), the stem-and-leaf display provides more detail than the histogram without being unwieldy. For instance, the stem-and-leaf display in Figure 2.11 clearly indicates that most states' rates are relatively evenly distributed from 6 to 15 (note that the five stem rows in the middle of the display have between 7 and 12 measurements "attached" to each stem). The stem-and-leaf display also clearly indicates the values of the individual measurements in the data set. For example, the lowest default rate (representing the measurement 2.7) is shown in the first stem row.

The histogram in Figure 2.10 also indicates a relatively even distribution of rates over the ten measurement classes with midpoints from 6 to 15. However, it's harder to determine the boundaries since we can't "see" the individual measurements in the display. Similarly, the two small bars over the outside classes with midpoints of 3 and 20 indicate the presence of some extreme measurements, but we can't see the exact number and location of these measurements. However, histograms are most useful for displaying very large data sets when the overall shape of the distribution of measurements is more important than the identification of individual measurements. _____ ■

Almost all statistical software packages can be used to generate histograms, stem-and-leaf displays, and dot plots. All three are useful tools for graphically describing data sets. We recommend that you generate and compare the displays whenever you can. You'll find that histograms are generally more useful for very large data sets, while stem-and-leaf displays and dot plots provide useful detail for smaller data sets.

# Using the TI-83 Graphing Calculator

## Making a Histogram on the TI-83

Step 1    *Enter the data*
Press **STAT  1**  for  **STAT edit**
Use the arrow and **ENTER** keys to enter the data set into **L1**.

Step 2    *Set up the histogram plot*
Press **2nd  Y=**    for  **STAT PLOT**
Press **1**        for  **Plot 1**

Use the arrow and **ENTER** keys to set up the screen as shown below.

## USING THE TI-83 GRAPHING CALCULATOR (*continued*)

Step 3    *Select your window settings*
Press WINDOW and adjust the settings as follows:

**Xmin** = lowest class boundary
**Xmax** = highest class boundary
**Xscl** = class width
**Ymin** = 0
**Ymax** = greatest class frequency
**Yscl** = approximately Ymax/10
**Xres** = 1

Step 4    *View the graph*
Press **GRAPH**

Optional    *Read class frequencies and class boundaries*

Step    You can press **TRACE** to read the class frequencies and class
boundaries.
Use the arrow keys to move between bars.

Example   The figures below show TI-83 window settings and histogram for
the following sample data:

86, 70, 62, 98, 73, 56, 53, 92, 86, 37, 62, 83, 78, 49, 78, 37, 67, 79, 57

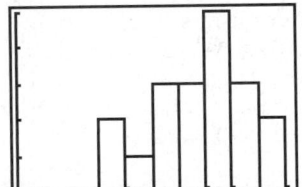

## Making a Histogram from a Frequency Table

Step 1    *Enter the data*
Enter a value from the middle of each class in **L1**
Enter the class frequencies or relative frequencies in **L2**

Step 2    *Set up the histogram plot*
Press   **2nd**   **Y=**   for   **STAT PLOT**
Press   **1**         for   **Plot1**

Use the arrow and **ENTER** keys to set up the screen as shown
below.

Steps 3-4   Follow steps 3-4 given above.

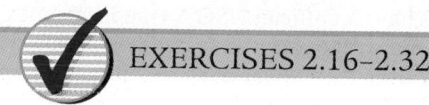 EXERCISES 2.16–2.32

## Learning the Mechanics

**2.16** Graph the relative frequency histogram for the 500 measurements summarized in the accompanying relative frequency table.

| Measurement Class | Relative Frequency |
|---|---|
| .5–2.5 | .10 |
| 2.5–4.5 | .15 |
| 4.5–6.5 | .25 |
| 6.5–8.5 | .20 |
| 8.5–10.5 | .05 |
| 10.5–12.5 | .10 |
| 12.5–14.5 | .10 |
| 14.5–16.5 | .05 |

**2.17** Refer to Exercise 2.16. Calculate the number of the 500 measurements falling into each of the measurement classes. Then graph a frequency histogram for these data.

**2.18** STATISTIX was used to generate the stem-and-leaf display shown below.
   **a.** How many observations were in the original data set?
   **b.** In the top row of the stem-and-leaf display, identify the stem, the leaves, and the numbers in the original data set represented by this stem and its leaves.
   **c.** Re-create all the numbers in the data set and construct a dot plot.

STATISTIX Output for Exercise 2.18

```
STEM AND LEAF PLOT OF X

LEAF DIGIT UNIT = 1        MINIMUM  0.0000
5  1   REPRESENTS 51.      MEDIAN   25.000
                          MAXIMUM  51.000
         STEM  LEAVES
      3    0   012
      3    0
      6    1   224
      7    1   8
     11    2   1134
     (3)   2   599
      9    3   0003
      5    3   6
      4    4   4
      3    4   57
      1    5   1

23 CASES INCLUDED     0 MISSING CASES
```

**2.19** MINITAB was used to generate the following histogram:

MINITAB Output for Exercise 2.19

   **a.** Is this a frequency histogram or a relative frequency histogram? Explain.
   **b.** How many measurement classes were used in the construction of this histogram?
   **c.** How many measurements are there in the data set described by this histogram?

## Applying the Concepts—Basic

**2.20** The graph below summarizes the scores obtained by 100 students on a questionnaire designed to measure aggressiveness. (Scores are integer values that range from 0 to 20. A high score indicates a high level of aggression.)
   **a.** Which measurement class contains the highest proportion of test scores?
   **b.** What is the proportion of scores that lie between 3.5 and 5.5?
   **c.** What is the proportion of scores that are higher than 11.5?
   **d.** How many students scored less than 5.5?

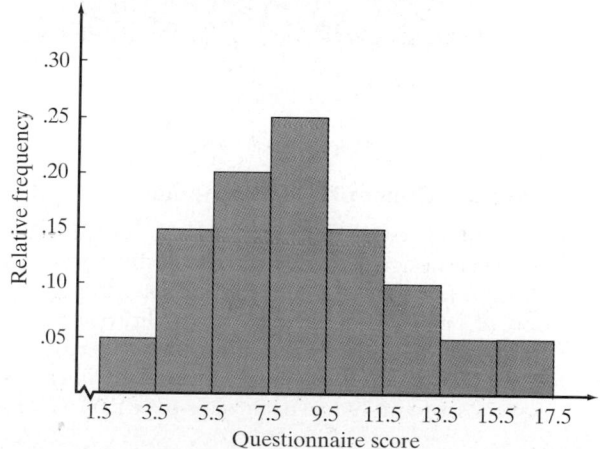

**2.21** The National Earthquake Information Center located 16,287 earthquakes worldwide in 1998. The magnitudes (on the Richter scale) of the earthquakes are summarized in the table. Display the data in a graph.

| Magnitude | Number of Earthquakes |
|---|---|
| 0 | 1,774 |
| 0.1–0.9 | 6 |
| 1.0–1.9 | 580 |
| 2.0–2.9 | 3,080 |
| 3.0–3.9 | 4,518 |
| 4.0–4.9 | 5,548 |
| 5.0–5.9 | 674 |
| 6.0–6.9 | 97 |
| 7.0–7.9 | 9 |
| 8.0–8.9 | 1 |

*Source:* National Earthquake Information Center, Department of Interior, U.S. Geological Survey.

**2.22** The United States Chess Federation (USCF) establishes a numerical rating for each competitive chess player. The USCF rating is a number between 0 and 4,000 that changes over time depending on the outcome of tournament games. The higher the rating, the better (more successful) the player. The graph describes the rating distribution of 27,563 players who were active competitors in 1997.
  **a.** What type of graph is displayed?
  **b.** Is the variable displayed on the graph quantitative or qualitative?
  **c.** What percentage of players has a USCF rating above 1,000?

*Source:* United States Chess Federation, Jan. 1998.

## Applying the Concepts—Intermediate

**2.23** The United States Golf Association (USGA) Handicap System is designed to allow golfers of differing abilities to enjoy fair competition. The handicap index is a measure of a player's potential scoring ability on an 18-hole golf course of standard difficulty. For example, on a par-72 course, a golfer with a handicap of 7 will typically have a score of 79 (seven strokes over par). Over 4.5 million golfers have an official USGA Handicap index. The handicap indexes for both male and female golfers were obtained from the USGA and are summarized in the two MINITAB histograms shown below.
  **a.** What percentage of male USGA golfers have a handicap greater than 20?
  **b.** What percentage of female USGA golfers have a handicap greater than 20?
  **c.** Identify the type of skewness (if any) present in the male handicap distribution.
  **d.** Identify the type of skewness (if any) present in the female handicap distribution.

**2.24** A group of University of Virginia biologists studied nectivory (nectar drinking) in crab spiders to determine if adult males were feeding on nectar to prevent fluid loss (*Animal Behavior*, June 1995). Nine male spiders were weighed and then placed on the flowers of Queen Anne's lace. One hour later, the spiders were removed and reweighed. The evaporative fluid loss (in milligrams) of each of the nine male spiders is given in the next table.

SPIDERS

| Male Spider | Fluid Loss |
|---|---|
| A | .018 |
| B | .020 |
| C | .017 |
| D | .024 |
| E | .020 |
| F | .024 |
| G | .003 |
| H | .001 |
| I | .009 |

*Source:* Pollard, S. D., *et al.* "Why do male crab spiders drink nectar?" *Animal Behavior*, Vol. 49, No. 6, June 1995, p.1445 (Table II).

**a.** Summarize the fluid losses of male crab spiders with a stem-and-leaf display.

**b.** Of the nine spiders, only three drank any nectar from the flowers of Queen Anne's lace. These three spiders are identified as G, H, and I in the table. Locate and circle these three fluid losses on the stem-and-leaf display. Does the pattern depicted in the graph give you any insight into whether feeding on flower nectar reduces evaporative fluid loss for male crab spiders? Explain.

**2.25** *Postmortem interval* (PMI) is defined as the elapsed time between death and an autopsy. Knowledge of PMI is considered essential when conducting medical research on human cadavers. The data in the table are the PMIs of 22 human brain specimens obtained at autopsy in a recent study (*Brain and Language*, June 1995). Graphically describe the PMI data with a dot plot. Based on the plot, make a summary statement about the PMI of the 22 human brain specimens.

BRAINPMI

**Postmortem Intervals for 22 Human Brain Specimens**

| | | | | | | | |
|---|---|---|---|---|---|---|---|
| 5.5 | 14.5 | 6.0 | 5.5 | 5.3 | 5.8 | 11.0 | 6.1 |
| 7.0 | 14.5 | 10.4 | 4.6 | 4.3 | 7.2 | 10.5 | 6.5 |
| 3.3 | 7.0 | 4.1 | 6.2 | 10.4 | 4.9 | | |

*Source:* Hayes. T. L., and Lewis, D. A. "Anatomical specialization of the anterior motor speech area: Hemispheric differences in magnopyramidal neurons." *Brain and Language*, Vol. 49, No. 3, June 1995, p. 292 (Table 1).

**2.26** It's not uncommon for hearing aids to cancel the desired signal. *IEEE Transactions on Speech and Audio Processing* (May 1995) reported on an audio processing system designed to limit the amount of signal cancellation that may occur. The system utilizes a mathematical equation that involves a variable, *V*, called a *sufficient norm constraint.* A histogram for realizations of *V*, produced using simulation, is shown here.

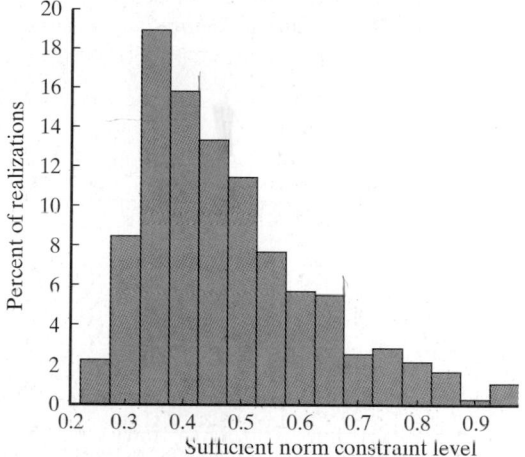

*Source:* Hoffman, M. W., and Buckley, K. M. "Robust time-domain processing of broadband microphone array data." *IEEE Transactions on Speech and Audio Processing*, Vol. 3, No. 3, May 1995, p. 199 (Figure 4).

**a.** Estimate the percentage of realizations of *V* with values ranging from .425 to .675.

**b.** Cancellation of the desired signal is limited by selecting a norm constraint *V* so that only 10% of the realizations have values below the selected level. Find this value of *V*.

**2.27** Data from a psychology experiment were reported and analyzed in *The American Statistician* (May 2001). Two samples of female students participated in the experiment. One sample consisted of 11 students known to suffer from the eating disorder bulimia; the other sample consisted of 14 students with normal eating habits. Each student completed a questionnaire from which a "fear of negative evaluation" (FNE) score was produced. (The higher the score, the greater the fear of negative evaluation.) The data are displayed in the table.

**a.** Construct a dot plot or stem-and-leaf display for the FNE scores of all 25 female students.

**b.** Highlight the bulimic students on the graph, part **a**. Does it appear that bulimics tend to have a greater fear of negative evaluation? Explain.

**c.** Why is it important to attach a measure of reliability to the inference made in part **b**?

BULIMIA

| Bulimic students: | 21 | 13 | 10 | 20 | 25 | 19 | 16 | 21 | 24 | 13 | 14 | | | |
|---|---|---|---|---|---|---|---|---|---|---|---|---|---|---|
| Normal students: | 13 | 6 | 16 | 13 | 8 | 19 | 23 | 18 | 11 | 19 | 7 | 10 | 15 | 20 |

*Source:* Randles, R. H. "On Neutral Responses (Zeros) in the Sign Test and Ties in the Wilcoxon-Mann-Whitney Test." *The American Statistician*, Vol. 55, No. 2, May 2001 (Figure 3).

**2.28** Neurological single-photon-emission computed tomography (neuroSPECT) is a relatively new method for detecting abnormalities of the brain that are not apparent on CT or MRI scans. Fourteen mild head injury patients with normal CT or MRI scans were examined using neuroSPECT compared to five age-matched hospital employees (controls) who had no history of head injury (*Journal of Head Trauma Rehabilitation*, June 1995). Density ratios for the right temporal lobes of the subjects are shown below.

NEUROSPECT

| Patients | | Controls |
|---|---|---|
| .852 | 1.004 | 1.012 |
| .901 | .942 | 1.111 |
| .849 | .977 | 1.094 |
| .779 | .891 | 1.216 |
| .842 | .898 | .989 |
| .960 | .866 | |
| .782 | 1.004 | |

*Source:* Varney, N. R., et al. "NeuroSPECT correlates of disabling mild head injury: Preliminary findings." *Journal of Head Trauma Rehabilitation*, Vol. 10, No. 3, June 1995, p. 24 (Table 2), Aspen Publishers, Inc.

a. Construct a dot plot for the neuroSPECT ratios of the 14 head injury patients.
b. Add the neuroSPECT ratios for the five controls to the dot plot, part **a**. Use a different-colored dot for these five measurements.
c. Do you detect a pattern in the dot plot, part **b**? Explain.

**2.29** Saturn has five satellites that rotate around the planet. *Astronomy* (Aug. 1995) lists 19 different events involving eclipses or occults of Saturnian satellites during the month of August. For each event, the percent of light lost by the eclipsed or occulted satellite at midevent is recorded in the table.

SATURN

| Date | Event | Light Loss (%) |
|---|---|---|
| Aug. 2 | Eclipse | 65 |
| 4 | Eclipse | 61 |
| 5 | Occult | 1 |
| 6 | Eclipse | 56 |
| 8 | Eclipse | 46 |
| 8 | Occult | 2 |
| 9 | Occult | 9 |
| 11 | Occult | 5 |
| 12 | Occult | 39 |
| 14 | Occult | 1 |
| 14 | Eclipse | 100 |
| 15 | Occult | 5 |
| 15 | Occult | 4 |
| 16 | Occult | 13 |
| 20 | Occult | 11 |
| 23 | Occult | 3 |
| 23 | Occult | 20 |
| 25 | Occult | 20 |
| 28 | Occult | 12 |

*Source:* *Astronomy* magazine, Aug. 1995, p. 60.

a. Construct a stem-and-leaf display for light loss percentage of the 19 events.
b. Locate on the stem-and-leaf plot, part **a**, the light losses associated with eclipses of Saturnian satellites. (Circle the light losses on the plot.)
c. Based on the marked stem-and-leaf display, part **b**, make an inference about which event type (eclipse or occult) is more likely to lead to a greater light loss.

## Applying the Concepts—Advanced

**2.30** Educators are constantly evaluating the efficacy of public schools in the education and training of American students. One quantitative assessment of change over time is the difference in scores on the SAT, which has been used for decades by colleges and universities as one criterion for admission. The following table shows the average SAT scores for each of the 50 states and District of Columbia for 1990 and 2000:

SAT

| State | 1990 | 2000 |
|---|---|---|
| Alabama | 1079 | 1114 |
| Alaska | 1015 | 1034 |
| Arizona | 1041 | 1044 |
| Arkansas | 1077 | 1117 |
| California | 1002 | 1015 |
| Colorado | 1067 | 1071 |
| Connecticut | 1002 | 1017 |
| Delaware | 1006 | 998 |
| D.C. | 950 | 980 |
| Florida | 988 | 998 |
| Georgia | 951 | 974 |
| Hawaii | 985 | 1007 |
| Idaho | 1066 | 1081 |
| Illinois | 1089 | 1154 |
| Indiana | 972 | 999 |
| Iowa | 1172 | 1189 |
| Kansas | 1129 | 1154 |
| Kentucky | 1089 | 1098 |
| Louisiana | 1088 | 1120 |
| Maine | 991 | 1004 |
| Maryland | 1008 | 1016 |
| Massachusetts | 1001 | 1024 |
| Michigan | 1063 | 1126 |
| Minnesota | 1110 | 1175 |
| Mississippi | 1090 | 1111 |
| Missouri | 1089 | 1149 |
| Montana | 1082 | 1089 |
| Nebraska | 1121 | 1131 |

| | | |
|---|---|---|
| Nevada | 1022 | 1027 |
| N. Hampshire | 1028 | 1039 |
| N. Jersey | 993 | 1011 |
| N. Mexico | 1100 | 1092 |
| New York | 985 | 1000 |
| N. Carolina | 948 | 988 |
| N. Dakota | 1157 | 1197 |
| Ohio | 1048 | 1072 |
| Oklahoma | 1095 | 1123 |
| Oregon | 1024 | 1054 |
| Pennsylvania | 987 | 995 |
| R. Island | 986 | 1005 |
| S. Carolina | 942 | 966 |
| S. Dakota | 1150 | 1175 |
| Tennessee | 1102 | 1116 |
| Texas | 979 | 993 |
| Utah | 1121 | 1139 |
| Vermont | 1000 | 1021 |
| Virginia | 997 | 1009 |
| Washington | 1024 | 1054 |
| W. Virginia | 1034 | 1037 |
| Wisconsin | 1111 | 1181 |
| Wyoming | 1072 | 1090 |

*Source:* College Entrance Examination Board, 2001.

**a.** Use graphs to display the two SAT score distributions. How have the distributions of average state scores changed over the last decade?

**b.** As another method of comparing the 1990 and 2000 average SAT scores, compute the **paired difference** by subtracting the 1990 score from the 2000 score for each state. Summarize these differences with a graph.

**c.** Interpret the graph, part **b**. How do your conclusions compare to those you reached when comparing the two graphs in part **a**?

**d.** What is the largest improvement in the average score between 1990 and 2000 as indicated on the graph, part **b**? With which state is this improvement associated in the table of data?

**2.31** The role that listener knowledge plays in the perception of imperfectly articulated speech was investigated in the *American Journal of Speech–Language Pathology* (Feb. 1995). Thirty female college students, randomly divided into three groups of ten, participated as listeners in the study. All subjects were required to listen to a 48-sentence audiotape of a Korean woman with cerebral palsy and asked to transcribe her entire speech. For the first group of students (the *control group*), the speaker used her normal manner of speaking. For the second group (the *treatment group*), the speaker employed a learned breathing pattern (called breath-group strategy) to improve speech efficiency. The third group (the *familiarity group*) also listened to the tape with the breath-group strategy, but only after they had practiced listening twice to another tape in which they were told exactly what the speaker was saying. At the end of the listening/transcribing session, two quantitative variables were measured for each listener: (1) the total number of words transcribed (called the rate of response) and (2) the percentage of words correctly transcribed (called the accuracy score). The data for all 30 subjects are provided in the table below.

**a.** MINITAB dot plots of the response rates (RESPRATE) for each listener group are displayed on p. 42. Use the information in the plots to describe the differences in the distributions of response rates among the three groups.

**b.** Use a graphical method to describe the distribution of accuracy scores for the three listener groups. Construct one graph for each group. Interpret the results.

**2.32** Mark McGwire of the St. Louis Cardinals and Sammy Sosa of the Chicago Cubs hit 70 and 66 home runs, respectively, during the 1998 Major League Baseball season, breaking a record held by Roger Maris (61 home runs) since 1961. J. S. Simonoff of New York University collected data on the number of runs scored by their respective teams in games in which McGwire and Sosa hit home runs (*Journal of Statistics Education*, Vol. 6, 1998).

LISTEN

| Control Group | | Treatment Group | | Familiarization Group | |
|---|---|---|---|---|---|
| **Rate of Response** | **Percent Correct** | **Rate of Response** | **Percent Correct** | **Rate of Response** | **Percent Correct** |
| 250 | 23.6 | 254 | 26.0 | 193 | 36.0 |
| 230 | 26.0 | 178 | 32.6 | 223 | 41.0 |
| 197 | 26.0 | 139 | 32.6 | 232 | 43.0 |
| 238 | 26.7 | 249 | 33.0 | 214 | 44.0 |
| 174 | 29.2 | 236 | 34.4 | 269 | 44.0 |
| 263 | 29.5 | 231 | 36.5 | 256 | 46.0 |
| 275 | 31.3 | 161 | 38.5 | 224 | 46.0 |
| 193 | 32.9 | 255 | 40.0 | 225 | 48.0 |
| 204 | 35.4 | 275 | 41.7 | 288 | 49.0 |
| 168 | 29.2 | 181 | 44.8 | 244 | 52.0 |

*Source:* Tjaden, K., and Liss, J. M. "The influence of familiarity on judgments of treated speech." *American Journal of Speech Language Pathology*, Vol. 4, No. 1, Feb. 1995, p. 43 (Table 1).

MINITAB Output for Exercise 2.31

The data is reproduced in the accompanyng table. (An asterisk indicates a game in which McGwire or Sosa hit multiple home runs.)

**a.** Construct a stem-and-leaf display for the number of runs scored by St. Louis during games when McGwire hit a home run.

**b.** Repeat part **a** for Chicago and Sosa.

**c.** Compare and contrast the two distributions.

**d.** On the stem-and-leaf display, circle the games in which McGwire or Sosa hit multiple home runs. Do you detect any patterns?

⚫ STLRUNS

| St. Louis Cardinals | | | | |
|---|---|---|---|---|
| 6 | 6 | 3 | 11 | 13 |
| 8 | 1 | 10 | 6 | 7 |
| 5 | 8 | 9 | 6* | 3 |
| 8 | 2 | 3 | 6* | 6 |
| 15* | 2 | 8* | 5 | 6 |
| 8 | 6 | 2 | 3 | 5 |
| 5 | 8 | 4 | 10* | |
| 8 | 9 | 4 | 4 | |
| 3 | 5 | 3 | 11* | |
| 5 | 7 | 6 | 1 | |
| 2 | 3 | 8 | 4 | |
| 6 | 2 | 7* | 6* | |
| 3 | 7 | 14* | 4* | |

⚫ CUBSRUNS

| Chicago Cubs | | | | |
|---|---|---|---|---|
| 3 | 6* | 8* | 6 | 7* |
| 4 | 5 | 2 | 7 | 2 |
| 1 | 8* | 6 | 13 | |
| 1 | 9* | 6 | 8 | |
| 4 | 2 | 3* | 10 | |
| 3 | 6 | 9 | 6* | |
| 5 | 4 | 10 | 5 | |
| 5 | 4 | 4 | 9 | |
| 2 | 9 | 5 | 7 | |
| 5* | 5 | 4 | 8 | |
| 5* | 6 | 5 | 15 | |
| 10* | 9 | 8 | 11* | |
| 5 | 3 | 11 | 6 | |

## 2.3   SUMMATION NOTATION

Now that we've examined some graphical techniques for summarizing and describing quantitative data sets, we turn to numerical methods for accomplishing this objective. Before giving the formulas for calculating numerical descriptive measures, let's look at some shorthand notation that will simplify our calculation instructions. Remember that such notation is used for one reason only—to avoid repeating the same verbal descriptions over and over. If you mentally substitute the verbal definition of a symbol each time you read it, you'll soon get used to it.

We denote the measurements of a quantitative data set as

$$x_1, x_2, x_3, \ldots, x_n$$

where $x_1$ is the first measurement in the data set, $x_2$ is the second measurement in the data set, $x_3$ is the third measurement in the data set, . . . , and $x_n$ is the $n$th (and last) measurement in the data set. Thus, if we have five measurements in a set of data, we will write $x_1, x_2, x_3, x_4, x_5$ to represent the measurements. If the actual numbers are 5, 3, 8, 5, and 4, we have $x_1 = 5, x_2 = 3, x_3 = 8, x_4 = 5$, and $x_5 = 4$.

Most of the formulas we use require a summation of numbers. For example, one sum we'll need to obtain is the sum of all the measurements in the data set, or $x_1 + x_2 + x_3 + \cdots + x_n$. To shorten the notation, we use the symbol $\sum$ for the summation. That is, $x_1 + x_2 + x_3 + \cdots + x_n = \sum_{i=1}^{n} x_i$. Verbally translate $\sum_{i=1}^{n} x_i$ as follows: "The sum of the measurements, whose typical member is $x_i$, beginning with the member $x_1$ and ending with the member $x_n$."

Suppose, as in our earlier example, $x_1 = 5$, $x_2 = 3$, $x_3 = 8$, $x_4 = 5$, and $x_5 = 4$. Then the sum of the five measurements, denoted $\sum_{i=1}^{5} x_i$ is obtained as follows:

$$\sum_{i=1}^{5} x_i = x_1 + x_2 + x_3 + x_4 + x_5$$
$$= 5 + 3 + 8 + 5 + 4 = 25$$

Another important calculation requires that we square each measurement and then sum the squares. The notation for this sum is $\sum_{i=1}^{n} x_i^2$. For the five measurements, we have

$$\sum_{i=1}^{5} x_i^2 = x_1^2 + x_2^2 + x_3^2 + x_4^2 + x_5^2$$
$$= 5^2 + 3^2 + 8^2 + 5^2 + 4^2$$
$$= 25 + 9 + 64 + 25 + 16 = 139$$

In general, the symbol following the summation sign $\sum$ represents the variable (or function of the variable) that is to be summed.

---

**The Meaning of Summation Notation** $\sum_{i=1}^{n} x_i$

---

Sum the measurements on the variable that appears to the right of the summation symbol, beginning with the 1st measurement and ending with the $n$th measurement.

---

 EXERCISES 2.33–2.36

**Learning the Mechanics**

*Note: In all exercises, $\sum$ represents $\sum_{i=1}^{n}$.*

**2.33** A data set contains the observations 5, 1, 3, 2, 1. Find:

  **a.** $\sum x$

  **b.** $\sum x^2$

  **c.** $\sum (x - 1)$

  **d.** $\sum (x - 1)^2$

  **e.** $\left( \sum x \right)^2$

**2.34** Suppose a data set contains the observations 3, 8, 4, 5, 3, 4, 6. Find:

  **a.** $\sum x$

  **b.** $\sum x^2$

  **c.** $\sum (x - 5)^2$

  **d.** $\sum (x - 2)^2$

  **e.** $\left( \sum x \right)^2$

**2.35** Refer to Exercise 2.33. Find:

  **a.** $\sum x^2 - \dfrac{\left( \sum x \right)^2}{5}$   **b.** $\sum (x - 2)^2$   **c.** $\sum x^2 - 10$

**2.36** A data set contains the observations 6, 0, −2, −1, 3. Find:

  **a.** $\sum x$   **b.** $\sum x^2$   **c.** $\sum x^2 - \dfrac{\left( \sum \right)^2}{5}$

## 2.4 NUMERICAL MEASURES OF CENTRAL TENDENCY

When we speak of a data set, we refer to either a sample or to a population. If statistical inference is our goal, we'll wish ultimately to use sample **numerical descriptive measures** to make inferences about the corresponding measures for a population.

As you'll see, a large number of numerical methods are available to describe quantitative data sets. Most of these methods measure one of two data characteristics:

1. The **central tendency** of the set of measurements—that is, the tendency of the data to cluster, or center, about certain numerical values (see Figure 2.12a).
2. The **variability** of the set of measurements—that is, the spread of the data (see Figure 2.12b).

**FIGURE 2.12**

*Numerical Descriptive Measures*

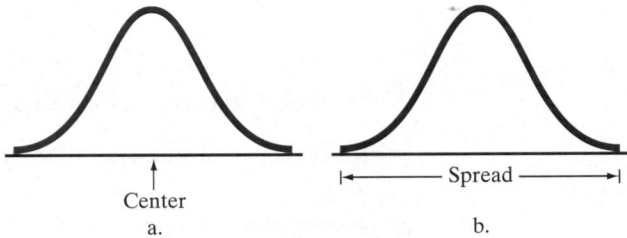

Center
a.

Spread
b.

In this section we concentrate on **measures of central tendency**. In the next section, we discuss measures of variability.

The most popular and best understood measure of central tendency for a quantitative data set is the *arithmetic mean* (or simply the mean) of a data set.

---

**DEFINITION 2.4**

The **mean** of a set of quantitative data is the sum of the measurements divided by the number of measurements contained in the data set.

---

In everyday terms, the mean is the average value of the data set and is often used to represent a "typical" value. We denote the **mean of a sample** of measurements by $\bar{x}$ (read "x-bar"), and represent the formula for its calculation as shown in the box:

---

**Calculating a Sample Mean**

$$\bar{x} = \frac{\sum\limits_{i=1}^{n} x_i}{n}$$

---

**EXAMPLE 2.3**

Calculate the mean of the following five sample measurements: 5, 3, 8, 5, 6.

### Solution

Using the definition of sample mean and the summation notation, we find

$$\bar{x} = \frac{\sum_{i=1}^{5} x_i}{5} = \frac{5 + 3 + 8 + 5 + 6}{5} = \frac{27}{5} = 5.4$$

Thus, the mean of this sample is 5.4.*

**EXAMPLE 2.4**  Calculate the sample mean for the 100 EPA mileages given in Table 2.3.

### Solution

The mean gas mileage for the 100 cars is denoted

$$\bar{x} = \frac{\sum_{i=1}^{100} x_i}{100}$$

Rather than compute $\bar{x}$ by hand (or calculator), we entered the data of Table 2.3 into a computer and employed SPSS statistical software to compute the mean. The SPSS printout is shown in Figure 2.13. The sample mean, highlighted on the printout, is $\bar{x} = 36.9940$.

**FIGURE 2.13**

*SPSS Printout of Numerical Descriptive Measures for 100 EPA Gas Mileages*

| | Descriptives | | Statistic | Std. Error |
|---|---|---|---|---|
| MPG | Mean | | 36.9940 | .2418 |
| | 95% Confidence Interval for Mean | Lower Bound | 36.5142 | |
| | | Upper Bound | 37.4738 | |
| | 5% Trimmed Mean | | 36.9922 | |
| | Median | | 37.0000 | |
| | Variance | | 5.846 | |
| | Std. Deviation | | 2.4179 | |
| | Minimum | | 30.00 | |
| | Maximum | | 44.90 | |
| | Range | | 14.90 | |
| | Interquartile Range | | 2.7500 | |
| | Skewness | | .051 | .241 |
| | Kurtosis | | .770 | .478 |

Given this information, you can visualize a distribution of gas mileage readings centered in the vicinity of $\bar{x} \approx 37$. An examination of the relative frequency histogram (Figure 2.8) confirms that $\bar{x}$ does in fact fall near the center of the distribution.

---

*In the examples given here, $\bar{x}$ is sometimes rounded to the nearest tenth, sometimes to the nearest hundredth, sometimes to the nearest thousandth, and so on. There is no specific rule for rounding when calculating $\bar{x}$ because $\bar{x}$ is specifically defined to be the sum of all measurements divided by $n$; that is, it is a specific fraction. When $\bar{x}$ is used for descriptive purposes, it is often convenient to round the calculated value of $\bar{x}$ to the number of significant figures used for the original measurements. When $\bar{x}$ is to be used in other calculations, however, it may be necessary to retain more significant figures.

The sample mean $\bar{x}$ will play an important role in accomplishing our objective of making inferences about populations based on sample information. For this reason we need to use a different symbol for the *mean of a population*—the mean of the set of measurements on every unit in the population. We use the Greek letter $\mu$ (mu) for the population mean.

---

### Symbols for the Sample Mean and the Population Mean

In this text, we adopt a general policy of using Greek letters to represent numerical descriptive measures for the population and Roman letters to represent corresponding descriptive measures for the sample. The symbols for the mean are

$\bar{x}$ = Sample mean                $\mu$ = Population mean

---

We'll often use the sample mean, $\bar{x}$, to estimate (make an inference about) the population mean, $\mu$. For example, the EPA mileages for the population consisting of *all* cars has a mean equal to some value, $\mu$. Our sample of 100 cars yielded mileages with a mean of $\bar{x} = 36.9940$. If, as is usually the case, we don't have access to the measurements for the entire population, we could use $\bar{x}$ as an estimator or approximator for $\mu$. Then we'd need to know something about the reliability of our inference. That is, we'd need to know how accurately we might expect $\bar{x}$ to estimate $\mu$. In Chapter 7, we'll find that this accuracy depends on two factors:

1. The *size of the sample*. The larger the sample, the more accurate the estimate will tend to be.
2. The *variability, or spread, of the data*. All other factors remaining constant, the more variable the data, the less accurate the estimate.

Another important measure of central tendency is the **median**.

---

**DEFINITION 2.5**
The **median** of a quantitative data set is the middle number when the measurements are arranged in ascending (or descending) order.

---

The median is of most value in describing large data sets. If the data set is characterized by a relative frequency histogram (Figure 2.14), the median is the point on the *x*-axis such that half the area under the histogram lies above the median and half lies below. [*Note*: In Section 2.2 we observed that the relative frequency associated with a particular interval on the *x*-axis is proportional to the amount of area under the histogram that lies above the interval.] We denote the *median* of a *sample* by *M*.

**FIGURE 2.14**

*Location of the Median*

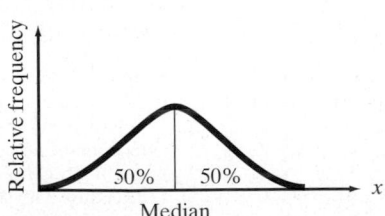

---

### Calculating a Sample Median, *M*

Arrange the *n* measurements from the smallest to the largest.

1. If *n* is odd, *M* is the middle number.
2. If *n* is even, *M* is the mean of the middle two numbers.

---

**EXAMPLE 2.5**

Consider the following sample of *n* = 7 measurements: 5, 7, 4, 5, 20, 6, 2

a. Calculate the median *M* of this sample.
b. Eliminate the last measurement (the 2) and calculate the median of the remaining *n* = 6 measurements.

**Solution**

a. The seven measurements in the sample are ranked in ascending order: 2, 4, 5, 5, 6, 7, 20. Because the number of measurements is odd, the median is the middle measurement. Thus, the median of this sample is *M* = 5.
b. After removing the 2 from the set of measurements, we rank the sample measurements in ascending order as follows: 4, 5, 5, 6, 7, 20. Now the number of measurements is even, so we average the middle two measurements. The median is *M* = (5 + 6)/2 = 5.5.

In certain situations, the median may be a better measure of central tendency than the mean. In particular, the median is less sensitive than the mean to extremely large or small measurements. Note, for instance, that all but one of the measurements in part **a** of Example 2.5 center about *x* = 5. The single relatively large measurement, *x* = 20, does not affect the value of the median, 5, but it causes the mean, $\bar{x}$ = 7, to lie to the right of most of the measurements.

As another example of data for which the central tendency is better described by the median than the mean, consider the household incomes of a community being studied by a sociologist. The presence of just a few households with very high incomes will affect the mean more than the median. Thus, the median will provide a more accurate picture of the typical income for the community. The mean could exceed the vast majority of the sample measurements (household incomes), making it a misleading measure of central tendency.

**EXAMPLE 2.6**

Calculate the median for the 100 EPA mileages given in Table 2.3. Compare the median to the mean computed in Example 2.4.

**Solution**

For this large data set, we again resort to a computer analysis. An SAS printout is displayed in Figure 2.15, with the median highlighted. You can see that the median

**FIGURE 2.15**

*SAS Printout of Numerical Descriptive Measures for 100 EPA Mileages*

| Basic Statistical Measures | | | |
|---|---|---|---|
| **Location** | | **Variability** | |
| Mean | 36.99400 | Std Deviation | 2.41790 |
| Median | 37.00000 | Variance | 5.84623 |
| Mode | 37.00000 | Range | 14.90000 |
| | | Interquartile Range | 2.70000 |

is 37.0. This value implies that half of the 100 mileages in the data set fall below 37.0 and half lie above 37.0. Note that the median, 37.0, and the mean, 36.9940, are almost equal. This fact indicates that the data form an approximately **symmetric distribution**. As indicated in the box on p. 49, a comparison of the mean and median gives an indication of the **skewness** (i.e, the tendency of the distribution to have elongated tails) of a data set. ∎

A third measure of central tendency is the **mode** of a set of measurements.

> **DEFINITION 2.6**
> The **mode** is the measurement that occurs most frequently in the data set.

Therefore, the mode shows where the data tend to concentrate.

**EXAMPLE 2.7**    Each of ten taste testers rated a new brand of barbecue sauce on a 10-point scale, where 1 = awful and 10 = excellent. Find the mode for the ten ratings shown below.

8    7    9    6    8    10    9    9    5    7

**Solution**

Since 9 occurs most often (three times), the mode of the ten taste ratings is 9. ∎

Note that the data in Example 2.7 are actually qualitative in nature (e.g., "awful," "excellent"). The mode is particularly useful for describing qualitative data. The modal category is simply the category (or class) that occurs most often. Because it emphasizes data concentration, the mode is also used with quantitative data sets to locate the region in which much of the data is concentrated. A retailer of men's clothing would be interested in the modal neck size and sleeve length of potential customers. The modal income class of the laborers in the United States is of interest to the Labor Department.

For some quantitative data sets, the mode may not be very meaningful. For example, consider the EPA mileage ratings in Table 2.3. A reexamination of the data reveals that the gas mileage of 37.0 occurs most often (four times). However, the mode of 37.0 (shown on the SAS printout, Figure 2.15) is not particularly useful as a measure of central tendency.

A more meaningful measure can be obtained from a relative frequency histogram for quantitative data. The measurement class containing the largest relative frequency is called the **modal class**. Several definitions exist for locating the position of the mode within a modal class, but the simplest is to define the mode as the midpoint of the modal class. For example, examine the frequency histogram for the EPA mileage ratings, in Figure 2.8 (p. 32). You can see that the modal class is the interval 36.0-37.5. The mode (the midpoint) is 36.75. This modal class (and the mode itself) identifies the area in which the data are most concentrated, and in that sense it is a measure of central tendency. However, for most applications involving quantitative data, the mean and median provide more descriptive information than the mode.

# Comparing the Mean and the Median

If the data set is skewed to the right, the mean is greater than (to the right of) the median.

If the data set is symmetric, the mean equals the median.

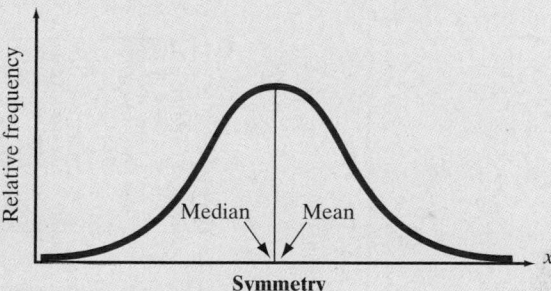

If the data set is skewed to the left, the mean is less than (to the left of) the median.

# Using the TI-83 Graphing Calculator

## Finding One-Variable Descriptive Statistics on the TI-83

Step 1    *Enter the data*
Press    **STAT   1**   for   **STAT Edit**
Enter the data into one of the lists.

Step 2    *Calculate descriptive statistics*
Press   **STAT**
Press the right arrow key to highlight **CALC**
Press   **ENTER**   for   **1-Var Stats**

Enter the name of the list containing your data.
Press   **2nd   1**   for   **L1**   (or **2nd 2** for   **L2** etc.)
Press   **ENTER**

You should see the statistics on your screen. Some of the statistics are off the bottom of the screen. Use the down arrow to scroll through to see the remaining statistics. Use the up arrow to scroll back up.

Example   The descriptive statistics for the sample data set

86, 70, 62, 98, 73, 56, 53, 92, 86, 37, 62, 83, 78, 49, 78, 37, 67, 79, 57

are shown below, as calculated by the TI-83.

Sorting    The descriptive statistics do not include the mode. To find the mode,
data       sort your data as follows:

Press   **STAT**
Press   **2**   for   **SORTA(**
Enter the name of the list your data is in. If your data is in **L1**,
press   **2nd 1**

Press    **ENTER**
The screen will say:   **DONE**
To see the sorted data, press   **STAT   1**   for   **STAT Edit**

  EXERCISES 2.37–2.53

## Learning the Mechanics

**2.37** Calculate the mode, mean, and median of the following data:

18    10    15    13    17    15    12    15    18    16    11

**2.38** Calculate the mean and median of the following grade point averages:

3.2    2.5    2.1    3.7    2.8    2.0

**2.39** Explain the difference between the calculation of the median for an odd and an even number of measurements.

Construct one data set consisting of five measurements and another consisting of six measurements for which the medians are equal.

**2.40** Explain how the relationship between the mean and median provides information about the symmetry or skewness of the data's distribution.

**2.41** Calculate the mean for samples where

a. $n = 10$, $\sum x = 85$

b. $n = 16$, $\sum x = 400$

c. $n = 45$, $\sum x = 35$

d. $n = 18$, $\sum x = 242$

**2.42** Calculate the mean, median, and mode for each of the following samples:

a. $7, -2, 3, 3, 0, 4$

b. $2, 3, 5, 3, 2, 3, 4, 3, 5, 1, 2, 3, 4$

c. $51, 50, 47, 50, 48, 41, 59, 68, 45, 37$

**2.43** Describe how the mean compares to the median for a distribution as follows:

a. Skewed to the left

b. Skewed to the right

c. Symmetric

### Applying the Concepts—Basic

**2.44** *Fortune* (Oct. 25, 1999) published a list of the 50 most powerful women in America. The data on age (in years) and title of each of these 50 women are stored in the WOMENPOWER file.

a. Find the mean, median, and modal age of these 50 women.

b. What do the mean and median indicate about the skewness of the age distribution?

c. Construct a relative frequency histogram for the age data. What is the modal age class?

### WOMENPOWER

| Rank | Name | Age | Company | Title |
|------|------|-----|---------|-------|
| 1 | Carly Fiorina | 45 | Hewlett-Packard | CEO |
| 2 | Heidi Miller | 46 | Citigroup | CFO |
| 3 | Mary Meeker | 40 | Morgan Stanley | Managing Director |
| 4 | Shelly Lazarus | 52 | Ogilvy & Mather | CEO |
| 5 | Meg Whitman | 43 | eBay | CEO |
| 6 | Debby Hopkins | 44 | Boeing | CFO |
| 7 | Marjorie Scardino | 52 | Pearson | CEO |
| 8 | Martha Stewart | 58 | Omnimedia | CEO |
| 9 | Nancy Peretsman | 45 | Allen & Co. | Ex. V.P. |
| 10 | Pat Russo | 47 | Lucent Technologies | Ex. V.P. |
| 11 | Patricia Dunn | 46 | Barclays Global Investors | Chairman |
| 12 | Abby Joseph Cohen | 47 | Goldman Sachs | Managing Director |
| 13 | Ann Livermore | 41 | Hewlett-Packard | CEO |
| 14 | Andrea Jung | 41 | Avon Products | COO |
| 15 | Sherry Lansing | 55 | Paramount Pictures | Chairman |
| 16 | Karen Katen | 50 | Pfizer | Ex. V.P. |
| 17 | Marilyn Carlson Nelson | 60 | Carlson Cos. | CEO |
| 18 | Judy McGrath | 47 | MTV & M2 | President |
| 19 | Lois Juliber | 50 | Colgate-Palmolive | COO |
| 20 | Gerry Laybourne | 52 | Oxygen Media | CEO |
| 21 | Judith Estrin | 44 | Cisco Systems | Sr. V.P. |
| 22 | Cathleen Black | 55 | Hearst Magazines | President |
| 23 | Linda Sandford | 46 | IBM | General Manager |
| 24 | Ann Moore | 49 | Time Inc. | President |
| 25 | Jill Barad | 48 | Mattel | CEO |
| 26 | Oprah Winfrey | 45 | Harpo Entertainment | Chairman |
| 27 | Judy Lewent | 50 | Merck | Sr. V.P. |
| 28 | Joy Covey | 36 | Amazon.com | COO |
| 29 | Rebecca Mark | 45 | Azurix | CEO |
| 30 | Deborah Willingham | 43 | Microsoft | V.P. |
| 31 | Dina Dubion | 46 | Chase Manhattan | Ex. V.P. |

*(continues on next page)*

WOMENPOWER (continued)

| Rank | Name | Age | Company | Title |
|------|------|-----|---------|-------|
| 32 | Patricia Woertz | 46 | Chevron | President |
| 33 | Lawton Fitt | 46 | Goldman Sachs | Man. Dir. |
| 34 | Ann Fudge | 48 | Kraft Foods | Ex. V.P. |
| 35 | Carolyn Ticknor | 52 | Hewlett-Packard | CEO |
| 36 | Dawn Lepore | 45 | Charles Schwab | CIO |
| 37 | Jeannine Rivet | 51 | United Healthcare | CEO |
| 38 | Jamie Gorelick | 49 | Fannie Mae | Vice Chairman |
| 39 | Jan Brandt | 48 | America Online | Mar. President |
| 40 | Bridget Macaskill | 51 | Oppenheimer Funds | CEO |
| 41 | Jeanne Jackson | 48 | Banana Republic | CEO |
| 42 | Cynthia Trudell | 46 | General Motors | V.P. |
| 43 | Nina DiSesa | 53 | McCann-Erickson | Chairman |
| 44 | Linda Wachner | 53 | Warnaco | Chairman |
| 45 | Darla Moore | 45 | Rainwater Inc. | President |
| 46 | Marion Sandler | 68 | Golden West | Co-CEO |
| 47 | Michelle Anthony | 42 | Sony Music | Ex. V.P. |
| 48 | Orit Gadlesh | 48 | Bain & Co. | Chairman |
| 49 | Charlotte Beers | 64 | J. Walter Thompson | Chairman |
| 50 | Abigail Johnson | 37 | Fidelity Investments | V.P. |

*Source: Fortune, Oct. 25, 1999.*

**2.45** Three-way catalytic converters have been installed in new vehicles in order to reduce pollutants from motor vehicle exhaust emissions. However, these converters unintentionally increase the level of ammonia in the air. *Environmental Science & Technology* (Sept. 1, 2000) published a study on the ammonia levels near the exit ramp of a San Francisco highway tunnel. The data in the table represent daily ammonia concentrations (parts per million) on eight randomly selected days during afternoon drive-time in the summer of 1999.

⚙ AMMONIA

| | | | | | | | |
|------|------|------|------|------|------|------|------|
| 1.53 | 1.50 | 1.37 | 1.51 | 1.55 | 1.42 | 1.41 | 1.48 |

a. Find the mean daily ammonia level in air in the tunnel.
b. Find the median ammonia level.
c. Interpret the values obtained in parts **a** and **b**.

**2.46** *The Condor* (May 1995) published a study of competition for nest holes among collared flycatchers, a bird species. The authors collected the data for the study by periodically inspecting nest boxes located on the island of Gotland in Sweden. The nest boxes were grouped into 14 discrete locations (called plots). The accompanying table gives the number of flycatchers killed and the number of flycatchers breeding at each plot.

⚙ CONDOR

| Plot Number | Number Killed | Number of Breeders |
|-------------|---------------|--------------------|
| 1 | 5 | 30 |
| 2 | 4 | 28 |
| 3 | 3 | 38 |
| 4 | 2 | 34 |
| 5 | 2 | 26 |
| 6 | 1 | 124 |
| 7 | 1 | 68 |
| 8 | 1 | 86 |
| 9 | 1 | 32 |
| 10 | 0 | 30 |
| 11 | 0 | 46 |
| 12 | 0 | 132 |
| 13 | 0 | 100 |
| 14 | 0 | 6 |

*Source: Merilä, J., and Wiggins, D. A. "Interspecific competition for nest holes causes adult mortality in the collared flycatcher." The Condor, Vol. 97, No. 2, May 1995, p. 447 (Table 4). Cooper Ornithological Society.*

a. Calculate the mean, median, and mode for the number of flycatchers killed at the 14 study plots.
b. Interpret the measures of central tendency, part **a**.
c. A MINITAB printout of descriptive statistics for the number of breeders at each plot is shown on p. 53. Locate the measures of central tendency on the printout and interpret these values.

```
MINITAB Output for Exercise 2.46
```

| Variable | N | Mean | Median | TrMean | StDev | SE Mean |
|----------|-----|------|--------|--------|-------|---------|
| Breeders | 14 | 55.7 | 36.0 | 53.5 | 39.6 | 10.6 |

| Variable | Minimum | Maximum | Q1 | Q3 |
|----------|---------|---------|------|------|
| Breeders | 6.0 | 132.0 | 29.5 | 89.5 |

**2.47** Applicants for an academic position (e.g., assistant professor) at a college or university are usually required to submit at least three letters of recommendation. A study of 148 applicants for an entry-level position in experimental psychology at the University of Alaska Anchorage revealed that many did not meet the three-letter requirement (*American Psychologist*, July 1995). Summary statistics for the number of recommendation letters in each application are given below. Interpret these summary measures.

Mean = 2.28    Median = 3    Mode = 3

## Applying the Concepts—Intermediate

**2.48** Platelet-activating factor (PAF) is a potent chemical that occurs in patients suffering from shock, inflammation, hypotension, respiratory, and cardiovascular disorders. A bioassay was undertaken to investigate the potential of 17 traditional Chinese herbal drugs in PAF inhibition (*Progress in Natural Science*, June 1995). The prevention of the PAF binding process, measured as a percentage, for each drug is provided in the accompanying table.

**a.** Construct a stem-and-leaf display for the data.

PAF

| Drug | PAF Inhibition (%) |
|------|-------------------|
| Hai-feng-teng (Fuji) | 77 |
| Hai-feng-teng (Japan) | 33 |
| Shan-ju | 75 |
| Zhang-yiz-hu-jiao | 62 |
| Shi-nan-teng | 70 |
| Huang-hua-hu-jiao | 12 |
| Hua-nan-hu-jiao | 0 |
| Xiao-yie-pa-ai-xiang | 0 |
| Mao-ju | 0 |
| Jia-ju | 15 |
| Xie-yie-ju | 25 |
| Da-yie-ju | 0 |
| Bian-yie-hu-jiao | 9 |
| Bi-bo | 24 |
| Duo-mai-hu-jiao | 40 |
| Yan-sen | 0 |
| Jiao-guo-hu-jiao | 31 |

*Source:* Guiqiu, H. "PAF receptor antagonistic principles from Chinese traditional drugs." *Progress in Natural Science*, Vol. 5, No. 3, June 1995, p. 301 (Table 1).

**b.** Compute the median inhibition percentage for the 17 herbal drugs. Interpret the result.

**c.** Compute the mean inhibition percentage for the 17 herbal drugs. Interpret the result.

**d.** Compute the mode of the 17 inhibition percentages. Interpret the result.

**e.** Locate the median, mean, and mode on the stem-and-leaf display, part **a**. Do these measures of central tendency appear to locate the center of the data?

**2.49** "The Training Game" is an activity used in psychology in which one person shapes an arbitrary behavior by selectively reinforcing the movements of another person. A group of 15 psychology students at Georgia Institute of Technology played "The Training Game" at Zoo Atlanta while participating in an experimental psychology laboratory in which they assisted in the training of animals (*Teaching of Psychology*, May 1998). At the end of the session, each student was asked to rate the statement: "'The Training Game' is a great way for students to understand the animal's perspective during training." Responses were recorded on a 7-point scale ranging from 1 (strongly disagree) to 7 (strongly agree). The 15 responses were summarized as follows:

mean = 5.87, mode = 6.

**a.** Interpret the measures of central tendency in the words of the problem.

**b.** What type of skewness (if any) is likely to be present in the distribution of student responses? Explain.

**2.50** Would you expect the data sets described below to possess relative frequency distributions that are symmetric, skewed to the right, or skewed to the left? Explain.

**a.** The salaries of all persons employed by a large university

**b.** The grades on an easy test

**c.** The grades on a difficult test

**d.** The amounts of time students in your class studied last week

**e.** The ages of automobiles on a used-car lot

**f.** The amounts of time spent by students on a difficult examination (maximum time is 50 minutes)

**2.51** Clinical observations suggest that specifically language-impaired (SLI) children have great difficulty with the proper use of pronouns. This phenomenon was investigated and reported in the *Journal of Communication Disorders* (Mar. 1995). Thirty children, all from low-income families, participated in the study. Ten were 5-year-old SLI children, ten were younger (3-year-old) normally developing (YND) children, and ten were older (5-year-old) normally developing (OND) children. The table contains the gender, deviation intelligence quotient (DIQ), and percentage of pronoun errors observed for each of the 30 subjects.

⊘ SLI

| Subject | Gender | Group | DIQ | Pronoun Errors (%) |
|---------|--------|-------|-----|--------------------|
| 1 | F | YND | 110 | 94.40 |
| 2 | F | YND | 92 | 19.05 |
| 3 | F | YND | 92 | 62.50 |
| 4 | M | YND | 100 | 18.75 |
| 5 | F | YND | 86 | 0 |
| 6 | F | YND | 105 | 55.00 |
| 7 | F | YND | 90 | 100.00 |
| 8 | M | YND | 96 | 86.67 |
| 9 | M | YND | 90 | 32.43 |
| 10 | F | YND | 92 | 0 |
| 11 | F | SLI | 86 | 60.00 |
| 12 | M | SLI | 86 | 40.00 |
| 13 | M | SLI | 94 | 31.58 |
| 14 | M | SLI | 98 | 66.67 |
| 15 | F | SLI | 89 | 42.86 |
| 16 | F | SLI | 84 | 27.27 |
| 17 | M | SLI | 110 | 33.33 |
| 18 | F | SLI | 107 | 0 |
| 19 | F | SLI | 87 | 0 |
| 20 | M | SLI | 95 | 0 |
| 21 | M | OND | 110 | 0 |
| 22 | M | OND | 113 | 0 |
| 23 | M | OND | 113 | 0 |
| 24 | F | OND | 109 | 0 |
| 25 | M | OND | 92 | 0 |
| 26 | F | OND | 108 | 0 |
| 27 | M | OND | 95 | 0 |
| 28 | F | OND | 87 | 0 |
| 29 | F | OND | 94 | 0 |
| 30 | F | OND | 98 | 0 |

*Source:* Moore, M. E. "Error analysis of pronouns by normal and language-impaired children." *Journal of Communication Disorders*, Vol. 28, No. 1, Mar. 1995, p. 62 (Table 2), p. 67 (Table 5).

**a.** Identify the variables in the data set as quantitative or qualitative.

**b.** Why is it nonsensical to compute numerical descriptive measures for qualitative variables?

**c.** Compute measures of central tendency for DIQ for the ten SLI children.

**d.** Compute measures of central tendency for DIQ for the ten YND children.

**e.** Compute measures of central tendency for DIQ for the ten OND children.

**f.** Use the results, parts **c-e**, to compare the DIQ central tendencies of the three groups of children.

**g.** An SPSS printout of descriptive statistics for the percentage of pronoun errors, by group, is shown at the bottom of the page. Locate the measures of central tendency on the printout and interpret their values.

**h.** Is it reasonable to use a single number (e.g., mean or median) to describe the center of the percentage of pronoun error distribution? Or should three "centers" be calculated, one for each of the three groups of children? Explain.

### Applying the Concepts—Advanced

**2.52** The conventional method of measuring the refractive status of an eye involves three quantities: (1) sphere power, (2) cylinder power, and (3) axis. Optometric researchers studied the variation in these three measures of refraction (*Optometry and Vision Science*, June 1995). Twenty-five successive refractive measurements were obtained on the eyes of over 100 university students. The cylinder power measurements for the left eye of one particular student (ID #11) are listed in the table. [*Note*: All measurements are negative values.] Numerical descriptive measures for the data set are provided in the SAS printout on p. 55.

⊘ LEFTEYE

| | | | | | | | | |
|------|------|------|------|------|------|------|------|------|
| .08 | .08 | 1.07 | .09 | .16 | .04 | .07 | .17 | .11 |
| .06 | .12 | .17 | .20 | .12 | .17 | .09 | .07 | .16 |
| .15 | .16 | .09 | .06 | .10 | .21 | .06 | | |

*Source:* Rubin, A., and Harris, W. F. "Refractive variation during autorefraction: Multivariate distribution of refractive status." *Optometry and Vision Science*, Vol. 72, No. 6, June 1995, p. 409 (Table 4).

**a.** Locate the measures of central tendency on the print-out and interpret their values.

**b.** Note that the data contains one unusually large (negative) cylinder power measurement relative to the other measurements in the data set. Find this

**SPSS Output for Exercise 2.51**

| PCTERROR | | | | | | | |
|----------|------|------|--------------|--------|----------|---------|---------|
| GROUP | Mean | N | Std.Deviation | Median | Variance | Minimum | Maximum |
| OND | .00 | 10 | .00 | .00 | .000 | 0 | 0 |
| SLI | 30.17 | 10 | 24.11 | 32.46 | 581.202 | 0 | 67 |
| YND | 46.88 | 10 | 38.21 | 43.72 | 1460.347 | 0 | 100 |
| Total | 25.68 | 30 | 31.98 | 9.38 | 10022.919 | 0 | 100 |

SAS Output for Exercise 2.52

The UNIVARIATE Procedure
Variable: CYLPOWER

Moments

| N | 25 | Sum Weights | 25 |
|---|---|---|---|
| Mean | 0.1544 | Sum Observations | 3.86 |
| Std Deviation | 0.19676721 | Variance | 0.03871733 |
| Skewness | 4.52208392 | Kurtosis | 21.7019614 |
| Uncorrected SS | 1.5252 | Corrected SS | 0.929216 |
| Coeff Variation | 127.4399 | Std Error Mean | 0.03935344 |

Basic Statistical Measures

| Location | | Variability | |
|---|---|---|---|
| Mean | 0.154400 | Std Deviation | 0.19677 |
| Median | 0.110000 | Variance | 0.03872 |
| Mode | 0.060000 | Range | 1.03000 |
| | | Interquartile Range | 0.08000 |

NOTE: The mode displayed is the smallest of 4 modes with a count of 3.

Extreme Observations

| -------Lowest------ | | -------Highest------- | |
|---|---|---|---|
| Value | Obs | Value | Obs |
| 0.04 | 6 | 0.17 | 12 |
| 0.06 | 25 | 0.17 | 15 |
| 0.06 | 22 | 0.20 | 13 |
| 0.06 | 10 | 0.21 | 24 |
| 0.07 | 17 | 1.07 | 3 |

measurement. (In Section 2.8, we call this value an **outlier**).

c. Delete the outlier, part **b**, from the data set and recalculate the measures of central tendency. Which measure is most affected by the deletion of the outlier?

**2.53** The salaries of superstar professional athletes receive much attention in the media. The multimillion-dollar long-term contract is now commonplace among this elite group. Nevertheless, rarely does a season pass without negotiations between one or more of the players' associations and team owners for additional salary and fringe benefits for *all* players in their particular sports.

a. If a players' association wanted to support its argument for higher "average" salaries, which measure of central tendency do you think it should use? Why?

b. To refute the argument, which measure of central tendency should the owners apply to the players' salaries? Why?

## 2.5 NUMERICAL MEASURES OF VARIABILITY

Measures of central tendency provide only a partial description of a quantitative data set. The description is incomplete without a **measure of the variability**, or **spread**, of the data set. Knowledge of the data set's variability along with its center can help us visualize the shape of a data set as well as its extreme values.

If you examine the two histograms in Figure 2.16, you'll notice that both hypothetical data sets are symmetric with equal modes, medians, and means. However, data set 1 (Figure 2.16a) has measurements spread with almost equal relative frequency over the measurement classes, while data set 2 (Figure 2.16b) has most of its measurements clustered about its center. Thus, data set 2 is *less variable* than data set 1. Consequently, you can see that we need a measure of variability as well as a measure of central tendency to describe a data set.

**FIGURE 2.16**
*Hypothetical Data Sets*

a. Data set 1                          b. Data set 2

Perhaps the simplest measure of the variability of a quantitative data set is its *range*.

---

**DEFINITION 2.7**
The **range** of a quantitative data set is equal to the largest measurement minus the smallest measurement.

---

The range is easy to compute and easy to understand, but it is a rather insensitive measure of data variation when the data sets are large. This is because two data sets can have the same range and be vastly different with respect to data variation. This phenomenon is demonstrated in Figure 2.16. Both distributions of data shown in the figure have the same range, but most of the measurements in data set 2 tend to concentrate near the center of the distribution. Consequently, the data are much less variable than the data in set 1. Thus, you can see that the range does not always detect differences in data variation for large data sets.

Let's see if we can find a measure of data variation that is more sensitive than the range. Consider the two samples in Table 2.6: Each has five measurements. (We have ordered the numbers for convenience.) Note that both samples have a mean of 3 and that we have also calculated the distance between each measurement and the mean. What information do these distances contain? If they tend to be large in magnitude, as in sample 1, the data are spread out, or highly variable. If the distances are mostly small, as in sample 2, the data are clustered around the mean, $\bar{x}$, and therefore do not exhibit much variability. You can see that these distances, displayed graphically in Figure 2.17, provide information about the variability of the sample measurements.

The next step is to condense the information in these distances into a single numerical measure of variability. Averaging the distances from $\bar{x}$ won't help because the negative and positive distances cancel; that is, the sum of the deviations (and thus the average deviation) is always equal to zero.

**TABLE 2.6   Two Hypothetical Data Sets**

|  | **Sample 1** | **Sample 2** |
|---|---|---|
| Measurements | $1, 2, 3, 4, 5$ | $2, 3, 3, 3, 4$ |
| Mean | $\bar{x} = \dfrac{1 + 2 + 3 + 4 + 5}{5} = \dfrac{15}{5} = 3$ | $\bar{x} = \dfrac{2 + 3 + 3 + 3 + 4}{5} = \dfrac{15}{5} = 3$ |
| Distances of measurement values from $\bar{x}$ | $(1 - 3), (2 - 3), (4 - 3),$ $(5 - 3)$ or $-2, -1, 0, 1, 2$ | $(2 - 3), (3 - 3), (3 - 3), (3 - 3),$ $(4 - 3)$ or $-1, 0, 0, 0, 1$ |

**FIGURE 2.17**

*Dot Plots for Two Data Sets*

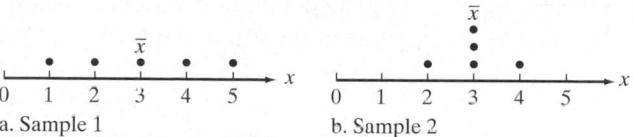

a. Sample 1    b. Sample 2

Two methods come to mind for dealing with the fact that positive and negative distances from the mean cancel. The first is to treat all the distances as though they were positive, ignoring the sign of the negative distances. We won't pursue this line of thought because the resulting measure of variability (the mean of the absolute values of the distances) presents analytical difficulties beyond the scope of this text. A second method of eliminating the minus signs associated with the distances is to square them. The quantity we can calculate from the squared distances will provide a meaningful description of the variability of a data set and presents fewer analytical difficulties in inference making.

To use the squared distances calculated from a data set, we first calculate the *sample variance*.

---

**DEFINITION 2.8**

The **sample variance** for a sample of $n$ measurements is equal to the sum of the squared distances from the mean divided by $(n - 1)$. In symbols, using $s^2$ to represent the sample variance,

$$s^2 = \frac{\sum_{i=1}^{n} (x_i - \bar{x})^2}{n - 1}$$

*Note*: A shortcut formula for calculating $s^2$ is

$$s^2 = \frac{\sum_{i=1}^{n} x_i^2 - \frac{\left(\sum_{i=1}^{n} x_i\right)^2}{n}}{n - 1}$$

---

Referring to the two samples in Table 2.6, you can calculate the variance for sample 1 as follows:

$$s^2 = \frac{(1 - 3)^2 + (2 - 3)^2 + (3 - 3)^2 + (4 - 3)^2 + (5 - 3)^2}{5 - 1}$$

$$= \frac{4 + 1 + 0 + 1 + 4}{4} = 2.5$$

The second step in finding a meaningful measure of data variability is to calculate the *standard deviation* of the data set:

---

**DEFINITION 2.9**

The **sample standard deviation**, $s$, is defined as the positive square root of the sample variance, $s^2$. Thus,

$$s = \sqrt{s^2}$$

The population variance, denoted by the symbol $\sigma^2$ (sigma squared), is the average of the squared distances of the measurements on *all* units in the population from the mean, $\mu$, and $\sigma$ (sigma) is the square root of this quantity.

---

### Symbols for Variance and Standard Deviation

$s^2$ = Sample variance
$s$ = Sample standard deviation
$\sigma^2$ = Population variance
$\sigma$ = Population standard deviation

---

Notice that, unlike the variance, the standard deviation is expressed in the original units of measurement. For example, if the original measurements are in dollars, the variance is expressed in the peculiar units "dollar squared," but the standard deviation is expressed in dollars.

You may wonder why we use the divisor $(n - 1)$ instead of $n$ when calculating the sample variance. Wouldn't using $n$ seem more logical, so that the sample variance would be the average squared distance from the mean? The trouble is, using $n$ tends to produce an underestimate of the population variance, $\sigma^2$. So we use $(n - 1)$ in the denominator to provide the appropriate correction for this tendency.* Since sample statistics like $s^2$ are primarily used to estimate population parameters like $\sigma^2$, $(n - 1)$ is preferred to $n$ when defining the sample variance.

**EXAMPLE 2.8**

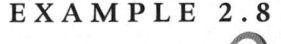

Calculate the variance and standard deviation of the following sample: 2, 3, 3, 3, 4.

### Solution

As the number of measurements increases, calculating $s^2$ and $s$ becomes very tedious. Fortunately, as we show in Example 2.9, we can use a statistical software package (or calculator) to find these values. If you must calculate these quantities by hand, it is advantageous to use the shortcut formula provided in Definition 2.8.

To do this, we need two summations: $\sum x$ and $\sum x^2$. These can easily be obtained from the following type of tabulation:

| $x$ | $x^2$ |
|---|---|
| 2 | 4 |
| 3 | 9 |
| 3 | 9 |
| 3 | 9 |
| 4 | 16 |
| $\sum x = 15$ | $\sum x^2 = 47$ |

---

*"Appropriate" here means that $s^2$ with a divisor of $(n - 1)$ is an *unbiased estimator* of $\sigma^2$. We define and discuss unbiasedness of estimators in Chapter 6.

**FIGURE 2.18**

*SAS Printout of Numerical Descriptive Measures for 100 EPA Mileages*

| | | | The MEANS Procedure | | |
| | | | Analysis Variable : MPG | | |
| N | Minimum | Maximum | Mean | Variance | Std Dev |
|---|---------|---------|------|----------|---------|
| 100 | 30.0000000 | 44.9000000 | 36.9940000 | 5.8462263 | 2.4178971 |

Then we use*

$$s^2 = \frac{\sum_{i=1}^{n} x_i^2 - \frac{\left(\sum_{i=1}^{n} x_i\right)^2}{n}}{n-1} = \frac{47 - \frac{(15)^2}{5}}{5-1} = \frac{2}{4} = .5$$

$$s = \sqrt{.5} = .71$$

**E X A M P L E  2 . 9**  Use the computer to find the sample variance $s^2$ and the sample standard deviation $s$ for the 100 gas mileage readings given in Table 2.3.

**Solution**

The SAS printout describing the gas mileage data is reproduced in Figure 2.18. The variance and standard deviation, highlighted on the printout, are (rounded) $s^2 = 5.84623$ and $s = 2.418$.

You now know that the standard deviation measures the variability of a set of data and how to calculate it. The larger the standard deviation, the more variable the data are. The smaller the standard deviation, the less variation in the data. But how can we practically interpret the standard deviation and use it to make inferences? This is the topic of Section 2.6.

---

*When calculating $s^2$, how many decimal places should you carry? Although there are no rules for the rounding procedure, it is reasonable to retain twice as many decimal places in $s^2$ as you ultimately wish to have in $s$. If you wish to calculate $s$ to the nearest hundredth (two decimal places), for example, you should calculate $s^2$ to the nearest ten-thousandth (four decimal places).

 EXERCISES 2.54–2.65

**Learning the Mechanics**

**2.54** Answer the following questions about variability of data sets:

**a.** What is the primary disadvantage of using the range to compare the variability of data sets?

**b.** Describe the sample variance using words rather than a formula. Do the same with the population variance.

**c.** Can the variance of a data set ever be negative? Explain. Can the variance ever be smaller than the standard deviation? Explain.

**2.55** Calculate the variance and standard deviation for samples where

**a.** $n = 10$, $\sum x^2 = 84$, $\sum x = 20$

**b.** $n = 40$, $\sum x^2 = 380$, $\sum x = 100$

**c.** $n = 20$, $\sum x^2 = 18$, $\sum x = 17$

**2.56** Calculate the range, variance, and standard deviation for the following samples:

**a.** 4, 2, 1, 0, 1

**b.** 1, 6, 2, 2, 3, 0, 3

**c.** 8, −2, 1, 3, 5, 4, 4, 1, 3

**d.** 0, 2, 0, 0, −1, 1, −2, 1, 0, −1, 1, −1, 0, −3, −2, −1, 0, 1

**2.57** Calculate the range, variance, and standard deviation for the following samples:

**a.** 39, 42, 40, 37, 41

**b.** 100, 4, 7, 96, 80, 3, 1, 10, 2

**c.** 100, 4, 7, 30, 80, 30, 42, 2

**2.58** Compute $\bar{x}$, $s^2$, and $s$ for each of the following data sets. If appropriate, specify the units in which your answer is expressed.
  **a.** 3, 1, 10, 10, 4
  **b.** 8 feet, 10 feet, 32 feet, 5 feet
  **c.** −1, −4, −3, 1, −4, −4
  **d.** 1/5 ounce, 1/5 ounce, 1/5 ounce, 2/5 ounce, 1/5 ounce, 4/5 ounce

**2.59** Using only integers between 0 and 10, construct two data sets with at least 10 observations each so that the two sets have the same mean but different variances. Construct dot plots for each of your data sets, and mark the mean of each data set on its dot plot.

**2.60** Using only integers between 0 and 10, construct two data sets with at least 10 observations each that have the same range but different means. Construct a dot plot for each of your data sets, and mark the mean of each data set on its dot plot.

**2.61** Consider the following sample of five measurements: 2, 1, 1, 0, 3.
  **a.** Calculate the range, $s^2$, and $s$.
  **b.** Add 3 to each measurement and repeat part **a**.
  **c.** Subtract 4 from each measurement and repeat part **a**.
  **d.** Considering your answers to parts **a**, **b**, and **c**, what seems to be the effect on the variability of a data set by adding the same number to or subtracting the same number from each measurement?

**2.62** Consider the following two samples:

  *Sample 1:*  10, 0, 1, 9, 10, 0, 8, 1, 1, 9
  *Sample 2:*  0, 5, 10, 5, 5, 5, 6, 5, 6, 5

  **a.** Examine both samples and identify the one that you believe has the greater variability.
  **b.** Calculate the range for each sample. Does the result agree with your answer to part **a**? Explain.
  **c.** Calculate the standard deviation for each sample. Does the result agree with your answer to part **a**? Explain.
  **d.** Which of the two, the range or the standard deviation, provides a better measure of variability?

## Applying the Concepts—Basic

**2.63** Refer to the *Environmental Science & Technology* (Sept. 1, 2000) study on the ammonia levels near the exit ramp of a San Francisco highway tunnel, Exercise 2.45 (p. 52). The data (in parts per million) for 8 days during afternoon drive-time are reproduced in the table.

  🖲 AMMONIA

| 1.53 | 1.50 | 1.37 | 1.51 | 1.55 | 1.42 | 1.41 | 1.48 |
|------|------|------|------|------|------|------|------|

  **a.** Find the range of the ammonia levels.
  **b.** Find the variance of the ammonia levels.
  **c.** Find the standard deviation of the ammonia levels.
  **d.** Suppose the standard deviation of the daily ammonia levels during morning drive-time at the exit ramp is

1.45 ppm. Which time, morning or afternoon drive-time, has more variable ammonia levels?

**2.64** Refer to *The Condor* (May 1995) study of collared flycatchers, Exercise 2.46 (p. 52). The data for the study is reproduced here.

  🖲 CONDOR

| Plot Number | Number Killed | Number of Breeders |
|-------------|---------------|--------------------|
| 1 | 5 | 30 |
| 2 | 4 | 28 |
| 3 | 3 | 38 |
| 4 | 2 | 34 |
| 5 | 2 | 26 |
| 6 | 1 | 124 |
| 7 | 1 | 68 |
| 8 | 1 | 86 |
| 9 | 1 | 32 |
| 10 | 0 | 30 |
| 11 | 0 | 46 |
| 12 | 0 | 132 |
| 13 | 0 | 100 |
| 14 | 0 | 6 |

  **a.** Find the range, variance, and standard deviation of the number of flycatchers killed at the 14 plots.
  **b.** Specify the units in which each of your answers to part **a** is expressed.
  **c.** Repeat parts **a** and **b** for the number of breeders.

## Applying the Concepts—Intermediate

**2.65** Refer to the *Fortune* (Oct. 25, 1999) ranking of the 50 most powerful women in America, Exercise 2.44 (p. 51). The distribution of the ages of the 50 women is summarized in the STATISTIX printout below.

**STATISTIX Output for Exercise 2.65**

```
         DESCRIPTIVE STATISTICS

                              AGE
         N                     50
         MEAN              48.160
         SD                6.0148
         VARIANCE          36.178
         MINIMUM           36.000
         MEDIAN            47.000
         MAXIMUM           68.000
```

  **a.** Locate the measures of variation on the printout.
  **b.** Give a value of the standard deviation that would make the age distribution more variable.
  **c.** Give a value of the standard deviation that would make the age distribution less variable.
  **d.** If the largest age in the data set (68 years) is omitted, would the standard deviation increase or decrease? Explain.

## 2.6 INTERPRETING THE STANDARD DEVIATION

We've seen that if we are comparing the variability of two samples selected from a population, the sample with the larger standard deviation is the more variable of the two. Thus, we know how to interpret the standard deviation on a relative or comparative basis, but we haven't explained how it provides a measure of variability for a single sample.

To understand how the standard deviation provides a measure of variability of a data set, consider a specific data set and answer the following questions: How many measurements are within 1 standard deviation of the mean? How many measurements are within 2 standard deviations? For example, look at the 100 mileage per gallon readings given in Table 2.3. Recall that $\bar{x} = 36.99$ and $s = 2.42$. Then

$$\bar{x} - s = 34.57 \qquad \bar{x} + s = 39.41$$
$$\bar{x} - 2s = 32.15 \qquad \bar{x} + 2s = 41.83$$

If we examine the data, we find that 68 of the 100 measurements, or 68%, are in the interval

$$\bar{x} - s \text{ to } \bar{x} + s$$

Similarly, we find that 96, or 96%, of the 100 measurements are in the interval

$$\bar{x} - 2s \text{ to } \bar{x} + 2s$$

We usually write these intervals as

$$(\bar{x} - s, \bar{x} + s) \text{ and } (\bar{x} - 2s, \bar{x} + 2s)$$

Such observations identify criteria for interpreting a standard deviation that apply to *any* set of data, whether a population or a sample. The criteria, expressed as a mathematical theorem and as a rule of thumb, are presented in Tables 2.7 and 2.8. In these tables we give two sets of answers to the questions of how many measurements fall within 1, 2, and 3 standard deviations of the mean. The first, which applies to *any* set of data, is derived from a theorem proved by the Russian mathematician P L. Chebyshev (1821–1894). The second, which applies to symmetric, **mound-shaped distributions** of data (where the mean, median, and mode are all about the same), is based upon empirical evidence that has accumulated over the years. However, the percentages given for the intervals in Table 2.8 provide remarkably good approximations even when the distribution of the data is slightly skewed or asymmetric.

### TABLE 2.7 Interpreting the Standard Deviation: Chebyshev's Rule

**Chebyshev's Rule** applies to any data set, regardless of the shape of the frequency distribution of the data.

a. No useful information is provided on the fraction of measurements that fall within 1 standard deviation of the mean, i.e., within the interval $(x - s, \bar{x} + s)$ for samples and $(\mu - \sigma, \mu + \sigma)$ for populations.

b. At least $\frac{3}{4}$ of the measurements will fall within 2 standard deviations of the mean, i.e., within the interval $(\bar{x} - 2s, \bar{x} + 2s)$ for samples and $(\mu - 2\sigma, \mu + 2\sigma)$ for populations.

c. At least $\frac{8}{9}$ of the measurements will fall within 3 standard deviations of the mean, i.e., within the interval $(\bar{x} - 3s, \bar{x} + 3s)$ for samples and $(\mu - 3\sigma, \mu + 3\sigma)$ for populations.

d. Generally, for any number $k$ greater than 1, at least $(1 - 1/k^2)$ of the measurements will fall within $k$ standard deviations of the mean, i.e., within the interval $(\bar{x} - ks, \bar{x} + ks)$ for samples and $(\mu - k\sigma, \mu + k\sigma)$ for populations.

---

**TABLE 2.8    Interpreting the Standard Deviation: The Empirical Rule**

The **Empirical Rule** is a rule of thumb that applies to data sets with frequency distributions that are mound-shaped and symmetric, as shown below.

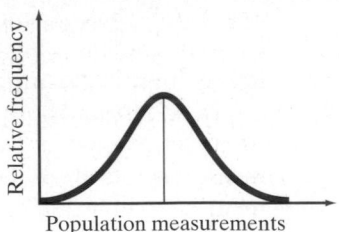

a. Approximately 68% of the measurements will fall within 1 standard deviation of the mean, i.e., within the interval $(\bar{x} - s, \bar{x} + s)$ for samples and $(\mu - \sigma, \mu + \sigma)$ for populations.
b. Approximately 95% of the measurements will fall within 2 standard deviations of the mean, i.e., within the interval $(\bar{x} - 2s, \bar{x} + 2s)$ for samples and $(\mu - 2\sigma, \mu + 2\sigma)$ for populations.
c. Approximately 99.7% (essentially all) of the measurements will fall within 3 standard deviations of the mean, i.e., within the interval $(\bar{x} - 3s, \bar{x} + 3s)$ for samples and $(\mu - 3\sigma, \mu + 3\sigma)$ for populations.

---

**EXAMPLE 2.10**

Thirty students in an experimental psychology class use various techniques to train a rat to move through a maze. At the end of the course, each student's rat is timed through the maze. The results (in minutes) are listed in Table 2.9. Determine the fraction of the 30 measurements in the intervals $\bar{x} \pm s$, $\bar{x} \pm 2s$, and $\bar{x} \pm 3s$, and compare the results with those predicted in Tables 2.7 and 2.8.

⊘ RATMAZE

**TABLE 2.9    Times (in Minutes) of 30 Rats Running Through a Maze**

| | | | | | | | | | |
|------|------|------|------|------|------|------|------|------|------|
| 1.97 | .60  | 4.02 | 3.20 | 1.15 | 6.06 | 4.44 | 2.02 | 3.37 | 3.65 |
| 1.74 | 2.75 | 3.81 | 9.70 | 8.29 | 5.63 | 5.21 | 4.55 | 7.60 | 3.16 |
| 3.77 | 5.36 | 1.06 | 1.71 | 2.47 | 4.25 | 1.93 | 5.15 | 2.06 | 1.65 |

**Solution**

First, we entered the data into the computer and used STATISTIX to produce summary statistics. The mean and standard deviation of the sample data, highlighted on the printout shown in Figure 2.19, are (rounded)

$$\bar{x} = 3.74 \text{ minutes} \quad s = 2.20 \text{ minutes}$$

**FIGURE 2.19**

*STATISTIX Printout for Example 2.10*

```
DESCRIPTIVE STATISTICS

                 RUNTIME
N                     30
MEAN              3.7443
SD                2.1982
VARIANCE          4.8323
MINIMUM           0.6000
MEDIAN            3.5100
MAXIMUM           9.7000
```

Now, we form the interval

$$(\bar{x} - s, \bar{x} + s) = (3.74 - 2.20, 3.74 + 2.20) = (1.54, 5.94)$$

A check of the measurements shows that 23 of the times are within this 1 standard deviation interval around the mean. This number represents 23/30, or $\approx 77\%$ of the sample measurements.

The next interval of interest is

$$(\bar{x} - 2s, \bar{x} + 2s) = (3.74 - 4.40, 3.74 + 4.40) = (-.66, 8.14)$$

All but two of the times are within this interval, so 28/30, or approximately 93%, are within 2 standard deviations of $\bar{x}$.

Finally, the 3-standard-deviation interval around $\bar{x}$ is

$$(\bar{x} - 3s, \bar{x} + 3s) = (3.74 - 6.60, 3.74 + 6.60) = (-2.86, 10.34)$$

All of the times fall within 3 standard deviations of the mean.

These 1-, 2-, and 3-standard-deviation percentages (77%, 93%, and 100%) agree fairly well with the approximations of 68%, 95%, and 100% given by the Empirical Rule (Table 2.8) for mound-shaped distributions. If you look at the STATISTIX frequency histogram for this data set in Figure 2.20, you'll note that the distribution is not really mound-shaped, nor is it extremely skewed. Thus, we get reasonably good results from the mound-shaped approximations. Of course, we know from Chebyshev's Rule (Table 2.7) that no matter what the shape of the distribution, we would expect at least 75% and 89% of the measurements to lie within 2 and 3 standard deviations of $\bar{x}$, respectively.

**FIGURE 2.20**

*STATISTIX Histogram for Times for Rats to Move Through Maze*

**EXAMPLE 2.11**

Chebyshev's Rule and the Empirical Rule (Tables 2.7 and 2.8) are useful as a check on the calculation of the standard deviation. For example, suppose we calculated the standard deviation for the gas mileage data (Table 2.3) to be 5.85. Are there any "clues" in the data that enable us to judge whether this number is reasonable?

**Solution**

The range of the mileage data in Table 2.3 is $44.0 - 30.0 = 14.9$. From Chebyshev's Rule and the Empirical Rule we know that most of the measurements (approximately 95% if the distribution is mound-shaped) will be within 2 standard deviations of the mean. And, regardless of the shape of the distribution and the number of measurements, almost all of them will fall within 3 standard deviations of the mean. Consequently, we would expect the range of the measurements to be between 4 (i.e., $\pm 2s$) and 6 (i.e., $\pm 3s$) standard deviations in length (see Figure 2.21). For the car mileage data, this means that $s$ should fall between

$$\frac{\text{Range}}{6} = \frac{14.9}{6} = 2.48 \quad \text{and} \quad \frac{\text{Range}}{4} = \frac{14.9}{4} = 3.73$$

In particular, the standard deviation should not be much larger than 1/4 of the range, particularly for the data set with 100 measurements. Thus, we have reason to believe that the calculation of 5.85 is too large. A check of our work reveals that 5.85 is the variance $s^2$, not the standard deviation $s$ (see Example 2.9). We "forgot" to take the square root (a common error); the correct value is $s = 2.42$. Note that this value is slightly smaller than the range divided by 6 (2.48). The larger the data set, the greater the tendency for very large or very small measurements (extreme values) to appear, and when they do, the range may exceed 6 standard deviations. ∎

**FIGURE 2.21**

*The Relation Between the Range and the Standard Deviation*

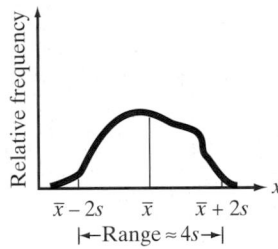

In examples and exercises we'll sometimes use $s \approx \text{range}/4$ to obtain a crude, and usually conservatively large, approximation for $s$. However, we stress that this is no substitute for calculating the exact value of $s$ when possible.

Finally, and most importantly, we will use the concepts in Chebyshev's Rule and the Empirical Rule to build the foundation for statistical inference-making. The method is illustrated in Example 2.12.

**EXAMPLE 2.12**

A manufacturer of automobile batteries claims that the average length of life for its grade A battery is 60 months. However, the guarantee on this brand is for just 36 months. Suppose the standard deviation of the life length is known to be 10 months, and the frequency distribution of the life-length data is known to be mound-shaped.

a. Approximately what percentage of the manufacturer's grade A batteries will last more than 50 months, assuming the manufacturer's claim is true?

b. Approximately what percentage of the manufacturer's batteries will last less than 40 months, assuming the manufacturer's claim is true?

c. Suppose your battery lasts 37 months. What could you infer about the manufacturer's claim?

### Solution

If the distribution of life length is assumed to be mound-shaped with a mean of 60 months and a standard deviation of 10 months, it would appear as shown in Figure 2.22. Note that we can take advantage of the fact that mound-shaped distributions are (approximately) symmetric about the mean, so that the percentages given by the Empirical Rule can be split equally between the halves of the distribution on each side of the mean. The approximations given in Figure 2.22 are more dependent on the assumption of a mound-shaped distribution than those given by the Empirical Rule (Table 2.8), because the approximations in Figure 2.22 depend on the (approximate) symmetry of the mound-shaped distribution. We saw in Example 2.10 that the Empirical Rule can yield good approximations even for skewed distributions. This will *not* be true of the approximations in Figure 2.22; the distribution *must* be mound-shaped and (approximately) symmetric.

**FIGURE 2.22**

*Battery Life-Length Distribution: Manufacturer's Claim Assumed True*

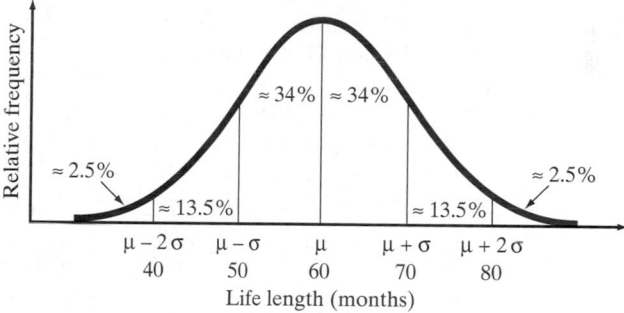

For example, since approximately 68% of the measurements will fall within 1 standard deviation of the mean, the distribution's symmetry implies that approximately $(1/2)(68\%) = 34\%$ of the measurements will fall between the mean and 1 standard deviation on each side. This concept is illustrated in Figure 2.22. The figure also shows that 2.5% of the measurements lie beyond 2 standard deviations in each direction from the mean. This result follows from the fact that if approximately 95% of the measurements fall within 2 standard deviations of the mean, then about 5% fall outside 2 standard deviations; if the distribution is approximately symmetric, then about 2.5% of the measurements fall beyond 2 standard deviations on each side of the mean.

a. It is easy to see in Figure 2.22 that the percentage of batteries lasting more than 50 months is approximately 34% (between 50 and 60 months) plus 50% (greater than 60 months). Thus, approximately 84% of the batteries should have life length exceeding 50 months.

b. The percentage of batteries that last less than 40 months can also be easily determined from Figure 2.22. Approximately 2.5% of the batteries should fail prior to 40 months, assuming the manufacturer's claim is true.

c. If you are so unfortunate that your grade A battery fails at 37 months, you can make one of two inferences: Either your battery was one of the approximately 2.5% that fail prior to 40 months, or something about the manufacturer's claim is not true. Because the chances are so small that a battery fails before 40 months, you would have good reason to have serious doubts about the manufacturer's claim. A mean smaller than 60 months and/or a standard deviation longer than 10 months would both increase the likelihood of failure prior to 40 months.* _____ ■

Example 2.12 is our initial demonstration of the statistical inference-making process. At this point you should realize that we'll use sample information (in Example 2.12, your battery's failure at 37 months) to make inferences about the population (in Example 2.12, the manufacturer's claim about the life length for the population of all batteries). We'll build on this foundation as we proceed.

 EXERCISES 2.66–2.81

## Learning the Mechanics

**2.66** To what kind of data sets can Chebyshev's Rule be applied? The Empirical Rule?

**2.67** The output from a statistical computer program indicates that the mean and standard deviation of a data set consisting of 200 measurements are $1,500 and $300, respectively.
  **a.** What are the units of measurement of the variable of interest? Based on the units, what type of data is this: quantitative or qualitative?
  **b.** What can be said about the number of measurements between $900 and $2,100? Between $600 and $2,400? Between $1,200 and $1,800? Between $1,500 and $2,100?

**2.68** For any set of data, what can be said about the percentage of the measurements contained in each of the following intervals?
  **a.** $\bar{x} - s$ to $\bar{x} + s$
  **b.** $\bar{x} - 2s$ to $\bar{x} + 2s$
  **c.** $\bar{x} - 3s$ to $\bar{x} + 3s$

**2.69.** For a set of data with a mound-shaped relative frequency distribution, what can be said about the percentage of the measurements contained in each of the intervals specified in Exercise 2.68?

**2.70** The following is a sample of 25 measurements:

| 7 | 6 | 6 | 11 | 8 | 9 | 11 | 9 | 10 | 8 | 7 | 7 | 5 |
| 9 | 10 | 7 | 7 | 7 | 7 | 9 | 12 | 10 | 10 | 8 | 6 | |

  **a.** Compute $\bar{x}$, $s^2$, and $s$ for this sample.
  **b.** Count the number of measurements in the intervals $\bar{x} \pm s$, $\bar{x} \pm 2s$, and $\bar{x} \pm 3s$. Express each count as a percentage of the total number of measurements.
  **c.** Compare the percentages found in part **b** to the percentages given by the Empirical Rule and Chebyshev's Rule.
  **d.** Calculate the range and use it to obtain a rough approximation for $s$. Does the result compare favorably with the actual value for $s$ found in part **a**?

**2.71** Given a data set with a largest value of 760 and a smallest value of 135, what would you estimate the standard deviation to be? Explain the logic behind the procedure you used to estimate the standard deviation. Suppose the standard deviation is reported to be 25. Is this feasible? Explain.

## Applying the Concepts—Basic

**2.72** To minimize the potential for gastrointestinal disease outbreaks, all passenger cruise ships arriving at U.S. ports are subject to unannounced sanitation inspections. Ships are rated on a 100-point scale by the Centers for Disease Control and Prevention. A score of 86 or higher indicates that the ship is providing an accepted standard of sanitation. The May 2001 sanitation scores for 151 cruise ships are listed in the accompanying table, followed by a MINITAB printout of descriptive statistics on p. 68.
  **a.** Locate the mean and standard deviation on the MINITAB printout.
  **b.** Calculate the intervals $\bar{x} \pm s$, $\bar{x} \pm 2s$, $\bar{x} \pm 3s$.
  **c.** Find the percentage of measurements in the data set that fall within each of the intervals, part **b**. Do these percentages agree with either Chebyshev's Theorem or the Empirical Rule?

---

*The assumption that the distribution is mound-shaped and symmetric may also be incorrect. However, if the distribution were skewed to the right, as life-length distributions often tend to be, the percentage of measurements more than 2 standard deviations below the mean would be even less than 2.5%.

SHIPSANIT

| Ship Name | Score | Ship Name | Score | Ship Name | Score |
| --- | --- | --- | --- | --- | --- |
| Grande Mariner | 95 | Crystal Symphony | 93 | Scotia Prince | 92 |
| Seabourn Sun | 90 | Statendam | 94 | Splendour of the Seas | 90 |
| Olympic Voyager | 95 | Texas Treasure | 96 | Silver Wind | 96 |
| Grandeur of the Seas | 94 | Elation | 95 | Clipper Adventurer | 92 |
| Jubilee | 95 | Delphin | 99 | Sea Bird | 96 |
| Club Med 2 | 94 | Sea Princess | 97 | Grande Caribe | 93 |
| Costa Victoria | 96 | Melody | 98 | Spirit of Columbia | 95 |
| Wind Spirit | 98 | Crystal Harmony | 96 | Yorktown Clipper | 90 |
| Norwegian Majesty | 94 | Millennium | 93 | Hanseatic | 97 |
| Galaxy | 98 | Zenith | 95 | Pacific Sky | 91 |
| Costa Atlantica | 100 | Legacy | 93 | Big Red Boat II | 94 |
| Volendam | 98 | Vision of the Seas | 92 | Pacific Princess | 88 |
| Norwegian Dream | 100 | Westerdam | 92 | Big Red Boat III | 92 |
| Maasdam | 94 | C. Columbus | 98 | Oceanic | 91 |
| Mercury | 96 | Queen Elizabeth 2 | 92 | Niagra Prince | 88 |
| Orient Venus | 87 | Fuji Maru | 91 | Sea Lion | 92 |
| Seven Seas Navigator | 100 | Pacific Venus | 82 | Regal Princess | 93 |
| Seabourn Pride | 93 | Universe Explorer | 95 | Dolphin IV | 88 |
| Infinity | 97 | The Emerald | 91 | Asuka | 94 |
| Seabourn Legend | 97 | Celebration | 91 | Enchanted Capri | 93 |
| Radisson Diamond | 96 | Inspiration | 98 | Horizon | 92 |
| Ocean Princess | 97 | Grand Princess | 94 | Rembrandt | 94 |
| Nordic Empress | 98 | Regal Empress | 89 | Enchanted Isle | 89 |
| Vistamar | 83 | Ocean Breeze | 95 | Maxim Gorky | 90 |
| Norwegian Wind | 92 | Palm Beach Princess | 92 | Enchanted Sun | 89 |
| Bolero | 96 | Regal Voyager | 87 | The Topaz | 77 |
| Oriana | 100 | Rotterdam | 97 | Costa Romantica | 90 |
| Nippon Maru | 99 | Dawn Princess | 96 | Royal Princess | 94 |
| Paradise | 95 | Noordam | 95 | Flamenco | 95 |
| Albatross | 93 | Enchantment of the Seas | 99 | Stella Solaris | 93 |
| Viking Serenade | 92 | Fascination | 95 | Contessa 1 | 93 |
| Holiday | 97 | Caronia | 95 | Aegean I | 52 |
| Island Adventure | 97 | Arcadia | 96 | Silver Cloud | 98 |
| Carnival Victory | 97 | Seabourn Goddess II | 94 | Deutschland | 91 |
| Norway | 93 | Explorer of the Seas | 91 | Spirit of Ninety Eight | 91 |
| Imagination | 94 | Voyager of the Sea | 92 | Legend of the Seas | 95 |
| Majesty of the Seas | 90 | Europa | 96 | Astor | 91 |
| Crown Princess | 96 | Discovery Sun | 95 | Paul Gauguin | 86 |
| Arkona | 90 | Norwegian Sea | 90 | Norwegian Star | 78 |
| Sun Princess | 98 | Tropicale | 86 | Black Watch | 86 |
| Monarch of the Seas | 93 | Ryndam | 96 | Spirit of Alaska | 93 |
| Carnival Destiny | 96 | Amsterdam | 93 | Aida | 90 |
| Seabourn Goddess I | 92 | Fantasy | 93 | Spirit of Discovery | 96 |
| Zaandam | 97 | Silver Shadow | 97 | Spirit of Glacier Bay | 95 |
| Aurora | 98 | Century | 93 | Seabourn Spirit | 92 |
| Disney Wonder | 99 | Ecstasy | 94 | Triton | 93 |
| Carnival Triumph | 98 | Norwegian Sky | 97 | Costa Allegra | 88 |
| Sensation | 90 | Nantucket Clipper | 88 | Bremen | 90 |
| Victoria | 94 | Rhapsody of the Seas | 94 | Costa Classica | 92 |
| Sovereign of the Seas | 98 | Veendam | 86 | | |
| Disney Magic | 96 | Le Levant | 91 | | |

*Source:* National Center for Environmental Health, Centers for Disease Control and Prevention, May 8, 2001.

**MINITAB Output for Exercise 2.72**

```
Descriptive Statistics: Score

Variable            N        Mean     Median     TrMean     StDev    SE Mean
Score             151      93.113     94.000     93.585     5.184      0.422

Variable      Minimum     Maximum         Q1         Q3
Score          52.000     100.000     91.000     96.000
```

**2.73** Refer to the *Marine Technology* (Jan. 1995) data on spillage amounts (in thousands of metric tons) for 50 major oil spills, Exercise 2.15 (p. 29). An SPSS histogram for the 50 spillage amounts is shown below.
  **a.** Interpret the histogram.
  **b.** Descriptive statistics for the 50 spillage amounts are also shown on the SPSS histogram. Use this information to form an interval that can be used to predict the spillage amount for the next major oil spill.

**2.74** Refer to the *American Psychologist* (July 1995) study of 148 applicants for a position in experimental psychology, Exercise 2.47 (p. 53). Recall that the mean number of recommendation letters included in each application packet was $\bar{x} = 2.28$. The standard deviation was also reported in the article; it's value was $s = 1.48$.
  **a.** Sketch the relative frequency distribution for the number of recommendation letters included in each application for the experimental psychology position. (Assume the distribution is mound-shaped and relatively symmetric.)

  **b.** Locate an interval on the distribution, part **a**, that captures approximately 95% of the measurements in the sample.
  **c.** Locate an interval on the distribution, part **a**, that captures almost all the sample measurements.

**2.75** For each day of last year, the number of vehicles passing through a certain intersection was recorded by a city engineer. One objective of this study was to determine the percentage of days that more than 425 vehicles used the intersection. Suppose the mean for the data was 375 vehicles per day and the standard deviation was 25 vehicles.
  **a.** What can you say about the percentage of days that more than 425 vehicles used the intersection? Assume you know nothing about the shape of the relative frequency distribution for the data.
  **b.** What is your answer to part **a** if you know that the relative frequency distribution for the data is mound-shaped?

**SPSS Output for Exercise 2.73**

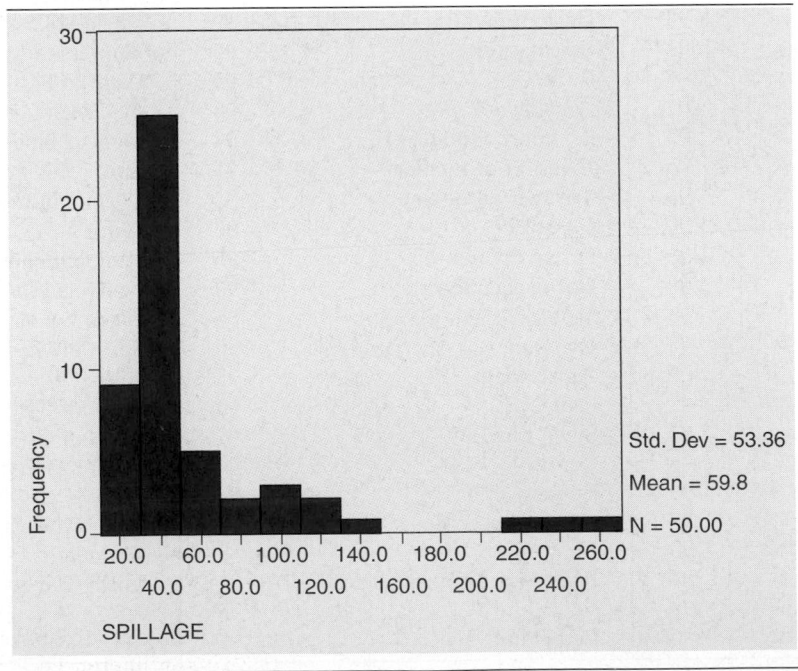

## Applying the Concepts—Intermediate

**2.76** A study published in *Applied Psycholinguistics* (June 1998) compared the language skills of young children (16–30 months old) from low income and middle income families. A total of 260 children—65 in the low income and 195 in the middle income group—completed the Communicative Development Inventory (CDI) exam. One of the variables measured on each child was sentence complexity score. Summary statistics for the scores of the two groups are reproduced in the table. Use this information to sketch a graph of the sentence complexity score distribution for each income group. (Assume the distributions are mound-shaped and symmetric.) Compare the distributions. What can you infer?

| | Low Income | Middle Income |
|---|---|---|
| Sample Size | 65 | 195 |
| Mean | 7.62 | 15.55 |
| Median | 4 | 14 |
| Standard Deviation | 8.91 | 12.24 |
| Minimum | 0 | 0 |
| Maximum | 36 | 37 |

*Source:* Arriaga, R. I. *et al.* "Scores on the MacArthur Communicative Development Inventory of children from low-income and middle-income families." *Applied Psycholinguistics*, Vol. 19, No. 2, June 1998, p. 217 (Table 7).

**2.77** The *American Rifleman* (June 1993) reported on the velocity of ammunition fired from the FEG P9R pistol, a 9mm gun manufactured in Hungary. Field tests revealed that Winchester bullets fired from the pistol had a mean velocity (at 15 feet) of 936 feet per second and a standard deviation of 10 feet per second. Tests were also conducted with Uzi and Black Hills ammunition.
**a.** Describe the velocity distribution of Winchester bullets fired from the FEG P9R pistol.
**b.** A bullet, brand unknown, is fired from the FEG P9R pistol. Suppose the velocity (at 15 feet) of the bullet is 1,000 feet per second. Is the bullet likely to be manufactured by Winchester? Explain.

**2.78** Astronomers theorize that cold dark matter (CDM) caused the formation of galaxies and clusters of galaxies in the universe. The theoretical CDM model requires an estimate of the velocity of light emitted from the galaxy cluster. *The Astronomical Journal* (July 1995) published a study of observed velocities for galaxies in four different galaxy clusters. Galaxy velocity was measured in kilometers per second (km/s) using a spectrograph and high-power telescope.
**a.** The observed velocities of 103 galaxies located in the cluster named A2142 are summarized in the accompanying histogram. Comment on whether the Empirical Rule is applicable for describing the velocity distribution for this cluster.

*Source:* Oegerle, W. R., Hill, J. M., and Fitchett, M. J. "Observations of high dispersion clusters of galaxies: Constraints on cold dark matter." *The Astronomical Journal*, Vol. 110, No. 1, July 1995, p. 37 (Figure 1).

**b.** The mean and standard deviation of the 103 velocities observed in galaxy cluster A2142 were reported as $\bar{x} = 27,117$ km/s and $s = 1,280$ km/s, respectively. Use this information to construct an interval that captures approximately 95% of the galaxy velocities in the cluster.
**c.** Recommend a single velocity value to be used in the CDM model for galaxy cluster A2142. Explain your reasoning

**2.79** The Locus of Control (LOC) is a measure of one's perception of control over factors affecting one's life. In one study, the LOC was measured for two groups of individuals undergoing weight-reduction for treatment of obesity (*Journal of Psychology*, Mar. 1991). The LOC mean and standard deviation for a sample of 46 adults were 6.45 and 2.89, respectively, while the LOC mean and standard deviation for a sample of 19 adolescents were 10.89 and 2.48, respectively. A lower score on the LOC scale indicates a perception that internal factors are in control, while a higher score indicates a perception that external factors are in control.
**a.** Calculate the 1- and 2-standard-deviation intervals around the means for each group. Plot these intervals on a line graph using different colors or symbols to represent each group.
**b.** Assuming that the distributions of LOC scores are approximately mound-shaped, estimate the numbers of individuals within each interval.
**c.** Based on your answers to parts **a** and **b**, do you think an inference can be made that *all* adults and adolescents undergoing weight-reduction treatment differ with respect to LOC? What factors did you consider in making this inference? (We'll reconsider this exercise in Chapter 9 to show how to measure the reliability of this inference.)

**Applying the Concepts—Advanced**

**2.80** A buyer for a lumber company must decide whether to buy a piece of land containing 5,000 pine trees. If 1,000 of the trees are at least 40 feet tall, the buyer will purchase the land; otherwise, he won't. The owner of the land reports that the height of the trees has a mean of 30 feet and a standard deviation of 3 feet. Based on this information, what is the buyer's decision?

**2.81** The National Education Longitudinal Survey (NELS) tracks a nationally representative sample of U.S. students from eighth grade through high school and college. Research published in *Chance* (Winter 2001) examined the Standardized Admission Test (SAT) scores of 265 NELS students who paid a private tutor to help them improve their scores. The table summarizes the changes in both the SAT-Mathematics and SAT-Verbal scores for these students.

| | SAT-Math | SAT-Verbal |
|---|---|---|
| Mean change in score | 19 | 7 |
| Standard deviation of score changes | 65 | 49 |

a. Suppose one of the 265 students who paid a private tutor is selected at random. Give an interval that is likely to contain this student's change in the SAT-Math score.

b. Repeat part **a** for the SAT-Verbal score.

c. Suppose the selected student's score increased on one of the SAT tests by 140 points. Which test, the SAT-Math or SAT-Verbal, is the one most likely to have the 140-point increase? Explain.

## 2.7 NUMERICAL MEASURES OF RELATIVE STANDING

We've seen that numerical measures of central tendency and variability describe the general nature of a quantitative data set (either a sample or a population). In addition, we may also be interested in describing the *relative* quantitative location of a particular measurement within a data set. Descriptive measures of the relationship of a measurement to the rest of the data are called **measures of relative standing**.

One measure of the relative standing of a measurement is its *percentile ranking*. For example, suppose you scored an 80 on a test and you want to know how you fared in comparison with others in your class. If the instructor tells you that you scored at the 90th percentile, it means that 90% of the grades were lower than yours and 10% were higher. Thus, if the scores were described by the relative frequency histogram in Figure 2.23, the 90th percentile would be located at a point such that 90% of the total area under the relative frequency histogram lies below the 90th percentile and 10% lies above. If the instructor tells you that you scored in the 50th percentile (the median of the data set), 50% of the test grades would be lower than yours and 50% would be higher.

**FIGURE 2.23**

*Location of 90th Percentile for Test Grades*

Percentile rankings are of practical value only for large data sets. Finding them involves a process similar to the one used in finding a median. The measurements are ranked in order and a rule is selected to define the location of each percentile. Since we are primarily interested in interpreting the percentile rankings of measurements (rather than finding particular percentiles for a data set), we define the *p*th *percentile* of a data set as shown in Definition 2.10.

> **DEFINITION 2.10**
>
> For any set of *n* measurements (arranged in ascending or descending order), the **ptb percentile** is a number such that *p*% of the measurements fall below the *p*th percentile and $(100 - p)$% fall above it.

**EXAMPLE 2.13**    Refer to the student default rates of the 50 states (and District of Columbia) in Table 2.5. An SAS printout describing the data is shown in Figure 2.24. Locate the 25th percentile and 95th percentile on the printout and interpret these values.

**FIGURE 2.24**

*SAS Descriptive Statistics for Student Default Rate Data*

The UNIVARIATE Procedure
Variable:  DEFAULT

Moments

| N | 51 | Sum Weights | 51 |
|---|---|---|---|
| Mean | 10.1960784 | Sum Observations | 520 |
| Std Deviation | 3.504395 | Variance | 12.2807843 |
| Skewness | 0.32759095 | Kurtosis | -0.0733636 |
| Uncorrected SS | 5916 | Corrected SS | 614.039216 |
| Coeff Variation | 34.3700279 | Std Error Mean | 0.49071345 |

Basic Statistical Measures

| Location | | Variability | |
|---|---|---|---|
| Mean | 10.19608 | Std Deviation | 3.50439 |
| Median | 9.70000 | Variance | 12.28078 |
| Mode | 8.80000 | Range | 17.00000 |
| | | Interquartile Range | 4.40000 |

Quantiles (Definition 5)

| Quantile | Estimate |
|---|---|
| 100% Max | 19.7 |
| 99% | 19.7 |
| 95% | 15.6 |
| 90% | 14.8 |
| 75% Q3 | 12.3 |
| 50% Median | 9.7 |
| 25% Q1 | 7.9 |
| 10% | 6.0 |
| 5% | 4.9 |
| 1% | 2.7 |
| 0% Min | 2.7 |

Extreme Observations

| ----Lowest---- | | ----Highest--- | |
|---|---|---|---|
| Value | Obs | Value | Obs |
| 2.7 | 51 | 15.2 | 44 |
| 4.8 | 35 | 15.5 | 34 |
| 4.9 | 28 | 15.6 | 25 |
| 5.5 | 42 | 16.6 | 21 |
| 5.7 | 17 | 19.7 | 2 |

**Solution**

Both the 25th percentile and 95th percentile are highlighted on the SAS printout, Figure 2.24. These values are 7.9 and 15.6, respectively. Our interpretations are as follows: 25% of the 51 default rates fall below 7.9 and 95% of the default rates fall below 15.6.

Another measure of relative standing in popular use is the *z-score*. As you can see in Definition 2.11, the *z*-score makes use of the mean and standard deviation of the data set in order to specify the relative location of the measurement.

---

**DEFINITION 2.11**

The **sample z-score** for a measurement $x$ is

$$z = \frac{x - \bar{x}}{s}$$

The **population z-score** for a measurement $x$ is

$$z = \frac{x - \mu}{\sigma}$$

---

Note that the *z*-score is calculated by subtracting $\bar{x}$ (or $\mu$) from the measurement $x$ and then dividing the result by $s$ (or $\sigma$). The final result, the **z-score**, represents the distance between a given measurement $x$ and the mean, expressed in standard deviations.

**EXAMPLE 2.14**

Suppose a sample of 2,000 high school seniors' verbal SAT scores is selected. The mean and standard deviation are

$$\bar{x} = 550 \quad s = 75$$

Suppose Joe Smith's score is 475. What is his sample *z*-score?

**Solution**

You can see that Joe Smith's score lies below the mean score of the 2,000 seniors:

| 325 | 475 | 550 | 775 |
|-----|-----|-----|-----|
| $\bar{x} - 3s$ | Joe Smith's score | $\bar{x}$ | $\bar{x} + 3s$ |

We compute

$$z = \frac{x - \bar{x}}{s} = \frac{475 - 550}{75} = -1.0$$

which tells us that Joe Smith's score is 1.0 standard deviation *below* the sample mean; in short, his sample *z*-score is $-1.0$.

The numerical value of the *z*-score reflects the relative standing of the measurement. A large positive *z*-score implies that the measurement is larger than almost all other measurements, whereas a large negative *z*-score indicates that the measurement is smaller than almost every other measurement. If a z-score is 0 or near 0, the measurement is located at or near the mean of the sample or population.

We can be more specific if we know that the frequency distribution of the measurements is mound-shaped. In this case, the following interpretation of the z-score can be given:

---

### Interpretation of z-Scores for Mound-Shaped Distributions of Data

1. Approximately 68% of the measurements will have a z-score between −1 and 1.
2. Approximately 95% of the measurements will have a z-score between −2 and 2.
3. Approximately 99.7% (almost all) of the measurements will have a z-score between −3 and 3.

---

Note that this interpretation of z-scores is identical to that given by the Empirical Rule for mound-shaped distributions (Table 2.8). The statement that a measurement falls in the interval $(\mu - \sigma)$ to $(\mu + \sigma)$ is equivalent to the statement that a measurement has a population z-score between −1 and 1, since all measurements between $(\mu - \sigma)$ and $(\mu + \sigma)$ are within 1 standard deviation of $\mu$. These z-scores are displayed in Figure 2.25.

**FIGURE 2.25**

*Population z-Scores for a Mound-Shaped Distribution*

## EXERCISES 2.82–2.93

### Learning the Mechanics

**2.82** Compute the z-score corresponding to each of the following values of x:

   **a.** $x = 40$, $\quad s = 5$, $\quad \bar{x} = 30$
   **b.** $x = 90$, $\quad \mu = 89$, $\quad \sigma = 2$
   **c.** $\mu = 50$, $\quad \sigma = 5$, $\quad x = 50$
   **d.** $s = 4$, $\quad x = 20$, $\quad \bar{x} = 30$

   **e.** In parts **a–d**, state whether the z-score locates x within a sample or a population.
   **f.** In parts **a–d**, state whether each value of x lies above or below the mean and by how many standard deviations.

**2.83** Give the percentage of measurements in a data set that are above and below each of the following percentiles:

**a.** 75th percentile
**b.** 50th percentile
**c.** 20th percentile
**d.** 84th percentile

**2.84** What is the 50th percentile of a quantitative data set called?

**2.85** Compare the $z$-scores to decide which of the following $x$ values lie the greatest distance above the mean and the greatest distance below the mean.

    **a.** $x = 100$,    $\mu = 50$,     $\sigma = 25$
    **b.** $x = 1$,      $\mu = 4$,      $\sigma = 1$
    **c.** $x = 0$,      $\mu = 200$,   $\sigma = 100$
    **d.** $x = 10$,    $\mu = 5$,      $\sigma = 3$

**2.86** Suppose that 40 and 90 are two elements of a population data set and that their $z$-scores are $-2$ and 3, respectively. Using only this information, is it possible to determine the population's mean and standard deviation? If so, find them. If not, explain why it's not possible.

## Applying the Concepts—Basic

**2.87** The distribution of scores on a nationally administered college achievement test has a median of 520 and a mean of 540.

    **a.** Explain why it is possible for the mean to exceed the median for this distribution of measurements.
    **b.** Suppose you are told that the 90th percentile is 660. What does this mean?
    **c.** Suppose you are told that you scored at the 94th percentile. Interpret this statement.

**2.88** According to the National Center for Education Statistics (2000), scores on a mathematics assessment test for United States eighth graders have a mean of 500, a 5th percentile of 356, a 25th percentile of 435, a 75th percentile of 563, and a 95th percentile of 653. Interpret each of these numerical descriptive measures.

**2.89** Refer to the sanitation levels of cruise ships, Exercise 2.72 (p. 66). The MINITAB descriptive statistics printout is reproduced below.

    **a.** Give a measure of relative standing for the Norwegian Star's score of 78. Interpret the result.
    **b.** Give a measure of relative standing for the Rotterdam's score of 97. Interpret the result.

## Applying the Concepts—Intermediate

**2.90** The US. Environmental Protection Agency (EPA) sets a limit on the amount of lead permitted in drinking water. The EPA *Action Level* for lead is .015 milligrams per liter (mg/L) of water. Under EPA guidelines, if 90% of a water system's study samples have a lead concentration less than .015 mg/L, the water is considered safe for drinking. I (co-author Sincich) received a report on a study of lead levels in the drinking water of homes in my subdivision. The 90th percentile of the study sample had a lead concentration of .00372 mg/L. Are water customers in my subdivision at risk of drinking water with unhealthy lead levels? Explain.

**2.91** Refer to the *Optometry and Vision Science* (June 1995) study of refractive variation in eyes, Exercise 2.52 (p. 54). A portion of the SAS descriptive statistics printout for the data set consisting of 25 cylinder power measurements is displayed on p. 75.

    **a.** Find the 10th percentile of cylinder power measurements on the printout. Interpret the result.
    **b.** Find the 95th percentile of cylinder power measurements on the printout. Interpret the result.
    **c.** Use the information on the SAS printout to calculate the $z$-score for the cylinder power measurement of 1.07. Interpret the result.

**2.92** In a study of how external clues influence performance, psychology professors at the University of Alberta and Pennsylvania State University gave two different forms of a midterm examination to a large group of introductory psychology students. The questions on the exam were identical and in the same order, but one exam was printed on blue paper and the other on red paper. (*Teaching Psychology*, May 1998.) Grading only the difficult questions on the exam, the researchers found that scores on the blue exam had a distribution with a mean of 53% and a standard deviation of 15%, while scores on the red exam had a distribution with a mean of 39% and a standard deviation of 12%. (Assume that both distributions are approximately mound-shaped and symmetric.)

    **a.** Give an interpretation of the standard deviation for the students who took the blue exam.
    **b.** Give an interpretation of the standard deviation for the students who took the red exam.
    **c.** Suppose a student is selected at random from the group of students who participated in the study and the student's score on the difficult questions is 20%. Which exam form is the student more likely to have taken, the blue or the red exam? Explain.

**2.93** At one university, the students are given $z$-scores at the end of each semester rather than the traditional GPAs. The mean and standard deviation of all students'

**MINITAB Output for Exercise 2.89**

| Descriptive Statistics: Score | | | | | |
|---|---|---|---|---|---|
| Variable | N | Mean | Median | TrMean | StDev | SE Mean |
| Score | 151 | 93.113 | 94.000 | 93.585 | 5.184 | 0.422 |
| | | | | | |
| Variable | Minimum | Maximum | Q1 | Q3 | |
| Score | 52.000 | 100.000 | 91.000 | 96.000 | |

cumulative GPAs, on which the $z$-scores are based, are 2.7 and .5, respectively.

**a.** Translate each of the following $z$-scores to corresponding GPA scores: $z = 2.0$, $z = -1.0$, $z = .5$, $z = -2.5$.

**b.** Students with $z$-scores below -1.6 are put on probation. What is the corresponding probationary GPA?

**c.** The president of the university wishes to graduate the top 16% of the students with *cum laude* honors and the top 2.5% with *summa cum laude* honors. Where (approximately) should the limits be set in terms of $z$ scores? In terms of GPAs? What assumption, if any, did you make about the distribution of the GPAs at the university?

**SAS Output for Exercise 2.91**

The UNIVARIATE Procedure
Variable: CYLPOWER

Basic Statistical Measures

| Location | | Variability | |
|---|---|---|---|
| Mean | 0.154400 | Std Deviation | 0.19677 |
| Median | 0.110000 | Variance | 0.03872 |
| Mode | 0.060000 | Range | 1.03000 |
| | | Interquartile Range | 0.08000 |

NOTE: The mode displayed is the smallest of 4 modes with a count of 3.

Quantiles (Definition 5)

| Quantile | Estimate |
|---|---|
| 100% Max | 1.07 |
| 99% | 1.07 |
| 95% | 0.21 |
| 90% | 0.20 |
| 75% Q3 | 0.16 |
| 50% Median | 0.11 |
| 25% Q1 | 0.08 |
| 10% | 0.06 |
| 5% | 0.06 |
| 1% | 0.04 |
| 0% Min | 0.04 |

# 2.8  METHODS FOR DETECTING OUTLIERS (OPTIONAL)

Sometimes it is important to identify inconsistent or unusual measurements in a data set. An observation that is unusually large or small relative to the data values we want to describe is called an **outlier**.

Outliers are often attributable to one of several causes. First, the measurement associated with the outlier may be invalid. For example, the experimental procedure used to generate the measurement may have malfunctioned, the experimenter may have misrecorded the measurement, or the data might have been coded incorrectly in the computer. Second, the outlier may be the result of a misclassified measurement. That is, the measurement belongs to a population different from that from which the rest of the sample was drawn. Finally, the measurement associated with the outlier may be recorded correctly and from the same population as the rest of the sample, but represents a rare (chance) event. Such outliers occur most often when the relative frequency distribution of the sample data is extremely skewed, because such a distribution has a tendency to include extremely large or small observations relative to the others in the data set.

## DEFINITION 2.12

An observation (or measurement) that is unusually large or small relative to the other values in a data set is called an **outlier**. Outliers typically are attributable to one of the following causes:

1. The measurement is observed, recorded, or entered into the computer incorrectly.
2. The measurement comes from a different population.
3. The measurement is correct, but represents a rare (chance) event.

**FIGURE 2.26**

*The Quartiles for a Data Set*

Two useful methods for detecting outliers, one graphical and one numerical, are **box plots** and $z$-scores. The box plot is based on the *quartiles* of a data set. **Quartiles** are values that partition the data set into four groups, each containing 25% of the measurements. The *lower quartile* $Q_L$ is the 25th percentile, the *middle quartile* is the median $M$ (the 50th percentile), and the *upper quartile* $Q_U$ is the 75th percentile (see Figure 2.26).

## DEFINITION 2.13

The **lower quartile** $Q_L$ is the 25th percentile of a data set. The **middle quartile** $M$ is the median. The **upper quartile** $Q_U$ is the 75th percentile.

A box plot is based on the *interquartile range (IQR)*, the distance between the lower and upper quartiles:

$$IQR = Q_U - Q_L$$

## DEFINITION 2.14

The **interquartile range (IQR)** is the distance between the lower and upper quartiles:

$$IQR = Q_U - Q_L$$

A vertical MINITAB box plot for the gas mileage data (Table 2.3) is shown in Figure 2.27.* Note that a rectangle (the *box*) is drawn, with the bottom and top of the rectangle (the **hinges**) drawn at the quartiles $Q_L$ and $Q_U$, respectively. By

---

*Although box plots can be generated by hand, the amount of detail required makes them particularly well suited for computer generation. We use computer software to generate the box plots in this section.

**FIGURE 2.27**

*MINITAB Box Plot for Gas Mileage Data*

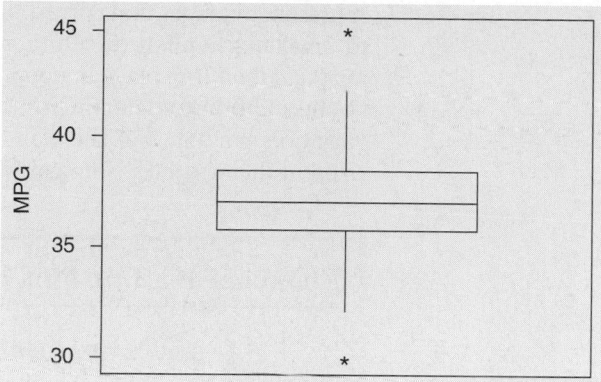

definition, then, the "middle" 50% of the observations—those between $Q_L$ and $Q_U$—fall inside the box. For the gas mileage data, these quartiles appear to be at (approximately) 35.5 and 38.5. Thus,

$$IQR = 38.5 - 35.5 = 3.0 \text{ (approximately)}$$

The median is shown at about 37 by a horizontal line within the box.

To guide the construction of the "tails" of the box plot, two sets of limits, called **inner fences** and **outer fences**, are used. Neither set of fences actually appears on the box plot. Inner fences are located at a distance of 1.5(IQR) from the hinges. Emanating from the hinges of the box are vertical lines called the **whiskers**. The two whiskers extend to the most extreme observation inside the inner fences. For example, the inner fence on the lower side of the gas mileage box plot is (approximately)

$$\begin{aligned}
\text{Lower inner fence} &= \text{Lower hinge} - 1.5(IQR) \\
&\approx 35.5 - 1.5(3.0) \\
&= 35.5 - 4.5 = 31.0
\end{aligned}$$

The smallest measurement *inside* this fence is the second smallest measurement, 31.8. Thus, the lower whisker extends to 31.8. Similarly, the upper whisker extends to 42.1, the largest measurement inside the upper inner fence at about 38.5 + 4.5 = 43.0.

Values that are beyond the inner fences are deemed *potential outliers* because they are extreme values that represent relatively rare occurrences. In fact, for mound-shaped distributions, fewer than 1% of the observations are expected to fall outside the inner fences. Two of the 100 gas mileage measurements, 30.0 and 44.9, fall beyond the inner fences, one on each end of the distribution. Each of these potential outliers is represented by the symbol * (an asterisk).

The other two imaginary fences, the outer fences, are defined at a distance 3(IQR) from each end of the box. Measurements that fall beyond the outer fences are represented by 0s (zeros) and are very extreme measurements that require special analysis. Since less than one-hundredth of 1% (.01% or .0001) of the measurements from mound-shaped distributions are expected to fall beyond the outer fences, these measurements are considered to be *outliers*.

Recall that outliers may be incorrectly recorded observations, members of a population different from the rest of the sample, or, at the least, very unusual measurements from the same population. The box plot of Figure 2.27 detected two potential outliers—the two gas mileage measurements beyond the inner fences. When we analyze these measurements, we find that they are correctly recorded. Perhaps they represent mileages that correspond to exceptional models of the car being tested or to unusual gas mixtures. Outlier analysis often reveals useful information of this kind and therefore plays an important role in the statistical inference-making process.

In addition to detecting outliers, box plots provide useful information on the variation in a data set. The elements (and nomenclature) of box plots are summarized in the next box. Some aids to the interpretation of box plots are also given.

---

### Elements of a Box Plot

1. A rectangle (the **box**) is drawn with the ends (the **hinges**) drawn at the lower and upper quartiles ($Q_L$ and $Q_U$). The median of the data is shown in the box, usually by a line.
2. The points at distances 1.5(IQR) from each hinge mark the **inner fences** of the data set. Lines (the **whiskers**) are drawn from each hinge to the most extreme measurement inside the inner fence.
3. A second pair of fences, the **outer fences**, appear at a distance of 3 interquartile ranges, 3(IQR), from the hinges. One symbol (usually "*") is used to represent measurements falling between the inner and outer fences, and another (usually "0") is used to represent measurements beyond the outer fences. Thus, outer fences are not shown unless one or more measurements lie beyond them.
4. The symbols used to represent the median and the extreme data points (those beyond the fences) will vary depending on the software you use to construct the box plot. (You may use your own symbols if you are constructing a box plot by hand.) You should consult the program's documentation to determine exactly which symbols are used.

### Aids to the Interpretation of Box Plots

1. Examine the length of the box. The IQR is a measure of the sample's variability and is especially useful for the comparison of two samples (see Example 2.16).
2. Visually compare the lengths of the whiskers. If one is clearly longer, the distribution of the data is probably skewed in the direction of the longer whisker.
3. Analyze any measurements that lie beyond the fences. Fewer than 5% should fall beyond the inner fences, even for very skewed distributions. Measurements beyond the outer fences are probably outliers, with one of the following explanations:
   a. The measurement is incorrect. It may have been observed, recorded, or entered into the computer incorrectly.
   b. The measurement belongs to a population different from the population that the rest of the sample was drawn from (see Example 2.16).
   c. The measurement is correct *and* from the same population as the rest. Generally, we accept this explanation only after carefully ruling out all others.

---

■ **EXAMPLE 2.15**    Use a statistical software package to draw a box plot for the student loan default data, Table 2.5. Identify any outliers in the data set.

### Solution

The MINITAB box plot for the student loan default rates is shown in Figure 2.28. Note that the median appears to be about 9.5, and, with the exception of a single extreme observation, the distribution appears to be symmetrically distributed between approximately 3% and 17%. The single outlier is beyond the inner fence

**FIGURE 2.28**

*MINITAB Box Plot for Student Default Rates*

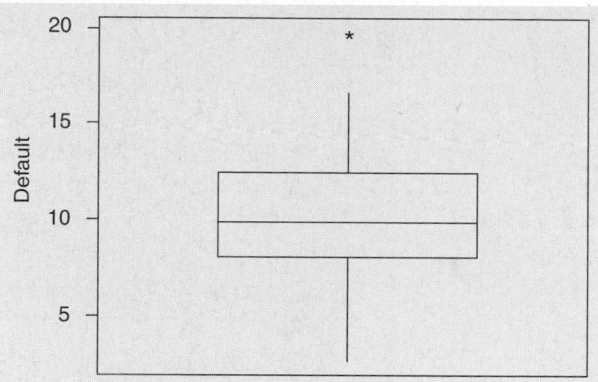

but inside the outer fence. Examination of the data reveals that this observation corresponds to Alaska's default rate of 19.7%.

E X A M P L E   2 . 16

A Ph.D. student in psychology conducted a stimulus reaction experiment as a part of her dissertation research. She subjected 50 subjects to a threatening stimulus and 50 to a nonthreatening stimulus. The reaction times of all 100 students, recorded to the nearest tenth of a second, are listed in Table 2.10. Box plots of the two resulting samples of reaction times, generated using SAS, are shown in Figure 2.29. Interpret the box plots.

🜨 REACTION

**TABLE 2.10   Reaction Times of Students**

| Nonthreatening Stimulus | | | | | | | | | |
|---|---|---|---|---|---|---|---|---|---|
| 2.0 | 1.8 | 2.3 | 2.1 | 2.0 | 2.2 | 2.1 | 2.2 | 2.1 | 2.1 |
| 2.0 | 2.0 | 1.8 | 1.9 | 2.2 | 2.0 | 2.2 | 2.4 | 2.1 | 2.0 |
| 2.2 | 2.1 | 2.2 | 1.9 | 1.7 | 2.0 | 2.0 | 2.3 | 2.1 | 1.9 |
| 2.0 | 2.2 | 1.6 | 2.1 | 2.3 | 2.0 | 2.0 | 2.0 | 2.2 | 2.6 |
| 2.0 | 2.0 | 1.9 | 1.9 | 2.2 | 2.3 | 1.8 | 1.7 | 1.7 | 1.8 |

| Threatening Stimulus | | | | | | | | | |
|---|---|---|---|---|---|---|---|---|---|
| 1.8 | 1.7 | 1.4 | 2.1 | 1.3 | 1.5 | 1.6 | 1.8 | 1.5 | 1.4 |
| 1.4 | 2.0 | 1.5 | 1.8 | 1.4 | 1.7 | 1.7 | 1.7 | 1.4 | 1.9 |
| 1.9 | 1.7 | 1.6 | 2.5 | 1.6 | 1.6 | 1.8 | 1.7 | 1.9 | 1.9 |
| 1.5 | 1.8 | 1.6 | 1.9 | 1.3 | 1.5 | 1.6 | 1.5 | 1.6 | 1.5 |
| 1.3 | 1.7 | 1.3 | 1.7 | 1.7 | 1.8 | 1.6 | 1.7 | 1.7 | 1.7 |

**Solution**

In SAS, the median is represented by the horizontal line through the box, while the asterisk (*) symbol represents the mean. Analysis of the box plots on the same numerical scale reveals that the distribution of times corresponding to the threatening stimulus lies below that of the nonthreatening stimulus. The implication is that the reaction times tend to be faster to the threatening stimulus. Note, too, that the upper whiskers of both samples are longer than the lower whiskers, indicating that the reaction times are positively skewed.

No observations in the two samples fall between the inner and outer fences. However, there is one outlier—the observation of 2.5 seconds corresponding to the threatening stimulus that is beyond the outer fence (denoted by the square symbol in SAS). When the researcher examined her notes from the experiments, she found that the

**FIGURE 2.29**

*SAS Box Plots for Reaction Time Data*

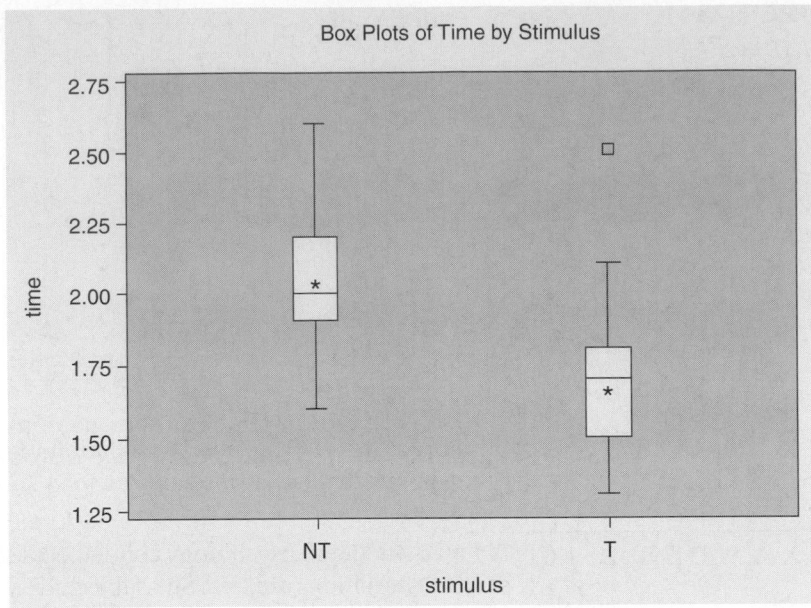

subject whose time was beyond the outer fence had mistakenly been given the nonthreatening stimulus. You can see in Figure 2.29 that his time would have been within the upper whisker if moved to the box plot corresponding to the nonthreatening stimulus. The box plots should be reconstructed since they will both change slightly when this misclassified reaction time is moved from one sample to the other.

The researcher concluded that the reactions to the threatening stimulus were faster than those to the nonthreatening stimulus. However, she was asked by her Ph.D. committee whether the results were *statistically significant*. Their question addresses the issue of whether the observed difference between the samples might be attributable to chance or sampling variation rather than to real differences between the populations. To answer their question, the researcher must use inferential statistics rather than graphical descriptions. We discuss how to compare two samples using inferential statistics in Chapter 9.    ∎

The following example illustrates how $z$-scores can be used to detect outliers and make inferences.

**EXAMPLE 2.17**

Suppose a female bank employee believes that her salary is low as a result of sex discrimination. To substantiate her belief, she collects information on the salaries of her male counterparts in the banking business. She finds that their salaries have a mean of $44,000 and a standard deviation of $2,000. Her salary is $37,000. Does this information support her claim of sex discrimination?

**Solution**

The analysis might proceed as follows: First, we calculate the $z$-score for the woman's salary with respect to those of her male counterparts. Thus,

$$z = \frac{\$37{,}000 - \$44{,}000}{\$2{,}000} = -3.5$$

The implication is that the woman's salary is 3.5 standard deviations *below* the mean of the male salary distribution. Furthermore, if a check of the male salary data shows that the frequency distribution is mound-shaped, we can infer that very few salaries in this distribution should have a $z$-score less than $-3$, as shown

**FIGURE 2.30**

*Male Salary Distribution*

in Figure 2.30. Clearly, a *z*-score of −3.5 represents an outlier—a measurement from a distribution different from the male salary distribution or a very unusual (highly improbable) measurement for the male salary distribution.

Which of the two situations do you think prevails? Do you think the woman's salary is simply unusually low in the distribution of salaries, or do you think her claim of sex discrimination is justified? Most people would probably conclude that her salary does not come from the male salary distribution. However, the careful investigator should require more information before inferring sex discrimination as the cause. We would want to know more about the data collection technique the woman used and more about her competence at her job. Also, perhaps other factors such as length of employment should be considered in the analysis.    ∎

Examples 2.16 and 2.17 exemplify an approach to statistical inference that might be called the **rare-event approach**. An experimenter hypothesizes a specific frequency distribution to describe a population of measurements. Then a sample of measurements is drawn from the population. If the experimenter finds it unlikely that the sample came from the hypothesized distribution, the hypothesis is concluded to be false. Thus, in Example 2.17 the woman believes her salary reflects discrimination. She hypothesizes that her salary should be just another measurement in the distribution of her male counterparts' salaries if no discrimination exists. However, it is so unlikely that the sample (in this case, her salary) came from the male frequency distribution that she rejects that hypothesis, concluding that the distribution from which her salary was drawn is different from the distribution for the men.

This rare-event approach to inference-making is discussed further in later chapters. Proper application of the approach requires a knowledge of probability, the subject of our next chapter.

We conclude this section with some rules of thumb for detecting outliers.

---

### Rules of Thumb for Detecting Outliers*

*Box plots:* Observations falling between the inner and outer fences are deemed **suspect outliers**. Observations falling beyond the outer fence are deemed **highly suspect outliers**.

*z-scores:* Observations with *z*-scores greater than 3 in absolute value are considered outliers. (For some highly skewed data sets, observations with *z*-scores greater than 1 in absolute value may be outliers.)

---

*The *z*-score and box plot methods both establish rule-of-thumb limits outside of which a measurement is deemed to be an outlier. Usually, the two methods produce similar results. However, the presence of one or more outliers in a data set can inflate the computed value of *s*. Consequently, it will be less likely that an errant observation would have a *z*-score larger than 3 in absolute value. In contrast, the values of the quartiles used to calculate the intervals for a box plot are not affected by the presence of outliers.

# Using the TI-83 Graphing Calculator

## Making a Box Plot on the TI-83

Step 1   *Enter the data*
Press   **STAT 1**   for   **STAT Edit**
Enter the data set into **L1**.

Step 2   *Set up the box plot*
Press   **2nd   Y =**   for   **STAT PLOT**
Press   **1**               for   **Plot 1**
Use the arrow and **ENTER** keys to set up the screen as shown below.

Step 3   *Select your window settings*
Press **WINDOW** and adjust the settings as follows:

**Xmin** = smallest data value (or smaller)
**Xmax** = largest data value (or larger)
**Xscl** = approximately (xmax − xmin)/10
**Ymin** = 0
**Ymax** = 10
**Yscl** = 1

Step 4   *View the graph*
Press   **GRAPH**

Optional   *Read the five number summary*
Step        Press   **TRACE**
Use the left and right arrow keys to move between minX, Q1, Med, Q3, and maxX.

Example:   Make a box plot for the given data.

86, 70, 62, 98, 73, 56, 53, 92, 86, 37, 62, 83, 78, 49, 78, 37, 67, 79, 57

The window settings and horizontal box plot are shown below.

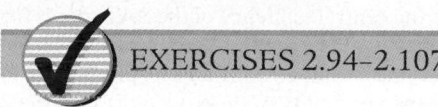

EXERCISES 2.94–2.107

## Learning the Mechanics

**2.94** A sample data set has a mean of 57 and a standard deviation of 11. Determine whether each of the following sample measurements are outliers.
  **a.** 65
  **b.** 21
  **c.** 72
  **d.** 98

**2.95** Define the 25th, 50th, and 75th percentiles of a data set. Explain how they provide a description of the data.

**2.96** Suppose a data set consisting of exam scores has a lower quartile $Q_L = 60$, a median $M = 75$, and an upper quartile $Q_U = 85$. The scores on the exam range from 18 to 100. Without having the actual scores available to you, construct as much of the box plot as possible.

**2.97** Consider the box plot at right:
  **a.** What is the median of the data set (approximately)?
  **b.** What are the upper and lower quartiles of the data set (approximately)?
  **c.** What is the interquartile range of the data set (approximately)?
  **d.** Is the data set skewed to the left, skewed to the right, or symmetric?
  **e.** What percentage of the measurements in the data set lie to the right of the median? To the left of the upper quartile?
  **f.** Identify any outliers in the data.

**2.98** Consider the following two sample data sets:

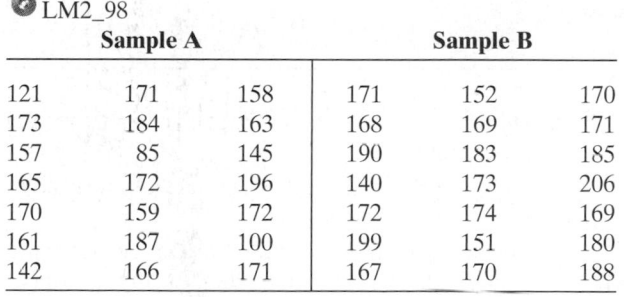

LM2_98

| Sample A | | | Sample B | | |
|---|---|---|---|---|---|
| 121 | 171 | 158 | 171 | 152 | 170 |
| 173 | 184 | 163 | 168 | 169 | 171 |
| 157 | 85 | 145 | 190 | 183 | 185 |
| 165 | 172 | 196 | 140 | 173 | 206 |
| 170 | 159 | 172 | 172 | 174 | 169 |
| 161 | 187 | 100 | 199 | 151 | 180 |
| 142 | 166 | 171 | 167 | 170 | 188 |

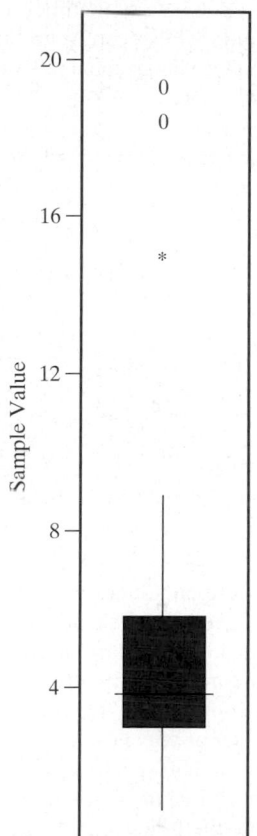

  **a.** Use a statistical software package to construct a box plot for each data set.
  **b.** Using information reflected in your box plots, describe the similarities and differences in the two data sets.
  **c.** Identify any outliers that may exist in the two data sets.

## Applying the Concepts—Basic

**2.99** Refer to the *Brain and Language* data on postmortem intervals (PMIs) of 22 human brain specimens, Exercise 2.25 (p. 39). An SAS descriptive statistics printout for the PMI variable is shown below.

SAS Output for Exercise 2.99

| | The MEANS Procedure | | | |
|---|---|---|---|---|
| | Analysis Variable : PMI | | | |
| N | Mean | Median | Variance | Std Dev |
| 22 | 7.3000000 | 6.1500000 | 10.1438095 | 3.1849348 |

a. Find the *z*-score for the PMI value of 3.3.

b. Is the PMI value of 3.3 considered an outlier? Explain.

**2.100** The table contains the top salary offer (in thousands of dollars) received by each member of a sample of 50 MBA students who recently graduated from the Graduate School of Management at Rutgers University.

⊘ MBASAL

| | | | | |
|---|---|---|---|---|
| 61.1 | 48.5 | 47.0 | 49.1 | 43.5 |
| 50.8 | 62.3 | 50.0 | 65.4 | 58.0 |
| 53.2 | 39.9 | 49.1 | 75.0 | 51.2 |
| 41.7 | 40.0 | 53.0 | 39.6 | 49.6 |
| 55.2 | 54.9 | 62.5 | 35.0 | 50.3 |
| 41.5 | 56.0 | 55.5 | 70.0 | 59.2 |
| 39.2 | 47.0 | 58.2 | 59.0 | 60.8 |
| 72.3 | 55.0 | 41.4 | 51.5 | 63.0 |
| 48.4 | 61.7 | 45.3 | 63.2 | 41.5 |
| 47.0 | 43.2 | 44.6 | 47.7 | 58.6 |

*Source:* Career Services Office, Graduate School of Management, Rutgers University.

a. The mean and standard deviation are 52.33 and 9.22, respectively. Find and interpret the *z*-score associated with the highest salary offer, the lowest salary offer, and the mean salary offer.

b. Would you consider the highest offer to be unusually high? Why or why not?

**2.101** Refer to Exercise 2.30 (p. 40) in which we compared states' average SAT scores in 1990 and 2000. SPSS was used to generate the following box plots for the states' SAT scores in 1990 and 2000.

SPSS Output for Exercise 2.101

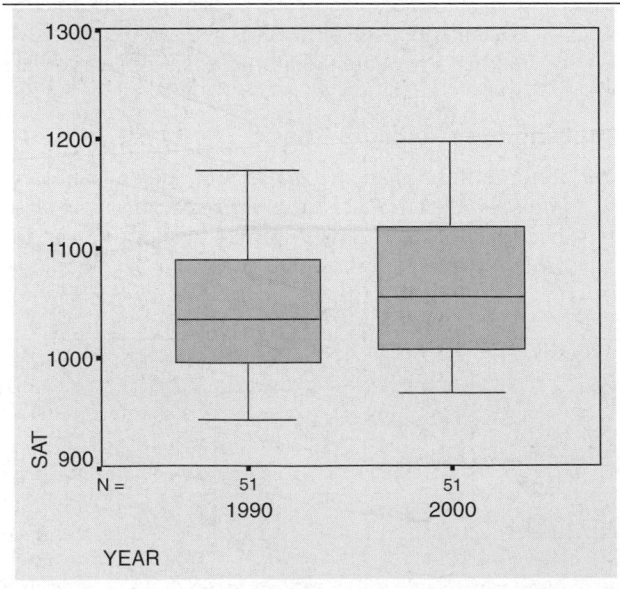

a. Compare the central tendency of the SAT scores for the 2 years. [*Note:* The median is represented by the horizontal line within each SPSS box plot.]

b. Compare the variability of the SAT scores for the 2 years.

c. Are any states' SAT scores outliers in either year? If so, identify them.

### Applying the Concepts—Intermediate

**2.102** Refer to the *American Journal of Speech–Language Pathology* (Feb. 1995) study, Exercise 2.31 (p. 41). Recall that three groups of college students listened to an audio tape of a woman with imperfect speech. The groups were called *control, treatment,* and *familiarity.* At the end of the session, each student transcribed the spoken words. Box plots, constructed using STATISTIX, for the percentage of words correctly transcribed (i.e., accuracy rate) by each group are shown here.

STATISTIX Output for Exercise 2.102

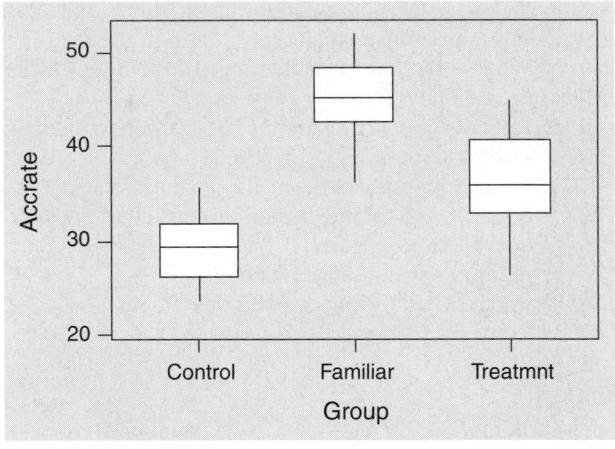

a. How do the median accuracy rates compare for the three groups?

b. How do the variabilities of the accuracy rates compare for the three groups?

c. The standard deviations of the accuracy rates are 3.56 for the control group, 5.45 for the treatment group, and 4.46 for the familiarity group. Do the standard deviations agree with the interquartile ranges (part **b**) with regard to the comparison of the variabilities of the accuracy rates?

d. Is there evidence of outliers in any of the three distributions?

**2.103** A manufacturer of minicomputers is investigating the down time of its systems. The 40 most recent customers were surveyed to determine the amount of down time (in hours) they had experienced during the previous month. These data are listed in the table on p. 85.

STATISTIX Output for Exercise 2.105

---

**DOWNTIME**

| Customer Number | Down Time | Customer Number | Down Time | Customer Number | Down Time |
|---|---|---|---|---|---|
| 230 | 12 | 244 | 2 | 258 | 28 |
| 231 | 16 | 245 | 11 | 259 | 19 |
| 232 | 5 | 246 | 22 | 260 | 34 |
| 233 | 16 | 247 | 17 | 261 | 26 |
| 234 | 21 | 248 | 31 | 262 | 17 |
| 235 | 29 | 249 | 10 | 263 | 11 |
| 236 | 38 | 250 | 4 | 264 | 64 |
| 237 | 14 | 251 | 10 | 265 | 19 |
| 238 | 47 | 252 | 15 | 266 | 18 |
| 239 | 0 | 253 | 7 | 267 | 24 |
| 240 | 24 | 254 | 20 | 268 | 49 |
| 241 | 15 | 255 | 9 | 269 | 50 |
| 242 | 13 | 256 | 22 | | |
| 243 | 8 | 257 | 18 | | |

  **a.** Use a statistical software package to construct a box plot for these data. Use the information reflected in the box plot to describe the frequency distribution of the data set. Your description should address central tendency, variation, and skewness.

  **b.** Use your box plot to determine which customers are having unusually lengthy down times.

  **c.** Find and interpret the $z$-scores associated with customers you identified in part **b**.

**2.104** Refer to the data on sanitation levels of cruise ships, Exercises 2.72 (p. 66) and 2.89 (p. 74).

  **a.** Use the box plot method to detect any outliers in the data.

  **b.** Use the $z$-score method to detect any outliers in the data.

  **c.** Do the two methods agree? If not, explain why.

**2.105** Refer to the *Fortune* (Oct. 25, 1999) ranking of the 50 most powerful women in America, Exercise 2.44 (p. 51). Use the STATISTIX box plots above to compare the ages of the women in three groups based on their position within the firm: Group 1 (CEO, CFO, CIO, or COO); Group 2 (Chairman, President, or Director); and Group 3 (Vice President, Vice Chairman, or General Manager).

### Applying the Concepts—Advanced

**2.106** A city librarian claims that books have been checked out an average of seven (or more) times in the last year. You suspect he has exaggerated the checkout rate (book usage) and that the mean number of checkouts per book per year is, in fact, less than seven. Using the computerized card catalog, you randomly select one book and find that it has been checked out four times in the last year. Assume that the standard deviation of the number of checkouts per book per year is approximately 1.

  **a.** If the mean number of checkouts per book per year really is seven, what is the $z$-score corresponding to four?

  **b.** Considering your answer to part **a**, do you have reason to believe that the librarian's claim is incorrect?

  **c.** If you knew that the distribution of the number of checkouts were mound-shaped, would your answer to part **b** change? Explain.

  **d.** If the standard deviation of the number of checkouts per book per year were 2 (instead of 1), would your answers to parts **b** and **c** change? Explain.

**2.107** A chemical company produces a substance composed of 98% cracked corn particles and 2% zinc phosphide for use in controlling rat populations in sugarcane fields. Production must be carefully controlled to maintain the 2% zinc phosphide because too much zinc phosphide will cause damage to the sugarcane and too little will be ineffective in controlling the rat population. Records from past production indicate that the distribution of the actual percentage of zinc phosphide present in the substance is approximately mound-shaped, with a mean of 2.0% and a standard deviation of .08%. Suppose one batch chosen randomly actually contains 1.80% zinc phosphide. Does this indicate that there is too little zinc phosphide in today's production? Explain your reasoning.

## 2.9 GRAPHING BIVARIATE RELATIONSHIPS (OPTIONAL)

The claim is often made that the crime rate and the unemployment rate are "highly correlated." Another popular belief is that smoking and lung cancer are "related." Some people even believe that the Dow Jones Industrial Average and the lengths of fashionable skirts are "associated." The words "correlated," "related," and "associated" imply a relationship between two variables—in the examples above, two *quantitative* variables.

One way to describe the relationship between two quantitative variables—called a **bivariate relationship**—is to plot the data in a **scattergram** (or **scatterplot**). A scattergram is a two-dimensional plot, with one variable's values plotted along the vertical axis and the other along the horizontal axis. For example, Figure 2.31 is a scattergram relating (1) the cost of mechanical work (heating, ventilating, and plumbing) to (2) the floor area of the building for a sample of 26 factory and warehouse buildings. Note that the scattergram suggests a general tendency for mechanical cost to increase as building floor area increases.

When an increase in one variable is generally associated with an increase in the second variable, we say that the two variables are "positively related" or "positively correlated."* Figure 2.31 implies that mechanical cost and floor area are positively correlated. Alternatively, if one variable has a tendency to decrease as the other increases, we say the variables are "negatively correlated." Figure 2.32 shows several hypothetical scattergrams that portray a positive bivariate relationship (Figure 2.32a), a negative bivariate relationship (Figure 2.32b), and a situation where the two variables are unrelated (Figure 2.32c).

**FIGURE 2.31**

*Scattergram of Cost vs. Floor Area*

---

*A formal definition of correlation is given in Chapter 11.

**FIGURE 2.32**

*Hypothetical Bivariate Relationship*

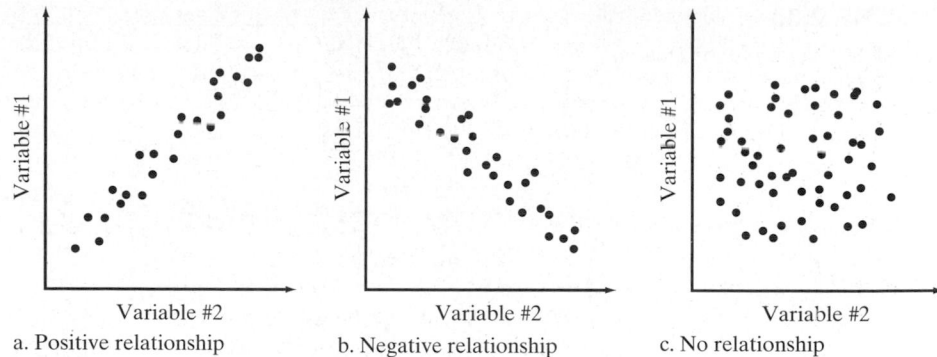

a. Positive relationship

b. Negative relationship

c. No relationship

EXAMPLE 2.18

A medical item used to administer to a hospital patient is called a *factor*. For example, factors can be intravenous (IV) tubing, IV fluid, needles, shave kits, bedpans, diapers, dressings, medications, and even code carts. The coronary care unit at Bayonet Point Hospital (St. Petersburg, Florida) recently investigated the relationship between the number of factors administered per patient and the patient's length of stay (in days). Data on these two variables for a sample of 50 coronary care patients are given in Table 2.11. Use a scattergram to describe the relationship between the two variables of interest, number of factors and length of stay.

⊙ MEDFACTORS

**TABLE 2.11    Data on Patient's Factors and Length of Stay**

| Number of Factors | Length of Stay (days) | Number of Factors | Length of Stay (days) |
|---|---|---|---|
| 231 | 9 | 354 | 11 |
| 323 | 7 | 142 | 7 |
| 113 | 8 | 286 | 9 |
| 208 | 5 | 341 | 10 |
| 162 | 4 | 201 | 5 |
| 117 | 4 | 158 | 11 |
| 159 | 6 | 243 | 6 |
| 169 | 9 | 156 | 6 |
| 55 | 6 | 184 | 7 |
| 77 | 3 | 115 | 4 |
| 103 | 4 | 202 | 6 |
| 147 | 6 | 206 | 5 |
| 230 | 6 | 360 | 6 |
| 78 | 3 | 84 | 3 |
| 525 | 9 | 331 | 9 |
| 121 | 7 | 302 | 7 |
| 248 | 5 | 60 | 2 |
| 233 | 8 | 110 | 2 |
| 260 | 4 | 131 | 5 |
| 224 | 7 | 364 | 4 |
| 472 | 12 | 180 | 7 |
| 220 | 8 | 134 | 6 |
| 383 | 6 | 401 | 15 |
| 301 | 9 | 155 | 4 |
| 262 | 7 | 338 | 8 |

*Source:* Bayonet Point Hospital, Coronary Care Unit.

**FIGURE 2.33**

*MINITAB Scatterplot of Data in*
*Table 2.11*

**Solution**

Rather than construct the plot by hand, we resort to a statistical software package. The MINITAB plot of the data in Table 2.11, with length of stay (LOS) on the horizontal axis and number of factors (FACTORS) on the vertical axis, is shown in Figure 2.33.

Although the plotted points exhibit a fair amount of variation, the scattergram clearly shows an increasing trend. It appears that a patient's length of stay is positively correlated with the number of factors administered to the patient. Hospital administrators may use this information to improve their forecasts of lengths of stay for future patients. ■

The scattergram is a simple but powerful tool for describing a bivariate relationship. However, keep in mind that it is only a graph. No measure of reliability can be attached to inferences made about bivariate populations based on scattergrams of sample data. The statistical tools that enable us to make inferences about bivariate relationships are presented in Chapter 11.

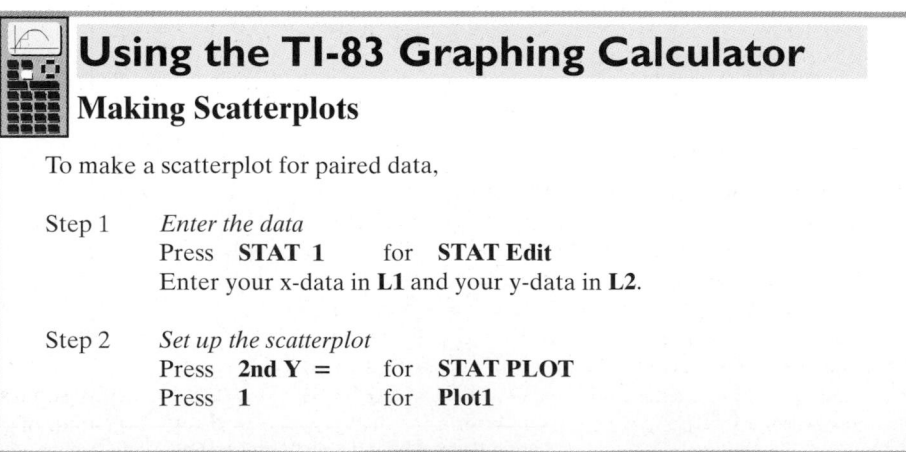

## Using the TI-83 Graphing Calculator

**Making Scatterplots**

To make a scatterplot for paired data,

Step 1     *Enter the data*
           Press   **STAT 1**     for   **STAT Edit**
           Enter your x-data in **L1** and your y-data in **L2**.

Step 2     *Set up the scatterplot*
           Press   **2nd Y =**     for   **STAT PLOT**
           Press   **1**           for   **Plot1**

## USING THE TI-83 GRAPHING CALCULATOR *(continued)*

Use the arrow and **ENTER** keys to set up the screen as shown below.

Step 3   *View the scatterplot*
Press  **ZOOM 9**  for  **ZoomStat**

Example  The figures below show a table of data entered on the TI-83 and the scatterplot of the data obtained using the steps given above.

---

## EXERCISES 2.108–2.115

### Learning the Mechanics

**2.108** Construct a scatterplot for the data in the following table. Do you detect a trend?

| Variable #1: | .5 | 1 | 1.5 | 2 | 2.5 | 3 | 3.5 | 4 | 4.5 | 5 |
|---|---|---|---|---|---|---|---|---|---|---|
| Variable #2: | 2 | 1 | 3 | 4 | 6 | 10 | 9 | 12 | 17 | 17 |

**2.109** Construct a scatterplot for the data in the following table. Do you detect a trend?

| Variable #1: | 5 | 3 | −1 | 2 | 7 | 6 | 4 | 0 | 8 |
|---|---|---|---|---|---|---|---|---|---|
| Variable #2: | 14 | 3 | 10 | 1 | 8 | 5 | 3 | 2 | 12 |

### Applying the Concepts—Basic

**2.110** Zoologists at the University of Western Australia conducted a study of the feeding behavior of blackbream fish spawned in aquariums (*Brain and Behavior Evolution*, April 2000). In one experiment, the zoologists recorded the number of aggressive strikes of two blackbream fish feeding at the bottom of the aquarium in the 10-minute period following the addition of food. The number of strikes and age of the fish (in days) was recorded approximately each week for nine weeks, as shown in the table.

⊘ BLACKBREAM

| Week | Number of Strikes | Age of Fish (days) |
|---|---|---|
| 1 | 85 | 120 |
| 2 | 63 | 136 |
| 3 | 34 | 150 |
| 4 | 39 | 155 |
| 5 | 58 | 162 |
| 6 | 35 | 169 |
| 7 | 57 | 178 |
| 8 | 12 | 184 |
| 9 | 15 | 190 |

*Source:* J. Shand, *et al.* "Variability in the location of the retinal ganglion cell area centralis is correlated with ontogenetic changes in feeding behavior in the Blackbream, Acanthopagrus 'butcher'," *Brain and Behavior*, Vol. 55, No. 4, April 2000 (Figure H).

**a.** Construct a scattergram for the data, with number of strikes on the *y*-axis and age of the fish on the *x*-axis.
**b.** Examine the scattergram of part **a**. Do you detect a trend?

**2.111** Baseball wisdom says if you can't hit, you can't win. But is the number of games won by a major league baseball team in a season related to the team's batting average?

The accompanying table shows the number of games won and the batting averages for the 14 teams in the American League for the 2000 season. Construct a two-dimensional plot for the data. Do you observe a trend?

🌐 ALWINS

| Team | Games Won | Team Batting Average |
|------|-----------|---------------------|
| Cleveland | 90 | .288 |
| New York | 87 | .277 |
| Boston | 85 | .267 |
| Toronto | 83 | .275 |
| Texas | 71 | .283 |
| Detroit | 79 | .275 |
| Minnesota | 69 | .270 |
| Baltimore | 74 | .272 |
| Anaheim | 82 | .280 |
| Tampa Bay | 69 | .257 |
| Seattle | 91 | .269 |
| Kansas City | 77 | .288 |
| Oakland | 91 | .270 |
| Chicago | 95 | .286 |

*Source:* Compiled by the Major League Baseball (MLB) Baseball Information System, 2001.

**2.112** *The Condor* data on number of flycatchers killed and number of flycatchers breeding at 14 nest sites, Exercise 2.46 (p. 52), are reproduced in the table. Graphically examine the relationship between the two variables. Do you detect a trend?

🌐 CONDOR

| Plot Number | Number Killed | Number of Breeders |
|-------------|---------------|--------------------|
| 1 | 5 | 30 |
| 2 | 4 | 28 |
| 3 | 3 | 38 |
| 4 | 2 | 34 |
| 5 | 2 | 26 |
| 6 | 1 | 124 |
| 7 | 1 | 68 |
| 8 | 1 | 86 |
| 9 | 1 | 32 |
| 10 | 0 | 30 |
| 11 | 0 | 46 |
| 12 | 0 | 132 |
| 13 | 0 | 100 |
| 14 | 0 | 6 |

*Source:* Merilä, J., and Wiggins, D. A. "Interspecific competition for nest holes causes adult mortality in the collared flycatcher." *The Condor,* Vol. 97, No. 2, May 1995, p. 447 (Table 4) Cooper Ornithological Society.

**2.113** A "self-avoiding walk" describes a path in which you never retrace your steps or cross your own path. An "unrooted walk" is a path in which it is impossible to distinguish between the starting point and ending point of the path. The *American Scientist* (July–Aug. 1998) investigated the relationship between self-avoiding and unrooted walks. The table gives the number of unrooted walks and possible number of self-avoiding walks of various lengths, where length is measured as number of steps.

🌐 WALK

| Walk Length (Number of Steps) | Unrooted Walks | Self-Avoiding Walks |
|-------------------------------|----------------|---------------------|
| 1 | 1 | 4 |
| 2 | 2 | 12 |
| 3 | 4 | 36 |
| 4 | 9 | 100 |
| 5 | 22 | 284 |
| 6 | 56 | 780 |
| 7 | 147 | 2,172 |
| 8 | 388 | 5,916 |

*Source:* Hayes, B. "How to avoid yourself." *American Scientist,* Vol. 86, No. 4, July–Aug.1998, p. 317 (Figure 5).

**a.** Construct a plot to investigate the relationship between total possible number of self-avoiding walks and walk length. What pattern (if any) do you observe?

**b.** Repeat part **a** for unrooted walks.

### Applying the Concepts—Intermediate

**2.114** Research published in the *American Journal of Psychiatry* (July 1995) attempted to establish a link between hippocampal (brain) volume and short-term verbal memory of patients with combat-related post-traumatic stress disorder (PTSD). A sample of 21 Vietnam veterans with a history of combat-related PTSD participated in the study. Magnetic resonance imaging was used to measure the volume of the right hippocampus (in cubic millimeters) of each subject. Verbal memory retention of each subject was measured by the percent retention subscale of the Wechsler Memory Scale. The data for the 21 patients is plotted in the scattergram. The researchers "hypothesized that smaller hippocampal volume would be associated with deficits in short-term verbal memory in patients with PTSD." Does the scattergram provide visual evidence to support this theory?

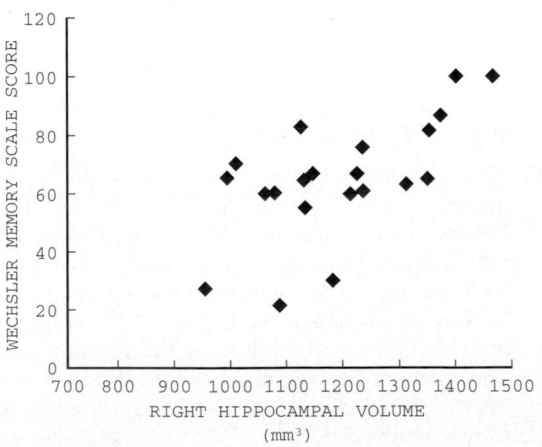

**2.115** The *Journal of Communication Disorders* data on deviation intelligence quotient (DIQ) and percent pronoun errors for 30 children, Exercise 2.51 (p. 53), are reproduced below.

**a.** Plot all the data to investigate a possible trend between DIQ and proper use of pronouns. What do you observe?

**b.** Plot the data for the ten SLI children only. Is there a trend between DIQ and proper use of pronouns?

SLI

| Subject | Gender | Group | DIQ | Pronoun Errors (%) |
|---------|--------|-------|-----|--------------------|
| 1 | F | YND | 110 | 94.40 |
| 2 | F | YND | 92 | 19.05 |
| 3 | F | YND | 92 | 62.50 |
| 4 | M | YND | 100 | 18.75 |
| 5 | F | YND | 86 | 0 |
| 6 | F | YND | 105 | 55.00 |
| 7 | F | YND | 90 | 100.00 |
| 8 | M | YND | 96 | 86.67 |
| 9 | M | YND | 90 | 32.43 |
| 10 | F | YND | 92 | 0 |
| 11 | F | SLI | 86 | 60.00 |
| 12 | M | SLI | 86 | 40.00 |
| 13 | M | SLI | 94 | 31.58 |
| 14 | M | SLI | 98 | 66.67 |
| 15 | F | SLI | 89 | 42.86 |
| 16 | F | SLI | 84 | 27.27 |
| 17 | M | SLI | 110 | 33.33 |
| 18 | F | SLI | 107 | 0 |
| 19 | F | SLI | 87 | 0 |
| 20 | M | SLI | 95 | 0 |
| 21 | M | OND | 110 | 0 |
| 22 | M | OND | 113 | 0 |
| 23 | M | OND | 113 | 0 |
| 24 | F | OND | 109 | 0 |
| 25 | M | OND | 92 | 0 |
| 26 | F | OND | 108 | 0 |
| 27 | M | OND | 95 | 0 |
| 28 | F | OND | 87 | 0 |
| 29 | F | OND | 94 | 0 |
| 30 | F | OND | 98 | 0 |

*Source:* Moore, M. E. "Error analysis of pronouns by normal and language-impaired children." *Journal of Communication Disorders*, Vol. 28, No. 1, Mar. 1995, p. 62 (Table 2), p. 67 (Table 5).

## 2.10  DISTORTING THE TRUTH WITH DESCRIPTIVE TECHNIQUES

A picture may be "worth a thousand words," but pictures can also color messages or distort them. In fact, the pictures in statistics—histograms, bar charts, and other graphical descriptions—are susceptible to distortion, so we have to examine each of them with care. We'll mention a few of the pitfalls to watch for when interpreting a chart or graph.

One common way to change the impression conveyed by a graph is to change the scale on the vertical axis, the horizontal axis, or both. For example, Figure 2.34 is a bar graph that shows the market share of sales for a company for each of the years 1997 to 2002. If you want to show that the change in firm A's market share over time is moderate, you should pack in a large number of units per inch on the vertical axis. That is, make the distance between successive units on the vertical scale small, as shown in Figure 2.34. You can see that a change in the firm's market share over time is barely apparent.

**FIGURE 2.34**

*Firm A's Market Share from 1997 to 2002: Packed Vertical Axis*

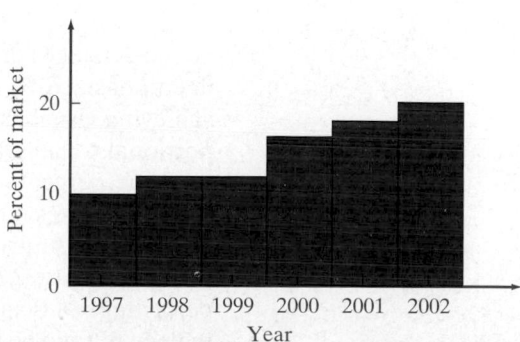

**FIGURE 2.35**

*Firm A's Market Share from 1997 to 2002: Stretched Vertical Axis*

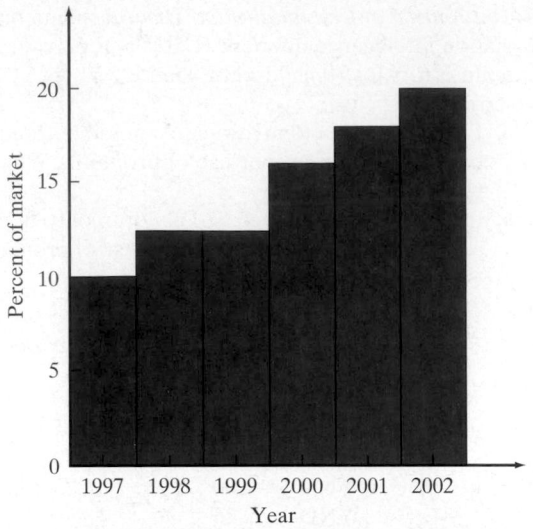

If you want to use the same data to make the changes in firm A's market share appear large, you should increase the distance between successive units on the vertical axis. That is, stretch the vertical axis by graphing only a few units per inch as in Figure 2.35. A telltale sign of stretching is a long vertical axis, but this is often hidden by starting the vertical axis at some point above 0. The same effect can be achieved by using a broken line—called a *scale break*—for the vertical axis, as shown in Figure 2.36.

**FIGURE 2.36**

*Firm A's Market Share from 1997 to 2002: Scale Break*

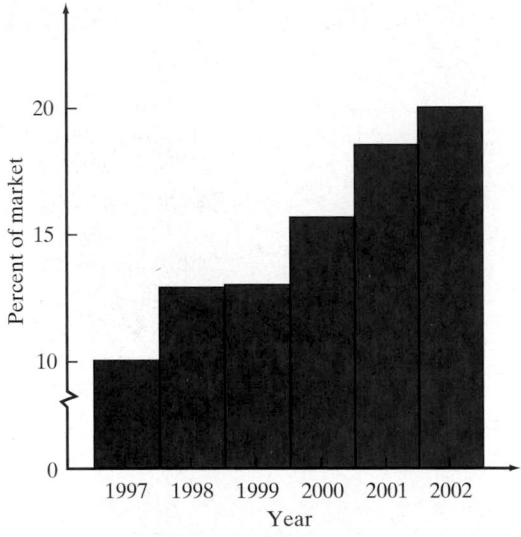

The changes in categories indicated by a bar graph can also be emphasized or deemphasized by stretching or shrinking the vertical axis. Another method of achieving visual distortion with bar graphs is by making the width of the bars proportional to height. For example, look at the bar chart in Figure 2.37a, which depicts the percentage of the total number of motor vehicle deaths in a year that occurred on each of four major highways. Now suppose we make both the width and the height grow as the percentage of fatal accidents grows. This change is shown in Figure 2.37b. The reader may tend to equate the *area* of the bars with the percentage of deaths occurring at each highway. But in fact, the true relative frequency of fatal accidents is proportional only to the *height* of the bars.

**FIGURE 2.37**

*Relative Frequency of Fatal Motor Vehicle Accidents on Each of Four Major Highways*

a. Bar chart

b. Width of bars grows with height

Although we've discussed only a few of the ways that graphs can be used to convey misleading pictures of phenomena, the lesson is clear. Look at all graphical descriptions of data with a critical eye. Particularly, check the axes and the size of the units on each axis. Ignore the visual changes and concentrate on the actual numerical changes indicated by the graph or chart.

The information in a data set can also be distorted by using numerical descriptive measures, as Example 2.19 indicates.

**EXAMPLE 2.19**     Suppose you're considering working for a small law firm—one that currently has a senior member and three junior members. You inquire about the salary you could expect to earn if you join the firm. Unfortunately, you receive two answers:

*Answer A:* The senior member tells you that an "average employee" earns $57,500

*Answer B:* One of the junior members later tells you that an "average employee" earns $45,000

Which answer can you believe?

**Solution**

The confusion exists because the phrase "average employee" has not been clearly defined. Suppose the four salaries paid are $45,000 for each of the three junior members and $95,000 for the senior member. Thus,

$$\bar{x} = \frac{3(\$45,000) + \$95,000}{4} = \frac{\$230,000}{4} = \$57,500$$

Median $= \$45,000$

You can now see how the two answers were obtained. The senior member reported the mean of the four salaries, and the junior member reported the median. The information you received was distorted because neither person stated which measure of central tendency was being used.

Another distortion of information in a sample occurs when *only* a measure of central tendency is reported. Both a measure of central tendency and a measure of variability are needed to obtain an accurate mental image of a data set.

Suppose you want to buy a new car and are trying to decide which of two models to purchase. Since energy and economy are both important issues, you decide to purchase model A because its EPA mileage rating is 32 miles per gallon in the city, whereas the mileage rating for model B is only 30 miles per gallon in the city.

**FIGURE 2.38**

*Mileage Distributions for Two Car Models*

However, you may have acted too quickly. How much variability is associated with the ratings? As an extreme example, suppose that further investigation reveals that the standard deviation for model A mileages is 5 miles per gallon, whereas that for model B is only 1 mile per gallon. If the mileages form a mound-shaped distribution, they might appear as shown in Figure 2.38. Note that the larger amount of variability associated with model A implies that more risk is involved in purchasing model A. That is, the particular car you purchase is more likely to have a mileage rating that will greatly differ from the EPA rating of 32 miles per gallon if you purchase model A, while a model B car is not likely to vary from the 30 miles per gallon rating by more than 2 miles per gallon.

We conclude this section with another example on distorting the truth with numerical descriptive measures.

**EXAMPLE 2.20**

*Children Out of School in America* is a report on delinquency of school-age children by the Children's Defense Fund (CDF). Consider the following three reported results of the CDF survey.

- Reported result: 25% of the 16- and 17-year-olds in the Portland, Maine, Bayside East Housing Project were out of school. Actual data: *Only eight children were surveyed; two were found to be out of school.*
- Reported result: Of all the secondary school students who had been suspended more than once in census tract 22 in Columbia, South Carolina, 33% had been suspended two times and 67% had been suspended three or more times. Actual data: *CDF found only three children in that entire census tract who had been suspended; one child was suspended twice and the other two children, three or more times.*
- Reported result: In the Portland Bayside East Housing Project, 50% of all the secondary school children who had been suspended more than once had been suspended three or more times. Actual data: *The survey found two secondary school children had been suspended in that area; one of them had been suspended three or more times.*
- Identify the potential distortions in the results reported by the CDF.

**Solution**

In each of these examples the reporting of percentages (i.e., relative frequencies) instead of the numbers themselves is misleading. No inference we might draw from the cited examples would be reliable. (We'll see how to measure the reliability of estimated percentages in Chapter 7.) In short, either the report should state the numbers alone instead of percentages, or, better yet, it should state that the numbers were too small to report by region. If several regions were combined, the numbers (and percentages) would be more meaningful.

## STATISTICS IN ACTION:
### The "Eye Cue" Test: Does Experience Improve Performance?

In 1948, famous child psychologist Jean Piaget devised a test of basic perceptual and conceptual skills dubbed the "water-level task." Subjects were shown a drawing of a glass being held perfectly still (at a 45° angle) by an invisible hand so that any water in it had to be at rest (see Figure 2.39).

**FIGURE 2.39**

*Drawing of the Water-Level Task*

The task for the subject was to draw a line representing the surface of the water—a line that would touch the black dot pictured on the right side of the glass. Piaget found that young children typically failed the test. Fifty years later, research psychologists still use the water-level task to test the perception of both adults and children. Surprisingly, about 40% of the adult population also fail. In addition, males tend to do better than females, and younger adults tend to do better than older adults.

Will people with experience handling liquid-filled containers perform better on this "eye cue" test? This question was the focus of research conducted by psychologists Heiko Hecht (NASA) and Dennis R. Proffitt (University of Virginia) and published in *Psychological Science* (Mar. 1995). The researchers presented the task to each of six different groups: (1) male college students, (2) female college students, (3) professional waitresses, (4) housewives, (5) male bartenders, and (6) male bus drivers. A total of 120 subjects (20 per group) participated in the study. Two of the groups, waitresses and bartenders, were assumed to have considerable experience handling liquid-filled glasses.

After each subject completed the drawing task, the researchers recorded the deviation* (in angle degrees) of the judged line from the true line. If the deviation was within 5° of the true water surface angle, the answer was considered correct. Deviations of more than 5° in either direction were considered incorrect answers.

The data for the water-level task (simulated based on summary results presented in the journal article) are provided in the file EYECUE on the data disk that accompanies this text. For each of the 120 subjects in the experiment, the following variables (in the order they appear on the data file) were measured:

### EYECUE

#### Variables

GENDER (F or M)
GROUP (Student, Waitres, Wife, Bartend, or Busdriv)
DEVIATION (angle, in degrees, of the judged line from the true line)
JUDGE (Within5, More5Above, More5Below)

### Focus

a. Use a graphical method to describe the qualitative responses for all 120 subjects.

b. Construct graphs that researchers could use to describe the effect of gender on the qualitative measure of task performance. Do the results tend to support or refute the prevailing theory that males do better than females?

c. Construct graphs that the researchers could use to describe the effect of age on the qualitative measure of task performance. [*Note*: The college students in the study were much younger than the subjects in the other four groups.] Do the results tend to support or refute the prevailing theory that younger adults do better than older adults?

d. Construct graphs that the researchers could use to describe the effect of experience in handling liquid-filled containers on the qualitative measure of task performance. Comment on the theory that experience improves task performance.

e. Refer to parts **a–d**. What are the drawbacks to using graphical displays of sample data to infer the nature of the population?

f. Repeat parts **a–e**, but use the quantitative measure of task performance, i.e., the actual deviations in water-line angles (in degrees) for each subject.

g. Discuss which graphs—those of parts **a–d** or those of part **f**—are more informative.

h. For each group of subjects, find and interpret numerical descriptive measures of the data for the quantitative measure of task performance. What do you conclude from this analysis?

*The true surface line is perfectly parallel to the table top.

> ➤➤➤ **QUICK REVIEW**

## Key Terms

*Note: Starred (*) terms are from the optional sections to this chapter*

Bar graph   22
Bivariate relationship*   86
Box plots*   76
Central tendency   44
Chebyshev's Rule   61
Class   20
Class frequency   20
Class relative frequency   20
Dot plot   30
Empirical Rule   62
Hinges*   76
Histogram   31
Inner fences*   77
Interquartile range*   76
Lower quartile*   76
Mean   44
Measurement classes   32

Measures of central
  tendency   44
Measures of relative
  standing   70
Measures of variation or
  spread   55
Median   46
Middle quartile*   76
Modal class   48
Mode   48
Mound-shaped distribution   61
Numerical descriptive
  measures   44
Outer fences*   77
Outlier*   75
Percentile   71
Pie chart   23

Quartiles*   76
Range   56
Rare-event approach*   81
Sample standard deviation   57
Sample variance   57
Scattergram*   86
Scatterplot*   86
Skewness   48
Standard deviation   57
Stem-and-leaf display   30
Symmetric distribution   48
Upper quartile*   76
Variability   44
Whiskers*   77
$z$-score   72

## Key Formulas

| | |
|---|---|
| $\dfrac{\text{Class frequency}}{n}$ | Class relative frequency   20 |
| $\bar{x} = \dfrac{\sum_{i=1}^{n} x_i}{n}$ | Sample mean   44 |
| $s^2 = \dfrac{\sum_{i=1}^{n}(x_i - \bar{x})^2}{n-1} = \dfrac{\sum_{i=1}^{n} x_i^2 - \dfrac{\left(\sum_{i=1}^{n} x_i\right)^2}{n}}{n-1}$ | Sample variance   57 |
| $s = \sqrt{s^2}$ | Sample standard deviation   57 |
| $z = \dfrac{x - \bar{x}}{s}$ | Sample $z$-score   72 |
| $z = \dfrac{x - \mu}{\sigma}$ | Population $z$-score   72 |
| $\text{IQR} = Q_U - Q_L$ * | Interquartile range   76 |

## Language Lab

| Symbol | Pronunciation | Description |
|---|---|---|
| $\sum$ | sum of | Summation notation; $\sum_{i=1}^{n} x_i$ represents the sum of the measurements $x_1, x_2, \ldots, x_n$, |
| $\mu$ | mu | Population mean |
| $\bar{x}$ | $x$-bar | Sample mean |
| $\sigma^2$ | sigma-squared | Population variance |
| $\sigma$ | sigma | Population standard deviation |
| $s^2$ | | Sample variance |
| $s$ | | Sample standard deviation |

| Symbol | Pronunciation | Description |
|--------|---------------|-------------|
| $z$ | | $z$-score for a measurement |
| $M$ | | Median (middle quartile) of a sample data set |
| $Q_L$ | | Lower quartile (25th percentile) |
| $Q_U$ | | Upper quartile (75th percentile) |
| IQR | | Interquartile range |

## SUPPLEMENTARY EXERCISES 2.116–2.140

*Note: Starred (*) exercises refer to the optional sections in this chapter.*

### Learning the Mechanics

**2.116** Construct a relative frequency histogram for the data summarized in the table below.

| Measurement Class | Relative Frequency | Measurement Class | Relative Frequency |
|-------------------|--------------------|-------------------|--------------------|
| .00–.75 | .02 | 5.25–6.00 | .15 |
| .75–1.50 | .01 | 6.00–6.75 | .12 |
| 1.50–2.25 | .03 | 6.75–7.50 | .09 |
| 2.25–3.00 | .05 | 7.50–8.25 | .05 |
| 3.00–3.75 | .10 | 8.25–9.00 | .04 |
| 3.75–4.50 | .14 | 9.00–9.75 | .01 |
| 4.50–5.25 | .19 | | |

**2.117** Consider the following three measurements: 50, 70, 80. Find the $z$-score for each measurement if they are from a population with a mean and standard deviation equal to
  **a.** $\mu = 60, \sigma = 10$
  **b.** $\mu = 60, \sigma = 5$
  **c.** $\mu = 40, \sigma = 10$
  **d.** $\mu = 40, \sigma = 100$

*__2.118__ Refer to Exercise 2.117. For parts **a–d**, determine whether the three measurements 50, 70, and 80 are outliers.

**2.119** Compute $s^2$ for data sets with the following characteristics:
  **a.** $\sum_{i=1}^{n} x_i^2 = 246, \sum_{i=1}^{n} x_i = 63, n = 22$
  **b.** $\sum_{i=1}^{n} x_i^2 = 666, \sum_{i=1}^{n} x_i = 106, n = 25$
  **c.** $\sum_{i=1}^{n} x_i^2 = 76, \sum_{i=1}^{n} x_i = 11, n = 7$

**2.120** If the range of a set of data is 20, find a rough approximation to the standard deviation of the data set.

**2.121** For each of the following data sets, compute $\bar{x}, s^2$, and $s$. If appropriate, specify the units in which your answers are expressed.
  **a.** 4, 6, 6, 5, 6, 7
  **b.** −$1, $4, −$3, $0, −$3, −$6
  **c.** 3/5%, 4/5%, 2/5%, 1/5%, 1/16%
  **d.** Calculate the range of each data set in parts **a–c**.

**2.122** For each of the following data sets, compute $\bar{x}, s^2$, and $s$:
  **a.** 13, 1, 10, 3, 3
  **b.** 13, 6, 6, 0
  **c.** 1, 0, 1, 10, 11, 11, 15
  **d.** 3, 3, 3, 3
  **e.** For each data set in parts **a–d**, form the interval $\bar{x} \pm 2s$ and calculate the percentage of the measurements that fall in the interval.

*__2.123__ Construct a scatterplot for the data listed here. Do you detect any trends?

| Variable #1: | 174 | 268 | 345 | 119 | 400 | 520 | 190 | 448 | 307 | 252 |
|--------------|-----|-----|-----|-----|-----|-----|-----|-----|-----|-----|
| Variable #2: | 8 | 10 | 15 | 7 | 22 | 31 | 15 | 20 | 11 | 9 |

**2.124** The following data sets have been invented to demonstrate that the lower bounds given by Chebyshev's Rule are appropriate. Notice that the data are contrived and would not be encountered in a real-life problem.
  **a.** Consider a data set that contains ten 0s, two 1s, and ten 2s. Calculate $\bar{x}, s^2$, and $s$. What percentage of the measurements are in the interval $\bar{x} \pm s$? Compare this result to Chebyshev's Rule.
  **b.** Consider a data set that contains five 0s, thirty-two 1s, and five 2s. Calculate $\bar{x}, s^2$, and $s$. What percentage of the measurements are in the interval $\bar{x} \pm 2s$? Compare this result to Chebyshev's Rule.
  **c.** Consider a data set that contains three 0s, fifty 1s, and three 2s. Calculate $\bar{x}, s^2$, and $s$. What percentage of the measurements are in the interval $\bar{x} \pm 3s$? Compare this result to Chebyshev's Rule.
  **d.** Draw a histogram for each of the data sets in parts **a, b**, and **c**. What do you conclude from these graphs and the answers to parts **a, b**, and **c**?

### Applying the Concepts—Basic

**2.125** Each year, the National Highway Traffic Safety Administration (NHTSA) crash tests new car models to determine how well they protect the driver and front-seat passenger in a head-on collision. The NHTSA has developed a "star" scoring system for the frontal crash test, with results ranging from one star (*) to five stars (*****). The more stars in the rating, the better the level

CRASH

of crash protection in a head-on collision. The NHTSA crash test results for 98 cars (model year 1997) are stored in the data file named CRASH. The driver-side star ratings for the 98 cars are summarized in the accompanying MINITAB printout. Use the information in the printout to form a pie chart. Interpret the graph.

**MINITAB Output for Exercise 2.125**

```
Tally for Discrete Variables: DrivStar

DrivStar  Count  Percent
       2      4     4.08
       3     17    17.35
       4     59    60.20
       5     18    18.37
     N=      98
```

2.126 Refer to Exercise 2.125 and the National Highway Traffic Safety Administration (NHTSA) crash test data. One quantitative variable recorded by the NHTSA is driver's severity of head injury (measured on a scale from 0 to 1,500). The mean and standard deviation for the 98 driver head-injury ratings in the CRASH file are displayed in the MINITAB printout below. Use these values to find the $z$-score for a driver head-injury rating of 408. Interpret the result.

2.127 Beanie Babies are toy stuffed animals that have become valuable collector's items. *Beanie World Magazine* provided the information on 50 Beanie Babies shown in the table on p. 99.
  **a.** Summarize the retired/current status of the 50 Beanie Babies with an appropriate graph. Interpret the graph.
  **b.** Summarize the values of the 50 Beanie Babies with an appropriate graph. Interpret the graph.
  *c. Use a graph to portray the relationship between a Beanie Baby's value and its age. Do you detect a trend?

2.128 A survey was conducted to investigate the impact of the mass media on the public's perception of mental illness (*Health Education Journal,* Sept. 1994). The media coverage for each of 562 items related to mental health in Scotland was classified into one of five categories: violence to others, sympathetic coverage, harm to self, comic images, and criticism of accepted defini-

tions of mental illness. A summary of the results is provided in the accompanying table.

| Media Coverage | Number of Items |
|---|---|
| Violence to others | 373 |
| Sympathetic | 102 |
| Harm to self | 71 |
| Comic images | 12 |
| Criticism of definitions | 4 |
| **Total** | **562** |

*Source:* Philo, G., *et al.* "The impact of the mass media on public images of mental illness: Media content and audience belief." *Health Education Journal,* Vol. 53, No. 3, Sept. 1994, p.274 (Table 1).

  **a.** Construct a relative frequency table for the data.
  **b.** Display the relative frequencies in a graph.
  **c.** Discuss the findings.

2.129 The extent of computer anxiety among secondary technical education teachers in West Virginia was examined in the *Journal of Studies in Technical Careers* (Vol. 15, 1995). Level of computer anxiety was measured by the Computer Anxiety Scale (COMPAS), with scores ranging from 10 to 50. (Lower scores reflect a higher degree of anxiety.) One objective was to compare the computer anxiety levels of male and female secondary technical education teachers in West Virginia. The data for the 116 teachers who participated in the survey are summarized in the following table.

| | Male Teachers | Female Teachers |
|---|---|---|
| $n$ | 68 | 48 |
| $\bar{x}$ | 26.4 | 24.5 |
| $s$ | 10.6 | 11.2 |

*Source:* Gordon, H. R. D. "Analysis of the computer anxiety levels of secondary technical education teachers in West Virginia." *Journal of Studies in Technical Careers*, Vol. 15, No. 2, 1995, pp. 26–27 (Table 1).

  **a.** Interpret the mean COMPAS score for male teachers.
  **b.** Repeat part **a** for female teachers.
  **c.** Give an interval that contains about 95% of the COMPAS scores for male teachers. (Assume the distribution of scores for males is symmetric and mound-shaped.)
  **d.** Repeat part **c** for female teachers.

**MINITAB Output for Exercise 2.126**

```
Descriptive Statistics: DrivHead

Variable      N      Mean   Median   TrMean   StDev   SE Mean
DrivHead     98     603.7    605.0    600.3   185.4      18.7

Variable   Minimum   Maximum      Q1       Q3
DrivHead     216.0    1240.0    475.0    724.3
```

● BEANIE

| Name | Age (Months) as of Sept. 1998 | Retired (R) Current (C) | Value ($) |
|---|---|---|---|
| 1. Ally the Alligator | 52 | R | 55.00 |
| 2. Batty the Bat | 12 | C | 12.00 |
| 3. Bongo the Brown Monkey | 28 | R | 40.00 |
| 4. Blackie the Bear | 52 | C | 10.00 |
| 5. Bucky the Beaver | 40 | R | 45.00 |
| 6. Bumble the Bee | 28 | R | 600.00 |
| 7. Crunch the Shark | 21 | C | 10.00 |
| 8. Congo the Gorilla | 28 | C | 10.00 |
| 9. Derby the Coarse Mained Horse | 28 | R | 30.00 |
| 10. Digger the Red Crab | 40 | R | 150.00 |
| 11. Echo the Dolphin | 17 | R | 20.00 |
| 12. Fetch the Golden Retriever | 5 | C | 15.00 |
| 13. Early the Robin | 5 | C | 20.00 |
| 14. Flip the White Cat | 28 | R | 40.00 |
| 15. Garcia the Teddy | 28 | R | 200.00 |
| 16. Happy the Hippo | 52 | R | 20.00 |
| 17. Grunt the Razorback | 28 | R | 175.00 |
| 18. Gigi the Poodle | 5 | C | 15.00 |
| 19. Goldie the Goldfish | 52 | R | 45.00 |
| 20. Iggy the Iguana | 10 | C | 10.00 |
| 21. Inch the Inchworm | 28 | R | 20.00 |
| 22. Jake the Mallard Duck | 5 | C | 20.00 |
| 23. Kiwi the Toucan | 40 | R | 165.00 |
| 24. Kuku the Cockatoo | 5 | C | 20.00 |
| 25. Mistic the Unicorn | 11 | R | 45.00 |
| 26. Mel the Koala Bear | 21 | C | 10.00 |
| 27. Nanook the Husky | 17 | C | 15.00 |
| 28. Nuts the Squirrel | 21 | C | 10.00 |
| 29. Peace the Tie Died Teddy | 17 | C | 25.00 |
| 30. Patty the Platypus | 64 | R | 800.00 |
| 31. Quacker the Duck | 40 | R | 15.00 |
| 32. Puffer the Penguin | 10 | C | 15.00 |
| 33. Princess the Bear | 12 | C | 65.00 |
| 34. Scottie the Scottie | 28 | R | 28.00 |
| 35. Rover the Dog | 28 | R | 15.00 |
| 36. Rex the Tyrannosaurus | 40 | R | 825.00 |
| 37. Sly the Fox | 28 | C | 10.00 |
| 38. Slither the Snake | 52 | R | 1,900.00 |
| 39. Skip the Siamese Cat | 21 | C | 10.00 |
| 40. Splash the Orca Whale | 52 | R | 150.00 |
| 41. Spooky the Ghost | 28 | R | 40.00 |
| 42. Snowball the Snowman | 12 | R | 40.00 |
| 43. Stinger the Scorpion | 5 | C | 15.00 |
| 44. Spot the Dog | 52 | R | 65.00 |
| 45. Tank the Armadillo | 28 | R | 85.00 |
| 46. Stripes the Tiger (Gold/Black) | 40 | R | 400.00 |
| 47. Teddy the 1997 Holiday Bear | 12 | R | 50.00 |
| 48. Tuffy the Terrier | 17 | C | 10.00 |
| 49. Tracker the Basset Hound | 5 | C | 15.00 |
| 50. Zip the Black Cat | 28 | R | 40.00 |

*Source: Beanie World Magazine, Sept. 1998.*

⊘TORNADO

| State | Number of Tornadoes per Year | Month of Peak Occurrence | State | Number of Tornadoes per Year | Month of Peak Occurrence |
|---|---|---|---|---|---|
| AL | 23 | April | MT | 5 | June |
| AK | 0 | — | NE | 36 | June |
| AZ | 4 | August | NV | 1 | June |
| AR | 21 | April | NH | 2 | July |
| CA | 4 | February | NJ | 3 | July |
| CO | 24 | June | NM | 9 | June |
| CT | 1 | July | NY | 5 | July |
| DE | 1 | July | NC | 14 | April |
| FL | 52 | June | ND | 20 | July |
| GA | 21 | April | OH | 16 | May |
| HI | 1 | January | OK | 47 | May |
| ID | 2 | July | OR | 1 | May |
| IL | 27 | May | PA | 10 | June |
| IN | 23 | May | RI | 0.23 | August |
| IA | 35 | May | SC | 10 | April |
| KS | 36 | May | SD | 28 | June |
| KY | 10 | May | TN | 12 | April |
| LA | 27 | April | TX | 137 | May |
| ME | 2 | July | UT | 2 | June |
| MD | 3 | May | VT | 1 | July |
| MA | 3 | August | VA | 6 | May |
| MI | 21 | June | WA | 1 | May |
| MN | 19 | June | WV | 2 | May |
| MS | 26 | April | WI | 18 | July |
| MO | 27 | May | WY | 11 | June |

*Source:* "Tornadoes ... nature's most violent storms." U.S. Dept. of Commerce, National Oceanic and Atmospheric Administration, Sept. 1992.

**2.130** A tornado is defined as a violently rotating column of air extending from a thunderstorm to the ground. The most violent tornadoes, with wind speeds of 250 mph or more, are capable of tremendous destruction. The accompanying table lists, for each state, the reported number of tornadoes per year and the month of peak tornado occurrence.

**a.** Identify the variables measured, number of tornadoes per year and month of peak occurrence, as quantitative or qualitative.

**b.** Use a graphical technique to summarize the month of peak tornado occurrence in the 50 states. Interpret the graph.

**2.131** Refer to Exercise 2.130. The descriptive statistics and stem-and-leaf display shown below were obtained by

**SPSS Output for Exercise 2.131**

Descriptives

| | | | Statistic | Std. Error |
|---|---|---|---|---|
| TORNUM | Mean | | 16.2000 | 3.0862 |
| | 95% Confidence | Lower Bound | 9.9980 | |
| | Interval for Mean | Upper Bound | 22.4020 | |
| | 5% Trimmed Mean | | 13.2667 | |
| | Median | | 10.0000 | |
| | Variance | | 476.245 | |
| | Std. Deviation | | 21.8230 | |
| | Minimum | | .00 | |
| | Maximum | | 137.00 | |
| | Range | | 137.00 | |
| | Interquartile Range | | 21.2500 | |
| | Skewness | | 3.692 | .337 |
| | Kurtosis | | 18.867 | .662 |

```
Frequency    Stem &  Leaf

  19.00       0 .  0011111112222233344
   4.00       0 .  5569
   6.00       1 .  000124
   3.00       1 .  689
   7.00       2 .  0111334
   5.00       2 .  67778
    .00       3 .
   3.00       3 .  566
    .00       4 .
   1.00       4 .  7
   1.00       5 .  2
   1.00    Extremes   (>=137)

Stem width:      10.00
Each leaf:        1 case(s)
```

using SPSS to analyze the reported number of tornadoes per year of the 50 states.

**a.** Use the stem-and-leaf display to give a verbal description of the data set. Do any of the measurements appear to be outliers?

**b.** Examine the computer output to determine the value of $\bar{x}$, the median, $s^2$, and $s$ for the number of tornadoes per year.

**c.** Based on the stem-and-leaf display and the relationship of the mean and median, is the distribution of these data approximately mound-shaped, or is it skewed to the right or left?

**d.** Based on your answer to part **c**, what percentage of the measurements do you expect to find in the intervals $\bar{x} \pm s, \bar{x} \pm 2s$, and $\bar{x} \pm 3s$?

**e.** Count the number of measurements that actually fall in each interval of part **d** and express each interval count as a percentage of the total number of measurements. Compare these results to your estimates from part **d**.

## Applying the Concepts—Intermediate

**2.132** Archaeologists gain insight into the social life of ancient tribes by measuring the distance (in centimeters) of each artifact found at a site from the "central hearth" (i.e., the middle of an artifact scatter). A graphical summary of these distances, in the form of a histogram, is called a *ring diagram* (*World Archaeology*, Oct. 1997). Ring diagrams for two archaeological sites (A and G) in Europe are shown at right.

**a.** Identify the type of skewness (if any) present in the data collected at the two sites.

**b.** Archaeologists have associated unimodal ring diagrams with open air hearths and multimodal ring diagrams with hearths inside dwellings. Can you identify the type of hearth (*open air or inside dwelling*) that was most likely present at each site?

**2.133** In 1961, Russian cosmonaut Yuri Gagarin was the first human to fly into space. Since then, more than 770 astronauts and cosmonauts have made space flights. The graphic at the top of p. 102, produced by the National Aeronautics and Space Administration (*NASA*), summarizes the annual number of people who flew into space between 1961 and 1998.

**a.** What type of graph is used to describe the number of humans in space each year?

**b.** In 1986, the space shuttle *Challenger* exploded, killing 7 American astronauts. Does the graph depict any trends that resulted from the disaster? Explain.

**2.134** One method used to control, trap, and kill predator species by the US. Fish and Wildlife Service is the steel-jaw leghold trap. Unfortunately, the traps often end up capturing and killing animals other than the intended predators (*UCLA Journal of Environmental Law & Policy*, Vol. 13, 1994/95). A study revealed that in one year, 44,982 animals were killed by steel traps intended for coyotes, but only 25,026 of these animals were

Histogram for Exercise 2.132

actually coyotes. The accompanying table lists the species of all animals captured by steel traps during that year and the number of each killed. Use a graphic method to summarize the data. Interpret the graph.

⊙ TRAPS

| Species | Number Killed by Steel Traps |
|---|---|
| Coyotes | 25,026 |
| Opossums | 2,698 |
| Porcupines | 1,367 |
| Racoons | 3,345 |
| Skunks | 6,348 |
| Beavers | 682 |
| Dogs | 273 |
| Cats | 73 |
| Goats | 49 |
| Foxes | 114 |
| Rabbits | 98 |
| Deer | 20 |
| Muskrats | 52 |
| Other | 4,837 |

**2.135** The *Journal of Agricultural, Biological, and Environmental Statistics* (Sept. 2000) published a study on the impact of the *Exxon Valdez* tanker oil spill on the seabird population in Prince William Sound, Alaska. A subset of the data analyzed is stored in the EVOS file. Data were collected on 96 shoreline locations (called transects) of constant width but variable length. For each transect, the number of seabirds found is recorded as well as the length (in kilometers) of the transect and whether or not the transect was in an oiled area. (The first five and last five observations in the EVOS file are listed in the table.)

**a.** Identify the variables measured as quantitative or qualitative.

**b.** Identify the experimental unit.

⊘ **EVOS**

| Transect | Seabirds | Length | Oil |
|----------|----------|--------|-----|
| 1 | 0 | 4.06 | No |
| 2 | 0 | 6.51 | No |
| 3 | 54 | 6.76 | No |
| 4 | 0 | 4.26 | No |
| 5 | 14 | 3.59 | No |
| ⋮ | ⋮ | ⋮ | ⋮ |
| 92 | 7 | 3.40 | Yes |
| 93 | 4 | 6.67 | Yes |
| 94 | 0 | 3.29 | Yes |
| 95 | 0 | 6.22 | Yes |
| 96 | 27 | 8.94 | Yes |

*Source:* McDonald, T. L., Erickson, W. P., and McDonald, L. L. "Analysis of count data from before–after control-impact studies," *Journal of Agricultural, Biological, and Environmental Statistics*, Vol. 5, No. 3, Sept. 2000, pp. 277–8 (Table A.1).

**c.** Use a pie chart to describe the percentage of transects in oiled and unoiled areas.

**d.** Use a graphical method to examine the relationship between observed number of seabirds and transect length.

**e.** Observed seabird density is defined as the observed count divided by the length of the transect. MINITAB descriptive statistics for seabird densities in unoiled and oiled transects are displayed in the printout shown below, followed by histograms for each type of transect on p. 103. Assess whether the distribution of seabird densities differs for transects in oiled and unoiled areas.

**f.** For unoiled transects, give an interval of values that is likely to contain at least 75% of the seabird densities.

**g.** For oiled transects, give an interval of values that is likely to contain at least 75% of the seabird densities.

**h.** Which type of transect, an oiled or unoiled one, is more likely to have a seabird density of 16? Explain.

**2.136** A study in the *Journal of Leisure Research* (Vol. 24, 1992) investigated the relationship between academic performance and leisure activities. Each in a sample of 159 high school students was asked to state how many leisure activities (sports, fishing, music, drama, photography, writing, watching TV, etc.) they participated in weekly. From this list, activities that involved reading, writing, or arithmetic were labeled "academic leisure activities." Some of the results of the study are presented below:

| | $\bar{x}$ | s |
|---|------|------|
| GPA | 2.96 | .71 |
| Number of leisure activities | 12.38 | 5.07 |
| Number of academic leisure activities | 2.77 | 1.97 |

```
MINITAB Output for Problem: 2.135

Descriptive Statistics: DENSITY by OIL

Variable   OIL            N      Mean    Median    TrMean    StDev
DENSITY    no            36      3.27      0.89      2.05      6.70
           yes           60     3.495     0.700     2.542     5.968

Variable   OIL       SE Mean   Minimum   Maximum        Q1        Q3
DENSITY    no           1.12      0.00     36.23      0.00      3.87
           yes         0.770     0.000    32.836     0.000     5.233
```

MINITAB Histograms for Exercise 2.135

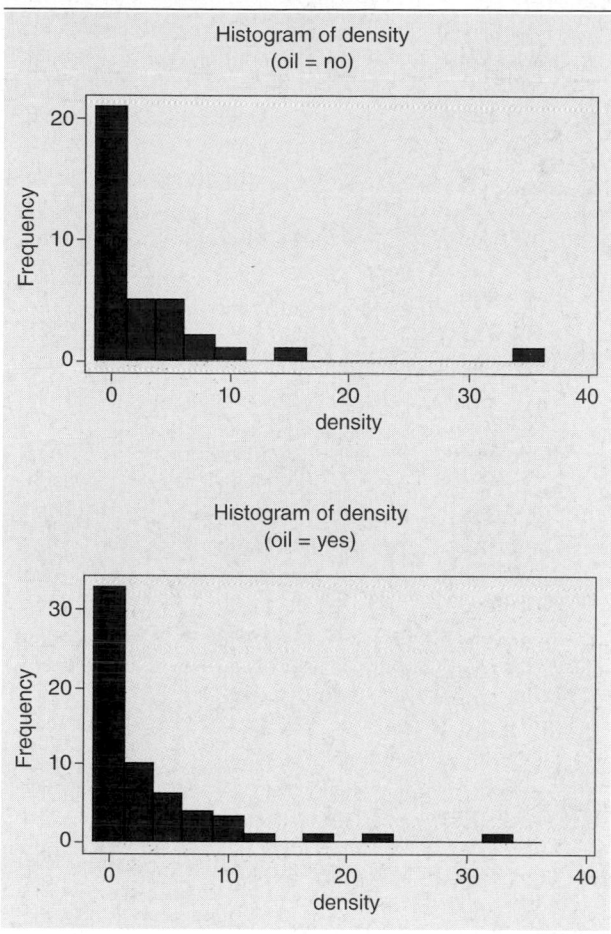

## Applying the Concepts—Advanced

**2.137** Refer to *The Astronomical Journal* study of galaxy velocities, Exercise 2.78 (p. 69). A second cluster of galaxies, named A1775, is thought to be a *double cluster*; that is, two clusters of galaxies in close proximity. Fifty-one velocity observations (in kilometers per second, km/s) from cluster A1775 are listed in the table.

GALAXY2

| | | | | | | |
|---|---|---|---|---|---|---|
| 22922 | 20210 | 21911 | 19225 | 18792 | 21993 | 23059 |
| 20785 | 22781 | 23303 | 22192 | 19462 | 19057 | 23017 |
| 20186 | 23292 | 19408 | 24909 | 19866 | 22891 | 23121 |
| 19673 | 23261 | 22796 | 22355 | 19807 | 23432 | 22625 |
| 22744 | 22426 | 19111 | 18933 | 22417 | 19595 | 23408 |
| 22809 | 19619 | 22738 | 18499 | 19130 | 23220 | 22647 |
| 22718 | 22779 | 19026 | 22513 | 19740 | 22682 | 19179 |
| 19404 | 22193 | | | | | |

*Source:* Oegerle, W. R., Hill, J. M., and Fitchett, M. J. "Observations of high dispersion clusters of galaxies: Constraints on cold dark matter." *The Astronomical Journal*, Vol. 110, No. 1, July 1995, p. 34 (Table 1), p. 37 (Figure 1).

**a.** Use a graphical method to describe the velocity distribution of galaxy cluster A1775.

**b.** Examine the graph, part **a**. Is there evidence to support the double cluster theory? Explain.

**c.** Calculate numerical descriptive measures (e.g., mean and standard deviation) for galaxy velocities in cluster A1775. Depending on your answer to part **b**, you may need to calculate two sets of numerical descriptive measures, one for each of the clusters (say, A1775A and A1775B) within the double cluster.

**d.** Suppose you observe a galaxy velocity of 20,000 km/s. Is this galaxy likely to belong to cluster A1775A or A1775B? Explain.

**2.138** *US News & World Report* (Mar. 5, 1990) reported on many factors contributing to the breakdown of public education. One study mentioned in the article found that over 90% of the nation's school districts reported that their students were scoring "above the national average" on standardized tests. Using your knowledge of measures of central tendency, explain why the schools' reports are incorrect. Does your analysis change if the term "average" refers to the mean? To the median? Explain what effect this misinformation might have on the perception of the nation's schools.

**2.139** Suicide is the leading cause of death of Americans incarcerated in correctional facilities. To determine what factors increase the risk of suicide in urban jails, a group of researchers collected data on all 37 suicides that occurred from 1967 to 1992 in the Wayne County Jail, Detroit, Michigan (*American Journal of Psychiatry*, July 1995). The data for each suicide victim is listed in the table shown on p. 104.

**a.** Identify the type (quantitative or qualitative) of each variable measured.

**a.** For GPA, calculate the intervals $\bar{x} \pm s$, $\bar{x} \pm 2s$, and $\bar{x} \pm 3s$. Based on these intervals, comment on the skewness or symmetry you would expect in these data. (Remember, the range of GPAs is 0 to 4.) Approximately what percentage of the students would you expect to find in each interval?

**b.** For number of leisure activities, calculate the intervals $\bar{x} \pm s$, $\bar{x} \pm 2s$, and $\bar{x} \pm 3s$. Based on these intervals, comment on the skewness or symmetry you would expect in these data. (Remember, the number of leisure activities cannot be negative.) Approximately what percentage of the students would you expect to find in each interval?

**c.** For number of academic leisure activities, calculate the intervals $\bar{x} \pm s$, $\bar{x} \pm 2s$, and $\bar{x} \pm 3s$. Based on these intervals, comment on the skewness or symmetry you would expect in these data. Approximately what percentage of the students would you expect to find in each interval?

**d.** Based on your answers, which of the variables' distributions would you expect to be most skewed? Why?

⊘SUICIDE

| Victim | Days in Jail Before Suicide | Marital Status | Race | Murder/ Manslaughter Charge | Time of Suicide | Year |
|---|---|---|---|---|---|---|
| 1 | 3 | Married | W | Yes | Night | 1972 |
| 2 | 4 | Single | W | Yes | Night | 1987 |
| 3 | 5 | Single | NW | Yes | Afternoon | 1975 |
| 4 | 7 | Widowed | NW | Yes | Night | 1981 |
| 5 | 10 | Single | NW | Yes | Afternoon | 1982 |
| 6 | 15 | Married | NW | Yes | Night | 1970 |
| 7 | 15 | Divorced | NW | Yes | Night | 1985 |
| 8 | 19 | Married | NW | Yes | Day | 1980 |
| 9 | 22 | Single | NW | Yes | Night | 1983 |
| 10 | 29 | Married | W | Yes | Night | 1981 |
| 11 | 31 | Single | NW | Yes | Night | 1970 |
| 12 | 85 | Single | NW | Yes | Afternoon | 1985 |
| 13 | 126 | Married | NW | Yes | Night | 1981 |
| 14 | 221 | Divorced | W | Yes | Night | 1970 |
| 15 | 14 | Single | NW | No | Night | 1980 |
| 16 | 22 | Single | W | No | Afternoon | 1970 |
| 17 | 42 | Single | W | No | Day | 1976 |
| 18 | 122 | Single | NW | No | Night | 1983 |
| 19 | 309 | Married | NW | No | Night | 1970 |
| 20 | 69 | Married | NW | No | Night | 1979 |
| 21 | 143 | Single | NW | No | Night | 1985 |
| 22 | 6 | Married | NW | No | Night | 1971 |
| 23 | 1 | Single | NW | No | Day | 1968 |
| 24 | 1 | Single | NW | No | Night | 1973 |
| 25 | 1 | Single | W | No | Night | 1989 |
| 26 | 2 | Single | W | No | Day | 1987 |
| 27 | 3 | Single | NW | No | Night | 1985 |
| 28 | 4 | Married | W | No | Afternoon | 1969 |
| 29 | 4 | Single | W | No | Afternoon | 1974 |
| 30 | 4 | Single | NW | No | Night | 1980 |
| 31 | 6 | Single | W | No | Day | 1968 |
| 32 | 6 | Married | NW | No | Night | 1975 |
| 33 | 10 | Single | W | No | Night | 1967 |
| 34 | 18 | Single | NW | No | Night | 1979 |
| 35 | 26 | Married | W | No | Night | 1976 |
| 36 | 41 | Single | NW | No | Night | 1985 |
| 37 | 86 | Married | W | No | Night | 1968 |

*Source:* DuRand, C. J., *et al.* "A quarter century of suicide in a major urban jail: Implications for community psychiatry." *American Journal of Psychiatry*, Vol. 152, No. 7, July 1995, p. 1078 (Table 1).

**b.** Are suicides at the jail more likely to be committed by inmates charged with murder/manslaughter or with lesser crimes? Illustrate with a graph.

**c.** Are suicides at the jail more likely to be committed at night? Illustrate with a graph.

**d.** What is the mean length of time an inmate is in jail before committing suicide? What is the median? Interpret these two numbers.

**e.** Is it likely that a future suicide at the jail will occur after 200 days? Explain.

**\*f.** Have suicides at the jail declined over the years? Support your answer with a graph.

**2.140** The final grades given by two professors in introductory statistics courses have been carefully examined. The students in the first professor's class had a grade point average of 3.0 and a standard deviation of .2. Those in the second professor's class had grade points with an average of 3.0 and a standard deviation of 1.0. If you had a choice, which professor would you take for this course? Explain.

## STUDENT PROJECT

We list here several sources of real-life data sets (many in the list have been obtained from Wasserman and Bernero's *Statistics Sources* and many can be accessed via the Internet). This index of data sources is very complete and is a useful reference for anyone interested in finding almost any type of data. First we list some almanacs:

*CBS News Almanac*
*Information Please Almanac*
*World Almanac and Book of Facts*

United States Government publications are also rich sources of data:

*Agricultural Statistics*
*Digest of Educational Statistics*
*Handbook of Labor Statistics*
*Housing and Urban Development Yearbook*
*Social Indicators*
*Uniform Crime Reports for the United States*
*Vital Statistics of the United States*
*Business Conditions Digest*
*Economic Indicators*
*Monthly Labor Review*
*Survey of Current Business*
*Bureau of the Census Catalog*
*Statistical Abstract of the United States*

Main data sources are published on an annual basis:

*Commodity Yearbook*
*Facts and Figures on Government Finance*
*Municipal Yearbook*
*Standard and Poor's Corporation, Trade and Securities: Statistics*

Some sources contain data that are international in scope:

*Compendium of Social Statistics*
*Demographic Yearbook*
*United Nations Statistical Yearbook*
*World Handbook of Political and Social Indicators*

Utilizing the data sources listed, sources suggested by your instructor, or your own resourcefulness, find one real-life quantitative data set that stems from an area of particular interest to you.

a. Describe the data set by using a relative frequency histogram, stem-and-leaf display, or dot plot.
b. Find the mean, median, variance, standard deviation, and range of the data set.
c. Use Chebyshev's Rule and the Empirical Rule to describe the distribution of this data set. Count the actual number of observations that fall within 1, 2, and 3 standard deviations of the mean of the data set and compare these counts with the description of the data set you developed in part **b**.

## REFERENCES

Hoff, D. *How to Lie with Statistics*. New York: Norton, 1954.

Mendenhall, W., Beaver, R. J., and Beaver, B. M. *Introduction to Probability and Statistics*, 10th ed. North Scituate, Mass.: Duxbury, 1999.

Tufte, E. R. *Envisioning Information*. Cheshire, Conn.: Graphics Press, 1990.

Tufte, E.R. *Visual Explanations*. Cheshire, Conn.: Graphics Press, 1997.

Tufte, E. R. *Visual Display of Quantitative Information*. Cheshire, Conn.: Graphics Press, 1983.

Tukey, J. *Exploratory Data Analysis*. Reading, Mass.: Addison-Wesley, 1977.

# Probability

## Contents

## Statistics in Action

Game Show Strategy: To Switch or Not to Switch?

### ✎ Where We've Been

We've identified inference, from a sample to a population, as the goal of statistics. And we've seen that to reach this goal, we must be able to describe a set of measurements. Thus, we explored the use of graphical and numerical methods for describing both quantitative and qualitative data sets and for phrasing inferences.

### ☞ Where We're Going

Now that we know how to phrase an inference about a population, we turn to the problem of making the inference. What is it that permits us to make the inferential jump from sample to population and then to give a measure of reliability for the inference? As you'll see, the answer is *probability*. This chapter is devoted to a study of probability—what it is and some of the basic concepts of the theory behind it.

Recall that one branch of statistics is concerned with decisions about a population based on sample information. You can see how this is accomplished more easily if you understand the relationship between population and sample—a relationship that becomes clearer if we reverse the statistical procedure of making inferences from sample to population. In this chapter we assume the population is *known* and calculate the chances of obtaining various samples from the population. Thus, we show that probability is the reverse of statistics: In probability, we use the population information to infer the probable nature of the sample.

Probability plays an important role in inference making. Suppose, for example, you have an opportunity to invest in an oil exploration company. Past records show that out of ten previous oil drillings (a sample of the company's experiences), all ten came up dry. What do you conclude? Do you think the chances are better than 50 : 50 that the company will hit a gusher? Should you invest in this company? Chances are, your answer to these questions will be an emphatic No. If the company's exploratory prowess is sufficient to hit a producing well 50% of the time, a record of ten dry wells out of ten drilled is an event that is just too improbable.

Or suppose you're playing poker with what your opponents assure you is a well-shuffled deck of cards. In three consecutive five-card hands, the person on your right is dealt four aces. Based on this sample of three deals, do you think the cards are being adequately shuffled? Again, your answer is likely to be No because dealing three hands of four aces is just too improbable if the cards were properly shuffled.

Note that the decisions concerning the potential success of the oil drilling company and the adequacy of card shuffling both involve knowing the chance—or probability—of a certain sample result. Both situations were contrived so that you could easily conclude that the probabilities of the sample results were small. Unfortunately, the probabilities of many observed sample results aren't so easy to evaluate intuitively. For these cases we need the assistance of a theory of probability.

## 3.1    EVENTS, SAMPLE SPACES, AND PROBABILITY

Let's begin our treatment of probability with straightforward examples that are easily described. With the aid of simple examples, we can introduce important definitions that will help us develop the notion of probability more easily.

Suppose a coin is tossed once and the up face is recorded. The result we see is called an *observation*, or *measurement*, and the process of making an observation is called an *experiment*. Notice that our definition of experiment is broader than the one used in the physical sciences, where you tend to picture test tubes, microscopes, and other laboratory equipment. Among other things, statistical experiments may include recording an Internet user's preference for a Web browser, recording a voter's opinion on an important political issue, measuring the amount of dissolved oxygen in a polluted river, observing the level of anxiety of a test taker, counting the number of errors in an inventory, and observing the fraction of insects killed by a new insecticide. The point is that a statistical experiment can be almost any act of observation as long as the outcome is uncertain.

**DEFINITION 3.1**
An **experiment** is an act or process of observation that leads to a single outcome that cannot be predicted with certainty.

Consider another simple experiment consisting of tossing a die and observing the number on the up face. The six possible outcomes to this experiment are:

1. Observe a 1
2. Observe a 2
3. Observe a 3
4. Observe a 4
5. Observe a 5
6. Observe a 6

Note that if this experiment is conducted once, *you can observe one and only one of these six basic outcomes, and the outcome cannot be predicted with certainty*. Also, these possibilities cannot be decomposed into more basic outcomes. Because observing the outcome of an experiment is similar to selecting a sample from a population, the basic possible outcomes to an experiment are called **sample points**.*

**DEFINITION 3.2**

A **sample point** is the most basic outcome of an experiment.

**EXAMPLE 3.1**

Two coins are tossed, and their up faces are recorded. List all the sample points for this experiment.

**Solution**

Even for a seemingly trivial experiment, we must be careful when listing the sample points. At first glance, we might expect three basic outcomes: Observe two heads; Observe two tails; or Observe one head and one tail. However, further reflection reveals that the last of these, Observe one head and one tail, can be decomposed into two outcomes: Head on coin 1, Tail on coin 2; and Tail on coin 1, Head on coin 2.† Thus, we have four sample points:

1. Observe *HH*
2. Observe *HT*
3. Observe *TH*
4. Observe *TT*

where *H* in the first position means "Head on coin 1," *H* in the second position means "Head on coin 2," and so on.

We often wish to refer to the collection of all the sample points of an experiment. This collection is called the *sample space* of the experiment. For example, there are six sample points in the sample space associated with the die-toss experiment. The sample spaces for the experiments discussed thus far are shown in Table 3.1.

**DEFINITION 3.3**

The **sample space** of an experiment is the collection of all its sample points.

---

*Alternatively, the term "simple event" can be used.
†Even if the coins are identical in appearance, there are, in fact, two distinct coins. Thus, the designation of one coin as coin 1 and the other as coin 2 is legitimate in any case.

**TABLE 3.1    Experiments and Their Sample Spaces**

*Experiment:*    Observe the up face on a coin.
*Sample Space:*    1. Observe a head
                    2. Observe a tail.

This sample space can be represented in set notation as a set containing two sample points:

$$S: \quad \{H, T\}$$

where $H$ represents the sample point Observe a head and $T$ represents the sample point Observe a tail.

*Experiment:*    Observe the up face on a die.
*Sample Space:*    1. Observe a 1
                    2. Observe a 2
                    3. Observe a 3
                    4. Observe a 4
                    5. Observe a 5
                    6. Observe a 6

This sample space can be represented in set notation as a set of six sample points:

$$S: \quad \{1, 2, 3, 4, 5, 6\}$$

*Experiment:*    Observe the up faces on two coins.
*Sample Space:*    1. Observe $HH$
                    2. Observe $HT$
                    3. Observe $TH$
                    4. Observe $TT$

This sample space can be represented in set notation as a set of four sample points:

$$S: \quad \{HH, HT, TH, TT\}$$

Just as graphs are useful in describing sets of data, a pictorial method for presenting the sample space will often be useful. Figure 3.1 shows such a representation for each of the experiments in Table 3.1. In each case, the sample space is shown as a closed figure, labeled **S**, containing all possible sample points. Each sample point is represented by a solid dot (i.e., a "point") and labeled accordingly. Such graphical representations are called **Venn diagrams**.

Now that we know that an experiment will result in *only one* basic outcome—called a sample point—and that the sample space is the collection of all possible sample points, we're ready to discuss the probabilities of the sample points. You have undoubtedly used the term *probability* and have some intuitive idea about its meaning. Probability is generally used synonymously with "chance," "odds," and similar concepts. For example, if a fair coin is tossed, we might reason that both the sample points, Observe a head and Observe a tail, have the same *chance* of occurring. Thus, we might state that "the probability of observing a head is 50%" or "the odds of seeing a head are 50:50." Both these statements are based on an informal knowledge of probability. We'll begin our treatment of probability by using such informal concepts and then solidify what we mean later.

*The probability of a sample point is a number between 0 and 1 that measures the likelihood that the outcome will occur when the experiment is performed.* This number is usually taken to be the relative frequency of the occurrence of a sample point in a very long series of repetitions of an experiment.* For example, if we are

---

*The result derives from an axiom in probability theory called the *Law of Large Numbers*. Phrased informally, the law states that the relative frequency of the number of times that an outcome occurs when an experiment is replicated over and over again (i.e., a large number of times) approaches the true (or theoretical) probability of the outcome.

**FIGURE 3.1**

*Venn Diagrams for the Three Experiments from Table 3.1*

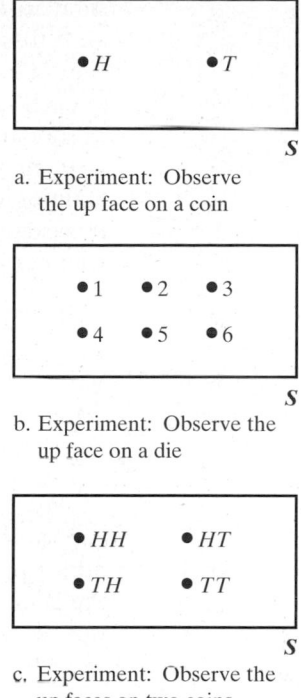

a. Experiment: Observe the up face on a coin

b. Experiment: Observe the up face on a die

c. Experiment: Observe the up faces on two coins

assigning probabilities to the two sample points in the coin-toss experiment (Observe a head and Observe a tail), we might reason that if we toss a balanced coin a very large number of times, the sample points Observe a head and Observe a tail will occur with the same relative frequency of .5.

Our reasoning is supported by Figure 3.2. The figure, generated using SAS, plots the relative frequency of the number of times that a head occurs when

**FIGURE 3.2**

*SAS Scatterplot Showing the Proportion of Heads in N Tosses of a Coin*

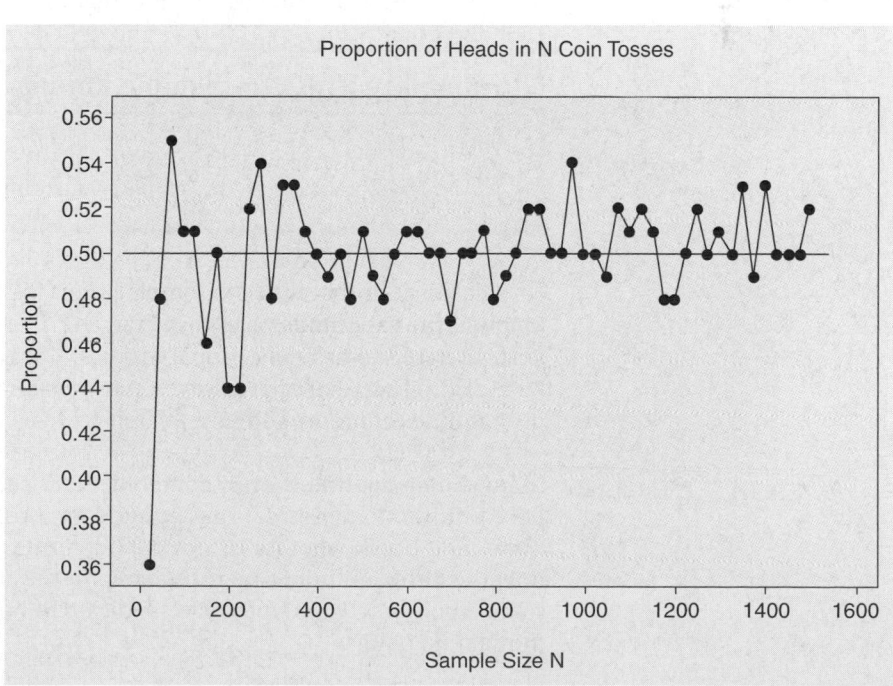

simulating (by computer) the toss of a coin $N$ times, where $N$ ranges from as few as 25 tosses to as many as 1,500 tosses of the coin. You can see that when $N$ is large (i.e., $N = 1{,}500$), the relative frequency is converging to .5. Thus, the probability of each sample point in the coin tossing experiment is .5.

For some experiments, we may have little or no information on the relative frequency of occurrence of the sample points; consequently, we must assign probabilities to the sample points based on general information about the experiment. For example, if the experiment is to invest in a business venture and to observe whether it succeeds or fails, the sample space would appear as in Figure 3.3.

### FIGURE 3.3

*Experiment: Invest in a*
*Business Venture and*
*Observe Whether It*
*Succeeds (S) or Fails (F)*

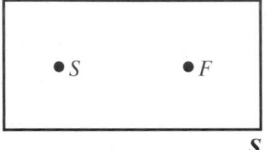

We are unlikely to be able to assign probabilities to the sample points of this experiment based on a long series of repetitions since unique factors govern each performance of this kind of experiment. Instead, we may consider factors such as the personnel managing the venture, the general state of the economy at the time, the rate of success of similar ventures, and any other information deemed pertinent. If we finally decide that the venture has an 80% chance of succeeding, we assign a probability of .8 to the sample point Success. This probability can be interpreted as a measure of our degree of belief in the outcome of the business venture; that is, it is a subjective probability. Notice, however, that such probabilities should be based on expert information that is carefully assessed. If not, we may be misled on any decisions based on these probabilities or based on any calculations in which they appear. [*Note:* For a text that deals in detail with the subjective evaluation of probabilities, see Winkler (1972) or Lindley (1985).]

No matter how you assign the probabilities to sample points, the probabilities assigned must obey two rules:

---

### Probability Rules for Sample Points

1. All sample point probabilities *must* lie between 0 and 1.
2. The probabilities of all the sample points within a sample space *must* sum to 1.

---

Assigning probabilities to sample points is easy for some experiments. For example, if the experiment is to toss a fair coin and observe the face, we would probably all agree to assign a probability of $\frac{1}{2}$ to the two sample points, Observe a head and Observe a tail. However, many experiments have sample points whose probabilities are more difficult to assign.

**EXAMPLE 3.2**

Many American hotels offer complimentary shampoo in their guest rooms. Suppose you randomly select one hotel from a registry of all hotels in the United States and check whether or not the hotel offers complimentary shampoo. Show how this problem might be formulated in the framework of an experiment with sample points and a sample space. Indicate how probabilities might be assigned to the sample points.

**Solution**

The experiment can be defined as the selection of an American hotel and the observation of whether or not complimentary shampoo is offered in the hotel's guest rooms. There are two sample points in the sample space corresponding to this experiment:

> *S*: {The hotel offers complimentary shampoo}
> *N*: {No complimentary shampoo is offered by the hotel}

The difference between this and the coin-toss experiment becomes apparent when we attempt to assign probabilities to the two sample points. What probability should we assign to the sample point *S*? If you answer .5, you are assuming that the events *S* and *N* should occur with equal likelihood, just like the sample points Heads and Tails in the coin-toss experiment. But assignment of sample point probabilities for the hotel-shampoo experiment is not so easy. In fact, a recent survey of American hotels found that 80% now offer complimentary shampoo to guests. Then it might be reasonable to approximate the probability of the sample point *S* as .8 and that of the sample point *N* as .2. Here we see that sample points are not always equally likely, so assigning probabilities to them can be complicated—particularly for experiments that represent real applications (as opposed to coin- and die-toss experiments).

Although the probabilities of sample points are often of interest in their own right, it is usually probabilities of collections of sample points that are important. Example 3.3 demonstrates this point.

**EXAMPLE 3.3**

A fair die is tossed, and the up face is observed. If the face is even, you win $1. Otherwise, you lose $1. What is the probability that you win?

**Solution**

Recall that the sample space for this experiment contains six sample points:

$$S: \{1, 2, 3, 4, 5, 6\}$$

Since the die is balanced, we assign a probability of $\frac{1}{6}$ to each of the sample points in this sample space. An even number will occur if one of the sample points, Observe a 2, Observe a 4, or Observe a 6, occurs. A collection of sample points such as this is called an *event*, which we denote by the letter *A*. Since the event *A* contains three sample points—all with probability $\frac{1}{6}$—and since no sample points can occur simultaneously, we reason that the probability of *A* is the sum of the probabilities of the sample points in *A*. Thus, the probability of *A* is $\frac{1}{6} + \frac{1}{6} + \frac{1}{6} = \frac{1}{2}$. This implies that, *in the long run*, you will win $1 half the time and lose $1 half the time.

Figure 3.4 is a Venn diagram depicting the sample space associated with a die-toss experiment and the event *A*, Observe an even number. The event *A* is represented by the closed figure inside the sample space *S*. This closed figure *A* contains all the sample points that comprise it.

To decide which sample points belong to the set associated with an event *A*, test each sample point in the sample space *S*. If event *A* occurs, then that sample point is in the event *A*. For example, the event *A*, Observe an even number, in the die-toss experiment will occur if the sample point Observe a 2 occurs. By the same reasoning, the sample points Observe a 4 and Observe a 6 are also in event *A*.

**FIGURE 3.4**

*Die-Toss Experiment with Event A: Observe an Even Number*

To summarize, we have demonstrated that an event can be defined in words or it can be defined as a specific set of sample points. This leads us to the following general definition of an *event*:

---

**DEFINITION 3.4**

An **event** is a specific collection of sample points.

---

**EXAMPLE 3.4**

Consider the experiment of tossing two *unbalanced* coins. Because the coins are *not* balanced, their outcomes (*H* or *T*) are not equiprobable. Suppose the correct probabilities associated with the sample points are given in the accompanying table. [*Note:* The necessary properties for assigning probabilities to sample points are satisfied.]

Consider the events

A: {Observe exactly one head}
B: {Observe at least one head}

Calculate the probability of *A* and the probability of *B*.

| Sample Point | Probability |
|---|---|
| *HH* | $\frac{4}{9}$ |
| *HT* | $\frac{2}{9}$ |
| *TH* | $\frac{2}{9}$ |
| *TT* | $\frac{1}{9}$ |

**Solution**

Event *A* contains the sample points *HT* and *TH*. Since two or more sample points cannot occur at the same time, we can easily calculate the probability of event *A* by summing the probabilities of the two sample points. Thus, the probability of observing exactly one head (event *A*), denoted by the symbol $P(A)$, is

$$P(A) = P(\text{Observe } HT) + P(\text{Observe } TH) = \frac{2}{9} + \frac{2}{9} = \frac{4}{9}$$

Similarly, since *B* contains the sample points *HH, HT,* and *TH,*

$$P(B) = \frac{4}{9} + \frac{2}{9} + \frac{2}{9} = \frac{8}{9}$$

The preceding example leads us to a general procedure for finding the probability of an event *A*:

---

### Probability of an Event

The probability of an event *A* is calculated by summing the probabilities of the sample points in the sample space for *A*.

---

Thus, we can summarize the steps for calculating the probability of any event, as indicated in the next box.

---

### Steps for Calculating Probabilities of Events

1. Define the experiment, that is, describe the process used to make an observation and the type of observation that will be recorded.
2. List the sample points.
3. Assign probabilities to the sample points.
4. Determine the collection of sample points contained in the event of interest.
5. Sum the sample point probabilities to get the event probability.

---

**EXAMPLE 3.5**

The American Association for Marriage and Family Therapy (AAMFT) is a group of professional therapists and family practitioners that treats many of the nation's couples and families. The AAMFT released the findings of a study that tracked the post-divorce history of 100 pairs of former spouses with children. Each divorced couple was classified into one of four groups, nicknamed "perfect pals (PP)," "cooperative colleagues (CC)," "angry associates (AA)," and "fiery foes (FF)." The proportions classified into each group are shown in Table 3.2.

Suppose one of the 100 couples is selected at random.

a. Define the experiment that generated the data in Table 3.2, and list the sample points.
b. Assign probabilities to the sample points.
c. What is the probability that the former spouses are "fiery foes"?
d. What is the probability that the former spouses have at least some conflict in their relationship?

**TABLE 3.2    Results of AAMFT Study of Divorced Couples**

| Group | Proportion |
|---|---|
| *Perfect Pals (PP)* <br> (Joint-custody parents who get along well) | .12 |
| *Cooperative Colleagues (CC)* <br> (Occasional conflict, likely to be remarried) | .38 |
| *Angry Associates (AA)* <br> (Cooperate on children-related issues only, conflicting otherwise) | .25 |
| *Fiery Foes (FF)* <br> (Communicate only through children, hostile toward each other) | .25 |

**FIGURE 3.5**

*Venn Diagram for AAMFT
Survey*

### Solution

a. The experiment is the act of classifying the randomly selected couple. The sample points, the simplest outcomes of the experiment, are the four groups (categories) listed in Table 3.2. They are shown in the Venn diagram in Figure 3.5.
b. If, as in Example 3.1, we were to assign equal probabilities in this case, each of the response categories would have a probability of one-fourth (1/4), or .25. But, by examining Table 3.2 you can see that equal probabilities are not reasonable in this case because the response percentages are not all the same in the four categories. It is more reasonable to assign a probability equal to the response proportion in each class, as shown in Table 3.3.*

**TABLE 3.3    Sample Point Probabilities for AAMFT Survey**

| Sample Point | Probability |
| --- | --- |
| *PP* | .12 |
| *CC* | .38 |
| *AA* | .25 |
| *FF* | .25 |

c. The event that the former spouses are "fiery foes" corresponds to the sample point *FF*. Consequently, the probability of the event is the probability of the sample point. From Table 3.3, we find $P(FF) = .25$. Therefore, there is a .25 probability (or one-fourth chance) that the couple we select are "fiery foes."
d. The event that the former spouses have at least some conflict in their relationship, call it event *C*, is not a sample point because it consists of more than one of the response classifications (the sample points). In fact, as shown in Figure 3.5, the event *C* consists of three sample points: *CC, AA*, and *FF*. The probability of *C* is defined to be the sum of the probabilities of the sample points in *C*:

$$P(C) = P(CC) + P(AA) + P(FF) = .38 + .25 + .25 = .88$$

Thus, the chance that we observe a divorced couple with some degree of conflict in their relationship is .88—a fairly high probability.    ∎

    For the experiments discussed so far, listing the sample points has been easy. For more complex experiments, the number of sample points may be so large that listing them is impractical. In solving probability problems for experiments with many sample points, we employ the same principles as for experiments with few sample points. The only difference is that we need *counting rules* for determining the number of sample points without actually enumerating all of them. In Section 3.8 (an optional section), we present several of the more useful counting rules.

---

*Since the response percentages were based on a sample of divorced comples, these assigned probabilities are estimates of the true population response percentages. You will learn how to measure the reliability of probability estimates in Chapter 7.

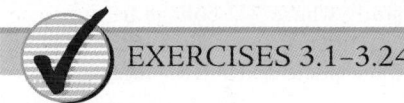 **EXERCISES 3.1–3.24**

### Learning the Mechanics

**3.1**   An experiment results in one of the following sample points: $E_1, E_2, E_3, E_4$, or $E_5$.
  **a.** Find $P(E_3)$ if $P(E_1) = .1, P(E_2) = .2, P(E_4) = .1$, and $P(E_5) = .1$
  **b.** Find $P(E_3)$ if $P(E_1) = P(E_3), P(E_2) = .1$, $P(E_4) = .2$, and $P(E_5) = .1$
  **c.** Find $P(E_3)$ if $P(E_1) = P(E_2) = P(E_4) = P(E_5) = .1$

**3.2**   The accompanying diagram describes the sample space of a particular experiment and events $A$ and $B$.

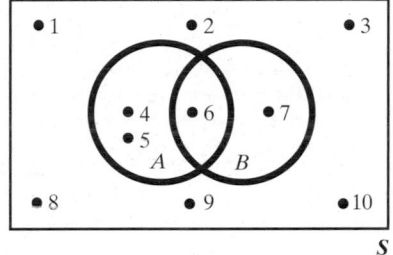

  **a.** What is this type of diagram called?
  **b.** Suppose the sample points are equally likely. Find $P(A)$ and $P(B)$
  **c.** Suppose $P(1) = P(2) = P(3) = P(4) = P(5) = \frac{1}{20}$ and $P(6) = P(7) = P(8) = P(9) = P(10) = \frac{3}{20}$. Find $P(A)$ and $P(B)$.

**3.3**   The sample space for an experiment contains five sample points with probabilities as shown in the table. Find the probability of each of the following events:

  $A$: {Either 1, 2, or 3 occurs}
  $B$: {Either 1, 3, or 5 occurs}
  $C$: {4 does not occur}

| Sample Points | Probabilities |
| --- | --- |
| 1 | .05 |
| 2 | .20 |
| 3 | .30 |
| 4 | .30 |
| 5 | .15 |

**3.4**   Consider the experiment of tossing a die and observing the up face.
  **a.** Draw a Venn diagram for the experiment. On your diagram indicate the event Observe a number greater than 4. Call this event $A$. Also indicate the event Observe an even number. Call this event $B$.
  **b.** We would all agree that the probability of observing a 3 on the toss of a "fair" die is $\frac{1}{6}$. Explain what it

means for a die to be fair (or balanced) and explain how knowing that a die is fair leads us to $P(3) = \frac{1}{6}$.
  **c.** For your Venn diagram of part **a**, assume the die is fair and find $P(A)$ and $P(B)$.
  **d.** If you knew that a particular die were unfair (i.e., "loaded"), how would you determine the probability of observing a 3?

**3.5**   The Venn diagram depicts an experiment with six sample points. The events $A$ and $B$ are also shown. The probabilities of the sample points are:

$$P(1) = P(2) = P(4) = \tfrac{2}{9}$$
$$P(3) = P(5) = P(6) = \tfrac{1}{9}$$

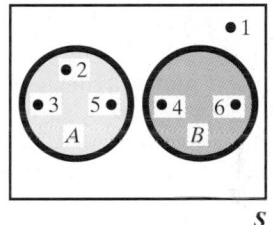

  **a.** Find $P(A)$.
  **b.** Find $P(B)$.
  **c.** Find the probability that the events $A$ and $B$ occur *simultaneously*.

**3.6**   Two fair dice are tossed, and the up face on each die is recorded.
  **a.** List the 36 sample points contained in the sample space.
  **b.** Find the probability of observing each of the following events:

  $A$: {A 3 appears on each of the two dice}
  $B$: {The sum of the numbers is even}
  $C$: {The sum of the numbers is equal to 7}
  $D$: {A 5 appears on at least one of the dice}
  $E$: {The sum of the numbers is 10 or more}

**3.7**   Consider the experiment composed of one roll of a fair die followed by one toss of a fair coin. List the sample points. Assign a probability to each sample point. Determine the probability of observing each of the following events:

  $A$: {6 on the die; $H$ on the coin}
  $B$: {Even number on the die; $T$ on the coin}
  $C$: {Even number on the die}
  $D$: {$T$ on the coin}

**3.8**   Two marbles are drawn at random and without replacement from a box containing two blue marbles and three red marbles. Determine the probability of observing each of the following events:

  $A$: {Two blue marbles are drawn}
  $B$: {A red and a blue marble are drawn}
  $C$: {Two red marbles are drawn}

**3.9**  Simulate the experiment described in Exercise 3.8 using any five identically shaped objects, two of which are one color and three, another. Mix the objects, draw two, record the results, and then replace the objects. Repeat the experiment a large number of times (at least 100). Calculate the proportion of times events *A, B,* and *C* occur. How do these proportions compare with the probabilities you calculated in Exercise 3.8? Should these proportions equal the probabilities? Explain.

### Applying the Concepts—Basic

**3.10**  According to *USA Today* (Sept. 19, 2000), there are 650 members of the International Nanny Association (INA). Of these, only three are men. Find the probability that a randomly selected member of the INA is a man.

**3.11**  *The Wall Street Journal* (Sept. 1, 2000) reported on an independent study of postal workers and violence at post offices. In a sample of 12,000 postal workers, 600 of them were physically assaulted on the job in the past year. Use this information to estimate the probability that a randomly selected postal worker will be physically assaulted on the job during the year.

**3.12**  Two species of rhinoceros native to Africa are black rhinos and white rhinos. The International Rhino Federation estimates that the African rhinoceros population consists of 2,600 white rhinos and 8,400 black rhinos. Suppose one rhino is selected at random from the African rhino population and its species (black or white) is observed.
  **a.** List the sample points for this experiment.
  **b.** Assign probabilities to the sample points based on the estimates made by the International Rhino Federation.

**3.13**  The United States Department of Agriculture (USDA) reports that, under its standard inspection system, one in every 100 slaughtered chickens pass inspection with fecal contamination (*Tampa Tribune*, Mar. 31, 2000).
  **a.** If a slaughtered chicken is selected at random, what is the probability that it passes inspection with fecal contamination?
  **b.** The probability of part **a** was based on a USDA study that found that 306 of 32,075 chicken carcasses passed inspection with fecal contamination. Do you agree with the USDA's statement about the likelihood of a slaughtered chicken passing inspection with fecal contamination?

**3.14**  Answer the following question posed in the *Atlanta Journal-Constitution* (Feb. 7, 2000): When a meteorologist says that "the probability of rain this afternoon is .4," does it mean that 40% of the time during the afternoon, it will be raining?

### Applying the Concepts—Intermediate

**3.15**  Refer to the *American Scientist* (July–Aug.1998) study of *Benford's Law*, Exercise 2.14 (p. 28). Recall that *Benford's Law* states that the integer 1 is more likely to occur as the first significant digit in a randomly selected number than the other integers. The table summarizes

the results of a study where 743 college freshmen selected a six-digit number at random. Assume that one of these 743 numbers is selected and the first significant digit is observed.

**DIGITS**

| First Digit | Number of Occurrences |
|---|---|
| 1 | 109 |
| 2 | 75 |
| 3 | 77 |
| 4 | 99 |
| 5 | 72 |
| 6 | 117 |
| 7 | 89 |
| 8 | 62 |
| 9 | 43 |
| **Total** | 743 |

*Source:* Hill, T. P "The First Digit Phenomenon." *American Scientist,* Vol. 86, No. 4, July-Aug. 1998, p. 363 (Figure 5).

  **a.** List the sample points for this experiment.
  **b.** Explain why the sample points are not equally likely to occur.
  **c.** Assign reasonable probabilities to the sample points. Verify that the probabilities sum to 1.
  **d.** What is the probability that the first significant digit is 1 or 2?
  **e.** What is the probability that the first significant digit is greater than 5?

**3.16**  The maximum score possible on the Standardized Admission Test (SAT) is 1600. According to the test developers, the chance that a student scores a perfect 1600 on the SAT is 5 in 10,000.
  **a.** Find the probability that a randomly selected student scores a 1600 on the SAT.
  **b.** In a recent year, 545 students scored a perfect 1600 on the SAT. How is this possible given the probability calculated in part **a**?

**3.17**  "Urban" and "rural" describe geographic areas upon which land zoning regulations, school district policy, and public service policy are often set. However, the characteristics of urban/rural areas are not clearly defined. Researchers at the University of Nevada (Reno) asked a sample of county commissioners to give their perception of the single most important factor in identifying urban counties (*Professional Geographer*, Feb. 2000). In all, five factors were mentioned by the commissioners: total population, agricultural change, presence of industry, growth, and population concentration. The survey results are displayed in the SPSS pie chart (p. 119). Suppose one of the commissioners is selected at random and the most important factor specified by the commissioner is recorded.
  **a.** List the sample points for this experiment.
  **b.** Assign reasonable probabilities to the sample points.
  **c.** Find the probability that the most important factor specified by the commissioner is population-related.

SPSS Output for Exercise 3.17

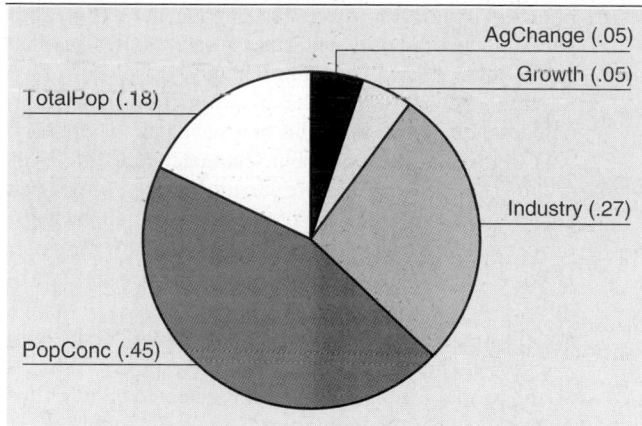

| Most Salient Role | Number |
|---|---|
| Spouse | 424 |
| Parent | 269 |
| Grandparent | 148 |
| Other relative | 59 |
| Friend | 73 |
| Homemaker | 59 |
| Provider | 34 |
| Volunteer, club, church member | 36 |
| **Total** | **1,102** |

*Source:* Krause, N., and Shaw, B. A. "Role-specific feelings of control and mortality." *Psychology and Aging,* Vol. 15, No. 4, Dec. 2000 (Table 2).

**3.18** Refer to the National Highway Transportation Safety Administration (NHTSA) study of requests for installation of on-off switches for passenger-side air bags, Exercise 2.5 (p. 26). The table summarizing the reasons for requesting the on-off switches for 1998–1999 is reproduced here. Suppose we select one of the 30,337 car owners who made such a request over this time period and record the reason for the request.

| Reason | Number of requests |
|---|---|
| Infant | 1,852 |
| Child | 17,148 |
| Medical | 8,377 |
| Infant & medical | 44 |
| Child & medical | 903 |
| Infant & child | 1,878 |
| Infant & child & medical | 135 |
| **Total** | **30,337** |

*Source:* National Highway Transportation Safety Administration, Sept. 2000.

    **a.** List the sample points for this experiment.

    **b.** Assign reasonable probabilities to the sample points.

    **c.** Find the probability that the car owner uses medical as a reason.

    **d.** Find the probability that the car owner does not use an infant as a reason.

**3.19** Refer to the *Psychology and Aging* (Dec. 2000) study of elderly people and their roles, Exercise 2.6 (p. 26). The table summarizing the most salient roles identified by each in a national sample of 1,102 adults, 65 years or older, is reproduced in the next column.

    **a.** What is the probability that a randomly selected elderly person feels his/her most salient role is a spouse?

    **b.** What is the probability that a randomly selected elderly person feels his/her most salient role is a parent or grandparent?

    **c.** What is the probability that a randomly selected elderly person feels his/her most salient role does not involve a spouse or relative of any kind?

**3.20** Scientists have already discovered two genes (on chromosomes 19 and 21) that mutate to cause the early onset of Alzheimer's disease. *Science News* (July 8, 1995) reported on a search for a third gene that causes Alzheimer's. An international team of scientists gathered genetic information from 21 families afflicted by the early-onset form of the disease. In six of these families, mutations in the S182 gene on chromosome 14 accounted for the early onset of Alzheimer's.

    **a.** Consider a family afflicted by the early onset of Alzheimer's disease. Find the approximate probability that the researchers will find mutations in the S182 gene on chromosome 14 for this family

    **b.** Discuss the reliability of the probability, part **a**. How could a more accurate estimate of the probability be obtained?

### Applying the Concepts—Advanced

**3.21** Three people play a game called "Odd Man Out." In this game, each player flips a fair coin until the outcome (heads or tails) for one of the players is not the same as the other two players. This player is then "the odd man out" and loses the game. Find the probability that the game ends (i.e., either exactly one of the coins will fall heads or exactly one of the coins will fall tails) after only one toss by each player. Suppose one of the players, hoping to reduce his chances of being the odd man, uses a two-headed coin. Will this ploy be successful? Solve by listing the sample points in the sample space.

**3.22** Consider the following question posed to Marilyn vos Savant in her weekly newspaper column "Ask Marilyn":

> I have two pairs of argyle socks, and they look nearly identical—one navy blue and the other black. [When doing the laundry] my wife matches the socks incorrectly much more often than she does correctly. ... If all four socks are in front of her, it seems to me that her chances are 50% for a wrong match and 50% for a right match. What do you think? *Source: Parade Magazine,* Feb. 27, 1994.

Use your knowledge of probability to answer this question. [*Hint:* List the sample points in the experiment.]

**3.23** Handicappers for horse races express their belief about the probabilities of each horse winning a race in terms of **odds**. If the probability of event $E$ is $P(E)$, then the *odds in favor of* $E$ are $P(E)$ to $1 - P(E)$. Thus, if a handicapper assesses a probability of .25 that Snow Chief will win the Belmont Stakes, the odds in favor of Snow Chief are $^{25}/_{100}$ to $^{75}/_{100}$, or 1 to 3. It follows that the *odds against* $E$ are $1 - P(E)$ to $P(E)$, or 3 to 1 against a win by Snow Chief. In general, if the odds in favor of event $E$ are $a$ to $b$, then $P(E) = a/(a + b)$.

**a.** A second handicapper assesses the probability of a win by Snow Chief to be $^1/_3$. According to the second handicapper, what are the odds in favor of a Snow Chief win?

**b.** A third handicapper assesses the odds in favor of Snow Chief to be 1 to 1. According to the third handicapper, what is the probability of a Snow Chief win?

**c.** A fourth handicapper assesses the odds against Snow Chief's winning to be 3 to 2. Find this handicapper's assessment of the probability that Snow Chief will win.

**3.24** An individual's genetic makeup is determined by the genes obtained from each parent. For every genetic trait, each parent possesses a gene pair; and each contributes one-half of this gene pair, with equal probability, to their offspring, forming a new gene pair. The offspring's traits (eye color, baldness, etc.) come from this new gene pair, where each gene in this pair possesses some characteristic.

For the gene pair that determines eye color, each gene trait may be one of two types: dominant brown ($B$) or recessive blue ($b$). A person possessing the gene pair $BB$ or $Bb$ has brown eyes, whereas the gene pair $bb$ produces blue eyes.

**a.** Suppose both parents of an individual are brown-eyed, each with a gene pair of the type $Bb$. What is the probability that a randomly selected child of this couple will have blue eyes?

**b.** If one parent has brown eyes, type $Bb$, and the other has blue eyes, what is the probability that a randomly selected child of this couple will have blue eyes?

**c.** Suppose one parent is brown-eyed, type $BB$. What is the probability that a child has blue eyes?

## 3.2 UNIONS AND INTERSECTIONS

An event can often be viewed as a composition of two or more other events. Such events, which are called **compound events**, can be formed (composed) in two ways, as defined and illustrated here.

**DEFINITION 3.5**

The **union** of two events $A$ and $B$ is the event that occurs if either $A$ or $B$ or both occur on a single performance of the experiment. We denote the union of events $A$ and $B$ by the symbol $A \cup B$. $A \cup B$ consists of all the sample points that belong to *A or B or both*. (See Figure 3.6a.)

**DEFINITION 3.6**

The **intersection** of two events $A$ and $B$ is the event that occurs if both $A$ and $B$ occur on a single performance of the experiment. We write $A \cap B$ for the intersection of $A$ and $B$. $A \cap B$ consists of all the sample points belonging to *both A and B*. (See Figure 3.6b.)

**FIGURE 3.6**

*Venn Diagrams for Union and Intersection*

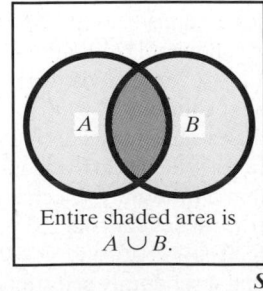

Entire shaded area is
$A \cup B$.

a. Union

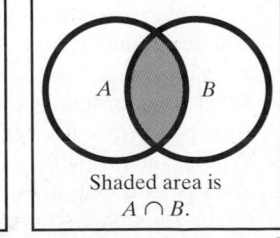

Shaded area is
$A \cap B$.

b. Intersection

**FIGURE 3.7**

*Venn Diagram for Die Toss*

**EXAMPLE 3.6**

Consider the die-toss experiment. Define the following events:

  *A*: {Toss an even number}
  *B*: {Toss a number less than or equal to 3}

a. Describe $A \cup B$ for this experiment.
b. Describe $A \cap B$ for this experiment.
c. Calculate $P(A \cup B)$ and $P(A \cap B)$ assuming the die is fair.

**Solution**

Draw the Venn diagram as shown in Figure 3.7.

a. The union of *A* and *B* is the event that occurs if we observe either an even number, a number less than or equal to 3, or both on a single throw of the die. Consequently, the sample points in the event $A \cup B$ are those for which *A* occurs, *B* occurs, or both *A* and *B* occur. Checking the sample points in the entire sample space, we find that the collection of sample points in the union of *A* and *B* is

$$A \cup B = \{1, 2, 3, 4, 6\}$$

b. The intersection of *A* and *B* is the event that occurs if we observe *both* an even number and a number less than or equal to 3 on a single throw of the die. Checking the sample points to see which imply the occurrence of *both* events *A* and *B*, we see that the intersection contains only one sample point:

$$A \cap B = \{2\}$$

In other words, the intersection of *A* and *B* is the sample point Observe a 2.

c. Recalling that the probability of an event is the sum of the probabilities of the sample points of which the event is composed, we have

$$P(A \cup B) = P(1) + P(2) + P(3) + P(4) + P(6)$$
$$= \frac{1}{6} + \frac{1}{6} + \frac{1}{6} + \frac{1}{6} + \frac{1}{6} = \frac{5}{6}$$

and

$$P(A \cap B) = P(2) = \frac{1}{6}$$

Unions and intersections can be defined for more than two events. For example, the event $A \cup B \cup C$ represents the union of three events—*A*, *B*, and *C*. This event, which includes the set of sample points in *A*, *B*, or *C*, will occur if any one or more of the events *A*, *B*, or *C* occurs. Similarly, the intersection $A \cap B \cap C$ is the event that all three of the events *A*, *B*, and *C* occur. Therefore, $A \cap B \cap C$ is the set of sample points that are in all three of the events *A*, *B*, and *C*.

EXAMPLE 3.7

Refer to Example 3.6 and define the event

    $C$: {Toss a number greater than 1}

    Find the sample points in:

a. $A \cup B \cup C$
b. $A \cap B \cap C$

where

    $A$: {Toss an even number}
    $B$: {Toss a number less than or equal to 3}

### Solution

a. Event $C$ contains the sample points corresponding to tossing a 2, 3, 4, 5, or 6; event $A$ contains the sample points 2, 4, and 6; and event $B$ contains the sample points 1, 2, and 3. Therefore, the event that either $A$, $B$, or $C$ occurs contains all six sample points in $S$—that is, those corresponding to tossing a 1, 2, 3, 4, 5, or 6.
b. You will observe all of the events $A$, $B$, and $C$ only if you observe a 2. Therefore, the intersection $A \cap B \cap C$ contains the single sample point Toss a 2.

EXAMPLE 3.8

*Family Planning Perspectives* (Nov./Dec. 1997) reported on a study of over 200,000 births in New Jersey over a recent two-year period. The study investigated the link between the mother's race and the age at which she gave birth (called maternal age). The percentages of the total number of births in New Jersey are given by the maternal age and race classifications in Table 3.4. Define the following events:

    $A$: {A New Jersey birth mother is white}
    $B$: {A New Jersey mother was a teenager when giving birth}

a. Find $P(A)$ and $P(B)$
b. Find $P(A \cup B)$
c. Find $P(A \cap B)$

### TABLE 3.4   Percentage of New Jersey Birth Mothers in Age–Race Classes

| Maternal Age (years) | Race | |
|---|---|---|
| | White | Black |
| ≤17 | 2% | 2% |
| 18–19 | 3% | 2% |
| 20–29 | 41% | 12% |
| ≥30 | 33% | 5% |

*Source:* Reichman, N. E., and Pagnini, D. L. "Maternal age and birth outcomes: Data from New Jersey." *Family Planning Perspectives*, Vol. 29, No. 6, Nov./Dec. 1997, p. 269 (adapted from Table 1).

### Solution

Following the steps for calculating probabilities of events, we first note that the objective is to characterize the race and maternal age distribution of New Jersey birth mothers. To accomplish this, we define the experiment to consist of selecting a birth mother from the collection of all New Jersey birth mothers during the two-year period and observing her race and maternal age class. The sample points are the eight different age-race classifications:

| | | | |
|---|---|---|---|
| $E_1$: | {≤17 yrs., white} | $E_5$: | {≤17 yrs., black} |
| $E_2$: | {18–19 yrs., white} | $E_6$: | {18–19 yrs., black} |
| $E_3$: | {20–29 yrs., white} | $E_7$: | {20–29 yrs., black} |
| $E_4$: | {≥30 yrs., white} | $E_8$: | {≥30 yrs., black} |

Next, we assign probabilities to the sample points. If we blindly select one of the birth mothers, the probability that she will occupy a particular age–race classification is just the proportion, or relative frequency, of birth mothers in the classification. These proportions (as percentages) are given in Table 3.4. Thus,

$P(E_1)$ = Relative frequency of birth mothers in age–race class {≤17 yrs., white} = .02
$P(E_2)$ = .03
$P(E_3)$ = .41
$P(E_4)$ = .33
$P(E_5)$ = .02
$P(E_6)$ = .02
$P(E_7)$ = .12
$P(E_8)$ = .05

You may verify that the sample point probabilities add to 1.

a. To find $P(A)$, we first determine the collection of sample points contained in event $A$. Since $A$ is defined as {white}, we see from Table 3.4 that $A$ contains the four sample points represented by the first column of the table. In words, the event $A$ consists of the race classification {white} in all four age classifications. The probability of $A$ is the sum of the probabilities of the sample points in $A$:

$$P(A) = P(E_1) + P(E_2) + P(E_3) + P(E_4) = .02 + .03 + .41 + .33 = .79$$

Similarly, $B$ = {teenage mother, age ≤ 19 years} consists of the four sample points in the first and second rows of Table 3.4:

$$P(B) = P(E_1) + P(E_2) + P(E_5) + P(E_6) = .02 + .03 + .02 + .02 = .09$$

b. The union of events $A$ and $B$, $A \cup B$, consists of all sample points in *either A or B or both*. That is, the union of $A$ and $B$ consists of all birth mothers who are white *or* who gave birth as a teenager. In Table 3.4 this is any sample point found in the first column *or* the first two rows. Thus,

$$P(A \cup B) = .02 + .03 + .41 + .33 + .02 + .02 = .83$$

c. The intersection of events $A$ and $B$, $A \cap B$, consists of all sample points in *both A and B*. That is, the intersection of $A$ and $B$ consists of all birth mothers who are white *and* who gave birth as a teenager. In Table 3.4 this is any sample point found in the first column *and* the first two rows. Thus,

$$P(A \cap B) = .02 + .03 = .05.$$

## 3.3 COMPLEMENTARY EVENTS

A very useful concept in the calculation of event probabilities is the notion of **complementary events**:

### DEFINITION 3.7

The **complement** of an event $A$ is the event that $A$ does *not* occur—that is, the event consisting of all sample points that are not in event $A$. We denote the complement of $A$ by $A^c$.

**FIGURE 3.8**

*Venn Diagram of
Complementary Events*

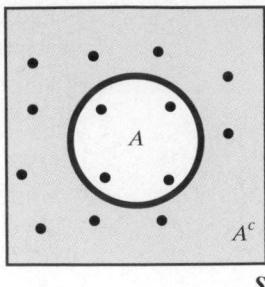

An event $A$ is a collection of sample points, and the sample points included in $A^c$ are those not in $A$. Figure 3.8 demonstrates this idea. Note from the figure that all sample points in $S$ are included in *either $A$ or $A^c$* and that *no* sample point is in both $A$ and $A^c$. This leads us to conclude that the probabilities of an event and its complement *must sum to 1:*

---

### Summing Probabilities of Complimentary Events

The sum of the probabilities of complementary events equals 1; that is,

$P(A) + P(A^c) = 1.$

---

In many probability problems calculating the probability of the complement of the event of interest is easier than calculating the event itself. Then, because

$$P(A) + P(A^c) = 1$$

we can calculate $P(A)$ by using the relationship

$$P(A) = 1 - P(A^c).$$

**EXAMPLE 3.9**

Consider the experiment of tossing two fair coins. (The sample space for this experiment is shown in Figure 3.9.) Calculate the probability of event $A$: {Observing at least one head}.

### Solution

We know that the event $A$: {Observing at least one head} consists of the sample points

   $A$: $\{HH, HT, TH\}$

The complement of $A$ is defined as the event that occurs when $A$ does not occur.

**FIGURE 3.9**

*Complementary Events in
the Toss of Two Coins*

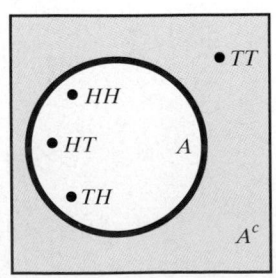

Therefore,

$A^c$: {Observe no heads} = $\{TT\}$

This complementary relationship is shown in Figure 3.9. Assuming the coins are balanced,

$$P(A^c) = P(TT) = \frac{1}{4}$$

and

$$P(A) = 1 - P(A^c) = 1 - \frac{1}{4} = \frac{3}{4}.$$

**EXAMPLE 3.10**

A fair coin is tossed 10 times, and the up face is recorded after each toss. What is the probability of event $A$: {Observe at least one head}?

**Solution**

We solve this problem by following the five steps for calculating probabilities of events (see Section 3.1).

**Step 1**  Define the experiment. The experiment is to record the results of the 10 tosses of the coin.

**Step 2**  List the sample points. A sample point consists of a particular sequence of ten heads and tails. Thus, one sample point is *HHTTTHTHTT*, which denotes head on first toss, head on second toss, tail on third toss, etc. Others are *HTHHHTTTTT* and *THHTHTHTTH*. Obviously, the number of sample points is very large—too many to list. It can be shown (see Section 3.8) that there are $2^{10} = 1{,}024$ sample points for this experiment.

**Step 3**  Assign probabilities. Since the coin is fair, each sequence of heads and tails has the same chance of occurring, and therefore all the sample points are equally likely. Then

$$P(\text{Each sample point}) = \frac{1}{1{,}024}$$

**Step 4**  Determine the sample points in event $A$. A sample point is in $A$ if at least one $H$ appears in the sequence of 10 tosses. However, if we consider the complement of $A$, we find that

$A^c$ = {No heads are observed in 10 tosses}

Thus, $A^c$ contains only one sample point:

$A^c$:  $\{TTTTTTTTTT\}$

and $P(A^c) = \dfrac{1}{1{,}024}$

**Step 5**  Now we use the relationship of complementary events to find $P(A)$:

$$P(A) = 1 - P(A^c) = 1 - \frac{1}{1{,}024} = \frac{1{,}023}{1{,}024} = .999$$

That is, we are virtually certain of observing at least one head in 10 tosses of the coin.

## 3.4   THE ADDITIVE RULE AND MUTUALLY EXCLUSIVE EVENTS

In Section 3.2 we saw how to determine which sample points are contained in a union and how to calculate the probability of the union by adding the probabilities of the sample points in the union. It is also possible to obtain the probability of the union of two events by using the *additive rule of probability*.

By studying the Venn diagram in Figure 3.10, you can see that the probability of the union of two events, $A$ and $B$, can be obtained by summing $P(A)$ and $P(B)$ and subtracting $P(A \cap B)$. We must subtract $P(A \cap B)$ because the sample point probabilities in $A \cap B$ have been included twice—once in $P(A)$ and once in $P(B)$.

**FIGURE 3.10**

*Venn Diagram of Union*

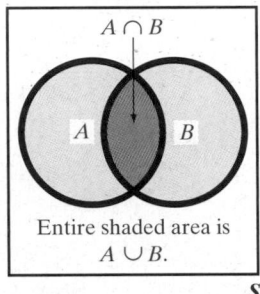

The formula for calculating the probability of the union of two events is given in the next box.

---

### Additive Rule of Probability

The probability of the union of events $A$ and $B$ is the sum of the probability of events $A$ and $B$ minus the probability of the intersection of events $A$ and $B$, that is

$$P(A \cup B) = P(A) + P(B) - P(A \cap B)$$

---

**EXAMPLE 3.11**

Hospital records show that 12% of all patients are admitted for surgical treatment, 16% are admitted for obstetrics, and 2% receive both obstetrics and surgical treatment. If a new patient is admitted to the hospital, what is the probability that the patient will be admitted either for surgery, obstetrics, or both?

**Solution**

Consider the following events:

   $A$: {A patient admitted to the hospital receives surgical treatment}
   $B$: {A patient admitted to the hospital receives obstetrics treatment}

Then, from the given information

$$P(A) = .12$$
$$P(B) = .16$$

and the probability of the event that a patient receives both obstetrics and surgical treatment is

$$P(A \cap B) = .02$$

The event that a patient admitted to the hospital receives either surgical treatment, obstetrics treatment, or both is the union $A \cup B$. The probability of $A \cup B$ is given by the additive rule of probability:

$$P(A \cup B) = P(A) + P(B) - P(A \cap B)$$
$$= .12 + .16 - .02 = .26$$

Thus, 26% of all patients admitted to the hospital receive either surgical treatment, obstetrics treatment, or both.

A very special relationship exists between events $A$ and $B$ when $A \cap B$ contains no sample points. In this case we call the events $A$ and $B$ *mutually exclusive events*.

### DEFINITION 3.8
Events $A$ and $B$ are **mutually exclusive** if $A \cap B$ contains no sample points, that is, if $A$ and $B$ have no sample points in common.

Figure 3.11 shows a Venn diagram of two mutually exclusive events. The events $A$ and $B$ have no sample points in common, that is, $A$ and $B$ cannot occur simultaneously and $P(A \cap B) = 0$. Thus, we have the important relationship given in the box.

### Probability of Union of Two Mutually Exclusive Events

If two events $A$ and $B$ are *mutually exclusive*, the probability of the union of $A$ and $B$ equals the sum of the probabilities of $A$ and $B$; that is, $P(A \cup B) = P(A) + P(B)$.

*Caution:* The formula shown above is *false* if the events are *not* mutually exclusive. In this case (i.e., two nonmutually exclusive events), you must apply the general additive rule of probability.

**FIGURE 3.11**

*Venn Diagram of Mutually Exclusive Events*

$S$

**EXAMPLE 3.12**  Consider the experiment of tossing two balanced coins. Find the probability of observing *at least* one head.

### Solution
Define the events

A: {Observe at least one head}
B: {Observe exactly one head}
C: {Observe exactly two heads}

**FIGURE 3.12**

*Venn Diagram for Coin Toss Experiment*

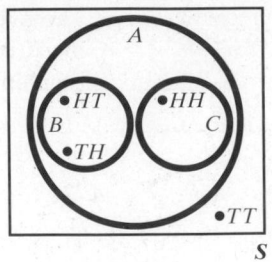

Note that

$$A = B \cup C$$

and that $B \cap C$ contains no sample points (see Figure 3.12). Thus, $B$ and $C$ are mutually exclusive, so that

$$P(A) = P(B \cup C) = P(B) + P(C) = \frac{1}{2} + \frac{1}{4} = \frac{3}{4}$$

Although Example 3.12 is very simple, it shows us that writing events with verbal descriptions that include the phrases "at least" or "at most" as unions of mutually exclusive events is very useful. This practice enables us to find the probability of the event by adding the probabilities of the mutually exclusive events.

 EXERCISES 3.25–3.39

**Learning the Mechanics**

**3.25** A fair coin is tossed three times and the events $A$ and $B$ are defined as follows:

   $A$: {At least one head is observed}
   $B$: {The number of heads observed is odd}

   **a.** Identify the sample points in the events $A$, $B$, $A \cup B$, $A^c$, and $A \cap B$.
   **b.** Find $P(A)$, $P(B)$, $P(A \cup B)$, $P(A^c)$, and $P(A \cap B)$ by summing the probabilities of the appropriate sample points.
   **c.** Find $P(A \cup B)$ using the additive rule. Compare your answer to the one you obtained in part **b**.
   **d.** Are the events $A$ and $B$ mutually exclusive? Why?

**3.26** A pair of fair dice is tossed. Define the following events:

   $A$: {You will roll a 7 (i.e., the sum of the numbers of dots on the upper faces of the two dice is equal to 7)}
   $B$: {At least one of the two dice is showing a 4}

   **a.** Identify the sample points in the events $A$, $B$, $A \cap B$, $A \cup B$, and $A^c$.
   **b.** Find $P(A)$, $P(B)$, $P(A \cap B)$, $P(A \cup B)$, and $P(A^c)$ by summing the probabilities of the appropriate sample points.

   **c.** Find $P(A \cup B)$ using the additive rule. Compare your answer to that for the same event in part **b**.
   **d.** Are $A$ and $B$ mutually exclusive? Why?

**3.27** Consider the Venn diagram, where $P(E_1) = P(E_2) = P(E_3) = \frac{1}{5}$, $P(E_4) = P(E_5) = \frac{1}{20}$, $P(E_6) = \frac{1}{10}$, and $P(E_7) = \frac{1}{5}$. Find each of the following probabilities.

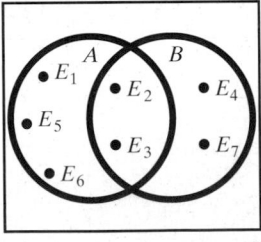

   **a.** $P(A)$
   **b.** $P(B)$
   **c.** $P(A \cup B)$
   **d.** $P(A \cap B)$
   **e.** $P(A^c)$
   **f.** $P(B^c)$
   **g.** $P(A \cup A^c)$
   **h.** $P(A^c \cap B)$

**3.28** Consider the accompanying Venn diagram, where $P(E_1) = .10$, $P(E_2) = .05$, $P(E_3) = P(E_4) = .2$, $P(E_5) = .06$, $P(E_6) = .3$, $P(E_7) = .06$, and $P(E_8) = .03$. Find the following probabilities:

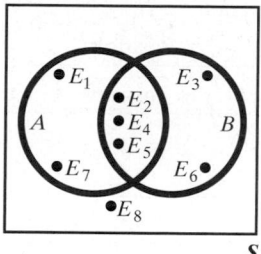

**a.** $P(A^c)$
**b.** $P(B^c)$
**c.** $P(A^c \cap B)$
**d.** $P(A \cup B)$
**e.** $P(A \cap B)$
**f.** $P(A^c \cup B^c)$
**g.** Are events $A$ and $B$ mutually exclusive? Why?

**3.29** The outcomes of two variables are (Low, Medium, High) and (On, Off), respectively. An experiment is conducted in which the outcomes of each of the two variables are observed. The probabilities associated with each of the six possible outcome pairs are given in the accompanying table.

|  | **Low** | **Medium** | **High** |
|---|---|---|---|
| **On** | .50 | .10 | .05 |
| **Off** | .25 | .07 | .03 |

Consider the following events:

$A$: {On}

$B$: {Medium or On}

$C$: {Off and Low}

**a.** Find $P(A)$.
**b.** Find $P(B)$.
**c.** Find $P(C)$.
**d.** Find $P(A^c)$.
**e.** Find $P(A \cup B)$.
**f.** Find $P(A \cap B)$.
**g.** Consider each pair of events ($A$ and $B$, $A$ and $C$, $B$ and $C$). List the pairs of events that are mutually exclusive. Justify your choices.

**3.30** Three fair coins are tossed. We wish to find the probability of the event $A$: {Observe at least one head}.
**a.** Express $A$ as the union of three mutually exclusive events. Find the probability of $A$ using this expression.
**b.** Express $A$ as the complement of an event. Find the probability of $A$ using this expression.

## Applying the Concepts—Basic

**3.31** According to the National Center for Education Statistics (2000), only 5% of United States eighth graders score above 653 on a mathematics assessment test. Make a probability statement about the event that a randomly selected eighth grader has a score of 653 or below on the mathematics assessment test.

**3.32** A study of binge alcohol drinking by college students was published in the *American Journal of Public Health* (July 1995). Suppose an experiment consists of randomly selecting one of the undergraduate students who participated in the study. Consider the following events:

$A$: {The student is a binge drinker}
$B$: {The student is a male}
$C$: {The student lives in a coed dorm}

Describe each of the following events in terms of unions, intersections, and complements ($A \cup B$, $A \cap B$, $A^c$, etc.):
**a.** The student is male and a binge drinker.
**b.** The student is not a binge drinker.
**c.** The student is male or lives in a coed dorm.
**d.** The student is female and not a binge drinker.

**3.33** *Roulette* is a very popular game in many American casinos. In Roulette, a ball spins on a circular wheel that is divided into 38 arcs of equal length, bearing the numbers 00, 0, 1, 2, ..., 35, 36. The number of the arc on which the ball stops is the outcome of one play of the game. The numbers are also colored in the manner shown in the table.

| **Red:** | 1, 3, 5, 7, 9, 12, 14, 16, 18, 19, 21, 23, 25, 27, 30, 32, 34, 36 |
|---|---|
| **Black:** | 2, 4, 6, 8, 10, 11, 13, 15, 17, 20, 22, 24, 26, 28, 29, 31, 33, 35 |
| **Green:** | 00, 0 |

Players may place bets on the table in a variety of ways, including bets on odd, even, red, black, high, low, etc. Consider the following events:

$A$: {Outcome is an odd number (00 and 0 are considered neither odd nor even)}
$B$: {Outcome is a black number}
$C$: {Outcome is a low number (1–18)}

**a.** Define the event $A \cap B$ as a specific set of sample points.
**b.** Define the event $A \cup B$ as a specific set of sample points.
**c.** Find $P(A)$, $P(B)$, $P(A \cap B)$, $P(A \cup B)$, and $P(C)$ by summing the probabilities of the appropriate sample points.
**d.** Define the event $A \cap B \cap C$ as a specific set of sample points.
**e.** Find $P(A \cup B)$ using the additive rule. Are events $A$ and $B$ mutually exclusive? Why?
**f.** Find $P(A \cap B \cap C)$ by summing the probabilities of the sample points given in part **d**.
**g.** Define the event $(A \cup B \cup C)$ as a specific set of sample points.
**h.** Find $P(A \cup B \cup C)$ by summing the probabilities of the sample points given in part **g**.

## Applying the Concepts—Intermediate

**3.34** *Ear and Hearing* (Apr. 1995) reported on a study of elderly hearing-impaired subjects who wear conventional analog hearing aids. Two of the variables measured for each hearing aid wearer were duration of experience with hearing aids and daily hearing aid use. The results are summarized in the table, where the numbers in the table represent the percentages of hearing aid wearers who fall into the respective categories. (Note that the percentages add to 100%.) Consider a randomly selected elderly hearing-impaired subject who wears a conventional analog hearing aid.

| Hearing Aid Experience | Daily Use (Hours) | | | | |
| --- | --- | --- | --- | --- | --- |
| | Less than 1 | 1–4 | 4–8 | 8–16 | Total |
| Less than 1 year | 4 | 6 | 7 | 15 | 32 |
| 1–10 years | 4 | 9 | 6 | 28 | 47 |
| More than 10 years | 0 | 3 | 6 | 12 | 21 |
| **Total** | 8 | 18 | 19 | 55 | 100 |

*Source:* Cox, R. M., and Alexander, G. C. "The abbreviated profile of hearing aid benefit." *Ear & Hearing*, Vol. 16, No. 2, Apr. 1995, p. 178 (Table 2).

**a.** List all the sample points for this experiment.

**b.** What is the set of all sample points called?

**c.** Let $C$ be the event that an elderly hearing-impaired subject has more than 10 years of experience wearing hearing aids. Find $P(C)$ by summing the probabilities of the sample points in $C$.

**d.** Let $G$ be the event that an elderly hearing-impaired subject uses hearing aids 8–16 hours per day. Find $P(G)$.

**e.** Let $A$ be the event that an elderly hearing-impaired subject has less than 1 year of experience wearing hearing aids. Find $P(A)$.

**f.** Let $D$ be the event that an elderly hearing-impaired subject uses hearing aids less than 1 hour per day. Find $P(D)$.

**g.** Let $E$ be the event that an elderly hearing-impaired subject uses hearing aids between 1 and 4 hours per day. Find $P(E)$.

**3.35** Refer to Exercise 3.34. Define the characteristics of an elderly hearing-impaired person portrayed by the following events, then find the probability of each. For each union, use the additive rule to find the probability. Also, determine whether the events are mutually exclusive.

**a.** $A \cap G$
**b.** $C \cup E$
**c.** $C \cap D$
**d.** $A \cup G$
**e.** $A \cup D$

**3.36** In professional sports, "stacking" is a term used to describe the practice of African-American players being excluded from certain positions because of race. To illustrate the stacking phenomenon, the *Sociology of Sport Journal* (Vol. 14, 1997) presented the table shown here. The table summarizes the race and positions of 368 National Basketball Association (NBA) players in 1993. Suppose an NBA player is selected at random from that year's player pool.

| | Position | | | |
| --- | --- | --- | --- | --- |
| | **Guard** | **Forward** | **Center** | **Totals** |
| **White** | 26 | 30 | 28 | 84 |
| **Black** | 128 | 122 | 34 | 284 |
| **Totals** | 154 | 152 | 62 | 368 |

**a.** What is the probability that the player is white?

**b.** What is the probability that the player is a center?

**c.** What is the probability that the player is African-American and plays guard?

**d.** What is the probability that the player is not a guard?

**e.** What is the probability that the player is white or a center?

**3.37** A Victoria University psychologist investigated whether Asian immigrants to Australia differed from Anglo-Australians in their attitudes toward mental illness (*Community Mental Health Journal*, Feb. 1999). Each in a sample of 139 Australian students was classified according to ethnic group and degree of contact with mentally ill people. The number in each category is displayed in the table at the top of p. 131. Suppose we randomly select one of the 139 students.

**a.** Find the probability that the student is an Anglo-Australian who has had little or no contact with mentally ill people.

**b.** Find the probability that the student has had close contact with mentally ill people.

**c.** Find the probability that the student is not an Anglo-Australian.

**d.** Find the probability that the student has had close contact with mentally ill people or is an Asian immigrant.

**3.38** The National Gang Crime Research Center (NGCRC) has developed a six-level gang classification system for both adults and juveniles. The NGCRC collected data on approximately 7,500 confined offenders and assigned each a score using the gang classification system (*Journal of Gang Research*, Winter 1997). One of several other variables measured by the NGCRC was whether or not the offender has ever carried a homemade weapon (e.g., knife) while in custody. The second table on p. 131 gives the number of confined offenders in each of the gang score and homemade weapon categories. Assume one of the confined offenders is randomly selected.

**a.** Find the probability that the offender has a gang score of 5.

**b.** Find the probability that the offender has carried a homemade weapon.

**c.** Find the probability that the offender has a gang score below 3.

| Mental Illness | Anglo-Australian | Short-Term Asian | Long-Term Asian | Other | Totals |
|---|---|---|---|---|---|
| Little or no contact | 30 | 17 | 20 | 15 | 82 |
| Some contact | 17 | 5 | 3 | 11 | 36 |
| Close contact | 16 | 0 | 1 | 4 | 21 |
| **Totals** | 63 | 22 | 24 | 30 | 139 |

*Source:* Fan, C. "A comparison of attitudes towards mental illness and knowledge of mental health services between Australian and Asian students." *Community Mental Health Journal,* Vol. 35, No. 1, Feb. 1999, p. 54 (Table 2).

**d.** Find the probability that the offender has a gang score of 5 and has carried a homemade weapon.

**e.** Find the probability that the offender has a gang score of 0 or has never carried a homemade weapon.

**f.** Are the events described in parts **a** and **b** mutually exclusive? Explain.

**g.** Are the events described in parts **a** and **c** mutually exclusive? Explain.

### Applying the Concepts—Advanced

**3.39** A buyer for a large metropolitan department store must choose two firms from the four available to supply the store's fall line of men's slacks. The buyer has not dealt with any of the four firms before and considers their products equally attractive. Unknown to the buyer, two of the four firms are having serious financial problems that may result in their not being able to deliver the fall line of slacks as soon as promised. The four firms are identified as $G_1$ and $G_2$ (firms in good financial condition) and $P_1$ and $P_2$ (firms in poor financial condition). If the buyer selects the firms at random, find

**a.** The probability that firm $P_1$ is selected.

**b.** The probability that at least one of the selected firms is in good financial condition.

| Gang Classification Score | Weapon | | Totals |
|---|---|---|---|
| | **Yes** | **No** | |
| 0 (Never joined a gang, no close friends in a gang) | 255 | 2,551 | 2,806 |
| 1 (Never joined a gang, 1–4 close friends in a gang) | 110 | 560 | 670 |
| 2 (Never joined a gang, 5 or more friends in a gang) | 151 | 636 | 787 |
| 3 (Inactive gang member) | 271 | 959 | 1,230 |
| 4 (Active gang member, no position of rank) | 175 | 513 | 688 |
| 5 (Active gang member, holds position of rank) | 476 | 831 | 1,307 |
| **Totals** | 1,438 | 6,050 | 7,488 |

*Source:* Knox, G.W., *et al.* "A gang classification system for corrections." *Journal of Gang Research*, Vol. 4, No. 2, Winter 1997, p. 54 (Table 4).

## 3.5  CONDITIONAL PROBABILITY

The event probabilities we've been discussing give the relative frequencies of the occurrences of the events when the experiment is repeated a very large number of times. Such probabilities are often called **unconditional probabilities** because no special conditions are assumed, other than those that define the experiment.

Often, however, we have additional knowledge that might affect the outcome of an experiment, so we may need to alter the probability of an event of interest. A probability that reflects such additional knowledge is called the **conditional probability** of the event. For example, we've seen that the probability of observing an even number (event $A$) on a toss of a fair die is $1/2$. But suppose we're given the information that on a particular throw of the die the result was a number less than or equal to 3 (event $B$). Would the probability of observing an even number

**FIGURE 3.13**

*Reduced Sample Space for the Die Toss Experiment: Given that Event B Has Occurred*

on that throw of the die still be equal to $^1/_2$? It can't be, because making the assumption that $B$ has occurred reduces the sample space from six sample points to three sample points (namely, those contained in event $B$). This reduced sample space is as shown in Figure 3.13.

Because the sample points for the die-toss experiment are equally likely, each of the three sample points in the reduced sample space is assigned an equal *conditional probability* of $^1/_3$. Since the only even number of the three in the reduced sample space $B$ is the number 2 and the die is fair, we conclude that the probability that $A$ occurs *given that $B$ occurs* is $^1/_3$. We use the symbol $P(A|B)$ to represent the probability of event $A$ given that event $B$ occurs. For the die-toss example

$$P(A|B) = \frac{1}{3}$$

To get the probability of event $A$ given that event $B$ occurs, we proceed as follows. We divide the probability of the part of $A$ that falls within the reduced sample space $B$, namely $P(A \cap B)$, by the total probability of the reduced sample space, namely, $P(B)$. Thus, for the die-toss example with event $A$: {Observe an even number} and event $B$: {Observe a number less than or equal to 3}, we find

$$P(A|B) = \frac{P(A \cap B)}{P(B)} = \frac{P(2)}{P(1) + P(2) + P(3)} = \frac{^1/_6}{^3/_6} = \frac{1}{3}$$

The formula for $P(A|B)$ is true in general:

---

### Formula for $P(A|B)$

To find the *conditional probability that event A occurs given that event B occurs*, divide the probability that *both $A$ and $B$ occur* by the probability that $B$ occurs, that is,

$$P(A|B) = \frac{P(A \cap B)}{P(B)} \qquad \text{[We assume that } P(B) \neq 0.]$$

---

This formula adjusts the probability of $A \cap B$ from its original value in the complete sample space $S$ to a conditional probability in the reduced sample space $B$. If the sample points in the complete sample space are equally likely, then the formula will assign equal probabilities to the sample points in the reduced sample space, as in the die-toss experiment. If, on the other hand, the sample points have unequal probabilities, the formula will assign conditional probabilities proportional to the probabilities in the complete sample space. This is illustrated by the following examples.

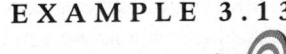

**EXAMPLE 3.13**    Many medical researchers have conducted experiments to examine the relationship between cigarette smoking and cancer. Consider an individual randomly selected from the adult male population. Let $A$ represent the event that the individual smokes, and let $A^c$ denote the complement of $A$ (the event that the individual does

**FIGURE 3.14**

*Sample Space for Example 3.13*

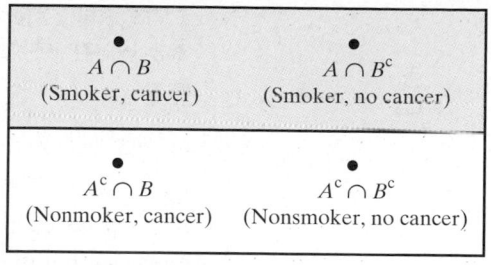

not smoke). Similarly, let $B$ represent the event that the individual develops cancer, and let $B^c$ be the complement of that event. Then the four sample points associated with the experiment are shown in Figure 3.14, and their probabilities for a certain section of the United States are given in Table 3.5. Use these sample point probabilities to examine the relationship between smoking and cancer.

**TABLE 3.5  Probabilities of Smoking and Developing Cancer**

|  | Develops Cancer | |
| --- | --- | --- |
| **Smoker** | **Yes, $B$** | **No, $B^c$** |
| **Yes, $A$** | .05 | .20 |
| **No, $A^c$** | .03 | .72 |

**Solution**

One method of determining whether these probabilities indicate that smoking and cancer are related is to compare the *conditional probability* that an adult male acquires cancer given that he smokes with the conditional probability that an adult male acquires cancer given that he does not smoke, i.e., compare $P(B|A)$ to $P(B|A^c)$.

First, we consider the reduced sample space $A$ corresponding to adult male smokers. This reduced sample space is highlighted in Figure 3.14. The two sample points $A \cap B$ and $A \cap B^c$ are contained in this reduced sample space, and the adjusted probabilities of these two sample points are the two conditional probabilities:

$$P(B|A) = \frac{P(A \cap B)}{P(A)} \quad \text{and} \quad P(B^c|A) = \frac{P(A \cap B^c)}{P(A)}$$

The probability of event $A$ is the sum of the probabilities of the sample points in $A$:

$$P(A) = P(A \cap B) + P(A \cap B^c) = .05 + .20 = .25$$

Then the values of the two conditional probabilities in the reduced sample space $A$ are:

$$P(B|A) = \frac{.05}{.25} = .20 \quad \text{and} \quad P(B^c|A) = \frac{.20}{.25} = .80$$

These two numbers represent the probabilities that an adult male smoker develops cancer and does not develop cancer, respectively. Notice that the conditional probabilities .80 and .20 are in the same 4 to 1 ratio as the original unconditional probabilities, .20 and .05. The conditional probability formula simply adjusts the unconditional probabilities so that they add to 1 in the reduced sample space, $A$, of adult male smokers.

In a like manner, the conditional probabilities of an adult male nonsmoker developing cancer and not developing cancer are

$$P(B|A^c) = \frac{P(A^c \cap B)}{P(A^c)} = \frac{.03}{.75} = .04$$

$$P(B^c|A^c) = \frac{P(A^c \cap B^c)}{P(A^c)} = \frac{.72}{.75} = .96$$

Observe that the conditional probabilities .96 and .04 are in the same 24 to 1 ratio as the unconditional probabilities .72 and .03.

Two of the conditional probabilities give some insight into the relationship between cancer and smoking: the probability of developing cancer given that the adult male is a smoker, and the probability of developing cancer given that the adult male is not a smoker. The conditional probability that an adult male smoker develops cancer (.20) is five times the probability that a nonsmoker develops cancer (.04). This does not imply that smoking *causes* cancer, but it does suggest a pronounced link between smoking and cancer.                              ■

**EXAMPLE 3.14**

The investigation of consumer product complaints by the Federal Trade Commission (FTC) has generated much interest by manufacturers in the quality of their products. A manufacturer of an electromechanical kitchen utensil conducted an analysis of a large number of consumer complaints and found that they fell into the six categories shown in Table 3.6. If a consumer complaint is received, what is the probability that the cause of the complaint was product appearance given that the complaint originated during the guarantee period?

**TABLE 3.6    Distribution of Product Complaints**

|  | **Reason For Complaint** | | | |
|---|---|---|---|---|
|  | **Electrical** | **Mechanical** | **Appearance** | **Totals** |
| **During Guarantee Period** | 18% | 13% | 32% | 63% |
| **After Guarantee Period** | 12% | 22% | 3% | 37% |
| **Totals** | 30% | 35% | 35% | 100% |

### Solution

Let $A$ represent the event that the cause of a particular complaint is product appearance, and let $B$ represent the event that the complaint occurred during the guarantee period. Checking Table 3.6, you can see that $(18 + 13 + 32)\% = 63\%$ of the complaints occur during the guarantee period. Hence, $P(B) = .63$. The percentage of complaints that were caused by appearance and occurred during the guarantee period (the event $A \cap B$) is 32%. Therefore, $P(A \cap B) = .32$.

Using these probability values, we can calculate the conditional probability $P(A|B)$ that the cause of a complaint is appearance given that the complaint occurred during the guarantee time:

$$P(A|B) = \frac{P(A \cap B)}{P(B)} = \frac{.32}{.63} = .51$$

Consequently, we can see that slightly more than half the complaints that occurred during the guarantee period were due to scratches, dents, or other imperfections in the surface of the kitchen devices.                              ■

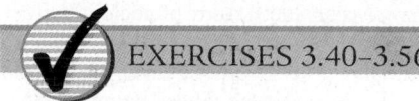

## EXERCISES 3.40–3.56

### Learning the Mechanics

**3.40** Consider the experiment defined by the accompanying Venn diagram, with the sample space $S$ containing five sample points. The sample points are assigned the following probabilities: $P(E_1) = .1$, $P(E_2) = .1$, $P(E_3) = .2$, $P(E_4) = .5$, $P(E_5) = .1$.

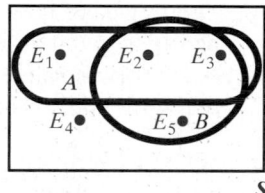

**a.** Calculate $P(A)$, $P(B)$, and $P(A \cap B)$.

**b.** Suppose we know event $A$ has occurred, so the reduced sample space consists of the three sample points in $A$: $E_1$, $E_2$, and $E_3$. Use the formula for conditional probability to determine the probabilities of these three sample points given that $A$ has occurred. Verify that the conditional probabilities are in the same ratio to one another as the original sample point probabilities.

**c.** Calculate the conditional probability $P(B|A)$, in two ways: First, add the adjusted (conditional) probabilities of the sample points in the intersection $A \cap B$, since these represent the event that $B$ occurs given that $A$ has occurred. Second, use the formula for conditional probability:

$$P(B|A) = \frac{P(A \cap B)}{P(A)}$$

Verify that the two methods yield the same result.

**3.41** Given that $P(A) = .3$, $P(B) = .6$, and $P(A \cap B) = .15$, find $P(A|B)$ and $P(B|A)$.

**3.42** A sample space contains six sample points and events $A$, $B$, and $C$ as shown in the Venn diagram. The probabilities of the sample points are $P(1) = .20$, $P(2) = .05$, $P(3) = .25$, $P(4) = .10$, $P(5) = .15$, $P(6) = .25$. Use the Venn diagram and the probabilities of the sample points to find:

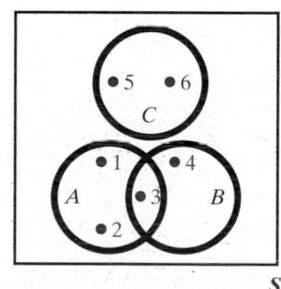

**a.** $P(A)$, $P(B)$, and $P(C)$.

**b.** $P(A \cap B)$, $P(A \cap C)$, and $P(B \cap C)$.

**c.** Suppose you know that event $A$ has occurred. Assign conditional probabilities to the three sample points contained in $A$. Verify that they add to 1 and are in the same ratio as the original (unconditional) probabilities.

**d.** Use the conditional probabilities from part **c** to calculate $P(B|A)$. Use the formula for $P(B|A)$ to verify your answer.

**e.** Use the formula for conditional probability to calculate $P(C|A)$ and $P(C|A^c)$. Verify the results by inspection of the Venn diagram, remembering that the "given event" is a reduced sample space for a conditional probability.

**3.43** Two fair coins are tossed and the events $A$ and $B$ are defined as follows:

$A$: {At least one head appears}
$B$: {Exactly one head appears}

**a.** Draw a Venn diagram for the experiment, labeling each sample point and showing events $A$ and $B$. Assign probabilities to the sample points.

**b.** Find $P(A)$, $P(B)$, and $P(A \cap B)$.

**c.** Use the formula for conditional probability to find $P(A|B)$ and $P(B|A)$. Verify your answers by inspecting the Venn diagram and using the concept of reduced sample spaces.

**3.44** A box contains two white, two red, and two blue poker chips. Two chips are randomly chosen without replacement and their colors are noted. Define the following events:

$A$: {Both chips are of the same color}
$B$: {Both chips are red}
$C$: {At least one chip is red or white}

Find $P(B|A)$, $P(B|A^c)$, $P(B|C)$, $P(A|C)$, and $P(C|A^c)$.

### Applying the Concepts—Basic

**3.45** To develop programs for business travelers staying at convention hotels, Hyatt Hotels Corp. commissioned a study of executives who play golf. The study revealed that 55% of the executives admitted they had cheated at golf. Also, 20% of the executives admitted they had cheated at golf and had lied in business. Given an executive had cheated at golf, what is the probability that the executive also had lied in business?

**3.46** *Forbes* (July 26, 1999) conducted a survey of the 20 largest non-domestic public companies in the world. Of these 20 companies, 6 were trading companies based in Japan. A total of 11 Japanese companies were on the top 20 list. Suppose we select one of these 20 companies at random. Given the company is based in Japan, what is the probability that it is a trading company?

**3.47** The New York Yankees, a member of the Eastern Division of the American League in Major League Baseball (MLB), recently won three consecutive World Series. The table summarizes the ten MLB World Series winners from 1990 to 2000 by division and league. (There was no World Series in 1994 due to a players' strike.) One of these ten World Series winners is to be chosen at random.

**a.** Given that the winner is a member of the American League, what is the probability that the winner plays in the Eastern Division?

**b.** If the winner plays in the Central Division, what is the probability that the winner is a member of the National League?

**c.** If the winner is a member of the National League, what is the probability that the winner plays in either the Central or Western Division?

| | | League | |
|---|---|---|---|
| | | **National** | **American** |
| | **Eastern** | 2 | 6 |
| **Division** | **Central** | 1 | 1 |
| | **Western** | 0 | 0 |

*Source:* Major League Baseball.

**3.48** Refer to the *Chance* (Fall 2000) study of ancient pottery found at the Greek settlement of Phylakopi, Exercise 2.8 (p. 26). Of the 837 pottery pieces uncovered at the excavation site, 183 were painted. These painted pieces included 14 painted in a curvilinear decoration, 165 painted in a geometric decoration, and 4 painted in a naturalistic decoration. Suppose one of the 837 pottery pieces is selected and examined.

**a.** What is the probability that the pottery piece is painted?

**b.** Given that the pottery piece is painted, what is the probability that it is painted in a curvilinear decoration?

**3.49** Refer to the *Forbes* (May 17, 1999) salary survey of chief executive officers, Exercise 2.9 (p. 27). The data for the top 25 best-paid CEOs are reproduced in the table. Suppose you randomly select 5 of the CEOs (without replacement) and record the highest degree obtained by each.

**a.** What is the probability that the highest degree obtained by the first CEO you select is a bachelor's degree?

**b.** Suppose the highest degree obtained by each of the first four CEOs you select is a bachelor's degree. What is the probability that the highest degree obtained by the fifth CEO you select is a bachelor's degree?

### CEODEGREES

| CEO Name | Company | Degree |
|---|---|---|
| 1. Michael D. Eisner | Walt Disney | Bachelors |
| 2. Mel Karmazin | CBS | Bachelors |
| 3. Stephen M. Case | America Online | Bachelors |
| 4. Stephen C. Hilbert | Conseco | none |
| 5. Craig R. Barrett | Intel | Doctorate |
| 6. Millard Drexler | GAP | Masters |
| 7. John F. Welch, Jr. | General Electric | Doctorate |
| 8. Thomas G. Stemberg | Staples | Masters |
| 9. Henry R. Silverman | Cendant | JD |
| 10. Reuben Mark | Colgate-Palmolive | Masters |
| 11. Philip J. Purcell | Morgan Stanley Dean Witter | Masters |
| 12. Scott G. McNealy | Sun Microsystems | Masters |
| 13. Margaret C. Whitman | eBay | Masters |
| 14. Louis V. Gerstner, Jr. | IBM | Masters |
| 15. John F. Gifford | Maxim Integrated Products | Bachelors |
| 16. Robert L. Waltrip | Service Corp. International | Bachelors |
| 17. M. Douglas Ivester | Coca-Cola | Bachelors |
| 18. Gordon M. Binder | Amgen | Masters |
| 19. Charles R. Schwab | Charles Schwab | Masters |
| 20. William R. Steere, Jr. | Pfizer | Bachelors |
| 21. Nolan D. Archibald | Black & Decker | Masters |
| 22. Charles A. Heimbold, Jr. | Bristol-Myers Squibb | LLB (law) |
| 23. William L. Larson | Network Associates | JD |
| 24. Maurice R. Greenberg | American International Group | LLB (law) |
| 25. Richard Jay Kogan | Schering-Plough | Masters |

*Source: Forbes*, May 17, 1999.

## Applying the Concepts—Intermediate

**3.50** The *Journal of the National Cancer Institute* (Feb. 16, 2000) published the results of a study that investigated the association between cigar smoking and death from tobacco-related cancers. Data were obtained for a national sample of 137,243 American men. The results are summarized in the table below. Each male in the study was classified according to his cigar-smoking status and whether or not he died from a tobacco-related cancer.

    **a.** Find the probability that a randomly selected man never smoked cigars and died from cancer.

    **b.** Find the probability that a randomly selected man was a former cigar smoker and died from cancer.

    **c.** Find the probability that a randomly selected man was a current cigar smoker and died from cancer.

    **d.** Given that a male was a current cigar smoker, find the probability that he died from cancer.

    **e.** Given that a male never smoked cigars, find the probability that he died from cancer.

| | **Died from Cancer** | | |
| **Cigars** | **Yes** | **No** | **Totals** |
| --- | --- | --- | --- |
| **Never Smoked** | 782 | 120,747 | 121,529 |
| **Former Smoker** | 91 | 7,757 | 7,848 |
| **Current Smoker** | 141 | 7,725 | 7,866 |
| **Totals** | 1,014 | 136,229 | 137,243 |

*Source:* Shapiro, J.A., Jacobs, E.J., and Thun, M.J. "Cigar Smoking in Men and Risk of Death from Tobacco-related Cancers." *Journal of the National Cancer Institute*, Vol. 92, No. 4, Feb. 16, 2000 (Table 2).

**3.51** Refer to the *Sociology of Sport Journal* (Vol. 14, 1997) study of "stacking" in the National Basketball Association (NBA), Exercise 3.36 (p. 130). Reconsider the table below, which summarizes the race and positions of 368 NBA players in 1993. Suppose an NBA player is selected at random from that year's player pool.

| | **Position** | | | |
| | **Guard** | **Forward** | **Center** | **Totals** |
| --- | --- | --- | --- | --- |
| **White** | 26 | 30 | 28 | 84 |
| **Black** | 128 | 122 | 34 | 284 |
| **Totals** | 154 | 152 | 62 | 368 |

    **a.** Given the player is white, what is the probability that he is a center?

    **b.** Given the player is African-American, what is the probability that he is a center?

    **c.** Are the events {White player} and {Center} independent?

    **d.** Recall that "stacking" refers to the practice of African-American players being excluded from certain positions because of race. Use your answers to parts **a–c** to make an inference about stacking in the NBA.

**3.52** Refer to the *Community Mental Health Journal* study of 139 Australian students' contact with mentally ill patients, Exercise 3.37 (p. 130). The results are reproduced in the table below. Suppose one of the 139 students is selected at random.

    **a.** Given the student is an Asian immigrant, what is the probability that he or she has had little or no contact with mentally ill patients?

    **b.** Given the student is Anglo-Australian, what is the probability that he or she has had little or no contact with mentally ill patients?

    **c.** Which ethnic group, Anglo-Australians or Asian immigrants, is most likely to have had little or no contact with mentally ill patients?

| | **Contact** | | | |
| **Mental Illness** | **Little/None** | **Some** | **Close** | **Totals** |
| --- | --- | --- | --- | --- |
| **Anglo-Australian** | 30 | 17 | 16 | 63 |
| **Short-Term Asian** | 17 | 5 | 0 | 22 |
| **Long-Term Asian** | 20 | 3 | 1 | 24 |
| **Other** | 15 | 11 | 4 | 30 |
| **Totals** | 82 | 36 | 21 | 139 |

*Source:* Fan, C. "A comparison of attitudes towards mental illness and knowledge of mental health services between Australian and Asian students." *Community Mental Health Journal,* Vol. 35, No. 1, Feb. 1999, p. 54 (Table 2).

**3.53** A British crime survey examined the relationship between the race of the attacker, the race of the victim, and the degree of the injury sustained in reported crimes (*The Howard Journal of Criminal Justice*, Aug. 1992). The table on p. 138 is cited in the article.

    **a.** What is the probability that a randomly selected reported crime involved a white attacker and a white victim?

    **b.** What is the probability that a randomly selected reported crime involved serious injuries?

    **c.** Given that both the attacker and victim were white, what is the probability that a randomly selected reported crime involved fatalities?

| Degree of Injury | Attacker/Victim | | | | |
| --- | --- | --- | --- | --- | --- |
| | White/ White | White/ Nonwhite | Nonwhite/ Nonwhite | Nonwhite/ White | Totals |
| Fatal | 183 | 18 | 18 | 18 | 237 |
| Serious | 580 | 49 | 111 | 141 | 881 |
| Slight | 4,336 | 440 | 594 | 1,656 | 7,026 |
| None | 1,422 | 136 | 214 | 1,801 | 3,573 |
| **Totals** | 6,521 | 643 | 937 | 3,616 | 11,717 |

*Source:* Shah, R., and Pease, K. "Crime, race and reporting to the police." *The Howard Journal of Criminal Justice*, Volume 31, No. 3, Aug. 1992.

**d.** If no injury was reported, what is the probability that a randomly selected reported crime involved a non-white attacker and a white victim?

**3.54** There are several methods of typing, or classifying, human blood. A method that is not as well known examines phosphoglucomutase (PGM) and classifies the blood into one of three main categories, 1-1, 2-1, or 2-2. Suppose a certain geographic region of the United States has the PGM percentages shown in the accompanying table. A person is to be chosen at random from this region.

**a.** What is the probability that a black person is chosen?

**b.** Given that a black person is chosen, what is the probability he or she is PGM type 1-1?

**c.** Given that a white person is chosen, what is the probability he or she is PGM type 1-1?

| Race | 1-1 | 2-1 | 2-2 |
| --- | --- | --- | --- |
| White | 46.3% | 39.2% | 4.0% |
| Black | 6.7% | 3.4% | .4% |

## Applying the Concepts—Advanced

**3.55** "Go" is one of the oldest and most popular strategic board games in the world, especially in Japan and Korea. The two-player game is played on a flat surface marked with 19 vertical and 19 horizontal lines. The objective is to control territory by placing pieces called "stones" on vacant points on the board. Players alternate placing their stones. The player using black stones goes first, followed by the player using white stones. [*Note:* The University of Virginia requires MBA students to learn Go to understand how the Japanese conduct business.] *Chance* (Summer 1995) published an article that investigated the advantage of playing first (i.e., using the black stones) in Go. The results of 577 games recently played by professional Go players were analyzed.

**a.** In the 577 games, the player with the black stones won 319 times and the player with the white stones won 258 times. Use this information to estimate the probability of winning when you play first in Go.

**b.** Professional Go players are classified by level. Group C includes the top-level players, followed by Group B (middle-level) and Group A (low-level) players.

The table describes the number of games won by the player with the black stones, categorized by level of the black player and level of the opponent. Estimate the probability of winning when you play first in Go for each combination of player and opponent level.

**c.** If the player with the black stones is ranked higher than the player with the white stones, what is the probability that black wins?

**d.** Given that the players are of the same level, what is the probability that the player with the black stones wins?

| Black Player Level | Opponent Level | Number of Wins | Number of Games |
| --- | --- | --- | --- |
| C | A | 34 | 34 |
| C | B | 69 | 79 |
| C | C | 66 | 118 |
| B | A | 40 | 54 |
| B | B | 52 | 95 |
| B | C | 27 | 79 |
| A | A | 15 | 28 |
| A | B | 11 | 51 |
| A | C | 5 | 39 |
| **Totals** | | 319 | 577 |

*Source:* J. Kim, and H. J. Kim, "The advantage of playing first in Go." *Chance*, Vol. 8, No. 3, Summer 1995, p. 26 (Table 3).

**3.56** Physicians and pharmacists sometimes fail to inform patients adequately about the proper application of prescription drugs and the precautions to take in order to avoid potential side effects. One method of increasing patients' awareness of the problem is for physicians to provide Patient Medication Instruction (PMI) sheets. The American Medical Association, however, has found that only 20% of the doctors who prescribe drugs frequently distribute PMI sheets to their patients. Assume that 20% of all patients receive the PMI sheet with their prescriptions and that 12% receive the PMI sheet and are hospitalized because of a drug-related problem. What is the probability that a person will be hospitalized for a drug-related problem given that the person has received the PMI sheet?

## 3.6 THE MULTIPLICATIVE RULE AND INDEPENDENT EVENTS

The probability of an intersection of two events can be calculated using the *multiplicative rule*, which employs the conditional probabilities we defined in the previous section. Actually, we've already developed the formula in another context. Recall that the conditional probability of $B$ given $A$ is

$$P(B|A) = \frac{P(A \cap B)}{P(A)}$$

Multiplying both sides of this equation by $P(A)$, we obtain a formula for the probability of the intersection of events $A$ and $B$. This is often called the **multiplicative rule of probability**.

---

### Multiplicative Rule of Probability

$P(A \cap B) = P(A)P(B|A)$,   or equivalently,   $P(A \cap B) = P(B)P(A|B)$

---

**EXAMPLE 3.15**

An agriculturist, who is interested in planting wheat next year, is concerned with the following events:

$B$: {The production of wheat will be profitable}

$A$: {A serious drought will occur}

Based on available information, the agriculturist believes that the probability is .01 that production of wheat will be profitable *assuming* a serious drought will occur in the same year and that the probability is .05 that a serious drought will occur. That is,

$$P(B|A) = .01 \quad \text{and} \quad P(A) = .05$$

Based on the information provided, what is the probability that a serious drought will occur *and* that a profit will be made? That is, find $P(A \cap B)$, the probability of the intersection of events $A$ and $B$.

**Solution**

We want to calculate $P(A \cap B)$. Using the formula for the multiplicative rule, we obtain

$$P(A \cap B) = P(A)P(B|A) = (.05)(.01) = .0005$$

The probability that a serious drought occurs *and* the production of wheat is profitable is only .0005. As we might expect, this intersection is a very rare event. ∎

Intersections often contain only a few sample points. In this case, the probability of an intersection is easy to calculate by summing the appropriate sample point probabilities. However, the formula for calculating intersection probabilities is invaluable when the intersection contains numerous sample points, as the next example illustrates.*

---

*The multiplicative rule of probability also plays an important role in an area of statistics known as *Bayesian statistics*. Consult the references at the end of the chapter for detailed discussions of Bayesian statistics.

**EXAMPLE 3.16**

A county welfare agency employs 10 welfare workers who interview prospective food stamp recipients. Periodically the supervisor selects, at random, the forms completed by two workers to audit for illegal deductions. Unknown to the supervisor, three of the workers have regularly been giving illegal deductions to applicants. What is the probability that both of the two workers chosen have been giving illegal deductions?

### Solution

Define the following two events:

  $A$: {First worker selected gives illegal deductions}
  $B$: {Second worker selected gives illegal deductions}

We want to find the probability of the event that both selected workers have been giving illegal deductions. This event can be restated as: {First worker gives illegal deductions *and* second worker gives illegal deductions}. Thus, we want to find the probability of the intersection, $A \cap B$. Applying the multiplicative rule, we have

$$P(A \cap B) = P(A)P(B|A)$$

To find $P(A)$ it is helpful to consider the experiment as selecting one worker from the 10. Then the sample space for the experiment contains 10 sample points (representing the 10 welfare workers), where the three workers giving illegal deductions are denoted by the symbol $I$ $(I_1, I_2, I_3)$, and the seven workers not giving illegal deductions are denoted by the symbol $N$ $(N_1, \ldots, N_7)$. The resulting Venn diagram is illustrated in Figure 3.15.

### FIGURE 3.15

*Venn Diagram for Finding*
$P(A)$

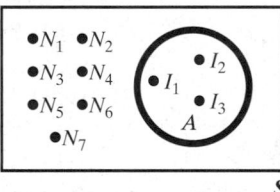

Since the first worker is selected at random from the 10, it is reasonable to assign equal probabilities to the 10 sample points. Thus, each sample point has a probability of $1/10$. The sample points in event $A$ are $\{I_1, I_2, I_3\}$—the three workers who are giving illegal deductions. Thus,

$$P(A) = P(I_1) + P(I_2) + P(I_3) = \frac{1}{10} + \frac{1}{10} + \frac{1}{10} = \frac{3}{10}$$

To find the conditional probability, $P(B|A)$, we need to alter the sample space $S$. Since we know $A$ has occurred—i.e., the first worker selected is giving illegal deductions—only two of the nine remaining workers in the sample space are giving illegal deductions. The Venn diagram for this new sample space is shown in Figure 3.16.

### FIGURE 3.16

*Venn Diagram for Finding*
$P(B|A)$

Each of these nine sample points is equally likely, so each is assigned a probability of $^1/_9$. Since the event $B|A$ contains the sample points $\{I_1, I_2\}$, we have

$$P(B|A) = P(I_1) + P(I_2) = \frac{1}{9} + \frac{1}{9} = \frac{2}{9}$$

Substituting $P(A) = ^3/_{10}$ and $P(B|A) = ^2/_9$ into the formula for the multiplicative rule, we find

$$P(A \cap B) = P(A)P(B|A) = \left(\frac{3}{10}\right)\left(\frac{2}{9}\right) = \frac{6}{90} = \frac{1}{15}$$

Thus, there is a 1 in 15 chance that both workers chosen by the supervisor have been giving illegal deductions to food stamp recipients. ∎

The sample space approach is only one way to solve the problem posed in Example 3.16. An alternative method employs the concept of a **tree diagram**. Tree diagrams are helpful for calculating the probability of an intersection.

To illustrate, a tree diagram for Example 3.16 is displayed in Figure 3.17. The tree begins at the far left with two branches. These branches represent the two possible outcomes $N$ (no illegal deductions) and $I$ (illegal deductions) for the first worker selected. The unconditional probability of each outcome is given (in parentheses) on the appropriate branch. That is, for the first worker selected, $P(N) = ^7/_{10}$ and $P(I) = ^3/_{10}$. (These can be obtained by summing sample point probabilities as in Example 3.16.)

The next level of the tree diagram (moving to the right) represents the outcomes for the second worker selected. The probabilities shown here are conditional probabilities since the outcome for the first worker is assumed to be known. For example, if the first worker is giving illegal deductions ($I$), the probability that the second worker is also giving illegal deductions ($I$) is $^2/_9$ since of the nine workers left to be selected, only two remain who are giving illegal deductions. This conditional probability, $^2/_9$, is shown in parentheses on the bottom branch of Figure 3.17.

**FIGURE 3.17**

*Tree Diagram for Example 3.16*

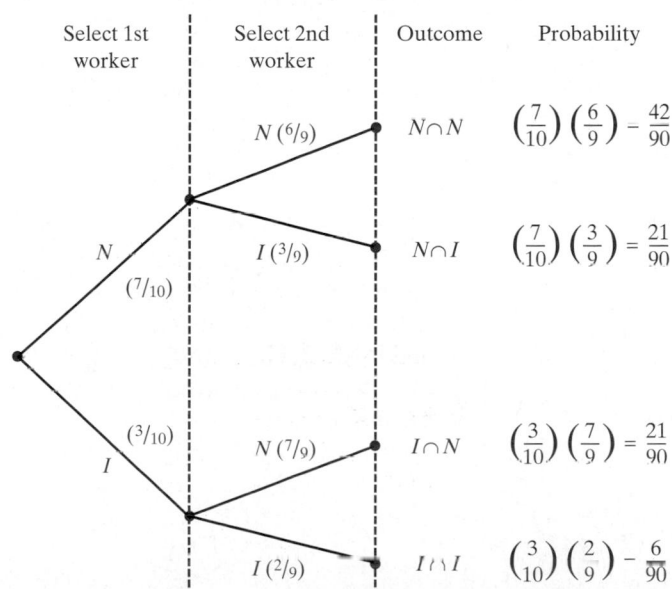

Finally, the four possible outcomes of the experiment are shown at the end of each of the four tree branches. These events are intersections of two events (outcome of first worker *and* outcome of second worker). Consequently, the multiplicative rule is applied to calculate each probability, as shown in Figure 3.17. You can see that the intersection $\{I \cap I\}$—the event that both workers selected are giving illegal deductions—has probability $^{6}/_{90} = ^{1}/_{15}$, which is the same value obtained in Example 3.16.

In Section 3.5 we showed that the probability of event $A$ may be substantially altered by the knowledge that an event $B$ has occurred. However, this will not always be the case. In some instances the assumption that event $B$ has occurred will *not* alter the probability of event $A$ at all. When this occurs, we say that the two events $A$ and $B$ are *independent events*.

---

**DEFINITION 3.9**

Events $A$ and $B$ are **independent events** if the occurrence of $B$ does not alter the probability that $A$ has occurred; that is, events $A$ and $B$ are independent if

$$P(A|B) = P(A)$$

When events $A$ and $B$ are independent, it is also true that

$$P(B|A) = P(B)$$

Events that are not independent are said to be **dependent**.

---

**EXAMPLE 3.17**

Consider the experiment of tossing a fair die and let

$A = \{\text{Observe an even number}\}$

$B = \{\text{Observe a number less than or equal to 4}\}$

Are events $A$ and $B$ independent?

**Solution**

The Venn diagram for this experiment is shown in Figure 3.18. We first calculate

$$P(A) = P(2) + P(4) + P(6) = \frac{1}{2}$$

$$P(B) = P(1) + P(2) + P(3) + P(4) = \frac{4}{6} = \frac{2}{3}$$

$$P(A \cap B) = P(2) + P(4) = \frac{2}{6} = \frac{1}{3}$$

**FIGURE 3.18**

*Venn Diagram for Die-Toss Experiment*

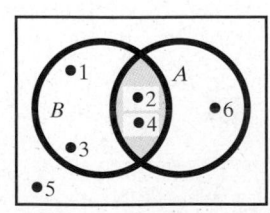

Now assuming $B$ has occurred, the conditional probability of $A$ given $B$ is

$$P(A|B) = \frac{P(A \cap B)}{P(B)} = \frac{1/3}{2/3} = \frac{1}{2} = P(A)$$

Thus, assuming that event $B$ occurs does not alter the probability of observing an even number—it remains $\frac{1}{2}$. Therefore, the events $A$ and $B$ are independent. Note that if we calculate the conditional probability of $B$ given $A$, our conclusion is the same:

$$P(B|A) = \frac{P(A \cap B)}{P(A)} = \frac{1/3}{1/2} = \frac{2}{3} = P(B)$$

**EXAMPLE 3.18**

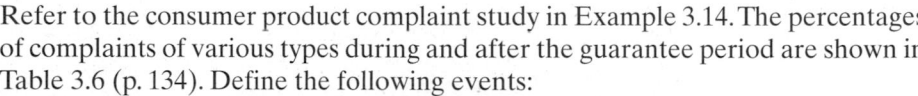

Refer to the consumer product complaint study in Example 3.14. The percentages of complaints of various types during and after the guarantee period are shown in Table 3.6 (p. 134). Define the following events:

$A$: {Cause of complaint is product appearance}

$B$: {Complaint occurred during the guarantee term}

Are $A$ and $B$ independent events?

**Solution**

Events $A$ and $B$ are independent if $P(A|B) = P(A)$. We calculated $P(A|B)$ in Example 3.14 to be .51, and from Table 3.6 we see that

$$P(A) = .32 + .03 = .35$$

Therefore, $P(A|B)$ is not equal to $P(A)$, and $A$ and $B$ are dependent events.

To gain an intuitive understanding of independence, think of situations in which the occurrence of one event does not alter the probability that a second event will occur. For example, new medical procedures are often tested on laboratory animals. The scientists conducting the tests generally try to perform the procedures on the animals so the results for one animal do not affect the results for the others. That is, the event that the procedure is successful on one animal is *independent* of the result for another. In this way, the scientists can get a more accurate idea of the efficacy of the procedure than if the results were dependent, with the success or failure for one animal affecting the results for other animals.

As a second example, consider an election poll in which 1,000 registered voters are asked their preference between two candidates. Pollsters try to use procedures for selecting a sample of voters so that the responses are independent. That is, the objective of the pollster is to select the sample so the event that one polled voter prefers candidate A does not alter the probability that a second polled voter prefers candidate A.

Now consider the world of sports. Do you think the results of a batter's successive trips to the plate in baseball, or of a basketball player's successive shots at the basket, are independent? If a basketball player makes two successive shots, is the probability of making the next shot altered from its value if the result of the first shot is not known? If a player makes two shots in a row, the probability of a third successful shot is likely to be different from what we would assign if we knew nothing about the first two shots. Why should this be so? Research has shown that many such results in sports tend to be *dependent* because players (and even teams) tend to get on "hot" and "cold" streaks, during which their probabilities of success may increase or decrease significantly.

We will make three final points about independence. The first is that the property of independence, unlike the mutually exclusive property, cannot be shown on or gleaned from a Venn diagram. This means *you can't trust your intuition*. In general, the only way to check for independence is by performing the calculations of the probabilities in the definition.

**FIGURE 3.19**

*Mutually Exclusive Events*
*are Dependent Events*

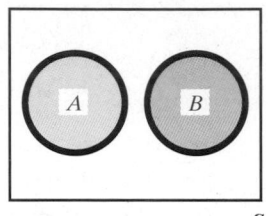

*S*

The second point concerns the relationship between the mutually exclusive and independence properties. Suppose that events *A* and *B* are mutually exclusive, as shown in Figure 3.19, and both events have nonzero probabilities. Are these events independent or dependent? That is, does the assumption that *B* occurs alter the probability of the occurrence of *A*? It certainly does, because if we assume that *B* has occurred, it is impossible for *A* to have occurred simultaneously. That is, $P(A|B) = 0$. Thus, *mutually exclusive events are dependent events* since $P(A) \neq P(A|B)$.

The third point is that the probability of the intersection of independent events is very easy to calculate. Referring to the formula for calculating the probability of an intersection, we find

$$P(A \cap B) = P(A)P(B|A)$$

Thus, since $P(B|A) = P(B)$ when *A* and *B* are independent, we have the following useful rule:

---

### Probability of Intersection of Two Independent Events

*If events A and B are independent*, the probability of the intersection of *A* and *B* equals the product of the probabilities of *A* and *B*; that is

$$P(A \cap B) = P(A)P(B)$$

The converse is also true: If $P(A \cap B) = P(A)P(B)$, then events *A* and *B* are independent.

---

In the die-toss experiment, we showed in Example 3.17 that the two events *A*: {Observe an even number} and *B*: {Observe a number less than or equal to 4} are independent if the die is fair. Thus,

$$P(A \cap B) = P(A)P(B) = \left(\frac{1}{2}\right)\left(\frac{2}{3}\right) = \frac{1}{3}$$

This agrees with the result that we obtained in the example:

$$P(A \cap B) = P(2) + P(4) = \frac{2}{6} = \frac{1}{3}$$

**EXAMPLE 3.19**

Refer to Example 3.5 (p. 115). Recall that the American Association for Marriage and Family Therapy (AAMFT) found that 25% of divorced couples are classified as "fiery foes," i.e., they communicate through their children and are hostile towards each other.

a. What is the probability that in a sample of two divorced couples, both are classified as "fiery foes"?

b. What is the probability that in a sample of ten divorced couples, all ten are classified as "fiery foes"?

**Solution**

a. Let $F_1$ represent the event that divorced couple 1 is classified as "fiery foes" and $F_2$ represent the event that divorced couple 2 are also "fiery foes." The event that *both* couples are "fiery foes" is the intersection of the two events, $F_1 \cap F_2$. Based on the AAMFT survey that found that 25% of divorced couples are "fiery foes," we could reasonably conclude that $P(F_1) = .25$ and $P(F_2) = .25$. However, in order to compute the probability of $F_1 \cap F_2$ from the multiplicative rule, we must make the assumption that the two events are independent. Since the classification of any divorced couple is not likely to affect the classification of another divorced couple, this assumption is reasonable. Assuming independence, we have

$$P(F_1 \cap F_2) = P(F_1)P(F_2) = (.25)(.25) = .0625$$

b. To see how to compute the probability that ten of ten divorced couples will all be classified as "fiery foes," first consider the event that three of three couples are "fiery foes." If $F_3$ represents the event that the third divorced couple are "fiery foes," then we want to compute the probability of the intersection $F_1 \cap F_2 \cap F_3$. Again assuming independence of the classifications, we have

$$P(F_1 \cap F_2 \cap F_3) = P(F_1)P(F_2)P(F_3) = (.25)(.25)(.25) = .015625$$

Similar reasoning leads us to the conclusion that the intersection of ten such events can be calculated as follows:

$$P(F_1 \cap F_2 \cap F_3 \cap \ldots \cap F_{10}) = P(F_1)P(F_2) \cdots P(F_{10}) = (.25)^{10} = .000001$$

Thus, the probability that ten of ten sampled divorced couples are all classified as "fiery foes" is 1 in 1 million, assuming the probability of each couple being classified as "fiery foes" is .25 and the classification decisions are independent. ∎

We conclude this section with an example that uses probability and statistics to make an inference.

**EXAMPLE 3.20**

Suppose a research psychologist is studying the hypothesis that trained rats will pass on at least part of their training to their offspring. To test the hypothesis, three offspring (no two with the same parents) of trained rats are randomly selected and subjected to a training test. From many previous experiments the psychologist knows that the relative frequency distribution of the scores for untrained rats is mound-shaped and symmetric with a mean ($\mu$) of 60 and a standard deviation ($\sigma$) of 10. Suppose all three of the trained rats' offspring score more than 70 on the test. Find the probability of this event. What can the research psychologist conclude?

**FIGURE 3.20**

*Relative Frequency Distribution of Training Test Scores, Example 3.20*

### Solution

The relative frequency distribution of the scores for untrained rats is shown in Figure 3.20. If the distribution is mound-shaped and approximately symmetric about the mean, we can conclude that approximately 16% of untrained rats will score more than 70 on the test (see Table 2.8).

Now define the events

$A_1$: {Offspring 1 scores more than 70}

$A_2$: {Offspring 2 scores more than 70}

$A_3$: {Offspring 3 scores more than 70}

We want to find $P(A_1 \cap A_2 \cap A_3)$, the probability that all three offspring score more than 70 on the training test. Since the offspring are selected so that they have different parents, it may be plausible to assume that the events $A_1$, $A_2$, and $A_3$ are independent. That is,

$$P(A_2|A_1) = P(A_2)$$

In words, knowing that the first offspring scores more than 70 on the test does not affect the probability that the second offspring scores more than 70. With the assumption of independence, we can calculate the probability of the intersection by multiplying the individual probabilities:

$$P(A_1 \cap A_2 \cap A_3) = P(A_1)P(A_2)P(A_3)$$
$$\approx (.16)(.16)(.16) = .004096$$

Thus, the probability that the research psychologist will observe all three offspring scoring more than 70 is only about .004 *if the offspring are untrained*. If this event were to occur, the psychologist might conclude that it lends credence to the theory that the offspring inherit some of the parents' training *since it is so unlikely to occur if they are untrained*. Such a conclusion would be an application of the rare-event approach to statistical inference.                                   ∎

 EXERCISES 3.57–3.75

### Learning the Mechanics

**3.57** Three fair coins are tossed and the following events are defined:

$A$: {Observe at least one head}
$B$: {Observe exactly two heads}

$C$: {Observe exactly two tails}
$D$: {Observe at most one head}

**a.** Sum the probabilities of the appropriate sample points to find: $P(A)$, $P(B)$, $P(C)$, $P(D)$, $P(A \cap B)$, $P(A \cap D)$, $P(B \cap C)$, and $P(B \cap D)$.

**b.** Use your answers to part **a** to calculate $P(B|A)$, $P(A|D)$, and $P(C|B)$.

**c.** Which pairs of events, if any, are independent? Why?

**3.58** An experiment results in one of five sample points with the following probabilities: $P(E_1) = .22$, $P(E_2) = .31$, $P(E_3) = .15$, $P(E_4) = .22$, and $P(E_5) = .1$. The following events have been defined:

$A: \{E_1, E_3\}$
$B: \{E_2, E_3, E_4\}$
$C: \{E_1, E_5\}$

Find each of the following probabilities:

**a.** $P(A)$
**b.** $P(B)$
**c.** $P(A \cap B)$
**d.** $P(A|B)$
**e.** $P(B \cap C)$
**f.** $P(C|B)$
**g.** Consider each pair of events: $A$ and $B$, $A$ and $C$, and $B$ and $C$. Are any of the pairs of events independent? Why?

**3.59** Two fair dice are tossed, and the following events are defined:

$A: \{\text{Sum of the numbers showing is odd}\}$
$B: \{\text{Sum of the numbers showing is } 9, 11, \text{ or } 12\}$

Are events $A$ and $B$ independent? Why?

**3.60** A sample space contains six sample points and events $A$, $B$, and $C$ as shown in the accompanying Venn diagram. The probabilities of the sample points are $P(1) = .20$, $P(2) = .05$, $P(3) = .30$, $P(4) = .10$, $P(5) = .10$, $P(6) = .25$.

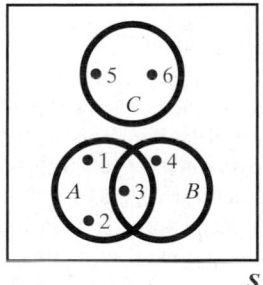

$S$

**a.** Which pairs of events, if any, are mutually exclusive? Why?

**b.** Which pairs of events, if any, are independent? Why?

**c.** Find $P(A \cup B)$ by adding the probabilities of the sample points and then by using the additive rule. Verify that the answers agree. Repeat for $P(A \cup C)$.

**3.61** Defend or refute each of the following statements:

**a.** Dependent events are always mutually exclusive.

**b.** Mutually exclusive events are always dependent.

**c.** Independent events are always mutually exclusive.

**3.62** For two events, $A$ and $B$, $P(A) = .4$ and $P(B) = .2$.

**a.** If $A$ and $B$ are independent, find $P(A \cap B)$, $P(A|B)$, and $P(A \cup B)$.

**b.** If $A$ and $B$ are dependent, with $P(A|B) = .6$, find $P(A \cap B)$ and $P(B|A)$.

## Applying the Concepts—Basic

**3.63** The National Center for Education Statistics (NCES) conducted a 1999 survey on the condition of America's public school facilities. The survey revealed the following information. The probability that a public school building has inadequate plumbing is .25. Of the buildings with inadequate plumbing, the probability that the school has plans for repairing the building is .38. Find the probability that a public school building has inadequate plumbing and will be repaired.

**3.64** According to the Children's Oncology Group Research Data Center at the University of Florida, 20% of children with neuroblastoma (a form of brain cancer) undergo surgery rather than the traditional treatment of chemotherapy or radiation. The surgery is successful in curing the disease 95% of the time (*Explore*, Spring 2001). Consider a child diagnosed with neuroblastoma. What are the chances that the child undergoes surgery and is cured?

**3.65** Consider an experiment in which 10 identical small boxes are placed side-by-side on a table. A crystal is placed, at random, inside one of the boxes. A self-professed "psychic" is asked to pick the box that contains the crystal.

**a.** If the "psychic" simply guesses, what is the probability that she picks the box with the crystal?

**b.** If the experiment is repeated seven times, what is the probability that the "psychic" guesses correctly at least once?

**c.** A group called the Tampa Bay Skeptics recently tested a self-proclaimed "psychic" using the above test. The "psychic" failed to pick the correct box all seven times (*Tampa Tribune*, Sept. 20, 1998). What would you infer about this person's psychic ability?

**3.66** Refer to *The American Association of Nurse Anesthetists Journal* (Feb. 2000) study on the use of herbal medicines before surgery, Exercise 1.18 (p. 16). Recall that 51% of surgical patients use herbal medicines against their doctor's advice prior to surgery.

**a.** What is the probability that a randomly selected surgical patient will use herbal medicines against his or her doctor's advice?

**b.** What is the probability that in a sample of two independently selected surgical patients, both will use herbal medicines against their doctor's advice?

**c.** What is the probability that in a sample of five independently selected surgical patients, all five will use herbal medicines against their doctor's advice?

**d.** Would you expect the event in part **c** to occur? Explain.

**3.67** "Channel One" is an education television network for which participating secondary schools are equipped with TV sets in every classroom. According to *Educational Technology* (May–June 1995), 40% of all U.S. secondary schools subscribe to the Channel One Communications Network (CCN). Of these subscribers, 5% never use the CCN broadcasts, while 20% use CCN more than five times per week.

**a.** Find the probability that a randomly selected U.S. secondary school subscribes to CCN and never uses the CCN broadcasts.

**b.** Find the probability that a randomly selected U.S. secondary school subscribes to CCN and uses the broadcasts more than five times per week.

## Applying the Concepts—Intermediate

**3.68** In Italy, all high school students must take a High School Diploma (HSD) exam and write a paper. In *Organizational Behavior and Human Decision Processes* (July 2000), University of Milan researcher L. Macchi provided the following information to a group of college undergraduates. *Fact 1*: In Italy, 360 out of every 1,000 students fail their HSD exam. *Fact 2*: Of those who fail the HSD, 75% also fail the written paper. *Fact 3*: Of those who pass the HSD, only 20% fail the written paper. Define events $A$ and $B$ as follows:

$A$ = {Student fails the HSD exam}
$B$ = {Student fails the written paper}

**a.** Write Fact 1 as a probability statement involving events $A$ and/or $B$.

**b.** Write Fact 2 as a probability statement involving events $A$ and/or $B$.

**c.** Write Fact 3 as a probability statement involving events $A$ and/or $B$.

**d.** State $P(A \cap B)$ in the words of the problem.

**e.** Find $P(A \cap B)$.

**3.69** A certain stretch of highway on the New Jersey Turnpike has a posted speed limit of 60 mph. A survey of drivers on this portion of the turnpike revealed the following facts (reported in *The Washington Post*, Aug.16, 1998):

**(1)** 14% of the drivers on this stretch of highway were African Americans

**(2)** 98% of the drivers were exceeding the speed limit by at least 5 mph (and, thus, subject to being stopped by the state police)

**(3)** Of these violators, 15% of the drivers were African Americans

**(4)** Of the drivers stopped for speeding by the New Jersey state police, 35% were African Americans

**a.** Convert each of the reported percentages above into a probability statement.

**b.** Find the probability that a driver on this stretch of highway is African American and exceeding the speed limit.

**c.** Given a driver on this stretch of highway is exceeding the speed limit, what is the probability the driver is not African American?

**d.** Given that a driver on this stretch of highway is stopped by the New Jersey state police, what is the probability that the driver is not African American?

**e.** For this stretch of highway, are the events {African American driver} and {Driver exceeding the speed limit} independent (to a close approximation)? Explain.

**f.** Use the probabilities in *The Washington Post* report to make a statement about whether African Americans are stopped more often for speeding than expected on this stretch of turnpike.

**3.70** Enterococci are bacteria that cause blood infections in hospitalized patients. One antibiotic used to battle enterococci is vancomycin. A study by the Federal Centers for Disease Control and Prevention revealed that 8% of all enterococci isolated in hospitals nationwide were resistant to vancomycin (*New York Times*, Sept.12, 1995). Consider a random sample of three patients with blood infections caused by the enterococci bacteria. Assume that all three patients are treated with the antibiotic vancomycin.

**a.** What is the probability that all three patients are successfully treated? What assumption did you make concerning the patients?

**b.** What is the probability that the bacteria resist the antibiotic for at least one patient?

**3.71** The genetic origin and properties of maize (modern-day corn) was investigated in *Economic Botany* (Jan.–Mar. 1995). Seeds from maize ears carry either single spikelets or paired spikelets, but not both. Progeny tests on approximately 600 maize ears revealed the following information. Forty percent of all seeds carry single spikelets, while 60% carry paired spikelets. A seed with single spikelets will produce maize ears with single spikelets 29% of the time and paired spikelets 71% of the time. A seed with paired spikelets will produce maize ears with single spikelets 26% of the time and paired spikelets 74% of the time.

**a.** Find the probability that a randomly selected maize ear seed carries a single spikelet and produces ears with single spikelets.

**b.** Find the probability that a randomly selected maize ear seed produces ears with paired spikelets.

**3.72** A new type of lie detector—called the Computerized Voice Stress Analyzer (CVSA)—has been developed. The manufacturer claims that the CVSA is 98% accurate, and, unlike a polygraph machine, will not be thrown off by drugs and medical factors. However, laboratory studies by the U.S. Defense Department found that the CVSA had an accuracy rate of 49.8%—slightly less than pure chance (*Tampa Tribune*, Jan. 10, 1999). Suppose the CVSA is used to test the veracity of four suspects. Assume the suspects' responses are independent.

**a.** If the manufacturer's claim is true, what is the probability that the CVSA will correctly determine the veracity of all four suspects?

**b.** If the manufacturer's claim is true, what is the probability that the CVSA will yield an incorrect result for at least one the four suspects?

**c.** Suppose that in a laboratory experiment conducted by the U.S. Defense Department on four suspects, the CVSA yielded incorrect results for two of the suspects. Use this result to make an inference about the true accuracy rate of the new lie detector.

## Applying the Concepts—Advanced

**3.73** One of the problems encountered in organ transplants is the body's rejection of the transplanted tissue. If the antigens attached to the tissue cells of the donor and receiver match, the body will accept the transplanted tissue. While the antigens in identical twins always match, the probability of a match in other siblings is .25 and that of a match in two people from the population at large is .001. Suppose you need a kidney, and you have two brothers and a sister.

**a.** If one of your three siblings offers a kidney, what is the probability that the antigens will match?

**b.** If all three siblings offer a kidney, what is the probability that all three antigens will match?

**c.** If all three siblings offer a kidney, what is the probability that none of the antigens will match?

**a.** Repeat parts **b** and **c**, this time assuming that the three donors were obtained from the population at large.

**3.74** In October 1994, a flaw was discovered in the Pentium microchip installed in personal computers. The chip produced an incorrect result when dividing two numbers. Intel, the manufacturer of the Pentium chip, initially announced that such an error would occur once in 9 billion divides, or "once in every 27,000 years" for a typical user; consequently, it did not immediately offer to replace the chip. Assume the probability of a divide error with the Pentium chip is, in fact, 1/9,000,000,000.

**a.** For a division performed using the flawed Pentium chip, what is the probability that no error will occur?

**b.** Consider two successive divisions performed using the flawed chip. What is the probability that neither result will be in error? (Assume that any one division has no impact on the result of any other division performed by the chip.)

**c.** Depending on the procedure, statistical software packages may perform an extremely large number of divisions to produce the required output. For heavy users of the software, 1 billion divisions over a short time frame is not unusual. Calculate the probability that 1 billion divisions performed using the flawed Pentium chip will result in no errors.

**d.** Use the result, part **c**, to compute the probability of at least one error in the 1 billion divisions. [*Note:* Two months after the flaw was discovered, Intel agreed to replace all Pentium chips free of charge.]

**3.75** In *Parade Magazine*'s (Nov. 26, 2000) column "Ask Marilyn," the following question was posed: "I have just tossed a [balanced] coin 10 times, and I ask you to guess which of the following three sequences was the result. One (and only one) of the sequences is genuine."

(1) H H H H H H H H H H
(2) H H T T H T T H H H
(3) T T T T T T T T T T

**a.** Demonstrate that prior to actually tossing the coins, the three sequences are equally likely to occur.

**b.** Find the probability that the 10 coin tosses result in all heads or all tails.

**c.** Find the probability that the 10 coin tosses result in a mix of heads and tails.

**d.** Marilyn's answer to the question posed was: "Though the chances of the three specific sequences occurring randomly are equal . . . it's reasonable for us to choose sequence (2) as the most likely genuine result." If you know that only one of the three sequences actually occurred, explain why Marilyn's answer is correct. [*Hint:* Compare the probabilities in parts **b** and **c**.]

## 3.7 RANDOM SAMPLING

How a sample is selected from a population is of vital importance in statistical inference because the probability of an observed sample will be used to infer the characteristics of the sampled population. To illustrate, suppose you deal yourself four cards from a deck of 52 cards and all four cards are aces. Do you conclude that your deck is an ordinary bridge deck, containing only four aces, or do you conclude that the deck is stacked with more than four aces? It depends on how the cards were drawn. If the four aces were always placed at the top of a standard bridge deck, drawing four aces is not unusual—it is certain. On the other hand, if the cards are thoroughly mixed, drawing four aces in a sample of four cards is highly improbable. The point, of course, is that in order to use the observed sample of four cards to draw inferences about the population (the deck of 52 cards), you need to know how the sample was selected from the deck.

One of the simplest and most frequently employed sampling procedures is implied in the previous examples and exercises. It produces what is known as a *random sample.* We learned in Section 1.5 (p. 10) that a random sample is likely to be *representative* of the population that it is selected from.

> **DEFINITION 3.10**
> If *n* elements are selected from a population in such a way that every set of *n* elements in the population has an equal probability of being selected, the *n* elements are said to be a **random sample**.*

**EXAMPLE 3.21**   Suppose a lottery consists of 10 tickets. (This number is small to simplify our example.) One ticket stub is to be chosen, and the corresponding ticket holder will receive a generous prize. How would you select this ticket stub so that the prize will be awarded fairly?

### Solution

If the prize is to be awarded fairly, it seems reasonable to require that each ticket stub have the same probability of being drawn. That is, each stub should have a probability of $\frac{1}{10}$ of being selected. A method to achieve the objective of equal selection probabilities is to *mix* the 10 stubs thoroughly and *blindly* pick one of the stubs. If this procedure were repeatedly used, each time replacing the selected stub, a particular stub should be chosen approximately $\frac{1}{10}$ of the time in a long series of draws. This method of sampling is known as **random sampling**.

If a population is not too large and the elements can be numbered on slips of paper, poker chips, etc., you can physically mix the slips of paper or chips and remove *n* elements from the total. The numbers that appear on the chips selected would indicate the population elements to be included in the sample. Since it is often difficult to achieve a thorough mix, such a procedure only provides an approximation to random sampling. Most researchers rely on **random number generators** to automatically generate the random sample. These random number generators are available in table form and they are built into most statistical software packages.

You can think of sampling as an experiment consisting of drawing *n* elements from a population of *N* elements, with each different sample representing a sample point of the experiment. Thus, in Example 3.21, in which a lottery consisted of drawing one of 10 tickets, there are 10 different sample points. If the drawing is held so that each sample point is equally likely with probability $\frac{1}{10}$, then the result of the experiment is a *random sample*.

Of course, for most applications the population will consist of more than $N = 10$ elements, and the sample will consist of more than $n = 1$ element. Often, the total number of possible samples will not be easy to visualize, so a method for counting the number of samples is needed. For example, suppose the lottery consists of drawing two tickets from ten. We can list the possible samples as in Table 3.7, where $T_1$ represents ticket 1, $T_2$ ticket 2, ..., and $T_{10}$ ticket 10. The systematic listing in Table 3.7 shows the 45 possible sample points of the experiment of sampling two elements from 10. However, the listing is tedious and only gets more so as the values of the population size *N* and the sample size *n* are increased.

A second method of determining the number of samples is to use **combinatorial mathematics**. The combinatorial symbol for the number of different ways of selecting *n* elements from *N* elements is $\binom{N}{n}$, which is read "the number of combinations of *N* elements taken *n* at a time." The formula** for calculating the number is

---

*Strictly speaking, this is a *simple random sample*. There are many different types of random samples. The simple random sample is the most common.
**For a more thorough discussion of the reasoning behind this and other counting rules, see (optional) Section 3.8.

**TABLE 3.7** **Listing of All Possible Samples of Two Tickets Drawn from 10 Tickets**

| $T_1, T_2$ | $T_2, T_3$ | $T_3, T_4$ | $T_4, T_5$ | $T_5, T_6$ | $T_6, T_7$ | $T_7, T_8$ | $T_8, T_9$ | $T_9, T_{10}$ |
|---|---|---|---|---|---|---|---|---|
| $T_1, T_3$ | $T_2, T_4$ | $T_3, T_5$ | $T_4, T_6$ | $T_5, T_7$ | $T_6, T_8$ | $T_7, T_9$ | $T_8, T_{10}$ | |
| $T_1, T_4$ | $T_2, T_5$ | $T_3, T_6$ | $T_4, T_7$ | $T_5, T_8$ | $T_6, T_9$ | $T_7, T_{10}$ | | |
| $T_1, T_5$ | $T_2, T_6$ | $T_3, T_7$ | $T_4, T_8$ | $T_5, T_9$ | $T_6, T_{10}$ | | | |
| $T_1, T_6$ | $T_2, T_7$ | $T_3, T_8$ | $T_4, T_9$ | $T_5, T_{10}$ | | | | |
| $T_1, T_7$ | $T_2, T_8$ | $T_3, T_9$ | $T_4, T_{10}$ | | | | | |
| $T_1, T_8$ | $T_2, T_9$ | $T_3, T_{10}$ | | | | | | |
| $T_1, T_9$ | $T_2, T_{10}$ | | | | | | | |
| $T_1, T_{10}$ | | | | | | | | |

$$\binom{N}{n} = \frac{N!}{n!(N-n)!}$$

where "!" is the factorial symbol and is a shorthand for the following multiplication:

$$n! = n(n-1)(n-2)\cdots(3)(2)(1)$$

Thus, for example, $5! = 5 \cdot 4 \cdot 3 \cdot 2 \cdot 1 = 120$. (The quantity $0!$ is defined to be 1.)

**EXAMPLE 3.22**

a. Use the combinatorial formula to count the number of different ways of drawing one lottery ticket from a total of 10 tickets.
b. Use the combinatorial formula to count the number of ways of drawing two lottery tickets from a total of 10 tickets.

**Solution**

a. Substituting $n = 1$ and $N = 10$ into the formula, we find

$$\binom{N}{n} = \binom{10}{1} = \frac{10!}{1!(10-1)!} = \frac{10!}{1!9!}$$
$$= \frac{10 \cdot 9 \cdot 8 \cdot 7 \cdot 6 \cdot 5 \cdot 4 \cdot 3 \cdot 2 \cdot 1}{(1)(9 \cdot 8 \cdot 7 \cdot 6 \cdot 5 \cdot 4 \cdot 3 \cdot 2 \cdot 1)} = 10$$

Thus, as we know intuitively, there are 10 different samples that can be selected when drawing one ticket from 10.

b. Substituting $n = 2$ and $N = 10$ into the formula, we find

$$\binom{N}{n} = \binom{10}{2} = \frac{10!}{2!(10-2)!} = \frac{10!}{2!8!}$$
$$= \frac{10 \cdot 9 \cdot 8 \cdot 7 \cdot 6 \cdot 5 \cdot 4 \cdot 3 \cdot 2 \cdot 1}{(2 \cdot 1)(8 \cdot 7 \cdot 6 \cdot 5 \cdot 4 \cdot 3 \cdot 2 \cdot 1)} = \frac{10 \cdot 9}{2 \cdot 1} = 45$$

This agrees with the number obtained by listing all possible samples in Table 3.7 but requires much less effort. And, as the next example illustrates, the combinatorial formula works long after listing has ceased to be practical.

**EXAMPLE 3.23**

Suppose you wish to randomly sample five households from a population of 100,000 households.

a. How many different samples can be selected?
b. Use a random number generator to select a random sample.

**Solution**

a. Using the combinatorial rule, we find

$$\binom{100,000}{5} = \frac{100,000!}{5!\,99,995!}$$
$$= \frac{100,000 \cdot 99,999 \cdot 99,998 \cdot 99,997 \cdot 99,996}{5 \cdot 4 \cdot 3 \cdot 2 \cdot 1}$$
$$= 8.33 \times 10^{22}$$

Thus, there are 83.3 billion trillion different samples of five households that can be selected from 100,000.

b. To ensure that each of the possible samples has an equal chance of being selected, as required for random sampling, we can employ a **random number table**, as provided in Table I of Appendix A. Random number tables are constructed in such a way that every number occurs with (approximately) equal probability. Furthermore, the occurrence of any one number in a position is independent of any of the other numbers that appear in the table. To use a table of random numbers, number the $N$ elements in the population from 1 to $N$.

Then turn to Table I and select a starting number in the table. Proceeding from this number either across the row or down the column, remove and record $n$ numbers from the table.

To illustrate, first we number the households in the population from 1 to 100,000. Then, we turn to a page of Table I, say the first page. (A partial reproduction of the first page of Table I is shown in Table 3.8.) Now, we arbitrarily select a starting number, say the random number appearing in the third row, second column. This number is 48,360. Then we proceed down the second column to obtain the remaining four random numbers. In this case we have selected five random numbers, which are highlighted in Table 3.8. Using the first five digits to represent households from 1 to 99,999 and the number 00000 to represent household 100,000, we can see that the households numbered

$$48,360 \qquad 93,093 \qquad 39,975 \qquad 6,907 \qquad 72,905$$

should be included in our sample.

**TABLE 3.8    Partial Reproduction of Table I in Appendix A**

| Row \ Column | 1 | 2 | 3 | 4 | 5 | 6 |
|---|---|---|---|---|---|---|
| 1 | 10480 | 15011 | 01536 | 02011 | 81647 | 91646 |
| 2 | 22368 | 46573 | 25595 | 85393 | 30995 | 89198 |
| 3 | 24130 | 48360 | 22527 | 97265 | 76393 | 64809 |
| 4 | 42167 | 93093 | 06243 | 61680 | 07856 | 16376 |
| 5 | 37670 | 39975 | 81837 | 16656 | 06121 | 91782 |
| 6 | 77921 | 06907 | 11008 | 42751 | 27756 | 53498 |
| 7 | 99562 | 72905 | 56420 | 69994 | 98872 | 31016 |
| 8 | 96301 | 91977 | 05463 | 07972 | 18876 | 20922 |
| 9 | 89579 | 14342 | 63661 | 10281 | 17453 | 18103 |
| 10 | 85475 | 36857 | 53342 | 53988 | 53060 | 59533 |
| 11 | 28918 | 69578 | 88231 | 33276 | 70997 | 79936 |
| 12 | 63553 | 40961 | 48235 | 03427 | 49626 | 69445 |
| 13 | 09429 | 93969 | 52636 | 92737 | 88974 | 33488 |

*Note:* Use only the necessary number of digits in each random number to identify the element to be included in the sample. If, in the course of recording the *n* numbers from the table, you select a number that has already been selected, simply discard the duplicate and select a replacement at the end of the sequence. Thus, you may have to record more than *n* numbers from the table to obtain a sample of *n* unique numbers.

Can we be sure that all 83.3 billion trillion samples have an equal chance of being selected? We can't; but to the extent that the random number table contains truly random sequences of digits, the sample should be very close to random.

Table I in Appendix A is just one example of a random number generator. For most scientific studies that require a large random sample, computers are used to generate the random sample. The SAS, MINITAB, STATISTIX, and SPSS statistical software packages all have easy-to-use random number generators.

For example, suppose we required a random sample of *n* = 50 households from the population of 100,000 households in Example 3.23. Here, we might employ the SAS random number generator. Figure 3.21 shows an SAS printout listing 50 random numbers (from a population of 100,000). The households with these identification numbers would be included in the random sample.

**FIGURE 3.21**

*SAS-Generated Random Sample of 50 Households*

| OBS | HOUSENUM | OBS | HOUSENUM | OBS | HOUSENUM | OBS | HOUSENUM |
|---|---|---|---|---|---|---|---|
| 1 | 47122 | 14 | 47271 | 27 | 17098 | 40 | 4260 |
| 2 | 94231 | 15 | 3642 | 28 | 23259 | 41 | 58140 |
| 3 | 95531 | 16 | 7611 | 29 | 30512 | 42 | 22903 |
| 4 | 41445 | 17 | 81646 | 30 | 91548 | 43 | 65959 |
| 5 | 80287 | 18 | 92158 | 31 | 7673 | 44 | 13962 |
| 6 | 11731 | 19 | 36667 | 32 | 68549 | 45 | 25819 |
| 7 | 47523 | 20 | 71811 | 33 | 85433 | 46 | 66497 |
| 8 | 84847 | 21 | 78988 | 34 | 5231 | 47 | 79559 |
| 9 | 69822 | 22 | 3819 | 35 | 13455 | 48 | 87017 |
| 10 | 18270 | 23 | 21873 | 36 | 71666 | 49 | 28483 |
| 11 | 52636 | 24 | 74938 | 37 | 66280 | 50 | 91806 |
| 12 | 21750 | 25 | 23635 | 38 | 66210 | | |
| 13 | 63363 | 26 | 35807 | 39 | 21998 | | |

EXERCISES 3.76–3.82

## Learning the Mechanics

**3.76** Suppose you wish to sample *n* = 3 elements from a total of *N* = 6 elements.
**a.** Count the number of different samples that can be drawn, first by listing them, and then by using combinatorial mathematics.
**b.** If random sampling is to be employed, what is the probability that any particular sample will be selected?
**c.** Show how to use the random number table, Table I in Appendix A, to select a random sample of three elements from a population of six elements. Perform the sampling procedure 20 times. Do any two of the samples contain the same three elements? Given your answer to part **b**, did you expect repeated samples?

**3.77** Suppose you wish to sample *n* = 3 elements from a total of *N* = 600 elements.
**a.** Count the number of different samples by using combinatorial mathematics.
**b.** If random sampling is to be employed, what is the probability that any particular sample will be selected?
**c.** Show how to use the random number table, Table I in Appendix A, to select a random sample of three elements from a population of 600 elements. Perform the sampling procedure 20 times. Do any two of the samples contain the same three elements? Given your answer to part **b**, did you expect repeated samples?
**d.** Use the computer to generate a random sample of three from the population of 600 elements.

**3.78** Suppose that a population contains $N = 200{,}000$ elements. Use a computer or Table I of Appendix A to select a random sample of $n = 10$ elements from the population. Explain how you selected your sample.

## Applying the Concepts—Basic

**3.79** To ascertain the effectiveness of their advertising campaigns, firms frequently conduct telephone interviews with consumers using *random-digit dialing*. With this method, a random number generator mechanically creates the sample of phone numbers to be called.
  **a.** Explain how the random number table (Table I of Appendix A) or a computer could be used to generate a sample of 7-digit telephone numbers.
  **b.** Use the procedure you described in part **a** to generate a sample of ten 7-digit telephone numbers.
  **c.** Use the procedure you described in part **a** to generate five 7-digit telephone numbers whose first three digits are 373.

**3.80** In addition to its decennial enumeration of the population, the U.S. Bureau of the Census regularly samples the population for demographic information such as income, family size, employment, and marital status. Suppose the Bureau plans to sample 1,000 households in a city that has a total of 534,322 households. Show how the Bureau could use the random number table in Appendix A or a computer to generate the sample. Select the first 10 households to be included in the sample.

## Applying the Concepts—Intermediate

**3.81** In auditing a firm's financial statements, an auditor is required to assess the operational effectiveness of the accounting system. In performing the assessment, the auditor frequently relies on a random sample of actual transactions (Stickney and Weil, *Financial Accounting: An Introduction to Concepts, Methods, and Uses*, 1994).

A particular firm has 5,382 customer accounts that are numbered from 0001 to 5382.
  **a.** One account is to be selected at random for audit. What is the probability that account number 3,241 is selected?
  **b.** Draw a random sample of 10 accounts and explain in detail the procedure you used.
  **c.** Refer to part **b**. The following are two possible random samples of size 10. Is one more likely to be selected than the other? Explain.

| Sample Number 1 | | | | |
|---|---|---|---|---|
| 5011 | 0082 | 0963 | 0772 | 3415 |
| 2663 | 1126 | 0008 | 0026 | 4189 |

| Sample Number 2 | | | | |
|---|---|---|---|---|
| 0001 | 0003 | 0005 | 0007 | 0009 |
| 0002 | 0004 | 0006 | 0008 | 0010 |

## Applying the Concepts—Advanced

**3.82** Archaeologists plan to perform test digs at a location they believe was inhabited several thousand years ago. The site is approximately 10,000 meters long and 5,000 meters wide. They first draw rectangular grids over the area, consisting of lines every 100 meters, creating a total of $100 \cdot 50 = 5{,}000$ intersections (not counting one of the outer boundaries). The plan is to randomly sample 50 intersection points and dig at the sampled intersections. Explain how you could use a random number generator to obtain a random sample of 50 intersections. Develop at least two plans: one that numbers the intersections from 1 to 5,000 prior to selection and another that selects the row and column of each sampled intersection (from the total of 100 rows and 50 columns).

# 3.8   SOME COUNTING RULES (OPTIONAL)

In Section 3.1 we pointed out that experiments sometimes have so many sample points that it is impractical to list them all. However, many of these experiments possess sample points with identical characteristics. If you can develop a **counting rule** to count the number of sample points for such an experiment, it can be used to aid in the solution of the problems.

■ **EXAMPLE 3.24**   A product can be shipped by four airlines and each airline can ship via three different routes. How many distinct ways exist to ship the product?

**Solution**

A pictorial representation of the different ways to ship the product will aid in counting them. The tree diagram (see Section 3.6) is useful for this purpose. Consider the tree diagram shown in Figure 3.22. At the starting point (stage 1), there are four choices—the different airlines—to begin the journey. Once we have chosen

**FIGURE 3.22**

*Tree Diagram for Shipping Problem*

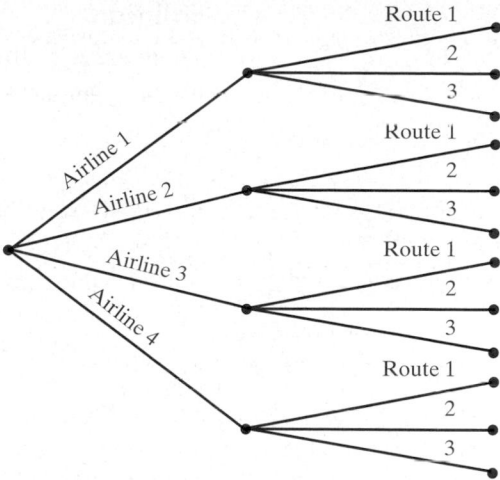

an airline (stage 2), there are three choices—the different routes—to complete the shipment and reach the final destination. Thus, the tree diagram clearly shows that there are $(4)(3) = 12$ distinct ways to ship the product.

The method of solving this example can be generalized to any number of stages with sets of different elements. This method is given by the *multiplicative rule*.

---

### The Multiplicative Rule

You have $k$ sets of elements, $n_1$ in the first set, $n_2$ in the second set,..., and $n_k$ in the $k$th set. Suppose you wish to form a sample of $k$ elements by *taking one element from each* of the $k$ sets. The number of different samples that can be formed is the product

$$n_1 n_2 n_3 \cdots n_k$$

---

**EXAMPLE 3.25**

There are 20 candidates for three different executive positions, $E_1$, $E_2$, and $E_3$. How many different ways could you fill the positions?

### Solution

For this example, there are $k = 3$ sets of elements corresponding to:

Set 1: Candidates available to fill position $E_1$
Set 2: Candidates remaining (after filling $E_1$) that are available to fill $E_2$
Set 3: Candidates remaining (after filling $E_1$ and $E_2$) that are available to fill $E_3$

The number of elements in the sets are $n_1 = 20, n_2 = 19, n_3 = 18$. Therefore, the number of different ways of filling the three positions is given by the multiplicative rule as $n_1 n_2 n_3 = (20)(19)(18) = 6,840$.

---

**EXAMPLE 3.26**

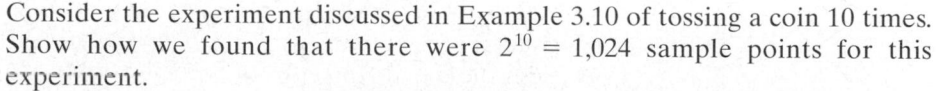

Consider the experiment discussed in Example 3.10 of tossing a coin 10 times. Show how we found that there were $2^{10} = 1,024$ sample points for this experiment.

### Solution

There are $k = 10$ sets of elements for this experiment. Each set contains two elements: a head and a tail. Thus, there are

$$(2)(2)(2)(2)(2)(2)(2)(2)(2)(2) = 2^{10} = 1,024$$

different outcomes (sample points) of this experiment.    ∎

**EXAMPLE 3.27**

Suppose there are five dangerous military missions, each requiring one soldier. In how many different ways can five soldiers from a squadron of 100 be assigned to these five missions?

### Solution

We can solve this problem by using the multiplicative rule. The entire set of 100 soldiers is available for the first mission, and after the selection of one soldier for that mission, 99 are available for the second mission, etc. Thus, the total number of different ways of choosing five soldiers for the five missions is

$$n_1 n_2 n_3 n_4 n_5 = (100)(99)(98)(97)(96) = 9,034,502,400$$

∎

The arrangement of elements in a distinct order is called a **permutation**. Thus, in Example 3.27 there are more than 9 billion different *permutations* of five elements (soldiers) drawn from a set of 100 elements!

---

### Permutations Rule

Given a *single set* of $N$ distinctly different elements, you wish to select $n$ elements from the $N$ and *arrange* them within $n$ positions. The number of different **permutations** of the $N$ elements taken $n$ at a time is denoted by $P_n^N$ and is equal to

$$P_n^N = N(N-1)(N-2) \cdots (N-n+1) = \frac{N!}{(N-n)!}$$

where $n! = n(n-1)(n-2) \cdots (3)(2)(1)$ and is called $n$ *factorial*. (Thus, for example, $5! = 5 \cdot 4 \cdot 3 \cdot 2 \cdot 1 = 120$.) The quantity $0!$ is defined to be 1.

---

**EXAMPLE 3.28**

You wish to drive, in sequence, from a starting point to each of five cities, and you wish to compare the distances—and ultimately the costs—of the different routes. How many different routes would you have to compare?

### Solution

Denote the cities as $C_1, C_2, \ldots, C_5$. Then a route moving from the starting point to $C_2$ to $C_1$ to $C_3$ to $C_4$ to $C_5$ would be represented as $C_2 C_1 C_3 C_4 C_5$. The total number of routes would equal the number of ways you could rearrange the $N = 5$ cities in $n = 5$ positions. This number is

$$P_n^N = P_5^5 = \frac{5!}{(5-5)!} = \frac{5!}{0!} = \frac{5 \cdot 4 \cdot 3 \cdot 2 \cdot 1}{1} = 120$$

(Recall that $0! = 1$.)

∎

**TABLE 3.9   Workers Assigned to Four Jobs, Example 3.29**

| Group 1 | Group 2 | Group 3 | Group 4 |
|---------|---------|---------|---------|
| ABCD | ABDC | ACDB | BCDA |
| ACBD | ADBC | ADCB | BDCA |
| BACD | BADC | CADB | CBDA |
| BCAD | BDAC | CDAB | CDBA |
| CABD | DABC | DACB | DBCA |
| CBAD | DBAC | DCAB | DCBA |

**EXAMPLE 3.29**

You are supervising four construction workers. You must assign three to job 1 and one to job 2. In how many different ways can you make this assignment?

**Solution**

To begin, suppose each worker is to be assigned to a distinct job. Then, using the multiplicative rule, you obtain $(4)(3)(2)(1) = 24$ ways of assigning the workers to four distinct jobs. The 24 ways are listed in four groups in Table 3.9 (where ABCD signifies that worker A was assigned the first distinct job, worker B the second, etc.).

Now suppose the first three positions represent job 1 and the last position represents job 2. You can now see that all the listings in group 1 represent the same outcome of the experiment of interest. That is, workers A, B, and C are assigned to job 1 and worker D to job 2. Similarly, group 2 listings are equivalent, as are group 3 and group 4 listings. Thus, there are only four different assignments of four workers to the two jobs. These are shown in Table 3.10.

**TABLE 3.10   Workers Assigned to Two Jobs, Example 3.29**

| Job 1 | Job 2 |
|-------|-------|
| ABC | D |
| ABD | C |
| ACD | B |
| BCD | A |

To generalize this result, we point out that the final result can be found by

$$\frac{(4)(3)(2)(1)}{(3)(2)(1)(1)} = 4$$

The $(4)(3)(2)(1)$ is the number of different ways (*permutations*) the workers could be assigned four distinct jobs. The division by $(3)(2)(1)$ is to remove the duplicated permutations resulting from the fact that three workers are assigned the same jobs. And the division by 1 is associated with the worker assigned to job 2. ∎

**Partitions Rule**

There exists a *single* set of $N$ distinctly different elements. You wish to partition them into $k$ sets, with the first set containing $n_1$ elements, the second containing $n_2$ elements, ..., and the $k$th set containing $n_k$ elements. The number of different partitions is

$$\frac{N!}{n_1!n_2!\cdots\cdot n_k!} \quad \text{where } n_1 + n_2 + n_3 + \cdots + n_k = N$$

**EXAMPLE 3.30**

You have 12 construction workers. You wish to assign three to job site 1, four to job site 2, and five to job site 3. In how many different ways can you make this assignment?

**Solution**

For this example, $k = 3$ (corresponding to the $k = 3$ different job sites), $N = 12$, and $n_1 = 3$, $n_2 = 4$, and $n_3 = 5$. Then the number of different ways to assign the workers to the job sites is

$$\frac{N!}{n_1!n_2!n_3!} = \frac{12!}{3!4!5!} = \frac{12 \cdot 11 \cdot 10 \cdot \cdots \cdot 3 \cdot 2 \cdot 1}{(3 \cdot 2 \cdot 1)(4 \cdot 3 \cdot 2 \cdot 1)(5 \cdot 4 \cdot 3 \cdot 2 \cdot 1)} = 27{,}720$$

**EXAMPLE 3.31**

How many samples of four college courses can you select from a list of 25 courses in a university catalog?

**Solution**

For this example, $k = 2$ (corresponding to the $n_1 = 4$ classes you *do* choose and the $n_2 = 21$ classes you *do not* choose) and $N = 25$. Then, the number of different ways to choose the four classes from 25 is

$$\frac{N!}{n_1!n_2!} = \frac{25!}{(4!)(21!)} = \frac{25 \cdot 24 \cdot 23 \cdot \cdots \cdot 3 \cdot 2 \cdot 1}{(4 \cdot 3 \cdot 2 \cdot 1)(21 \cdot 20 \cdot \cdots \cdot 2 \cdot 1)} = 12{,}650$$

(Now you know why there's so much confusion surrounding registration time at many colleges.)

This special application of the partitions rule—partitioning a set of $N$ elements into $k = 2$ groups (the elements that appear in a sample and those that do not)—is very common. Therefore, we give a different name to the rule for counting the number of different ways of partitioning a set of elements into two parts: the **combinations rule**.

---

## Combinations Rule

A sample of $n$ elements is to be chosen from a set of $N$ elements. Then the number of different samples of $n$ elements can be selected from $N$ denoted by $\binom{N}{n}$ and is equal to

$$\binom{N}{n} = \frac{N!}{n!(N-n)!}$$

Note that the order in which the $n$ elements are drawn is not important.

---

**EXAMPLE 3.32**

Five soldiers from a squadron of 100 are to be chosen for a dangerous mission. In how many ways can groups of five be formed?

**Solution**

This problem is equivalent to sampling $n = 5$ elements from a set of $N = 100$ elements. Thus, the number of ways is the number of possible combinations of five

soldiers selected from 100, or

$$\binom{100}{5} = \frac{100!}{(5!)(95!)} = \frac{100 \cdot 99 \cdot 98 \cdot 97 \cdot 96 \cdot 95 \cdot 94 \cdots 2 \cdot 1}{(5 \cdot 4 \cdot 3 \cdot 2 \cdot 1)(95 \cdot 94 \cdots 2 \cdot 1)}$$
$$= \frac{100 \cdot 99 \cdot 98 \cdot 97 \cdot 96}{5 \cdot 4 \cdot 3 \cdot 2 \cdot 1} = 75,287,520$$

Compare this result with that of Example 3.27, where we found that the number of permutations of five elements drawn from 100 was more than 9 billion. Because the order of the elements does not affect combinations, there are *fewer* combinations than permutations.

When working a probability problem, you should carefully examine the experiment to determine whether you can use one or more of the rules we have discussed in this section. Let's see how these rules (summarized in the next box) can help solve a probability problem.

---

## Summary of Counting Rules

1. *Multiplicative rule.* If you are drawing *one element from each of k sets of elements*, with the sizes of the sets $n_1, n_2, \ldots, n_k$, the number of different results is

   $$n_1 n_2 n_3 \cdots n_k$$

2. *Permutations rule.* If you are drawing *n elements from a set of N elements and arranging the n elements in a distinct order*, the number of different results is

   $$P_n^N = \frac{N!}{(N-n)!}$$

3. *Partitions rule.* If you are partitioning the *elements of a set of N elements into k groups consisting of* $n_1, n_2, \ldots, n_k$ *elements* $(n_1 + \cdots + n_k = N)$, the number of different results is

   $$\frac{N!}{n_1! n_2! \cdots n_k!}$$

4. *Combinations rule.* If you are drawing *n elements from a set of N elements without regard to the order of the n elements*, the number of different results is

   $$\binom{N}{n} = \frac{N!}{n!(N-n)!}$$

   [*Note:* The combinations rule is a special case of the partitions rule when $k = 2$.]

---

**EXAMPLE 3.33**  A consumer testing service is commissioned to rank the top three brands of laundry detergent. A total of 10 brands are to be included in the study.

a. In how many different ways can the consumer testing service arrive at the final ranking?

b. If the testing service can distinguish no difference among the brands and therefore arrives at the final ranking by random selection, what is the probability that company Z's brand is ranked first? In the top three?

### Solution

a. Since the testing service is drawing three elements (brands) from a set of 10 elements and arranging the three elements in a distinct order, we use the permutations rule to find the number of different results:

$$P_3^{10} = \frac{10!}{(10-3)!} = 10 \cdot 9 \cdot 8 = 720$$

b. The steps for calculating the probability of interest are as follows:

**Step 1** The experiment is to select and rank three brands of detergent from 10 brands.

**Step 2** There are too many sample points to list. However, we know from part **a** that there are 720 different outcomes (i.e., sample points) of this experiment.

**Step 3** If we assume the testing service determines the rankings at random, each of the 720 sample points should have an equal probability of occurrence. Thus,

$$P(\text{Each sample point}) = \frac{1}{720}$$

**Step 4** One event of interest to company Z is that their brand receives top ranking. We will call this event $A$. The list of sample points that result in the occurrence of event $A$ is long, but the *number* of sample points contained in event $A$ is determined by breaking event $A$ into two parts:

Thus, event $A$ can occur in 72 different ways.

Now define $B$ as the event that company Z's brand is ranked in the top three. Since event $B$ specifies only that brand Z appear in the top three, we repeat the calculations above, fixing brand Z in position 2 and then in position 3. We conclude that the number of sample points contained in event $B$ is $3(72) = 216$.

**Step 5** The final step is to calculate the probabilities of events $A$ and $B$. Since the 720 sample points are equally likely to occur, we find

$$P(A) = \frac{\text{Number of sample points in } A}{\text{Total number of sample points}} = \frac{72}{720} = \frac{1}{10}$$

Similarly,

$$P(B) = \frac{216}{720} = \frac{3}{10}$$

∎

**E X A M P L E   3 . 3 4**   Refer to Example 3.33. Suppose the consumer testing service is to choose the top three laundry detergents from the group of 10 *but is not to rank the three.*

a. In how many different ways can the testing service choose the three to be designated as top detergents?
b. Assuming that the testing service makes its choice by a random selection and that company X has two brands in the group of 10, what is the probability that exactly one of the company X brands is selected in the top three? At least one?

### Solution

a. The testing service is selecting three elements (brands) from a set of 10 elements *without regard to order*, so we can apply the combinations rule to determine the number of different results:

$$\binom{10}{3} = \frac{10!}{3!(10-3)!} = \frac{10 \cdot 9 \cdot 8}{3 \cdot 2 \cdot 1} = 120$$

b. The steps for calculating the probability of interest are as follows:

**Step 1**   The experiment is to select (*but not rank*) three brands from 10.
**Step 2**   There are 120 sample points for this experiment.
**Step 3**   Since the selection is made at random,

$$P(\text{Each sample point}) = \frac{1}{120}$$

**Step 4**   Define events $A$ and $B$ as follows:

$A$: {Exactly one company X brand is selected}

$B$: {At least one company X brand is selected}

Since each of the sample points is equally likely to occur, we need only know the number of sample points in $A$ and $B$ to determine their probabilities.

For event $A$ to occur, exactly one company X brand must be selected, along with two of the remaining eight brands. We thus break $A$ into two parts:

Note that the one company X brand can be selected in two ways, whereas the two other brands can be selected in 28 ways. (We use the combinations rule because the order of selection is not important.) Then, we use the multiplicative rule to combine one of the two ways to select a company X brand with one of the 28 ways to select two other brands, yielding a total of 56 sample points for event $A$.

To count the sample points contained in $B$, we first note that

$$B = A \cup C$$

where $A$ is defined as before and

   $C$: {Both company X brands are selected}

We have determined that $A$ contains 56 sample points. Using the same method for event $C$, we find:

Thus, event $C$ contains eight sample points and $P(C) = {}^{8}/_{120}$.

**Step 5**    Since all the sample points are equally likely,

$$P(A) = \frac{\text{Number of sample points in } A}{\text{Total number of sample points}} = \frac{56}{120} = \frac{7}{15}$$

$$P(B) = P(A \cup C) = P(A) + P(C) - P(A \cap C)$$

where $P(A \cap C) = 0$ because the testing service cannot select exactly one *and* exactly two of the company X brands; that is, $A$ and $C$ are mutually exclusive events. Thus,

$$P(B) = \frac{56}{120} + \frac{8}{120} = \frac{64}{120} = \frac{8}{15}$$

■

Learning how to decide whether a particular counting rule applies to an experiment takes patience and practice. If you wish to develop this skill, attempt to use the rules to solve the following exercises and especially the optional exercises (marked with an asterisk) at the end of this chapter. Proofs of the counting rules can be found in the text by W. Feller (1968, Chapter 3).

 EXERCISES 3.83–3.101

**Learning the Mechanics**

**3.83** Use the multiplicative rule to determine the number of sample points in the sample space corresponding to the experiment of tossing a coin the following number of times:

    **a.** 2 times
    **b.** 3 times
    **c.** 5 times
    **d.** $n$ times

**3.84** Find the numerical values of

a. $\binom{6}{3}$

b. $P_2^5$

c. $P_2^4$

d. $\binom{100}{98}$

e. $\binom{50}{50}$

f. $\binom{50}{0}$

g. $P_3^5$

h. $P_0^{10}$

**3.85** An experiment consists of choosing objects without regard to order. How many sample points are there if you choose the following?

a. 3 objects from 7

b. 2 objects from 6

c. 2 objects from 30

d. 8 objects from 10

e. $r$ objects from $q$

**3.86** Determine the number of sample points contained in the sample space when you toss the following:

a. 1 die

b. 2 dice

c. 4 dice

d. $n$ dice

**3.87** Suppose you are managing 10 employees, and you need to form three teams to work on different projects. Assume that each employee may serve on any team. In how many different ways can the teams be formed if the number of members on each project team are as follows:

a. 3, 3, 4

b. 2, 3, 5

c. 1, 4, 5

d. 2, 4, 4

## Applying the Concepts—Basic

**3.88** Flying into New York from a certain city, you can choose one of three airlines and can travel either first class or coach. How many travel options do you have?

**3.89** Suppose you are to choose a basketball team (five players) from eight available athletes.

a. How many ways can you choose a team (ignoring positions)?

b. How many ways can you choose a team composed of two guards, two forwards, and a center?

c. How many ways can you choose a team composed of point guard, shooting guard, power forward, small forward, and center?

**3.90** The 9-digit Zip code has become an integral part of the U.S. postal system.

a. How many different Zip codes are available for use by the postal service? (Assume any 9-digit number can be used as a Zip code.)

b. The first three digits of a Chicago Zip code are 606. If no other city in the United States has these first three digits as part of its Zip, how many different Zip codes may possibly exist in Chicago?

**3.91** A salesperson living in city A wishes to visit four cities, B, C, D, and E.

a. If the cities are all connected by airlines, how many different travel plans could be constructed to visit each city exactly once and then return home?

b. Suppose all cities are connected except that B and C are not directly connected. How many different flight plans would be available to the salesperson?

## Applying the Concepts—Intermediate

**3.92** Mathematicians at the University of Florida solved a 30-year-old math problem using the theory of partitions (*Explore*, Fall 2000). In math terminology, a partition is a representation of an integer as a sum of positive integers. (For example, the number 3 has three possible partitions: 3, 2 + 1, and 1 + 1 + 1.) The researchers solved the problem by using "colored partitions" of a number, where the colors correspond to the four suits —red hearts, red diamonds, black spades, and black clubs —in a standard 52-card bridge deck. Consider forming colored partitions of an integer.

a. How many colored partitions of the number 3 are possible? [*Hint:* One partition is 3♥; another is 2♦ + 1♣.]

b. How many colored partitions of the number 5 are possible?

**3.93** Successful companies such as Amway, Herbalife, and NuSkin employ multilevel marketing (MLM) schemes to distribute their products. In MLM, each distributor recruits several additional distributors to work at the first level. Each of these first-level distributors, in turn, recruits distributors to work at the second level. Distributors at any level continue to recruit distributors, forming additional levels. A distributor at any particular level makes a 5% commission on the sales of all distributors at lower levels in his or her "group" (*Success*, Dec. 1995).

a. Elaine is a distributor for a MLM company. Elaine recruits six distributors to work for her at the first level. Each of these distributors recruits an additional five distributors to work at the second level. How many distributors are under Elaine?

b. Refer to part **a**. Suppose each distributor at Elaine's second level recruits seven third-level distributors, and each of these third-level distributors recruits five level-four distributors. Now, how many distributors are under Elaine? (Your answer should provide insight into why some MLM distributors can earn $100,000 in commissions per month!)

**3.94** In the mid-1980s the state of Florida ran out of combinations of letters and numbers for its license plates. Then, each license plate contained three letters of the alphabet followed by three digits selected from the 10 digits 0, 1, 2, ..., 9.

a. How many different license plates did this system allow?

b. New Florida tags were obtained by reversing the procedure, starting with three digits followed by three letters. How many new tags did the new system provide?

c. Since the new tag numbers were added to the old numbers, what is the total number of licenses available for registration in Florida?

**3.95** The National Resident Matching Program (NRMP) is a service provided by the Association of American Medical Colleges to match graduating medical students with residency appointments at hospitals. After students and hospitals have evaluated each other, they submit rank-order lists of their preferences to the NRMP. Using a matching algorithm, the NRMP then generates final, nonnegotiable assignments of students to the residency programs of hospitals (*Academic Medicine*, June 1995). Assume that three graduating medical students (#1, #2, and #3) have applied for positions at three different hospitals (A, B, and C), where each hospital has one and only one resident opening.

a. How many different assignments of medical students to hospitals are possible? List them.

b. Suppose student #1 prefers hospital B. If the NRMP algorithm is entirely random, what is the probability that the student is assigned to hospital B?

**3.96** A company sells midlevel models of automobiles in five different styles. A buyer can get an automobile in one of eight colors with either standard or automatic transmission. Would it be reasonable to expect a dealer to stock at least one automobile in every style, color, and transmission combination? At a minimum, how many automobiles would the dealer have to stock?

**3.97** The Republican governor of a state with no income tax is appointing a committee of five members to consider changes in the income tax law. There are 15 state representatives available for appointment to the committee, seven Democrats and eight Republicans. Assume that the governor selects the committee of five members randomly from the 15 representatives.

a. In how many different ways can the committee members be selected?

b. What is the probability that no Democrat is appointed to the committee? If this were to occur, what would you conclude about the assumption that the appointments were made at random? Why?

c. What is the probability that the majority of the committee members are Republican? If this were to occur, would you have reason to doubt that the governor made the selections randomly? Why?

**3.98** Intercollegiate volleyball rules require that after the opposing team has lost its serve, each of the six members of the serving team must rotate into new positions on the court. Hence, each player must be able to play all six different positions. How many different team combinations by player and position are possible during a volleyball game? If players are initially assigned to the positions in a random manner, find the probability that the best server on the team is in the serving position.

**3.99** A college professor hands out a list of 10 questions, five of which will appear on the final examination for the

course. One of the students taking the course is pressed for time and can prepare for only seven of the 10 questions. Suppose the professor chooses the five questions at random from the 10.

a. What is the probability that the student will be prepared for all five questions that appear on the final examination?

b. What is the probability that the student will be prepared for fewer than three questions?

c. What is the probability that the student will be prepared for exactly four questions?

## Applying the Concepts—Advanced

**3.100** Consider five-card poker hands dealt from a standard 52-card bridge deck. Two important events are:

*A*: {You draw a flush}
*B*: {You draw a straight}

[*Note:* A *flush* consists of any five cards of the same suit. A *straight* consists of any five cards with values in sequence. In a straight, the cards may be of any suit and an ace may be considered as having a value of 1 or a value higher than a king.]

a. How many different five-card hands can be dealt from a 52-card bridge deck?

b. Find $P(A)$.

c. Find $P(B)$.

c. The event that both *A* and *B* occur, that is $A \cap B$, is called a *straight flush*. Find $P(A \cap B)$.

**3.101** Many states run lotteries to generate revenue for education, road construction, etc. A popular game is the Pick-6 Lotto. To play Pick-6 Lotto, you select six numbers of your choice from a field of numbers ranging from 1 to *N*, where *N* depends on which state's game you are playing. For example, Florida's Lotto game involves picking six numbers ranging from 1 to 53 (denoted 6/53). Delaware's Lotto is a 6/30 game and Pennsylvania's is a 6/40 game. The cost of a ticket is $1 and the payoff, if your six numbers match the winning numbers drawn at the end of each week, is $6 million or more, depending on the number of tickets purchased.

a. Consider Florida's 6/53 Lotto game. Determine the number of possible ways in which you can choose the six numbers from the 53 available. If you purchase a single $1 ticket, what is the probability that you win the grand prize (i.e., match all six numbers)?

b. Repeat part **a** for Delaware's 6/30 game.

c. Repeat part **a** for Pennsylvania's 6/40 game.

d. Since you can play any state's Lotto game, which of the three, Florida, Delaware, or Pennsylvania, would you choose to play? Why?

e. One strategy used to increase your odds of winning a Lotto is to employ a *wheeling system*. In a complete wheeling system, you select more than six numbers, say, seven, and play every combination of six of those seven numbers. Suppose you choose to "wheel" the following seven numbers in a 6/40 game: 2, 7, 18, 23, 30, 32, 39. How many tickets would you need to purchase to have every possible com-

bination of the seven numbers? List the six numbers on each of these tickets.

**f.** Refer to part **e**. What is the probability of winning the 6/40 Lotto when you wheel seven numbers? Does the strategy, in fact, increase your odds of winning?

---

## STATISTICS IN ACTION
### Game Show Strategy: To Switch or Not to Switch?

Marilyn vos Savant, who is listed in *Guinness Book of World Records Hall of Fame* for "Highest IQ," writes a weekly column in the Sunday newspaper supplement, *Parade Magazine*. Her column, "Ask Marilyn," is devoted to games of skill, puzzles, and mind-bending riddles. In one issue, vos Savant posed the following question:

Suppose you're on a game show, and you're given a choice of three doors. Behind one door is a car; behind the others, goats. You pick a door—say, #1—and the host, who knows what's behind the doors, opens another door—say #3—which has a goat. He then says to you, "Do you want to pick door #2?" Is it to your advantage to switch your choice?

Marilyn's answer: "Yes, you should switch. The first door has a $\frac{1}{3}$ chance of winning [the car], but the second has a $\frac{2}{3}$ chance [of winning the car]." Predictably, vos Savant's surprising answer elicited thousands of critical letters, many of them from Ph.D. mathematicians, who disagreed with her. Some of the more interesting and critical letters, which were printed in her next column (*Parade Magazine*, Feb. 24, 1991) are condensed below:

"May I suggest you obtain and refer to a standard textbook on probability before you try to answer a question of this type again?" (University of Florida)

"Your logic is in error, and I am sure you will receive many letters on this topic from high school and college students. Perhaps you should keep a few addresses for help with future columns." (Georgia State University)

"You are utterly incorrect about the game-show question, and I hope this controversy will call some public attention to the serious national crisis in mathematical education. If you can admit your error you will have contributed constructively toward the solution of a deplorable situation. How many irate mathematicians are needed to get you to change your mind?" (Georgetown University)

"I am in shock that after being corrected by at least three mathematicians, you still do not see your mistake." (Dickinson State University)

"You are the goat!" (Western State University)

"You're wrong, but look on the positive side. If all the Ph.D.'s were wrong, the country would be in serious trouble." (U.S. Army Research Institute)

The logic employed by those who disagree with vos Savant is as follows: Once the host shows you door #3 (a goat), only two doors remain. The probability of the car being behind door #1 (your door) is $\frac{1}{2}$; similarly, the probability is $\frac{1}{2}$ for door #2. Therefore, in the long run (i.e., over a long series of trials) it doesn't matter whether you switch to door #2 or keep door #1. Approximately 50% of the time you'll win a car, and 50% of the time you'll get the goat.

Who is correct, the Ph.D.'s or Marilyn? By answering the following series of questions, you'll arrive at the correct solution.

### Focus

**a.** Before taping of the game show, the host randomly decides the door behind which to put the car; then the goats go behind the remaining two doors. List the sample points for this experiment.

**b.** Suppose you choose at random door #1. Now, form a new sample space for this conditional event as follows: For each sample point in part **a**, eliminate one of the remaining two doors that hides a goat. (This is the door that the host shows the contestant—always a goat.)

**c.** Refer to the altered sample points in part **b**. Assume your strategy is to keep door #1. Count the number of sample points for which this is a "winning" strategy (i.e., you win the car). Assuming equally likely sample points, what is the probability that you win the car?

**d.** Repeat part **c**, but assume your strategy is to always switch doors.

**e.** Based on the probabilities of parts **c** and **d**, is it to your advantage to switch your choice?

**f.** Repeat parts **b–e**, but assume you select door #2 at random.

**g.** Repeat parts **b–e**, but assume you select door #3 at random.

**h.** Demonstrate that your choice of doors does not impact your *long-run* strategy.

## QUICK REVIEW

### Key Terms

*Note: Starred (*) terms are from the optional section in this chapter.*

Additive rule of probability   126
Combinations rule*   158
Combinatorial mathematics   150
Complementary event   123
Compound event   120
Conditional probability   131
Counting rule*   154
Dependent events   142
Event   114
Experiment   108
Independent events   142

Intersection   120
Multiplicative counting rule*   155
Multiplicative rule
  of probability   139
Mutually exclusive events   127
Odds   120
Partitions rule*   157
Permutation*   156
Permutations rule*   156
Probability rules for simple
  points   112

Random number
  generator   150
Random number
  table   152
Random sample   150
Sample point   109
Sample space   109
Tree diagram   141
Unconditional probabilities   131
Union   120
Venn diagram   110

### Key Formulas

| | |
|---|---|
| $P(A) + P(A^c) = 1$ | Complementary events   124 |
| $P(A \cup B) = P(A) + P(B) - P(A \cap B)$ | Additive rule   126 |
| $P(A \cap B) = 0$ | Mutually exclusive events   127 |
| $P(A \cup B) = P(A) + P(B)$ | Additive rule for mutually exclusive events   127 |
| $P(A|B) = \dfrac{P(A \cap B)}{P(B)}$ | Conditional probability   132 |
| $P(A \cap B) = P(A)P(B|A) = P(B)P(A|B)$ | Multiplicative rule   139 |
| $P(A|B) = P(A)$ | Independent events   142 |
| $P(A \cap B) = P(A)P(B)$ | Multiplicative rule for independent events   144 |
| $\dbinom{N}{n} = \dfrac{N!}{n!(N-n)!}$ where $N! = N(N-1)(N-2)\cdots(2)(1)$ | Combinatorial rule   158 |

[*Note*: For a summary of formulas for counting rules, see page 159.]

### Language Lab

| Symbol | Pronunciation | Description |
|---|---|---|
| $S$ | | Sample space |
| $S$: $\{1,2,3,4,5\}$ | | Set of sample points, 1,2,3,4,5, in sample space |
| $A$: $\{1,2\}$ | | Set of sample points, 1,2, in event $A$ |
| $P(A)$ | Probability of $A$ | Probability that event $A$ occurs |
| $A \cup B$ | $A$ union $B$ | Union of events $A$ and $B$ (either $A$ or $B$ or both occur) |
| $A \cap B$ | $A$ intersect $B$ | Intersection of events $A$ and $B$ (both $A$ and $B$ occur) |
| $A^c$ | $A$ complement | Complement of event $A$ (the event that $A$ does not occur) |
| $P(A|B)$ | Probability of $A$ given $B$ | Conditional probability that event $A$ occurs given that event $B$ occurs |
| $\dbinom{N}{n}$ | $N$ choose $n$ | Number of combinations of $N$ elements taken $n$ at a time |
| $N!$ | $N$ factorial | Multiply $N(N-1)(N-2)\cdots(2)(1)$ |
| $P_n^N$ | $N$ permute $n$ | Number of different permutations of $N$ elements taken $n$ at a time |

## SUPPLEMENTARY EXERCISES 3.102–3.135

*Note: Starred (\*) exercises refer to the optional section in this chapter.*

### Learning the Mechanics

**3.102** The accompanying Venn diagram illustrates a sample space containing six sample points and three events, $A$, $B$, and $C$. The probabilities of the sample points are: $P(1) = .3$, $P(2) = .2$, $P(3) = .1$, $P(4) = .1$, $P(5) = .1$, and $P(6) = .2$.

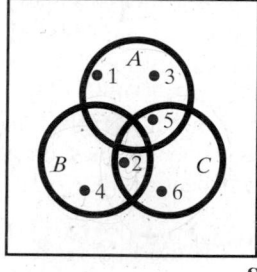

**a.** Find $P(A \cap B)$, $P(B \cap C)$, $P(A \cup C)$, $P(A \cup B \cup C)$, $P(B^c)$, $P(A^c \cap B)$, $P(B|C)$, and $P(B|A)$.

**b.** Are $A$ and $B$ independent? Mutually exclusive? Why?

**c.** Are $B$ and $C$ independent? Mutually exclusive? Why?

**3.103** A fair die is tossed and the up face is noted. If the number is even, the die is tossed again; if the number is odd, a fair coin is tossed. Define the events:

$A$: {A head appears on the coin}
$B$: {The die is tossed only one time}

**a.** List the sample points in the sample space.

**b.** Give the probability for each of the sample points.

**c.** Find $P(A)$ and $P(B)$.

**d.** Identify the sample points in $A^c$, $B^c$, $A \cap B$, and $A \cup B$.

**e.** Find $P(A^c)$, $P(B^c)$, $P(A \cap B)$, $P(A \cup B)$, $P(A|B)$, and $P(B|A)$.

**f.** Are $A$ and $B$ mutually exclusive events? Independent events? Why?

**3.104** A balanced die is thrown once. If a 4 appears, a ball is drawn from urn 1; otherwise, a ball is drawn from urn 2. Urn 1 contains four red, three white, and three black balls. Urn 2 contains six red and four white balls.

**a.** Find the probability that a red ball is drawn.

**b.** Find the probability that urn 1 was used given that a red ball was drawn.

**3.105** Which of the following pairs of events are mutually exclusive? Justify your response.

**a.** St. Louis Cardinals win the World Series. Mark McGwire, Cardinals first baseman, hits 70 home runs.

**b.** An IBM notebook computer is purchased. An Apple notebook computer is purchased.

**c.** Subject A in a psychology experiment responds to a stimulus within 5 seconds. Subject A in a psychology experiment records the fastest response to a stimulus (2.3 seconds).

**3.106** A random sample of five students is to be selected from 50 sociology majors for participation in a special program.

**a.** In how many different ways can the sample be drawn?

**b.** Show how the random number table, Table I of Appendix A, can be used to select the sample of students.

**3.107** Two events, $A$ and $B$, are independent, with $P(A) = .3$ and $P(B) = .1$.

**a.** Are $A$ and $B$ mutually exclusive? Why?

**b.** Find $P(A|B)$ and $P(B|A)$.

**c.** Find $P(A \cup B)$.

**\*3.108** Find the numerical value of:

**a.** $6!$

**b.** $\binom{10}{9}$

**c.** $\binom{10}{1}$

**d.** $P_2^6$

**e.** $\binom{6}{3}$

**f.** $0!$

**g.** $P_4^{10}$

**h.** $P_2^{50}$

### Applying the Concepts—Basic

**3.109** Refer to the National Highway Traffic Safety Administration (NHTSA) crash tests of new car models, Exercise 2.125 (p. 98). Recall that the NHTSA has developed a "star" scoring system, with results ranging from one star (\*) to five stars (\*\*\*\*\*). The more stars in the rating, the better the level of crash protection in a head-on collision. A summary of the driver-side star ratings for 98 cars is reproduced in the accompanying MINITAB printout. Assume that one of the 98 cars is selected at random. State whether each of the following is true or false.

**a.** The probability that the car has a rating of two stars is 4.

**b.** The probability that the car has a rating of four or five stars is .7857.

**c.** The probability that the car has a rating of one star is 0.

**d.** The car has a better chance of having a 2-star rating than of having a 5-star rating.

**MINITAB Output for Exercise 3.109**

Tally for Discrete Variables: DrivStar

| DrivStar | Count | Percent |
|---|---|---|
| 2 | 4 | 4.08 |
| 3 | 17 | 17.35 |
| 4 | 59 | 60.20 |
| 5 | 18 | 18.37 |
| N= | 98 | |

**3.110** Refer to the *Journal of Agricultural, Biological, and Environmental Statistics* (Sept. 2000) study of the impact of the *Exxon Valdez* tanker oil spill on the seabird population in Prince William Sound, Alaska, Exercise 2.135 (p. 103). Recall that data were collected on 96 shoreline locations (called transects), and it was determined whether or not the transect was in an oiled area. The data are stored in the file called EVOS. From the data, estimate the probability that a randomly selected transect in Prince William Sound is contaminated with oil.

EVOS

**3.111** According to a national survey conducted for CACI Marketing Systems, 25% of American adults smoke cigarettes. Of these smokers, 13% attempted (but failed) to quit smoking during the past year. Define the following events:

$A$: {An American adult smokes}
$B$: {A smoker attempted to quit smoking last year}

**a.** Find $P(A)$.
**b.** Find $P(B|A)$.
**c.** Find $P(A^c)$. State this probability in the words of the problem.
**d.** Find $P(A \cap B)$. State this probability in the words of the problem.

**3.112** The *American Journal of Public Health* (July 1995) published a study on unintentional carbon monoxide (CO) poisoning of Colorado residents. A total of 981 cases of CO poisoning were reported during a six-year period. Each case was classified as fatal or nonfatal and by source of exposure. The number of cases occurring in each of the categories is shown in the accompanying table. Assume that one of the 981 cases of unintentional CO poisoning is randomly selected.

| Source of Exposure | Fatal | Nonfatal | Total |
|---|---|---|---|
| Fire | 63 | 53 | 116 |
| Auto exhaust | 60 | 178 | 238 |
| Furnace | 18 | 345 | 363 |
| Kerosene or spaceheater | 9 | 18 | 27 |
| Appliance | 9 | 63 | 72 |
| Other gas-powered motor | 3 | 73 | 76 |
| Fireplace | 0 | 16 | 16 |
| Other | 3 | 19 | 22 |
| Unknown | 9 | 42 | 51 |
| **Total** | 174 | 807 | 981 |

*Source:* Cook, M. C., Simon P. A., and Hoffman, R. E., "Unintentional carbon monoxide poisoning in Colorado, 1986 through 1991." *American Journal of Public Health,* Vol. 85, No. 7, July 1995, p. 989 (Table 1).

**a.** List all sample points for this experiment.
**b.** What is the set of all sample points called?
**c.** Let $A$ be the event that the CO poisoning is caused by fire. Find $P(A)$.
**d.** Let $B$ be the event that the CO poisoning is fatal. Find $P(B)$.
**e.** Let $C$ be the event that the CO poisoning is caused by auto exhaust. Find $P(C)$.

**f.** Let $D$ be the event that the CO poisoning is caused by auto exhaust and is fatal. Find $P(D)$.
**g.** Let $E$ be the event that the CO poisoning is caused by fire but is nonfatal. Find $P(E)$.
**h.** Given that the source of the poisoning is fire, what is the probability that the case is fatal?
**i.** Given that the case is nonfatal, what is the probability that it is caused by auto exhaust?
**j.** If the case is fatal, what is the probability that the source is unknown?
**k.** If the case is nonfatal, what is the probability that the source is not fire or a fireplace?

**3.113** Refer to the *Health Education Journal* (Sept. 1994) study of media coverage of stories involving mental illness in Scotland, Exercise 2.128 (p. 99). The media coverage of each of 562 stories was classified by type, with the results shown in the table. Assume that one of the 562 stories on mental illness is selected and the type of media coverage is noted.

| Media Coverage | Number of Items |
|---|---|
| Violence to others | 373 |
| Sympathetic | 102 |
| Harm to self | 71 |
| Comic images | 12 |
| Criticism of definitions | 4 |
| **Total** | 562 |

*Source:* Philo, G. *et al.* "The impact of the mass media on public images of mental illness: Media content and audience belief." *Health Education Journal,* Vol. 53, No. 3, Sept. 1994, p. 274 (Table 1).

**a.** List the sample points for this experiment.
**b.** Explain why the sample points, part **a**, are not equally likely to occur.
**c.** Assign reasonable probabilities to the sample points.
**d.** Find the probability that a randomly selected story on mental health is portrayed sympathetically or comically by the Scottish media.

**3.114** The characteristics of families with young children were examined in *Children and Youth Services Review* (Vol. 17, 1995). Using data obtained from the National Child Care Survey, the income distribution and employment status of these families was summarized as shown in the table on p. 169.

**a.** Find the probability that a randomly selected family with young children has an income above the poverty line, but less than $25,000.
**b.** Find the probability that a randomly selected family with young children has unemployed parents or no parents.
**c.** Find the probability that a randomly selected family with young children has an income below the poverty line.

**3.115** An entomologist is studying the effect of a chemical sex attractant (pheromone) on insects. Several insects

| Income Characteristic | Percentage |
|---|---|
| No parent | 1 |
| Below poverty line; not employed | 7 |
| Below poverty line; employed | 7 |
| Above poverty line, but less than $25,000; not employed | 2 |
| Above poverty line, but less than $25,000; employed | 22 |
| $25,000 or more | 61 |
| **Total** | 100 |

are released at a site equidistant from the pheromone under study and a control substance. If the pheromone has an effect, more insects will travel toward it rather than toward the control. Otherwise, the insects are equally likely to travel in either direction. Suppose the pheromone under study has no effect, so that it is equally likely that an insect will move toward either the pheromone or the control. If five insects are released, what is the probability that

**a.** All five travel toward the pheromone?

**b.** Exactly four?

**c.** What inference would you make if the event in part **b** actually occurs? Explain.

## Applying the Concepts—Intermediate

**3.116** Psychologists tend to believe that there is a relationship between aggressiveness and order of birth. To test this belief, a psychologist chose 500 elementary school students at random and administered each a test designed to measure the student's aggressiveness. Each student was classified according to one of four categories. The percentages of students falling in the four categories are shown here.

| | **Firstborn** | **Not Firstborn** |
|---|---|---|
| **Aggressive** | 15% | 15% |
| **Not Aggressive** | 25% | 45% |

**a.** If one student is chosen at random from the 500, what is the probability that the student is firstborn?

**b.** What is the probability that the student is aggressive?

**c.** What is the probability that the student is aggressive, given the student was firstborn?

**d.** If

$A$: {Student chosen is aggressive}
$B$: {Student chosen is firstborn}

are $A$ and $B$ independent? Explain.

**3.117** In college basketball games a player may be afforded the opportunity to shoot two consecutive foul shots (free throws).

**a.** Suppose a player who makes (i.e., scores on) 80% of his foul shots has been awarded two free throws. If the two throws are considered independent, what is the probability that the player makes both shots? Exactly one? Neither shot?

**b.** Suppose a player who makes 80% of his first attempted foul shots has been awarded two free throws, and the outcome on the second shot is dependent on the outcome of the first shot. In fact, if this player makes the first shot, he makes 90% of the second shots; and if he misses the first shot, he makes 70% of the second shots. In this case, what is the probability that the player makes on both shots? Exactly one? Neither shot?

**c.** In parts **a** and **b**, we considered two ways of *modeling* the probability a basketball player makes two consecutive foul shots. Which model do you think is a more realistic attempt to explain the outcome of shooting foul shots; that is, do you think two consecutive foul shots are independent or dependent? Explain.

**3.118** The probability that an Avon salesperson sells beauty products to a prospective customer on the first visit to the customer is .4. If the salesperson fails to make the sale on the first visit, the probability that the sale will be made on the second visit is .65. The salesperson never visits a prospective customer more than twice. What is the probability that the salesperson will make a sale to a particular customer?

**3.119** Sandoz, a pharmaceutical firm, reports that kidney transplant patients who receive the drug cyclosporine have an 80% chance of surviving the first year. Suppose a hospital performs four kidney transplants, and all patients receive the drug cyclosporine. Assuming one patient's survival is independent of another's, what is the probability that

**a.** All four patients are alive at the end of 1 year?

**b.** None of the four patients is alive at the end of 1 year?

**c.** At least one of the patients is alive at the end of 1 year?

**3.120** The system shown in the schematic (lower left) operates properly only if all three components operate properly. (The three components are said to operate *in series*.) The probability of failure for each component is listed in the table below. Assume the components operate independently of each other.

**Figure for Exercise 3.120**

A System Comprised of Three Components in Series

Input ———→ (#1) ———→ (#2) ———→ (#3) ———→ ► Output

| Component | Probability of Failure |
|---|---|
| 1 | .12 |
| 2 | .09 |
| 3 | .11 |

**Figure for Exercise 3.121**

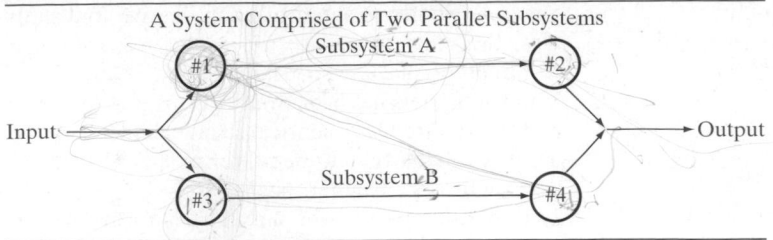

A System Comprised of Two Parallel Subsystems

a. Find the probability that the system operates properly.

b. What is the probability that at least one of the components will fail and therefore that the system will fail?

**3.121** The schematic above is a representation of a system that comprises two subsystems said to operate *in parallel*. Each subsystem has two components that operate in series (refer to Exercise 3.120). The system will operate properly as long as at least one of the subsystems functions properly. The probability of failure for each component in the system is .1. Assume the components operate independently of each other.

a. Find the probability that the system operates properly.

b. Find the probability that exactly one subsystem fails.

c. Find the probability that the system fails to operate properly.

d. How many parallel subsystems like the two shown here would be required to guarantee that the system would operate properly at least 99% of the time?

**3.122** A small brewery has two bottling machines. Machine A produces 75% of the bottles and machine B produces 25%. One out of every 20 bottles filled by A is rejected for some reason, while one out of every 30 bottles from B is rejected.

a. What proportion of bottles is rejected?

b. What is the probability that a randomly selected bottle comes from machine A, given that it is accepted?

**3.123** Refer to *The Howard Journal of Criminal Justice* study of the relationship between the race of the attacker, the race of the victim, and the degree of injury sustained in reported crimes, Exercise 3.53 (p. 137). The information is reproduced below.

a. What is the probability that a randomly selected reported crime involved a nonwhite attacker?

b. What is the probability that a randomly selected reported crime involved no injury?

c. What is the probability that a randomly selected reported crime involved both a nonwhite attacker and no injury?

d. Are the race of the attacker and the degree of injury independent?

*****3.124** Suppose a manufacturer of big-screen television sets is prepared to send 20 new sets to three retail dealers. Dealer A is to get ten of the sets, dealer B six of the sets, and dealer C four of the sets.

a. In how many different ways can the 20 sets be divided among the dealers in the prescribed numbers?

b. Assuming the sets are randomly divided among the dealers and there are three defective sets among the 20, what is the probability that dealer A gets all three defective sets?

c. Making the assumptions given in part **b**, what is the probability that dealer A gets two defective sets and dealer B the other defective set?

**3.125** A clinical psychologist is asked to view tapes in which each of six experimental subjects is discussing his or her recent dreams. Three of the six subjects have previously been classified as "high-anxiety" individuals and the other three as "low-anxiety." The psychologist is told only that there are three of each type and is asked to select the three high-anxiety subjects.

a. List all possible outcomes (sample points) for this experiment.

b. Assuming that the psychologist guesses at the classifications of the subjects, assign probabilities to the sample points.

c. Find the probability that the psychologist guesses all classifications correctly.

d. Find the probability that the psychologist guesses at least two of the three high-anxiety subjects correctly.

| | Attacker/Victim | | | | |
|---|---|---|---|---|---|
| **Degree of Injury** | **White/White** | **White/Nonwhite** | **Nonwhite/Nonwhite** | **Nonwhite/White** | **Totals** |
| Fatal | 183 | 18 | 18 | 18 | 237 |
| Serious | 580 | 49 | 111 | 141 | 881 |
| Slight | 4,336 | 440 | 594 | 1,656 | 7,026 |
| None | 1,422 | 136 | 214 | 1,801 | 3,573 |
| **Totals** | 6,521 | 643 | 937 | 3,616 | 11,717 |

*Source:* Shah, R., and Pease, K. "Crime, race and reporting to the police." *The Howard Journal of Criminal Justice*, Volume 31, No. 3, Aug. 1992.

**\*3.126** Refer to Exercise 3.125. Suppose the clinical psychologist is given tapes of 20 subjects and is told only that 10 are high-anxiety individuals and 10 are low-anxiety individuals. The psychologist is asked to select the 10 high anxiety subjects.

**a.** Using the appropriate counting rule from Section 3.8, count the number of sample points for this experiment.

**b.** Assuming that the psychologist is guessing, assign probabilities to each of the sample points.

**c.** Find the probability that the psychologist guesses all classifications correctly.

**d.** Find the probability that the psychologist guesses at least nine of the 10 high-anxiety subjects correctly.

**\*3.127** Suppose there are 500 applicants for five equivalent positions at a factory and the company is able to narrow the field to 30 equally qualified applicants. Seven of the finalists are minority candidates. Assume that the five who are chosen are selected at random from this final group of 30.

**a.** In how many different ways can the selection be made?

**b.** What is the probability that none of the minority candidates is hired?

**c.** What is the probability that no more than one minority candidate is hired?

## Applying the Concepts—Advanced

**3.128** Pneumovax is an antipneumonia vaccine designed especially for elderly or debilitated patients who are usually the most vulnerable to bacterial pneumonia. The vaccine is 90% effective in stimulating the production of antibodies to pneumonia-producing bacteria (i.e., it is 90% successful in preventing a person exposed to pneumonia-producing bacteria from acquiring the disease). Suppose the probability that an elderly or debilitated person is exposed to these bacteria is .40 (whether inoculated or not) and, after being exposed, the probability that the person will contract bacterial pneumonia if not inoculated with the vaccine is .95. Find the probability that an elderly or debilitated person inoculated with this new vaccine acquires pneumonia. What is the probability if this person has not been inoculated?

**3.129** In the game of Parcheesi each player rolls a pair of dice on each turn. In order to begin the game, you must throw a 5 on at least one of the dice, or a total of 5 on the two dice. What is the probability that you can begin the game on your first turn? The second turn? The third turn? The $n$th turn?

**3.130** Blackjack, a favorite game of gamblers, is played by a dealer and at least one opponent. At the outset of the game, two cards of a 52-card bridge deck are dealt to the player and two cards to the dealer. Drawing an ace and a face card is called *blackjack*. If the dealer does not draw a blackjack and the player does, the player wins. If both the dealer and player draw blackjack, a "push" (i.e., a tie) occurs.

**a.** What is the probability that the dealer will draw a blackjack?

**b.** What is the probability that the player wins with a blackjack?

**3.131** Seventy-five percent of all women who submit to pregnancy tests are really pregnant. A certain pregnancy test gives a *false positive* result with probability .02 and a *valid positive result* with probability .99. If a particular woman's test is positive, what is the probability that she really is pregnant? [*Hint:* If $A$ is the event that a woman is pregnant and $B$ is the event that the pregnancy test is positive, then $B$ is the union of the two mutually exclusive events $A \cap B$ and $A^c \cap B$. Also, the probability of a false positive result may be written as $P(B|A^c) = .02$.]

**3.132** Children who develop unexpected difficulties with acquisition of spoken language are often diagnosed with specific language impairment (SLI). A study published in the *Journal of Speech, Language, and Hearing Research* (Dec. 1997) investigated the incidence of SLI in kindergarten children. As an initial screen, each in a national sample of over 7,000 children was given a test for language performance. The percentages of children who passed and failed the screen were 73.8% and 26.2%, respectively. All children who failed the screen were tested clinically for SLI. About one-third of those who passed the screen were randomly selected and also tested for SLI. The percentage of children diagnosed with SLI in the "failed screen" group was 20.5%; the percentage diagnosed with SLI in the "pass screen" group was 2.8%.

**a.** For this problem, let "pass" represent a child who passed the language performance screen, "fail" represent a child who failed the screen, and "SLI" represent a child diagnosed with SLI. Now find each of the following probabilities: $P(\text{Pass})$, $P(\text{Fail})$, $P(\text{SLI}|\text{Pass})$, and $P(\text{SLI}|\text{Fail})$.

**b.** Use the probabilities, part **a**, to find $P(\text{Pass} \cap \text{SLI})$ and $P(\text{Fail} \cap \text{SLI})$. What probability law did you use to calculate these probabilities?

**c.** Use the probabilities, part **b**, to find $P(\text{SLI})$. What probability law did you use to calculate this probability?

**\*3.133** According to a morning news program, a very rare event recently occurred in Dubuque, Iowa. Each of four women playing bridge was astounded to note that she had been dealt a perfect bridge hand. That is, one woman was dealt all 13 spades, another all 13 hearts, another all the diamonds, and another all the clubs. What is the probability of this rare event?

**3.134** The National Center for Health Statistics compiles data on suicide rates by race and gender. These data were used to estimate the probability of committing suicide during a lifetime in *Geriatric Psychiatry* (Vol. 27, 1994). The probabilities of lifetime suicide for four race/gender cohorts are given in the next table. For example, a black female selected at random has a $^1/_{535}$ chance of committing suicide during her lifetime.

| Cohort | Probability of Suicide |
|---|---|
| Black female | $^1/_{535}$ |
| White female | $^1/_{196}$ |
| Black male | $^1/_{133}$ |
| White male | $^1/_{66}$ |

**a.** Are the probabilities listed in the table conditional or unconditional probabilities? Explain.

**b.** Consider the following statement: "The probability that a randomly selected male commits suicide is obtained by summing the probabilities for black male and white male: $^1/_{133} + ^1/_{66}$." Defend or refute this argument.

**3.135** A version of the dice game "craps" is played in the following manner: A player starts by rolling two dice. If the result is a 7 or 11, the player wins. For most other sums appearing on the dice, the player continues to roll the dice until that sums recurs (in which case the player loses). But if on any roll the outcome is 2 or 3 (called "craps"), the game is over, and the player loses.

**a.** What is the probability that a player wins the game on the first roll of the dice? (Assume the dice are balanced.)

**b.** What is the probability that a player loses the game on the first roll of the dice?

**c.** If the player throws a total of 4 on the first roll, what is the probability that the game ends on the next roll?

## STUDENT PROJECTS

Obtain a standard deck of 52 playing cards (the kind commonly used for bridge, poker, or solitaire). An experiment will consist of drawing one card at random from the deck of cards and recording which card was observed. This random drawing will be simulated by shuffling the deck thoroughly and observing the top card. Consider the following two events:

*A:* {Card observed is a heart}
*B:* {Card observed is an ace, king, queen, or jack}

a. Find $P(A)$, $P(B)$, $P(A \cap B)$, and $P(A \cup B)$.

b. Conduct the experiment 10 times and record the observed card each time. Be sure to return the observed card each time and thoroughly shuffle the deck before making the draw. After you've observed 10 cards, calculate the proportion of observations that satisfy event *A*, event *B*, event $A \cap B$, and event $A \cup B$. Compare the observed proportions with the true probabilities calculated in part **a**.

c. Conduct the experiment 40 more times to obtain a total of 50 observed cards. Now calculate the proportion of observations that satisfy event *A*, event *B*, event $A \cap B$, and event $A \cup B$. Compare these proportions with those found in part **b** and the true probabilities found in part **a**.

d. Conduct the experiment 50 more times to obtain a total of 100 observations. Compare the observed proportions for the 100 trials with those found previously. What comments do you have concerning the different proportions found in parts **b**, **c**, and **d** as compared to the true probabilities found in part **a**? How do you think the observed proportions and true probabilities would compare if the experiment were conducted 1,000 times? 1 million times?

## REFERENCES

Epstein, R. A. *The Theory of Gambling and Statistical Logic*, rev. ed. New York: Academic Press, 1977.

Feller, W. *An Introduction to Probability Theory and Its Applications*, 3d ed., Vol. 1. New York: Wiley, 1968.

Lindley, D. V. *Making Decisions*, 2d ed. London: Wiley, 1985.

Mosteller, F., Rourke, R., and Thomas, G. *Probability with Statistical Applications*, 2d ed. Reading, Mass.: Addison-Wesley, 1970.

Parzen, E. *Modern Probability Theory and Its Applications*. New York: Wiley, 1960.

Wackerly, D., Mendenhall, W., and Scheaffer, R. *Mathematical Statistics with Applications*, 6th ed. Boston: Duxbury, 2002.

Williams, B. *A Sampler on Sampling*. New York: Wiley, 1978.

Winkler, R. L. *An Introduction to Bayesian Inference and Decision*. New York: Holt, Rinehart and Winston, 1972.

Wright, G., and Ayton, P., eds. *Subjective Probability*. New York: Wiley, 1994.

# Discrete Random Variables

## Contents

## Statistics in Action

Probability in a Reverse Cocaine Sting

### 🛎 Where We've Been

We saw by illustration in Chapter 3 how probability would be used to make an inference about a population from data contained in an observed sample. We also noted that probability would be used to measure the reliability of the inference.

### ☞ Where We're Going

Most of the experimental events we encountered in Chapter 3 were events described in words and denoted by capital letters. In real life, most sample observations are numerical—in other words, they are numerical data. In this chapter, we learn that data are observed values of random variables. We study three important random variables and learn how to find the probabilities of specific numerical outcomes.

You may have noticed that many of the examples of experiments in Chapter 3 generated quantitative (numerical) observations. The unemployment rate, the percentage of voters favoring a particular candidate, the cost of textbooks for a school term, and the amount of pesticide in the discharge waters of a chemical plant are all examples of numerical measurements of some phenomenon. Thus, most experiments have sample points that correspond to values of some numerical variable.

**FIGURE 4.1**

*Venn Diagram for Coin Tossing Experiment*

```
┌─────────────────────────────┐
│  HH            HT           │
│  ●             ●            │
│  (2)           (1)          │
│                             │
│  TT            TH           │
│  ●             ●            │
│  (0)           (1)          │
└─────────────────────────────┘
                            S
```

To illustrate, consider the coin-tossing experiment of Chapter 3. Figure 4.1 is a Venn diagram showing the sample points when two coins are tossed and the up faces (heads or tails) of the coins are observed. One possible numerical outcome is the total number of heads observed. These values (0, 1, or 2) are shown in parentheses on the Venn diagram, one numerical value associated with each sample point. In the jargon of probability, the variable "total number of heads observed in two tosses of a coin" is called a *random variable*.

**DEFINITION 4.1**

A **random variable** is a variable that assumes numerical values associated with the random outcomes of an experiment, where one (and only one) numerical value is assigned to each sample point.

The term *random variable* is more meaningful than the term *variable* alone because the adjective *random* indicates that the coin-tossing experiment may result in one of the several possible values of the variable—0, 1, and 2—according to the *random* outcome of the experiment, *HH*, *HT*, *TH*, and *TT*. Similarly, if the experiment is to count the number of customers who use the drive-up window of a bank each day, the random variable (the number of customers) will vary from day to day, partly because of the random phenomena that influence whether customers use the drive-up window. Thus, the possible values of this random variable range from 0 to the maximum number of customers the window could possibly serve in a day.

We define two different types of random variables, *discrete* and *continuous*, in Section 4.1. Then we spend the remainder of this chapter discussing specific types of discrete random variables and the aspects that make them important to the statistician. We discuss continuous random variables in Chapter 5.

## 4.1 TWO TYPES OF RANDOM VARIABLES

Recall that the sample point probabilities corresponding to an experiment must sum to 1. Dividing one unit of probability among the sample points in a sample space and consequently assigning probabilities to the values of a random variable is not always as easy as the examples in Chapter 3 might lead you to believe. If the number of sample points can be completely listed, the job is straightforward. But if the experiment results in an infinite number of sample points that are impossible to list, the task of assigning probabilities to the sample points is impossible without the aid of a probability model. The next three examples demonstrate the need for different probability models depending on the number of values that a random variable can assume.

---

**EXAMPLE 4.1**

A panel of 10 experts for the *Wine Spectator* (a national publication) is asked to taste a new white wine and assign a rating of 0, 1, 2, or 3. A score is then obtained by adding together the ratings of the 10 experts. How many values can this random variable assume?

### Solution

A sample point is a sequence of 10 numbers associated with the rating of each expert. For example, one sample point is

$$\{1, 0, 0, 1, 2, 0, 0, 3, 1, 0\}.$$

The random variable assigns a score to each one of these sample points by adding the 10 numbers together. Thus, the smallest score is 0 (if all 10 ratings are 0) and the largest score is 30 (if all 10 ratings are 3). Since every integer between 0 and 30 is a possible score, the random variable denoted by the symbol $x$ can assume 31 values. Note that the value of the random variable for the sample point above is $x = 8$.*

This is an example of a *discrete random variable*, since there is a finite number of distinct possible values. Whenever all the possible values a random variable can assume can be listed (or *counted*), the random variable is *discrete*. ∎

---

**EXAMPLE 4.2**

Suppose the Environmental Protection Agency (EPA) takes readings once a month on the amount of pesticide in the discharge water of a chemical company. If the amount of pesticide exceeds the maximum level set by the EPA, the company is forced to take corrective action and may be subject to penalty. Consider the following random variable:

Number, $x$, of months before the company's discharge exceeds the EPA's maximum level

What values can $x$ assume?

### Solution

The company's discharge of pesticide may exceed the maximum allowable level on the first month of testing, the second month of testing, etc. It is possible that the company's discharge will *never* exceed the maximum level. Thus, the set of possible values for the number of months until the level is first exceeded is the set of all positive integers 1, 2, 3, 4, ... .

If we can list the values of a random variable $x$, even though the list is never-ending, we call the list *countable* and the corresponding random variable *discrete*. Thus, the number of months until the company's discharge first exceeds the limit is a *discrete random variable*. ∎

---

*The standard mathematical convention is to use a capital letter (e.g., $X$) to denote the theoretical random variable. The possible values (or realizations) of the random variable are typically denoted with a lowercase letter (e.g., $x$). Thus, in Example 4.1, the random variable $X$ can take on the values $x = 0, 1, 2, \ldots, 30$. Since this notation can be confusing for introductory statistics students, we simplify the notation by using the lowercase $x$ to represent the random variable throughout.

**EXAMPLE 4.3**

Refer to Example 4.2. A second random variable of interest is the amount $x$ of pesticide (in milligrams per liter) found in the monthly sample of discharge waters from the chemical company. What values can this random variable assume?

**Solution**

Unlike the *number* of months before the company's discharge exceeds the EPA's maximum level, the set of all possible values for the *amount* of discharge *cannot be listed*—i.e., is not countable. The possible values for the amount $x$ of pesticide would correspond to the points on the interval between 0 and the largest possible value the amount of the discharge could attain, the maximum number of milligrams that could occupy 1 liter of volume. (Practically, the interval would be much smaller, say, between 0 and 500 milligrams per liter.) When the values of a random variable are not countable but instead correspond to the points on some interval, we call it a *continuous random variable*. Thus, the *amount* of pesticide in the chemical plant's discharge waters is a *continuous random variable*. ∎

> **DEFINITION 4.2**
> Random variables that can assume a *countable* number of values are called **discrete**.
>
> **DEFINITION 4.3**
> Random variables that can assume values corresponding to any of the points contained in one or more intervals are called **continuous**.

The following are examples of discrete random variables:

1. The number of seizures an epileptic patient has in a given week: $x = 0, 1, 2, \ldots$
2. The number of voters in a sample of 500 who favor impeachment of the president: $x = 0, 1, 2, \ldots, 500$
3. The number of students applying to medical schools this year: $x = 0, 1, 2, \ldots$
4. The number of errors on a page of an accountant's ledger: $x = 0, 1, 2, \ldots$
5. The number of customers waiting to be served in a restaurant at a particular time: $x = 0, 1, 2, \ldots$

Note that each of the examples of discrete random variables begins with the words "The number of . . . ." This wording is very common, since the discrete random variables most frequently observed are counts. The following are examples of continuous random variables:

1. The length of time (in seconds) between arrivals at a hospital clinic: $0 \leq x \leq \infty$ (infinity)
2. The length of time (in minutes) it takes a student to complete a one-hour exam: $0 \leq x \leq 60$
3. The amount (in ounces) of carbonated beverage loaded into a 12-ounce can in a can-filling operation: $0 \leq x \leq 12$
4. The depth (in feet) at which a successful oil drilling venture first strikes oil: $0 \leq x \leq c$, where $c$ is the maximum depth obtainable
5. The weight (in pounds) of a food item bought in a supermarket: $0 \leq x \leq 500$ [*Note:* Theoretically, there is no upper limit on $x$, but it is unlikely that it would exceed 500 pounds.]

Discrete random variables and their probability distributions are discussed in this chapter. Continuous random variables and their probability distributions are the topic of Chapter 5.

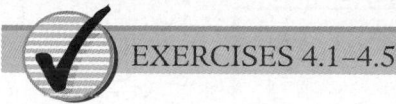 EXERCISES 4.1–4.5

**Applying the Concepts—Basic**

**4.1** What is a random variable?

**4.2** How do discrete and continuous random variables differ?

**4.3** Classify the following random variables according to whether they are discrete or continuous:
  **a.** The number of words spelled correctly by a student on a spelling test
  **b.** The amount of water flowing through the Hoover Dam in a day
  **c.** The length of time an employee is late for work
  **d.** The number of bacteria in a particular cubic centimeter of drinking water
  **e.** The amount of carbon monoxide produced per gallon of unleaded gas
  **f.** Your weight

**4.4** Identify the following random variables as discrete or continuous:
  **a.** The amount of flu vaccine in a syringe

  **b.** The heart rate (number of beats per minute) of an American male
  **c.** The time it takes a student to complete an examination
  **d.** The barometric pressure at a given location
  **e.** The number of registered voters who vote in a national election
  **f.** Your score on the Scholastic Assessment Test (SAT)

**4.5** Identify the following variables as discrete or continuous:
  **a.** The reaction time difference to the same stimulus before and after training
  **b.** The number of violent crimes committed per month in your community
  **c.** The number of commercial aircraft near-misses per month
  **d.** The number of winners each week in a state lottery
  **e.** The number of free throws made per game by a basketball team
  **f.** The distance traveled by a school bus each day

## 4.2 PROBABILITY DISTRIBUTIONS FOR DISCRETE RANDOM VARIABLES

A complete description of a discrete random variable requires that we *specify the possible values the random variable can assume* and *the probability associated with each value*. To illustrate, consider Example 4.4.

**EXAMPLE 4.4**  Recall the experiment of tossing two coins (Section 4.1), and let $x$ be the number of heads observed. Find the probability associated with each value of the random variable $x$, assuming the two coins are fair.

**Solution**

The sample space and sample points for this experiment are reproduced in Figure 4.2. Note that the random variable $x$ can assume values $0, 1, 2$. Recall (from Chapter 3) that the probability associated with each of the four sample points is $\frac{1}{4}$. Then, identifying the probabilities of the sample points associated with each of these values of $x$, we have

$$P(x = 0) = P(TT) = \frac{1}{4}$$

$$P(x = 1) = P(TH) + P(HT) = \frac{1}{4} + \frac{1}{4} = \frac{1}{2}$$

$$P(x = 2) = P(HH) = \frac{1}{4}$$

Thus, we now know the values the random variable can assume $(0, 1, 2)$ and how the probability is *distributed over* these values $\left(\frac{1}{4}, \frac{1}{2}, \frac{1}{4}\right)$. This completely describes

**FIGURE 4.2**

*Venn Diagram for the Two-Coin-Toss Experiment*

the random variable and is referred to as the *probability distribution*, denoted by the symbol $p(x)$.* The probability distribution for the coin-toss example is shown in tabular form in Table 4.1 and in graphic form in Figure 4.3. Since the probability distribution for a discrete random variable is concentrated at specific points (values of $x$), the graph in Figure 4.3a represents the probabilities as the heights of vertical lines over the corresponding values of $x$. Although the representation of the probability distribution as a histogram, as in Figure 4.3b, is less precise (since the probability is spread over a unit interval), the histogram representation will prove useful when we approximate probabilities of certain discrete random variables in Section 4.4.

**Table 4.1    Probability Distribution for Coin-Toss Experiment: Tabular Form**

| $x$ | $p(x)$ |
|:---:|:---:|
| 0 | $1/4$ |
| 1 | $1/2$ |
| 2 | $1/4$ |

**FIGURE 4.3**

*Probability Distribution for Coin-Toss Experiment: Graphical Form*

a. Point representation of $p(x)$

b. Histogram representation of $p(x)$

We could also present the probability distribution for $x$ as a formula, but this would unnecessarily complicate a very simple example. We give the formulas for the probability distributions of some common discrete random variables later in this chapter.  ∎

**DEFINITION 4.4**

The **probability distribution** of a discrete random variable is a graph, table, or formula that specifies the probability associated with each possible value the random variable can assume.

---

*In standard mathematical notation, the probability that a random variable $X$ takes on a value $x$ is denoted $P(X = x) = p(x)$. Thus, $P(X = 0) = p(0)$, $P(X = 1) = p(1)$, etc. In this text, we adopt the simpler $p(x)$ notation.

Two requirements must be satisfied by all probability distributions for discrete random variables.

---

**Requirements for the Probability Distribution of a Discrete Random Variable $x$**

(1)  $p(x) \geq 0$  for all values of $x$
(2)  $\Sigma p(x) = 1$

where the summation of $p(x)$ is over all possible values of $x$.*

---

Example 4.4 illustrates how the probability distribution for a discrete random variable can be derived, but for many practical situations the task is much more difficult. Fortunately, many experiments and associated discrete random variables observed in nature possess identical characteristics. Thus, you might observe a random variable in a psychology experiment that would possess the same probability distribution as a random variable observed in an engineering experiment or a social sample survey. We classify random variables according to type of experiment, derive the probability distribution for each of the different types, and then use the appropriate probability distribution when a particular type of random variable is observed in a practical situation. The probability distributions for most commonly occurring discrete random variables have already been derived. This fact simplifies the problem of finding the probability distributions for random variables.

In Sections 4.4 and 4.5, we describe two important types of discrete random variables, give their probability distributions, and explain where and how they can be applied in practice. (Mathematical derivations of the probability distributions will be omitted, but these details can be found in the chapter references.)

But first, in Section 4.3, we discuss some descriptive measures of these sometimes complex probability distributions. Since probability distributions are analogous to the relative frequency distributions of Chapter 2, it should be no surprise that the mean and standard deviation are useful descriptive measures.

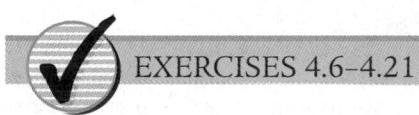 EXERCISES 4.6–4.21

**Learning the Mechanics**

**4.6**  Consider the following probability distribution:

| $x$ | −4 | 0 | 1 | 3 |
|---|---|---|---|---|
| $p(x)$ | .1 | .2 | .4 | .3 |

**a.** List the values that $x$ may assume.
**b.** What value of $x$ is most probable?
**c.** What is the probability that $x$ is greater than 0?
**d.** What is the probability that $x = -2$?

**4.7**  A discrete random variable $x$ can assume five possible values: 2, 3, 5, 8, and 10. Its probability distribution is shown here:

| $x$ | 2 | 3 | 5 | 8 | 10 |
|---|---|---|---|---|---|
| $p(x)$ | .15 | .10 | | .25 | .25 |

**a.** What is $p(5)$?
**b.** What is the probability that $x$ equals 2 or 10?
**c.** What is $P(x \leq 8)$?

---

*Unless otherwise indicated, summations will always be over all possible values of $x$.

**4.8** Explain why each of the following is or is not a valid probability distribution for a discrete random variable $x$:

**a.**

| $x$ | 0 | 1 | 2 | 3 |
|-----|---|---|---|---|
| $p(x)$ | .2 | .3 | .3 | .2 |

**b.**

| $x$ | −2 | −1 | 0 |
|-----|----|----|---|
| $p(x)$ | .25 | .50 | .20 |

**c.**

| $x$ | 4 | 9 | 20 |
|-----|---|---|----|
| $p(x)$ | −.3 | 1.0 | .3 |

**d.**

| $x$ | 2 | 3 | 5 | 6 |
|-----|---|---|---|---|
| $p(x)$ | .15 | .20 | .40 | .35 |

**4.9** The random variable $x$ has the following discrete probability distribution:

| $x$ | 10 | 11 | 12 | 13 | 14 |
|-----|----|----|----|----|----|
| $p(x)$ | .2 | .3 | .2 | .1 | .2 |

Since the values that $x$ can assume are mutually exclusive events, the event $\{x \leq 12\}$ is the union of three mutually exclusive events:

$$\{x = 10\} \cup \{x = 11\} \cup \{x = 12\}$$

**a.** Find $P(x \leq 12)$.
**b.** Find $P(x > 12)$.
**c.** Find $P(x \leq 14)$.
**d.** Find $P(x = 14)$.
**e.** Find $P(x \leq 11$ or $x > 12)$.

**4.10** The random variable $x$ has the discrete probability distribution shown here.

| $x$ | −2 | −1 | 0 | 1 | 2 |
|-----|----|----|---|---|---|
| $p(x)$ | .10 | .15 | .40 | .30 | .05 |

**a.** Find $P(x \leq 0)$.
**b.** Find $P(x > -1)$.
**c.** Find $P(-1 \leq x \leq 1)$.
**d.** Find $P(x < 2)$.
**e.** Find $P(-1 < x < 2)$.
**f.** Find $P(x < 1)$.

**4.11** Toss three fair coins and let $x$ equal the number of heads observed.
**a.** Identify the sample points associated with this experiment and assign a value of $x$ to each sample point.
**b.** Calculate $p(x)$ for each value of $x$.
**c.** Construct a probability histogram for $p(x)$.
**d.** What is $P(x = 2$ or $x = 3)$?

## Applying the Concepts—Basic

**4.12** According to the American Dental Association, 60% of all dentists use nitrous oxide ("laughing gas") in their practice (*New York Times*, June 20, 1995). If $x$ equals the number of dentists in a random sample of five dentists who use laughing gas in practice, then the probability distribution of $x$ is

| $x$ | 0 | 1 | 2 | 3 | 4 | 5 |
|-----|---|---|---|---|---|---|
| $p(x)$ | .0102 | .0768 | .2304 | .3456 | .2592 | .0778 |

Find the probability that the number of dentists using laughing gas in the sample of five is
**a.** 4
**b.** less than 2
**c.** greater than or equal to 3

**4.13** A dust mite allergen level that exceeds 2 micrograms per gram ($\mu g/g$) of dust has been associated with the development of allergies. Consider a random sample of four homes and let $x$ be the number of homes with a dust mite level that exceeds $2\,\mu g/g$. The probability distribution for $x$, based on a May 2000 study by the National Institute of Environmental Health Sciences, is shown in the following table.

| $x$ | 0 | 1 | 2 | 3 | 4 |
|-----|---|---|---|---|---|
| $p(x)$ | .09 | .30 | .37 | .20 | .04 |

**a.** Verify that the sum of the probabilities for $x$ in the table sum to 1.
**b.** Find the probability that three or four of the homes in the sample have a dust mite level that exceeds $2\,\mu g/g$.
**c.** Find the probability that fewer than two homes in the sample have a dust mite level that exceeds $2\,\mu g/g$.

**4.14** The probability distribution of $x$, the number of customer arrivals per 15-minute period, at a Wendy's Restaurant in New Jersey is shown below.

| $x$ | 5 | 6 | 7 | 8 | 9 | 10 | 11 | 12 | 13 | 14 | 15 |
|-----|---|---|---|---|---|----|----|----|----|----|----|
| $p(x)$ | .01 | .02 | .03 | .05 | .08 | .09 | .11 | .13 | .12 | .10 | .08 |

| $x$ | 16 | 17 | 18 | 19 | 20 | 21 |
|-----|----|----|----|----|----|----|
| $p(x)$ | .06 | .05 | .03 | .02 | .01 | .01 |

*Source:* Ford, R., Roberts, D., and Saxton, P. *Queuing Models.* Graduate School of Management, Rutgers University, 1992.

**a.** Does this distribution meet the two requirements for the probability distribution of a discrete random variable? Justify your answer.

**b.** What is the probability that exactly 16 customers enter the restaurant in the next 15 minutes?

**c.** Find $P(x \leq 10)$.

**d.** Find $P(5 \leq x \leq 15)$.

## Applying the Concepts—Intermediate

**4.15** In Exercise 3.16 (p. 118) you learned that five in every 10,000 students who take the Scholastic Assessment Test (SAT) score a perfect 1600. Consider a random sample of three students who take the SAT. Let $x$ equal the number who score a perfect 1600.

**a.** Find $p(x)$ for $x = 0, 1, 2, 3$.

**b.** Graph $p(x)$.

**c.** Find $P(x \leq 1)$.

**4.16** A weapons manufacturer uses a liquid propellant to produce gun cartridges. During the manufacturing process, the propellant can get mixed with another liquid to produce a contaminated cartridge. A University of South Florida statistician, hired by the company to investigate the level of contamination in the stored cartridges, found that 23% of the cartridges in a particular lot were contaminated. Suppose you randomly sample (without replacement) gun cartridges from this lot until you find a contaminated one. Let $x$ be the number of cartridges sampled until a contaminated one is found. It is known that the probability distribution for $x$ is given by the formula:

$$p(x) = (.23)(.77)^{x-1}, \qquad x = 1, 2, 3, \ldots$$

**a.** Find $p(1)$. Interpret this result.

**b.** Find $p(5)$. Interpret this result.

**c.** Find $P(x \geq 2)$. Interpret this result.

**4.17** In his article "American Football" (*Statistics in Sport*, 1998), Iowa State University statistician Hal Stern evaluates winning strategies of teams in the National Football League (NFL). In a section on estimating the probability of winning a game, Stern used actual NFL play-by-play data to approximate the probabilities associated with certain outcomes (e.g., running plays, short pass plays, and long pass plays). The table below gives the probability

| $x$, Yards | Probability | $x$, Yards | Probability |
|---|---|---|---|
| −4 | .020 | 6 | .090 |
| −2 | .060 | 8 | .060 |
| 1 | .070 | 10 | .050 |
| 0 | .150 | 15 | .085 |
| 1 | .130 | 30 | .010 |
| 2 | .110 | 50 | .004 |
| 3 | .090 | 99 | .001 |
| 4 | .070 | | |

distribution for the yardage gained, $x$, on a running play. (A negative gain represents a loss of yards on the play.)

**a.** Find the probability of gaining 10 yards or more on a running play.

**b.** Find the probability of losing yardage on a running play.

**4.18** A team of Chinese university professors investigated the reliability of several capacitated-flow networks in the journal *Networks* (May 1995). One network examined in the article and illustrated here is a bridge network with arcs $a_1, a_2, a_3, a_4, a_5,$ and $a_6$. The probability distribution of the capacity $x$ for each of the six arcs is provided in the table.

| Arc | Capacity $x$ | $p(x)$ | Arc | Capacity $x$ | $p(x)$ |
|---|---|---|---|---|---|
| $a_1$ | 3 | .60 | $a_4$ | 1 | .90 |
| | 2 | .25 | | 0 | .10 |
| | 1 | .10 | | | |
| | 0 | .05 | | | |
| $a_2$ | 2 | .60 | $a_5$ | 1 | .90 |
| | 1 | .30 | | 0 | .10 |
| | 0 | .10 | | | |
| $a_3$ | 1 | .90 | $a_6$ | 2 | .70 |
| | 0 | .10 | | 1 | .25 |
| | | | | 0 | .05 |

*Source:* Lin, J., et al. "On reliability evaluation of capacitated-flow network in terms of minimal pathsets." *Networks*, Vol. 25, No. 3, May 1995, p. 135 (Table 1).

**a.** Verify that the properties of discrete probability distributions are satisfied for each arc capacity distribution.

**b.** Find the probability that the capacity for arc $a_1$ will exceed 1.

**c.** Repeat part **b** for each of the remaining five arcs.

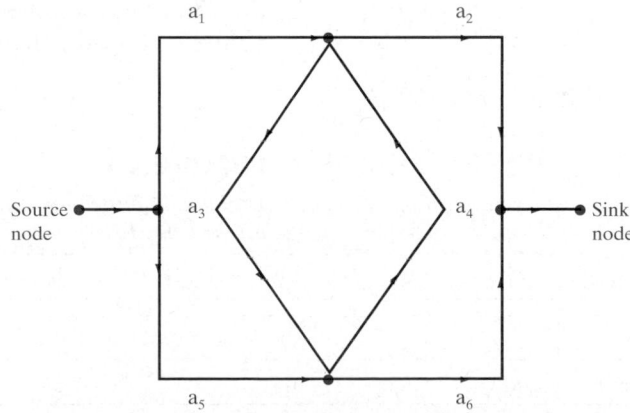

**Applying the Concepts—Advanced**

**4.19** Refer to Exercise 4.18.

    **a.** One path from the source node to the sink node is through arcs $a_1$ and $a_2$. Find the probability that the system maintains a capacity of more than 1 through the $a_1$–$a_2$ path. (Assume that the arc capacities are independent.)

    **b.** Another path from the source node to the sink node is through $a_1$, $a_3$, and $a_6$. Find the probability that the system maintains a capacity of 1 through the $a_1$–$a_3$–$a_6$ path.

**4.20** Geneticists use a grid—called a *Punnett square*—to display all possible gene combinations in genetic crosses. (The grid is named for Reginald Punnett, a British geneticist who developed the method in the early 1900s.) The accompanying figure is a Punnett square for a cross involving human earlobes. In humans, free earlobes (E) are dominant over attached earlobes (e). Consequently, the gene pairs EE and Ee will result in free earlobes, while the gene pair ee results in attached earlobes. Consider a couple with genes as shown in the Punnett square. Suppose the couple has seven children. Let $x$ represent

the number of children with attached earlobes (i.e., with the gene pair ee). Find the probability distribution of $x$.

**4.21** Every human possesses two sex chromosomes. A copy of one or the other (equally likely) is contributed to an offspring. Males have one X chromosome and one Y chromosome. Females have two X chromosomes. If a couple has three children, what is the probability that they have at least one boy?

## 4.3 EXPECTED VALUES OF DISCRETE RANDOM VARIABLES

If a discrete random variable $x$ were observed a very large number of times and the data generated were arranged in a relative frequency distribution, the relative frequency distribution would be indistinguishable from the probability distribution for the random variable. Thus, the probability distribution for a random variable is a theoretical model for the relative frequency distribution of a population. To the extent that the two distributions are equivalent (and we will assume they are), the probability distribution for $x$ possesses a mean $\mu$ and a variance $\sigma^2$ that are identical to the corresponding descriptive measures for the population. This section explains how you can find the mean value for a random variable. We illustrate the procedure with an example.

Examine the probability distribution for $x$ (the number of heads observed in the toss of two fair coins) in Figure 4.4. Try to locate the mean of the distribution intuitively. We may reason that the mean $\mu$ of this distribution is equal to 1 as follows: In a large number of experiments, $1/4$ should result in $x = 0$, $1/2$ in $x = 1$, and $1/4$ in $x = 2$ heads. Therefore, the average number of heads is

$$\mu = 0\left(\tfrac{1}{4}\right) + 1\left(\tfrac{1}{2}\right) + 2\left(\tfrac{1}{4}\right) = 0 + \tfrac{1}{2} + \tfrac{1}{2} = 1$$

**FIGURE 4.4**

*Probability Distribution for a Two-Coin Toss*

Note that to get the population mean of the random variable $x$, we multiply each possible value of $x$ by its probability $p(x)$, and then we sum this product over all possible values of $x$. The *mean of $x$* is also referred to as the *expected value of $x$*, denoted $E(x)$.

---

**DEFINITION 4.5**

The **mean**, or **expected value**, of a discrete random variable $x$ is

$$\mu = E(x) = \Sigma x p(x)$$

---

The term *expected* is a mathematical term and should not be interpreted as it is typically used. Specifically, a random variable might never be equal to its "expected value." Rather, the expected value is the mean of the probability distribution, or a measure of its central tendency. You can think of $\mu$ as the mean value of $x$ in a *very large* (actually, *infinite*) number of repetitions of the experiment, where the values of $x$ occur in proportions equivalent to the probabilities of $x$.

**EXAMPLE 4.5**

Suppose you work for an insurance company, and you sell a $10,000 1-year term insurance policy at an annual premium of $290. Actuarial tables show that the probability of death during the next year for a person of your customer's age, sex, health, etc., is .001. What is the expected gain (amount of money made by the company) for a policy of this type?

**Solution**

The experiment is to observe whether the customer survives the upcoming year. The probabilities associated with the two sample points, Live and Die, are .999 and .001, respectively. The random variable you are interested in is the gain $x$, which can assume the values shown in the following table.

| Gain $x$ | Sample Point | Probability |
|---|---|---|
| $290 | Customer lives | .999 |
| $-$ $9,710 | Customer dies | .001 |

If the customer lives, the company gains the $290 premium as profit. If the customer dies, the gain is negative because the company must pay $10,000, for a net "gain" of $ (290 - 10,000) = -$ $9,710$. The expected gain is therefore

$$\mu = E(x) = \Sigma x p(x)$$
$$= (290)(.999) + (-9,710)(.001) = \$ 280$$

In other words, if the company were to sell a very large number of 1-year $10,000 policies to customers possessing the characteristics described above, it would (on the average) net $280 per sale in the next year.

Example 4.5 illustrates that the expected value of a random variable $x$ need not equal a possible value of $x$. That is, the expected value is $280, but $x$ will equal either $290 or $-$ $9,710 each time the experiment is performed (a policy is sold and a year elapses). The expected value is a measure of central tendency—and in this case represents the average over a very large number of 1-year policies—but is not a possible value of $x$.

We learned in Chapter 2 that the mean and other measures of central tendency tell only part of the story about a set of data. The same is true about probability distributions. We need to measure variability as well. Since a probability distribution can be viewed as a representation of a population, we will use the population variance to measure its variability.

The *population variance* $\sigma^2$ is defined as the average of the squared distance of $x$ from the population mean $\mu$. Since $x$ is a random variable, the squared distance, $(x - \mu)^2$, is also a random variable. Using the same logic used to find the mean value of $x$, we find the mean value of $(x - \mu)^2$ by multiplying all possible values of $(x - \mu)^2$ by $p(x)$ and then summing over all possible $x$ values.* This quantity

$$E[(x - \mu)^2] = \sum_{\text{all } x} (x - \mu)^2 p(x)$$

is also called the *expected value of the squared distance from the mean; that is,* $\sigma^2 = E[(x - \mu)^2]$. The standard deviation of $x$ is defined as the square root of the variance $\sigma^2$.

---

**DEFINITION 4.6**

The **variance** of a random variable $x$ is

$$\sigma^2 = E[(x - \mu)^2] = \Sigma(x - \mu)^2 p(x)$$

**DEFINITION 4.7**

The **standard deviation** of a discrete random variable is equal to the square root of the variance, i.e., $\sigma = \sqrt{\sigma^2}$.

---

Knowing the mean $\mu$ and standard deviation $\sigma$ of the probability distribution of $x$, in conjunction with Chebyshev's Rule (Table 2.7) and the Empirical Rule (Table 2.8), we can make statements about the likelihood of values of $x$ falling within the intervals $\mu \pm \sigma, \mu \pm 2\sigma$, and $\mu \pm 3\sigma$. These probabilities are given in the box.

---

## Chebyshev's Rule and Empirical Rule for a Discrete Random Variable

Let $x$ be a discrete random variable with probability distribution $p(x)$, mean $\mu$, and standard deviation $\sigma$. Then, depending on the shape of $p(x)$, the following probability statements can be made:

| | Chebyshev's Rule | Empirical Rule |
|---|---|---|
| | Applies to any probability distribution (see Figure 4.5a) | Applies to probability distributions that are mound-shaped and symmetric (see Figure 4.5b) |
| $P(\mu - \sigma < x < \mu + \sigma)$ | $\geq 0$ | $\approx .68$ |
| $P(\mu - 2\sigma < x < \mu + 2\sigma)$ | $\geq \frac{3}{4}$ | $\approx .95$ |
| $P(\mu - 3\sigma < x < \mu + 3\sigma)$ | $\geq \frac{8}{9}$ | $\approx 1.00$ |

---

*It can be shown that $E[(x - \mu)^2] = E(x^2) - \mu^2$, where $E(x^2) = \Sigma x^2 p(x)$. Note the similarity between this expression and the shortcut formula $\Sigma(x - \bar{x})^2 = \Sigma x^2 - (\Sigma x)^2/n$ given in Chapter 2.

**FIGURE 4.5**

*Shapes of Two Probability Distributions for a Discrete Random Variable x*

a. Skewed distribution

b. Mound-shaped, symmetric

**EXAMPLE 4.6**

Medical research has shown that a certain type of chemotherapy is successful 70% of the time when used to treat skin cancer. Suppose five skin cancer patients are treated with this type of chemotherapy and let $x$ equal the number of successful cures out of the five. The probability distribution for the number $x$ of successful cures out of five is given in the table:

| $x$ | 0 | 1 | 2 | 3 | 4 | 5 |
|------|------|------|------|------|------|------|
| $p(x)$ | .002 | .029 | .132 | .309 | .360 | .168 |

a. Find $\mu = E(x)$. Interpret the result.
b. Find $\sigma = \sqrt{E[(x - \mu)^2]}$, Interpret the result.
c. Graph $p(x)$. Locate $\mu$ and the interval $\mu \pm 2\sigma$ on the graph. Use either Chebyshev's Rule or the Empirical Rule to approximate the probability that $x$ falls in this interval. Compare this result with the actual probability.
d. Would you expect to observe fewer than two successful cures out of five?

**Solution**

a. Applying the formula,

$$\mu = E(x) = \Sigma x p(x)$$
$$= 0(.002) + 1(.029) + 2(.132) + 3(.309) + 4(.360) + 5(.168) = 3.50$$

On average, the number of successful cures out of five skin cancer patients treated with chemotherapy will equal 3.5. Remember that this expected value only has meaning when the experiment—treating five skin cancer patients with chemotherapy—is repeated a large number of times.

b. Now we calculate the variance of $x$:

$$\sigma^2 = E[(x - \mu)^2] = \Sigma(x - \mu)^2 p(x)$$
$$= (0 - 3.5)^2(.002) + (1 - 3.5)^2(.029) + (2 - 3.5)^2(.132)$$
$$+ (3 - 3.5)^2(.309) + (4 - 3.5)^2(.360) + (5 - 3.5)^2(.168)$$
$$= 1.05$$

Thus, the standard deviation is

$$\sigma = \sqrt{\sigma^2} = \sqrt{1.05} = 1.02$$

This value measures the spread of the probability distribution of $x$, the number of successful cures out of five. A more useful interpretation is obtained by answering parts **c** and **d**.

**FIGURE 4.6**

*Graph of $p(x)$ for Example 4.6*

c. The graph of $p(x)$ is shown in Figure 4.6 with the mean $\mu$ and the interval $\mu \pm 2\sigma = 3.50 \pm 2(1.02) = 3.50 \pm 2.04 = (1.46, 5.54)$ shown on the graph. Note particularly that $\mu = 3.5$ locates the center of the probability distribution. Since this distribution is a theoretical relative frequency distribution that is moderately mound-shaped (see Figure 4.6), we expect (from Chebyshev's Rule) at least 75% and, more likely (from the Empirical Rule), approximately 95% of observed $x$ values to fall between 1.46 and 5.54. You can see from Figure 4.6 that the actual probability that $x$ falls in the interval $\mu \pm 2\sigma$ includes the sum of $p(x)$ for the values $x = 2$, $x = 3$, $x = 4$, and $x = 5$. This probability is $p(2) + p(3) + p(4) + p(5) = .132 + .309 + .360 + .168 = .969$. Therefore, 96.9% of the probability distribution lies within 2 standard deviations of the mean. This percentage is consistent with both Chebyshev's Rule and the Empirical Rule.

d. Fewer than two successful cures out of five implies that $x = 0$ or $x = 1$. Since both these values of $x$ lie outside the interval $\mu \pm 2\sigma$, we know from the Empirical Rule that such a result is unlikely (approximate probability of .05). The exact probability, $P(x \le 1)$, is $p(0) + p(1) = .002 + .029 = .031$. Consequently, in a single experiment where five skin cancer patients are treated with chemotherapy, we would not expect to observe fewer than two successful cures. _____ ∎

   EXERCISES 4.22–4.34

**Learning the Mechanics**

**4.22** Consider the probability distribution for the random variable $x$ shown here.

| $x$ | 10 | 20 | 30 | 40 | 50 | 60 |
|---|---|---|---|---|---|---|
| $p(x)$ | .05 | .20 | .30 | .25 | .10 | .10 |

**a.** Find $\mu, \sigma^2$, and $\sigma$.

**b.** Graph $p(x)$.

**c.** Locate $\mu$ and the interval $\mu \pm 2\sigma$ on your graph. What is the probability that $x$ will fall within the interval $\mu \pm 2\sigma$?

**4.23** Consider the probability distribution shown here.

| $x$ | 1 | 2 | 4 | 10 |
|-----|-----|-----|-----|-----|
| $p(x)$ | .2 | .4 | .2 | .2 |

**a.** Find $\mu = E(x)$.

**b.** Find $\sigma^2 = E[(x - \mu)^2]$.

**c.** Find $\sigma$.

**d.** Interpret the value you obtained for $\mu$.

**e.** In this case, can the random variable $x$ ever assume the value $\mu$? Explain.

**f.** In general, can a random variable ever assume a value equal to its expected value? Explain.

**4.24** Consider the probability distribution shown here.

| $x$ | −4 | −3 | −2 | −1 | 0 | 1 | 2 | 3 | 4 |
|-----|-----|-----|-----|-----|-----|-----|-----|-----|-----|
| $p(x)$ | .02 | .07 | .10 | .15 | .30 | .18 | .10 | .06 | .02 |

**a.** Calculate $\mu, \sigma^2$, and $\sigma$.

**b.** Graph $p(x)$. Locate $\mu, \mu - 2\sigma$, and $\mu + 2\sigma$ on the graph.

**c.** What is the probability that $x$ will fall in the interval $\mu \pm 2\sigma$?

**4.25** Consider the probability distributions shown here.

| $x$ | 0 | 1 | 2 |
|-----|-----|-----|-----|
| $p(x)$ | .3 | .4 | .3 |

| $y$ | 0 | 1 | 2 |
|-----|-----|-----|-----|
| $p(y)$ | .1 | .8 | .1 |

**a.** Use your intuition to find the mean for each distribution. How did you arrive at your choice?

**b.** Which distribution appears to be more variable? Why?

**c.** Calculate $\mu$ and $\sigma^2$ for each distribution. Compare these answers to your answers in parts **a** and **b**.

## Applying the Concepts—Basic

**4.26** Refer to Exercise 4.12 (p. 180). The probability distribution for the number $x$ out of five randomly selected dentists who use "laughing gas" is reproduced here. Find the mean number of dentists out of the five who use "laughing gas." Interpret this value.

| $x$ | 0 | 1 | 2 | 3 | 4 | 5 |
|-----|-----|-----|-----|-----|-----|-----|
| $p(x)$ | .0102 | .0768 | .2304 | .3456 | .2592 | .0778 |

**4.27** Exercise 4.13 (p. 180) gives the probability distribution for the number $x$ of homes with high dust mite levels. The probability distribution is reproduced here.

| Number of Homes, $x$ | $p(x)$ |
|-----|-----|
| 0 | .09 |
| 1 | .30 |
| 2 | .37 |
| 3 | .20 |
| 4 | .04 |

**a.** Find $E(x)$. Give a meaningful interpretation of the result.

**b.** Find $\sigma$.

**c.** Find the exact probability that $x$ is in the interval $\mu \pm 2\sigma$. Compare to Chebyshev's Rule and the Empirical Rule.

**4.28** A capacitated-bridge network with six arcs (*Networks*, May 1995) was presented in Exercise 4.18 (p. 181). The probability distributions for the capacities of the six arcs are reproduced here.

| Arc | Capacity $x$ | $p(x)$ | | Arc | Capacity $x$ | $p(x)$ |
|-----|-----|-----|---|-----|-----|-----|
| $a_1$ | 3 | .60 | | $a_4$ | 1 | .90 |
|  | 2 | .25 | |  | 0 | .10 |
|  | 1 | .10 | |  |  |  |
|  | 0 | .05 | |  |  |  |
| $a_2$ | 2 | .60 | | $a_5$ | 1 | .90 |
|  | 1 | .30 | |  | 0 | .10 |
|  | 0 | .10 | |  |  |  |
| $a_3$ | 1 | .90 | | $a_6$ | 2 | .70 |
|  | 0 | .10 | |  | 1 | .25 |
|  |  |  | |  | 0 | .05 |

*Source:* Lin, J., *et al.* "On reliability evaluation of capacitated-flow network in terms of minimal pathsets." *Networks*, Vol. 25, No. 3, May 1995, p. 135 (Table 1).

**a.** Compute the mean capacity of each arc and interpret its value.

**b.** Compute $\sigma$ for each arc and interpret its value.

## Applying the Concepts—Intermediate

**4.29** The National Weather Service issues precipitation forecasts that indicate the likelihood of measurable precipitation ($\geq .01$ inch) at a specific point—called the official rain gauge—during a given time period. Suppose that if a measurable amount of rain falls during the next 24 hours, an Ohio river will reach flood stage and a business will incur damages of $300,000. The National Weather Service has indicated that there is a 30% chance of a measurable amount of rain during the next 24 hours.

**a.** Construct the probability distribution that describes the potential flood damages.

**b.** Find the firm's expected loss due to flood damage. Interpret the result.

**4.30** A patient complaining of severe stomach pains checked into a local hospital. After a series of tests the doctors narrowed their diagnosis to four possible ailments. They believe there is a 40% chance that the patient has hepatitis; a 10% chance that she has cirrhosis; a 45% chance of gallstones; and a 5% chance of cancer of the pancreas. The doctors are certain that the patient has only one of the diseases, but will not know which disease until further tests are performed. The cost associated with treating each disease is given in the table:

| Disease | Hepatitis | Cirrhosis | Gallstones | Pancreatic Cancer |
|---------|-----------|-----------|------------|-------------------|
| Cost | $700 | $1,100 | $3,320 | $16,450 |

**a.** Construct the probability distribution for the cost of treating the patient.
**b.** Calculate the mean of the probability distribution you constructed in part **a**. What does this number represent?
**c.** Further testing reveals that the patient has either hepatitis or cirrhosis. Given this information, construct the new probability distribution for the cost of treating the patient.
**d.** Calculate the mean of the probability distribution you constructed in part **c**. What does this number represent?

**4.31** The chance of winning Florida's Pick-6 Lotto game is 1 in approximately 23 million. Suppose you buy a $1 Lotto ticket in anticipation of winning the $7 million grand prize. Calculate your expected net winnings. Interpret the result.

**4.32** On the popular television game show, *The Price is Right*, contestants can play "The Showcase Showdown." The game involves a large wheel with twenty nickel values, 5, 10, 15, 20, ..., 95, 100, marked on it. Contestants spin the wheel once or twice, with the objective of obtaining the highest total score *without going over a dollar (100)*. [According to the *American Statistician* (Aug. 1995), the optimal strategy for the first spinner in a three-player game is to spin a second time only if the value of the initial spin is 65 or less.] Let $x$ represent a single contestant playing "The Showcase Showdown." Assume a "fair" wheel, i.e., a wheel with equally likely outcomes. If the total of the player's spins exceeds 100, the total score is set to 0.
**a.** If the player is permitted only one spin of the wheel, find the probability distribution for $x$.
**b.** Refer to part **a**. Find $E(x)$ and interpret this value.
**c.** Refer to part **a**. Give a range of values within which $x$ is likely to fall.
**d.** Suppose the player will spin the wheel twice, no matter what the outcome of the first spin. Find the probability distribution for $x$.

**e.** What assumption did you make to obtain the probability distribution, part **d**? Is it a reasonable assumption?
**f.** Find $\mu$ and $\sigma$ for the probability distribution, part **d**, and interpret the results.
**g.** Refer to part **d**. What is the probability that in two spins the player's total score exceeds a dollar (i.e., is set to 0)?
**h.** Suppose the player obtains a 20 on the first spin and decides to spin again. Find the probability distribution for $x$.
**i.** Refer to part **h**. What is the probability that the player's total score exceeds a dollar?
**j.** Given the player obtains a 65 on the first spin and decides to spin again, find the probability that the player's total score exceeds a dollar.
**k.** Repeat part **j** for different first-spin outcomes. Use this information to suggest a strategy for the one-player game.

### Applying the Concepts—Advanced

**4.33** The number of training units that must be passed before a complex computer software program is mastered varies from one to five, depending on the student. After much experience, the software manufacturer has determined the probability distribution that describes the fraction of users mastering the software after each number of training units:

| Number of Units | 1 | 2 | 3 | 4 | 5 |
|-----------------|---|---|---|---|---|
| Probability of Mastery | .1 | .25 | .4 | .15 | .1 |

**a.** Calculate the mean number of training units necessary to master the program. Calculate the median. Interpret each.
**b.** If the firm wants to ensure that at least 75% of the students master the program, what is the minimum number of training units that must be administered? At least 90%?
**c.** Suppose the firm develops a new training program that increases the probability that only one unit of training is needed from .1 to .25, increases the probability that only two units are needed to .35, leaves the probability that three units are needed at .4, and completely eliminates the need for four or five units. How do your answers to parts **a** and **b** change for this new program?

**4.34** Odds makers try to predict which professional and college football teams will win and by how much (the *spread*). If the odds makers do this accurately, adding the spread to the underdog's score should make the final score a tie. Suppose a bookie will give you $6 for every $1 you risk if you pick the winners in three ballgames (adjusted by the spread) on a "parlay" card. What is the bookie's expected earnings per dollar wagered? Interpret this value.

## 4.4  THE BINOMIAL RANDOM VARIABLE

Many experiments result in *dichotomous* responses—i.e., responses for which there exist two possible alternatives, such as Yes–No, Pass–Fail, Defective–Nondefective, or Male–Female. A simple example of such an experiment is the coin-toss experiment. A coin is tossed a number of times, say 10. Each toss results in one of two outcomes, Head or Tail, and the probability of observing each of these two outcomes remains the same for each of the 10 tosses. Ultimately, we are interested in the probability distribution of $x$, the number of heads observed. Many other experiments are equivalent to tossing a coin (either balanced or unbalanced) a fixed number $n$ of times and observing the number $x$ of times that one of the two possible outcomes occurs. Random variables that possess these characteristics are called **binomial random variables**.

Public opinion and consumer preference polls (e.g., the CNN, Gallup, and Harris polls) frequently yield observations on binomial random variables. For example, suppose a sample of 100 students is selected from a large student body and each person is asked whether he or she favors (a Head) or opposes (a Tail) a certain campus issue. Suppose we are interested in $x$, the number of students in the sample who favor the issue. Sampling 100 students is analogous to tossing the coin 100 times. Thus, you can see that opinion polls that record the number of people who favor a certain issue are real-life equivalents of coin-toss experiments. We have been describing a **binomial experiment**; it is identified by the following characteristics.

---

### Characteristics of a Binomial Random Variable

1. The experiment consists of $n$ identical trials.
2. There are only two possible outcomes on each trial. We will denote one outcome by $S$ (for Success) and the other by $F$ (for Failure).
3. The probability of $S$ remains the same from trial to trial. This probability is denoted by $p$, and the probability of $F$ is denoted by $q$. Note that $q = 1 - p$.
4. The trials are independent.
5. The binomial random variable $x$ is the number of $S$'s in $n$ trials.

---

**EXAMPLE 4.7**

For the following examples, decide whether $x$ is a binomial random variable.

a. A university scholarship committee must select two students to receive a scholarship for the next academic year. The committee receives 10 applications for the scholarships—six from male students and four from female students. Suppose the applicants are all equally qualified, so that the selections are randomly made. Let $x$ be the number of female students who receive a scholarship.

b. Before marketing a new product on a large scale, many companies will conduct a consumer-preference survey to determine whether the product is likely to be successful. Suppose a company develops a new diet soda and then conducts a taste-preference survey in which 100 randomly chosen consumers state their preferences among the new soda and the two leading sellers. Let $x$ be the number of the 100 who choose the new brand over the two others.

c. Some surveys are conducted by using a method of sampling other than simple random sampling (defined in Chapter 3). For example, suppose a television cable company plans to conduct a survey to determine the fraction of households in the city that would use the cable television service. The sampling method is to choose a city block at random and then survey every household on that block. This sampling technique is called *cluster sampling*. Suppose 10 blocks are so sampled, producing a total of 124 household responses. Let $x$ be the number of the 124 households that would use the television cable service.

### Solution

a. In checking the binomial characteristics, a problem arises with independence (characteristic 4 in the preceding box). Given that the first student selected is female, the probability that the second chosen is female is $3/9$. On the other hand, given that the first selection is a male student, the probability that the second is female is $4/9$. Thus, the conditional probability of a Success (choosing a female student to receive a scholarship) on the second trial (selection) depends on the outcome of the first trial, and the trials are therefore dependent. Since the trials are *not independent*, this is not a binomial random variable. (This variable is actually a *hypergeometric* random variable, the topic of Optional Section 4.6.)

b. Surveys that produce dichotomous responses and use random sampling techniques are classic examples of binomial experiments. In our example, each randomly selected consumer either states a preference for the new diet soda or does not. The sample of 100 consumers is a very small proportion of the totality of potential consumers, so the response of one would be, for all practical purposes, independent of another.* Thus, $x$ is a binomial random variable.

c. This example is a survey with dichotomous responses (Yes or No to the cable service), but the sampling method is not simple random sampling. Again, the binomial characteristic of independent trials would probably not be satisfied. The responses of households within a particular block would be dependent, since households within a block tend to be similar with respect to income, level of education, and general interests. Thus, the binomial model would not be satisfactory for $x$ if the cluster sampling technique were employed. ∎

**EXAMPLE 4.8** The Heart Association claims that only 10% of U.S. adults over 30 can pass the President's Physical Fitness Commission's minimum requirements. Suppose four adults are randomly selected, and each is given the fitness test.

a. Use the steps given in Chapter 3 (box on p. 115) to find the probability that none of the four adults passes the test.
b. Find the probability that three of the four adults pass the test.
c. Let $x$ represent the number of the four adults who pass the fitness test. Explain why $x$ is a binomial random variable.
d. Use the answers to parts **a** and **b** to derive a formula for $p(x)$, the probability distribution of the binomial random variable $x$.

### Solution

a. 1. The first step is to define the experiment. Here we are interested in observing the fitness test results of each of the four adults: pass $(S)$ or fail $(F)$.
   2. Next, we list the sample points associated with the experiment. Each sample point consists of the test results of the four adults. For example, *SSSS* represents the sample point that all four adults pass, while *FSSS* represents the sample point that adult 1 fails, while adults 2, 3, and 4 pass the test. The 16 sample points are listed in Table 4.2.
   3. We now assign probabilities to the sample points. Note that each sample point can be viewed as the intersection of four adults' test results and, as-

---

*In most real-life applications of the binomial distribution, the population of interest has a finite number of elements (trials), denoted $N$. When $N$ is large and the sample size $n$ is small relative to $N$, say $n/N \leq .05$, the sampling procedure, for all practical purposes, satisfies the conditions of a binomial experiment.

**Table 4.2  Sample Points for Fitness Test of Example 4.8**

| SSSS | FSSS | FFSS | SFFF | FFFF |
|------|------|------|------|------|
|      | SFSS | FSFS | FSFF |      |
|      | SSFS | FSSF | FFSF |      |
|      | SSSF | SFFS | FFFS |      |
|      |      | SFSF |      |      |
|      |      | SSFF |      |      |

suming the results are independent, the probability of each sample point can be obtained using the multiplicative rule, as follows:

$$P(SSSS) = P[(\text{adult 1 passes}) \cap (\text{adult 2 passes})$$
$$\cap (\text{adult 3 passes}) \cap (\text{adult 4 passes})]$$
$$= P(\text{adult 1 passes}) \times P(\text{adult 2 passes})$$
$$\times P(\text{adult 3 passes}) \times P(\text{adult 4 passes})$$
$$= (.1)(.1)(.1)(.1) = (.1)^4 = .0001$$

All other sample point probabilities are calculated using similar reasoning. For example,

$$P(FSSS) = (.9)(.1)(.1)(.1) = .0009$$

You can check that this reasoning results in sample point probabilities that add to 1 over the 16 points in the sample space.

4. Finally, we add the appropriate sample point probabilities to obtain the desired event probability. The event of interest is that all four adults fail the fitness test. In Table 4.2 we find only one sample point, *FFFF*, contained in this event. All other sample points imply that at least one adult passes. Thus,

$$P(\text{All four adults fail}) = P(FFFF) = (.9)^4 = .6561$$

b. The event that three of the four adults pass the fitness test consists of the four sample points in the second column of Table 4.2: *FSSS*, *SFSS*, *SSFS*, and *SSSF*. To obtain the event probability we add the sample point probabilities:

$$P(3 \text{ of 4 adults pass}) = P(FSSS) + P(SFSS) + P(SSFS) + P(SSSF)$$
$$= (.1)^3(.9) + (.1)^3(.9) + (.1)^3(.9) + (.1)^3(.9)$$
$$= 4(.1)^3(.9) = .0036$$

Note that each of the four sample point probabilities is the same, because each sample point consists of three $S$'s and one $F$; the order does not affect the probability because the adults' test results are (assumed) independent.

c. We can characterize the experiment as consisting of four identical trials—the four test results. There are two possible outcomes to each trial, $S$ or $F$, and the probability of passing, $p = .1$, is the same for each trial. Finally, we are assuming that each adult's test result is independent of all others, so that the four trials are independent. Then it follows that $x$, the number of the four adults who pass the fitness test, is a binomial random variable.

d. The event probabilities in parts **a** and **b** provide insight into the formula for the probability distribution $p(x)$. First, consider the event that three adults pass (part **b**). We found that

$$P(x = 3) = (\text{Number of sample points for which } x = 3) \times (.1)^{\text{Number of successes}} \times (.9)^{\text{Number of failures}}$$
$$= 4(.1)^3(.9)^1$$

In general, we can use combinatorial mathematics to count the number of sample points. For example,

Number of sample points for which $x = 3$
= Number of different ways of selecting 3 successes of the 4 trials
$$= \binom{4}{3} = \frac{4!}{3!(4 - 3)!} = \frac{4 \cdot 3 \cdot 2 \cdot 1}{(3 \cdot 2 \cdot 1) \cdot 1} = 4$$

The formula that works for any value of $x$ can be deduced as follows:

$$P(x = 3) = \binom{4}{3}(.1)^3(.9)^1 = \binom{4}{x}(.1)^x(.9)^{4-x}$$

The component $\binom{4}{x}$ counts the number of sample points with $x$ successes and the component $(.1)^x(.9)^{4-x}$ is the probability associated with each sample point having $x$ successes.

For the general binomial experiment, with $n$ trials and probability of Success $p$ on each trial, the probability of $x$ successes is

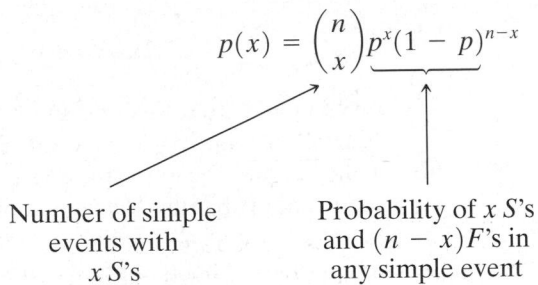

$$
\begin{array}{c}
\text{Number of simple} \\
\text{events with} \\
x \text{ } S\text{'s}
\end{array}
\qquad
\begin{array}{c}
\text{Probability of } x \text{ } S\text{'s} \\
\text{and } (n - x)F\text{'s in} \\
\text{any simple event}
\end{array}
$$

In theory, you could always resort to the principles developed in Example 4.8 to calculate binomial probabilities; list the sample points and sum their probabilities. However, as the number of trials ($n$) increases, the number of sample points grows very rapidly (the number of sample points is $2^n$). Thus, we prefer the formula for calculating binomial probabilities, since its use avoids listing sample points.

The binomial distribution* is summarized in the box.

---

### The Binomial Probability Distribution

$$p(x) = \binom{n}{x}p^x q^{n-x} \qquad (x = 0, 1, 2, \ldots, n)$$

where

$p$ = Probability of a success on a single trial
$q = 1 - p$
$n$ = Number of trials
$x$ = Number of successes in $n$ trials
$\binom{n}{x} = \dfrac{n!}{x!(n - x)!}$

---

*The binomial distribution is so named because the probabilities, $p(x)$, $x = 0, 1, \ldots, n$, are terms of the binomial expansion, $(q + p)^n$.

As noted in Chapter 3, the symbol 5! means $5 \cdot 4 \cdot 3 \cdot 2 \cdot 1 = 120$. Similarly, $n! = n(n-1)(n-2) \cdots 3 \cdot 2 \cdot 1$; remember, $0! = 1$.

**EXAMPLE 4.9**

Refer to Example 4.8. Use the formula for a binomial random variable to find the probability distribution of $x$, where $x$ is the number of adults who pass the fitness test. Graph the distribution.

### Solution

For this application, we have $n = 4$ trials. Since a success $S$ is defined as an adult who passes the test, $p = P(S) = .1$ and $q = 1 - p = .9$. Substituting $n = 4$, $p = .1$, and $q = .9$ into the formula for $p(x)$, we obtain

$$p(0) = \frac{4!}{0!(4-0)!}(.1)^0(.9)^{4-0} = \frac{4 \cdot 3 \cdot 2 \cdot 1}{(1)(4 \cdot 3 \cdot 2 \cdot 1)}(.1)^0(.9)^4 = 1(1)^0(.9)^4 = .6561$$

$$p(1) = \frac{4!}{1!(4-1)!}(.1)^1(.9)^{4-1} = \frac{4 \cdot 3 \cdot 2 \cdot 1}{(1)(3 \cdot 2 \cdot 1)}(.1)^1(.9)^3 = 4(.1)(.9)^3 = .2916$$

$$p(2) = \frac{4!}{2!(4-2)!}(.1)^2(.9)^{4-2} = \frac{4 \cdot 3 \cdot 2 \cdot 1}{(2 \cdot 1)(2 \cdot 1)}(.1)^2(.9)^2 = 6(.1)^2(.9)^2 = .0486$$

$$p(3) = \frac{4!}{3!(4-3)!}(.1)^3(.9)^{4-3} = \frac{4 \cdot 3 \cdot 2 \cdot 1}{(3 \cdot 2 \cdot 1)(1)}(.1)^3(.9)^1 = 4(.1)^3(.9) = .0036$$

$$p(4) = \frac{4!}{4!(4-4)!}(.1)^4(.9)^{4-4} = \frac{4 \cdot 3 \cdot 2 \cdot 1}{(4 \cdot 3 \cdot 2 \cdot 1)(1)}(.1)^4(.9)^0 = 1(.1)^4(.9) = .0001$$

Note that these probabilities, listed in Table 4.3, sum to 1. A graph of this probability distribution is shown in Figure 4.7.

**Table 4.3**   **Probability Distribution for Physical Fitness Example: Tabular Form**

| $x$ | $p(x)$ |
|---|---|
| 0 | .6561 |
| 1 | .2916 |
| 2 | .0486 |
| 3 | .0036 |
| 4 | .0001 |

**FIGURE 4.7**

*Probability Distribution for Physical Fitness Example: Graphical Form*

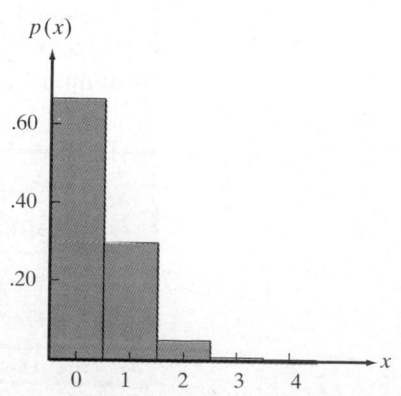

**EXAMPLE 4.10**

Refer to Examples 4.8 and 4.9. Calculate $\mu$ and $\sigma$, the mean and standard deviation, respectively, of the number of the four adults who pass the test. Interpret the results.

**Solution**

From Section 4.3 we know that the mean of a discrete probability distribution is

$$\mu = \Sigma x p(x)$$

Referring to Table 4.3, the probability distribution for the number $x$ who pass the fitness test, we find

$$\mu = 0(.6561) + 1(.2916) + 2(.0486) + 3(.0036) + 4(.0001) = .4$$
$$= 4(.1) = np$$

Thus, in the long run, the average number of adults (out of four) who pass the test is only .4. [*Note: The relationship $\mu = np$ holds in general for a binomial random variable.*]

The variance is

$$\sigma^2 = \Sigma(x - \mu)^2 p(x) = \Sigma(x - .4)^2 p(x)$$
$$= (0 - .4)^2(.6561) + (1 - .4)^2(.2916) + (2 - .4)^2(.0486)$$
$$\quad + (3 - .4)^2(.0036) + (4 - .4)^2(.0001)$$
$$= .104976 + .104976 + .124416 + .024336 + .001296$$
$$= .36 = 4(.1)(.9) = npq$$

[*Note: The relationship $\sigma^2 = npq$ holds in general for a binomial random variable.*] Finally, the standard deviation of the number who pass the fitness test is

$$\sigma = \sqrt{\sigma^2} = \sqrt{.36} = .6$$

Applying the Empirical Rule, we know that approximately 95% of the $x$ values will fall in the interval $\mu \pm 2\sigma = .4 \pm 2(.6) = (-.8, 1.6)$. Since $x$ cannot be negative, we expect (i.e., in the long run) the number of adults out of four who pass the fitness test to be less than 1.6.　■

We emphasize that you need not use the expectation summation rules to calculate $\mu$ and $\sigma^2$ for a binomial random variable. You can find them easily using the formulas $\mu = np$ and $\sigma^2 = npq$.

---

## Mean, Variance, and Standard Deviation for a Binomial Random Variable

Mean: $\mu = np$
Variance: $\sigma^2 = npq$
Standard deviation: $\sigma = \sqrt{npq}$

## Using Binomial Tables

Calculating binomial probabilities becomes tedious when $n$ is large. For some values of $n$ and $p$ the binomial probabilities have been tabulated in Table II of Appendix A. Part of Table II is shown in Table 4.4; a graph of the binomial probability distribution for $n = 10$ and $p = .10$ is shown in Figure 4.8.

**Table 4.4**   **Reproduction of Part of Table II of Appendix A: Binomial Probabilities for $n = 10$**

| $k$ \\ $p$ | .01 | .05 | .10 | .20 | .30 | .40 | .50 | .60 | .70 | .80 | .90 | .95 | .99 |
|---|---|---|---|---|---|---|---|---|---|---|---|---|---|
| 0 | .904 | .599 | .349 | .107 | .028 | .006 | .001 | .000 | .000 | .000 | .000 | .000 | .000 |
| 1 | .996 | .914 | .736 | .376 | .149 | .046 | .011 | .002 | .000 | .000 | .000 | .000 | .000 |
| 2 | 1.000 | .988 | .930 | .678 | .383 | .167 | .055 | .012 | .002 | .000 | .000 | .000 | .000 |
| 3 | 1.000 | .999 | .987 | .879 | .650 | .382 | .172 | .055 | .011 | .001 | .000 | .000 | .000 |
| 4 | 1.000 | 1.000 | .998 | .967 | .850 | .633 | .377 | .166 | .047 | .006 | .000 | .000 | .000 |
| 5 | 1.000 | 1.000 | 1.000 | .994 | .953 | .834 | .623 | .367 | .150 | .033 | .002 | .000 | .000 |
| 6 | 1.000 | 1.000 | 1.000 | .999 | .989 | .945 | .828 | .618 | .350 | .121 | .013 | .001 | .000 |
| 7 | 1.000 | 1.000 | 1.000 | 1.000 | .998 | .988 | .945 | .833 | .617 | .322 | .070 | .012 | .000 |
| 8 | 1.000 | 1.000 | 1.000 | 1.000 | 1.000 | .988 | .989 | .954 | .851 | .624 | .264 | .086 | .004 |
| 9 | 1.000 | 1.000 | 1.000 | 1.000 | 1.000 | 1.000 | .999 | .994 | .972 | .893 | .651 | .401 | .096 |

Table II actually contains a total of nine tables, labeled (**a**) through (**i**), one each corresponding to $n = 5, 6, 7, 8, 9, 10, 15, 20,$ and 25, respectively. In each of these tables the columns correspond to values of $p$, and the rows correspond to values of the random variable $x$. The entries in the table represent **cumulative binomial probabilities**. Thus, for example, the entry in the column corresponding to $p = .10$ and the row corresponding to $x = 2$ is .930 (highlighted), and its interpretation is

$$P(x \leq 2) = P(x = 0) + P(x = 1) + P(x = 2) = .930$$

This probability is also highlighted in the graphical representation of the binomial distribution with $n = 10$ and $p = .10$ in Figure 4.8.

You can also use Table II to find the probability that $x$ equals a specific value. For example, suppose you want to find the probability that $x = 2$ in the binomial distribution with $n = 10$ and $p = .10$. This is found by subtraction as follows:

$$P(x = 2) = [P(x = 0) + P(x = 1) + P(x = 2)] - [P(x = 0) + P(x = 1)]$$
$$= P(x \leq 2) - P(x \leq 1) = .930 - .736 = .194$$

The probability that a binomial random variable exceeds a specified value can be found using Table II and the notion of complementary events. For example, to find the probability that $x$ exceeds 2 when $n = 10$ and $p = .10$, we use

$$P(x > 2) = 1 - P(x \leq 2) = 1 - .930 = .070$$

**FIGURE 4.8**

*Binomial Probability Distribution for $n = 10$ and $p = .10$; $P(x \leq 2)$ Highlighted*

Note that this probability is represented by the unhighlighted portion of the graph in Figure 4.8.

All probabilities in Table II are rounded to three decimal places. Thus, although none of the binomial probabilities in the table is exactly zero, some are small enough (less than .0005) to round to .000. For example, using the formula to find $P(x = 0)$ when $n = 10$ and $p = .6$, we obtain

$$P(x = 0) = \binom{10}{0}(.6)^0(.4)^{10-0} = .4^{10} = .00010486$$

but this is rounded to .000 in Table II of Appendix A (see Table 4.4).

Similarly, none of the table entries is exactly 1.0, but when the cumulative probabilities exceed .9995, they are rounded to 1.000. The row corresponding to the largest possible value for $x$, $x = n$, is omitted, because all the cumulative probabilities in that row are equal to 1.0 (exactly). For example, in Table 4.4 with $n = 10$, $P(x \le 10) = 1.0$, no matter what the value of $p$.

The following example further illustrates the use of Table II.

**EXAMPLE 4.11**

Suppose a poll of 20 voters is taken in a large city. The purpose is to determine $x$, the number who favor a certain candidate for mayor. Suppose that 60% of all the city's voters favor the candidate.

a. Find the mean and standard deviation of $x$.
b. Use Table II of Appendix A to find the probability that $x \le 10$.
c. Use Table II to find the probability that $x > 12$.
d. Use Table II to find the probability that $x = 11$.
e. Graph the probability distribution of $x$ and locate the interval $\mu \pm 2\sigma$ on the graph.

**Solution**

a. The number of voters polled is presumably small compared with the total number of eligible voters in the city. Thus, we may treat $x$, the number of the 20 who favor the mayoral candidate, as a binomial random variable. The value of $p$ is the fraction of the total voters who favor the candidate; i.e., $p = .6$. Therefore, we calculate the mean and variance:

$$\mu = np = 20(.6) = 12$$
$$\sigma^2 = npq = 20(.6)(.4) = 4.8$$
$$\sigma = \sqrt{4.8} = 2.19$$

b. Looking in the $k = 10$ row and the $p = .6$ column of Table II (Appendix A) for $n = 20$, we find the value of .245. Thus,

$$P(x \le 10) = .245$$

c. To find the probability

$$P(x > 12) = \sum_{x=13}^{20} p(x)$$

we use the fact that for all probability distributions,

$$\sum_{\text{all } x} p(x) = 1.$$

Therefore,

$$P(x > 12) = 1 - P(x \leq 12) = 1 - \sum_{x=0}^{12} p(x)$$

Consulting Table II, we find the entry in row $k = 12$, column $p = .6$ to be .584. Thus,

$$P(x > 12) = 1 - .584 = .416$$

d. To find the probability that exactly 11 voters favor the candidate, recall that the entries in Table II are cumulative probabilities and use the relationship

$$P(x = 11) = [p(0) + p(1) + \cdots + p(11)] - [p(0) + p(1) + \cdots + p(10)]$$
$$= P(x \leq 11) - P(x \leq 10)$$

Then

$$P(x = 11) = .404 - .245 = .159$$

e. The probability distribution for $x$ is shown in Figure 4.9. Note that

$$\mu - 2\sigma = 12 - 2(2.2) = 7.6 \qquad \mu + 2\sigma = 12 + 2(2.2) = 16.4$$

The interval $\mu - 2\sigma$ to $\mu + 2\sigma$ is shown in Figure 4.9. The probability that $x$ falls in the interval $\mu + 2\sigma$ is $P(x = 8, 9, 10, \ldots, 16) = P(x \leq 16) - P(x \leq 7) = .984 - .021 = .963$. Note that this probability is very close to the .95 given by the Empirical Rule. Thus, we expect the number of voters in the sample of 20 who favor the mayoral candidate to be between 8 and 16.    ■

**FIGURE 4.9**

*The Binomial Probability Distribution for x in Example 4.11: n = 20 and p = .6*

# Using the TI-83 Graphing Calculator

## Binomial Probabilities on the TI-83

### $P(x = k)$

To compute $P(x = k)$, the probability of **k** successes in **n** trials, where the **p** is probability of success for each trial, use the **binompdf(** command. Binom**p**df stands for "binomial probability density function." This command is under the **DISTR**ibution menu and has the format **binompdf(n, p, k)**.

Example:  Compute the probability of 5 successes in 8 trials where the probability of success for a single trial is 40%.

In this example, $n = 8$, $p = .4$, and $k = 5$

Press  **2nd**  **VARS**  for  **DISTR**
Press the down arrow key until  **0:binompdf**  is highlighted
Press  **ENTER**
After  **binompdf(**, type  **8, .4, 5)**
Press  **ENTER**

You should see

```
binompdf(8,.4,5)
        .12386304
```

Thus, $P(x = 5)$ is about 12.4%.

### $P(x \le k)$

To compute $P(x \le k)$, the probability of **k** or fewer successes in **n** trials, where the **p** is probability of success for each trial, use the **binomcdf(** command. Binom**c**df stands for "binomial *cumulative* probability density function." This command is under the **DISTR**ibution menu and has the format **binomcdf(n, p, k)**.

Example:  Compute the probability of less than 5 successes in 8 trials where the probability of success for a single trial is 40%.

In this example, $n = 8$, $p = .4$, and $k = 5$

Press  **2nd**  **VARS**  for  **DISTR**
Press down the arrow key until  **A:binomcdf**  is highlighted
Press  **ENTER**
After  **binomcdf(**, type  **8, .4, 5)**
Press  **ENTER**

---

**USING THE TI-83 GRAPHING CALCULATOR (*continued*)**

You should see

```
binomcdf(8,.4,5)
           .95019264
```

Thus, $P(x \leq 5)$ is about 95%.

**$P(x < k), P(x > k), P(x \geq k)$**

To find the probability of less than $k$ successes $P(x < k)$, more than $k$ successes $P(x > k)$, or at least $k$ successes $P(x \geq k)$, variations of the **binomcdf(** command must be used as shown below.

| | | |
|---|---|---|
| $P(x < k)$ | use | **binomcdf($n, p, k - 1$)** |
| $P(x > k)$ | use | **$1 - $ binomcdf($n, p, k$)** |
| $P(x \geq k)$ | use | **$1 - $ binomcdf($n, p, k - 1$)** |

where $n = $ number of trials and $p = $ probability of success in each trial.

---

 EXERCISES 4.35–4.56

**Learning the Mechanics**

**4.35** Compute the following:

  **a.** $\dfrac{6!}{2!(6 - 2)!}$

  **b.** $\dbinom{5}{2}$

  **c.** $\dbinom{7}{0}$

  **d.** $\dbinom{6}{6}$

  **e.** $\dbinom{4}{3}$

**4.36** Consider the following probability distribution:

$$p(x) = \binom{5}{x}(.7)^x(.3)^{5-x} \quad (x = 0, 1, 2, \ldots, 5)$$

  **a.** Is $x$ a discrete or a continuous random variable? Explain.

  **b.** What is the name of this probability distribution?

  **c.** Graph the probability distribution.

  **d.** Find the mean and standard deviation of $x$.

  **e.** Show the mean and the 2-standard-deviation interval on each side of the mean on the graph you drew in part **c**.

**4.37** If $x$ is a binomial random variable, compute $p(x)$ for each of the following cases:

  **a.** $n = 5, x = 1, p = .2$

  **b.** $n = 4, x = 2, q = .4$

  **c.** $n = 3, x = 0, p = .7$

  **d.** $n = 5, x = 3, p = .1$

  **e.** $n = 4, x = 2, q = .6$

  **f.** $n = 3, x = 1, p = .9$

**4.38** Suppose $x$ is a binomial random variable with $n = 3$ and $p = .3$.

  **a.** Calculate the value of $p(x)$, $x = 0, 1, 2, 3$, using the formula for a binomial probability distribution.

  **b.** Using your answers to part **a**, give the probability distribution for $x$ in tabular form.

**4.39** If $x$ is a binomial random variable, calculate $\mu$, $\sigma^2$, and $\sigma$ for each of the following:

  **a.** $n = 25, p = .5$

  **b.** $n = 80, p = .2$

  **c.** $n = 100, p = .6$

  **d.** $n = 70, p = .9$

  **e.** $n = 60, p = .8$

  **f.** $n = 1,000, p = .04$

**4.40** If $x$ is a binomial random variable, use Table II in Appendix A to find the following probabilities:

  **a.** $P(x = 2)$ for $n = 10, p = .4$

  **b.** $P(x \leq 5)$ for $n = 15, p = .6$

c. $P(x > 1)$ for $n = 5, p = .1$

d. $P(x < 10)$ for $n = 25, p = .7$

e. $P(x \geq 10)$ for $n = 15, p = .9$

f. $P(x = 2)$ for $n = 20, p = .2$

**4.41** Suppose $x$ is a binomial random variable with $n = 5$ and $p = .5$. Compute $p(x)$ for $x = 0, 1, 2, 3, 4$, and 5 using the following two methods:

a. List the sample points (using $S$ for Success and $F$ for Failure on each trial) corresponding to each value of $x$, assign probabilities to each sample point, and obtain $p(x)$ by adding sample point probabilities.

b. Use the formula for the binomial probability distribution to obtain $p(x)$.

**4.42** The binomial probability distribution is a family of probability distributions with each single distribution depending on the values of $n$ and $p$. Assume that $x$ is a binomial random variable with $n = 4$.

a. Determine a value of $p$ such that the probability distribution of $x$ is symmetric.

b. Determine a value of $p$ such that the probability distribution of $x$ is skewed to the right.

c. Determine a value of $p$ such that the probability distribution of $x$ is skewed to the left.

d. Graph each of the binomial distributions you obtained in parts **a**, **b**, and **c**. Locate the mean for each distribution on its graph.

e. In general, for what values of $p$ will a binomial distribution be symmetric? Skewed to the right? Skewed to the left?

## Applying the Concepts—Basic

**4.43** According to the University of South Florida's Tobacco Research and Intervention Program, only 5% of the nation's cigarette smokers ever enter into a treatment program to help them quit smoking (*USF Magazine*, Spring 2000). In a random sample of 200 smokers, let $x$ be the number who enter into a treatment program.

a. Explain why $x$ is a binomial random variable (to a reasonable degree of approximation).

b. What is the value of $p$? Interpret this value.

c. What is the expected value of $x$? Interpret this value.

**4.44** The American College of Obstetricians and Gynecologists report that 22% of all births in the U.S. take place by Caesarian section each year (*USA Today*, Sept. 19, 2000).

a. In a random sample of 1,000 births, how many, on average will take place by Caesarian section?

b. What is the standard deviation of the number of Caesarian section births in a sample of 1,000 births?

c. Use your answers to parts **a** and **b** to form an interval that is likely to contain the number of Caesarian section births in a sample of 1,000 births.

**4.45** A national poll conducted by *The New York Times* (May 7, 2000) revealed that 80% of Americans believe that after you die, some part of you lives on, either in a next life on earth or in heaven. Consider a random sample of 10 Americans and count $x$, the number who believe in life after death.

a. Find $P(x = 3)$.

b. Find $P(x \leq 7)$.

c. Find $P(x > 4)$.

**4.46** According to a nationwide survey, 60% of parents with young children condone spanking their child as a regular form of punishment (*Tampa Tribune*, Oct. 5, 2000). Consider a random sample of three people, each of whom is a parent with young children. Assume that $x$, the number in the sample that condone spanking, is a binomial random variable.

a. What is the probability that none of the three parents condones spanking as a regular form of punishment for their children?

b. What is the probability that at least one condones spanking as a regular form of punishment?

c. Give the mean and standard deviation of $x$. Interpret the results.

**4.47** A Federal Trade Commission (FTC) study of the pricing accuracy of electronic checkout scanners at stores found that one of every 30 items is priced incorrectly (*Price Check II: A Follow-Up Report on the Accuracy of Checkout Scanner Prices*, Dec. 16, 1998). Suppose the FTC randomly selects five items at a retail store and checks the accuracy of the scanner price of each. Let $x$ represent the number of the five items that is priced incorrectly.

a. Show that $x$ is (approximately) a binomial random variable.

b. Use the information in the FTC study to estimate $p$ for the binomial experiment.

c. What is the probability that exactly one of the five items is priced incorrectly by the scanner?

d. What is the probability that at least one of the five items is priced incorrectly by the scanner?

## Applying the Concepts—Intermediate

**4.48** The degree to which democratic and non-democratic countries attempt to control the news media was examined in the *Journal of Peace Research* (Nov. 1997). Between 1948 and 1996, 80% of all democratic regimes allowed a free press. In contrast, over the same time period, 10% of all non-democratic regimes allowed a free press.

a. In a random sample of 50 democratic regimes, how many would you expect to allow a free press? Give a range that is highly likely to include the number of democratic regimes with a free press.

b. In a random sample of 50 non-democratic regimes, how many would you expect to allow a free press? Give a range that is highly likely to include the number of non-democratic regimes with a free press.

**4.49** The *American Journal of Public Health* (July 1995) published a study of the relationship between passive smoking and nasal allergies in Japanese female students. The study revealed that 80% of students from heavy-smoking families showed signs of nasal allergies on physical examinations. Consider a sample of 25 Japanese female students exposed daily to heavy smoking.

**a.** What is the probability that fewer than 20 of the students will have nasal allergies?

**b.** What is the probability that more than 15 of the students will have nasal allergies?

**c.** On average, how many of the 25 students would you expect to show signs of nasal allergies?

**4.50** The United States Department of Agriculture (USDA) reports that, under its standard inspection system, one in every 100 slaughtered chickens passes inspection with fecal contamination (*Tampa Tribune*, Mar. 31, 2000). In Exercise 3.13 (p. 118), you found the probability that a randomly selected slaughtered chicken passes inspection with fecal contamination. Now find the probability that, in a random sample of five slaughtered chickens, at least one passes inspection with fecal contamination.

**4.51** According to researchers at Johns Hopkins University School of Medicine, one in every three women has been a victim of domestic abuse (*Annals of Internal Medicine*, Nov. 1995). This probability was obtained from a survey of nearly 2,000 adult women residing in Baltimore, Maryland. Suppose we randomly sample 15 women and find that four have been abused.

**a.** What is the probability of observing four or more abused women in a sample of 15 if the proportion $p$ of women who are victims of domestic abuse is really $p = \frac{1}{3}$?

**b.** Many experts on domestic violence believe that the proportion of women who are domestically abused is closer to $p = .10$. Calculate the probability of observing four or more abused women in a sample of 15 if $p = .10$.

**c.** Why might your answers to parts **a** and **b** lead you to believe that $p = \frac{1}{3}$?

**4.52** Refer to Exercise 3.65 (p. 147) and the experiment conducted by the Tampa Bay Skeptics to see whether an acclaimed psychic has extrasensory perception (ESP). Recall that a crystal is placed, at random, inside one of 10 identical boxes lying side-by-side on a table. The experiment was repeated seven times, and $x$, the number of correct decisions, was recorded. (Assume that the seven trials are independent.)

**a.** If the psychic is guessing—i.e., if the psychic does *not* possess ESP—what is the value of $p$, the probability of a correct decision on each trial?

**b.** If the psychic is guessing, what is the expected number of correct decisions in seven trials?

**c.** If the psychic is guessing, what is the probability of no correct decisions in seven trials?

**d.** Now suppose the psychic has ESP and $p = .5$. What is the probability that the psychic guesses incorrectly in all seven trials?

**e.** Refer to part **d**. Recall that the psychic failed to select the box with the crystal on all seven trials. Is this evidence against the psychic having ESP? Explain.

## Applying the Concepts—Advanced

**4.53** According to the U.S. Golf Association (USGA), "The weight of the [golf] ball shall not be greater than 1.620 ounces avoirdupois (45.93 grams).... The diameter of the ball shall not be less than 1.680 inches.... The velocity of the ball shall not be greater than 250 feet per second" (USGA, 2001). The USGA periodically checks the specifications of golf balls sold in the United States by randomly sampling balls from pro shops around the country. Two dozen of each kind are sampled, and if more than three do not meet size and/or velocity requirements, that kind of ball is removed from the USGA's approved-ball list.

**a.** What assumptions must be made and what information must be known in order to use the binomial probability distribution to calculate the probability that the USGA will remove a particular kind of golf ball from its approved-ball list?

**b.** Suppose 10% of all balls produced by a particular manufacturer are less than 1.680 inches in diameter, and assume that the number of such balls, $x$, in a sample of two dozen balls can be adequately characterized by a binomial probability distribution. Find the mean and standard deviation of the binomial distribution.

**c.** Refer to part **b**. If $x$ has a binomial distribution, then so does the number, $y$, of balls in the sample that meet the USGA's minimum diameter. [*Note:* $x + y = 24$.] Describe the distribution of $y$. In particular, what are $p$, $q$, and $n$? Also, find $E(y)$ and the standard deviation of $y$.

**4.54** Ataxia-telangiectasia (A-T) is a neurological disorder that weakens immune systems and causes premature aging. According to *Science News* (June 24, 1995), when both members of a couple carry the A-T gene, their children have a one in five chance of developing the disease.

**a.** Consider 15 couples in which both members of each couple carry the A-T gene. What is the probability that more than 8 of 15 couples have children that develop the neurological disorder?

**b.** Consider 10,000 couples in which both members of each couple carry the A-T gene. Is it likely that fewer than 3,000 will have children that develop the disease? [*Hint:* Calculate the mean and standard deviation of the distribution.]

**4.55** Suppose you are a purchasing officer for a large company. You have purchased 5 million electrical switches and your supplier has guaranteed that the shipment will contain no more than .1% defectives. To check the shipment, you randomly sample 500 switches, test them, and find that four are defective. Based on this evidence, do you think the supplier has complied with the guarantee? Explain.

**4.56** A literature professor decides to give a 20-question true-false quiz to determine who has read an assigned novel. She wants to choose the passing grade such that the probability of passing a student who guesses on every question is less than .05. What score should she set as the lowest passing grade?

## 4.5    THE POISSON RANDOM VARIABLE (OPTIONAL)

A type of probability distribution that is often useful in describing the number of events that will occur in a specific period of time or in a specific area or volume is the **Poisson distribution** (named after the 18th-century physicist and mathematician, Siméon Poisson). Typical examples of random variables for which the Poisson probability distribution provides a good model are

1. The number of traffic accidents per month at a busy intersection
2. The number of noticeable surface defects (scratches, dents, etc.) found by quality inspectors on a new automobile
3. The parts per million of some toxin found in the water or air emission from a manufacturing plant
4. The number of diseased trees per acre of a certain woodland
5. The number of death claims received per day by an insurance company
6. The number of unscheduled admissions per day to a hospital

---

### Characteristics of a Poisson Random Variable

1. The experiment consists of counting the number of times a certain event occurs during a given unit of time or in a given area or volume (or weight, distance, or any other unit of measurement).
2. The probability that an event occurs in a given unit of time, area, or volume is the same for all the units.
3. The number of events that occur in one unit of time, area, or volume is independent of the number that occur in other units.
4. The mean (or expected) number of events in each unit is denoted by the Greek letter lambda, $\lambda$.

---

The characteristics of the Poisson random variable are usually difficult to verify for practical examples. The examples given satisfy them well enough that the Poisson distribution provides a good model in many instances.* As with all probability models, the real test of the adequacy of the Poisson model is in whether it provides a reasonable approximation to reality—that is, whether empirical data support it.

The probability distribution, mean, and variance for a Poisson random variable are shown in the next box.

---

### Probability Distribution, Mean, and Variance for a Poisson Random Variable

$$p(x) = \frac{\lambda^x e^{-\lambda}}{x!} \quad (x = 0, 1, 2, \dots)$$

$$\mu = \lambda \qquad \sigma^2 = \lambda$$

where

$\lambda$ = Mean number of events during given unit of time, area, volume, etc.

---

*The Poisson probability distribution also provides a good approximation to a binomial probability distribution with mean $\lambda = np$ when $n$ is large and $p$ is small (say, $np \leq 7$).

The calculation of Poisson probabilities is made easier by the use of Table III in Appendix A, which gives the cumulative probabilities $P(x \leq k)$ for various values of $\lambda$. The use of Table III is illustrated in Example 4.12.

**EXAMPLE 4.12**

Ecologists often use the number of reported sightings of a rare species of animal to estimate the remaining population size. For example, suppose the number, $x$, of reported sightings per week of blue whales is recorded. Assume that $x$ has (approximately) a Poisson probability distribution. Furthermore, assume that the average number of weekly sightings is 2.6.

a. Find the mean and standard deviation of $x$, the number of blue whale sightings per week. Interpret the results.
b. Use Table III to find the probability that fewer than two sightings are made during a given week.
c. Use Table III to find the probability that more than five sightings are made during a given week.
d. Use Table III to find the probability that exactly five sightings are made during a given week.

### Solution

a. The mean and variance of a Poisson random variable are both equal to $\lambda$. Thus, for this example,

$$\mu = \lambda = 2.6$$
$$\sigma^2 = \lambda = 2.6$$

Then the standard deviation of $x$ is

$$\sigma = \sqrt{2.6} = 1.61$$

Remember that the mean measures the central tendency of the distribution and does not necessarily equal a possible value of $x$. In this example, the mean is 2.6 sightings, and although there cannot be 2.6 sightings during a given week, the average number of weekly sightings is 2.6. Similarly, the standard deviation of 1.61 measures the variability of the number of sightings per week. Perhaps a more helpful measure is the interval $\mu \pm 2\sigma$, which in this case stretches from $-.62$ to $5.82$. We expect the number of sightings to fall in this interval most of the time—with at least 75% relative frequency (according to Chebyshev's Rule) and probably with more than 90% relative frequency (the Empirical Rule). The mean and the 2-standard-deviation interval around it are shown in Figure 4.10.

**FIGURE 4.10**

*Probability Distribution for Number of Blue Whale Sightings*

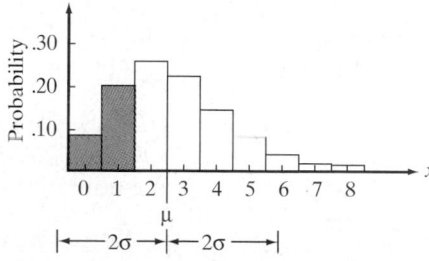

b. A partial reproduction of Table III is shown in Table 4.5. The rows of the table correspond to different values of $\lambda$, and the columns correspond to different values $k$ of the Poisson random variable $x$. The entries in the table (like the binomial probabilities in Table II) give the cumulative probability $P(x \leq k)$. To find the probability that fewer than two sightings are made during a given week, we first note that

$$P(x < 2) = P(x \leq 1)$$

**Table 4.5  Reproduction of Part of Table III in Appendix A**

| λ \ k | 0 | 1 | 2 | 3 | 4 | 5 | 6 | 7 | 8 | 9 |
|---|---|---|---|---|---|---|---|---|---|---|
| 2.2 | .111 | .355 | .623 | .819 | .928 | .975 | .993 | .998 | 1.000 | 1.000 |
| 2.4 | .091 | .308 | .570 | .779 | .904 | .964 | .988 | .997 | .999 | 1.000 |
| 2.6 | .074 | .267 | .518 | .736 | .877 | .951 | .983 | .995 | .999 | 1.000 |
| 2.8 | .061 | .231 | .469 | .692 | .848 | .935 | .976 | .992 | .998 | .999 |
| 3.0 | .050 | .199 | .423 | .647 | .815 | .916 | .966 | .988 | .996 | .999 |
| 3.2 | .041 | .171 | .380 | .603 | .781 | .895 | .955 | .983 | .994 | .998 |
| 3.4 | .033 | .147 | .340 | .558 | .744 | .871 | .942 | .977 | .992 | .997 |
| 3.6 | .027 | .126 | .303 | .515 | .706 | .844 | .927 | .969 | .988 | .996 |
| 3.8 | .022 | .107 | .269 | .473 | .668 | .816 | .909 | .960 | .984 | .994 |
| 4.0 | .018 | .092 | .238 | .433 | .629 | .785 | .889 | .949 | .979 | .992 |
| 4.2 | .015 | .078 | .210 | .395 | .590 | .753 | .867 | .936 | .972 | .989 |
| 4.4 | .012 | .066 | .185 | .359 | .551 | .720 | .844 | .921 | .964 | .985 |
| 4.6 | .010 | .056 | .163 | .326 | .513 | .686 | .818 | .905 | .955 | .980 |
| 4.8 | .008 | .048 | .143 | .294 | .476 | .651 | .791 | .887 | .944 | .975 |
| 5.0 | .007 | .040 | .125 | .265 | .440 | .616 | .762 | .867 | .932 | .968 |
| 5.2 | .006 | .034 | .109 | .238 | .406 | .581 | .732 | .845 | .918 | .960 |
| 5.4 | .005 | .029 | .095 | .213 | .373 | .546 | .702 | .822 | .903 | .951 |
| 5.6 | .004 | .024 | .082 | .191 | .342 | .512 | .670 | .797 | .886 | .941 |
| 5.8 | .003 | .021 | .072 | .170 | .313 | .478 | .638 | .771 | .867 | .929 |
| 6.0 | .002 | .017 | .062 | .151 | .285 | .446 | .606 | .744 | .847 | .916 |

This probability is a cumulative probability and therefore is the entry in Table III in the row corresponding to $\lambda = 2.6$ and the column corresponding to $k = 1$. The entry is .267, shown highlighted in Table 4.5. This probability corresponds to the highlighted area in Figure 4.10 and may be interpreted as meaning that there is a 26.7% chance that fewer than two sightings will be made during a given week.

c. To find the probability that more than five sightings are made during a given week, we consider the complementary event

$$P(x > 5) = 1 - P(x \le 5) = 1 - .951 = .049$$

where .951 is the entry in Table III corresponding to $\lambda = 2.6$ and $k = 5$ (see Table 4.5). Note from Figure 4.10 that this is the area in the interval $\mu \pm 2\sigma$, or $-.62$ to 5.82. Then the number of sightings should exceed 5—or, equivalently, should be more than 2 standard deviations from the mean—during only about 4.9% of all weeks. Note that this percentage agrees remarkably well with that given by the Empirical Rule for mound-shaped distributions, which tells us to expect approximately 5% of the measurements (values of the random variable) to lie farther than 2 standard deviations from the mean.

d. To use Table III to find the probability that *exactly* five sightings are made during a given week, we must write the probability as the difference between two cumulative probabilities:

$$P(x = 5) = P(x \le 5) - P(x \le 4) = .951 - .877 = .074$$

∎

Note that the probabilities in Table III are all rounded to three decimal places. Thus, although in theory a Poisson random variable can assume infinitely large values, the values of $k$ in Table III are extended only until the cumulative probability is 1.000. This does not mean that $x$ *cannot* assume larger values, but only that the likelihood is less than .001 (in fact, less than .0005) that it will do so.

Finally, you may need to calculate Poisson probabilities for values of $\lambda$ not found in Table III. You may be able to obtain an adequate approximation by interpolation, but if not, consult more extensive tables for the Poisson distribution.

## Using the TI-83 Graphing Calculator

### Poisson Probabilities on the TI-83

**$P(x = k)$**

To compute $P(x = k)$, the probability of exactly **k** successes in a specified interval where $\lambda$ is the mean number of successes in the interval, use the **poissonpdf(** command. Poisson**p**df stands for "Poisson probability density function." This command is under the **DISTR**ibution menu and has the format **poissonpdf($\lambda$, k)**.

Example: Suppose that the number, $x$, of reported sightings per week of blue whales is recorded. Assume that $x$ has approximately a Poisson probability distribution, and that the average number of weekly sightings is 2.6.

Compute the probability that exactly five sightings are made during a given week.

In this example, $\lambda = 2.5$ and $k = 5$

> Press **2nd** **VARS** for **DISTR**
> Press the down arrow key until **B:poissonpdf** is highlighted
> Press **ENTER**
> After **poissonpdf(**, type **2.6, 5)**
> Press **ENTER**

You should see

```
poissonpdf(2.6,5
)
        .0735393591
```

Thus, the $P(x = 5)$ is about 7.4%.

**$P(x \le k)$**

To compute $P(x \le k)$, the probability of **k** or fewer successes in a specified interval, where $\lambda$ is the mean number of successes in the interval, use the **poissoncdf(** command. Poisson**c**df stands for "Poisson cumulative probability density function." This command is under the **DISTR**ibution menu and has the format **poissoncdf($\lambda$, k)**.

Example: In the example given above, compute the probability 5 or fewer sightings are made during a given week.

*(continues on next page)*

## USING TI-83 GRAPHING CALCULATOR (*continued*)

In this example, $\lambda = 2.6$ and $k = 5$

Press **2nd** **VARS** for **DISTR**
Press the down arrow key until **C:poissoncdf** is highlighted
Press **ENTER**
After **poissoncdf(**, type **2.6, 5**
Press **ENTER**

You should see

```
poissoncdf(2.6,5
)
        .9509628481
```

Thus, the $P(x \le 5)$ is about 95.1%.

**$P(x < k), P(x > k), P(x \ge k)$**
To find the probability of less than $k$ successes $P(x < k)$, more than $k$ successes $P(x > k)$, or at least $k$ successes $P(x \ge k)$, variations of **poissoncdf(** command must be used as shown below.

| | | |
|---|---|---|
| **$P(x < k)$** | use | **poissoncdf($\lambda$, $k$-1)** |
| **$P(x > k)$** | use | **1 − poissoncdf($\lambda$, $k$)** |
| **$P(x \ge k)$** | use | **1 − poissoncdf($\lambda$, $k$-1)** |

where $\lambda$ is the mean number of successes in the specified interval.

## EXERCISES 4.57–4.71

### Learning the Mechanics

**4.57** Consider the probability distribution shown here:

$$p(x) = \frac{3^x e^{-3}}{x!} \qquad (x = 0, 1, 2, \dots)$$

   **a.** Is $x$ a discrete or continuous random variable? Explain.
   **b.** What is the name of this probability distribution?
   **c.** Graph the probability distribution.
   **d.** Find the mean and standard deviation of $x$.
   **e.** Find the mean and standard deviation of the probability distribution.

**4.58** Given that $x$ is a random variable for which a Poisson probability distribution provides a good approximation, use Table III to compute the following:

   **a.** $P(x \le 2)$ when $\lambda = 1$
   **b.** $P(x \le 2)$ when $\lambda = 2$
   **c.** $P(x \le 2)$ when $\lambda = 3$
   **d.** What happens to the probability of the event $\{x \le 2\}$ as $\lambda$ increases from 1 to 3? Is this intuitively reasonable?

**4.59** Assume that $x$ is a random variable having a Poisson probability distribution with a mean of 1.5. Use Table III to find the following probabilities:

   **a.** $P(x \le 3)$
   **b.** $P(x \ge 3)$
   **c.** $P(x = 3)$
   **d.** $P(x = 0)$
   **e.** $P(x > 0)$
   **f.** $P(x > 6)$

**4.60** Suppose $x$ is a random variable for which a Poisson probability distribution with $\lambda = 1$ provides a good characterization.
  **a.** Graph $p(x)$ for $x = 0, 1, 2, \ldots, 9$.
  **b.** Find $\mu$ and $\sigma$ for $x$, and locate $\mu$ and the interval $\mu \pm 2\sigma$ on the graph.
  **c.** What is the probability that $x$ will fall within the interval $\mu \pm 2\sigma$?

**4.61** Suppose $x$ is a random variable for which a Poisson probability distribution with $\lambda = 3$ provides a good characterization.
  **a.** Graph $p(x)$ for $x = 0, 1, 2, \ldots, 9$.
  **b.** Find $\mu$ and $\sigma$ for $x$, and locate $\mu$ and the interval $\mu \pm 2\sigma$ on the graph.
  **c.** What is the probability that $x$ will fall within the interval $\mu \pm 2\sigma$?

**4.62** As mentioned in Section 4.5, when $n$ is large, $p$ is small, and $np \leq 7$, the Poisson probability distribution provides a good approximation to the binomial probability distribution. Since we provide exact binomial probabilities (Table II in Appendix A) for relatively small values of $n$, you can investigate the adequacy of the approximation for $n = 25$. Use Table II to find $p(0)$, $p(1)$, and $p(2)$ for $n = 25$ and $p = .05$. Calculate the corresponding Poisson approximations using $\lambda = \mu = np$. [*Note:* These approximations are reasonably good for $n$ as small as 25, but to use the approximation in a practical situation we would prefer to have $n \geq 100$.]

## Applying the Concepts—Basic

**4.63** Cognitive scientists at the University of Massachusetts designed an experiment to measure $x$, the number of times a reader's eye fixated on a single word before moving past that word (*Memory and Cognition*, Sept. 1997). For this experiment, $x$ was found to have a mean of 1. Suppose one of the readers in the experiment is randomly selected and assume that $x$ has a Poisson distribution.
  **a.** Find $P(x = 0)$.
  **b.** Find $P(x > 1)$.
  **c.** Find $P(x < 2)$.

**4.64** A can company reports that the number of breakdowns per 8-hour shift on its machine-operated assembly line follows a Poisson distribution with a mean of 1.5.
  **a.** What is the probability of exactly two breakdowns on the midnight shift?
  **b.** What is the probability of fewer than two breakdowns on the afternoon shift?
  **c.** What is the probability that more than two breakdowns occur on the midnight shift?
  **d.** What is the probability of no breakdowns during three consecutive 8-hour shifts? (Assume that the machine operates independently across shifts.)

**4.65** U.S. airlines average about 3.8 fatalities per month (*Statistical Abstract of the United States: 2000*). Assume the probability distribution for $x$, the number of fatalities per month, can be approximated by a Poisson probability distribution.
  **a.** What is the probability that no fatalities will occur during any given month?
  **b.** What is the probability that one fatality will occur during any given month?
  **c.** Find $E(x)$ and the standard deviation of $x$.

**4.66** Economists at the University of New Mexico modeled the number of casualties (deaths or missing persons) experienced by a deep-draft U.S. flag vessel over a three-year period as a Poisson random variable, $x$. The researchers estimated $E(x)$ to be .03 (*Management Science*, Jan. 1999).
  **a.** Find the variance of $x$.
  **b.** Discuss the conditions that would make the researchers' Poisson assumption plausible.
  **c.** What is the probability that a deep-draft U.S. flag vessel will have no casualties in a three-year time period?

## Applying the Concepts—Intermediate

**4.67** The Environmental Protection Agency (EPA) limits the amount of vinyl chloride in plant air emissions to no more than 10 parts per million. Suppose the mean emission of vinyl chloride for a particular plant is 4 parts per million. Assume that the number of parts per million of vinyl chloride in air samples, $x$, follows a Poisson probability distribution.
  **a.** What is the standard deviation of $x$ for the plant?
  **b.** Is it likely that a sample of air from the plant would yield a value of $x$ that would exceed the EPA limit? Explain.
  **c.** Discuss conditions that would make the Poisson assumption plausible.

**4.68** Davey Crockett, a U.S. congressman during the 1830s, published *A Narrative of the Life of Davey Crockett, Written by Himself*. In a *Chance* (Spring 1999) article, researchers determined that $x$, the number of times (per 1,000 words) that the word "though" appears in the *Life* narrative, has a mean of .25. Assume that $x$ has an approximate Poisson distribution.
  **a.** Find the variance of $x$.
  **b.** Calculate the interval $\mu \pm 2\sigma$. What proportion of the time will $x$ fall in this interval?
  **c.** In 1836, Davey Crockett died at the Alamo defending Texans from an attack by the Mexican general, Santa Anna. The following year, the narrative *Colonel Crockett's Exploits and Adventures in Texas* was published. Some historians have doubts whether Crockett was the actual author of this narrative. In the *Texas* narrative, "though" appears twice in the first 1,000 words. Use this information and your answer to part **b** to make an inference about whether Crockett actually wrote the *Texas* narrative. (Assume the *Life* narrative was really written by Davey Crockett.)

**4.69** As part of a project targeted at improving the services of a local bakery, a management consultant (L. Lei of Rutgers University) monitored customer arrivals for several Saturdays and Sundays. Using the arrival data, she estimated the average number of customer arrivals per 10-minute period on Saturdays to be 6.2. She assumed that arrivals per 10-minute interval followed the Poisson distribution shown below, some of whose values are missing.
  a. Compute the missing probabilities.
  b. Graph the distribution.
  c. Find $\mu$ and $\sigma$ and show the intervals $\mu \pm \sigma, \mu \pm 2\sigma$, and $\mu \pm 3\sigma$ on your graph of part **b**.
  d. The owner of the bakery claims that more than 75 customers per hour enter the store on Saturdays. Based on the consultant's data, is this likely? Explain.

**4.70** In many cities, neighborhood Crime Watch groups are formed in an attempt to reduce the amount of criminal activity. Suppose one neighborhood that has experienced an average of 10 crimes per year organizes such a group. During the first year following the creation of the group, three crimes are committed in the neighborhood.
  a. Use the Poisson distribution to calculate the probability that three or fewer crimes are committed in a year assuming the average number is still 10 crimes per year.
  b. Do you think this event provides some evidence that the Crime Watch group has been effective in this neighborhood?

**Applying the Concepts—Advanced**

**4.71** A certain automatic car wash takes exactly 5 minutes to wash a car. On the average, 10 cars per hour arrive at the car wash. Suppose that, 30 minutes before closing time, five cars are in line. If the car wash is in continuous use until closing time, is it likely anyone will be in line at closing time?

| x | 0 | 1 | 2 | 3 | 4 | 5 | 6 | 7 | 8 | 9 | 10 | 11 | 12 | 13 |
|---|---|---|---|---|---|---|---|---|---|---|----|----|----|----|
| p(x) | .002 | .013 | — | .081 | .125 | .155 | — | .142 | .110 | .076 | — | .026 | .014 | .007 |

*Source:* Lei, L. *Dorsi's Bakery: Modeling Service Operations.* Graduate School of Management, Rutgers University, 1993.

## 4.6 THE HYPERGEOMETRIC RANDOM VARIABLE (OPTIONAL)

The **hypergeometric probability distribution** provides a realistic model for some types of enumerative (countable) data. The characteristics of the hypergeometric distribution are listed in the box.

---

### Characteristics of a Hypergeometric Random Variable

1. The experiment consists of randomly drawing $n$ elements without replacement from a set of $N$ elements, $r$ of which are $S$'s (for Success) and $(N - r)$ of which are $F$'s (for Failure).
2. The hypergeometric random variable $x$ is the number of $S$'s in the draw of $n$ elements.

---

Note that both the hypergeometric and binomial characteristics stipulate that each draw, or trial, results in one of two outcomes. The basic difference between these random variables is that the hypergeometric trials are dependent, while the binomial trials are independent. The draws are dependent because the probability of drawing an $S$ (or an $F$) is dependent on what occurred on preceding draws.

To illustrate the dependence between trials, we note that the probability of drawing an $S$ on the first draw is $r/N$. Then, the probability of drawing an $S$ on the second draw depends on the outcome of the first. It will be either $(r - 1)/(N - 1)$, or $r/(N - 1)$, depending on whether the first draw was an $S$ or an $F$. Consequently, the results of the draws represent dependent events.

For example, suppose we define $x$ as the number of women hired in a random selection of three applicants from a total of six men and four women. This random variable satisfies the characteristics of a hypergeometric random variable with $N = 10$ and $n = 3$. The possible outcomes on each trial are either selection of a female ($S$) or selection of a male ($F$). Another example of a hypergeometric random variable is the number, $x$, of defective large-screen television picture tubes in a random selection of $n = 4$ from a shipment of $N = 8$ tubes. And, as a third example, suppose $n = 5$ stocks are randomly selected from a list of $N = 15$ stocks. Then, the number $x$ of the five selected companies that pay regular dividends to stockholders is a hypergeometric random variable.

The hypergeometric probability distribution is summarized in the box.

---

**Probability Distribution, Mean, and Variance of the Hypergeometric Random Variable**

$$p(x) = \frac{\binom{r}{x}\binom{N-r}{n-x}}{\binom{N}{n}} \quad [x = \text{Maximum}\,[0, n - (N - r)], \dots, \text{Minimum}\,(r, n)]$$

$$\mu = \frac{nr}{N} \qquad \sigma^2 = \frac{r(N-r)n(N-n)}{N^2(N-1)}$$

where

$N$ = Total number of elements
$r$ = Number of $S$'s in the $N$ elements
$n$ = Number of elements drawn
$x$ = Number of $S$'s drawn in the $n$ elements

---

**EXAMPLE 4.13**

Suppose a professor randomly selects three new teaching assistants from a total of 10 applicants, six male and four female students. Let $x$ be the number of females who are hired.

a. Find the mean and standard deviation of $x$.
b. Find the probability that no females are hired.

**Solution**

a. Since $x$ is a hypergeometric random variable with $N = 10, n = 3$, and $r = 4$, the mean and variance are

$$\mu = \frac{nr}{N} = \frac{(3)(4)}{10} = 1.2$$

$$\sigma^2 = \frac{r(N-r)n(N-n)}{N^2(N-1)} = \frac{4(10-4)3(10-3)}{(10)^2(10-1)}$$

$$= \frac{(4)(6)(3)(7)}{(100)(9)} = .56$$

The standard deviation is

$$\sigma = \sqrt{.56} = .75$$

b. The probability that no female students are hired by the professor, assuming the selection is truly random, is

$$P(x = 0) = p(0) = \frac{\binom{4}{0}\binom{10-4}{3-0}}{\binom{10}{3}}$$

$$= \frac{\dfrac{4!}{0!(4-0)!}\dfrac{6!}{3!(6-3)!}}{\dfrac{10!}{3!(10-3)!}} = \frac{(1)(20)}{120} = \frac{1}{6}$$

The entire probability distribution for $x$ is shown in Figure 4.11. The mean $\mu = 1.2$ and the interval $\mu \pm 2\sigma = (-.3, 2.7)$ are indicated. You can see that if this random variable were to be observed over and over again a large number of times, most of the values of $x$ would fall within the interval $\mu \pm 2\sigma$. ∎

**FIGURE 4.11**

*Probability Distribution for*
*x in Example 4.13*

## EXERCISES 4.72–4.88

### Learning the Mechanics

**4.72** Explain the difference between sampling with replacement and sampling without replacement.

**4.73** How do binomial and hypergeometric random variables differ? In what respects are they similar?

**4.74** Given that $x$ is a hypergeometric random variable with $N = 8, n = 3$, and $r = 5$, compute the following:
   **a.** $P(x = 1)$
   **b.** $P(x = 0)$
   **c.** $P(x = 3)$
   **d.** $P(x \ge 4)$

**4.75** Given that $x$ is a hypergeometric random variable, compute $p(x)$ for each of the following cases:
   **a.** $N = 5, n = 3, r = 3, x = 1$
   **b.** $N = 9, n = 5, r = 3, x = 3$

   **c.** $N = 4, n = 2, r = 2, x = 2$
   **d.** $N = 4, n = 2, r = 2, x = 0$

**4.76** Given that $x$ is a hypergeometric random variable with $N = 10, n = 5$, and $r = 7$:
   **a.** Display the probability distribution for $x$ in tabular form.
   **b.** Compute the mean and variance of $x$.
   **c.** Graph $p(x)$ and locate $\mu$ and the interval $\mu \pm 2\sigma$ on the graph.
   **d.** What is the probability that $x$ will fall within the interval $\mu \pm 2\sigma$?

**4.77** Given that $x$ is a hypergeometric random variable with $N = 12, n = 8$, and $r = 6$:
   **a.** Display the probability distribution for $x$ in tabular form.

**b.** Compute $\mu$ and $\sigma$ for $x$.

**c.** Graph $p(x)$ and locate $\mu$ and the interval $\mu \pm 2\sigma$ on the graph.

**d.** What is the probability that $x$ will fall within the interval $\mu \pm 2\sigma$?

**4.78** Use the results of Exercise 4.77 to find the following probabilities:

**a.** $P(x = 1)$

**b.** $P(x = 4)$

**c.** $P(x \leq 4)$

**d.** $P(x \geq 5)$

**e.** $P(x < 3)$

**f.** $P(x \geq 8)$

**4.79** Suppose you plan to sample 10 items from a population of 100 items and would like to determine the probability of observing four defective items in the sample. Which probability distribution should you use to compute this probability under the following conditions? Justify your answers.

**a.** The sample is drawn without replacement.

**b.** The sample is drawn with replacement.

## Applying the Concepts—Basic

**4.80** According to *USA Today* (Sept. 19, 2000), there are 650 members of the International Nanny Association (INA). Of these, only three are men. In Exercise 3.10 (p. 118), you found the probability that a randomly selected member of the INA is a man. Now find the probability that in a random sample of four INA members, at least one is a man.

**4.81** The Resource Conservation and Recovery Act mandates the tracking and disposal of hazardous waste produced at U.S. facilities. *Professional Geographer* (Feb. 2000) reported the hazardous waste generation and disposal characteristics of 209 facilities. Only eight of these facilities treated hazardous waste on-site.

**a.** In a random sample of 10 of the 209 facilities, what is the expected number in the sample that treat hazardous waste on-site? Interpret this result.

**b.** Find the probability that four of the eight selected facilities treat hazardous waste on-site.

**4.82** Imagine you are purchasing small lots of a manufactured product. If it is very costly to test a single item, it may be desirable to test a sample of items from the lot instead of testing every item in the lot. Suppose each lot contains 10 items. You decide to sample four items per lot and reject the lot if you observe one or more defectives.

**a.** If the lot contains one defective item, what is the probability that you will accept the lot?

**b.** What is the probability that you will accept the lot if it contains two defective items?

**4.83** Refer to the investigation of contaminated gun cartridges at a weapons manufacturer, Exercise 4.16 (p. 181). In a sample of 158 cartridges from a certain lot, 36 were found to be contaminated and 122 were "clean." If you randomly select 5 of these 158 cartridges, what is the probability that all 5 will be "clean"?

## Applying the Concepts—Intermediate

**4.84** "Hotspots" are species-rich geographical areas. A *Nature* (Sept. 1993) study estimated the probability of a bird species in Great Britain inhabiting a butterfly hotspot at .70. Consider a random sample of four British bird species selected from a total of 10 tagged species. Assume that seven of the 10 tagged species inhabit a butterfly hotspot.

**a.** What is the probability that exactly half of the four bird species sampled inhabit a butterfly hotspot?

**b.** What is the probability that at least one of the four bird species sampled inhabit a butterfly hotspot?

**4.85** Suppose you are purchasing cases of wine (twelve bottles per case) and that, periodically, you select a test case to determine the adequacy of the bottles' seals. To do this, you randomly select and test three bottles in the case. If a case contains one spoiled bottle of wine, what is the probability that this bottle will turn up in your sample?

**4.86** The cornrake is a European species of bird in danger of worldwide extinction. A census revealed that 12 cornrakes inhabit mainland Scotland (*Journal of Applied Ecology*, 1993). Suppose that two of these Scottish cornrakes are captured for mating purposes. Let $x$ be the number of these captured cornrakes that are capable of mating. If exactly four of the original 12 cornrakes inhabiting Scotland are infertile and hence unable to mate, find the probability distribution for $x$.

## Applying the Concepts—Advanced

**4.87** A curious event was described in the *Minneapolis Star and Tribune*. The Minneapolis Community Development Agency (MCDA) makes home improvement grants each year to homeowners in depressed city neighborhoods. Of the $708,000 granted one year, $233,000 was awarded by the city council using a "random selection" of 140 homeowners' applications from among a total of 743 applications: 601 from the north side and 142 from the south side of Minneapolis. Oddly, all 140 grants awarded were from the north side—clearly a highly improbable outcome, if, in fact, the 140 winners were randomly selected from among the 743 applicants.

**a.** Suppose the 140 winning applications were randomly selected from among the total of 743, and let $x$ equal the number in the sample from the north side. Find the mean and standard deviation of $x$.

**b.** Use the results of part **a** to support a contention that the grant winners were not randomly selected.

**4.88** The *Journal of Business & Economic Statistics* (July, 2000) presented a case in which a charge of gender discrimination was filed against the U.S. Postal Service. At the time, there were 302 U.S. Postal Service employees (229 men and 73 women) who applied for promotion. Of the 72 employees who were awarded promotion, 5 were female. Make an inference about whether or not females at the U.S. Postal Service were promoted fairly.

## STATISTICS IN ACTION
### Probability in a Reverse Cocaine Sting

*The American Statistician* (May 1991) described an interesting application of a discrete probability distribution in a case involving illegal drugs. It all started with a "bust" in a mid-sized Florida city. During the bust, police seized 496 foil packets of a white, powdery substance, presumably cocaine. However, it is not a crime to buy or sell nonnarcotic cocaine look-alikes (e.g., inert powders), so detectives had to prove that the packets contained genuine cocaine in order to convict their suspects of drug trafficking. When the police laboratory randomly selected and chemically tested four of the packets, all four tested positive for cocaine. This finding led to the conviction of the traffickers.

### Focus

a. Of the 496 packets confiscated, suppose 331 contained genuine cocaine and 165 contained an inert (legal) powder. Find the probability that four randomly selected packets will test positive for cocaine.

After the conviction, the police decided to use the 492 remaining foil packets (i.e., those not tested) in reverse sting operations. Two of the 492 packets were randomly selected and sold by undercover officers to a buyer. Between the sale and the arrest, however, the buyer disposed of the evidence. The key question is: Beyond a reasonable doubt, did the defendant really purchase cocaine?

In court, the defendant's attorney argued that his client should not be convicted because the police could not prove that the missing foil packets contained cocaine. The police contended, however, that since four of the original 496 packets tested positive for cocaine, the two packets sold in the reverse sting were also highly likely to contain cocaine.

### Focus

b. Given that four of the original packets tested positive for cocaine, what is the probability that two packets sold in the reverse sting did not contain cocaine? Assume the information provided in part **a** is correct.

c. Use the results, parts **a** and **b**, to compute the probability that of the original 496 foil packets, the first four selected (at random) would test positive for cocaine and the next two selected (at random) would test negative.

A statistician hired as an expert witness, Professor Jonathan Shuster of the University of Florida, showed that the probability of the event described in part **c** is maximized when 331 of the 496 packets contain cocaine and 165 do not. Consequently, the probability, part **c**, represents the best possible case for the defendant. Nevertheless, the defense was satisfied that the probability of part **c** was large enough to establish reasonable doubt in a criminal trial.

### Focus

d. Recalculate the probabilities, parts **a–c**, assuming that 400 of the original 496 packets contain cocaine. Comment on the reasonable-doubt claim using this probability.

To the surprise of the defense, the prosecution revealed that the remaining 490 packets had not been used in any other reverse sting operations and offered to test a sample of them. On the advice of the statistician, the defense requested that an additional 20 packets be tested. All 20 tested positive for cocaine! As a consequence, the defendant was ultimately convicted. The case represented Florida's first cocaine-possession conviction without the actual physical evidence.

### Focus

e. Assuming the best possible case for the defendant (i.e., 331 of the original 496 packets contain cocaine), find the probability that the first four test positive, the next two test negative, and the next 20 test positive. [*Hint:* $P(ABC) = P(C|AB) \cdot P(B|A) \cdot P(A)$]. Comment on the reasonable-doubt claim.

[*Note:* If you did not cover optional Section 4.6, use the binomial probability distribution to approximate the probabilities in parts **a–e**. If you did cover Section 4.6, show that the binomial probabilities give good approximations to the true probabilities in parts **a–e**.]

## QUICK REVIEW

### Key Terms

*Note: Starred (\*) terms are from the optional sections in this chapter.*

## Key Formulas

*Note: Starred (\*) formulas are from the optional sections in this chapter.*

| Probability Distribution | Mean ($\mu$) | Variance ($\sigma^2$) | | |
|---|---|---|---|---|
| $p(x)$ | $\sum_{\text{all } x} x p(x)$ | $\sum_{\text{all } x} (x - \mu)^2 p(x)$ | General discrete random variable | 183,184 |
| $\binom{n}{x} p^x q^{n-x}$ | $np$ | $npq$ | Binomial random variable | 192,194 |
| $\dfrac{\lambda^x e^{-\lambda}}{x!}$ | $\lambda$ | $\lambda$ | Poisson random variable* | 202 |
| $\dfrac{\binom{r}{x}\binom{N-r}{n-x}}{\binom{N}{n}}$ | $\dfrac{nr}{N}$ | $\dfrac{r(N-r)n(N-n)}{N^2(N-1)}$ | Hypergeometric random variable* | 209 |

## Language Lab

| Symbol | Pronunciation | Description |
|---|---|---|
| $p(x)$ | | Probability distribution of the random variable $x$ |
| $S$ | | The outcome of a binomial trial denoted a "success" |
| $F$ | | The outcome of a binomial trial denoted a "failure" |
| $p$ | | The probability of success ($S$) in a binomial trial |
| $q$ | | The probability of failure ($F$) in a binomial trial, where $q = 1 - p$ |
| $\lambda$ | lambda | The mean (or expected) number of events for a Poisson random variable |
| $e$ | | A constant used in the Poisson probability distribution, where $e = 2.71828\ldots$ |

 SUPPLEMENTARY EXERCISES 4.89–4.112

*Note: Starred (\*) exercises refer to the optional sections in this chapter.*

### Learning the Mechanics

**4.89** Which of the following describe discrete random variables, and which describe continuous random variables?

**a.** The length of time that an exercise physiologist's program takes to elevate her client's heart rate to 140 beats per minute

**b.** The number of crimes committed on a college campus per year

**c.** The number of square feet of vacant office space in a large city

**d.** The number of voters who favor a new tax proposal

**4.90** For each of the following examples, decide whether $x$ is a binomial random variable and explain your decision:

**a.** A manufacturer of computer chips randomly selects 100 chips from each hour's production in order to estimate the proportion defectives. Let $x$ represent the number of defectives in the 100 sampled chips.

**b.** Of five applicants for a job, two will be selected. Although all applicants appear to be equally qualified, only three have the ability to fulfill the expectations of the company. Suppose that the two selections are made at random from the five applicants, and let $x$ be the number of qualified applicants selected.

**c.** A software developer establishes a support hotline for customers to call in with questions regarding use of the software. Let $x$ represent the number of calls received on the support hotline during a specified workday.

**d.** Florida is one of a minority of states with no state income tax. A poll of 1,000 registered voters is conducted to determine how many would favor a state income tax in light of the state's current fiscal condition. Let $x$ be the number in the sample who would favor the tax.

**4.91** Identify the type of random variable—binomial, Poisson, or hypergeometric—described by each of the following probability distributions:

**\*a.** $p(x) = \dfrac{.5^x e^{-.5}}{x!}$ $(x = 0, 1, 2, \ldots)$

**b.** $p(x) = \binom{6}{x}(.2)^x(.8)^{6-x}$ $(x = 0, 1, 2, \ldots, 6)$

**c.** $p(x) = \dfrac{10!}{x!(10-x)!}(.9)^x(.1)^{10-x}$ $(x = 0, 1, 2, \ldots, 10)$

**\*4.92** Given that $x$ is a hypergeometric random variable, compute $p(x)$ for each of the following cases:

**a.** $N = 8, n = 5, r = 3, x = 2$

**b.** $N = 6, n = 2, r = 2, x = 2$

**c.** $N = 5, n = 4, r = 4, x = 3$

**4.93** Suppose $x$ is a binomial random variable. Find $p(x)$ for each of the following combinations of $x, n$, and $p$.

**a.** $x = 1, n = 3, p = .1$

**b.** $x = 4, n = 20, p = .3$

**c.** $x = 0, n = 2, p = .4$

**d.** $x = 4, n = 5, p = .5$

**e.** $n = 15, x = 12, p = .9$

**f.** $n = 10, x = 8, p = .6$

**4.94** Consider the discrete probability distribution shown here.

| $x$ | 10 | 12 | 18 | 20 |
|------|------|------|------|------|
| $p(x)$ | .2 | .3 | .1 | .4 |

**a.** Calculate $\mu, \sigma^2$, and $\sigma$.

**b.** What is $P(x < 15)$?

**c.** Calculate $\mu \pm 2\sigma$.

**d.** What is the probability that $x$ is in the interval $\mu \pm 2\sigma$?

*4.95 Suppose $x$ is a Poisson random variable. Compute $p(x)$ for each of the following cases:

**a.** $\lambda = 2, x = 3$

**b.** $\lambda = 1, x = 4$

**c.** $\lambda = .5, x = 2$

**4.96** Suppose $x$ is a binomial random variable with $n = 20$ and $p = .7$.

**a.** Find $P(x = 14)$.

**b.** Find $P(x \leq 12)$.

**c.** Find $P(x > 12)$.

**d.** Find $P(9 \leq x \leq 18)$.

**e.** Find $P(8 < x < 18)$.

**f.** Find $\mu, \sigma^2$, and $\sigma$.

**g.** What is the probability that $x$ is in the interval $\mu \pm 2\sigma$?

## Applying the Concepts—Basic

**4.97** According to *Ladies Home Journal* (June 1988), 80% of married women would marry their current husbands again, if given the chance. Consider the number $x$ in a random sample of ten married women who would marry their husbands again. Identify the discrete probability distribution that best models the distribution of $x$. Explain.

*4.98 An emergency rescue vehicle is used an average of 1.3 times daily. Use the Poisson distribution to find:

**a.** The probability that the vehicle will be used exactly twice tomorrow.

**b.** The probability that it will be used more than twice.

**c.** The probability that it will be used exactly three times.

**4.99** In Exercise 2.11 (p. 27) you read about a study of book reviews in American history, geography, and area studies published in *Choice* magazine (*Library Acquisitions:*

*Practice and Theory*, Vol. 19, 1995). The overall rating stated in each review was ascertained and recorded as follows: 1 = would not recommend, 2 = cautious or very little recommendation, 3 = little or no preference, 4 = favorable/recommended, 5 = outstanding/significant contribution. Based on a sample of 375 book reviews, the probability distribution of rating, $x$, is

| Book Rating $x$ | $p(x)$ |
|------|------|
| 1 | .051 |
| 2 | .099 |
| 3 | .093 |
| 4 | .635 |
| 5 | .122 |

**a.** Is this a valid probability distribution?

**b.** What is the probability that a book reviewed in *Choice* has a rating of 1?

**c.** What is the probability that a book reviewed in *Choice* has a rating of at least 4?

**d.** What is the probability that a book reviewed in *Choice* has a rating of 2 or 3?

**e.** Find $E(x)$.

**f.** Interpret the result, part **e**.

*4.100 A nursery advertises that it has 10 elm trees for sale. Unknown to the nursery, three of the trees have already been infected with Dutch elm disease and will die within a year.

**a.** If a buyer purchases two trees, what is the probability that both trees will be healthy?

**b.** Refer to part **a**. What is the probability that at least one of the trees is infected?

## Applying the Concepts—Intermediate

**4.101** The Food and Drug Administration (FDA) produces a quarterly report called *Total Diet Study*. The FDA's report covers more than 200 food items, each of which is analyzed for dangerous chemical compounds. A recent *Total Diet Study* reported that no pesticides at all were found on 65% of the domestically produced food samples (*Consumer's Research*, June 1995). Consider a random sample of 800 food items analyzed for the presence of pesticides.

**a.** Compute $\mu$ and $\sigma$ for the random variable $x$, the number of food items found without any trace of pesticide.

**b.** Based on a sample of 800 food items, is it likely you would observe less than half without any traces of pesticide? Explain.

**4.102** Refer to the *Professional Geographer* (Feb. 2000) study of urban and rural counties, Exercise 3.17 (p. 118). Forty-five percent of county commissioners in Nevada feel that "population concentration" is the single most important factor in identifying urban counties. Suppose 10 county commissioners are selected at random to serve on a Nevada review board that will consider re-defining urban and rural areas. What is the probability that more than half of the commissioners will specify "population concentration" as the single most important factor in identifying urban counties?

**4.103** *Pediatric Exercise Science* (Feb. 1995) published an article on the behavior of parents at competitive youth gymnastic meets. Based on a survey of the gymnasts, the researchers estimated the probability of a parent "yelling" at their child before, during, or after the meet as .05. In a random sample of 20 parents attending a gymnastic meet, find the probability that at least one parent yells at their child before, during, or after the meet.

**\*4.104** By mistake, a manufacturer of CD mini-rack systems includes three defective systems in a shipment of 10 going out to a small retailer. The retailer has decided to accept the shipment of CD systems only if none is found to be defective. Upon receipt of the shipment, the retailer examines only five of the CD systems.

**a.** What is the probability that the shipment will be rejected?

**b.** If the retailer inspects six of the CD systems, what is the probability the shipment will be accepted?

**4.105** The state highway patrol has determined that one out of every six calls for help originating from roadside call boxes is a hoax. Five calls for help have come in and five tow trucks have been dispatched.

**a.** What is the probability that none of the calls was a hoax?

**b.** What is the probability that only three of the callers really needed assistance?

**c.** What assumptions do you have to make in order to solve this problem?

**d.** If the highway patrol answers 10,000 calls for help next year and each call costs the patrol about $30 (labor, gas, etc.), approximately how much money will be wasted answering false alarms?

**\*4.106** The number $x$ of people who arrive at a cashier's counter in a bank during a specified period of time often exhibits (approximately) a Poisson probability distribution. If we know the mean arrival rate $\lambda$, the Poisson probability distribution can be used to aid in the design of the customer service facility. Suppose you estimate that the mean number of arrivals per minute for cashier service at a bank is one person per minute.

**a.** What is the probability that in a given minute the number of arrivals will equal three or more?

**b.** Can you tell the bank manager that the number of arrivals will rarely exceed two per minute?

**4.107** The efficacy of insecticides is often measured by the dose necessary to kill a certain percentage of insects. Suppose a certain dose of a new insecticide is supposed to kill 80% of the insects that receive it. To test the claim, 25 insects are exposed to the insecticide.

**a.** If the insecticide really kills 80% of the exposed insects, what is the probability that fewer than 15 die?

**b.** If you observed such a result, what would you conclude about the new insecticide? Explain your logic.

**4.108** The owner of construction company A makes bids on jobs so that if awarded the job, company A will make a $10,000 profit. The owner of construction company B makes bids on jobs so that if awarded the job, company B will make a $15,000 profit. Each company describes the probability distribution of the number of jobs $x$ the company is awarded per year as shown in the table.

| Company A | | Company B | |
|---|---|---|---|
| $x$ | $P(x)$ | $x$ | $P(x)$ |
| 2 | .05 | 2 | .15 |
| 3 | .15 | 3 | .30 |
| 4 | .20 | 4 | .30 |
| 5 | .35 | 5 | .20 |
| 6 | .25 | 6 | .05 |

**a.** Find the expected number of jobs each will be awarded in a year.

**b.** What is the expected profit for each company?

**c.** Find the variance and standard deviation of the distribution of number of jobs awarded per year for each company.

**d.** Graph $p(x)$ for both companies A and B. For each company, what proportion of the time will $x$ fall in the interval $\mu \pm 2\sigma$?

### Applying the Concepts—Advanced

**4.109** The probability that a person responds to a mailed questionnaire is .4.

**a.** What is the probability that of 20 questionnaires, more than 12 will be returned?

**b.** How many questionnaires should be mailed if you want to be reasonably certain that at least 100 will be returned?

**\*4.110** On January 28, 1986, the space shuttle *Challenger* exploded, killing all seven astronauts aboard. An investigation concluded that the explosion was caused by the failure of the O-ring seal in the joint between the two lower segments of the right solid rocket booster. In a report made one year prior to the catastrophe, the National Aeronautics and Space Administration (NASA) claimed that the probability of such a failure was about $1/60,000$, or about once in every 60,000 flights. But a risk-assessment study conducted for the Air Force at about the same time assessed the probability to be $1/35$, or about once in every 35 missions.

**a.** The shuttle had flown 24 successful missions prior to the disaster. Assuming NASA's failure-rate estimate was accurate, compute the probability that no disasters would have occurred during 25 shuttle missions.

**b.** Repeat part **a**, but use the Air Force's failure rate estimate.

**c.** What conditions must exist for the probabilities, parts **a** and **b**, to be valid?

**d.** Given the events of January 28, 1986, which risk assessment—NASA's or the Air Force's—appears to be more appropriate? [*Hint:* Consider the complement of the events, parts **a** and **b**.]

**4.111** For a space vehicle to gain reentry into the earth's atmosphere, a particular system must work properly. One component of the system operates successfully only 85% of the time. To increase the reliability of the system, four of these components are installed in such a way that the system will operate successfully if at least one component is working. What is the probability that the system will fail? Assume the components operate independently.

**\*4.112** The mean number of patients admitted per day to the emergency room of a small hospital is 2.5. If, on a given day, there are only four beds available for new patients, what is the probability the hospital will not have enough beds to accommodate its newly admitted patients?

## STUDENT PROJECTS

Consider the following random variables:

**1.** The number $x$ of people who recover from a certain disease out of a sample of five patients

**2.** The number $x$ of voters in a sample of five who favor a method of tax reform

**3.** The number $x$ of hits a baseball player gets in five official times at bat

In each case, $x$ is a binomial random variable (or approximately so) with $n = 5$ trials. Assume that in each case the probability of Success is .3. (What is a Success in each of the examples?) If this were true, the probability distribution for $x$ would be the same for each of the three examples. To obtain a relative frequency histogram for $x$, conduct the following experiment: Place 10 poker chips (pennies, marbles, or any 10 *identical* items) in a bowl and mark three of the 10 Success—the remaining seven will represent Failure. Randomly select a chip from the 10, observing whether it was a Success or Failure. Then return the chip and randomly select a second chip from the 10 available chips, and record this outcome. Repeat this process until a total of five trials has been conducted. Count the number $x$ of Successes observed in the five trials. Repeat the entire process 100 times to obtain 100 observed values of $x$.

**a.** Use the 100 values of $x$ obtained from the simulation to construct a relative frequency histogram for $x$. Note that this histogram is an approximation to $p(x)$.

**b.** Calculate the exact values of $p(x)$ for $n = 5$ and $p = .3$ and compare these values with the approximations found in part **a**.

**c.** If you were to repeat the simulation an extremely large number of times (say 100,000), how do you think the relative frequency histogram and true probability distribution would compare?

## REFERENCES

Hogg, R. V., and Craig, A. *Introduction to Mathematical Statistics*, 5th ed. Upper Saddle River, N.J.: Prentice Hall, 1995.

Wackerly, D., Mendenhall, W., and Scheaffer, R.L. *Mathematical Statistics with Applications*, 6th ed. North Scituate, Mass.: Duxbury, 2002.

# Continuous Random Variables

## Contents

## Statistics in Action

IQ, Economic Mobility, and the Bell Curve

## 🖐 Where We've Been

We've seen that because numerical data represent observed values of random variables, we need to find the probabilities associated with specific sample observations. The probability theory of Chapter 3 provided the mechanism for finding the probabilities associated with discrete random variables. Finding and describing this set of probabilities—the probability distribution for a discrete random variable—was the subject of Chapter 4.

## ☞ Where We're Going

Data may be derived from observations on continuous as well as discrete random variables. Thus, we need to know about probability distributions associated with continuous random variables and also how to use the mean and standard deviation to describe these distributions. Chapter 5 addresses this problem and, in particular, introduces the *normal probability distribution*. As you'll subsequently see, the normal probability distribution is one of the most useful distributions in statistics.

In this chapter we'll consider some continuous random variables that are commonly encountered. Recall that a continuous random variable is one that can assume any value within some interval or intervals. For example, the length of time between a person's visits to a doctor, the thickness of sheets of steel produced in a rolling mill, and the yield of wheat per acre of farmland are all continuous random variables. The methodology we employ to describe continuous random variables will necessarily be somewhat different from that used to describe discrete random variables. We first discuss the general form of **continuous probability distributions**, and then we explore three specific types that are used in making statistical decisions. The normal probability distribution, which plays a basic and important role in both the theory and application of statistics, is essential to the study of most of the subsequent chapters in this book. The other types have practical applications, but a study of these topics is optional.

## 5.1 CONTINUOUS PROBABILITY DISTRIBUTIONS

The graphical form of the probability distribution for a continuous random variable $x$ is a smooth curve that might appear as shown in Figure 5.1. This curve, a function of $x$, is denoted by the symbol $f(x)$ and is variously called a **probability density function**, a **frequency function**, or a **probability distribution**.

The areas under a probability distribution correspond to probabilities for $x$. For example, the area $A$ beneath the curve between the two points $a$ and $b$, as shown in Figure 5.1, is the probability that $x$ assumes a value between $a$ and $b(a < x < b)$. Because there is no area over a point, say $x = a$, it follows that (according to our model) the probability associated with a particular value of $x$ is equal to 0; that is, $P(x = a) = 0$ and hence $P(a < x < b) = P(a \le x \le b)$. In other words, the probability is the same regardless of whether or not you include the endpoints of the interval. Also, because areas over intervals represent probabilities, it follows that the total area under a probability distribution, the probability assigned to all values of $x$, should equal 1. Note that probability distributions for continuous random variables possess different shapes depending on the relative frequency distributions of real data that the probability distributions are supposed to model.

**FIGURE 5.1**

*A Probability Distribution $f(x)$ for a Continuous Random Variable $x$*

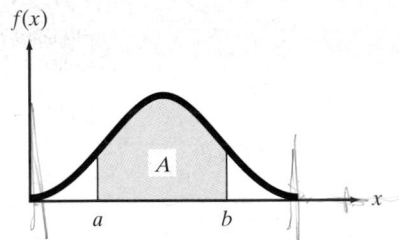

The areas under most probability distributions are obtained by using calculus or numerical methods.* Because these methods often involve difficult procedures, we will give the areas for some of the most common probability distributions in tabular form in Appendix A. Then, to find the area between two values of $x$, say $x = a$ and $x = b$, you simply have to consult the appropriate table.

For each of the continuous random variables presented in this chapter, we will give the formula for the probability distribution along with its mean $\mu$ and standard deviation $\sigma$. These two numbers will enable you to make some approximate

---

*Students with knowledge of calculus should note that the probability that $x$ assumes a value in the interval $a < x < b$ is $P(a < x < b) = \int_a^b f(x)\, dx$, assuming the integral exists. Similar to the requirements for a discrete probability distribution, we require $f(x) \ge 0$ and $\int_{-\infty}^{\infty} f(x)\, dx = 1$.

probability statements about a random variable even when you do not have access to a table of areas under the probability distribution.

## 5.2   THE UNIFORM DISTRIBUTION

All the probability problems discussed in Chapter 3 had sample spaces that contained a finite number of sample points. In many of these problems, the sample points were assigned equal probabilities—for example, the die toss or the coin toss. For continuous random variables, there is an infinite number of values in the sample space, but in some cases the values may appear to be equally likely. For example, if a short exists in a 5-meter stretch of electrical wire, it may have an equal probability of being in any particular 1-centimeter segment along the line. Or if a safety inspector plans to choose a time at random during the four afternoon work hours to pay a surprise visit to a certain area of a plant, then each 1-minute time interval in this 4-work-hour period will have an equally likely chance of being selected for the visit.

Continuous random variables that appear to have equally likely outcomes over their range of possible values possess a **uniform probability distribution**, perhaps the simplest of all continuous probability distributions. Suppose the random variable $x$ can assume values only in an interval $c \le x \le d$. Then the uniform frequency function has a rectangular shape, as shown in Figure 5.2. Note that the possible values of $x$ consist of all points in the interval between point $c$ and point $d$. The height of $f(x)$ is constant in that interval and equals $1/(d - c)$. Therefore, the total area under $f(x)$ is given by

$$\text{Total area of rectangle} = (\text{Base})(\text{Height}) = (d - c)\left(\frac{1}{d - c}\right) = 1$$

The uniform probability distribution provides a model for continuous random variables that are *evenly distributed* over a certain interval. That is, a uniform random variable is one that is just as likely to assume a value in one interval as it is to assume a value in any other interval of equal size. There is no clustering of values around any value; instead, there is an even spread over the entire region of possible values.

**FIGURE 5.2**

*The Uniform Probability Distribution*

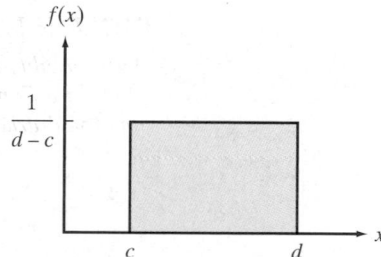

The uniform distribution is sometimes referred to as the **randomness distribution**, since one way of generating a uniform random variable is to perform an experiment in which a point is *randomly selected* on the horizontal axis between the points $c$ and $d$. If we were to repeat this experiment infinitely often, we would create a uniform probability distribution like that shown in Figure 5.2. The random selection of points in an interval can also be used to generate random numbers such as those in Table I of Appendix A. Recall that random numbers are selected in such a way that every number would have an equal probability of selection. Therefore, random numbers are realizations of a uniform random variable. (Random numbers were used to draw random samples in Section 3.7.) The formulas for the uniform probability distribution, its mean, and its standard deviation are shown in the box.

---

### Probability Distribution, Mean, and Standard Deviation of a Uniform Random Variable $x$

$$f(x) = \frac{1}{d - c} \qquad (c \le x \le d)$$

$$\mu = \frac{c + d}{2} \qquad \sigma = \frac{d - c}{\sqrt{12}}$$

---

Suppose the interval $a < x < b$ lies within the domain of $x$; that is, it falls within the larger interval $c \le x \le d$. Then the probability that $x$ assumes a value within the interval $a < x < b$ is equal to the area of the rectangle over the interval, namely, $(b - a)/(d - c)$.*

**EXAMPLE 5.1**  An unprincipled used car dealer sells a car to an unsuspecting buyer, even though the dealer knows that the car will have a major breakdown within the next 6 months. The dealer provides a warranty of 45 days on all cars sold. Let $x$ represent the length of time until the breakdown occurs. Assume that $x$ is a uniform random variable with values between 0 and 6 months.

a. Calculate and interpret the mean and standard deviation of $x$.
b. Graph the probability distribution of $x$, and show the mean on the horizontal axis. Also show 1- and 2-standard-deviation intervals around the mean.
c. Calculate the probability that the breakdown occurs while the car is still under warranty.

### Solution

a. To calculate the mean and standard deviation for $x$, we substitute 0 and 6 months for $c$ and $d$, respectively, in the formulas for uniform random variables. Thus,

$$\mu = \frac{c + d}{2} = \frac{0 + 6}{2} = 3 \text{ months}$$

$$\sigma = \frac{d - c}{\sqrt{12}} = \frac{6 - 0}{\sqrt{12}} = \frac{6}{3.464} = 1.73 \text{ months}$$

Our interpretations of $\mu$ and $\sigma$ follow:
The average length of time until breakdown, $x$, for all similar used cars is $\mu = 3$ months. From Chebyshev's Theorem (Table 2.7, p. 61), we know that at least 75% of the values of $x$ in the distribution will fall in the interval

$$\mu \pm 2\sigma = 3 \pm 2(1.73)$$
$$= 3 \pm 3.46$$

or, between $-.46$ and 6.46 months. Consequently, we expect the length of time until breakdown to be less than 6.46 months at least 75% of the time.

b. The uniform probability distribution is

$$f(x) = \frac{1}{d - c} = \frac{1}{6 - 0} = \frac{1}{6} \qquad (0 \le x \le 6)$$

---

*The student with knowledge of calculus should note that

$$P(a < x < b) = \int_a^b f(x)\, d(x) = \int_a^b [1/(d - c)]\, dx = (b - a)/(d - c)$$

## FIGURE 5.3

*Distribution for x in Example 5.1*

The graph of this function is shown in Figure 5.3. The mean and 1- and 2-standard-deviation intervals around the mean are shown on the horizontal axis. Note that the entire distribution of $x$ lies within the $\mu \pm 2\sigma$ interval. (This demonstrates, once again, the conservativeness of Chebyshev's Theorem.)

c. To find the probability that the car is still under warranty when it breaks down, we must find the probability that $x$ is less than 45 days, or (about) 1.5 months. As indicated in Figure 5.4, we need to calculate the area under the frequency function $f(x)$ between the points $x = 0$ and $x = 1.5$. This is the area of a rectangle with base $1.5 - 0 = 1.5$ and height $-\frac{1}{6}$. The probability that the unsuspecting car buyer will be able to have the car repaired at the dealer's expense is then $P(0 < x < 1.5) = (\text{Base})(\text{Height}) = 1.5(\frac{1}{6}) = .25$ ∎

## FIGURE 5.4

*Probability that Car Breaks Down Within 1.5 Months of Purchase*

 EXERCISES 5.1–5.14

### Learning the Mechanics

**5.1** Suppose $x$ is a random variable best described by a uniform probability distribution with $c = 10$ and $d = 30$.
   **a.** Find $f(x)$.
   **b.** Find the mean and standard deviation of $x$.
   **c.** Graph $f(x)$ and locate $\mu$ and the interval $\mu \pm 2\sigma$ on the graph. Note that the probability that $x$ assumes a value within the interval $\mu \pm 2\sigma$ is equal to 1.

**5.2** Refer to Exercise 5.1. Find the following probabilities:
   **a.** $P(10 \le x \le 25)$
   **b.** $P(20 < x < 30)$
   **c.** $P(x \ge 25)$
   **d.** $P(x \le 10)$
   **e.** $P(x \le 25)$
   **f.** $P(20.5 \le x \le 25.5)$

**5.3** Suppose $x$ is a random variable best described by a uniform probability distribution with $c = 2$ and $d = 4$.
   **a.** Find $f(x)$.
   **b.** Find the mean and standard deviation of $x$.
   **c.** Find $P(\mu - \sigma \le x \le \mu + \sigma)$.
   **d.** Find $P(x > 2.78)$.
   **e.** Find $P(2.4 \le x \le 3.7)$.
   **f.** Find $P(x < 2)$.

**5.4** Refer to Exercise 5.3. Find the value of $a$ that makes each of the following probability statements true.
   **a.** $P(x \ge a) = .5$
   **b.** $P(x \le a) = .2$
   **c.** $P(x \le a) = 0$
   **d.** $P(2.5 \le x \le a) = .5$

**5.5** The random variable $x$ is best described by a uniform probability distribution with $c = 100$ and $d = 200$. Find the probability that $x$ assumes a value
   **a.** More than 2 standard deviations from $\mu$.
   **b.** Less than 3 standard deviations from $\mu$.
   **c.** Within 2 standard deviations of $\mu$.

**5.6**    The random variable $x$ is best described by a uniform probability distribution with mean 50 and standard deviation 5. Find $c, d$, and $f(x)$. Graph the probability distribution.

## Applying the Concepts—Basic

**5.7**    Records indicate that the daily high January temperatures on a tropical island tend to have a uniform distribution over the interval from 75°F to 90°F. A tourist arrives on the island on a randomly selected day in January.
   **a.** What is the probability that the temperature will be above 80°F?
   **b.** What is the probability that the temperature will be between 80°F and 85°F?
   **c.** What is the expected temperature?

**5.8**    Researchers at the University of California-Berkeley have designed, built, and tested a switched-capacitor circuit for generating random signals (*International Journal of Circuit Theory and Applications*, May–June 1990). The circuit's trajectory was shown to be uniformly distributed on the interval $(0, 1)$.
   **a.** Give the mean and variance of the circuit's trajectory.
   **b.** Compute the probability that the trajectory falls between .2 and .4.
   **c.** Would you expect to observe a trajectory that exceeds .995? Explain.

**5.9**    The data set listed here was created using the MINITAB random number generator. Construct a relative frequency histogram for the data. Except for the expected variation in relative frequencies among the class intervals, does your histogram suggest that the data are observations on a uniform random variable with $c = 0$ and $d = 100$? Explain.

**⦿ RANUNI**

| | | | | |
|---|---|---|---|---|
| 38.8759 | 98.0716 | 64.5788 | 60.8422 | .8413 |
| 99.3734 | 31.8792 | 32.9847 | .7434 | 93.3017 |
| 12.4337 | 11.7828 | 87.4506 | 94.1727 | 23.0892 |
| 47.0121 | 43.3629 | 50.7119 | 88.2612 | 69.2875 |
| 62.6626 | 55.6267 | 78.3936 | 28.6777 | 71.6829 |
| 44.0466 | 57.8870 | 71.8318 | 28.9622 | 23.0278 |
| 35.6438 | 38.6584 | 46.7404 | 11.2159 | 96.1009 |
| 95.3660 | 21.5478 | 87.7819 | 12.0605 | 75.1015 |

## Applying the Concepts—Intermediate

**5.10**    The frequency distribution shown in the table depicts the property and marine losses incurred by a large oil company over a 2-year period. In the insurance business, each "loss" interval is called a *layer*. *Research Review* (Summer 1998) demonstrated that analysts often treat the actual loss value within a layer as a uniform random variable.

| Layer | Marine Losses ($ thousands) | Relative Frequency |
|---|---|---|
| 1 | 0–10 | .927 |
| 2 | 10–50 | .053 |
| 3 | 50–100 | .010 |
| 4 | 100–250 | .006 |
| 5 | 250–500 | .003 |
| 6 | 500–1,000 | .001 |
| 7 | 1,000–2,500 | .000 |

*Source:* Cozzolino, J. M., and Mikola, P. J. "Applications of the Piecewise Constant Pareto Distribution." *Research Review*, Summer 1998.

   **a.** Use the uniform distribution to find the mean loss amount in layer 2.
   **b.** Use the uniform distribution to find the mean loss amount in layer 6.
   **c.** If a loss occurs in layer 2, what is the probability that it exceeds $30,000?
   **d.** If a loss occurs in layer 6, what is the probability that it is between $750,000 and $800,000?

**5.11**    The manager of a local soft-drink bottling company believes that when a new beverage-dispensing machine is set to dispense 7 ounces, it in fact dispenses an amount $x$ at random anywhere between 6.5 and 7.5 ounces inclusive. Suppose $x$ has a uniform probability distribution.
   **a.** Is the amount dispensed by the beverage machine a discrete or a continuous random variable? Explain.
   **b.** Graph the frequency function for $x$, the amount of beverage the manager believes is dispensed by the new machine when it is set to dispense 7 ounces.
   **c.** Find the mean and standard deviation for the distribution graphed in part **b**, and locate the mean and the interval $\mu \pm 2\sigma$ on the graph.
   **d.** Find $P(x \geq 7)$.
   **e.** Find $P(x < 6)$.
   **f.** Find $P(6.5 \leq x \leq 7.25)$.
   **g.** What is the probability that each of the next six bottles filled by the new machine will contain more than 7.25 ounces of beverage? Assume that the amount of beverage dispensed in one bottle is independent of the amount dispensed in another bottle.

**5.12**    A bus is scheduled to stop at a certain bus stop every half hour on the hour and the half hour. At the end of the day, buses still stop after every 30 minutes, but because delays often occur earlier in the day, the bus is never early and likely to be late. The director of the bus line claims that the length of time a bus is late is uniformly distributed and the maximum time that a bus is late is 20 minutes.
   **a.** If the director's claim is true, what is the expected number of minutes a bus will be late?
   **b.** If the director's claim is true, what is the probability that the last bus on a given day will be more than 19 minutes late?
   **c.** If you arrive at the bus stop at the end of a day at exactly half-past the hour and must wait more than 19

minutes for the bus, what would you conclude about the director's claim? Why?

### Applying the Concepts—Advanced

**5.13** A tool-and-die machine shop produces extremely high-tolerance spindles. The spindles are 18-inch slender rods used in a variety of military equipment. A piece of equipment used in the manufacture of the spindles malfunctions on occasion and places a single gouge somewhere on the spindle. However, if the spindle can be cut so that it has 14 consecutive inches without a gouge, then the spindle can be salvaged for other purposes. Assuming that the location of the gouge along the spindle is best described by a uniform distribution, what is the probability that a defective spindle can be salvaged?

**5.14** The *reliability* of a piece of equipment is frequently defined to be the probability, $p$, that the equipment performs its intended function successfully for a given period of time under specific conditions (Render and Heizer, *Principles of Operations Management*, 1995).

Because $p$ varies from one point in time to another, some reliability analysts treat $p$ as if it were a random variable. Suppose an analyst characterizes the uncertainty about the reliability of a particular robotic device used in an automobile assembly line using the following distribution:

$$f(p) = \begin{cases} 1 & 0 \le p \le 1 \\ 0 & \text{otherwise} \end{cases}$$

**a.** Graph the analyst's probability distribution for $p$.
**b.** Find the mean and variance of $p$.
**c.** According to the analyst's probability distribution for $p$, what is the probability that $p$ is greater than .95? Less than .95?
**d.** Suppose the analyst receives the additional information that $p$ is definitely between .90 and .95, but that there is complete uncertainty about where it lies between these values. Describe the probability distribution the analyst should now use to describe $p$.

## 5.3   THE NORMAL DISTRIBUTION

One of the most commonly observed continuous random variables has a **bell-shaped** probability distribution as shown in Figure 5.5. It is known as a **normal random variable** and its probability distribution is called a **normal distribution**.

**FIGURE 5.5**

*A Normal Probability Distribution*

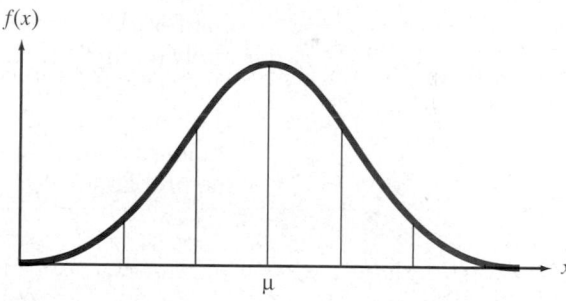

The normal distribution plays a very important role in the science of statistical inference. Moreover, many phenomena generate random variables with probability distributions that are very well approximated by a normal distribution. For example, the error made in measuring a person's blood pressure may be a normal random variable, and the probability distribution for the yearly rainfall in a certain region might be approximated by a normal probability distribution. You can determine the adequacy of the normal approximation to an existing population of data by comparing the relative frequency distribution of a large sample of the data to the normal probability distribution. Methods to detect disagreement between a set of data and the assumption of normality are presented in Section 5.4.

The normal distribution is perfectly symmetric about its mean $\mu$, as can be seen in the examples in Figure 5.6. Its spread is determined by the value of its standard deviation $\sigma$.

The formula for the normal probability distribution is shown in the box. When plotted, this formula yields a curve like that shown in Figure 5.5.

**FIGURE 5.6**

*Several Normal Distributions with Different Means and Standard Deviations*

---

**Probability Distribution for a Normal Random Variable $x$**

$$f(x) = \frac{1}{\sigma\sqrt{2\pi}}e^{-(1/2)[(x-\mu)/\sigma]^2}$$

where

$\mu$ = Mean of the normal random variable $x$

$\sigma$ = Standard deviation

$\pi$ = 3.1416...

$e$ = 2.71828...

---

Note that the mean $\mu$ and standard deviation $\sigma$ appear in this formula, so that no separate formulas for $\mu$ and $\sigma$ are necessary. To graph the normal curve we have to know the numerical values of $\mu$ and $\sigma$.

Computing the area over intervals under the normal probability distribution is a difficult task.* Consequently, we will use the computed areas listed in Table IV of Appendix A. Although there are an infinitely large number of normal curves— one for each pair of values for $\mu$ and $\sigma$—we have formed a single table that will apply to any normal curve.

Table IV is based on a normal distribution with mean $\mu = 0$ and standard deviation $\sigma = 1$, called a *standard normal distribution*. A random variable with a standard normal distribution is typically denoted by the symbol $z$. The formula for the probability distribution of $z$ is given by

$$f(z) = \frac{1}{\sqrt{2\pi}}e^{-(1/2)z^2}$$

Figure 5.7 shows the graph of a standard normal distribution.

**FIGURE 5.7**

*Standard Normal Distribution: $\mu = 0, \sigma = 1$*

---

*The student with knowledge of calculus should note that there is not a closed-form expression for $P(a < x < b) = \int_a^b f(x)\,dx$ for the normal probability distribution. The value of this definite integral can be obtained to any desired degree of accuracy by numerical approximation procedures. For this reason, it is tabulated for the user.

**Table 5.1    Reproduction of Part of Table IV in Appendix A**

| z | .00 | .01 | .02 | .03 | .04 | .05 | .06 | .07 | .08 | .09 |
|------|------|------|------|------|------|------|------|------|------|------|
| .0 | .0000 | .0040 | .0080 | .0120 | .0160 | .0199 | .0239 | .0279 | .0319 | .0359 |
| .1 | .0398 | .0438 | .0478 | .0517 | .0557 | .0596 | .0636 | .0675 | .0714 | .0753 |
| .2 | .0793 | .0832 | .0871 | .0910 | .0948 | .0987 | .1026 | .1064 | .1103 | .1141 |
| .3 | .1179 | .1217 | .1255 | .1293 | .1331 | .1368 | .1406 | .1443 | .1480 | .1517 |
| .4 | .1554 | .1591 | .1628 | .1664 | .1700 | .1736 | .1772 | .1808 | .1844 | .1879 |
| .5 | .1915 | .1950 | .1985 | .2019 | .2054 | .2088 | .2123 | .2157 | .2190 | .2224 |
| .6 | .2257 | .2291 | .2324 | .2357 | .2389 | .2422 | .2454 | .2486 | .2517 | .2549 |
| .7 | .2580 | .2611 | .2642 | .2673 | .2704 | .2734 | .2764 | .2794 | .2823 | .2852 |
| .8 | .2881 | .2910 | .2939 | .2967 | .2995 | .3023 | .3051 | .3078 | .3106 | .3133 |
| .9 | .3159 | .3186 | .3212 | .3238 | .3264 | .3289 | .3315 | .3340 | .3365 | .3389 |
| 1.0 | .3413 | .3438 | .3461 | .3485 | .3508 | .3531 | .3554 | .3577 | .3599 | .3621 |
| 1.1 | .3643 | .3665 | .3686 | .3708 | .3729 | .3749 | .3770 | .3790 | .3810 | .3830 |
| 1.2 | .3849 | .3869 | .3888 | .3907 | .3925 | .3944 | .3962 | .3980 | .3997 | .4015 |
| 1.3 | .4032 | .4049 | .4066 | .4082 | .4099 | .4115 | .4131 | .4147 | .4162 | .4177 |
| 1.4 | .4192 | .4207 | .4222 | .4236 | .4251 | .4265 | .4279 | .4292 | .4306 | .4319 |
| 1.5 | .4332 | .4345 | .4357 | .4370 | .4382 | .4394 | .4406 | .4418 | .4429 | .4441 |

**DEFINITION 5.1**

The **standard normal distribution** is a normal distribution with $\mu = 0$ and $\sigma = 1$. A random variable with a standard normal distribution, denoted by the symbol $z$, is called a **standard normal random variable**.

Since we will ultimately convert all normal random variables to standard normal in order to use Table IV to find probabilities, it is important that you learn to use Table IV well. A partial reproduction of Table IV is shown in Table 5.1. Note that the values of the standard normal random variable $z$ are listed in the left-hand column. The entries in the body of the table give the area (probability) between 0 and $z$. Examples 5.2–5.5 illustrate the use of the table.

**EXAMPLE 5.2**

Find the probability that the standard normal random variable $z$ falls between $-1.33$ and $+1.33$.

**Solution**

The standard normal distribution is shown again in Figure 5.8. Since all probabilities associated with standard normal random variables can be depicted as areas under the standard normal curve, you should always draw the curve and then equate the desired probability to an area.

**FIGURE 5.8**

*Areas Under the Standard Normal Curve for Example 5.2*

In this example we want to find the probability that $z$ falls between $-1.33$ and $+1.33$, which is equivalent to the area between $-1.33$ and $+1.33$, shown highlighted in Figure 5.8. Table IV provides the area between $z = 0$ and any value of $z$, so that if we look up $z = 1.33$ (the value in the 1.3 row and .03 column), we find that the area between $z = 0$ and $z = 1.33$ is .4082. This is the area labeled $A_1$ in Figure 5.8. To find the area $A_2$ located between $z = 0$ and $z = -1.33$, we note that the symmetry of the normal distribution implies that the area between $z = 0$ and any point to the left is equal to the area between $z = 0$ and the point equidistant to the right. Thus, in this example the area between $z = 0$ and $z = -1.33$ is equal to the area between $z = 0$ and $z = +1.33$. That is,

$$A_1 = A_2 = .4082$$

The probability that $z$ falls between $-1.33$ and $+1.33$ is the sum of the areas of $A_1$ and $A_2$. We summarize in probabilistic notation:

$$P(-1.33 < z < +1.33) = P(-1.33 < z < 0) + P(0 < z \le 1.33)$$
$$= A_1 + A_2 = .4082 + .4082 = .8164$$

Remember that " $<$ " and " $\le$ " are equivalent in events involving $z$, because the inclusion (or exclusion) of a single point does not alter the probability of an event involving a continuous random variable.

**EXAMPLE 5.3**

Find the probability that a standard normal random variable exceeds 1.64; that is, find $P(z > 1.64)$.

### Solution

The area under the standard normal distribution to the right of 1.64 is the highlighted area labeled $A_1$ in Figure 5.9. This area represents the desired probability that $z$ exceeds 1.64. However, when we look up $z = 1.64$ in Table IV, we must remember that the probability given in the table corresponds to the area between $z = 0$ and $z = 1.64$ (the area labeled $A_2$ in Figure 5.9). From Table IV we find that $A_2 = .4495$. To find the area $A_1$ to the right of 1.64, we make use of two facts:

1. The standard normal distribution is symmetric about its mean $z = 0$.
2. The total area under the standard normal probability distribution equals 1.

**FIGURE 5.9**

*Areas Under the Standard Normal Curve for Example 5.3*

Taken together, these two facts imply that the areas on either side of the mean $z = 0$ equal .5; thus, the area to the right of $z = 0$ in Figure 5.9 is $A_1 + A_2 = .5$. Then

$$P(z > 1.64) = A_1 = .5 - A_2 = .5 - .4495 = .0505$$

To attach some practical significance to this probability, note that the implication is that the chance of a standard normal random variable exceeding 1.64 is only about .05.

**E X A M P L E   5 . 4**   Find the probability that a standard normal random variable lies to the left of .67.

**Solution**

The event is shown as the highlighted area in Figure 5.10. We want to find $P(z < .67)$. We divide the highlighted area into two parts: the area $A_1$ between $z = 0$ and $z = .67$, and the area $A_2$ to the left of $z = 0$. We must always make such a division when the desired area lies on both sides of the mean ($z = 0$) because Table IV contains areas between $z = 0$ and the point you look up. We look up $z = .67$ in Table IV to find that $A_1 = .2486$. The symmetry of the standard normal distribution also implies that half the distribution lies on each side of the mean, so the area $A_2$ to the left of $z = 0$ is .5. Then,

$$P(z < .67) = A_1 + A_2 = .2486 + .5 = .7486$$

Note that this probability is approximately .75. Thus, about 75% of the time the standard normal random variable $z$ will fall below .67. This implies that $z = .67$ represents the approximate 75th percentile for the distribution.

**FIGURE 5.10**

*Areas Under the Standard Normal Curve for Example 5.4*

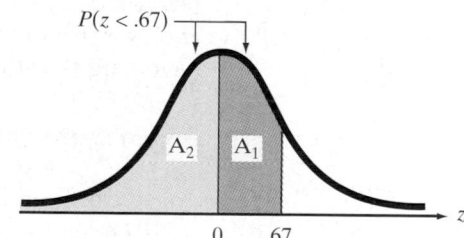

**E X A M P L E   5 . 5**   Find the probability that a standard normal random variable exceeds 1.96 in absolute value.

**Solution**

The event is shown highlighted in Figure 5.11. We want to find

$$P(|z| > 1.96) = P(z < -1.96 \text{ or } z > 1.96)$$

**FIGURE 5.11**

*Areas Under the Standard Normal Curve for Example 5.5*

Note that the total highlighted area is the sum of two areas, $A_1$ and $A_2$—areas that are equal because of the symmetry of the normal distribution.

We look up $z = 1.96$ and find the area between $z = 0$ and $z = 1.96$ to be .4750. Then the area to the right of 1.96, $A_2$, is $.5 - .4750 = .0250$ so that

$$P(|z| > 1.96) = A_1 + A_2 = .0250 + .0250 = .05$$                          ■

To apply Table IV to a normal random variable $x$ with any mean $\mu$ and any standard deviation $\sigma$, we must first convert the value of $x$ to a $z$-score. The population $z$-score for a measurement was defined (in Section 2.6) as the *distance* between the measurement and the population mean, divided by the population standard deviation. Thus, the $z$-score gives the distance between a measurement and the mean in units equal to the standard deviation. In symbolic form, the $z$-score for the measurement $x$ is

$$z = \frac{x - \mu}{\sigma}$$

Note that when $x = \mu$, we obtain $z = 0$.

An important property of the normal distribution is that if $x$ is normally distributed with any mean and any standard deviation, $z$ is *always* normally distributed with mean 0 and standard deviation 1. That is, $z$ is a standard normal random variable.

---

### Property of Normal Distributions

If $x$ is a normal random variable with mean $\mu$ and standard deviation $\sigma$, then the random variable $z$ defined by the formula

$$z = \frac{x - \mu}{\sigma}$$

has a standard normal distribution. The value $z$ describes the number of standard deviations between $x$ and $\mu$.

---

Recall from Example 5.5 that $P(|z| > 1.96) = .05$. This probability coupled with our interpretation of $z$ implies that any normal random variable lies more than 1.96 standard deviations from its mean only 5% of the time. Compare this to the Empirical Rule (Table 2.8, p. 62), which tells us that about 5% of the measurements in mound-shaped distributions will lie beyond 2 standard deviations from the mean. The normal distribution actually provides the model on which the Empirical Rule is based, along with much "empirical" experience with real data that often approximately obey the rule, whether drawn from a normal distribution or not.

■
**EXAMPLE 5.6**

Assume that the length of time, $x$, between charges of a cellular phone is normally distributed with a mean of 10 hours and a standard deviation of 1.5 hours. Find the probability that the cell phone will last between 8 and 12 hours between charges.

### Solution

The normal distribution with mean $\mu = 10$ and $\sigma = 1.5$ is shown in Figure 5.12. The desired probability that the cell phone lasts between 8 and 12 hours is highlighted. In order to find the probability, we must first convert the distribution to standard normal, which we do by calculating the $z$-score:

**FIGURE 5.12**

*Areas Under the Normal Curve for Example 5.6*

$$z = \frac{x - \mu}{\sigma}$$

The $z$-scores corresponding to the important values of $x$ are shown beneath the $x$ values on the horizontal axis in Figure 5.12. Note that $z = 0$ corresponds to the mean of $\mu = 10$ hours, whereas the $x$ values 8 and 12 yield $z$-scores of 1.33 and +1.33, respectively. Thus, the event that the cell phone lasts between 8 and 12 hours is equivalent to the event that a standard normal random variable lies between $-1.33$ and $+1.33$. We found this probability in Example 5.2 (see Figure 5.8) by doubling the area corresponding to $z = 1.33$ in Table IV. That is,

$$P(8 \le x \le 12) = P(-1.33 \le z \le 1.33) = 2(.4082) = .8164$$

The steps to follow when calculating a probability corresponding to a normal random variable are shown in the box.

---

## Steps for Finding a Probability Corresponding to a Normal Random Variable

1. Sketch the normal distribution and indicate the mean of the random variable $x$. Then shade the area corresponding to the probability you want to find.
2. Convert the boundaries of the shaded area from $x$ values to standard normal random variable $z$ values using the formula

   $$z = \frac{x - \mu}{\sigma}$$

   Show the $z$ values under the corresponding $x$ values on your sketch.
3. Use Table IV in Appendix A to find the areas corresponding to the $z$ values. If necessary, use the symmetry of the normal distribution to find areas corresponding to negative $z$ values and the fact that the total area on each side of the mean equals .5 to convert the areas from Table IV to the probabilities of the event you have shaded.

---

**E X A M P L E   5 . 7**

Suppose an automobile manufacturer introduces a new model that has an advertised mean in-city mileage of 27 miles per gallon. Although such advertisements seldom report any measure of variability, suppose you write the manufacturer for the details of the tests, and you find that the standard deviation is 3 miles per gallon. This information leads you to formulate a probability model for the random variable $x$, the in-city mileage for this car model. You believe that the probability distribution of $x$ can be approximated by a normal distribution with a mean of 27 and a standard deviation of 3.

# Using the TI-83 Graphing Calculator

## How to Graph the Area under the Standard Normal

Start from the home screen.

Step 1    Set the viewing window. (*Recall Standard Normal uses z-values, not the actual data.*)

Step 2    Access the Distributions menu.
*Press*   **WINDOW**
*Set*    Xmin = −5
         Xmax = 5
         Xscl = 1
         Ymin = 0
         Ymax = .5
         Yscl = 0
         Xres = 1

> Recall the negative sign is the gray key, not the blue key.

Step 3    *View Graph*
The graph will be displayed along with the area, lower limit, and upper limit.
*Press*   **2ⁿᵈ VARS**
Arrow right to    **DRAW**
*Press*   **Enter**
Enter your lower limit
*Press* **comma**
Enter your upper limit
*Press* **) Press Enter**

Example    What is the probability that *z* is less than 1.5 under the Standard Normal curve?
Step 1    Set your window. (See screen below.)

```
WINDOW
 Xmin=-5
 Xmax=5
 Xscl=1
 Ymin=0
 Ymax=.5
 Yscl=0
 Xres=1
```

Step 2    Enter your values (see lower left screen). Press **ENTER**.

Step 3    You will see display (see lower right screen).

Step 4    Clear the screen for the next problem. Return to the home screen.
*Press* **2ⁿᵈ PRGM ENTER CLEAR CLEAR**.

a. If you were to buy this model of automobile, what is the probability that you would purchase one that averages less than 20 miles per gallon for in-city driving? In other words, find $P(x < 20)$.

b. Suppose you purchase one of these new models and it does get less than 20 miles per gallon for in-city driving. Should you conclude that your probability model is incorrect?

### Solution

a. The probability model proposed for $x$, the in-city mileage, is shown in Figure 5.13. We are interested in finding the area $A$ to the left of 20 since this area corresponds to the probability that a measurement chosen from this distribution falls below 20. In other words, if this model is correct, the area $A$ represents the fraction of cars that can be expected to get less than 20 miles per gallon for in-city driving. To find $A$, we first calculate the $z$ value corresponding to $x = 20$. That is,

$$z = \frac{x - \mu}{\sigma} = \frac{20 - 27}{3} = \frac{7}{3} = -2.33$$

Then

$$P(x < 20) = P(z < -2.33)$$

as indicated by the highlighted area in Figure 5.13. Since Table IV gives only areas to the right of the mean (and because the normal distribution is symmetric about its mean), we look up 2.33 in Table IV and find that the corresponding area is .4901. This is equal to the area between $z = 0$ and $z = -2.33$, so we find

$$P(x < 20) = A = .5 - .4901 = .0099 \approx .01$$

According to this probability model, you should have only about a 1% chance of purchasing a car of this make with an in-city mileage under 20 miles per gallon.

### FIGURE 5.13

*Area Under the Normal Curve for Example 5.7*

b. Now you are asked to make an inference based on a sample—the car you purchased. You are getting less than 20 miles per gallon for in-city driving. What do you infer? We think you will agree that one of two possibilities is true:

1. The probability model is correct. You simply were unfortunate to have purchased one of the cars in the 1% that get less than 20 miles per gallon in the city.
2. The probability model is incorrect. Perhaps the assumption of a normal distribution is unwarranted, or the mean of 27 is an overestimate, or the standard deviation of 3 is an underestimate, or some combination of these errors was made. At any rate, the form of the actual probability model certainly merits further investigation.

   You have no way of knowing with certainty which possibility is correct, but the evidence points to the second one. We are again relying on the rare-event approach to statistical inference that we introduced earlier. The sample (one measurement in this case) was so unlikely to have been drawn from the proposed probability model that it casts serious doubt on the model. We would be inclined to believe that the model is somehow in error. ∎

Occasionally you will be given a probability and will want to find the values of the normal random variable that correspond to the probability. For example, suppose the scores on a college entrance examination are known to be normally distributed, and a certain prestigious university will consider for admission only those applicants whose scores exceed the 90th percentile of the test score distribution. To determine the minimum score for admission consideration, you will need to be able to use Table IV in reverse, as demonstrated in the following example.

**EXAMPLE 5.8**

Find the value of $z$, call it $z_0$, in the standard normal distribution that will be exceeded only 10% of the time. That is, find $z_0$ such that $P(z \geq z_0) = .10$.

### Solution

In this case we are given a probability, or an area, and asked to find the value of the standard normal random variable that corresponds to the area. Specifically, we want to find the value $z_0$ such that only 10% of the standard normal distribution exceeds $z_0$ (see Figure 5.14).

**FIGURE 5.14**

*Areas Under the Standard Normal Curve for Example 5.8*

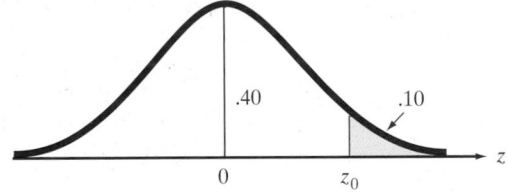

We know that the total area to the right of the mean $z = 0$ is .5, which implies that $z_0$ must lie to the right of (above) 0. To pinpoint the value, we use the fact that the area to the right of $z_0$ is .10, which implies that the area between $z = 0$ and $z_0$ is $.5 - .1 = .4$. But areas between $z = 0$ and some other $z$ value are exactly the types given in Table IV. Therefore, we look up the area .4000 in the body of Table IV and find that the corresponding $z$ value is (to the closest approximation) $z_0 = 1.28$. The implication is that the point 1.28 standard deviations above the mean is the 90th percentile of a normal distribution. ∎

**EXAMPLE 5.9**

Find the value of $z_0$ such that 95% of the standard normal $z$ values lie between $-z_0$ and $+z_0$—that is, $P(-z_0 \leq z \leq z_0) = .95$.

### Solution

Here we wish to move an equal distance $z_0$ in the positive and negative directions from the mean $z = 0$ until 95% of the standard normal distribution is enclosed. This means that the area on each side of the mean will be equal to $\frac{1}{2}(.95) = .475$, as shown in Figure 5.15. Since the area between $z = 0$ and $z_0$ is .475, we look up

**FIGURE 5.15**

*Areas Under the Standard Normal Curve for Example 5.9*

.475 in the body of Table IV to find the value $z_0 = 1.96$. Thus, as we found in the reverse order in Example 5.5, 95% of a normal distribution lies between $+1.96$ and $-1.96$ standard deviations of the mean.  ■

Now that you have learned to use Table IV to find a standard normal $z$ value that corresponds to a specified probability, we demonstrate a practical application in Example 5.10.

**EXAMPLE 5.10**

Suppose the scores, $x$, on a college entrance examination are normally distributed with a mean of 550 and a standard deviation of 100. A certain prestigious university will consider for admission only those applicants whose scores exceed the 90th percentile of the distribution. Find the minimum score an applicant must achieve in order to receive consideration for admission to the university.

### Solution

In this example, we want to find a score, $x_0$, such that 90% of the scores ($x$ values) in the distribution fall below $x_0$ and only 10% fall above $x_0$. That is,

$$P(x \leq x_0) = .90$$

Converting $x$ to a standard normal random variable, where $\mu = 550$ and $\sigma = 100$, we have

$$P(x \leq x_0) = P\left(z \leq \frac{x_0 - \mu}{\sigma}\right)$$
$$= P\left(z \leq \frac{x_0 - 550}{100}\right) = .90$$

In Example 5.8 (see Figure 5.14) we found the 90th percentile of the standard normal distribution to be $z_0 = 1.28$. That is, we found $P(z \leq 1.28) = .90$. Consequently, we know the minimum test score $x_0$ corresponds to a $z$-score of 1.28; that is,

$$\frac{x_0 - 550}{100} = 1.28$$

If we solve this equation for $x_0$, we find

$$x_0 = 550 + 1.28(100) = 550 + 128 = 678$$

This $x$ value is shown in Figure 5.16. Thus, the 90th percentile of the test score distribution is 678. An applicant must score at least 678 on the entrance exam to receive consideration for admission by the university.  ■

**FIGURE 5.16**

*Area Under the Normal Curve for Example 5.10*

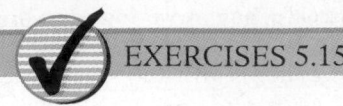

EXERCISES 5.15–5.42

## Learning the Mechanics

**5.15** Find the area under the standard normal probability distribution between the following pairs of $z$-scores:
  **a.** $z = 0$ and $z = 2.00$
  **b.** $z = 0$ and $z = 1.00$
  **c.** $z = 0$ and $z = 3$
  **d.** $z = 0$ and $z = .58$
  **e.** $z = -2.00$ and $z = 0$
  **f.** $z = -1.00$ and $z = 0$
  **g.** $z = -1.69$ and $z = 0$
  **h.** $z = -.58$ and $z = 0$

**5.16** Find the following probabilities for the standard normal random variable $z$:
  **a.** $P(z > 1.46)$
  **b.** $P(z < -1.56)$
  **c.** $P(.67 \leq z \leq 2.41)$
  **d.** $P(-1.96 \leq z < -.33)$
  **e.** $P(z \geq 0)$
  **f.** $P(-2.33 < z < 1.50)$
  **g.** $P(z \geq -2.33)$
  **h.** $P(z < 2.33)$

**5.17** Find each of the following probabilities for a standard normal random variable $z$:
  **a.** $P(z = 1)$
  **b.** $P(z \leq 1)$
  **c.** $P(z < 1)$
  **d.** $P(z > 1)$
  **e.** $P(-1 \leq z \leq 1)$
  **f.** $P(-2 \leq z \leq 2)$
  **g.** $P(-2.16 \leq z \leq .55)$
  **h.** $P(-.42 < z < 1.96)$

**5.18** Find each of the following probabilities for the standard normal random variable $z$:
  **a.** $P(-1 \leq z \leq 1)$
  **b.** $P(-1.96 \leq z \leq 1.96)$
  **c.** $P(-1.645 \leq z \leq 1.645)$
  **d.** $P(-2 \leq z \leq 2)$

**5.19** Find a value of the standard normal random variable $z$, call it $z_0$, such that
  **a.** $P(z \geq z_0) = .05$
  **b.** $P(z \geq z_0) = .025$
  **c.** $P(z \leq z_0) = .025$
  **d.** $P(z \geq z_0) = .10$
  **e.** $P(z > z_0) = .10$

**5.20** Find a value of the standard normal random variable $z$, call it $z_0$, such that
  **a.** $P(z \leq z_0) = .0401$
  **b.** $P(-z_0 \leq z \leq z_0) = .95$
  **c.** $P(-z_0 \leq z \leq z_0) = .90$
  **d.** $P(-z_0 \leq z \leq z_0) = .8740$
  **e.** $P(-z_0 \leq z \leq 0) = .2967$
  **f.** $P(-2 < z < z_0) = .9710$

  **g.** $P(z \geq z_0) = .5$
  **h.** $P(z \geq z_0) = .0057$

**5.21** Give the $z$-score for a measurement from a normal distribution for the following:
  **a.** 1 standard deviation above the mean
  **b.** 1 standard deviation below the mean
  **c.** Equal to the mean
  **d.** 2.5 standard deviations below the mean
  **e.** 3 standard deviations above the mean

**5.22** Suppose the random variable $x$ is best described by a normal distribution with $\mu = 25$ and $\sigma = 5$. Find the $z$-score that corresponds to each of the following $x$ values:
  **a.** $x = 25$
  **b.** $x = 30$
  **c.** $x = 37.5$
  **d.** $x = 10$
  **e.** $x = 50$
  **f.** $x = 32$

**5.23** The random variable $x$ has a normal distribution with $\mu = 300$ and $\sigma = 30$.
  **a.** Find the probability that $x$ assumes a value more than 2 standard deviations from its mean. More than 3 standard deviations from $\mu$.
  **b.** Find the probability that $x$ assumes a value within 1 standard deviation of its mean. Within 2 standard deviations of $\mu$.
  **c.** Find the value of $x$ that represents the 80th percentile of this distribution. The 10th percentile.

**5.24** Suppose $x$ is a normally distributed random variable with $\mu = 11$ and $\sigma = 2$. Find each of the following:
  **a.** $P(10 \leq x \leq 12)$
  **b.** $P(6 \leq x \leq 10)$
  **c.** $P(13 \leq x \leq 16)$
  **d.** $P(7.8 \leq x \leq 12.6)$
  **e.** $P(x \geq 13.24)$
  **f.** $P(x \geq 7.62)$

**5.25** Suppose $x$ is a normally distributed random variable with $\mu = 30$ and $\sigma = 8$. Find a value of the random variable, call it $x_0$, such that
  **a.** $P(x \geq x_0) = .5$
  **b.** $P(x < x_0) = .025$
  **c.** $P(x > x_0) = .10$
  **d.** $P(x > x_0) = .95$
  **e.** 10% of the values of $x$ are less than $x_0$
  **f.** 80% of the values of $x$ are less than $x_0$
  **g.** 1% of the values of $x$ are greater than $x_0$

**5.26** Suppose $x$ is a normally distributed random variable with mean 100 and standard deviation 8. Draw a rough graph of the distribution of $x$. Locate $\mu$ and the interval $\mu \pm 2\sigma$ on the graph. Find the following probabilities:
  **a.** $P(\mu - 2\sigma \leq x \leq \mu + 2\sigma)$
  **b.** $P(x \geq \mu + 2\sigma)$
  **c.** $P(x \leq 92)$

   **d.** $P(92 \leq x \leq 116)$
   **e.** $P(92 \leq x \leq 96)$
   **f.** $P(76 \leq x \leq 124)$

**5.27** The random variable $x$ has a normal distribution with standard deviation 25. It is known that the probability that $x$ exceeds 150 is .90. Find the mean $\mu$ of the probability distribution.

## Applying the Concepts—Basic

**5.28** Psychology students at Wittenberg University completed the Dental Anxiety Scale questionnaire (*Psychological Reports*, Aug. 1997). Scores on the scale range from 0 (no anxiety) to 20 (extreme anxiety). The mean score was 11 and the standard deviation was 3.5. Assume that the distribution of all scores on the Dental Anxiety Scale is normal with $\mu = 11$ and $\sigma = 3.5$.
   **a.** Suppose you score a 16 on the Dental Anxiety Scale. Find the $z$ value for this score.
   **b.** Find the probability that someone scores between a 10 and a 15 on the Dental Anxiety Scale.
   **c.** Find the probability that someone scores above a 17 on the Dental Anxiety Scale.

**5.29** The alkalinity level of water specimens collected from the Han River in Seoul, Korea, has a mean of 50 milligrams per liter and a standard deviation of 3.2 milligrams per liter (*Environmental Science & Engineering*, Sept. 1, 2000). Assume the distribution of alkalinity levels is approximately normal and find the probability that a water specimen collected from the river has an alkalinity level
   **a.** exceeding 45 milligrams per liter.
   **b.** below 55 milligrams per liter.
   **c.** between 51 and 52 milligrams per liter.

**5.30** Refer to the *Chance* (Winter 2001) study of students who paid a private tutor to help them improve their Standardized Admission Test (SAT) scores, Exercise 2.81 (p. 70). The table summarizing the changes in both the SAT-Mathematics and SAT-Verbal scores for these students is reproduced here. Assume that both distributions of SAT score changes are approximately normal.

|                                      | SAT-Math | SAT-Verbal |
| ------------------------------------ | -------- | ---------- |
| Mean change in score                 | 19       | 7          |
| Standard deviation of score changes  | 65       | 49         |

   **a.** What is the probability that a student increases his or her score on the SAT-Math test by at least 50 points?
   **b.** What is the probability that a student increases his or her score on the SAT-Verbal test by at least 50 points?

**5.31** A baseball player's batting average is determined by dividing the player's total number of hits by the total number of official batting attempts. The batting averages of all professional baseball players are compiled each year by Major League Baseball. The distribution of batting averages for players in the American League (which uses a designated hitter) and National League (which does not use a designated hitter) are summarized in the table. Both distributions are approximately normal.

| American League | National League |
| --------------- | --------------- |
| $\mu = .268$    | $\mu = .262$    |
| $\sigma = .031$ | $\sigma = .039$ |

*Source:* Major League Baseball, 1998.

   **a.** Find the probability that an American League player has a batting average of .300 or higher.
   **b.** Find the probability that a National League player has a batting average of .300 or higher.

**5.32** A physical-fitness association is including the mile run in its secondary-school fitness test for students. The time for this event for students in secondary school is approximately normally distributed with a mean of 450 seconds and a standard deviation of 40 seconds. If the association wants to designate the fastest 10% as "excellent," what time should the association set for this criterion?

**5.33** *Fisheries Science* (Feb. 1995) published a study of the length distributions of sardines inhabiting Japanese waters. At two years of age, fish have a length distribution that is approximately normal with $\mu = 20.20$ centimeters (cm) and $\sigma = .65$ cm.
   **a.** Find the probability that a two-year-old sardine inhabiting Japanese waters is between 20 and 21 cm long.
   **b.** Find the probability that a two-year-old sardine in Japanese waters is less than 19.84 cm long.
   **c.** Find the probability that a two-year-old sardine in Japanese waters is greater than 22.01 cm long.

## Applying the Concepts—Intermediate

**5.34** The physical fitness of a patient is often measured by the patient's maximum oxygen uptake (recorded in milliliters per kilogram, ml/kg). The mean maximum oxygen uptake for cardiac patients who regularly participate in sports or exercise programs was found to be 24.1 with a standard deviation of 6.30 (*Adapted Physical Activity Quarterly*, Oct. 1997). Assume this distribution is approximately normal.
   **a.** What is the probability that a cardiac patient who regularly participates in sports has a maximum oxygen uptake of at least 20 ml/kg?
   **b.** What is the probability that a cardiac patient who regularly exercises has a maximum oxygen uptake of 10.5 ml/kg or lower?
   **c.** Consider a cardiac patient with a maximum oxygen uptake of 10.5. Is it likely that this patient participates regularly in sports or exercise programs? Explain.

**5.35** The *Journal of Visual Impairment & Blindness* (May–June 1997) published a study of the lifestyles of visually impaired students. Using diaries, the students kept track of several variables, including number of hours of sleep obtained in a typical day. These visually impaired students had a mean of 9.06 hours and a standard deviation of 2.11 hours. Assume that the distribution of the number of hours of sleep for this group of students is approximately normal.

**a.** Find the probability that a visually impaired student obtains less than 6 hours of sleep on a typical day.

**b.** Find the probability that a visually impaired student gets between 8 and 10 hours of sleep on a typical day.

**c.** Twenty percent of all visually impaired students obtain less than how many hours of sleep on a typical day?

**5.36** A group of Florida State University psychologists examined the effects of alcohol on the reactions of people to a threat (*Journal of Abnormal Psychology*, Vol. 107, 1998). After obtaining a specified blood alcohol level, experimental subjects were placed in a room and threatened with electric shocks. Using sophisticated equipment to monitor the subjects' eye movements, the startle response (measured in milliseconds) was recorded for each subject. The mean and standard deviation of the startle responses were 37.9 and 12.4, respectively. Assume that the startle response $x$ for a person with the specified blood alcohol level is approximately normally distributed.

**a.** Find the probability that $x$ is between 40 and 50 milliseconds.

**b.** Find the probability that $x$ is less than 30 milliseconds.

**c.** Give an interval for $x$, centered around 37.9 milliseconds, so that the probability that $x$ falls in the interval is .95.

**d.** Ten percent of the experimental subjects have startle responses above what value?

**5.37** *Ecological Applications* (May 1995) published a study on the development of forests following wildfires in the Pacific Northwest. One variable of interest to the researcher was tree diameter at breast height 110 years after the fire. The population of Douglas fir trees was shown to have an approximately normal diameter distribution, with $\mu = 50$ centimeters (cm) and $\sigma = 12$ cm. Find the diameter, $d$, such that 30% of the Douglas fir trees in the population have diameters that exceed $d$.

**5.38** Based on data from the National Center for Health Statistics, N. Wetzel used the normal distribution to model the length of gestation for pregnant U.S. women (*Chance*, Spring 2001). Gestation length has a mean of 280 days with a standard deviation of 20 days.

**a.** Find the probability that gestation length is between 275.5 and 276.5 days. (This estimates the probability that a women has her baby 4 days earlier than the "average" due date.)

**b.** Find the probability that gestation length is between 258.5 and 259.5 days. (This estimates the probability that a women has her baby 21 days earlier than the "average" due date.)

**c.** Find the probability that gestation length is between 254.5 and 255.5 days. (This estimates the probability

that a women has her baby 25 days earlier than the "average" due date.)

**d.** The *Chance* article referenced a newspaper story about three sisters who all gave birth on the same day (March 11, 1998). Karralee had her baby 4 days early; Marrianne had her baby 21 days early; and Jennifer had her baby 25 days early. Use the results, parts **a–c**, to estimate the probability that three women have their babies 4, 21, and 25 days early, respectively. Assume the births are independent events.

**5.39** A study of serum cholesterol levels in psychiatric patients of a maximum-security forensic hospital revealed that cholesterol level is approximately normally distributed with a mean of 208 milligrams per deciliter (mg/dL) and a standard deviation of 25 mg/dL (*Journal of Behavioral Medicine*, Feb. 1995). Prior research has shown that patients who exhibit violent behavior have cholesterol levels below 200 mg/dL.

**a.** What is the probability of observing a psychiatric patient with a cholesterol level below 200 mg/dL?

**b.** For three randomly selected patients, what is the probability that at least one will have a cholesterol level below 200 mg/dL?

## Applying the Concepts—Advanced

**5.40** The characteristics of an industrial filling process in which an expensive liquid is injected into a container were investigated in *Journal of Quality Technology* (July 1999). The quantity injected per container is approximately normally distributed with mean 10 units and standard deviation .2 units. Each unit of fill costs $20. If a container contains less than 10 units (i.e., is underfilled), it must be reprocessed at a cost of $10. A properly filled container sells for $230.

**a.** Find the probability that a container is underfilled.

**b.** A container is initially underfilled and must be reprocessed. Upon refilling it contains 10.6 units. How much profit will the company make on this container?

**c.** The operations manager adjusts the mean of the filling process upward to 10.5 units in order to make the probability of underfilling approximately zero. Under these conditions, what is the expected profit per container?

**5.41** What relationship exists between the standard normal distribution and the box-plot methodology (optional Section 2.8) for describing distributions of data using quartiles? The answer depends on the true underlying probability distribution of the data. Assume for the remainder of this exercise that the distribution is normal.

**a.** Calculate the values of the standard normal random variable $z$, call them $z_L$ and $z_U$, that correspond to the hinges of the box plot—i.e., the lower and upper quartiles, $Q_L$ and $Q_U$—of the probability distribution.

**b.** Calculate the $z$ values that correspond to the inner fences of the box plot for a normal probability distribution.

**c.** Calculate the $z$ values that correspond to the outer fences of the box plot for a normal probability distribution.

**d.** What is the probability that an observation lies beyond the inner fences of a normal probability distribution? The outer fences?

**e.** Can you better understand why the inner and outer fences of a box plot are used to detect outliers in a distribution? Explain.

**5.42** A machine used to regulate the amount of dye dispensed for mixing shades of paint can be set so that it discharges an average of $\mu$ milliliters (mL) of dye per can of paint. The amount of dye discharged is known to have a normal distribution with a standard deviation of .4 mL. If more than 6 mL of dye are discharged when making a certain shade of blue paint, the shade is unacceptable. Determine the setting for $\mu$ so that only 1% of the cans of paint will be unacceptable.

## 5.4 DESCRIPTIVE METHODS FOR ASSESSING NORMALITY

In the chapters that follow, we learn how to make inferences about the population based on information in the sample. Several of these techniques are based on the assumption that the population is approximately normally distributed. Consequently, it will be important to determine whether the sample data come from a normal population before we can properly apply these techniques.

Several descriptive methods can be used to check for normality. In this section, we consider the three methods summarized in the box.

---

### Determining Whether the Data Are From an Approximately Normal Distribution

1. Construct either a histogram or stem-and-leaf display for the data and note the shape of the graph. If the data are approximately normal, the shape of the histogram or stem-and-leaf display will be similar to the normal curve, Figure 5.5 (i.e., mound-shaped and symmetric about the mean).

2. Compute the intervals $\bar{x} \pm s$, $\bar{x} \pm 2s$, and $\bar{x} \pm 3s$, and determine the percentage of measurements falling in each. If the data are approximately normal, the percentages will be approximately equal to 68%, 95%, and 100%, respectively.

3. Find the interquartile range, IQR, and standard deviation, $s$, for the sample, then calculate the ratio IQR/$s$. If the data are approximately normal, then IQR/$s \approx 1.3$.

4. Construct a normal probability plot for the data. If the data are approximately normal, the points will fall (approximately) on a straight line.

---

The first two methods come directly from the properties of a normal distribution established in Section 5.3. Method 3 is based on the fact that for normal distributions, the $z$ values corresponding to the 25th and 75th percentiles are $-.67$ and $.67$, respectively (see Example 5.4). Since $\sigma = 1$ for a standard normal distribution,

$$\frac{\text{IQR}}{\sigma} = \frac{Q_U - Q_L}{\sigma} = \frac{.67 - (-.67)}{1} = 1.34$$

The final descriptive method for checking normality is based on a *normal probability plot*. In such a plot, the observations in a data set are ordered from smallest to largest and then plotted against the expected $z$-scores of observations calculated under the assumption that the data come from a normal distribution. When the data are, in fact, normally distributed, a linear (straight-line) trend will result. A nonlinear trend in the plot suggests that the data are non-normal.

**DEFINITION 5.2**
A **normal probability plot** for a data set is a scatterplot with the ranked data values on one axis and their corresponding expected $z$-scores from a standard normal distribution on the other axis. [*Note:* Computation of the expected standard normal $z$-scores are beyond the scope of this text. Therefore, we will rely on available statistical software packages to generate a normal probability plot.]

**EXAMPLE 5.11**

The EPA mileage ratings on 100 cars, first presented in Chapter 2 (p. 30), are reproduced in Table 5.2. Numerical and graphical descriptive measures for the data are shown on the MINITAB and SPSS printouts, Figure 5.17a–c. Determine whether the EPA mileage ratings are from an approximate normal distribution.

EPAGAS

**Table 5.2    EPA Gas Mileage Ratings for 100 Cars (miles per gallon)**

| | | | | | | | | | |
|---|---|---|---|---|---|---|---|---|---|
| 36.3 | 41.0 | 36.9 | 37.1 | 44.9 | 36.8 | 30.0 | 37.2 | 42.1 | 36.7 |
| 32.7 | 37.3 | 41.2 | 36.6 | 32.9 | 36.5 | 33.2 | 37.4 | 37.5 | 33.6 |
| 40.5 | 36.5 | 37.6 | 33.9 | 40.2 | 36.4 | 37.7 | 37.7 | 40.0 | 34.2 |
| 36.2 | 37.9 | 36.0 | 37.9 | 35.9 | 38.2 | 38.3 | 35.7 | 35.6 | 35.1 |
| 38.5 | 39.0 | 35.5 | 34.8 | 38.6 | 39.4 | 35.3 | 34.4 | 38.8 | 39.7 |
| 36.3 | 36.8 | 32.5 | 36.4 | 40.5 | 36.6 | 36.1 | 38.2 | 38.4 | 39.3 |
| 41.0 | 31.8 | 37.3 | 33.1 | 37.0 | 37.6 | 37.0 | 38.7 | 39.0 | 35.8 |
| 37.0 | 37.2 | 40.7 | 37.4 | 37.1 | 37.8 | 35.9 | 35.6 | 36.7 | 34.5 |
| 37.1 | 40.3 | 36.7 | 37.0 | 33.9 | 40.1 | 38.0 | 35.2 | 34.8 | 39.5 |
| 39.9 | 36.9 | 32.9 | 33.8 | 39.8 | 34.0 | 36.8 | 35.0 | 38.1 | 36.9 |

**FIGURE 5.17a**
*MINITAB Histogram for Mileage Data*

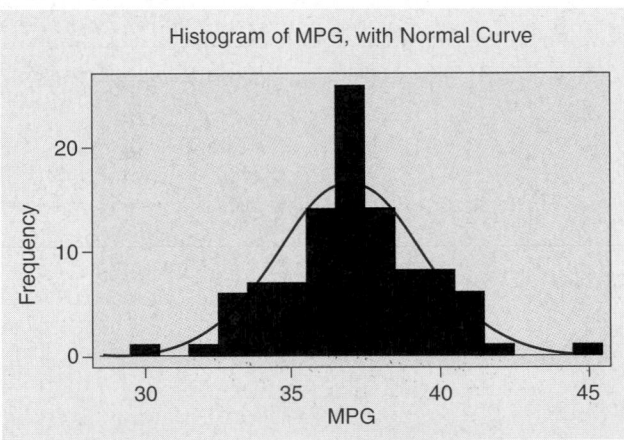

**FIGURE 5.17b**
*MINITAB Descriptive Statistics for Mileage Data*

Descriptive Statistics: MPG

| Variable | N | Mean | Median | TrMean | StDev | SE Mean |
|---|---|---|---|---|---|---|
| MPG | 100 | 36.994 | 37.000 | 36.992 | 2.418 | 0.242 |

| Variable | Minimum | Maximum | Q1 | Q3 |
|---|---|---|---|---|
| MPG | 30.000 | 44.900 | 35.625 | 38.375 |

**FIGURE 5.17c**

*SPSS Normal Probability Plot
for Mileage Data*

## Solution

As a first check, we examine the MINITAB histogram of the data shown in Figure 5.17a. Clearly, the mileages fall in an approximately mound-shaped, symmetric distribution centered around the mean of approximately 37 mpg. Note that a normal curve is superimposed on the figure. Therefore, using check #1 in the box, the data appear to be approximately normal.

To apply check #2, we obtain $\bar{x} = 37$ and $s = 2.4$ from the MINITAB printout, Figure 5.17b. The intervals $\bar{x} \pm s, \bar{x} \pm 2s$, and $\bar{x} \pm 3s$, are shown in Table 5.3, as well as the percentage of mileage ratings that fall in each interval. (We obtained these results in Section 2.6, p. 61.) These percentages agree almost exactly with those from a normal distribution.

**Table 5.3   Describing the 100 EPA Mileage Ratings**

| Interval | Percentage in Interval |
|---|---|
| $\bar{x} \pm \ \ s = (34.6, 39.4)$ | 68 |
| $\bar{x} \pm 2s = (32.2, 41.8)$ | 96 |
| $\bar{x} \pm 3s = (29.8, 44.2)$ | 99 |

Check #3 in the box requires that we find the ratio IQR/$s$. From Figure 5.17b, the 25th percentile (labeled $Q_1$ by MINITAB) is $Q_L = 35.625$ and the 75th percentile (labeled $Q_3$ by MINITAB) is $Q_U = 38.375$. Then, IQR = $Q_U - Q_L = 2.75$ and the ratio is

$$\frac{\text{IRQ}}{s} = \frac{2.75}{2.4} = 1.15$$

Since this value is approximately equal to 1.3, we have further confirmation that the data are approximately normal.

A fourth descriptive method is to interpret a normal probability plot. An SPSS normal probability plot for the mileage data is shown in Figure 5.17c. Notice that the ordered mileage values (shown on the horizontal axis) fall reasonably close to a straight line when plotted against the expected values from a normal distribution. Thus, check #4 also suggests that the EPA mileage data are likely to be approximately normally distributed.

∎

The checks for normality given in the box are simple, yet powerful, techniques to apply, but they are only descriptive in nature. It is possible (although unlikely) that the data are nonnormal even when the checks are reasonably satisfied. Thus, we should be careful not to claim that the 100 EPA mileage ratings of Example 5.11 are, in fact, normally distributed. We can only state that it is reasonable to believe that the data are from a normal distribution.*

As we will learn in the next chapter, several inferential methods of analysis require the data to be approximately normal. If the data are clearly nonnormal, inferences derived from the method may be invalid. Therefore, it is advisable to check the normality of the data prior to conducting the analysis.

---

# Using the TI-83 Graphing Calculator
## How to Graph a Normal Probability Plot

Start from the home screen.

Step 1   Enter your data into **List 1** (*Recall the Lists are accessed by pressing* **STAT ENTER**. *Always begin from a "clear" List*).

Step 2   Access the "Stat Plot" menu.
*Press*   **2ⁿᵈ ENTER**
*Press*   **ENTER**
*Press*   **ENTER**
Arrow down and right to last graph.
*Press*   **ENTER**
*Set*   Data List: **L1**
Data Axis: **X**

Step 3   *Press* **Zoom 9**
Your data will be displayed against the residuals. If you see a "generally" linear relationship your data are near normal.

Example   Using a Normal Probability Plot test whether or not the data are normally distributed.

| 9.7 | 93.1 | 33.0 | 21.2 | 81.4 | 51.1 |
|------|------|------|------|------|------|
| 43.5 | 10.6 | 12.8 | 7.8 | 18.1 | 12.7 |

Step 1   Enter Data into List1 (see screen below).

```
L1      L2     L3      1
51.1
43.5
10.6
12.8
7.8
18.1
12.7
L1(12) =12.7
```

---

*Statistical tests of normality that provide a measure of reliability for the inference are available. However, these tests tend to be very sensitive to slight departures from normality, i.e., they tend to reject the hypothesis of normality for any distribution that is not perfectly symmetrical and mound shaped. Consult the references (see Ramsey & Ramsey, 1990) if you want to learn more about these tests.

**USING THE TI-83 GRAPHING CALCULATOR (*continued*)**

Step 2    Access the STAT PLOT Menu. *Press* 2$^{nd}$ = **ENTER ENTER**.
(You will see the screen below left after you change your setting to match.)

Step 3    View Display. *Press* **Zoom 9** (see screen above right).
There is a noticeable curve. The data are not normally distributed.

Step 4    Clear the screen for the next problem. Return to the home screen.
*Press* 2$^{nd}$Y = **ENTER**. Arrow right. *Press* **ENTER CLEAR**.

 EXERCISES 5.43–5.53

### Learning the Mechanics

**5.43** If a population data set is normally distributed, what is the proportion of measurements you would expect to fall within the following intervals?
  **a.** $\mu \pm \sigma$
  **b.** $\mu \pm 2\sigma$
  **c.** $\mu \pm 3\sigma$

**5.44** Consider a sample data set with the following summary statistics:   $s = 95, Q_L = 72, Q_U = 195$.
  **a.** Calculate IQR.
  **b.** Calculate IQR/$s$.
  **c.** Is the value of IQR/$s$ approximately equal to 1.3? What does this imply?

**5.45** Normal probability plots for three data sets are shown here. Which plot indicates that the data are approximately normally distributed?

**5.46** Examine the sample data in the table.

  ⊘ LM5_46

| 5.9 | 5.3 | 1.6 | 7.4 | 8.6 | 1.2 | 2.1 |
|-----|-----|-----|-----|-----|-----|-----|
| 4.0 | 7.3 | 8.4 | 8.9 | 6.7 | 4.5 | 6.3 |
| 7.6 | 9.7 | 3.5 | 1.1 | 4.3 | 3.3 | 8.4 |
| 1.6 | 8.2 | 6.5 | 1.1 | 5.0 | 9.4 | 6.4 |

  **a.** Construct a stem-and-leaf plot to assess whether the data are from an approximately normal distribution.
  **b.** Compute $s$ for the sample data.
  **c.** Find the values of $Q_L$ and $Q_U$ and the value of $s$ from part **b** to assess whether the data come from an approximately normal distribution.
  **d.** Generate a normal probability plot for the data and use it to assess whether the data are approximately normal.

**Plots for Exercise 5.45**

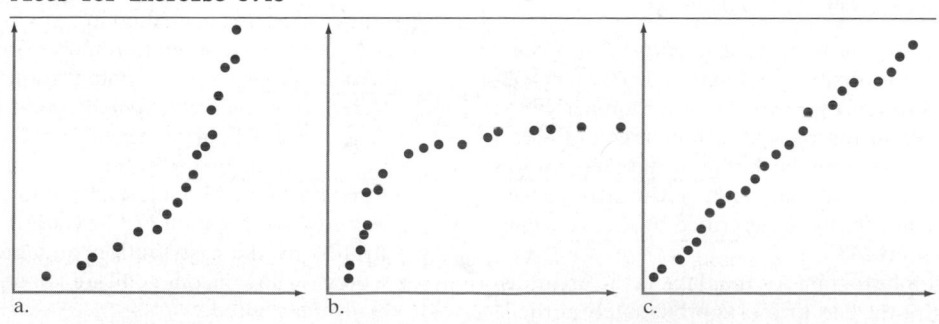

a.    b.    c.

MINITAB Output for Exercise 5.47

| Variable | N | Mean | Median | Tr Mean | StDev | SE Mean |
|---|---|---|---|---|---|---|
| Age | 50 | 48.160 | 47.000 | 47.795 | 6.015 | 0.851 |

| Variable | Min | Max | Q1 | Q3 |
|---|---|---|---|---|
| Age | 36.000 | 68.000 | 45.000 | 51.250 |

## Applying the Concepts—Basic

WOMENPOWER

**5.47** Refer to Exercise 2.44 (p. 51) and the ages of the 50 most powerful women in corporate America by *Fortune* (Oct. 25, 1999). A MINITAB printout with summary statistics for the age distribution is reproduced below.
  **a.** Use the relevant statistics on the printout to determine whether the age distribution is approximately normal.
  **b.** In Exercise 2.44d you constructed a relative frequency histogram for the age data. Use this graph to support your conclusion in part a.

**5.48** Refer to the *Astronomical Journal* (July 1995) study of galaxy velocities, Exercise 2.78 (p. 69). A histogram of the velocity for 103 galaxies located in a particular cluster named A2142 is reproduced here. Comment on whether or not the galaxy velocities are approximately normally distributed.

*Source:* Oegerle, W. R., Hill, J. M., and Fitchett, M. J. "Observations of high dispersion clusters of galaxies: Constraints on cold dark matter." *The Astronomical Journal*, Vol. 110, No. 1, July 1995, p. 37 (Figure 1).

**5.49** How accurate are you at the game of darts? Researchers at Iowa State University attempted to develop a probability model for dart throws (*Chance*, Summer 1997). For each of 590 throws made at certain targets on a dart board, the distance from the dart to the target point was measured (to the nearest millimeter). The error distribution for the dart throws is described by the frequency table shown on p. 243.
  **a.** Construct a histogram for the data. Is the error distribution for the dart throws approximately normal?

**b.** Descriptive statistics for the distances from the target for the 590 throws are given below. Use this information to decide whether the error distribution is approximately normal.

$$\bar{x} = 24.4 \text{ mm}$$
$$s = 12.8 \text{ mm}$$
$$Q_L = 14 \text{ mm}$$
$$Q_U = 34 \text{ mm}$$

**c.** Construct a normal probability plot for the data and use it to assess whether the error distribution is approximately normal.

## Applying the Concepts—Intermediate

**5.50** Refer to the distribution of batting averages for Major League Baseball players, Exercise 5.31 (p. 235). The batting averages for American League players with at least 100 official at bats are stored in the data file ALBATAVG, while the batting averages for National League players with at least 100 official at bats are stored in the data file NLBATAVG. Determine whether each distribution is approximately normal.

**5.51** In Exercise 2.32 (p. 41) you read about a *Journal of Statistics Education* study of team performance on games in which Mark McGwire (of the St. Louis Cardinals) and Sammy Sosa (of the Chicago Cubs) hit home runs during their record-breaking 1998 Major League Baseball season. The data on number of runs scored by their respective teams in these games are reproduced in the table at the bottom of p. 243.
  **a.** Determine whether the data on number of runs scored by the St. Louis Cardinals are approximately normal.
  **b.** Repeat part **a** for the Chicago Cubs.

**5.52** Refer to the data on May 2001 sanitation scores for 151 cruise ships, first presented in Exercise 2.72 (p. 67). Assess whether the sanitation scores (saved in the SHIP-SANIT file) are approximately normally distributed.

**5.53** Refer to the *Applied Psycholinguistics* (June 1998) study of language skills in young children, Exercise 2.76 (p. 69). Recall that the mean sentence complexity score (measured on a 0 to 48 point scale) of low income children was 7.62 with a standard deviation of 8.91. Demonstrate why the distribution of sentence complexity scores for low income children is unlikely to be normally distributed.

## DARTS

| Distance from Target (mm) | Frequency | Distance from Target (mm) | Frequency | Distance from Target (mm) | Frequency |
|---|---|---|---|---|---|
| 0 | 3 | 21 | 15 | 41 | 7 |
| 2 | 2 | 22 | 19 | 42 | 7 |
| 3 | 5 | 23 | 13 | 43 | 11 |
| 4 | 10 | 24 | 11 | 44 | 7 |
| 5 | 7 | 25 | 9 | 45 | 5 |
| 6 | 3 | 26 | 21 | 46 | 3 |
| 7 | 12 | 27 | 16 | 47 | 1 |
| 8 | 14 | 28 | 18 | 48 | 3 |
| 9 | 9 | 29 | 9 | 49 | 1 |
| 10 | 11 | 30 | 14 | 50 | 4 |
| 11 | 14 | 31 | 13 | 51 | 4 |
| 12 | 13 | 32 | 11 | 52 | 3 |
| 13 | 32 | 33 | 11 | 53 | 4 |
| 14 | 19 | 34 | 11 | 54 | 1 |
| 15 | 20 | 35 | 16 | 55 | 1 |
| 16 | 9 | 36 | 13 | 56 | 4 |
| 17 | 18 | 37 | 5 | 57 | 3 |
| 18 | 23 | 38 | 9 | 58 | 2 |
| 19 | 25 | 39 | 6 | 61 | 1 |
| 20 | 21 | 40 | 5 | 62 | 2 |
|  |  |  |  | 66 | 1 |

*Source:* Stern, H. S., and Wilcox, W. "Shooting Darts." *Chance*, Vol. 10, No. 3, Summer 1997, p. 17 (adapted from Figure 2).

## STLRUNS

### St. Louis Cardinals

| | | | |
|---|---|---|---|
| 6 | 6 | 3 | 11 |
| 8 | 1 | 10 | 6 |
| 5 | 8 | 9 | 6 |
| 8 | 2 | 3 | 6 |
| 15 | 2 | 8 | |
| 8 | 6 | 2 | |
| 5 | 8 | 4 | |
| 8 | 9 | 4 | |
| 3 | 5 | 3 | |
| 5 | 7 | 6 | |
| 2 | 3 | 8 | |
| 6 | 2 | 7 | |
| 3 | 7 | 14 | |
| 5 | 1 | 7 | |
| 3 | 4 | 3 | |
| 10 | 6 | 6 | |
| 4 | 4 | 6 | |
| 11 | 13 | 5 | |

## CUBSRUNS

### Chicago Cubs

| | | |
|---|---|---|
| 3 | 6 | 8 |
| 4 | 5 | 2 |
| 1 | 8 | 6 |
| 1 | 9 | 6 |
| 4 | 2 | 3 |
| 3 | 6 | 9 |
| 5 | 4 | 10 |
| 5 | 4 | 4 |
| 2 | 9 | 5 |
| 5 | 5 | 4 |
| 5 | 6 | 5 |
| 10 | 9 | 8 |
| 5 | 3 | 11 |
| 6 | 6 | 15 |
| 7 | 5 | 11 |
| 13 | 9 | 6 |
| 8 | 7 | 7 |
| 10 | 8 | 2 |

## 5.5    APPROXIMATING A BINOMIAL DISTRIBUTION WITH A NORMAL DISTRIBUTION (OPTIONAL)

When the discrete binomial random variable (Section 4.4) can assume a large number of values, the calculation of its probabilities may become very tedious. To contend with this problem, we provide tables in Appendix A to give the probabilities for some values of $n$ and $p$, but these tables are by necessity incomplete. Recall that the binomial probability table (Table II) can be used only for $n = 5, 6, 7, 8, 9, 10, 15, 20$, or $25$. To deal with this limitation, we seek approximation procedures for calculating the probabilities associated with a binomial probability distribution.

When $n$ is large, a normal probability distribution may be used to provide a good approximation to the probability distribution of a binomial random variable. To show how this approximation works, we refer to Example 4.11, in which we used the binomial distribution to model the number $x$ of 20 eligible voters who favor a candidate. We assumed that 60% of all the eligible voters favored the candidate. The mean and standard deviation of $x$ were found to be $\mu = 12$ and $\sigma = 2.2$. The binomial distribution for $n = 20$ and $p = .6$ is shown in Figure 5.18, and the approximating normal distribution with mean $\mu = 12$ and standard deviation $\sigma = 2.2$ is superimposed.

**FIGURE 5.18**

*Binomial Distribution for $n = 20$, $p = .6$ and Normal Distribution with $\mu = 12$, $\sigma = 2.2$*

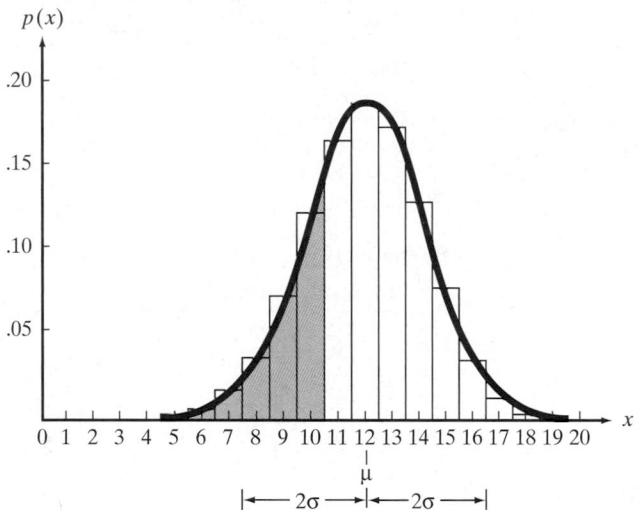

As part of Example 4.11, we used Table II to find the probability that $x \leq 10$. This probability, which is equal to the sum of the areas contained in the rectangles (shown in Figure 5.18) that correspond to $p(0), p(1), p(2), \ldots, p(10)$, was found to equal .245. The portion of the approximating normal curve that would be used to approximate the area $p(0) + p(1) + p(2) + \cdots + p(10)$ is highlighted in Figure 5.18. Note that this highlighted area lies to the left of 10.5 (not 10), so we may include all of the probability in the rectangle corresponding to $p(10)$. Because we are approximating a discrete distribution (the binomial) with a continuous distribution (the normal), we call the use of 10.5 (instead of 10 or 11) a **correction for continuity**. That is, we are correcting the discrete distribution so that it can be approximated by the continuous one. The use of the correction for continuity leads to the calculation of the following standard normal $z$ value:

$$z = \frac{x - \mu}{\sigma} = \frac{10.5 - 12}{2.2} = -.68$$

Using Table IV, we find the area between $z = 0$ and $z = .68$ to be .2517. Then the probability that $x$ is less than or equal to 10 is approximated by the area under the normal distribution to the left of 10.5, shown highlighted in Figure 5.18. That is,

$$P(x \leq 10) \approx P(z \leq -.68) - .5 - P(-.68 < z < 0) = .5 - .2517 = .2483$$

The approximation differs only slightly from the exact binomial probability, .245. Of course, when tables of exact binomial probabilities are available, we will use the exact value rather than a normal approximation.

Use of the normal distribution will not always provide a good approximation for binomial probabilities. The following is a useful rule of thumb to determine when $n$ is large enough for the approximation to be effective: *The interval $\mu \pm 3\sigma$ should lie within the range of the binomial random variable $x$ (i.e., 0 to n) in order for the normal approximation to be adequate.* The rule works well because almost all of the normal distribution falls within 3 standard deviations of the mean, so if this interval is contained within the range of $x$ values, there is "room" for the normal approximation to work.

As shown in Figure 5.19a for the preceding example with $n = 20$ and $p = .6$, the interval $\mu \pm 3\sigma = 12 + 3(2.19) = (5.43, 18.57)$ lies within the range 0 to 20. However, if we were to try to use the normal approximation with $n = 10$ and $p = .1$, the interval $\mu \pm 3\sigma$ is $1 \pm 3(.95)$, or $(-1.85, 3.85)$. As shown in Figure 5.19b, this interval is not contained within the range of $x$ since $x = 0$ is the lower bound for a binomial random variable. Note in Figure 5.19b that the normal distribution will not "fit" in the range of $x$, and therefore it will not provide a good approximation to the binomial probabilities.

**FIGURE 5.19**

*Rule of Thumb for Normal Approximation to Binomial Probabilities*

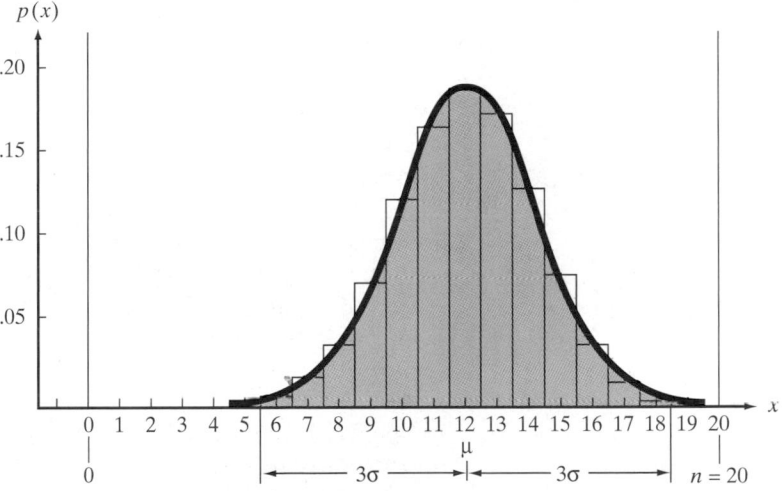

a. $n = 20, p = .6$: Normal approximation is good

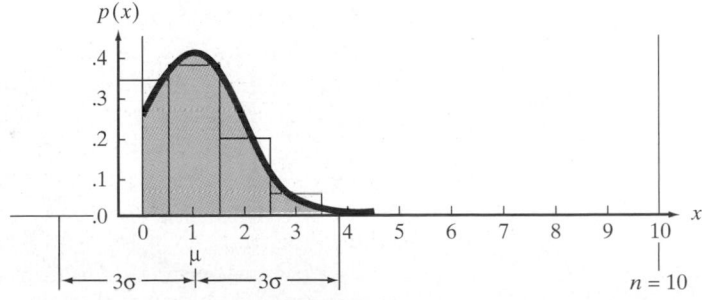

b. $n = 10, p = .1$: Normal approximation is poor

**EXAMPLE 5.12**  One problem with any product that is mass-produced (e.g., a graphing calculator) is quality control. The process must be monitored or audited to be sure the output of the process conforms to requirements. One monitoring method is *lot acceptance sampling*, in which items being produced are sampled at various stages of the production process and are carefully inspected. The lot of items from which the sample is drawn is then accepted or rejected, based on the number of defectives in the sample. Lots that are accepted may be sent forward for further processing or may be shipped to customers; lots that are rejected may be reworked or scrapped. For example, suppose a manufacturer of calculators chooses 200 stamped circuits from the day's production and determines $x$, the number of defective circuits in the sample. Suppose that up to a 6% rate of defectives is considered acceptable for the process.

a. Find the mean and standard deviation of $x$, assuming the defective rate is 6%.
b. Use the normal approximation to determine the probability that 20 or more defectives are observed in the sample of 200 circuits (i.e., find the approximate probability that $x \geq 20$).

**Solution**

a. The random variable $x$ is binomial with $n = 200$ and the fraction defective $p = .06$. Thus,

$$\mu = np = 200(.06) = 12$$
$$\sigma = \sqrt{npq} = \sqrt{200(.06)(.94)} = \sqrt{11.28} = 3.36$$

We first note that

$$\mu \pm 3\sigma = 12 \pm 3(3.36) = 12 \pm 10.08 = (1.92, 22.08)$$

lies completely within the range from 0 to 200. Therefore, a normal probability distribution should provide an adequate approximation to this binomial distribution.

b. Using the rule of complements, $P(x \geq 20) = 1 - P(x \leq 19)$. To find the approximating area corresponding to $x \leq 19$, refer to Figure 5.20. Note that we want to include all the binomial probability histograms from 0 to 19, inclusive. Since the event is of the form $x \leq a$, the proper correction for continuity is $a + .5 = 19 + .5 = 19.5$. Thus, the $z$ value of interest is

$$z = \frac{(a + .5) - \mu}{\sigma} = \frac{19.5 - 12}{3.36} = 2.23$$

**FIGURE 5.20**

*Normal Approximation to the Binomial Distribution with $n = 200$, $p = .06$*

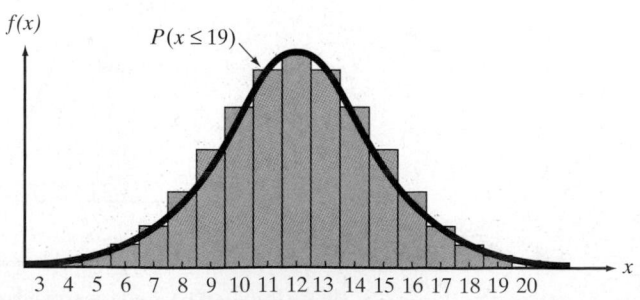

**FIGURE 5.21**

*Standard Normal Distribution*

Referring to Table IV in Appendix A, we find that the area to the right of the mean 0 corresponding to $z = 2.23$ (see Figure 5.21) is .4871. So the area $A = P(z \leq 2.23)$ is

$$A = .5 + .4871 = .9871$$

Thus, the normal approximation to the binomial probability we seek is

$$P(x \geq 20) = 1 - P(x \leq 19) \approx 1 - .9871 = .0129$$

In other words, the probability is extremely small that 20 or more defectives will be observed in a sample of 200 circuits—*if in fact the true fraction of defectives is* .06. If the manufacturer observes $x \geq 20$, the likely reason is that the process is producing more than the acceptable 6% defectives. The lot acceptance sampling procedure is another example of using the rare-event approach to make inferences. ■

The steps for approximating a binomial probability by a normal probability are given in the accompanying box.

---

## Using a Normal Distribution to Approximate Binomial Probabilities

1. After you have determined $n$ and $p$ for the binomial distribution, calculate the interval

    $$\mu \pm 3\sigma = np \pm 3\sqrt{npq}$$

    If the interval lies in the range 0 to $n$, the normal distribution will provide a reasonable approximation to the probabilities of most binomial events.

2. Express the binomial probability to be approximated in the form $P(x \leq a)$ or $P(x \leq b) - P(x \leq a)$. For example,

    $$P(x < 3) = P(x \leq 2)$$
    $$P(x \geq 5) = 1 - P(x \leq 4)$$
    $$P(7 \leq x \leq 10) = P(x \leq 10) - P(x \leq 6)$$

3. For each value of interest $a$, the correction for continuity is $(a + .5)$, and the corresponding standard normal $z$ value is

    $$z = \frac{(a + .5) - \mu}{\sigma} \quad \text{(see Figure 5.22)}$$

4. Sketch the approximating normal distribution and shade the area corresponding to the probability of the event of interest, as in Figure 5.22. Verify that the rectangles you have included in the shaded area correspond to the event probability you wish to approximate. Using Table IV and the $z$ value(s) you calculated in step 3, find the shaded area. This is the approximate probability of the binomial event.

**FIGURE 5.22**

*Approximating Binomial Probabilities by Normal Probabilities*

   EXERCISES 5.54–5.70

## Learning the Mechanics

**5.54** Assume that $x$ is a binomial random variable with $n$ and $p$ as specified in parts **a–f**. For which cases would it be appropriate to use a normal distribution to approximate the binomial distribution?

    **a.** $n = 100, p = .01$
    **b.** $n = 20, p = .6$
    **c.** $n = 10, p = .4$
    **d.** $n = 1,000, p = .05$
    **e.** $n = 100, p = .8$
    **f.** $n = 35, p = .7$

**5.55** Suppose $x$ is a binomial random variable with $p = .4$ and $n = 25$.

    **a.** Would it be appropriate to approximate the probability distribution of $x$ with a normal distribution? Explain.
    **b.** Assuming that a normal distribution provides an adequate approximation to the distribution of $x$, what are the mean and variance of the approximating normal distribution?
    **c.** Use Table II of Appendix A to find the exact value of $P(x \geq 9)$.
    **d.** Use the normal approximation to find $P(x \geq 9)$.

**5.56** Assume that $x$ is a binomial random variable with $n = 25$ and $p = .5$. Use Table II of Appendix A and the normal approximation to find the exact and approximate values, respectively, for the following probabilities:

    **a.** $P(x \leq 11)$
    **b.** $P(x \geq 16)$
    **c.** $P(8 \leq x \leq 16)$

**5.57** Assume that $x$ is a binomial random variable with $n = 100$ and $p = .40$. Use a normal approximation to find the following:

    **a.** $P(x \leq 35)$
    **b.** $P(40 \leq x \leq 50)$
    **c.** $P(x \geq 38)$

**5.58** Assume that $x$ is a binomial random variable with $n = 1000$ and $p = .50$. Find each of the following probabilities:

    **a.** $P(x > 500)$
    **b.** $P(490 \leq x < 500)$
    **c.** $P(x > 550)$

## Applying the Concepts—Basic

**5.59** In Exercise 4.43 (p. 200), you learned that only 5% of the nation's cigarette smokers ever enter into a treatment program to help them quit smoking (*USF Magazine*, Spring 2000). In a random sample of 200 smokers, let $x$ be the number who enter into a treatment program.

    **a.** Find the mean of $x$. (This value should agree with your answer to Exercise 4.43**c**.)
    **b.** Find the standard deviation of $x$.
    **c.** Find the $z$-score for the value $x = 10.5$.
    **d.** Find the approximate probability that the number of smokers in a sample of 200 who will enter into a treatment program is 11 or more.

**5.60** In Exercise 4.44 (p. 200), you learned that 22% of all births in the U.S. occur by Caesarian section each year (*USA Today*, Sept. 19, 2000). In a random sample of 1,000 births this year, let $x$ be the number that occur by Caesarian section.

    **a.** Find the mean of $x$. (This value should agree with your answer to Exercise 4.44**a**.)
    **b.** Find the standard deviation of $x$. (This value should agree with your answer to Exercise 4.44**b**.)
    **c.** Find the $z$-score for the value $x = 200.5$.
    **d.** Find the approximate probability that the number of Caesarian sections in a sample of 1,000 births is less than or equal to 200.

**5.61** According to the *American Cancer Society*, melanoma, a form of skin cancer, kills 60% of Americans who suffer from the disease each year. Consider a sample of 10,000 melanoma patients.

    **a.** What are the expected value and variance of $x$, the number of the 10,000 melanoma patients who die of the affliction this year?
    **b.** Find the probability that $x$ will exceed 6,100 patients per year.
    **c.** Would you expect $x$, the number of patients dying of melanoma, to exceed 6,500 in any single year? Explain.

**5.62** According to *Time* (Oct. 11, 1999), 1% of all patients who undergo laser surgery to correct their vision have serious post-laser vision problems. In a sample of 100,000 patients, what is the approximate probability that fewer than 950 will experience serious post-laser vision problems?

## Applying the Concepts—Intermediate

**5.63** The computer chips in notebook and laptop computers are produced from semiconductor wafers. Certain semiconductor wafers are exposed to an environment that generates up to 100 possible defects per wafer. The number of defects per wafer, $x$, was found to follow a binomial distribution if the manufacturing process is stable and generates defects that are randomly distributed on the wafers (*IEEE Transactions on Semiconductor Manufacturing*, May 1995). Let $p$ represent the probability that a defect occurs at any one of the 100 points of the wafer. For each of the following cases, determine whether the normal approximation can be used to characterize $x$.
 **a.** $p = .01$
 **b.** $p = .50$
 **c.** $p = .90$

**5.64** Refer to the FTC study of the pricing accuracy of supermarket electronic scanners, Exercise 4.47 (p. 200). Recall that the probability that a scanned item is priced incorrectly is $\frac{1}{30} = .033$.
 **a.** Suppose 10,000 supermarket items are scanned. What is the approximate probability that you observe at least 100 items with incorrect prices?
 **b.** Suppose 100 items are scanned and you are interested in the probability that fewer than five are incorrectly priced. Explain why the approximate method of part **a** may not yield an accurate estimate of the probability.

**5.65** In Exercise 4.51 (p. 201) you learned that some researchers believe that one in every three women are victims of domestic abuse (*Annals of Internal Medicine*, Nov. 1995).
 **a.** For a random sample of 150 women, what is the approximate probability that more than half are victims of domestic abuse?
 **b.** For a random sample of 150 women, what is the approximate probability that fewer than 50 are victims of domestic abuse?
 **c.** Would you expect to observe fewer than 30 domestically abused women in a sample of 150? Explain.

**5.66** In Exercise 4.110 (p. 215), the number of shuttle catastrophes caused by booster failure in $n$ missions was treated as a binomial random variable. Using the binomial distribution and the probability of catastrophe determined by the Air Force's risk assessment study $(\frac{1}{35})$, you determined the probability of at least one shuttle catastrophe in 25 missions to be .5155.
 **a.** Based on the guidelines presented in this section, would it have been advisable to approximate this probability using the normal approximation to the binomial distribution? Explain.
 **b.** Regardless of your answer to part **a**, use the normal distribution to approximate the binomial probability.

Comment on the difference between the exact and approximate probabilities.
 **c.** Refer to part **a**. Would the normal approximation be advisable if $n = 100$? If $n = 500$? If $n = 1,000$?
 **d.** Approximate the probability that more than 25 catastrophes occur in 1,000 flights, assuming that the probability of a catastrophe in any given flight remains $\frac{1}{35}$.

**5.67** A manufacturer of CD-ROMs claims that 99.4% of its CDs are defect-free. A large software company that buys and uses large numbers of the CDs wants to verify this claim, so it selects 1,600 CDs to be tested. The tests reveal 12 CDs to be defective. Assuming that the manufacturer's claim is correct, what is the probability of finding 12 or more defective CDs in a sample of 1,600? Does your answer cast doubt on the manufacturer's claim? Explain.

## Applying the Concepts—Advanced

**5.68** The percentage of fat in the bodies of American men is an approximately normal random variable with mean equal to 15% and standard deviation equal to 2%.
 **a.** If these values were used to describe the body fat of men in the United States Army and if 20% or more body fat is characterized as obese, what is the approximate probability that a random sample of 10,000 soldiers will contain fewer than 50 who would actually be characterized as obese?
 **b.** If the army actually were to check the percentage of body fat for a random sample of 10,000 men and if only 30 contained 20% (or higher) body fat, would you conclude that the army was successful in reducing the percentage of obese men below the percentage in the general population? Explain your reasoning.

**5.69** According to *New Jersey Business* (Feb. 1996), Newark International Airport's new terminal handles an average of 3,000 international passengers an hour, but is capable of handling twice that number. Also, 80% of arriving international passengers pass through without their luggage being inspected and the remainder are detained for inspection. The inspection facility can handle 600 passengers an hour without unreasonable delays for the travelers.
 **a.** When international passengers arrive at the rate of 1,500 per hour, what is the expected number of passengers who will be detained for luggage inspection?
 **b.** In the future, it is expected that as many as 4,000 international passengers will arrive per hour. When that occurs, what is the expected number of passengers who will be detained for luggage inspection?
 **c.** Refer to part **b**. Find the approximate probability that more than 600 international passengers will be detained for luggage inspection. (This is also the probability that travelers will experience unreasonable luggage inspection delays.)

**5.70** The median time a patient waits to see a doctor in a large clinic is 20 minutes. On a day when 150 patients visit the clinic, what is the approximate probability that
  **a.** More than half will wait more than 20 minutes?

  **b.** More than 85 will wait more than 20 minutes?
  **c.** More than 60 but fewer than 90 will wait more than 20 minutes?

## 5.6   THE EXPONENTIAL DISTRIBUTION (OPTIONAL)

The length of time between emergency arrivals at a hospital, the length of time between breakdowns of manufacturing equipment, the length of time between catastrophic events (floods, earthquakes, etc.), and the distance traveled by a wildlife ecologist between sightings of an endangered species are all random phenomena that we might want to describe probabilistically. The amount of time or distance between occurrences of random events like these can often be described by the **exponential probability distribution**. For this reason, the exponential distribution is sometimes called the **waiting time distribution**. The formula for the exponential probability distribution is shown in the box along with its mean and standard deviation.

---

### Probability Distribution, Mean, and Standard Deviation for an Exponential Random Variable $x$

$$f(x) = \frac{1}{\theta} e^{-x/\theta} \qquad (x > 0)$$

$$\mu = \theta \qquad \sigma = \theta$$

---

Unlike the normal distribution, which has a shape and location determined by the values of the two quantities $\mu$ and $\sigma$, the shape of the exponential distribution is governed by a single quantity, $\theta$. Further, it is a probability distribution with the property that its mean equals its standard deviation. Exponential distributions corresponding to $\theta = .5$, $1$, and $2$ are shown in Figure 5.23.

To calculate probabilities for exponential random variables, we need to be able to find areas under the exponential probability distribution. Suppose we want to find the area $A$ to the right of some number $a$, as shown in Figure 5.24.

**FIGURE 5.23**

*Exponential Distributions*

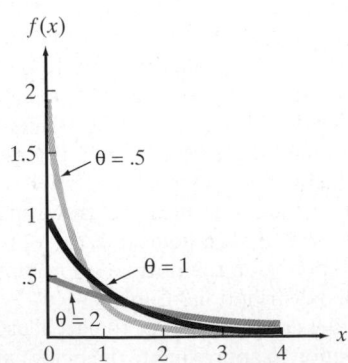

**FIGURE 5.24**

*The Area A to the Right of a Number a for an Exponential Distribution*

This area can be calculated by using the formula shown in the box. Use Table V in Appendix A or a pocket calculator with an exponential function to find the value of $e^{-a/\theta}$ after substituting the appropriate numerical values for $\theta$ and $a$.

---

**Finding the Area $A$ to the Right of a Number $a$ for an Exponential Distribution\***

$$A = P(x \geq a) = e^{-a/\theta}$$

---

**EXAMPLE 5.13**

Suppose the length of time (in hours) between emergency arrivals at a certain hospital is modeled as an exponential distribution with $\theta = 2$. What is the probability that more than 5 hours pass without an emergency arrival?

**Solution**

The probability we want is the area $A$ to the right of $a = 5$ in Figure 5.25. To find this probability, use the formula given for area:

$$A = e^{-a/\theta} = e^{-(5/2)} = e^{-2.5}$$

Referring to Table V (or using a calculator), we find

$$A = e^{-2.5} = .082085$$

Our exponential model indicates that the probability that more than 5 hours pass between emergency arrivals is about .08 for this hospital.

**FIGURE 5.25**

*Area to the Right of* a = 5 *for Example 5.13*

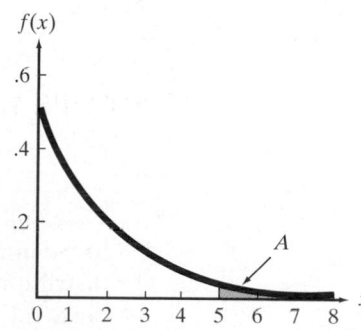

---

\*For students with a knowledge of calculus, the highlighted area in Figure 5.24 corresponds to the integral

$$\int_{a}^{\infty} \frac{1}{\theta} e^{-x/\theta} dx = -e^{-x/\theta} \Big|_{a}^{\infty} = e^{-a/\theta}.$$

**EXAMPLE 5.14**

A microwave oven manufacturer is trying to determine the length of warranty period it should attach to its magnetron tube, the most critical component in the oven. Preliminary testing has shown that the length of life (in years), $x$, of a magnetron tube has an exponential probability distribution with $\theta = 6.25$.

a. Find the mean and standard deviation of $x$.
b. Suppose a warranty period of 5 years is attached to the magnetron tube. What fraction of tubes must the manufacturer plan to replace, assuming that the exponential model with $\theta = 6.25$ is correct?
c. Find the probability that the length of life of a magnetron tube will fall within the interval $\mu - 2\sigma$ to $\mu + 2\sigma$.

**Solution**

a. Since $\theta = \mu = \sigma$, both $\mu$ and $\sigma$ equal 6.25.
b. To find the fraction of tubes that will have to be replaced before the 5-year warranty period expires, we need to find the area between 0 and 5 under the distribution. This area, $A$, is shown in Figure 5.26. To find the required probability, we recall the formula

$$P(x > a) = e^{-a/\theta}$$

Using this formula, we can find

$$P(x > 5) = e^{-a/\theta} = e^{-5/6.25} = e^{-.80} = .449329$$

(see Table V). To find the area $A$, we use the complementary relationship:

$$P(x \le 5) = 1 - P(x > 5) = 1 - .449329 = .550671$$

So approximately 55% of the magnetron tubes will have to be replaced during the 5-year warranty period.

**FIGURE 5.26**

*Area to the Left of* a = 5
*for Example 5.14*

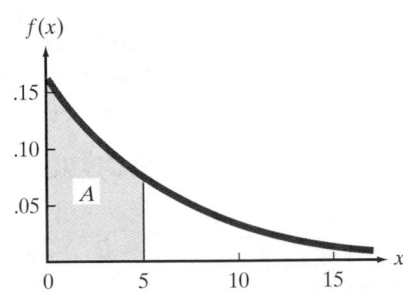

c. We would expect the probability that the life of a magnetron tube, $x$, falls within the interval $\mu - 2\sigma$ to $\mu + 2\sigma$ to be quite large. A graph of the exponential distribution showing the interval $\mu - 2\sigma$ to $\mu + 2\sigma$ is given in Figure 5.27. Since the point $\mu - 2\sigma$ lies below $x = 0$, we need to find only the area between $x = 0$ and $x = \mu + 2\sigma = 6.25 + 2(6.25) = 18.75$. This area, $P$, which is highlighted in Figure 5.27, is

$$P = 1 - P(x > 18.75) = 1 - e^{-18.75/\theta} = 1 - e^{-18.75/6.25} = 1 - e^{-3}$$

## FIGURE 5.27

*Area in the Interval $\mu \pm 2\sigma$
for Example 5.14*

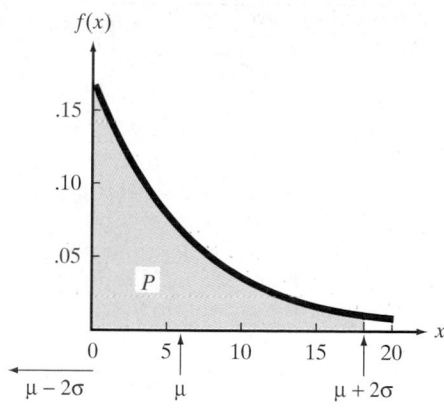

Checking Table V for the value of $e^{-3}$, we find $e^{-3} = .049787$. Therefore, the probability that the life $x$ of a magnetron tube will fall within the interval $\mu - 2\sigma$ to $\mu + 2\sigma$ is

$$P = 1 - e^{-3} = 1 - .049787 = .950213$$

You can see that this probability agrees very well with the interpretation of a standard deviation given by the Empirical Rule (Table 2.8, p. 62) even though this probability distribution is not mound-shaped. (It is strongly skewed to the right.) ■

## EXERCISES 5.71–5.85

### Learning the Mechanics

**5.71** The random variables $x$ and $y$ have exponential distributions with $\theta = 3$ and $\theta = 1$, respectively. Using Table V in Appendix A, carefully plot both distributions on the same set of axes.

**5.72** Use Table V in Appendix A to determine the value of $e^{-a/\theta}$ for each of the following cases.
  **a.** $\theta = 1, a = 1$
  **b.** $\theta = 1, a = 2.5$
  **c.** $\theta = .4, a = 3$
  **d.** $\theta = .2, a = .3$

**5.73** Suppose $x$ has an exponential distribution with $\theta = 1$. Find the following probabilities:
  **a.** $P(x > 1)$
  **b.** $P(x \le 3)$
  **c.** $P(x > 1.5)$
  **d.** $P(x \le 5)$

**5.74** Suppose $x$ has an exponential distribution with $\theta = 2.5$. Find the following probabilities:
  **a.** $P(x \le 4)$
  **b.** $P(x > 5)$
  **c.** $P(x \le 2)$
  **d.** $P(x > 3)$

**5.75** Suppose the random variable $x$ has an exponential probability distribution with $\theta = 1$. Find the mean and standard deviation of $x$. Find the probability that $x$ will assume a value within the interval $\mu \pm 2\sigma$.

**5.76** The random variable $x$ can be adequately approximated by an exponential probability distribution with $\theta = 2$. Find the probability that $x$ assumes a value
  **a.** More than 3 standard deviations from $\mu$.
  **b.** Less than 2 standard deviations from $\mu$.
  **c.** Within .5 standard deviation of $\mu$.

### Applying the Concepts—Basic

**5.77** Based on data collected from metal shredders across the nation, the amount $x$ of extractable lead in metal shredder residue has an approximate exponential distribution with mean $\theta = 2.5$ milligrams per liter (Florida Shredder's Association).
  **a.** Find the probability that $x$ is greater than 2 milligrams per liter.
  **b.** Find the probability that $x$ is less than 5 milligrams per liter.

**5.78** After bacteria are subjected to a certain drug, the length of time until the bacteria die follows an exponential distribution with a mean of 30 minutes.

a. How long will it take for half the bacteria to die after the drug is administered?

b. What proportion of bacteria will die between 12 minutes and 60 minutes?

**5.79** Refer to the *Ecological Applications* (May 1995) study of forest development 110 years following a wildfire, Exercise 5.37 (p. 236). Another species of tree, western hemlock, was found to have a breast height diameter distribution that resembled an exponential distribution with $\theta = 30$ centimeters. Find the probability that a western hemlock tree growing in the forest damaged by wildfire 110 years ago has a diameter that exceeds 25 centimeters.

## Applying the Concepts—Intermediate

**5.80** In the National Hockey League (NHL), games that are tied at the end of three periods are sent to "sudden-death" overtime. In overtime, the team to score the first goal wins. An analysis of NHL overtime games showed that the length of time elapsed before the winning goal is scored has an exponential distribution with mean 9.15 minutes (*Chance*, Winter 1995).

a. For a randomly selected overtime NHL game, find the probability that the winning goal is scored in 3 minutes or less.

b. In the NHL, each period (including overtime) lasts 20 minutes. If neither team scores a goal in overtime, the game is considered a tie. What is the probability of an NHL game ending in a tie?

**5.81** An article in *IEEE Transactions* (Mar. 1990) gave an example of a flexible manufacturing system (FMS) with four machines operating independently. The repair rates for the machines (i.e., the time, in hours, it takes to repair a failed machine) are exponentially distributed with means $\mu_1 = 1, \mu_2 = 2, \mu_3 = .5$, and $\mu_4 = .5$, respectively.

a. Find the probability that the repair time for machine 1 exceeds 1 hour.

b. Repeat part **a** for machine 2.

c. Repeat part **a** for machines 3 and 4.

d. If all four machines fail simultaneously, find the probability that the repair time for the entire system exceeds 1 hour.

**5.82** An infestation of a certain species of caterpillar, the spruce budworm, can cause extensive damage to the timberlands of the northern United States. It is known that an outbreak of this type of infestation occurs, on the average, every 30 years. Assuming that this phenomenon obeys an exponential probability law, what is the probability that catastrophic outbreaks of spruce budworm infestation will occur within 6 years of each other?

**5.83** Lack of port facilities or shallow water may require cargo on a large ship to be transferred to a pier using smaller craft. This process may require the smaller craft to cycle back and forth from ship to shore many times.

Researchers G. Horne (Center for Naval Analysis) and T. Irony (George Washington University) developed models of this transfer process that provide estimates of ship-to-shore transfer times (*Naval Research Logistics*, Vol. 41, 1994). They modeled the time between arrivals of the smaller craft at the pier using an exponential distribution.

a. Assume the mean time between arrivals at the pier is 17 minutes. Give the value of $\theta$ for this exponential distribution. Graph the distribution.

b. Suppose there is only one unloading zone at the pier available for the small craft to use. If the first craft docks at 10:00 am and doesn't finish unloading until 10:15 am, what is the probability that the second craft will arrive at the unloading zone and have to wait before docking?

**5.84** The importance of modeling machine downtime correctly in simulation studies was discussed in *Industrial Engineering* (Aug. 1990). The paper presented simulation results for a single-machine-tool system with the following properties:

(1) The interarrival times of jobs are exponentially distributed with a mean of 1.25 minutes

(2) The amount of time the machine operates before breaking down is exponentially distributed with a mean of 540 minutes

a. Find the probability that two jobs arrive for processing at most 1 minute apart.

b. Find the probability that the machine operates for at least 720 minutes (12 hours) before breaking down.

## Applying the Concepts—Advanced

**5.85** For a certain type of halogen light bulb, an old bulb that has been in use for a while tends to have a longer lifelength than a new bulb. Let $x$ represent the lifelength (in hours) of a new halogen light bulb and assume that $x$ has an exponential distribution with mean $\theta = 250$ hours. According to *Microelectronics and Reliability* (Jan. 1986), the "life" distribution of $x$ is considered *new better than used* (NBU) if

$$P(x > a + b) \leq P(x > a)P(x > b)$$

Alternatively, a "life" distribution is considered *new worse than used* (NWU) if

$$P(x > a + b) \geq P(x > a)P(x > b)$$

a. Show that when $a = 300$ and $b = 200$ the exponential distribution is both NBU and NWU.

b. Choose any two positive numbers $a$ and $b$, where $a > 0$ and $b > 0$, and repeat part **a**.

c. Show that, in general, for any $a$ and $b$ ($a > 0$ and $b > 0$), the exponential distribution with mean $\theta$ is both NBU and NWU. Such a "life" distribution is said to be *new same as used* or *memoryless*. Explain why.

## STATISTICS IN ACTION
### IQ, Economic Mobility, and the Bell Curve

In their controversial book *The Bell Curve* (Free Press, 1994), Professors Richard J. Herrnstein (a Harvard psychologist who died while the book was in production) and Charles Murray (a political scientist at MIT) explore, as the subtitle states, "intelligence and class structure in American life." *The Bell Curve* heavily employs statistical analyses in an attempt to support the authors' positions. Since the book's publication, many expert statisticians have raised doubts about the authors' statistical methods and the inferences drawn from them. (See, for example, "Wringing *The Bell Curve*: A cautionary tale about the relationships among race, genes, and IQ," *Chance*, Summer 1995.) In the Statistics in Action of Chapter 12, we explore a few of these problems.

One of the many controversies sparked by the book is the authors' tenet that level of intelligence (or lack thereof) is a cause of a wide range of intractable social problems, including constrained economic mobility. "America has taken great pride in the mobility of generations," state Herrnstein and Murray, "but this mobility has its limits. . . . The son of a father whose earnings are in the bottom five percent of the [income] distribution has something like one chance in twenty (or less) of rising to the top fifth of the income distribution and almost a fifty-fifty chance of staying in the bottom fifth. He has less than one chance in four of rising above even the median income. . . . Most people at present are stuck near where their parents were on the income distribution in part because [intelligence], which has become a major predictor of income, passes on sufficiently from one generation to the next to constrain economic mobility."

The measure of intelligence chosen by the authors is the well known Intelligent Quotient (IQ). Numerous tests have been developed to measure IQ; Herrnstein and Murray use the Armed Forces Qualification Test (AFQT), originally designed to measure the cognitive ability of military recruits.

Psychologists traditionally treat IQ as a random variable having a normal distribution with mean $\mu = 100$ and standard deviation $\sigma = 15$. This distribution, or **bell curve**, is shown in Figure 5.28.

In their book, Herrnstein and Murray refer to five cognitive classes of people defined by percentiles of the normal distribution. Class I ("very bright") consists of those with IQs above the 95th percentile; Class II ("bright") are those with IQs between the 75th and 95th percentiles; Class III ("normal") includes IQs between the 25th and 75th percentiles; Class IV ("dull") are those with IQs between the 5th and 25th percentiles; and Class V ("very dull") are IQs below the 5th percentile. These classes are also illustrated in Figure 5.28.

### Focus

a. Assuming that the distribution of IQ is accurately represented by the bell curve in Figure 5.28, determine the proportion of people with IQs in each of the five cognitive classes defined by Herrnstein and Murray.

b. Although the cognitive classes above are defined in terms of percentiles, the authors stress that IQ scores should be compared with $z$-scores, not percentiles. In other words, it is more informative to give the difference in $z$-scores for two IQ scores than it is to give the difference in percentiles. To demonstrate this point, calculate the difference in $z$-scores for IQs at the 50th and 55th percentiles. Do the same for IQs at the 94th and 99th percentiles. What do you observe?

c. Researchers have found that scores on many intelligence tests are decidedly nonnormal. Some distributions are skewed toward higher scores, others toward lower scores. How would the proportions in the five cognitive classes differ for an IQ distribution that is skewed right? Skewed left?

## FIGURE 5.28
*The Distribution of IQ*

# QUICK REVIEW

## Key Terms

*Note: Starred (\*) terms are from the optional sections in this chapter.*

Bell curve   255
Bell-shaped distribution   223
Continuous probability
   distribution   218
Correction for continuity\*   244
Exponential probability
   distribution\*   250
Exponential random variable\*   250

Frequency function   218
Normal distribution   223
Normal probability plot   238
Normal random variable   223
Probability density
   function   218
Probability distribution   218
Randomness distribution   219

Standard normal
   distribution   225
Standard normal random
   variable   225
Uniform probability
   distribution   219
Uniform random variable   220
Waiting time distribution\*   250

## Key Formulas

*Note: Starred (\*) formulas are from the optional sections in this chapter.*

| Density Function | Mean | Standard Deviation | Random Variable |
|---|---|---|---|
| $f(x) = \dfrac{1}{d-c}(c \le x \le d)$ | $\mu = \dfrac{c+d}{2}$ | $\sigma = \dfrac{d-c}{\sqrt{12}}$ | Uniform, $x$   220 |
| $f(x) = \dfrac{1}{\sigma\sqrt{2\pi}}e^{-(1/2)[(x-\mu)/\sigma]^2}$ | $\mu$ | $\sigma$ | Normal, $x$   224 |
| $f(z) = \dfrac{1}{\sqrt{2\pi}}e^{-(1/2)z^2}$ | $\mu = 0$ | $\sigma = 1$ | Standard Normal, $z = \left(\dfrac{x-\mu}{\sigma}\right)$   224 |
| $f(x) = \dfrac{1}{\theta}e^{-x/\theta}(x > 0)$ | $\mu = \theta$ | $\sigma = \theta$ | \*Exponential, $x$   250 |
| \*Normal Approximation to Binomial: $P(x \le a) = P\left[z \le \dfrac{(a+.5)-\mu}{\sigma}\right]$   247 | | | |

## Language Lab

| Symbol | Pronunciation | Description |
|---|---|---|
| $f(x)$ | $f$ of $x$ | Probability density function for a continuous random variable $x$ |
| $\theta$ | theta | Mean of exponential random variable |

 ## SUPPLEMENTARY EXERCISES 5.86–5.115

*Note: Starred (\*) exercises refer to the optional sections in this chapter.*

### Learning the Mechanics

**5.86** Assume that $x$ is a random variable best described by a uniform distribution with $c = 40$ and $d = 70$.
   **a.** Find $f(x)$.
   **b.** Find the mean and standard deviation of $x$.
   **c.** Graph the probability distribution for $x$ and locate its mean and the interval $\mu \pm 2\sigma$ on the graph.
   **d.** Find $P(x \le 45)$.
   **e.** Find $P(x \ge 58)$.
   **f.** Find $P(x \le 100)$.
   **g.** Find $P(\mu - \sigma \le x \le \mu + \sigma)$.
   **h.** Find $P(x > 60)$.

**5.87** Find the following probabilities for the standard normal random variable $z$:
   **a.** $P(z \le 2.1)$
   **b.** $P(z \ge 2.1)$
   **c.** $P(z \ge -1.65)$
   **d.** $P(-2.13 \le z \le -.41)$
   **e.** $P(-1.45 \le z \le 2.15)$
   **f.** $P(z \le -1.43)$

**5.88** Find a $z$-score, call it $z_0$, such that
   **a.** $P(z \le z_0) = .8708$
   **b.** $P(z \ge z_0) = .0526$
   **c.** $P(z \le z_0) = .5$
   **d.** $P(-z_0 \le z \le z_0) = .8164$

**e.** $P(z \geq z_0) = .8023$
**f.** $P(z \geq z_0) = .0041$

**5.89** The random variable $x$ has a normal distribution with $\mu = 70$ and $\sigma = 10$. Find the following probabilities:
**a.** $P(x \leq 75)$
**b.** $P(x \geq 90)$
**c.** $P(60 \leq x \leq 75)$
**d.** $P(x > 75)$
**e.** $P(x = 75)$
**f.** $P(x \leq 95)$

**5.90** The random variable $x$ has a normal distribution with $\mu = 40$ and $\sigma^2 = 36$. Find a value of $x$, call it $x_0$, such that
**a.** $P(x \geq x_0) = .5$
**b.** $P(x \leq x_0) = .9911$
**c.** $P(x \leq x_0) = .0028$
**d.** $P(x \geq x_0) = .0228$
**e.** $P(x \leq x_0) = .1003$
**f.** $P(x \geq x_0) = .7995$

**\*5.91** Assume that $x$ is a binomial random variable with $n = 100$ and $p = .5$. Use the normal probability distribution to approximate the following probabilities:
**a.** $P(x \leq 48)$
**b.** $P(50 \leq x \leq 65)$
**c.** $P(x \geq 70)$
**d.** $P(55 \leq x \leq 58)$
**e.** $P(x = 62)$
**f.** $P(x \leq 49 \text{ or } x \geq 72)$

**\*5.92** Assume that $x$ has an exponential distribution with $\theta = 3$. Find
**a.** $P(x \leq 1)$
**b.** $P(x > 1)$
**c.** $P(x = 1)$
**d.** $P(x \leq 6)$
**e.** $P(2 \leq x \leq 10)$

### Applying the Concepts—Basic

**5.93** All Florida high schools require their students to demonstrate competence in mathematics by scoring 70% or above on the FCAT mathematics achievement test. The FCAT math scores of those students taking the test for the first time are normally distributed with a mean of 77% and a standard deviation of 7.3%. What percentage of students who take the test for the first time will pass the test?

**5.94** In baseball, a "no-hitter" is a regulation 9-inning game in which the pitcher yields no hits to the opposing batters. *Chance* (Summer 1994) reported on a study of no-hitters in Major League Baseball (MLB). The initial analysis focused on the total number of hits yielded per game per team for all 9-inning MLB games played between 1989 and 1993. The distribution of hits/9-innings is approximately normal with mean 8.72 and standard deviation 1.10.
**a.** What percentage of 9-inning MLB games result in fewer than 6 hits?
**b.** Demonstrate, statistically, why a no-hitter is considered an extremely rare occurrence in MLB.

**\*5.95** Sixteen percent of the African-American population is known to suffer from sickle-cell anemia. If 1,000 African

Americans are sampled at random, what is the approximate probability that:
**a.** More than 175 have the disease?
**b.** Fewer than 140 have the disease?
**c.** The number of people in the sample with the disease is between 130 and 180 inclusive.

**\*5.96** The *Statistical Abstract of the United States* reports that 25% of the country's households are composed of one person. If 1,000 randomly selected homes are to participate in a Nielsen survey to determine television ratings, find the approximate probability that no more than 240 of these homes are one-person households.

**\*5.97** On the average, the main chute fails in one of every 1,000 parachute jumps. Suppose during a lifetime a professional parachutist makes 4,000 jumps, and let $x$ equal the number of times the main chute fails.
**a.** Show that $x$ has a binomial distribution.
**b.** What is the approximate probability that the parachutist's main chute fails on at least one jump?
**c.** Because $n$ is so large and $p$ is so small, the Poisson probability distribution will also provide a good approximation to this probability. (See Exercise 4.62, p. 207.) If you covered Section 4.5, find the Poisson approximation to $P(x > 0)$.

**5.98** The Council on Environmental Quality (CEQ) considers a river with a dissolved oxygen content of less than 5 milligrams per liter (mg/L) of water to be undesirable because it is unlikely to support aquatic life. Suppose an industrial plant discharges its waste into a river and the downstream daily oxygen content measurements are normally distributed with a mean equal to 6.3 mg/L and a standard deviation of .6 mg/L.
**a.** What percentage of the days would the dissolved oxygen content in the river be considered undesirable by the CEQ?
**b.** Within what limits would we expect the dissolved oxygen content to fall?

**\*5.99** An article in *The International Journal of Sports Psychology* (July–Sept. 1990) evaluated the relationship between physical fitness and stress. The research revealed that white-collar workers in good physical condition have only a 10% probability of developing a stress-related health problem. What is the probability that more than 60 in a random sample of 400 white-collar employees in good physical condition will develop stress-related illnesses?

**\*5.100** The length of time between arrivals at a hospital clinic has an approximately exponential probability distribution. Suppose the mean time between arrivals for patients at a clinic is 4 minutes.
**a.** What is the probability that a particular interarrival time (the time between the arrival of two patients) is less than 1 minute?
**b.** What is the probability that the next four interarrival times are all less than 1 minute?
**c.** What is the probability that an interarrival time will exceed 10 minutes?

## Applying the Concepts—Intermediate

**5.101** The manager of a large department store with three floors reports that the time a customer on the second floor must wait for an elevator has a uniform distribution ranging from 0 to 4 minutes. If it takes the elevator 15 seconds to go from floor to floor, find the probability that a hurried customer can reach the first floor in less than 1.5 minutes after pushing the second-floor elevator button.

**5.102** Refer to the *Teaching Psychology* (May 1998) study of how external clues influence performance, Exercise 2.92 (p. 74). Recall that two different forms of a midterm psychology examination were given—one printed on blue paper and the other printed on red paper. Grading only the difficult questions, scores on the blue exam had a distribution with a mean of 53% and a standard deviation of 15%, while scores on the red exam had a distribution with a mean of 39% and a standard deviation of 12%. Assuming that both distributions are approximately normal, on which exam form is a student more likely to score below 20% on the difficult questions, the blue or the red exam? (Compare your answer to 2.92**c**.)

**5.103** Recall *The Astronomical Journal* (July 1995) study of galaxy velocity, Exercise 5.48 (p. 242). The observed velocity of a galaxy located in galaxy cluster A2142 was found to have a normal distribution with mean 27,117 kilometers per second (km/s) and standard deviation 1,280 km/s. A galaxy with a velocity of 24,350 km/s is observed. Comment on the likelihood of this galaxy being located in cluster A2142.

**5.104** It is well known that children with developmental delays (i.e., mild mental retardation) are slower, cognitively, than normally developing children. Are their social skills also lacking? A study compared the social interactions of the two groups of children in a controlled playground environment (*American Journal on Mental Retardation*, Jan. 1992). One variable of interest was the number of intervals of "no play" by each child. Children with developmental delays had a mean of 2.73 intervals of "no play" and a standard deviation of 2.58 intervals. Based on this information, is it possible for the variable of interest to be normally distributed? Explain.

**5.105** The extract of a plant native to Taiwan (Republic of China) is being tested as a possible treatment for leukemia. The chemical properties of the extract were examined in *Planta Medica* (Apr. 1995). A particular compound produced from the plant extract was analyzed for the protein collagen. The collagen amount $x$ for the compound was found to be normally distributed with $\mu = 76.8$ and $\sigma = 9.2$ grams per milliliter.

**a.** What percentage of compounds formed from the plant extract will have a collagen amount greater than 90 grams per milliliter?

**b.** Give an interval around the mean $\mu$ that contains the collagen amount for 80% of the compounds formed from the plant extract.

**5.106** A bakery has determined that the number of loaves of its white bread demanded daily has a normal distribution with mean 7,200 loaves and standard deviation 300 loaves. Based on cost considerations, the company has decided that its best strategy is to produce a sufficient number of loaves so that it will fully supply demand on 94% of all days.

**a.** How many loaves of bread should the company produce?

**b.** Based on the production in part **a**, on what percentage of days will the company be left with more than 500 loaves of unsold bread?

**5.107** A chemical analysis of metamorphic rock in western Turkey was reported in *Geological Magazine* (May 1995). The trace amount (in parts per million) of the element rubidium was measured for 20 rock specimens. The data are listed here. Assess whether the sample data come from a normal population.

### 🜨 RUBIDIUM

| | | | | | | | | | |
|---|---|---|---|---|---|---|---|---|---|
| 164 | 286 | 355 | 308 | 277 | 330 | 323 | 370 | 241 | 402 |
| 301 | 200 | 202 | 341 | 327 | 285 | 277 | 247 | 213 | 424 |

*Source:* Bozkurt, E. *et al.* "Geochemistry and the tectonic significance of augen gneisses from the southern Menderes Massif (West Turkey)." *Geological Magazine*, Vol. 132, No. 3, May 1995, p. 291 (Table 1).

**5.108** Geologists have successfully used statistical models to evaluate the nature of sedimentary deposits (called *facies*) in reservoirs. One of the model's key parameters is the proportion $P$ of facies bodies in a reservoir. An article in *Mathematical Geology* (Apr. 1995) demonstrated that the number of facies bodies that must be sampled to satisfactorily estimate $P$ is approximately normally distributed with $\mu = 99$ and $\sigma = 4.3$. How many facies bodies are required to satisfactorily estimate $P$ for 99% of the reservoirs evaluated?

**5.109** A. K. Shah published a simple approximation for areas under the normal curve in *The American Statistician* (Feb. 1985). Shah showed that the area $A$ under the standard normal curve between 0 and $z$ is

$$A \approx \begin{cases} z(4.4 - z)/10 & \text{for } 0 \le z \le 2.2 \\ .49 & \text{for } 2.2 < z < 2.6 \\ .50 & \text{for } z \ge 2.6 \end{cases}$$

**a.** Use the approximation to find

   **(i)**    $P(0 < z < 1.2)$

   **(ii)**   $P(0 < z < 2.5)$

   **(iii)**  $P(z > .8)$

   **(iv)**  $P(z < 1.0)$

**b.** Find the exact probabilities in part **a**.

**c.** Shah showed that the approximation has a maximum absolute error of .0052. Verify this for the approximations in part **a**.

**\*5.110** Based on sample data collected in the Denver area, a study found that in some cases the exponential distribution is an adequate approximation for the distribution of the time (in weeks) an individual is unemployed (*Journal of Economics*, Vol. 28, 1985). In particular, the study found the exponential distribution to be appropriate for white and African-American workers but not for Hispanics. Use $\theta = 13$ to answer the following questions.

   **a.** What is the mean time workers are unemployed according to the exponential distribution?

   **b.** Find the probability that a white worker who just lost her job will be unemployed for at least 2 weeks. For more than 6 weeks.

   **c.** What is the probability that an unemployed worker will find a new job within 12 weeks?

**5.111** A professor believes that if a class is allowed to work on an examination as long as desired, the times spent by the students would be approximately normal with mean 40 minutes and standard deviation 6 minutes. Approximately how long should be allotted for the examination if the professor wants almost all (say, 97.5%) of the class to finish?

**\*5.112** In an article published in the *European Journal of Operational Research* (Vol. 21, 1985), the vehicle-dispatching decisions of an airport-based taxi service were investigated. In modeling the system, the authors assumed travel times of successive taxi trips to and from the terminal to be independent exponential random variables. Assume $\theta = 20$ minutes.

   **a.** What is the mean trip time for the taxi service?

   **b.** What is the probability that a particular trip will take more than 30 minutes?

   **c.** Two taxis have just been dispatched. What is the probability that both will be gone for more than 30 minutes? That at least one of the taxis will return within 30 minutes?

## Applying the Concepts—Advanced

**5.113** A team of soil scientists investigated the water retention properties of soil cores sampled from an uncropped field consisting of silt loam (*Soil Science*, Jan. 1995). At a pressure of .1 megapascal (MPa), the water content of the soil (measured in cubic meters of water per cubic meter of soil) was determined to be approximately normally distributed with $\mu = .27$ and $\sigma = .04$. In addition to water content readings at a pressure of .1 MPa, measurements were obtained at pressures 0, .005, .01, .03, and 1.5 MPa. Consider a soil core with a water content reading of .14. Is it likely that this reading was obtained at a pressure of .1 MPa? Explain.

**\*5.114** The net weight per package of a certain brand of corn chips is listed as 10 ounces. The weight actually delivered to each package by an automated machine is a normal random variable with mean 10.5 ounces and standard deviation .2 ounce. Suppose 1,500 packages are chosen at random and the net weights are ascertained. Let $x$ be the number of the 1,500 selected packages that contain at least 10 ounces of corn chips. Then $x$ is a binomial random variable with $n = 1,500$ and $p = $ probability that a randomly selected package contains at least 10 ounces. What is the probability that they all contain at least 10 ounces of corn chips? What is the probability that at least 90% of the packages contain 10 ounces or more?

**\*5.115** The number of serious accidents in a manufacturing plant has (approximately) a Poisson probability distribution with a mean of two serious accidents per month. If $x$, the number of events per unit time, has a Poisson distribution with mean $\lambda$, then it can be shown that the time between two successive events has an exponential probability distribution with mean $\theta = \frac{1}{\lambda}$.

   **a.** If an accident occurs today, what is the probability that the next serious accident will not occur within the next month?

   **b.** What is the probability that more than one serious accident will occur within the next month?

## STUDENT PROJECTS

For large values of $n$ the computational effort involved in working with the binomial probability distribution is considerable. Fortunately, in many instances the normal distribution provides a good approximation to the binomial distribution. This project was designed to enable you to demonstrate to yourself how well the normal distribution approximates the binomial distribution.

a. Let the random variable $x$ have a binomial probability with $n = 10$ and $p = .5$. Using the binomial distribution, find the probability that $x$ takes on a value in each of the following intervals: $\mu \pm \sigma$, $\mu \pm 2\sigma$, and $\mu \pm 3\sigma$.

b. Find the probabilities requested in part **a** by using a normal approximation to the given binomial distribution.

c. Determine the magnitude of the difference between each of the three probabilities as determined by the binomial distribution and by the normal approximation.

d. Letting $x$ have a binomial distribution with $n = 20$ and $p = .5$, repeat parts **a**, **b**, and **c**. Notice that the probability estimates provided by the normal distribution are more accurate for $n = 20$ than for $n = 10$.

e. Letting $x$ have a binomial distribution with $n = 20$ and $p = .01$, repeat parts **a**, **b**, and **c**. Notice that the probability estimates provided by the normal distribution are very poor in this case. Explain why this occurs.

## REFERENCES

Hogg, R. V., and Craig, A. T. *Introduction to Mathematical Statistics*, 5th ed. Upper Saddle River, N.J.: Prentice Hall, 1995.

Ramsey, P. P., and Ramsey, P. H. "Simple tests of normality in small samples." *Journal of Quality Technology*, Vol. 22, 1990, pp. 299–309.

Wackerly, D., Mendenhall, W., and Scheaffer, R. L. *Mathematical Statistics with Applications*, 6th ed. North Scituate, Mass.: Duxbury, 1999.

# Sampling Distributions

## Contents

## Statistics in Action

The Insomnia Pill

### ✊ Where We've Been

We've learned that the objective of most statistical investigations is inference—that is, making decisions or predictions about a population based on information in a sample. To actually make the decision, we use the sample data to compute sample statistics, such as the sample mean or variance. The knowledge of random variables and their probability distributions enables us to construct theoretical models of populations.

### ☛ Where We're Going

Because sample measurements are observed values of random variables, the value for a sample statistic will vary in a random manner from sample to sample. In other words, since sample statistics are random variables, they possess probability distributions that are either discrete or continuous. In this chapter, we'll explore these probability distributions, which are called *sampling distributions*. You'll learn why many sampling distributions tend to be approximately normal, and you'll see how sampling distributions can be used to evaluate the reliability of inferences made using the statistics.

In Chapters 4 and 5 we assumed that we knew the probability distribution of a random variable, and using this knowledge we were able to compute the mean, variance, and probabilities associated with the random variable. However, in most practical applications, this information is not available. To illustrate, in Example 4.11 (p. 196) we calculated the probability that the binomial random variable $x$, the number of 20 polled voters who favor a certain mayoral candidate, assumed specific values. To do this, it was necessary to assume some value for $p$, the proportion of all voters who favor the candidate. Thus, for the purposes of illustration, we assumed $p = .6$ when, in all likelihood, the exact value of $p$ would be unknown. In fact, the probable purpose of taking the poll is to estimate $p$. Similarly, when we modeled the in-city gas mileage of a certain automobile model, we used the normal probability distribution with an *assumed* mean and standard deviation of 27 and 3 miles per gallon, respectively. In most situations, the true mean and standard deviation are unknown quantities that would have to be estimated. Numerical quantities that describe probability distributions are called *parameters*. Thus, $p$, the probability of a success in a binomial experiment, and $\mu$ and $\sigma$, the mean and standard deviation of a normal distribution, are examples of parameters.

> **DEFINITION 6.1**
>
> A **parameter** is a numerical descriptive measure of a population. Because it is based on the observations in the population, its value is almost always unknown.

We have also discussed the sample mean $\bar{x}$, sample variance $s^2$, sample standard deviation $s$, etc., which are numerical descriptive measures calculated from the sample. We will often use the information contained in these *sample statistics* to make inferences about the parameters of a population.

> **DEFINITION 6.2**
>
> A **sample statistic** is a numerical descriptive measure of a sample. It is calculated from the observations in the sample.

Note that the term *statistic* refers to a *sample* quantity and the term *parameter* refers to a *population* quantity.

Before we can show you how to use sample statistics to make inferences about population parameters, we need to be able to evaluate their properties. Does one sample statistic contain more information than another about a population parameter? On what basis should we choose the "best" statistic for making inferences about a parameter? The purpose of this chapter is to answer these questions.

## 6.1    WHAT IS A SAMPLING DISTRIBUTION?

If we want to estimate a parameter of a population—say, the population mean $\mu$—we could use a number of sample statistics for our estimate. Two possibilities are the sample mean $\bar{x}$ and the sample median $M$. Which of these do you think will provide a better estimate of $\mu$?

Before answering this question, consider the following example: Toss a fair die, and let $x$ equal the number of dots showing on the up face. Suppose the die is

**FIGURE 6.1**

*Comparing the Sample Mean*
*($\bar{x}$) and Sample Median ($M$) as*
*Estimators of the Population*
*Mean ($\mu$)*

a. Sample 1:   $\bar{x}$ is closer than $M$ to $\mu$     b. Sample 2:   $M$ is closer than $\bar{x}$ to $\mu$

tossed three times, producing the sample measurements 2, 2, 6. The sample mean is $\bar{x} = 3.33$ and the sample median is $M = 2$. Since the population mean of $x$ is $\mu = 3.5$, you can see that for this sample of three measurements, the sample mean $\bar{x}$ provides an estimate that falls closer to $\mu$ than does the sample median (see Figure 6.1a). Now suppose we toss the die three more times and obtain the sample measurements 3, 4, 6. The mean and median of this sample are $\bar{x} = 4.33$ and $M = 4$, respectively. This time $M$ is closer to $\mu$ (see Figure 6.1b).

This simple example illustrates an important point: Neither the sample mean nor the sample median will *always* fall closer to the population mean. Consequently, we cannot compare these two sample statistics, or, in general, any two sample statistics, on the basis of their performance for a single sample. Instead, we need to recognize that sample statistics are themselves random variables, because different samples can lead to different values for the sample statistics. As random variables, sample statistics must be judged and compared on the basis of their probability distributions, i.e., the *collection* of values and associated probabilities of each statistic that would be obtained if the sampling experiment were repeated a *very large number of times*. We will illustrate this concept with another example.

Suppose it is known that in a certain part of Canada the daily high temperature recorded for all past months of January has a mean of $\mu = 10°F$ and a standard deviation of $\sigma = 5°F$. Consider an experiment consisting of randomly selecting 25 daily high temperatures from the records of past months of January and calculating the sample mean $\bar{x}$. If this experiment were repeated a very large number of times, the value of $\bar{x}$ would vary from sample to sample. For example, the first sample of 25 temperature measurements might have a mean $\bar{x} = 9.8$, the second sample a mean $\bar{x} = 11.4$, the third sample a mean $\bar{x} = 10.5$, etc. If the sampling experiment were repeated a very large number of times, the resulting histogram of sample means would be approximately the probability distribution of $\bar{x}$. If $\bar{x}$ is a good estimator of $\mu$, we would expect the values of $\bar{x}$ to cluster around $\mu$ as shown in Figure 6.2. This probability distribution is called a *sampling distribution* because it is generated by repeating a sampling experiment a very large number of times.

**DEFINITION 6.3**

The **sampling distribution** of a sample statistic calculated from a sample of $n$ measurements is the probability distribution of the statistic.

**FIGURE 6.2**

*Sampling Distribution for $\bar{x}$*
*Based on a Sample of*
*$n = 25$ Measurements*

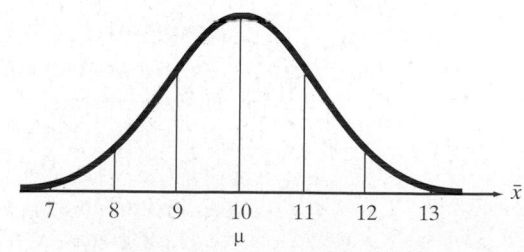

In actual practice, the sampling distribution of a statistic is obtained mathematically or (at least approximately) by simulating the sample on a computer using a procedure similar to that just described.

If $\bar{x}$ has been calculated from a sample of $n = 25$ measurements selected from a population with mean $\mu = 10$ and standard deviation $\sigma = 5$, the sampling distribution (Figure 6.2) provides information about the behavior of $\bar{x}$ in repeated sampling. For example, the probability that you will draw a sample of 25 measurements and obtain a value of $\bar{x}$ in the interval $9 \leq \bar{x} \leq 10$ will be the area under the sampling distribution over that interval.

Since the properties of a statistic are typified by its sampling distribution, it follows that to compare two sample statistics you compare their sampling distributions. For example, if you have two statistics, $A$ and $B$, for estimating the same parameter (for purposes of illustration, suppose the parameter is the population variance $\sigma^2$) and if their sampling distributions are as shown in Figure 6.3, you would choose statistic $A$ in preference to statistic $B$. You would make this choice because the sampling distribution for statistic $A$ centers over $\sigma^2$ and has less spread (variation) than the sampling distribution for statistic $B$. When you draw a single sample in a practical sampling situation, the probability is higher that statistic $A$ will fall nearer $\sigma^2$.

**FIGURE 6.3**

*Two Sampling Distributions for Estimating the Population Variance, $\sigma^2$*

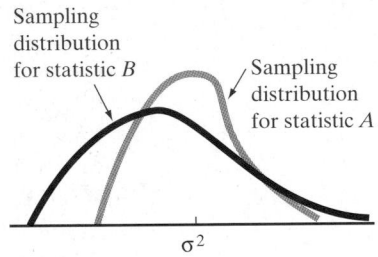

Sampling distribution for statistic $B$

Sampling distribution for statistic $A$

$\sigma^2$

Remember that in practice we will not know the numerical value of the unknown parameter $\sigma^2$ so we will not know whether statistic $A$ or statistic $B$ is closer to $\sigma^2$ for a sample. We have to rely on our knowledge of the theoretical sampling distributions to choose the best sample statistic and then use it sample after sample. The procedure for finding the sampling distribution for a statistic is demonstrated in Example 6.1.

**EXAMPLE 6.1**

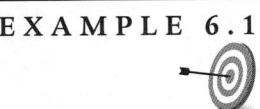

Consider a population consisting of the measurements 0, 3, and 12 and described by the probability distribution shown here. A random sample of $n = 3$ measurements is selected from the population.

| $x$ | 0 | 3 | 12 |
|---|---|---|---|
| $p(x)$ | $\frac{1}{3}$ | $\frac{1}{3}$ | $\frac{1}{3}$ |

a. Find the sampling distribution of the sample mean $\bar{x}$.
b. Find the sampling distribution of the sample median $M$.

**Solution**

Every possible sample of $n = 3$ measurements is listed in Table 6.1 along with the sample mean and median. Also, because any one sample is as likely to be selected as any other (random sampling), the probability of observing any particular sample is $\frac{1}{27}$. The probability is also listed in Table 6.1.

a. From Table 6.1 you can see that $\bar{x}$ can assume the values 0, 1, 2, 3, 4, 5, 6, 8, 9, and 12. Because $\bar{x} = 0$ occurs in only one sample, $P(\bar{x} = 0) = \frac{1}{27}$. Similarly, $\bar{x} = 1$

**TABLE 6.1**

| Possible Samples | $\bar{x}$ | M | Probability |
|---|---|---|---|
| 0, 0, 0 | 0 | 0 | $1/27$ |
| 0, 0, 3 | 1 | 0 | $1/27$ |
| 0, 0, 12 | 4 | 0 | $1/27$ |
| 0, 3, 0 | 1 | 0 | $1/27$ |
| 0, 3, 3 | 2 | 3 | $1/27$ |
| 0, 3, 12 | 5 | 3 | $1/27$ |
| 0, 12, 0 | 4 | 0 | $1/27$ |
| 0, 12, 3 | 5 | 3 | $1/27$ |
| 0, 12, 12 | 8 | 12 | $1/27$ |
| 3, 0, 0 | 1 | 0 | $1/27$ |
| 3, 0, 3 | 2 | 3 | $1/27$ |
| 3, 0, 12 | 5 | 3 | $1/27$ |
| 3, 3, 0 | 2 | 3 | $1/27$ |
| 3, 3, 3 | 3 | 3 | $1/27$ |
| 3, 3, 12 | 6 | 3 | $1/27$ |
| 3, 12, 0 | 5 | 3 | $1/27$ |
| 3, 12, 3 | 6 | 3 | $1/27$ |
| 3, 12, 12 | 9 | 12 | $1/27$ |
| 12, 0, 0 | 4 | 0 | $1/27$ |
| 12, 0, 3 | 5 | 3 | $1/27$ |
| 12, 0, 12 | 8 | 12 | $1/27$ |
| 12, 3, 0 | 5 | 3 | $1/27$ |
| 12, 3, 3 | 6 | 3 | $1/27$ |
| 12, 3, 12 | 9 | 12 | $1/27$ |
| 12, 12, 0 | 8 | 12 | $1/27$ |
| 12, 12, 3 | 9 | 12 | $1/27$ |
| 12, 12, 12 | 12 | 12 | $1/27$ |

occurs in three samples: $(0,0,3)$, $(0,3,0)$, and $(3,0,0)$. Therefore, $P(\bar{x} = 1) = 3/27 = 1/9$. Calculating the probabilities of the remaining values of $\bar{x}$ and arranging them in a table, we obtain the probability distribution shown here.

| $\bar{x}$ | 0 | 1 | 2 | 3 | 4 | 5 | 6 | 8 | 9 | 12 |
|---|---|---|---|---|---|---|---|---|---|---|
| $p(\bar{x})$ | $1/27$ | $3/27$ | $3/27$ | $1/27$ | $3/27$ | $6/27$ | $3/27$ | $3/27$ | $3/27$ | $1/27$ |

This is the sampling distribution for $\bar{x}$ because it specifies the probability associated with each possible value of $\bar{x}$.

b. In Table 6.1 you can see that the median $M$ can assume one of the three values: 0, 3, or 12. The value $M = 0$ occurs in seven different samples. Therefore, $P(M = 0) = 7/27$. Similarly, $M = 3$ occurs in 13 samples and $M = 12$ occurs in seven samples. Therefore, the probability distribution (i.e., the sampling distribution) for the median $M$ is as shown below.

| $M$ | 0 | 3 | 12 |
|---|---|---|---|
| $p(M)$ | $7/27$ | $13/27$ | $7/27$ |

Example 6.1 demonstrates the procedure for finding the exact sampling distribution of a statistic when the number of different samples that could be selected from the population is relatively small. In the real world, populations often consist of a large number of different values, making samples difficult (or impossible) to enumerate. When this situation occurs, we may choose to obtain the approximate sampling distribution for a statistic by simulating the sampling over and over again and recording the proportion of times different values of the statistic occur. Example 6.2 illustrates this procedure.

**EXAMPLE 6.2**

Suppose we perform the following experiment over and over again: Take a sample of 11 measurements from the uniform distribution shown in Figure 6.4. Calculate the two sample statistics

$$\bar{x} = \text{Sample mean} = \frac{\Sigma x}{11}$$

$M = \text{Median} = $ Sixth sample measurement when the 11 measurements are arranged in ascending order

Obtain approximations to the sampling distributions of $\bar{x}$ and $M$.

**FIGURE 6.4**

*Uniform Distribution from 0 to 1*

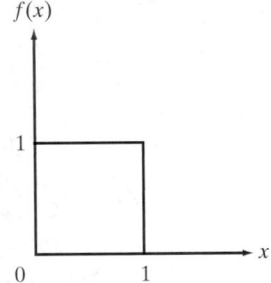

**Solution**

We use a computer to generate 1,000 samples, each with $n = 11$ observations. Then we compute $\bar{x}$ and $M$ for each sample. Our goal is to obtain approximations to the sampling distributions of $\bar{x}$ and $M$ to find out which sample statistic ($\bar{x}$ or $M$) contains more information about $\mu$. (Note that, in this particular example, we *know* the population mean is $\mu = .5$.) The first 10 of the 1,000 samples generated are presented in Table 6.2. For instance, the first computer-generated sample from the

**TABLE 6.2    First 10 Samples of $n = 11$ Measurements from a Uniform Distribution**

| Sample | Measurements | | | | | | | | | | |
|--------|------|------|------|------|------|------|------|------|------|------|------|
| *1* | .217 | .786 | .757 | .125 | .139 | .919 | .506 | .771 | .138 | .516 | .419 |
| *2* | .303 | .703 | .812 | .650 | .848 | .392 | .988 | .469 | .632 | .012 | .065 |
| *3* | .383 | .547 | .383 | .584 | .098 | .676 | .091 | .535 | .256 | .163 | .390 |
| *4* | .218 | .376 | .248 | .606 | .610 | .055 | .095 | .311 | .086 | .165 | .665 |
| *5* | .144 | .069 | .485 | .739 | .491 | .054 | .953 | .179 | .865 | .429 | .648 |
| *6* | .426 | .563 | .186 | .896 | .628 | .075 | .283 | .549 | .295 | .522 | .674 |
| *7* | .643 | .828 | .465 | .672 | .074 | .300 | .319 | .254 | .708 | .384 | .534 |
| *8* | .616 | .049 | .324 | .700 | .803 | .399 | .557 | .975 | .569 | .023 | .072 |
| *9* | .093 | .835 | .534 | .212 | .201 | .041 | .889 | .728 | .466 | .142 | .574 |
| *10* | .957 | .253 | .983 | .904 | .696 | .766 | .880 | .485 | .035 | .881 | .732 |

a. Sampling distribution for $\bar{x}$ (based on 1,000 samples of $n = 11$ measurements)

b. Sampling distribution for $M$ (based on 1,000 samples of $n = 11$ measurements)

**FIGURE 6.5**

*Relative Frequency Histograms for $\bar{x}$ and M, Example 6.2*

uniform distribution (arranged in ascending order) contained the following measurements: .125, .138, .139, .217, .419, .506, .516, .757, .771, .786, and .919. The sample mean $\bar{x}$ and median $M$ computed for this sample are

$$\bar{x} = \frac{.125 + .138 + \cdots + .919}{11} = .481$$

$$M = \text{Sixth ordered measurement} = .506$$

The relative frequency histograms for $\bar{x}$ and $M$ for the 1,000 samples of size $n = 11$ are shown in Figure 6.5.

You can see that the values of $\bar{x}$ tend to cluster around $\mu$ to a greater extent than do the values of $M$. Thus, on the basis of the observed sampling distributions, we conclude that $\bar{x}$ contains more information about $\mu$ than $M$ does—at least for samples of $n = 11$ measurements from the uniform distribution.

As noted earlier, many sampling distributions can be derived mathematically, but the theory necessary to do this is beyond the scope of this text. Consequently, when we need to know the properties of a statistic, we will present its sampling distribution and simply describe its properties. Several of the important properties we look for in sampling distributions are discussed in the next section.

 EXERCISES 6.1–6.7

**Learning the Mechanics**

**6.1** The probability distribution shown here describes a population of measurements that can assume values of 0, 2, 4, and 6, each of which occurs with the same relative frequency:

| $x$ | 0 | 2 | 4 | 6 |
|---|---|---|---|---|
| $p(x)$ | $\frac{1}{4}$ | $\frac{1}{4}$ | $\frac{1}{4}$ | $\frac{1}{4}$ |

**a.** List all the different samples of $n = 2$ measurements that can be selected from this population.

**b.** Calculate the mean of each different sample listed in part **a**.

**c.** If a sample of $n = 2$ measurements is randomly selected from the population, what is the probability that a specific sample will be selected?

**d.** Assume that a random sample of $n = 2$ measurements is selected from the population. List the different values of $\bar{x}$ found in part **b**, and find the probability of each. Then give the sampling distribution of the sample mean $\bar{x}$ in tabular form.

**e.** Construct a probability histogram for the sampling distribution of $\bar{x}$.

6.2 Simulate sampling from the population described in Exercise 6.1 by marking the values of $x$, one on each of four identical coins (or poker chips, etc.). Place the coins (marked 0, 2, 4, and 6) into a bag, randomly select one, and observe its value. Replace this coin, draw a second coin, and observe its value. Finally, calculate the mean $\bar{x}$ for this sample of $n = 2$ observations randomly selected from the population (Exercise 6.1, part **b**). Replace the coins, mix, and using the same procedure, select a sample of $n = 2$ observations from the population. Record the numbers and calculate $\bar{x}$ for this sample. Repeat this sampling process until you acquire 100 values of $\bar{x}$. Construct a relative frequency distribution for these 100 sample means. Compare this distribution to the exact sampling distribution of $\bar{x}$ found in part **e** of Exercise 6.1. [*Note:* The distribution obtained in this exercise is an approximation to the exact sampling distribution. But, if you were to repeat the sampling procedure, drawing two coins not 100 times but 10,000 times, the relative frequency distribution for the 10,000 sample means would be almost identical to the sampling distribution of $\bar{x}$ found in Exercise 6.1, part **e**.]

6.3 Consider the population described by the probability distribution shown here.

| $x$ | 1 | 2 | 3 | 4 | 5 |
|-----|---|---|---|---|---|
| $p(x)$ | .2 | .3 | .2 | .2 | .1 |

The random variable $x$ is observed twice. If these observations are independent, verify that the different samples of size 2 and their probabilities are as shown at the top of the next column.

a. Find the sampling distribution of the sample mean $\bar{x}$.
b. Construct a probability histogram for the sampling distribution of $\bar{x}$.
c. What is the probability that $\bar{x}$ is 4.5 or larger?
d. Would you expect to observe a value of $\bar{x}$ equal to 4.5 or larger? Explain.

6.4 Refer to Exercise 6.3 and find $E(x) = \mu$. Then use the sampling distribution of $\bar{x}$ found in Exercise 6.3 to find the expected value of $\bar{x}$. Note that $E(\bar{x}) = \mu$.

6.5 Refer to Exercise 6.3. Assume that a random sample of $n = 2$ measurements is randomly selected from the population.

| Sample | Probability | Sample | Probability |
|--------|-------------|--------|-------------|
| 1, 1 | .04 | 3, 4 | .04 |
| 1, 2 | .06 | 3, 5 | .02 |
| 1, 3 | .04 | 4, 1 | .04 |
| 1, 4 | .04 | 4, 2 | .06 |
| 1, 5 | .02 | 4, 3 | .04 |
| 2, 1 | .06 | 4, 4 | .04 |
| 2, 2 | .09 | 4, 5 | .02 |
| 2, 3 | .06 | 5, 1 | .02 |
| 2, 4 | .06 | 5, 2 | .03 |
| 2, 5 | .03 | 5, 3 | .02 |
| 3, 1 | .04 | 5, 4 | .02 |
| 3, 2 | .06 | 5, 5 | .01 |
| 3, 3 | .04 | | |

a. List the different values that the sample median $M$ may assume and find the probability of each. Then give the sampling distribution of the sample median.
b. Construct a probability histogram for the sampling distribution of the sample median and compare it with the probability histogram for the sample mean (Exercise 6.3, part **b**).

6.6 In Example 6.2 we use the computer to generate 1,000 samples, each containing $n = 11$ observations, from a uniform distribution over the interval from 0 to 1. For this exercise, use the computer to generate 500 samples, each containing $n = 15$ observations, from this population.
a. Calculate the sample mean for each sample. To approximate the sampling distribution of $\bar{x}$, construct a relative frequency histogram for the 500 values of $\bar{x}$.
b. Repeat part **a** for the sample median. Compare this approximate sampling distribution with the approximate sampling distribution of $\bar{x}$ found in part **a**.

6.7 Consider a population that contains values of $x$ equal to 00, 01, 02, 03, ..., 96, 97, 98, 99. Assume that these values of $x$ occur with equal probability. Use the computer to generate 500 samples, each containing $n = 25$ measurements, from this population. Calculate the sample mean $\bar{x}$ and sample variance $s^2$ for each of the 500 samples.
a. To approximate the sampling distribution of $\bar{x}$, construct a relative frequency histogram for the 500 values of $\bar{x}$.
b. Repeat part **a** for the 500 values of $s^2$.

## 6.2 PROPERTIES OF SAMPLING DISTRIBUTIONS: UNBIASEDNESS AND MINIMUM VARIANCE (OPTIONAL)

The simplest type of statistic used to make inferences about a population parameter is a *point estimator*. A point estimator is a rule or formula that tells us how to use the sample data to calculate a single number that is intended to estimate the value of some population parameter. For example, the sample mean $\bar{x}$ is a point estimator of the population mean $\mu$. Similarly, the sample variance $s^2$ is a point estimator of the population variance $\sigma^2$.

---

**DEFINITION 6.4**

A **point estimator** of a population parameter is a rule or formula that tells us how to use the sample data to calculate a single number that can be used as an *estimate* of the population parameter.

---

Often, many different point estimators can be found to estimate the same parameter. Each will have a sampling distribution that provides information about the point estimator. By examining the sampling distribution, we can determine how large the difference between an estimate and the true value of the parameter (called the **error of estimation**) is likely to be. We can also tell whether an estimator is more likely to overestimate or to underestimate a parameter.

**EXAMPLE 6.3**

Suppose two statistics, $A$ and $B$, exist to estimate the same population parameter, $\theta$ (theta). (Note that $\theta$ could be any parameter, $\mu$, $\sigma^2$, $\sigma$, etc.) Suppose the two statistics have sampling distributions as shown in Figure 6.6. Based on these sampling distributions, which statistic is more attractive as an estimator of $\theta$?

**FIGURE 6.6**

*Sampling Distributions of Unbiased and Biased Estimators*

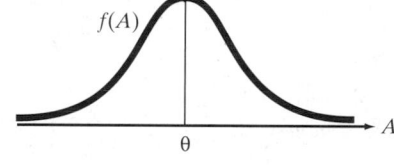

a. Unbiased sample statistic
   for the parameter $\theta$

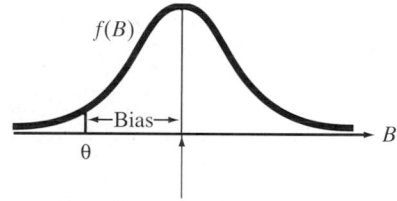

Mean of sampling distribution

b. Biased sample statistic
   for the parameter $\theta$

### Solution

As a first consideration, we would like the sampling distribution to center over the value of the parameter we wish to estimate. One way to characterize this property is in terms of the mean of the sampling distribution. Consequently, we say that a statistic is *unbiased* if the mean of the sampling distribution is equal to the parameter it is intended to estimate. This situation is shown in Figure 6.6a, where the mean $\mu_A$ of statistic $A$ is equal to $\theta$. If the mean of a sampling distribution is not equal to the parameter it is intended to estimate, the statistic is said to be *biased*. The sampling distribution for a biased statistic is shown in Figure 6.6b. The mean $\mu_B$ of the sampling distribution for statistic $B$ is not equal to $\theta$; in fact, it is shifted to the right of $\theta$.

You can see that biased statistics tend either to overestimate or to underestimate a parameter. Consequently, when other properties of statistics tend to be equivalent, we will choose an unbiased statistic to estimate a parameter of interest.*

---

*Unbiased statistics do not exist for all parameters of interest, but they do exist for the parameters considered in this text.

> **DEFINITION 6.5**
>
> If the sampling distribution of a sample statistic has a mean equal to the population parameter the statistic is intended to estimate, the statistic is said to be an **unbiased estimate** of the parameter.
>
> If the mean of the sampling distribution is not equal to the parameter, the statistic is said to be a **biased estimate** of the parameter.

The standard deviation of a sampling distribution measures another important property of statistics—the spread of these estimates generated by repeated sampling. Suppose two statistics, $A$ and $B$, are both unbiased estimators of the population parameter. Since the means of the two sampling distributions are the same, we turn to their standard deviations in order to decide which will provide estimates that fall closer to the unknown population parameter we are estimating. Naturally, we will choose the sample statistic that has the smaller standard deviation. Figure 6.7 depicts sampling distributions for $A$ and $B$. Note that the standard deviation of the distribution of $A$ is smaller than the standard deviation for $B$, indicating that over a large number of samples, the values of $A$ cluster more closely around the unknown population parameter than do the values of $B$. Stated differently, the probability that $A$ is close to the parameter value is higher than the probability that $B$ is close to the parameter value.

**FIGURE 6.7**

*Sampling Distributions for Two Unbiased Estimators*

In summary, to make an inference about a population parameter, we use the sample statistic with a sampling distribution that is unbiased and has a small standard deviation (usually smaller than the standard deviation of other unbiased sample statistics). The derivation of this sample statistic will not concern us, because the "best" statistic for estimating specific parameters is a matter of record. We will simply present an unbiased estimator with its standard deviation for each population parameter we consider. [*Note:* The standard deviation of the sampling distribution of a statistic is also called the **standard error of the statistic**.]

**EXAMPLE 6.4**

In Example 6.1, we found the sampling distributions of the sample mean $\bar{x}$ and the sample median $M$ for random samples of $n = 3$ measurements from a population defined by the probability distribution shown below.

| $x$ | 0 | 3 | 12 |
|---|---|---|---|
| $p(x)$ | $1/3$ | $1/3$ | $1/3$ |

The sampling distributions of $\bar{x}$ and $M$ were found to be as shown here.

| $\bar{x}$ | 0 | 1 | 2 | 3 | 4 | 5 | 6 | 8 | 9 | 12 |
|---|---|---|---|---|---|---|---|---|---|---|
| $p(\bar{x})$ | $1/27$ | $3/27$ | $3/27$ | $1/27$ | $3/27$ | $6/27$ | $3/27$ | $3/27$ | $3/27$ | $1/27$ |

| M | 0 | 3 | 12 |
|---|---|---|---|
| p(M) | $\frac{7}{27}$ | $\frac{13}{27}$ | $\frac{7}{27}$ |

a. Show that $\bar{x}$ is an unbiased estimator of $\mu$ in this situation.
b. Show that $M$ is a biased estimator of $\mu$ in this situation.

**Solution**

a. The expected value of a discrete random variable $x$ (see Section 4.3) is defined to be $E(x) = \Sigma xp(x)$, where the summation is over all values of $x$. Then

$$E(x) = \mu = \sum xp(x) = (0)\left(\tfrac{1}{3}\right) + (3)\left(\tfrac{1}{3}\right) + (12)\left(\tfrac{1}{3}\right) = 5$$

The expected value of the discrete random variable $\bar{x}$ is

$$E(\bar{x}) = \sum (\bar{x})p(\bar{x})$$

summed over all values of $\bar{x}$. Or

$$E(\bar{x}) = (0)\left(\tfrac{1}{27}\right) + (1)\left(\tfrac{3}{27}\right) + 2\left(\tfrac{3}{27}\right) + \cdots + (12)\left(\tfrac{1}{27}\right) = 5$$

Since $E(\bar{x}) = \mu$, we see that $\bar{x}$ is an unbiased estimator of $\mu$.
b. The expected value of the sample median $M$ is

$$E(M) = \sum Mp(M) = (0)\left(\tfrac{7}{27}\right) + (3)\left(\tfrac{13}{27}\right) + (12)\left(\tfrac{7}{27}\right) = 4.56$$

Since the expected value of $M$ is not equal to $\mu(\mu = 5)$, the sample median $M$ is a biased estimator of $\mu$.

**EXAMPLE 6.5**

Refer to Example 6.4 and find the standard deviations of the sampling distributions of $\bar{x}$ and $M$. Which statistic would appear to be a better estimator of $\mu$?

**Solution**

The variance of the sampling distribution of $\bar{x}$ (we denote it by the symbol $\sigma_{\bar{x}}^2$) is found to be

$$\sigma_{\bar{x}}^2 = E\{[\bar{x} - E(\bar{x})]^2\} = \sum (\bar{x} - \mu)^2 p(\bar{x})$$

where, from Example 6.4,

$$E(\bar{x}) = \mu = 5$$

Then

$$\sigma_{\bar{x}}^2 = (0 - 5)^2\left(\tfrac{1}{27}\right) + (1 - 5)^2\left(\tfrac{3}{27}\right) + (2 - 5)^2\left(\tfrac{3}{27}\right)$$
$$+ \cdots + (12 - 5)^2\left(\tfrac{1}{27}\right) = 8.6667$$

and

$$\sigma_{\bar{x}} = \sqrt{8.6667} = 2.94$$

Similarly, the variance of the sampling distribution of $M$ (we denote it by $\sigma_M^2$) is

$$\sigma_M^2 = E\{[M - E(M)]^2\}$$

where, from Example 6.4, the expected value of $M$ is $E(M) = 4.56$. Then

$$\sigma_M^2 = E\{[M - E(M)]^2\} = \sum [M - E(M)]^2 p(M)$$
$$= (0 - 4.56)^2 \left(\frac{7}{27}\right) + (3 - 4.56)^2 \left(\frac{13}{27}\right) + (12 - 4.56)^2 \left(\frac{7}{27}\right) = 20.9136$$

and

$$\sigma_M = \sqrt{20.9136} = 4.57$$

Which statistic appears to be the better estimator for the population mean $\mu$: the sample mean $\bar{x}$ or the median $M$? To answer this question, we compare the sampling distributions of the two statistics. The sampling distribution of the sample median $M$ is biased (i.e., it is located to the left of the mean $\mu$) and its standard deviation $\sigma_M = 4.57$ is much larger than the standard deviation of the sampling distribution of $\bar{x}$, $\sigma_{\bar{x}} = 2.94$. Consequently, the sample mean $\bar{x}$ would be a better estimator of the population mean $\mu$, for the population in question, than would the sample median $M$.    ∎

## EXERCISES 6.8–6.14

**Learning the Mechanics**

**6.8**  Consider the probability distribution shown here.

| $x$ | 0 | 1 | 4 |
|-----|------|------|------|
| $p(x)$ | $\frac{1}{3}$ | $\frac{1}{3}$ | $\frac{1}{3}$ |

**a.** Find $\mu$ and $\sigma^2$.

**b.** Find the sampling distribution of the sample mean $\bar{x}$ for a random sample of $n = 2$ measurements from this distribution.

**c.** Show that $\bar{x}$ is an unbiased estimator for $\mu$. [*Hint:*

Show that $E(\bar{x}) = \sum \bar{x}p(\bar{x}) = \mu$.]

**d.** Find the sampling distribution of the sample variance $s^2$ for a random sample of $n = 2$ measurements from this distribution.

**e.** Show that $s^2$ is an unbiased estimator for $\sigma^2$.

**6.9**  Consider the probability distribution shown here.

| $x$ | 2 | 4 | 9 |
|-----|------|------|------|
| $p(x)$ | $\frac{1}{3}$ | $\frac{1}{3}$ | $\frac{1}{3}$ |

**a.** Calculate $\mu$ for this distribution.

**b.** Find the sampling distribution of the sample mean $\bar{x}$ for a random sample of $n = 3$ measurements from this distribution, and show that $\bar{x}$ is an unbiased estimator of $\mu$.

**c.** Find the sampling distribution of the sample median $M$ for a random sample of $n = 3$ measurements from this distribution, and show that the median is a biased estimator of $\mu$.

**d.** If you wanted to estimate $\mu$ using a sample of three measurements from this population, which estimator would you use? Why?

**6.10**  Consider the probability distribution shown here.

| $x$ | 0 | 1 | 2 |
|-----|------|------|------|
| $p(x)$ | $\frac{1}{3}$ | $\frac{1}{3}$ | $\frac{1}{3}$ |

**a.** Find $\mu$.

**b.** For a random sample of $n = 3$ observations from this distribution, find the sampling distribution of the sample mean.

**c.** Find the sampling distribution of the median of a sample of $n = 3$ observations from this population.

**d.** Refer to parts **b** and **c** and show that both the mean and median are unbiased estimators of $\mu$ for this population.

**e.** Find the variances of the sampling distributions of the sample mean and the sample median.

**f.** Which estimator would you use to estimate $\mu$? Why?

**6.11** Use the computer to generate 500 samples, each containing $n = 25$ measurements, from a population that contains values of $x$ equal to $1, 2, \ldots, 48, 49, 50$. Assume that these values of $x$ are equally likely. Calculate the sample mean $\bar{x}$ and median $M$ for each sample. Construct relative frequency histograms for the 500 values of $\bar{x}$ and the 500 values of $M$. Use these approximations to the sampling distributions of $\bar{x}$ and $M$ to answer the following questions:

**a.** Does it appear that $\bar{x}$ and $M$ are unbiased estimators of the population mean? [*Note:* $\mu = 25.5.$]

**b.** Which sampling distribution displays greater variation?

**6.12** Refer to Exercise 6.3.

**a.** Show that $\bar{x}$ is an unbiased estimator of $\mu$.

**b.** Find $\sigma_{\bar{x}}^2$

**c.** Find the probability that $\bar{x}$ will fall within $2\sigma_{\bar{x}}$ of $\mu$.

**6.13** Refer to Exercise 6.3.

**a.** Find the sampling distribution of $s^2$.

**b.** Find the population variance $\sigma^2$.

**c.** Show that $s^2$ is an unbiased estimator of $\sigma^2$.

**d.** Find the sampling distribution of the sample standard deviation $s$.

**e.** Show that $s$ is a biased estimator of $\sigma$.

**6.14** Refer to Exercise 6.5, where we found the sampling distribution of the sample median. Is the median an unbiased estimator of the population mean $\mu$?

## 6.3   THE CENTRAL LIMIT THEOREM

Estimating the mean useful life of automobiles, the mean number of crimes per month in a large city, and the mean yield per acre of a new soybean hybrid are practical problems with something in common. In each case we are interested in making an inference about the mean $\mu$ of some population. As we mentioned in Chapter 2, the sample mean $\bar{x}$ is, in general, a good estimator of $\mu$. We now develop pertinent information about the sampling distribution for this useful statistic.

**EXAMPLE 6.6**

Suppose a population has the uniform probability distribution given in Figure 6.8. The mean and standard deviation of this probability distribution are $\mu = .5$ and $\sigma = .29$. (See Section 5.2 for the formulas for $\mu$ and $\sigma$.) Now suppose a sample of 11 measurements is selected from this population. Describe the sampling distribution of the sample mean $\bar{x}$ based on the 1,000 sampling experiments discussed in Example 6.2.

**FIGURE 6.8**

*Sampled Uniform Population*

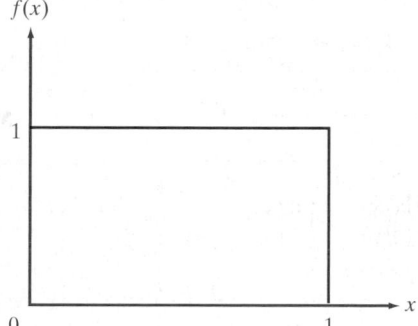

**Solution**

You will recall that in Example 6.2 we generated 1,000 samples of $n = 11$ measurements each. The relative frequency histogram for the 1,000 sample means is shown in Figure 6.9 with a normal probability distribution superimposed. You can see that this normal probability distribution approximates the computer-generated sampling distribution very well.

To fully describe a normal probability distribution, it is necessary to know its mean and standard deviation. Inspection of Figure 6.9 indicates that the mean of

**FIGURE 6.9**

*Relative Frequency Histogram for x̄ in 1,000 Samples of n = 11 Measurements with Normal Distribution Superimposed*

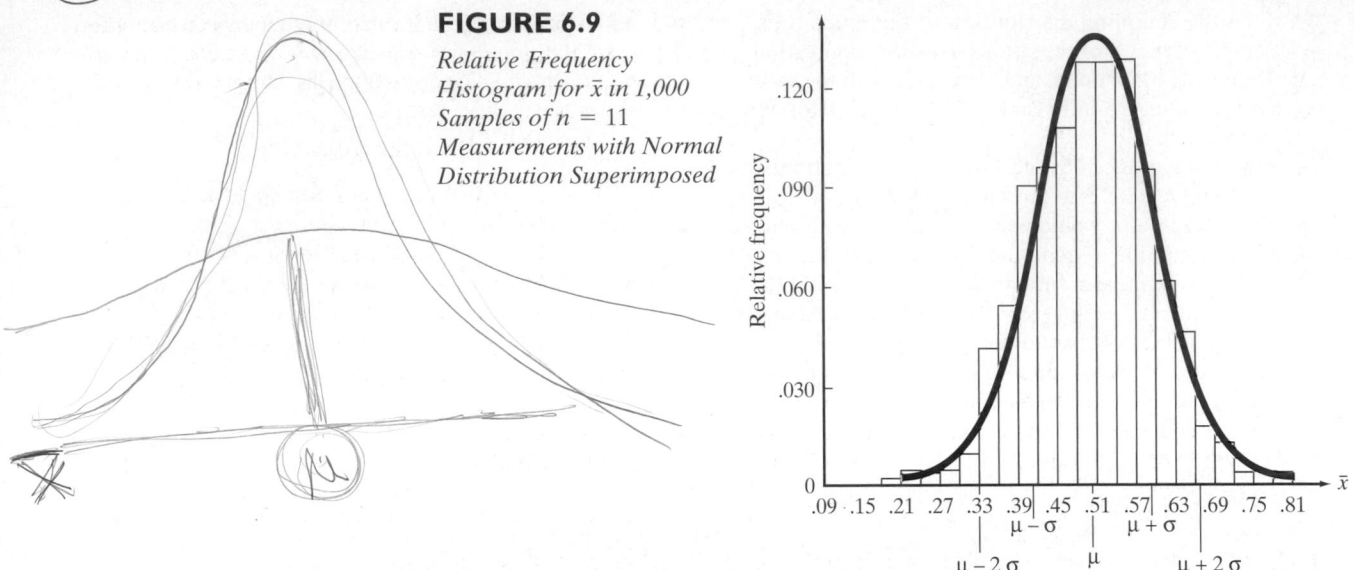

the distribution of $\bar{x}$, $\mu_{\bar{x}}$, appears to be very close to .5, the mean of the sampled uniform population. Furthermore, for a mound-shaped distribution such as that shown in Figure 6.9, almost all the measurements should fall within 3 standard deviations of the mean. Since the number of values of $\bar{x}$ is very large (1,000), the range of the observed $\bar{x}$'s divided by 6 (rather than 4) should give a reasonable approximation to the standard deviation of the sample means, $\sigma_{\bar{x}}$. The values of $\bar{x}$ range from about .2 to .8, so we calculate

$$\sigma_{\bar{x}} \approx \frac{\text{Range of } \bar{x}\text{'s}}{6} = \frac{.8 - .2}{6} = .1$$

To summarize our findings based on 1,000 samples, each consisting of 11 measurements from a uniform population, the sampling distribution of $\bar{x}$ appears to be approximately normal with a mean of about .5 and a standard deviation of <u>about .1.</u>  ∎

The sampling distribution of $\bar{x}$ has the properties given in the next box, assuming only that a random sample of $n$ observations has been selected from *any* population.

---

### Properties of the Sampling Distribution of $\bar{x}$

1. Mean of sampling distribution equals mean of sampled population. That is, $\mu_{\bar{x}} = E(\bar{x}) = \mu$.
2. Standard deviation of sampling distribution equals

$$\frac{\text{Standard deviation of sampled population}}{\text{Square root of sample size}}$$

That is, $\sigma_{\bar{x}} = \sigma / \sqrt{n}$.*
The standard deviation $\sigma_{\bar{x}}$ is often referred to as the **standard error of the mean**.

---

*If the sample size $n$ is large relative to the number $N$ of elements in the population (e.g., 5% or more), $\sigma/\sqrt{n}$ must be multiplied by a finite population correction factor, $\sqrt{(N-n)/(N-1)}$. For most sampling situations, this correction factor will be close to 1 and can be ignored.

You can see that our approximation to $\mu_{\bar{x}}$ in Example 6.6 was precise, since property 1 assures us that the mean is the same as that of the sampled population: .5. Property 2 tells us how to calculate the standard deviation of the sampling distribution of $\bar{x}$. Substituting $\sigma = .29$, the standard deviation of the sampled uniform distribution, and the sample size $n = 11$ into the formula for $\sigma_{\bar{x}}$, we find

$$\sigma_{\bar{x}} = \frac{\sigma}{\sqrt{n}} = \frac{.29}{\sqrt{11}} = .09$$

Thus, the approximation we obtained in Example 6.6, $\sigma_{\bar{x}} \approx .1$, is very close to the exact value, $\sigma_{\bar{x}} = .09$.

A third property, applicable when the sample size $n$ is large, is contained in one of the most important theoretical results in statistics: the *Central Limit Theorem*.

---

### Central Limit Theorem

Consider a random sample of $n$ observations selected from a population (*any* population) with mean $\mu$ and standard deviation $\sigma$. Then, when $n$ is sufficiently large, the sampling distribution of $\bar{x}$ will be approximately a normal distribution with mean $\mu_{\bar{x}} = \mu$ and standard deviation $\sigma_{\bar{x}} = \sigma/\sqrt{n}$. The larger the sample size, the better will be the normal approximation to the sampling distribution of $\bar{x}$.*

---

Thus, for sufficiently large samples the sampling distribution of $\bar{x}$ is approximately normal. How large must the sample size $n$ be so that the normal distribution provides a good approximation for the sampling distribution of $\bar{x}$? The answer depends on the shape of the distribution of the sampled population, as shown by Figure 6.10. Generally speaking, the greater the skewness of the sampled population distribution, the larger the sample size must be before the normal distribution is an adequate approximation for the sampling distribution of $\bar{x}$. For most sampled populations, sample sizes of $n \geq 30$ will suffice for the normal approximation to be reasonable. We will use the normal approximation for the sampling distribution of $\bar{x}$ when the sample size is at least 30.

**EXAMPLE  6.7**

Suppose we have selected a random sample of $n = 25$ observations from a population with mean equal to 80 and standard deviation equal to 5. It is known that the population is not extremely skewed.

a. Sketch the relative frequency distributions for the population and for the sampling distribution of the sample mean, $\bar{x}$.
b. Find the probability that $\bar{x}$ will be larger than 82.

### Solution

a. We do not know the exact shape of the population relative frequency distribution, but we do know that it should be centered about $\mu = 80$, its spread should be measured by $\sigma = 5$, and it is not highly skewed. One possibility is shown in Figure 6.11a. From the Central Limit Theorem, we know that the

---

*Moreover, because of the Central Limit Theorem, the sum of a random sample of $n$ observations, $\Sigma x$, will possess a sampling distribution that is approximately normal for large samples. This distribution will have a mean equal to $n\mu$ and a variance equal to $n\sigma^2$. Proof of the Central Limit Theorem is beyond the scope of this book, but it can be found in many mathematical statistics texts.

**FIGURE 6.10**

*Sampling Distributions of $\bar{x}$ for Different Populations and Different Sample Sizes*

a. Population relative frequency distribution

b. Sampling distribution of $\bar{x}$

**FIGURE 6.11**

*A Population Relative Frequency Distribution and the Sampling Distribution for $\bar{x}$*

sampling distribution of $\bar{x}$ will be approximately normal since the sampled population distribution is not extremely skewed. We also know that the sampling distribution will have mean and standard deviation

$$\mu_{\bar{x}} = \mu = 80 \quad \text{and} \quad \sigma_{\bar{x}} = \frac{\sigma}{\sqrt{n}} = \frac{5}{\sqrt{25}} = 1$$

The sampling distribution of $\bar{x}$ is shown in Figure 6.11b.

b. The probability that $\bar{x}$ will exceed 82 is equal to the highlighted area in Figure 6.12. To find this area, we need to find the $z$ value corresponding to $\bar{x} = 82$. Recall that the standard normal random variable $z$ is the difference between any normally distributed random variable and its mean, expressed in units of its standard deviation. Since $\bar{x}$ is a normally distributed random variable with mean $\mu_{\bar{x}} = \mu$ and standard deviation $\sigma_{\bar{x}} = \sigma/\sqrt{n}$, it follows that the standard normal $z$ value corresponding to the sample mean, $\bar{x}$, is

$$z = \frac{(\text{Normal random variable}) - (\text{Mean})}{\text{Standard Deviation}} = \frac{\bar{x} - \mu_{\bar{x}}}{\sigma_{\bar{x}}}$$

Therefore, for $\bar{x} = 82$, we have

$$z = \frac{\bar{x} - \mu_{\bar{x}}}{\sigma_{\bar{x}}} = \frac{82 - 80}{1} = 2$$

The area $A$ in Figure 6.12 corresponding to $z = 2$ is given in the table of areas under the normal curve (see Table IV of Appendix A) as .4772. Therefore, the tail area corresponding to the probability that $\bar{x}$ exceeds 82 is

$$P(\bar{x} > 82) = P(z > 2) = .5 - .4772 = .0228$$

**FIGURE 6.12**

*The Sampling Distribution of $\bar{x}$*

**EXAMPLE 6.8**

A manufacturer of automobile batteries claims that the distribution of the lengths of life of its best battery has a mean of 54 months and a standard deviation of 6 months. Suppose a consumer group decides to check the claim by purchasing a sample of 50 of these batteries and subjecting them to tests that determine battery life.

a. Assuming that the manufacturer's claim is true, describe the sampling distribution of the mean lifetime of a sample of 50 batteries.

b. Assuming that the manufacturer's claim is true, what is the probability the consumer group's sample has a mean life of 52 or fewer months?

**Solution**

a. Even though we have no information about the shape of the probability distribution of the lives of the batteries, we can use the Central Limit Theorem to deduce that the sampling distribution for a sample mean lifetime of 50 batteries is approximately normally distributed. Furthermore, the mean of this sampling distribution is the same as the mean of the sampled population, which is $\mu = 54$ months according to the manufacturer's claim. Finally, the standard deviation of the sampling distribution is given by

$$\sigma_{\bar{x}} = \frac{\sigma}{\sqrt{n}} = \frac{6}{\sqrt{50}} = .85 \text{ month}$$

Note that we used the claimed standard deviation of the sampled population, $\sigma = 6$ months. Thus, if we assume that the claim is true, the sampling distribution for the mean life of the 50 batteries sampled is as shown in Figure 6.13.

**FIGURE 6.13**

*Sampling Distribution of $\bar{x}$ in Example 6.8 for $n = 50$*

$P(\bar{x} \le 52)$

52   54

b. If the manufacturer's claim is true, the probability that the consumer group observes a mean battery life of 52 or fewer months for their sample of 50 batteries, $P(\bar{x} \le 52)$, is equivalent to the highlighted area in Figure 6.13. Since the sampling distribution is approximately normal, we can find this area by computing the standard normal $z$ value:

$$z = \frac{\bar{x} - \mu_{\bar{x}}}{\sigma_{\bar{x}}} = \frac{\bar{x} - \mu}{\sigma_{\bar{x}}} = \frac{52 - 54}{.85} = -2.35$$

where $\mu_{\bar{x}}$, the mean of the sampling distribution of $\bar{x}$, is equal to $\mu$, the mean of the lifetimes of the sampled population, and $\sigma_{\bar{x}}$ is the standard deviation of the sampling distribution of $\bar{x}$. Note that $z$ is the familiar standardized distance ($z$-score) of Section 2.6 and, since $\bar{x}$ is approximately normally distributed, it will possess the standard normal distribution of Section 5.3.

The area $A$ shown in Figure 6.13 between $\bar{x} = 52$ and $\bar{x} = 54$ (corresponding to $z = -2.35$) is found in Table IV of Appendix A to be .4906. Therefore, the area to the left of $\bar{x} = 52$ is

$$P(\bar{x} \le 52) = .5 - A = .5 - .4906 = .0094$$

Thus, the probability the consumer group will observe a sample mean of 52 or less is only .0094 if the manufacturer's claim is true. If the 50 tested batteries do exhibit a mean of 52 or fewer months, the consumer group will have strong evidence that the manufacturer's claim is untrue, because such an event is very unlikely to occur if the claim is true. (This is still another application of the *rare-event approach* to statistical inference.) ■

We conclude this section with two final comments on the sampling distribution of $\bar{x}$. First, from the formula $\sigma_{\bar{x}} = \sigma/\sqrt{n}$, we see that the standard deviation of the sampling distribution of $\bar{x}$ gets smaller as the sample size $n$ gets larger. For example, we computed $\sigma_{\bar{x}} = .85$ when $n = 50$ in Example 6.8. However, for $n = 100$ we obtain $\sigma_{\bar{x}} = \sigma/\sqrt{n} = 6/\sqrt{100} = .60$. This relationship will hold true for most of the sample statistics encountered in this text. That is: *The standard deviation of the sampling distribution decreases as the sample size increases.* Consequently, the larger the sample size, the more accurate the sample statistic (e.g., $\bar{x}$) is in estimating a population parameter (e.g., $\mu$). We will use this result in Chapter 7 to help us determine the sample size needed to obtain a specified accuracy of estimation.

Our second comment concerns the Central Limit Theorem. In addition to providing a very useful approximation for the sampling distribution of a sample mean, the Central Limit Theorem offers an explanation for the fact that many relative frequency distributions of data possess mound-shaped distributions. Many of the measurements we take in various areas of research are really means or sums of a large number of small phenomena. For example, a year's growth of a pine seedling is the total of the many individual components that affect the plant's growth. Similarly, we can view the length of time a construction company takes to build a house as the total of the times taken to complete a multitude of distinct jobs, and we can regard the monthly demand for blood at a hospital as the total of the many individual patients' needs. Whether or not the observations entering into these sums satisfy the assumptions basic to the Central Limit Theorem is open to question. However, it is a fact that many distributions of data in nature are mound-shaped and possess the appearance of normal distributions.

## EXERCISES 6.15–6.33

### Learning the Mechanics

**6.15** Suppose a random sample of $n$ measurements is selected from a population with mean $\mu = 100$ and variance $\sigma^2 = 100$. For each of the following values of $n$, give the mean and standard deviation of the sampling distribution of the sample mean $\bar{x}$.
  **a.** $n = 4$
  **b.** $n = 25$
  **c.** $n = 100$
  **d.** $n = 50$
  **e.** $n = 500$
  **f.** $n = 1,000$

**6.16** Suppose a random sample of $n = 25$ measurements is selected from a population with mean $\mu$ and standard deviation $\sigma$. For each of the following values of $\mu$ and $\sigma$, give the values of $\mu_{\bar{x}}$ and $\sigma_{\bar{x}}$.
  **a.** $\mu = 10, \sigma = 3$
  **b.** $\mu = 100, \sigma = 25$
  **c.** $\mu = 20, \sigma = 40$
  **d.** $\mu = 10, \sigma = 100$

**6.17** Consider the probability distribution shown here.

| $x$ | 1 | 2 | 3 | 8 |
|-----|-----|-----|-----|-----|
| $p(x)$ | .1 | .4 | .4 | .1 |

  **a.** Find $\mu, \sigma^2$, and $\sigma$.
  **b.** Find the sampling distribution of $\bar{x}$ for random samples of $n = 2$ measurements from this distribution by listing all possible values of $\bar{x}$, and find the probability associated with each.
  **c.** Use the results of part **b** to calculate $\mu_{\bar{x}}$ and $\sigma_{\bar{x}}$. Confirm that $\mu_{\bar{x}} = \mu$ and that $\sigma_{\bar{x}} = \sigma/\sqrt{n} = \sigma/\sqrt{2}$.

**6.18** Will the sampling distribution of $\bar{x}$ always be approximately normally distributed? Explain.

**6.19** A random sample of $n = 64$ observations is drawn from a population with a mean equal to 20 and standard deviation equal to 16.
  **a.** Give the mean and standard deviation of the (repeated) sampling distribution of $\bar{x}$.
  **b.** Describe the shape of the sampling distribution of $\bar{x}$. Does your answer depend on the sample size?
  **c.** Calculate the standard normal $z$-score corresponding to a value of $\bar{x} = 15.5$.
  **d.** Calculate the standard normal $z$-score corresponding to $\bar{x} = 23$.

**6.20** Refer to Exercise 6.19. Find the probability that
  **a.** $\bar{x}$ is less than 16
  **b.** $\bar{x}$ is greater than 23
  **c.** $\bar{x}$ is greater than 25
  **d.** $\bar{x}$ falls between 16 and 22
  **e.** $\bar{x}$ is less than 14

**6.21** A random sample of $n = 100$ observations is selected from a population with $\mu = 30$ and $\sigma = 16$. Approximate the following probabilities:
  **a.** $P(\bar{x} \geq 28)$
  **b.** $P(22.1 \leq \bar{x} \leq 26.8)$
  **c.** $P(\bar{x} \leq 28.2)$
  **d.** $P(\bar{x} \geq 27.0)$

**6.22** A random sample of $n = 900$ observations is selected from a population with $\mu = 100$ and $\sigma = 10$.
  **a.** What are the largest and smallest values of $\bar{x}$ that you would expect to see?
  **b.** How far, at the most, would you expect $\bar{x}$ to deviate from $\mu$?
  **c.** Did you have to know $\mu$ to answer part **b**? Explain.

**6.23** Consider a population that contains values of $x$ equal to 0, 1, 2, ..., 97, 98, 99. Assume that the values of $x$ are equally likely. For each of the following values of $n$, use the computer to generate 500 random samples and calculate $\bar{x}$ for each sample. For each sample size, construct a relative frequency histogram of the 500 values of $\bar{x}$. What changes occur in the histograms as the value of $n$ increases? What similarities exist? Use $n = 2, n = 5, n = 10, n = 30,$ and $n = 50$.

## Applying the Concepts—Basic

**6.24** According to the American Automobile Association (AAA), the average daily meal and lodging costs for a family of four is $213 (*Travel News*, May 11, 1999). Assume that the standard deviation of such costs is $15. Consider a random sample of 30 families of four and their travel expenses.
 **a.** Describe the sampling distribution of $\bar{x}$, the average daily meal and lodging costs for the sample of families. In particular, how is $\bar{x}$ distributed and what are the mean and variance of $\bar{x}$?
 **b.** Find the probability that $\bar{x}$ is greater than $200.

**6.25** The ocean quahog is a type of clam found in the coastal waters of the mid-Atlantic states. A federal survey of offshore ocean quahog harvesting in New Jersey revealed an average catch per unit effort (CPUE) of 89.34 clams. The CPUE standard deviation was 7.74 (*Journal of Shellfish Research*, June 1995). Let $\bar{x}$ represent the mean CPUE for a sample of 35 attempts to catch ocean quahogs off the New Jersey shore.
 **a.** Compute $\mu_{\bar{x}}$ and $\sigma_{\bar{x}}$. Interpret their values.
 **b.** Sketch the sampling distribution of $\bar{x}$.
 **c.** Find $P(\bar{x} > 88)$.
 **d.** Find $P(\bar{x} < 87)$.

**6.26** Refer to *The American Statistician* (May 2001) study of female students who suffer from bulimia, Exercise 2.27 (p. 39). Recall that each student completed a questionnaire from which a "fear of negative evaluation" (FNE) score was produced. (The higher the score, the greater the fear of negative evaluation.) Suppose the FNE scores of bulimic students have a distribution with mean $\mu = 18$ and standard deviation $\sigma = 5$. Now consider a random sample of 45 female students with bulimia.
 **a.** What is the probability that the sample mean FNE score is greater than 17.5?
 **b.** What is the probability that the sample mean FNE score is between 18 and 18.5?
 **c.** What is the probability that the sample mean FNE score is less than 18.5?

**6.27** Refer to the *Chance* (Winter 2001) examination of Standardized Admission Test (SAT) scores of students who pay a private tutor to help them improve their results, Exercise 2.81 (p. 70). On the SAT-Mathematics test, these students had a mean score change of +19 points, with a standard deviation of 65 points. In a random sam-

ple of 100 students who pay a private tutor to help them improve their results, what is the likelihood that the sample mean score change is less than 10 points?

## Applying the Concepts—Intermediate

**6.28** The *College Student Journal* (Dec. 1992) investigated differences in traditional and nontraditional students, where nontraditional students are generally defined as those 25 years old or older. Based on the study results, we can assume that the population mean and standard deviation for the GPA of all nontraditional students is $\mu = 3.5$ and $\sigma = .5$. Suppose that a random sample of $n = 100$ nontraditional students is selected from the population of all nontraditional students, and the GPA of each student is determined. Then $\bar{x}$, the sample mean, will be approximately normally distributed (because of the Central Limit Theorem).
 **a.** Calculate $\mu_{\bar{x}}$ and $\sigma_{\bar{x}}$.
 **b.** What is the approximate probability that the nontraditional student sample has a mean GPA between 3.40 and 3.60?
 **c.** What is the approximate probability that the sample of 100 nontraditional students has a mean GPA that exceeds 3.62?
 **d.** How would the sampling distribution of $\bar{x}$ change if the sample size $n$ were doubled from 100 to 200? How do your answers to parts **b** and **c** change when the sample size is doubled?

**6.29** Interpersonal violence (e.g., rape) generally leads to psychological stress for the victim. *Clinical Psychology Review* (Vol. 15, 1995) reported on the results of all recently published studies of the relationship between interpersonal violence and psychological stress. The distribution of the time elapsed between the violent incident and the initial sign of stress has a mean of 5.1 years and a standard deviation of 6.1 years. Consider a random sample of $n = 150$ victims of interpersonal violence. Let $\bar{x}$ represent the mean time elapsed between the violent act and the first sign of stress for the sampled victims.
 **a.** Give the mean and standard deviation of the sampling distribution of $\bar{x}$.
 **b.** Will the sampling distribution of $\bar{x}$ be approximately normal? Explain.
 **c.** Find $P(\bar{x} > 5.5)$.
 **d.** Find $P(4 < \bar{x} < 5)$.

**6.30** The Computer-Assisted Hypnosis Scale (CAHS) is designed to measure a person's susceptibility to hypnosis. In computer-assisted hypnosis, the computer serves as a facilitator of hypnosis by using digitized speech processing coupled with interactive involvement with the hypnotic subject. CAHS scores range from 0 (no susceptibility) to 12 (extremely high susceptibility). A study in *Psychological Assessment* (Mar. 1995) reported a mean CAHS score of 4.59 and a standard deviation of 2.95 for University of Tennessee undergraduates. Assume that $\mu = 4.29$ and $\sigma = 2.95$ for this population.

Suppose a psychologist uses CAHS to test a random sample of 50 subjects.

**a.** Would you expect to observe a sample mean CAHS score of $\bar{x} = 6$ or higher? Explain.

**b.** Suppose the psychologist actually observes $\bar{x} = 6.2$. Based on your answer to part **a**, make an inference about the population from which the sample was selected.

**6.31** Last year a company began a program to compensate its employees for unused sick days, paying each employee a bonus of one-half the usual wage earned for each unused sick day. The question that naturally arises is: "Did this policy motivate employees to use fewer sick days?" *Before* last year, the number of sick days used by employees had a distribution with a mean of 7 days and a standard deviation of 2 days.

**a.** Assuming that these parameters did not change last year, find the approximate probability that the sample mean number of sick days used by 100 employees chosen at random was less than or equal to 6.4 last year.

**b.** How would you interpret the result if the sample mean for the 100 employees was 6.4?

## Applying the Concepts—Advanced

**6.32** The Test of Knowledge About Epilepsy (KAE), which is designed to measure attitudes toward persons with epilepsy, uses 20 multiple-choice items where all of the choices are incorrect. For each person, two scores (ranging from 0 to 20) are obtained, an attitude score (KAE-A) and a general knowledge score (KAE-GK). Based on a large-scale study of college students, the distribution of KAE-A scores has a mean of $\mu = 11.92$ and a standard deviation of $\sigma = 2.95$ while the distribution of KAE-GK scores has a mean of $\mu = 6.35$ and a standard deviation of $\sigma = 2.12$ (*Rehabilitative Psychology*, Spring 1995). Consider a random sample of 100 college students, and suppose you observe a sample mean KAE score of 6.5. Is this result more likely to be the mean of the attitude scores (KAE-A) or the general knowledge scores (KAE-GK)? Explain.

**6.33** A soft drink bottler purchases glass bottles from a vendor. The bottles are required to have an internal pressure of at least 150 pounds per square inch (psi). A prospective bottle vendor claims that its production process yields bottles with a mean internal pressure of 157 psi and a standard deviation of 3 psi. The bottler strikes an agreement with the vendor that permits the bottler to sample from the vendor's production process to verify the vendor's claim. The bottler randomly selects 40 bottles from the last 10,000 produced, measures the internal pressure of each, and finds the mean pressure for the sample to be 1.3 psi below the process mean cited by the vendor.

**a.** Assuming the vendor's claim to be true, what is the probability of obtaining a sample mean this far or farther below the process mean? What does your answer suggest about the validity of the vendor's claim?

**b.** If the process standard deviation were 3 psi as claimed by the vendor, but the mean were 156 psi, would the observed sample result be more or less likely than in part **a**? What if the mean were 158 psi?

**c.** If the process mean were 157 psi as claimed, but the process standard deviation were 2 psi, would the sample result be more or less likely than in part **a**? What if instead the standard deviation were 6 psi?

---

## QUICK REVIEW

*Note: Starred (\*) items are from the optional section in this chapter.*

## Key Terms

## Key Formulas

| | Mean | Standard Deviation | z-score | |
|---|---|---|---|---|
| Sampling distribution of $\bar{x}$ | $\mu_{\bar{x}} = \mu$ | $\sigma_{\bar{x}} = \dfrac{\sigma}{\sqrt{n}}$ | $z = \dfrac{\bar{x} - \mu_{\bar{x}}}{\sigma_{\bar{x}}} = \dfrac{\bar{x} - \mu}{\sigma/\sqrt{n}}$ | 274 |

## STATISTICS IN ACTION
### The Insomnia Pill

A research report published in the *Proceedings of the National Academy of Sciences* (Mar. 1994) brought encouraging news to insomniacs and long-distance travelers. Neuroscientists at the Massachusetts Institute of Technology (MIT) have been experimenting with melatonin—a hormone secreted by the pineal gland in the brain—as a sleep-inducing hormone. Since the hormone is naturally produced, it is considered nonaddictive. The researchers believe melatonin may also be effective in treating jet lag—the body's response to rapid travel across many time zones so that a daylight–darkness change disrupts sleep patterns.

In the MIT study, young male volunteers were given various doses of melatonin or a placebo (no dosage of melatonin). Then they were placed in a dark room at midday and told to close their eyes for 30 minutes. The variable of interest was the time (in minutes) elapsed before each volunteer fell asleep.

According to the lead investigator, Professor Richard Wurtman, "Our volunteers fall asleep in five or six minutes on melatonin, while those on placebo take about 15 minutes." Wurtman warns, however, that uncontrolled doses of melatonin could cause serious mood-altering side effects. (Melatonin is sold in some health food stores. However, sales are unregulated, and the purity and strength of the hormone are often uncertain.)

### Focus

With the placebo (i.e., no hormone) the researchers found that the mean time to fall asleep was 15 minutes. Assume that with the placebo treatment $\mu = 15$ and $\sigma = 5$. Now, consider a random sample of 40 young males, each of whom is given a dose of the sleep-inducing hormone, melatonin. The times (in minutes) to fall asleep for these 40 males are listed in Table 6.3. Use the data to make an inference about the true value of $\mu$ for those taking the melatonin. (The data are stored in the file named INSOMNIA.) Does melatonin appear to be an effective drug against insomnia?

⊘ INSOMNIA

**TABLE 6.3    Times (in Minutes) for 40 Male Volunteers to Fall Asleep***

| 6.4 | 6.0 | 3.2 | 4.4 | 6.2 | 1.7 | 5.1 |
|-----|-----|-----|------|-----|-----|-----|
| 5.9 | 1.6 | 4.4 | 16.2 | 4.8 | 8.3 | 7.5 |
| 4.8 | 3.3 | 4.0 | 6.2 | 6.3 | 5.0 | 6.3 |
| 5.1 | 6.4 | 15.6 | 3.4 | 3.1 | 6.1 | 6.0 |
| 5.0 | 1.8 | 6.1 | 4.5 | 4.5 | 1.5 | 4.7 |
| 7.6 | 8.2 | 4.9 | 6.1 | 3.9 | | |

*These are simulated sleep times based on summary information provided in the MIT study.

## Language Lab

| Symbol | Pronunciation | Description |
|--------|---------------|-------------|
| $\theta$ | theta | Population parameter (general) |
| $\mu_{\bar{x}}$ | mu of *x*-bar | True mean of sampling distribution of $\bar{x}$ |
| $\sigma_{\bar{x}}$ | sigma of *x*-bar | True standard deviation of sampling distribution of $\bar{x}$ |

## SUPPLEMENTARY EXERCISES 6.34–6.52

*Note: Starred (\*) exercises refer to the optional section in this chapter.*

### Learning the Mechanics

**\*6.34** The standard deviation (or, as it is usually called, the *standard error*) of the sampling distribution for the sample mean, $\bar{x}$, is equal to the standard deviation of the population from which the sample was selected divided by the square root of the sample size. That is,

$$\sigma_{\bar{x}} = \frac{\sigma}{\sqrt{n}}$$

**a.** As the sample size is increased, what happens to the standard error of $\bar{x}$? Why is this property considered important?

**b.** Suppose that a sample statistic has a standard error that is not a function of the sample size. In other words, the standard error remains constant as $n$ changes. What would this imply about the statistic as an estimator of a population parameter?

**c.** Suppose another unbiased estimator (call it $A$) of the population mean is a sample statistic with a standard error equal to

$$\sigma_A = \frac{\sigma}{\sqrt[3]{n}}$$

Which of the sample statistics, $\bar{x}$ or $A$, is preferable as an estimator of the population mean? Why?

**d.** Suppose that the population standard deviation $\sigma$ is equal to 10 and that the sample size is 64. Calculate the standard errors of $\bar{x}$ and $A$. Assuming that the sampling distribution of $A$ is approximately normal, interpret the standard errors. Why is the assumption of (approximate) normality unnecessary for the sampling distribution of $\bar{x}$?

**6.35** Consider a sample statistic $A$. As with all sample statistics, $A$ is computed by utilizing a specified function (formula) of the sample measurements. (For example, if $A$ were the sample mean, the specified formula would be to sum the measurements and divide by the number of measurements.)

  **a.** Describe what we mean by the phrase "the sampling distribution of the sample statistic $A$."

  *__b.__ Suppose $A$ is to be used to estimate a population parameter $\alpha$. What is meant by the assertion that $A$ is an unbiased estimator of $\alpha$?

  *__c.__ Consider another sample statistic, $B$. Assume that $B$ is also an unbiased estimator of the population parameter $\alpha$. How can we use the sampling distributions of $A$ and $B$ to decide which is the better estimator of $\alpha$?

  **d.** If the sample sizes on which $A$ and $B$ are based are large, can we apply the Central Limit Theorem and assert that the sampling distributions of $A$ and $B$ are approximately normal? Why or why not?

**6.36** A random sample of 40 observations is to be drawn from a large population of measurements. It is known that 30% of the measurements in the population are 1's, 20% are 2's, 20% are 3's, and 30% are 4's.

  **a.** Give the mean and standard deviation of the (repeated) sampling distribution of $\bar{x}$, the sample mean of the 40 observations.

  **b.** Describe the shape of the sampling distribution of $\bar{x}$. Does your answer depend on the sample size?

**6.37** A random sample of $n = 68$ observations is selected from a population with $\mu = 19.6$ and $\sigma = 3.2$. Approximate each of the following probabilities.

  **a.** $P(\bar{x} \le 19.6)$

  **b.** $P(\bar{x} \le 19)$

  **c.** $P(\bar{x} \ge 20.1)$

  **d.** $P(19.2 \le \bar{x} \le 20.6)$

**6.38** Use a statistical software package to generate 100 random samples of size $n = 2$ from a population characterized by a normal probability distribution with mean 100 and standard deviation 10. Compute $\bar{x}$ for each sample and plot a frequency distribution for the 100 values of $\bar{x}$. Repeat this process for $n = 5, 10, 30,$ and 50. How does the fact that the sampled population is normal affect the sampling distribution of $\bar{x}$?

**6.39** Use a statistical software package to generate 100 random samples of size $n = 2$ from a population characterized by a uniform probability distribution with $c = 0$ and $d = 10$. Compute $\bar{x}$ for each sample, and plot a frequency distribution for the 100 $\bar{x}$ values. Repeat this process for $n = 5, 10, 30,$ and 50. Explain how your plots illustrate the Central Limit Theorem.

**6.40** Suppose $x$ equals the number of heads observed when a single coin is tossed; that is, $x = 0$ or $x = 1$. The population corresponding to $x$ is the set of 0's and 1's generated when the coin is tossed repeatedly a large number of times. Suppose we select $n = 2$ observations from this population. (That is, we toss the coin twice and observe two values of $x$.)

  **a.** List the three different samples (combinations of 0's and 1's) that could be obtained.

  **b.** Calculate the value of $\bar{x}$ for each of the samples.

  **c.** List the values that $\bar{x}$ can assume, and find the probabilities of observing these values.

  **d.** Construct a graph of the sampling distribution of $\bar{x}$.

**6.41** A random sample of size $n$ is to be drawn from a large population with mean 100 and standard deviation 10, and the sample mean $\bar{x}$ is to be calculated. To see the effect of different sample sizes on the standard deviation of the sampling distribution of $\bar{x}$, plot $\sigma/\sqrt{n}$ against $n$ for $n = 1, 5, 10, 20, 30, 40,$ and 50.

## Applying the Concepts—Basic

**6.42** The distribution of the number of barrels of oil produced by a certain oil well each day for the past 3 years has a mean of 400 and a standard deviation of 75.

  **a.** Describe the sampling distribution of the mean number of barrels produced per day for samples of 40 production days drawn from the past 3 years.

  **b.** What is the approximate probability that the sample mean will be greater than 425?

  **c.** What is the approximate probability that the sample mean will be less than 400?

  **d.** What assumptions did you have to make in order to answer parts **a–c**? Justify the assumptions.

**6.43** The distribution of violent crimes per day in a certain city possesses a mean equal to 1.3 and a standard deviation equal to 1.7. A random sample of 50 days is observed, and the daily mean number of crimes for this sample, $\bar{x}$, is calculated.

  **a.** Give the mean and standard deviation of the sampling distribution of $\bar{x}$.

  **b.** Will the sampling distribution of $\bar{x}$ be approximately normal? Explain.

  **c.** Find an approximate value of $P(\bar{x} < 1)$.

  **d.** Find an approximate value of $P(\bar{x} > 1.9)$.

**6.44** To determine whether a metal lathe that produces machine bearings is properly adjusted, a random sample of 25 bearings is collected and the diameter of each is measured.

  **a.** If the standard deviation of the diameters of the bearings measured over a long period of time is .001 inch, what is the approximate probability that the mean diameter $\bar{x}$ of the sample of 25 bearings will lie within .0001 inch of the population mean diameter of the bearings?

  **b.** If the population of diameters has an extremely skewed distribution, how will your approximation in part **a** be affected?

**6.45** This past year, an elementary school began using a new method to teach arithmetic to first graders. A standardized test, administered at the end of the year, was used to measure the effectiveness of the new method. The distribution of past scores on the standardized test produced a mean of 75 and a standard deviation of 10.

a. If the new method is no different from the old method, what is the approximate probability that the mean score $\bar{x}$ of a random sample of 36 students will be greater than 79?

b. What assumptions must be satisfied to make your answer valid?

## Applying the Concepts—Intermediate

**6.46** Researchers at the Terry College of Business at the University of Georgia sampled 344 business students and asked them this question: "Over the course of your lifetime, what is the maximum number of years you expect to work for any one employer?" The sample resulted in $\bar{x} = 19.1$ years. Assume the sample of students was randomly selected from the 6,000 undergraduate students at the Terry College and that $\sigma = 6$ years.

a. Describe the sampling distribution of $\bar{x}$.

b. If the mean for the 6,000 undergraduate students is $\mu = 18.5$ years, find $P(\bar{x} > 19.1)$.

c. If the mean for the 6,000 undergraduate students is $\mu = 19.5$ years, find $P(\bar{x} > 19.1)$.

d. If $P(\bar{x} > 19.1) = .5$, what is $\mu$?

e. If $P(\bar{x} > 19.1) = .2$, is $\mu$ greater than or less than 19.1 years? Explain.

**6.47** The *Florida Employer Opinion Survey* gives the results of an extensive survey of employer opinions in Florida. Each employer was asked to rate his or her satisfaction with the preparation of employees by the public education system. Responses were 1, 1.5, or 2, representing very dissatisfied, neither satisfied nor dissatisfied, and very satisfied, respectively. A sample of 651 employers was selected. Assume that the mean for all employers in Florida is 1.50 (the "dividing line" between satisfied and dissatisfied) and the standard deviation is .45.

a. Which type of distribution describes the individual survey responses, continuous or discrete?

b. Describe the distribution that best approximates the sample mean response of 651 employers. What are the mean and standard deviation of this distribution? What assumptions, if any, are necessary to ensure the validity of your answers?

c. What is the approximate probability that the sample mean will be 1.45 or less?

d. The mean of the sample of 651 employers surveyed in a recent year was 1.36. Given this result, do you think it is likely that all Florida employers' opinions were evenly divided on the effectiveness of public education? That is, do you think the assumption that the population mean is 1.50 is correct? Why or why not?

**6.48** Many species of terrestrial tree frogs that hibernate at or near the ground surface can survive prolonged exposure to low winter temperatures. In freezing conditions, the frog's body temperature, called its *supercooling*

*temperature*, remains relatively higher because of an accumulation of glycerol in its body fluids. Studies have shown that the supercooling temperature of terrestrial frogs frozen at $-6°C$ has a relative frequency distribution with a mean of $-2.18°C$ and a standard deviation of $.32°C$ (*Science*, May 1983). Consider the mean supercooling temperature, $\bar{x}$, of a random sample of $n = 42$ terrestrial frogs frozen at $-6°C$.

a. Find the probability that $\bar{x}$ exceeds $-2.05°C$.

b. Find the probability that $\bar{x}$ falls between $-2.20°C$ and $-2.10°C$.

**6.49** Over the past 20 years, the *Journal of Abnormal Psychology* has published numerous studies on the relationship between juvenile delinquency and poor verbal abilities. Assume that scores on a verbal IQ test have a population mean $\mu = 107$ and a population standard deviation $\sigma = 15$.

a. What shape would you expect the sampling distribution of $\bar{x}$ for $n = 84$ juveniles to have? Does your answer depend on the shape of the distribution of verbal IQ scores for all juveniles?

b. Assuming that the population mean and standard deviation for juveniles with no record of delinquency are the same as those for all juveniles, approximate the probability that the sample mean verbal IQ for $n = 84$ juveniles will be 110 or more. State any assumptions you make.

c. In one cited study, the researchers found that a sample of $n = 84$ juveniles with no record of delinquency had a mean verbal IQ of $\bar{x} = 110$. Considering your answer to part **b**, do you think that the population mean and standard deviation for nondelinquent juveniles are the same as those for all juveniles? Explain.

**6.50** Refer to Exercise 6.44. The mean diameter of the bearings produced by the machine is supposed to be .5 inch. The company decides to use the sample mean (from Exercise 6.44) to decide whether the process is in control; i.e., whether it is producing bearings with a mean diameter of .5 inch. The machine will be considered out of control if the mean of the sample of $n = 25$ diameters is less than .4994 inch or larger than .5006 inch. If the true mean diameter of the bearings produced by the machine is .501 inch, what is the approximate probability that the test will imply that the process is out of control?

## Applying the Concepts—Advanced

**6.51** [*Note:* This exercise refers to an optional section in Chapter 4.] A building contractor has decided to purchase a load of factory-reject aluminum siding as long as the average number of flaws per piece of siding in a sample of size 35 from the factory's reject pile is 2.1 or less. If it is known that the number of flaws per piece of siding in the factory's reject pile has a Poisson probability distribution with a mean of 2.5, find the approximate probability that the contractor will not purchase a load of siding. [*Hint:* If $x$ is a Poisson random variable with mean $\lambda$, then $\sigma_x^2$ also equals $\lambda$.]

**6.52** [*Note:* This exercise refers to an optional section in Chapter 5.] An article in *Industrial Engineering* (Aug. 1990) discussed the importance of modeling machine downtime correctly in simulation studies. As an illustration, the researcher considered a single-machine-tool system with repair times (in minutes) that can be modeled by an exponential distribution with $\theta = 60$ (see optional Section 5.6). Of interest is the mean repair time, $\bar{x}$, of a sample of 100 machine breakdowns.

**a.** Find $E(\bar{x})$ and the variance of $\bar{x}$.

**b.** What probability distribution provides the best model of the sampling distribution of $\bar{x}$? Why?

**c.** Calculate the probability that the mean repair time, $\bar{x}$, is no longer than 30 minutes.

## STUDENT PROJECTS

To understand the Central Limit Theorem and sampling distribution, consider the following experiment: Toss four identical coins, and record the number of heads observed. Then repeat this experiment four more times, so that you end up with a total of five observations for the random variable $x$, the number of heads when four coins are tossed.

Now derive and graph the probability distribution for $x$, assuming the coins are balanced. Note that the mean of this distribution is $\mu - 2$ and the standard deviation is $\sigma = 1$. This probability distribution represents the one from which you are drawing a random sample of five measurements.

Next, calculate the mean $\bar{x}$ of the five measurements—that is, calculate the mean number of heads you observed in five repetitions of the experiment. Although you have repeated the basic experiment five times, you have only one observed value of $\bar{x}$. To derive the probability distribution or sampling distribution of $\bar{x}$ empirically, you have to repeat the entire process (of tossing four coins five times) many times. Do it 100 times.

The approximate sampling distribution of $\bar{x}$ can be derived theoretically by making use of the Central Limit Theorem. We expect at least an approximate normal probability distribution with a mean $\mu = 2$ and a standard deviation

$$\sigma_{\bar{x}} = \frac{\sigma}{\sqrt{n}} = \frac{1}{\sqrt{5}} = .45$$

Count the number of your 100 $\bar{x}$'s that fall in each of the intervals in the accompanying figure. Use the normal probability distribution with $\mu = 2$ and $\sigma_{\bar{x}} = .45$ to calculate the expected number of the 100 $\bar{x}$'s in each of the intervals. How closely does the theory describe your experimental results?

| Interval | Interval | Interval | Interval | Interval | Interval |
|---|---|---|---|---|---|
| 1 | 2 | 3 | 4 | 5 | 6 |

| 1.10 | 1.55 | 2 | 2.45 | 2.90 |
|---|---|---|---|---|
| $\mu - 2\sigma_{\bar{x}}$ | $\mu - \sigma_{\bar{x}}$ | $\mu$ | $\mu + \sigma_{\bar{x}}$ | $\mu + 2\sigma_{\bar{x}}$ |

## REFERENCES

Hogg, R. V., and Craig, A. T. *Introduction to Mathematical Statistics*, 5th ed. Upper Saddle River, N. J.: Prentice Hall, 1995.

Larsen, R. J., and Marx, M. L. *An Introduction to Mathematical Statistics and Its Applications*, 3rd ed. Upper Saddle River, N. J.: Prentice Hall, 2001.

Wackerly, D., Mendenhall, W., and Scheaffer, R. L. *Mathematical Statistics with Applications*, 6th ed. North Scituate, Mass.: Duxbury, 1999.

# CHAPTER 7

# Inferences Based on a Single Sample:

*ESTIMATION WITH CONFIDENCE INTERVALS*

## Contents

## Statistics in Action

Scallops, Sampling, and the Law

### 🞂 Where We've Been

We've learned that populations are characterized by numerical descriptive measures (called *parameters*) and that decisions about their values are based on sample statistics computed from sample data. Since statistics vary in a random manner from sample to sample, inferences based on them will be subject to uncertainty. This property is reflected in the sampling (probability) distribution of a statistic.

### ☞ Where We're Going

In this chapter, we'll put all the preceding material into practice; that is, we'll estimate population means and proportions based on a single sample selected from the population of interest. Most important, we use the sampling distribution of a sample statistic to assess the reliability of an estimate.

The estimation of the mean gas mileage for a new car model, the estimation of the expected life of a computer monitor, and the estimation of the mean yearly sales for companies in the steel industry are problems with a common element. In each case, we're interested in estimating the mean of a population of measurements. This important problem constitutes the primary topic of this chapter.

You'll see that different techniques are used for estimating a mean, depending on whether a sample contains a large or small number of measurements. Nevertheless, our objectives remain the same: We want to use the sample information to estimate the mean and to assess the reliability of the estimate.

First, we consider a method of estimating a population mean using a large sample (Section 7.1) and a small sample (Section 7.2). Then, we consider estimation of population proportions (Section 7.3). Finally, we see how to determine the sample sizes necessary for reliable estimates (Section 7.4).

# 7.1 LARGE-SAMPLE CONFIDENCE INTERVAL FOR A POPULATION MEAN

Suppose a large hospital wants to estimate the average length of time patients remain in the hospital. To accomplish this objective, the hospital administrators plan to randomly sample 100 of all previous patients' records and to use the sample mean, $\bar{x}$, of the lengths of stay to estimate $\mu$, the mean of *all* patients' visits. The sample mean $\bar{x}$ represents a *point estimator* of the population mean $\mu$ (Definition 6.4). How can we assess the accuracy of this large-sample point estimator?

According to the Central Limit Theorem, the sampling distribution of the sample mean is approximately normal for large samples, as shown in Figure 7.1. Let us calculate the interval

$$\bar{x} \pm 2\sigma_{\bar{x}} = \bar{x} \pm \frac{2\sigma}{\sqrt{n}}$$

That is, we form an interval 4 standard deviations wide—from 2 standard deviations below the sample mean to 2 standard deviations above the mean. Prior to drawing the sample, what are the chances that this interval will enclose $\mu$, the population mean?

**FIGURE 7.1**

*Sampling Distribution of $\bar{x}$*

To answer this question, refer to Figure 7.1. If the 100 measurements yield a value of $\bar{x}$ that falls between the two lines on either side of $\mu$—i.e., within 2 standard deviations of $\mu$—then the interval $\bar{x} \pm 2\sigma_{\bar{x}}$ will contain $\mu$; if $\bar{x}$ falls outside these boundaries, the interval $\bar{x} \pm 2\sigma_{\bar{x}}$ will not contain $\mu$. Since the area under the normal curve (the sampling distribution of $\bar{x}$) between these boundaries is about .95 (more precisely, from Table IV in Appendix A the area is .9544), we know that the interval $\bar{x} \pm 2\sigma_{\bar{x}}$ will contain $\mu$ with a probability approximately equal to .95.

For instance, consider the lengths of time spent in the hospital for 100 patients shown in Table 7.1. An SAS printout of summary statistics for the sample of 100 lengths of stay is shown in Figure 7.2.

HOSPLOS
**TABLE 7.1 Lengths of Stay (in Days) for 100 Patients**

| 2 | 3 | 8 | 6 | 4 | 4 | 6 | 4 | 2 | 5 |
|---|---|---|---|---|---|---|---|---|---|
| 8 | 10 | 4 | 4 | 4 | 2 | 1 | 3 | 2 | 10 |
| 1 | 3 | 2 | 3 | 4 | 3 | 5 | 2 | 4 | 1 |
| 2 | 9 | 1 | 7 | 17 | 9 | 9 | 9 | 4 | 4 |
| 1 | 1 | 1 | 3 | 1 | 6 | 3 | 3 | 2 | 5 |
| 1 | 3 | 3 | 14 | 2 | 3 | 9 | 6 | 6 | 3 |
| 5 | 1 | 4 | 6 | 11 | 22 | 1 | 9 | 6 | 5 |
| 2 | 2 | 5 | 4 | 3 | 6 | 1 | 5 | 1 | 6 |
| 17 | 1 | 2 | 4 | 5 | 4 | 4 | 3 | 2 | 3 |
| 3 | 5 | 2 | 3 | 3 | 2 | 10 | 2 | 4 | 2 |

**FIGURE 7.2**

*SAS Summary Statistics for the Lengths of 100 Hospital Stays*

The MEANS Procedure

Analysis Variable : LOS

| N | Mean | Std Dev | Minimum | Maximum |
|---|------|---------|---------|---------|
| 100 | 4.5300000 | 3.6775458 | 1.0000000 | 22.0000000 |

From the printout, we find $\bar{x} = 4.53$ days and $s = 3.68$ days. To achieve our objective, we must construct the interval

$$\bar{x} \pm 2\sigma_{\bar{x}} = 4.53 \pm 2\frac{\sigma}{\sqrt{100}}$$

But now we face a problem. You can see that without knowing the standard deviation $\sigma$ of the original population—that is, the standard deviation of the lengths of stay of *all* patients—we cannot calculate this interval. However, since we have a large sample ($n = 100$ measurements), we can approximate the interval by using the sample standard deviation $s$ to approximate $\sigma$. Thus,

$$\bar{x} \pm 2\frac{\sigma}{\sqrt{100}} \approx \bar{x} \pm 2\frac{s}{\sqrt{100}} = 4.53 \pm 2\left(\frac{3.68}{10}\right) = 4.53 \pm .74$$

That is, we estimate the mean length of stay in the hospital for all patients to fall in the interval 3.79 to 5.27 days.

Can we be sure that $\mu$, the true mean, is in the interval 3.79 to 5.27? We cannot be certain, but we can be reasonably confident that it is. This confidence is derived from the knowledge that if we were to draw repeated random samples of 100 measurements from this population and form the interval $\bar{x} \pm 2\sigma_{\bar{x}}$ each time, approximately 95% of the intervals would contain $\mu$. We have no way of knowing (without looking at all the patients' records) whether our sample interval is one of the 95% that contain $\mu$ or one of the 5% that do not, but the odds certainly favor its containing $\mu$. Consequently, the interval 3.79 to 5.27 provides a reliable estimate of the mean length of patient stay in the hospital.

The formula that tells us how to calculate an interval estimate based on sample data is called an *interval estimator*, or *confidence interval*. The probability, .95, that measures the confidence we can place in the interval estimate is called a *confidence coefficient*. The percentage, 95%, is called the *confidence level* for the interval estimate. It is not usually possible to assess precisely the reliability of point estimators because they are single points rather than intervals. So, because we prefer to use estimators for which a measure of reliability can be calculated, we will generally use interval estimators.

---

**DEFINITION 7.1**

An **interval estimator** (or **confidence interval**) is a formula that tells us how to use sample data to calculate an interval that estimates a population parameter.

**DEFINITION 7.2**

The **confidence coefficient** is the probability that an interval estimator encloses the population parameter—that is, the relative frequency with which the interval estimator encloses the population parameter when the estimator is used repeatedly a very large number of times. The **confidence level** is the confidence coefficient expressed as a percentage.

---

Now we have seen how an interval can be used to estimate a population mean. When we use an interval estimator, we can usually calculate the probability that the estimation *process* will result in an interval that contains the true value of the population mean. That is, the probability that the interval contains the parameter in repeated usage is usually known. Figure 7.3 shows what happens when 10 different samples are drawn from a population, and a confidence interval for $\mu$ is calculated from each. The location of $\mu$ is indicated by the vertical line in the figure. Ten confidence intervals, each based on one of 10 samples, are shown as horizontal line segments. Note that the confidence intervals move from sample to sample—sometimes containing $\mu$ and other times missing $\mu$. *If our confidence level is 95%, in the long run, 95% of our sample confidence intervals will contain $\mu$.*

**FIGURE 7.3**

*Confidence Intervals for $\mu$:*
*10 Samples*

Suppose you wish to choose a confidence coefficient other than .95. Notice in Figure 7.1 that the confidence coefficient .95 is equal to the total area under the sampling distribution, less .05 of the area, which is divided equally between the two tails. Using this idea, we can construct a confidence interval with any desired confidence coefficient by increasing or decreasing the area (call it $\alpha$) assigned to the tails of the sampling distribution (see Figure 7.4). For example, if we place area $\alpha/2$ in each tail and if $z_{\alpha/2}$ is the $z$ value such that the area $\alpha/2$ will lie to its right, then the confidence interval with confidence coefficient $(1 - \alpha)$ is

$$\bar{x} \pm z_{\alpha/2}\sigma_{\bar{x}}$$

**FIGURE 7.4**

*Locating $z_{\alpha/2}$ on the Standard Normal Curve*

**FIGURE 7.5**

*The z Value ($z_{.05}$) Corresponding to an Area Equal to .05 in the Upper Tail of the z-Distribution*

$z_{.05} = 1.645$

To illustrate, for a confidence coefficient of .90 we have $(1 - \alpha) = .90, \alpha = .10$, and $\alpha/2 = 0.5$; $z_{.05}$ is the $z$ value that locates area .05 in the upper tail of the sampling distribution. Recall that Table IV in Appendix A gives the areas between the mean and a specified $z$ value. Since the total area to the right of the mean is .5, we find that $z_{.05}$ will be the $z$ value corresponding to an area of $.5 - .05 = .45$ to the right of the mean (see Figure 7.5). This $z$ value is $z_{.05} = 1.645$.

Confidence coefficients used in practice usually range from .90 to .99. The most commonly used confidence coefficients with corresponding values of $\alpha$ and $z_{\alpha/2}$ are shown in Table 7.2.

**TABLE 7.2    Commonly Used Values of $z_{\alpha/2}$**

| Confidence Level $100(1 - \alpha)$ | $\alpha$ | $\alpha/2$ | $z_{\alpha/2}$ |
|---|---|---|---|
| 90% | .10 | .05 | 1.645 |
| 95% | .05 | .025 | 1.96 |
| 99% | .01 | .005 | 2.575 |

---

### Large-Sample $100(1 - \alpha)\%$ Confidence Interval for $\mu$

$$\bar{x} \pm z_{\alpha/2}\sigma_{\bar{x}} = \bar{x} \pm z_{\alpha/2}\frac{\sigma}{\sqrt{n}}$$

where $z_{\alpha/2}$ is the $z$ value with an area $\alpha/2$ to its right (see Figure 7.4) and $\sigma_{\bar{x}} = \sigma/\sqrt{n}$. The parameter $\sigma$ is the standard deviation of the sampled population and $n$ is the sample size.

*Note:* When $\sigma$ is unknown (as is almost always the case) and $n$ is large (say, $n \geq 30$), the confidence interval is approximately equal to

$$\bar{x} \pm \left(z_{\alpha/2}\right)\left(\frac{s}{\sqrt{n}}\right)$$

where $s$ is the sample standard deviation.

*Assumptions:* None, since the Central Limit Theorem guarantees that the sampling distribution of $\bar{x}$ is approximately normal.

| Variable | N | Mean | StDev | SE Mean | 90.0% CI |
|----------|-----|--------|-------|---------|-------------------|
| NOSEATS | 225 | 11.596 | 4.103 | 0.274 | ( 11.144, 12.047) |

**E X A M P L E   7.1**

AIRNOSHOWS

Unoccupied seats on flights cause airlines to lose revenue. Suppose a large airline wants to estimate its average number of unoccupied seats per flight over the past year. To accomplish this, the records of 225 flights are randomly selected, and the number of unoccupied seats is noted for each of the sampled flights. Descriptive statistics for the data are displayed in the MINITAB printout, Figure 7.6.

Estimate $\mu$, the mean number of unoccupied seats per flight during the past year, using a 90% confidence interval.

### Solution

The general form of the 90% confidence interval for a population mean is

$$\bar{x} \pm z_{\alpha/2}\sigma_{\bar{x}} = \bar{x} \pm z_{.05}\sigma_{\bar{x}} = \bar{x} \pm 1.645\left(\frac{\sigma}{\sqrt{n}}\right)$$

From Figure 7.6, we find (after rounding) $\bar{x} = 11.6$. Since we do not know the value of $\sigma$ (the standard deviation of the number of unoccupied seats per flight for all flights of the year), we use our best approximation—the sample standard deviation, $s = 4.1$, shown on the MINITAB printout. Then the 90% confidence interval is, approximately,

$$11.6 \pm 1.645\left(\frac{4.1}{\sqrt{225}}\right) = 11.6 \pm .45$$

or from 11.15 to 12.05. That is, at the 90% confidence level, we estimate the mean number of unoccupied seats per flight to be between 11.15 and 12.05 during the sampled year. This result is verified (except for rounding) on the right side of the MINITAB printout in Figure 7.6.

We stress that the confidence level for this example, 90%, refers to the procedure used. If we were to apply this procedure repeatedly to different samples, approximately 90% of the intervals would contain $\mu$. Although we do not know for sure whether this particular interval (11.15, 12.05) is one of the 90% that contain $\mu$ or one of the 10% that do not, our knowledge of probability gives us "confidence" that the interval contains $\mu$.

The interpretation of confidence intervals for a population mean is summarized in the box on p. 293.

Sometimes, the estimation procedure yields a confidence interval that is too wide for our purposes. In this case, we will want to reduce the width of the interval to obtain a more precise estimate of $\mu$. One way to accomplish this is to decrease the confidence coefficient, $1 - \alpha$. For example, consider the problem of estimating the mean length of stay, $\mu$, for hospital patients. Recall that for a sample of 100 patients, $\bar{x} = 4.53$ days and $s = 3.68$ days. A 90% confidence interval for $\mu$ is

$$\bar{x} \pm 1.645(\sigma/\sqrt{n}) \approx 4.53 \pm (1.645)(3.68/\sqrt{100}) = 4.53 \pm .61$$

or (3.92, 5.14). You can see that this interval is narrower than the previously calculated 95% confidence interval, (3.79, 5.27). Unfortunately, we also have "less confidence" in the 90% confidence interval. An alternative method used to decrease the width of an interval without sacrificing "confidence" is to increase the sample size $n$. We demonstrate this method in Section 7.4.

## Interpretation of a Confidence Interval for a Population Mean

When we form a $100(1 - \alpha)\%$ confidence interval for $\mu$, we usually express our confidence in the interval with a statement such as, "We can be $100(1 - \alpha)\%$ confident that $\mu$ lies between the lower and upper bounds of the confidence interval," where for a particular application, we substitute the appropriate numerical values for the level of confidence and for the lower and upper bounds. *The statement reflects our confidence in the estimation process rather than in the particular interval that is calculated from the sample data.* We know that repeated application of the same procedure will result in different lower and upper bounds on the interval. Furthermore, we know that $100(1 - \alpha)\%$ of the resulting intervals will contain $\mu$. There is (usually) no way to determine whether any particular interval is one of those that contain $\mu$, or one that does not. However, unlike point estimators, confidence intervals have some measure of reliability, the confidence coefficient, associated with them. For that reason they are generally preferred to point estimators.

# Using the TI-83 Graphing Calculator

## Confidence Interval for a Population Mean
## (known $\sigma$ or $n \geq 30$)

Say we want to develop a **95%** confidence interval for the population mean from a sample size of **35** where we know the sample mean is **100** and the **population** deviation is **12**. For this problem we want **Z-Interval** because we are given sigma. On the calculator we will choose **"7:Zinterval"** from the **"TESTS"** menu.

Step 1    *Access the Statistical Tests Menu*
Press    **STAT** (this screen will appear).
Arrow right to **"TEST"**
Arrow down to **"7:ZInterval"**
Press    **ENTER**

```
EDIT CALC TESTS
1:Edit…
2:SortA(
3:SortD(
4:ClrList
5:SetUpEditor
```

Step 2    *Choose* **"Data"** *or* **"Stats"**

*If you choose "Data" the calculator will read the data from LIST.*

*If you choose "Stats" then you will need to enter the settings yourself.*

```
EDIT CALC TESTS
1:Z-Test…
2:T-Test…
3:2-SampZTest…
4:2-SampTTest…
5:1-PropZTest…
6:2-PropZTest…
7↓ZInterval…
```

Highlight **"Stats"**
Press **ENTER**

```
ZInterval
Inpt:Data Stats
σ:12
x̄:100
n:35
C-Level:.95
Calculate
```

You must set:
**$\sigma$: 12**
**$\bar{x}$: 100**
**$n$: 35**
**C-Level: .95** (don't forget the decimal).

Step 3    *View Display*
Next, highlight **"Calculate"**
and Press **ENTER**

As you can see our confidence interval is **(96.024, 103.98)**.

```
ZInterval
(96.024,103.98)
x̄:100
n:35
```

## EXERCISES 7.1–7.20

### Learning the Mechanics

**7.1** Find $z_{\alpha/2}$ for each of the following:
   **a.** $\alpha = .10$
   **b.** $\alpha = .01$
   **c.** $\alpha = .05$
   **d.** $\alpha = .20$

**7.2** What is the confidence level of each of the following confidence intervals for $\mu$?

   **a.** $\bar{x} \pm 1.96\left(\dfrac{\sigma}{\sqrt{n}}\right)$

   **b.** $\bar{x} \pm 1.645\left(\dfrac{\sigma}{\sqrt{n}}\right)$

   **c.** $\bar{x} \pm 2.575\left(\dfrac{\sigma}{\sqrt{n}}\right)$

   **d.** $\bar{x} \pm 1.282\left(\dfrac{\sigma}{\sqrt{n}}\right)$

   **e.** $\bar{x} \pm .99\left(\dfrac{\sigma}{\sqrt{n}}\right)$

**7.3** A random sample of 64 observations from a population produced the following summary statistics:
$$\Sigma x_i = 700 \qquad \Sigma(x_i - \bar{x})^2 = 4{,}238$$
   **a.** Find a 95% confidence interval for $\mu$.
   **b.** Interpret the confidence interval you found in part **a**.

**7.4** A random sample of 90 observations produced a mean $\bar{x} = 25.9$ and a standard deviation $s = 2.7$.
   **a.** Find a 95% confidence interval for the population mean $\mu$.
   **b.** Find a 90% confidence interval for $\mu$.
   **c.** Find a 99% confidence interval for $\mu$.

**7.5** A random sample of 100 observations from a normally distributed population possesses a mean equal to 83.2 and a standard deviation equal to 6.4.
   **a.** Find a 95% confidence interval for $\mu$.
   **b.** What do you mean when you say that a confidence coefficient is .95?
   **c.** Find a 99% confidence interval for $\mu$.
   **d.** What happens to the width of a confidence interval as the value of the confidence coefficient is increased while the sample size is held fixed?
   **e.** Would your confidence intervals of parts **a** and **c** be valid if the distribution of the original population was not normal? Explain.

**7.6** A random sample of $n$ measurements was selected from a population with unknown mean $\mu$ and standard deviation $\sigma$. Calculate a 95% confidence interval for $\mu$ for each of the following situations:
   **a.** $n = 75$, $\bar{x} = 28$, $s^2 = 12$
   **b.** $n = 200$, $\bar{x} = 102$, $s^2 = 22$
   **c.** $n = 100$, $\bar{x} = 15$, $s = .3$

   **d.** $n = 100$, $\bar{x} = 4.05$, $s = .83$
   **e.** Is the assumption that the underlying population of measurements is normally distributed necessary to ensure the validity of the confidence intervals in parts **a–d**? Explain.

**7.7** Explain the difference between an interval estimator and a point estimator for $\mu$.

**7.8** Explain what is meant by the statement, "We are 95% confident that an interval estimate contains $\mu$."

**7.9** Will a large-sample confidence interval be valid if the population from which the sample is taken is not normally distributed? Explain.

**7.10** The mean and standard deviation of a random sample of $n$ measurements are equal to 33.9 and 3.3, respectively.
   **a.** Find a 95% confidence interval for $\mu$ if $n = 100$.
   **b.** Find a 95% confidence interval for $\mu$ if $n = 400$.
   **c.** Find the widths of the confidence intervals found in parts **a** and **b**. What is the effect on the width of a confidence interval of quadrupling the sample size while holding the confidence coefficient fixed?

### Applying the Concepts—Basic

**7.11** The *Journal of the American Medical Association* (Apr. 21, 1993) reported on the results of a National Health Interview Survey designed to determine the prevalence of smoking among U.S. adults. Current smokers (more than 11,000 adults in the survey) were asked: "On the average, how many cigarettes do you now smoke a day?" The results yielded a mean of 20.0 cigarettes per day with an associated 95% confidence interval of (19.7, 20.3).
   **a.** Interpret the 95% confidence interval.
   **b.** State any assumptions about the target population of current cigarette smokers that must be satisfied for inferences derived from the interval to be valid.
   **c.** A tobacco industry researcher claims that the mean number of cigarettes smoked per day by regular cigarette smokers is less than 15. Comment on this claim.

**7.12** Refer to the *Applied Psycholinguistics* (June 1998) study of language skills of low-income children, Exercise 2.76 (p. 69). Each in a sample of 65 low-income children was administered the Communicative Development Inventory (CDI) exam. The sentence complexity scores had a mean of 7.62 and a standard deviation of 8.91.
   **a.** Construct a 90% confidence interval for the mean sentence complexity score of all low-income children.
   **b.** Interpret the interval, part **a**, in the words of the problem.
   **c.** Suppose we know that the true mean sentence complexity score of middle-income children is 15.55. Is there evidence that the true mean for low-income children differs from 15.55? Explain.

**7.13** Refer to the *Chance* (Winter 2001) and National Education Longitudinal Survey (NELS) study of 265 students who paid a private tutor to help them improve their SAT scores, Exercise 2.81 (p. 70). The changes in both the SAT-Mathematics and SAT-Verbal scores for these students are reproduced in the table.

|                                      | SAT-Math | SAT-Verbal |
| ------------------------------------ | :------: | :--------: |
| Mean change in score                 |    19    |     7      |
| Standard deviation of score changes  |    65    |     49     |

a. Construct and interpret a 95% confidence interval for the population mean change in SAT-Mathematics score for students who pay a private tutor.

b. Repeat part a for the population mean change in SAT-Verbal score.

c. Suppose the true population mean change in score on one of the SAT tests for all students who paid a private tutor is 15. Which of the two tests, SAT-Mathematics or SAT-Verbal, is most likely to have this mean change? Explain.

7.14 Donations to tax-exempt organizations such as the Red Cross, the Salvation Army, the YMCA, and the American Cancer Society not only go to the stated charitable purpose, but are used to cover fundraising expenses and overhead. For a sample of 30 charities, the table lists their charitable commitment, i.e., the percentage of their expenses that go toward the stated charitable purpose.

a. Give a point estimate for the mean charitable commitment of tax-exempt organizations.

b. Construct a 98% confidence interval for the mean charitable commitment.

c. What assumption(s) must hold for the method of estimation used in part b to be appropriate?

d. Why is the confidence interval of part b a better estimator of the mean charitable commitment than the point estimator of part a?

⊙ CHARITY

| Organization                            | Charitable Commitment |
| --------------------------------------- | :-------------------: |
| American Cancer Society                 |          62%          |
| American National Red Cross             |          91           |
| Big Brothers Big Sisters of America     |          77           |
| Boy Scouts of America National Council  |          81           |
| Boys & Girls Clubs of America           |          81           |
| CARE                                    |          91           |
| Covenant House                          |          15           |
| Disabled American Veterans              |          65           |
| Ducks Unlimited                         |          78           |
| Feed The Children                       |          90           |
| Girl Scouts of the USA                  |          83           |
| Goodwill Industries International        |          89           |
| Habitat for Humanity International       |          81           |
| Mayo Foundation                         |          26           |

| Mothers Against Drunk Drivers            | 71 |
| Multiple Sclerosis Association of America | 56 |
| Museum of Modern Art                     | 79 |
| Nature Conservancy                       | 77 |
| Paralyzed Veterans of America            | 50 |
| Planned Parenthood Federation            | 81 |
| Salvation Army                           | 84 |
| Shriners Hospital for Children           | 95 |
| Smithsonian Institution                  | 87 |
| Special Olympics                         | 72 |
| Trust for Public Land                    | 88 |
| United Jewish Appeal/Federation - NY     | 75 |
| United States Olympic Committee          | 78 |
| United Way of New York City              | 85 |
| WGBH Educational Foundation              | 81 |
| YMCA of the USA                          | 80 |

*Source:* "Look Before You Give," *Forbes*, Dec. 27, 1999, pp. 206–216.

## Applying the Concepts—Intermediate

7.15 Animal behaviorists have discovered that the more domestic chickens peck at objects placed in their environment, the healthier the chickens seem to be. White string has been found to be a particularly attractive pecking stimulus. In one experiment, 72 chickens were exposed to a string stimulus. Instead of white string, blue-colored string was used. The number of pecks each chicken took at the blue string over a specified time interval was recorded. Summary statistics for the 72 chickens were: $\bar{x} = 1.13$ pecks, $s = 2.21$ pecks (*Applied Animal Behaviour Science*, Oct. 2000).

a. Estimate the population mean number of pecks made by chickens pecking at blue string using a 99% confidence interval. Interpret the result.

b. Previous research has shown that $\mu = 7.5$ pecks if chickens are exposed to white string. Based on the results, part a, is there evidence that chickens are more apt to peck at white string than blue string? Explain.

7.16 Refer to *The Astronomical Journal* (July 1995) study of the velocity of light emitted from a galaxy in the universe, Exercise 2.78 (p. 69). A sample of 103 galaxies located in the galaxy cluster A2142 had a mean velocity of $\bar{x} = 27,117$ kilometers per second (km/s) and a standard deviation of $s = 1,280$ km/s. Suppose your goal is to make an inference about the population mean light velocity of galaxies in cluster A2142.

a. In part b of Exercise 2.78, you constructed an interval that captured approximately 95% of the galaxy velocities in the cluster. Explain why this interval is inappropriate for this inference.

b. Construct a 95% confidence interval for the mean light velocity emitted from all galaxies in cluster A2142. Interpret the result.

7.17 The relationship between an employee's participation in the performance appraisal process and subsequent subordinate reactions toward the appraisal was investigated

MINITAB Output for Exercise 7.17

| Variable | N | Mean | StDev | SE Mean | 95.0% CI |
|---|---|---|---|---|---|
| CORR | 34 | 0.4224 | 0.1998 | 0.0343 | ( 0.3526, 0.4921) |

in the *Journal of Applied Psychology* (Aug. 1998). In Chapter 11 we will discuss a quantitative measure of the relationship between two variables, called the coefficient of correlation $r$. The researchers obtained $r$ for a sample of 34 studies that examined the relationship between appraisal participation and a subordinate's satisfaction with the appraisal. These correlations are listed in the table. (Values of $r$ near $+1$ reflect a strong positive relationship between the variables.) A MINITAB printout showing a 95% confidence interval for the mean of the data is provided above. Locate the 95% confidence interval on the printout and interpret it in the words of the problem.

⊚ CORR34

| | | | | | | | | | | | |
|---|---|---|---|---|---|---|---|---|---|---|---|
| .50 | .58 | .71 | .46 | .63 | .66 | .31 | .35 | .51 | .06 | .35 | .19 |
| .40 | .63 | .43 | .16 | −.08 | .51 | .59 | .43 | .30 | .69 | .25 | .20 |
| .39 | .20 | .51 | .68 | .74 | .65 | .34 | .45 | .31 | .27 | | |

*Source:* Cawley, B. D., Keeping, L. M., and Levy, P. E. "Participation in the performance appraisal process and employee reactions: A meta-analytic review of field investigations." *Journal of Applied Psychology*, Vol. 83, No. 4, Aug. 1998, pp. 632–633 (Appendix).

**7.18** Psychologists have found that twins, in their early years, tend to have lower IQs and pick up language more slowly than nontwins. The slower intellectual growth of most twins may be caused by benign parental neglect. Suppose it is desired to estimate the mean attention time given to twins per week by their parents. A sample of 50 sets of $2\frac{1}{2}$-year-old twin boys is taken, and at the end of 1 week the attention time given to each pair is recorded. The data (in hours) are listed in the accompanying table. An SAS printout gives summary statistics. Use this

⊚ ATTIMES

| | | | | |
|---|---|---|---|---|
| 20.7 | 16.7 | 22.5 | 12.1 | 2.9 |
| 23.5 | 6.4 | 1.3 | 39.6 | 35.6 |
| 10.9 | 7.1 | 46.0 | 23.4 | 29.4 |
| 44.1 | 13.8 | 24.3 | 9.3 | 3.4 |

SAS Output for Exercise 7.18

| Analysis Variable : ATTIME | | | | |
|---|---|---|---|---|
| Mean | Std Dev | N | Minimum | Maximum |
| 20.8480000 | 13.4138253 | 50 | 0.8000000 | 48.2000000 |

| | | | | |
|---|---|---|---|---|
| 15.7 | 46.6 | 10.6 | 6.7 | 5.4 |
| 14.0 | 20.7 | 48.2 | 7.7 | 22.2 |
| 20.3 | 34.0 | 44.5 | 23.8 | 20.0 |
| 43.1 | 14.3 | 21.9 | 17.5 | 9.6 |
| 36.4 | 0.8 | 1.1 | 19.3 | 14.6 |
| 32.5 | 19.1 | 36.9 | 27.9 | 14.0 |

information to find a 90% confidence interval for the mean attention time given to all twin boys by their parents. Interpret the confidence interval.

### Applying the Concepts—Advanced

**7.19** Research reported in the *Journal of Psychology and Aging* (May 1992) studied the role that the age of workers has in determining their level of job satisfaction. The researcher hypothesized that both younger and older workers would have a higher job satisfaction rating than middle-age workers. Each of a sample of 1,686 adults was given a job satisfaction score based on answers to a series of questions. Higher job satisfaction scores indicate higher levels of job satisfaction. The data, arranged by age group, are summarized below.

| | Age Group | | |
|---|---|---|---|
| | **Younger 18–24** | **Middle-Age 25–44** | **Older 45–64** |
| $\bar{x}$ | 4.17 | 4.04 | 4.31 |
| $s$ | .75 | .81 | .82 |
| $n$ | 241 | 768 | 677 |

**a.** Construct 95% confidence intervals for the mean job satisfaction scores of each age group. Carefully interpret each interval.

**b.** In the construction of three 95% confidence intervals, is it more or less likely that at least one of them will *not* contain the population mean it is intended to estimate than it is for a single confidence interval to miss the population mean? [*Hint:* Assume the three intervals are independent, and calculate the probability that at least one of them will not contain the population mean it estimates. Compare this probability to the probability that a single interval fails to enclose the mean.]

c. Based on these intervals, does it appear that the researcher's hypothesis is supported? [*Caution:* We'll learn how to use sample information to compare population means in Chapter 9, and we'll return to this exercise at that time. Here, simply base your opinion on the individual confidence intervals you constructed in part **a**.]

7.20 According to scientists, the cockroach has had 300 million years to develop a resistance to destruction. In a study conducted by researchers for S. C. Johnson & Son, Inc. (manufacturers of Raid and Off), 5,000 roaches (the expected number in a roach-infested house) were released in the Raid test kitchen. One week later the kitchen was fumigated and 16,298 dead roaches were counted, a gain of 11,298 roaches for the 1-week period. Assume that none of the original roaches died during the 1-week period and that the standard deviation of $x$, the number of roaches produced per roach in a 1-week period, is 1.5. Use the number of roaches produced by the sample of 5,000 roaches to find a 95% confidence interval for the mean number of roaches produced per week for each roach in a typical roach-infested house.

## 7.2 SMALL-SAMPLE CONFIDENCE INTERVAL FOR A POPULATION MEAN

Federal legislation requires pharmaceutical companies to perform extensive tests on new drugs before they can be marketed. Initially, a new drug is tested on animals. If the drug is deemed safe after this first phase of testing, the pharmaceutical company is then permitted to begin human testing on a limited basis. During this second phase, inferences must be made about the safety of the drug based on information in very small samples.

Suppose a pharmaceutical company must estimate the average increase in blood pressure of patients who take a certain new drug. Assume that only six patients (randomly selected from the population of all patients) can be used in the initial phase of human testing. The use of a *small sample* in making an inference about $\mu$ presents two immediate problems when we attempt to use the standard normal $z$ as a test statistic.

### PROBLEM I

The shape of the sampling distribution of the sample mean $\bar{x}$ (and the $z$ statistic) now depends on the shape of the population that is sampled. We can no longer assume that the sampling distribution of $\bar{x}$ is approximately normal, because the Central Limit Theorem ensures normality only for samples that are sufficiently large.

#### Solution to Problem 1

The sampling distribution of $\bar{x}$ (and $z$) is exactly normal even for relatively small samples if the sampled population is normal. It is approximately normal if the sampled population is approximately normal.

### PROBLEM 2

The population standard deviation $\sigma$ is almost always unknown. Although it is still true that $\sigma_{\bar{x}} = \sigma/\sqrt{n}$, the sample standard deviation $s$ may provide a poor approximation for $\sigma$ when the sample size is small.

#### Solution to Problem 2

Instead of using the standard normal statistic

$$z = \frac{\bar{x} - \mu}{\sigma_{\bar{x}}} = \frac{\bar{x} - \mu}{\sigma/\sqrt{n}}$$

which requires knowledge of or a good approximation to $\sigma$, we define and use the statistic

$$t = \frac{\bar{x} - \mu}{s/\sqrt{n}}$$

in which the sample standard deviation, $s$, replaces the population standard deviation, $\sigma$.

The distribution of the *t* **statistic** in repeated sampling was discovered by W. S. Gosset, a chemist in the Guinness brewery in Ireland, who published his discovery in 1908 under the pen name of Student. The main result of Gosset's work is that if we are sampling from a normal distribution, the *t* statistic has a sampling distribution very much like that of the *z* statistic: mound-shaped, symmetric, with mean 0. The primary difference between the sampling distributions of *t* and *z* is that the *t* statistic is more variable than the *z*, which follows intuitively when you realize that *t* contains two random quantities ($\bar{x}$ and $s$), whereas *z* contains only one ($\bar{x}$).

The actual amount of variability in the sampling distribution of *t* depends on the sample size *n*. A convenient way of expressing this dependence is to say that the *t* statistic has $(n - 1)$ **degrees of freedom (df)**. Recall that the quantity $(n - 1)$ is the divisor that appears in the formula for $s^2$. This number plays a key role in the sampling distribution of $s^2$ and appears in discussions of other statistics in later chapters. Particularly, the smaller the number of degrees of freedom associated with the *t* statistic, the more variable will be its sampling distribution.

In Figure 7.7 we show both the sampling distribution of *z* and the sampling distribution of a *t* statistic with 4 df. You can see that the increased variability of the *t* statistic means that the *t* value, $t_\alpha$, that locates an area $\alpha$ in the upper tail of the *t*-distribution is larger than the corresponding value $z_\alpha$. For any given value of $\alpha$, the *t* value $t_\alpha$ increases as the number of degrees of freedom (df) decreases. Values of *t* that will be used in forming small-sample confidence intervals of $\mu$ are given in Table VI of Appendix A. A partial reproduction of this table is shown in Table 7.3.

**FIGURE 7.7**

*Standard Normal (z)*
*Distribution and t-Distribution*
*with 4 df*

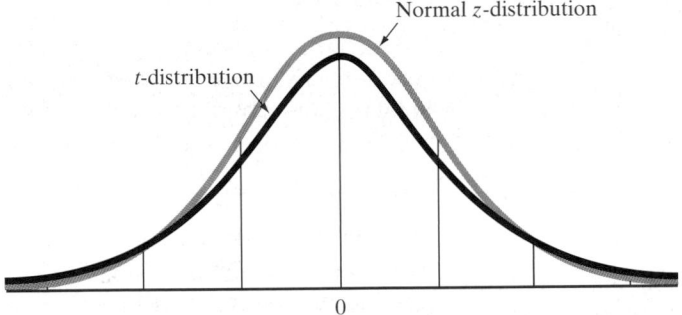

Note that $t_\alpha$ values are listed for various degrees of freedom where $\alpha$ refers to the tail area under the *t*-distribution to the right of $t_\alpha$. For example, if we want the *t* value with an area of .025 to its right and 4 df, we look in the table under the column $t_{.025}$ for the entry in the row corresponding to 4 df. This entry is $t_{.025} = 2.776$, as shown in Figure 7.8. The corresponding standard normal *z*-score is $z_{.025} = 1.96$.

Note that the last row of Table VI, where df = $\infty$ (infinity), contains the standard normal *z* values. This follows from the fact that as the sample size *n* grows very large, *s* becomes closer to $\sigma$ and thus *t* becomes closer in distribution to *z*. In fact, when df = 29, there is little difference between corresponding tabulated values of *z* and *t*. Thus, we choose the arbitrary cutoff of $n = 30$ (df = 29) to distinguish between the large-sample and small-sample inferential techniques.

**TABLE 7.3   Reproduction of Part of Table VI in Appendix A**

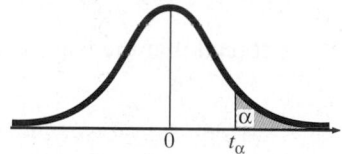

| Degrees of Freedom | $t_{.100}$ | $t_{.050}$ | $t_{.025}$ | $t_{.010}$ | $t_{.005}$ | $t_{.001}$ | $t_{.0005}$ |
|---|---|---|---|---|---|---|---|
| 1 | 3.078 | 6.314 | 12.706 | 31.821 | 63.657 | 318.31 | 636.62 |
| 2 | 1.886 | 2.920 | 4.303 | 6.965 | 9.925 | 22.326 | 31.598 |
| 3 | 1.638 | 2.353 | 3.182 | 4.541 | 5.841 | 10.213 | 12.924 |
| 4 | 1.533 | 2.132 | 2.776 | 3.747 | 4.604 | 7.173 | 8.610 |
| 5 | 1.476 | 2.015 | 2.571 | 3.365 | 4.032 | 5.893 | 6.869 |
| 6 | 1.440 | 1.943 | 2.447 | 3.143 | 3.707 | 5.208 | 5.959 |
| 7 | 1.415 | 1.895 | 2.365 | 2.998 | 3.499 | 4.785 | 5.408 |
| 8 | 1.397 | 1.860 | 2.306 | 2.896 | 3.355 | 4.501 | 5.041 |
| 9 | 1.383 | 1.833 | 2.262 | 2.821 | 3.250 | 4.297 | 4.781 |
| 10 | 1.372 | 1.812 | 2.228 | 2.764 | 3.169 | 4.144 | 4.587 |
| 11 | 1.363 | 1.796 | 2.201 | 2.718 | 3.106 | 4.025 | 4.437 |
| 12 | 1.356 | 1.782 | 2.179 | 2.681 | 3.055 | 3.930 | 4.318 |
| 13 | 1.350 | 1.771 | 2.160 | 2.650 | 3.012 | 3.852 | 4.221 |
| 14 | 1.345 | 1.761 | 2.145 | 2.624 | 2.977 | 3.787 | 4.140 |
| 15 | 1.341 | 1.753 | 2.131 | 2.602 | 2.947 | 3.733 | 4.073 |
| ⋮ | ⋮ | ⋮ | ⋮ | ⋮ | ⋮ | ⋮ | ⋮ |
| ∞ | 1.282 | 1.645 | 1.960 | 2.326 | 2.576 | 3.090 | 3.291 |

**FIGURE 7.8**

*The $t_{.025}$ Value in a t-Distribution with 4 df and the Corresponding $z_{.025}$ Value*

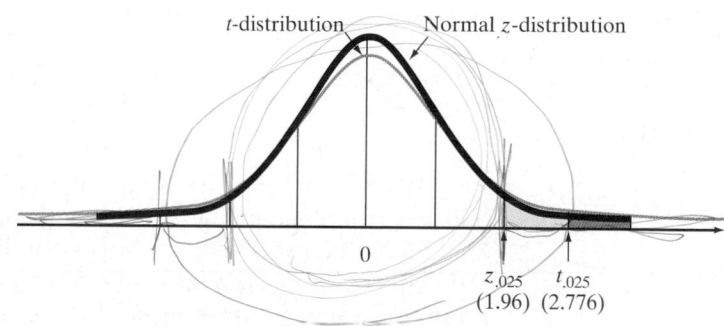

Returning to the example of testing a new drug, suppose that the six test patients have blood pressure increases of 1.7, 3.0, .8, 3.4, 2.7, and 2.1 points. How can we use this information to construct a 95% confidence interval for $\mu$, the mean increase in blood pressure associated with the new drug for all patients in the population?

First, we know that we are dealing with a sample too small to assume that the sample mean $\bar{x}$ is approximately normally distributed by the Central Limit Theorem. That is, we do not get the normal distribution of $\bar{x}$ "automatically" from the Central Limit Theorem when the sample size is small. Instead, we must assume that the measured variable, in this case the increase in blood pressure, is normally distributed in order for the distribution of $\bar{x}$ to be normal.

Second, unless we are fortunate enough to know the population standard deviation $\sigma$, which in this case represents the standard deviation of *all* the patients' increases in blood pressure when they take the new drug, we cannot use the standard normal $z$ statistic to form our confidence interval for $\mu$. Instead, we must use the $t$-distribution, with $(n - 1)$ degrees of freedom.

In this case, $n - 1 = 5$ df, and the $t$ value is found in Table 7.3 to be

$$t_{.025} = 2.571 \text{ with 5 df}$$

Recall that the large-sample confidence interval would have been of the form

$$\bar{x} \pm z_{\alpha/2}\sigma_{\bar{x}} = \bar{x} \pm z_{\alpha/2}\frac{\sigma}{\sqrt{n}} = \bar{x} \pm z_{.025}\frac{\sigma}{\sqrt{n}}$$

where 95% is the desired confidence level. To form the interval for a small sample from *a normal distribution, we simply substitute t for z and s for σ in the preceding formula:*

$$\bar{x} \pm t_{\alpha/2}\frac{s}{\sqrt{n}}$$

A STATISTIX printout showing descriptive statistics for the six blood pressure increases is displayed in Figure 7.9. After rounding, we have $\bar{x} = 2.283$ and $s = .950$.

Substituting these numerical values into the confidence interval formula, we get

$$2.283 \pm (2.571)\left(\frac{.950}{\sqrt{6}}\right) = 2.283 \pm .997$$

or 1.286 to 3.280 points. Note that this interval agrees (except for rounding) with the confidence interval highlighted on the STATISTIX printout in Figure 7.9.

**FIGURE 7.9**

*STATISTIX Analysis of Six Blood Pressure Increases*

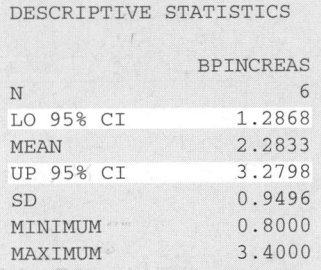

```
DESCRIPTIVE STATISTICS

                    BPINCREAS
N                           6
LO 95% CI              1.2868
MEAN                   2.2833
UP 95% CI              3.2798
SD                     0.9496
MINIMUM                0.8000
MAXIMUM                3.4000
```

We interpret the interval as follows: We can be 95% confident that the mean increase in blood pressure associated with taking this new drug is between 1.286 and 3.28 points. As with our large-sample interval estimates, our confidence is in the process, not in this particular interval. We know that if we were to repeatedly use this estimation procedure, 95% of the confidence intervals produced would contain the true mean $\mu$, *assuming that the probability distribution of changes in blood pressure from which our sample was selected is normal.* The latter assumption is necessary for the small-sample interval to be valid.

What price did we pay for having to utilize a small sample to make the inference? First, we had to assume the underlying population is normally distributed, and if the assumption is invalid, our interval might also be invalid.* Second, we had to form the interval using a $t$ value of 2.571 rather than a $z$ value of 1.96, resulting in a wider interval to achieve the same 95% level of confidence. If the interval from 1.286 to 3.28 is too wide to be of much use, we know how to remedy the situation: increase the number of patients sampled in order to decrease the interval width (on average).

The procedure for forming a small-sample confidence interval is summarized in the next box.

---

*By *invalid*, we mean that the probability that the procedure will yield an interval that contains $\mu$ is not equal to $(1 - \alpha)$. Generally, if the underlying population is approximately normal, the confidence coefficient will approximate the probability that the interval contains $\mu$.

> ## Small-Sample Confidence Interval* for $\mu$
>
> $$\bar{x} \pm t_{\alpha/2}\left(\frac{s}{\sqrt{n}}\right)$$
>
> where $t_{\alpha/2}$ is based on $(n - 1)$ degrees of freedom
> *Assumption:*  A random sample is selected from a population with a relative frequency distribution that is approximately normal.

**EXAMPLE 7.2**

Some quality control experiments require *destructive sampling* (i.e., the test to determine whether the item is defective destroys the item) in order to measure some particular characteristic of the product. The cost of destructive sampling often dictates small samples. Suppose a manufacturer of printers for personal computers wishes to estimate the mean number of characters printed before the printhead fails. The printer manufacturer tests $n = 15$ printheads and records the number of characters printed until failure for each. These 15 measurements (in millions of characters) are listed in Table 7.4, followed by an SPSS summary statistics printout in Figure 7.10.

a. Form a 99% confidence interval for the mean number of characters printed before the printhead fails. Interpret the result.

b. What assumption is required for the interval, part **a**, to be valid? Is it reasonably satisfied?

⊙ PRINTHEAD

**TABLE 7.4   Number of Characters (in Millions) for $n = 15$ Printhead Tests**

| | | | | | | | | | | | | | | |
|---|---|---|---|---|---|---|---|---|---|---|---|---|---|---|
| 1.13 | 1.55 | 1.43 | .92 | 1.25 | 1.36 | 1.32 | .85 | 1.07 | 1.48 | 1.20 | 1.33 | 1.18 | 1.22 | 1.29 |

**FIGURE 7.10**

*SPSS Summary Statistics Printout for Data in Table 7.4*

| Descriptives | | | Statistic | Std. Error |
|---|---|---|---|---|
| NUMBER | Mean | | 1.2387 | 4.987E-02 |
| | 99% Confidence | Lower Bound | 1.0902 | |
| | Interval for Mean | Upper Bound | 1.3871 | |
| | | | | |
| | 5% Trimmed Mean | | 1.2430 | |
| | Median | | 1.2500 | |
| | Variance | | 3.731E-02 | |
| | Std. Deviation | | .1932 | |
| | Minimum | | .85 | |
| | Maximum | | 1.55 | |
| | Range | | .70 | |
| | Interquartile Range | | .2300 | |
| | Skewness | | -.491 | .580 |
| | Kurtosis | | .064 | 1.121 |

---

*The procedure given in the box assumes that the population standard deviation $\sigma$ is unknown, which is almost always the case. If $\sigma$ is known, we can form the small-sample confidence interval just as we would a large-sample confidence interval using a standard normal $z$ value instead of $t$. However, we must still assume that the underlying population is approximately normal.

**Solution**

a. For this small sample ($n = 15$), we use the $t$ statistic to form the confidence interval. We use a confidence coefficient of .99 and $n - 1 = 14$ degrees of freedom to find $t_{\alpha/2}$ in Table VI:

$$t_{\alpha/2} = t_{.005} = 2.977$$

[*Note:* The small sample forces us to extend the interval almost 3 standard deviations (of $\bar{x}$) on each side of the sample mean in order to form the 99% confidence interval.] From the SPSS printout, Figure 7.10, we find $\bar{x} = 1.24$ and $s = .19$. Substituting these (rounded) values into the confidence interval formula, we obtain:

$$\bar{x} \pm t_{.005}\left(\frac{s}{\sqrt{n}}\right) = 1.24 \pm 2.977\left(\frac{.19}{\sqrt{15}}\right)$$
$$= 1.24 \pm .15 \quad \text{or} \quad (1.09, 1.39)$$

This interval is highlighted on Figure 7.10. Thus, the manufacturer can be 99% confident that the printhead has a mean life of between 1.09 and 1.39 million characters. If the manufacturer were to advertise that the mean life of its printheads is (at least) 1 million characters, the interval would support such a claim. Our confidence is derived from the fact that 99% of the intervals formed in repeated applications of this procedure will contain $\mu$.

b. Since $n$ is small, we must assume that the number of characters printed before printhead failure is a random variable from a normal distribution. That is, we assume that the population from which the sample of 15 measurements is selected is distributed normally. One way to check this assumption is to graph the distribution of data in Table 7.4. If the sample data is approximately normal, then the population from which the sample is selected is very likely to be normal. A MINITAB stem-and-leaf plot for the sample data is displayed in Figure 7.11. The distribution is mound-shaped and nearly symmetric. Therefore, the assumption of normality appears to be reasonably satisfied. ⎯⎯⎯⎯⎯■

**FIGURE 7.11**

*MINITAB Stem-and-Leaf Display of Data in Table 7.4*

```
Stem-and-leaf of number    N = 15
Leaf Unit = 0.010

    1       8  5
    2       9  2
    3      10  7
    5      11  38
   (4)     12  0259
    6      13  236
    3      14  38
    1      15  5
```

We have emphasized throughout this section that an assumption that the population is approximately normally distributed is necessary for making small-sample inferences about $\mu$ when using the $t$ statistic. Although many phenomena do have approximately normal distributions, it is also true that many random phenomena have distributions that are not normal or even mound-shaped. Empirical evidence acquired over the years has shown that the *t*-distribution is rather insensitive to moderate departures from normality. That is, use of the *t* statistic when sampling from slightly skewed mound-shaped populations generally produces credible results; however, for cases in which the distribution is distinctly nonnormal, we must either take a large sample or use a *nonparametric method* (the topic of Chapter 14).

---

### What Do You Do When the Population Relative Frequency Distribution Departs Greatly from Normality?

*Answer:*   Use the nonparametric statistical methods of Chapter 14.

---

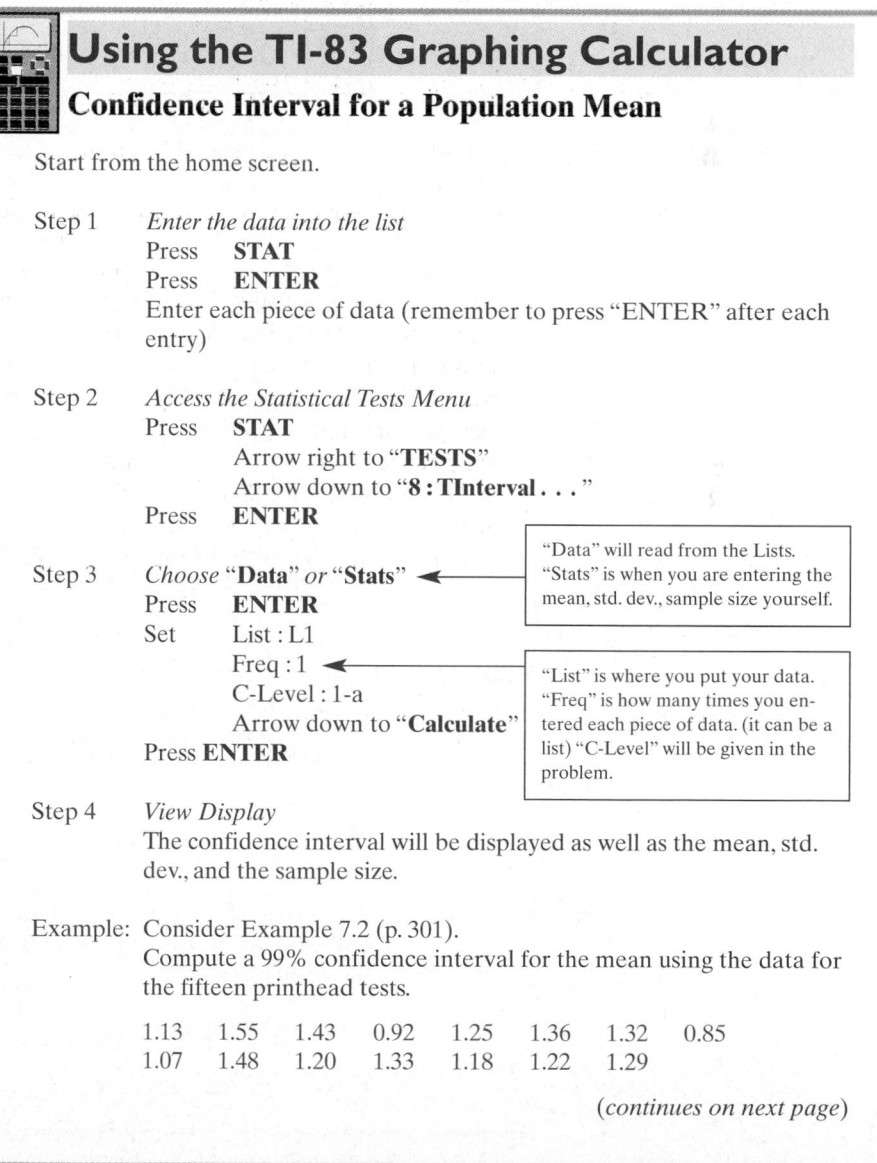

# Using the TI-83 Graphing Calculator
## Confidence Interval for a Population Mean

Start from the home screen.

Step 1   *Enter the data into the list*
　　　　Press   **STAT**
　　　　Press   **ENTER**
　　　　Enter each piece of data (remember to press "ENTER" after each entry)

Step 2   *Access the Statistical Tests Menu*
　　　　Press   **STAT**
　　　　　　　Arrow right to "**TESTS**"
　　　　　　　Arrow down to "**8 : TInterval . . .**"
　　　　Press   **ENTER**

Step 3   *Choose "**Data**" or "**Stats**"*   ◄――   "Data" will read from the Lists. "Stats" is when you are entering the mean, std. dev., sample size yourself.
　　　　Press   **ENTER**
　　　　Set   List : L1
　　　　　　　Freq : 1   ◄――   "List" is where you put your data. "Freq" is how many times you entered each piece of data. (it can be a list) "C-Level" will be given in the problem.
　　　　　　　C-Level : 1-a
　　　　　　　Arrow down to "**Calculate**"
　　　　Press **ENTER**

Step 4   *View Display*
　　　　The confidence interval will be displayed as well as the mean, std. dev., and the sample size.

Example:   Consider Example 7.2 (p. 301).
　　　　Compute a 99% confidence interval for the mean using the data for the fifteen printhead tests.

　　　　1.13   1.55   1.43   0.92   1.25   1.36   1.32   0.85
　　　　1.07   1.48   1.20   1.33   1.18   1.22   1.29

*(continues on next page)*

### USING THE TI-83 GRAPHING CALCULATOR *(continued)*

Step 1   Enter the data into List1
Press   **STAT**
Press   **ENTER**
Enter each piece of data.

Step 2   Press   **STAT**
Arrow right to "**TESTS**"
Arrow down to "**8 : TInterval . . .**"
(you should be looking at this screen).

```
EDIT CALC TESTS
2↑T-Test…
3:2-SampZTest…
4:2-SampTTest…
5:1-PropZTest…
6:2-PropZTest…
7:ZInterval…
8↓TInterval…
```

Press   **ENTER**

Step 3   Highlight "**Data**" then press **ENTER** (you will see this screen).

```
TInterval
 Inpt:Data Stats
 List:L₁
 Freq:1
 C-Level:.99
 Calculate
```

Match the settings on this screen then arrow down, highlight
"**Calculate**," and press **ENTER** (you will see this screen).

```
TInterval
 (1.0902,1.3871)
 x̄=1.238666667
 Sx=.1931641296
 n=15
▮
```

Step 4   As you can see from the screen our 99% confidence interval is
**(1.0902, 1.3871)**. You will also notice it gives the mean, std. dev., and
the sample size.

Step 5   Return to the home screen.

Press   **CLEAR   ENTER**

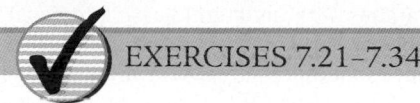

## EXERCISES 7.21–7.34

### Learning the Mechanics

**7.21** Explain the differences in the sampling distributions of $\bar{x}$ for large and small samples under the following assumptions.
   **a.** The variable of interest, $x$, is normally distributed.
   **b.** Nothing is known about the distribution of the variable $x$.

**7.22** Suppose you have selected a random sample of $n = 7$ measurements from a normal distribution. Compare the standard normal $z$ values with the corresponding $t$ values if you were forming the following confidence intervals.
   **a.** 80% confidence interval
   **b.** 90% confidence interval
   **c.** 95% confidence interval
   **d.** 98% confidence interval
   **e.** 99% confidence interval
   **f.** Use the table values you obtained in parts **a–e** to sketch the $z$- and $t$-distributions. What are the similarities and differences?

**7.23** Let $t_0$ be a specific value of $t$. Use Table VI in Appendix A to find $t_0$ values such that the following statements are true.
   **a.** $P(t \geq t_0) = .025$ where df $= 10$
   **b.** $P(t \geq t_0) = .01$ where df $= 17$
   **c.** $P(t \leq t_0) = .005$ where df $= 6$
   **d.** $P(t \leq t_0) = .05$ where df $= 13$

**7.24** Let $t_0$ be a particular value of $t$. Use Table VI of Appendix A to find $t_0$ values such that the following statements are true.
   **a.** $P(-t_0 < t < t_0) = .95$ where df $= 16$
   **b.** $P(t \leq -t_0 \text{ or } t \geq t_0) = .05$ where df $= 16$
   **c.** $P(t \leq t_0) = .05$ where df $= 16$
   **d.** $P(t \leq -t_0 \text{ or } t \geq t_0) = .10$ where df $= 12$
   **e.** $P(t \leq -t_0 \text{ or } t \geq t_0) = .01$ where df $= 8$

**7.25** The following random sample was selected from a normal distribution: $4, 6, 3, 5, 9, 3$.
   **a.** Construct a 90% confidence interval for the population mean $\mu$.
   **b.** Construct a 95% confidence interval for the population mean $\mu$.
   **c.** Construct a 99% confidence interval for the population mean $\mu$.
   **d.** Assume that the sample mean $\bar{x}$ and sample standard deviation $s$ remain exactly the same as those you just calculated but that they are based on a sample of $n = 25$ observations rather than $n = 6$ observations. Repeat parts **a–c**. What is the effect of increasing the sample size on the width of the confidence intervals?

**7.26** The following sample of 16 measurements was selected from a population that is approximately normally distributed:

| 91 | 80 | 99 | 110 | 95 | 106 | 78 | 121 | 106 | 100 |
|----|----|----|-----|----|-----|----|-----|-----|-----|
| 97 | 82 | 100 | 83 | 115 | 104 | | | | |

   **a.** Construct an 80% confidence interval for the population mean.
   **b.** Construct a 95% confidence interval for the population mean and compare the width of this interval with that of part **a**.
   **c.** Carefully interpret each of the confidence intervals and explain why the 80% confidence interval is narrower.

### Applying the Concepts—Basic

**7.27** Periodically, the Hillsborough County (Florida) Water Department tests the drinking water of homeowners for contaminants such as lead and copper. The lead and copper levels in water specimens collected in 1998 for a sample of 10 residents of the Crystal Lakes Manors subdivision are shown below.

🌐 **LEADCOPP**

| Lead ($\mu$g/L) | Copper (mg/L) |
|-----------------|---------------|
| 1.32 | .508 |
| 0 | .279 |
| 13.1 | .320 |
| .919 | .904 |
| .657 | .221 |
| 3.0 | .283 |
| 1.32 | .475 |
| 4.09 | .130 |
| 4.45 | .220 |
| 0 | .743 |

*Source:* Hillsborough County Water Department Environmental Laboratory, Tampa, Florida.

   **a.** Construct a 99% confidence interval for the mean lead level in water specimens from Crystal Lake Manors.
   **b.** Construct a 99% confidence interval for the mean copper level in water specimens from Crystal Lake Manors.
   **c.** Interpret the intervals, parts **a** and **b**, in the words of the problem.
   **d.** Discuss the meaning of the phrase, "99% confident."

**7.28** A group of Harvard University School of Public Health researchers studied the impact of cooking on the size of indoor air particles (*Environmental Science & Technology*, Sept. 1, 2000). The decay rate (measured as $\mu$m/hour) for fine particles produced from oven cooking or toasting was recorded on six randomly selected days. These six measurements are

### ⊙ DECAY

| .95 | .83 | 1.20 | .89 | 1.45 | 1.12 |
|-----|-----|------|-----|------|------|

*Source:* Abt, E., *et al.* "Relative contribution of outdoor and indoor particle sources to indoor concentrations." *Environmental Science & Technology*, Vol. 34, No. 17, Sept. 1, 2000 (Table 3).

**a.** Find and interpret a 95% confidence interval for the true average decay rate of fine particles produced from oven cooking or toasting.

**b.** Explain what the phrase "95% confident" implies in the interpretation of part **a**.

**c.** What must be true about the distribution of the population of decay rates for the inference to be valid?

**7.29** The *Australian Journal of Zoology* (Vol. 43, 1995) reported on a study of the diets and water requirements of spinifex pigeons. Sixteen pigeons were captured in the desert and the crop (i.e., stomach) contents of each examined. The accompanying table reports the weight (in grams) of dry seed in the crop of each pigeon. Use the accompanying SAS printout to find a 99% confidence interval for the average weight of dry seeds in the crops of spinifex pigeons inhabiting the Western Australian desert. Interpret the result.

### ⊙ PIGEONS

| .457 | 3.751 | .238 | 2.967 | 2.509 | 1.384 | 1.454 | .818 |
|------|-------|------|-------|-------|-------|-------|------|
| .335 | 1.436 | 1.603 | 1.309 | .201 | .530 | 2.144 | .834 |

*Source:* Excerpted from Williams, J.B., Bradshaw, D., and Schmidt, L. "Field metabolism and water requirements of spinifex pigeons *(Geophaps plumifera)* in Western Australia." *Australian Journal of Zoology,* Vol. 43, No. 1, 1995, p. 7 (Table 2).

**SAS Output for Exercise 7.29**

Sample Statistics for SEEDWT

| N | Mean | Std. Dev. | Std. Error |
|---|------|-----------|------------|
| 16 | 1.37 | 1.03 | 0.26 |

99 % Confidence Interval for the Mean

Lower Limit:  0.61
Upper Limit:  2.13

### Applying the Concepts—Intermediate

**7.30** Patients with normal hearing in one ear and unaided sensorineural hearing loss in the other are characterized as suffering from unilateral hearing loss. In a study reported in the *American Journal of Audiology* (Mar.

1995), eight patients with unilateral hearing loss were fitted with a special wireless hearing aid in the "bad" ear. The absolute sound pressure level (SPL) was then measured near the eardrum of the ear when noise was produced at a frequency of 500 hertz. The SPLs of the eight patients, recorded in decibels, are listed below.

### ⊙ SOUND

| 73.0 | 80.1 | 82.8 | 76.8 | 73.5 | 74.3 | 76.0 | 68.1 |
|------|------|------|------|------|------|------|------|

Construct and interpret a 90% confidence interval for the true mean SPL of unilateral hearing loss patients when noise is produced at a frequency of 500 hertz.

**7.31** It is customary practice in the United States to base roadway design on the 30th highest hourly volume in a year. Thus, all roadway facilities are expected to operate at acceptable levels of service for all but 29 hours of the year. The Florida Department of Transportation (DOT), however, has shifted from the 30th highest hour to the 100th highest hour as the basis for level-of-service determinators. Florida Atlantic University researcher Reid Ewing investigated whether this shift was warranted in the *Journal of STAR Research* (July 1994). The accompanying table gives the traffic counts at the 30th highest hour and the 100th highest hour of a recent year for 20 randomly selected DOT permanent count stations. MINITAB stem-and-leaf plots for the two variables as well as summary statistics and 95% confidence interval printouts are provided on p. 307.

### ⊙ TRAFFIC

| Station | Type of Route | 30th Highest Hour | 100th Highest Hour |
|---------|---------------|-------------------|--------------------|
| 0117 | small city | 1,890 | 1,736 |
| 0087 | recreational | 2,217 | 2,069 |
| 0166 | small city | 1,444 | 1,345 |
| 0013 | rural | 2,105 | 2,049 |
| 0161 | urban | 4,905 | 4,815 |
| 0096 | urban | 2,022 | 1,958 |
| 0145 | rural | 594 | 548 |
| 0149 | rural | 252 | 229 |
| 0038 | urban | 2,162 | 2,048 |
| 0118 | rural | 1,938 | 1,748 |
| 0047 | rural | 879 | 811 |
| 0066 | urban | 1,913 | 1,772 |
| 0094 | rural | 3,494 | 3,403 |
| 0105 | small city | 1,424 | 1,309 |
| 0113 | small city | 4,571 | 4,425 |
| 0151 | urban | 3,494 | 3,359 |
| 0159 | rural | 2,222 | 2,137 |
| 0160 | small city | 1,076 | 989 |
| 0164 | recreational | 2,167 | 2,039 |
| 0165 | recreational | 3,350 | 3,123 |

*Source:* Ewing, R. "Roadway levels of service in an era of growth management." *Journal of STAR Research.* Vol. 3, July 1994, p. 103 (Table 2).

```
MINITAB Output for Exercise 7.31

Stem-and-leaf of HOUR30     N  = 20
Leaf Unit = 100

    1    0  2
    3    0  58
    6    1  044
    9    1  899
   (6)   2  011122
    5    2
    5    3  344
    2    3
    2    4
    2    4  59

Stem-and-leaf of HOUR100    N  = 20
Leaf Unit = 100

    1    0  2
    4    0  589
    6    1  33
   10    1  7779
   10    2  00001
    5    2
    5    3  134
    2    3
    2    4  4
    1    4  8

Variable          N     Mean    StDev   SE Mean       95.0% CI
HOUR30           20     2206     1224       274  (  1633,    2779)
HOUR100          20     2096     1203       269  (  1533,    2659)
```

a. Describe the population from which the sample data is selected.

b. Does the sample appear to be representative of the population? Explain.

c. Locate and interpret the 95% confidence interval for the mean traffic count at the 30th highest hour.

d. What assumption is necessary for the confidence interval to be valid? Does it appear to be satisfied? Explain.

e. Repeat parts **c** and **d** for the 100th highest hour.

7.32 In Exercise 2.25 (p. 39) you learned that postmortem interval (PMI) is the elapsed time between death and the performance of an autopsy on the cadaver. *Brain and Language* (June 1995) reported on the PMIs of 22 randomly selected human brain specimens obtained at autopsy. The data are reproduced in the table above.

a. Construct a 95% confidence interval for the true mean PMI of human brain specimens obtained at autopsy.

⊘ BRAINPMI

| | | | | | | | |
|---|---|---|---|---|---|---|---|
| 5.5 | 14.5 | 6.0 | 5.5 | 5.3 | 5.8 | 11.0 | 6.4 |
| 7.0 | 14.5 | 10.4 | 4.6 | 4.3 | 7.2 | 10.5 | 6.5 |
| 3.3 | 7.0 | 4.1 | 6.2 | 10.4 | 4.9 | | |

*Source:* Hayes, T. L., and Lewis, D. A. "Anatomical specialization of the anterior motor speech area: Hemispheric differences in magnopyramidal neurons." *Brain and Language.* Vol. 49, No. 3, June 1995, p. 292 (Table 1).

b. Interpret the interval, part **a**.

c. What assumption is required for the interval, part **a**, to be valid? Is this assumption satisfied? Explain.

d. What is meant by the phrase "95% confidence"?

7.33 The "fear of negative evaluation" (FNE) scores for 11 bulimic female students and 14 normal female students, first presented in Exercise 2.27 (p. 39) are reproduced below. (Recall that the higher the score, the greater the fear of negative evaluation.)

⊘ BULIMIA

| | | | | | | | | | | | | | |
|---|---|---|---|---|---|---|---|---|---|---|---|---|---|
| **Bulimic students** | 21 | 13 | 10 | 20 | 25 | 19 | 16 | 21 | 24 | 13 | 14 | | |
| **Normal students** | 13 | 6 | 16 | 13 | 8 | 19 | 23 | 18 | 11 | 19 | 7 | 10 | 15 | 20 |

*Source:* Randles, R. H. "On neutral responses (zeros) in the sign test and ties in the Wilcoxon-Mann-Whitney Test." *The American Statistician,* Vol. 55, No. 2, May 2001 (Figure 3).

a. Construct a 95% confidence interval for the mean FNE score of the population of bulimic female students. Interpret the result.

b. Construct a 95% confidence interval for the mean FNE score of the population of normal female students. Interpret the result.

c. What assumptions are required for the intervals of parts **a** and **b** to be statistically valid? Are these assumptions reasonably satisfied? Explain.

## Applying the Concepts—Advanced

**7.34** Scientists have discovered increased levels of the hormone adrenocorticotropin in people just before they awake from sleeping (*Nature*, Jan. 7, 1999). In the study, 15 subjects were monitored during their sleep after being told they would be woken at a particular time. One hour prior to the designated wake-up time, the adrenocorticotropin level (pg/ml) was measured in each, with the following results:

$$\bar{x} = 37.3 \qquad s = 13.9$$

a. Estimate the true mean adrenocorticotropin level of sleepers one hour prior to waking using a 95% confidence interval.

b. Interpret the interval, part **a**, in the words of the problem.

c. The researchers also found that if the subjects were woken three hours earlier than they anticipated, the average adrenocorticotropin level was 25.5 pg/ml. Assume that $\mu = 25.5$ for all sleepers who are woken three hours earlier than expected. Use the interval, part **a**, to make an inference about the mean adrenocorticotropin level of sleepers under two conditions: one hour before anticipated wake-up time and three hours before anticipated wake-up time.

## 7.3 LARGE-SAMPLE CONFIDENCE INTERVAL FOR A POPULATION PROPORTION

The number of public opinion polls has grown at an astounding rate in recent years. Almost daily, the news media report the results of some poll. Pollsters regularly determine the percentage of people in favor of the president's welfare-reform program, the fraction of voters in favor of a certain candidate, the fraction of customers who favor a particular brand of wine, and the proportion of people who smoke cigarettes. In each case, we are interested in estimating the percentage (or proportion) of some group with a certain characteristic. In this section we consider methods for making inferences about population proportions when the sample is large.

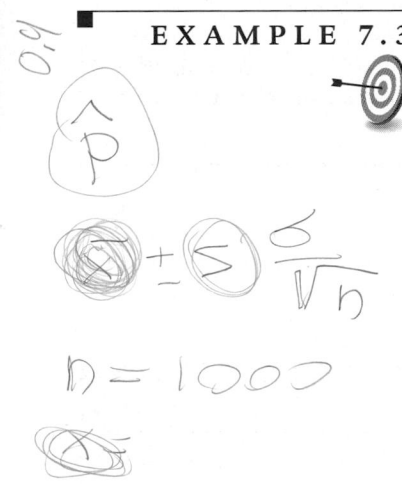

**EXAMPLE 7.3** Public opinion polls are conducted regularly to estimate the fraction of U.S. citizens who trust the president. Suppose 1,000 people are randomly chosen and 637 answer that they trust the president. How would you estimate the true fraction of *all* U.S. citizens who trust the president?

### Solution

What we have really asked is how you would estimate the probability $p$ of success in a binomial experiment, where $p$ is the probability that a person chosen trusts the president. One logical method of estimating $p$ for the population is to use the proportion of successes in the sample. That is, we can estimate $p$ by calculating

$$\hat{p} = \frac{\text{Number of people sampled who trust the president}}{\text{Number of people sampled}}$$

where $\hat{p}$ is read "$p$ hat." Thus, in this case,

$$\hat{p} = \frac{637}{1,000} = .637$$

To determine the reliability of the estimator $\hat{p}$, we need to know its sampling distribution. That is, if we were to draw samples of 1,000 people over and over again, each time calculating a new estimate $\hat{p}$, what would be the frequency distribution of all the $\hat{p}$ values? The answer lies in viewing $\hat{p}$ as the average, or mean, number of successes per trial over the $n$ trials. If each success is assigned a value equal to 1 and a failure is assigned a value of 0, then the sum of all $n$ sample observations is $x$, the total number of successes, and $\hat{p} = x/n$ is the average, or mean, number of successes per trial in the $n$ trials. The Central Limit Theorem tells us that the relative frequency distribution of the sample mean for any population is approximately normal for sufficiently large samples.

The repeated sampling distribution of $\hat{p}$ has the characteristics listed in the next box and shown in Figure 7.12.

**FIGURE 7.12**

*Sampling Distribution of $\hat{p}$*

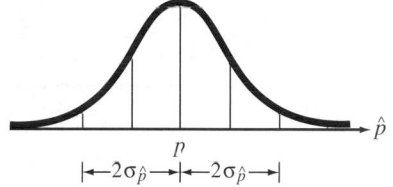

---

### Sampling Distribution of $\hat{p}$

1. The mean of the sampling distribution of $\hat{p}$ is $p$; that is, $\hat{p}$ is an unbiased estimator of $p$.
2. The standard deviation of the sampling distribution of $\hat{p}$ is $\sqrt{pq/n}$; that is, $\sigma_p = \sqrt{pq/n}$, where $q = 1 - p$.
3. For large samples, the sampling distribution of $\hat{p}$ is approximately normal. A sample size is considered large if the interval $\hat{p} \pm 3\sigma_{\hat{p}}$ does not include 0 or 1. [*Note:* This requirement is almost equivalent to that given in Section 5.5 for approximating a binomial distribution with a normal one. The difference is that we assumed $p$ to be known in Section 5.5; now we are trying to make inferences about an unknown $p$, so we use $\hat{p}$ to estimate $p$ in checking the adequacy of the normal approximation.]

---

The fact that $\hat{p}$ is a "sample mean fraction of successes" allows us to form confidence intervals about $p$ in a manner that is completely analogous to that used for large-sample estimation of $\mu$.

---

### Large-Sample Confidence Interval for $p$

$$\hat{p} \pm z_{\alpha/2}\sigma_{\hat{p}} = \hat{p} \pm z_{\alpha/2}\sqrt{\frac{pq}{n}} \approx \hat{p} \pm z_{\alpha/2}\sqrt{\frac{\hat{p}\hat{q}}{n}}$$

where $\hat{p} = \dfrac{x}{n}$ and $\hat{q} = 1 - \hat{p}$

*Note:* When $n$ is large, $\hat{p}$ can approximate the value of $p$ in the formula for $\sigma_{\hat{p}}$.

Thus, if 637 of 1,000 U.S. citizens say they trust the president, a 95% confidence interval for the proportion of *all* U.S. citizens who trust the president is

$$\hat{p} \pm z_{\alpha/2}\sigma_{\hat{p}} = .637 \pm 1.96\sqrt{\frac{pq}{1,000}}$$

where $q = 1 - p$. Just as we needed an approximation for $\sigma$ in calculating a large-sample confidence interval for $\mu$, we now need an approximation for $p$. As Table 7.5 shows, the approximation for $p$ does not have to be especially accurate, because the value of $\sqrt{pq}$ needed for the confidence interval is relatively insensitive to changes in $p$. Therefore, we can use $\hat{p}$ to approximate $p$. Keeping in mind that $\hat{q} = 1 - \hat{p}$, we substitute these values into the formula for the confidence interval:

$$\hat{p} \pm 1.96\sqrt{pq/1,000} \approx \hat{p} \pm 1.96\sqrt{\hat{p}\hat{q}/1,000}$$
$$= .637 \pm 1.96\sqrt{(.637)(.363)/1,000} = .637 \pm .030$$
$$= (.607, .667)$$

Then we can be 95% confident that the interval from 60.7% to 66.7% contains the true percentage of *all* U.S. citizens who trust the president. That is, in repeated construction of confidence intervals, approximately 95% of all samples would produce confidence intervals that enclose $p$. Note that the guidelines for interpreting a confidence interval about $\mu$ also apply to interpreting a confidence interval for $p$ because $p$ is the "population mean fraction of successes" in a binomial experiment.

**TABLE 7.5    Values of *pq* for Several Different Values of *p***

| *p* | *pq* | $\sqrt{pq}$ |
|---|---|---|
| .5 | .25 | .50 |
| .6 or .4 | .24 | .49 |
| .7 or .3 | .21 | .46 |
| .8 or .2 | .16 | .40 |
| .9 or .1 | .09 | .30 |

**EXAMPLE 7.4**

Many public polling agencies conduct surveys to determine the current consumer sentiment concerning the state of the economy. For example, the Bureau of Economic and Business Research (BEBR) at the University of Florida conducts quarterly surveys to gauge consumer sentiment in the Sunshine State. Suppose that BEBR randomly samples 484 consumers and finds that 257 are optimistic about the state of the economy. Use a 90% confidence interval to estimate the proportion of all consumers in Florida who are optimistic about the state of the economy. Based on the confidence interval, can BEBR infer that the majority of Florida consumers are optimistic about the economy?

**Solution**

The number, $x$, of the 484 sampled consumers who are optimistic about the Florida economy is a binomial random variable if we can assume that the sample was randomly selected from the population of Florida consumers and that the poll was conducted identically for each sampled consumer.

The point estimate of the proportion of Florida consumers who are optimistic about the economy is

$$\hat{p} = \frac{x}{n} = \frac{257}{484} = .531$$

We first check to be sure that the sample size is sufficiently large that the normal distribution provides a reasonable approximation for the sampling distribution of $\hat{p}$. We check the 3-standard-deviation interval around $\hat{p}$:

$$\hat{p} \pm 3\sigma_{\hat{p}} \approx \hat{p} \pm 3\sqrt{\frac{\hat{p}\hat{q}}{n}}$$

$$= .531 \pm 3\sqrt{\frac{(.531)(.469)}{484}} = .531 \pm .068 = (.463, .599)$$

Since this interval is wholly contained in the interval $(0, 1)$, we may conclude that the normal approximation is reasonable.

We now proceed to form the 90% confidence interval for $p$, the true proportion of Florida consumers who are optimistic about the state of the economy:

$$\hat{p} \pm z_{\alpha/2}\sigma_{\hat{p}} = \hat{p} \pm z_{\alpha/2}\sqrt{\frac{pq}{n}} \approx \hat{p} \pm z_{\alpha/2}\sqrt{\frac{\hat{p}\hat{q}}{n}}$$

$$= .531 \pm 1.645\sqrt{\frac{(.531)(.469)}{484}} = .531 \pm .037 = (.494, .568)$$

Thus, we can be 90% confident that the proportion of all Florida consumers who are confident about the economy is between .494 and .568. As always, our confidence stems from the fact that 90% of all similarly formed intervals will contain the true proportion $p$ and not from any knowledge about whether this particular interval does.

Can we conclude that the majority of Florida consumers are optimistic about the economy based on this interval? If we wished to use this interval to infer that a majority is optimistic, the interval would have to support the inference that $p$ exceeds .5—that is, that more than 50% of the Florida consumers are optimistic about the economy. Note that the interval contains some values below .5 (as low as .494) as well as some above .5 (as high as .568). Therefore, we cannot conclude that the true value of $p$ exceeds .5 based on this 90% confidence interval. ∎

*Caution:* Unless $n$ is extremely large, the large-sample procedure presented in this section performs poorly when $p$ is near 0 or 1. For example, suppose you want to estimate the proportion of people who die from a bee sting. This proportion is likely to be near 0 (say, $p \approx .001$). Confidence intervals for $p$ based on a sample of size $n = 50$ will probably be misleading.

To overcome this potential problem, an *extremely* large sample size is required. Since the value of $n$ required to satisfy "extremely large" is difficult to determine, statisticians (see Agresti & Coull, 1998) have proposed an alternative method, based on the Wilson (1927) point estimator of $p$. The procedure is outlined in the box. Researchers have shown that this confidence interval works well for any $p$ even when the sample size $n$ is very small.

---

### Adjusted $(1 - \alpha)$ 100% Confidence Interval for a Population Proportion, $p$

$$p^* \pm z_{\alpha/2}\sqrt{\frac{p^*(1 - p^*)}{n + 4}}$$

where $p^* = \dfrac{x + 2}{n + 4}$ is the adjusted sample proportion of observations with the characteristic of interest, $x$ is the number of successes in the sample, and $n$ is the sample size.

**EXAMPLE 7.5**

According to *True Odds: How Risk Affects Your Everyday Life* (Walsh, 1997), the probability of being the victim of a violent crime is less than .01. Suppose that in a random sample of 200 Americans, 3 were victims of a violent crime. Estimate the true proportion of Americans who were victims of a violent crime using a 95% confidence interval.

### Solution

Let $p$ represent the true proportion of Americans who were victims of a violent crime. Since $p$ is near 0, an "extremely large" sample is required to estimate its value using the usual large-sample method. Since we are unsure whether the sample size of 200 is large enough, we will apply Wilson's adjustment outlined in the box.

The number of "successes" (i.e., number of violent crime victims) in the sample is $x = 3$. Therefore, the adjusted sample proportion is

$$p^* = \frac{x + 2}{n + 4} = \frac{3 + 2}{200 + 4} = \frac{5}{204} = .025$$

Note that this adjusted sample proportion is obtained by adding a total of four observations—two "successes" and two "failures"—to the sample data. Substituting $p^* = .025$ into the equation for a 95% confidence interval, we obtain

$$p^* \pm 1.96\sqrt{\frac{p^*(1 - p^*)}{n + 4}} = .025 \pm 1.96\sqrt{\frac{(.025)(.975)}{204}}$$
$$= .025 \pm .021$$

or (.004, .046). Consequently, we are 95% confident that the true proportion of Americans who are victims of a violent crime falls between .004 and .046. ∎

---

 **Using the TI-83 Graphing Calculator**

**Confidence Intervals for Proportions**

You can find confidence intervals for proportions using the **1-PropZInt** command in the **STAT TESTS** Menu.

Step 1    *Select the appropriate type of confidence interval*
Press **STAT** and highlight **TESTS**
Use the arrow keys to highlight **A:1-PropZInt**
Press **ENTER**

Step 2    *Determine your values for x, n, and confidence level*
Enter these values for **x, n,** and **C-Level**
where   **x** = number of successes
**n** = sample size
**C-Level** = level of confidence

Step 3    *Calculate the confidence interval*
Highlight **Calculate**
Press **ENTER**

---

**USING THE TI-83 GRAPHING CALCULATOR (continued)**

Example  Suppose that 1100 U.S. citizens are randomly chosen and 532 answer that they favor a flat income tax rate. Use a 95% confidence interval to estimate the true proportion of citizens who favor a flat income tax rate.

In this example, $x = 532, n = 1100$, and C-Level $= .95$

The TI-83 screens for this example are shown below.

Thus, we can be 95% confident that the interval from 45.4% to 51.3% contains the true percentage of all U.S. citizens who favor a flat income tax rate.

---

 EXERCISES 7.35–7.50

### Learning the Mechanics

**7.35** Describe the sampling distribution of $\hat{p}$ based on large samples of size $n$. That is, give the mean, the standard deviation, and the (approximate) shape of the distribution of $\hat{p}$ when large samples of size $n$ are (repeatedly) selected from the binomial distribution with probability of success $p$.

**7.36** Explain the meaning of the phrase "$\hat{p}$ is an unbiased estimator of $p$."

**7.37** A random sample of size $n = 196$ yielded $\hat{p} = .64$.
  **a.** Is the sample size large enough to use the methods of this section to construct a confidence interval for $p$? Explain.
  **b.** Construct a 95% confidence interval for $p$.
  **c.** Interpret the 95% confidence interval.
  **d.** Explain what is meant by the phrase "95% confidence interval."

**7.38** A random sample of size $n = 144$ yielded $\hat{p} = .76$.
  **a.** Is the sample size large enough to use the methods of this section to construct a confidence interval for $p$? Explain.
  **b.** Construct a 90% confidence interval for $p$.
  **c.** What assumption is necessary to ensure the validity of this confidence interval?

**7.39** For the binomial sample information summarized in each part, indicate whether the sample size is large enough to use the methods of this chapter to construct a confidence interval for $p$.
  **a.** $n = 500, \hat{p} = .05$

  **b.** $n = 100, \hat{p} = .05$
  **c.** $n = 10, \hat{p} = .5$
  **d.** $n = 10, \hat{p} = .3$

**7.40** A random sample of 50 consumers taste-tested a new snack food. Their responses were coded (0: do not like; 1: like; 2: indifferent) and recorded as follows:

🔵 SNACK

| | | | | | | | | | |
|---|---|---|---|---|---|---|---|---|---|
| 1 | 0 | 0 | 1 | 2 | 0 | 1 | 1 | 0 | 0 |
| 0 | 1 | 0 | 2 | 0 | 2 | 2 | 0 | 0 | 1 |
| 1 | 0 | 0 | 0 | 0 | 1 | 0 | 2 | 0 | 0 |
| 0 | 1 | 0 | 0 | 1 | 0 | 0 | 1 | 0 | 1 |
| 0 | 2 | 0 | 0 | 1 | 1 | 0 | 0 | 0 | 1 |

  **a.** Use an 80% confidence interval to estimate the proportion of consumers who like the snack food.
  **b.** Provide a statistical interpretation for the confidence interval you constructed in part **a**.

**7.41** A poll is taken in which 278 of 500 randomly sampled voters indicate their preference for a certain candidate. Use a 95% confidence interval to estimate the true fraction, $p$, of voters who prefer the candidate.

### Applying the Concepts—Basic

**7.42** According to a University of Michigan study, many adults have experienced lingering "fright" effects from a scary movie or TV show they saw as a teenager (*Tampa Tribune*, Mar. 10, 1999). In a survey of 150 college students, 39 said they still experience "residual anxiety" from a scary TV show or movie.

a. Give a point estimate, $\hat{p}$, for the true proportion of college students who experience "residual anxiety" from a scary TV show or movie.

b. Find a 95% confidence interval for $p$.

c. Interpret the interval, part **b**.

**7.43** Refer to the *Chance* (Fall, 2000) study of 837 pottery pieces found at the ancient Greek settlement at Phylakopi, Exercise 2.8 (p. 26). Of these, 183 pieces were painted with either a curvilinear, geometric, or naturalistic decoration. Find a 90% confidence interval for the population proportion of all pottery artifacts at Phylakopi that are painted. Interpret the resulting interval.

**7.44** The *American Journal of Orthopsychiatry* (July 1995) published an article on the prevalence of homelessness in the United States. A sample of 487 U.S. adults were asked to respond to the question: "Was there ever a time in your life when you did not have a place to live?" A 95% confidence interval for the true proportion who were/are homeless, based on the survey data, was determined to be (.045, .091). Fully interpret the result.

## Applying the Concepts—Intermediate

**7.45** Researchers at the University of Florida's Department of Exercise and Sport Sciences conducted a study of variety in exercise workouts (*Journal of Sport Behavior*, 2001). A sample of 114 men and women were randomly divided into three groups, 38 people per group. Group 1 members varied their exercise routine in workouts, group 2 members performed the same exercise at each workout, and group 3 members had no set schedule or regulations for their workouts.

a. By the end of the study, 14 people had dropped out of the first exercise group. Estimate the dropout rate for exercisers who vary their routine in workouts. Use a 90% confidence interval and interpret the result.

b. By the end of the study, 23 people had dropped out of the third exercise group. Estimate the dropout rate for exercisers who have no set schedule for their workouts. Use a 90% confidence interval and interpret the result.

**7.46** The Gallup Organization surveyed 1,252 debit cardholders in the U.S. and found that 180 had used the debit card to purchase a product or service on the Internet (*Card Fax*, November 12, 1999).

a. Describe the population of interest to the Gallup Organization.

b. Is the sample size large enough to construct a valid confidence interval for the proportion of debit cardholders who have used their card in making purchases over the Internet?

c. Estimate the proportion referred to in part **b** using a 99% confidence interval. Interpret your result in the context of the problem.

d. If you had constructed a 90% confidence interval instead, would it be wider or narrower?

**7.47** By law, all new cars must be equipped with both driver-side and passenger-side safety air bags. There is concern, however, over whether air bags pose a danger for children sitting on the passenger side. In a National Highway Traffic Safety Administration (NHTSA) study of 55 people killed by the explosive force of air bags, 35 were children seated on the front-passenger side (*Wall Street Journal*, Jan. 22, 1997). This study led some car owners with children to disconnect the passenger-side air bag. Consider all fatal automobile accidents in which it is determined that air bags were the cause of death. Let $p$ represent the true proportion of these accidents involving children seated on the front-passenger side.

a. Use the data from the NHTSA study to estimate $p$.

b. Construct a 99% confidence interval for $p$.

c. Interpret the interval, part **b**, in the words of the problem.

d. NHTSA investigators determined that 24 of 35 children killed by the air bags were not wearing seat belts or were improperly restrained. How does this information impact your assessment of the risk of an air bag fatality?

## OILSPILL

**7.48** Refer to the *Marine Technology* (Jan. 1995) study of the causes of fifty recent major oil spills from tankers and carriers, Exercise 2.15 (p. 29). The data are stored in the OILSPILL file.

a. Give a point estimate for the proportion of major oil spills that are caused by hull failure.

b. Form a 95% confidence interval for the estimate, part **a**. Interpret the result.

**7.49** In Exercise 2.13 (p. 28) you learned about the signature whistles of bottlenose dolphins. Marine scientists categorize signature whistles by type—type a, type b, type c, etc. For one study of a sample of 185 whistles emitted from bottlenose dolphins in captivity, 97 were categorized as type a whistles (*Ethology*, July 1995).

a. Estimate the true proportion of bottlenose dolphin signature whistles that are type a whistles. Use a 99% confidence interval.

b. Interpret the interval, part **a**.

**7.50** Refer to the Federal Trade Commission (FTC) 1998 "Price Check" study of electronic checkout scanners, Exercise 4.47 (p. 200). The FTC inspected 1,669 scanners at retail stores and supermarkets by scanning a sample of items at each store and determining if the scanned price was accurate. The FTC gives a store a "passing" grade if 98% or more of the items are priced accurately. Of the 1,669 stores in the study, 1,185 passed inspection.

a. Find a 90% confidence interval for the true proportion of retail stores and supermarkets with electronic scanners that pass the FTC price-check test. Interpret the result.

b. In 1996, the FTC found that 45% of the stores passed inspection. Use the interval, part **a**, to determine whether the proportion of stores that pass inspection in 1998 exceeds .45.

## 7.4 DETERMINING THE SAMPLE SIZE

Recall (Section 1.5) that one way to collect the relevant data for a study used to make inferences about the population is to implement a designed (planned) experiment. Perhaps the most important design decision faced by the analyst is to determine the size of the sample. We show in this section that the appropriate sample size for making an inference about a population mean or proportion depends on the desired reliability.

### Estimating a Population Mean

Consider the example from Section 7.1 in which we estimated the mean length of stay for patients in a large hospital. A sample of 100 patients' records produced the 95% confidence interval $\bar{x} \pm 2\sigma_{\bar{x}} \approx 4.53 \pm .74$. Consequently, our estimate $\bar{x}$ was within .74 day of the true mean length of stay $\mu$ for all the hospital's patients at the 95% confidence level. That is, the 95% confidence interval for $\mu$ was $2(.74) = 1.48$ days wide when 100 accounts were sampled. This is illustrated in Figure 7.13a.

Now suppose we want to estimate $\mu$ to within .25 day with 95% confidence. That is, we want to narrow the width of the confidence interval from 1.48 days to .50 day, as shown in Figure 7.13b. How much will the sample size have to be increased to accomplish this? If we want the estimator $\bar{x}$ to be within .25 day of $\mu$, we must have

$$2\sigma_{\bar{x}} = .25 \qquad \text{or, equivalently,} \qquad 2\left(\frac{\sigma}{\sqrt{n}}\right) = .25$$

The necessary sample size is obtained by solving this equation for $n$. To do this we need an approximation for $\sigma$. We have an approximation from the initial sample of 100 patients' records—namely, the sample standard deviation, $s = 3.68$. Thus,

$$2\left(\frac{\sigma}{\sqrt{n}}\right) \approx 2\left(\frac{s}{\sqrt{n}}\right) = 2\left(\frac{3.68}{\sqrt{n}}\right) = .25$$

$$\sqrt{n} = \frac{2(3.68)}{.25} = 29.44$$

$$n = (29.44)^2 = 866.71 \approx 867$$

a. $n = 100$

b. $n = 867$

**FIGURE 7.13**

*Relationship Between Sample Size and Width of Confidence Interval: Hospital Stay Example*

Approximately 867 patients' records will have to be sampled to estimate the mean length of stay $\mu$ to within .25 day with (approximately) 95% confidence. The confidence interval resulting from a sample of this size will be approximately .50 day wide (see Figure 7.13b).

In general, we express the reliability associated with a confidence interval for the population mean $\mu$ by specifying the **bound, B,** within which we want to estimate $\mu$ with $100(1 - \alpha)\%$ confidence. The bound $B$ then is equal to the half-width of the confidence interval, as shown in Figure 7.14.

The procedure for finding the sample size necessary to estimate $\mu$ to within a given bound $B$ is given in the box.

**FIGURE 7.14**

*Specifying the Bound B as the Half-Width of a Confidence Interval*

$\mu - z_{\alpha/2}\sigma_{\bar{x}}$ $\mu$ $\mu + z_{\alpha/2}\sigma_{\bar{x}}$

$B$ $B$

---

### Sample Size Determination for $100(1 - \alpha)\%$ Confidence Intervals for $\mu$

In order to estimate $\mu$ to within a bound $B$ with $100(1 - \alpha)\%$ confidence, the required sample size is found as follows:

$$z_{\alpha/2}\left(\frac{\sigma}{\sqrt{n}}\right) = B$$

The solution can be written in terms of $B$ as follows:

$$n = \frac{(z_{\alpha/2})^2\sigma^2}{B^2}$$

The value of $\sigma$ is usually unknown. It can be estimated by the standard deviation $s$ from a prior sample. Alternatively, we may approximate the range $R$ of observations in the population, and (conservatively) estimate $\sigma \approx R/4$. In any case, you should round the value of $n$ obtained *upward* to ensure that the sample size will be sufficient to achieve the specified reliability.

---

**EXAMPLE 7.6**

Suppose the manufacturer of official NFL footballs uses a machine to inflate the new balls to a pressure of 13.5 pounds. When the machine is properly calibrated, the mean inflation pressure is 13.5 pounds, but uncontrollable factors cause the pressures of individual footballs to vary randomly from about 13.3 to 13.7 pounds. For quality control purposes, the manufacturer wishes to estimate the mean inflation pressure to within .025 pound of its true value with a 99% confidence interval. What sample size should be specified for the experiment?

### Solution

We desire a 99% confidence interval that estimates $\mu$ to within $B = .025$ pound of its true value. For a 99% confidence interval, we have $z_{\alpha/2} = z_{.005} = 2.575$. To estimate $\sigma$, we note that the range of observations is $R = 13.7 - 13.3 = .4$ and use

$\sigma \approx R/4 = .1$. Now we use the formula derived in the box to find the sample size $n$:

$$n = \frac{(z_{\alpha/2})^2 \sigma^2}{B^2} \approx \frac{(2.575)^2(.1)^2}{(.025)^2} = 106.09$$

We round this up to $n = 107$. Realizing that $\sigma$ was approximated by $R/4$, we might even advise that the sample size be specified as $n = 110$ to be more certain of attaining the objective of a 99% confidence interval with bound $B = .025$ pound or less. ∎

Sometimes the formula will lead to a solution that indicates a small sample size is sufficient to achieve the confidence interval goal. Unfortunately, the procedures and assumptions for small samples differ from those for large samples, as we discovered in Section 7.2. Therefore, if the formulas yield a small sample size ($n < 30$), one simple strategy is to select a sample size $n = 30$.

## Estimating a Population Proportion

The method outlined above is easily applied to a population proportion $p$. For example, in Section 7.3 a pollster used a sample of 1,000 U.S. citizens to calculate a 95% confidence interval for the proportion who trust the president, obtaining the interval $.637 \pm .03$. Suppose the pollster wishes to estimate more precisely the proportion who trust the president, say, to within .015 with a 95% confidence interval.

The pollster wants a confidence interval with a bound $B$ on the estimate of $p$ of $B = .015$. The sample size required to generate such an interval is found by solving the following equation for $n$:

$$z_{\alpha/2}\sigma_{\hat{p}} = B \qquad \text{or} \qquad z_{\alpha/2}\sqrt{\frac{pq}{n}} = .015 \qquad \text{(see Figure 7.15)}$$

Since a 95% confidence interval is desired, the appropriate $z$ value is $z_{\alpha/2} = z_{.025} = 1.96$. We must approximate the value of the product $pq$ before we can solve the equation for $n$. As shown in Table 7.5, the closer the values of $p$ and $q$ to .5, the larger the product $pq$. Thus, to find a conservatively large sample size that will generate a confidence interval with the specified reliability, we generally choose an approximation of $p$ close to .5. In the case of the proportion of U.S. citizens who trust the president, however, we have an initial sample estimate of $\hat{p} = .637$. A conservatively large estimate of $pq$ can therefore be obtained by using, say, $p = .60$. We now substitute into the equation and solve for $n$:

$$1.96\sqrt{\frac{(.60)(.40)}{n}} = .015$$

$$n = \frac{(1.96)^2(.60)(.40)}{(.015)^2} = 4,097.7 \approx 4,098$$

**FIGURE 7.15**

*Specifying the Bound B of a Confidence Interval for a Population Proportion p*

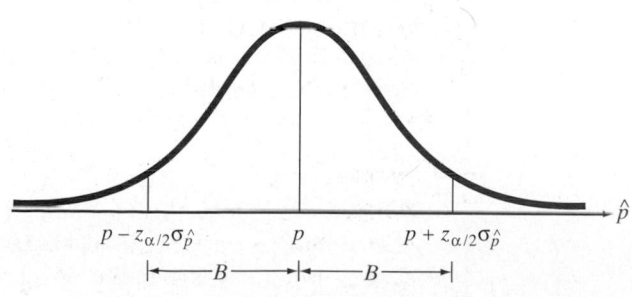

The pollster must sample about 4,098 U.S. citizens to estimate the percentage who trust the president with a confidence interval of width .03.

The procedure for finding the sample size necessary to estimate a population proportion $p$ to within a given bound $B$ is given in the box.

---

### Sample Size Determination for $100(1 - \alpha)\%$ Confidence Interval for $p$

In order to estimate a binomial probability $p$ to within a bound $B$ with $100(1 - \alpha)\%$ confidence, the required sample size is found by solving the following equation for $n$:

$$z_{\alpha/2}\sqrt{\frac{pq}{n}} = B$$

The solution can be written in terms of B:

$$n = \frac{(z_{\alpha/2})^2(pq)}{B^2}$$

Since the value of the product $pq$ is unknown, it can be estimated by using the sample fraction of successes, $\hat{p}$, from a prior sample. Remember (Table 7.5) that the value of $pq$ is at its maximum when $p$ equals .5, so that you can obtain conservatively large values of $n$ by approximating $p$ by .5 or values close to .5. In any case, you should round the value of $n$ obtained *upward* to ensure that the sample size will be sufficient to achieve the specified reliability.

---

**EXAMPLE 7.7**

Suppose a large telephone manufacturer that entered the post-regulation market quickly has an initial problem with excessive customer complaints and consequent returns of the phones for repair or replacement. The manufacturer wants to estimate the magnitude of the problem in order to design a quality control program. How many telephones should be sampled and checked in order to estimate the fraction defective, $p$, to within .01 with 90% confidence?

### Solution

In order to estimate $p$ to within a bound of .01, we set the half-width of the confidence interval equal to $B = .01$, as shown in Figure 7.16.

The equation for the sample size $n$ requires an estimate of the product $pq$. We could most conservatively estimate $pq = .25$ (i.e., use $p = .5$), but this may be overly conservative when estimating a fraction defective. A value of .1, corresponding to 10% defective, will probably be conservatively large for this application. The solution is therefore

**FIGURE 7.16**

*Specified Reliability for Estimate of Fraction Defective in Example 7.7*

$$n = \frac{(z_{\alpha/2})^2(pq)}{B^2} = \frac{(1.645)^2(.1)(.9)}{(.01)^2} = 2,435.4 \approx 2,436$$

Thus, the manufacturer should sample 2,436 telephones in order to estimate the fraction defective, $p$, to within .01 with 90% confidence. Remember that this answer depends on our approximation for $pq$, where we used .09. If the fraction defective is closer to .05 than .10, we can use a sample of 1,286 telephones (check this) to estimate $p$ to within .01 with 90% confidence. ∎

The cost of sampling will also play an important role in the final determination of the sample size to be selected to estimate either $\mu$ or $p$. Although more complex formulas can be derived to balance the reliability and cost considerations, we will solve for the necessary sample size and note that the sampling budget may be a limiting factor. Consult the references for a more complete treatment of this problem.

## STATISTICS IN ACTION
### Scallops, Sampling, and the Law

Arnold Bennett, a Sloan School of Management professor at the Massachusetts Institute of Technology (MIT), describes a recent legal case in which he served as a statistical "expert" in *Interfaces* (Mar.–Apr. 1995). The case involved a ship that fishes for scallops off the coast of New England. In order to protect baby scallops from being harvested, the U.S. Fisheries and Wildlife Service requires that "the average meat per scallop weigh at least $\frac{1}{36}$ of a pound." The ship was accused of violating this weight standard. Bennett lays out the scenario:

The vessel arrived at a Massachusetts port with 11,000 bags of scallops, from which the harbormaster randomly selected 18 bags for weighing. From each such bag, his agents took a large scoopful of scallops; then, to estimate the bag's average meat per scallop, they divided the total weight of meat in the scoopful by the number of scallops it contained. Based on the 18 [numbers] thus generated, the harbormaster estimated that each of the ship's scallops possessed an average $\frac{1}{39}$ of a pound of meat (that is, they were about seven percent lighter than the minimum requirement). Viewing this outcome as conclusive evidence that the weight standard had been violated, federal authorities at once confiscated *95 percent* of the catch (which they then sold at auction). The fishing voyage was thus transformed into a financial catastrophe for its participants.

Bennett provided the actual scallop weight measurements for each of the 18 sampled bags in the article. The data (saved in the SCALLOPS file) are listed in Table 7.6. For ease of exposition, Bennett expressed each number as a multiple of $\frac{1}{36}$ of a pound, the minimum permissible average weight per scallop. Consequently, numbers below one indicate individual bags that do not meet the standard.

The ship's owner filed a lawsuit against the federal government, declaring that his vessel had fully complied with the weight standard. A Boston law firm was hired to represent the owner in legal proceedings and Bennett was retained by the firm to provide statistical litigation support and, if necessary, expert witness testimony.

### Focus

a. Recall that the harbormaster sampled only 18 of the ship's 11,000 bags of scallops. One of the questions the lawyers asked Bennett was: "Can a reliable estimate of the mean weight of all the scallops be obtained from a sample of size 18?" Give your opinion on this issue.

b. As stated in the article, the government's decision rule is to confiscate a scallop catch if the sample mean weight of the scallops is less than $\frac{1}{36}$ of a pound. Do you see any flaws in this rule?

c. Develop your own procedure for determining whether a ship is in violation of the minimum weight restriction. Apply your rule to the data in Table 7.6. Draw a conclusion about the ship in question.

**TABLE 7.6   Scallop Weight Measurements for 18 Bags Sampled**

⊙ SCALLOPS

| | | | | | | | | |
|---|---|---|---|---|---|---|---|---|
| .93 | .88 | .85 | .91 | .91 | .84 | .90 | .98 | .88 |
| .89 | .98 | .87 | .91 | .92 | .99 | 1.14 | 1.06 | .93 |

*Source:* Bennett, A. "Misapplications review: Jail terms." *Interfaces*, Vol. 25, No. 2, Mar.–Apr. 1995, p. 20.

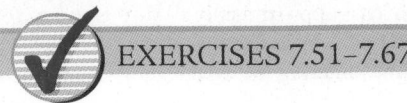

EXERCISES 7.51–7.67

## Learning the Mechanics

**7.51** If you wish to estimate a population mean to within a bound $B = .2$ using a 95% confidence interval and you know from prior sampling that $\sigma^2$ is approximately equal to 5.4, how many observations would you have to include in your sample?

**7.52** If nothing is known about $p$, .5 can be substituted for $p$ in the sample-size formula for a population proportion. But when this is done, the resulting sample size may be larger than needed. Under what circumstances will using $p = .5$ in the sample-size formula yield a sample size larger than needed to construct a confidence interval for $p$ with a specified bound and a specified confidence level?

**7.53** Suppose you wish to estimate a population mean correct to within a bound $B = .15$ with probability equal to .90. You do not know $\sigma^2$, but you know that the observations will range in value between 31 and 39.

    **a.** Find the approximate sample size that will produce the desired accuracy of the estimate. You wish to be conservative to ensure that the sample size will be ample to achieve the desired accuracy of the estimate. [*Hint:* Using your knowledge of data variation from Section 2.5, assume that the range of the observations will equal $4\sigma$.]

    **b.** Calculate the approximate sample size making the less conservative assumption that the range of the observations is equal to $6\sigma$.

**7.54** In each case, find the approximate sample size required to construct a 95% confidence interval for $p$ that has bound $B = .06$.

    **a.** Assume $p$ is near .3

    **b.** Assume you have no prior knowledge about $p$, but you wish to be certain that your sample is large enough to achieve the specified accuracy for the estimate.

**7.55** The following is a 90% confidence interval for $p$: $(.26, .54)$. How large was the sample used to construct this interval?

**7.56** It costs you $10 to draw a sample of size $n = 1$ and measure the attribute of interest. You have a budget of $1,200.

    **a.** Do you have sufficient funds to estimate the population mean for the attribute of interest with a 95% confidence interval 4 units in width? Assume $\sigma = 12$.

    **b.** If a 90% confidence level were used, would your answer to part **a** change? Explain.

**7.57** Suppose you wish to estimate the mean of a normal population using a 95% confidence interval, and you know from prior information that $\sigma^2 \approx 1$.

    **a.** To see the effect of the sample size on the width of the confidence interval, calculate the width of the confidence interval for $n = 16, 25, 49, 100,$ and $400$.

    **b.** Plot the width as a function of sample size $n$ on graph paper. Connect the points by a smooth curve and note how the width decreases as $n$ increases.

## Applying the Concepts—Basic

**7.58** A gigantic warehouse located in Tampa, Florida, stores approximately 60 million empty aluminum beer and soda cans. Recently, a fire occurred at the warehouse. The smoke from the fire contaminated many of the cans with blackspot, rendering them unusable. A University of South Florida statistician was hired by the insurance company to estimate $p$, the true proportion of cans in the warehouse that were contaminated by the fire. How many aluminum cans should be randomly sampled to estimate the true proportion to within .02 with 90% confidence?

**7.59** The EPA wants to test a randomly selected sample of $n$ water specimens and estimate $\mu$, the mean daily rate of pollution produced by a mining operation. If the EPA wants a 95% confidence interval estimate with a bound on the error of 1 milligram per liter (mg/L), how many water specimens are required in the sample? Assume prior knowledge indicates that pollution readings in water samples taken during a day are approximately normally distributed with a standard deviation equal to 5 (mg/L).

## Applying the Concepts—Intermediate

**7.60** Refer to Exercise 2.49 (p. 53) and the *Teaching of Psychology* (May 1998) study in which students assisted in the training of zoo animals. A sample of 15 psychology students rated "The Training Game" as a "great" method of understanding the animal's perspective during training on a 7-1 point scale (where $1 =$ strongly disagree and $7 =$ strongly agree). The mean response was 5.87 with a standard deviation of 1.51.

    **a.** Construct a 95% confidence interval for the true mean response of the students.

    **b.** Suppose you want to reduce the width of the 95% confidence interval to half the size obtained in part **a**. How many students are required in the sample to obtain the desired confidence interval width?

**7.61** Is the bottled water you drink safe? According to an article in *U.S. News & World Report* (April 12, 1999), the Natural Resources Defense Council warns that the bottled water you are drinking may contain more bacteria and other potentially carcinogenic chemicals than allowed by state and federal regulations. Of the more than 1,000 bottles studied, nearly one-third exceeded government levels. Suppose that the Natural Resources Defense Council wants an updated estimate of the population proportion of bottled water that violates at least one government standard. Determine the sample size (number of bottles) needed to estimate this proportion to within ±0.01 with 99% confidence.

**7.62** The chemical benzalkonium chloride (BAC) is an antibacterial agent that is added to some asthma medications to prevent contamination. Researchers at the University of Florida College of Pharmacy have discovered that adding BAC to asthma drugs can cause airway constriction in patients. In a sample of 18 asthmatic patients, each of whom received a heavy dose of BAC, 10 experienced a significant drop in breathing capacity (*Journal of Allergy and Clinical Immunology*, Jan. 2001). Based on this information, a 95% confidence interval for the true percentage of asthmatic patients who experience breathing difficulties after taking BAC is (.326, .785).

**a.** Why might the confidence interval lead to an erroneous inference?

**b.** How many asthma patients must be included in the study to estimate the true percentage who experience a significant drop in breathing capacity to within 4% using a 95% confidence interval?

**7.63** Refer to Exercise 7.28 (p. 305). Suppose that we want to estimate the average decay rate of fine particles produced from oven cooking or toasting to within .04 with 95% confidence. How large a sample should be selected? Use the standard deviation obtained in Exercise 7.28.

**7.64** According to a Food and Drug Administration (FDA) study, a cup of coffee contains an average of 115 milligrams (mg) of caffeine, with the amount per cup ranging from 60 to 180 mg. Suppose you want to repeat the FDA experiment to obtain an estimate of the mean caffeine content in a cup of coffee correct to within 5 mg with 95% confidence. How many cups of coffee would have to be included in your sample?

**7.65** The United States Golf Association (USGA) tests all new brands of golf balls to ensure that they meet USGA specifications. One test conducted is intended to measure the average distance traveled when the ball is hit by a machine called "Iron Byron." Suppose the USGA wishes to estimate the mean distance for a new brand to within 1 yard with 90% confidence. Assume that past tests have indicated that the standard deviation of the distances Iron

Byron hits golf balls is approximately 10 yards. How many golf balls should be hit by Iron Byron to achieve the desired accuracy in estimating the mean?

### Applying the Concepts—Advanced

**7.66** Destructive sampling, in which the test to determine whether an item is defective destroys the item, is generally expensive, and the high costs involved often prohibit large sample sizes. For example, suppose the National Highway Traffic Safety Administration (NHTSA) wishes to determine the proportion of new tires that will fail when subjected to hard braking at a speed of 60 miles per hour. NHTSA can obtain the tires for $25 (wholesale price) each. Suppose the budget for the experiment is $10,000, and NHTSA wishes to estimate the percentage that will fail to within .02 with 95% confidence. Assuming that the entire $10,000 can be spent on tires (ignoring other costs) and that the true fraction that will fail is approximately .05, can NHTSA attain its goal while staying within the budget? Explain.

**7.67** It costs more to produce defective items—since they must be scrapped or reworked—than it does to produce nondefective items. This simple fact suggests that manufacturers should ensure the quality of their products by perfecting their production processes instead of depending on inspection of finished products (Deming, 1986). In order to better understand a particular metal stamping process, a manufacturer wishes to estimate the mean length of items produced by the process during the past 24 hours.

**a.** How many parts should be sampled in order to estimate the population mean to within .1 millimeter (mm) with 90% confidence? Previous studies of this machine have indicated that the standard deviation of lengths produced by the stamping operation is about 2 mm.

**b.** Time permits the use of a sample size no larger than 100. If a 90% confidence interval for $\mu$ is constructed using $n = 100$, will it be wider or narrower than would have been obtained using the sample size determined in part **a**? Explain.

**c.** If management requires that $\mu$ be estimated to within .1 mm and that a sample size of no more than 100 be used, what is (approximately) the maximum confidence level that could be attained for a confidence interval that meets management's specifications?

---

![](QUICK REVIEW)

## Key Terms

Bound on the error of estimation   316
Confidence coefficient   290
Confidence interval   290
Confidence level   290

Degrees of freedom   298
Interval estimator   290
*t* statistic   298
Wilson adjustment for estimating *p*   311

## Key Formulas

$\hat{\theta} \pm (z_{\alpha/2})\sigma_{\hat{\theta}}$    Large-sample confidence interval for population parameter $\theta$ where $\hat{\theta}$ and $\sigma_{\hat{\theta}}$ are obtained from the table below.

| Parameter $\theta$ | Estimator $\hat{\theta}$ | Standard Error $\sigma_{\hat{\theta}}$ | |
|---|---|---|---|
| Mean, $\mu$ | $\bar{x}$ | $\sigma/\sqrt{n}$ | 291 |
| Proportion, $p$ | $\hat{p}$ | $\sqrt{\dfrac{pq}{n}}$ | 309 |

$\bar{x} \pm t_{\alpha/2}\left(\dfrac{s}{\sqrt{n}}\right)$    Small-sample confidence interval for population mean $\mu$   301

$p^* = \dfrac{x+2}{n+4}$    Adjusted estimator of $p$   311

$n = \dfrac{(z_{\alpha/2})^2\sigma^2}{B^2}$    Determining the sample size $n$ for estimating $\mu$   316

$n = \dfrac{(z_{\alpha/2})^2(pq)}{B^2}$    Determining the sample size $n$ for estimating $p$   318

## Language Lab

| Symbol | Pronunciation | Description |
|---|---|---|
| $\theta$ | theta | General population parameter |
| $\mu$ | mu | Population mean |
| $p$ | | Population proportion |
| $B$ | | Bound on error of estimation |
| $\alpha$ | alpha | $(1-\alpha)$ represents the confidence coefficient |
| $z_{\alpha/2}$ | z of alpha over 2 | $z$ value used in a $100(1-\alpha)\%$ large-sample confidence interval |
| $t_{\alpha/2}$ | t of alpha over 2 | $t$ value used in a $100(1-\alpha)\%$ small-sample confidence interval |
| $\bar{x}$ | x-bar | Sample mean; point estimate of $\mu$ |
| $\hat{p}$ | p-hat | Sample proportion; point estimate of $p$ |
| $\sigma$ | sigma | Population standard deviation |
| $s$ | | Sample standard deviation; point estimate of $\sigma$ |
| $\sigma_{\bar{x}}$ | sigma of $\bar{x}$ | Standard deviation of sampling distribution of $\bar{x}$ |
| $\sigma_{\hat{p}}$ | sigma of $\hat{p}$ | Standard deviation of sampling distribution of $\hat{p}$ |
| $p^*$ | | Wilson's adjusted estimate of $p$ |

 SUPPLEMENTARY EXERCISES 7.68–7.90

*Note: List the assumptions necessary for the valid implementation of the statistical procedures you use in solving all these exercises.*

### Learning the Mechanics

**7.68** Let $t_0$ represent a particular value of $t$ from Table VI of Appendix A. Find the table values such that the following statements are true.

   **a.** $P(t \leq t_0) = .05$ where df $= 17$

   **b.** $P(t \geq t_0) = .005$ where df $= 14$

   **c.** $P(t \leq -t_0 \text{ or } t \geq t_0) = .10$ where df $= 6$

   **d.** $P(t \leq -t_0 \text{ or } t \geq t_0) = .01$ where df $= 22$

**7.69** In each of the following instances, determine whether you would use a $z$ or $t$ statistic (or neither) to form a 95% confidence interval; then look up the appropriate $z$ or $t$ value.

   **a.** Random sample of size $n = 21$ from a normal distribution with unknown mean $\mu$ and standard deviation $\sigma$.

   **b.** Random sample of size $n = 175$ from a normal distribution with unknown mean $\mu$ and standard deviation $\sigma$

**c.** Random sample of size $n = 12$ from a normal distribution with unknown mean $\mu$ and standard deviation $\sigma = 5$

**d.** Random sample of size $n = 65$ from a distribution about which nothing is known

**e.** Random sample of size $n = 8$ from a distribution about which nothing is known

**7.70** A random sample of 225 measurements is selected from a population, and the sample mean and standard deviation are $\bar{x} = 32.5$ and $s = 30.0$, respectively.

**a.** Use a 99% confidence interval to estimate the mean of the population, $\mu$.

**b.** How large a sample would be needed to estimate $\mu$ to within .5 with 99% confidence?

**c.** What is meant by the phrase "99% confidence" as it is used in this exercise?

### Applying the Concepts—Basic

**7.71** The Centers for Disease Control and Prevention (CDCP) in Atlanta, Georgia, conducts an annual survey of the general health of the U.S. population as part of its Behavioral Risk Factor Surveillance System (*New York Times*, Mar. 29, 1995). Using random-digit dialing, the CDCP telephones U.S. citizens over 18 years of age and asks them the following four questions:

**(1)** Is your health generally excellent, very good, good, fair, or poor?

**(2)** How many days during the previous 30 days was your physical health not good because of injury or illness?

**(3)** How many days during the previous 30 days was your mental health not good because of stress, depression, or emotional problems?

**(4)** How many days during the previous 30 days did your physical or mental health prevent you from performing your usual activities?

Identify the parameter of interest for each question.

**7.72** Refer to Exercise 7.71. According to the CDCP, 89,582 of 102,263 adults interviewed stated their health was good, very good, or excellent.

**a.** Use a 99% confidence interval to estimate the true proportion of U.S. adults who believe their health to be good to excellent. Interpret the interval.

**b.** Why might the estimate, part **a**, be overly optimistic (i.e., biased high)?

**7.73** Refer to the *Journal of the American Medical Association* (Apr. 21, 1993) report on the prevalence of cigarette smoking among U.S. adults, Exercise 7.11 (p. 294). Of the 43,732 survey respondents, 11,239 indicated that they were current smokers and 10,539 indicated they were former smokers.

**a.** Construct and interpret a 90% confidence interval for the percentage of U.S. adults who currently smoke cigarettes.

**b.** Construct and interpret a 90% confidence interval for the percentage of U.S. adults who are former cigarette smokers.

**7.74** Refer to the *Environmental Science & Technology* (Sept. 1, 2000) study on the ammonia levels near the exit ramp of a San Francisco highway tunnel, Exercise 2.45 (p. 52). The ammonia concentration (parts per million) data for eight randomly selected days during afternoon drive-time are reproduced in the table. Find a 99% confidence interval for the population mean daily ammonia level in air in the tunnel. Interpret your result.

### AMMONIA

| 1.53 | 1.50 | 1.37 | 1.51 | 1.55 | 1.42 | 1.41 | 1.48 |

**7.75** A health researcher wishes to estimate the mean number of cavities per child for children under the age of 12 who live in a specified environment. The number of cavities per child for a random sample of 35 children under the age of 12 has a mean of 2 and a standard deviation of 1.7. Construct a 90% confidence interval for the mean number of cavities per child under the age of 12 who lives in the sampled environment.

### Applying the Concepts—Intermediate

**7.76** Pulse rate is an important measure of the fitness of a person's cardiovascular system. The mean pulse rate for all U.S. adult males is approximately 72 heart beats per minute. A random sample of 21 U.S. adult males who jog at least 15 miles per week had a mean pulse rate of 52.6 beats per minute and a standard deviation of 3.22 beats per minute.

**a.** Find a 95% confidence interval for the mean pulse rate of all U.S. adult males who jog at least 15 miles per week.

**b.** Interpret the interval found in part **a**. Does it appear that jogging at least 15 miles per week reduces the mean pulse rate for adult males?

**c.** What assumptions are required for the validity of the confidence interval and the inference, part **b**?

**7.77** In a study reported in *The Wall Street Journal* (April 4, 1999), the Tupperware Corporation surveyed 1,007 U.S. workers. Of the people surveyed, 665 indicated that they take their lunch to work with them. Of these 665 taking their lunch, 200 reported that they take it in brown bags.

**a.** Find a 95% confidence interval estimate of the population proportion of U.S. workers who take their lunch to work with them. Interpret the interval.

**b.** Consider the population of U.S. workers who take their lunch to work with them. Find a 95% confidence interval estimate of the population proportion who take brown-bag lunches. Interpret the interval.

**7.78** Refer to the *Psychology and Aging* (Dec. 2000) study of the roles that elderly people feel are the most important to them, Exercise 2.6 (p. 26). Recall that in a national sample of 1,102 adults, 65 years or older, 424 identified "spouse" as their most salient role. Use a 95% confidence interval to estimate the true percentage of adults, 65 years or older, who feel being a spouse is their most salient role. Give a practical interpretation of this interval.

**7.79** Tropical swarm-founding wasps rely on female workers to raise their offspring. One possible explanation for this strange behavior is inbreeding, which increases relatedness among the wasps, presumably making it easier for the workers to pick out their closest relatives as propagators of their own genetic material. To test this theory, 197 swarm-founding wasps were captured in Venezuela, frozen at $-70°$ C, and then subjected to a series of genetic tests (*Science*, Nov. 1988). The data were used to generate an inbreeding coefficient, $x$, for each wasp specimen, with the following results: $\bar{x} = .044$ and $s = .884$.

   **a.** Construct a 99% confidence interval for the mean inbreeding coefficient of this species of wasp.

   **b.** A coefficient of 0 implies that the wasp has no tendency to inbreed. Use the confidence interval, part **a**, to make an inference about the tendency for this species of wasp to inbreed.

**7.80** A company is interested in estimating $\mu$, the mean number of days of sick leave taken by all its employees. The firm's statistician selects at random 100 personnel files and notes the number of sick days taken by each employee. The following sample statistics are computed: $\bar{x} = 12.2$ days, $s = 10$ days.

   **a.** Estimate $\mu$ using a 90% confidence interval.

   **b.** How many personnel files would the statistician have to select in order to estimate $\mu$ to within 2 days with a 99% confidence interval?

**7.81** *IEEE Transactions* (June 1990) presented a hybrid algorithm for solving a polynomial zero-one mathematical programming problem. The algorithm incorporates a mixture of pseudo-Boolean concepts and time-proven implicit enumeration procedures. Fifty-two random problems were solved using the hybrid algorithm; the times to solution (CPU time in seconds) are listed in the table. Use the accompanying SAS printout to estimate, with 95% confidence, the mean solution time for the hybrid algorithm. Interpret the result.

**MATHCPU**

| | | | | | | | | |
|---|---|---|---|---|---|---|---|---|
| .045 | 1.055 | .136 | 1.894 | .379 | .136 | .336 | .258 | 1.070 |
| .506 | .088 | .242 | 1.639 | .912 | .412 | .361 | 8.788 | .579 |
| 1.267 | .567 | .182 | .036 | .394 | .209 | .445 | .179 | .118 |
| .333 | .554 | .258 | .182 | .070 | 3.985 | .670 | 3.888 | .136 |
| .091 | .600 | .291 | .327 | .130 | .145 | 4.170 | .227 | .064 |
| .194 | .209 | .258 | 3.046 | .045 | .049 | .079 | | |

*Source:* Snyder, W. S., and Chrissis, J. W. "A hybrid algorithm for solving zero-one mathematical programming problems." *IEEE Transactions,* Vol. 22, No. 2, June 1990, p. 166 (Table 1). The Institute of Industrial Engineers, 25 Technology Park/Atlanta, Norcross, GA 30092. (770) 449-0461.

**7.82** A meteorologist wishes to estimate the mean amount of snowfall per year in Spokane, Washington. A random sample of the recorded snowfalls for 20 years produces a sample mean equal to 54 inches and a standard deviation of 9.59 inches.

**SAS Output for Exercise 7.81**

Sample Statistics for soltime

| N | Mean | Std. Dev. | Std. Error |
|---|---|---|---|
| 52 | 0.81 | 1.50 | 0.21 |

Hypothesis Test

Null hypothesis:     Mean of soltime = 0
Alternative:        Mean of soltime ^= 0

| t Statistic | Df | Prob > t |
|---|---|---|
| 3.892 | 51 | 0.0003 |

95 % Confidence Interval for the Mean

Lower Limit:          0.39
Upper Limit:         1.23

   **a.** Estimate the true mean amount of snowfall in Spokane using a 99% confidence interval.

   **b.** If you were purchasing snow-removal equipment for a city, what numerical descriptive measure of the distribution of depth of snowfall would be of most interest to you? Would it be the mean?

**7.83** A company purchases large quantities of naphtha in 50-gallon drums. Because the purchases are ongoing, small shortages in the drums can represent a sizable loss to the company. The weights of the drums vary slightly from drum to drum, so the weight of the naphtha is determined by removing it from the drums and measuring it. Suppose the company samples the contents of 20 drums, measures the naphtha in each, and calculates $\bar{x} = 49.70$ gallons and $s = .32$ gallon. Find a 95% confidence interval for the mean number of gallons of naphtha per drum. What assumptions are necessary to assure the validity of the confidence interval?

**7.84** Before approval is given for the use of a new insecticide, the United States Department of Agriculture (USDA) requires that several tests be performed to see how the substance will affect wildlife. In particular, the USDA would like to know the proportion of starlings that will die after being exposed to the insecticide. A random sample of 80 starlings were caught and fed a diet of regular food that had been treated with the substance. After 10 days, 10 starlings had died. Use a 99% confidence interval to estimate the true proportion of starlings that will be killed by the substance.

**7.85** In Exercise 2.139 (p. 105) you considered data on suicides in correctional facilities, published in the *American Journal of Psychiatry* (July 1995). The researchers wanted to know what factors increase the risk of suicide by those incarcerated in urban jails. The data on 37 suicides that occurred from 1967 to 1992 in Wayne County Jail, Detroit, Michigan, are reproduced in the table. Answer each of the questions posed below, using a 95% confidence interval for the target parameter implied in the question. Fully interpret each confidence interval.

**a.** What proportion of suicides at the jail are committed by inmates charged with murder/manslaughter?

**b.** What proportion of suicides at the jail are committed at night?

**c.** What is the average length of time an inmate is in jail before committing suicide?

**d.** What percentage of suicides are committed by white inmates?

**7.86** Refer to the *American Psychologist* (July 1995) study of applications for university positions in experimental psychology, Exercise 2.47 (p. 53). One of the objectives of the analysis was to identify problems and errors (e.g., not submitting at least three letters of recommendation) in the application packages. In a sample of 148 applications for the positions, 30 applicants failed to provide *any* letters of recommendation. Estimate, with 90% confidence, the true fraction of applicants for a university position in experimental psychology that fail to provide letters of recommendation. Interpret the result.

**7.87** Recently, a case of salmonella (bacterial) poisoning was traced to a particular brand of ice cream bar, and the

⊘ SUICIDE

| Victim | Days in Jail Before Suicide | Marital Status | Race | Murder/Manslaughter Charge | Time of Suicide |
|---|---|---|---|---|---|
| 1 | 3 | Married | W | Yes | 11 pm–7 am |
| 2 | 4 | Single | W | Yes | 11 pm–7 am |
| 3 | 5 | Single | NW | Yes | 3 pm–11 pm |
| 4 | 7 | Widowed | NW | Yes | 11 pm–7 am |
| 5 | 10 | Single | NW | Yes | 3 pm–11 am |
| 6 | 15 | Married | NW | Yes | 11 pm–7 am |
| 7 | 15 | Divorced | NW | Yes | 11 pm–7 am |
| 8 | 19 | Married | NW | Yes | 7 am–3 pm |
| 9 | 22 | Single | NW | Yes | 11 pm–7 am |
| 10 | 29 | Married | W | Yes | 11 pm–7 am |
| 11 | 31 | Single | NW | Yes | 11 pm–7 am |
| 12 | 85 | Single | NW | Yes | 3 pm–11 pm |
| 13 | 126 | Married | NW | Yes | 11 pm–7 am |
| 14 | 221 | Divorced | W | Yes | 11 pm–7 am |
| 15 | 14 | Single | NW | No | 11 pm–7 am |
| 16 | 22 | Single | W | No | 3 pm–11 pm |
| 17 | 42 | Single | W | No | 7 am–3 pm |
| 18 | 122 | Single | NW | No | 11 pm–7 am |
| 19 | 309 | Married | NW | No | 11 pm–7 am |
| 20 | 69 | Married | NW | No | 11 pm–7 am |
| 21 | 143 | Single | NW | No | 11 pm–7 am |
| 22 | 6 | Married | NW | No | 11 pm–7 am |
| 23 | 1 | Single | NW | No | 7 am–3 pm |
| 24 | 1 | Single | NW | No | 11 pm–7 am |
| 25 | 1 | Single | W | No | 11 pm–7 am |
| 26 | 2 | Single | W | No | 7 am–3 pm |
| 27 | 3 | Single | NW | No | 11 pm–7 am |
| 28 | 4 | Married | W | No | 3 pm–11 pm |
| 29 | 4 | Single | W | No | 3 pm–11 pm |
| 30 | 4 | Single | NW | No | 11 pm–7 am |
| 31 | 6 | Single | W | No | 7 am–3 pm |
| 32 | 6 | Married | NW | No | 11 pm–7 am |
| 33 | 10 | Single | W | No | 11 pm–7 am |
| 34 | 18 | Single | NW | No | 11 pm–7 am |
| 35 | 26 | Married | W | No | 11 pm–7 am |
| 36 | 41 | Single | NW | No | 11 pm–7 am |
| 37 | 86 | Married | W | No | 11 pm–7 am |

*Source:* DuRand, C. J., *et al.* "A quarter century of suicide in a major urban jail: Implications for community psychiatry." *American Journal of Psychiatry,* Vol. 152, No. 7, July 1995, p. 1078 (Table 1).

manufacturer removed the bars from the market. Despite this response, many consumers refused to purchase *any* brand of ice cream bars for some period of time after the event (McClave, personal consulting). One manufacturer conducted a survey of consumers 6 months after the poisoning. A sample of 244 ice cream bar consumers were contacted, and 23 of them indicated that they would not purchase ice cream bars because of the potential for food poisoning.

**a.** What is the point estimate of the true fraction of the entire market who refuse to purchase bars 6 months after the poisoning?

**b.** Is the sample size large enough to use the normal approximation for the sampling distribution of the estimator of the binomial probability? Justify your response.

**c.** Construct a 95% confidence interval for the true proportion of the market who still refuse to purchase ice cream bars 6 months after the event.

**d.** Interpret both the point estimate and confidence interval in terms of this application.

**7.88** Refer to Exercise 7.87. Suppose it is now 1 year after the poisoning was traced to ice cream bars. The manufacturer wishes to estimate the proportion who still will not purchase bars to within .02 using a 95% confidence interval. How many consumers should be sampled?

**7.89** A common hazardous compound found in contaminated soil is benzo(a)pyrene [B(a)p]. An experiment was conducted to determine the effectiveness of a method designed to remove B(a)p from soil (*Journal of Hazardous Materials*, June 1995). Three soil specimens contaminated with a known amount of B(a)p were treated with a toxin that inhibits microbial growth. After 95 days of incubation, the percentage of B(a)p removed from each soil specimen was measured. The experiment produced the following summary statistics: $\bar{x} = 49.3$ and $s = 1.5$.

**a.** Use a 99% confidence interval to estimate the mean percentage of B(a)p removed from a soil specimen in which the toxin was used.

**b.** Interpret the interval in terms of this application.

**c.** What assumption is necessary to ensure the validity of this confidence interval?

**d.** How many soil specimens must be sampled to estimate the mean percentage removed to within .5 using a 99% confidence interval?

**7.90** The accounting firm Price Waterhouse periodically monitors the U.S. Postal Service's performance. One parameter of interest is the percentage of mail delivered on time. In a sample of 332,000 mailed items, Price Waterhouse determined that 282,200 items were delivered on time (*Tampa Tribune*, Mar. 26, 1995). Use this information to estimate with 99% confidence the true percentage of items delivered on time by the U.S. Postal Service. Interpret the result.

---

## STUDENT PROJECTS

Choose a population pertinent to your major area of interest—a population that has an unknown mean or, if the population is binomial, that has an unknown probability of success. For example, a marketing major may be interested in the proportion of consumers who prefer a certain product. A sociology major may be interested in estimating the proportion of people in a certain socioeconomic group or the mean income of people living in a particular part of a city. A political science major may wish to estimate the proportion of an electorate in favor of a certain candidate, a certain amendment, or a certain presidential policy. A pre-med student might want to find the average length of time patients stay in the hospital or the average number of people treated daily in the emergency room. We could continue with examples, but the point should be clear—choose something of interest to you.

Define the parameter you want to estimate and conduct a *pilot study* to obtain an initial estimate of the parameter of interest and, more importantly, an estimate of the variability associated with the estimator. A pilot study is a small experiment (perhaps 20 to 30 observations) used to gain some information about the population of interest. The purpose of the study is to help plan more elaborate future experiments. Using the results of your pilot study, determine the sample size necessary to estimate the parameter to within a reasonable bound (of your choice) with a 95% confidence interval.

---

## REFERENCES

Agresti, A., and Coull, B. A. "Approximate is better than 'exact' for interval estimation of binomial proportions." *The American Statistician*, Vol. 52, No. 2, May 1998, pp. 119–126.

Deming, W. E. *Out of the Crisis*. Cambridge, Mass.: M.I.T. Center for Advanced Study of Engineering, 1986.

Mendenhall, W., Beaver, R. J., and Beaver, B. *Introduction to Probability and Statistics*, 10th ed. North Scituate, Mass.: Duxbury, 1999.

Wilson, E. B. "Probable inference, the law of succession, and statistical inference." *Journal of the American Statistical Association*, Vol. 22, 1927, pp. 209–212.

# CHAPTER 8

# Inferences Based on a Single Sample:

*TESTS OF HYPOTHESIS*

## Contents

## Statistics in Action

March Madness: Handicapping the NCAA Basketball Tourney

---

### 👍 Where We've Been

We saw how to use sample information to estimate population parameters in Chapter 7. The sampling distribution of a statistic is used to assess the reliability of an estimate, which we express in terms of a confidence interval.

### ☞ Where We're Going

We'll see how to utilize sample information to test what the value of a population parameter may be. This type of inference is called a *test of hypothesis*. We'll also see how to conduct a test of hypothesis about a population mean $\mu$ and a population proportion $p$. And, just as with estimation, we'll stress the measurement of the reliability of the inference. An inference without a measure of reliability is little more than a guess.

Suppose you wanted to determine whether the mean level of a driver's blood alcohol exceeds the legal limit after two drinks, or whether the majority of registered voters approve of the president's performance. In both cases you are interested in making an inference about how the value of a parameter relates to a specific numerical value. Is it less than, equal to, or greater than the specified number? This type of inference, called a **test of hypothesis**, is the subject of this chapter.

We introduce the elements of a test of hypothesis in Section 8.1. We then show how to conduct a large-sample test of hypothesis about a population mean in Sections 8.2 and 8.3. In Section 8.4 we utilize small samples to conduct tests about means. Large-sample tests about binomial probabilities are the subject of Section 8.5, and some advanced methods for determining the reliability of a test are covered in optional Section 8.6. Finally, we show how to conduct a test about a population variance in optional Section 8.7.

## 8.1    THE ELEMENTS OF A TEST OF HYPOTHESIS

Suppose building specifications in a certain city require that the average breaking strength of residential sewer pipe be more than 2,400 pounds per foot of length (i.e., per linear foot). Each manufacturer who wants to sell pipe in this city must demonstrate that its product meets the specification. Note that we are again interested in making an inference about the mean $\mu$ of a population. However, in this example we are less interested in estimating the value of $\mu$ than we are in testing a *hypothesis* about its value. That is, *we want to decide whether the mean breaking strength of the pipe exceeds 2,400 pounds per linear foot.*

The method used to reach a decision is based on the rare-event concept explained in earlier chapters. We define two hypotheses: (1) The **null hypothesis** is that which represents the status quo to the party performing the sampling experiment—the hypothesis that will be accepted unless the data provide convincing evidence that it is false. (2) The **alternative**, or **research**, **hypothesis** is that which will be accepted only if the data provide convincing evidence of its truth. From the point of view of the city conducting the tests, the null hypothesis is that the manufacturer's pipe does *not* meet specifications unless the tests provide convincing evidence otherwise. The null and alternative hypotheses are therefore

> **Null hypothesis** ($H_0$): $\mu \leq 2,400$
> (i.e., the manufacturer's pipe does not meet specifications)
>
> **Alternative (research) hypothesis** ($H_a$): $\mu > 2,400$
> (i.e., the manufacturer's pipe meets specifications)

How can the city decide when enough evidence exists to conclude that the manufacturer's pipe meets specifications? Since the hypotheses concern the value of the population mean $\mu$, it is reasonable to use the sample mean $\bar{x}$ to make the inference, just as we did when forming confidence intervals for $\mu$ in Sections 7.1 and 7.2. The city will conclude that the pipe meets specifications only when the sample mean $\bar{x}$ convincingly indicates that the population mean exceeds 2,400 pounds per linear foot.

"Convincing" evidence in favor of the alternative hypothesis will exist when the value of $\bar{x}$ exceeds 2,400 by an amount that cannot be readily attributed to sampling variability. To decide, we compute a **test statistic**, which is the $z$ value that measures the distance between the value of $\bar{x}$ and the value of $\mu$ specified in the null hypothesis. When the null hypothesis contains more than one value of $\mu$, as in this case ($H_0$: $\mu \leq 2,400$), we use the value of $\mu$ closest to the values specified in the alternative hypothesis. The idea is that if the hypothesis that $\mu$ *equals* 2,400 can be rejected in favor of $\mu > 2,400$, then $\mu$ *less than or equal to* 2,400 can certainly be rejected. Thus, the test statistic is

$$z = \frac{\bar{x} - 2,400}{\sigma_{\bar{x}}} = \frac{\bar{x} - 2,400}{\sigma/\sqrt{n}}$$

Note that a value of $z = 1$ means that $\bar{x}$ is 1 standard deviation above $\mu = 2,400$; a value of $z = 1.5$ means that $\bar{x}$ is 1.5 standard deviations above $\mu = 2,400$, etc. How large must $z$ be before the city can be convinced that the null hypothesis can be rejected in favor of the alternative and conclude that the pipe meets specifications?

If you examine Figure 8.1, you will note that the chance of observing $\bar{x}$ more than 1.645 standard deviations above 2,400 is only .05—*if in fact the true mean $\mu$ is 2,400*. Thus, if the sample mean is more than 1.645 standard deviations above 2,400, either $H_0$ is true and a relatively rare event has occurred (.05 probability) or $H_a$ is true and the population mean exceeds 2,400. Since we would most likely reject the notion that a rare event has occurred, we would reject the null hypothesis ($\mu \leq 2,400$) and conclude that the alternative hypothesis ($\mu > 2,400$) is true. What is the probability that this procedure will lead us to an incorrect decision?

**FIGURE 8.1**

*The Sampling Distribution of $\bar{x}$, Assuming $\mu = 2,400$*

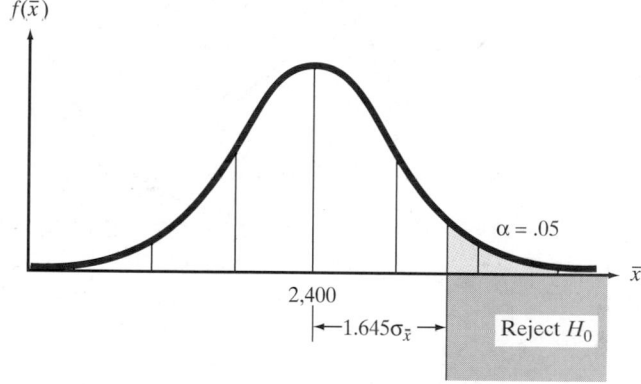

Such an incorrect decision—deciding that the null hypothesis is false when in fact it is true—is called a **Type I decision error**. As indicated in Figure 8.1, the risk of making a Type I error is denoted by the symbol $\alpha$. That is,

$$\alpha = P(\text{Type I error})$$
$$= P(\text{Rejecting the null hypothesis when in fact the null hypothesis is true})$$

In our example

$$\alpha = P(z > 1.645 \text{ when in fact } \mu = 2,400) = .05$$

We now summarize the elements of the test:

$H_0: \mu \leq 2,400$

$H_a: \mu > 2,400$

*Test statistic*: $z = \dfrac{\bar{x} - 2,400}{\sigma_{\bar{x}}}$

*Rejection region*: $z > 1.645$, which corresponds to $\alpha = .05$

Note that the **rejection region** refers to the values of the test statistic for which we will *reject the null hypothesis*.

To illustrate the use of the test, suppose we test 50 sections of sewer pipe and find the mean and standard deviation for these 50 measurements to be:

$$\bar{x} = 2{,}460 \text{ pounds per linear foot} \qquad s = 200 \text{ pounds per linear foot}$$

As in the case of estimation, we can use $s$ to approximate $\sigma$ when $s$ is calculated from a large set of sample measurements.

The test statistic is

$$z = \frac{\bar{x} - 2{,}400}{\sigma_{\bar{x}}} = \frac{\bar{x} - 2{,}400}{\sigma/\sqrt{n}} \approx \frac{\bar{x} - 2{,}400}{s/\sqrt{n}}$$

Substituting $\bar{x} = 2{,}460$, $n = 50$, and $s = 200$, we have

$$z \approx \frac{2{,}460 - 2{,}400}{200/\sqrt{50}} = \frac{60}{28.28} = 2.12$$

Therefore, the sample mean lies $2.12\sigma_{\bar{x}}$ above the hypothesized value of $\mu = 2{,}400$ as shown in Figure 8.2. Since this value of $z$ exceeds 1.645, it falls in the rejection region. That is, we reject the null hypothesis that $\mu = 2{,}400$ and conclude that $\mu > 2{,}400$. Thus, it appears that the company's pipe has a mean strength that exceeds 2,400 pounds per linear foot.

**FIGURE 8.2**

*Location of the Test Statistic for a Test of the Hypothesis $H_0: \mu = 2{,}400$*

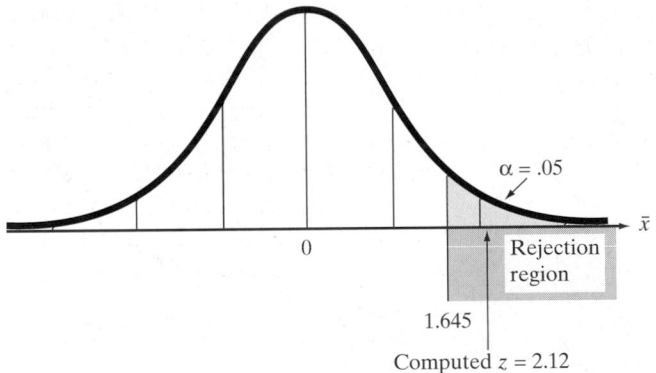

How much faith can be placed in this conclusion? What is the probability that our statistical test could lead us to reject the null hypothesis (and conclude that the company's pipe meets the city's specifications) when in fact the null hypothesis is true? The answer is $\alpha = .05$. That is, we selected the level of risk, $\alpha$, of making a Type I error when we constructed the test. Thus, the chance is only 1 in 20 that our test would lead us to conclude the manufacturer's pipe satisfies the city's specifications when in fact the pipe does *not* meet specifications.

Now, suppose the sample mean breaking strength for the 50 sections of sewer pipe turned out to be $\bar{x} = 2{,}430$ pounds per linear foot. Assuming that the sample standard deviation is still $s = 200$, the test statistic is

$$z = \frac{2{,}430 - 2{,}400}{2/\sqrt{50}} = \frac{30}{28.28} = 1.06$$

Therefore, the sample mean $\bar{x} = 2{,}430$ is only 1.06 standard deviations above the null hypothesized value of $\mu = 2{,}400$. As shown in Figure 8.3, this value does not fall into the rejection region ($z > 1.645$). Therefore, we know that we cannot reject $H_0$ using $\alpha = .05$. Even though the sample mean exceeds the city's specification of 2,400 by 30 pounds per linear foot, it does not exceed the specification by enough to provide *convincing* evidence that the *population mean* exceeds 2,400.

**FIGURE 8.3**

*Location of the Test Statistic when $\bar{x} = 2,430$*

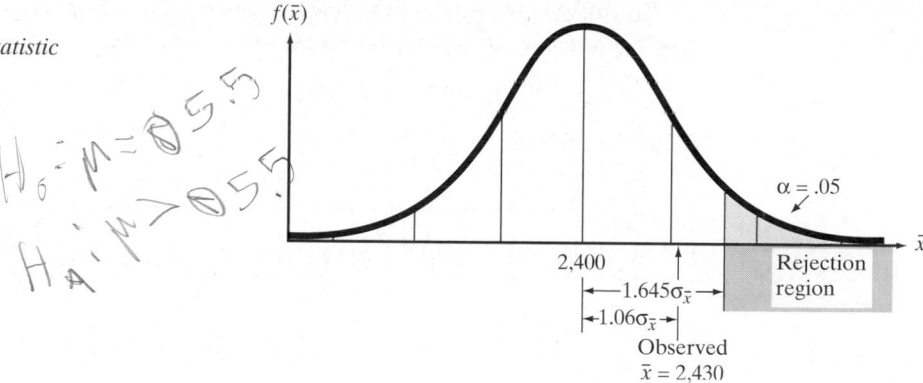

Should we accept the null hypothesis $H_0: \mu \leq 2,400$ and conclude that the manufacturer's pipe does not meet specifications? To do so would be to risk a **Type II error**—that of concluding that the null hypothesis is true (the pipe does not meet specifications) when in fact it is false (the pipe does meet specifications). We denote the probability of committing a Type II error by $\beta$, and we show in optional Section 8.6 that $\beta$ is often difficult to determine precisely. Rather than make a decision (accept $H_0$) for which the probability of error ($\beta$) is unknown, we avoid the potential Type II error by avoiding the conclusion that the null hypothesis is true. Instead, we will simply state that *the sample evidence is insufficient to reject $H_0$ at* $\alpha = .05$. Since the null hypothesis is the "status-quo" hypothesis, the effect of not rejecting $H_0$ is to maintain the status quo. In our pipe-testing example, the effect of having insufficient evidence to reject the null hypothesis that the pipe does not meet specifications is probably to prohibit the utilization of the manufacturer's pipe unless and until there is sufficient evidence that the pipe does meet specifications. That is, until the data indicate convincingly that the null hypothesis is false, we usually maintain the status quo implied by its truth.

**TABLE 8.1  Conclusions and Consequences for a Test of Hypothesis**

| | True State of Nature | |
|---|---|---|
| **Conclusion** | $H_0$ **True** | $H_a$ **True** |
| Accept $H_0$ (Assume $H_0$ True) | Correct decision | Type II error (probability $\beta$) |
| Reject $H_0$ (Assume $H_a$ True) | Type I error (probability $\alpha$) | Correct decision |

Table 8.1 summarizes the four possible outcomes of a test of hypothesis. The "true state of nature" columns in Table 8.1 refer to the fact that either the null hypothesis $H_0$ is true or the alternative hypothesis $H_a$ is true. Note that the true state of nature is unknown to the researcher conducting the test. The "conclusion" rows in Table 8.1 refer to the action of the researcher, assuming that he or she will conclude either that $H_0$ is true or that $H_a$ is true, based on the results of the sampling experiment. Note that a Type I error can be made *only* when the alternative hypothesis is accepted (equivalently, when the null hypothesis is rejected), and a Type II error can be made *only* when the null hypothesis is accepted. Our policy will be to make a decision only when we know the probability of making the error that corresponds to that decision. Since $\alpha$ is usually specified by the analyst (typically .05 in research), we will generally be able to reject $H_0$ (accept $H_a$) when the sample evidence supports that decision. However, since $\beta$ is usually not specified, *we will generally avoid the decision to accept $H_0$, preferring instead to state that the sample evidence is insufficient to reject $H_0$ when the test statistic is not in the rejection region.*

The elements of a test of hypothesis are summarized in the following box. Note that the first four elements are all specified *before* the sampling experiment is performed. In no case will the results of the sample be used to determine the hypotheses—the data are collected to test the predetermined hypotheses, not to formulate them.

---

### Elements of a Test of Hypothesis

1. *Null hypothesis* ($H_0$): A theory about the values of one or more population parameters. The theory generally represents the status quo, which we accept until it is proven false.

2. *Alternative (research) hypothesis* ($H_a$): A theory that contradicts the null hypothesis. The theory generally represents that which we will accept only when sufficient evidence exists to establish its truth.

3. *Test statistic:* A sample statistic used to decide whether to reject the null hypothesis.

4. *Rejection region:* The numerical values of the test statistic for which the null hypothesis will be rejected. The rejection region is chosen so that the probability is $\alpha$ that it will contain the test statistic when the null hypothesis is true, thereby leading to a Type I error. The value of $\alpha$ is usually chosen to be small (e.g., .01, .05, or .10), and is referred to as the **level of significance** of the test.

5. *Assumptions:* Clear statement(s) of any assumptions made about the population(s) being sampled.

6. *Experiment and calculation of test statistic:* Performance of the sampling experiment and determination of the numerical value of the test statistic.

7. *Conclusion:*

   a. If the numerical value of the test statistic falls in the rejection region, we reject the null hypothesis and conclude that the alternative hypothesis is true. We know that the hypothesis-testing process will lead to this conclusion incorrectly (Type I error) only $100\alpha\%$ of the time when $H_0$ is true.

   b. If the test statistic does not fall in the rejection region, we do not reject $H_0$. Thus, we reserve judgment about which hypothesis is true. We do not conclude that the null hypothesis is true because we do not (in general) know the probability $\beta$ that our test procedure will lead to an incorrect acceptance of $H_0$ (Type II error).

---

 EXERCISES 8.1–8.16

### Learning the Mechanics

**8.1** Which hypothesis, the null or the alternative, is the status-quo hypothesis? Which is the research hypothesis?

**8.2** Which element of a test of hypothesis is used to decide whether to reject the null hypothesis in favor of the alternative hypothesis?

**8.3** What is the level of significance of a test of hypothesis?

**8.4** What is the difference between Type I and Type II errors in hypothesis testing? How do $\alpha$ and $\beta$ relate to Type I and Type II errors?

**8.5** List the four possible results of the combinations of decisions and true states of nature for a test of hypothesis.

**8.6** We (generally) reject the null hypothesis when the test statistic falls in the rejection region, but we do not accept the null hypothesis when the test statistic does not fall in the rejection region. Why?

**8.7** If you test a hypothesis and reject the null hypothesis in favor of the alternative hypothesis, does your test prove that the alternative hypothesis is correct? Explain.

## Applying the Concepts—Basic

**8.8** American Express Consulting reported in *USA Today* (June 15, 2001) that 80% of U.S. companies have formal, written travel and entertainment policies for their employees. Give the null hypothesis for testing the claim made by American Express Consulting.

**8.9** According to the *Journal of Advanced Nursing* (Jan. 2001), 45% of senior women (i.e., women over the age of 65) use herbal therapies to prevent or treat health problems. Also, of senior women who use herbal therapies, they use an average of 2.5 herbal products in a year.
  **a.** Give the null hypothesis for testing the first claim by the journal.
  **b.** Give the null hypothesis for testing the second claim by the journal.

**8.10** *Science* (Jan. 1, 1999) reported that the mean listening time of 7-month-old infants exposed to a three-syllable sentence (e.g., "ga ti ti") is 9 seconds. Set up the null and alternative hypotheses for testing the claim.

**8.11** Infants exposed to cocaine in their mother's womb are thought to be at a high risk for major birth defects. However, according to a University of Florida study, more than 75% of all cocaine-exposed babies suffer no major problems (*Explore*, Fall 1998). Set up the null and alternative hypotheses if you want to support the findings.

**8.12** The national student loan default rate has dropped steadily over the last decade. *USF Magazine* (Spring, 1999) reported the default rate (i.e., the proportion of college students who default on their loans) at .10 in fiscal year 1996. Set up the null and alternative hypotheses if you want to determine if the student loan default rate in 2000 is less than .10.

## Applying the Concepts—Intermediate

**8.13** According to a University of Florida wildlife ecology and conservation researcher, the average level of mercury uptake in wading birds in the Everglades has declined over the past several years (*UF News*, Dec. 15, 2000). In 1994, the average level was 15 parts per million.
  **a.** Give the null and alternative hypotheses for testing whether the average level in 2000 was less than 15 ppm.
  **b.** Describe a Type I error for this test.
  **c.** Describe a Type II error for this test.

**8.14** The Computer-Assisted Hypnosis Scale (CAHS) is designed to measure a person's susceptibility to hypnosis. CAHS scores range from 0 (no susceptibility to hypnosis) to 12 (extremely high susceptibility to hypnosis). *Psychological Assessment* (Mar. 1995) reported that University of Tennessee undergraduates had a mean CAHS score of $\mu = 4.6$. Suppose you want to test whether undergraduates at your college or university are more susceptible to hypnosis than University of Tennessee undergraduates.
  **a.** Set up $H_0$ and $H_a$ for the test.
  **b.** Describe a Type I error for this test.
  **c.** Describe a Type II error for this test.

## Applying the Concepts—Advanced

**8.15** A group of physicians subjected the *polygraph* (or *lie detector*) to the same careful testing given to medical diagnostic tests. They found that if 1,000 people were subjected to the polygraph and 500 told the truth and 500 lied, the polygraph would indicate that approximately 185 of the truth-tellers were liars and that approximately 120 of the liars were truth-tellers (*Discover*, 1986).
  **a.** In the application of a polygraph test, an individual is presumed to be a truth-teller ($H_0$) until "proven" a liar ($H_a$). In this context, what is a Type I error? A Type II error?
  **b.** According to the study, what is the probability (approximately) that a polygraph test will result in a Type I error? A Type II error?

**8.16** Sometimes, the outcome of a jury trial defies the "commonsense" expectations of the general public (e.g., the O. J. Simpson verdict in the "Trial of the Century"). Such a verdict is more acceptable if we understand that the jury trial of an accused murderer is analogous to the statistical hypothesis-testing process. The null hypothesis in a jury trial is that the accused is innocent. (The status-quo hypothesis in the U.S. system of justice is innocence, which is assumed to be true until proven *beyond a reasonable doubt*.) The alternative hypothesis is guilt, which is accepted only when sufficient evidence exists to establish its truth. If the vote of the jury is unanimous in favor of guilt, the null hypothesis of innocence is rejected and the court concludes that the accused murderer is guilty. Any vote other than a unanimous one for guilt results in a "not guilty" verdict. The court never accepts the null hypothesis; that is, the court never declares the accused "innocent." A "not guilty" verdict (as in the O. J. Simpson case) implies that the court could not find the defendant guilty *beyond a reasonable doubt*.
  **a.** Define Type I and Type II errors in a murder trial.
  **b.** Which of the two errors is the more serious? Explain.
  **c.** The court does not, in general, know the values of $\alpha$ and $\beta$; but ideally, both should be small. One of these probabilities is assumed to be smaller than the other in a jury trial. Which one, and why?
  **d.** The court system relies on the belief that the value of $\alpha$ is made very small by requiring a unanimous vote before guilt is concluded. Explain why this is so.
  **e.** For a jury prejudiced against a guilty verdict as the trial begins, will the value of $\alpha$ increase or decrease? Explain.
  **f.** For a jury prejudiced against a guilty verdict as the trial begins, will the value of $\beta$ increase or decrease? Explain.

## 8.2 LARGE-SAMPLE TEST OF HYPOTHESIS ABOUT A POPULATION MEAN

In Section 8.1 we learned that the null and alternative hypotheses form the basis for a test of hypothesis inference. The null and alternative hypotheses may take one of several forms. In the sewer pipe example we tested the null hypothesis that the population mean strength of the pipe is less than or equal to 2,400 pounds per linear foot against the alternative hypothesis that the mean strength exceeds 2,400. That is, we tested

$$H_0: \mu \leq 2{,}400$$
$$H_a: \mu > 2{,}400$$

This is a **one-tailed** (or **one-sided**) **statistical test** because the alternative hypothesis specifies that the population parameter (the population mean $\mu$, in this example) is strictly greater than a specified value (2,400, in this example). If the null hypothesis had been $H_0: \mu \geq 2{,}400$ and the alternative hypothesis had been $H_a: \mu < 2{,}400$, the test would still be one-sided, because the parameter is still specified to be on "one side" of the null hypothesis value. Some statistical investigations seek to show that the population parameter is *either larger or smaller* than some specified value. Such an alternative hypothesis is called a **two-tailed** (or **two-sided**) **hypothesis**.

While alternative hypotheses are always specified as strict inequalities, such as $\mu < 2{,}400$, $\mu > 2{,}400$, or $\mu \neq 2{,}400$, null hypotheses are usually specified as equalities, such as $\mu = 2{,}400$. *Even when the null hypothesis is an inequality, such as $\mu \leq 2{,}400$, we specify $H_0: \mu = 2{,}400$, reasoning that if sufficient evidence exists to show that $H_a: \mu > 2{,}400$ is true when tested against $H_0: \mu = 2{,}400$, then surely sufficient evidence exists to reject $\mu < 2{,}400$ as well.* Therefore, the null hypothesis is specified as the value of $\mu$ closest to a one-sided alternative hypothesis and as the only value *not* specified in a two-tailed alternative hypothesis. The steps for selecting the null and alternative hypotheses are summarized in the box on p. 335.

The rejection region for a two-tailed test differs from that for a one-tailed test. When we are trying to detect departure from the null hypothesis in *either* direction, we must establish a rejection region in both tails of the sampling distribution of the test statistic. Figures 8.4a and 8.4b show the one-tailed rejection regions for lower- and upper-tailed tests, respectively. The two-tailed rejection region is illustrated in Figure 8.4c. Note that a rejection region is established in each tail of the sampling distribution for a two-tailed test.

The rejection regions corresponding to typical values selected for $\alpha$ are shown in Table 8.2 for one-and two-tailed tests. Note that the smaller $\alpha$ you select, the more evidence (the larger $z$ ) you will need before you can reject $H_0$.

**FIGURE 8.4**

*Rejection Regions Corresponding to One- and Two-Tailed Tests*

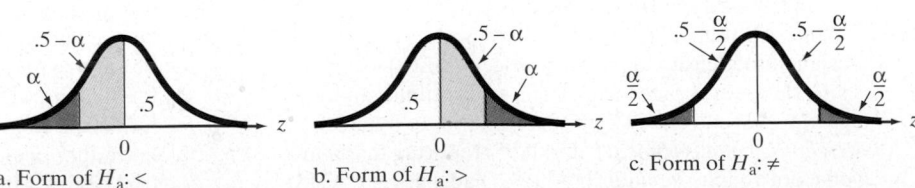

a. Form of $H_a$: $<$    b. Form of $H_a$: $>$    c. Form of $H_a$: $\neq$

## Steps for Selecting the Null and Alternative Hypotheses

1. Select the *alternative hypothesis* as that which the sampling experiment is intended to establish. The alternative hypothesis will assume one of three forms:

   a. One-tailed, upper-tailed   *Example: $H_a$:   $\mu > 2{,}400$*

   b. One-tailed, lower-tailed   *Example: $H_a$:   $\mu < 2{,}400$*

   c. Two-tailed   *Example: $H_a$:   $\mu \neq 2{,}400$*

2. Select the *null hypothesis* as the status quo, that which will be presumed true unless the sampling experiment conclusively establishes the alternative hypothesis. The null hypothesis will be specified as that parameter value closest to the alternative in one-tailed tests, and as the complementary (or only unspecified) value in two-tailed tests.

   *Example:*   $H_0: \mu = 2{,}400$

**TABLE 8.2   Rejection Regions for Common Values of $\alpha$**

|  | Alternative Hypotheses | | |
|---|---|---|---|
|  | **Lower-Tailed** | **Upper-Tailed** | **Two-Tailed** |
| $\alpha = .10$ | $z < -1.28$ | $z > 1.28$ | $z < -1.645$ or $z > 1.645$ |
| $\alpha = .05$ | $z < -1.645$ | $z > 1.645$ | $z < -1.96$  or $z > 1.96$ |
| $\alpha = .01$ | $z < -2.33$ | $z > 2.33$ | $z < -2.575$ or $z > 2.575$ |

**EXAMPLE 8.1**

The effect of drugs and alcohol on the nervous system has been the subject of considerable research. Suppose a research neurologist is testing the effect of a drug on response time by injecting 100 rats with a unit dose of the drug, subjecting each to a neurological stimulus, and recording its response time. The neurologist knows that the mean response time for rats not injected with the drug (the "control" mean) is 1.2 seconds. She wishes to test whether the mean response time for drug-injected rats differs from 1.2 seconds. Set up the test of hypothesis for this experiment, using $\alpha = .01$.

### Solution

Since the neurologist wishes to detect whether the mean response time, $\mu$, for drug-injected rats differs from the control mean of 1.2 seconds in *either* direction—that is, $\mu < 1.2$ or $\mu > 1.2$—we conduct a two-tailed statistical test. Following the procedure for selecting the null and alternative hypotheses, we specify as the alternative hypothesis that the mean differs from 1.2 seconds, since determining whether the drug-injected mean differs from the control mean is the purpose of the experiment. The null hypothesis is the presumption that drug-injected rats have the same mean response time as control rats unless the research indicates otherwise. Thus,

$$H_0: \mu = 1.2$$
$$H_a: \mu \neq 1.2 \qquad (\text{i.e., } \mu < 1.2 \text{ or } \mu > 1.2)$$

The test statistic measures the number of standard deviations between the observed value of $\bar{x}$ and the null hypothesized value $\mu = 1.2$:

$$\textit{Test statistic:} \quad z = \frac{\bar{x} - 1.2}{\sigma_{\bar{x}}}$$

The rejection region must be designated to detect a departure from $\mu = 1.2$ in *either* direction, so we will reject $H_0$ for values of $z$ that are either too small (negative) or too large (positive). To determine the precise values of $z$ that comprise the rejection region, we first select $\alpha$, the probability that the test will lead to incorrect rejection of the null hypothesis. Then we divide $\alpha$ equally between the lower and upper tail of the distribution of $z$, as shown in Figure 8.5. In this example, $\alpha = .01$, so $\alpha/2 = .005$ is placed in each tail. The areas in the tails correspond to $z = -2.575$ and $z = 2.575$, respectively (from Table 8.2):

$$\text{Rejection region}: z < -2.575 \text{ or } z > 2.575 \quad \text{(see Figure 8.5)}$$

*Assumptions:*   Since the sample size of the experiment is large enough ($n > 30$), the Central Limit Theorem will apply, and no assumptions need be made about the population of response time measurements. The sampling distribution of the sample mean response of 100 rats will be approximately normal regardless of the distribution of the individual rats' response times.

**FIGURE 8.5**

*Two-Tailed Rejection Region:* $\alpha = .01$

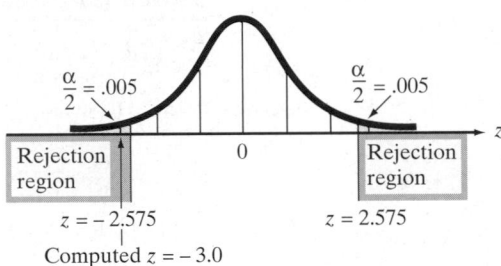

Note that the test in Example 8.1 is set up *before* the sampling experiment is conducted. The data are not used to develop the test. Evidently, the neurologist wants to conclude that the mean response time for the drug-injected rats differs from the control mean only when the evidence is very convincing, because the value of $\alpha$ has been set quite low at .01. If the experiment results in the rejection of $H_0$, she can be 99% confident that the mean response time of the drug-injected rats differs from the control mean.

Once the test is set up, she is ready to perform the sampling experiment and conduct the test. The test is performed in Example 8.2.

**EXAMPLE 8.2**

Refer to the neurological response-time test set up in Example 8.1. Suppose the sampling experiment is conducted with the following results:

$$n = 100 \text{ response times for drug-injected rats}$$
$$\bar{x} = 1.05 \text{ seconds} \qquad s = .5 \text{ second}$$

Use the results of the sampling experiment to conduct the test of hypothesis.

**Solution**

Since the test is completely specified in Example 8.1, we simply substitute the sample statistics into the test statistic:

$$z = \frac{\bar{x} - 1.2}{\sigma_{\bar{x}}} = \frac{\bar{x} - 1.2}{\sigma/\sqrt{n}} = \frac{1.05 - 1.2}{\sigma/\sqrt{100}}$$
$$\approx \frac{1.05 - 1.2}{s/10} = \frac{-.15}{.5/10} = -3.0$$

The implication is that the sample mean, 1.05, is (approximately) 3 standard deviations below the null hypothesized value of 1.2 in the sampling distribution of $\bar{x}$. You can see in Figure 8.5 that this value of $z$ is in the lower-tail rejection region, which consists of all values of $z < -2.575$. This sampling experiment provides sufficient evidence to reject $H_0$ and conclude, at the $\alpha = .01$ level of significance, that the mean response time for drug-injected rats differs from the control mean of 1.2 seconds. It appears that the rats receiving an injection of this drug have a mean response time that is less than 1.2 seconds.

■

The setup of a large-sample test of hypothesis about a population mean is summarized in the following box. Both the one- and two-tailed tests are shown.

---

**Large-Sample Test of Hypothesis About $\mu$**

**ONE-TAILED TEST**

$H_0: \mu = \mu_0$

$H_a: \mu < \mu_0$
  (or $H_a: \mu > \mu_0$)

*Test statistic:* $z = \dfrac{\bar{x} - \mu_0}{\sigma_{\bar{x}}}$

*Rejection region:* $z < -z_\alpha$
  (or $z > z_\alpha$ when $H_a: \mu > \mu_0$)

where $z_\alpha$ is chosen so that
  $P(z > z_\alpha) = \alpha$

**TWO-TAILED TEST**

$H_0: \mu = \mu_0$

$H_a: \mu \neq \mu_0$

*Test statistics:* $z = \dfrac{\bar{x} - \mu_0}{\sigma_{\bar{x}}}$

*Rejection region:* $z < -z_{\alpha/2}$  or  $z > z_{\alpha/2}$

where $z_{\alpha/2}$ is chosen so that
  $P(z > z_{\alpha/2}) = \alpha/2$

*Assumptions:* No assumptions need to be made about the probability distribution of the population because the Central Limit Theorem assures us that, for large samples, the test statistic will be approximately normally distributed regardless of the shape of the underlying probability distribution of the population.

*Note:* $\mu_0$ is the symbol for the numerical value assigned to $\mu$ under the null hypothesis.

---

Once the test has been set up, the sampling experiment is performed and the test statistic is calculated. The next box contains possible conclusions for a test of hypothesis, depending on the result of the sampling experiment.

Two final points about the test of hypothesis in Example 8.2 apply to all statistical tests:

1. Since $z$ is less than $-2.575$, it is tempting to state our conclusion at a significance level lower than $\alpha = .01$. We resist the temptation because the level of $\alpha$ is determined *before* the sampling experiment is performed. If we decide that we are willing to tolerate a 1% Type I error rate, the result of the sampling experiment should have no effect on that decision. *In general, the same data should not be used both to set up and to conduct the test.*

2. When we state our conclusion at the .01 level of significance, we are referring to the failure rate of the *procedure*, not the result of this particular test. We know that the test procedure will lead to the rejection of the null hypothesis only 1% of the time when in fact $\mu = 1.2$. *Therefore, when the test statistic falls in the rejection region, we infer that the alternative $\mu \neq 1.2$ is true and express our confidence in the procedure by quoting the $\alpha$ level of significance, or the $100(1 - \alpha)\%$ confidence level.*

---

### Possible Conclusions for a Test of Hypothesis

1. If the calculated test statistic falls in the rejection region, reject $H_0$ and conclude that the alternative hypothesis $H_a$ is true. State that you are rejecting $H_0$ at the $\alpha$ level of significance. Remember that the confidence is in the testing *process*, not the particular result of a single test.

2. If the test statistic does not fall in the rejection region, conclude that the sampling experiment does not provide sufficient evidence to reject $H_0$ at the $\alpha$ level of significance. [Generally, we will not "accept" the null hypothesis unless the probability $\beta$ of a Type II error has been calculated (see optional Section 8.6).]

---

 EXERCISES 8.17–8.29

### Learning the Mechanics

**8.17** For each of the following rejection regions, sketch the sampling distribution for $z$ and indicate the location of the rejection region.
  **a.** $z > 1.96$
  **b.** $z > 1.645$
  **c.** $z > 2.575$
  **d.** $z < -1.28$
  **e.** $z < -1.645$ or $z > 1.645$
  **f.** $z < -2.575$ or $z > 2.575$
  **g.** For each of the rejection regions specified in parts **a–f**, what is the probability that a Type I error will be made?

**8.18** Suppose you are interested in conducting the statistical test of $H_0: \mu = 200$ against $H_a: \mu > 200$, and you have decided to use the following decision rule: Reject $H_0$ if the sample mean of a random sample of 100 items is more than 215. Assume that the standard deviation of the population is 80.
  **a.** Express the decision rule in terms of $z$.
  **b.** Find $\alpha$, the probability of making a Type I error, by using this decision rule.

**8.19** A random sample of 100 observations from a population with standard deviation 60 yielded a sample mean of 110.
  **a.** Test the null hypothesis that $\mu = 100$ against the alternative hypothesis that $\mu > 100$ using $\alpha = .05$. Interpret the results of the test.
  **b.** Test the null hypothesis that $\mu = 100$ against the alternative hypothesis that $\mu \neq 100$ using $\alpha = .05$. Interpret the results of the test.
  **c.** Compare the results of the two tests you conducted. Explain why the results differ.

**8.20** A random sample of 64 observations produced the following summary statistics: $\bar{x} = .323$ and $s^2 = .034$.
  **a.** Test the null hypothesis that $\mu = .36$ against the alternative hypothesis that $\mu < .36$ using $\alpha = .10$.
  **b.** Test the null hypothesis that $\mu = .36$ against the alternative hypothesis that $\mu \neq .36$ using $\alpha = .10$. Interpret the result.

### Applying the Concepts—Basic

**8.21** In Exercise 5.29 (p. 235) you learned that the mean alkalinity level of water specimens collected from the Han River in Seoul, Korea, is 50 milligrams per liter (*Environmental Science & Engineering*, Sept. 1, 2000). Consider a random sample of 100 water specimens collected from a tributary of the Han River. Suppose the mean and standard deviation of the alkalinity levels for the sample are $\bar{x} = 67.8$ mpl and $s = 14.4$ mpl. Is there sufficient evidence (at $\alpha = .01$) to indicate that the population mean alkalinity level of water in the tributary exceeds 50 mpl?

**8.22** A group of St. Louis University researchers administered the Vividness of Visual Imagery Questionnaire (VVIQ) to a sample of 51 college students on varsity sports teams (American Statistical Association Joint Meetings, Aug. 1996). The mean and standard deviation of the VVIQ scores were 65.60 and 19.38, respectively. Scores range from 16 to 80, with a lower score representing greater vividness. In the general population, the mean VVIQ score is 67. Conduct a test to determine whether the mean VVIQ score of college athletes is less than 67. Test using $\alpha = .01$.

**8.23** A study reported in the *Journal of Occupational and Organizational Psychology* (Dec. 1992) investigated the relationship of employment status to mental health. Each of a sample of 49 unemployed men was given a mental health examination using the General Health Questionnaire (GHQ). The GHQ is a widely recognized measure of present mental health, with lower values indicating better mental health. The mean and standard deviation of the GHQ scores were $\bar{x} = 10.94$ and $s = 5.10$, respectively.
  **a.** Specify the appropriate null and alternative hypotheses if we wish to test the research hypothesis that the mean GHQ score for all unemployed men exceeds 10. Is the test one-tailed or two-tailed? Why?
  **b.** If we specify $\alpha = .05$, what is the appropriate rejection region for this test?

**c.** Conduct the test, and state your conclusion clearly in the language of this exercise.

## Applying the Concepts—Intermediate

**8.24** During the National Football League (NFL) season, Las Vegas oddsmakers establish a point spread on each game for betting purposes. For example, the Baltimore Ravens were established as 3-point favorites over the New York Giants in the 2001 Super Bowl. The final scores of NFL games were compared against the final point spreads established by the oddsmakers in *Chance* (Fall 1998). The difference between the game outcome and point spread (called a point-spread error) was calculated for 240 NFL games. The mean and standard deviation of the point-spread errors are $\bar{x} = -1.6$ and $s = 13.3$. Use this information to test the hypothesis that the true mean point-spread error for all NFL games is 0. Conduct the test at $\alpha = .01$ and interpret the result.

**8.25** *Psychological Assessment* (Mar. 1995) published the results of a study of World War II aviators who were captured by German forces after they were shot down. Having located a total of 239 World War II aviator POW survivors, the researchers asked each veteran to participate in the study; 33 responded to the letter of invitation. Each of the 33 POW survivors were administered the Minnesota Multiphasic Personality Inventory, one component of which measures level of post-traumatic stress disorder (PTSD). [*Note:* The higher the score, the higher the level of PTSD.] The aviators produced a mean PTSD score of $\bar{x} = 9.00$ and a standard deviation of $s = 9.32$.

**a.** Set up the null and alternative hypotheses for determining whether the true mean PTSD score of all World War II aviator POWs is less than 16. [*Note:* The value, 16, represents the mean PTSD score established for Vietnam POWs.]

**b.** Conduct the test, part **a**, using $\alpha = .10$. What are the practical implications of the test?

**c.** Discuss the representativeness of the sample used in the study and its ramifications.

**8.26** Current technology uses X-rays and lasers for inspection of solder-joint defects on printed circuit boards (PCBs) (*Quality Congress Transactions*, 1986). A particular manufacturer of laser-based inspection equipment claims that its product can inspect on average at least 10 solder joints per second when the joints are spaced .1 inch apart. The equipment was tested by a potential buyer on 48 different PCBs. In each case, the equipment was operated for exactly 1 second. The number of solder joints inspected on each run follows:

**⬤ PCB**

| | | | | | | | | | | | |
|---|---|---|---|---|---|---|---|---|---|---|---|
| 10 | 9 | 10 | 10 | 11 | 9 | 12 | 8 | 8 | 9 | 6 | 10 |
| 7 | 10 | 11 | 9 | 9 | 13 | 9 | 10 | 11 | 10 | 12 | 8 |
| 9 | 9 | 9 | 7 | 12 | 6 | 9 | 10 | 10 | 8 | 7 | 9 |
| 11 | 12 | 10 | 0 | 10 | 11 | 12 | 9 | 7 | 9 | 9 | 10 |

*37 40 40 26 42 39 42 37 36 36 34 37*

**a.** The potential buyer wants to know whether the sample data refute the manufacturer's claim. Specify the null and alternative hypotheses that the buyer should test.

**b.** In the context of this exercise, what is a Type I error? A Type II error?

**c.** Conduct the hypothesis test you described in part **a**, and interpret the test's results in the context of this exercise. Use $\alpha = .05$ and the SPSS descriptive statistics printout at the bottom of the page.

**8.27** Humerus bones from the same species of animal tend to have approximately the same length-to-width ratios. When fossils of humerus bones are discovered, archeologists can often determine the species of animal by examining the length-to-width ratios of the bones. It is known that species A exhibits a mean ratio of 8.5. Suppose 41 fossils of humerus bones were unearthed at an archeological site in East Africa, where species A is believed to have lived. (Assume that the unearthed bones were all from the same unknown species.) The length-to-width ratios of the bones were calculated and listed, as shown in the table. Summary statistics for the data set are shown in the STATISTIX printout on p. 340.

**a.** Test whether the population mean ratio of all bones of this particular species differs from 8.5. Use $\alpha = .01$.

**b.** What are the practical implications of the test, part **a**?

**⬤ BONES**

| | | | | | | | |
|---|---|---|---|---|---|---|---|
| 10.73 | 8.89 | 9.07 | 9.20 | 10.33 | 9.98 | 9.84 | 9.59 |
| 8.48 | 8.71 | 9.57 | 9.29 | 9.94 | 8.07 | 8.37 | 6.85 |
| 8.52 | 8.87 | 6.23 | 9.41 | 6.66 | 9.35 | 8.86 | 9.93 |
| 8.91 | 11.77 | 10.48 | 10.39 | 9.39 | 9.17 | 9.89 | 8.17 |
| 8.93 | 8.80 | 10.02 | 8.38 | 11.67 | 8.30 | 9.17 | 12.00 |
| 9.38 | | | | | | | |

`SPSS Output for Exercise 8.26`

| Descriptive Statistics | | | | | |
|---|---|---|---|---|---|
| | N | Minimum | Maximum | Mean | Std. Deviation |
| NUMBER | 48 | .00 | 13.00 | 9.2917 | 2.1033 |
| Valid N (listwise) | 48 | | | | |

STATISTIX Output for Exercise 8.27

| DESCRIPTIVE STATISTICS | |
| --- | --- |
| | LWRATIO |
| N | 41 |
| MEAN | 9.2576 |
| SD | 1.2036 |
| MINIMUM | 6.2300 |
| MEDIAN | 9.2000 |
| MAXIMUM | 12.000 |

**8.28** *Environmental Science & Technology* (Oct. 1993) reported on a study of contaminated soil in The Netherlands. Seventy-two 400-gram soil specimens were sampled, dried, and analyzed for the contaminant cyanide. The cyanide concentration [in milligrams per kilogram (mg/kg) of soil] of each soil specimen was determined using an infrared microscopic method. The sample resulted in a mean cyanide level of $\bar{x} = 84$ mg/kg and a standard deviation of $s = 80$ mg/kg.

**a.** Test the hypothesis that the true mean cyanide level in soil in The Netherlands exceeds 100 mg/kg. Use $\alpha = .10$.

**b.** Would you reach the same conclusion as in part **a** using $\alpha = .05$? Using $\alpha = .01$? Why can the conclusion of a test change when the value of $\alpha$ is changed?

**Applying the Concepts—Advanced**

**8.29** The *Community Mental Health Journal* (Aug. 2000) presented the results of a survey of over 6,000 clients of the Department of Mental Health and Addiction Services (DMHAS) in Connecticut. One of the many variables measured for each mental health patient was frequency of social interaction (on a 5-point scale, where 1 = very infrequently, 3 = occasionally, and 5 = very frequently). The 6,681 clients who were evaluated had a mean social interaction score of 2.95 with a standard deviation of 1.10.

**a.** Conduct a hypothesis test (at $\alpha = .01$) to determine if the true mean social interaction score of all Connecticut mental health patients differs from 3.

**b.** Examine the results of the study from a practical view, then discuss why "statistical significance" does not always imply "practical significance."

**c.** Because the variable of interest is measured on a 5-point scale, it is unlikely that the population of ratings will be normally distributed. Consequently, some analysts may perceive the test, part **a**, to be invalid and search for alternative methods of analysis. Defend or refute this position.

## 8.3 OBSERVED SIGNIFICANCE LEVELS: *p*-VALUES

According to the statistical test procedure described in Section 8.2, the rejection region and, correspondingly, the value of $\alpha$ are selected prior to conducting the test, and the conclusions are stated in terms of rejecting or not rejecting the null hypothesis. A second method of presenting the results of a statistical test is one that reports the extent to which the test statistic disagrees with the null hypothesis and leaves to the reader the task of deciding whether to reject the null hypothesis. This measure of disagreement is called the *observed significance level* (or *p-value*) for the test.

> ### DEFINITION 8.1
> The **observed significance level**, or ***p*-value**, for a specific statistical test is the probability (assuming $H_0$ is true) of observing a value of the test statistic that is at least as contradictory to the null hypothesis, and supportive of the alternative hypothesis, as the actual one computed from the sample data.

For example, the value of the test statistic computed for the sample of $n = 50$ sections of sewer pipe was $z = 2.12$. Since the test is one-tailed—i.e., the alternative (research) hypothesis of interest is $H_a: \mu > 2,400$—values of the test statistic even more contradictory to $H_0$ than the one observed would be values larger than $z = 2.12$. Therefore, the observed significance level (*p*-value) for this test is

$$p\text{-value} = P(z \geq 2.12)$$

**FIGURE 8.6**

*Finding the p-Value for an
Upper-Tailed Test when
z = 2.12*

or, equivalently, the area under the standard normal curve to the right of $z = 2.12$ (see Figure 8.6).

The area $A$ in Figure 8.6 is given in Table IV in Appendix A as .4830. Therefore, the upper-tail area corresponding to $z = 2.12$ is

$$p\text{-value} = .5 - .4830 = .0170$$

Consequently, we say that these test results are "very significant"; that is, they disagree rather strongly with the null hypothesis, $H_0: \mu = 2,400$, and favor $H_a: \mu > 2,400$. The probability of observing a $z$ value as large as 2.12 is only .0170, if in fact the true value of $\mu$ is 2,400.

If you are inclined to select $\alpha = .05$ for this test, then you would reject the null hypothesis because the $p$-value for the test, .0170, is less than .05. In contrast, if you choose $\alpha = .01$, you would not reject the null hypothesis because the $p$-value for the test is larger than .01. Thus, the use of the observed significance level is identical to the test procedure described in the preceding sections except that the choice of $\alpha$ is left to you.

The steps for calculating the $p$-value corresponding to a test statistic for a population mean are given in the box.

---

**Steps for Calculating the *p*-Value for a Test of Hypothesis**

---

1. Determine the value of the test statistic $z$ corresponding to the result of the sampling experiment.

2. a. If the test is one-tailed, the $p$-value is equal to the tail area beyond $z$ in the same direction as the alternative hypothesis. Thus, if the alternative hypothesis is of the form $>$, the $p$-value is the area to the right of, or above, the observed $z$ value. Conversely, if the alternative is of the form $<$, the $p$-value is the area to the left of, or below, the observed $z$ value. (See Figure 8.7.)

   b. If the test is two-tailed, the $p$-value is equal to twice the tail area beyond the observed $z$ value in the direction of the sign of $z$. That is, if $z$ is positive, the $p$-value is twice the area to the right of, or above, the observed $z$ value. Conversely, if $z$ is negative, the $p$-value is twice the area to the left of, or below, the observed $z$ value. (See Figure 8.8.)

---

**FIGURE 8.7**

*Finding the p-Value for a
One-Tailed Test*

a. Lower–tailed test, $H_a: \mu < \mu_0$      b. Upper–tailed test, $H_a: \mu > \mu_0$

## FIGURE 8.8

*Finding the p-Value for a Two-Tailed Test: p-Value = 2(p/2)*

a. Test statistic *z* negative

b. Test statistic *z* positive

**EXAMPLE 8.3** Find the observed significance level for the test of the mean response time for drug-injected rats in Examples 8.1 and 8.2.

**Solution**

Example 8.1 presented a two-tailed test of the hypothesis

$$H_0: \mu = 1.2 \text{ seconds}$$

against the alternative hypothesis

$$H_a: \mu \neq 1.2 \text{ seconds}$$

The observed value of the test statistic in Example 8.2 was $z = -3.0$, and any value of $z$ less than $-3.0$ or greater than $+3.0$ (because this is a two-tailed test) would be even more contradictory to $H_0$. Therefore, the observed significance level for the test is

$$p\text{-value} = P(z < -3.0 \text{ or } z > +3.0)$$

Thus, we calculate the area below the observed $z$ value, $z = -3.0$, and double it. Consulting Table IV in Appendix A, we find that $P(z < -3.0) = .5 - .4987 = .0013$. Therefore, the $p$-value for this two-tailed test is

$$2P(z < -3.0) = 2(.0013) = .0026$$

We can interpret this $p$-value as a strong indication that the mean reaction time of drug-injected rats differs from the control mean ($\mu \neq 1.2$), since we would observe a test statistic this extreme or more extreme only 26 in 10,000 times if the drug-injected mean were equal to the control mean ($\mu = 1.2$). The extent to which the mean differs from 1.2 could be better determined by calculating a confidence interval for $\mu$.

When publishing the results of a statistical test of hypothesis in journals, case studies, reports, etc., many researchers make use of $p$-values. Instead of selecting $a$ beforehand and then conducting a test, as outlined in this chapter, the researcher computes (usually with the aid of a statistical software package) and reports the value of the appropriate test statistic and its associated $p$-value. It is left to the reader of the report to judge the significance of the result—i.e., the reader must determine whether to reject the null hypothesis in favor of the alternative hypothesis, based on the reported $p$-value. Usually, the null hypothesis is rejected if the observed significance level is *less than* the fixed significance level, $\alpha$, chosen by the reader. The inherent advantage of reporting test results in this manner is twofold: (1) Readers are permitted to select the maximum value of $\alpha$ that they would be willing to tolerate if they actually carried out a standard test of hypothesis in the manner outlined in this chapter, and (2) a measure of the degree of significance of the result (i.e., the $p$ value) is provided.

---

### Reporting Test Results as *p*-Values: How to Decide Whether to Reject $H_0$

1. Choose the maximum value of $\alpha$ that you are willing to tolerate.
2. If the observed significance level (*p*-value) of the test is less than the chosen value of $\alpha$, reject the null hypothesis. Otherwise, do not reject the null hypothesis.

---

**E X A M P L E   8 . 4**

The lengths of stay (in days) for 100 randomly selected hospital patients, first presented in Table 7.1, are reproduced in Table 8.3. Suppose we want to test the hypothesis that the true mean length of stay (LOS) at the hospital is less than 5 days, that is,

$$H_0\colon \mu = 5$$
$$H_a\colon \mu < 5$$

Assuming $\sigma = 3.68$, use the data in the table to conduct the test at $\alpha = .05$.

**⊘ HOSPLOS**

**TABLE 8.3   Lengths of Stay for 100 Hospital Patients**

| | | | | | | | | | |
|---|---|---|---|---|---|---|---|---|---|
| 2 | 3 | 8 | 6 | 4 | 4 | 6 | 4 | 2 | 5 |
| 8 | 10 | 4 | 4 | 4 | 2 | 1 | 3 | 2 | 10 |
| 1 | 3 | 2 | 3 | 4 | 3 | 5 | 2 | 4 | 1 |
| 2 | 9 | 1 | 7 | 17 | 9 | 9 | 9 | 4 | 4 |
| 1 | 1 | 1 | 3 | 1 | 6 | 3 | 3 | 2 | 5 |
| 1 | 3 | 3 | 14 | 2 | 3 | 9 | 6 | 6 | 3 |
| 5 | 1 | 4 | 6 | 11 | 22 | 1 | 9 | 6 | 5 |
| 2 | 2 | 5 | 4 | 3 | 6 | 1 | 5 | 1 | 6 |
| 17 | 1 | 2 | 4 | 5 | 4 | 4 | 3 | 2 | 3 |
| 3 | 5 | 2 | 3 | 3 | 2 | 10 | 2 | 4 | 2 |

### Solution

Instead of conducting the test by hand, we will use a statistical software package. The data were entered into a computer and MINITAB was used to conduct the analysis. The MINITAB printout for the lower-tailed test is displayed in Figure 8.9. Both the

**FIGURE 8.9**

*MINITAB Printout for the Lower-Tailed Test in Example 8.4*

```
Test of mu = 5 vs mu < 5
The assumed sigma = 3.68

Variable            N       Mean     StDev     SE Mean
LOS               100      4.530     3.678       0.368

Variable      95.0% Upper Bound          Z           P
LOS                       5.135      -1.28       0.101
```

test statistic, $z = -1.28$, and *p*-value of the test, $p = .101$, are highlighted on the MINITAB printout. Since the *p*-value exceeds our selected $\alpha$ value, $\alpha = .05$, we cannot reject the null hypothesis. Hence, there is insufficient evidence (at $\alpha = .05$) to conclude that the true mean LOS at the hospital is less than 5 days.  ▄

*Note:* MINITAB, SAS, and STATISTIX all provide an option for selecting one-tailed or two-tailed tests and report the appropriate *p*-value. Some statistical software packages such as SPSS will conduct only two-tailed tests of hypothesis. For these packages, you obtain the *p*-value for a one-tailed test as follows:

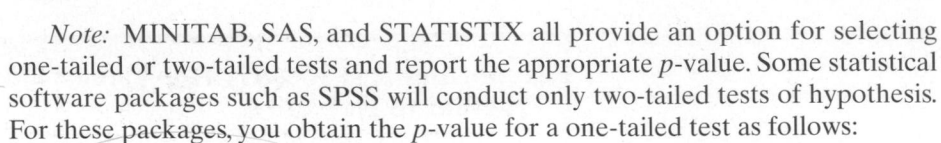

$$p = \frac{\text{Reported } p\text{-value}}{2} \quad \text{if} \begin{cases} H_{\text{a}} \text{ is of form } > \text{ and } z \text{ is positive} \\ H_{\text{a}} \text{ is of form } < \text{ and } z \text{ is negative} \end{cases}$$

$$p = 1 - \left( \frac{\text{Reported } p\text{-value}}{2} \right) \quad \text{if} \begin{cases} H_{\text{a}} \text{ is of form } > \text{ and } z \text{ is negative} \\ H_{\text{a}} \text{ is of form } < \text{ and } z \text{ is positive} \end{cases}$$

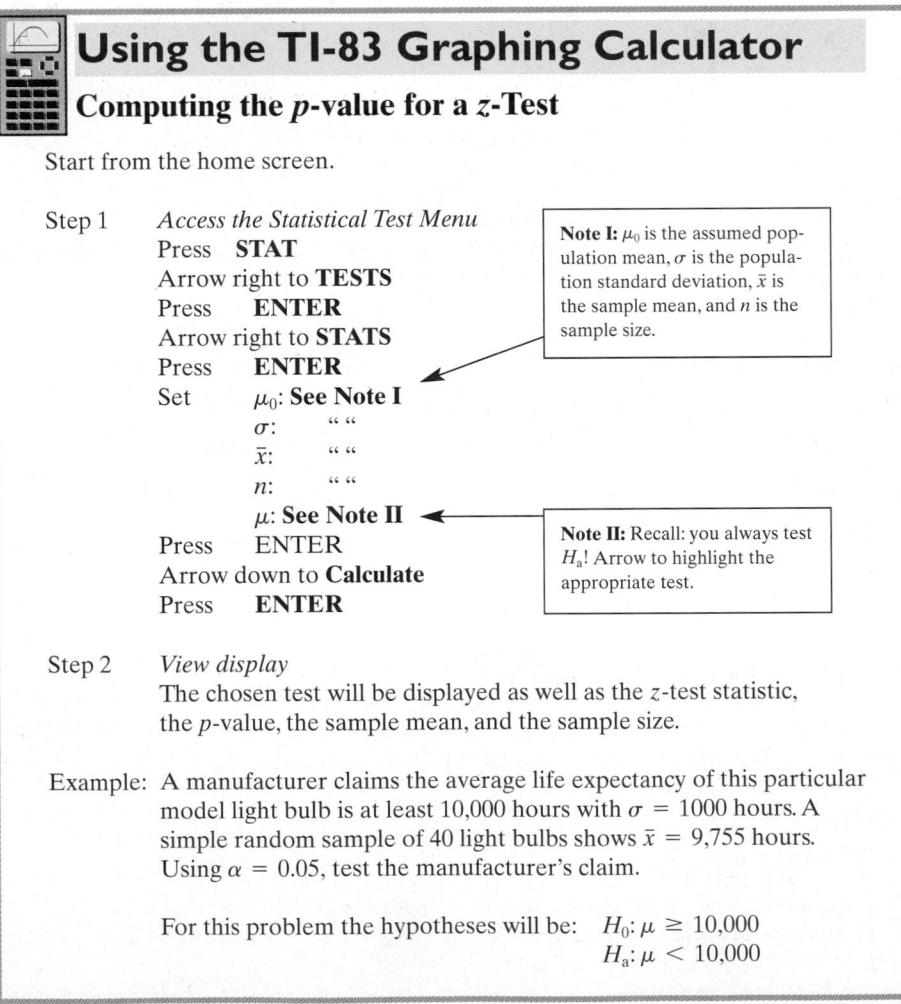

# Using the TI-83 Graphing Calculator

## Computing the *p*-value for a *z*-Test

Start from the home screen.

Step 1   *Access the Statistical Test Menu*
Press   **STAT**
Arrow right to **TESTS**
Press   **ENTER**
Arrow right to **STATS**
Press   **ENTER**
Set   $\mu_0$: **See Note I**
      $\sigma$:   " "
      $\bar{x}$:   " "
      $n$:   " "
      $\mu$: **See Note II**
Press   ENTER
Arrow down to **Calculate**
Press   **ENTER**

**Note I:** $\mu_0$ is the assumed population mean, $\sigma$ is the population standard deviation, $\bar{x}$ is the sample mean, and $n$ is the sample size.

**Note II:** Recall: you always test $H_{\text{a}}$! Arrow to highlight the appropriate test.

Step 2   *View display*
The chosen test will be displayed as well as the *z*-test statistic, the *p*-value, the sample mean, and the sample size.

Example:   A manufacturer claims the average life expectancy of this particular model light bulb is at least 10,000 hours with $\sigma = 1000$ hours. A simple random sample of 40 light bulbs shows $\bar{x} = 9,755$ hours. Using $\alpha = 0.05$, test the manufacturer's claim.

For this problem the hypotheses will be:   $H_0: \mu \geq 10,000$
                                            $H_{\text{a}}: \mu < 10,000$

**USING THE TI-83 GRAPHING CALCULATOR** (*continued*)

Step 1     *Access the Statistical Test Menu*
Enter your setting for this problem (see the screen below left).

Arrow down to Calculate
Press **ENTER** (above right screen will be displayed).

Step 2     *View display*
As you can see, the *p*-value is 0.061. Recall we reject $H_0$ only if $p < \alpha$. Therefore *do not* reject $H_0$.

Step 3     Clear the screen for the next problem. Return to the home screen. Press **CLEAR**

 EXERCISES 8.30–8.44

**Learning the Mechanics**

**8.30** If a hypothesis test were conducted using $\alpha = .05$, for which of the following *p*-values would the null hypothesis be rejected?
    **a.** .06
    **b.** .10
    **c.** .01
    **d.** .001
    **e.** .251
    **f.** .042

**8.31** For each $\alpha$ and observed significance level (*p*-value) pair, indicate whether the null hypothesis would be rejected.
    **a.** $\alpha = .05$, *p*-value $= .10$
    **b.** $\alpha = .10$, *p*-value $= .05$
    **c.** $\alpha = .01$, *p*-value $= .001$
    **d.** $\alpha = .025$, *p*-value $= .05$
    **e.** $\alpha = .10$, *p*-value $= .45$

**8.32** An analyst tested the null hypothesis $\mu \geq 20$ against the alternative hypothesis $\mu < 20$. The analyst reported a *p*-value of .06. What is the smallest value of $\alpha$ for which the null hypothesis would be rejected?

**8.33** In a test of $H_0: \mu = 100$ against $H_a: \mu > 100$, the sample data yielded the test statistic $z = 2.17$. Find the *p*-value for the test.

**8.34** In a test of $H_0: \mu = 100$ against $H_a: \mu \neq 100$, the sample data yielded the test statistic $z = 2.17$. Find the *p*-value for the test.

**8.35** In a test of the hypothesis $H_0: \mu = 50$ versus $H_a: \mu > 50$, a sample of $n = 100$ observations possessed mean $\bar{x} = 49.4$ and standard deviation $s = 4.1$. Find and interpret the *p*-value for this test.

**8.36** In a test of the hypothesis $H_0: \mu = 10$ versus $H_a: \mu \neq 10$, a sample of $n = 50$ observations possessed mean $\bar{x} = 10.7$ and standard deviation $s = 3.1$. Find and interpret the *p*-value for this test.

**8.37** Consider a test of $H_0: \mu = 75$ performed using the computer. SPSS reports a two-tailed *p*-value of .1032. Make the appropriate conclusion for each of the following situations:
    **a.** $H_a: \mu < 75$, $z = -1.63$, $\alpha = .05$
    **b.** $H_a: \mu < 75$, $z = 1.63$, $\alpha = .10$
    **c.** $H_a: \mu > 75$, $z = 1.63$, $\alpha = .10$
    **d.** $H_a: \mu \neq 75$, $z = -1.63$, $\alpha = .01$

## Applying the Concepts—Basic

**8.38** Refer to the *Psychological Assessment* study of World War II aviator POWs, Exercise 8.25 (p. 339). You tested whether the true mean post-traumatic stress disorder score of World War II aviator POWs is less than 16. Recall that $\bar{x} = 9.00$ and $s = 9.32$ for a sample of $n = 33$ POWs.

  **a.** Compute the *p*-value of the test.

  **b.** In general, do large *p*-values or small *p*-values support the alternative hypothesis? Explain.

**8.39** According to *USA Today* (Dec. 30, 1999), the average age of viewers of MSNBC cable television news programming is 50 years. A random sample of 50 U.S. households that receive cable television programming yielded the following additional information about the ages of MSNBC news viewers: $\bar{x} = 51.3$ years and $s = 7.1$ years. Consider a test to determine whether the average age of all MSNBC's viewers is greater than 50 years.

  **a.** Find the observed significance level (*p*-value) of the test and interpret its value in the context of the problem.

  **b.** Refer to part **a**. Would the *p*-value have been larger or smaller if $\bar{x}$ had been larger? Explain.

**8.40** Refer to Exercise 8.23, (p. 338), in which a random sample of 49 unemployed men were administered the General Health Questionnaire (GHQ). The sample mean and standard deviation were 10.94 and 5.10, respectively. Denoting the population mean GHQ for unemployed workers by $\mu$, we wish to test the null hypothesis $H_0$: $\mu = 10$ versus the one-tailed alternative $H_a$: $\mu > 10$. When the data are analyzed in MINITAB, the results are as shown in the printout below. What conclusion would you reach about the test based on the MINITAB analysis?

**8.41** In Exercise 8.27 (p. 339), you tested $H_0$: $\mu = 8.5$ versus $H_a$: $\mu \neq 8.5$, where $\mu$ is the population mean length-to-width ratio of humerus bones of a particular species of animal. A STATISTIX printout for the hypothesis test is provided (below, right). Locate the *p*-value on the printout and interpret its value.

## Applying the Concepts—Intermediate

**8.42** Refer to the *Applied Animal Behaviour Science* (Oct. 2000) study of domestic chickens exposed to a pecking stimulus, Exercise 7.15 (p. 295). Recall that the average number of pecks a chicken takes at a white string over a specified time interval is known to be $\mu = 7.5$ pecks.

In an experiment where 72 chickens were exposed to blue string, the average number of pecks was $\bar{x} = 1.13$ pecks, with a standard deviation of $s = 2.21$ pecks.

  **a.** On average, are chickens more apt to peck at white string than at blue string? Conduct the appropriate test of hypothesis using $\alpha = .05$.

  **b.** Compare your answer to part **a** with your answer to Exercise 7.15**b**.

  **c.** Find the *p*-value for the test and interpret its value.

**8.43** Research published in *Nature* (Aug. 27, 1998) revealed that people are more attracted to "feminized" faces, regardless of gender. In one experiment, 50 human subjects viewed both a Japanese female and Caucasian male face on a computer. Using special computer graphics, each subject could morph the faces (by making them more feminine or more masculine) until they attained the "most attractive" face. The level of feminization $x$ (measured as a percentage) was measured.

  **a.** For the Japanese female face, $\bar{x} = 10.2\%$ and $s = 31.3\%$. The researchers used this sample information to test the null hypothesis of a mean level of feminization equal to 0%. Verify that the test statistic is equal to 2.3.

  **b.** Refer to part **a**. The researchers reported the *p*-value of the test as $p \approx .02$. Verify and interpret this result.

  **c.** For the Caucasian male face, $\bar{x} = 15.0\%$ and $s = 25.1\%$. The researchers reported the test statistic (for the test of the null hypothesis stated in part **a**) as 4.23 with an associated *p*-value of approximately 0. Verify and interpret these results.

**8.44** In Exercise 8.26 (p. 339), you tested $H_0$: $\mu \geq 10$ versus $H_a$: $\mu < 10$, where $\mu$ is the average number of solder joints inspected per second when the joints are spaced .1 inch apart. An SPSS printout of the hypothesis test is shown on p. 347.

  **a.** Locate the two-tailed *p*-value of the test shown on the printout.

  **b.** Adjust the *p*-value for the one-tailed test (if necessary) and interpret its value.

**STATISTIX Output for Exercise 8.41**

```
ONE-SAMPLE T TEST FOR LWRATIO

NULL HYPOTHESIS: MU =   8.5
ALTERNATIVE HYP: MU <> 8.5

MEAN          9.2576
STD ERROR     0.1880
MEAN - H0     0.7576
LO 95% CI     0.3777
UP 95% CI     1.1375
T                4.03
DF                 40
P             0.0002

CASES INCLUDED 41    MISSING CASES 0
```

**MINITAB Output for Exercise 8.40**

```
Test of mu = 10 vs mu > 10

Variable        N      Mean     StDev    SE Mean
GHQ            49     10.94      5.10       0.73

Variable   95.0% Lower Bound       T        P
GHQ                    8.776     1.29    .1016
```

SPSS Output for Exercise 8.44

**One-Sample Test**

| | | | | | 95% Confidence Interval of the Difference | |
|---|---|---|---|---|---|---|
| | | | Test Value = 10 | | | |
| | t | df | Sig. (2-tailed) | Mean Difference | Lower | Upper |
| NUMBER | -2.333 | 47 | .024 | -.7083 | -1.3191 | -9.76E-02 |

## 8.4 SMALL-SAMPLE TEST OF HYPOTHESIS ABOUT A POPULATION MEAN

Most water-treatment facilities monitor the quality of their drinking water on an hourly basis. One variable monitored is pH, which measures the degree of alkalinity or acidity in the water. A pH below 7.0 is acidic, one above 7.0 is alkaline, and a pH of 7.0 is neutral. One water-treatment plant has a target pH of 8.5 (most try to maintain a slightly alkaline level). The mean and standard deviation of 1 hour's test results, based on 17 water samples at this plant, are

$$\bar{x} = 8.42 \qquad s = .16$$

Does this sample provide sufficient evidence that the mean pH level in the water differs from 8.5?

This inference can be placed in a test of hypothesis framework. We establish the target pH as the null hypothesized value and then utilize a two-tailed alternative that the true mean pH differs from the target:

$$H_0: \mu = 8.5$$
$$H_a: \mu \neq 8.5$$

Recall from Section 7.3 that when we are faced with making inferences about a population mean using the information in a small sample, two problems emerge:

1. The normality of the sampling distribution for $\bar{x}$ does not follow from the Central Limit Theorem when the sample size is small. We must assume that the distribution of measurements from which the sample was selected is approximately normally distributed in order to ensure the approximate normality of the sampling distribution of $\bar{x}$.
2. If the population standard deviation $\sigma$ is unknown, as is usually the case, then we cannot assume that $s$ will provide a good approximation for $\sigma$ when the sample size is small. Instead, we must use the $t$-distribution rather than the standard normal $z$-distribution to make inferences about the population mean $\mu$.

Therefore, as the test statistic of a small-sample test of a population mean, we use the $t$ statistic:

$$Test\ statistic: t = \frac{\bar{x} - \mu_0}{s/\sqrt{n}} = \frac{\bar{x} - 8.5}{s/\sqrt{n}}$$

where $\mu_0$ is the null hypothesized value of the population mean, $\mu$. In our example, $\mu_0 = 8.5$.

To find the rejection region, we must specify the value of $\alpha$, the probability that the test will lead to rejection of the null hypothesis when it is true, and then

**FIGURE 8.10**

*Two-Tailed Rejection Region for Small-Sample t-Test*

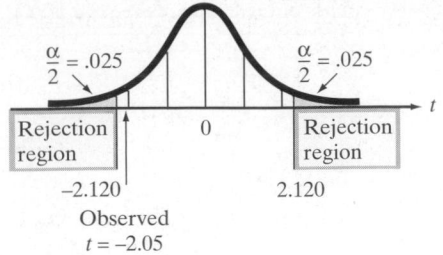

consult the *t* table (Table VI of Appendix A). Using $\alpha = .05$, the two-tailed rejection region is

$$Rejection\ region: t_{\alpha/2} = t_{.025} = 2.120 \text{ with } n - 1 = 16 \text{ degrees of freedom}$$
$$Reject\ H_0 \text{ if } t < -2.120 \text{ or } t > 2.120$$

The rejection region is shown in Figure 8.10.

We are now prepared to calculate the test statistic and reach a conclusion:

$$t = \frac{\bar{x} - \mu_0}{s/\sqrt{n}} = \frac{8.42 - 8.50}{.16/\sqrt{17}} = \frac{-.08}{.039} = -2.05$$

Since the calculated value of *t* does not fall in the rejection region (Figure 8.10), we cannot reject $H_0$ at the $\alpha = .05$ level of significance. Thus, the water-treatment plant should not conclude that the mean pH differs from the 8.5 target based on the sample evidence.

It is interesting to note that the calculated *t* value, $-2.05$, is *less than* the .05 level *z* value, $-1.96$. The implication is that if we had *incorrectly* used a *z* statistic for this test, we would have rejected the null hypothesis at the .05 level, concluding that the mean pH level differs from 8.5. The important point is that the statistical procedure to be used must always be closely scrutinized and all the assumptions understood. Many statistical lies are the result of misapplications of otherwise valid procedures.

The technique for conducting a small-sample test of hypothesis about a population mean is summarized in the following box.

---

## Small-Sample Test of Hypothesis About $\mu$

**ONE-TAILED TEST**

$H_0: \mu = \mu_0$

$H_a: \mu < \mu_0$ (or $H_a: \mu > \mu_0$)

Test statistic: $t = \dfrac{\bar{x} - \mu_0}{s/\sqrt{n}}$

Rejection region: $t < -t_\alpha$
  (or $t > t_\alpha$ when $H_a: \mu > \mu_0$)

**TWO-TAILED TEST**

$H_0: \mu = \mu_0$

$H_a: \mu \neq \mu_0$

Test statistic: $t = \dfrac{\bar{x} - \mu_0}{s/\sqrt{n}}$

Rejection region: $t < -t_{\alpha/2}$ or $t > t_{\alpha/2}$

where $t_\alpha$ and $t_{\alpha/2}$ are based on $(n - 1)$ degrees of freedom

*Assumption:* A random sample is selected from a population with a relative frequency distribution that is approximately normal.

**EXAMPLE 8.5**

A major car manufacturer wants to test a new engine to determine whether it meets new air-pollution standards. The mean emission $\mu$ of all engines of this type must be less than 20 parts per million of carbon. Ten engines are manufactured for testing purposes, and the emission level of each is determined. The data (in parts per million) are listed below:

⊘ EMISSIONS

| 15.6 | 16.2 | 22.5 | 20.5 | 16.4 | 19.4 | 16.6 | 17.9 | 12.7 | 13.9 |

Do the data supply sufficient evidence to allow the manufacturer to conclude that this type of engine meets the pollution standard? Assume that the manufacturer is willing to risk a Type I error with probability $\alpha = .01$.

### Solution

The manufacturer wants to support the research hypothesis that the mean emission level $\mu$ for all engines of this type is less than 20 parts per million. The elements of this small-sample one-tailed test are

$$H_0: \mu = 20$$
$$H_a: \mu < 20$$
$$\text{Test statistic: } t = \frac{\bar{x} - 20}{s/\sqrt{n}}$$

*Assumption:* The relative frequency distribution of the population of emission levels for all engines of this type is approximately normal.

*Rejection region:* For $\alpha = .01$ and df $= n - 1 = 9$, the one-tailed rejection region (see Figure 8.11) is $t < -t_{.01} = -2.821$.

To calculate the test statistic, we entered the data into a computer and analyzed it using SAS. The SAS printout is shown in Figure 8.12. From the printout, we obtain $\bar{x} = 17.17$ and $s = 2.98$. Substituting these values into the test statistic formula, we get (rounding)

$$t = \frac{\bar{x} - 20}{s/\sqrt{n}} = \frac{17.17 - 20}{2.98/\sqrt{10}} = -3.00$$

Since the calculated $t$ falls in the rejection region (see Figure 8.11), the manufacturer concludes that $\mu < 20$ parts per million and the new engine type meets the pollution standard. Are you satisfied with the reliability associated with this inference? The probability is only $\alpha = .01$ that the test would support the research hypothesis if in fact it were false. ∎

**FIGURE 8.11**

*A t-Distribution with 9 df and the Rejection Region for Example 8.5*

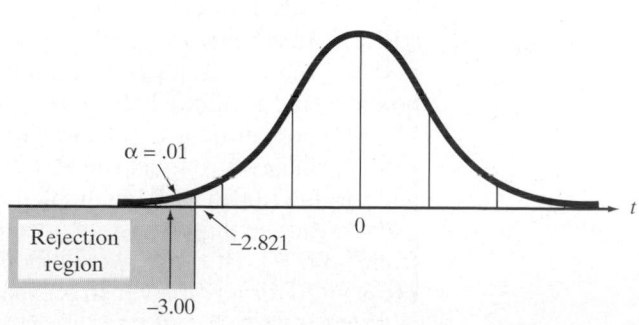

## FIGURE 8.12

*SAS Analysis of 10 Emission Levels*

### One Sample t-test for a Mean

Sample Statistics for EMIT

| N | Mean | Std. Dev. | Std. Error |
|---|------|-----------|------------|
| 10 | 17.17 | 2.98 | 0.94 |

Hypothesis Test

Null hypothesis:    Mean of EMIT => 20
Alternative:        Mean of EMIT <  20

| t Statistic | Df | Prob > t |
|-------------|----|-----------| 
| -3.002 | 9 | 0.0075 |

---

**EXAMPLE 8.6**

Find the observed significance level for the test described in Example 8.5. Interpret the result.

### Solution

The test of Example 8.5 was a lower-tailed test: $H_0: \mu = 20$ versus $H_a: \mu < 20$. Since the value of $t$ computed from the sample data was $t = -3.00$, the observed significance level (or $p$-value) for the test is equal to the probability that $t$ would assume a value less than or equal to $-3.00$ if in fact $H_0$ were true. This is equal to the area in the lower tail of the $t$-distribution (highlighted in Figure 8.13).

## FIGURE 8.13

*The Observed Significance Level for the Test of Example 8.5*

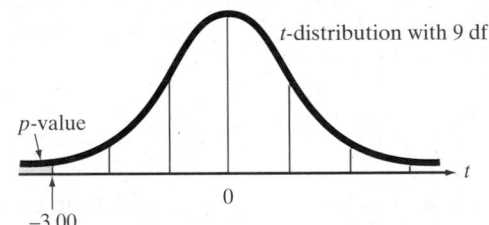

One way to find this area—i.e., the $p$-value for the test—is to consult the $t$-table (Table VI in Appendix A). Unlike the table of areas under the normal curve, Table VI gives only the $t$ values corresponding to the areas .100, .050, .025, .010, .005, .001, and .0005. Therefore, we can only approximate the $p$-value for the test. Since the observed $t$ value was based on 9 degrees of freedom, we use the df = 9 row in Table VI and move across the row until we reach the $t$ values that are closest to the observed $t = -3.00$. [*Note:* We ignore the minus sign.] The $t$ values corresponding to $p$-values of .010 and .005 are 2.821 and 3.250, respectively. Since the observed $t$ value falls between $t_{.010}$ and $t_{.005}$, the $p$-value for the test lies between .005 and .010. In other words, $.005 < p\text{-value} < .01$. Thus, we would reject the null hypothesis, $H_0: \mu = 20$ parts per million, for any value of $\alpha$ larger than .01 (the upper bound of the $p$-value).

A second, more accurate, way to obtain the $p$-value is to use a statistical software package to conduct the test of hypothesis. Both the test statistic ($-3.002$) and $p$-value ($.0075$) are highlighted on the SAS printout, Figure. 8.12.

You can see that the actual $p$-value of the test falls within the bounds obtained from Table VI. Thus, the two methods agree; we will reject $H_0: \mu = 20$ in favor of $H_a: \mu < 20$ for any $\alpha$ level larger than .01.

Small-sample inferences typically require more assumptions and provide less information about the population parameter than do large-sample inferences. Nevertheless, the *t*-test is a method of testing a hypothesis about a population mean of a normal distribution when only a small number of observations is available.

---

### What Can Be Done if the Population Relative Frequency Distribution Departs Greatly from Normal?

*Answer:*   Use one of the nonparametric statistical methods of Chapter 14.

---

 EXERCISES 8.45–8.60

### Learning the Mechanics

**8.45** Under what circumstances should you use the *t*-distribution in testing a hypothesis about a population mean?

**8.46** In what ways are the distributions of the *z* statistic and *t* statistic alike? How do they differ?

**8.47** For each of the following rejection regions, sketch the sampling distribution of *t*, and indicate the location of the rejection region on your sketch:
   **a.** $t > 1.440$ where df $= 6$
   **b.** $t < -1.782$ where df $= 12$
   **c.** $t < -2.060$ or $t > 2.060$ where df $= 25$

**8.48** For each of the rejection regions defined in Exercise 8.47, what is the probability that a Type I error will be made?

**8.49** A random sample of *n* observations is selected from a normal population to test the null hypothesis that $\mu = 10$. Specify the rejection region for each of the following combinations of $H_a$, $\alpha$, and *n*:
   **a.** $H_a: \mu \neq 10; \alpha = .05; n = 14$
   **b.** $H_a: \mu > 10; \alpha = .01; n = 24$
   **c.** $H_a: \mu > 10; \alpha = .10; n = 9$
   **d.** $H_a: \mu < 10; \alpha = .01; n = 12$
   **e.** $H_a: \mu \neq 10; \alpha = .10; n = 20$
   **f.** $H_a: \mu < 10; \alpha = .05; n = 4$

**8.50** The following sample of six measurements was randomly selected from a normally distributed population: $1, 3, -1, 5, 1, 2$.

   **a.** Test the null hypothesis that the mean of the population is 3 against the alternative hypothesis, $\mu < 3$. Use $\alpha = .05$.
   **b.** Test the null hypothesis that the mean of the population is 3 against the alternative hypothesis, $\mu \neq 3$. Use $\alpha = .05$.
   **c.** Find the observed significance level for each test.

**8.51** A sample of five measurements, randomly selected from a normally distributed population, resulted in the following summary statistics: $\bar{x} = 4.8, s = 1.3$.
   **a.** Test the null hypothesis that the mean of the population is 6 against the alternative hypothesis, $\mu < 6$. Use $\alpha = .05$.
   **b.** Test the null hypothesis that the mean of the population is 6 against the alternative hypothesis, $\mu \neq 6$. Use $\alpha = .05$.
   **c.** Find the observed significance level for each test.

**8.52** MINITAB is used to conduct a *t*-test for the null hypothesis $H_0: \mu = 1,000$ versus the alternative hypothesis $H_a: \mu > 1,000$ based on a sample of 17 observations. The software's output is shown below.
   **a.** What assumptions are necessary for the validity of this procedure?
   **b.** Interpret the results of the test.
   **c.** Suppose the alternative hypothesis had been the two-tailed $H_a: \mu \neq 1,000$. If the *t* statistic were unchanged, what would the *p*-value be for this test? Interpret the *p*-value for the two-tailed test.

**MINITAB Output for Exercise 8.52**

```
Test of mu = 1000 vs mu > 1000
```

| Variable | N | Mean | StDev | SE Mean |
|----------|-----|------|-------|---------|
| X | 17 | 1020 | 43.54 | 10.56 |

| Variable | 95.0% Lower Bound | T | P |
|----------|-------------------|------|-------|
| X | 981.62 | 1.89 | .0382 |

## Applying the Concepts—Basic

**8.53** Refer to the *Psychological Reports* (Aug. 1997) study of college students who completed the Dental Anxiety Scale, Exercise 5.28 (p. 235). Recall that scores range from 0 (no anxiety) to 20 (extreme anxiety). Summary statistics for the scores of the 27 students who completed the questionnaire follow: $\bar{x} = 10.7$, $s = 3.6$. Conduct a test of hypothesis to determine if the mean Dental Anxiety Scale score for the population of college students differs from $\mu = 11$. Use $\alpha = .05$.

**8.54** The Cleveland Casting Plant produces iron automotive castings for Ford Motor Company. When the process is stable, the target pouring temperature of the molten iron is 2,550 degrees (*Quality Engineering*, Vol. 7, 1995). The pouring temperatures (in degrees Fahrenheit) for a random sample of 10 crankshafts produced at the plant are listed in the table. Conduct a test to determine whether the true mean pouring temperature differs from the target setting. Test using $\alpha = .01$. Use the accompanying SAS printout to carry out the test.

### ⊘ IRONTEMP

| 2,543 | 2,541 | 2,544 | 2,620 | 2,560 | 2,559 | 2,562 |
|-------|-------|-------|-------|-------|-------|-------|
| 2,553 | 2,552 | 2,553 | | | | |

*Source:* Price, B., & Barth, B. "A structural model relating process inputs and final product characteristics." *Quality Engineering*, Vol. 7, No. 4, 1995, p. 696 (Table 2).

**SAS Output for Exercise 8.54**

One Sample t-test for a Mean

Sample Statistics for IRONTEMP

| N | Mean | Std. Dev. | Std. Error |
|-----|---------|-----------|------------|
| 10 | 2558.70 | 22.75 | 7.19 |

Hypothesis Test

Null hypothesis:    Mean of IRONTEMP = 2550
Alternative:        Mean of IRONTEMP ^= 2550

| t Statistic | Df | Prob > t |
|-------------|-----|----------|
| 1.210 | 9 | 0.2573 |

**8.55** In Exercise 7.29 (p. 306), you analyzed data from a study of the diets of spinifex pigeons. The data, extracted from the *Australian Journal of Zoology*, are reproduced in the table. Each measurement in the data set represents the weight (in grams) of the crop content of a spinifex pigeon. MINITAB was used to conduct a test of the null hypothesis $H_0: \mu = 1$, where $\mu$ represents the mean weight in the crops of all spinifex pigeons. The resulting printout is shown at the bottom of the page.

**a.** Identify the alternative hypothesis specified on the printout.

### ⊘ PIGEONS

| .457 | 3.751 | .238 | 2.967 | 2.509 | 1.384 | 1.454 | .818 |
|------|-------|------|-------|-------|-------|-------|------|
| .335 | 1.436 | 1.603 | 1.309 | .201 | .530 | 2.144 | .834 |

**b.** Locate the value of the test statistic and give its value.

**c.** Find and interpret the *p*-value of the test shown on the printout.

## Applying the Concepts—Intermediate

**8.56** A study was conducted to evaluate the effectiveness of a new mosquito repellent designed by the U.S. Army to be applied as camouflage face paint (*Journal of the Mosquito Control Association*, June 1995). The repellent was applied to the forearms of five volunteers and then they were exposed to fifteen active mosquitos for a ten-hour period. The percentage of the forearm surface area protected from bites (called percent repellency) was calculated for each of the five volunteers. For one color of paint (loam), the following summary statistics were obtained:

$$\bar{x} = 83\% \quad s = 15\%$$

**a.** The new repellent is considered effective if it provides a percent repellency of at least 95. Conduct a test to determine whether the mean repellency percentage of the new mosquito repellent is less than 95. Test using $\alpha = .10$.

**b.** What assumptions are required for the hypothesis test in part **a** to be valid?

**8.57** The Mississippi Department of Transportation collected data on the number of cracks (called *crack intensity*) in an undivided two-lane highway using van-mounted state-of-the-art video technology (*Journal of Infrastructure Systems*, Mar. 1995). The mean number of cracks found in a sample of eight 50-meter sections of the highway was

**MINITAB Output for Exercise 8.55**

Test of mu = 1 vs mu not = 1

| Variable | N | Mean | StDev | SE Mean |
|----------|-----|-------|-------|---------|
| WEIGHT | 16 | 1.373 | 1.034 | 0.258 |

| Variable | 95.0% CI | T | P |
|----------|------------------|------|-------|
| WEIGHT | ( 0.822, 1.924) | 1.44 | 0.169 |

$\bar{x} = .210$, with a variance of $s^2 = .011$. Suppose the American Association of State Highway and Transportation Officials (AASHTO) recommends a maximum mean crack intensity of .100 for safety purposes. Test the hypothesis that the true mean crack intensity of the Mississippi highway exceeds the AASHTO recommended maximum. Use $\alpha = .01$.

**8.58** One of the most feared predators in the ocean is the great white shark. It is known that the white shark grows to a mean length of 21 feet; however, one marine biologist believes that great white sharks off the Bermuda coast grow much longer owing to unusual feeding habits. To test this claim, some full-grown great white sharks were captured off the Bermuda coast, measured, and then set free. However, because the capture of sharks is difficult, costly, and very dangerous, only three specimens were sampled. Their lengths were 24, 20, and 22 feet.

**a.** Do the data provide sufficient evidence to support the marine biologist's claim? Use $\alpha = .10$.

**b.** Give the approximate observed significance level for the test in part **a**, and interpret its value.

**c.** What assumptions must be made in order to carry out the test?

**d.** Do you think these assumptions are likely to be satisfied in this sampling situation?

**8.59** The SCL-90-R is a 90-item symptom inventory checklist designed to reflect the psychological status of an individual. Each symptom (e.g., obsessive-compulsive behavior) is scored on a scale of 0 (none) to 4 (extreme). The total of these scores yields an individual's Positive Symptom Total (PST). "Normal" individuals are known to have a mean PST of about 40. The *Journal of Head Trauma Rehabilitation* (Apr. 1995) reported that a sample of 23 patients diagnosed with mild to moderate traumatic brain injury had a mean PST score of $\bar{x} = 48.43$ and a standard deviation of $s = 20.76$. Is there sufficient evidence to claim that the true mean PST score of all patients with mild to moderate traumatic brain injury exceeds the "normal" value of 40? Test using $\alpha = .05$.

### Applying the Concepts—Advanced

**8.60** "Hot Tamales" are chewy, cinnamon-flavored candies. A bulk vending machine is known to dispense, on average, 15 Hot Tamales per bag. *Chance* (Fall 2000) published an article on a classroom project in which students were required to purchase bags of Hot Tamales from the machine and count the number of candies per bag. One student group claimed they purchased five bags that had the following candy counts: 25, 23, 21, 21, and 20. There was some question as to whether the students had fabricated the data. Use a hypothesis test to gain insight into whether or not the data collected by the students are fabricated. Use a level of significance that gives the benefit of the doubt to the students.

## 8.5 LARGE-SAMPLE TEST OF HYPOTHESIS ABOUT A POPULATION PROPORTION

Inferences about population proportions (or percentages) are often made in the context of the probability, $p$, of "success" for a binomial distribution. We saw how to use large samples from binomial distributions to form confidence intervals for $p$ in Section 7.3. We now consider tests of hypotheses about $p$.

Consider, for example, a method currently used by doctors to screen women for possible breast cancer. The method fails to detect cancer in 20% of the women who actually have the disease. Suppose a new method has been developed that researchers hope will detect cancer more accurately. This new method was used to screen a random sample of 140 women known to have breast cancer. Of these, the new method failed to detect cancer in 12 women. Does this sample provide evidence that the failure rate of the new method differs from the one currently in use?

We first view this as a binomial experiment with 140 screened women as the trials and failure to detect breast cancer as "Success" (in binomial terminology). Let $p$ represent the probability that the new method fails to detect the breast cancer. If the new method is no better than the current one, then the failure rate is $p = .2$. On the other hand, if the new method is either better or worse than the current method, then the failure rate is either smaller or larger than 20%; that is, $p \neq .2$.

We can now place the problem in the context of a test of hypothesis:

$$H_0: p = .2$$
$$H_a: p \neq .2$$

Recall that the sample proportion, $\hat{p}$, is really just the sample mean of the outcomes of the individual binomial trials and, as such, is approximately normally distributed (for large samples) according to the Central Limit Theorem. Thus, for large samples we can use the standard normal $z$ as the test statistic:

$$\textit{Test statistic:} \quad z = \frac{\text{Sample proportion} - \text{Null hypothesized proportion}}{\text{Standard deviation of sample proportion}}$$

$$= \frac{\hat{p} - p_0}{\sigma_{\hat{p}}}$$

where we use the symbol $p_0$ to represent the null hypothesized value of $p$.

*Rejection region:* We use the standard normal distribution to find the appropriate rejection region for the specified value of $\alpha$. Using $\alpha = .05$, the two-tailed rejection region is

$$z < -z_{\alpha/2} = -z_{.025} = -1.96 \quad \text{or} \quad z > z_{\alpha/2} = z_{.025} = 1.96$$

See Figure 8.14.

**FIGURE 8.14**

*Rejection Region for Breast Cancer Example*

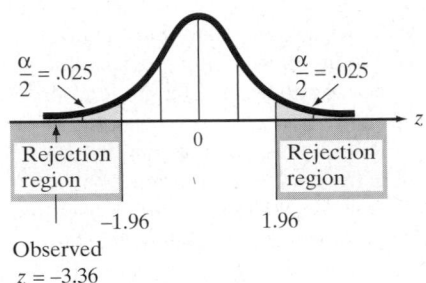

We are now prepared to calculate the value of the test statistic. Before doing so, we want to be sure that the sample size is large enough to ensure that the normal approximation for the sampling distribution of $\hat{p}$ is reasonable. To check this, we calculate a 3-standard-deviation interval around the null hypothesized value, $p_0$, which is assumed to be the true value of $p$ until our test procedure proves otherwise. Recall that $\sigma_{\hat{p}} = \sqrt{pq/n}$ and that we need an estimate of the product $pq$ in order to calculate a numerical value of the test statistic $z$. Since the null hypothesized value is generally the accepted-until-proven-otherwise value, we use the value of $p_0 q_0$ (where $q_0 = 1 - p_0$) to estimate $pq$ in the calculation of $z$. Thus,

$$\sigma_{\hat{p}} = \sqrt{\frac{pq}{n}} \approx \sqrt{\frac{p_0 q_0}{n}} = \sqrt{\frac{(.2)(.8)}{140}} = .034$$

and the 3-standard-deviation interval around $p_0$ is

$$p_0 \pm 3\sigma_{\hat{p}} \approx .2 \pm 3(.034) = (.166, .234)$$

As long as this interval does not contain 0 or 1 (i.e., is completely contained in the interval 0 to 1), as is the case here, the normal distribution will provide a reasonable approximation for the sampling distribution of $\hat{p}$.

Returning to the hypothesis test at hand, the proportion of the screenings that failed to detect breast cancer is

$$\hat{p} = \frac{12}{140} = .086$$

Finally, we calculate the number of standard deviations (the $z$ value) between the sampled and hypothesized value of the binomial proportion:

$$z = \frac{\hat{p} - p_0}{\sigma_{\hat{p}}} = \frac{\hat{p} - p_0}{\sqrt{p_0 q_0 / n}} - \frac{.086 - .2}{.034} = \frac{-.114}{.034} = -3.36$$

The implication is that the observed sample proportion is (approximately) 3.36 standard deviations below the null hypothesized proportion .2 (Figure 8.14). Therefore, we reject the null hypothesis, concluding at the .05 level of significance that the true failure rate of the new method for detecting breast cancer differs from .20. Since $\hat{p} = .086$, it appears that the new method is better (i.e., has a smaller failure rate) than the method currently in use. (To estimate the magnitude of the failure rate for the new method, a confidence interval can be constructed.)

The test of hypothesis about a population proportion $p$ is summarized in the next box. Note that the procedure is entirely analogous to that used for conducting large-sample tests about a population mean.

---

**Large-Sample Test of Hypothesis About $p$**

**ONE-TAILED TEST**

$H_0: p = p_0$

$H_a: p < p_0$
  (or $H_a: p > p_0$)

Test statistic:  $z = \dfrac{\hat{p} - p_0}{\sigma_{\hat{p}}}$

**TWO-TAILED TEST**

$H_0: p = p_0$

$H_a: p \neq p_0$

Test statistic:  $z = \dfrac{\hat{p} - p_0}{\sigma_{\hat{p}}}$

where, $p_0 =$ hypothesized value of $p$, $\sigma_{\hat{p}} = \sqrt{p_0 q_0 / n}$ and $q_0 = 1 - p_0$

*Rejection region:*  $z < -z_\alpha$
  (or $z > z_\alpha$ when $H_a: p > p_0$)

*Rejection region:*  $z < -z_{\alpha/2}$ or $z > z_{\alpha/2}$

*Assumption:*   The experiment is binomial, and the sample size is large enough that the interval $p_0 \pm 3\sigma_{\hat{p}}$ does not include 0 or 1.

---

**EXAMPLE 8.7**

The reputations (and hence sales) of many businesses can be severely damaged by shipments of manufactured items that contain a large percentage of defectives. For example, a manufacturer of alkaline batteries may want to be reasonably certain that fewer than 5% of its batteries are defective. Suppose 300 batteries are randomly selected from a very large shipment; each is tested and 10 defective batteries are found. Does this provide sufficient evidence for the manufacturer to conclude that the fraction defective in the entire shipment is less than .05? Use $\alpha = .01$.

**Solution**

Before conducting the test of hypothesis, we check to determine whether the sample size is large enough to use the normal approximation for the sampling distribution of $\hat{p}$. The criterion is tested by the interval

$$p_0 \pm 3\sigma_{\hat{p}} = p_0 \pm 3\sqrt{\frac{p_0 q_0}{n}} = .05 \pm 3\sqrt{\frac{(.05)(.95)}{300}}$$
$$= .05 \pm .04 \quad \text{or} \quad (.01, .09)$$

**FIGURE 8.15**

*Rejection Region for*
*Example 8.7*

Since the interval lies within the interval $(0, 1)$, the normal approximation will be adequate.

The objective of the sampling is to determine whether there is sufficient evidence to indicate that the fraction defective, $p$, is less than .05. Consequently, we will test the null hypothesis that $p = .05$ against the alternative hypothesis that $p < .05$. The elements of the test are

$$H_0: p = .05$$
$$H_a: p < .05$$

*Test statistic:* $\quad z = \dfrac{\hat{p} - p_0}{\sigma_{\hat{p}}}$

*Rejection region:* $\quad z < -z_{.01} = -2.33$ $\qquad$ (see Figure 8.15)

We now calculate the test statistic:

$$z = \frac{\hat{p} - .05}{\sigma_{\hat{p}}} = \frac{(10/300) - .05}{\sqrt{p_0 q_0 / n}} = \frac{.033 - .05}{\sqrt{p_0 q_0 / 300}}$$

Notice that we use $p_0$ to calculate $\sigma_{\hat{p}}$ because, in contrast to calculating $\sigma_{\hat{p}}$ for a confidence interval, the test statistic is computed on the assumption that the null hypothesis is true—that is, $p = p_0$. Therefore, substituting the values for $\hat{p}$ and $p_0$ into the $z$ statistic, we obtain

$$z \approx \frac{-.017}{\sqrt{(.05)(.95)/300}} = \frac{-.017}{.0126} = -1.35$$

As shown in Figure 8.15, the calculated $z$ value does not fall in the rejection region. Therefore, there is insufficient evidence at the .01 level of significance to indicate that the shipment contains fewer than 5% defective batteries. ∎

**E X A M P L E   8 . 8**

In Example 8.7 we found that we did not have sufficient evidence, at the $\alpha = .01$ level of significance, to indicate that the fraction defective $p$ of alkaline batteries was less than $p = .05$. How strong was the weight of evidence favoring the alternative hypothesis ($H_a: p < .05$)? Find the observed significance level (*p*-value) for the test.

**Solution**

The computed value of the test statistic $z$ was $z = -1.35$. Therefore, for this lower-tailed test, the observed significance level is

$$p\text{-value} = P(z \leq -1.35)$$

This lower-tail area is shown in Figure 8.16. The area between $z = 0$ and $z = 1.35$ is given in Table IV in Appendix A as .4115. Therefore, the observed significance

**FIGURE 8.16**

*The Observed Significance Level for Example 8.8*

level is $.5 - .4115 = .0885$. Note that this probability is quite small. Although we did not reject $H_0$: $p = .05$ at $\alpha = .01$, the probability of observing a $z$ value as small as or smaller than $-1.35$ is only $.0885$ if in fact $H_0$ is true. Therefore, we would reject $H_0$ if we choose $\alpha = .10$ (since the observed significance level is less than $.10$), and we would not reject $H_0$ (the conclusion of Example 8.7) if we choose $\alpha = .05$ or $\alpha = .01$

■

Small-sample test procedures are also available for $p$, although most surveys use samples that are large enough to employ the large-sample tests presented in this section. A test of proportions that can be applied to small samples is discussed in Chapter 13.

 EXERCISES 8.61–8.75

### Learning the Mechanics

**8.61** For the binomial sample sizes and null hypothesized values of $p$ in each part, determine whether the sample size is large enough to use the normal approximation methodology presented in this section to conduct a test of the null hypothesis $H_0$: $p = p_0$.
  **a.** $n = 500$, $p_0 = .05$
  **b.** $n = 100$, $p_0 = .99$
  **c.** $n = 50$, $p_0 = .2$
  **d.** $n = 20$, $p_0 = .2$
  **e.** $n = 10$, $p_0 = .4$

**8.62** Suppose a random sample of 100 observations from a binomial population gives a value of $\hat{p} = .69$ and you wish to test the null hypothesis that the population parameter $p$ is equal to $.75$ against the alternative hypothesis that $p$ is less than $.75$.
  **a.** Noting that $\hat{p} = .69$, what does your intuition tell you? Does the value of $\hat{p}$ appear to contradict the null hypothesis?
  **b.** Use the large-sample $z$-test to test $H_0$: $p = .75$ against the alternative hypothesis $H_a$: $p < .75$. Use $\alpha = .05$. How do the test results compare with your intuitive decision from part **a**?
  **c.** Find and interpret the observed significance level of the test you conducted in part **b**.

**8.63** Suppose the sample in Exercise 8.62 has produced $\hat{p} = .84$ and we wish to test $H_0$: $p = .9$ against the alternative $H_a$: $p < .9$.
  **a.** Calculate the value of the $z$ statistic for this test.
  **b.** Note that the numerator of the $z$ statistic ($\hat{p} - p_0 = .84 - .90 = -.06$) is the same as for Exercise 8.62.

Considering this, why is the absolute value of $z$ for this exercise larger than that calculated in Exercise 8.62?
  **c.** Complete the test using $\alpha = .05$ and interpret the result.
  **d.** Find the observed significance level for the test and interpret its value.

**8.64** A random sample of 100 observations is selected from a binomial population with unknown probability of success $p$. The computed value of $\hat{p}$ is equal to $.74$.
  **a.** Test $H_0$: $p = .65$ against $H_a$: $p > .65$. Use $\alpha = .01$.
  **b.** Test $H_0$: $p = .65$ against $H_a$: $p > .65$. Use $\alpha = .10$.
  **c.** Test $H_0$: $p = .90$ against $H_a$: $p \neq .90$. Use $\alpha = .05$.
  **d.** Form a 95% confidence interval for $p$.
  **e.** Form a 99% confidence interval for $p$.

**8.65** Refer to Exercise 7.40 (p. 313), in which 50 consumers taste-tested a new snack food.
  **a.** Test $H_0$: $p = .5$ against $H_a$: $p > .5$, where $p$ is the proportion of customers who do not like the snack food. Use $\alpha = .10$.
  **b.** Report the observed significance level of your test.

**8.66** A statistics student will use statistical software to test the null hypothesis $H_0$: $p = .5$ against the one-tailed alternative, $H_a$: $p > .5$. A sample of 500 observations are input into the computer, which returns the following result: $z = .44$, one-tailed $p$-value $= .3300$.
  **a.** The student concludes, based on the $p$-value, that there is a 33% chance that the alternative hypothesis is true. Do you agree? If not, correct the interpretation.
  **b.** How would the $p$-value change if the alternative hypothesis were two-tailed, $H_a$: $p \neq .5$? Interpret this $p$-value.

## Applying the Concepts—Basic

**8.67** Examining data collected on 835 males from the National Youth Survey (a longitudinal survey of a random sample of U.S. households), researchers at Carnegie Mellon University found that 401 of the male youths were raised in a single-parent family (*Sociological Methods & Research*, Feb. 2001). Does this information allow you to conclude that more than 45% of male youths are raised in a single-parent family? Test at $\alpha = .05$.

**8.68** Pond's Age-Defying Complex, a cream with alpha-hydroxy acid, advertises that it can reduce wrinkles and improve the skin. In a study published in *Archives of Dermatology* (June 1996), 33 women over age 40 used a cream with alpha-hydroxy acid for twenty-two weeks. At the end of the study period, 23 of the women exhibited skin improvement (as judged by a dermatologist).
  **a.** Is this evidence that the cream will improve the skin of more than 60% of women over age 40? Test using $\alpha = .05$.
  **b.** Find and interpret the *p*-value of the test.

**8.69** According to *Forbes* (Dec. 13, 1999), Creative Good, a New York consulting firm, claims that 40% of shoppers fail in their attempts to purchase merchandise online because Web sites are too complex. Another consulting firm asked a random sample of 60 online shoppers to each test a different randomly selected e-commerce Web site. Only 15 reported sufficient frustration with their sites to deter making a purchase. Do these data provide sufficient evidence to reject the claim made by Creative Good? Conduct the appropriate hypothesis test using $\alpha = .01$.

## Applying the Concepts—Intermediate

**8.70** Any sentence that contains an animate noun as a direct object and another noun as a second object (e.g., "Sue offered Ann a cookie.") is termed a double object dative (DOD). The connection between certain verbs and a DOD was investigated in *Applied Psycholinguistics* (June 1998). The subjects were 35 native English speakers who were enrolled in an introductory English composition course at a Hawaiian community college. After viewing pictures of a family, each subject was asked to write a sentence about the family using a specified verb. Of the 35 sentences using the verb "buy," 10 had a DOD structure. Conduct a test to determine if the true fraction of sentences with the verb "buy" that are DODs is less than $\frac{1}{3}$. Use $\alpha = .05$.

**8.71** Refer to the *Nature* (Aug. 27, 1998) study of facial characteristics that are deemed attractive, Exercise 8.43 (p. 346). In another experiment, 67 human subjects viewed side-by-side an image of a Caucasian male face and the same image 50% masculinized. Each subject was asked to select the facial image that they deemed more attractive. Fifty-eight of the 67 subjects felt that masculinization of face shape decreased attractiveness of the male face. The researchers used this sample infor-

mation to test whether the subjects showed preference for either the unaltered or morphed male face.
  **a.** Set up the null and alternative hypotheses for this test.
  **b.** Compute the test statistic.
  **c.** The researchers reported *p*-value $\approx 0$ for the test. Do you agree?
  **d.** Make the appropriate conclusion in the words of the problem. Use $\alpha = .01$.

**8.72** *Meteoritics* (Mar. 1995) reported the results of a study of lunar soil evolution. Data were obtained from the Apollo 16 mission to the moon, during which a 62-cm core was extracted from the soil near the landing site. Monomineralic grains of lunar soil were separated out and examined for coating with dust and glass fragments. Each grain was then classified as coated or uncoated. Of interest is the "coat index," that is, the proportion of grains that are coated. According to soil evolution theory, the coat index will exceed .5 at the top of the core, equal .5 in the middle of the core, and fall below .5 at the bottom of the core. Use the summary data in the accompanying table to test each part of the 3-part theory. Use $\alpha = .05$ for each test.

| | Location (depth) | | |
|---|---|---|---|
| | **Top (4.25 cm)** | **Middle (28.1 cm)** | **Bottom (54.5 cm)** |
| **Number of grains sampled** | 84 | 73 | 81 |
| **Number coated** | 64 | 35 | 29 |

*Source:* Basu, A., and McKay, D. S. "Lunar soil evolution processes and Apollo 16 core 60013/60014." *Meteoritics*, Vol. 30, No. 2, Mar. 1995, p. 166 (Table 2).

**8.73** The *placebo effect* describes the phenomenon of improvement in the condition of a patient taking a placebo—a pill that looks and tastes real but contains no medically active chemicals. Physicians at a clinic in La Jolla, California, gave what they thought were drugs to 7,000 asthma, ulcer, and herpes patients. Although the doctors later learned that the drugs were really placebos, 70% of the patients reported an improved condition (*Forbes*, May 22, 1995). Use this information to test (at $\alpha = .05$) the placebo effect at the clinic. Assume that if the placebo is ineffective, the probability of a patient's condition improving is .5.

## Applying the Concepts—Advanced

**8.74** In Exercise 4.46 (p. 200) you read about a nationwide survey that claimed that 60% of parents with young children condone spanking their child as a regular form of punishment (*Tampa Tribune*, Oct. 5, 2000). In a random sample of 100 parents with young children, how many parents would need to say they condone spanking as a form of punishment in order to refute the claim?

**8.75** To get their names on the ballot of a local election, political candidates often must obtain petitions bearing the signatures of a minimum number of registered voters. In Pinellas County, Florida, a certain political candidate obtained petitions with 18,200 signatures (*St. Petersburg Times*, Apr. 7, 1992). To verify that the names on the petitions were signed by actual registered voters, election officials randomly sampled 100 of the names and checked each for authenticity. Only two were invalid signatures.

**a.** Is 98 out of 100 verified signatures sufficient to believe that more than 17,000 of the total 18,200 signatures are valid?

**b.** Repeat part **a** if only 16,000 valid signatures are required.

## 8.6 CALCULATING TYPE II ERROR PROBABILITIES: MORE ABOUT $\beta$ (OPTIONAL)

In our introduction to hypothesis testing in Section 8.1, we showed that the probability of committing a Type I error, $\alpha$, can be controlled by the selection of the rejection region for the test. Thus, when the test statistic falls in the rejection region and we make the decision to reject the null hypothesis, we do so knowing the error rate for incorrect rejections of $H_0$. The situation corresponding to accepting the null hypothesis, and thereby risking a Type II error, is not generally as controllable. For that reason, we adopted a policy of nonrejection of $H_0$ when the test statistic does not fall in the rejection region, rather than risking an error of unknown magnitude.

To see how $\beta$, the probability of a Type II error, can be calculated for a test of hypothesis, recall the example in Section 8.1 in which a city tests a manufacturer's pipe to see whether it meets the requirement that the mean strength exceed 2,400 pounds per linear foot. The setup for the test is as follows:

$$H_0: \mu = 2,400$$
$$H_a: \mu > 2,400$$
$$\text{Test statistic: } z = \frac{\bar{x} - 2,400}{\sigma/\sqrt{n}}$$
$$\text{Rejection region: } z > 1.645 \qquad \text{for } \alpha = .05$$

Figure 8.17a shows the rejection region for the **null distribution**—that is, the distribution of the test statistic assuming the null hypothesis is true. The area in the rejection region is .05, and this area represents $\alpha$, the probability that the test statistic leads to rejection of $H_0$ when in fact $H_0$ is true.

The Type II error probability $\beta$ is calculated assuming that the null hypothesis is false, because it is defined as the *probability of accepting $H_0$ when it is false*. Since $H_0$ is false for any value of $\mu$ exceeding 2,400, one value of $\beta$ exists for each possible value of $\mu$ greater than 2,400 (an infinite number of possibilities). Figures 8.17b, 8.17c, and 8.17d show three of the possibilities, corresponding to alternative hypothesis values of $\mu$ equal to 2,425, 2,450, and 2,475, respectively. Note that $\beta$ is the area in the *nonrejection* (or *acceptance*) *region* in each of these distributions and that $\beta$ decreases as the true value of $\mu$ moves farther from the null hypothesized value of $\mu = 2,400$. This is sensible because the probability of incorrectly accepting the null hypothesis should decrease as the distance between the null and alternative values of $\mu$ increases.

In order to calculate the value of $\beta$ for a specific value of $\mu$ in $H_a$, we proceed as follows:

1. Calculate the value of $\bar{x}$ that corresponds to the border between the acceptance and rejection regions. For the sewer pipe example, this is the value of $\bar{x}$ that lies 1.645 standard deviations above $\mu = 2,400$ in the sampling distribution

**FIGURE 8.17**

*Values of α and β for Various Values of μ*

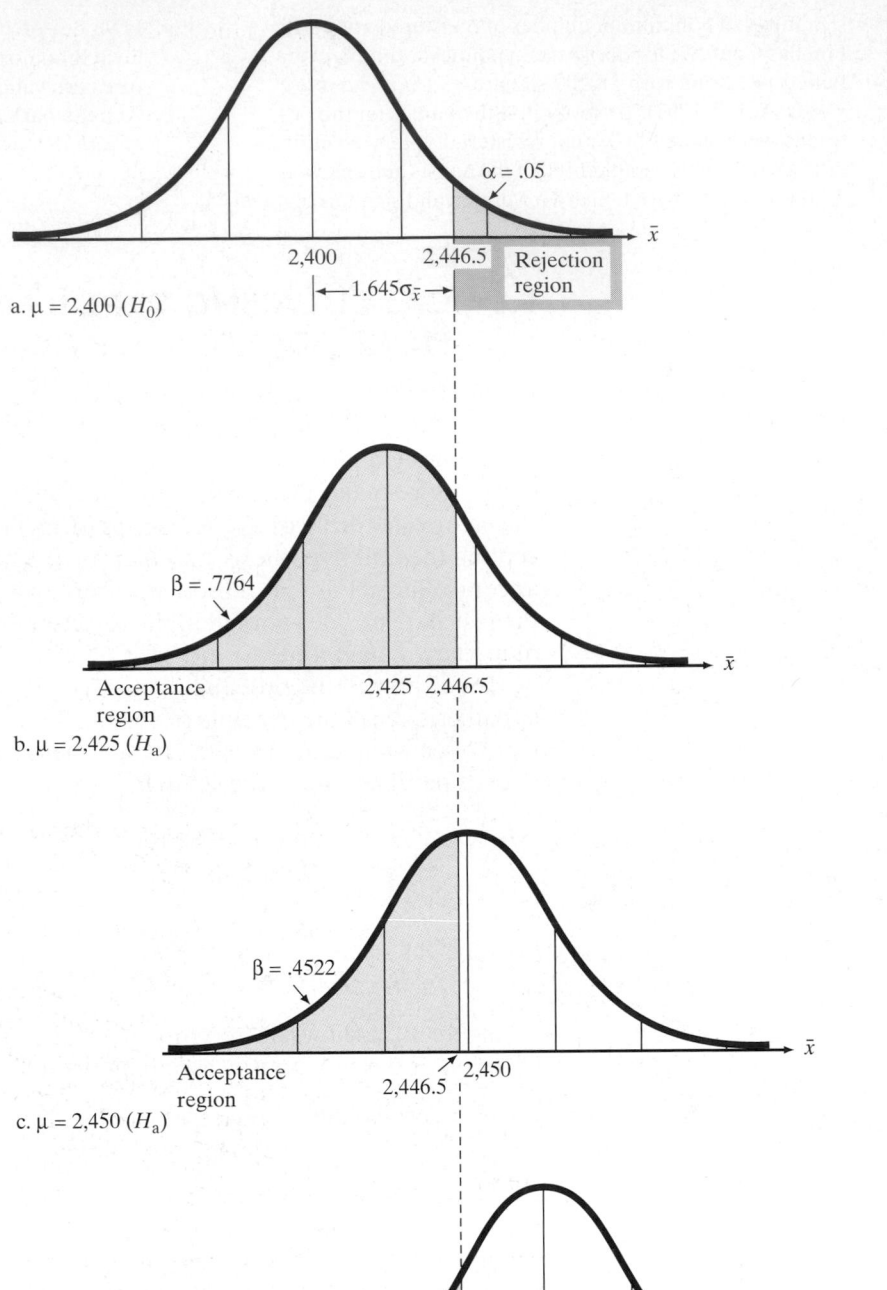

a. μ = 2,400 ($H_0$)

b. μ = 2,425 ($H_a$)

c. μ = 2,450 ($H_a$)

d. μ = 2,475 ($H_a$)

of $\bar{x}$. Denoting this value by $\bar{x}_0$, corresponding to the largest value of $\bar{x}$ that supports the null hypothesis, we find (recalling that $s = 200$ and $n = 50$)

$$\bar{x}_0 = \mu_0 + 1.645\sigma_x = 2,400 + 1.645\left(\frac{\sigma}{\sqrt{n}}\right)$$

$$\approx 2,400 + 1.645\left(\frac{s}{\sqrt{n}}\right) = 2,400 + 1.645\left(\frac{200}{\sqrt{50}}\right)$$

$$= 2,400 + 1.645(28.28) = 2,446.5$$

2. For a particular alternative distribution corresponding to a value of $\mu$ denoted by $\mu_a$, we calculate the $z$ value corresponding to $\bar{x}_0$, the border between the rejection and acceptance regions. We then use this $z$ value and Table IV of Appendix A to determine the area in the *acceptance region* under the alternative distribution. This area is the value of $\beta$ corresponding to the particular alternative $\mu_a$. For example, for the alternative $\mu_a = 2,425$, we calculate

$$z = \frac{\bar{x}_0 - 2,425}{\sigma_{\bar{x}}} = \frac{\bar{x}_0 - 2,425}{\sigma/\sqrt{n}}$$

$$\approx \frac{\bar{x}_0 - 2,425}{s/\sqrt{n}} = \frac{2,446.5 - 2,425}{28.28} = .76$$

Note in Figure 8.17b that the area in the acceptance region is the area to the left of $z = .76$. This area is

$$\beta = .5 + .2764 = .7764$$

Thus, the probability that the test procedure will lead to an incorrect acceptance of the null hypothesis $\mu = 2,400$ when in fact $\mu = 2,425$ is about .78. As the average strength of the pipe increases to 2,450, the value of $\beta$ decreases to .4522 (Figure 8.17c). If the mean strength is further increased to 2,475, the value of $\beta$ is further decreased to .1562 (Figure 8.17d). Thus, even if the true mean strength of the pipe exceeds the minimum specification by 75 pounds per linear foot, the test procedure will lead to an incorrect acceptance of the null hypothesis (rejection of the pipe) approximately 16% of the time. The upshot is that the pipe must be manufactured so that the mean strength well exceeds the minimum requirement if the manufacturer wants the probability of its acceptance by the city to be large (i.e., $\beta$ to be small).

The steps for calculating $\beta$ for a large-sample test about a population mean are summarized in the next box.

---

## Steps for Calculating $\beta$ for a Large-Sample Test About $\mu$

1. Calculate the value(s) of $\bar{x}$ corresponding to the border(s) of the rejection region. There will be one border value for a one-tailed test, and two for a two-tailed test. The formula is one of the following, corresponding to a test with level of significance $\alpha$:

*Upper-tailed test:* $\quad \bar{x}_0 = \mu_0 + z_\alpha\sigma_{\bar{x}} \approx \mu_0 + z_\alpha\left(\frac{s}{\sqrt{n}}\right)$

*Lower-tailed test:* $\quad \bar{x}_0 = \mu_0 - z_\alpha\sigma_{\bar{x}} \approx \mu_0 - z_\alpha\left(\frac{s}{\sqrt{n}}\right)$

*(continues on next page)*

*Two-tailed test:* $\bar{x}_{0,\mathrm{L}} = \mu_0 - z_{\alpha/2}\sigma_{\bar{x}} \approx \mu_0 - z_{\alpha/2}\left(\dfrac{s}{\sqrt{n}}\right)$

$$\bar{x}_{0,\mathrm{U}} = \mu_0 + z_{\alpha/2}\sigma_{\bar{x}} \approx \mu_0 + z_{\alpha/2}\left(\dfrac{s}{\sqrt{n}}\right)$$

2. Specify the value of $\mu_a$ in the alternative hypothesis for which the value of $\beta$ is to be calculated. Then convert the border value(s) of $\bar{x}_0$ to $z$ value(s) using the alternative distribution with mean $\mu_a$. The general formula for the $z$ value is

$$z = \frac{\bar{x}_0 - \mu_a}{\sigma_{\bar{x}}}$$

Sketch the alternative distribution (centered at $\mu_a$), and shade the area in the acceptance (nonrejection) region. Use the $z$ statistic(s) and Table IV of Appendix A to find the shaded area, which is $\beta$.

Following the calculation of $\beta$ for a particular value of $\mu_a$, you should interpret the value in the context of the hypothesis testing application. It is often useful to interpret the value of $1 - \beta$, which is known as the *power of the test* corresponding to a particular alternative, $\mu_a$. Since $\beta$ is the probability of accepting the null hypothesis when the alternative hypothesis is true with $\mu = \mu_a$, $1 - \beta$ is the probability of the complementary event, or the probability of rejecting the null hypothesis when the alternative $H_a: \mu = \mu_a$ is true. That is, the power $(1 - \beta)$ measures the likelihood that the test procedure will lead to the correct decision (reject $H_0$) for a particular value of the mean in the alternative hypothesis.

**DEFINITION 8.2**

The **power of a test** is the probability that the test will correctly lead to the rejection of the null hypothesis for a particular value of $\mu$ in the alternative hypothesis. The power is equal to $(1 - \beta)$ for the particular alternative considered.

For example, in the sewer pipe example we found that $\beta = .7764$ when $\mu = 2,425$. This is the probability that the test leads to the (incorrect) acceptance of the null hypothesis when $\mu = 2,425$. Or, equivalently, the power of the test is $1 - .7764 = .2236$, which means that the test will lead to the (correct) rejection of the null hypothesis only 22% of the time when the pipe exceeds specifications by 25 pounds per linear foot. When the manufacturer's pipe has a mean strength of 2,475 (that is, 75 pounds per linear foot in excess of specifications), the power of the test increases to $1 - .1562 = .8438$. That is, the test will lead to the acceptance of the manufacturer's pipe 84% of the time if $\mu = 2,475$.

**EXAMPLE 8.9**

Recall the drug experiment in Examples 8.1 and 8.2, in which we tested to determine whether the mean response time for rats injected with a drug differs from the control mean response time of $\mu = 1.2$ seconds. The test setup is repeated here:

$$H_0: \mu = 1.2$$
$$H_a: \mu \neq 1.2 \qquad \text{(i.e., } \mu < 1.2 \text{ or } \mu > 1.2\text{)}$$
$$\textit{Test statistic:} \quad z = \frac{\bar{x} - 1.2}{\sigma_{\bar{x}}}$$

$$\text{Rejection region:} \quad z < -1.96 \quad \text{or} \quad z > 1.96 \quad \text{for } \alpha = .05$$
$$z < -2.575 \quad \text{or} \quad z > 2.575 \quad \text{for } \alpha = .01$$

Note that two rejection regions have been specified corresponding to values of $\alpha = .05$ and $\alpha = .01$, respectively. Assume that $n = 100$ and $s = .5$.

a. Suppose drug-injected rats have a mean response time of 1.1 seconds, that is, $\mu = 1.1$. Calculate the values of $\beta$ corresponding to the two rejection regions. Discuss the relationship between the values of $\alpha$ and $\beta$.

b. Calculate the power of the test for each of the rejection regions when $\mu = 1.1$.

### Solution

a. We first consider the rejection region corresponding to $\alpha = .05$. The first step is to calculate the border values of $\bar{x}$ corresponding to the two-tailed rejection region, $z < -1.96$ or $z > 1.96$:

$$\bar{x}_{0,\,L} = \mu_0 - 1.96\sigma_{\bar{x}} \approx \mu_0 - 1.96\left(\frac{s}{\sqrt{n}}\right) = 1.2 - 1.96\left(\frac{.5}{10}\right) = 1.102$$

$$\bar{x}_{0,\,U} = \mu_0 + 1.96\sigma_x \approx \mu_0 + 1.96\left(\frac{s}{\sqrt{n}}\right) = 1.2 + 1.96\left(\frac{.5}{10}\right) = 1.298$$

These border values are shown in Figure 8.18.

Next, we convert these values to $z$ values in the alternative distribution with $\mu_a = 1.1$:

$$z_L = \frac{\bar{x}_{0,\,L} - \mu_a}{\sigma_{\bar{x}}} \approx \frac{1.102 - 1.1}{.05} = .04$$

$$z_U = \frac{\bar{x}_{0,\,U} - \mu_a}{\sigma_{\bar{x}}} \approx \frac{1.298 - 1.1}{.05} = 3.96$$

These $z$ values are shown in Figure 8.18b. You can see that the acceptance (or nonrejection) region is the area between them. Using Table IV of Appendix A, we find that the area between $z = 0$ and $z = .04$ is .0160, and the area between $z = 0$ and $z = 3.96$ is (approximately) .5 (since $z = 3.96$ is off the scale of Table IV). Then the area between $z = .04$ and $z = 3.96$ is, approximately,

$$\beta = .5 - .0160 = .4840$$

Thus, the test with $\alpha = .05$ will lead to a Type II error about 48% of the time when the mean reaction time for drug-injected rats is .1 second less than the control mean response time.

For the rejection region corresponding to $\alpha = .01$, $z < -2.575$ or $z > 2.575$, we find

$$\bar{x}_{0,\,L} = 1.2 - 2.575\left(\frac{.5}{10}\right) = 1.0712$$

$$\bar{x}_{0,\,U} = 1.2 + 2.575\left(\frac{.5}{10}\right) = 1.3288$$

These border values of the rejection region are shown in Figure 8.18c.

Converting these to $z$ values in the alternative distribution with $\mu_a = 1.1$, we find $z_L = -.58$ and $z_U = 4.58$. The area between these values is, approximately,

$$\beta = .2190 + .5 = .7190$$

**FIGURE 8.18**

*Calculation of β for Drug-Injected Rats (Example 8.9)*

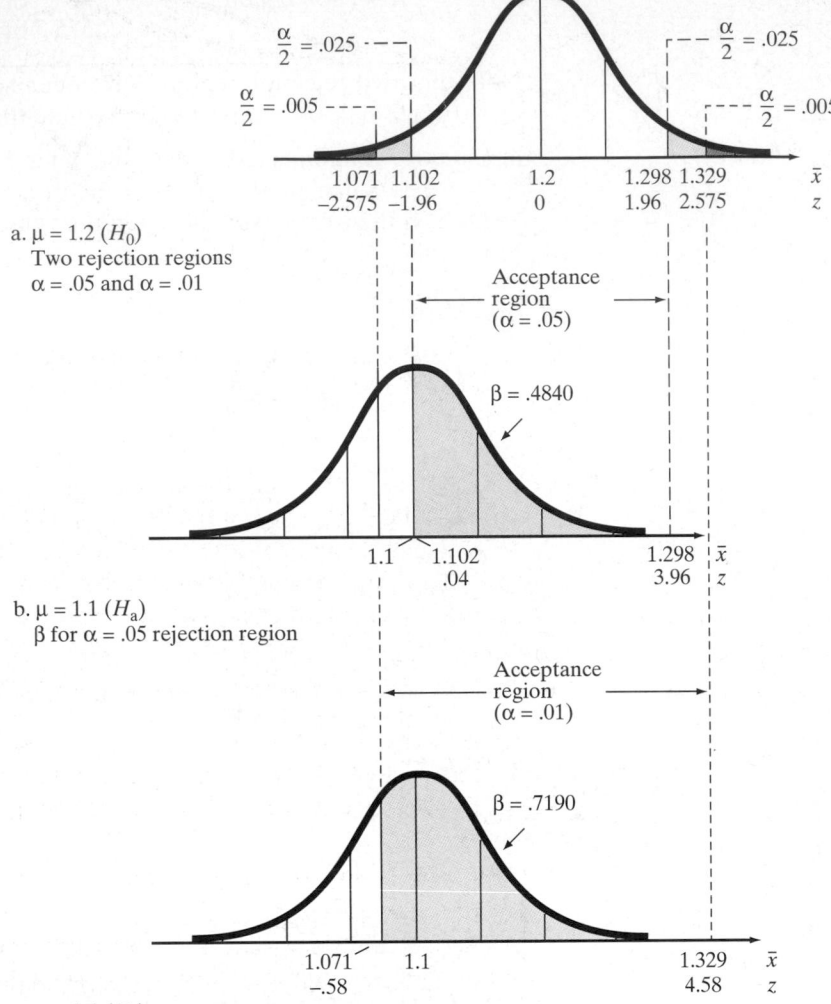

a. $\mu = 1.2$ ($H_0$)
   Two rejection regions
   $\alpha = .05$ and $\alpha = .01$

b. $\mu = 1.1$ ($H_a$)
   β for $\alpha = .05$ rejection region

c. $\mu = 1.1$ ($H_a$)
   β for $\alpha = .01$ rejection region

Thus, the chance that the test procedure with $\alpha = .01$ will lead to an incorrect acceptance of $H_0$ is about 72%.

Note that the value of β increases from .4840 to .7190 when we decrease the value of $\alpha$ from .05 to .01. This is a general property of the relationship between $\alpha$ and β: *as $\alpha$ is decreased (increased), β is increased (decreased).*

b. The power is defined to be the probability of (correctly) rejecting the null hypothesis when the alternative is true. When $\mu = 1.1$ and $\alpha = .05$, we find

$$\text{Power} = 1 - \beta = 1 - .4840 = .5160$$

When $\mu = 1.1$ and $\alpha = .01$, we find

$$\text{Power} = 1 - \beta = 1 - .7190 = .2810$$

You can see that the power of the test is decreased as the level of $\alpha$ is decreased. This means that as the probability of incorrectly rejecting the null hypothesis is decreased, the probability of correctly accepting the null hypothesis for a given alternative is also decreased. Thus, the value of $\alpha$ must be selected carefully,

**FIGURE 8.19**

*Relationship Between α, β, and n*

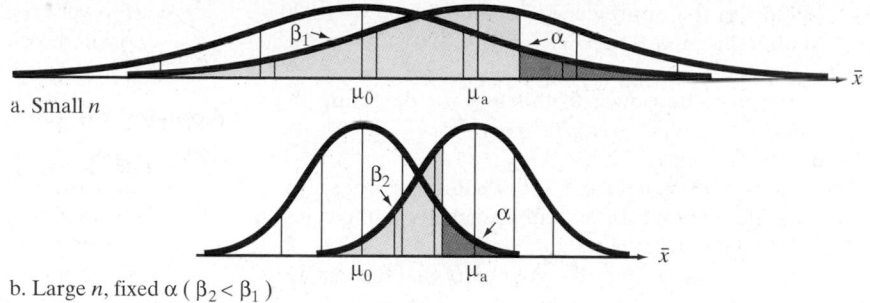

a. Small $n$

b. Large $n$, fixed α ( $\beta_2 < \beta_1$ )

with the realization that a test is made less capable of detecting departures from the null hypothesis when the value of α is decreased.

We have shown that the probability of committing a Type II error, $\beta$, is inversely related to α (Example 8.9), and that the value of $\beta$ decreases as the value of $\mu_a$ moves farther from the null hypothesis value (sewer pipe example). The sample size $n$ also affects $\beta$. Remember that the standard deviation of the sampling distribution of $\bar{x}$ is inversely proportional to the square root of the sample size $(\sigma_{\bar{x}} = \sigma/\sqrt{n})$. Thus, as illustrated in Figure 8.19, the variability of both the null and alternative sampling distributions is decreased as $n$ is increased. If the value of α is specified and remains fixed, the value of $\beta$ decreases as $n$ increases, as illustrated in Figure 8.19. Conversely, the power of the test for a given alternative hypothesis is increased as the sample size is increased.

The properties of $\beta$ and power are summarized in the following box.

---

### Properties of $\beta$ and Power

1. For fixed $n$ and α, the value of $\beta$ decreases and the power increases as the distance between the specified null value $\mu_0$ and the specified alternative value $\mu_a$ increases (see Figure 8.17).
2. For fixed $n$ and values of $\mu_0$ and $\mu_a$, the value of $\beta$ increases and the power decreases as the value of α is decreased (see Figure 8.18).
3. For fixed α and values of $\mu_0$ and $\mu_a$, the value of $\beta$ decreases and the power increases as the sample size $n$ is increased (see Figure 8.19).

---

 EXERCISES 8.76–8.85

**Learning the Mechanics**

**8.76** Consider testing $H_0: \mu = 125$ against $H_a: \mu \neq 125$.
   **a.** What is the relationship between $\beta$, the probability of committing a Type II error, and the power of the test?
   **b.** List three factors that will increase the power of the test.

**8.77** Suppose you want to test $H_0: \mu = 1,000$ against $H_a: \mu > 1,000$ using α = .05. The population in question is normally distributed with standard deviation 120. A random sample of size $n = 36$ will be used.

**a.** Sketch the sampling distribution of $\bar{x}$ assuming that $H_0$ is true.

**b.** Find the value of $\bar{x}_0$, that value of $\bar{x}$ above which the null hypothesis will be rejected. Indicate the rejection region on your graph of part **a**. Shade the area above the rejection region and label it α.

**c.** On your graph of part **a**, sketch the sampling distribution of $\bar{x}$ if $\mu = 1,020$. Shade the area under this distribution that corresponds to the probability that

$\bar{x}$ falls in the nonrejection region when $\mu = 1,020$. Label this area $\beta$.

d. Find $\beta$.

e. Compute the power of this test for detecting the alternative $H_a: \mu = 1,020$.

**8.78** Refer to Exercise 8.77.

a. If $\mu = 1,040$ instead of 1,020, what is the probability that the hypothesis test will incorrectly fail to reject $H_0$? That is, what is $\beta$?

b. If $\mu = 1,040$, what is the probability that the test will correctly reject the null hypothesis? That is, what is the power of the test?

c. Compare $\beta$ and the power of the test when $\mu = 1,040$ to the values you obtained in Exercise 8.77 for $\mu = 1,020$. Explain the differences.

**8.79** It is desired to test $H_0: \mu = 50$ against $H_a: \mu < 50$ using $\alpha = .10$. The population in question is uniformly distributed with standard deviation 20. A random sample of size 64 will be drawn from the population.

a. Describe the (approximate) sampling distribution of $\bar{x}$ under the assumption that $H_0$ is true.

b. Describe the (approximate) sampling distribution of $\bar{x}$ under the assumption that the population mean is 45.

c. If $\mu$ were really equal to 45, what is the probability that the hypothesis test would lead the investigator to commit a Type II error?

d. What is the power of this test for detecting the alternative $H_a: \mu = 45$?

**8.80** Refer to Exercise 8.79.

a. Find $\beta$ for each of the following values of the population mean: 49, 47, 45, 43, and 41.

b. Plot each value of $\beta$ you obtained in part **a** against its associated population mean. Show $\beta$ on the vertical axis and $\mu$ on the horizontal axis. Draw a curve through the five points on your graph.

c. Use your graph of part **b** to find the approximate probability that the hypothesis test will lead to a Type II error when $\mu = 48$.

d. Convert each of the $\beta$ values you calculated in part **a** to the power of the test at the specified value of $\mu$. Plot the power on the vertical axis against $\mu$ on the horizontal axis. Compare the graph of part **b** to the *power curve* of this part.

e. Examine the graphs of parts **b** and **d**. Explain what they reveal about the relationships among the distance between the true mean $\mu$ and the null hypothesized mean $\mu_0$, the value of $\beta$, and the power.

**8.81** Suppose you want to conduct the two-tailed test of $H_0: \mu = 10$ against $H_a: \mu \neq 10$ using $\alpha = .05$. A random sample of size 100 will be drawn from the population in question. Assume the population has a standard deviation equal to 1.0.

a. Describe the sampling distribution of $\bar{x}$ under the assumption that $H_0$ is true.

b. Describe the sampling distribution of $\bar{x}$ under the assumption that $\mu = 9.9$.

c. If $\mu$ were really equal to 9.9, find the value of $\beta$ associated with the test.

d. Find the value of $\beta$ for the alternative $H_a: \mu = 10.1$.

## Applying the Concepts—Intermediate

**8.82** Refer to Exercise 8.25 (p. 339), in which the null hypothesis that the mean PTSD score of World War II POWs is 16 is tested against the alternative hypothesis that the mean is less than 16. Recall that 33 POWs were included in the sample. Assume that the resulting standard deviation of $s = 9.32$ is a good estimate of the true standard deviation.

a. Calculate the power of the test for the mean values of 15.5, 15.0, 14.5, 14.0, and 13.5.

b. Plot the power of the test on the vertical axis against the mean on the horizontal axis. Draw a curve through the points.

c. Use the power curve of part **b** to estimate the power for the mean value $\mu = 14.25$. Calculate the power for this value of $\mu$, and compare it to your approximation.

d. Use the power curve to approximate the power of the test when $\mu = 12.10$. If the true value of the mean PTSD score for WWII POWs is really 10, what (approximately) are the chances that the test will fail to reject the null hypothesis that the mean is 16?

**8.83** Refer to Exercise 8.82. Show what happens to the power curve when the sample size is increased from $n = 33$ to $n = 100$. Assume that the standard deviation is $\sigma = 9.32$.

**8.84** In 1999, the average size of single-family homes built in the U.S. was 2,230 square feet, an increase of over 200 square feet from a decade earlier (*Wall Street Journal Interactive Edition*, Jan. 7, 2000). A random sample of 100 new homes sold in California in late 1999 yielded the following size information: $\bar{x} = 2,347$ square feet and $s = 257$ square feet.

a. Assume the average size of U.S. homes is known with certainty. Do the sample data provide sufficient evidence to conclude that the mean size of all California homes built in late 1999 exceeds the national average? Test using $\alpha = .01$.

b. Suppose the actual mean size of new California homes was $\mu = 2,330$ square feet. What is the power of the test in part **a** to detect this 100-square-feet difference?

c. If the California mean were actually $\mu = 2,280$ square feet, what is the power of the test in part **a** to detect this 50-square-feet difference?

**8.85** Refer to Exercise 8.26 (p. 339), in which the performance of a particular type of laser-based inspection equipment was investigated. Assume that the standard deviation of the number of solder joints inspected on each run is 1.2. If $\alpha = .05$ is used in conducting the hypothesis test of interest using a sample of 48 circuit boards, and if the true mean number of solder joints that can be inspected is really equal to 9.5, what is the probability that the test will result in a Type II error?

# 8.7 TEST OF HYPOTHESIS ABOUT A POPULATION VARIANCE (OPTIONAL)

Although many practical problems involve inferences about a population mean (or proportion), it is sometimes of interest to make an inference about a population variance, $\sigma^2$. To illustrate, a quality control supervisor in a cannery knows that the exact amount each can contains will vary since there are certain uncontrollable factors that affect the amount of fill. The mean fill per can is important, but equally important is the variation of fill. If $\sigma^2$, the variance of the fill, is large, some cans will contain too little and others too much. Suppose regulatory agencies specify that the standard deviation of the amount of fill in 16 ounce cans should be less than .1 ounce. To determine whether the process is meeting this specification, the supervisor randomly selects 10 cans and weighs the contents of each. The results are given in the following table:

### ● FILLAMOUNTS

| | | | | |
|---|---|---|---|---|
| 16.00 | 15.95 | 16.10 | 16.02 | 15.99 |
| 16.06 | 16.04 | 16.05 | 16.03 | 16.02 |

Do these data provide sufficient evidence to indicate that the variability is as small as desired? To answer this question, we need a procedure for testing a hypothesis about $\sigma^2$.

Intuitively it seems that we should compare the sample variance $\sigma^2$ to the hypothesized value of $\sigma^2$ (or $s$ to $\sigma$) in order to make a decision about the population's variability. The quantity

$$\frac{(n-1)s^2}{\sigma^2}$$

has been shown to have a sampling distribution called a **chi-square** $(\chi^2)$ **distribution** when the population from which the sample is taken is *normally distributed*. Several chi-square distributions are shown in Figure 8.20.

The upper-tail areas for this distribution have been tabulated and are given in Table VII of Appendix A, a portion of which is reproduced in Table 8.4. The table gives the values of $\chi^2$, denoted as $\chi^2_\alpha$, that locate an area of $\alpha$ in the upper tail of the chi-square distribution; that is, $P(\chi^2 > \chi^2_\alpha) = \alpha$. In this case, as with the $t$ statistic, the shape of the chi-square distribution depends on the degrees of freedom associated with $s^2$, namely $(n-1)$. Thus, for $n = 10$ and an upper-tail value $\alpha = .05$, you will have $n - 1 = 9$ df and $\chi^2_{.05} = 16.9190$ (highlighted area in Table 8.4). To further illustrate the use of Table VII, we return to the can-filling example.

## FIGURE 8.20

*Several $\chi^2$ Probability Distributions*

**TABLE 8.4    Reproduction of Part of Table VII in Appendix A: Critical Values of Chi-Square**

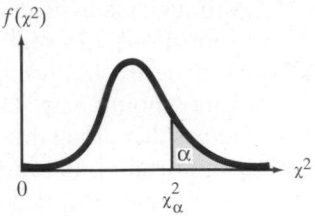

| Degrees of Freedom | $\chi^2_{.100}$ | $\chi^2_{.050}$ | $\chi^2_{.025}$ | $\chi^2_{.010}$ | $\chi^2_{.005}$ |
|---|---|---|---|---|---|
| 1 | 2.70554 | 3.84146 | 5.02389 | 6.63490 | 7.87944 |
| 2 | 4.60517 | 5.99147 | 7.37776 | 9.21034 | 10.5966 |
| 3 | 6.25139 | 7.81473 | 9.34840 | 11.3449 | 12.8381 |
| 4 | 7.77944 | 9.48773 | 11.1433 | 13.2767 | 14.8602 |
| 5 | 9.23635 | 11.0705 | 12.8325 | 15.0863 | 16.7496 |
| 6 | 10.6446 | 12.5916 | 14.4494 | 16.8119 | 18.5476 |
| 7 | 12.0170 | 14.0671 | 16.0128 | 18.4753 | 20.2777 |
| 8 | 13.3616 | 15.5073 | 17.5346 | 20.0902 | 21.9550 |
| 9 | 14.6837 | 16.9190 | 19.0228 | 21.6660 | 23.5893 |
| 10 | 15.9871 | 18.3070 | 20.4831 | 23.2093 | 25.1882 |
| 11 | 17.2750 | 19.6751 | 21.9200 | 24.7250 | 26.7569 |
| 12 | 18.5494 | 21.0261 | 23.3367 | 26.2170 | 28.2995 |
| 13 | 19.8119 | 22.3621 | 24.7356 | 27.6883 | 29.8194 |
| 14 | 21.0642 | 23.6848 | 26.1190 | 29.1413 | 31.3193 |
| 15 | 22.3072 | 24.9958 | 27.4884 | 30.5779 | 32.8013 |
| 16 | 23.5418 | 26.2862 | 28.8454 | 31.9999 | 34.2672 |
| 17 | 24.7690 | 27.5871 | 30.1910 | 33.4087 | 35.7185 |
| 18 | 25.9894 | 28.8693 | 31.5264 | 34.8053 | 37.1564 |
| 19 | 27.2036 | 30.1435 | 32.8523 | 36.1908 | 38.5822 |

**EXAMPLE 8.10**

Refer to the data saved in the FILLAMOUNTS file. Is there sufficient evidence to conclude that the true standard deviation $\sigma$ of the fill measurements of 16 ounce cans is less than .1 ounce?

**Solution**

Here, we want to test whether $\sigma < .1$. Since the null and alternative hypotheses must be stated in terms of $\sigma^2$ (rather than $\sigma$), we want to test the null hypothesis that $\sigma^2 = (.1)^2 = .01$ against the alternative that $\sigma^2 < .01$. Therefore, the elements of the test are

$$H_0: \sigma^2 = .01$$
$$H_a: \sigma^2 < .01$$

$$\text{Test statistic: } \chi^2 = \frac{(n-1)s^2}{\sigma^2}$$

*Assumption:*    The distribution of the amounts of fill is approximately normal.

*Rejection region:*    The smaller the value of $s^2$ we observe, the stronger the evidence in favor of $H_a$. Thus, we reject $H_0$ for "small values" of the test statistic. With $\alpha = .05$ and 9 df, the $\chi^2$ value for rejection is found in Table VII and pictured in Figure 8.21. We will reject $H_0$ if $\chi^2 < 3.32511$.

**FIGURE 8.21**

*Rejection Region for*
*Example 8.10*

(Remember that the area given in Table VII is the area to the *right* of the numerical value in the table. Thus, to determine the lower-tail value, which has $\alpha = .05$ to its *left*, we used the $\chi^2_{.95}$ column in Table VII.)

An SAS printout of the analysis is displayed in Figure 8.22. The value of $s$ (highlighted on the printout) is $s = .0412$. Substituting into the formula for the test statistic, we have

$$\chi^2 = \frac{(n-1)s^2}{\sigma^2} = \frac{9(.0412)^2}{.01} = 1.524$$

Since the value of the test statistic is less than 3.32511, the supervisor can conclude that the variance $\sigma^2$ of the population of all amounts of fill is less than $.01$(i.e., $\sigma < .1$) with probability of a Type I error equal to $\alpha = .05$. If this procedure is repeatedly used, it will incorrectly reject $H_0$ only 5% of the time. Thus, the quality control supervisor is confident in the decision that the cannery is operating within the desired limits of variability.

Note that both the test statistic (1.524) and one-tailed $p$-value of the test (.0030) are shown on the SAS printout. Also note that the $p$-value is less than $\alpha = .05$, thus confirming our conclusion to reject $H_0$. ∎

**FIGURE 8.22**

*SAS Analysis of Ten Fill*
*Amounts*

One Sample Chi-square Test for a Variance

Sample Statistics for fill

| N | Mean | Std. Dev. | Variance |
|---|------|-----------|----------|
| 10 | 16.026 | 0.0412 | 0.0017 |

Hypothesis Test

Null hypothesis:    Variance of fill => 0.01
Alternative:      Variance of fill <  0.01

| Chi-square | Df | Prob |
|------------|-----|------|
| 1.524 | 9 | 0.0030 |

One-tailed and two-tailed tests of hypothesis for $\sigma^2$ are given in the following box.*

---

### Test of Hypothesis About $\sigma^2$

**ONE-TAILED TEST**

$H_0: \sigma^2 = \sigma_0^2$

$H_a: \sigma^2 < \sigma_0^2$
  (or $H_a: \sigma^2 > \sigma_0^2$)

*Test statistic:* $\chi^2 = \dfrac{(n-1)s^2}{\sigma_0^2}$

*Rejection region:* $\chi^2 < \chi_{(1-\alpha)}^2$
  (or $\chi^2 > \chi_\alpha^2$ when $H_a: \sigma^2 > \sigma_0^2$)

**TWO-TAILED TEST**

$H_0: \sigma^2 = \sigma_0^2$

$H_a: \sigma^2 \neq \sigma_0^2$

*Test statistic:* $\chi^2 = \dfrac{(n-1)s^2}{\sigma_0^2}$

*Rejection region:* $\chi^2 < \chi_{(1-\alpha/2)}^2$
  or $\chi^2 > \chi_{(\alpha/2)}^2$

where $\sigma_0^2$ is the hypothesized variance and the distribution of $\chi^2$ is based on $(n-1)$ degrees of freedom.
*Assumption:* The population from which the sample is drawn is approximately normal.

---

### Caution

The procedure for conducting a hypothesis test for $\sigma^2$ in the above examples requires an assumption regardless of whether the sample size $n$ is large or small. We must assume that the population from which the sample is selected has an approximate normal distribution. Unlike small sample tests for $\mu$ based on the $t$ statistic, *slight to moderate departures from normality will render the $\chi^2$-test invalid.*

---

 EXERCISES 8.86–8.96

### Learning the Mechanics

**8.86** Let $\chi_0^2$ be a particular value of $\chi^2$. Find the value of $\chi_0^2$ such that:
  **a.** $P(\chi^2 > \chi_0^2) = .10$ for $n = 12$
  **b.** $P(\chi^2 > \chi_0^2) = .05$ for $n = 9$
  **c.** $P(\chi^2 > \chi_0^2) = .025$ for $n = 5$

**8.87** A random sample of $n$ observations is selected from a normal population to test the null hypothesis that $\sigma^2 = 25$. Specify the rejection region for each of the following combinations of $H_a$, $\alpha$, and $n$:
  **a.** $H_a: \sigma^2 \neq 25; \alpha = .05; n = 16$
  **b.** $H_a: \sigma^2 > 25; \alpha = .01; n = 23$
  **c.** $H_a: \sigma^2 > 25; \alpha = .10; n = 15$

  **d.** $H_a: \sigma^2 < 25; \alpha = .01; n = 13$
  **e.** $H_a: \sigma^2 \neq 25; \alpha = .10; n = 7$
  **f.** $H_a: \sigma^2 < 25; \alpha = .05; n = 25$

**8.88** A random sample of seven measurements gave $\bar{x} = 9.4$ and $s^2 = 4.84$.
  **a.** What assumptions must you make concerning the population in order to test a hypothesis about $\sigma^2$?
  **b.** Suppose the assumptions in part **a** are satisfied. Test the null hypothesis, $\sigma^2 = 1$, against the alternative hypothesis, $\sigma^2 > 1$. Use $\alpha = .05$.
  **c.** Test the null hypothesis that $\sigma^2 = 1$ against the alternative hypothesis that $\sigma^2 \neq 1$. Use $\alpha = .05$.

**8.89** Refer to Exercise 8.88. Suppose we had $n = 100$, $\bar{x} = 9.4$, and $s^2 = 4.84$.

---

*A confidence interval for $\sigma^2$ can also be formed using the $\chi^2$-distribution with $(n-1)$ degrees of freedom. A $(1-\alpha)100\%$ confidence interval is

$$\frac{(n-1)s^2}{\chi_{\alpha/2}^2} < \sigma^2 < \frac{(n-1)s^2}{\chi_{(1-\alpha/2)}^2}$$

**a.** Test the null hypothesis, $H_0: \sigma^2 = 1$, against the alternative hypothesis, $H_a: \sigma^2 > 1$.

**b.** Compare your test result with that of Exercise 8.88.

**8.90** A random sample of $n = 7$ observations from a normal population produced the following measurements: 4, 0, 6, 3, 3, 5, 9. Do the data provide sufficient evidence to indicate that $\sigma^2 < 1$? Test using $\alpha = .05$.

## Applying the Concepts—Basic

**8.91** Refer to the *Chance* (Fall 1998) study of point-spread errors in NFL games, Exercise 8.24 (p. 339). Recall that the difference between the actual game outcome and the point spread established by oddsmakers—the point-spread error—was calculated for 240 NFL games. The results are summarized as follows: $\bar{x} = -1.6$, $s = 13.3$. Suppose the researcher wants to know whether the true standard deviation of the point-spread errors exceeds 15.

**a.** Set up $H_0$ and $H_a$ for the researcher.

**b.** Compute the value of the test statistic.

**c.** Conduct the test at $\alpha = .10$. Interpret the results in the words of the problem.

**8.92** A group of researchers at the University of Texas-Houston conducted a comprehensive study of pregnant cocaine-dependent women (*Journal of Drug Issues*, Summer 1997). All the women in the study used cocaine on a regular basis (at least three times a week) for more than a year. One of the many variables measured was birthweight (in grams) of the baby delivered. For a sample of 16 cocaine-dependent women, the mean birthweight was 2,971 grams and the standard deviation was 410 grams. Test (at $\alpha = .01$) to determine whether the variance in birthweights of babies delivered by cocaine-dependent women is less than 200,000 grams$^2$.

## Applying the Concepts—Intermediate

**8.93** A new technique for extracting toxic organic compounds was evaluated in *Chromatographia* (Mar. 1995). Uncontaminated fish fillets were injected with a toxic substance and the method was used to extract the toxicant. To test the precision of the new method, 7 measurements were obtained on a single fish fillet injected with a toxin. Summary statistics on percent of the toxin recovered are given as follows: $\bar{x} = 99$, $s = 9$. Determine whether the standard deviation of the percent recovery using the new method differs from 15. (Use $\alpha = .10$.)

**8.94** It is essential in the manufacture of machinery to utilize parts that conform to specifications. In the past, diameters of the ball-bearings produced by a certain manufacturer had a variance of .00156. To cut costs, the manufacturer instituted a less expensive production method. The variance of the diameters of 100 randomly sampled bearings produced by the new process was .00211. Do the data provide sufficient evidence to indicate that diameters of ball-bearings produced by the new process are more variable than those produced by the old process?

**8.95** Geologists use laser Raman microprobe (LRM) spectroscopy to analyze *fluid inclusions* (pockets of gas or liquid) in rock. A chip of natural Brazilian quartz was artificially injected with several fluid inclusions of liquid carbon dioxide ($CO_2$) and then subjected to LRM spectroscopy (*Applied Spectroscopy*, Feb. 1986). The amount of $CO_2$ present in the inclusion was recorded for the same inclusion on four different days. The data (in mole percent) follow:

CO2

| | | | |
|---|---|---|---|
| 86.6 | 84.6 | 85.5 | 85.9 |

**a.** Do the data indicate that the variation in the $CO_2$ concentration measurements using the LRM method differs from 1?

**b.** What assumption is required for the test to be valid?

## Applying the Concepts—Advanced

**8.96** To improve the signal-to-noise ratio (SNR) in the electrical activity of the brain, neurologists repeatedly stimulate subjects and average the responses—a procedure that assumes that signal responses are homogeneous. A study was conducted to test the homogeneous signal theory (*IEEE Engineering in Medicine and Biology Magazine*, Mar. 1990). For this study, the variance of the SNR readings of subjects will equal .54 under the homogeneous signal theory. If the SNR variance exceeds .54, the researchers will conclude that the signals are nonhomogeneous. Signal-to-noise ratios recorded for a sample of 41 normal children ranged from .03 to 3.0. Use this information to test the theory at $\alpha = .10$.

# QUICK REVIEW

## Key Terms

*Note: Starred (\*) terms are from the optional sections in this chapter.*

## STATISTICS IN ACTION
### *March Madness: Handicapping the NCAA Basketball Tourney*

For three weeks each March, the National Collegiate Athletic Association (NCAA) holds its annual men's basketball championship tournament. The 64 best college basketball teams in the nation play a single-elimination tournament—a total of 63 games—to determine the NCAA champion. Due to its extreme popularity (all 63 games are televised by CBS), the media refers to the tournament as "March Madness."

The NCAA groups the 64 teams into four regions (East, South, Midwest, and West) of 16 teams each. The teams in each region are ranked (seeded) from 1 to 16 based on their performance during the regular season. The NCAA considers such factors as overall record, strength of schedule, average margin of victory, wins on the road, and conference affiliation to determine a team's seeding. In the first round, the number one seed plays the number 16 seed, the second seed plays the fifteenth seed, the third seed plays the fourteenth seed, etc. (See Figure 8.23.) Winners continue playing until the champion is determined.

Tournament followers, from hardcore gamblers to the casual fan who enters the office betting pool, have a strong interest in handicapping the games. Obviously, predicting the eventual champion is of prime interest. However, knowing who will win each game and the margin of victory may be just as important. To provide insight into this phenomenon, statisticians Hal Stern and Barbara Mock analyzed data from 13 NCAA tournaments and published their results in *Chance* (Winter 1998). The results of first-round games are summarized in Table 8.5.

#### *Focus*

a. A common perception among fans, media, and gamblers is that the higher seeded team has a better than 50-50 chance of winning a first-round game. Is there evidence to support this perception? Conduct the appropriate test for each matchup. What trends do you observe?

b. Is there evidence to support the claim that a 1-, 2-, 3-, or 4-seeded team will win by an average of more than 10 points in first-round games? Conduct the appropriate test for each matchup.

c. Is there evidence to support the claim that a 5-, 6-, 7-, or 8-seeded team will win by an average of less than five-points in first-round games? Conduct the appropriate test for each matchup.

**FIGURE 8.23**

*The Design of a 16-Team NCAA Regional Tournament*

## STATISTICS IN ACTION (*continued*)

d. (Optional) For each matchup, test the null hypothesis that the standard deviation of the victory margin is 11 points.

e. The researchers also calculated the difference between the game outcome (victory margin, in points) and point spread established by Las Vegas oddsmakers for a sample of 360 recent NCAA tournament games. The mean dif-

ference is .7 and the standard deviation of the difference is 11.3. If the true mean difference is 0, then the point spread can be considered a good predictor of the game outcome. Use this sample information to test the hypothesis that the point spread, on average, is a good predictor of the victory margin in NCAA tournament games.

**TABLE 8.5   Summary of First-Round NCAA Tournament Games, 1985–1997**

| Matchup (Seeds) | Number of Games | Number Won by Favorite (Higher Seed) | Margin of Victory (Points) | |
|---|---|---|---|---|
| | | | Mean | Standard Deviation |
| 1 vs 16 | 52 | 52 | 22.9 | 12.4 |
| 2 vs 15 | 52 | 49 | 17.2 | 11.4 |
| 3 vs 14 | 52 | 41 | 10.6 | 12.0 |
| 4 vs 13 | 52 | 42 | 10.0 | 12.5 |
| 5 vs 12 | 52 | 37 | 5.3 | 10.4 |
| 6 vs 11 | 52 | 36 | 4.3 | 10.7 |
| 7 vs 10 | 52 | 35 | 3.2 | 10.5 |
| 8 vs 9 | 52 | 22 | −2.1 | 11.0 |

*Source:* Stern, H. S., & Mock, B. "College basketball upsets: Will a 16-seed ever beat a 1-seed?" *Chance*, Vol. 11, No. 1, Winter 1998, p. 29 (Table 3).

## Key Formulas

*Note: Starred (\*) formulas are from the optional sections in this chapter.*

For testing $H_0: \theta = \theta_0$, the **large-sample test statistic** is

$$z = \frac{\hat{\theta} - \theta_0}{\sigma_{\hat{\theta}}}$$

where $\hat{\theta}$, $\theta_0$, and $\sigma_{\hat{\theta}}$ are obtained from the table below:

| Parameter, $\theta$ | Hypothesized Parameter Value, $\theta_0$ | Estimator, $\hat{\theta}$ | Standard Error of Estimator, $\sigma_{\hat{\theta}}$ | |
|---|---|---|---|---|
| $\mu$ | $\mu_0$ | $\bar{x}$ | $\dfrac{\sigma}{\sqrt{n}}$ | 337 |
| $p$ | $p_0$ | $\hat{p}$ | $\sqrt{\dfrac{p_0 q_0}{n}}$ | 355 |

For testing $H_0: \mu = \mu_0$, the **small-sample test statistic** is

$$t = \frac{\bar{x} - \mu_0}{s/\sqrt{n}} \quad 348$$

\*$\beta = P(\hat{\theta}$ falls in acceptance region $|\theta = \theta_a)$   361

\*Power $= 1 - \beta$   362

\*For testing $H_0: \sigma^2 = \sigma_0^2$, the test statistic is

$$\chi^2 = \frac{(n-1)s^2}{\sigma_0^2} \quad 370$$

## Language Lab

| Symbol | Pronunciation | Description |
|---|---|---|
| $H_0$ | *H*-oh | Null hypothesis |
| $H_a$ | *H*-a | Alternative hypothesis |
| $\alpha$ | alpha | Probability of Type I error |
| $\beta$ | beta | Probability of Type II error |
| $\chi^2$ | chi-square | Sampling distribution of $s^2$ when data are from a normal population |

 SUPPLEMENTARY EXERCISES 8.97–8.125

*Note: List the assumptions necessary for the valid implementation of the statistical procedures you use in solving all these exercises. Starred (*) exercises refer to the optional sections in this chapter.*

### Learning the Mechanics

**8.97** *Complete the following statement:* The smaller the *p*-value associated with a test of hypothesis, the stronger the support for the _____ hypothesis. Explain your answer.

**8.98** Specify the differences between a large-sample and small-sample test of hypothesis about a population mean $\mu$. Focus on the assumptions and test statistics.

**8.99** Which of the elements of a test of hypothesis can and should be specified *prior* to analyzing the data that are to be utilized to conduct the test?

**8.100** If the rejection of the null hypothesis of a particular test would cause your firm to go out of business, would you want $\alpha$ to be small or large? Explain.

**8.101** *Complete the following statement:* The larger the *p*-value associated with a test of hypothesis, the stronger the support for the _____ hypothesis. Explain your answer.

**8.102** A random sample of 20 observations selected from a normal population produced $\bar{x} = 72.6$ and $s^2 = 19.4$.
  **a.** Form a 90% confidence interval for the population mean.
  **b.** Test $H_0: \mu = 80$ against $H_a: \mu < 80$. Use $\alpha = .05$.
  **c.** Test $H_0: \mu = 80$ against $H_a: \mu \neq 80$. Use $\alpha = .01$.
  **d.** Form a 99% confidence interval for $\mu$.
  **e.** How large a sample would be required to estimate $\mu$ to within 1 unit with 95% confidence?

**8.103** A random sample of $n = 200$ observations from a binomial population yields $\hat{p} = .29$.
  **a.** Test $H_0: p = .35$ against $H_a: p < .35$. Use $\alpha = .05$.
  **b.** Test $H_0: p = .35$ against $H_a: p \neq .35$. Use $\alpha = .05$.
  **c.** Form a 95% confidence interval for *p*.

  **d.** Form a 99% confidence interval for *p*.
  **e.** How large a sample would be required to estimate *p* to within .05 with 99% confidence?

**8.104** A random sample of 175 measurements possessed a mean $\bar{x} = 8.2$ and a standard deviation $s = .79$.
  **a.** Form a 95% confidence interval for $\mu$.
  **b.** Test $H_0: \mu = 8.3$ against $H_a: \mu \neq 8.3$. Use $\alpha = .05$.
  **c.** Test $H_0: \mu = 8.4$ against $H_a: \mu \neq 8.4$. Use $\alpha = .05$.

**8.105** A random sample of 41 observations from a normal population possessed a mean $\bar{x} = 88$ and a standard deviation $s = 6.9$.
  **a.** Test $H_0: \sigma^2 = 30$ against $H_a: \sigma^2 > 30$. Use $\alpha = .05$.
  **b.** Test $H_0: \sigma^2 = 30$ against $H_a: \sigma^2 \neq 30$. Use $\alpha = .05$.

**8.106** A *t*-test is conducted for the null hypothesis $H_0: \mu = 10$ versus the alternative $H_a: \mu > 10$ for a random sample of $n = 17$ observations. The data are analyzed using MINITAB, with the results shown below.
  **a.** Interpret the *p*-value.
  **b.** What assumptions are necessary for the validity of this test?
  **c.** Calculate and interpret the *p*-value assuming the alternative hypothesis was instead $H_a: \mu \neq 10$.

### Applying the Concepts—Basic

**8.107** When a new drug is formulated, the pharmaceutical company must subject it to lengthy and involved testing before receiving the necessary permission from the Food and Drug Administration (FDA) to market the drug. The FDA requires the pharmaceutical company to provide substantial evidence that the new drug is safe for potential consumers.

---

```
MINITAB Output for Exercise 8.106

Test of mu = 10 vs mu > 10

Variable          N       Mean      StDev    SE Mean
X                17      12.50       8.78       2.13

Variable          95.0% CI              T          P
X            (  8.00,   17.00)       1.17      .1288
```

**a.** If the new drug testing were to be placed in a test of hypothesis framework, would the null hypothesis be that the drug is safe or unsafe? The alternative hypothesis?

**b.** Given the choice of null and alternative hypotheses in part **a**, describe Type I and Type II errors in terms of this application. Define $\alpha$ and $\beta$ in terms of this application.

**c.** If the FDA wants to be very confident that the drug is safe before permitting it to be marketed, is it more important that $\alpha$ or $\beta$ be small? Explain.

**8.108** An article in the *Annals of the Association of American Geographers* (June 1992) revealed that only 133 of 337 randomly selected residences in Los Angeles County were protected by earthquake insurance.

**a.** What are the appropriate null and alternative hypotheses to test the research hypothesis that less than 40% of the residents of Los Angeles County were protected by earthquake insurance?

**b.** Do the data provide sufficient evidence to support the research hypothesis? Use $\alpha = .10$.

**c.** Calculate and interpret the *p*-value for the test.

**8.109** In a British study, 12 healthy college students, deprived of one night's sleep, received an array of tests intended to measure thinking time, fluency, flexibility, and originality of thought. The overall test scores of the sleep-deprived students were compared to the average score expected from students who received their accustomed sleep (*Sleep*, Jan. 1989). Suppose the overall scores of the 12 sleep-deprived students had a mean of $\bar{x} = 63$ and a standard deviation of 17. (Lower scores are associated with a decreased ability to think creatively.)

**a.** Test the hypothesis that the true mean score of sleep-deprived subjects is less than 80, the mean score of subjects who received sleep prior to taking the test. Use $\alpha = .05$.

**b.** What assumption is required for the hypothesis test of part **a** to be valid?

**8.110** "Take the Pepsi Challenge" was a marketing campaign used by the Pepsi-Cola Company. Coca-Cola drinkers participated in a blind taste test in which they tasted unmarked cups of Pepsi and Coke and were asked to select their favorite. Pepsi claimed that "in recent blind taste tests, more than half the Diet Coke drinkers surveyed said they preferred the taste of Diet Pepsi" (*Consumer's Research*, May 1993). Suppose 100 Diet Coke drinkers took the Pepsi Challenge and 56 preferred the taste of Diet Pepsi. Test the hypothesis that more than half of all Diet Coke drinkers will select Diet Pepsi in a blind taste test. Use $\alpha = .05$.

**8.111** A sporting goods manufacturer who produces both white and yellow golf balls claims that more than 75% of all golf balls sold are white. A marketing study of the purchases of white and yellow golf balls at a number of stores showed that of 470 balls sold, 410 were white and 60 were yellow.

**a.** Is there sufficient evidence to support the manufacturer's claim? Test using $\alpha = .01$.

**b.** Find the observed significance level for the test and interpret its value.

**\*c.** Calculate the probability, $\beta$, of a Type II error if in fact 80% of the golf balls sold are white.

**\*8.112** Refer to the *IEEE Transactions* (June 1990) study of a new hybrid algorithm for solving polynomial 0/1 mathematical programs, Exercise 7.81 (p. 324). An SAS printout giving descriptive statistics for the sample of 52 solution times is reproduced below. Use this information to determine whether the variance of the solution times differs from 2. Use $\alpha = .05$

**8.113** Golf-course designers are concerned that old courses are becoming obsolete because new equipment enables golfers to hit the ball so far: In effect, golf courses are "shrinking." One designer believes that new courses need to be built with the expectation that players will be able to hit the ball more than 250 yards (with their drivers), on average. Suppose a sample of 135 golfers is tested, and their mean driving distance is 256.3 yards, with a standard deviation of 43.4 yards.

**a.** What are appropriate null and alternative hypotheses to test the designer's research hypothesis?

**b.** Calculate and interpret the *p*-value for this test.

**c.** If you were to make this a two-tailed test, how would your answer to part **b** change?

## Applying the Concepts—Intermediate

**8.114** Medical tests have been developed to detect many serious diseases. A medical test is designed to minimize the probability that it will produce a "false positive" or a "false negative." A false positive is a positive test result for an individual who does not have the disease, whereas a false negative is a negative test result for an individual who does have the disease.

**a.** If we treat a medical test for a disease as a statistical test of hypothesis, what are the null and alternative hypotheses for the medical test?

SAS Output for Exercise 8.112

| Analysis Variable : CPU | | | | | |
|---|---|---|---|---|---|
| Mean | Std Dev | Variance | N | Minimum | Maximum |
| 0.8121923 | 1.5047603 | 2.2643035 | 52 | 0.0360000 | 8.7880000 |

b. What are the Type I and Type II errors for the test? Relate each to false positives and false negatives.

c. Which of these errors has graver consequences? Considering this error, is it more important to minimize $\alpha$ or $\beta$? Explain.

**8.115** In order to be effective, a certain mechanical component used in a spacecraft must have a mean length of life greater than 1,100 hours. Owing to the prohibitive cost of this component, only three were tested under simulated space conditions. The lifetimes (hours) of the components were recorded and the computed statistics were $\bar{x} = 1,173.6$ and, $s = 36.3$. These data were analyzed using MINITAB, as shown below.

a. Verify that the software has correctly calculated the *t* statistic, and use Table VI of Appendix A to determine whether the *p*-value is in the appropriate range.

b. Interpret the *p*-value.

c. What assumptions are necessary for the validity of this test?

d. Which type of error, I or II, is of greater concern for this test? Explain.

e. Would you recommend that this component be passed as meeting specifications?

**8.116** A consumer protection group is concerned that a ketchup manufacturer is filling its 20-ounce family-size containers with less than 20 ounces of ketchup. The group purchases 10 family-size bottles of this ketchup, weighs the contents of each, and finds that the mean weight is equal to 19.86 ounces, and the standard deviation is equal to .22 ounce.

a. Do the data provide sufficient evidence for the consumer group to conclude that the mean fill per family-size bottle is less than 20 ounces? Test using $\alpha = .05$.

b. If the test in part **a** were conducted on a periodic basis by the company's quality control department, is the consumer group more concerned about making a Type I error or a Type II error? (The probability of making this type of error is called the *consumer's risk*.)

c. The ketchup company is also interested in the mean amount of ketchup per bottle. It does not wish to overfill them. For the test conducted in part **a**, which type of error is more serious from the company's point of view—a Type I error or a Type II error? (The probability of making this type of error is called the *producer's risk*.)

**8.117** Refer to the *Science* (Nov. 1988) study of inbreeding in tropical swarm-founding wasps, Exercise 7.79 (p. 324). A sample of 197 wasps, captured, frozen, and subjected to a series of genetic tests, yielded a sample mean inbreeding coefficient of $\bar{x} = .044$ with a standard deviation of $s = .884$. Recall that if the wasp has no tendency to inbreed, the true mean inbreeding coefficient $\mu$ for the species will equal 0.

a. Test the hypothesis that the true mean inbreeding coefficient $\mu$ for this species of wasp exceeds 0. Use $\alpha = .05$.

b. Compare the inference, part **a**, to the inference obtained in Exercise 7.79 using a confidence interval. Do the inferences agree? Explain.

*__8.118__ A marine biologist wishes to use parrotfish for experimental purposes due to the belief that their weight is fairly stable (i.e., the variability in weights among parrotfish is small). The biologist randomly samples 10 parrotfish and finds that their mean weight is 4.3 pounds and the standard deviation is 1.4 pounds. The biologist will use the parrotfish only if there is evidence that the variance of their weights is less than 4.

a. Is there sufficient evidence for the biologist to claim that the variability in weights among parrotfish is small enough to justify their use in the experiment? Test at $\alpha = .05$.

b. State any assumptions that are needed for the test, part **a**, to be valid.

**8.119** The EPA sets a limit of 5 parts per million (ppm) on PCB (a dangerous substance) in water. A major manufacturing firm producing PCB for electrical insulation discharges small amounts from the plant. The company management, attempting to control the PCB in its discharge, has given instructions to halt production if the mean amount of PCB in the effluent exceeds 3 ppm. A random sample of 50 water specimens produced the following statistics: $\bar{x} = 3.1$ ppm and $s = .5$ ppm.

a. Do these statistics provide sufficient evidence to halt the production process? Use $\alpha = .01$.

b. If you were the plant manager, would you want to use a large or a small value for $\alpha$ for the test in part **a**?

*__8.120__ Refer to Exercise 8.119.

a. In the context of the problem, define a Type II error.

b. Calculate $\beta$ for the test described in part **a** of Exercise 8.119 assuming that the true mean is $\mu = 3.1$ ppm.

---

**MINITAB Output for Exercise 8.115**

```
Test of mu = 1100 vs mu > 1100

Variable         N        Mean      StDev     SE Mean
X                3      1173.6       36.3       20.96


Variable            95.0% CI              T          P
X             ( 1083.5, 1263.7)        3.51      .0362
```

**c.** What is the power of the test to detect the effluent's departure from the standard of 3.0 ppm when the mean is 3.1 ppm?

**d.** Repeat parts **b** and **c** assuming that the true mean is 3.2 ppm. What happens to the power of the test as the plant's mean PCB departs farther from the standard?

*8.121 Refer to Exercises 8.119 and 8.120.

**a.** Suppose an $\alpha$ value of .05 is used to conduct the test. Does this change favor the manufacturer? Explain.

**b.** Determine the value of $\beta$ and the power for the test when $\alpha = .05$ and $\mu = 3.1$.

**c.** What happens to the power of the test when $\alpha$ is increased?

8.122 Research reported in the *Journal of Psychology* (Mar. 1991) studied the personality characteristics of obese individuals. One variable, the locus of control (LOC), measures the individual's degree of belief that he or she has control over situations. High scores on the LOC scale indicate less perceived control. For one sample of 19 obese adolescents, the mean LOC score was 10.89 with a standard deviation of 2.48. Suppose we wish to test whether the mean LOC score for all obese adolescents exceeds 10, the average for "normal" individuals.

**a.** Specify the null and alternative hypotheses for this test.

**b.** The data are analyzed by computer, and the following STATISTIX output is obtained. Check the calculation of the $t$ statistic, and determine whether the $p$-value is in the correct range according to Table VI of Appendix A.

**STATISTIX Output for Exercise 8.122**

```
ONE-SAMPLE T TEST FOR LOC

NULL HYPOTHESIS: MU =   10
ALTERNATIVE HYP: MU >   10

MEAN          10.89
STD ERROR      0.569
MEAN - HO      0.89
T              1.564
DF               18
P             0.0676

CASES INCLUDED 19    MISSING CASES 0
```

8.123 The National Science Foundation, in a survey of 2,237 engineering graduate students who earned their Ph.D. degrees, found that 607 were U.S. citizens; the majority (1,630) of the Ph.D. degrees were awarded to foreign nationals (*Science*, Sept. 24, 1993). Conduct a test to determine whether the true percentage of engineering Ph.D. degrees awarded to foreign nationals exceeds 50%. Use $\alpha = .01$.

*8.124 Ophthalmologists require an instrument that can rapidly measure interocular pressure for glaucoma patients. The device now in general use is known to yield readings of this pressure with a variance of 10.3. The variance of five pressure readings on the same eye by a newly developed instrument is equal to 9.8. Does this sample variance provide sufficient evidence to indicate that the new instrument is more reliable than the instrument currently in use? (Use $\alpha = .05$.)

### Applying the Concepts—Advanced

*8.125 If a manufacturer (the vendee) buys all items of a particular type from a particular vendor, the manufacturer is practicing *sole sourcing* (Schonberger and Knod, *Operations Managment*, 1994). As part of a sole-sourcing arrangement, a vendor agrees to periodically supply its vendee with sample data from its production process. The vendee uses the data to investigate whether the mean length of rods produced by the vendor's production process is truly 5.0 millimeters (mm) or more, as claimed by the vendor and desired by the vendee.

**a.** If the production process has a standard deviation of .01 mm, the vendor supplies $n = 100$ items to the vendee, and the vendee uses $\alpha = .05$ in testing $H_0: \mu = 5.0$ mm against $H_a: \mu < 5.0$ mm, what is the probability that the vendee's test will fail to reject the null hypothesis when in fact $\mu = 4.9975$ mm? What is the name given to this type of error?

**b.** Refer to part **a**. What is the probability that the vendee's test will reject the null hypothesis when in fact $\mu = 5.0$? What is the name given to this type of error?

**c.** What is the power of the test to detect a departure of .0025 mm below the specified mean rod length of 5.0 mm?

## STUDENT PROJECTS

The "efficient market" theory postulates that the best predictor of a stock's price at some point in the future is the current price of the stock (with some adjustments for inflation and transaction costs, which we shall assume to be negligible for the purpose of this exercise). To test this theory, select a random sample of 25 stocks on the New York Stock Exchange and record the closing prices on the last days of two recent consecutive months. Calculate the increase or decrease in the stock price over the 1-month period.

a. Define $\mu$ as the mean change in price of all stocks over a 1-month period. Set up the appropriate null and alternative hypotheses in terms of $\mu$.

b. Use the sample of the 25 stock price differences to conduct the test of hypothesis established in part a. Use $\alpha = .05$.

c. Tabulate the number of rejections and nonrejections of the null hypothesis in your class. If the null hypothesis were true, how many rejections of the null hypothesis would you expect among those in your class? How does this expectation compare with the actual number of nonrejections? What does the result of this exercise indicate about the efficient market theory?

## REFERENCES

Snedecor, G. W., and Cochran, W. G. *Statistical Methods*, 7th ed. Ames: Iowa State University Press, 1980.

Wackerly, D., Mendenhall, W., and Scheaffer, R. *Mathematical Statistics with Applications*, 6th ed. North Scituate, Mass.: Duxbury, 2002.

# CHAPTER 9

# Inferences Based on Two Samples:

## CONFIDENCE INTERVALS AND TESTS OF HYPOTHESES

### Contents

### Statistics in Action

Detection of Rigged Milk Prices

---

### 🐾 Where We've Been

We explored two methods for making statistical inferences, confidence intervals and tests of hypotheses based on single samples, in Chapters 7 and 8. In particular, we gave confidence intervals and tests of hypotheses concerning a population mean $\mu$ and a population proportion $p$, and we learned how to select the sample size necessary to obtain a specified amount of information concerning a parameter. (Inferences about a population variance $\sigma^2$ were included in an optional section.)

### ☞ Where We're Going

Now that we've learned to make inferences about a single population, we'll learn how to compare two populations. For example, we may wish to compare the mean gas mileages for two models of automobiles, or the mean reaction times of men and women to a visual stimulus, or the proportions of subjects in a psychological experiment that respond to two different stimuli. In this chapter we'll see how to decide whether differences exist and how to estimate the differences between population means and proportions.

Many experiments involve a comparison of two populations. For instance, a sociologist may want to estimate the difference in mean life expectancy between inner-city and suburban residents. A consumer group may want to test whether two major brands of food freezers differ in the mean amount of electricity they use. A political candidate may wish to estimate the difference in the proportions of voters in two districts who favor his or her candidacy. A professional golfer may wish to compare the variability in the distance that two competing brands of golf balls travel when struck with the same club. In this chapter we consider techniques for using two samples to compare the populations from which they were selected.

## 9.1 COMPARING TWO POPULATION MEANS: INDEPENDENT SAMPLING

Many of the same procedures that are used to estimate and test hypotheses about a single parameter can be modified to make inferences about two parameters. Both the $z$ and $t$ statistics may be adapted to make inferences about the difference between two population means.

In this section we develop both large-sample and small-sample methodologies for comparing two population means. In the large-sample case we use the $z$ statistic, while in the small-sample case we use the $t$ statistic.

### Large Samples

**EXAMPLE 9.1**

A dietitian has developed a diet that is low in fats, carbohydrates, and cholesterol. Although the diet was initially intended to be used by people with heart disease, the dietitian wishes to examine the effect this diet has on the weights of obese people. Two random samples of 100 obese people each are selected, and one group of 100 is placed on the low-fat diet. The other 100 are placed on a diet that contains approximately the same quantity of food but is not as low in fats, carbohydrates, and cholesterol. For each person, the amount of weight lost (or gained) in a 3-week period is recorded. The data, saved in the DIETSTUDY file, are listed in Table 9.1. Form a 95% confidence interval for the difference between the population mean weight losses for the two diets. Interpret the result.

### Solution

Recall that the general form of a large-sample confidence interval for a single mean $\mu$ is $\bar{x} \pm z_{\alpha/2}\sigma_{\bar{x}}$. That is, we add and subtract $z_{\alpha/2}$ standard deviations of the sample estimate, $\bar{x}$, to the value of the estimate. We employ a similar procedure to form the confidence interval for the difference between two population means.

Let $\mu_1$ represent the mean of the conceptual population of weight losses for all obese people who could be placed on the low-fat diet. Let $\mu_2$ be similarly defined for the other diet. We wish to form a confidence interval for $(\mu_1 - \mu_2)$. An intuitively appealing estimator for $(\mu_1 - \mu_2)$ is the difference between the sample means, $(\bar{x}_1 - \bar{x}_2)$. Thus, we will form the confidence interval of interest by

$$(\bar{x}_1 - \bar{x}_2) \pm z_{\alpha/2}\sigma_{(\bar{x}_1 - \bar{x}_2)}$$

Assuming the two samples are independent, the standard deviation of the difference between the sample means is

$$\sigma_{(\bar{x}_1 - \bar{x}_2)} = \sqrt{\frac{\sigma_1^2}{n_1} + \frac{\sigma_2^2}{n_2}} \approx \sqrt{\frac{s_1^2}{n_1} + \frac{s_2^2}{n_2}}$$

⊘ DIETSTUDY

**TABLE 9.1   Diet Study Data, Example 9.1**

**Weight Losses for Lowfat Diet**

| 8 | 10 | 10 | 12 | 9 | 3 | 11 | 7 | 9 | 2 |
|---|---|---|---|---|---|---|---|---|---|
| 21 | 8 | 9 | 2 | 2 | 20 | 14 | 11 | 15 | 6 |
| 13 | 8 | 10 | 12 | 1 | 7 | 10 | 13 | 14 | 4 |
| 8 | 12 | 8 | 10 | 11 | 19 | 0 | 9 | 10 | 4 |
| 11 | 7 | 14 | 12 | 11 | 12 | 4 | 12 | 9 | 2 |
| 4 | 3 | 3 | 5 | 9 | 9 | 4 | 3 | 5 | 12 |
| 3 | 12 | 7 | 13 | 11 | 11 | 13 | 12 | 18 | 9 |
| 6 | 14 | 14 | 18 | 10 | 11 | 7 | 9 | 7 | 2 |
| 16 | 16 | 11 | 11 | 3 | 15 | 9 | 5 | 2 | 6 |
| 5 | 11 | 14 | 11 | 6 | 9 | 4 | 17 | 20 | 10 |

**Weight Losses for Regular Diet**

| 6 | 6 | 5 | 5 | 2 | 6 | 10 | 3 | 9 | 11 |
|---|---|---|---|---|---|---|---|---|---|
| 14 | 4 | 10 | 13 | 3 | 8 | 8 | 13 | 9 | 3 |
| 4 | 12 | 6 | 11 | 12 | 9 | 8 | 5 | 8 | 7 |
| 6 | 2 | 6 | 8 | 5 | 7 | 16 | 18 | 6 | 8 |
| 13 | 1 | 9 | 8 | 12 | 10 | 6 | 1 | 0 | 13 |
| 11 | 2 | 8 | 16 | 14 | 4 | 6 | 5 | 12 | 9 |
| 11 | 6 | 3 | 9 | 9 | 14 | 2 | 10 | 4 | 13 |
| 8 | 1 | 1 | 4 | 9 | 4 | 1 | 1 | 5 | 6 |
| 14 | 0 | 7 | 12 | 9 | 5 | 9 | 12 | 7 | 9 |
| 8 | 9 | 8 | 10 | 5 | 8 | 0 | 3 | 4 | 8 |

**FIGURE 9.1**

*SPSS Summary Statistics for Diet Study*

| Group Statistics | | | | | |
|---|---|---|---|---|---|
| | DIET | N | Mean | Std. Deviation | Std. Error Mean |
| WTLOSS | LOWFAT | 100 | 9.31 | 4.67 | .47 |
| | REGULAR | 100 | 7.40 | 4.04 | .40 |

Summary statistics for the diet data are displayed in the SPSS printout, Figure 9.1. Note that $\bar{x}_1 = 9.31$, $\bar{x}_2 = 7.40$, $s_1 = 4.67$, and $s_2 = 4.04$. Using these values and noting that $\alpha = .05$ and $z_{.025} = 1.96$, we find that the 95% confidence interval is, approximately,

$$(9.31 - 7.40) \pm 1.96 \sqrt{\frac{(4.67)^2}{100} + \frac{(4.04)^2}{100}} = 1.91 \pm (1.96)(.62) = 1.91 \pm 1.22$$

or (.69, 3.13). Using this estimation procedure over and over again for different samples, we know that approximately 95% of the confidence intervals formed in this manner will enclose the difference in population means $(\mu_1 - \mu_2)$. Therefore, we are highly confident that the mean weight loss for the low-fat diet is between .69 and 3.13 pounds more than the mean weight loss for the other diet. With this information, the dietitian better understands the potential of the low-fat diet as a weight-reducing diet.

∎

**FIGURE 9.2**

*Sampling Distribution of*
$(\bar{x}_1 - \bar{x}_2)$

The justification for the procedure used in Example 9.1 to estimate $(\mu_1 - \mu_2)$ relies on the properties of the sampling distribution of $(\bar{x}_1 - \bar{x}_2)$. The performance of the estimator in repeated sampling is pictured in Figure 9.2, and its properties are summarized in the next box.

---

## Properties of the Sampling Distribution of $(\bar{x}_1 - \bar{x}_2)$

1. The mean of the sampling distribution of $(\bar{x}_1 - \bar{x}_2)$ is $(\mu_1 - \mu_2)$.
2. If the two samples are independent, the standard deviation of the sampling distribution is

$$\sigma_{(\bar{x}_1 - \bar{x}_2)} = \sqrt{\frac{\sigma_1^2}{n_1} + \frac{\sigma_2^2}{n_2}}$$

where $\sigma_1^2$ and $\sigma_2^2$ are the variances of the two populations being sampled and $n_1$ and $n_2$ are the respective sample sizes. We also refer to $\sigma_{(\bar{x}_1 - \bar{x}_2)}$ as the **standard error** of the statistic $(\bar{x}_1 - \bar{x}_2)$.

3. The sampling distribution of $(\bar{x}_1 - \bar{x}_2)$ is approximately normal for *large samples* by the Central Limit Theorem.

---

In Example 9.1, we noted the similarity in the procedures for forming a large-sample confidence interval for one population mean and a large-sample confidence interval for the difference between two population means. When we are testing hypotheses, the procedures are again very similar. The general large-sample procedures for forming confidence intervals and testing hypotheses about $(\mu_1 - \mu_2)$ are summarized in the next two boxes.

---

## Large Sample Confidence Interval for $(\mu_1 - \mu_2)$

$$(\bar{x}_1 - \bar{x}_2) \pm z_{\alpha/2}\sigma_{(\bar{x}_1 - \bar{x}_2)} = (\bar{x}_1 - \bar{x}_2) \pm z_{\alpha/2}\sqrt{\frac{\sigma_1^2}{n_1} + \frac{\sigma_2^2}{n_2}}$$

*Assumptions:*  The two samples are randomly selected in an independent manner from the two populations. The sample sizes, $n_1$ and $n_2$, are large enough so that $\bar{x}_1$ and $\bar{x}_2$ both have approximately normal sampling distributions and so that $s_1^2$ and $s_2^2$ provide good approximations to $\sigma_1^2$ and $\sigma_2^2$. This will be true if $n_1 \geq 30$ and $n_2 \geq 30$.

### Large-Sample Test of Hypothesis for ($\mu_1 - \mu_2$)

**ONE-TAILED TEST**

$H_0: (\mu_1 - \mu_2) = D_0$

$H_a: (\mu_1 - \mu_2) < D_0$
$\quad [\text{or } H_a: (\mu_1 - \mu_2) > D_0]$

**TWO-TAILED TEST**

$H_0: (\mu_1 - \mu_2) = D_0$

$H_a: (\mu_1 - \mu_2) \neq D_0$

where $D_0$ = Hypothesized difference between the means (this difference is often hypothesized to be equal to 0)

*Test statistic:*

$$z = \frac{(\bar{x}_1 - \bar{x}_2) - D_0}{\sigma_{(\bar{x}_1 - \bar{x}_2)}} \quad \text{where} \quad \sigma_{(\bar{x}_1 - \bar{x}_2)} = \sqrt{\frac{\sigma_1^2}{n_1} + \frac{\sigma_2^2}{n_2}}$$

*Rejection region:* $z < -z_\alpha$
$\quad [\text{or } z > z_\alpha \text{ when}$
$\quad H_a: (\mu_1 - \mu_2) > D_0]$

*Rejection region:* $|z| > z_{\alpha/2}$

*Assumptions:* Same as for the large-sample confidence interval.

---

**EXAMPLE 9.2**

Refer to the study of obese people on a low-fat diet and a regular diet, Example 9.1. Another way to compare the mean weight losses for the two different diets is to conduct a test of hypothesis. Use the information on the SPSS printout, Figure 9.1, to conduct the test. Use $\alpha = .05$.

### Solution

Again, we let $\mu_1$ and $\mu_2$ represent the population mean weight losses of obese people on the low-fat diet and regular diet, respectively. If one diet is more effective in reducing the weights of obese people, then either $\mu_1 < \mu_2$ or $\mu_2 < \mu_1$; that is, $\mu_1 \neq \mu_2$.

Thus, the elements of the test are as follows:

$H_0: (\mu_1 - \mu_2) = 0 \quad$ (i.e., $\mu_1 = \mu_2$; note that $D_0 = 0$ for this hypothesis test)

$H_a: (\mu_1 - \mu_2) \neq 0 \quad$ (i.e., $\mu_1 \neq \mu_2$)

*Test statistic:* $z = \dfrac{(\bar{x}_1 - \bar{x}_2) - D_0}{\sigma_{(\bar{x}_1 - \bar{x}_2)}} = \dfrac{\bar{x}_1 - \bar{x}_2 - 0}{\sigma_{(\bar{x}_1 - \bar{x}_2)}}$

*Rejection region:* $z < -z_{\alpha/2} = -1.96$ or $z > z_{\alpha/2} = 1.96 \quad$ (see Figure 9.3)

Substituting the summary statistics given in Figure 9.1 into the test statistic, we obtain

$$z = \frac{(\bar{x}_1 - \bar{x}_2) - 0}{\sigma_{(\bar{x}_1 - \bar{x}_2)}} = \frac{9.31 - 7.40}{\sqrt{\dfrac{\sigma_1^2}{n_1} + \dfrac{\sigma_2^2}{n_2}}}$$

$$\approx \frac{1.91}{\sqrt{\dfrac{s_1^2}{n_1} + \dfrac{s_2^2}{n_2}}} = \frac{1.91}{\sqrt{\dfrac{(4.67)^2}{100} + \dfrac{(4.04)^2}{100}}} = \frac{1.91}{.617} = 3.09$$

As you can see in Figure 9.3, the calculated $z$ value clearly falls in the rejection region. Therefore, the samples provide sufficient evidence, at $\alpha = .05$, for the dietitian to conclude that the mean weight losses for the two diets differ.

**FIGURE 9.3**

*Rejection Region for*
*Example 9.2*

This conclusion agrees with the inference drawn from the 95% confidence interval in Example 9.1. However, the confidence interval provides more information on the mean weight losses. From the hypothesis test, we know only that the two means differ; that is, $\mu_1 \neq \mu_2$. From the confidence interval in Example 9.1, we found that the mean weight loss $\mu_1$ of the low-fat diet was between .69 and 3.13 pounds more than the mean weight loss $\mu_2$ of the regular diet. In other words, the test tells us that the means differ, but the confidence interval tells us how large the difference is. Both inferences are made with the same degree of reliability—namely, with 95% confidence (or, at $\alpha = .05$.)

**EXAMPLE 9.3**

Find the observed significance level for the test in Example 9.2. Interpret the result.

**Solution**

The alternative hypothesis in Example 9.2, $H_a: \mu_1 - \mu_2 \neq 0$, required a two-tailed test using

$$z = \frac{\bar{x}_1 - \bar{x}_2}{\sigma_{(\bar{x}_1 - \bar{x}_2)}}$$

as a test statistic. Since the $z$ value calculated from the sample data was 3.09, the observed significance level (*p*-value) for the two-tailed test is the probability of observing a value of $z$ at least as contradictory to the null hypothesis as $z = 3.09$; that is,

$$p\text{-value} = 2 \cdot P(z \geq 3.09)$$

This probability is computed assuming $H_0$ is true and is equal to the highlighted area shown in Figure 9.4.

The tabulated area corresponding to $z = 3.09$ in Table IV of Appendix A is .4990. Therefore,

$$P(z \geq 3.09) = .5 - .4990 = .0010$$

and the observed significance level for the test is

$$p\text{-value} = 2(.001) = .002$$

**FIGURE 9.4**

*The Observed Significance*
*Level for Example 9.2*

**FIGURE 9.5**

*MINITAB printout for the Hypothesis Test of Example 9.2*

```
Two-sample T for LOWFAT vs REGULAR

              N      Mean     StDev    SE Mean
LOWFAT      100      9.31      4.67      0.47
REGULAR     100      7.40      4.04      0.40


Difference = mu LOWFAT - mu REGULAR
Estimate for difference:  1.910
95% CI for difference: (0.693, 3.127)
T-Test of difference = 0 (vs not =): T-Value = 3.10   P-Value = 0.002
DF = 193
```

Since our selected $\alpha$ value, .05, exceeds this *p*-value, we have sufficient evidence to reject $H_0: \mu_1 - \mu_2 = 0$.

The *p*-value of the test is more easily obtained from a statistical software package. A MINITAB printout for the hypothesis test is displayed in Figure 9.5. The *p*-value, highlighted on the printout, is .002. This value agrees with our calculated *p*-value. ∎

## Small Samples

When comparing two population means with small samples (say, $n_1 < 30$ and $n_2 < 30$), the methodology of the previous three examples is invalid. The reason? When the sample sizes are small, estimates of $\sigma_1^2$ and $\sigma_2^2$ are unreliable and the Central Limit Theorem (which guarantees that the *z* statistic is normal) can no longer be applied. But as in the case of a single mean (Section 8.4), we use the familiar Student's *t*-distribution described in Chapter 7.

*To use the t-distribution, both sampled populations must be approximately normally distributed with equal population variances, and the random samples must be selected independently of each other.* The normality and equal variances assumptions imply relative frequency distributions for the populations that would appear as shown in Figure 9.6

**FIGURE 9.6**

*Assumptions for the Two-Sample t: (1) Normal Populations, (2) Equal Variances*

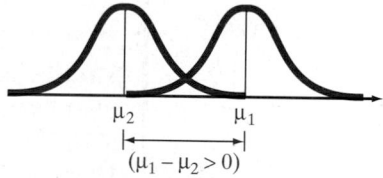

Since we assume the two populations have equal variances ($\sigma_1^2 = \sigma_2^2 = \sigma^2$), it is reasonable to use the information contained in both samples to construct a **pooled sample estimator $\sigma^2$** for use in confidence intervals and test statistics. Thus, if $s_1^2$ and $s_2^2$ are the two sample variances (both estimating the variance $\sigma^2$ common to both populations), the pooled estimator of $\sigma^2$, denoted as $s_p^2$, is

$$s_p^2 = \frac{(n_1 - 1)s_1^2 + (n_2 - 1)s_2^2}{(n_1 - 1) + (n_2 - 1)} = \frac{(n_1 - 1)s_1^2 + (n_2 - 1)s_2^2}{n_1 + n_2 - 2}$$

or

$$s_p^2 = \frac{\overbrace{\sum (x_1 - \bar{x}_1)^2}^{\text{From sample 1}} + \overbrace{\sum (x_2 - \bar{x}_2)^2}^{\text{From sample 2}}}{n_1 + n_2 - 2}$$

where $x_1$ represents a measurement from sample 1 and $x_2$ represents a measurement from sample 2. Recall that the term *degrees of freedom* was defined in Section 7.2 as 1 less than the sample size. Thus, in this case, we have $(n_1 - 1)$ degrees of freedom for sample 1 and $(n_2 - 1)$ degrees of freedom for sample 2. Since we are pooling the information on $\sigma^2$ obtained from both samples, the degrees of freedom associated with the pooled variance $s_p^2$ is equal to the sum of the degrees of freedom for the two samples, namely, the denominator of $s_p^2$; that is, $(n_1 - 1) + (n_2 - 1) = n_1 + n_2 - 2$.

Note that the second formula given for $s_p^2$ shows that the pooled variance is simply a *weighted average* of the two sample variances $s_1^2$ and $s_2^2$. The weight given each variance is proportional to its degrees of freedom. If the two variances have the same number of degrees of freedom (i.e., if the sample sizes are equal), then the pooled variance is a simple average of the two sample variances. The result is an average or "pooled" variance that is a better estimate of $\sigma^2$ than either $s_1^2$ or $s_2^2$ alone.

Both the confidence interval and the test of hypothesis procedures for comparing two population means with small samples are summarized in the accompanying box.

---

### Small-Sample Confidence Interval for $(\mu_1 - \mu_2)$: Independent Samples

$$(\bar{x}_1 - \bar{x}_2) \pm t_{\alpha/2} \sqrt{s_p^2 \left( \frac{1}{n_1} + \frac{1}{n_2} \right)}$$

where $s_p^2 = \dfrac{(n_1 - 1)s_1^2 + (n_2 - 1)s_2^2}{n_1 + n_2 - 2}$

and $t_{\alpha/2}$ is based on $(n_1 + n_2 - 2)$ degrees of freedom.

*Assumptions:*    1.  Both sampled populations have relative frequency distributions that are approximately normal.

                 2.  The population variances are equal.

                 3.  The samples are randomly and independently selected from the populations.

---

### Small-Sample Test of Hypothesis for $(\mu_1 - \mu_2)$: Independent Samples

**ONE-TAILED TEST**

$H_0: (\mu_1 - \mu_2) = D_0$

$H_a: (\mu_1 - \mu_2) < D_0$

[or $H_a: (\mu_1 - \mu_2) > D_0$]

*Test statistic:*

**TWO-TAILED TEST**

$H_0: (\mu_1 - \mu_2) = D_0$

$H_a: (\mu_1 - \mu_2) \neq D_0$

$$t = \frac{(\bar{x}_1 - \bar{x}_2) - D_0}{\sqrt{s_p^2 \left( \dfrac{1}{n_1} + \dfrac{1}{n_2} \right)}}$$

*Rejection region:* $t < -t_\alpha$

[or $t > t_\alpha$ when

$H_a: (\mu_1 - \mu_2) > D_0$]

*Rejection region:* $|t| > t_{\alpha/2}$

where $t_\alpha$ and $t_{\alpha/2}$ are based on $(n_1 + n_2 - 2)$ degrees of freedom.

*Assumptions:* Same as for the small-sample confidence interval for $(\mu_1 - \mu_2)$ in the previous box.

**EXAMPLE 9.4**

Suppose you wish to compare a new method of teaching reading to "slow learners" to the current standard method. You decide to base this comparison on the results of a reading test given at the end of a learning period of 6 months. Of a random sample of 22 slow learners, 10 are taught by the new method and 12 are taught by the standard method. All 22 children are taught by qualified instructors under similar conditions for a 6-month period. The results of the reading test at the end of this period are given in Table 9.2.

⊘ READING

**TABLE 9.2   Reading Test Scores for Slow Learners**

| New Method | | | | Standard Method | | | |
|---|---|---|---|---|---|---|---|
| 80 | 80 | 79 | 81 | 79 | 62 | 70 | 68 |
| 76 | 66 | 71 | 76 | 73 | 76 | 86 | 73 |
| 70 | 85 | | | 72 | 68 | 75 | 66 |

a. Use the data in the table to estimate the true mean difference between the test scores for the new method and the standard method. Use a 95% confidence interval.
b. Interpret the interval, part **a**.
c. What assumptions must be made in order that the estimate be valid? Are they reasonably satisfied?

**Solution**

a. For this experiment, let $\mu_1$ and $\mu_2$ represent the mean reading test scores of slow learners taught with the new and standard methods, respectively. Then, the objective is to obtain a 95% confidence interval for $(\mu_1 - \mu_2)$.

The first step in constructing the confidence interval is to obtain summary statistics (e.g., $\bar{x}$ and $s$) on reading test scores for each method. The data of Table 9.2 was entered into a computer, and SAS used to obtain these descriptive statistics. The SAS printout appears in Figure 9.7. Note that $\bar{x}_1 = 76.4$, $s_1 = 5.8348$, $\bar{x}_2 = 72.333$, and $s_2 = 6.3437$

**FIGURE 9.7**

*SAS Output for Example 9.4*

Two Sample t-test for the Means of SCORE within METHOD

Sample Statistics

| Group | N | Mean | Std. Dev. | Std. Error |
|---|---|---|---|---|
| NEW | 10 | 76.4 | 5.8348 | 1.8451 |
| STD | 12 | 72.33333 | 6.3437 | 1.8313 |

Hypothesis Test

Null hypothesis:    Mean 1 - Mean 2 =  0
Alternative:        Mean 1 - Mean 2 ^= 0

| If Variances Are | t statistic | Df | Pr > t |
|---|---|---|---|
| Equal | 1.552 | 20 | 0.1364 |
| Not Equal | 1.564 | 19.77 | 0.1336 |

95% Confidence Interval for the Difference between Two Means

| Lower Limit | Upper Limit |
|---|---|
| -1.40 | 9.53 |

Next, we calculate the pooled estimate of variance:

$$s_p^2 = \frac{(n_1 - 1)s_1^2 + (n_2 - 1)s_2^2}{n_1 + n_2 - 2}$$

$$= \frac{(10 - 1)(5.8348)^2 + (12 - 1)(6.3437)^2}{10 + 12 - 2} = 37.45$$

where $s_p^2$ is based on $(n_1 + n_2 - 2) = (10 + 12 - 2) = 20$ degrees of freedom. Also, we find $t_{\alpha/2} = t_{.025} = 2.086$ (based on 20 degrees of freedom) from Table VI of Appendix A.

Finally, the 95% confidence interval for $(\mu_1 - \mu_2)$, the difference between mean test scores for the two methods, is

$$(\bar{x}_1 - \bar{x}_2) \pm t_{\alpha/2}\sqrt{s_p^2\left(\frac{1}{n_1} + \frac{1}{n_2}\right)} = (76.4 - 72.33) \pm t_{.025}\sqrt{37.45\left(\frac{1}{10} + \frac{1}{12}\right)}$$

$$= 4.07 \pm (2.086)(2.62)$$
$$= 4.07 \pm 5.47$$

or $(-1.4, 9.54)$. This interval agrees (except for rounding) with the one shown at the bottom of the SAS printout, Figure 9.7.

b. The interval can be interpreted as follows. With a confidence coefficient equal to .95, we estimate the difference in mean test scores between using the new method of teaching and using the standard method to fall in the interval $-1.4$ to 9.54. In other words, we estimate (with 95% confidence) the mean test score for the new method to be anywhere from 1.4 points less than to 9.54 points more than the mean test score for the standard method. Although the sample means seem to suggest that the new method is associated with a higher mean test score, there is insufficient evidence to indicate that $(\mu_1 - \mu_2)$ differs from 0 because the interval includes 0 as a possible value for $(\mu_1 - \mu_2)$. To show a difference in mean test scores (if it exists), you could increase the sample size and thereby narrow the width of the confidence interval for $(\mu_1 - \mu_2)$. An alternative is to design the experiment differently. This possibility is discussed in the next section.

c. To properly use the small-sample confidence interval, the following assumptions must be satisfied:
   (1) The samples are randomly and independently selected from the populations of slow learners taught by the new method and the standard method.
   (2) The test scores are normally distributed for both teaching methods.
   (3) The variance of the test scores are the same for the two populations, that is, $\sigma_1^2 = \sigma_2^2$.

The first assumption is satisfied, based on the information provided about the sampling procedure in the problem description. To check the plausibility of the remaining two assumptions, we resort to graphical methods. Figure 9.8 is a portion of the SPSS printout that gives normal probability plots for the test scores of the two samples of slow learners. The near straight-line trends on both plots indicate that the score distributions are approximately mound-shaped and symmetric. Consequently, each sample data set appears to come from a population that is approximately normal.

One way to check assumption #3 is to test the null hypothesis $H_0: \sigma_1^2 = \sigma_2^2$. This test is covered in Section 9.5. Another approach is to examine box plots for the sample data. Figure 9.9 is an SPSS printout that shows side-by-side vertical box plots for the test scores in the two samples. Recall, from Section 2.9, that the box plot represents the "spread" of a data set. The two box plots appear to have about the same spread; thus, the samples appear to come from populations with approximately the same variance.

**FIGURE 9.8**

*SPSS Normal Probability Plots for Example 9.4*

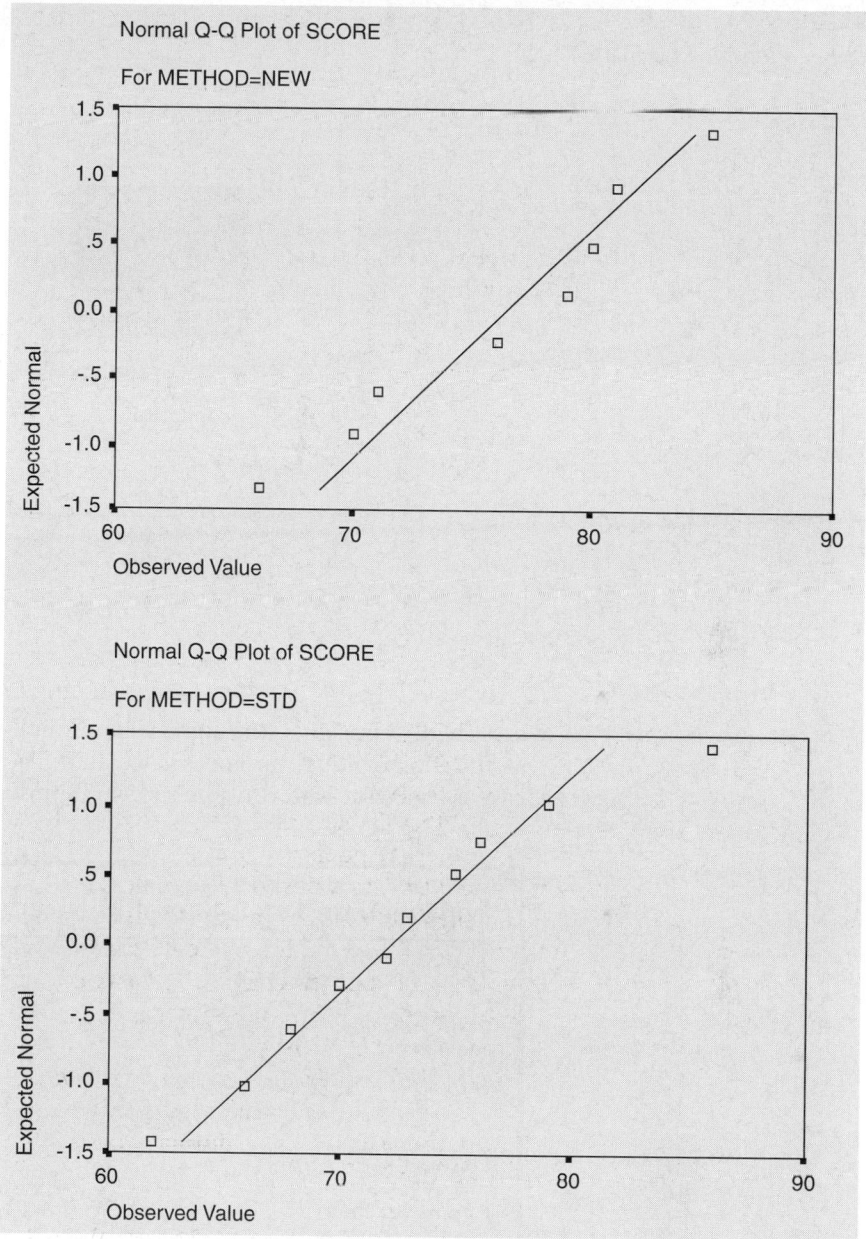

All three assumptions, then, appear to be reasonably satisfied for this application of the small-sample confidence interval. ∎

The two-sample $t$ statistic is a powerful tool for comparing population means when the assumptions are satisfied. It has also been shown to retain its usefulness when the sampled populations are only approximately normally distributed. And when the sample sizes are equal, the assumption of equal population variances can be relaxed. That is, if $n_1 = n_2$, then $\sigma_1^2$ and $\sigma_2^2$ can be quite different and the test statistic will still possess, approximately, a Student's $t$-distribution. In the case where $\sigma_1^2 \neq \sigma_2^2$ and $n_1 \neq n_2$, an approximate small-sample confidence interval or test can be obtained by modifying the degrees of freedom associated with the $t$-distribution.

**FIGURE 9.9**

*SPSS Box Plots for Example 9.4*

The next box gives the approximate small-sample procedures to use when the assumption of equal variances is violated. The test for the "unequal sample sizes" case is based on Satterthwaite's (1946) approximation.

---

## Approximate Small-Sample Procedures when $\sigma_1^2 \neq \sigma_2^2$

**EQUAL SAMPLE SIZES ($n_1 = n_2 = n$)**

*Confidence interval:*          $(\bar{x}_1 - \bar{x}_2) \pm t_{\alpha/2}\sqrt{(s_1^2 + s_2^2)/n}$

*Test statistic for $H_0$: $(\mu_1 - \mu_2) = 0$:*          $t = (\bar{x}_1 - \bar{x}_2)/\sqrt{(s_1^2 + s_2^2)/n}$
where $t$ is based on $\nu = n_1 + n_2 - 2 = 2(n - 1)$ degrees of freedom.

**UNEQUAL SAMPLE SIZES ($n_1 \neq n_2$)**

*Confidence interval:*          $(\bar{x}_1 - \bar{x}_2) \pm t_{\alpha/2}\sqrt{(s_1^2/n_1) + (s_2^2/n_2)}$

*Test statistic for $H_0$: $(\mu_1 - \mu_2) = 0$:*          $t = (\bar{x}_1 - \bar{x}_2)/\sqrt{(s_1^2/n_1) + (s_2^2/n_2)}$
where $t$ is based on degrees of freedom equal to

$$\nu = \frac{(s_1^2/n_1 + s_2^2/n_2)^2}{\dfrac{(s_1^2/n_1)^2}{n_1 - 1} + \dfrac{(s_2^2/n_2)^2}{n_2 - 1}}$$

*Note:* The value of $\nu$ will generally not be an integer. Round $\nu$ down to the nearest integer to use the $t$ table.

---

When the assumptions are not clearly satisfied, you can select larger samples from the populations or you can use other available statistical tests (nonparametric statistical tests, which are described in Chapter 14).

---

# What Should You Do If the Assumptions Are Not Satisfied?

*Answer:* If you are concerned that the assumptions are not satisfied, use the Wilcoxon rank sum test for independent samples to test for a shift in population distributions. See Chapter 14.

---

 EXERCISES 9.1–9.26

## Learning the Mechanics

**9.1** In order to compare the means of two populations, independent random samples of 400 observations are selected from each population, with the following results:

| Sample 1 | Sample 2 |
|---|---|
| $\bar{x}_1 = 5,275$ | $\bar{x}_2 = 5,240$ |
| $s_1 = 150$ | $s_2 = 200$ |

**a.** Use a 95% confidence interval to estimate the difference between the population means $(\mu_1 - \mu_2)$. Interpret the confidence interval.

**b.** Test the null hypothesis $H_0: (\mu_1 - \mu_2) = 0$ versus the alternative hypothesis $H_a: (\mu_1 - \mu_2) \neq 0$. Give the significance level of the test, and interpret the result.

**c.** Suppose the test in part **b** was conducted with the alternative hypothesis $H_a: (\mu_1 - \mu_2) > 0$. How would your answer to part **b** change?

**d.** Test the null hypothesis $H_0: (\mu_1 - \mu_2) = 25$ versus the alternative $H_a: (\mu_1 - \mu_2) \neq 25$. Give the significance level, and interpret the result. Compare your answer to the test conducted in part **b**.

**e.** What assumptions are necessary to ensure the validity of the inferential procedures applied in parts **a–d**?

**9.2** The purpose of this exercise is to compare the variability of $\bar{x}_1$ and $\bar{x}_2$ with the variability of $(\bar{x}_1 - \bar{x}_2)$.

**a.** Suppose the first sample is selected from a population with mean $\mu_1 = 150$ and variance $\sigma_1^2 = 900$. Within what range should the sample mean vary about 95% of the time in repeated samples of 100 measurements from this distribution? That is, construct an interval extending 2 standard deviations of $\bar{x}_1$ on each side of $\mu_1$.

**b.** Suppose the second sample is selected independently of the first from a second population with mean $\mu_2 = 150$ and variance $\sigma_2^2 = 1,600$. Within what range should the sample mean vary about 95% of the time in repeated samples of 100 measurements from this distribution? That is, construct an interval extending 2 standard deviations of $\bar{x}_2$ on each side of $\mu_2$.

**c.** Now consider the difference between the two sample means $(\bar{x}_1 - \bar{x}_2)$. What are the mean and standard deviation of the sampling distribution of $(\bar{x}_1 - \bar{x}_2)$?

**d.** Within what range should the difference in sample means vary about 95% of the time in repeated independent samples of 100 measurements each from the two populations?

**e.** What, in general, can be said about the variability of the difference between independent sample means relative to the variability of the individual sample means?

**9.3** Independent random samples of 100 observations each are chosen from two normal populations with the following means and standard deviations:

| Population 1 | Population 2 |
|---|---|
| $\mu_1 = 14$ | $\mu_2 = 10$ |
| $\sigma_1 = 4$ | $\sigma_2 = 3$ |

Let $\bar{x}_1$ and $\bar{x}_2$ denote the two sample means.

**a.** Give the mean and standard deviation of the sampling distribution of $\bar{x}_1$.

**b.** Give the mean and standard deviation of the sampling distribution of $\bar{x}_2$.

**c.** Suppose you were to calculate the difference $(\bar{x}_1 - \bar{x}_2)$ between the sample means. Find the mean and standard deviation of the sampling distribution of $(\bar{x}_1 - \bar{x}_2)$.

**d.** Will the statistic $(\bar{x}_1 - \bar{x}_2)$ be normally distributed? Explain.

**9.4** To use the $t$ statistic to test for a difference between the means of two populations, what assumptions must be made about the two populations? About the two samples?

**9.5** Two populations are described in each of the following cases. In which cases would it be appropriate to apply the small-sample $t$-test to investigate the difference between the population means?

**a.** Population 1: Normal distribution with variance $\sigma_1^2$ Population 2: Skewed to the right with variance $\sigma_2^2 = \sigma_1^2$

**b.** Population 1: Normal distribution with variance $\sigma_1^2$ Population 2: Normal distribution with variance $\sigma_2^2 \neq \sigma_1^2$

**c.** Population 1: Skewed to the left with variance $\sigma_1^2$ Population 2: Skewed to the left with variance $\sigma_2^2 = \sigma_1^2$

**d.** Population 1: Normal distribution with variance $\sigma_1^2$ Population 2: Normal distribution with variance $\sigma_2^2 = \sigma_1^2$

**e.** Population 1: Uniform distribution with variance $\sigma_1^2$
Population 2: Uniform distribution with variance $\sigma_2^2 = \sigma_1^2$

**9.6** Assume that $\sigma_1^2 = \sigma_2^2 = \sigma^2$. Calculate the pooled estimator of $\sigma^2$ for each of the following cases:

**a.** $s_1^2 = 200, s_2^2 = 180, n_1 = n_2 = 25$
**b.** $s_1^2 = 25, s_2^2 = 40, n_1 = 20, n_2 = 10$
**c.** $s_1^2 = .20, s_2^2 = .30, n_1 = 8, n_2 = 12$
**d.** $s_1^2 = 2,500, s_2^2 = 1,800, n_1 = 16, n_2 = 17$
**e.** Note that the pooled estimate is a weighted average of the sample variances. To which of the variances does the pooled estimate fall nearer in each of the above cases?

**9.7** Independent random samples from normal populations produced the results shown below:

LM9_7

| Sample 1 | Sample 2 |
|---|---|
| 1.2 | 4.2 |
| 3.1 | 2.7 |
| 1.7 | 3.6 |
| 2.8 | 3.9 |
| 3.0 | |

**a.** Calculate the pooled estimate of $\sigma^2$.
**b.** Do the data provide sufficient evidence to indicate that $\mu_2 > \mu_1$? Test using $\alpha = .10$.
**c.** Find a 90% confidence interval for $(\mu_1 - \mu_2)$.
**d.** Which of the two inferential procedures, the test of hypothesis in part **b** or the confidence interval in part **c**, provides more information about $(\mu_1 - \mu_2)$?

**9.8** Two independent random samples have been selected, 100 observations from population 1 and 100 from population 2. Sample means $\bar{x}_1 = 70$ and $\bar{x}_2 = 50$ were obtained. From previous experience with these populations, it is known that the variances are $\sigma_1^2 = 100$ and $\sigma_2^2 = 64$.

**a.** Find $\sigma_{(\bar{x}_1 - \bar{x}_2)}$.
**b.** Sketch the approximate sampling distribution $(\bar{x}_1 - \bar{x}_2)$ assuming $(\mu_1 - \mu_2) = 5$.

**c.** Locate the observed value of $(\bar{x}_1 - \bar{x}_2)$ on the graph you drew in part **b**. Does it appear that this value contradicts the null hypothesis $H_0: (\mu_1 - \mu_2) = 5$?
**d.** Use the $z$-table to determine the rejection region for the test of $H_0: (\mu_1 - \mu_2) = 5$ against $H_a: (\mu_1 - \mu_2) \neq 5$. Use $\alpha = .05$.
**e.** Conduct the hypothesis test of part **d** and interpret your result.
**f.** Construct a 95% confidence interval for $(\mu_1 - \mu_2)$. Interpret the interval.
**g.** Which inference provides more information about the value of $(\mu_1 - \mu_2)$—the test of hypothesis in part **e** or the confidence interval in part **f**?

**9.9** Independent random samples are selected from two populations and used to test the hypothesis $H_0: (\mu_1 - \mu_2) = 0$ against the alternative $H_a: (\mu_1 - \mu_2) \neq 0$. A total of 233 observations from population 1 and 312 from population 2 are analyzed by MINITAB, with the result shown below.

**a.** Interpret the results of the computer analysis.
**b.** If the alternative hypothesis had been $H_a: (\mu_1 - \mu_2) < 0$, how would the $p$-value change? Interpret the $p$-value for this one-tailed test.

**9.10** Independent random samples selected from two normal populations produced the sample means and standard deviations shown below:

| Sample 1 | Sample 2 |
|---|---|
| $n_1 = 17$ | $n_2 = 12$ |
| $\bar{x}_1 = 5.4$ | $\bar{x}_2 = 7.9$ |
| $s_1 = 3.4$ | $s_2 = 4.8$ |

**a.** The test $H_0: (\mu_1 - \mu_2) = 0$ against $H_a: (\mu_1 - \mu_2) \neq 0$ was conducted using SAS, with the results shown on p. 393. Check and interpret the results.
**b.** Check and interpret the 95% confidence interval for $(\mu_1 - \mu_2)$ shown on the printout.

## Applying the Concepts—Basic

**9.11** Are children who repeat a grade in elementary school shorter on average, than their peers? To answer this

---

**MINITAB Output for Exercise 9.9**

```
Two-sample T for density

              N      Mean     StDev    SE Mean
X1          233      473        84      15.26
X2          312      485        93      17.66

Difference = mu (X1) - mu (X2)
Estimate for difference: -8.0
95% CI for difference: (-22.93, 6.93)
T-Test of difference = 0 (vs not =): T-Value = -1.58  P-Value = 0.115
DF = 543
```

SAS Output for Exercise 9.10

Two Sample t-test for the Means of X within SAMPLE

Sample Statistics

| Group | N | Mean | Std. Dev. | Std. Error |
|-------|-----|------|-----------|------------|
| SAM1 | 17 | 5.4 | 3.4 | 4.123 |
| SAM2 | 12 | 7.9 | 4.8 | 3.464 |

Hypothesis Test

Null hypothesis:    Mean 1 - Mean 2 = 0
Alternative:    Mean 1 - Mean 2 ^= 0

| If Variances Are | t statistic | Df | Pr > t |
|------------------|-------------|-------|--------|
| Equal | -1.646 | 27 | 0.1114 |
| Not Equal | -1.550 | 18.57 | 0.1386 |

95% Confidence Interval for the Difference between Two Means

| Lower Limit | Upper Limit |
|-------------|-------------|
| -5.48 | 0.48 |

question, researchers compared the heights of Australian school children who repeated a grade to those who did not (*The Archives of Disease in Childhood*, Apr. 2000). All height measurements were standardized using $z$-scores. A summary of the results, by gender, is shown in the table.

| | **Never Repeated** | **Repeated a Grade** |
|------|--------------------|----------------------|
| **Boys** | $n = 1,349$ | $n = 86$ |
| | $\bar{x} = .30$ | $\bar{x} = -.04$ |
| | $s = .97$ | $s = 1.17$ |
| **Girls** | $n = 1,366$ | $n = 43$ |
| | $\bar{x} = .22$ | $\bar{x} = .26$ |
| | $s = 1.04$ | $s = .94$ |

*Source:* Wake, M., Coghlan, D., and Hesketh, K. "Does height influence progression through primary school grades?" *The Archives of Disease in Childhood*, Vol. 82, Apr. 2000 (Table 3).

**a.** Conduct a test of hypothesis to determine whether the average height of Australian boys who repeated a grade is less than the average height of boys who never repeated. Use $\alpha = .05$.

**b.** Repeat part **a** for Australian girls.

**c.** Summarize the results of the hypothesis tests in the words of the problem.

**9.12** A group of University of Florida psychologists investigated the effects of age and gender on the short-term memory of adults (*Cognitive Aging Conference,* Apr. 1996). Each in a sample of 152 adults was asked to place 20 common household items (e.g., eyeglasses, keys, hat, hammer) into the rooms of a computer-image house. After performing some unrelated activities, each subject was asked to recall the locations of the objects they had placed. The number of correct responses (out of 20) was recorded.

**a.** The researchers theorized that women will have a higher mean recall score than men. Set up the null and alternative hypotheses to test this theory.

**b.** Refer to part **a.** The 43 men in the study had a mean recall score of 13.5, while the 109 women had a mean recall score of 14.4. The observed significance level for comparing these two means was found to be $p = .0001$. Interpret this value.

**c.** The researchers also hypothesized that younger adults would have a higher mean recall score than older adults. Set up $H_0$ and $H_a$ to test this theory.

**d.** The observed significance level for the test of part **c** was reported as $p = .0001$. Interpret this result.

**9.13** The "fear of negative evaluation" (FNE) scores for 11 female students known to suffer from the eating disorder bulimia and 14 female students with normal eating habits, first presented in Exercise 2.27 (p. 39), are reproduced on p. 394. (Recall that the higher the score, the greater the fear of negative evaluation.)

**a.** Find a 95% confidence interval for the difference between the population means of the FNE scores for bulimic and normal female students. Interpret the result.

## BULIMIA

| Bulimic students | 21 | 13 | 10 | 20 | 25 | 19 | 16 | 21 | 24 | 13 | 14 | | | |
|---|---|---|---|---|---|---|---|---|---|---|---|---|---|---|
| Normal students | 13 | 6 | 16 | 13 | 8 | 19 | 23 | 18 | 11 | 19 | 7 | 10 | 15 | 20 |

*Source:* Randles, R. H. "On neutral responses (zeros) in the sign test and ties in the Wilcoxon-Mann-Whitney test." *The American Statistician*, Vol. 55, No. 2, May 2001 (Figure 3).

**b.** What assumptions are required for the interval of part **a** to be statistically valid? Are these assumptions reasonably satisfied? Explain.

**9.14** Do the early questions in an exam affect your performance on later questions? Research reported in the *Journal of Psychology* (Jan. 1991) investigated this relationship by conducting the following experiment: 140 college students were randomly assigned to two groups. Group A took an exam that contained 15 difficult questions followed by five moderate questions. Group B took an exam that contained 15 easy questions followed by the same five moderate questions. One of the goals of the study was to compare the average score on the moderate questions for the two groups.

**a.** The researchers hypothesized that the students who had the easy questions would score better than the students who had the difficult questions. Set up the appropriate null and alternative hypotheses to test their research hypothesis, defining any symbols you use.

**b.** Suppose the *p*-value for this test was reported as .3248. What conclusion would you reach based on this *p*-value?

**c.** What assumptions are necessary to ensure the validity of this conclusion?

**9.15** On average, do males outperform females in mathematics? To answer this question, psychologists at the University of Minnesota compared the scores of male and female eighth-grade students who took a basic skills mathematics achievement test (*American Educational Research Journal,* Fall 1998). One form of the test consisted of 68 multiple-choice questions. A summary of the test scores is displayed in the table.

| | **Males** | **Females** |
|---|---|---|
| Sample size | 1,764 | 1,739 |
| Mean | 48.9 | 48.4 |
| Standard deviation | 12.96 | 11.85 |

*Source:* Bielinski, J., and Davison, M. L. "Gender differences by item difficulty interactions in multiple-choice mathematics items." *American Educational Research Journal,* Vol. 35, No. 3, Fall 1998, p. 464 (Table 1).

**a.** Is there evidence of a difference between the true mean mathematics test scores of male and female eighth-graders?

**b.** Use a 90% confidence interval to estimate the true difference in mean test scores between males and females. Does the confidence interval support the result of the test you conducted in part **a**?

**c.** What assumptions about the distributions of the populations of test scores are necessary to ensure the validity of the inferences you made in parts **a** and **b**?

**d.** What is the observed significance level of the test you conducted in part **a**?

### Applying the Concepts—Intermediate

## EVOS

**9.16** Refer to the *Journal of Agricultural, Biological, and Environmental Statistics* (Sept. 2000) impact study of a tanker oil spill on the seabird population in Alaska, Exercise 2.135 (p. 102). Recall that for each of 96 shoreline locations (called transects), the number of seabirds found, the length (in kilometers) of the transect, and whether or not the transect was in an oiled area were recorded. (The data are saved in the EVOS file.) Observed seabird density is defined as the observed count divided by the length of the transect. A comparison of the mean densities of oiled and unoiled transects is displayed in the following MINITAB printout on p. 395. Use this information to make an inference about the difference in the population mean seabird densities of oiled and unoiled transects.

**9.17** Teachers Involve Parents in Schoolwork (TIPS) is an interactive homework process designed to improve the quality of homework assignments for elementary, middle, and high school students. TIPS homework assignments require students to conduct interactions with family partners (parents, guardians, etc.) while completing the homework. Frances Van Voorhis (Johns Hopkins University) conducted a study to investigate the effects of TIPS in science, mathematics, and language arts homework assignments (April, 2001). Each in a sample of 128 middle school students was assigned to complete TIPS homework assignments, while 98 students in a second sample were assigned traditional, non-interactive, homework assignments (called ATIPS). At the end of the study, all students reported on the level of family involvement in their homework on a 4-point scale (0 = never, 1 = rarely, 2 = sometimes, 3 = frequently, 4 = always). Three scores were recorded for each student: one for science homework, one for math homework, and one for language arts homework. The data for the study are saved in the HWSTUDY file. (The first 5 and last 5 observations in the data set are listed in the table on p. 395)

● EVOS
**MINITAB Output for Problem 9.16**

```
Two-sample T for DENSITY

OIL        N      Mean     StDev    SE Mean
no        36      3.27     6.70        1.1
yes       60      3.50     5.97       0.77


Difference = mu (no) - mu (yes)
Estimate for difference: -0.22
95% CI for difference: (-2.93, 2.49)
T-Test of difference = 0 (vs not =): T-Value = -0.16   P-Value = 0.871
DF = 67
```

● HWSTUDY (First and last 5 observations)

| Homework Condition | Science | Math | Language |
|---|---|---|---|
| ATIPS | 1 | 0 | 0 |
| ATIPS | 0 | 1 | 1 |
| ATIPS | 0 | 1 | 0 |
| ATIPS | 1 | 2 | 0 |
| ATIPS | 1 | 1 | 2 |
| TIPS | 2 | 3 | 2 |
| TIPS | 1 | 4 | 2 |
| TIPS | 2 | 4 | 2 |
| TIPS | 4 | 0 | 3 |
| TIPS | 2 | 0 | 1 |

*Source:* Van Voorhis, F. L., "Teachers' use of interactive homework and its effects on family involvement and science achievement of middle grade students." Paper presented at the annual meeting of the American Educational Research Association, Seattle, April 2001.

a. Conduct an analysis to compare the mean level of family involvement in science homework assignments of TIPS and ATIPS students. Use $\alpha = .05$. Make a practical conclusion.

b. Repeat part a for mathematics homework assignments.

c. Repeat part a for language arts homework assignments.

d. What assumptions are necessary for the inferences of parts a–c to be valid? Are they reasonably satisfied?

9.18 Two methods of castrating male piglets were investigated in *Applied Animal Behaviour Science* (Nov. 1, 2000). Method 1 involved an incision of the spermatic cords while Method 2 involved pulling and severing the cords. Forty-nine male piglets were randomly allocated to one of the two methods. During castration, the researchers measured the number of high frequency vocal responses (squeals) per second over a 5-second period. The data are summarized in the table. Conduct a test of hypothesis to determine whether the population mean number of high frequency vocal responses differs for piglets castrated by the two methods. Use $\alpha = .05$.

|  | Method 1 | Method 2 |
|---|---|---|
| Sample size | 24 | 25 |
| Mean number of squeals | .74 | .70 |
| Standard deviation | .09 | .09 |

*Source:* Taylor, A. A., and Weary, D. M. "Vocal responses of piglets to castration: Identifying procedural sources of pain." *Applied Animal Behaviour Science*, Vol. 70, No. 1, November 1, 2000.

9.19 Each week, *Billboard* magazine ranks the top 200 selling albums (or CDs) in the commercial music industry. Prior to 1991, *Billboard*'s charting methodology was based on reported album sales at a sample of record stores. In 1991, *Billboard* began using the SoundScan system to chart album sales. With SoundScan, record sales were pulled directly form barcode-reading sales registers across the country. The impact of SoundScan on *Billboard* magazine's ratings was investigated in *Organizational Science* (May–June, 2000). Based on summary statistics provided in the article, the table lists the (simulated) number of new artists entering the weekly *Billboard* chart for 12 weeks prior to the introduction of SoundScan and for 12 weeks after the introduction of SoundScan.

● BILLBOARD

| **Pre-SoundScan** | 1 | 1 | 0 | 2 | 3 | 4 | 3 | 4 | 5 | 5 | 2 | 3 |
|---|---|---|---|---|---|---|---|---|---|---|---|---|
| **Post-SoundScan** | 0 | 2 | 3 | 1 | 3 | 1 | 0 | 1 | 4 | 2 | 2 | 1 |

a. Find a 90% confidence interval for the difference between the mean number of new artists entering the *Billboard* chart per week prior to SoundScan and the mean number of new artists per week after SoundScan was introduced. Interpret the result.

b. *Organizational Science* reported the *p*-value for testing the difference between the two means as $p < .1$. Find the actual *p*-value of a two-tailed test and interpret it. Does the result agree with the reported value?

**9.20** To help students organize global information about people, places, and environments, geographers encourage them to develop "mental maps" of the world. A series of lessons was designed to aid students in the development of mental maps (*Journal of Geography,* May/June 1997). In one experiment, a class of 24 seventh-grade geography students was given mental map lessons while a second class of 20 students received traditional instruction. All students were asked to sketch a map of the world, and each portion of the map was evaluated for accuracy using a 5-point scale (1 = low accuracy, 5 = high accuracy).

**a.** The mean accuracy scores of the two groups of seventh-graders were compared using a test of hypothesis. State $H_0$ and $H_a$ for a test to determine whether the mental map lessons improve a student's ability to sketch a world map.

**b.** What assumptions (if any) are required for the test to be statistically valid?

**c.** The observed significance level of the test for comparing the mean accuracy scores for continents drawn is .0507. Interpret this result.

**d.** The observed significance level of the test for comparing the mean accuracy scores for labeling oceans is .7371. Interpret the result.

**e.** The observed significance level of the test for comparing the mean accuracy scores for the entire map is .0024. Interpret the result.

**9.21** Bear gallbladder is used in Chinese medicine to treat inflamation. Due to the difficulty of obtaining bear gallbladder, Chinese medical researchers are searching for a more readily available source of animal bile. A study in the *Journal of Ethnopharmacology* (June 1995) examined pig gallbladder as an effective substitute for bear gallbladder. Twenty male mice were divided randomly into two groups: 10 were given a dosage of bear bile and 10 were given a dosage of pig bile. All the mice then received an injection of croton oil in the left ear lobe to induce inflammation. Four hours later, both the left and right ear lobes were weighed, with the difference (in milligrams) representing the degree of swelling. Summary statistics on the degree of swelling are provided in the following table.

| Bear Bile | Pig Bile |
|-----------|----------|
| $n_1 = 10$ | $n_2 = 10$ |
| $\bar{x}_1 = 9.19$ | $\bar{x}_2 = 9.71$ |
| $s_1 = 4.17$ | $s_2 = 3.33$ |

**a.** Compare the mean degree of swelling for mice treated with bear and pig bile with a 95% confidence interval. Interpret the result.

**b.** What assumptions are necessary for the inference, part **a**, to be valid?

**c.** Another group of 10 mice (the control group) was given normal saline instead of bear or pig bile. The

degree of swelling for this control group produced the following summary statistics: $\bar{x} = 18.30$, $s = 3.38$. Compare the mean degree of swelling for the control group to the mice treated with pig bile using a 95% confidence interval. Interpret the results.

**9.22** Hermaphrodites are animals that possess the reproductive organs of both sexes. *Genetical Research* (June 1995) published a study of the mating systems of hermaphroditic snail species. The mating habits of the snails were classified into two groups: (1) self-fertilizing (selfing) snails that mate with snails of the same sex and (2) cross-fertilizing (outcrossing) snails that mate with snails of the opposite sex. One variable of interest in the study was the effective population size of the snail species. The means and standard deviations of the effective population size for independent random samples of 17 outcrossing snail species and 5 selfing snail species are given in the table. Compare the mean effective population sizes of the two types of snail species with a 90% confidence interval. Interpret the result.

| Snail Mating System | Sample Size | Effective Population Size Mean | Standard Deviation |
|---------------------|-------------|------|-----------|
| Outcrossing | 17 | 4,894 | 1,932 |
| Selfing | 5 | 4,133 | 1,890 |

*Source:* Jarne, P. "Mating system, bottlenecks, and genetic polymorphism in hermaphroditic animals." *Genetical Research,* Vol. 65, No. 3, June 1995, p. 197 (Table 4).

**9.23** Refer to the *Journal of Communication Disorders* (Mar. 1995) study of specifically language-impaired (SLI) children, Exercise 2.51 (p. 53). The data on deviation intelligence quotient (DIQ) for 10 SLI children and 10 younger, normally developing (YND) children are reproduced in the table. Use the methodology of this section to compare the mean DIQ of the two groups of children. (Use $\alpha = .10$.) What do you conclude?

SLI

| SLI Children | | | YND Children | | |
|------|------|------|------|------|------|
| 86 | 87 | 84 | 110 | 90 | 105 |
| 94 | 86 | 107 | 92 | 92 | 96 |
| 89 | 98 | 95 | 86 | 100 | 92 |
| 110 | | | 90 | | |

**9.24** Do cocaine abusers have radically different personalities than nonabusing college students? This was one of the questions researched in *Psychological Assessment* (June 1995). A personality questionnaire (ZKPQ) was administered to a sample of 450 cocaine abusers and a sample of 589 college students. The ZKPQ yields scores (measured on a 20-point scale) on each of five dimensions:

impulsive–sensation seeking, sociability, neuroticism–anxiety, aggression–hostility, and activity. The results are summarized in the table. Compare the mean ZKPQ scores of the two groups on each dimension using a statistical test of hypothesis. Interpret the results at $\alpha = .01$.

| ZKPQ Dimension | Cocaine Abusers (n = 450) | | College Students (n = 589) | |
|---|---|---|---|---|
| | **Mean** | **Std. Dev.** | **Mean** | **Std. Dev.** |
| Impulsive–sensation seeking | 9.4 | 4.4 | 9.5 | 4.4 |
| Sociability | 10.4 | 4.3 | 12.5 | 4.0 |
| Neuroticism–anxiety | 8.6 | 5.1 | 9.1 | 4.6 |
| Aggression–hostility | 8.6 | 3.9 | 7.3 | 4.1 |
| Activity | 11.1 | 3.4 | 8.0 | 4.1 |

*Source:* Ball, S. A. "The validity of an alternative five-factor measure of personality in cocaine abusers." *Psychological Assessment,* Vol. 7, No. 2, June 1995, p. 150 (Table 1).

### Applying the Concepts—Advanced

**9.25** An investigation of ethnic differences in reports of pain perception was presented at the annual meeting of the American Psychosomatic Society (March 2001). A sample of 55 blacks and 159 whites participated in the study. Subjects rated (on a 13-point scale) the intensity and unpleasantness of pain felt when a bag of ice was placed on their foreheads for two minutes. (Higher ratings correspond to higher pain intensity.) A summary of the results is provided in the accompanying table.

| | Blacks | Whites |
|---|---|---|
| Sample size | 55 | 159 |
| Mean pain intensity | 8.2 | 6.9 |

**a.** Why is it dangerous to draw a statistical inference from the summarized data? Explain.

**b.** Give values of the missing sample standard deviations that would lead you to conclude (at $\alpha = .05$) that blacks, on average, have a higher pain intensity rating than whites.

**c.** Give values of the missing sample standard deviations that would lead you to an inconclusive decision (at $\alpha = .05$) regarding whether blacks or whites have a higher mean intensity rating.

**9.26** The *Florida Scientist* (Summer/Autumn 1991) reported on a study of the feeding habits of sea urchins. A sample of 20 urchins were captured from Biscayne Bay (Miami), placed in marine aquaria, then starved for 48 hours. Each sea urchin was then fed a 5-cm blade of turtle grass. Ten of the urchins received only green blades, while the other half received only decayed blades. (Assume that the two samples of sea urchins—10 urchins per sample—were randomly and independently selected.) The ingestion time, measured from the time the blade first made contact with the urchin's mouth to the time the urchin had finished ingesting the blade, was recorded. A summary of the results is provided in the next table.

| | Green Blades | Decayed Blades |
|---|---|---|
| Number of sea urchins | 10 | 10 |
| Mean ingestion time (hours) | 3.35 | 2.36 |
| Standard deviation (hours) | .79 | .47 |

*Source:* Montague, J. R., *et al.* "Laboratory measurement of ingestion rate for the sea urchin *Lytechinus variegatus*." *Florida Scientist,* Vol. 54, Nos. 3/4, Summer/Autumn, 1991 (Table 1).

According to the researchers, "the difference in rates at which the urchins ingested the blades suggest that green, unblemished turtle grass may not be a particularly palatable food compared with decayed turtle grass. If so, urchins in the field may find it more profitable to selectively graze on decayed portions of the leaves." Do the results support this conclusion?

## 9.2 COMPARING TWO POPULATION MEANS: PAIRED DIFFERENCE EXPERIMENTS

In Example 9.4, we compared two methods of teaching reading to slow learners by means of a 95% confidence interval. Suppose it is possible to measure the slow learners' "reading IQs" *before* they are subjected to a teaching method. Eight pairs of slow learners with similar reading IQs are found, and one member of each pair is randomly assigned to the standard teaching method while the other is assigned to the new method. The data are given in Table 9.3. Do the data support the hypothesis that the population mean reading test score for slow learners taught by the new method is greater than the mean reading test score for those taught by the standard method?

⊘ PAIREDSCORES

**TABLE 9.3   Reading Test Scores for Eight Pairs of Slow Learners**

| Pair | New Method (1) | Standard Method (2) |
|------|---------------|---------------------|
| 1 | 77 | 72 |
| 2 | 74 | 68 |
| 3 | 82 | 76 |
| 4 | 73 | 68 |
| 5 | 87 | 84 |
| 6 | 69 | 68 |
| 7 | 66 | 61 |
| 8 | 80 | 76 |

**FIGURE 9.10**

*STATISTIX Analysis of Data in Table 9.3*

```
TWO-SAMPLE T TESTS FOR NEW VS STANDARD


                            SAMPLE
VARIABLE        MEAN        SIZE        S.D.        S.E.
----------      -------     -------     --------    --------
NEW             76.000      8           6.9282      2.4495
STANDARD        71.625      8           7.0089      2.4780
DIFFERENCE      4.3750


NULL HYPOTHESIS:   DIFFERENCE = 0
ALTERNATIVE HYP:   DIFFERENCE > 0


ASSUMPTION           T        DF         P       95% CI FOR DIFFERENCE
------------------   -----    ------    ------   ---------------------
EQUAL VARIANCES      1.26     14        0.1149   (-3.0982,  11.848)
UNEQUAL VARIANCES    1.26     14.0      0.1149   (-3.0983,  11.848)


                     F        NUM DF    DEN DF    P
TESTS FOR EQUALITY   -------  ------    ------    ------
      OF VARIANCES   1.02     7         7         0.4882


CASES INCLUDED 16       MISSING CASES 0
```

We want to test

$$H_0: (\mu_1 - \mu_2) = 0$$
$$H_a: (\mu_1 - \mu_2) > 0$$

One way to conduct this test is to use the $t$ statistic for two independent samples (Section 9.1). The analysis is shown on the STATISTIX printout, Figure 9.10. The test statistic, $t = 1.26$, is highlighted on the printout as well as the $p$-value of the test, $p = .1149$. At $\alpha = .10$, the $p$-value exceeds $\alpha$. Thus, from *this* analysis we might conclude that we do not have sufficient evidence to infer a difference in the mean test scores for the two methods.

If you carefully examine the data in Table 9.3, however, you will find this result difficult to accept. The test score of the new method is larger than the corresponding test score for the standard method *for every one of the eight pairs of slow learners*. This, in itself, seems to provide strong evidence to indicate that $\mu_1$ exceeds $\mu_2$. Why, then, did the $t$-test fail to detect this difference? The answer is: *The independent samples t-test is not a valid procedure to use with this set of data.*

The $t$-test is inappropriate because the assumption of independent samples is invalid. We have randomly chosen *pairs of test scores,* and thus, once we have chosen the sample for the new method, we have *not* independently chosen the sample for the standard method. The dependence between observations within pairs can be seen by examining the pairs of test scores, which tend to rise and fall together as we go from pair to pair. This pattern provides strong visual evidence of a violation of the assumption of independence required for the two-sample $t$-test of Section 9.1. Note, also, that

$$s_p^2 = \frac{(n_1 - 1)s_1^2 + (n_2 - 1)s_2^2}{n_1 + n_2 - 2} = \frac{(8 - 1)(6.93)^2 + (8 - 1)(7.01)^2}{8 + 8 - 2} = 48.55$$

Thus, there is a *large variation within samples* (reflected by the large value of $s_p^2$) in comparison to the relatively *small difference between the sample means*. Because $s_p^2$ is so large, the $t$-test of Section 9.1 is unable to detect a difference between $\mu_1$ and $\mu_2$.

**TABLE 9.4   Differences in Reading Test Scores**

| Pair | New Method | Standard Method | Difference (New Method–Standard Method) |
|------|-----------|-----------------|------------------------------------------|
| 1 | 77 | 72 | 5 |
| 2 | 74 | 68 | 6 |
| 3 | 82 | 76 | 6 |
| 4 | 73 | 68 | 5 |
| 5 | 87 | 84 | 3 |
| 6 | 69 | 68 | 1 |
| 7 | 66 | 61 | 5 |
| 8 | 80 | 76 | 4 |

We now consider a valid method of analyzing the data of Table 9.3. In Table 9.4 we add the column of differences between the test scores of the pairs of slow learners. We can regard these differences in test scores as a random sample of differences for all pairs (matched on reading IQ) of slow learners, past and present. Then we can use this sample to make inferences about the mean of the population of differences, $\mu_D$, which is equal to the difference $(\mu_1 - \mu_2)$. That is, the mean of the population (and sample) of differences equals the difference between the population (and sample) means. Thus, our test becomes

$$H_0: \mu_D = 0 \quad (\mu_1 - \mu_2 = 0)$$
$$H_a: \mu_D > 0 \quad (\mu_1 - \mu_2 > 0)$$

The test statistic is a one-sample $t$ (Section 8.4), since we are now analyzing a single sample of differences for small $n$:

$$\textit{Test statistic:} \quad t = \frac{\bar{x}_D - 0}{s_D / \sqrt{n_D}}$$

where   $\bar{x}_D$ = Sample mean difference
   $s_D$ = Sample standard deviation of differences
   $n_D$ = Number of differences = Number of pairs

*Assumptions:* The population of differences in test scores is approximately normally distributed. The sample differences are randomly selected from the population differences. [*Note:* We do not need to make the assumption that $\sigma_1^2 = \sigma_2^2$.]
*Rejection region:* At significance level $\alpha = .05$, we will reject $H_0$: if $t > t_{.05}$, where $t_{.05}$ is based on $(n_D - 1)$ degrees of freedom.

Referring to Table VI in Appendix A, we find the $t$ value corresponding to $\alpha = .05$ and $n_D - 1 = 8 - 1 = 7$ df to be $t_{.05} = 1.895$. Then we will reject the null hypothesis if $t > 1.895$ (see Figure 9.11). Note that the number of degrees of freedom

**FIGURE 9.11**

*Rejection Region for
Example 9.4*

Because this value of *t* falls in the rejection region, we conclude (at $\alpha = .05$) that the population mean test score for slow learners taught by the new method exceeds the population mean score for those taught by the standard method. We can reach the same conclusion by noting that the *p*-value of the test, highlighted in Figure 9.12, is much smaller than $\alpha = .05$.

decreases from $n_1 + n_2 - 2 = 14$ to 7 when we use the paired difference experiment rather than the two independent random samples design.

Summary statistics for the $n = 8$ differences are shown in the MINITAB printout, Figure 9.12. Note that $\bar{x}_D = 4.375$ and $s_D = 1.685$. Substituting these values into the formula for the test statistic, we have

$$t = \frac{\bar{x}_D - 0}{s_D/\sqrt{n_D}} = \frac{4.375}{1.685/\sqrt{8}} = 7.34$$

**FIGURE 9.12**

*MINITAB Paired Difference
Analysis of Data in Table 9.4*

```
Paired T for NEW - STANDARD

              N       Mean      StDev      SE Mean
NEW           8      76.00       6.93         2.45
STANDARD      8      71.63       7.01         2.48
Difference    8      4.375       1.685        0.596


95% lower bound for mean difference: 3.246
T-Test of mean difference = 0 (vs > 0): T-Value = 7.34   P-Value = 0.000
```

This kind of experiment, in which observations are paired and the differences are analyzed, is called a **paired difference experiment**. In many cases, a paired difference experiment can provide more information about the difference between population means than an independent samples experiment. The idea is to compare population means by comparing the differences between pairs of experimental units (objects, people, etc.) that were very similar prior to the experiment. The differencing removes sources of variation that tend to inflate $\sigma^2$. For example, when two children are taught to read by two different methods, the observed difference in achievement may be due to a difference in the effectiveness of the two teaching methods *or* it may be due to differences in the initial reading levels and IQs of the two children (random error). To reduce the effect of differences in the children on the observed differences in reading achievement, the two methods of reading are imposed on two children who are more likely to possess similar intellectual capacity, namely children with nearly equal IQs. The effect of this pairing is to remove the larger source of variation that would be present if children with different abilities were randomly assigned to the two samples. Making comparisons within groups of similar experimental units is called **blocking**, and the paired difference experiment is a simple example of a **randomized block experiment**. In our example, pairs of children with matching IQ scores represent the blocks.

Some other examples for which the paired difference experiment might be appropriate are the following:

1. Suppose you want to estimate the difference $(\mu_1 - \mu_2)$ in mean price per gallon between two major brands of premium gasoline. If you choose two independent random samples of stations for each brand, the variability in price due to geographic location may be large. To eliminate this source of variability you could choose pairs of stations of similar size, one station for each brand, in close geographic proximity and use the sample of differences between the prices of the brands to make an inference about $(\mu_1 - \mu_2)$.

2. Suppose a college placement center wants to estimate the difference $(\mu_1 - \mu_2)$ in mean starting salaries for men and women graduates who seek jobs through the center. If it independently samples men and women, the starting salaries may vary because of their different college majors and differences in grade point averages. To eliminate these sources of variability, the placement center could match male and female job-seekers according to their majors and grade point averages. Then the differences between the starting salaries of each pair in the sample could be used to make an inference about $(\mu_1 - \mu_2)$.

3. Suppose you wish to estimate the difference $(\mu_1 - \mu_2)$ in mean absorption rate into the bloodstream for two drugs that relieve pain. If you independently sample people, the absorption rates might vary because of age, weight, sex, blood pressure, etc. In fact, there are many possible sources of nuisance variability, and pairing individuals who are similar in all the possible sources would be quite difficult. However, it may be possible to obtain two measurements *on the same person*. First, we administer one of the two drugs and record the time until absorption. After a sufficient amount of time, the other drug is administered and a second measurement on absorption time is obtained. The differences between the measurements for each person in the sample could then be used to estimate $(\mu_1 - \mu_2)$. This procedure would be advisable only if the amount of time allotted between drugs is sufficient to guarantee little or no carryover effect. Otherwise, it would be better to use different people matched as closely as possible on the factors thought to be most important.

The hypothesis-testing procedures and the method of forming confidence intervals for the difference between two means using a paired difference experiment are summarized in the next two boxes for both large and small $n$.

---

### Paired Difference Confidence Interval for $\mu_D = \mu_1 - \mu_2$

**LARGE SAMPLE**

$$\bar{x}_D + z_{\alpha/2}\frac{\sigma_D}{\sqrt{n_D}} \approx \bar{x}_D \pm z_{\alpha/2}\frac{s_D}{\sqrt{n_D}}$$

*Assumption:* The sample differences are randomly selected from the population of differences.

**SMALL SAMPLE**

$$\bar{x}_D \pm t_{\alpha/2}\frac{s_D}{\sqrt{n_D}}$$

where $t_{\alpha/2}$ is based on $(n_D - 1)$ degrees of freedom
*Assumptions:*

1. The relative frequency distribution of the population of differences is normal.
2. The sample differences are randomly selected from the population of differences.

**Paired Difference Test of Hypothesis for $\mu_D = \mu_1 - \mu_2$**

**ONE-TAILED TEST**

$H_0: \mu_D = D_0$

$H_a: \mu_D < D_0$
[or $H_a: \mu_D > D_0$]

**TWO-TAILED TEST**

$H_0: \mu_D = D_0$

$H_a: \mu_D \neq D_0$

**LARGE SAMPLE**

*Test statistic:* $\quad z = \dfrac{\bar{x}_D - D_0}{\sigma_D/\sqrt{n_D}} \approx \dfrac{\bar{x}_D - D_0}{s_D/\sqrt{n_D}}$

*Rejection region:* $z < -z_\alpha$
[or $z > z_\alpha$ when $H_a: \mu_D > D_0$]

*Rejection region:* $|z| > z_{\alpha/2}$

*Assumption:* The differences are randomly selected from the population of differences.

**SMALL SAMPLE**

*Test statistic:* $\quad t = \dfrac{\bar{x}_D - D_0}{s_D/\sqrt{n_D}}$

*Rejection region:* $t < -t_\alpha$
[or $t > t_\alpha$ when $H_a: \mu_D > D_0$]

*Rejection region:* $|t| > t_{\alpha/2}$

where $t_\alpha$ and $t_{\alpha/2}$ are based on $(n_D - 1)$ degrees of freedom

*Assumptions:*

1. The relative frequency distribution of the population of differences is normal.

2. The differences are randomly selected from the population of differences.

**EXAMPLE 9.5**

An experiment is conducted to compare the starting salaries of male and female college graduates who find jobs. Pairs are formed by choosing a male and a female with the same major and similar grade point averages (GPA). Suppose a random sample of 10 pairs is formed in this manner and the starting annual salary of each person is recorded. The results are shown in Table 9.5. Compare the mean starting salary, $\mu_1$, for males to the mean starting salary, $\mu_2$, for females using a 95% confidence interval. Interpret the results.

**TABLE 9.5  Data on Annual Salaries for Matched Pairs of College Graduates**

| Pair | Male | Female | Difference (Male – Female) | Pair | Male | Female | Difference (Male – Female) |
|------|------|--------|---------------------------|------|------|--------|---------------------------|
| 1 | $29,300 | $28,800 | $ 500 | 6 | $37,800 | $38,000 | $ −200 |
| 2 | 41,500 | 41,600 | −100 | 7 | 69,500 | 69,200 | 300 |
| 3 | 40,400 | 39,800 | 600 | 8 | 41,200 | 40,100 | 1,100 |
| 4 | 38,500 | 38,500 | 0 | 9 | 38,400 | 38,200 | 200 |
| 5 | 43,500 | 42,600 | 900 | 10 | 59,200 | 58,500 | 700 |

**FIGURE 9.13a**

*SAS Descriptive Statistics for Differences in Table 9.5*

| | The MEANS Procedure | | |
|---|---|---|---|
| | Analysis Variable : DIFF | | |
| N | Mean | Std Dev | Minimum | Maximum |
| 10 | 400.0000000 | 434.6134937 | -200.0000000 | 1100.00 |

**FIGURE 9.13b**

*SAS Paired Difference Analysis of Annual Salaries*

```
        Two Sample Paired t-test for the Means of MALE and FEMALE

Sample Statistics

         Group        N       Mean     Std. Dev.    Std. Error
    -------------------------------------------------------------------
         MALE         10      43930      11665        3688.8
         FEMALE       10      43530      11617        3673.6


Hypothesis Test

         Null hypothesis:      Mean of (MALE - FEMALE)  =  0
         Alternative:          Mean of (MALE - FEMALE) ^= 0


         t  Statistic        Df         Prob > t
    -------------------------------------------------
             2.910            9          0.0173


95% Confidence Interval for the Difference between Two Paired Means


             Lower Limit         Upper Limit
             ------------         ------------
               89.10               710.90
```

**Solution**

Since the data on annual salary are collected in pairs of males and females matched on GPA and major, a paired difference experiment is performed. To conduct the analysis, we first compute the differences between the salaries, as shown in Table 9.5. Summary statistics for these $n = 10$ differences are displayed in the SAS printout, Figure 9.13a.

The 95% confidence interval for $\mu_D = (\mu_1 - \mu_2)$ for this small sample is

$$\bar{x}_D \pm t_{\alpha/2} \frac{s_D}{\sqrt{n_D}}$$

where $t_{\alpha/2} = t_{.025} = 2.262$ (obtained from Table VI, Appendix A) is based on $n - 1 = 9$ degrees of freedom. Substituting the values of $\bar{x}_D$ and $s_D$ shown on the printout, we obtain

$$\bar{x}_D \pm 2.262 \frac{s_D}{\sqrt{n_D}} = 400 \pm 2.262 \left( \frac{434.613}{\sqrt{10}} \right)$$

$$= 400 \pm 310.88 \approx 400 \pm 311 = (\$89, \$711)$$

[*Note:* This interval is also shown highlighted on the SAS printout, Figure 9.13b.] Our interpretation is that the true mean difference between the starting salaries of males and females falls between $89 and $711, with 95% confidence. Since the interval falls above 0, we infer that $\mu_1 - \mu_2 > 0$; that is, the mean salary for males exceeds the mean salary for females.

**FIGURE 9.14**

*SAS Analysis of Data in Table 9.5, Assuming Independent Samples*

Two Sample t-test for the Means of MALE and FEMALE

Sample Statistics

| Group | N | Mean | Std. Dev. | Std. Error |
|-------|---|------|-----------|------------|
| MALE | 10 | 43930 | 11665 | 3688.8 |
| FEMALE | 10 | 43530 | 11617 | 3673.6 |

Hypothesis Test

Null hypothesis:    Mean 1 - Mean 2 =  0
Alternative:        Mean 1 - Mean 2 ^= 0

| If Variances Are | t statistic | Df | Pr > t |
|------------------|-------------|-----|--------|
| Equal | 0.077 | 18 | 0.9396 |
| Not Equal | 0.077 | 18.00 | 0.9396 |

95% Confidence Interval for the Difference between Two Means

| Lower Limit | Upper Limit |
|-------------|-------------|
| -10537.5 | 11337.50 |

To measure the amount of information about $(\mu_1 - \mu_2)$ gained by using a paired difference experiment in Example 9.5 rather than an independent samples experiment, we can compare the relative widths of the confidence intervals obtained by the two methods. A 95% confidence interval for $(\mu_1 - \mu_2)$ using the paired difference experiment is, from Example 9.5, ($89, $711). If we analyzed the same data as though this were an independent samples experiment,* we would first obtain the descriptive statistics shown in the SAS printout, Figure 9.14. Then we substitute the sample means and standard deviations shown on the printout into the formula for a 95% confidence interval for $(\mu_1 - \mu_2)$ using independent samples:

$$(\bar{x}_1 - \bar{x}_2) \pm t_{.025} \sqrt{s_p^2 \left( \frac{1}{n_1} + \frac{1}{n_2} \right)}$$

where

$$s_p^2 = \frac{(n_1 - 1)s_1^2 + (n_2 - 1)s_2^2}{n_1 + n_2 - 2}$$

SAS performed these calculations and obtained the interval ($-10,537.50, $11,337.50). This interval is highlighted on Figure 9.14.

Notice that the independent samples interval includes 0. Consequently, if we were to use this interval to make an inference about $(\mu_1 - \mu_2)$, we would incorrectly conclude that the mean starting salaries of males and females do not differ! You can see that the confidence interval for the independent sampling experiment is about five times wider than for the corresponding paired difference confidence interval. Blocking out the variability due to differences in majors and grade point

---

*This is done only to provide a measure of the increase in the amount of information obtained by a paired design in comparison to an unpaired design. Actually, if an experiment is designed using pairing, an unpaired analysis would be invalid because the assumption of independent samples would not be satisfied.

averages significantly increases the information about the difference in male and female mean starting salaries by providing a much more accurate (smaller confidence interval for the same confidence coefficient) estimate of $(\mu_1 - \mu_2)$.

You may wonder whether conducting a paired difference experiment is always superior to an independent samples experiment. The answer is: Most of the time, but not always. We sacrifice half the degrees of freedom in the $t$ statistic when a paired difference design is used instead of an independent samples design. This is a loss of information, and unless this loss is more than compensated for by the reduction in variability obtained by blocking (pairing), the paired difference experiment will result in a net loss of information about $(\mu_1 - \mu_2)$. Thus, we should be convinced that the pairing will significantly reduce variability before performing the paired difference experiment. Most of the time this will happen.

*One final note:* The pairing of the observations is determined *before* the experiment is performed (that is, by the *design* of the experiment). A paired difference experiment is *never* obtained by pairing the sample observations after the measurements have been acquired.

---

**What Do You Do When the Assumption of a Normal Distribution for the Population of Differences Is Not Satisfied?**

---

*Answer:* Use the Wilcoxon signed rank test for the paired difference design (Chapter 14).

---

 EXERCISES 9.27–9.42

## Learning the Mechanics

**9.27** A paired difference experiment yielded $n_D$ pairs of observations. In each case, what is the rejection region for testing $H_0: \mu_D = 2$ against $H_a: \mu_D > 2$?
**a.** $n_D = 10, \alpha = .05$
**b.** $n_D = 20, \alpha = .10$
**c.** $n_D = 5, \alpha = .025$
**d.** $n_D = 9, \alpha = .01$

**9.28** A paired difference experiment produced the following data:

$$n_D = 16 \quad \bar{x}_1 = 143 \quad \bar{x}_2 = 150 \quad \bar{x}_D = -7 \quad s_D^2 = 64$$

**a.** Determine the values of $t$ for which the null hypothesis, $\mu_1 - \mu_2 = 0$, would be rejected in favor of the alternative hypothesis, $\mu_1 - \mu_2 < 0$. Use $\alpha = .10$.
**b.** Conduct the paired difference test described in part **a**. Draw the appropriate conclusions.
**c.** What assumptions are necessary so that the paired difference test will be valid?
**d.** Find a 90% confidence interval for the mean difference $\mu_D$.
**e.** Which of the two inferential procedures, the confidence interval of part **d** or the test of hypothesis of

part **b**, provides more information about the difference between the population means?

**9.29** The data for a random sample of six paired observations are shown in the table.

LM9_29

| Pair | Sample from Population 1 | Sample from Population 2 |
|------|:---:|:---:|
| 1 | 7 | 4 |
| 2 | 3 | 1 |
| 3 | 9 | 7 |
| 4 | 6 | 2 |
| 5 | 4 | 4 |
| 6 | 8 | 7 |

**a.** Calculate the difference between each pair of observations by subtracting observation 2 from observation 1. Use the differences to calculate $\bar{x}_D$ and $s_D^2$.
**b.** If $\mu_1$ and $\mu_2$ are the means of populations 1 and 2, respectively, express $\mu_D$ in terms of $\mu_1$ and $\mu_2$.
**c.** Form a 95% confidence interval for $\mu_D$.
**d.** Test the null hypothesis $H_0: \mu_D = 0$ against the alternative hypothesis $H_a: \mu_D \neq 0$. Use $\alpha = .05$.

**9.30** The data for a random sample of 10 paired observations are shown in the next table.

◉ LM9_30

| Pair | Population 1 | Population 2 |
|------|--------------|--------------|
| 1 | 19 | 24 |
| 2 | 25 | 27 |
| 3 | 31 | 36 |
| 4 | 52 | 53 |
| 5 | 49 | 55 |
| 6 | 34 | 34 |
| 7 | 59 | 66 |
| 8 | 47 | 51 |
| 9 | 17 | 20 |
| 10 | 51 | 55 |

**a.** If you wish to test whether these data are sufficient to indicate that the mean for population 2 is larger than that for population 1, what are the appropriate null and alternative hypotheses? Define any symbols you use.

**STATISTIX Output for Exercise 9.30**

```
PAIRED T TEST FOR X1 - X2

NULL HYPOTHESIS: DIFFERENCE =  0
ALTERNATIVE HYP: DIFFERENCE <> 0

MEAN          -3.7000
STD ERROR      0.7000
LO 95% CI     -5.2835
UP 95% CI     -2.1165
T             -5.29
DF             9
P              0.0005

CASES INCLUDED 10    MISSING CASES 0
```

**b.** The data are analyzed using STATISTIX, with the output shown above. Interpret the test results.
**c.** The STATISTIX output also includes a confidence interval. Interpret this interval.
**d.** What assumptions are necessary to ensure the validity of this analysis?

**9.31** A paired difference experiment yielded the data shown in the table.

◉ LM9_31

| Pair | x | y | Pair | x | y |
|------|-----|-----|------|-----|-----|
| 1 | 55 | 44 | 5 | 75 | 62 |
| 2 | 68 | 55 | 6 | 52 | 38 |
| 3 | 40 | 25 | 7 | 49 | 31 |
| 4 | 55 | 56 | | | |

**a.** Test $H_0: \mu_D = 10$ against $H_a: \mu_D \neq 10$, where $\mu_D = (\mu_1 - \mu_2)$. Use $\alpha = .05$.
**b.** Report the *p*-value for the test you conducted in part a. Interpret the *p*-value.

### Applying the Concepts—Basic

**9.32** Does winning an Academy of Motion Picture Arts and Sciences award lead to long-term mortality for movie actors? In an article in the *Annals of Internal Medicine* (May 15, 2001), researchers identified 762 Academy Award winners and matched each one with another actor of the same sex who was in the same winning film and was born in the same era. The life expectancy (age) of each pair of actors was compared.
**a.** Explain why the data should be analyzed as a paired difference experiment.
**b.** Set up the null hypothesis for a test to compare the mean life expectancies of Academy Award winners and non-winners.
**c.** The sample mean life expectancies of Academy Award winners and non-winners were reported as 79.7 years and 75.8 years, respectively. The *p*-value for comparing the two population means was reported as $p = .003$. Interpret this value in the context of the problem.

**9.33** Hypersexual behavior caused by traumatic brain injury (TBI) is often treated with the drug Depo-Provera. In one clinical study, eight young male TBI patients who exhibited hypersexual behavior were treated weekly with 400 milligrams of Depo-Provera for six months (*Journal of Head Trauma Rehabilitation*, June 1995). The testosterone levels (in nanograms per deciliter) of the patients both prior to treatment and at the end of the 6-month treatment period are given in the table below.

◉ DEPOPROV

| Patient | Pretreatment | After 6 months on Depo-Provera |
|---------|--------------|-------------------------------|
| 1 | 849 | 96 |
| 2 | 903 | 41 |
| 3 | 890 | 31 |
| 4 | 1,092 | 124 |
| 5 | 362 | 46 |
| 6 | 900 | 53 |
| 7 | 1,006 | 113 |
| 8 | 672 | 174 |

*Source:* Emory, L. E., Cole, C. M., and Meyer, W. J. "Use of Depo-Provera to control sexual aggression in persons with traumatic brain injury." *Journal of Head Trauma Rehabilitation*, Vol. 10, No. 3, June 1995, p. 52 (Table 2).

**a.** Construct a 99% confidence interval for the true mean difference between the pretreatment and after-treatment testosterone levels of young male TBI patients.

**b.** Use the interval, part **a**, to make an inference about the effectiveness of Depo-Provera in reducing testosterone levels of young male TBI patients.

**9.34** Researchers at the University of South Alabama compared the attitudes of male college students toward their fathers with their attitudes toward their mothers (*Journal of Genetic Psychology*, March 1998). Each of a sample of 13 males was asked to complete the following statement about each of their parents: My relationship with my father (mother) can best be described as: (1) Awful, (2) Poor, (3) Average, (4) Good, or (5) Great. The following data were obtained:

**⊘ FMATTITUDES**

| Student | Attitude toward Father | Attitude toward Mother |
|---|---|---|
| 1 | 2 | 3 |
| 2 | 5 | 5 |
| 3 | 4 | 3 |
| 4 | 4 | 5 |
| 5 | 3 | 4 |
| 6 | 5 | 4 |
| 7 | 4 | 5 |
| 8 | 2 | 4 |
| 9 | 4 | 5 |
| 10 | 5 | 4 |
| 11 | 4 | 5 |
| 12 | 5 | 4 |
| 13 | 3 | 3 |

*Source:* Adapted from Vitulli, W. F., and Richardson, D. K. "College student's attitudes toward relationships with parents: A five-year comparative analysis." *Journal of Genetic Psychology*, Vol. 159, No. 1, (March 1998), pp. 45–52.

**a.** Specify the appropriate hypotheses for testing whether male students' attitudes toward their fathers differ from their attitudes toward their mothers, on average.

**b.** Conduct the test of part **a** at $\alpha = .05$. Interpret the results in the context of the problem.

**Applying the Concepts—Intermediate**

**9.35** According to *Webster's New World Dictionary*, a tongue-twister is "a phrase that is hard to speak rapidly." Do tongue-twisters have an effect on the length of time it takes to read silently? To answer this question, 42 undergraduate psychology students participated in a reading experiment (*Memory & Cognition*, Sept. 1997). Two lists, each comprised of 600 words, were constructed. One list contained a series of tongue-twisters and the other list (called the *control*) did not contain any tongue-twisters. Each student read both lists and the length of time (in minutes) required to complete the lists was recorded. The researchers used a test of hypothesis to compare the mean reading response times for the tongue-twister and control lists.

**a.** Set up the null hypothesis for the test.

**b.** Use the information in the accompanying table to find the test statistic and *p*-value of the test.

**c.** Give the appropriate conclusion. Use $\alpha = .05$.

| List Type | Response Time (minutes) Mean | Standard Deviation |
|---|---|---|
| Tongue-twister | 6.59 | 1.94 |
| Control | 6.34 | 1.92 |
| Difference | .25 | .78 |

*Source:* Robinson, D. H., and Katayama, A. D. "At-lexical, articulatory interference in silent reading: The 'upstream' tongue-twister effect." *Memory & Cognition*, Vol. 25, No. 5, Sept. 1997, p. 663.

**9.36** When searching for an item (e.g., a roadside traffic sign, a lost earring, or a tumor in a mammogram), common sense dictates that you will not re-examine items previously rejected. However, researchers at Harvard Medical School found that a visual search has no memory (*Nature*, Aug. 6, 1998). In their experiment, nine subjects searched for the letter "T" mixed among several letters "L." Each subject conducted the search under two conditions: random and static. In the random condition, the location of the letters were changed every 111 milliseconds; in the static condition, the location of the letters remained unchanged. In each trial, the reaction time (i.e., the amount of time it took the subject to locate the target letter) was recorded in milliseconds.

**a.** One goal of the research is to compare the mean reaction times of subjects in the two experimental conditions. Explain why the data should be analyzed as a paired-difference experiment.

**b.** If a visual search has no memory, then the main reaction times in the two conditions will not differ. Specify $H_0$ and $H_a$ for testing the "no memory" theory.

**c.** The test statistic was calculated as $t = 1.52$ with *p*-value = .15. Make the appropriate conclusion.

**9.37** A study published in *Clinical Kinesiology* (Spring 1995) was designed to examine the metabolic and cardiopulmonary responses during exercise of persons diagnosed with multiple sclerosis (MS). Leg cycling and arm cranking exercises were performed by 10 MS patients and 10 healthy (non-MS) control subjects. Each member of the control group was selected based on gender, age, height, and weight to match (as closely as possible) with one member of the MS group. Consequently, the researchers compared the MS and non-MS groups by matched-pairs *t*-tests on such outcome variables as oxygen uptake, carbon dioxide output, and peak aerobic power. The data on the matching variables used in the experiment are shown in the shown in the table on p. 408.

**a.** Visually examine the data table. Does it appear that the researchers have successfully matched the MS and non-MS subjects?

⊘ MSSTUDY

| Matched Pair | MS Subjects | | | | Non-MS Subjects | | | |
|:---:|:---:|:---:|:---:|:---:|:---:|:---:|:---:|:---:|
| | Gender | Age (years) | Height (cm) | Weight (kg) | Gender | Age (years) | Height (cm) | Weight (kg) |
| 1 | M | 48 | 171.0 | 80.8 | M | 45 | 173.0 | 76.3 |
| 2 | F | 34 | 158.5 | 75.0 | F | 34 | 158.0 | 75.6 |
| 3 | F | 34 | 167.6 | 55.5 | F | 34 | 164.5 | 57.7 |
| 4 | M | 38 | 167.0 | 71.3 | M | 34 | 161.3 | 70.0 |
| 5 | M | 45 | 182.5 | 90.9 | M | 39 | 179.0 | 96.0 |
| 6 | F | 42 | 166.0 | 72.4 | F | 42 | 167.0 | 77.8 |
| 7 | M | 32 | 172.0 | 70.5 | M | 34 | 165.8 | 74.7 |
| 8 | F | 35 | 166.5 | 55.3 | F | 43 | 165.1 | 71.4 |
| 9 | F | 33 | 166.5 | 57.9 | F | 31 | 170.1 | 60.4 |
| 10 | F | 46 | 175.0 | 79.9 | F | 43 | 175.0 | 77.9 |

*Source:* Ponichtera-Mulcare, J. A., *et al.* "Maximal aerobic exercise of individuals with multiple sclerosis using three modes of ergometry." *Clinical Kinesiology,* Vol. 49, No. 1, Spring 1995, p. 7 (Table 1).

**b.** Conduct a test to determine whether the mean ages of the MS and non-MS groups are significantly different. Use $\alpha = .05$.

**c.** SPSS printouts for comparing the mean heights and weights of the two groups as shown below. Interpret the results.

**d.** Based on the results, parts **b** and **c**, comment on the validity of the *t*-tests performed by the researchers on the outcome variables.

**9.38** Twins, because they share an identical genotype, make ideal subjects for investigating the degree to which various environmental conditions impact personality. The classical method of studying this phenomenon and the subject of an interesting book by Susan Farber (*Identical Twins Reared Apart,* New York: Basic Books, 1981) is the study of identical twins separated early in life and reared apart. Much of Farber's discussion focuses on a comparison of IQ scores. The data for this analysis appear in the table on p. 409. One member (A) of each of the $n = 32$ pairs of twins was reared by a natural parent; the other member (B) was reared by a relative or some other person. Is there a significant difference between the average IQ scores of identical twins, where one member of the pair is reared by the natural parents and the other member of the pair is not? Use $\alpha = .05$ to make your conclusion.

**SPSS Output for Exercise 9.37**

Paired Samples Statistics

| | | Mean | N | Std. Deviation | Std. Error Mean |
|:---|:---|:---:|:---:|:---:|:---:|
| Pair 1 | MSHT | 169.2600 | 10 | 6.3969 | 2.0229 |
| | NONMSHT | 167.8800 | 10 | 6.4175 | 2.0294 |
| Pair 2 | MSWT | 70.9500 | 10 | 11.7751 | 3.7236 |
| | NONMSWT | 73.7800 | 10 | 10.5141 | 3.3248 |

Paired Samples Test

| | | Paired Differences | | | | | | | |
|:---|:---|:---:|:---:|:---:|:---:|:---:|:---:|:---:|:---:|
| | | | | | 95% Confidence Interval of the Difference | | | | |
| | | Mean | Std. Deviation | Std. Error Mean | Lower | Upper | t | df | Sig. (2-tailed) |
| Pair 1 | MSHT - NONMSHT | 1.3800 | 3.2303 | 1.0215 | -.9308 | 3.6908 | 1.351 | 9 | .210 |
| Pair 2 | MSWT - NONMSWT | -2.8300 | 5.6698 | 1.7930 | -6.8859 | 1.2259 | -1.578 | 9 | .149 |

⊘TWINSIQ

| Pair ID | Twin A | Twin B | Pair ID | Twin A | Twin B |
|---------|--------|--------|---------|--------|--------|
| 112 | 113 | 109 | 228 | 100 | 88 |
| 114 | 94 | 100 | 232 | 100 | 104 |
| 126 | 99 | 86 | 236 | 93 | 84 |
| 132 | 77 | 80 | 306 | 99 | 95 |
| 136 | 81 | 95 | 308 | 109 | 98 |
| 148 | 91 | 106 | 312 | 95 | 100 |
| 170 | 111 | 117 | 314 | 75 | 86 |
| 172 | 104 | 107 | 324 | 104 | 103 |
| 174 | 85 | 85 | 328 | 73 | 78 |
| 180 | 66 | 84 | 330 | 88 | 99 |
| 184 | 111 | 125 | 338 | 92 | 111 |
| 186 | 51 | 66 | 342 | 108 | 110 |
| 202 | 109 | 108 | 344 | 88 | 83 |
| 216 | 122 | 121 | 350 | 90 | 82 |
| 218 | 97 | 98 | 352 | 79 | 76 |
| 220 | 82 | 94 | 416 | 97 | 98 |

*Source:* Adapted from *Identical Twins Reared Apart,* by Susan L. Farber 1981 by Basic Books, Inc.

**9.39** A study reported in the *Journal of Psychology* (Mar. 1991) measures the change in female students' self-concepts as they move from high school to college. A sample of 133 Boston College female students were selected for the study. Each was asked to evaluate several aspects of her life at two points in time: at the end of her senior year of high school, and during her sophomore year of college. Each student was asked to evaluate where she believed she stood on a scale that ranged from top 10% of class (1) to lowest 10% of class (5). The results for three of the traits evaluated are reported in the table below.

| Trait | $n$ | Senior Year of High School $\bar{x}$ | Sophomore Year of College $\bar{x}$ |
|-------|-----|-----------------------------|-----------------------------|
| Leadership | 133 | 2.09 | 2.33 |
| Popularity | 133 | 2.48 | 2.69 |
| Intellectual self-confidence | 133 | 2.29 | 2.55 |

   **a.** What null and alternative hypotheses would you test to determine whether the mean self-concept of females decreases between the senior year of high school and the sophomore year of college as measured by each of these three traits?
   **b.** Are these traits more appropriately analyzed using an independent samples test or a paired difference test? Explain.
   **c.** Noting the size of the sample, what assumptions are necessary to ensure the validity of the tests?
   **d.** The article reports that the leadership test results in a *p*-value greater than .05, while the tests for popularity and intellectual self-confidence result in *p*-values less than .05. Interpret these results.

**9.40** Merck Research Labs conducted an experiment to evaluate the effect of a new drug using the single-T swim maze. Nineteen impregnated dam rats were allocated a dosage of 12.5 milligrams of the drug. One male and one female rat pup were randomly selected from each resulting litter to perform in the swim maze. Each pup was placed in the water at one end of the maze and allowed to swim until it escaped at the opposite end. If the pup failed to escape after a certain period of time, it was placed at the beginning of the maze and given another chance. The experiment was repeated until each pup accomplished three successful escapes. The accompanying table reports the number of swims required by each pup to perform three successful escapes. Is there sufficient evidence of a difference between the mean number of swims required by male and female pups? Use the accompanying MINITAB printout to conduct the test (at $\alpha = .10$). Comment on the assumptions required for the test to be valid.

⊘RATPUPS

| Litter | Male | Female | Litter | Male | Female |
|--------|------|--------|--------|------|--------|
| 1 | 8 | 5 | 11 | 6 | 5 |
| 2 | 8 | 4 | 12 | 6 | 3 |
| 3 | 6 | 7 | 13 | 12 | 5 |
| 4 | 6 | 3 | 14 | 3 | 8 |
| 5 | 6 | 5 | 15 | 3 | 4 |
| 6 | 6 | 3 | 16 | 8 | 12 |
| 7 | 3 | 8 | 17 | 3 | 6 |
| 8 | 5 | 10 | 18 | 6 | 4 |
| 9 | 4 | 4 | 19 | 9 | 5 |
| 10 | 4 | 4 | | | |

*Source:* Thomas E. Bradstreet, Merck Research Labs, BL 3–2, West Point, PA 19486.

**MINITAB Output for Exercise 9.40**

```
Paired T for male - female

                N      Mean     StDev    SE Mean
male           19     5.895     2.378     0.546
female         19     5.526     2.458     0.564
Difference     19     0.368     3.515     0.806

95% CI for mean difference: (-1.326, 2.063)
T-Test of mean difference = 0 (vs not = 0): T-Value = 0.46   P-Value = 0.653
```

## Applying the Concepts—Advanced

**9.41** A *homophone* is a word whose pronunciation is the same as that of another word having a different meaning and spelling (e.g., *nun* and *none, doe* and *dough,* etc.). *Brain and Language* (Apr. 1995) reported on a study of homophone spelling in patients with Alzheimer's disease. Twenty Alzheimer's patients were asked to spell 24 homophone pairs given in random order, then the number of homophone confusions (e.g., spelling *doe* given the context, *bake bread dough*) was recorded for each patient. One year later, the same test was given to the same patients. The data for the study are provided in the table. The researchers posed the following question: "Do Alzheimer's patients show a significant increase in mean homophone confusion errors over time?" Perform an analysis of the data to answer the researchers' question. Use the relevant information in the SAS printout. What assumptions are necessary for the procedure used to be valid? Are they satisfied?

**SAS output for Exercise 9.41**

Two Sample Paired t-test for the Means of TIME1 and TIME2

Sample Statistics

| Group | N | Mean | Std. Dev. | Std. Error |
|-------|---|------|-----------|-----------|
| TIME1 | 20 | 4.15 | 3.4985 | 0.7823 |
| TIME2 | 20 | 5.8 | 4.2128 | 0.942 |

Hypothesis Test

Null hypothesis:     Mean of (TIME1 - TIME 2) => 0
Alternative:        Mean of (TIME1 - TIME 2) < 0

| t Statistic | Df | Prob > t |
|-------------|-----|----------|
| -2.306 | 19 | 0.0163 |

95% Confidence Interval for the Difference between Two Paired Means

| Lower Limit | Upper Limit |
|-------------|-------------|
| -3.15 | -0.15 |

🧭 HOMOPHON

| Patient | Time 1 | Time 2 |
|---------|--------|--------|
| 1 | 5 | 5 |
| 2 | 1 | 3 |
| 3 | 0 | 0 |
| 4 | 1 | 1 |
| 5 | 0 | 1 |
| 6 | 2 | 1 |
| 7 | 5 | 6 |
| 8 | 1 | 2 |
| 9 | 0 | 9 |
| 10 | 5 | 8 |
| 11 | 7 | 10 |
| 12 | 0 | 3 |
| 13 | 3 | 9 |
| 14 | 5 | 8 |
| 15 | 7 | 12 |
| 16 | 10 | 16 |
| 17 | 5 | 5 |
| 18 | 6 | 3 |
| 19 | 9 | 6 |
| 20 | 11 | 8 |

*Source:* Neils, J., Roeltgen, D. P., and Constantinidou, F. "Decline in homophone spelling associated with loss of semantic influence on spelling in Alzheimer's disease." *Brain and Language,* Vol. 49, No. 1, Apr. 1995, p. 36 (Table 3).

**9.42** Inositol is a complex cyclic alcohol found to be effective against clinical depression. Medical researchers believe inositol may also be used to treat panic disorder. To test this theory, a double-blind, placebo-controlled study of 21 patients diagnosed with panic disorder was conducted (*American Journal of Psychiatry,* July 1995). Patients completed daily panic diaries recording the occurrence of their panic attacks. The data for a week in which patients received a glucose placebo and for a week when they were treated with inositol are provided in the table. [*Note:* Neither the patients nor the treating physicians knew which week the placebo was given.] Analyze the data and interpret the results. Comment on the validity of the assumptions.

🧭 INOSITOL

| Patient | Placebo | Inositol |
|---------|---------|----------|
| 1 | 0 | 0 |
| 2 | 2 | 1 |
| 3 | 0 | 3 |
| 4 | 1 | 2 |
| 5 | 0 | 0 |
| 6 | 10 | 5 |
| 7 | 2 | 0 |
| 8 | 6 | 4 |
| 9 | 1 | 1 |
| 10 | 1 | 0 |
| 11 | 1 | 3 |
| 12 | 0 | 2 |
| 13 | 3 | 1 |
| 14 | 3 | 1 |
| 15 | 3 | 4 |
| 16 | 4 | 2 |
| 17 | 6 | 4 |
| 18 | 15 | 21 |
| 19 | 28 | 8 |
| 20 | 30 | 0 |
| 21 | 13 | 0 |

*Source:* Benjamin, J., *et al.* "Double-blind, placebo-controlled, crossover trial of inositol treatment for panic disorder." *American Journal of Psychiatry,* Vol. 152, No. 7, July 1995, p. 1085 (Table 1).

# 9.3 COMPARING TWO POPULATION PROPORTIONS: INDEPENDENT SAMPLING

Suppose a presidential candidate wants to compare the preference of registered voters in the northeastern United States (NE) to those in the southeastern United States (SE). Such a comparison would help determine where to concentrate campaign efforts. The candidate hires a professional pollster to randomly choose 1,000 registered voters in the northeast and 1,000 in the southeast and interview each to learn her or his voting preference. The objective is to use this sample information to make an inference about the difference $(p_1 - p_2)$ between the proportion $p_1$ of *all* registered voters in the northeast and the proportion $p_2$ of *all* registered voters in the southeast who plan to vote for the presidential candidate.

The two samples represent independent binomial experiments. (See Section 4.4 for the characteristics of binomial experiments.) The binomial random variables are the numbers $x_1$ and $x_2$ of the 1,000 sampled voters in each area who indicate they will vote for the candidate. The results are summarized below.

| NE | SE |
|---|---|
| $n_1 = 1,000$ | $n_2 = 1,000$ |
| $x_1 = 546$ | $x_2 = 475$ |

We can now calculate the sample proportions $\hat{p}_1$ and $\hat{p}_2$ of the voters in favor of the candidate in the northeast and southeast, respectively:

$$\hat{p} = \frac{x_1}{n_1} = \frac{546}{1,000} = .546 \quad \hat{p}_2 = \frac{x_2}{n_2} = \frac{475}{1,000} = .475$$

The difference between the sample proportions $(\hat{p}_1 - \hat{p}_2)$ makes an intuitively appealing point estimator of the difference between the population $(p_1 - p_2)$. For our example, the estimate is

$$(\hat{p}_1 - \hat{p}_2) = .546 - .475 = .071$$

To judge the reliability of the estimator $(\hat{p}_1 - \hat{p}_2)$, we must observe its performance in repeated sampling from the two populations. That is, we need to know the sampling distribution of $(\hat{p}_1 - \hat{p}_2)$. The properties of the sampling distribution are given in the following box. Remember that $\hat{p}_1$ and $\hat{p}_2$ can be viewed as means of the number of successes per trial in the respective samples, so the Central Limit Theorem applies when the sample sizes are large.

---

## Properties of the Sampling Distribution of $(\hat{p}_1 - \hat{p}_2)$

1. The mean of the sampling distribution of $(\hat{p}_1 - \hat{p}_2)$ is $(p_1 - p_2)$ that is,

   $$E(\hat{p}_1 - \hat{p}_2) = p_1 - p_2$$

   Thus, $(\hat{p}_1 - \hat{p}_2)$ is an unbiased estimator of $(p_1 - p_2)$.

2. The standard deviation of the sampling distribution of $(\hat{p}_1 - \hat{p}_2)$ is

   $$\sigma_{(\hat{p}_1 - \hat{p}_2)} = \sqrt{\frac{p_1 q_1}{n_1} + \frac{p_2 q_2}{n_2}}$$

3. If the sample sizes $n_1$ and $n_2$ are large (see Section 7.3 for a guideline), the sampling distribution of $(\hat{p}_1 - \hat{p}_2)$ is approximately normal.

Since the distribution of $(\hat{p}_1 - \hat{p}_2)$ in repeated sampling is approximately normal, we can use the $z$ statistic to derive confidence intervals for $(p_1 - p_2)$ or to test a hypothesis about $(p_1 - p_2)$.

For the voter example, a 95% confidence interval for the difference $(p_1 - p_2)$ is

$$(\hat{p}_1 - \hat{p}_2) \pm 1.96\sigma_{(\hat{p}_1 - \hat{p}_2)} \quad \text{or} \quad (\hat{p}_1 - \hat{p}_2) \pm 1.96\sqrt{\frac{p_1 q_1}{n_1} + \frac{p_2 q_2}{n_2}}$$

The quantities $p_1 q_1$ and $p_2 q_2$ must be estimated in order to complete the calculation of the standard deviation, $\sigma_{(\hat{p}_1 - \hat{p}_2)}$, and hence the calculation of the confidence interval. In Section 7.3 we showed that the value of $pq$ is relatively insensitive to the value chosen to approximate $p$. Therefore, $\hat{p}_1 \hat{q}_1$ and $\hat{p}_2 \hat{q}_2$ will provide satisfactory estimates to approximate $p_1 q_1$ and $p_2 q_2$, respectively. Then

$$\sqrt{\frac{p_1 q_1}{n_1} + \frac{p_2 q_2}{n_2}} \approx \sqrt{\frac{\hat{p}_1 \hat{q}_1}{n_1} + \frac{\hat{p}_2 \hat{q}_2}{n_2}}$$

and we will approximate the 95% confidence interval by

$$(\hat{p}_1 - \hat{p}_2) \pm 1.96\sqrt{\frac{\hat{p}_1 \hat{q}_1}{n_1} + \frac{\hat{p}_2 \hat{q}_2}{n_2}}$$

Substituting the sample quantities yields

$$(.546 - .475) \pm 1.96\sqrt{\frac{(.546)(.454)}{1,000} + \frac{(.475)(.525)}{1,000}}$$

or $.071 \pm .044$. Thus, we are 95% confident that the interval from .027 to .115 contains $(p_1 - p_2)$.

We infer that there are between 2.7% and 11.5% more registered voters in the northeast than in the southeast who plan to vote for the presidential candidate. It seems that the candidate should direct a greater campaign effort in the southeast compared to the northeast.

The general form of a confidence interval for the difference $(p_1 - p_2)$ between population proportions is given in the following box.

---

### Large-Sample $100(1 - \alpha)\%$ Confidence Interval for $(p_1 - p_2)$

$$(\hat{p}_1 - \hat{p}_2) \pm z_{\alpha/2}\sigma_{(\hat{p}_1 - \hat{p}_2)} = (\hat{p}_1 - \hat{p}_2) \pm z_{\alpha/2}\sqrt{\frac{p_1 q_1}{n_1} + \frac{p_2 q_2}{n_2}}$$

$$\approx (\hat{p}_1 - \hat{p}_2) \pm z_{\alpha/2}\sqrt{\frac{\hat{p}_1 \hat{q}_1}{n_1} + \frac{\hat{p}_2 \hat{q}_2}{n_2}}$$

*Assumption:* The two samples are independent random samples. Both samples should be large enough that the normal distribution provides an adequate approximation to the sampling distribution of $\hat{p}_1$ and $\hat{p}_2$ (see Section 8.5).

---

The $z$ statistic,

$$z = \frac{(\hat{p}_1 - \hat{p}_2) - (p_1 - p_2)}{\sigma_{(\hat{p}_1 - \hat{p}_2)}}$$

is used to test the null hypothesis that $(p_1 - p_2)$ equals some specified difference, say $D_0$. For the special case where $D_0 = 0$, that is, where we want to test the null hypothesis $H_0: (p_1 - p_2) = 0$ (or, equivalently, $H_0: p_1 = p_2$), the best estimate of $p_1 = p_2 = p$ is obtained by dividing the total number of successes $(x_1 + x_2)$ for the two samples by the total number of observations $(n_1 + n_2)$; that is,

$$\hat{p} = \frac{x_1 + x_2}{n_1 + n_2} \quad \text{or} \quad \hat{p} = \frac{n_1\hat{p}_1 + n_2\hat{p}_2}{n_1 + n_2}$$

The second equation shows that $\hat{p}$ is a weighted average of $\hat{p}_1$ and $\hat{p}_2$, with the larger sample receiving more weight. If the sample sizes are equal, then $\hat{p}$ is a simple average of the two sample proportions of successes.

We now substitute the weighted average $\hat{p}$ for both $p_1$ and $p_2$ in the formula for the standard deviation of $(\hat{p}_1 - \hat{p}_2)$:

$$\sigma_{(\hat{p}_1-\hat{p}_2)} = \sqrt{\frac{p_1 q_1}{n_1} + \frac{p_2 q_2}{n_2}} \approx \sqrt{\frac{\hat{p}\hat{q}}{n_1} + \frac{\hat{p}\hat{q}}{n_2}} = \sqrt{\hat{p}\hat{q}\left(\frac{1}{n_1} + \frac{1}{n_2}\right)}$$

The test is summarized in the next box.

---

## Large-Sample Test of Hypothesis About ($p_1 - p_2$)

**ONE-TAILED TEST**

$H_0: (p_1 - p_2) = 0*$

$H_a: (p_1 - p_2) < 0$
  $[\text{or } H_a: (p_1 - p_2) > 0]$

*Test statistic:*

*Rejection region:* $z < -z_\alpha$
  $[\text{or } z > z_\alpha \text{ when } H_a: (p_1 - p_2) > 0]$

**TWO-TAILED TEST**

$H_0: (p_1 - p_2) = 0$

$H_a: (p_1 - p_2) \neq 0$

$$z = \frac{(\hat{p}_1 - \hat{p}_2)}{\sigma_{(\hat{p}_1-\hat{p}_2)}}$$

*Rejection region:* $|z| > z_{\alpha/2}$

*Note:* $\sigma_{(\hat{p}_1-\hat{p}_2)} = \sqrt{\dfrac{p_1 q_1}{n_1} + \dfrac{p_2 q_2}{n_2}} \approx \sqrt{\hat{p}\hat{q}\left(\dfrac{1}{n_1} + \dfrac{1}{n_2}\right)}$   where   $\hat{p} = \dfrac{x_1 + x_2}{n_1 + n_2}$

*Assumption:* Same as for large-sample confidence interval for $(p_1 - p_2)$. (See previous box.)

---

**EXAMPLE  9.6**

In the past decade intensive antismoking campaigns have been sponsored by both federal and private agencies. Suppose the American Cancer Society randomly sampled 1,500 adults in 1992 and then sampled 1,750 adults in 2002 to determine whether there was evidence that the percentage of smokers had decreased. The results of the two sample surveys are shown in the table, where $x_1$ and $x_2$ represent the numbers of smokers in the 1992 and 2002 samples, respectively. Do these data indicate that the fraction of smokers decreased over this 10-year period? Use $\alpha = .05$.

| 1992 | 2002 |
|------|------|
| $n_1 = 1,500$ | $n_2 = 1,750$ |
| $x_1 = 555$ | $x_2 = 578$ |

---

*The test can be adapted to test for a difference $D_0 \neq 0$. Because most applications call for a comparison of $p_1$ and $p_2$, implying $D_0 = 0$, we will confine our attention to this case.

### Solution

If we define $p_1$ and $p_2$ as the true proportions of adult smokers in 1992 and 2002, the elements of our test are

$$H_0: (p_1 - p_2) = 0$$
$$H_a: (p_1 - p_2) > 0$$

(The test is one-tailed since we are interested only in determining whether the proportion of smokers *decreased*.)

$$\text{Test statistic: } z = \frac{(\hat{p}_1 - \hat{p}_2) - 0}{\sigma_{(\hat{p}_1 - \hat{p}_2)}}$$

*Rejection region using* $\alpha = .05$:
$$z > z_\alpha = z_{.05} = 1.645 \qquad \text{(see Figure 9.15)}$$

**FIGURE 9.15**

*Rejection Region for Example 9.6*

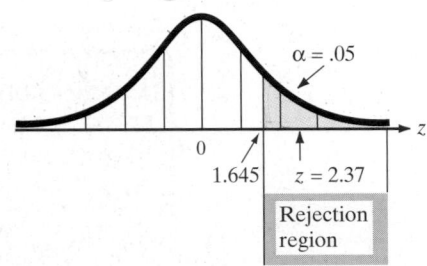

We now calculate the sample proportions of smokers

$$\hat{p}_1 = \frac{555}{1,500} = .37 \qquad \hat{p}_2 = \frac{578}{1,750} = .33$$

Then

$$z = \frac{(\hat{p}_1 - \hat{p}_2) - 0}{\sigma_{(\hat{p}_1 - \hat{p}_2)}} \approx \frac{(\hat{p}_1 - \hat{p}_2)}{\sqrt{\hat{p}\hat{q}\left(\dfrac{1}{n_1} + \dfrac{1}{n_2}\right)}}$$

where

$$\hat{p} = \frac{x_1 + x_2}{n_1 + n_2} = \frac{555 + 578}{1,500 + 1,750} = .349$$

Note that $\hat{p}$ is a weighted average of $\hat{p}_1$ and $\hat{p}_2$, with more weight given to the larger (2002) sample.

Thus, the computed value of the test statistic is

$$z = \frac{.37 - .33}{\sqrt{(.349)(.651)\left(\dfrac{1}{1,500} + \dfrac{1}{1,750}\right)}} = \frac{.040}{.0168} = 2.37$$

There is sufficient evidence at the $\alpha = .05$ level to conclude that the proportion of adults who smoke has decreased over the 1992–2002 period. We could place a confidence interval on $(p_1 - p_2)$ if we were interested in estimating the extent of the decrease. ∎

**EXAMPLE 9.7**

Use a statistical software package to conduct the test in Example 9.6.

**Solution**

We entered the sample sizes ($n_1$ and $n_2$) and numbers of successes ($x_1$ and $x_2$) into MINITAB and obtained the printout shown in Figure 9.16. The test statistic for this one-tailed test, $z = 2.37$, is shaded on the printout, as well as the *p*-value of the test. Note that *p*-value = .009 is smaller than $\alpha = .05$. Consequently, we have strong evidence to reject $H_0$ and conclude that $p_1$ exceeds $p_2$.

**FIGURE 9.16**

*MINITAB Output for Test of Two Proportions*

```
Test and CI for Two Proportions

Sample       X       N   Sample p
1          555    1500   0.370000
2          578    1750   0.330286

Estimate for p(1) - p(2):  0.0397143
95% lower bound for p(1) - p(2): 0.0121024
Test for p(1) - p(2) = 0 (vs > 0):  Z = 2.37   P-Value = 0.009
```

 **EXERCISES 9.43–9.60**

## Learning the Mechanics

**9.43** Explain why the Central Limit Theorem is important in finding an approximate distribution for $(\hat{p}_1 - \hat{p}_2)$.

**9.44** In each case, determine whether the sample sizes are large enough to conclude that the sampling distribution of $(\hat{p}_1 - \hat{p}_2)$ is approximately normal.
  **a.** $n_1 = 10, n_2 = 12, \hat{p}_1 = .50, \hat{p}_2 = .50$
  **b.** $n_1 = 10, n_2 = 12, \hat{p}_1 = .10, \hat{p}_2 = .08$
  **c.** $n_1 = n_2 = 30, \hat{p}_1 = .20, \hat{p}_2 = .30$
  **d.** $n_1 = 100, n_2 = 200, \hat{p}_1 = .05, \hat{p}_2 = .09$
  **e.** $n_1 = 100, n_2 = 200, \hat{p}_1 = .95, \hat{p}_2 = .91$

**9.45** For each of the following values of $\alpha$, find the values of $z$ for which $H_0: (p_1 - p_2) = 0$ would be rejected in favor of $H_a: (p_1 - p_2) < 0$.
  **a.** $\alpha = .01$
  **b.** $\alpha = .025$
  **c.** $\alpha = .05$
  **d.** $\alpha = .10$

**9.46** Independent random samples, each containing 800 observations, were selected from two binomial populations. The samples from populations 1 and 2 produced 320 and 400 successes, respectively.
  **a.** Test $H_0: (p_1 - p_2) = 0$ against $H_a: (p_1 - p_2) \neq 0$. Use $\alpha = .05$.
  **b.** Test $H_0: (p_1 - p_2) = 0$ against $H_a: (p_1 - p_2) \neq 0$. Use $\alpha = .01$.
  **c.** Test $H_0: (p_1 - p_2) = 0$ against $H_a: (p_1 - p_2) < 0$. Use $\alpha = .01$.
  **d.** Form a 90% confidence interval for $(p_1 - p_2)$.

**9.47** Construct a 95% confidence interval for $(p_1 - p_2)$ in each of the following situations:
  **a.** $n_1 = 400, \hat{p}_1 = .65; n_2 = 400, \hat{p}_2 = .58$
  **b.** $n_1 = 180, \hat{p}_1 = .31; n_2 = 250, \hat{p}_2 = .25$
  **c.** $n_1 = 100, \hat{p}_1 = .46; n_2 = 120, \hat{p}_2 = .61$

**9.48** Sketch the sampling distribution of $(\hat{p}_1 - \hat{p}_2)$ based on independent random samples of $n_1 = 100$ and $n_2 = 200$ observations from two binomial populations with success probabilities $p_1 = .1$ and $p_2 = .5$, respectively.

**9.49** Random samples of size $n_1 = 50$ and $n_2 = 60$ were drawn from populations 1 and 2, respectively. The samples yielded $\hat{p}_1 = .4$ and $\hat{p}_2 = .2$. Test $H_0: (p_1 - p_2) = .1$ against $H_a: (p_1 - p_2) > .1$ using $\alpha = .05$.

## Applying the Concepts—Basic

**9.50** Refer to the *Journal of Sport Behavior* (2001) study of variety in exercise workouts, Exercise 7.45 (p. 314). One group of 38 people varied their exercise routine in workouts while a second group of 38 exercisers had no set schedule or regulations for their workouts. By the end of the study, 14 people had dropped out of the first exercise group and 23 had dropped out of the second group.
  **a.** Find the dropout rates (i.e., the percentage of exercisers who had dropped out of the exercise group) for each of the two groups of exercisers.
  **b.** Find a 90% confidence interval for the difference between the dropout rates of the two groups of exercisers.

**c.** Give a practical interpretation of the confidence interval, part **c**.

**9.51** The *Journal of the American Medical Association* (April 18, 2001) published a study of the effectiveness of using extracts of the herbal medicine, St. John's wort, in treating major depression. In an 8-week randomized, controlled trial, 200 patients diagnosed with major depression were divided into two groups: one group ($n_1 = 98$) received St. John's wort extract while the other group ($n_2 = 102$) received a placebo (no drug). At the end of the study period, 14 of the St. John's wort patients were in remission compared to 5 of the placebo patients.

   **a.** Compute the proportion of the St. John's wort patients who were in remission.

   **b.** Compute the proportion of the placebo patients who were in remission.

   **c.** If St. John's wort is effective in treating major depression, then the proportion of St. John's wort patients in remission will exceed the proportion of placebo patients in remission. At $\alpha = .01$, is St. John's wort effective in treating major depression?

   **d.** Repeat part **c**, but use $\alpha = .10$.

   **e.** Explain why the choice of $\alpha$ is critical for this study.

**9.52** The *Journal of Fish Biology* (Aug. 1990) reported on a study to compare the incidence of parasites (tapeworms) in species of Mediterranean and Atlantic fish. In the Mediterranean Sea, 588 brill were captured and dissected, and 211 were found to be infected by the parasite. In the Atlantic Ocean, 123 brill were captured and dissected, and 26 were found to be infected. Compare the proportions of infected brill at the two capture sites using a 90% confidence interval. Interpret the interval.

## Applying the Concepts—Intermediate

**9.53** A University of South Florida biologist conducted an experiment to determine whether increased levels of carbon dioxide kill leaf-eating moths (*USF Magazine,* Winter 1999). Moth larvae were placed in open containers filled with oak leaves. Half the containers had normal carbon dioxide levels while the other half had double the normal level of carbon dioxide. Ten percent of the larvae in the containers with high carbon dioxide levels died, compared to 5 percent in the containers with normal levels. Assume that 80 moth larvae were placed, at random, in each of the two types of containers. Do the experimental results demonstrate that an increased level of carbon dioxide is effective in killing a higher percentage of leaf-eating moth larvae? Test using $\alpha = .01$.

**9.54** When female undergraduates switch from science, mathematics, and engineering (SME) majors into disciplines that are not science-based, are their reasons different from those of their male counterparts? This question was investigated in *Science Education* (July 1995). A sample of 335 junior/senior undergraduates—172 females and 163 males—at two large research universities were identified as "switchers," that is, they left a declared SME major for a non-SME major. Each student listed one or more factors that contributed to the switching decision.

   **a.** Of the 172 females in the sample, 74 listed lack or loss of interest in SME (i.e., "turned off" by science) as a major factor, compared to 72 of the 163 males. Conduct a test (at $\alpha = .10$) to determine whether the proportion of female switchers who give "lack of interest in SME" as a major reason for switching differs from the corresponding proportion of males.

   **b.** Thirty-three of the 172 females in the sample indicated that they were discouraged or lost confidence because of low grades in SME during their early years, compared to 44 of 163 males. Construct a 90% confidence interval for the difference between the proportions of female and male switchers who lost confidence due to low grades in SME. Interpret the result.

**9.55** Trying to think of a word you know, but can't instantly retrieve, is called the "tip of the tongue" phenomenon. *Psychology and Aging* (Sept. 2001) published a study of this phenomenon in senior citizens. The researchers compared 40 people between 60 and 72 years of age with 40 between 73 and 83 years of age. When primed with the initial syllable of a missing word (e.g., seeing the word "include" to help recall the word "incisor"), the younger seniors had a higher recall rate. Suppose 31 of the 40 seniors in the younger group could recall the word when primed with the initial syllable, while only 22 of the 40 seniors could recall the word. The recall rates of the two groups were compared, with the results shown in the accompanying MINITAB printout. Interpret these results. Does one group of elderly people have a significantly higher recall rate than the other?

---

**MINITAB Output for Exercise 9.55**

```
Test and CI for Two Proportions


Sample      X      N   Sample p
1          31     40   0.775000
2          22     40   0.550000


Estimate for p(1) - p(2):  0.225
95% CI for p(1) - p(2): (0.0237157, 0.426284)
Test for p(1) - p(2) = 0 (vs not = 0): Z = 2.13   P-Value = 0.033
```

**9.56** Geneticists at Duke University Medical Center have identified the E2F1 transcription factor as an important component of cell proliferation control (*Nature,* Sept. 23, 1993). The researchers induced DNA synthesis in two batches of serum-starved cells. Each cell in one batch was microinjected with the E2F1 gene, while the cells in the second batch (the controls) were not exposed to E2F1. After 30 hours, the number of cells in each batch that exhibited altered growth was determined. The results of the experiment are summarized in the table.

|  | Control | E2F1-Treated Cells |
|---|---|---|
| Total number of cells | 158 | 92 |
| Number of growth-altered cells | 15 | 41 |

*Source:* Johnson, D. G., *et al.* "Expression of transcription factor E2F1 induces quiescent cells to enter S phase." *Nature,* Vol. 365, No. 6444, Sept. 23, 1993, p. 351 (Table 1).

**a.** Compare the percentages of cells exhibiting altered growth in the two batches with a 90% confidence interval.

**b.** Use the interval, part **a,** to make an inference about the ability of the E2F1 transcription factor to induce cell growth.

**9.57** Scientists have linked a catastrophic decline in the number of frogs inhabiting the world to ultraviolet radiation (*Tampa Tribune,* Mar. 1, 1994). The Pacific tree frog, however, is not believed to be in decline, possibly because it manufactures an enzyme that protects its eggs from ultraviolet radiation. Researchers at Oregon State University compared the hatching rates of two groups of Pacific tree frog eggs. One group of eggs was shielded with ultraviolet-blocking sun shades, while the second group was not. The number of eggs successfully hatched in each group is provided in the following table. Compare the hatching rates of the two groups of Pacific tree frog eggs with a test of hypothesis. Use $\alpha = .01$.

|  | Sun-Shaded Eggs | Unshaded Eggs |
|---|---|---|
| Total number | 70 | 80 |
| Number hatched | 34 | 31 |

**9.58** Can stress management help heart patients reduce their risk of heart attacks or heart surgery? In a study published in the *Archives of Internal Medicine* (Oct. 27, 1997), 107 heart patients were randomly divided into three groups. One group participated in a stress management program, the second group underwent an exercise program, and the third group received usual heart care from their doctors. After three years, the number in each group suffering "cardiac events" (e.g., heart attack, bypass surgery, or angioplasty) was recorded. The results are shown in the next table.

|  | Management | Exercise | Usual Care |
|---|---|---|---|
| Number of heart patients | 33 | 34 | 40 |
| Number who suffer a "cardiac event" | 3 | 7 | 12 |

**a.** Use a 99% confidence interval to compare the proportion of stress management patients who suffer a cardiac event to the proportion of exercise patients who suffer a cardiac event. What inference can you make?

**b.** Repeat part **a,** but compare the stress management group to the usual care group.

**c.** Repeat part **a,** but compare the exercise group to the usual care group.

## Applying the Concepts—Advanced

**9.59** Do you have an insatiable craving for chocolate or some other food? Since many people apparently do, psychologists are designing scientific studies to examine the phenomenon. According to the *New York Times* (Feb. 22, 1995), one of the largest studies of food cravings involved a survey of 1,000 McMaster University (Canada) students. The survey revealed that 97% of the women in the study acknowledged specific food cravings while only 67% of the men did. Assume that 600 of the respondents were women and 400 were men.

**a.** Is there sufficient evidence to claim that the true proportion of women who acknowledge having food cravings exceeds the corresponding proportion of men? Test using $\alpha = .01$.

**b.** Why is it dangerous to conclude from the study that women have a higher incidence of food cravings than men?

**9.60** Can taking an antidepressant drug help cigarette smokers kick their habit? The *New England Journal of Medicine* (Oct. 23, 1997) published a study in which 615 smokers (all who wanted to give up smoking) were randomly assigned to receive either Zyban (an antidepressant) or a placebo (a dummy pill) for six weeks. Of the 309 patients who received Zyban, 71 were not smoking one year later. Of the 306 patients who received a placebo, 37 were not smoking one year later. Conduct a test of hypothesis (at $\alpha = .05$) to answer the research question posed above.

## 9.4   DETERMINING THE SAMPLE SIZE

You can find the appropriate sample size to estimate the difference between a pair of parameters with a specified degree of reliability by using the method described in Section 7.4. That is, to estimate the difference between a pair of parameters correct to within $B$ units with probability $(1 - \alpha)$, let $z_{\alpha/2}$ standard deviations of the sampling distribution of the estimator equal $B$. Then solve for the sample size. To do this, you have to solve the problem for a specific ratio between $n_1$ and $n_2$. Most often, you will want to have equal sample sizes, that is, $n_1 = n_2 = n$. We will illustrate the procedure with two examples.

**EXAMPLE 9.8**

New fertilizer compounds are often advertised with the promise of increased crop yields. Suppose we want to compare the mean yield $\mu_1$ of wheat when a new fertilizer is used to the mean yield $\mu_2$ with a fertilizer in common use. The estimate of the difference in mean yield per acre is to be correct to within .25 bushel with a confidence coefficient of .95. If the sample sizes are to be equal, find $n_1 = n_2 = n$, the number of 1-acre plots of wheat assigned to each fertilizer.

### Solution

To solve the problem, you need to know something about the variation in the bushels of yield per acre. Suppose from past records you know the yields of wheat possess a range of approximately 10 bushels per acre. You could then approximate $\sigma_1 = \sigma_2 = \sigma$ by letting the range equal $4\sigma$. Thus,

$$4\sigma \approx 10 \text{ bushels}$$
$$\sigma \approx 2.5 \text{ bushels}$$

The next step is to solve the equation

$$z_{\alpha/2}\sigma_{(\bar{x}_1 - \bar{x}_2)} = B \quad \text{or} \quad z_{\alpha/2}\sqrt{\frac{\sigma_1^2}{n_1} + \frac{\sigma_2^2}{n_2}} = B$$

for $n$, where $n = n_1 = n_2$. Since we want the estimate to lie within $B = .25$ of $(\mu_1 - \mu_2)$ with confidence coefficient equal to .95, we have $z_{\alpha/2} = z_{.025} = 1.96$. Then, letting $\sigma_1 = \sigma_2 = 2.5$ and solving for $n$, we have

$$1.96\sqrt{\frac{(2.5)^2}{n} + \frac{(2.5)^2}{n}} = .25$$

$$1.96\sqrt{\frac{2(2.5)^2}{n}} = .25$$

$$n = 768.32 \approx 769 \text{ (rounding up)}$$

Consequently, you will have to sample 769 acres of wheat for each fertilizer to estimate the difference in mean yield per acre to within .25 bushel. Since this would necessitate extensive and costly experimentation, you might decide to allow a larger bound (say, $B = .50$ or $B = 1$) in order to reduce the sample size, or you might decrease the confidence coefficient. The point is that we can obtain an idea of the experimental effort necessary to achieve a specified precision in our final estimate by determining the approximate sample size *before* the experiment is begun. ∎

**EXAMPLE 9.9**

A production supervisor suspects that a difference exists between the proportions $p_1$ and $p_2$ of defective items produced by two different machines. Experience has shown that the proportion defective for each of the two machines is in the neighborhood of .03. If the supervisor wants to estimate the difference in the proportions to within .005

using a 95% confidence interval, how many items must be randomly sampled from the production of each machine? (Assume that the supervisor wants $n_1 = n_2 = n$.)

### Solution

In this sampling problem, $B = .005$, and for the specified level of reliability, $z_{\alpha/2} = z_{.025} = 1.96$. Then, letting $p_1 = p_2 = .03$ and $n_1 = n_2 = n$, we find the required sample size per machine by solving the following equation for $n$:

$$z_{\alpha/2}\sigma_{(\hat{p}_1 - \hat{p}_2)} = B$$

or

$$z_{\alpha/2}\sqrt{\frac{p_1 q_1}{n_1} + \frac{p_2 q_2}{n_2}} = B$$

$$1.96\sqrt{\frac{(.03)(.97)}{n} + \frac{(.03)(.97)}{n}} = .005$$

$$1.96\sqrt{\frac{2(.03)(.97)}{n}} = .005$$

$$n = 8,943.2$$

You can see that this may be a tedious sampling procedure. If the supervisor insists on estimating $(p_1 - p_2)$ correct to within .005 with 95% confidence, approximately 9,000 items will have to be inspected for each machine. ∎

You can see from the calculations in Example 9.9 that $\sigma_{(\hat{p}_1 - \hat{p}_2)}$ (and hence the solution, $n_1 = n_2 = n$) depends on the actual (but unknown) values of $p_1$ and $p_2$. In fact, the required sample size $n_1 = n_2 = n$ is largest when $p_1 = p_2 = .5$. Therefore, if you have no prior information on the approximate values of $p_1$ and $p_2$, use $p_1 = p_2 = .5$ in the formula for $\sigma_{(\hat{p}_1 - \hat{p}_2)}$. If $p_1$ and $p_2$ are in fact close to .5, then the values of $n_1$ and $n_2$ that you have calculated will be correct. If $p_1$ and $p_2$ differ substantially from .5, then your solutions for $n_1$ and $n_2$ will be larger than needed. Consequently, using $p_1 = p_2 = .5$ when solving for $n_1$ and $n_2$ is a conservative procedure because the sample sizes $n_1$ and $n_2$ will be at least as large as (and probably larger than) needed.

The procedures for determining sample sizes necessary for estimating $(\mu_1 - \mu_2)$ or $(p_1 - p_2)$ for the case $n_1 = n_2$ are given in the next boxes.

---

### Determination of Sample Size for Estimating $\mu_1 - \mu_2$

To estimate $(\mu_1 - \mu_2)$ to within a given bound $B$ with probability $(1 - \alpha)$, use the following formula to solve for equal sample sizes that will achieve the desired reliability:

$$n_1 = n_2 = \frac{(z_{\alpha/2})^2(\sigma_1^2 + \sigma_2^2)}{B^2}$$

You will need to substitute estimates for the values of $\sigma_1^2$ and $\sigma_2^2$ before solving for the sample size. These estimates might be sample variances $s_1^2$ and $s_2^2$ from prior sampling (e.g., a pilot study), or from an educated (and conservatively large) guess based on the range—that is, $s \approx R/4$.

---

## Determination of Sample Size for Estimating $p_1 - p_2$

To estimate $(p_1 - p_2)$ to within a given bound $B$ with probability $(1 - \alpha)$, use the following formula to solve for equal sample sizes that will achieve the desired reliability:

$$n_1 = n_2 = \frac{(z_{\alpha/2})^2(p_1q_1 + p_2q_2)}{B^2}$$

You will need to substitute estimates for the values of $p_1$ and $p_2$ before solving for the sample size. These estimates might be based on prior samples, obtained from educated guesses or, most conservatively, specified as $p_1 = p_2 = .5$.

---

## EXERCISES 9.61–9.71

### Learning the Mechanics

**9.61** Suppose you want to estimate the difference between two population means correct to within 2.2 with probability .95. If prior information suggests that the population variances are approximately equal to $\sigma_1^2 = \sigma_2^2 = 15$ and you want to select independent random samples of equal size from the populations, how large should the sample sizes, $n_1$ and $n_2$, be?

**9.62** Find the appropriate values of $n_1$ and $n_2$ (assume $n_1 = n_2$) needed to estimate $(\mu_1 - \mu_2)$ with:
  **a.** A bound on the error of estimation equal to 3.2 with 95% confidence. From prior experience it is known that $\sigma_1 \approx 15$ and $\sigma_2 \approx 17$.
  **b.** A bound on the error of estimation equal to 8 with 99% confidence. The range of each population is 60.
  **c.** A 90% confidence interval of width 1.0. Assume that $\sigma_1^2 \approx 5.8$ and $\sigma_2^2 \approx 7.5$.

**9.63** Assuming that $n_1 = n_2$, find the sample sizes needed to estimate $(p_1 - p_2)$ for each of the following situations:
  **a.** Bound = .01 with 99% confidence. Assume that $p_1 \approx .4$ and $p_2 \approx .7$.
  **b.** A 90% confidence interval of width .05. Assume there is no prior information available to obtain approximate values of $p_1$ and $p_2$.
  **c.** Bound = .03 with 90% confidence. Assume that $p_1 \approx .2$ and $p_2 \approx .3$.

**9.64** Enough money has been budgeted to collect independent random samples of size $n_1 = n_2 = 100$ from populations 1 and 2 in order to estimate $(p_1 - p_2)$. Prior information indicates that $p_1 = p_2 \approx .6$. Have sufficient funds been allocated to construct a 90% confidence interval for $(p_1 - p_2)$ of width .1 or less? Justify your answer.

### Applying the Concepts—Basic

**9.65** Refer to *The American Statistician* (May 2001) study comparing the "fear of negative evaluation" (FNE) scores for bulimic and normal female students, Exercise 9.13 (p. 393). Suppose you want to estimate $(\mu_B - \mu_N)$, the difference between the population means of the FNE scores for bulimic and normal female students, using a 95% confidence interval with a bound on the error of estimation of two points. Find the sample sizes required to obtain such an estimate. Assume equal sample sizes of $\sigma_B^2 = \sigma_N^2 = 25$.

**9.66** Refer to the *Journal of Genetic Psychology* (Mar. 1998) study comparing male students' attitudes toward their fathers and mothers, Exercise 9.34 (p. 407). Suppose you want to estimate $\mu_D = (\mu_F - \mu_M)$, the difference between the population mean attitudes toward fathers and mothers, using a 90% confidence interval with a bound on the error of estimation of .3. Find the number of male students required to obtain such an estimate. Assume the variance of the paired differences is $\sigma_D^2 = 1$.

**9.67** Refer to the *Journal of Sport Behavior* (2001) study comparing the dropout rates of two groups of exercisers, Exercise 9.50 (p. 415). Suppose you want to reduce the bound on the error of estimating $(p_1 - p_2)$ with a 90% confidence interval to .1. Determine the number of exercisers to be sampled from each group in order to obtain such an estimate. Assume equal sample sizes, $p_1 \approx .4$, and $p_2 \approx .6$.

### Applying the Concepts—Intermediate

**9.68** All cable television companies carry at least one home shopping channel. Who uses these home shopping services? Are the shoppers primarily men or women? Suppose you want to estimate the difference in the proportions of men and women who say they have used or expect to use televised home shopping using an 80% confidence interval of width .06 or less.
  **a.** Approximately how many people should be included in your samples?
  **b.** Suppose you want to obtain individual estimates for the two proportions of interest. Will the sample size found in part **a** be large enough to provide estimates

of each proportion correct to within .02 with probability equal to .90? Justify your response.

**9.69** Is housework hazardous to your health? A study in the *Public Health Reports* (July–Aug. 1992) compares the life expectancies of 25-year-old white women in the labor force to those who are housewives. How large a sample would have to be taken from each group in order to be 95% confident that the estimate of difference in mean life expectancies for the two groups is within 1 year of the true difference? Assume that equal sample sizes will be selected from the two groups, and that the standard deviation for both groups is approximately 15 years.

**9.70** Poisons are used to prevent rat damage in sugarcane fields. The U.S. Department of Agriculture is investigating whether the rat poison should be located in the middle of the field or on the outer perimeter. One way to answer this question is to determine where the greater amount of damage occurs. If damage is measured by the proportion of cane stalks that have been damaged by rats, how many stalks from each section of the field should be sampled in order to estimate the true difference between proportions of stalks damaged in the two sections to within .02 with 95% confidence?

**9.71** In seeking a free agent NFL running back, a general manager is looking for a player with high mean yards gained per carry and a small standard deviation. Suppose the GM wishes to compare the mean yards gained per carry for two free agents based on independent random samples of their yards gained per carry. Data from last year's pro football season indicate that $\sigma_1 = \sigma_2 \approx 5$ yards. If the GM wants to estimate the difference in means correct to within 1 yard with probability equal to .9, how many runs would have to be observed for each player? (Assume equal sample sizes.)

## 9.5 COMPARING TWO POPULATION VARIANCES: INDEPENDENT SAMPLING (OPTIONAL)

Many times it is of practical interest to use the techniques developed in this chapter to compare the means or proportions of two populations. However, there are also important instances when we wish to compare two population variances. For example, when two devices are available for producing precision measurements (scales, calipers, thermometers, etc.), we might want to compare the variability of the measurements of the devices before deciding which one to purchase. Or when two standardized tests can be used to rate job applicants, the variability of the scores for both tests should be taken into consideration before deciding which test to use.

For problems like these we need to develop a statistical procedure to compare population variances. The common statistical procedure for comparing population variances, $\sigma_1^2$ and $\sigma_2^2$, makes an inference about the ratio $\sigma_1^2/\sigma_2^2$. In this section, we will show how to test the null hypothesis that the ratio $\sigma_1^2/\sigma_2^2$ equals 1 (the variances are equal) against the alternative hypothesis that the ratio differs from 1 (the variances differ):

$$H_0: \frac{\sigma_1^2}{\sigma_2^2} = 1 \quad (\sigma_1^2 = \sigma_2^2)$$

$$H_a: \frac{\sigma_1^2}{\sigma_2^2} \neq 1 \quad (\sigma_1^2 \neq \sigma_2^2)$$

To make an inference about the ratio $\sigma_1^2/\sigma_2^2$, it seems reasonable to collect sample data and use the ratio of the sample variances, $s_1^2/s_2^2$. We will use the test statistic

$$F = \frac{s_1^2}{s_2^2}$$

To establish a rejection region for the test statistic, we need to know the sampling distribution of $s_1^2/s_2^2$. As you will subsequently see, the sampling distribution of $s_1^2/s_2^2$ is based on two of the assumptions already required for the *t*-test:

1. The two sampled populations are normally distributed.
2. The samples are randomly and independently selected from their respective populations.

**FIGURE 9.17**

*An F-Distribution with*
*7 Numerator and*
*9 Denominator Degrees*
*of Freedom*

When these assumptions are satisfied and when the null hypothesis is true (that is, $\sigma_1^2 = \sigma_2^2$), the sampling distribution of $F = s_1^2/s_2^2$ is the **F-distribution** with $(n_1 - 1)$ numerator degrees of freedom and $(n_2 - 1)$ denominator degrees of freedom, respectively. The shape of the $F$-distribution depends on the degrees of freedom associated with $s_1^2$ and $s_2^2$—that is, on $(n_1 - 1)$ and $(n_2 - 1)$. An $F$-distribution with 7 and 9 df is shown in Figure 9.17. As you can see, the distribution is skewed to the right, since $s_1^2/s_2^2$ cannot be less than 0 but can increase without bound.

We need to be able to find $F$ values corresponding to the tail areas of this distribution in order to establish the rejection region for our test of hypothesis because we expect the ratio $F$ of the sample variances to be either very large or very small when two population variances are unequal. The upper-tail $F$ values for $\alpha = .10, .05, .025,$ and .01 can be found in Tables VIII, IX, X, and XI of Appendix A. Table IX is partially reproduced in Table 9.6. It gives $F$ values that correspond to $\alpha = .05$

**TABLE 9.6**    **Reproduction of Part of Table IX in Appendix A: Percentage Points of the F-distribution, $\alpha = .05$**

| $\nu_1$ | **Numerator Degrees of Freedom** | | | | | | | | |
|---|---|---|---|---|---|---|---|---|---|
| $\nu_2$ | 1 | 2 | 3 | 4 | 5 | 6 | 7 | 8 | 9 |
| 1 | 161.4 | 199.5 | 215.7 | 224.6 | 230.2 | 234.0 | 236.8 | 238.9 | 240.5 |
| 2 | 18.51 | 19.00 | 19.16 | 19.25 | 19.30 | 19.33 | 19.35 | 19.37 | 19.38 |
| 3 | 10.13 | 9.55 | 9.28 | 9.12 | 9.01 | 8.94 | 8.89 | 8.85 | 8.81 |
| 4 | 7.71 | 6.94 | 6.59 | 6.39 | 6.26 | 6.16 | 6.09 | 6.04 | 6.00 |
| 5 | 6.61 | 5.79 | 5.41 | 5.19 | 5.05 | 4.95 | 4.88 | 4.82 | 4.77 |
| 6 | 5.99 | 5.14 | 4.76 | 4.53 | 4.39 | 4.28 | 4.21 | 4.15 | 4.10 |
| 7 | 5.59 | 4.74 | 4.35 | 4.12 | 3.97 | 3.87 | 3.79 | 3.73 | 3.68 |
| 8 | 5.32 | 4.46 | 4.07 | 3.84 | 3.69 | 3.58 | 3.50 | 3.44 | 3.39 |
| 9 | 5.12 | 4.26 | 3.86 | 3.63 | 3.48 | 3.37 | 3.29 | 3.23 | 3.18 |
| 10 | 4.96 | 4.10 | 3.71 | 3.48 | 3.33 | 3.22 | 3.14 | 3.07 | 3.02 |
| 11 | 4.84 | 3.98 | 3.59 | 3.36 | 3.20 | 3.09 | 3.01 | 2.95 | 2.90 |
| 12 | 4.75 | 3.89 | 3.49 | 3.25 | 3.11 | 3.00 | 2.91 | 2.85 | 2.80 |
| 13 | 4.67 | 3.81 | 3.41 | 3.18 | 3.03 | 2.92 | 2.83 | 2.77 | 2.71 |
| 14 | 4.60 | 3.74 | 3.34 | 3.11 | 2.96 | 2.85 | 2.76 | 2.70 | 2.65 |

*Denominator Degrees of Freedom*

**FIGURE 9.18**

*An F-Distribution for $\nu_1 = 7$ and $\nu_2 = 9$ df; $\alpha = .05$*

upper-tail areas for different degrees of freedom; the columns correspond to the degrees of freedom $\nu_1$ for the numerator sample variance, $s_1^2$, whereas the rows correspond to the degrees of freedom $\nu_2$ for the denominator sample variance, $s_2^2$. Thus, if the numerator degrees of freedom is $\nu_1 = 7$ and the denominator degrees of freedom is $\nu_2 = 9$, we look in the seventh column and ninth row to find $F_{.05} = 3.29$. As shown in Figure 9.18, $\alpha = .05$ is the tail area to the right of 3.29 in the $F$-distribution with 7 and 9 df. That is, if $\sigma_1^2 = \sigma_2^2$, then the probability that the $F$ statistic will exceed 3.29 is $\alpha = .05$.

**EXAMPLE 9.10**

An experimenter wants to compare the metabolic rates of white mice subjected to different drugs. The weights of the mice may affect their metabolic rates, and thus the experimenter wishes to obtain mice that are relatively homogeneous with respect to weight. Five hundred mice will be needed to complete the study. Currently, 18 mice from supplier 1 and another 13 mice from supplier 2 are available for comparison. The experimenter weighs these mice and obtains the data shown in Table 9.7. Do these data provide sufficient evidence to indicate a difference in the variability of weights of mice obtained from the two suppliers? (Use $\alpha = .10$.) Using the results of this analysis, what would you suggest to the experimenter?

⦿ MICEWTS

**TABLE 9.7   Weights (in ounces) of Experimental Mice**

**Supplier 1**

| | | | | | |
|---|---|---|---|---|---|
| 4.14 | 4.26 | 4.05 | 4.11 | 4.31 | 4.12 |
| 4.17 | 4.35 | 4.25 | 4.21 | 4.05 | 4.28 |
| 4.15 | 4.20 | 4.32 | 4.25 | 4.02 | 4.14 |

**Supplier 2**

| | | | | | |
|---|---|---|---|---|---|
| 4.23 | 4.35 | 4.05 | 3.75 | 4.41 | 4.37 |
| 4.01 | 4.06 | 4.15 | 4.19 | 4.52 | 4.21 |
| 4.29 | | | | | |

**Solution**

Let

$$\sigma_1^2 = \text{Population variance of weights of white mice from supplier 1}$$
$$\sigma_2^2 = \text{Population variance of weights of white mice from supplier 2}$$

The hypotheses of interest are then

$$H_0: \frac{\sigma_1^2}{\sigma_2^2} = 1 \quad (\sigma_1^2 = \sigma_2^2)$$

$$H_a: \frac{\sigma_1^2}{\sigma_2^2} \neq 1 \quad (\sigma_1^2 \neq \sigma_2^2)$$

The nature of the *F*-tables given in Appendix A affects the form of the test statistic. To form the rejection region for a two-tailed *F*-test, we want to make certain that the upper tail is used, because only the upper-tail values of *F* are shown in Tables VIII, IX, X, and XI. To accomplish this, *we will always place the larger sample variance in the numerator of the F-test statistic.* This has the effect of doubling the tabulated value for $\alpha$, since we double the probability that the *F*-ratio will fall in the upper tail by always placing the larger sample variance in the numerator. That is, we establish a one-tailed rejection region by putting the larger variance in the numerator rather than establishing rejection regions in both tails.

Thus, for our example, we have a denominator $s_1^2$ with df $= \nu_2 = n_1 - 1 = 17$ and a numerator $s_2^2$ with df $= \nu_1 = n_2 - 1 = 12$. Therefore, the test statistic will be

$$F = \frac{\text{Larger sample variance}}{\text{Smaller sample variance}} = \frac{s_2^2}{s_1^2}$$

and we will reject $H_0: \sigma_1^2 = \sigma_2^2$ for $\alpha = .10$ when the calculated value of $F$ exceeds the tabulated value:

$$F_{\alpha/2} = F_{.05} = 2.38 \quad \text{(see Figure 9.19)}$$

To calculate the value of the test statistic, we require the sample variances. These are shown on the STATISTIX printout, Figure 9.20. The sample standard deviations are $s_1 = .0982$ and $s_2 = .2021$. Therefore,

$$F = \frac{s_2^2}{s_1^2} = \frac{(.2021)^2}{(.0982)^2} = 4.24$$

When we compare this result to the rejection region shown in Figure 9.19, we see that $F = 4.24$ falls in the rejection region. Therefore, the data provide sufficient evidence to indicate that the population variances differ. It appears that the weights of mice obtained from supplier 1 tend to be more homogeneous than the weights of mice obtained from supplier 2. On the basis of this evidence, we would advise the experimenter to purchase the mice from supplier 1. ∎

**FIGURE 9.19**

*Rejection Region for Example 9.10*

**FIGURE 9.20**

*STATISTIX Analysis of Mice
Weights*

```
TWO-SAMPLE T TESTS FOR SUPP1 VS SUPP2

                      SAMPLE
VARIABLE     MEAN      SIZE       S.D.        S.E.
----------  -------   -------    --------    --------
SUPP1        4.1878     18       0.0982      0.0231
SUPP2        4.1992     13       0.2021      0.0561
DIFFERENCE  -0.0115


NULL HYPOTHESIS:   DIFFERENCE =  0
ALTERNATIVE HYP:   DIFFERENCE <> 0


ASSUMPTION              T       DF       P      95% CI FOR DIFFERENCE
------------------    -----    -----    ------  ---------------------
EQUAL VARIANCES       -0.21     29      0.8355  (-0.1233,  0.1004)
UNEQUAL VARIANCES     -0.19    16.1     0.8526  (-0.1400,  0.1171)


                       F      NUM DF   DEN DF     P
TESTS FOR EQUALITY  -------   ------   ------   ------
      OF VARIANCES    4.24      12       17     0.0036


CASES INCLUDED 31      MISSING CASES 5
```

What would you have concluded if the value of $F$ calculated from the samples had not fallen in the rejection region? Would you conclude that the null hypothesis of equal variances is true? No, because then you risk the possibility of a Type II error (accepting $H_0$ if $H_a$ is true) without knowing the value of $\beta$, the probability of accepting $H_0: \sigma_1^2 = \sigma_2^2$ if in fact it is false. Since we will not consider the calculation of $\beta$ for specific alternatives in this text, when the $F$ statistic does not fall in the rejection region we simply conclude that insufficient sample evidence exists to refute the null hypothesis that $\sigma_1^2 = \sigma_2^2$.

The $F$-test for equal population variances is summarized in the next box.*

**EXAMPLE 9.11**

Find the $p$-value for the test in Example 9.10 using the $F$-tables in Appendix A. Compare this to the exact $p$-value obtained from a computer printout.

**Solution**

Since the observed value of the $F$ statistic in Example 9.10 was 4.24, the observed significance level of the test would equal the probability of observing a value of $F$ at least as contradictory to $H_0: \sigma_1^2 = \sigma_2^2$ as $F = 4.24$, if in fact $H_0$ is true. Since we give the $F$-tables in Appendix A only for values of $\alpha$ equal to .10, .05, .025, and .01, we can only approximate the observed significance level. Checking Table XI, we find $F_{.01} = 3.46$. Since the observed value of $F$ is greater than $F_{.01}$, the observed significance level for the test is less than $2(.01) = .02$. (Note that we double the $\alpha$ value shown in Table XI because this is a two-tailed test.) The exact $p$-value, $p = .0072$, is twice the highlighted one-tailed $p$-value (.0036) given at the bottom of the STATISTIX printout, Figure 9.20.

---

*A $100(1 - \alpha)\%$ confidence interval for $\sigma_1^2/\sigma_2^2$ can also be obtained using the $F$-distribution. The interval takes the form:

$$\left(\frac{s_1^2}{s_2^2}\right)\left(\frac{1}{F_{L,\alpha/2}}\right) < \frac{\sigma_1^2}{\sigma_2^2} < \left(\frac{s_1^2}{s_2^2}\right)F_{U,\alpha/2}$$

where $F_{L,\alpha/2}$ is the value of $F$ that places an area $\alpha/2$ in the upper tail of an $F$-distribution with $\nu_1 = (n_1 - 1)$ numerator and $\nu_2 = (n_2 - 1)$ denominator degrees of freedom, and $F_{U,\alpha/2}$ is the upper-tailed value of $F$ based on $\nu_1 = (n_2 - 1)$ numerator and $\nu_2 = (n_1 - 1)$ denominator degrees of freedom.

## *F*-Test for Equal Population Variances*

**ONE-TAILED TEST**

$H_0: \sigma_1^2 = \sigma_2^2$

$H_a: \sigma_1^2 < \sigma_2^2$

   (or $H_a: \sigma_1^2 > \sigma_2^2$)

*Test statistic:*

$$F = \frac{s_2^2}{s_1^2}$$

$$\left(\text{or } F = \frac{s_1^2}{s_2^2} \text{ when } H_a: \sigma_1^2 > \sigma_2^2\right)$$

*Rejection region:*

$F > F_\alpha$

**TWO-TAILED TEST**

$H_0: \sigma_1^2 = \sigma_2^2$

$H_a: \sigma_1^2 \neq \sigma_2^2$

*Test statistic:*

$$F = \frac{\text{Larger sample variance}}{\text{Smaller sample variance}}$$

$$= \frac{s_1^2}{s_2^2} \text{ when } s_1^2 > s_2^2$$

$$\left(\text{or } \frac{s_2^2}{s_1^2} \text{ when } s_2^2 > s_1^2\right)$$

*Rejection region:*

$F > F_{\alpha/2}$

where $F_\alpha$ and $F_{\alpha/2}$ are based on $\nu_1$ = numerator degrees of freedom and $\nu_2$ = denominator degrees of freedom; $\nu_1$ and $\nu_2$ are the degrees of freedom for the numerator and denominator sample variances, respectively.

*Assumptions:*

1.  Both sampled populations are normally distributed.
2.  The samples are random and independent.

To conclude this section, we consider the comparison of population variances as a check of the assumption $\sigma_1^2 = \sigma_2^2$ needed for the two-sample *t*-test. Rejection of the null hypothesis $\sigma_1^2 = \sigma_2^2$ would indicate that the assumption is invalid. [*Note:* Nonrejection of the null hypothesis does *not* imply that the assumption is valid.]

■ **EXAMPLE 9.12**

In Example 9.4 (Section 9.1) we used the two-sample *t* statistic to compare the mean reading scores of two groups of slow learners who had been taught to read using two different methods. The data are repeated in Table 9.8 for convenience. The use of the *t* statistic was based on the assumption that the population variances of the test scores were equal for the two methods. Conduct a test of hypothesis to check this assumption at $\alpha = .10$.

⦿ READING

**TABLE 9.8    Reading Test Scores for Slow Learners**

| New Method | | | | Standard Method | | | |
|---|---|---|---|---|---|---|---|
| 80 | 80 | 79 | 81 | 79 | 62 | 70 | 68 |
| 76 | 66 | 71 | 76 | 73 | 76 | 86 | 73 |
| 70 | 85 | | | 72 | 68 | 75 | 66 |

---

*Although a test of a hypothesis of equality of variances is the most common application of the *F*-test, it can also be used to test a hypothesis that the ratio between the population variances is equal to some specified value, $H_0: \sigma_1^2/\sigma_2^2 = k$. The test is conducted in exactly the same way as specified in the box, except that we use the test statistic

$$F = \left(\frac{s_1^2}{s_2^2}\right)\left(\frac{1}{k}\right)$$

**FIGURE 9.21**

*SAS F-Test for the Data in Table 9.8*

Two Sample Test for Variances of SCORE within METHOD

Sample Statistics

| METHOD Group | N | Mean | Std. Dev. | Variance |
|---|---|---|---|---|
| NEW | 10 | 76.4 | 5.8348 | 34.04444 |
| STD | 12 | 72.33333 | 6.3437 | 40.24242 |

Hypothesis Test

Null hypothesis:     Variance 1 / Variance 2  =  1
Alternative:         Variance 1 / Variance 2  ^= 1

|   | - Degrees of Freedom - | | |
| F | Numer. | Denom. | Pr > F |
|---|---|---|---|
| 0.85 | 9 | 11 | 0.8148 |

### Solution

We want to test

$$H_0: \sigma_1^2 = \sigma_2^2$$
$$H_a: \sigma_1^2 \neq \sigma_2^2$$

The data were entered into SAS, and the SAS printout shown in Figure 9.21 was obtained. Both the test statistic, $F = .85$, and two-tailed $p$-value, .8148, are highlighted on the printout. Since $\alpha = .10$ is less than the $p$-value, we do not reject the null hypothesis that the population variances of the reading test scores are equal. It is here that the temptation to misuse the $F$-test is strongest. *We cannot conclude that the data justify the use of the t statistic.* This is equivalent to accepting $H_0$, and we have repeatedly warned against this conclusion because the probability of a Type II error, $\beta$, is unknown. The $\alpha$ level of .10 protects us only against rejecting $H_0$ if it is true. This use of the $F$-test may prevent us from abusing the $t$ procedure when we obtain a value of $F$ that leads to a rejection of the assumption that $\sigma_1^2 = \sigma_2^2$. But when the $F$ statistic does not fall in the rejection region, we know little more about the validity of the assumption than before we conducted the test.

---

### What Do You Do If the Assumption of Normal Population Distributions Is Not Satisfied?

*Answer:* The $F$-test is much less robust (i.e., much more sensitive) to departures from normality than the $t$-test for comparing the population means (Section 9.1). If you have doubts about the normality of the population frequency distributions, use a *nonparametric method* for comparing the two population variances. A method can be found in the nonparametric statistics texts listed in the references.

EXERCISES 9.72–9.84

## Learning the Mechanics

**9.72** Use Tables VIII, IX, X, and XI of Appendix A to find each of the following $F$ values:
- **a.** $F_{.05}$ where $\nu_1 = 8$ and $\nu_2 = 5$
- **b.** $F_{.01}$ where $\nu_1 = 20$ and $\nu_2 = 14$
- **c.** $F_{.025}$ where $\nu_1 = 10$ and $\nu_2 = 5$
- **d.** $F_{.10}$ where $\nu_1 = 20$ and $\nu_2 = 5$

**9.73** Given $\nu_1$ and $\nu_2$, find the following probabilities:
- **a.** $\nu_1 = 2, \nu_2 = 30, P(F \geq 4.18)$
- **b.** $\nu_1 = 24, \nu_2 = 14, P(F < 1.94)$
- **c.** $\nu_1 = 9, \nu_2 = 1, P(F \leq 6,022.0)$
- **d.** $\nu_1 = 30, \nu_2 = 30, P(F > 1.84)$

**9.74** For each of the following cases, identify the rejection region that should be used to test $H_0: \sigma_1^2 = \sigma_2^2$ against $H_a: \sigma_1^2 > \sigma_2^2$. Assume $\nu_1 = 20$ and $\nu_2 = 30$.
- **a.** $\alpha = .10$
- **b.** $\alpha = .05$
- **c.** $\alpha = .025$
- **d.** $\alpha = .01$

**9.75** For each of the following cases, identify the rejection region that should be used to test $H_0: \sigma_1^2 = \sigma_2^2$ against $H_a: \sigma_1^2 \neq \sigma_2^2$. Assume $\nu_1 = 8$ and $\nu_2 = 40$.
- **a.** $\alpha = .20$
- **b.** $\alpha = .10$
- **c.** $\alpha = .05$
- **d.** $\alpha = .02$

**9.76** Specify the appropriate rejection region for testing against $H_0: \sigma_1^2 = \sigma_2^2$ in each of the following situations:
- **a.** $H_a: \sigma_1^2 > \sigma_2^2; \alpha = .05, n_1 = 25, n_2 = 20$
- **b.** $H_a: \sigma_1^2 < \sigma_2^2; \alpha = .05, n_1 = 10, n_2 = 15$
- **c.** $H_a: \sigma_1^2 \neq \sigma_2^2; \alpha = .10, n_1 = 21, n_2 = 31$
- **d.** $H_a: \sigma_1^2 < \sigma_2^2; \alpha = .01, n_1 = 31, n_2 = 41$
- **e.** $H_a: \sigma_1^2 \neq \sigma_2^2; \alpha = .05, n_1 = 7, n_2 = 16$

**9.77** Independent random samples were selected from each of two normally distributed populations, $n_1 = 16$ from population 1 and $n_2 = 25$ from population 2. The means and variances for the two samples are shown in the table.

| Sample 1 | Sample 2 |
|---|---|
| $n_1 = 16$ | $n_2 = 25$ |
| $\bar{x}_1 = 22.5$ | $\bar{x}_2 = 28.2$ |
| $s_1^2 = 2.87$ | $s_2^2 = 9.85$ |

- **a.** Test the null hypothesis $H_0: \sigma_1^2 = \sigma_2^2$ against the alternative hypothesis $H_a: \sigma_1^2 \neq \sigma_2^2$. Use $\alpha = .05$.
- **b.** Test $H_0: \sigma_1^2 = \sigma_2^2$ against $H_a: \sigma_1^2 < \sigma_2^2$. Use $\alpha = .05$.

**9.78** Independent random samples were selected from each of two normally distributed populations, $n_1 = 6$ from population 1 and $n_2 = 4$ from population 2. The data are shown in the table.

LM9_78

| Sample 1 | Sample 2 |
|---|---|
| 3.1 | 2.3 |
| 4.3 | 1.4 |
| 1.2 | 3.7 |
| 1.7 | 8.9 |
| .6 | |
| 3.4 | |

- **a.** Test $H_0: \sigma_1^2 = \sigma_2^2$ against $H_a: \sigma_1^2 < \sigma_2^2$. Use $\alpha = .01$.
- **b.** Test $H_0: \sigma_1^2 = \sigma_2^2$ against $H_a: \sigma_1^2 \neq \sigma_2^2$. Use $\alpha = .10$.

## Applying the Concepts—Basic

**9.79** Refer to Exercise 9.13 (p. 393). The "fear of negative evaluation" (FNE) scores for the 11 bulimic females and 14 females with normal eating habits are reproduced in the table. The confidence interval you constructed in Exercise 9.13 requires that the variance of the FNE scores of bulimic females is equal to the variance of the FNE scores of normal females. Conduct a test (at $\alpha = .05$) to determine the validity of this assumption.

 BULIMIA

| **Bulimic students** | 21 | 13 | 10 | 20 | 25 | 19 | 16 | 21 | 24 | 13 | 14 | | |
|---|---|---|---|---|---|---|---|---|---|---|---|---|---|
| **Normal students** | 13 | 6 | 16 | 13 | 8 | 19 | 23 | 18 | 11 | 19 | 7 | 10 | 15 | 20 |

*Source:* Randles, R. H. "On neutral responses (zeros) in the sign test and ties in the Wilcoxon-Mann-Whitney test." *The American Statistician,* Vol. 55, No. 2, May 2001 (Figure 3).

### BILLBOARD

| | | | | | | | | | | | | |
|---|---|---|---|---|---|---|---|---|---|---|---|---|
| **Pre-SoundScan** | 1 | 1 | 0 | 2 | 3 | 4 | 3 | 4 | 5 | 5 | 2 | 3 |
| **Post-SoundScan** | 0 | 2 | 3 | 1 | 3 | 1 | 0 | 1 | 4 | 2 | 2 | 1 |

**9.80** Refer to Exercise 9.19 (p. 395) and the *Organizational Science* (May–June 2000) analysis of the weekly number of new artists entering the *Billboard* Top 200 chart prior to and after *Billboard* began using the SoundScan system. The data are reproduced in the table above. The confidence interval and hypothesis test you performed in Exercise 9.19 requires that the variance of the weekly number of new artists pre-SoundScan is equal to the variance of the weekly number of new artists post-SoundScan. Conduct a test (at $\alpha = .05$) to determine the validity of this assumption.

**9.81** Refer to the *American Educational Research Journal* (Fall 1998) study to compare the mathematics achievement test scores of male and female students, Exercise 9.15 (p. 394). The researchers hypothesized that the distribution of test scores for males is more variable than the corresponding distribution for females. Use the summary information reproduced in the table to test this claim at $\alpha = .01$.

| | **Males** | **Females** |
|---|---|---|
| Sample size | 1,764 | 1,739 |
| Mean | 48.9 | 48.4 |
| Standard deviation | 12.96 | 11.85 |

*Source:* Bielinski, J., and Davison, M. L. "Gender differences by item difficulty interactions in multiple-choice mathematics items." *American Educational Research Journal,* Vol. 35, No. 3, Fall 1998, p. 464 (Table 1).

### Applying the Concepts—Intermediate

**9.82** Tests of product quality using human inspectors can lead to serious inspection error problems (*Journal of Quality Technology,* Apr. 1986). To evaluate the performance of inspectors in a new company, a quality manager had a sample of 12 novice inspectors evaluate 200 finished products. The same 200 items were evaluated by 12 experienced inspectors. The quality of each item—whether defective or nondefective—was known to the manager. The following table lists the number of inspection errors (classifying a defective item as nondefective or vice versa) made by each inspector. A SAS printout comparing the two types of inspectors is also shown.

### INSPECT

| **Novice Inspectors** | | | | **Experienced Inspectors** | | | |
|---|---|---|---|---|---|---|---|
| 30 | 35 | 26 | 40 | 31 | 15 | 25 | 19 |
| 36 | 20 | 45 | 31 | 28 | 17 | 19 | 18 |
| 33 | 29 | 21 | 48 | 24 | 10 | 20 | 21 |

**a.** Prior to conducting this experiment, the manager believed the variance in inspection errors was lower for experienced inspectors than for novice inspectors. Do the sample data support her belief? Test using $\alpha = .05$.

**b.** What is the appropriate *p*-value of the test you conducted in part **a**?

SAS Output for Exercise 9.82

Two Sample Test for Variances of ERRORS within INSPECT

Sample Statistics

| INSPECT Group | N | Mean | Std. Dev. | Variance |
|---|---|---|---|---|
| 1NOVICE | 12 | 32.83333 | 8.6427 | 74.69697 |
| 2EXPER | 12 | 20.58333 | 5.7439 | 32.99242 |

Hypothesis Test

Null hypothesis:     Variance 1 / Variance 2  <= 1
Alternative:          Variance 1 / Variance 2  >  1

| | - Degrees of Freedom - | | |
|---|---|---|---|
| F | Numer. | Denom. | Pr > F |
| 2.26 | 11 | 11 | 0.0955 |

**9.83** Refer to the *Genetical Research* (June 1995) study of the mating habits of hermaphroditic snails, Exercise 9.22 (p.396). Recall that the snails were identified as either selfers or outcrossers and that the mean effective population sizes of the two groups were compared. The data for the study are reproduced in the table. Geneticists are often more interested in comparing the variation in population size of the two types of mating systems. Conduct this analysis for the researcher. Interpret the result.

**Effective Population Size**

| Snail Mating System | Sample Size | Mean | Standard Deviation |
|---|---|---|---|
| Outcrossing | 17 | 4,894 | 1,932 |
| Selfing | 5 | 4,133 | 1,890 |

*Source:* Jarne, P. "Mating system, bottlenecks, and genetic polymorphism in hermaphroditic animals." *Genetical Research,* Vol. 65, No. 3, June 1995, p. 197 (Table 4).

**9.84** Following the Persian Gulf War, the Pentagon changed its logistics processes to be more corporate-like. The extravagant "just-in-case" mentality was replaced with "just-in-time" systems. Emulating Federal Express and United Parcel Service, deliveries from factories to foxholes are now expedited using bar codes, laser cards,

⬤ ORDTIMES

| Persian Gulf (1991) | Bosnia (1995) |
|---|---|
| 28.0 | 15.1 |
| 20.0 | 6.4 |
| 26.5 | 5.0 |
| 10.6 | 11.4 |
| 9.1 | 6.5 |
| 35.2 | 6.5 |
| 29.1 | 3.0 |
| 41.2 | 7.0 |
| 27.5 | 5.5 |

*Source:* Adapted from Crock, S. "The Pentagon goes to B-school." *Business Week,* Dec. 11, 1995, p. 98.

radio tags, and databases to track supplies. The table contains order-to-delivery times (in days) for a sample of shipments from the U.S. to the Persian Gulf in 1991 and a sample of shipments to Bosnia in 1995.

**a.** Determine whether the variances in order-to-delivery times for Persian Gulf and Bosnia shipments are equal. Use $\alpha = .05$.

**b.** Given your answer to part **a,** is it appropriate to construct a confidence interval for the difference between the mean order-to-delivery times? Explain.

---

**STATISTICS IN ACTION**
*Detection of Rigged Milk Prices*

Many products and services are purchased by government agencies, cities, states, and businesses on the basis of sealed bids, and contracts are awarded to the lowest bidders. This process works extremely well in competitive markets, but it has the potential to increase the cost of purchasing if the markets are noncompetitive or if bidders are engaged in collusive practices. The latter may have occurred recently in the Kentucky milk market.

Each year, the state of Kentucky invites bids from dairies to supply half-pint containers of fluid milk products for its school districts. In several districts in northern Kentucky, the suppliers (dairies) were accused of "price-fixing," that is, conspiring to set the price of half-pint milk containers above the "fair," or competitive, price. A market analysis disclosed that two dairies—Meyer Dairy and Trauth Dairy—were the only two bidders on the milk contracts in the school districts in the "tri-county market" of Boone, Campbell, and Kenton Counties between 1983 and 1991. Consequently, these two companies were awarded all the milk contracts in the market. Speculation is that Meyer and Trauth conspired to rig their bids in the tri-county market.

Numerous economic and statistical methods exist for detecting the possibility of collusive practices among bidders. These typically involve (1) market-shares analyses, (2) an analysis of incumbency rates (i.e., the percentage of markets that are won by the same vendor who won the previous year), (3) an examination of bid levels and their dispersion, (4) price-distance analyses, (5) bid-sequence analyses, (6) comparison of mean winning prices, and (7) econometric modeling of the winning price. For this case, we'll focus on a comparison of the mean winning bid prices; however, a complete examination would consist of all seven types of analyses.

Consider two similar markets, one in which the bids are rigged and the other in which bids are made competitively. In theory, the mean winning price in the "rigged" market will be higher than the mean winning price in the competitive market. Recall that in this case, the speculation is that from 1983 to 1991 the tri-county milk market (where Meyer and Trauth dairies supplied most of the milk) was rigged while the school districts in northern Kentucky that surround the tri-county market (the "surrounding market") was competitive. Consequently, you will want to compare the mean winning price in the tri-county market with that of the surrounding market for each of the years in question.

The state of Kentucky has granted permission for you to analyze its milk data. The data, saved in the MILK file, consist of the prices (in dollars) of all winning bids in the tri-county and surrounding markets from 1983 to 1991 for each of three products: whole white milk, low-fat white milk, and low-fat chocolate milk. ⬤ MILK

## STATISTICS IN ACTION *(continued)*

### *Focus*

a. The mean winning bid prices for half-pint containers of low-fat white milk in the two markets are plotted by year on the same graph in Figure 9.22. Do you detect a pattern in the graph that supports the bid-collusion theory? Explain.

b. Why is Figure 9.22 insufficient to claim, with a high degree of confidence, that the true population mean winning price of low-fat white milk in the tri-county market exceeds the population mean in the surrounding market?

c. For each year, test the bid-collusion theory by comparing the mean winning bid of low-fat white milk in the

tri-county market to the corresponding mean in the surrounding market. Give your conclusions and recommendations in a written report intended for the state of Kentucky.

d. (Optional question for those covering Section 9.5.) For each test conducted in part c, test whether the assumption about equal population variances is satisfied. After performing these tests, adjust (if necessary) the conclusions you made in part c.

e. Repeat the analysis in parts a–d for both whole milk and low-fat chocolate milk. Summarize your conclusions and make recommendations in a written report.

### FIGURE 9.22

*Plot of Average Winning Low-Fat White Milk Price*

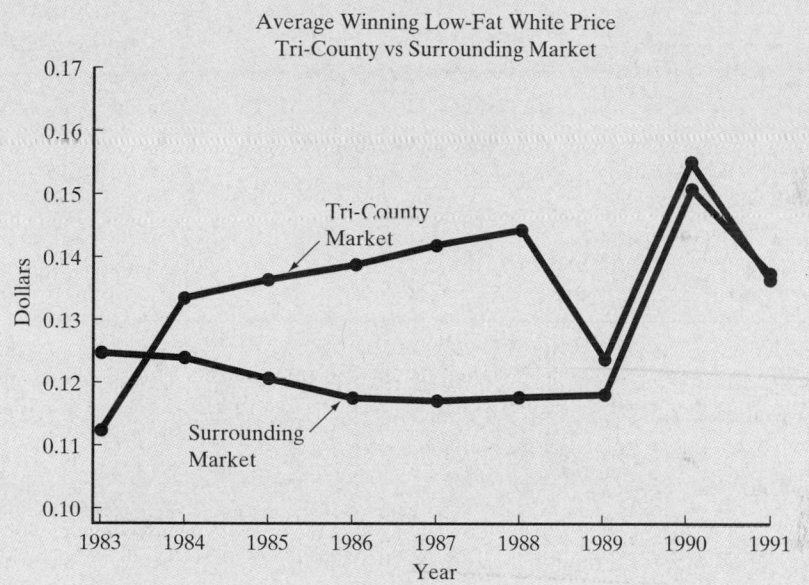

## QUICK REVIEW

### Key Terms

*Note: Starred (\*) terms are from the optional section in this chapter.*

Blocking   400
*F*-distribution\*   422
Paired difference experiment   400

Pooled sample estimator $\sigma^2$   385
Randomized block experiment   400
Standard error   382

### Key Formulas

*Note: Starred (\*) formulas are from the optional section in this chapter.*

$(1 - \alpha)100\%$ confidence interval for $\theta$: (see table)

Large samples:   $\hat{\theta} \pm z_{\alpha/2}\sigma_{\hat{\theta}}$

Small samples:   $\hat{\theta} \pm t_{\alpha/2}\sigma_{\hat{\theta}}$

For testing $H_0: \theta = D_0$: (see table)

Large samples: $z = \dfrac{\hat{\theta} - D_0}{\sigma_{\hat{\theta}}}$

Small samples: $t = \dfrac{\hat{\theta} - D_0}{\sigma_{\hat{\theta}}}$

| Parameter, $\theta$ | Estimator, $\hat{\theta}$ | Standard Error of Estimator, $\sigma_{\hat{\theta}}$ | Estimated Standard Error | |
|---|---|---|---|---|
| $\mu_1 - \mu_2$ (independent samples) | $\bar{x}_1 - \bar{x}_2$ | $\sqrt{\dfrac{\sigma_1^2}{n_1} + \dfrac{\sigma_2^2}{n_2}}$ | Large $n$: $\sqrt{\dfrac{s_1^2}{n_1} + \dfrac{s_2^2}{n_2}}$ | 382, 383 |
| | | | Small $n$: $\sqrt{s_p^2\left(\dfrac{1}{n_1} + \dfrac{1}{n_2}\right)}$ | 386 |
| $\mu_D$ (paired sample) | $\bar{x}_D$ | $\dfrac{\sigma_D}{\sqrt{n_D}}$ | $\dfrac{s_D}{\sqrt{n_D}}$ | 401, 402 |
| $p_1 - p_2$ (large, independent samples) | $\hat{p}_1 - \hat{p}_2$ | $\sqrt{\dfrac{p_1 q_1}{n_1} + \dfrac{p_2 q_2}{n_2}}$ | Confidence intervals: $\sqrt{\dfrac{\hat{p}_1 \hat{q}_1}{n_1} + \dfrac{\hat{p}_2 \hat{q}_2}{n_2}}$ | 411, 412 |
| | | | Hypothesis tests: $\sqrt{\hat{p}\hat{q}\left(\dfrac{1}{n_1} + \dfrac{1}{n_2}\right)}$ | 413 |

| | | |
|---|---|---|
| Pooled sample variance: | $s_p^2 = \dfrac{(n_1 - 1)s_1^2 + (n_2 - 1)s_2^2}{n_1 + n_2 - 2}$ | 386 |
| Pooled sample proportion: | $\hat{p} = \dfrac{x_1 + x_2}{n_1 + n_2}, \qquad \hat{q} = 1 - \hat{p}$ | 413 |
| Determining the sample size for estimating $\mu_1 - \mu_2$: | $n_1 = n_2 = \dfrac{(z_{\alpha/2})^2(\sigma_1^2 + \sigma_2^2)}{B^2}$ | 419 |
| Determining the sample size for estimating $p_1 - p_2$: | $n_1 = n_2 = \dfrac{(z_{\alpha/2})^2(p_1 q_1 + p_2 q_2)}{B^2}$ | 420 |
| *Test statistic for testing $H_0: \dfrac{\sigma_1^2}{\sigma_2^2}$ | $F = \dfrac{\text{larger } s^2}{\text{smaller } s^2} \quad \text{if} \quad H_a: \dfrac{\sigma_1^2}{\sigma_2^2} \neq 1$ | 425–426 |
| | $F = \dfrac{s_1^2}{s_2^2} \quad \text{if} \quad H_a: \dfrac{\sigma_1^2}{\sigma_2^2} > 1$ | |

## Language Lab

| Symbol | Pronunciation | Description |
|---|---|---|
| $\mu_1 - \mu_2$ | mu-1 minus mu-2 | Difference between population means |
| $\bar{x}_1 - \bar{x}_2$ | x-bar-1 minus x-bar-2 | Difference between sample means |
| $\sigma_{(\bar{x}_1 - \bar{x}_2)}$ | sigma of x-bar-1 minus x-bar-2 | Standard deviation of the sampling distribution of $(\bar{x}_1 - \bar{x}_2)$ |
| $s_p^2$ | s-p squared | Pooled sample variance |
| $D_0$ | D naught | Hypothesized value of difference |
| $\mu_D$ | mu D | Difference between population means, paired data |
| $\bar{x}_D$ | x-bar D | Mean of sample differences |
| $s_D$ | s-D | Standard deviation of sample differences |
| $n_D$ | n-D | Number of differences in sample |

| Symbol | Pronunciation | Description |
|--------|---------------|-------------|
| $p_1 - p_2$ | p-1 minus p-2 | Difference between population proportions |
| $\hat{p}_1 - \hat{p}_2$ | p-1 hat minus p-2 hat | Difference between sample proportions |
| $\sigma_{(\hat{p}_1 - \hat{p}_2)}$ | sigma of p-1 hat minus p-2 hat | Standard deviation of the sampling distribution of $(\hat{p}_1 - \hat{p}_2)$ |
| $F_\alpha$ | F-alpha | Critical value of $F$ associated with tail area $\alpha$ |
| $\nu_1$ | nu-1 | Numerator degrees of freedom for $F$ statistic |
| $\nu_2$ | nu-2 | Denominator degrees of freedom for $F$ statistic |
| $\dfrac{\sigma_1^2}{\sigma_2^2}$ | sigma-1 squared over sigma-2 squared | Ratio of two population variances |

# ✓ SUPPLEMENTARY EXERCISES 9.85–9.115

*Note: Starred (*) exercises refer to the optional section in this chapter.*

## Learning the Mechanics

**9.85** Independent random samples were selected from two normally distributed populations with means $\mu_1$ and $\mu_2$, respectively. The sample sizes, means, and variances are shown in the following table.

| Sample 1 | Sample 2 |
|----------|----------|
| $n_1 = 12$ | $n_2 = 14$ |
| $\bar{x}_1 = 17.8$ | $\bar{x}_2 = 15.3$ |
| $s_1^2 = 74.2$ | $s_2^2 = 60.5$ |

**a.** Test $H_0: (\mu_1 - \mu_2) = 0$ against $H_a: (\mu_1 - \mu_2) > 0$. Use $\alpha = .05$.

**b.** Form a 99% confidence interval for $(\mu_1 - \mu_2)$.

**c.** How large must $n_1$ and $n_2$ be if you wish to estimate $(\mu_1 - \mu_2)$ to within 2 units with 99% confidence? Assume that $n_1 = n_2$.

**9.86** Two independent random samples were selected from normally distributed populations with means and variances $(\mu_1, \sigma_1^2)$ and $(\mu_2, \sigma_2^2)$, respectively. The sample sizes, means, and variances are shown in the table.

| Sample 1 | Sample 2 |
|----------|----------|
| $n_1 = 20$ | $n_2 = 15$ |
| $\bar{x}_1 = 123$ | $\bar{x}_2 = 116$ |
| $s_1^2 = 31.3$ | $s_2^2 = 120.1$ |

***a.** Test $H_0: \sigma_1^2 = \sigma_2^2$ against $H_a: \sigma_1^2 \neq \sigma_2^2$. Use $\alpha = .05$.

**b.** Would you be willing to use a *t*-test to test the null hypothesis $H_0: (\mu_1 - \mu_2) = 0$ against the alternative hypothesis $H_a: (\mu_1 - \mu_2) \neq 0$? Why?

**9.87** Two independent random samples are taken from two populations. The results of these samples are summarized in the following table.

| Sample 1 | Sample 2 |
|----------|----------|
| $n_1 = 135$ | $n_2 = 148$ |
| $\bar{x}_1 = 12.2$ | $\bar{x}_2 = 8.3$ |
| $s_1^2 = 2.1$ | $s_2^2 = 3.0$ |

**a.** Form a 90% confidence interval for $(\mu_1 - \mu_2)$.

**b.** Test $H_0: (\mu_1 - \mu_2) = 0$ against $H_a: (\mu_1 - \mu_2) \neq 0$. Use $\alpha = .01$.

**c.** What sample size would be required if you wish to estimate $(\mu_1 - \mu_2)$ to within .2 with 90% confidence? Assume that $n_1 = n_2$.

**9.88** Independent random samples were selected from two binomial populations. The size and number of observed successes for each sample are shown in the table.

| Sample 1 | Sample 2 |
|----------|----------|
| $n_1 = 200$ | $n_2 = 200$ |
| $x_1 = 110$ | $x_2 = 130$ |

**a.** Test $H_0: (p_1 - p_2) = 0$ against $H_a: (p_1 - p_2) < 0$. Use $\alpha = .10$.

**b.** Form a 95% confidence interval for $(p_1 - p_2)$.

**c.** What sample sizes would be required if we wish to use a 95% confidence interval of width .01 to estimate $(p_1 - p_2)$?

**9.89** A random sample of five pairs of observations were selected, one observation from a population with mean $\mu_1$, the other from a population with mean $\mu_2$. The data are shown in the accompanying table.

🌐 LM9_89

| Pair | Value from Population 1 | Value from Population 2 |
|------|------------------------|-------------------------|
| 1 | 28 | 22 |
| 2 | 31 | 27 |
| 3 | 24 | 20 |
| 4 | 30 | 27 |
| 5 | 22 | 20 |

**a.** Test the null hypothesis $H_0: \mu_D = 0$ against $H_a: \mu_D \neq 0$, where $\mu_D = \mu_1 - \mu_2$. Use $\alpha = .05$.

**b.** Form a 95% confidence interval for $\mu_D$.

**c.** When are the procedures you used in parts **a** and **b** valid?

**9.90** List the assumptions necessary for each of the following inferential techniques:
  **a.** Large-sample inferences about the difference $(\mu_1 - \mu_2)$ between population means using a two-sample $z$ statistic
  **b.** Small-sample inferences about $(\mu_1 - \mu_2)$ using an independent samples design and a two-sample $t$ statistic
  **c.** Small-sample inferences about $(\mu_1 - \mu_2)$ using a paired difference design and a single-sample $t$ statistic to analyze the differences
  **d.** Large-sample inferences about the differences $(p_1 - p_2)$ between binomial proportions using a two-sample $z$ statistic
  *__e.__ Inferences about the ratio $\sigma_1^2 / \sigma_2^2$ of two population variances using an $F$ test.

**9.91** Suppose you use the SPSS statistical program package to test the null hypothesis $H_0 : (\mu_1 - \mu_2) = 0$ against $H_a : (\mu_1 - \mu_2) \neq 0$ with independent samples of size 12 and 10, respectively. Assuming you are using $\alpha = .05$, what conclusion would you reach in each of the following instances of observed significance level reported by the program? (Recall that SPSS reports the two-tailed $p$-value.)
  **a.** .0429
  **b.** .1984
  **c.** .0001
  **d.** .0344
  **e.** .0545
  **f.** .9633
  **g.** What assumptions are necessary to ensure the validity of this test?

## Applying the Concepts—Basic

⊘ OILSPILL

**9.92** Refer to the *Marine Technology* (Jan. 1995) study of major oil spills from tankers and carriers, Exercise 2.15 (p. 29). The data for the 50 spills are saved in the OILSPILL file.
  **a.** Construct a 90% confidence interval for the difference between the mean spillage amount of accidents caused by collision and the mean spillage amount of accidents caused by fire/explosion. Interpret the result.
  **b.** Conduct a test of hypothesis to compare the mean spillage amount of accidents caused by grounding to the corresponding mean of accidents caused by hull failure. Use $\alpha = .05$.
  **c.** Refer to parts **a** and **b**. State any assumptions required for the inferences derived from the analyses to be valid. Are these assumptions reasonably satisfied?
  *__d.__ Conduct a test of hypothesis to compare the variation in spillage amounts for accidents caused by collision and accidents caused by grounding. Use $\alpha = .02$.

**9.93** Nontraditional university students are generally defined as those at least 25 years old. A study reported in the *College Student Journal* (Dec. 1992) compared tradi-

tional and nontraditional students on a number of factors, including grade point average (GPA). The table summarizes the information from the sample.

|  | GPA | |
|---|---|---|
|  | **Traditional Students** | **Nontraditional Students** |
| $n$ | 94 | 73 |
| $\bar{x}$ | 2.90 | 3.50 |
| $s$ | .50 | .50 |

  **a.** What are the appropriate null and alternative hypotheses if we want to test whether the mean GPAs of traditional and nontraditional students differ?
  **b.** Conduct the test using $\alpha = .01$, and interpret the result.
  **c.** What assumptions are necessary to ensure the validity of the test?

**9.94** Does marriage hinder the career progression of women to a greater degree than that of men? According to *The Journal of Applied Psychology* (Vol. 77, 1992), in a random sample of 795 male managers and 223 female managers from 20 Fortune 500 corporations, 86% of the male managers and 45% of the female managers were married.
  **a.** Use a 95% confidence interval to estimate the difference between the proportions of men and women managers who are married.
  **b.** Interpret the interval. State your conclusion in terms of the populations from which the samples were drawn.

**9.95** Dr. Philip Lieberman, a neuroscientist at Brown University, recently conducted a field experiment to gauge the effect of high altitude on a person's ability to think critically (*New York Times*, Aug. 23, 1995). The subjects of the experiment were five male members of an American expedition climbing Mount Everest. At the base camp, Lieberman read sentences to the climbers while they looked at simple pictures in a book. The length of time (in seconds) it took for each climber to match the picture with a sentence was recorded. Using a radio, Lieberman repeated the task when the climbers reached a camp 5 miles above sea level. At this altitude, he noted that the climbers took 50% longer to complete the task.
  **a.** What is the variable measured for this experiment?
  **b.** What are the experimental units?
  **c.** Discuss how the data should be analyzed.

*__9.96__ A medical researcher wants to compare the variability of hemoglobin in the blood of men and women who are between the ages of 20 and 30. In random samples of 25 women and 20 men in this age group, the researcher recorded the amount of hemoglobin in the blood of each. (The quantity of the hemoglobin is measured as a percentage of the total volume of blood.) The results are summarized in the table on p. 435. Use a hypothesis test to compare the variance of the amount of hemoglobin in women's blood to the corresponding variance for men.

| Women | Men |
|---|---|
| $n_1 = 25$ | $n_2 = 20$ |
| $\bar{x}_1 = 42.7$ | $\bar{x}_2 = 41.8$ |
| $s_1^2 = 18.3$ | $s_2^2 = 8.5$ |

**9.97** A plant that purifies its liquid waste discharges the water into a local river. An EPA inspector has collected six water specimens of the discharge of the plant and six water specimens in the river upstream from the plant. The bacteria counts for each of the 12 specimens are reported in the accompanying table.

🔘 BACTERIA

| Plant Discharge | | | Upstream | | |
|---|---|---|---|---|---|
| 30 | 36 | 33 | 29 | 30 | 26 |
| 28 | 29 | 34 | 27 | 31 | 32 |

  **a.** Why might the bacteria counts shown tend to be approximately normally distributed?
  **b.** What are the appropriate null and alternative hypotheses to test whether the mean bacteria count for the plant discharge exceeds that for the upstream location? Be sure to define any symbols you use.
  **c.** The data are analyzed using SPSS. Carefully interpret the SPSS output shown below.
  **d.** What assumptions are necessary to ensure the validity of this test?

**9.98** Some power plants are located near rivers or oceans so that the available water can be used for cooling the condensers. Suppose that, as part of an environmental impact study, a power company wants to estimate the difference in mean water temperature between the discharge of its plant and the offshore waters. How many sample measurements must be taken at each site in order to estimate the true difference between means to within .2°C with 95% confidence? Assume that the range in readings will be about 4°C at each site and the same number of readings will be taken at each site.

**9.99** A study in the *Journal of Psychology & Marketing* (Jan. 1992) investigated the degree to which consumers are concerned about product tampering. Large random samples of male and female consumers were asked to rate their concern about product tampering on a scale of 1 (little or no concern) to 9 (very concerned).
  **a.** What are the appropriate null and alternative hypotheses to determine whether a difference exists in the mean level of concern about product tampering between men and women? Define any symbols you use.
  **b.** The statistics reported include those shown in the MINITAB printout on p. 436. Interpret these results.
  **c.** What assumptions are necessary to ensure the validity of this test?

**9.100** Consider a study conducted to compare the abilities of men and women to perform the strenuous tasks required of a shipboard firefighter (*Human Factors*, Vol. 24, 1982). The researchers measured the pulling force (in newtons) that a firefighter was able to exert in pulling the starter cord of a P-250 water pump.

**SPSS Output for Exercise 9.97**

**Group Statistics**

| | LOCATION | N | Mean | Std. Deviation | Std. Error Mean |
|---|---|---|---|---|---|
| BACTERIA | PLANT | 6 | 31.67 | 3.14 | 1.28 |
| | UPSTREAM | 6 | 29.17 | 2.32 | .95 |

**Independent Samples Test**

| | | Levene's Test for Equality of Variances | | t-test for Equality of Means | | | | | 95% Confidence Interval of the Difference | |
|---|---|---|---|---|---|---|---|---|---|---|
| | | F | Sig. | t | df | Sig. (2-tailed) | Mean Difference | Std. Error Difference | Lower | Upper |
| BACTERIA | Equal variances assumed | 1.563 | .240 | 1.569 | 10 | .148 | 2.50 | 1.59 | -1.05 | 6.05 |
| | Equal variances not assumed | | | 1.569 | 9.197 | .150 | 2.50 | 1.59 | -1.09 | 6.09 |

**MINITAB Output for Exercise 9.99**

```
Two-sample T for TAMPER

                N       Mean      StDev     SE Mean
MSCORE        200        3.21       2.33       0.165
FSCORE        200        3.92       2.94       0.208


Difference = mu (MSCORE) - mu (FSCORE)
Estimate for difference: -0.71
95% CI for difference: (-0.85, -0.57)
T-Test of difference = 0 (vs not =): T-Value = -2.69   P-Value = 0.007
DF = 398
```

Firefighters were matched in pairs according to weight, thus producing data for a matched pairs (or paired difference) experiment, as shown below.

⬤ PULLFORCE

| Pair | Female | Male |
|---|---|---|
| 1 | 40.03 | 84.51 |
| 2 | 75.62 | 80.06 |
| 3 | 53.38 | 102.30 |
| 4 | 62.27 | 88.96 |

*Source:* Phillips, M.D., and Pepper, R.L., "Shipboard firefighting performance of females and males." *Human Factors*, Vol. 24, No. 3, 1982.

a. Do the data provide sufficient evidence to indicate a difference in mean pulling force between female and male firefighters? Test using $\alpha = .05$

b. Find a 95% confidence interval for the difference in mean pulling force between female and male firefighters. Interpret the interval.

9.101 Does the time of day during which one works affect job satisfaction? A study in *The Journal of Occupational Psychology* (Sept. 1991) examined differences in job satisfaction between day-shift and night-shift nurses. Nurses' satisfaction with their hours of work, free time away from work, and breaks during work were measured. The table shows the mean scores for each measure of job satisfaction (higher scores indicate greater satisfaction), along with the observed significance level comparing the means for the day-shift and night-shift samples:

| | Day Shift | Night Shift | *p*-value |
|---|---|---|---|
| Mean Satisfaction with: | | | |
| Hours of work | 3.91 | 3.56 | .813 |
| Free time | 2.55 | 1.72 | .047 |
| Breaks | 2.53 | 3.75 | .0073 |

a. Specify the null and alternative hypotheses if we wish to test whether a difference in job satisfaction exists between day-shift and night-shift nurses on each of the three measures. Define any symbols you use.

b. Interpret the *p*-value for each of the tests. (Each of the *p*-values in the table is two-tailed.)

c. Assume that each of the tests is based on small samples of nurses from each group. What assumptions are necessary for the tests to be valid?

**Applying the Concepts—Intermediate**

9.102 A pupillometer is a device used to observe changes in an individual's pupil dilations as he or she is exposed to different visual stimuli. The Design and Market Research Laboratories of the Container Corporation of America used a pupillometer to evaluate consumer reaction to different silverware patterns for one of its clients. Suppose 15 consumers were chosen at random, and each was shown two different silverware patterns. The pupillometer readings (in millimeters) for each consumer are shown below.

⬤ PUPILL

| Consumer | Pattern 1 | Pattern 2 |
|---|---|---|
| 1 | 1.00 | .80 |
| 2 | .97 | .66 |
| 3 | 1.45 | 1.22 |
| 4 | 1.21 | 1.00 |
| 5 | .77 | .81 |
| 6 | 1.32 | 1.11 |
| 7 | 1.81 | 1.30 |
| 8 | .91 | .32 |
| 9 | .98 | .91 |
| 10 | 1.46 | 1.10 |
| 11 | 1.85 | 1.60 |
| 12 | .33 | .21 |
| 13 | 1.77 | 1.50 |
| 14 | .85 | .65 |
| 15 | .15 | .05 |

a. Which type of experiment does this represent—independent samples or paired difference? Explain.

b. Use a 90% confidence interval to estimate the difference in mean pupil dilation per consumer for silverware patterns 1 and 2. Interpret the confidence interval, assuming that the pupillometer indeed measures consumer interest.

**c.** Test the hypothesis that the mean dilation differs for the two patterns. Use $\alpha = .10$. Does the test conclusion support your interpretation of the confidence interval in part **b**?

**d.** What assumptions are necessary to ensure the validity of the inferences in parts **b** and **c**?

**9.103** A traditional approach to teaching basic nursing skills was compared with an innovative approach in the *Journal of Nursing Education* (Jan. 1992). The innovative approach utilizes Vee heuristics and concept maps to link theoretical concepts with practical skills. Forty-two students enrolled in an upper-division nursing course participated in the study. Half (21) were randomly assigned to labs that utilized the innovative approach. After completing the course, all students were given short-answer questions about scientific principles underlying each of 10 nursing skills. The objective of the research is to compare the mean scores of the two groups of students.

**a.** What is the appropriate test to use to compare the two groups?

**b.** Are any assumptions required for the test?

**c.** One question dealt with the use of clean/sterile gloves. The mean scores for this question were 3.28 (traditional) and 3.40 (innovative). Is there sufficient information to perform the test?

**d.** Refer to part **c**. The *p*-value for the test was reported as $p = .79$. Interpret this result.

**e.** Another question concerned the choice of a stethoscope. The mean scores of the two groups were 2.55 (traditional) and 3.60 (innovative) with an associated *p*-value of .02. Interpret these results.

**9.104** *Arctic and Alpine Research* (Vol. 14, 1982) investigated the relationship between the mean daily air temperature and the cocoon temperature of wooly-bear caterpillars of the High Arctic. The data in the table indicate that the caterpillar's body temperature (inside the cocoon) is higher than the outside air temperature. Estimate the mean difference in temperature between the cocoon and the outside air. Use a 95% confidence interval. Interpret the result.

**⊘ WOOLYCAT**

| Day | Air | Cocoon[a] | Day | Air | Cocoon[a] |
|---|---|---|---|---|---|
| 1 | 10.4 | 15.1 | 7 | 1.7 | 3.6 |
| 2 | 9.2 | 14.6 | 8 | 2.0 | 5.3 |
| 3 | 2.2 | 6.8 | 9 | 3.0 | 7.0 |
| 4 | 2.6 | 6.8 | 10 | 3.5 | 7.1 |
| 5 | 4.1 | 8.0 | 11 | 4.5 | 9.6 |
| 6 | 3.7 | 8.7 | 12 | 4.4 | 9.5 |

[a] Each cocoon temperature is the average of the temperatures of two cocoons.
*Source:* Kevan, P. G., Jensen, T. S., and Shorthouse, J. D. "Body temperatures and behavioral thermoregulation of High Arctic wooly-bear caterpillars and pupae (*Gynaephora rossii*, Lymantridae: Lepidoptera) and the importance of sunshine." *Arctic and Alpine Research*, 1982, 14.

**9.105** Can a person control certain body functions if that person is trained in a program of *biofeedback* exercises? An experiment is conducted to show that blood pressure levels can be consciously reduced in people trained in this program. The blood pressure measurements (in millimeters of mercury) listed in the table represent readings before and after the biofeedback training of six subjects.

**⊘ BIOFEED**

| Subject | Before | After |
|---|---|---|
| 1 | 136.9 | 130.2 |
| 2 | 201.4 | 180.7 |
| 3 | 166.8 | 149.6 |
| 4 | 150.0 | 153.2 |
| 5 | 173.2 | 162.6 |
| 6 | 169.3 | 160.1 |

**a.** If we want to test whether the mean blood pressure decreases after the training, what are the appropriate null and alternative hypotheses? Define any symbols you use.

**b.** Conduct the test of part **a** at $\alpha = .05$. Interpret these results.

**c.** Find a 95% confidence interval for the parameter of interest. Interpret this interval.

**9.106** Excess postexercise oxygen consumption (EPOC) describes energy expended during the body's recovery period immediately following aerobic exercise. *The Journal of Sports Medicine and Physical Fitness* (Dec. 1994) published a study designed to investigate the effect of fitness level on the magnitude and duration of EPOC. Ten healthy young adult males volunteered for the study. Five of these were endurance-trained and comprised the fit group; the other five were not engaged in any systematic training and comprised the sedentary group. Each volunteer engaged in a weight-supported exercise on a cycle ergometer until 300 kilocalories were expended. The magnitude (in kilocalories) and duration (in minutes) of the EPOC of each exerciser were measured. The study results are summarized in the table below.

| Variable | | Fit $(n = 5)$ | Sedentary $(n = 5)$ | *p*-value |
|---|---|---|---|---|
| Magnitude (kcal) | Mean | 12.2 | 12.2 | .998 |
| | Std. dev. | 3.1 | 4.3 | |
| Duration (min) | Mean | 16.6 | 20.4 | .344 |
| | Std. dev. | 3.1 | 7.8 | |

*Source:* Sedlock, D. A. "Fitness levels and postexercise energy expenditure." *The Journal of Sports Medicine and Physical Fitness*, Vol. 34, No. 4, Dec. 1994, p. 339 (Table III).

**a.** Conduct a test of hypothesis to determine whether the true mean magnitude of EPOC differs for fit and sedentary young adult males. Use $\alpha = .10$.

**b.** The *p*-value for the test, part **a**, is given in the table. Interpret this value.

**c.** Conduct a test of hypothesis to determine whether the true mean duration of EPOC differs for fit and sedentary young adult males. Use $\alpha = .10$.

**d.** The *p*-value for the test, part **c**, is given in the table. Interpret this value.

**9.107** Inositol—a complex alcohol nutrient found in breast milk—has been found to reduce the risk of lung and eye damage in premature infants. The study, published in the *New England Journal of Medicine* (May 7, 1992), involved 220 premature infants. The infants were randomly divided into two groups; half (110) of the infants received an intravenous feeding of inositol, while the other half received a standard diet. The researchers found that 14 of the inositol-fed infants suffered retinopathy of prematurity, an eye injury that results from the high oxygen levels needed to compensate for poorly developed lungs. In contrast, 29 of the infants on the standard diet developed this disease. Analyze the results with a test of hypothesis. Use $\alpha = .01$.

**9.108** Operation Crossroads was a 1946 military exercise in which atomic bombs were detonated over empty target ships in the Pacific Ocean. The Navy assigned sailors to wash down the test ships immediately after the atomic blasts. The National Academy of Science reported "that the overall death rate among Operation Crossroads sailors was 4.6% higher than among a comparable group of sailors. . . . However, this increase was not statistically significant." (*Tampa Tribune*, Oct. 30, 1996.)

**a.** Describe the parameter of interest in the National Academy of Science study.

**b.** Interpret the statement: "This increase was not statistically significant."

**9.109** Students enrolled in music classes at the University of Texas (Austin) participated in a study to compare the observations and teacher evaluations of music education majors and non-music majors (*Journal of Research in Music Education*, Winter 1991). Independent random samples of 100 music majors and 100 nonmajors rated the overall performance of their teacher using a 6-point scale, where 1 = lowest rating and 6 = highest rating. Use the information in the table to compare the mean teacher ratings of the two groups of music students with a 95% confidence interval. Interpret the result.

| | **Music Majors** | **Non-Music Majors** |
|---|---|---|
| Sample size | 100 | 100 |
| Mean "overall" rating | 4.26 | 4.59 |
| Standard deviation | .81 | .78 |

*Source:* Duke, R. A., and Blackman, M. D. "The relationship between observers' recorded teacher behavior and evaluation of music instruction." *Journal of Research in Music Education*, Vol. 39, No. 4, Winter 1991 (Table 2).

**9.110** For each of the following studies, give the parameter of interest and state any assumptions that are necessary for the inferences to be invalid.

**a.** To investigate a possible link between jet lag and memory impairment, a University of Bristol (England) neurologist recruited 20 female flight attendants who worked flights across several time zones. Half of the attendants had only a short recovery time between flights and half had a long recovery time between flights. The average size of the right temporal lobe of the brain for the short-recovery group was significantly smaller than the average size of the right temporal lobe of the brain for the long-recovery group (*Tampa Tribune*, May 23, 2001).

**b.** In a study presented at the March, 2001, meeting of the Association for the Advancement of Applied Sport Psychology, researchers revealed that the proportion of athletes who have a good self-image of their body is 20% higher than the corresponding proportion of non-athletes.

**c.** A University of Florida animal sciences professor has discovered that feeding chickens corn oil causes them to produce larger eggs (*UF News*, April 11, 2001). The weight of eggs produced by each of a sample of chickens on a regular feed diet was recorded. Then, the same chickens were fed a diet supplemented by corn oil, and the weight of eggs produced by each was recorded. The mean weight of the eggs produced with corn oil was 3 grams heavier than the mean weight produced with the regular diet.

**\*9.111** The quality control department of a paper company measures the brightness (a measure of reflectance) of finished paper on a periodic basis throughout the day. Two instruments that are available to measure the paper specimens are subject to error, but they can be adjusted so that the mean readings for a control paper specimen are the same for both instruments. Suppose you are concerned about the precision of the two instruments and want to compare the variability in the readings of instrument 1 to those of instrument 2. Five brightness measurements were made on a single paper specimen using each of the two instruments. The data are shown in the table.

🖉 BRIGHT

| **Instrument 1** | **Instrument 2** |
|---|---|
| 29 | 26 |
| 28 | 34 |
| 30 | 30 |
| 28 | 32 |
| 30 | 28 |

**a.** Is the variance of the measurements obtained by instrument 1 significantly different from the variance of the measurements obtained using instrument 2?

**b.** What assumptions must be satisfied for the test in part **a** to be valid?

| | Contra Costa | Santa Clara | Los Angeles | San Bernardino |
|---|---|---|---|---|
| Sample size | 521 | 556 | 337 | 372 |
| Number with earthquake insurance | 117 | 222 | 133 | 109 |

**9.112** An article in the *Annals of the Association of American Geographers* (June 1992) explored some factors that are considered when California homeowners purchase earthquake insurance, including the proximity to a major earthquake fault. Surveys were mailed to residents in four California counties. A summary of the data collected is shown in the table above.

   **a.** Los Angeles County is the closest of the four to a major earthquake fault. Calculate 95% confidence intervals for the difference in the proportions of earthquake-insured residents in Los Angeles County and each of the other counties.

   **b.** Do these results support the contention that closer proximities to major earthquake faults result in higher proportions of earthquake-insured residents?

   **c.** How large would the samples from Los Angeles and San Bernardino counties have to be in order to estimate the difference between earthquake-insured proportions to within .03 with 95% confidence?

**9.113** Doctors have theorized that renegade antibodies in blood may be a major underlying cause of stroke. To investigate the causal link between antibodies and stroke, medical researchers examined two groups of patients for renegade antibodies (*Tampa Tribune*, Feb. 2, 1992). One group of 255 patients had suffered a first stroke while the other group of 257 patients were hospitalized for other reasons. Twenty-five of the stroke patients had antibodies in their blood compared to 11 of the other patients. Analyze the data for the medical researchers. What conclusions can you draw from the analysis?

**9.114** Marine biochemists at the University of Tokyo studied the properties of crustacean striated muscles (*The Journal of Experimental Zoology*, Sept. 1993). The main purpose of the experiment was to compare the biochemical properties of fast and slow muscles of crayfish. Using crayfish obtained from a local supplier, the researchers excised twelve fast-muscle fiber bundles and tested each fiber bundle for uptake of calcium. Twelve slow-muscle fiber bundles were excised from a second sample of crayfish, and calcium uptake was measured. The results of the experiment are summarized here. (All calcium measurements are in moles per milligram.) Analyze the data using a 95% confidence interval. Make an inference about the difference between the calcium uptake means of fast and slow muscles.

| Fast Muscle | Slow Muscle |
|---|---|
| $n_1 = 12$ | $n_2 = 12$ |
| $\bar{x}_1 = .57$ | $\bar{x}_2 = .37$ |
| $s_1 = .104$ | $s_2 = .035$ |

*Source:* Ushio, H., and Watabe, S. "Ultrastructural and biochemical analysis of the sarcoplasmic reticulum from crayfish fast and slow striated muscles." *The Journal of Experimental Zoology*, Vol. 267, Sept. 1993, p. 16 (Table 1). Copyright © 1993 John Wiley & Sons, Inc. Reprinted with permission.

## Applying the Concepts—Advanced

**9.115** Researchers at Rochester Institute of Technology investigated the use of isolation timeout as a behavioral management technique (*Exceptional Children*, Feb. 1995). Subjects for the study were 155 emotionally disturbed students enrolled in a special education facility. The students were randomly assigned to one of two types of classrooms—Option II classrooms (one teacher, one paraprofessional, and a maximum of 12 students) and Option III classrooms (one teacher, one paraprofessional, and a maximum of 6 students). Over the academic year the number of behavioral incidents resulting in an isolation timeout was recorded for each student. Summary statistics for the two groups of students are shown in the following table.

| | Option II | Option III |
|---|---|---|
| Number of students | 100 | 55 |
| Mean number of timeout incidents | 78.67 | 102.87 |
| Standard deviation | 59.08 | 69.33 |

*Source:* Costenbader, V., and Reading-Brown, M. "Isolation timeout used with students with emotional disturbance." *Exceptional Children*, Vol. 61, No.4, Feb. 1995, p. 359 (Table 3).

   Do you agree with the following statement: "On average, students in Option III classrooms had significantly more timeout incidents than students in Option II classrooms?"

## STUDENT PROJECTS

We have now discussed two methods of collecting data to compare two population means. In many experimental situations a decision must be made either to collect two independent samples or to conduct a paired difference experiment.

The importance of this decision cannot be overemphasized, since the amount of information obtained and the cost of the experiment are both directly related to the method of experimentation that is chosen.

Choose two populations (pertinent to your major area) which have unknown means and for which you could both collect two independent samples and collect paired observations. Before conducting the experiment, state which method of sampling you think will provide more information (and why). Compare the two methods, first performing the independent sampling procedure by collecting 10 observations from each population (a total of 20 measurements), then performing the paired difference experiment by collecting 10 pairs of observations.

Construct two 95% confidence intervals, one for each experiment you conduct. Which method provides the narrower confidence interval and hence more information on this performance of the experiment? Does this result agree with your preliminary expectations?

## REFERENCES

Mendenhall, W., Beaver, R. J., and Beaver, B. M. *Introduction to Probability and Statistics*, 10th ed. North Scituate, Mass.: Duxbury, 1999.

Satterthwaite, F. W. "An approximate distribution of estimates of variance components." *Biometrics Bulletin*, Vol. 2, 1946, pp. 110–114.

Snedecor, G. W., and Cochran, W. *Statistical Methods*, 7th ed. Ames: Iowa State University Press, 1980.

Steel, R. G. D., and Torrie, J. H. *Principles and Procedures of Statistics*, 2nd ed. New York: McGraw-Hill, 1980.

# CHAPTER 10

# Analysis of Variance:

## COMPARING MORE THAN TWO MEANS

### Contents

### Statistics in Action

On the Trail of the Cockroach

### ✊ Where We've Been

As we've seen in preceding chapters, the solutions of many practical problems are based on inferences about population means. Methods for estimating and testing hypotheses about a single mean and the comparison of two means were presented in Chapters 7–9.

### ☞ Where We're Going

In this chapter we extend the methodology of Chapters 7–9 in two important ways. First, we discuss the critical elements in the *design* of a sampling experiment. Then we see how to *analyze* the experiment in order to compare more than two populations. We'll look at several of the more popular experimental designs, and we'll see how to use the computer to analyze designed experiments using analysis of variance software.

Most of the data analyzed in previous chapters were collected in *observational* sampling experiments rather than *designed* sampling experiments. In *observational experiments* the analyst has little or no control over the variables under study and merely observes their values. In contrast, *designed experiments* are those in which the analyst attempts to control the levels of one or more variables to determine their effect on a variable of interest. Although many practical situations do not present the opportunity for such control, it is instructive, even for observational experiments, to have a working knowledge of the analysis and interpretation of data that result from designed experiments and to know the basics of how to design experiments when the opportunity arises.

We first present the basic elements of an experimental design in Section 10.1. We then discuss two of the simpler, and more popular, experimental designs in Sections 10.2 and 10.4. Slightly more complex experiments are discussed in Section 10.5.

## 10.1 ELEMENTS OF A DESIGNED EXPERIMENT

Certain elements are common to almost all designed experiments, regardless of the specific area of application. For example, the *response* is the variable of interest in the experiment. The response might be the SAT scores of a high school senior, the total sales of a firm last year, or the total income of a particular household this year. We will also refer to the response as the *dependent variable*.

> **DEFINITION 10.1**
> The **response variable** is the variable of interest to be measured in the experiment. We also refer to the response as the **dependent variable**.

The intent of most statistical experiments is to determine the effect of one or more variables on the response. These variables are usually referred to as the *factors* in a designed experiment. Factors are either *quantitative* or *qualitative*, depending on whether the variable is measured on a numerical scale or not. For example, we might want to explore the effect of the qualitative factor Gender on the response SAT score. In other words, we want to compare the SAT scores of male and female high school seniors. Or, we might wish to determine the effect of the quantitative factor Number of sales people on the response Total sales for retail firms. Often two or more factors are of interest. For example, we might want to determine the effect of the quantitative factor Number of wage earners and the qualitative factor Location on the response Household income.

> **DEFINITION 10.2**
> **Factors** are those variables whose effect on the response is of interest to the experimenter. **Quantitative factors** are measured on a numerical scale, whereas **qualitative factors** are those that are not (naturally) measured on a numerical scale.

*Levels* are the values of the factors that are utilized in the experiment. The levels of qualitative factors are usually nonnumerical. For example, the levels of Gender are Male and Female, and the levels of Location might be North, East, South, and West.* The levels of quantitative factors are the numerical values of the variable utilized in the experiment. The Number of sales people for each of a set of companies, the Number of wage earners in each of a set of households, and the GPAs for a set of high school seniors all represent levels of the respective quantitative factors.

> **DEFINITION 10.3**
> **Factor levels** are the values of the factor utilized in the experiment.

When a *single factor* is employed in an experiment, the *treatments* of the experiment are the levels of the factor. For example, if the effect of the factor Gender on the response SAT score is being investigated, the treatments of the experiment are the two levels of Gender—Female and Male. Or, if the effect of the Number of wage earners on Household income is the subject of the experiment, the numerical values assumed by the quantitative factor Number of wage earners are the treatments. If *two or more factors* are utilized in an experiment, the treatments are the factor–level combinations used. For example, if the effects of the factors Gender and GPA on the response SAT score are being investigated, the treatments are the combinations of the levels of Gender and GPA used; thus (Female, 2.61), (Male, 3.43), and (Female, 3.82) would all be treatments.

> **DEFINITION 10.4**
> The **treatments** of an experiment are the factor–level combinations utilized.

The objects on which the response variable and factors are observed are the *experimental units*. For example, SAT score, High school GPA, and Gender are all variables that can be observed on the same experimental unit—a high school senior. Or, the Total sales, the Earnings per share, and the Number of salespeople can be measured on a particular firm in a particular year, and the firm–year combination is the experimental unit. The Total income, the Number of female wage earners, and the Location can be observed for a household at a particular point in time, and the household–time combination is the experimental unit. Every experiment, whether observational or designed, has experimental units on which the variables are observed. However, the identification of the experimental units is more important in designed experiments, when the experimenter must actually sample the experimental units and measure the variables.

> **DEFINITION 10.5**
> An **experimental unit** is the object on which the response and factors are observed or measured.[†]

---

*The levels of a qualitative variable may bear numerical labels. For example, the Locations could be numbered 1, 2, 3, and 4. However, in such cases the numerical labels for a qualitative variable will usually be codes representing non-numerical levels.

[†] Recall (Chapter 1) that the set of all experimental units is the population.

When the specification of the treatments and the method of assigning the experimental units to each of the treatments is controlled by the analyst, the experiment is said to be *designed*. In contrast, if the analyst is just an observer of the treatments on a sample of experimental units, the experiment is *observational*. For example, if you specify the number of female and male high school students within each GPA range to be randomly selected in order to evaluate the effect of gender and GPA on SAT scores, you are designing the experiment. If, on the other hand, you simply observe the SAT scores, gender, and GPA for all students who took the SAT test last month at a particular high school, the experiment is observational.

---

**DEFINITION 10.6**

A **designed experiment** is one for which the analyst controls the specification of the treatments and the method of assigning the experimental units to each treatment. An **observational experiment** is one for which the analyst simply observes the treatments and the response on a sample of experimental units.

---

The diagram in Figure 10.1 provides an overview of the experimental process and a summary of the terminology introduced in this section. Note that the experimental unit is at the core of the process. The method by which the sample of experimental units is selected from the population determines the type of experiment. The level of every factor (the treatment) and the response are all variables that are observed or measured on each experimental unit.

**EXAMPLE 10.1**

The USGA (United States Golf Association) regularly tests golf equipment to ensure that it conforms to USGA standards. Suppose it wishes to compare the mean distance traveled by four different brands of golf balls when struck by a driver

**FIGURE 10.1**

*Sampling Experiment: Process and Terminology*

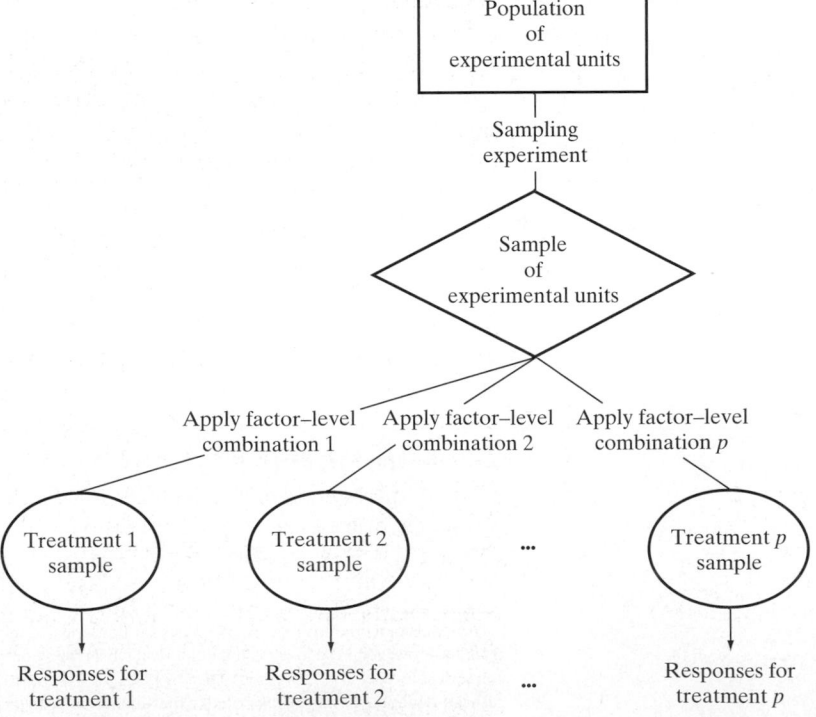

(the club used to maximize distance). The following experiment is conducted: 10 balls of each brand are randomly selected. Each is struck by "Iron Byron" (the USGA's golf robot named for the famous golfer, Byron Nelson) using a driver, and the distance traveled is recorded. Identify each of the following elements in this experiment: response, factors, factor types, levels, treatments, and experimental units.

### Solution

The response is the variable of interest, Distance traveled. The only factor being investigated is Brand of golf ball, and it is nonnumerical and therefore qualitative. The four brands (say A, B, C, and D) represent the levels of this factor. Since only one factor is utilized, the treatments are the four levels of this factor—that is, the four brands. The experimental unit is a golf ball; more specifically, it is a golf ball at a particular position in the striking sequence, since the distance traveled can be recorded only when the ball is struck, and we would expect the distance to be different (due to random factors such as wind resistance, landing place, and so forth) if the same ball is struck a second time. Note that 10 experimental units are sampled for each treatment, generating a total of 40 observations.

    This experiment, like many real applications, is a blend of designed and observational: The analyst cannot control the assignment of the brand to each golf ball (observational), but he or she can control the assignment of each ball to the position in the striking sequence (designed). ∎

**EXAMPLE 10.2**

Suppose the USGA is also interested in comparing the mean distances the four brands of golf balls travel when struck by a five-iron and by a driver. Ten balls of each brand are randomly selected, five to be struck by the driver, and five by the five-iron. Identify the elements of the experiment, and construct a schematic diagram similar to Figure 10.1 to provide an overview of this experiment.

### Solution

The response is the same as in Example 10.1—Distance traveled. The experiment now has two factors, Brand of golf ball and Club utilized. There are four levels of Brand (A, B, C, and D) and two of Club (driver and five-iron, or 1 and 5). Treatments are factor–level combinations, so there are $4 \times 2 = 8$ treatments in this experiment: (A, 1), (A, 5), (B, 1), (B, 5), (C, 1), (C, 5), (D, 1), and (D, 5). The experimental units are still the combinations of golf ball and hitting position. Note that five experimental units are sampled per treatment, generating 40 observations. The experiment is summarized in Figure 10.2. ∎

    Our objective in designing an experiment is usually to maximize the amount of information obtained about the relationship between the treatments and the response. Of course, we are almost always subject to constraints on budget, time, and even the availability of experimental units. Nevertheless, designed experiments are generally preferred to observational experiments. Not only do we have better control of the amount and quality of the information collected, but we also avoid the biases inherent in observational experiments in the selection of the experimental units representing each treatment. Inferences based on observational experiments always carry the implicit assumption that the sample has no hidden bias that was not considered in the statistical analysis. Better understanding of the potential problems with observational experiments is a by-product of our study of experimental design in the remainder of this chapter.

**FIGURE 10.2**

*Two-Factor Golf Experiment
Summary: Example 10.2*

---

 EXERCISES 10.1–10.10

### Learning the Mechanics

**10.1** What are the treatments for a designed experiment that utilizes one qualitative factor with four levels—A, B, C, and D?

**10.2** What are the treatments for a designed experiment with two factors, one qualitative with two levels (A and B) and one quantitative with five levels (50, 60, 70, 80, and 90)?

**10.3** What is the difference between an observational and a designed experiment?

### Applying the Concepts—Basic

**10.4** What are the experimental units on which each of the following responses are observed?
   **a.** GPA of a college student
   **b.** Household income
   **c.** Your time in running the 100-yard dash
   **d.** A patient's reaction to a new drug

**10.5** Brief descriptions of a number of experiments are given next. Determine whether each is observational or designed, and explain your reasoning.
   **a.** An economist obtains the unemployment rate and gross state product for a sample of states over the past 10 years, with the objective of examining the relationship between the unemployment rate and the gross state product by census region.

   **b.** A psychologist tests the effects of three different feedback programs by randomly assigning five rats to each program and recording their response times at specified intervals during the program.

   **c.** A marketer of computers runs ads in each of four national publications for one quarter, and keeps track of the number of sales that are attributable to each publication's ad.

   **d.** An electric utility engages a consultant to monitor the discharge from its smokestack on a monthly basis over a 1-year period in order to relate the level of sulfur dioxide in the discharge to the load on the facility's generators.

   **e.** Intrastate trucking rates are compared before and after governmental deregulation of prices charged, with the comparison also taking into account distance of haul, goods hauled, and the price of diesel fuel.

   **f.** An agriculture student compares the amount of rainfall in four different states over the past 5 years.

**10.6** Suppose the mean cholesterol levels are to be compared for adults in four socioeconomic classes—poverty, low income, middle income, and high income. If random samples of each socioeconomic class are selected to make the comparison, identify each of the following elements of the experiment.
  **a.** Response
  **b.** Factor(s) and factor type(s)
  **c.** Treatments
  **d.** Experimental units

## Applying the Concepts—Intermediate

**10.7** *Applied Animal Behaviour Science* (Oct. 2000) published a study of the taste preferences of caged cockatiels. A sample of birds bred at the University of California, Davis, were randomly divided into three experimental groups. Group 1 was fed purified water in bottles on both sides of the cage. Group 2 was fed water on one side and a liquid sucrose (sweet) mixture on the opposite side of the cage. Group 3 was fed water on one side and a liquid sodium chloride (salty) mixture on the opposite side of the cage. One variable of interest to the researchers was total consumption of liquid by each cockatiel.
  **a.** What is the experimental unit for this study?
  **b.** Is the study a designed experiment? Why?
  **c.** What are the factors in the study?
  **d.** Give the levels of each factor.
  **e.** How many treatments are in the study? Identify them.
  **f.** What is the response variable?

**10.8** In *Teaching of Psychology* (Aug. 1998), a study investigated whether final exam performance is affected by whether or not students take a practice test. Students in an introductory psychology class at Pennsylvania State University were initially divided into three groups based on their class standing: Low, Medium, or High. Within each group, the students were randomly assigned to either attend a review session or take a practice test prior to the final exam. Thus, six groups were formed: (Low, Review), (Low, Practice exam), (Medium, Review), (Medium, Practice exam), (High, Review), and (High, Practice exam). One goal of the study was to compare the mean final exam scores of the six groups of students.
  **a.** What is the experimental unit for this study?
  **b.** Is the study a designed experiment? Why?
  **c.** What are the factors in the study?
  **d.** Give the levels of each factor.
  **e.** How many treatments are in the study? Identify them.
  **f.** What is the response variable?

## Applying the Concepts—Advanced

**10.9** Within marketing, the area of personal sales has long suffered from a poor ethical image, particularly in the eyes of college students. An article in *Journal of Business Ethics* (Vol. 15, 1996) investigated whether such opinions by college students are a function of the type of sales jobs (high tech versus low tech) and/or the sales task (new account development versus account maintenance). Four different samples of college students were confronted with the four different situations (new account development in a high-tech sales task; new account development in a low-tech sales task; account maintenance in a high-tech sales task; and account maintenance in a low-tech sales task), and were asked to evaluate the ethical behavior of the salesperson on a 7-point scale ranging from 1 (not a serious ethical violation) to 7 (a very serious ethical violation). Identify each of the following elements of the experiment:
  **a.** Response
  **b.** Factor(s) and factor level(s)
  **c.** Treatments
  **d.** Experimental units

**10.10** A quality control supervisor measures the quality of a steel ingot on a scale from 0 to 10. He designs an experiment in which three different temperatures (ranging from 1,100 to 1,200°F) and five different pressures (ranging from 500 to 600 psi) are utilized, with 20 ingots produced at each Temperature–Pressure combination. Identify the following elements of the experiment:
  **a.** Response
  **b.** Factor(s) and factor type(s)
  **c.** Treatments
  **d.** Experimental units

# 10.2 THE COMPLETELY RANDOMIZED DESIGN

The simplest experimental design, a *completely randomized design*, consists of the *independent random selection* of experimental units representing each treatment. For example, we could independently select random samples of 20 female and 15 male high school seniors to compare their mean SAT scores. Or, we could independently select random samples of 30 households from each of four census districts to compare the mean income per household among the districts. In both examples our objective is to compare treatment means by selecting random, independent samples for each treatment.

> **DEFINITION 10.7**
> A **completely randomized design** is a design for which independent random samples of experimental units are selected for each treatment.*

The objective of a completely randomized design is usually to compare the treatment means. If we denote the true, or population, means of the $p$ treatments as $\mu_1, \mu_2, \ldots, \mu_p$, then we will test the null hypothesis that the treatment means are all equal against the alternative that at least two of the treatment means differ:

$$H_0: \mu_1 = \mu_2 = \cdots = \mu_p$$
$$H_a: \text{At least two of the } p \text{ treatment means differ}$$

The $\mu$'s might represent the means of *all* female and male high school seniors' SAT scores or the means of *all* households' income in each of four census regions.

To conduct a statistical test of these hypotheses, we will use the means of the independent random samples selected from the treatment populations using the completely randomized design. That is, we compare the $p$ sample means $\bar{x}_1, \bar{x}_2, \ldots, \bar{x}_p$.

For example, suppose you select independent random samples of five female and five male high school seniors and obtain sample mean SAT scores of 550 and 590, respectively. Can we conclude that males score 40 points higher, on average, than females? To answer this question, we must consider the amount of sampling variability among the experimental units (students). If the scores are as depicted in the dot plot shown in Figure 10.3, then the difference between the means is small relative to the sampling variability of the scores within the treatments, Female and Male. We would be inclined not to reject the null hypothesis of equal population means in this case.

In contrast, if the data are as depicted in the dot plot of Figure 10.4, then the sampling variability is small relative to the difference between the two means. We would be inclined to favor the alternative hypothesis that the population means differ in this case.

**FIGURE 10.3**

*Dot Plot of SAT Scores: Difference Between Means Dominated by Sampling Variability*

**FIGURE 10.4**

*Dot Plot of SAT Scores: Difference Between Means Large Relative to Sampling Variability*

---

*We use *completely randomized design* to refer to both designed and observational experiments. Thus, the only requirement is that the experimental units to which treatments are applied (designed) or on which treatments are observed (observational) are independently selected for each treatment.

You can see that the key is to compare the difference between the treatment means to the amount of sampling variability. To conduct a formal statistical test of the hypothesis requires numerical measures of the difference between the treatment means and the sampling variability within each treatment. The variation between the treatment means is measured by the **Sum of Squares for Treatments** (SST), which is calculated by squaring the distance between each treatment mean and the overall mean of *all* sample measurements, multiplying each squared distance by the number of sample measurements for the treatment, and adding the results over all treatments:

$$\text{SST} = \sum_{i=1}^{p} n_i(\bar{x}_i - \bar{x})^2 = 5(550 - 570)^2 + 5(590 - 570)^2 = 4,000$$

where we use $\bar{x}$ to represent the overall mean response of all sample measurements, that is, the mean of the combined samples. The symbol $n_i$ is used to denote the sample size for the $i$th treatment. You can see that the value of SST is 4,000 for the two samples of five female and five male SAT scores depicted in Figures 10.3 and 10.4.

Next, we must measure the sampling variability within the treatments. We call this the **Sum of Squares for Error** (SSE) because it measures the variability around the treatment means that is attributed to sampling error. Suppose the 10 measurements in the first dot plot (Figure 10.3) are 490, 520, 550, 580, and 610 for females and 530, 560, 590, 620, and 650 for males. Then the value of SSE is computed by summing the squared distance between each response measurement and the corresponding treatment mean, and then adding the squared differences over all measurements in the entire sample:

$$\text{SSE} = \sum_{j=1}^{n_1} \left(x_{1j} - \bar{x}_1\right)^2 + \sum_{j=1}^{n_2} \left(x_{2j} - \bar{x}_2\right)^2 + \ldots \sum_{j=1}^{n_p} \left(x_{pj} - \bar{x}_p\right)^2$$

where the symbol $x_{1j}$ is the $j$th measurement in sample 1, $x_{2j}$ is the $j$th measurement in sample 2, and so on. This rather complex-looking formula can be simplified by recalling the formula for the sample variance, $s^2$, given in Chapter 2:

$$s^2 = \sum_{i=1}^{n} \frac{(x_i - \bar{x})^2}{n - 1}$$

Note that each sum in SSE is simply the numerator of $s^2$ for that particular treatment. Consequently, we can rewrite SSE as

$$\text{SSE} = (n_1 - 1)s_1^2 + (n_2 - 1)s_2^2 + \cdots + (n_p - 1)s_p^2$$

where $s_1^2, s_2^2, \cdots, s_p^2$ are the sample variances for the $p$ treatments. For our samples of SAT scores, we find $s_1^2 = 2,250$ (for females) and $s_2^2 = 2,250$ (for males); then we have

$$\text{SSE} = (5 - 1)(2,250) + (5 - 1)(2,250) = 18,000$$

To make the two measurements of variability comparable, we divide each by the degrees of freedom to convert the sums of squares to mean squares. First, the **Mean Square for Treatments** (MST), which measures the variability *among* the treatment means, is equal to

$$\text{MST} = \frac{\text{SST}}{p - 1} = \frac{4,000}{2 - 1} = 4,000$$

where the number of degrees of freedom for the $p$ treatments is $(p - 1)$. Next, the **Mean Square for Error** (MSE), which measures the sampling variability *within* the treatments, is

$$\text{MSE} = \frac{\text{SSE}}{n - p} = \frac{18,000}{10 - 2} = 2,250$$

Finally, we calculate the ratio of MST to MSE—an **F statistic**:

$$F = \frac{\text{MST}}{\text{MSE}} = \frac{4{,}000}{2{,}250} = 1.78$$

Values of the $F$ statistic near 1 indicate that the two sources of variation, between treatment means and within treatments, are approximately equal. In this case, the difference between the treatment means may well be attributable to sampling error, which provides little support for the alternative hypothesis that the population treatment means differ. Values of $F$ well in excess of 1 indicate that the variation among treatment means well exceeds that within means and therefore support the alternative hypothesis that the population treatment means differ.

When does $F$ exceed 1 by enough to reject the null hypothesis that the means are equal? This depends on the degrees of freedom for treatments and for error, and on the value of $\alpha$ selected for the test. We compare the calculated $F$ value to a table $F$ value (Tables VIII–XI of Appendix A) with $\nu_1 = (p - 1)$ degrees of freedom in the numerator and $\nu_2 = (n - p)$ degrees of freedom in the denominator and corresponding to a Type I error probability of $\alpha$. For the SAT score example, the $F$ statistic has $\nu_1 = (2 - 1)$ numerator degree of freedom and $\nu_2 = (10 - 2) = 8$ denominator degrees of freedom. Thus, for $\alpha = .05$ we find (Table IX of Appendix A)

$$F_{.05} = 5.32$$

The implication is that MST would have to be 5.32 times greater than MSE before we could conclude at the .05 level of significance that the two population treatment means differ. Since the data yielded $F = 1.78$, our initial impressions for the dot plot in Figure 10.3 are confirmed—there is insufficient information to conclude that the mean SAT scores differ for the populations of female and male high school seniors. The rejection region and the calculated $F$ value are shown in Figure 10.5.

**FIGURE 10.5**

*Rejection Region and Calculated F Values for SAT Score Samples*

In contrast, consider the dot plot in Figure 10.4. Since the means are the same as in the first example, 550 and 590, respectively, the variation between the means is the same, MST $= 4{,}000$. But the variation within the two treatments appears to be considerably smaller. The observed SAT scores are 540, 545, 550, 555, and 560 for females and 580, 585, 590, 595, and 600 for males. These values yield $s_1^2 = 62.5$ and $s_2^2 = 62.5$. Thus, the variation within the treatments is measured by

$$\text{SSE} = (5 - 1)(62.5) + (5 - 1)(62.5)$$
$$= 500$$

$$\text{MSE} = \frac{\text{SSE}}{n - p} = \frac{500}{8} = 62.5$$

Then the $F$-ratio is

$$F = \frac{\text{MST}}{\text{MSE}} = \frac{4,000}{62.5} = 64.0$$

Again, our visual analysis of the dot plot is confirmed statistically: $F = 64.0$ well exceeds the tabled $F$ value, 5.32, corresponding to the .05 level of significance. We would therefore reject the null hypothesis at that level and conclude that the SAT mean score of males differs from that of females.

Recall that we performed a hypothesis test for the difference between two means in Section 9.1 using a two-sample $t$ statistic for two independent samples. When two independent samples are being compared, the $t$- and $F$-tests are equivalent. To see this, recall the formula

$$t = \frac{\bar{x}_1 - \bar{x}_2}{\sqrt{s_p^2 \left( \frac{1}{n_1} + \frac{1}{n_2} \right)}} = \frac{590 - 550}{\sqrt{(62.5)\left( \frac{1}{5} + \frac{1}{5} \right)}} = \frac{40}{5} = 8$$

where we used the fact that $s_p^2 = \text{MSE}$, which you can verify by comparing the formulas. Note that the calculated $F$ for these samples ($F = 64$) equals the square of the calculated $t$ for the same samples ($t = 8$). Likewise, the tabled $F$ value (5.32) equals the square of the tabled $t$ value at the two-sided .05 level of significance ($t_{.025} = 2.306$ with 8 df). Since both the rejection region and the calculated values are related in the same way, the tests are equivalent. Moreover, the assumptions that must be met to ensure the validity of the $t$- and $F$-tests are the same:

1. The probability distributions of the populations of responses associated with each treatment must all be normal.
2. The probability distributions of the populations of responses associated with each treatment must have equal variances.
3. The samples of experimental units selected for the treatments must be random and independent.

In fact, the only real difference between the tests is that the $F$-test can be used to compare *more than two* treatment means, whereas the $t$-test is applicable to two samples only. The **$F$-test** is summarized in the accompanying box.

---

## Test to Compare $p$ Treatment Means for a Completely Randomized Design

$H_0: \mu_1 = \mu_2 = \cdots = \mu_p$
$H_a$: At least two treatment means differ

*Test statistic:* $F = \dfrac{\text{MST}}{\text{MSE}}$

*Assumptions:* 1. Samples are selected randomly and independently from the respective populations.
2. All $p$ population probability distributions are normal.
3. The $p$ population variances are equal.

*Rejection region:* $F > F_\alpha$, where $F_\alpha$ is based on $\nu_1 = (p - 1)$ numerator degrees of freedom (associated with MST) and $\nu_2 = (n - p)$ denominator degrees of freedom (associated with MSE).

Computational formulas for MST and MSE are given in Appendix C. We will rely on statistical software to compute the $F$ statistic, concentrating on the interpretation of the results rather than their calculations.

**EXAMPLE 10.3**

Suppose the USGA wants to compare the mean distances associated with four different brands of golf balls when struck with a driver. A completely randomized design is employed, with Iron Byron, the USGA's robotic golfer, using a driver to hit a random sample of 10 balls of each brand in a random sequence. The distance is recorded for each hit, and the results are shown in Table 10.1, organized by brand.

a. Set up the test to compare the mean distances for the four brands. Use $\alpha = .10$.
b. Use SAS to obtain the test statistic and $p$-value. Interpret the results.

⊘ GOLFCRD
**TABLE 10.1    Results of Completely Randomized Design: Iron Byron Driver**

|  | **Brand A** | **Brand B** | **Brand C** | **Brand D** |
|---|---|---|---|---|
|  | 251.2 | 263.2 | 269.7 | 251.6 |
|  | 245.1 | 262.9 | 263.2 | 248.6 |
|  | 248.0 | 265.0 | 277.5 | 249.4 |
|  | 251.1 | 254.5 | 267.4 | 242.0 |
|  | 260.5 | 264.3 | 270.5 | 246.5 |
|  | 250.0 | 257.0 | 265.5 | 251.3 |
|  | 253.9 | 262.8 | 270.7 | 261.8 |
|  | 244.6 | 264.4 | 272.9 | 249.0 |
|  | 254.6 | 260.6 | 275.6 | 247.1 |
|  | 248.8 | 255.9 | 266.5 | 245.9 |
| Sample means | 250.8 | 261.1 | 270.0 | 249.3 |

**Solution**

a. To compare the mean distances of the four brands, we first specify the hypotheses to be tested. Denoting the population mean of the $i$th brand by $\mu_i$, we test

$H_0: \mu_1 = \mu_2 = \mu_3 = \mu_4$
$H_a$: The mean distances differ for at least two of the brands.

The test statistic compares the variation among the four treatment (Brand) means to the sampling variability within each of the treatments.

$$\textit{Test statistic:} \quad F = \frac{\text{MST}}{\text{MSE}}$$

$$\textit{Rejection region:} \quad F > F_\alpha = F_{.10} \text{ with } \nu_1 = (p-1) = 3 \text{ df}$$
$$\text{and } \nu_2 = (n-p) = 36 \text{ df}$$

From Table VIII of Appendix A, we find $F_{.10} \approx 2.25$ for 3 and 36 df. Thus, we will reject $H_0$ if $F > 2.25$. (See Figure 10.7.)

The assumptions necessary to ensure the validity of the test are as follows:
1. The samples of 10 golf balls for each brand are selected randomly and independently.
2. The probability distributions of the distances for each brand are normal.

**FIGURE 10.6**

*SAS Analysis of Variance Printout for Golf Ball Distance Data: Completely Randomized Design*

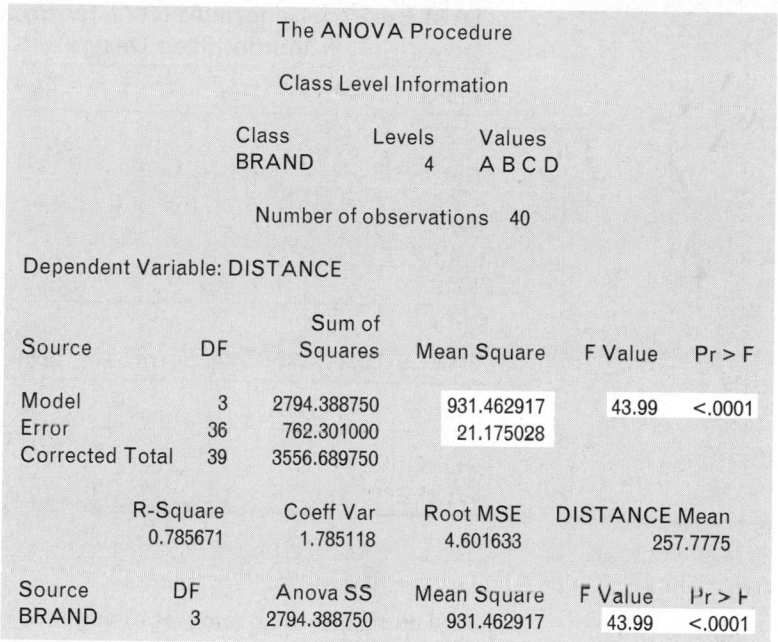

The ANOVA Procedure

Class Level Information

| Class | Levels | Values |
|-------|--------|--------|
| BRAND | 4 | A B C D |

Number of observations   40

Dependent Variable: DISTANCE

| Source | DF | Sum of Squares | Mean Square | F Value | Pr > F |
|--------|-----|----------------|-------------|---------|--------|
| Model | 3 | 2794.388750 | 931.462917 | 43.99 | <.0001 |
| Error | 36 | 762.301000 | 21.175028 | | |
| Corrected Total | 39 | 3556.689750 | | | |

| R-Square | Coeff Var | Root MSE | DISTANCE Mean |
|----------|-----------|----------|---------------|
| 0.785671 | 1.785118 | 4.601633 | 257.7775 |

| Source | DF | Anova SS | Mean Square | F Value | Pr > F |
|--------|-----|----------|-------------|---------|--------|
| BRAND | 3 | 2794.388750 | 931.462917 | 43.99 | <.0001 |

3. The variances of the distance probability distributions for each brand are equal.

b. The SAS printout for the data in Table 10.1 resulting from this completely randomized design is given in Figure 10.6. The Total Sum of Squares is designated the **Corrected Total**, and it is partitioned into the **Model** and **Error Sums of Squares**. The bottom part of the printout further partitions the **Model** component into the factors that comprise the model. In this single-factor experiment, the Model and Brand sums of squares are the same. The **Sum of Squares** column is headed **ANOVA SS**.

The values of the mean squares, MST and MSE (highlighted on the printout), are 931.46 and 21.18, respectively. The *F*-ratio, 43.99, also highlighted on the printout, exceeds the tabled value of 2.25. We therefore reject the null hypothesis at the .10 level of significance, concluding that at least two of the brands differ with respect to mean distance traveled when struck by the driver.

The observed significance level of the *F*-test is also highlighted on the printout: *p*-value < .0001. This is the area to the right of the calculated *F* value and it implies that we would reject the null hypothesis that the means are equal at any $\alpha$ level of .0001 or greater. The result of the test is displayed in Figure 10.7. ■

**FIGURE 10.7**

*F-Test for Completely Randomized Design: Golf Ball Experiment*

**TABLE 10.2** **General ANOVA Summary Table for a Completely Randomized Design**

| Source | df | SS | MS | F |
|---|---|---|---|---|
| Treatments | $p - 1$ | SST | $\text{MST} = \dfrac{\text{SST}}{p - 1}$ | $\dfrac{\text{MST}}{\text{MSE}}$ |
| Error | $n - p$ | SSE | $\text{MSE} = \dfrac{\text{SSE}}{n - p}$ | |
| Total | $n - 1$ | SS(Total) | | |

**TABLE 10.3** **ANOVA Summary Table for Example 10.3**

| Source | df | SS | MS | F | p-Value |
|---|---|---|---|---|---|
| Brands | 3 | 2,794.39 | 931.46 | 43.99 | .0001 |
| Error | 36 | 762.30 | 21.18 | | |
| Total | 39 | 3,556.69 | | | |

The results of an **analysis of variance (ANOVA)** can be summarized in a simple tabular format similar to that obtained from the SAS program in Example 10.3. The general form of the table is shown in Table 10.2, where the symbols df, SS, and MS stand for degrees of freedom, Sum of Squares, and Mean Square, respectively. Note that the two sources of variation, Treatments and Error, add to the Total Sum of Squares, SS(Total). The ANOVA summary table for Example 10.3 is given in Table 10.3, and the partitioning of the Total Sum of Squares into its two components is illustrated in Figure 10.8.

**FIGURE 10.8**

*Partitioning of the Total Sum of Squares for the Completely Randomized Design*

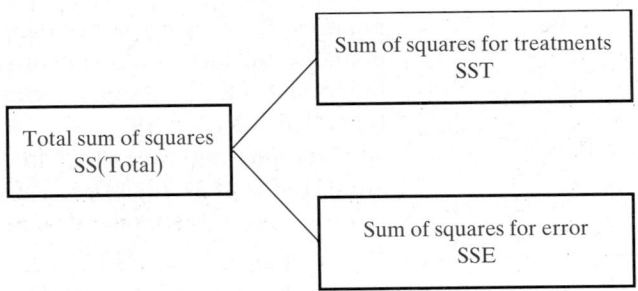

Suppose the *F*-test results in a rejection of the null hypothesis that the treatment means are equal. Is the analysis complete? Usually, the conclusion that at least two of the treatment means differ leads to other questions. Which of the means differ, and by how much? For example, the *F*-test in Example 10.3 leads to the conclusion that at least two of the brands of golf balls have different mean distances traveled when struck with a driver. Now the question is, which of the brands differ? How are the brands ranked with respect to mean distance?

One way to obtain this information is to construct a confidence interval for the difference between the means of any pair of treatments using the method of Section 9.1. For example, if a 95% confidence interval for $\mu_A - \mu_C$ in Example 10.3 is found to be $(-24, -13)$, we are confident that the mean distance for Brand C exceeds the mean for Brand A (since all differences in the interval are negative). Constructing these confidence intervals for all possible brand pairs allows you to rank the brand means. A method for conducting these *multiple comparisons—* one that controls for Type I errors—is presented in Section 10.3.

# Using the TI-83 Graphing Calculator

## "One-Way ANOVA" on the TI-83

Start from the home screen.

Step 1     Enter each data set into its own list. (i.e., sample 1 into List1, sample 2 into List2, sample 3 into List3, etc.)

Step 2     *Access the "Statistical Test" Menu*
Press **STAT**
Arrow right to **TEST**
Arrow up to **F: ANOVA(**
Press **ENTER**

Step 3     Enter each list, separated by commas, for which you want to perform the analysis. (e.g., L1, L2, L3, L4)
Press **ENTER**

Step 4     *View Display*
The calculator will display the $F$-test statistic, as well as the $p$-value, the factor degrees of freedom, sum of squares, mean square, and by arrowing down the error degrees of freedom, sum of squares, mean square, and the pooled standard deviation.

Example    Below are four different samples. At the $\alpha = 0.05$ level of significance, test whether the four population means are equal. The null hypothesis will be $H_0: \mu_1 = \mu_2 = \mu_3 = \mu_4$. The alternative hypothesis is $H_a$: At least one mean is different.

| SAMPLE 1 | SAMPLE 2 | SAMPLE 3 | SAMPLE 4 |
|:---:|:---:|:---:|:---:|
| 60 | 59 | 55 | 58 |
| 61 | 52 | 55 | 58 |
| 56 | 51 | 52 | 55 |

Step 1     Enter the data into the lists.

Step 2     Access the Statistical Test Menu.

Step 3     Enter each list for which you wish to perform the analysis (see screen below left).

Step 4     Press **ENTER** and view screen (see screen above right.) As you can see from the screen the $p$-value is 0.1598 which is *not less than* 0.05 therefore we should *not reject* $H_0: \mu_1 = \mu_2 = \mu_3 = \mu_4$. The differences are not significant.

Step 5     Clear the screen for the next problem. Return to the home screen Press **CLEAR.**

**EXAMPLE 10.4**

Refer to the completely randomized design ANOVA conducted in Example 10.3. Are the assumptions required for the test approximately satisfied?

**Solution**

The assumptions for the test are repeated below.

1. The samples of golf balls for each brand are selected randomly and independently.
2. The probability distributions of the distances for each brand are normal.
3. The variances of the distance probability distributions for each brand are equal.

Since the sample consisted of 10 randomly selected balls of each brand and the robotic golfer Iron Byron was used to drive all the balls, the first assumption of independent random samples is satisfied. To check the next two assumptions, we will employ two graphical methods presented in Chapter 2: stem-and-leaf displays and dot plots. A MINITAB stem-and-leaf display for the sample distances of each brand of golf ball is shown in Figure 10.9, and a MINITAB dot plot in Figure 10.10.

The normality assumption can be checked by examining the stem-and-leaf displays in Figure 10.9. With only 10 sample measurements for each brand, however, the displays are not very informative. More data would need to be collected for each brand before we could assess whether the distances come from normal distributions. Fortunately, analysis of variance has been shown to be a very **robust method** when the assumption of normality is not satisfied exactly: That is, *moderate departures from normality do not have much effect on the significance level of the ANOVA F-test or on confidence coefficients.* Rather than spend the time, energy, or money to collect additional data for this experiment in order to verify the normality assumption, we will rely on the robustness of the ANOVA methodology.

Dot plots are a convenient way to obtain a rough check on the assumption of equal variances. With the exception of a possible outlier for Brand D, the dot plots in Figure 10.10 show that the spread of the distance measurements is about the same for each brand. Since the sample variances appear to be the same, the assumption of equal population variances for the brands is probably satisfied. Although robust with respect to the normality assumption, ANOVA is *not robust* with respect to the equal variances assumption. Departures from the assumption of equal population variances can affect the associated measures of reliability (e.g., *p*-values and confidence levels). Fortunately, the effect is slight when the sample sizes are equal, as in this experiment. ∎

Although graphs can be used to check the ANOVA assumptions, as in Example 10.4, no measures of reliability can be attached to these graphs. When you have a plot that is unclear as to whether or not an assumption is satisfied, you can use formal statistical tests that are beyond the scope of this text. Consult the references at the end of the chapter for information on these tests. When the validity of the ANOVA assumptions is in doubt, nonparametric statistical methods are useful.

---

**What Do You Do When the Assumptions Are Not Satisfied for the Analysis of Variance for a Completely Randomized Design?**

---

*Answer:* Use a nonparametric statistical method such as the Kruskal-Wallis *H*-Test of Section 14.4.

**FIGURE 10.9**

*MINITAB Stem-and-Leaf Displays for Distance Data: Completely Randomized Design*

```
Stem-and-leaf of BrandA     N = 10
Leaf Unit = 1.0

    2    24  45
    2    24
    4    24  88
   (3)   25  011
    3    25  3
    2    25  4
    1    25
    1    25
    1    26  0


Stem-and-leaf of BrandB     N = 10
Leaf Unit = 1.0

    2    25  45
    3    25  7
    3    25
    4    26  0
   (3)   26  223
    3    26  445


Stem-and-leaf of BrandC     N = 10
Leaf Unit = 1.0

    1    26  3
    2    26  5
    4    26  67
    5    26  9
    5    27  00
    3    27  2
    2    27  5
    1    27  7


Stem-and-leaf of BrandD     N = 10
Leaf Unit = 1.0

    1    24  2
    2    24  5
    4    24  67
   (3)   24  899
    3    25  11
    1    25
    1    25
    1    25
    1    25
    1    26  1
```

**FIGURE 10.10**

*MINITAB Dot Plots for Distance Data: Completely Randomized Design*

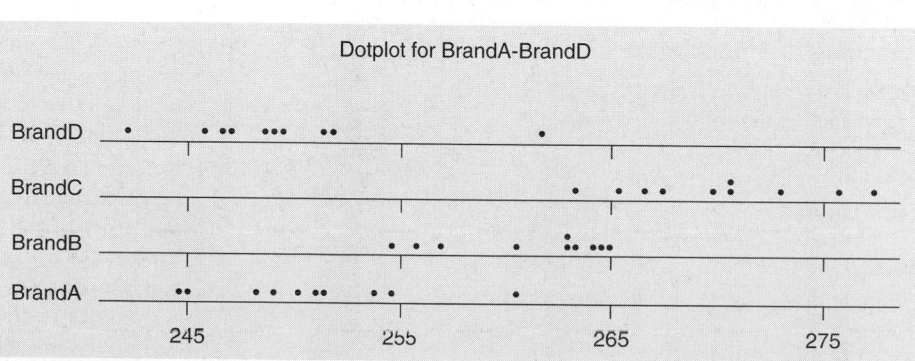

Dotplot for BrandA-BrandD

The procedure for conducting an analysis of variance for a completely randomized design is summarized in the following box. Remember that the hallmark of this design is independent random samples of experimental units associated with each treatment. We discuss a design with dependent samples in Section 10.4.

---

### Steps for Conducting an ANOVA for a Completely Randomized Design

1. Be sure the design is truly completely randomized, with independent random samples for each treatment.
2. Check the assumptions of normality and equal variances.
3. Create an ANOVA summary table that specifies the variability attributable to treatments and error, making sure that it leads to the calculation of the $F$ statistic for testing the null hypothesis that the treatment means are equal in the population. Use a statistical software program to obtain the numerical results. If no such package is available, use the calculation formulas in Appendix C.
4. If the $F$-test leads to the conclusion that the means differ:
   a. Conduct a multiple comparisons procedure for as many of the pairs of means as you wish to compare (see Section 10.3). Use the results to summarize the statistically significant differences among the treatment means.
   b. If desired, form confidence intervals for one or more individual treatment means.
5. If the $F$-test leads to the nonrejection of the null hypothesis that the treatment means are equal, consider the following possibilities:
   a. The treatment means are equal—that is, the null hypothesis is true.
   b. The treatment means really differ, but other important factors affecting the response are not accounted for by the completely randomized design. These factors inflate the sampling variability, as measured by MSE, resulting in smaller values of the $F$ statistic. Either increase the sample size for each treatment or use a different experimental design (as in Section 10.4) that accounts for the other factors affecting the response.

[*Note:* Be careful not to automatically conclude that the treatment means are equal since the possibility of a Type II error must be considered if you accept $H_0$.]

---

## EXERCISES 10.11–10.29

### Learning the Mechanics

**10.11** Use Tables VIII, IX, X, and XI of Appendix A to find each of the following $F$ values:
  **a.** $F_{.05}, v_1 = 3, v_2 = 4$
  **b.** $F_{.01}, v_1 = 3, v_2 = 4$
  **c.** $F_{.10}, v_1 = 20, v_2 = 40$
  **d.** $F_{.025}, v_1 = 12, v_2 = 9$

**10.12** Find the following probabilities:
  **a.** $P(F \leq 3.48)$ for $v_1 = 5, v_2 = 9$
  **b.** $P(F > 3.09)$ for $v_1 = 15, v_2 = 20$
  **c.** $P(F > 2.40)$ for $v_1 = 15, v_2 = 15$
  **d.** $P(F \leq 1.83)$ for $v_1 = 8, v_2 = 40$

**10.13** Consider dot plots a. and b. shown on p. 459. In which dot plot is the difference between the sample means small relative to the variability within the sample observations? Justify your answer.

**10.14** Refer to Exercise 10.13. Assume that the two samples represent independent, random samples corresponding to two treatments in a completely randomized design.
  **a.** Calculate the treatment means, i.e., the means of samples 1 and 2, for both dot plots.
  **b.** Use the means to calculate the Sum of Squares for Treatments (SST) for each dot plot.

**Plots for Exercise 10.13**

• Sample 1
□ Sample 2

**c.** Calculate the sample variance for each sample and use these values to obtain the Sum of Squares for Error (SSE) for each dot plot.

**d.** Calculate the Total Sum of Squares [SS(Total)] for the two dot plots by adding the Sums of Squares for Treatment and Error. What percentage of SS(Total) is accounted for by the treatments—that is, what percentage of the Total Sum of Squares is the Sum of Squares for Treatment—in each case?

**e.** Convert the Sum of Squares for Treatment and Error to mean squares by dividing each by the appropriate number of degrees of freedom. Calculate the $F$-ratio of the Mean Square for Treatment (MST) to the Mean Square for Error (MSE) for each dot plot.

**f.** Use the $F$-ratios to test the null hypothesis that the two samples are drawn from populations with equal means. Use $\alpha = .05$.

**g.** What assumptions must be made about the probability distributions corresponding to the responses for each treatment in order to ensure the validity of the $F$-tests conducted in part **f**?

**10.15** Refer to Exercises 10.13 and 10.14. Conduct a two-sample $t$-test (Section 9.1) of the null hypothesis that the two treatment means are equal for each dot plot. Use $\alpha = .05$ and two-tailed tests. In the course of the test, compare each of the following with the $F$-tests in Exercise 10.14:

**a.** The pooled variances and the MSEs

**b.** The $t$- and the $F$-test statistics

**c.** The tabled values of $t$ and $F$ that determine the rejection regions

**d.** The conclusions of the $t$- and $F$-tests

**e.** The assumptions that must be made in order to ensure the validity of the $t$- and $F$-tests

**10.16** Refer to Exercises 10.13 and 10.14. Complete the following ANOVA table for each of the two dot plots:

| Source | df | SS | MS | F |
|---|---|---|---|---|
| Treatments | | | | |
| Error | | | | |
| Total | | | | |

**10.17** Suppose the Total Sum of Squares for a completely randomized design with $p = 5$ treatments and $n = 30$ total measurements (six per treatment) is equal to 500. In each of the following cases, conduct an $F$-test of the null hypothesis that the mean responses for the five treatments are the same. Use $\alpha = .10$.

**a.** Sum of Squares for Treatment (SST) is 20% of SS(Total)

**b.** SST is 50% of SS(Total)

**c.** SST is 80% of SS(Total)

**d.** What happens to the $F$-ratio as the percentage of the Total Sum of Squares attributable to treatments is increased?

**10.18** A partially completed ANOVA table for a completely randomized design is shown here:

| Source | df | SS | MS | F |
|---|---|---|---|---|
| Treatments | 6 | 18.4 | | |
| Error | | | | |
| Total | 41 | 45.2 | | |

**a.** Complete the ANOVA table.

**b.** How many treatments are involved in the experiment?

**c.** Do the data provide sufficient evidence to indicate a difference among the population means? Test using $\alpha = .10$.

**d.** Find the approximate observed significance level for the test in part **c**, and interpret it.

**10.19** Below is a MINITAB printout for an experiment utilizing a completely randomized design.

**a.** How many treatments are involved in the experiment? What is the total sample size?

**MINITAB Output for Exercise 10.19**

```
Analysis of Variance for RESPONSE

Source    DF       SS         MS       F       P
FACTOR     3     57258      19086    14.80   0.000
Error     34     43836       1289
Total     37    101094
```

**b.** Conduct a test of the null hypothesis that the treatment means are equal. Use $\alpha = .01$ .

**c.** What additional information is needed in order to be able to compare specific pairs of treatment means?

**10.20** The data in the table below resulted from an experiment that utilized a completely randomized design.

⊘ LM10_20

| Treatment 1 | Treatment 2 | Treatment 3 |
|---|---|---|
| 3.9 | 5.4 | 1.3 |
| 1.4 | 2.0 | .7 |
| 4.1 | 4.8 | 2.2 |
| 5.5 | 3.8 | |
| 2.3 | 3.5 | |

**a.** Use statistical software (or the formulas in Appendix C) to complete the following ANOVA table:

| Source | df | SS | MS | F |
|---|---|---|---|---|
| Treatments | | | | |
| Error | | | | |
| Total | | | | |

**b.** Test the null hypothesis that $\mu_1 = \mu_2 = \mu_3$, where $\mu_i$ represents the true mean for treatment $i$, against the alternative that at least two of the means differ. Use $\alpha = .01$.

## Applying the Concepts—Basic

**10.21** Robotics researchers investigated whether robots could be trained to behave like ants in an ant colony (*Nature*, Aug. 2000). Robots were trained and randomly assigned to "colonies" (i.e., groups) consisting of 3, 6, 9, or 12 robots. The robots were assigned the task of foraging for "food" and to recruit another robot when they identified a resource-rich area. One goal of the experiment was to compare the mean energy expended (per robot) of the four different colony sizes.

**a.** What type of experimental design was employed?

**b.** Identify the treatments and the dependent variable.

**c.** Set up the null and alternative hypotheses of the test.

**d.** The following ANOVA results were reported: $F = 7.70$, numerator df $= 3$, denominator df $= 56$, $p$-value $< .001$. Conduct the test at a significance level of $\alpha = .05$ and interpret the result.

**10.22** An article in the *American Journal of Political Science* (Jan. 1998) examined the attitudes of three groups of professionals that influence U.S. policy. Random samples of 100 scientists, 100 journalists, and 100 government officials were asked about the safety of nuclear power plants. Responses were made on a seven-point scale, where 1 = very unsafe and 7 = very safe. The mean safety scores for the groups are: scientists, 4.1; journalists, 3.7; government officials, 4.2.

**a.** Identify the response variable for this study.

**b.** How many treatments are included in this study? Describe them.

**c.** Specify the null and alternative hypotheses that should be used to investigate whether there are differences in the attitudes of scientists, journalists, and government officials regarding the safety of nuclear power plants.

**d.** The MSE for the sample data is 2.355. At least how large must MST be in order to reject the null hypothesis of the test of part **a** using $\alpha = .05$?

**e.** If the MST $= 11.280$, what is the approximate $p$-value of the test of part **a**?

**10.23** Refer to *The Archives of Disease in Childhood* (Apr. 2000) study of whether height influences a child's progression through elementary school, Exercise 9.11 (p. 393). Within each grade, Australian school children were divided into equal thirds (tertiles) based on age (youngest third, middle third, and oldest third). The researchers compared the average heights of the three groups using an analysis of variance. (All height measurements were standardized using $z$-scores.) A summary of the results for all grades combined, by gender, is shown in the table below.

**a.** What is the null hypothesis for the ANOVA of the boys' data?

**b.** Interpret the results of the test, part **a**. Use $\alpha = .05$.

**c.** Repeat parts **a** and **b** for the girls' data.

**d.** Summarize the results of the hypothesis tests in the words of the problem.

| | Sample Size | Youngest Tertile Mean Height | Middle Tertile Mean Height | Oldest Tertile Mean Height | F Value | p-Value |
|---|---|---|---|---|---|---|
| Boys | 1439 | 0.33 | 0.33 | 0.16 | 4.57 | 0.01 |
| Girls | 1409 | 0.27 | 0.18 | 0.21 | 0.85 | 0.43 |

*Source:* Wake, M., Coghlan, D., and Hesketh, K. "Does height influence progression through primary school grades?" *The Archives of Disease in Childhood*, Vol. 82, Apr. 2000 (Table 2).

**10.24** The *Journal of Hazardous Materials* (July 1995) published the results of a study of the chemical properties of three different types of hazardous organic solvents used to clean metal parts: aromatics, chloroalkanes, and esters. One variable studied was sorption rate, measured as mole percentage. Independent samples of solvents from each type were tested and their sorption rates were recorded, as shown in the table. A MINITAB analysis of variance of the data is provided.

🔴 SORPRATE

| Aromatics | | Chloroalkanes | | Esters | | |
|-----------|-----|------|------|-----|-----|-----|
| 1.06 | .95 | 1.58 | 1.12 | .29 | .43 | .06 |
| .79  | .65 | 1.45 | .91  | .06 | .51 | .09 |
| .82  | 1.15| .57  | .83  | .44 | .10 | .17 |
| .89  | 1.12| 1.16 | .43  | .61 | .34 | .60 |
| 1.05 |     |      |      | .55 | .53 | .17 |

*Source:* Reprinted from *Journal of Hazardous Materials*, Vol. 42, No. 2, J. D. Ortego *et al.*, "A review of polymeric geosynthetics used in hazardous waste facilities." p. 142 (Table 9), July 1995, Elsevier Science-NL, Sara Burgerhartstraat 25, 1055 KV Amsterdam, The Netherlands.

**MINITAB Output for Exercise 10.24**

```
Analysis of Variance for SORPRATE

Source    DF      SS        MS        F       P
SOLVENT    2   3.3054    1.6527    24.51   0.000
Error     29   1.9553    0.0674
Total     31   5.2608
```

**a.** Construct an ANOVA table from the MINITAB printout.

**b.** Is there evidence of differences among the mean sorption rates of the three organic solvent types? Test using $\alpha = .10$.

### Applying the Concepts—Intermediate

**10.25** Studies conducted at the University of Melbourne (Australia) indicate that there may be a difference between the pain thresholds of blondes and brunettes. Men and women of various ages were divided into four categories according to hair color: light blond, dark blond, light brunette, and dark brunette. The purpose of the experiment was to determine whether hair color is related to the amount of pain evoked by common types of mishaps and assorted types of trauma. Each person in the experiment was given a pain threshold score based on his or her performance in a pain sensitivity test (the higher the score, the higher the person's pain tolerance). SAS was used to conduct the analysis of variance of the data listed in the table. The SAS printout is also provided.

🔴 HAIRPAIN

| Light Blond | Dark Blond | Light Brunette | Dark Brunette |
|-------------|------------|----------------|---------------|
| 62 | 63 | 42 | 32 |
| 60 | 57 | 50 | 39 |
| 71 | 52 | 41 | 51 |
| 55 | 41 | 37 | 30 |
| 48 | 43 |    | 35 |

**SAS Output for Exercise 10.25**

The ANOVA Procedure

Class Level Information

| Class | Levels | Values |
|-------|--------|--------|
| COLOR | 4 | DkBlond DkBrunet LtBlond LtBrunet |

Number of observations   19

Dependent Variable: PAIN

| Source | DF | Sum of Squares | Mean Square | F Value | Pr > F |
|--------|----|----------------|-------------|---------|--------|
| Model | 3 | 1360.726316 | 453.575439 | 6.79 | 0.0041 |
| Error | 15 | 1001.800000 | 66.786667 | | |
| Corrected Total | 18 | 2362.526316 | | | |

| R-Square | Coeff Var | Root MSE | PAIN Mean |
|----------|-----------|----------|-----------|
| 0.575962 | 17.08184 | 8.172311 | 47.84211 |

| Source | DF | Anova SS | Mean Square | F Value | Pr > F |
|--------|----|----------|-------------|---------|--------|
| COLOR | 3 | 1360.726316 | 453.575439 | 6.79 | 0.0041 |

a. Based on the given information, what type of experimental design appears to have been employed?

b. Using the SAS printout, conduct a test to determine whether the mean pain thresholds differ among people possessing the four types of hair color. Use $\alpha = .05$.

c. What is the observed significance level for the test in part **b**? Interpret it.

d. What assumptions must be met in order to ensure the validity of the inferences you made in part **b**?

**10.26** Psychologists at Lancaster University (United Kingdom) evaluated three methods of name retrieval in a controlled setting (*Journal of Experimental Psychology-Applied*, June 2000.) A sample of 139 students was randomly divided into three groups, and each group of students used a different method to learn the names of the other students in the group. Group 1 used the "simple name game," where the first student states his/her full name, the second student announces his/her name and the name of the first student, the third student says his/her name and the names of the first two students, etc. Group 2 used the "elaborate name game," a modification of the simple name game where the students not only state their names but also their favorite activity (e.g., sports). Group 3 used "pairwise introductions," where students are divided into pairs and each student must introduce the other member of the pair. One year later, all subjects were sent pictures of the students in their group and asked to state the full name of each. The researchers measured the percentage of names recalled for each student respondent. The data (simulated based on summary statistics provided in the research article) are shown in the table. Conduct an analysis of variance to determine whether the mean percentages of names recalled differ for the three name retrieval methods. Use $\alpha = .05$.

**10.27** What do people infer from facial expressions of emotion? This was the research question of interest in an article published in the *Journal of Nonverbal Behavior* (Fall 1996). A sample of 36 introductory psychology students was randomly divided into six groups. Each group was assigned to view one of six slides showing a person making a facial expression.* The six expressions were (1) angry, (2) disgusted, (3) fearful, (4) happy, (5) sad, and (6) neutral faces. After viewing the slides, the students rated the degree of dominance they inferred from the facial expression (on a scale ranging from $-15$ to $+15$). The data (simulated from summary information provided in the article) are listed in the table. Conduct an analysis of variance to determine whether the mean dominance ratings differ among the six facial expressions. Use $\alpha = .10$.

**FACES**

| Angry | Disgusted | Fearful | Happy | Sad | Neutral |
|-------|-----------|---------|-------|------|---------|
| 2.10 | .40 | .82 | 1.71 | .74 | 1.69 |
| .64 | .73 | $-2.93$ | $-.04$ | $-1.26$ | $-.60$ |
| .47 | $-.07$ | $-.74$ | 1.04 | $-2.27$ | $-.55$ |
| .37 | $-.25$ | .79 | 1.44 | $-.39$ | .27 |
| 1.62 | .89 | $-.77$ | 1.37 | $-2.65$ | $-.57$ |
| $-.08$ | 1.93 | $-1.60$ | .59 | $-.44$ | $-2.16$ |

**OILSPILL**

**10.28** Refer to the *Marine Technology* (Jan. 1995) study of major ocean oil spills by tanker vessels, Exercise 2.15 (p. 29). The spillage amounts (thousands of metric tons) and cause of accident for 48 tankers are saved in the OILSPILL file. (*Note:* Delete the two tankers with oil spills of unknown causes.) Conduct an analysis of variance (at $\alpha = .01$) to compare the mean spillage amounts for the four accident types: (1) collision, (2) grounding, (3) fire/explosion, and (4) hull failure. Interpret your results.

**Applying the Concepts—Advanced**

**10.29** Do you experience episodes of excessive eating accompanied by being overweight? If so, you may suffer from binge eating disorder. Cognitive-behavioral therapy (CBT), in which patients are taught how to make changes in specific behavior patterns (e.g., exercise, eat only low-fat foods), can be effective in treating the disorder. A group of Stanford University researchers investigated the effectiveness of interpersonal therapy (IPT) as a second level of treatment for binge eaters

**NAMEGAME**
**Simple Name Game**

| | | | | | | | | | | | | | |
|--|--|--|--|--|--|--|--|--|--|--|--|--|--|
| 24 | 43 | 38 | 65 | 35 | 15 | 44 | 44 | 18 | 27 | 0 | 38 | 50 | 31 |
| 7 | 46 | 33 | 31 | 0 | 29 | 0 | 0 | 52 | 0 | 29 | 42 | 39 | 26 |
| 51 | 0 | 42 | 20 | 37 | 51 | 0 | 30 | 43 | 30 | 99 | 39 | 35 | 19 |
| 24 | 34 | 3 | 60 | 0 | 29 | 40 | 40 | | | | | | |

**Elaborate Name Game**

| | | | | | | | | | | | | | |
|--|--|--|--|--|--|--|--|--|--|--|--|--|--|
| 39 | 71 | 9 | 86 | 26 | 45 | 0 | 38 | 5 | 53 | 29 | 0 | 62 | 0 |
| 1 | 35 | 10 | 6 | 33 | 48 | 9 | 26 | 83 | 33 | 12 | 5 | 0 | 0 |
| 25 | 36 | 39 | 1 | 37 | 2 | 13 | 26 | 7 | 35 | 3 | 8 | 55 | 50 |

**Pairwise Introductions**

| | | | | | | | | | | | | | |
|--|--|--|--|--|--|--|--|--|--|--|--|--|--|
| 5 | 21 | 22 | 3 | 32 | 29 | 32 | 0 | 4 | 41 | 0 | 27 | 5 | 9 |
| 66 | 54 | 1 | 15 | 0 | 26 | 1 | 30 | 2 | 13 | 0 | 2 | 17 | 14 |
| 5 | 29 | 0 | 45 | 35 | 7 | 11 | 4 | 9 | 23 | 4 | 0 | 8 | 2 |
| 18 | 0 | 5 | 21 | 14 | | | | | | | | | |

*Source:* Morris, P.E., and Fritz, C.O. "The name game: Using retrieval practice to improve the learning of names." *Journal of Experimental Psychology-Applied*, Vol. 6, No. 2, June 2000 (data simulated from Figure 1).

---

*In the actual experiment, each group viewed all six facial expression slides and the design employed was a Latin Square (beyond the scope of this text).

(*Journal of Consulting and Clinical Psychology*, June 1995). The researchers employed a design that randomly assigned a sample of 41 overweight individuals diagnosed with binge eating disorder to either a treatment group (30 subjects) or a control group (11 subjects). Subjects in the treatment group received 12 weeks of cognitive-behavior therapy, then were subdivided into two groups. Those who responded successfully to CBT (17 subjects) were assigned to a weight loss therapy (WLT) program for the next 12 weeks. Those CBT subjects who did not respond to treatment (13 subjects) received 12 weeks of IPT. The subjects in the control group received no therapy of any type. Thus, the study ultimately consisted of three groups of overweight binge eaters: the CBT-WLT group, the CBT-IPT group, and the control group. One outcome (response) variable measured for each subject was the number of binge eating episodes per week, $x$. Summary statistics for each of the three groups at the end of the 24-week period are shown in the table. The data were analyzed as a completely randomized design with

|  | CBT-WLT | CBT-IPT | Control |
|---|---|---|---|
| Sample size | 17 | 13 | 11 |
| Mean number of binges per week | 0.2 | 1.9 | 2.9 |
| Standard deviation | 0.4 | 1.7 | 2.0 |

*Source:* Agras, W. S., *et al.* "Does interpersonal therapy help patients with binge eating disorder who fail to respond to cognitive-behavioral therapy?" *Journal of Consulting and Clinical Psychology*, Vol. 63, No. 3, June 1995, p. 358 (Table 1).

three treatments (CBT-WLT, CBT-IPT, and Control). Although the ANOVA tables were not provided in the article, sufficient information is provided in the table to reconstruct them.

**a.** Compute SST for the ANOVA using the formula on p. 449:

$$SST = \sum_{i=1}^{3} n_i(\bar{x}_i - \bar{x})^2$$

where $\bar{x}$ is the overall mean number of binges per week of all 41 subjects. [*Hint:* $\bar{x} = (\sum_{i=1}^{3} n_i\bar{x}_i)/41$].

**b.** Recall that SSE for the ANOVA can be written as

$$SSE = (n_1 - 1)s_1^2 + (n_2 - 1)s_2^2 + (n_3 - 1)s_3^2$$

where $s_1^2, s_2^2$, and $s_3^2$ are the sample variances associated with the three treatments. Compute SSE for the ANOVA.

**c.** Use the results, parts **a** and **b,** to construct the ANOVA table.

**d.** Is there sufficient evidence (at $\alpha = .01$) of differences among the mean number of binges experienced per week by the subjects in the three groups?

**e.** Comment on the validity of the ANOVA assumptions. How might this impact the results of the study?

**f.** Comment on the use of a completely randomized design for this study. Have subjects been randomly and independently assigned to each group? How might this impact the results of the study?

## 10.3   MULTIPLE COMPARISONS OF MEANS

Consider a completely randomized design with three treatments, A, B, and C. Suppose we determine that the treatment means are statistically different via the ANOVA $F$-test of Section 10.2. To complete the analysis, we want to rank the three treatment means. As mentioned in Section 10.2, we start by placing confidence intervals on the difference between various pairs of treatment means in the experiment. In the three-treatment experiment, for example, we would construct confidence intervals for the following differences: $\mu_A - \mu_B, \mu_A - \mu_C$, and $\mu_B - \mu_C$.

In general, if there are $p$ treatment means, there are

$$c = p(p - 1)/2$$

pairs of means that can be compared. If we want to have $100(1 - \alpha)\%$ confidence that each of the $c$ confidence intervals contains the true difference it is intended to estimate, we must use a smaller value of $\alpha$ for each individual confidence interval than we would use for a single interval. For example, suppose we want to rank the means of the three treatments, A, B, and C, with 95% confidence that all three confidence intervals comparing the means contain the true differences between the treatment means. Then each individual confidence interval will need to be constructed using a level of significance smaller than $\alpha = .05$ in order to have 95% confidence that the three intervals collectively include the true differences.*

To make **multiple comparisons** of a set of treatment means we can use a number of procedures that, under various assumptions, ensure that the overall confidence level associated with all the comparisons remains at or above the specified $100(1 - \alpha)\%$ level. Three widely used techniques are the Bonferroni, Scheffé, and Tukey methods. For each of these procedures, the risk of making a Type I error applies to the comparisons of the treatment means in the experiment; thus, the value of $\alpha$ selected is called an **experimentwise error rate** (in contrast to a **comparisonwise error rate**).

The choice of a multiple comparisons method in ANOVA will depend on the type of experimental design used and the comparisons of interest to the analyst. For example, **Tukey** (1949) developed his procedure specifically for pairwise comparisons when the sample sizes of the treatments are equal. The **Bonferroni** method (see Miller, 1981), like the Tukey procedure, can be applied when pairwise comparisons are of interest; however, Bonferroni's method does not require equal sample sizes. **Scheffé** (1953) developed a more general procedure for comparing all possible linear combinations of treatment means (called *contrasts*). Consequently, when making pairwise comparisons, the confidence intervals produced by Scheffé's method will generally be wider than the Tukey or Bonferroni confidence intervals.

The formulas for constructing confidence intervals for differences between treatment means using the Tukey, Bonferroni, or Scheffé method are beyond the scope of this text. However, these procedures (and many others) are available in the ANOVA programs of most statistical software packages. The programs generate a confidence interval for the difference between two treatment means for all possible pairs of treatments, based on the experimentwise error rate ($\alpha$) selected by the analyst.

**EXAMPLE 10.5**

Refer to the completely randomized design of Example 10.3, in which we concluded that at least two of the four brands of golf balls are associated with different mean distances traveled when struck with a driver.

a. Use Tukey's multiple comparisons procedure to rank the treatment means with an overall confidence level of 95%.
b. Estimate the mean distance traveled for balls manufactured by the brand with the highest rank.

### Solution

a. To rank the treatment means with an overall confidence level of .95, we require the experimentwise error rate of $\alpha = .05$. The confidence intervals generated by Tukey's method appear at the bottom of the SAS ANOVA printout, Figure 10.11. Note that for any pair of means, $\mu_i$ and $\mu_j$, SAS computes two confidence intervals—one for $(\mu_i - \mu_j)$ and one for $(\mu_j - \mu_i)$. Only one of these intervals is necessary to decide whether the means differ significantly.

---

*The reason each interval must be formed at a higher confidence level than that specified for the collection of intervals can be demonstrated as follows:

$P$ {At least one of $c$ intervals fails to contain the true difference}

$= 1 - P$ {All $c$ intervals contain the true differences}

$= 1 - (1 - \alpha)^c \geq \alpha$

Thus, to make this probability of at least one failure equal to $\alpha$, we must specify the individual levels of significance to be less than $\alpha$.

**FIGURE 10.11**

*SAS Printout for Example 10.5*

The ANOVA Procedure

Dependent Variable: DISTANCE

| Source | DF | Sum of Squares | Mean Square | F Value | Pr > F |
|---|---|---|---|---|---|
| Model | 3 | 2794.388750 | 931.462917 | 43.99 | <.0001 |
| Error | 36 | 762.301000 | 21.175028 | | |
| Corrected Total | 39 | 3556.689750 | | | |

| R-Square | Coeff Var | Root MSE | DISTANCE Mean |
|---|---|---|---|
| 0.785671 | 1.785118 | 4.601633 | 257.7775 |

| Source BRAND | DF | Anova SS | Mean Square | F Value | Pr > F |
|---|---|---|---|---|---|
| | 3 | 2794.388750 | 931.462917 | 43.99 | <.0001 |

Tukey's Studentized Range (HSD) Test for distance

NOTE: This test controls the Type I experimentwise error rate.

| | |
|---|---|
| Alpha | 0.05 |
| Error Degrees of Freedom | 36 |
| Error Mean Square | 21.17503 |
| Critical Value of Studentized Range | 3.80880 |
| Minimum Significant Difference | 5.5424 |

Comparisons significant at the 0.05 level are indicated by ***.

| BRAND Comparison | Difference Between Means | Simultaneous 95% Confidence Limits | | |
|---|---|---|---|---|
| C - B | 8.890 | 3.348 | 14.432 | *** |
| C - A | 19.170 | 13.628 | 24.712 | *** |
| C - D | 20.630 | 15.088 | 26.172 | *** |
| | | | | |
| B - C | -8.890 | -14.432 | -3.348 | *** |
| B - A | 10.280 | 4.738 | 15.822 | *** |
| B - D | 11.740 | 6.198 | 17.282 | *** |
| | | | | |
| A - C | -19.170 | -24.712 | -13.628 | *** |
| A - B | -10.280 | -15.822 | -4.738 | *** |
| A - D | 1.460 | -4.082 | 7.002 | |
| | | | | |
| D - C | -20.630 | -26.172 | -15.088 | *** |
| D - B | -11.740 | -17.282 | -6.198 | *** |
| D - A | -1.460 | -7.002 | 4.082 | |

In this example, we have $p = 4$ brand means to compare. Consequently, the number of relevant pairwise comparisons—that is, the number of nonredundant confidence intervals—is $c = 4(3)/2 = 6$. These six intervals, highlighted in Figure 10.11, are given below.

| Brand Comparison | Confidence Interval |
|---|---|
| $(\mu_A - \mu_B)$ | $(-15.822, -4.738)$ |
| $(\mu_A - \mu_C)$ | $(-24.712, -13.628)$ |
| $(\mu_A - \mu_D)$ | $(-4.082, 7.002)$ |
| $(\mu_B - \mu_C)$ | $(-14.432, -3.348)$ |
| $(\mu_B - \mu_D)$ | $(6.198, 17.282)$ |
| $(\mu_C - \mu_D)$ | $(15.088, 26.172)$ |

We are 95% confident that the intervals *collectively* contain all the differences between the true brand mean distances. Note that intervals that contain 0, such as the (Brand A–Brand D) interval from −4.082 to 7.002, do not support a conclusion that the true brand mean distances differ. If both endpoints of the interval are positive, as with the (Brand B–Brand D) interval from 6.198 to 17.282, the implication is that the first brand (B) mean distance exceeds the second (D). Conversely, if both endpoints of the interval are negative, as with the (Brand A–Brand C) interval from −24.712 to −13.628, the implication is that the second brand (C) mean distance exceeds the first brand (A) mean distance. A convenient summary of the results of the Tukey multiple comparisons is a listing of the brand means from highest to lowest, with a solid line connecting those that are *not* significantly different. This summary is shown in Figure 10.12. The interpretation is that brand C's mean distance exceeds all others; brand B's mean exceeds that of brands A and D; and the means of brands A and D do not differ significantly. All these inferences are made with 95% confidence, the overall confidence level of the Tukey multiple comparisons.

**FIGURE 10.12**

*Summary of Tukey Multiple Comparisons*

| BRAND | MEAN |
|-------|-------|
| C | 270.0 |
| B | 261.1 |
| A | 250.8 |
| D | 249.3 |

b. Brand C is ranked highest; thus, we want a confidence interval for $\mu_C$. Since the samples were selected independently in a completely randomized design, a confidence interval for an individual treatment mean is obtained with the one-sample $t$ confidence interval of Section 7.2, using the standard deviation, $s = \sqrt{\text{MSE}}$, as the measure of sampling variability for the experiment. A 95% confidence interval on the mean distance traveled by brand C (apparently the "longest ball" of those tested), is

$$\bar{x}_C \pm t_{.025}s\sqrt{1/n}$$

where $n = 10, t_{.025} \approx 2$ (based on 36 degrees of freedom), and $s = 4.601633$ (obtained from the SAS printout, Figure 10.11). Substituting, we obtain

$$270.0 \pm (2)(4.60)(\sqrt{.1})$$
$$270.0 \pm 2.9 \quad \text{or} \quad (267.1, 272.9)$$

Thus, we are 95% confident that the true mean distance traveled for brand C is between 267.1 and 272.9 yards, when hit with a driver by Iron Byron. ∎

Many of the available statistical software packages that have multiple comparisons routines will also produce the rankings shown in Figure 10.12. For example, the SAS printout in Figure 10.13 displays the ranking of the mean distances for the four golf ball brands using Bonferroni's method.

The experimentwise error rate (.05), MSE value (21.17503), and critical $t$ value (2.79197) used in the analysis are highlighted on the printout. The Bonferroni rankings of the means are displayed at the bottom of Figure 10.13. Instead of a solid line, note that SAS uses a **Bon Grouping** letter to connect means that are not significantly different. You can see that the mean for brand C is ranked highest, followed by the mean for brand B. These two brand means are significantly different since they are associated with a different Bon Grouping letter. Brands A and D are ranked lowest and their means are not significantly different (since they have the same Bon Grouping letter).

**FIGURE 10.13**

*SAS Bonferroni Analysis for Four Golf Ball Brands*

The ANOVA Procedure

Bonferroni (Dunn) t Tests for DISTANCE

NOTE: This test controls the Type I experimentwise error rate, but it generally has a higher Type II error rate than REGWQ.

| Alpha | 0.05 |
|---|---|
| Error Degrees of Freedom | 36 |
| Error Mean Square | 21.17503 |
| Critical Value of t | 2.79197 |
| Minimum Significant Difference | 5.7456 |

Means with the same letter are not significantly different.

| Bon Grouping | Mean | N | BRAND |
|---|---|---|---|
| A | 269.950 | 10 | C |
| B | 261.060 | 10 | B |
| C | 250.780 | 10 | A |
| C | | | |
| C | 249.320 | 10 | D |

Remember that the Tukey and Bonferroni methods are just two of numerous multiple comparisons procedures available. Another technique may be more appropriate for the experimental design you employ. Consult the references for details on these other methods and when they should be applied.

 EXERCISES 10.30–10.40

## Learning the Mechanics

**10.30** Consider a completely randomized design with $p$ treatments. Assume all pairwise comparisons of treatment means are to be made using a multiple comparisons procedure. Determine the total number of treatment means to be compared for the following values of $p$.
 **a.** $p = 3$
 **b.** $p = 5$
 **c.** $p = 4$
 **d.** $p = 10$

**10.31** Define an experimentwise error rate and a comparisonwise error rate.

**10.32** Consider a completely randomized design with 5 treatments, A, B, C, D, and E. The ANOVA $F$-test revealed significant differences among the means. A multiple comparisons procedure was used to compare all possible pairs of treatment means at $\alpha = .05$. The ranking of the 5 treatment means is summarized below. Identify which pairs of means are significantly different.
 **a.** $\overline{A \quad C \quad E \quad B \quad D}$
 **b.** $\overline{A \quad C \quad E} \quad \overline{B \quad D}$
 **c.** $A \quad \overline{C \quad E \quad B \quad D}$
 **d.** $\overline{A \quad C \quad E \quad B \quad D}$

## Applying the Concepts—Basic

**10.33** Refer to the *Nature* (Aug. 2000) study of robots trained to behave like ants, Exercise 10.21 (p. 460). Multiple comparisons of mean energy expended for the four colony sizes were conducted using an experimentwise error rate of .05. The results are summarized below.

| Sample mean: | .97 | .95 | .93 | .80 |
|---|---|---|---|---|
| Group size: | 3 | 6 | 9 | 12 |

 **a.** How many pairwise comparisons are conducted in this analysis?
 **b.** Interpret the results shown in the table.

**10.34** Refer to the *American Journal of Political Science* (Jan. 1998) study of the attitudes of three groups of professionals (scientists, journalists, and federal government policymakers) regarding the safety of nuclear power plants, Exercise 10.22 (p. 460). The mean safety scores for the groups were:

| Government officials | 4.2 |
|---|---|
| Scientists | 4.1 |
| Journalists | 3.7 |

a. Determine the number of pairwise comparisons of treatment means that can be made in this study.

b. Using an experimentwise error rate of $\alpha = .05$, Tukey's minimum significant difference for comparing means is .23. Use this information to conduct a multiple comparisons of the safety score means. Fully interpret the results.

**10.35** Refer to *The Archives of Disease in Childhood* (Apr. 2000) comparison of the average heights of the three groups of Australian school children based on age, Exercise 10.23 (p. 460). The three height means for boys were ranked using the Bonferroni method at $\alpha = .05$. The results are summarized below. (Recall that all height measurements were standardized using z-scores.)

| Sample mean: | 0.16 | 0.33 | 0.33 |
|---|---|---|---|
| Age group: | Oldest | Youngest | Middle |

a. Is there a significant difference between the standardized height means for the oldest and youngest boys?

b. Is there a significant difference between the standardized height means for the oldest and middle-aged boys?

c. Is there a significant difference between the standardized height means for the youngest and middle-aged boys?

d. What is the experimentwise error rate for the inferences made in parts **a–c**? Interpret this value.

e. The researchers did not perform a Bonferroni analysis of the height means for the three groups of girls. Explain why not.

**10.36** Refer to the *Journal of Hazardous Materials* study of mean sorption rates for three types of organic solvents, Exercise 10.24 (p. 461). SAS was used to produce Bonferroni confidence intervals for all pairs of treatment means. The SAS printout is shown above right.

a. Find and interpret the experimentwise error rate shown on the printout.

b. Locate and interpret the confidence interval for the difference between the mean sorption rates of aromatics and esters.

c. Use the confidence intervals to determine which pairs of treatment means are significantly different.

### Applying the Concepts—Intermediate

**10.37** Does recalling a traumatic dental experience increase your level of anxiety at the dentist's office? In a study published in *Psychological Reports* (Aug. 1997), researchers at Wittenberg University randomly assigned 74 undergraduate psychology students to one of three experimental conditions. Subjects in the "Slide" condition viewed ten slides of scenes from a dental office. Subjects in the "Questionnaire" condition completed a full dental history questionnaire; one of the questions asked them to describe their worst dental experience. Subjects in the "Control" condition received no actual

**SAS Output for Exercise 10.36**

```
                The ANOVA Procedure

        Bonferroni (Dunn) t Tests for SORPRATE

NOTE: This test controls the Type I experimentwise error rate, but it
      generally has a higher Type II error rate than Tukey's for all
      pairwise comparisons.

              Alpha                        0.05
              Error Degrees of Freedom       29
              Error Mean Square         0.067426
              Critical Value of t        2.54091

Comparisons significant at the 0.05 level are indicated by ***.

                            Difference
            SOLVENT          Between      Simultaneous 95%
          Comparison          Means      Confidence Limits

        CHLOR - AROMA        0.06403    -0.25657   0.38462
        CHLOR - ESTER        0.67625     0.38740   0.96510   ***
        AROMA - CHLOR       -0.06403    -0.38462   0.25657
        AROMA - ESTER        0.61222     0.33403   0.89041   ***
        ESTER - CHLOR       -0.67625    -0.96510  -0.38740   ***
        ESTER - AROMA       -0.61222    -0.89041  -0.33403   ***
```

treatment. All students then completed the Dental Fear Scale, with scores ranging from 27 (no fear) to 135 (extreme fear). The sample dental fear means for the Slide, Questionnaire, and Control groups were reported as 43.1, 53.8, and 41.8, respectively.

a. A completely randomized design ANOVA was carried out on the data, with the following results: $F = 4.43$, p-value $< .05$. Interpret these results.

b. According to the article, a Bonferroni ranking of the three dental fear means (at $\alpha = .05$) "indicated a significant difference between the mean scores on the Dental Fear Scale for the Control and Questionnaire groups, but not for the means between the Control and Slide groups." Summarize these results in a chart similar to Figure 10.12 (p. 466).

**10.38** Refer to the *Journal of Nonverbal Behavior* study of students' evaluations of facial expressions, Exercise 10.27 (p. 462). Use a statistical software package with Tukey's method to rank the dominance rating means of the six facial expressions. (Use an experimentwise error rate of $\alpha = .10$.)

**10.39** Refer to the *Marine Technology* study of major ocean spills by tankers, Exercise 10.28 (p. 462). Use a statistical software package with Bonferroni's method to rank the spillage amount means of the four accident types. (Use an experimentwise error rate of $\alpha = .01$).

### Applying the Concepts—Advanced

**10.40** Refer to the *Journal of Consulting and Clinical Psychology* study of binge eaters, Exercise 10.29 (p. 462). You found evidence of a difference among the treatment means for the three treatments, CBT-WT, CBT-IPT, and Control. Conduct a Bonferroni analysis to rank the three treatment means. Use an experiment-wise error rate of $\alpha = .03$. Interpret the results for the researchers. [*Hint:* The Bonferroni formula for a confidence interval for the difference $(\mu_i - \mu_j)$ is:

$$(\bar{x}_i - \bar{x}_j) \pm t_{\alpha*/2}(s)\sqrt{(1/n_i) + (1/n_j)}$$

where $\alpha* = 2\alpha/[(p)(p - 1)]$, $\alpha$ is the experimentwise error rate, and $p$ is the total number of treatment means compared.]

## 10.4  THE RANDOMIZED BLOCK DESIGN

If the completely randomized design results in nonrejection of the null hypothesis that the treatment means differ because the sampling variability (as measured by MSE) is large, we may want to consider an experimental design that better controls the variability. In contrast to the selection of independent samples of experimental units specified by the completely randomized design, the *randomized block design* utilizes experimental units that are *matched sets*, assigning one from each set to each treatment. The matched sets of experimental units are called *blocks*. The theory behind the randomized block design is that the sampling variability of the experimental units in each block will be reduced, in turn reducing the measure of error, MSE.

> **DEFINITION 10.8**
> The **randomized block design** consists of a two-step procedure:
>
> 1. Matched sets of experimental units, called **blocks**, are formed, each block consisting of $p$ experimental units (where $p$ is the number of treatments). The $b$ blocks should consist of experimental units that are as similar as possible.
> 2. One experimental unit from each block is randomly assigned to each treatment, resulting in a total of $n = bp$ responses.

For example, if we wish to compare SAT scores of female and male high school seniors, we could select independent random samples of five females and five males, and analyze the results of the completely randomized design as outlined in Section 10.2. Or, we could select matched pairs of females and males according to their scholastic records, and analyze the SAT scores of the pairs. For instance, we could select pairs of students with approximately the same GPAs from the same high school. Five such pairs (blocks) are depicted in Table 10.4. Note that this is just a *paired difference experiment*, first discussed in Section 9.2.

As before, the variation between the treatment means is measured by squaring the distance between each treatment mean and the overall mean, multiplying

**TABLE 10.4  Randomized Block Design: SAT Score Comparison**

| Block | Female SAT Score | Male SAT Score | Block Mean |
|---|---|---|---|
| 1 (School A, 2.75 GPA) | 540 | 530 | 535 |
| 2 (School B, 3.00 GPA) | 570 | 550 | 560 |
| 3 (School C, 3.25 GPA) | 590 | 580 | 585 |
| 4 (School D, 3.50 GPA) | 640 | 620 | 630 |
| 5 (School E, 3.75 GPA) | 690 | 690 | 690 |
| Treatment mean | 606 | 594 | |

each squared distance by the number of measurements for the treatment, and summing over treatments:

$$SST = \sum_{i=1}^{p} b(\bar{x}_{T_i} - \bar{x})^2$$
$$= 5(606 - 600)^2 + 5(594 - 600)^2 = 360$$

where $\bar{x}_{T_i}$ represents the sample mean for the $i$th treatment, $b$ (the number of blocks) is the number of measurements for each treatment, and $p$ is the number of treatments.

The blocks also account for some of the variation among the different responses. That is, just as SST measures the variation between the female and male means, we can calculate a measure of variation among the five block means representing different schools and scholastic abilities. Analogous to the computation of SST, we sum the squares of the differences between each block mean and the overall mean, multiplying each squared difference by the number of measurements for each block, and sum over blocks to calculate the **Sum of Squares for Blocks** (SSB):

$$SSB = \sum_{i=1}^{b} p(\bar{x}_{B_i} - \bar{x})^2$$
$$= 2(535 - 600)^2 + 2(560 - 600)^2 + 2(585 - 600)^2 +$$
$$2(630 - 600)^2 + 2(690 - 600)^2$$
$$= 30{,}100$$

where $\bar{x}_{B_i}$ represents the sample mean for the $i$th block and $p$ (the number of treatments) is the number of measurements in each block. As we expect, the variation in SAT scores attributable to Schools and Levels of scholastic achievement is apparently large.

As before, we want to compare the variability attributed to treatments with that which is attributed to sampling variability. In a randomized block design, the sampling variability is measured by subtracting that portion attributed to treatments and blocks from the Total Sum of Squares, SS(Total). The total variation is the sum of squared differences of each measurement from the overall mean:

$$SS(\text{Total}) = \sum_{i=1}^{n} (x_i - \bar{x})^2$$
$$= (540 - 600)^2 + (530 - 600)^2 + (570 - 600)^2 + (550 - 600)^2 +$$
$$\cdots + (690 - 600)^2$$
$$= 30{,}600$$

Then the variation attributable to sampling error is found by subtraction:

$$SSE = SS(\text{Total}) - SST - SSB = 30{,}600 - 360 - 30{,}100 = 140$$

In summary, the Total Sum of Squares, 30,600, is divided into three components: 360 attributed to treatments (Gender), 30,100 attributed to blocks (Scholastic ability), and 140 attributed to sampling error.

The mean squares associated with each source of variability are obtained by dividing the sum of squares by the appropriate number of degrees of freedom. The partitioning of the Total Sum of Squares and the total degrees of freedom for a randomized block experiment is summarized in Figure 10.14.

**FIGURE 10.14**

*Partitioning of the Total Sum of Squares for the Randomized Block Design*

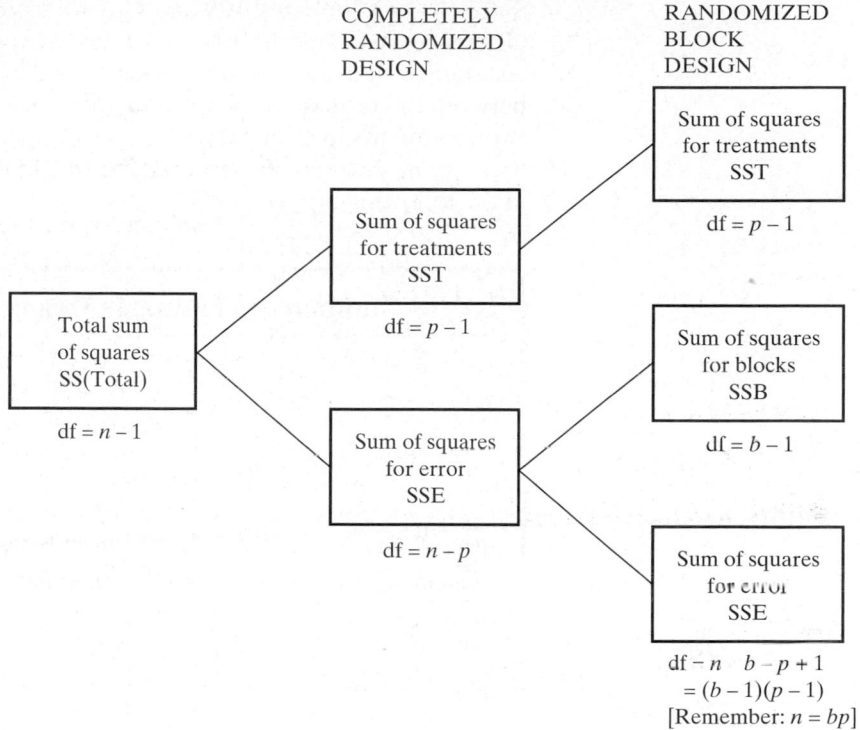

To determine whether we can reject the null hypothesis that the treatment means are equal in favor of the alternative that at least two of them differ, we calculate

$$\text{MST} = \frac{\text{SST}}{p - 1} = \frac{360}{2 - 1} = 360$$

$$\text{MSE} = \frac{\text{SSE}}{n - b - p + 1} = \frac{140}{10 - 5 - 2 + 1} = 35$$

The *F*-ratio that is used to test the hypothesis is

$$F = \frac{360}{35} = 10.29$$

Comparing this ratio to the tabled *F* value corresponding to $\alpha = .05$, $\nu_1 = (p - 1) = 1$ degree of freedom in the numerator, and $\nu_2 = (n - b - p + 1) = 4$ degrees of freedom in the denominator, we find that

$$F = 10.29 > F_{.05} = 7.71$$

which indicates that we should reject the null hypothesis and conclude that the mean SAT scores differ for females and males.

If you review Section 9.2, you will find that the analysis of a paired difference experiment results in a one-sample *t*-test on the differences between the treatment responses within each block. Applying the procedure to the differences between female and male scores in Table 10.4, we find

$$t = \frac{\bar{x}_D}{s_D / \sqrt{n_D}} = \frac{12}{\sqrt{70} / \sqrt{5}} = 3.207$$

At the .05 level of significance with $(n_D - 1) = 4$ degrees of freedom,

$$t = 3.207 > t_{.025} = 2.776$$

Since $t^2 = (3.207)^2 = 10.29$ and $t_{.025}^2 = (2.776)^2 = 7.71$, we find that the paired difference $t$-test and the ANOVA $F$-test are equivalent, with both the calculated test statistics and the rejection region related by the formula $F = t^2$. The difference between the tests is that the paired difference $t$-test can be used to compare only two treatments in a randomized block design, whereas the $F$-test can be applied to *two or more* treatments in a randomized block design. The $F$-test is summarized in the following box.

---

### Test to Compare $p$ Treatment Means: Randomized Block Design

$H_0: \mu_1 = \mu_2 = \cdots = \mu_p$
$H_a:$ At least two treatment means differ

*Test statistic:* $\quad F = \dfrac{\text{MST}}{\text{MSE}}$

*Rejection region:* $F > F_\alpha$, where $F_\alpha$ is based on $(p - 1)$ numerator degrees of freedom and $(n - b - p + 1)$ denominator degrees of freedom.

*Assumptions:* 1. The probability distributions of observations corresponding to all the block–treatment combinations are normal.
           2. The variances of all the probability distributions are equal.

---

Note that the assumptions concern the probability distributions associated with each block–treatment combination. The experimental unit selected for each combination is assumed to have been randomly selected from all possible experimental units for that combination, and the response is assumed to be normally distributed with the same variance for each of the block–treatment combinations. For example, the $F$-test comparing female and male SAT score means requires the scores for each combination of gender and scholastic ability (e.g., females with 3.25 GPA) to be normally distributed with the same variance as the other combinations employed in the experiment.

The calculation formulas for randomized block designs are given in Appendix C. We will rely on statistical software packages to analyze randomized block designs and to obtain the necessary ingredients for testing the null hypothesis that the treatment means are equal.

**EXAMPLE 10.6**

Refer to Examples 10.3–10.5. Suppose the USGA wants to compare the mean distances associated with the four brands of golf balls when struck by a driver, but wishes to employ human golfers rather than the robot, Iron Byron. Assume that 10 balls of each brand are to be utilized in the experiment.

a. Explain how a completely randomized design could be employed.
b. Explain how a randomized block design could be employed.
c. Which design is likely to provide more information about the differences among the brand mean distances?

### Solution

a. Since the completely randomized design calls for independent samples, we can employ such a design by randomly selecting 40 golfers and then randomly assigning 10 golfers to each of the four brands. Finally, each golfer will strike the

# FIGURE 10.15

*Illustration of Completely
Randomized Design and
Randomized Block Design:
Comparison of Four Golf Ball
Brands*

a. Completely randomized design

b. Randomized block design

ball of the assigned brand, and the distance will be recorded. The design is illustrated in Figure 10.15a.

b. The randomized block design employs blocks of relatively homogeneous experimental units. For example, we could randomly select 10 golfers, and permit each golfer to hit four balls, one of each brand, in a random sequence. Then each golfer is a block, with each treatment (brand) assigned to each block (golfer). The design is summarized in Figure 10.15b.

c. Because we expect much more variability among distances generated by "real" golfers than by Iron Byron, we would expect the randomized block design to control the variability better than the completely randomized design. That is, with 40 different golfers, we would expect the sampling variability among the measured distances within each brand to be greater than that among the four distances generated by each of 10 golfers hitting one ball of each brand. ∎

**EXAMPLE 10.7** Refer to Example 10.6. Suppose the randomized block design of part b is employed, utilizing a random sample of 10 golfers with each golfer using a driver to hit four balls, one of each brand, in a random sequence.

a. Set up a test of the research hypothesis that the brand mean distances differ. Use $\alpha = .05$.

b. The data for the experiment are given in Table 10.5. Use statistical software to analyze the data, and conduct the test set up in part a.

⊘ GOLFRBD
**TABLE 10.5   Distance Data for Randomized Block Design**

| Golfer (Block) | Brand A | Brand B | Brand C | Brand D |
|:---:|:---:|:---:|:---:|:---:|
| 1 | 202.4 | 203.2 | 223.7 | 203.6 |
| 2 | 242.0 | 248.7 | 259.8 | 240.7 |
| 3 | 220.4 | 227.3 | 240.0 | 207.4 |
| 4 | 230.0 | 243.1 | 247.7 | 226.9 |
| 5 | 191.6 | 211.4 | 218.7 | 200.1 |
| 6 | 247.7 | 253.0 | 268.1 | 244.0 |
| 7 | 214.8 | 214.8 | 233.9 | 195.8 |
| 8 | 245.4 | 243.6 | 257.8 | 227.9 |
| 9 | 224.0 | 231.5 | 238.2 | 215.7 |
| 10 | 252.2 | 255.2 | 265.4 | 245.2 |
| Sample means | 227.0 | 233.2 | 245.3 | 220.7 |

### Solution

a. We want to test whether the data in Table 10.5 provide sufficient evidence to conclude that the brand mean distances differ. Denoting the population mean of the $i$th brand by $\mu_i$, we test

$$H_0: \mu_1 = \mu_2 = \mu_3 = \mu_4$$
$$H_a: \text{The mean distances differ for at least two of the brands}$$

The test statistic compares the variation among the four treatment (brand) means to the sampling variability within each of the treatments.

$$\textit{Test statistic:} \quad F = \frac{\text{MST}}{\text{MSE}}$$

*Rejection region:* $F > F_\alpha = F_{.05}$, with $\nu_1 = (p - 1) = 3$ numerator degrees of freedom and $\nu_2 = (n - p - b + 1) = 27$ denominator degrees of freedom. From Table IX of Appendix A, we find $F_{.05} = 2.96$. Thus, we will reject $H_0$ if $F > 2.96$.

The assumptions necessary to ensure the validity of the test are as follows: (1) The probability distributions of the distances for each brand–golfer combination are normal. (2) The variances of the distance probability distributions for each brand–golfer combination are equal.

b. MINITAB was used to analyze the data in Table 10.5, and the result is shown in Figure 10.16. The values of MST and MSE (highlighted on the printout) are 1,099.6 and 20.2, respectively. The $F$-ratio for Brand (also highlighted on the printout) is $F = 54.31$, which exceeds the tabled value of 2.96. We therefore

**FIGURE 10.16**

*MINITAB Printout for Randomized Block Design: Golf Ball Brand Comparison*

```
Analysis of Variance for DISTANCE

Source      DF        SS        MS        F         P
BRAND        3     3298.7    1099.6    54.31     0.000
GOLFER       9    12073.9    1341.5    66.26     0.000
Error       27      546.6      20.2
Total       39    15919.2
```

reject the null hypothesis at the $\alpha = .05$ level of significance, concluding that at least two of the brands differ with respect to mean distance traveled when struck by the driver. (This result is confirmed by noting that the observed significance level of the test, highlighted on the printout, is $p \approx 0.$) ∎

The results of an ANOVA can be summarized in a simple tabular format, similar to that utilized for the completely randomized design in Section 10.2. The general form of the table is shown in Table 10.6, and that for Example 10.7 is given in Table 10.7. Note that the randomized block design is characterized by three sources of variation—Treatments, Blocks, and Error—which sum to the Total Sum of Squares. We hope that employing blocks of experimental units will reduce the error variability, thereby making the test for comparing treatment means more powerful.

When the *F*-test results in the rejection of the null hypothesis that the treatment means are equal, we will usually want to compare the various pairs of treatment means to determine which specific pairs differ. We can employ a multiple comparisons procedure as in Section 10.3. The number of pairs of means to be compared will again be $c = p(p - 1)/2$, where $p$ is the number of treatment means. In Example 10.7, $c = 4(3)/2 = 6$; that is, there are six pairs of golf ball brand means to be compared.

**TABLE 10.6  General ANOVA Summary Table for a Randomized Block Design**

| Source | df | SS | MS | F |
|---|---|---|---|---|
| Treatment | $p - 1$ | SST | MST | MST/MSE |
| Block | $b - 1$ | SSB | MSB | |
| Error | $n - p - b + 1$ | SSE | MSE | |
| Total | $n - 1$ | SS(Total) | | |

**TABLE 10.7  ANOVA Table for Example 10.7**

| Source | df | SS | MS | F | p |
|---|---|---|---|---|---|
| Treatment (Brand) | 3 | 3,298.7 | 1,099.6 | 54.31 | .000 |
| Block (Golfer) | 9 | 12,073.9 | 1,341.5 | | |
| Error | 27 | 546.6 | 20.2 | | |
| Total | 39 | 15,919.2 | | | |

**EXAMPLE 10.8**

Bonferroni's procedure is used to compare the mean distances of the four golf ball brands in Example 10.7. The resulting confidence intervals, with an experimentwise error rate of $\alpha = .05$, are shown in the SPSS printout, Figure 10.17. Interpret the results.

### Solution

Note that 12 confidence intervals are shown in Figure 10.17 rather than 6. SPSS (like SAS) computes intervals for both $\mu_i - \mu_j$ and $\mu_j - \mu_i$, $i \neq j$. Only half of these are necessary to conduct the analysis, and these are highlighted on the printout. The intervals (rounded) are summarized below:

$$(\mu_A - \mu_B): \quad (-11.9, -.4)$$
$$(\mu_A - \mu_C): \quad (-24.0, -12.6)$$
$$(\mu_A - \mu_D): \quad (.6, 12.0)$$

**FIGURE 10.17**

*SPSS Printout of Bonferroni*
*Confidence Intervals for*
*Randomized Block Design*

**Multiple Comparisons**

Dependent Variable: DISTANCE
Bonferroni

| (I) BRAND | (J) BRAND | Mean Difference (I-J) | Std. Error | Sig. | 95% Confidence Interval Lower Bound | Upper Bound |
|---|---|---|---|---|---|---|
| A | B | -6.1300* | 2.0122 | .031 | -11.8586 | -.4014 |
|   | C | -18.2800* | 2.0122 | .000 | -24.0086 | -12.5514 |
|   | D | 6.3200* | 2.0122 | .024 | .5914 | 12.0486 |
| B | A | 6.1300* | 2.0122 | .031 | .4014 | 11.8586 |
|   | C | -12.1500* | 2.0122 | .000 | -17.8786 | -6.4214 |
|   | D | 12.4500* | 2.0122 | .000 | 6.7214 | 18.1786 |
| C | A | 18.2800* | 2.0122 | .000 | 12.5514 | 24.0086 |
|   | B | 12.1500* | 2.0122 | .000 | 6.4214 | 17.8786 |
|   | D | 24.6000* | 2.0122 | .000 | 18.8714 | 30.3286 |
| D | A | -6.3200* | 2.0122 | .024 | -12.0486 | -.5914 |
|   | B | -12.4500* | 2.0122 | .000 | -18.1786 | -6.7214 |
|   | C | -24.6000* | 2.0122 | .000 | -30.3286 | -18.8714 |

Based on observed means.
\*. The mean difference is significant at the .05 level.

$$(\mu_B - \mu_C): \quad (-17.9, -6.4)$$
$$(\mu_B - \mu_D): \quad (6.7, 18.2)$$
$$(\mu_C - \mu_D): \quad (18.9, 30.3)$$

Note that we are 95% confident that all the brand means differ because none of the intervals contains 0. The listing of the brand means in Figure 10.18 has no lines connecting them because there are no nonsignificant differences at the .05 level. ■

**FIGURE 10.18**

*Listing of Brand Means for*
*Randomized Block Design*
*[Note: All differences are*
*statistically significant.]*

| Brand | Mean |
|---|---|
| C | 245.3 |
| B | 233.2 |
| A | 227.0 |
| D | 220.7 |

Unlike the completely randomized design, the randomized block design cannot, in general, be used to estimate individual treatment means. Whereas the completely randomized design employs a random sample for each treatment, the randomized block design does not necessarily employ a random sample of experimental units for each treatment. The experimental units within the blocks are assumed to be randomly selected, but the blocks themselves may not be randomly selected.

We can, however, test the hypothesis that the block means are significantly different. We simply compare the variability attributable to differences among the block means to that associated with sampling variability. The ratio of MSB to MSE is an $F$-ratio similar to that formed in testing treatment means. The $F$ statistic is compared to a tabled value for a specific value of $\alpha$, with numerator degrees of freedom $(b - 1)$ and denominator degrees of freedom $(n - p - b + 1)$. The test is usually given on the same printout as the test for treatment means. Refer to the MINITAB printout in Figure 10.16, and note that the test statistic for comparing the block means is

$$F = \frac{MSB}{MSE} = \frac{MS(\text{Golfers})}{MS(\text{Error})} = \frac{1{,}341.5}{20.2} = 66.26$$

**TABLE 10.8** **ANOVA Table for Randomized Block Design: Test for Blocks Included**

| Source | df | SS | MS | F | p |
|---|---|---|---|---|---|
| Treatments (Brands) | 3 | 3,298.7 | 1,099.6 | 54.31 | .000 |
| Blocks (Golfers) | 9 | 12,073.9 | 1,341.5 | 66.26 | .000 |
| Error | 27 | 546.6 | 20.2 | | |
| Total | 39 | 15,919.2 | | | |

with a *p*-value of .000. Since $\alpha = .05$ exceeds this *p*-value, we conclude that the block means are different. The results of the test are summarized in Table 10.8.

In the golf example, the test for block means confirms our suspicion that the golfers vary significantly; therefore, the use of the block design was a good decision. However, be careful not to conclude that the block design was a mistake if the *F*-test for blocks does not result in rejection of the null hypothesis that the block means are the same. Remember that the possibility of a Type II error exists, and we are not controlling its probability as we are the probability $\alpha$ of a Type I error. If the experimenter believes that the experimental units are more homogeneous within blocks than between blocks, then he or she should use the randomized block design regardless of the results of a single test comparing the block means.

The procedure for conducting an analysis of variance for a randomized block design is summarized in the next box. Remember that the hallmark of this design is the utilization of blocks of homogeneous experimental units in which each treatment is represented.

---

## Steps for Conducting an ANOVA for a Randomized Block Design

1. Be sure the design consists of blocks (preferably, blocks of homogeneous experimental units) and that each treatment is randomly assigned to one experimental unit in each block.

2. If possible, check the assumptions of normality and equal variances for all block–treatment combinations. [*Note:* This may be difficult to do since the design will likely have only one observation for each block–treatment combination.]

3. Create an ANOVA summary table that specifies the variability attributable to Treatments, Blocks, and Error, and that leads to the calculation of the *F* statistic to test the null hypothesis that the treatment means are equal in the population. Use a statistical software package or the calculation formulas in Appendix C to obtain the necessary numerical ingredients.

4. If the *F*-test leads to the conclusion that the means differ, use the Bonferroni, Tukey, or similar procedure to conduct multiple comparisons of as many of the pairs of means as you wish. Use the results to summarize the statistically significant differences among the treatment means. Remember that, in general, the randomized block design cannot be used to form confidence intervals for individual treatment means.

5. If the *F*-test leads to the nonrejection of the null hypothesis that the treatment means are equal, several possibilities exist:

   a. The treatment means are equal—that is, the null hypothesis is true.

*(continues on next page)*

## Steps for Conducting an ANOVA for a Randomized Block Design *(continued)*

b. The treatment means really differ, but other important factors affecting the response are not accounted for by the randomized block design. These factors inflate the sampling variability, as measured by MSE, resulting in smaller values of the $F$ statistic. Either increase the sample size for each treatment, or conduct an experiment that accounts for the other factors affecting the response (as in Section 10.5). Do not automatically reach the former conclusion, since the possibility of a Type II error must be considered if you accept $H_0$.

6. If desired, conduct the $F$-test of the null hypothesis that the block means are equal. Rejection of this hypothesis lends statistical support to the utilization of the randomized block design.

*Note:* It is often difficult to check whether the assumptions for a randomized block design are satisfied. When you feel these assumptions are likely to be violated, a nonparametric procedure is advisable.

## What Do You Do When the Assumptions Are Not Satisfied for the Analysis of Variance for a Completely Randomized Design?

*Answer:* Use a nonparametric statistical method such as the Friedman $F_r$ test of Section 14.5.

 EXERCISES 10.41–10.52

### Learning the Mechanics

**10.41** A randomized block design yielded the ANOVA table shown here.

| Source | df | SS | MS | F |
|---|---|---|---|---|
| Treatments | 4 | 501 | 125.25 | 9.109 |
| Blocks | 2 | 225 | 112.50 | 8.182 |
| Error | 8 | 110 | 13.75 | |
| Total | 14 | 836 | | |

a. How many blocks and treatments were used in the experiment?
b. How many observations were collected in the experiment?
c. Specify the null and alternative hypotheses you would use to compare the treatment means.

d. What test statistic should be used to conduct the hypothesis test of part **c**?
e. Specify the rejection region for the test of parts **c** and **d**. Use $\alpha = .01$.
f. Conduct the test of parts **c–e**, and state the proper conclusion.
g. What assumptions are necessary to ensure the validity of the test you conducted in part **f**?

**10.42** An experiment was conducted using a randomized block design. The data from the experiment are displayed in the following table.

LM10_42

| Treatment | Block | | |
|---|---|---|---|
| | 1 | 2 | 3 |
| 1 | 2 | 3 | 5 |
| 2 | 8 | 6 | 7 |
| 3 | 7 | 6 | 5 |

**a.** Fill in the missing entries in the ANOVA table.

| Source | df | SS | MS | F |
|---|---|---|---|---|
| Treatments | | 21.5555 | | |
| Blocks | | | | |
| Error | | | | |
| Total | | 30.2222 | | |

**b.** Specify the null and alternative hypotheses you would use to investigate whether a difference exists among the treatment means.

**c.** What test statistic should be used in conducting the test of part **b**?

**d.** Describe the Type I and Type II errors associated with the hypothesis test of part **b**.

**e.** Conduct the hypothesis test of part **b** using $\alpha = .05$.

**10.43** A randomized block design was used to compare the mean responses for three treatments. Four blocks of three homogeneous experimental units were selected, and each treatment was randomly assigned to one experimental unit within each block. The data are shown in the table, followed by the SPSS ANOVA printout for this experiment.

**LM10_43**

| Treatment | Block | | | |
|---|---|---|---|---|
| | 1 | 2 | 3 | 4 |
| A | 3.4 | 5.5 | 7.9 | 1.3 |
| B | 4.4 | 5.8 | 9.6 | 2.8 |
| C | 2.2 | 3.4 | 6.9 | .3 |

**a.** Use the printout to fill in the entries in the following ANOVA table.

| Source | df | SS | MS | F |
|---|---|---|---|---|
| Treatments | | | | |
| Blocks | | | | |
| Error | | | | |
| Total | | | | |

**SPSS Output for Exercise 10.43**

**Tests of Between-Subjects Effects**

Dependent Variable: RESPONSE

| Source | Type III Sum of Squares | df | Mean Square | F | Sig. |
|---|---|---|---|---|---|
| Corrected Model | 83.781[a] | 5 | 16.756 | 141.935 | .000 |
| Intercept | 238.521 | 1 | 238.521 | 2020.412 | .000 |
| TRTMENT | 12.032 | 2 | 6.016 | 50.958 | .000 |
| BLOCK | 71.749 | 3 | 23.916 | 202.586 | .000 |
| Error | .708 | 6 | .118 | | |
| Total | 323.010 | 12 | | | |
| Corrected Total | 84.489 | 11 | | | |

a. R Squared = .992 (Adjusted R Squared = .985)

**Multiple Comparisons**

Dependent Variable: RESPONSE
Tukey HSD

| (I) TRTMENT | (J) TRTMENT | Mean Difference (I-J) | Std. Error | Sig. | 95% Confidence Interval | |
|---|---|---|---|---|---|---|
| | | | | | Lower Bound | Upper Bound |
| A | B | -1.125* | .243 | .009 | -1.870 | -.380 |
| | C | 1.325* | .243 | .004 | .580 | 2.070 |
| B | A | 1.125* | .243 | .009 | .380 | 1.870 |
| | C | 2.450* | .243 | .000 | 1.705 | 3.195 |
| C | A | -1.325* | .243 | .004 | -2.070 | -.580 |
| | B | -2.450* | .243 | .000 | -3.195 | -1.705 |

Based on observed means.

*. The mean difference is significant at the .05 level.

**b.** Do the data provide sufficient evidence to indicate that the treatment means differ? Use $\alpha = .05$.

**c.** Do the data provide sufficient evidence to indicate that blocking was effective in reducing the experimental error? Use $\alpha = .05$.

**d.** Use the printout to rank the treatment means at $\alpha = .05$.

**e.** What assumptions are necessary to ensure the validity of the inferences made in parts **b**, **c**, and **d**?

**10.44** Suppose an experiment utilizing a randomized block design has four treatments and nine blocks, for a total of $4 \times 9 = 36$ observations. Assume that the Total Sum of Squares for the response is SS(Total) = 500. For each of the following partitions of SS(Total), test the null hypothesis that the treatment means are equal, and the null hypothesis that the block means are equal. Use $\alpha = .05$ for each test.

**a.** The Sum of Squares for Treatments (SST) is 20% of SS(Total), and the Sum of Squares for Blocks (SSB) is 30% of SS(Total).

**b.** SST is 50% of SS(Total), and SSB is 20% of SS(Total).

**c.** SST is 20% of SS(Total), and SSB is 50% of SS(Total).

**d.** SST is 40% of SS(Total), and SSB is 40% of SS(Total).

**e.** SST is 20% of SS(Total), and SSB is 20% of SS(Total).

## Applying the Concepts—Basic

**10.45** Rugel's pawpaw (yellow squirrel banana) is an endangered species of a dwarf shrub. Biologists from Stetson University conducted an experiment to determine the effects of fire on the shrub's growth (*Florida Scientist*, Spring 1997). Twelve experimental plots of land were selected in a pasture where the shrub is abundant. Within each plot, three pawpaws were randomly selected and treated as follows: one shrub was subjected to fire, another to clipping, and the third was left unmanipulated (a control). After five months, the number of flowers produced by each of the 36 shrubs was determined. The objective of the study was to compare the mean number of flowers produced by pawpaws for the three treatments (fire, clipping, and control).

**a.** Identify the type of experimental design employed, including the treatments, response variable, and experimental units.

**b.** Illustrate the layout of the design using a graphic similar to Figure 10.15.

**c.** The ANOVA of the data resulted in a test statistic of $F = 5.42$ for treatments with $p$-value $= .009$. Interpret this result.

**d.** The three treatment means were compared using Tukey's method at $\alpha = .05$. Interpret the results shown below.

| Mean number of flowers: | 1.17 | 10.58 | 17.08 |
|---|---|---|---|
| Treatment: | Control | Clipping | Burning |

**10.46** Using decoys is a common method of hunting waterfowl. A study in the *Journal of Wildlife Management* (July 1995) compared the effectiveness of three different decoy types—taxidermy-mounted decoys, plastic shell decoys, and full-bodied plastic decoys—in attracting Canada geese to sunken pit blinds. In order to account for an extraneous source of variation, three pit blinds were used as blocks in the experiment. Thus, a randomized block design with three treatments (decoy types) and three blocks (pit blinds) was employed. The response variable was the percentage of a goose flock to approach within 46 meters of the pit blind on a given day. The data are given in the table,* followed by a MINITAB printout of the analysis.

⬤ DECOY

| Blind | Shell | Full-Bodied | Taxidermy-Mounted |
|---|---|---|---|
| 1 | 7.3 | 13.6 | 17.8 |
| 2 | 12.6 | 10.4 | 17.0 |
| 3 | 16.4 | 23.4 | 13.6 |

*Source:* Harrey, W. F., Hindman, L. J., and Rhodes, W. E. "Vulnerability of Canada geese to taxidermy-mounted decoys." *Journal of Wildlife Management*, Vol. 59, No. 3, July 1995, p. 475 (Table 1).

**a.** Find and interpret the $F$ statistic for comparing the response means of the three decoy types.

**MINITAB Output for Exercise 10.46**

```
Analysis of Variance for PERCENT

Source    DF        SS        MS        F         P
DECOY      2      30.1      15.0      0.61     0.589
BLIND      2      44.1      22.1      0.89     0.479
Error      4      99.3      24.8
Total      8     173.6
```

*The actual design employed in the study was more complex than the randomized block design shown here. In the actual study, each number in the table represents the mean daily percentage of goose flocks attracted to the blind, averaged over 13–17 days.

**b.** What assumptions are necessary for the validity of the inference made in part **a**?

**c.** Why is it not necessary to conduct multiple comparisons of the response means for the three decoy types?

**10.47** A study was conducted to investigate the effect of prompting in a walking program (*Health Psychology*, Mar. 1995). Five groups of walkers—27 in each group—agreed to participate by walking for 20 minutes at least one day per week over a 24-week period. The participants were prompted to walk each week via telephone calls, but different prompting schemes were used for each group. Walkers in the control group received no prompting phone calls; walkers in the "frequent/low" group received a call once a week with low structure (i.e., "just touching base"); walkers in the "frequent/high" group received a call once a week with high structure (i.e., goals are set); walkers in the "infrequent/low" group received a call once every 3 weeks with low structure; and walkers in the "infrequent/high" group received a call once every 3 weeks with high structure. The table lists the number of participants in each group who actually walked the minimum requirement each week for weeks 1, 4, 8, 12, 16, and 24. The data were subjected to an analysis of variance for a randomized block design, with the five walker groups representing the treatments and the six time periods (weeks) representing the blocks. A STATISTIX printout of the ANOVA is provided below.

⊘ WALKERS

| Week | Control | Frequent/Low | Frequent/High | Infrequent/Low | Infrequent/High |
|------|---------|--------------|---------------|----------------|-----------------|
| 1 | 7 | 23 | 25 | 21 | 19 |
| 4 | 2 | 19 | 25 | 10 | 12 |
| 8 | 2 | 18 | 19 | 9 | 9 |
| 12 | 2 | 7 | 20 | 8 | 2 |
| 16 | 2 | 18 | 18 | 8 | 7 |
| 24 | 1 | 17 | 17 | 7 | 6 |

*Source:* Lombard, D. N., *et al.* "Walking to meet health guidelines: The effect of prompting frequency and prompt structure." *Health Psychology*, Vol. 14, No. 2, Mar. 1995, p. 167 (Table 2).

**STATISTIX Output for Exercise 10.47**

```
ANALYSIS OF VARIANCE TABLE FOR WALKERS

SOURCE         DF      SS          MS         F         P
-----------    ----    ----------  ---------  -------   -------
PROMPT (A)      4      1185.00     296.250    39.87     0.0000
WEEK (B)        5       386.400     77.2800   10.40     0.0001
A*B            20       148.600      7.43000
-----------    ----    ----------
TOTAL          29      1720.00

TUKEY (HSD) COMPARISON OF MEANS OF WALKERS BY PROMPT

                        HOMOGENEOUS
   PROMPT     MEAN      GROUPS
   --------   -------   -----------
   FREQ-HI    20.667    I
   FREQ-LOW   17.000    I
   INF-LOW    10.500    .. I
   INF-HI      9.1667   .. I
   CONTROL     2.6667   .... I

THERE ARE 3 GROUPS IN WHICH THE MEANS ARE
NOT SIGNIFICANTLY DIFFERENT FROM ONE ANOTHER.

CRITICAL Q VALUE                 4.232    REJECTION LEVEL     0.050
CRITICAL VALUE FOR COMPARISON    4.7095
STANDARD ERROR FOR COMPARISON    1.5737

ERROR TERM USED: PROMPT*WEEK, 20 DF
```

[*Note:* On a STATISTIX randomized block design printout, A*B represents experimental error.]

a. What is the purpose of blocking on weeks in this study?

b. Construct an ANOVA summary table using the printout.

c. Is there sufficient evidence of a difference in the mean number of walkers per week among the five walker groups? Use $\alpha = .05$.

d. Tukey's technique was used to compare all pairs of treatment means with an experimentwise error rate of $\alpha = .05$. The rankings are shown at the bottom of the printout. Interpret these results.

e. What assumptions must hold to ensure the validity of the inferences in parts **c** and **d**?

## Applying the Concepts—Intermediate

**10.48** Eight amateur boxers participated in an experiment to investigate the effect of massage on boxing performance (*British Journal of Sports Medicine*, Apr. 2000). The punching power of each boxer (measured in Newtons) was recorded in the round following each of four different interventions: (M1) in round 1 following a pre-bout sports massage, (R1) in round 1 following a pre-bout period of rest, (M5) in round 5 following a sports massage between rounds, and (R5) in round 5 following a period of rest between rounds. Based on information provided in the article, the data in the table (p. 483) were obtained. The main goal of the experiment is to compare the

**SAS Output for Exercise 10.48**

### Class Level Information

| Class | Levels | Values |
|---|---|---|
| INTERVEN | 4 | M1 M5 R1 R5 |
| BOXER | 8 | 1 2 3 4 5 6 7 8 |

Number of observations   32

Dependent Variable: POWER

| Source | DF | Sum of Squares | Mean Square | F Value | Pr > F |
|---|---|---|---|---|---|
| Model | 10 | 132798.3125 | 13279.8313 | 10.51 | <.0001 |
| Error | 21 | 26524.9062 | 1263.0908 | | |
| Corrected Total | 31 | 159323.2188 | | | |

| R-Square | Coeff Var | Root MSE | POWER Mean |
|---|---|---|---|
| 0.833515 | 2.835614 | 35.53999 | 1253.344 |

| Source | DF | Type III SS | Mean Square | F Value | Pr > F |
|---|---|---|---|---|---|
| INTERVEN | 3 | 15754.3438 | 5251.4479 | 4.16 | 0.0185 |
| BOXER | 7 | 117043.9688 | 16720.5670 | 13.24 | <.0001 |

Tukey's Studentized Range (HSD) Test for POWER

NOTE: This test controls the Type I experimentwise error rate, but it generally has a higher Type II error rate than REGWQ.

| Alpha | 0.05 |
|---|---|
| Error Degrees of Freedom | 21 |
| Error Mean Square | 1263.091 |
| Critical Value of Studentized Range | 3.94188 |
| Minimum Significant Difference | 49.531 |

Means with the same letter are not significantly different.

| Tukey Grouping | | Mean | N | INTERVEN |
|---|---|---|---|---|
| | A | 1277.75 | 8 | M5 |
| | A | | | |
| B | A | 1271.50 | 8 | R5 |
| B | A | | | |
| B | A | 1240.38 | 8 | M1 |
| B | | | | |
| B | | 1223.75 | 8 | R1 |

punching power means of the four interventions. Draw the appropriate inferences based on the SAS printout (p. 482). What conclusions can you draw regarding the effect of massage on punching power?

⚫ BOXING

| Boxer | Intervention | | | |
|---|---|---|---|---|
| | **M1** | **R1** | **M5** | **R5** |
| 1 | 1243 | 1244 | 1291 | 1262 |
| 2 | 1147 | 1053 | 1169 | 1177 |
| 3 | 1247 | 1375 | 1309 | 1321 |
| 4 | 1274 | 1235 | 1290 | 1285 |
| 5 | 1177 | 1139 | 1233 | 1238 |
| 6 | 1336 | 1313 | 1366 | 1362 |
| 7 | 1238 | 1279 | 1275 | 1261 |
| 8 | 1261 | 1152 | 1289 | 1266 |

*Source:* Hemmings, B., Smith, M., Graydon, J., and Dyson, R. "Effects of massage on physiological restoration, perceived recovery, and repeated sports performance." *British Journal of Sports Medicine*, Vol. 34, No. 2, Apr. 2000 (adapted from Table 3).

**10.49** Plant therapists believe that plants can reduce the stress levels of humans. A Kansas State University study was conducted to investigate this phenomenon. Two weeks prior to final exams, 10 undergraduate students took part in an experiment to determine what effect the presence of a live plant, a photo of a plant, or absence of a plant has on the student's ability to relax while isolated in a dimly lit room. Each student participated in three sessions—one with a live plant, one with a plant photo, and one with no plant (control).* During each session, finger temperature was measured at one-minute intervals for 20 minutes. Since increasing finger temperature indicates an increased level of relaxation, the maximum temperature (in degrees) was used as the response variable. The data for the experiment are provided in the table. Conduct an ANOVA and make the proper inferences at $\alpha = .10$.

⚫ PLANTS

| Student | Live Plant | Plant Photo | No Plant (control) |
|---|---|---|---|
| 1 | 91.4 | 93.5 | 96.6 |
| 2 | 94.9 | 96.6 | 90.5 |
| 3 | 97.0 | 95.8 | 95.4 |
| 4 | 93.7 | 96.2 | 96.7 |
| 5 | 96.0 | 96.6 | 93.5 |
| 6 | 96.7 | 95.5 | 94.8 |
| 7 | 95.2 | 94.6 | 95.7 |
| 8 | 96.0 | 97.2 | 96.2 |
| 9 | 95.6 | 94.8 | 96.0 |
| 10 | 95.6 | 92.6 | 96.6 |

*Source:* Elizabeth Schreiber, Department of Statistics, Kansas State University, Manhattan, Kansas.

**10.50** Two drugs, A and B, used for the treatment of glaucoma (an eye disease) were tested for effectiveness on 10 diseased dogs. Drug A was administered to one eye (chosen randomly) of each dog and drug B to the other eye. Pressure measurements were taken 1 hour later on both eyeballs of each dog. The 10 diseased dogs serve as the blocks for comparing the two treatments, drugs A and B. Pressure measurements are given in the table. (The smaller the measurement, the less serious the eye disease.)

⚫ EYEDOGS

| Dog | Drug A | Drug B |
|---|---|---|
| 1 | .17 | .15 |
| 2 | .20 | .18 |
| 3 | .14 | .13 |
| 4 | .18 | .18 |
| 5 | .23 | .19 |
| 6 | .19 | .12 |
| 7 | .12 | .07 |
| 8 | .10 | .09 |
| 9 | .16 | .14 |
| 10 | .13 | .08 |

**a.** Perform an analysis of variance for these data. Do the data provide sufficient evidence to indicate a difference in mean pressure readings for the two treatments (i.e., is one of the glaucoma drugs better than the other)? Use $\alpha = .05$.

**b.** What is the purpose of using the dogs as blocks in this experiment?

**c.** Recall that a randomized block design with $p = 2$ treatments is a paired difference experiment (Chapter 9). Analyze the data as a paired difference experiment using a $t$-test to compare the treatment means. Use $\alpha = .05$.

**d.** Compare the computed $F$ and $t$ values from parts **a** and **c**, and verify that $F = t^2$. Also verify that for the rejection region values of $F$ and $t$, $F_\alpha = t^2_{\alpha/2}$.

**e.** Find the approximate observed significance level for the test in part **a** and interpret its value.

**10.51** A species of Caribbean mosquito is known to be resistant against certain insecticides. The effectiveness of five different types of insecticides—temephos, malathion, fenitrothion, fenthion, and chlorpyrifos—in controlling this mosquito species was investigated in the *Journal of the American Mosquito Control Association* (Mar. 1995). Mosquito larvae were collected from each of seven Caribbean locations. In a laboratory, the larvae from each location were divided into five batches and each batch exposed to one of the five insecticides. The dosage of insecticide required to kill 50% of the larvae was recorded and divided by the known dosage for a susceptible mosquito strain. The resulting

*The experiment is simplified for this exercise. The actual experiment involved 30 students who participated in 12 sessions.

● MOSQUITO

| Location | Insecticide | | | | |
|---|---|---|---|---|---|
| | **Temephos** | **Malathion** | **Fenitrothion** | **Fenthion** | **Chlorpyrifos** |
| Anguilla | 4.6 | 1.2 | 1.5 | 1.8 | 1.5 |
| Antigua | 9.2 | 2.9 | 2.0 | 7.0 | 2.0 |
| Dominica | 7.8 | 1.4 | 2.4 | 4.2 | 4.1 |
| Guyana | 1.7 | 1.9 | 2.2 | 1.5 | 1.8 |
| Jamaica | 3.4 | 3.7 | 2.0 | 1.5 | 7.1 |
| St. Lucia | 6.7 | 2.7 | 2.7 | 4.8 | 8.7 |
| Suriname | 1.4 | 1.9 | 2.0 | 2.1 | 1.7 |

*Source:* Rawlins, S. C., and Oh Hing Wan, J. "Resistance in some Caribbean population of *Aedes aegypti* to several insecticides." *Journal of the American Mosquito Control Association*, Vol. 11, No. 1, Mar. 1995 (Table 1).

value is called the resistance ratio. (The higher the ratio, the more resistant the mosquito species is to the insecticide relative to the susceptible mosquito strain.) The resistance ratios for the study are listed in the table above. The researchers want to compare the mean resistance ratios of the five insecticides.

**a.** Explain why the experimental design is a randomized block design. Identify the treatments and the blocks.

**b.** Conduct a complete analysis of the data. Are any of the insecticides more effective than any of the others?

### Applying the Concepts—Advanced

**10.52** *Science* (Jan. 1, 1999) reported on the ability of 7-month-old infants to learn an unfamiliar language. In one experiment, 16 infants were trained in an artificial language. Then, each infant was presented with two 3-word sentences that consisted entirely of new words (e.g., "wo fe wo"). One sentence was consistent (i.e., constructed from the same grammar as in the training session) and one sentence was inconsistent (i.e., constructed from grammar in which the infant was not trained). The variable measured in each trial was the time (in seconds) the infant spent listening to the speaker, with the goal to compare the mean listening times of consistent and inconsistent sentences.

**a.** The data were analyzed as a randomized block design with the 16 infants representing the blocks and the two sentence types (consistent and inconsistent) representing the treatments. Do you agree with this data analysis method? Explain.

**b.** Refer to part **a.** The test statistic for testing treatments was $F = 25.7$ with an associated observed significance level of $p < .001$. Interpret this result.

**c.** Explain why the data could also be analyzed as a paired difference experiment, with a test statistic of $t = 5.07$.

**d.** The mean listening times and standard deviations for the two treatments are given here. Use this information to calculate the $F$ statistic for comparing the treatment means in an ANOVA for a completely randomized design. Explain why this test statistic provides weaker evidence of a difference between treatment means than the test in part **b.**

| | Consistent Sentences | Inconsistent Sentences |
|---|---|---|
| Mean | 6.3 | 9.0 |
| Standard deviation | 2.6 | 2.16 |

**e.** Explain why there is no need to control the experimentwise error rate when ranking the treatment means for this experiment.

## 10.5 FACTORIAL EXPERIMENTS

All the experiments discussed in Sections 10.2 and 10.4 were **single-factor experiments**. The treatments were levels of a single factor, with the sampling of experimental units performed using either a completely randomized or randomized block design. However, most responses are affected by more than one factor, and we will therefore often wish to design experiments involving more than one factor.

Consider an experiment in which the effects of two factors on the response are being investigated. Assume that factor $A$ is to be investigated at $a$ levels, and factor $B$ at $b$ levels. Recalling that treatments are factor–level combinations, you can see that the experiment has, potentially, $ab$ treatments that could be included in the experiment. A *complete factorial experiment* is one in which all possible *ab* treatments are utilized.

---

**DEFINITION 10.9**

A **complete factorial experiment** is one in which every factor–level combination is utilized. That is, the number of treatments in the experiment equals the total number of factor–level combinations.

---

For example, suppose the USGA wants to determine not only the relationship between distance and brand of golf ball, but also between distance and the club used to hit the ball. If they decide to use four brands and two clubs (say, driver and five-iron) in the experiment, then a complete factorial would call for utilizing all $4 \times 2 = 8$ Brand–Club combinations. This experiment is referred to more specifically as a *complete* $4 \times 2$ *factorial*. A layout for a two-factor factorial experiment (we are henceforth referring to a *complete factorial* when we use the term *factorial*) is given in Table 10.9. The factorial experiment is also referred to as a **two-way classification** because it can be arranged in the row–column format exhibited in Table 10.9.

**TABLE 10.9   Schematic Layout of Two-Factor Factorial Experiment**

| | Level | \multicolumn{5}{c}{Factor *B* at *b* Levels} |
| | | **1** | **2** | **3** | $\cdots$ | **b** |
|---|---|---|---|---|---|---|
| | 1 | Trt. 1 | Trt. 2 | Trt. 3 | $\cdots$ | Trt. *b* |
| **Factor A** | 2 | Trt. $b + 1$ | Trt. $b + 2$ | Trt. $b + 3$ | $\cdots$ | Trt. 2*b* |
| **at *a* levels** | 3 | Trt. $2b + 1$ | Trt. $2b + 2$ | Trt. $2b + 3$ | $\cdots$ | Trt. 3*b* |
| | $\vdots$ | $\vdots$ | $\vdots$ | $\vdots$ | $\cdots$ | $\vdots$ |
| | *a* | Trt. $(a - 1)b + 1$ | Trt. $(a - 1)b + 2$ | Trt. $(a - 1)b + 3$ | $\cdots$ | Trt. *ab* |

In order to complete the specification of the experimental design, the treatments must be assigned to the experimental units. If the assignment of the *ab* treatments in the factorial experiment is random and independent, the design is completely randomized. For example, if the machine Iron Byron is used to hit 80 golf balls, 10 for each of the eight Brand–Club combinations, in a random sequence, the design would be completely randomized. On the other hand, if the assignment is made within homogeneous blocks of experimental units, then the design is a randomized block. For example, if ten golfers are employed to hit each of the eight golf balls, and each golfer hits all eight Brand–Club combinations in a random sequence, then the design is randomized block, with the golfers serving as blocks. In the remainder of this section, we confine our attention to factorial experiments utilizing completely randomized designs.

If we utilize a completely randomized design to conduct a factorial experiment with *ab* treatments, we can proceed with the analysis in exactly the same way as we did in Section 10.2. That is, we calculate (or let the computer calculate) the measure of treatment mean variability (MST) and the measure of sampling variability (MSE) and use the *F*-ratio of these two quantities to test the null hypothesis that the treatment means are equal. However, if this hypothesis is rejected, so that we conclude some differences exist among the treatment means, important questions remain. Are both factors affecting the response, or only one? If both, do they affect the response independently, or do they interact to affect the response?

**FIGURE 10.19**

*Illustration of Possible Treatment Effects: Factorial Experiment*

For example, suppose the distance data indicate that at least two of the eight treatment (Brand–Club combinations) means differ in the golf experiment. Does the brand of ball (factor *A*) or the club utilized (factor *B*) affect mean distance, or do both affect it? Several possibilities are shown in Figure 10.19. In Figure 10.19a, the brand means are equal (only three are shown for the purpose of illustration), but the distances differ for the two levels of factor *B* (Club). Thus, there is no effect of Brand on distance, but a Club main effect is present. In Figure 10.19b, the Brand means differ, but the Club means are equal for each Brand. Here a Brand main effect is present, but no effect of Club is present.

Figures 10.19c and 10.19d illustrate cases in which both factors affect the response. In Figure 10.19c, the mean distance between clubs does not change for the three Brands, so that the effect of Brand on distance is independent of Club. That is, the two factors Brand and Club *do not interact*. In contrast, Figure 10.19d shows that the difference between mean distances between clubs varies with Brand. Thus, the effect of Brand on distance depends on Club; therefore, the two factors *do interact*.

In order to determine the nature of the treatment effect, if any, on the response in a factorial experiment, we need to break the treatment variability into three components: Interaction Between Factors *A* and *B*, Main Effect of Factor *A*, and Main Effect of Factor *B*. The **Factor Interaction** component is used to test whether the factors combine to affect the response, while the **Factor Main Effect** components are used to determine whether the factors separately affect the response.

The partitioning of the Total Sum of Squares into its various components is illustrated in Figure 10.20. Notice that at stage 1 the components are identical to those in the one-factor, completely randomized designs of Section 10.2; the Sums of Squares for Treatment and Error sum to the Total Sum of Squares. The degrees of freedom for treatments is equal to $(ab - 1)$, one less than the number of treatments.

**FIGURE 10.20**

*Partitioning the Total Sum of Squares for a Two-Factor Factorial*

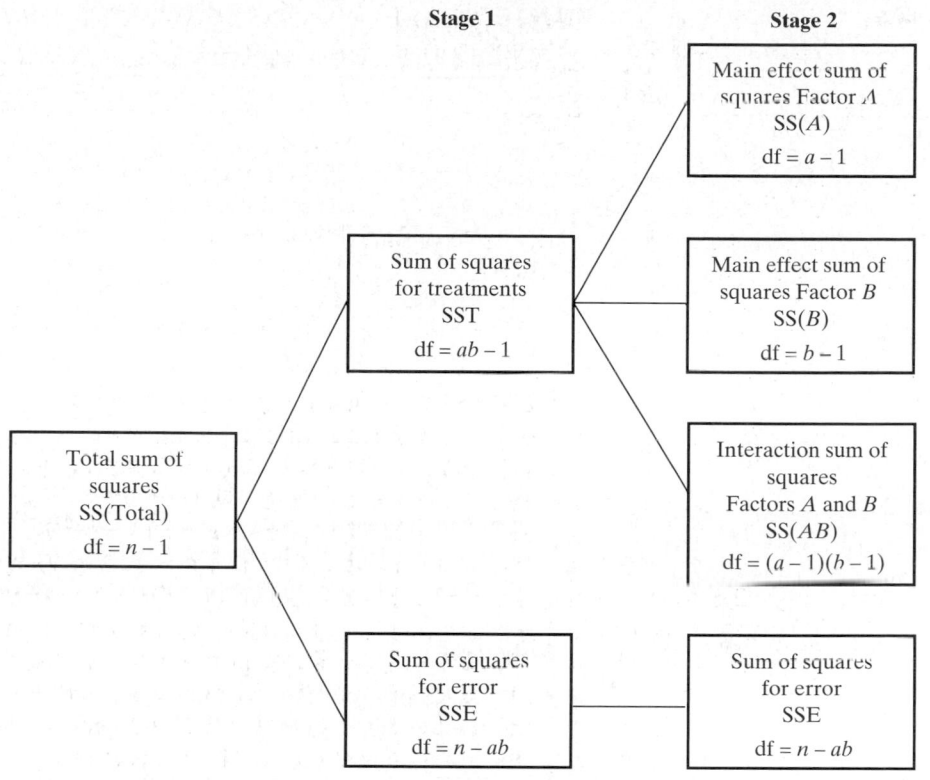

The degrees of freedom for error is equal to $(n - ab)$, the total sample size minus the number of treatments. Only at stage 2 of the partitioning does the factorial experiment differ from those previously discussed. Here we divide the Treatment Sum of Squares into its three components: Interaction and the two Main Effects. These components can then be used to test the nature of the differences, if any, among the treatment means.

There are a number of ways to proceed in the testing and estimation of factors in a factorial experiment. We present one approach in the box on p. 49.

We assume the completely randomized design is a **balanced design**, meaning that the same number of observations are made for each treatment. That is, we assume that $r$ experimental units are randomly and independently selected for each treatment. The numerical value of $r$ must exceed 1 in order to have any degrees of freedom with which to measure the sampling variability. [Note that if $r = 1$, then $n = ab$, and the degrees of freedom associated with Error (Figure 10.20) is df $= n - ab = 0$.] The value of $r$ is often referred to as the number of **replicates** of the factorial experiment since we assume that all $ab$ treatments are repeated, or replicated, $r$ times. Whatever approach is adopted in the analysis of a factorial experiment, several tests of hypotheses are usually conducted. The tests are summarized in the box on page 491.

**EXAMPLE 10.9**

Suppose the USGA tests four different brands (A, B, C, D) of golf balls and two different clubs (driver, five-iron) in a $4 \times 2$ factorial design. Each of the eight Brand–Club combinations (treatments) is randomly and independently assigned to four experimental units, each experimental unit consisting of a specific position in the sequence of hits by Iron Byron. The distance response is recorded for each of the 32 hits, and the results are shown in Table 10.10.

⊘ GOLFFAC1
**TABLE 10.10    Distance Data for 4 × 2 Factorial Golf Experiment**

| Club | Brand | | | |
|------|-------|---|---|---|
| | **A** | **B** | **C** | **D** |
| **Driver** | 226.4 | 238.3 | 240.5 | 219.8 |
| | 232.6 | 231.7 | 246.9 | 228.7 |
| | 234.0 | 227.7 | 240.3 | 232.9 |
| | 220.7 | 237.2 | 244.7 | 237.6 |
| **Five-iron** | 163.8 | 184.4 | 179.0 | 157.8 |
| | 179.4 | 180.6 | 168.0 | 161.8 |
| | 168.6 | 179.5 | 165.2 | 162.1 |
| | 173.4 | 186.2 | 156.5 | 160.3 |

a. Use statistical software to partition the Total Sum of Squares into the components necessary to analyze this 4 × 2 factorial experiment.

b. Follow the steps for analyzing a two-factor factorial experiment and interpret the results of your analysis. Use $\alpha = .10$ for the tests you conduct.

### Solution

a. The SAS printout that partitions the Total Sum of Squares for this factorial experiment is given in Figure 10.21. The partitioning takes place in two stages: First, the Total Sum of Squares is partitioned into the Model (Treatment) and Error Sums of Squares at the top of the printout. Note that SST is 33,659.8 with 7 degrees of freedom, and SSE is 822.2 with 24 degrees of freedom, adding to 34,482.0 and 31 degrees of freedom. In the second stage of partitioning, the Treatment Sum of Squares is further divided into the Main Effect and Interaction Sums of Squares. At the bottom of the printout we see that SS(Club) is 32,093.1 with 1 degree of freedom, SS(Brand) is 800.7 with 3 degrees of freedom, and SS(Club × Brand) is 766.0 with 3 degrees of freedom, adding to 33,659.8 and 7 degrees of freedom.

b. Once partitioning is accomplished, our first test is

$H_0$: The eight treatment means are equal
$H_a$: At least two of the eight means differ

*Test statistic:* $F = \dfrac{\text{MST}}{\text{MSE}} = 140.35$     (top of printout)

*Observed significance level:* $p < .0001$     (top of printout)

Since $\alpha = .10$ exceeds $p$, we reject this null hypothesis and conclude that at least two of the Brand-Club combinations differ in mean distance.

After accepting the hypothesis that the treatment means differ, and therefore that the factors Brand and/or Club somehow affect the mean distance, we want to determine how the factors affect the mean response. We begin with a test of interaction between Brand and Club:

$H_0$: The factors Brand and Club do not interact to affect the mean response
$H_a$: Brand and Club interact to affect mean response

*Test statistic:* $F = \dfrac{\text{MS}(AB)}{\text{MSE}} = \dfrac{\text{MS}(\text{Brand} \times \text{Club})}{\text{MSE}}$

$= \dfrac{255.32}{34.26} = 7.45$     (bottom of printout)

*Observed significance level:* $p = .0011$     (bottom of printout)

**FIGURE 10.21**

*SAS Printout for Factorial ANOVA of Golf Experiment Data*

Analysis of Variance Procedure

Dependent Variable: DISTANCE

| Source | DF | Sum of Squares | Mean Square | F Value | Pr > F |
|--------|----|----------------|-------------|---------|--------|
| Model | 7 | 33659.80875 | 4808.54411 | 140.35 | <.0001 |
| Error | 24 | 822.24000 | 34.26000 | | |
| Corrected Total | 31 | 34482.04875 | | | |

| R-Square | Coeff Var | Root MSE | DISTANCE Mean |
|----------|-----------|----------|---------------|
| 0.976155 | 2.896461 | 5.853204 | 202.0813 |

| Source | DF | Type III SS | Mean Square | F Value | Pr > F |
|--------|----|-------------|-------------|---------|--------|
| CLUB | 1 | 32093.11125 | 32093.11125 | 936.75 | <.0001 |
| BRAND | 3 | 800.73625 | 266.91208 | 7.79 | 0.0008 |
| CLUB*BRAND | 3 | 765.96125 | 255.32042 | 7.45 | 0.0011 |

Since $\alpha = .10$ exceeds the $p$-value, we conclude that the factors Brand and Club interact to affect mean distance.

Because the factors interact, we do not test the main effects for Brand and Club. Instead, we compare the treatment means in an attempt to learn the nature of the interaction. Rather than compare all $8(7)/2 = 28$ pairs of treatment means, we test for differences only between pairs of brands within each club. That differences exist *between* clubs can be assumed. Therefore, only $4(3)/2 = 6$ pairs of means need to be compared for each club, or a total of 12 comparisons for the two clubs. The results of these comparisons using Tukey's method with an experimentwise error rate of $\alpha = .10$ for each club are displayed in the SAS printout, Figure 10.22. For each club, the brand means are listed in descending order in Figure 10.22, and those not significantly different are connected by the same letter in the **Tukey Grouping** column.

As shown in Figure 10.22, the picture is unclear with respect to Brand means. For the five-iron (top of Figure 10.22), the brand B mean significantly exceeds all other brands. However, when hit with a driver (bottom of Fig. 10.22), brand B's mean is not significantly different from any of the other brands. Note the nontransitive nature of the multiple comparisons. For example, for the driver the brand C mean can be "the same" as the brand B mean, and the brand B mean can be "the same" as the brand D mean, and yet the brand C mean can significantly exceed the brand D mean. The reason lies in the definition of "the same"—we must be careful not to conclude two means are equal simply because they are connected by a line or letter. The line indicates only that *the connected means are not significantly different*. You should conclude (at the overall $\alpha$ level of significance) only that means *not* connected are different, while withholding judgment on those that are connected. The picture of which means differ and by how much will become clearer as we increase the number of replicates of the factorial experiment.

The Club × Brand interaction can be seen in the plot of means in Figure 10.23. Note that the difference between the mean distances of the two clubs (driver and five-iron) varies depending on brand. The biggest difference appears for Brand C while the smallest difference is for Brand B.

**FIGURE 10.22**

*SAS Ranking of Treatment Means for Factorial Golf Experiment*

```
--------------------------------- CLUB=5IRON ---------------------------------
          Tukey's Studentized Range (HSD) Test for DISTANCE

NOTE: This test controls the Type I experimentwise error rate, but it
      generally has a higher Type II error rate than REGWQ.

          Alpha                                       0.1
          Error Degrees of Freedom                     12
          Error Mean Square                      36.10792
          Critical Value of Studentized Range     3.62071
          Minimum Significant Difference           10.878

      Means with the same letter are not significantly different.

          Tukey Grouping        Mean        N       BRAND

                      A       182.675        4        B

                      B       171.300        4        A
                      B
                      B       167.175        4        C
                      B
                      B       160.500        4        D

--------------------------------- CLUB=DRIVER --------------------------------
          Tukey's Studentized Range (HSD) Test for DISTANCE

NOTE: This test controls the Type I experimentwise error rate, but it
      generally has a higher Type II error rate than REGWQ.

          Alpha                                       0.1
          Error Degrees of Freedom                     12
          Error Mean Square                      32.41208
          Critical Value of Studentized Range     3.62071
          Minimum Significant Difference           10.307

      Means with the same letter are not significantly different.

          Tukey Grouping        Mean        N       BRAND

                      A       243.100        4        C
                      A
                  B   A       233.725        4        B
                  B
                  B           229.750        4        D
                  B
                  B           228.425        4        A
```

**FIGURE 10.23**

*Sample Mean Plot for Factorial Golf Experiment*

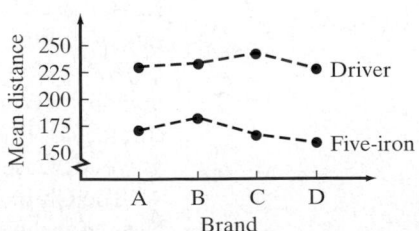

## Procedure for Analysis of Two-Factor Factorial Experiment

1. Partition the Total Sum of Squares into the Treatment and Error components (stage 1 of Figure 10.20). Use either a statistical software package or the calculation formulas in Appendix C to accomplish the partitioning.

2. Use the *F*-ratio of Mean Square for Treatments to Mean Square for Error to test the null hypothesis that the treatment means are equal.*

   a. If the test results in nonrejection of the null hypothesis, consider refining the experiment by increasing the number of replications or introducing other factors. Also consider the possibility that the response is unrelated to the two factors.

   b. If the test results in rejection of the null hypothesis, then proceed to step 3.

3. Partition the Treatment Sum of Squares into the Main Effect and Interaction Sum of Squares (stage 2 of Figure 10.20). Use either a statistical software package or the calculation formulas in Appendix C to accomplish the partitioning.

4. Test the null hypothesis that factors *A* and *B* do not interact to affect the response by computing the *F*-ratio of the Mean Square for Interaction to the Mean Square for Error.

   a. If the test results in nonrejection of the null hypothesis, proceed to step 5.

   b. If the test results in rejection of the null hypothesis, conclude that the two factors interact to affect the mean response. Then proceed to step 6a.

5. Conduct tests of two null hypotheses that the mean response is the same at each level of factor *A* and factor *B*. Compute two *F*-ratios by comparing the Mean Square for each Factor Main Effect to the Mean Square for Error.

   a. If one or both tests result in rejection of the null hypothesis, conclude that the factor affects the mean response. Proceed to step 6b.

   b. If both tests result in nonrejection, an apparent contradiction has occurred. Although the treatment means apparently differ (step 2 test), the interaction (step 4) and main effect (step 5) tests have not supported that result. Further experimentation is advised.

6. Compare the means:

   a. If the test for interaction (step 4) is significant, use a multiple comparisons procedure to compare any or all pairs of the treatment means.

   b. If the test for one or both main effects (step 5) is significant, use a multiple comparisons procedure to compare the pairs of means corresponding to the levels of the significant factor(s).

## Tests Conducted in Analyses of Factorial Experiments: Completely Randomized Design, *r* Replicates per Treatment

**TEST FOR TREATMENT MEANS**

$H_0$: No difference among the *ab* treatment means
$H_a$: At least two treatment means differ

*Test statistic:* $F = \dfrac{\text{MST}}{\text{MSE}}$

*Rejection region:* $F \geq F_\alpha$, based on $(ab - 1)$ numerator and $(n - ab)$ denominator degrees of freedom [*Note:* $n = abr$.]          *(continues on next page)*

---

*Some analysts prefer to proceed directly to test the interaction and main effect components, skipping the test of treatment means. We begin with this test to be consistent with our approach in the one-factor completely randomized design.

## Tests Conducted in Analyses of Factorial Experiments: Completely Randomized Design, *r* Replicates per Treatment *(continued)*

**TEST FOR FACTOR INTERACTION**

$H_0$: Factors $A$ and $B$ do not interact to affect the response mean

$H_a$: Factors $A$ and $B$ do interact to affect the response mean

*Test statistic:* $F = \dfrac{\text{MS}(AB)}{\text{MSE}}$

*Rejection region:* $F \geq F_\alpha$, based on $(a - 1)(b - 1)$ numerator and $(n - ab)$ denominator degrees of freedom

**TEST FOR MAIN EFFECT OF FACTOR A**

$H_0$: No difference among the $a$ mean levels of factor $A$

$H_a$: At least two factor $A$ mean levels differ

*Test statistic:* $F = \dfrac{\text{MS}(A)}{\text{MSE}}$

*Rejection region:* $F \geq F_\alpha$, based on $(a - 1)$ numerator and $(n - ab)$ denominator degrees of freedom

**TEST FOR MAIN EFFECT OF FACTOR B**

$H_0$: No difference among the $b$ mean levels of factor $B$

$H_a$: At least two factor $B$ mean levels differ

*Test statistic:* $F = \dfrac{\text{MS}(B)}{\text{MSE}}$

*Rejection region:* $F \geq F_\alpha$, based on $(b - 1)$ numerator and $(n - ab)$ denominator degrees of freedom

**ASSUMPTIONS FOR ALL TESTS**

1. The response distribution for each factor–level combination (treatment) is normal.

2. The response variance is constant for all treatments.

3. Random and independent samples of experimental units are associated with each treatment.

**EXAMPLE 10.10**

Refer to Example 10.9. Suppose the same factorial experiment is performed on four other brands (E, F, G, and H), and the results are as shown in Table 10.11. Repeat the factorial analysis and interpret the results.

GOLFFAC2

**TABLE 10.11    Distance Data for Second Factorial Golf Experiment**

| Club | Brand | | | |
|---|---|---|---|---|
| | **E** | **F** | **G** | **H** |
| **Driver** | 238.6 | 261.4 | 264.7 | 235.4 |
| | 241.9 | 261.3 | 262.9 | 239.8 |
| | 236.6 | 254.0 | 253.5 | 236.2 |
| | 244.9 | 259.9 | 255.6 | 237.5 |
| **Five-iron** | 165.2 | 179.2 | 189.0 | 171.4 |
| | 156.9 | 171.0 | 191.2 | 159.3 |
| | 172.2 | 178.0 | 191.3 | 156.6 |
| | 163.2 | 182.7 | 180.5 | 157.4 |

**FIGURE 10.24**

*SPSS Printout for Second Factorial ANOVA*

Tests of Between-Subjects Effects

Dependent Variable: DISTANCE

| Source | Type III Sum of Squares | df | Mean Square | F | Sig. |
|---|---|---|---|---|---|
| Corrected Model | 49959.375[a] | 7 | 7137.054 | 290.120 | .000 |
| Intercept | 1423532.828 | 1 | 1423532.828 | 57866.453 | .000 |
| CLUB | 46443.900 | 1 | 46443.900 | 1887.939 | .000 |
| BRAND | 3410.316 | 3 | 1136.772 | 46.210 | .000 |
| CLUB*BRAND | 105.158 | 3 | 35.053 | 1.425 | .260 |
| Error | 590.407 | 24 | 24.600 | | |
| Total | 1474082.610 | 32 | | | |
| Corrected Total | 50549.782 | 31 | | | |

a. R Squared = .988 (Adjusted R Squared = .985)

### Solution

An SPSS printout for the second factorial experiment is shown in Figure 10.24. The *F*-ratio for Treatments is $F = 290.1$ (highlighted at the top of the printout), which exceeds the tabled value of $F_{.10} = 1.98$ for 7 numerator and 24 denominator degrees of freedom. (Note that the same rejection regions will apply in this example as in Example 10.9 since the factors, treatments, and replicates are the same.) We conclude that at least two of the Brand–Club combinations have different mean distances.

We next test for interaction between Brand and Club. The *F* value (highlighted on the SPSS printout) is

$$F = \frac{\text{MS(Brand} \times \text{Club)}}{\text{MSE}} = 1.425$$

Since this *F*-ratio does *not* exceed the tabled value of $F_{.10} = 2.33$ with 3 and 24 df, we cannot conclude at the .10 level of significance that the factors interact. In fact, note that the observed significance level (highlighted on the SPSS printout) for the test of interaction is .260. Thus, at any level of significance lower than $\alpha = .26$, we could not conclude that the factors interact. We therefore test the main effects for Brand and Club.

We first test the Brand main effect:

$H_0$: No difference exists among the true Brand mean distances
$H_a$: At least two Brand mean distances differ

*Test statistic:*
$$F = \frac{\text{MS(Brand)}}{\text{MSE}} = \frac{1,136.772}{24.600} = 46.210 \quad \text{(highlighted on printout)}$$
*Observed significance level*: $p = .000$   (highlighted on printout)

Since $\alpha = .10$ exceeds the *p*-value, we conclude that at least two of the brand means differ. We will subsequently determine which brand means differ using Tukey's multiple comparisons procedure. But first, we want to test the Club main effect:

$H_0$: No differences exist between the Club mean distances
$H_a$: The Club mean distances differ

*Test statistic:*
$$F = \frac{\text{MS(Club)}}{\text{MSE}} = \frac{46,443.900}{24.600} = 1,887.939 \quad \text{(highlighted on printout)}$$
*Observed significance level:* $p = .000$   (highlighted on printout)

**FIGURE 10.25**

*SPSS Ranking of Golf Ball Brand Means for Second Factorial Experiment*

DISTANCE

Tukey HSD[a,b]

| BRAND | N | Subset 1 | Subset 2 |
|-------|---|----------|----------|
| H | 8 | 199.2000 | |
| E | 8 | 202.4375 | |
| F | 8 | | 218.4375 |
| G | 8 | | 223.5875 |
| Sig. | | .568 | .189 |

Means for groups in homogeneous subsets are displayed.
Based on Type III Sum of Squares
The error term is Mean Square(Error) = 24.600.
a. Uses Harmonic Mean Sample Size = 8.000.
b. Alpha = .10.

Since $\alpha = .10$ exceeds the *p*-value, we conclude that the two clubs are associated with different mean distances. Since only two levels of Club were utilized in the experiment, this *F*-test leads to the inference that the mean distance differs for the two clubs. It is no surprise (to golfers) that the mean distance for balls hit with the driver is significantly greater than the mean distance for those hit with the five-iron.

To determine which of the Brands' mean distances differ, we wish to compare the four Brand means using Tukey's method at $\alpha = .10$. The results of these multiple comparisons are displayed in the SPSS printout, Figure 10.25. The Brand means are shown, grouped by subset, in Figure 10.25, with means that are not significantly different in the same subset. Brands G and F (subset 2) are apparently associated with significantly greater mean distances than brands E and H (subset 1). However, since G and F are in the same Tukey subset and E and H are in the same subset, we cannot distinguish between brands G and F or between brands E and H using these data. Since the interaction between Brand and Club was not significant, we conclude that this difference among brands applies to both clubs. The sample means for all Club–Brand combinations are shown in Figure 10.26 and appear to support the conclusions of the tests and comparisons. Note that the Brand means maintain their relative positions for each Club—brands F and G dominate brands E and H for both the driver and five-iron. ∎

**FIGURE 10.26**

*Sample Mean Plot for Second Factorial Golf Experiment*

The analysis of factorial experiments can become complex if the number of factors is increased. Even the two-factor experiment becomes more difficult to analyze if some factor combinations have different numbers of observations than others. We have provided an introduction to these important experiments using two-factor factorials with equal numbers of observations for each treatment. Although similar principles apply to most factorial experiments, you should consult the references at the end of this chapter if you need to design and analyze more complex factorials.

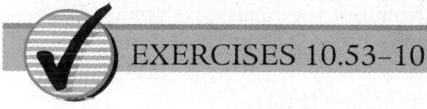 EXERCISES 10.53–10.68

## Learning the Mechanics

**10.53** Suppose you conduct a 3 × 5 factorial experiment.
   **a.** How many factors are used in the experiment?
   **b.** Can you determine the factor type(s)—qualitative or quantitative—from the information given? Explain.
   **c.** Can you determine the number of levels used for each factor? Explain.
   **d.** Describe a treatment for this experiment, and determine the number of treatments used.
   **e.** What problem is caused by using a single replicate of this experiment? How is the problem solved?

**10.54** The partially complete ANOVA table given here is for a two-factor factorial experiment.

| Source | df | SS | MS | F |
|--------|----|------|-----|---|
| *A* | 3 | | .75 | |
| *B* | 1 | .95 | | |
| *AB* | | | .30 | |
| Error | | | | |
| Total | 23 | 6.5 | | |

   **a.** Give the number of levels for each factor.
   **b.** How many observations were collected for each factor–level combination?
   **c.** Complete the ANOVA table.
   **d.** Test to determine whether the treatment means differ. Use $\alpha = .10$.
   **e.** Conduct the tests of factor interaction and mean effects, each at the $\alpha = .10$ level of significance. Which of the tests are warranted as part of the factorial analysis? Explain.

**10.55** The partially completed ANOVA table for a 3 × 4 factorial experiment with two replications is shown in the next column.

| Source | df | SS | MS | F |
|--------|----|------|-----|---|
| *A* | | .8 | | |
| *B* | | 5.3 | | |
| *AB* | | 9.6 | | |
| Error | | | | |
| Total | | 18.1 | | |

   **a.** Complete the ANOVA table.
   **b.** Which sums of squares are combined to find the Sum of Squares for Treatment? Do the data provide sufficient evidence to indicate that the treatment means differ? Use $\alpha = .05$.
   **c.** Does the result of the test in part **b** warrant further testing? Explain.
   **d.** What is meant by factor interaction, and what is the practical implication if it exists?
   **e.** Test to determine whether these factors interact to affect the response mean. Use $\alpha = .05$, and interpret the result.
   **f.** Does the result of the interaction test warrant further testing? Explain.

**10.56** The following two-way table gives data for a 2 × 3 factorial experiment with two observations for each factor–level combination.

⊘ LM10_56

| | | Factor *B* | | |
|--------|-------|-----|-----|-----|
| | Level | 1 | 2 | 3 |
| Factor *A* | 1 | 3.1, 4.0 | 4.6, 4.2 | 6.4, 7.1 |
| | 2 | 5.9, 5.3 | 2.9, 2.2 | 3.3, 2.5 |

   **a.** Identify the treatments for this experiment. Calculate and plot the treatment means, using the response variable as the *y*-axis and the levels of factor *B* as the *x*-axis. Use the levels of factor *A* as plotting symbols. Do the treatment means appear to differ? Do the factors appear to interact?

```
MINITAB Output for Exercise 10.56

Amalysis of Variance for RESPONSE

Source    DF        SS        MS        F       P
A          1    4.4408    4.4408    18.06   0.005
B          2    4.1267    2.0633     8.39   0.018
A*B        2   18.0067    9.0033    36.62   0.000
Error      6    1.4750    0.2458
Total     11   28.0492
```

**b.** The MINITAB ANOVA printout for this experiment is shown at the bottom of p. 495. Test to determine whether the treatment means differ at the $\alpha = .05$ level of significance. Does the test support your visual interpretation from part **a**?

**c.** Does the result of the test in part **b** warrant a test for interaction between the two factors? If so, perform it using $\alpha = .05$.

**d.** Do the results of the previous tests warrant tests of the two factor main effects? If so, perform them using $\alpha = .05$.

**e.** Interpret the results of the tests. Do they support your visual interpretation from part **a**?

**10.57** Suppose a $3 \times 3$ factorial experiment is conducted with three replications. Assume that SS(Total) $= 1,000$. For each of the following scenarios, form an ANOVA table, conduct the appropriate tests, and interpret the results.

**a.** The Sum of Squares of factor $A$ main effect [SS($A$)] is 20% of SS(Total), the Sum of Squares for factor $B$ main effect [SS($B$)] is 10% of SS(Total), and the Sum of Squares for interaction [SS($AB$)] is 10% of SS(Total).

**b.** SS($A$) is 10%, SS($B$) is 10%, and SS($AB$) is 50% of SS(Total).

**c.** SS($A$) is 40%, SS($B$) is 10%, and SS($AB$) is 20% of SS(Total).

**d.** SS($A$) is 40%, SS($B$) is 40%, and SS($AB$) is 10% of SS(Total).

**10.58** The two-way table below gives data for a $2 \times 2$ factorial experiment with two observations per factor–level combination.

⊘ LM10_58

| | | Factor *B* | |
|---|---|---|---|
| | **Level** | **1** | **2** |
| **Factor *A*** | 1 | 29.6, 35.2 | 47.3, 42.1 |
| | 2 | 12.9, 17.6 | 28.4, 22.7 |

**a.** Identify the treatments for this experiment. Calculate and plot the treatment means, using the response variable as the *y*-axis and the levels of factor *B* as the *x*-axis. Use the levels of factor *A* as plotting symbols. Do the treatment means appear to differ? Do the factors appear to interact?

**b.** Construct an ANOVA table for this experiment.

**c.** Test to determine whether the treatment means differ at the $\alpha = .05$ level of significance. Does the test support your visual interpretation from part **a**?

**d.** Does the result of the test in part **b** warrant a test for interaction between the two factors? If so, perform it using $\alpha = .05$.

**e.** Do the results of the previous tests warrant tests of the two factor main effects? If so, perform them using $\alpha = .05$.

**f.** Interpret the results of the tests. Do they support your visual interpretation from part **a**?

**g.** Given the results of your tests, which pairs of means, if any, should be compared?

**Applying the Concepts—Basic**

**10.59** A coagulation–microfiltration process for removing bacteria from water was investigated in *Environmental Science & Engineering* (Sept. 1, 2000). Chemical engineers at Seoul National University performed a designed experiment to estimate the effect of both the level of the coagulant and acidity (pH) level on the coagulation efficiency of the process. Six levels of coagulant (5, 10, 20, 50, 100, and 200 milligrams per liter) and six pH levels (4.0, 5.0, 6.0, 7.0, 8.0, and 9.0) were employed. Water specimens collected from the Han River in Seoul, Korea, were placed in jars, and each jar randomly assigned to receive one of the $6 \times 6 = 36$ combinations of coagulant level and pH level.

**a.** What type of experimental design was applied in this study?

**b.** Give the factors, factor levels, and treatments for the study.

**10.60** Parapsychologists define "lucky" people as individuals who report that seemingly chance events consistently tend to work out in their favor. A team of British psychologists designed a study to examine the effects of luckiness and competition on performance in a guessing task (*The Journal of Parapsychology*, Mar. 1997). Each in a sample of 56 college students was classified as lucky, unlucky, or uncertain based on their responses to a Luckiness Questionnaire. In addition, the participants were randomly assigned to either a competitive or noncompetitive condition. All students were then asked to guess the outcomes of 50 flips of a coin. The response variable measured was percentage of coin-flips correctly guessed.

**a.** An ANOVA for a $2 \times 3$ factorial design was conducted on the data. Identify the factors and their levels for this design.

**b.** The results of the ANOVA are summarized in the table. Fully interpret the results.

| Source | df | *F* | *p*-Value |
|---|---|---|---|
| Luckiness (L) | 2 | 1.39 | .26 |
| Competition (C) | 1 | 2.84 | .10 |
| L × C | 2 | 0.72 | .72 |
| Error | 50 | | |
| Total | 55 | | |

**10.61** A study published in *Teaching Psychology* (May, 1998) examined how external clues influence student performance. Introductory psychology students were randomly assigned to one of four different midterm examinations. Form 1 was printed on blue paper and contained difficult questions, while form 2 was also printed on blue paper but contained simple questions. Form 3 was printed on red paper, with difficult questions;

form 4 was printed on red paper with simple questions. The researchers were interested in the impact that Color (red or blue) and Question (simple or difficult) had on mean exam score.

**a.** What experimental design was employed in this study? Identify the factors and treatments.

**b.** The researchers conducted an ANOVA and found a significant interaction between Color and Question ($p$-value $< .03$). Interpret this result.

**c.** The sample mean scores (percentage correct) for the four exam forms are listed below. Plot the four means on a graph to illustrate the Color $\times$ Question interaction.

| Form | Color | Question | Mean Score |
|------|-------|----------|------------|
| 1 | Blue | Difficult | 53.3 |
| 2 | Blue | Simple | 80.0 |
| 3 | Red | Difficult | 39.3 |
| 4 | Red | Simple | 73.6 |

**10.62** A computer lab at the University of Oklahoma is open 24 hours a day, seven days a week. In *Production and Inventory Management Journal* (3rd Quarter, 1999), S. Barman investigated whether computer usage differed significantly (1) among the seven days of the week and (2) among the 24 hours of the day. Using student log-on records, data on hourly student loads (number of users per hour) were collected during a 7-week period. A factorial ANOVA was employed with the results presented in the table below.

**a.** Is this an observational or a designed experiment? Explain.

**b.** What are the two factors of the experiment and how many levels of each factor are used?

**c.** This is an $a \times b$ factorial experiment. What are $a$ and $b$?

**d.** Specify the null and alternative hypotheses that should be used to test for an interaction effect between the two factors of the study.

**e.** Conduct the test of part **d** using $\alpha = .05$. Interpret your result in the context of the problem.

**f.** If appropriate, conduct main effects tests for both day and time. Use $\alpha = .05$. Interpret your results in the context of the problem.

**Applying the Concepts—Intermediate**

**10.63** Do women enjoy the thrill of a close basketball game as much as men? To answer this question, male and female undergraduate students were recruited to participate in an experiment (*Journal of Sport & Social Issues*, Feb. 1997). The students watched one of eight live televised games of a recent NCAA basketball tournament. (None of the games involved a home team to which the students could be considered emotionally committed.) The "suspense" of each game was classified into one of four categories according to the closeness of scores at the game's conclusions: minimal (15 point or greater differential), moderate (10–14 point differential), substantial (5–9 point differential), and extreme (1–4 point differential). After the game, each student rated his or her enjoyment on an 11-point scale ranging from 0 (not at all) to 10 (extremely). The enjoyment rating data were analyzed as a $4 \times 2$ factorial design, with suspense (four levels) and gender (two levels) as the two factors. The $4 \times 2 = 8$ treatment means are shown in the accompanying table.

| Suspense | Gender | |
|----------|--------|--------|
| | **Male** | **Female** |
| Minimal | 1.77 | 2.73 |
| Moderate | 5.38 | 4.34 |
| Substantial | 7.16 | 7.52 |
| Extreme | 7.59 | 4.92 |

*Source:* Gan, Su-lin, *et al.* "The thrill of a close game: Who enjoys it and who doesn't." *Journal of Sport & Social Issues*, Vol. 21, No. 1, Feb. 1997, pp. 59–60.

**a.** Plot the treatment means in a graph similar to Figure 10.23. Does the pattern of means suggest interaction between suspense and gender? Explain.

| Source | df | Sum of Squares | Mean Square |
|--------|-----|----------------|-------------|
| Model | 167 | 191047.21 | 1143.99 |
| Error | 1004 | 45829.97 | 45.65 |
| Total | 1171 | 236877.18 | |

$R = .8065$   $F$ value $= 25.06$   Pr $> F = .001$

| Source | df | Sum of Squares | Mean Square | *F* Value | Pr $> F$ |
|--------|-----|----------------|-------------|-----------|----------|
| Day | 6 | 18732.13 | 3122.02 | 68.39 | .0001 |
| Time | 23 | 164629.86 | 7157.82 | 156.80 | .0001 |
| Day $\times$ Time | 138 | 7685.22 | 55.69 | 1.22 | .0527 |

*Source:* Barman, S. "A statistical analysis of the attendance pattern of a computer laboratory." *Production and Inventory Management Journal*, 3rd Quarter, 1999, pp. 26–30.

**b.** The ANOVA *F*-test for interaction yielded the following results: numerator df = 3, denominator df = 68, *F* = 4.42, *p*-value = .007. What can you infer from these results?

**c.** Based on the test, part **b,** is the difference between the mean enjoyment levels of males and females the same, regardless of the suspense level of the game?

**10.64** An experiment was conducted to examine the effects of alcohol on the marital interactions of husbands and wives (*Journal of Abnormal Psychology*, Nov. 1998). A total of 135 couples participated in the experiment. The husband in each couple was classified as aggressive (60 husbands) or nonaggressive (75 husbands), based on an interview and his response to a questionnaire. Before the marital interactions of the couples were observed, each husband was randomly assigned to three groups: receive no alcohol, receive several alcoholic mixed drinks, or receive placebos (nonalcoholic drinks disguised as mixed drinks). Consequently, a 2 × 3 factorial design was employed, with husband's aggressiveness at 2 levels (aggressive or nonaggressive) and husband's alcohol condition at 3 levels (no alcohol, alcohol, and placebo). The response variable observed during the marital interaction was severity of conflict (measured on a 100-point scale).

**a.** A partial ANOVA table is shown below. Fill in the missing degrees of freedom.

| Source | df | F |
|---|---|---|
| Aggressiveness (A) | – | 16.43 ($p < .001$) |
| Alcohol Condition (C) | – | 6.00 ($p < .01$) |
| A × C | – | – |
| Error | 129 | |
| Total | – | |

**b.** Interpret the *p*-value of the *F*-test for Aggressiveness.

**c.** Interpret the *p*-value of the *F*-test for Alcohol Condition.

**d.** The *F*-test for interaction was omitted from the article. Discuss the dangers of making inferences based on the tests, parts **a** and **b,** without knowing the result of the interaction test.

**10.65** In the *Journal of Nutrition* (July 1995), University of Georgia researchers examined the impact of a vitamin-B supplement (nicotinamide) on the kidney. The experimental "subjects" were 28 Zucker rats—a species that tends to develop kidney problems. Half of the rats were classified as obese and half as lean. Within each group, half were randomly assigned to receive a vitamin-B supplemented diet and half were not. Thus, a 2 × 2 factorial experiment was conducted with seven rats assigned to each of the four combinations of size (lean or obese) and diet (supplemental or not). One of the response variables measured was weight (in grams)

of the kidney at the end of a 20-week feeding period. The data (simulated from summary information provided in the journal article) are shown in the table.

**⊘ VITAMINB**

| | | Diet | | | |
|---|---|---|---|---|---|
| | | Regular | | Vitamin-B Supplement | |
| **Rat Size** | **Lean** | 1.62 | 1.47 | 1.51 | 1.63 |
| | | 1.80 | 1.37 | 1.65 | 1.35 |
| | | 1.71 | 1.71 | 1.45 | 1.66 |
| | | 1.81 | | 1.44 | |
| | **Obese** | 2.35 | 2.84 | 2.93 | 2.63 |
| | | 2.97 | 2.05 | 2.72 | 2.61 |
| | | 2.54 | 2.82 | 2.99 | 2.64 |
| | | 2.93 | | 2.19 | |

**a.** Conduct an analysis of variance on the data. Summarize the results in an ANOVA table.

**b.** Conduct the appropriate ANOVA *F*-tests at $\alpha = .01$. Interpret the results.

**10.66** Most people are right-handed due to the propensity of the left hemisphere of the brain to control sequential movement. Similarly, the fact that some tasks are performed better with the left hand is likely due to the superiority of the right hemisphere of the brain to process the necessary information. Does such cerebral specialization in spatial processing occur in adults with Down syndrome? A 2 × 2 factorial experiment was conducted to answer this question (*American Journal on Mental Retardation*, May 1995). A sample of adults with Down syndrome was compared to a control group of normal individuals of a similar age. Thus, one factor was Group at two levels (Down syndrome and control) and the second factor was the Handedness (left or right) of the subject. All the subjects performed a task that typically yields a left-hand advantage. The response variable was "laterality index," measured on a −100 to 100 point scale. (A large positive index indicates a right-hand advantage, while a large negative index indicates a left-hand advantage.)

**a.** Identify the treatments in this experiment.

**b.** Construct a graph that would support a finding of no interaction between the two factors.

**c.** Construct a graph that would support a finding of interaction between the two factors.

**d.** The *F*-test for factor interaction yielded an observed significance level of $p < .05$. Interpret this result.

**e.** Multiple comparisons of all pairs of treatment means yielded the rankings shown here. Interpret the results.

| Mean laterality index: | −30 | −4 | −.5 | + .5 |
|---|---|---|---|---|
| Group/Handed: | Down/Left | Control/Right | Control/Left | Down/Right |

**f.** The experimentwise error rate for the analysis, part **e**, was $\alpha = .05$. Interpret this value.

**10.67** Refer to the *Teaching of Psychology* (Aug. 1998) study of whether a practice test helps students prepare for a final exam, Exercise 10.8 (p. 447). Recall that students in an introductory psychology class were grouped according to their class standing and whether they attended a review session or took a practice test prior to the final exam. The experimental design was a $3 \times 2$ factorial design, with Class Standing at 3 levels (low, medium, or high) and Exam Preparation at 2 levels (practice exam or review session). There were 22 students in each of the $3 \times 2 = 6$ treatment groups. After completing the final exam, each student rated her or his exam preparation on an 11-point scale ranging from 0 (not helpful at all) to 10 (extremely helpful). The data for this experiment (simulated from summary statistics provided in the article) are saved in the PRACEXAM file. The first five and last five observations in the data set are listed below. Conduct a complete analysis of variance of the helpfulness ratings data, including (if warranted) multiple comparisons of means. Do your findings support the research conclusion that "students at all levels of academic ability benefit from a . . . practice exam"?

🌐 PRACEXAM

| Exam Preparation | Class Standing | Helpfulness Rating |
|---|---|---|
| PRACTICE | LOW | 6 |
| PRACTICE | LOW | 7 |
| PRACTICE | LOW | 7 |
| PRACTICE | LOW | 5 |
| PRACTICE | LOW | 3 |
| ⋮ | ⋮ | ⋮ |
| REVIEW | HI | 5 |
| REVIEW | HI | 2 |
| REVIEW | HI | 5 |
| REVIEW | HI | 4 |
| REVIEW | HI | 3 |

*Source:* Balch, W. R. "Practice versus review exams and final exam performance." *Teaching of Psychology*, Vol. 25, No. 3, Aug. 1998 (adapted from Table 1).

## Applying the Concepts—Advanced

**10.68** How do women compare with men in their ability to perform laborious tasks that require strength? Some information on this question is provided in a study of the firefighting ability of men and women (*Human Factors*, 1982). The researchers conducted a $2 \times 2$ factorial experiment to investigate the effect of the factor Sex (male or female) and the factor Weight (light or heavy) on the length of time required for a person to perform a particular firefighting task. Eight persons were selected for each of the $2 \times 2 = 4$ Sex–Weight categories of the $2 \times 2$ factorial experiment, and the length of time needed to complete the task was recorded for each of the 32 persons. The means and standard deviations of the four samples are shown in the following table.

| | Light | | Heavy | |
|---|---|---|---|---|
| | **Mean** | **Standard Deviation** | **Mean** | **Standard Deviation** |
| Female | 18.30 | 6.81 | 14.50 | 2.93 |
| Male | 13.00 | 5.04 | 12.25 | 5.70 |

*Source:* M. D. Phillips and R. L. Pepper, "Shipboard firefighting performance of females and males." Reprinted with permission from *Human Factors*, Vol. 24, No. 3, 1982. Copyright 1982 by the Human Factors and Ergonomics Society. All rights reserved.

**a.** Calculate the total of the $n = 8$ time measurements for each of the four categories of the $2 \times 2$ factorial experiment.

**b.** Calculate the correction for mean, CM. (See Appendix C for computational formulas.)

**c.** Use the results of parts **a** and **b** to calculate the sums of squares for Sex, Weight, and for the Sex $\times$ Weight interaction.

**d.** Calculate each sample variance. Then calculate the sum of squares of deviations *within* each sample for each of the four samples.

**e.** Calculate SSE. [*Hint:* SSE is the pooled sum of squares for the deviations calculated in part **d**.]

**f.** Now that you know SS(Sex), SS(Weight), SS(Sex $\times$ Weight), and SSE, find SS(Total).

**g.** Summarize the calculations in an ANOVA table.

**h.** Conduct a complete analysis of these data. Use $\alpha = .05$ for any inferential techniques you employ. Interpret your conclusions graphically.

**i.** What assumptions are necessary to ensure the validity of the inferential techniques you utilized? State them in terms of this experiment.

## STATISTICS IN ACTION
### On the Trail of the Cockroach

Entomologists have long established that insects such as ants, bees, caterpillars, and termites use chemical or "odor" trails for navigation. These trails are used as highways between sources of food and the insect nest. Until recently, however, "bug" researchers believed that navigational behavior of cockroaches scavenging for food was random and not linked to a chemical trail.

One of the first researchers to challenge the "random-walk" theory for cockroaches was professor and entomologist Dini Miller of Virginia Tech University. According to Miller, "the idea that roaches forage randomly means that they would have to come out of their hiding places every night and bump into food and water by accident. But roaches never seem to go hungry." Since cockroaches had never before been evaluated for trail following behavior, Miller designed an experiment to test a cockroach's ability to follow a trail to their fecal material (*Explore*, Research at the University of Florida, Fall 1998).

First, Dr. Miller developed a methanol extract from roach feces—called a pheromone. She theorized that "pheromones are communication devices between cockroaches. If you have an infestation and have a lot of fecal material around, it advertises, 'Hey, this is a good cockroach place.'" Then, she created a chemical trail with the pheromone on a strip of white chromatography paper and placed the paper at the bottom of a plastic, V-shaped container, 122 square centimeters in size. German cockroaches were released into the container at the beginning of the trail, one at a time, and a video surveillance camera was used to monitor the roach's movements.

In addition to the trail containing the fecal extract (the treatment), a trail using methanol only was created. This second trail served as a "control" to compare back against the treated trail. Since Dr. Miller also wanted to determine if trail following ability differed among cockroaches of different age, sex, and reproductive stage, four roach groups were utilized in the experiment: adult males, adult females, gravid (pregnant) females, and nymphs (immatures). Twenty roaches of each type were randomly assigned to the treatment trail and 10 of each type were randomly assigned to the control trail. Thus, a total of 120 roaches were used in the experiment. A layout of the design is illustrated in Figure 10.27.

The movement pattern of each cockroach tested was translated into *xy* coordinates every one-tenth of a second by the Dynamic Animal Movement Analyzer (DAMA) program. Miller measured the perpendicular distance of each *xy* coordinate from the trail and then averaged these distances, or deviations, for each cockroach. The average trail deviations (measured in pixels*) for 120 cockroaches in the study are listed in Table 10.12 and saved in the ROACH file.

### Focus

Conduct the appropriate analysis of variance of the data. Use the results to answer the following research questions (not necessarily in the order presented):

(1) Is there evidence that cockroaches within a group exhibit the ability to follow fecal extract trails?
(2) Does trail following ability differ among cockroaches of different age, sex, and reproductive status?

Write up the results of your analysis in a report and present it to your class.

---

*1 pixel ≈ 2 centimeters

**FIGURE 10.27** **Layout of Experimental Design for Cockroach Study**

| | | Roach Type | | | |
|---|---|---|---|---|---|
| | | **Adult Male** | **Adult Female** | **Gravid Female** | **Nymph** |
| **Trail** | **Extract** | $n = 20$ | $n = 20$ | $n = 20$ | $n = 20$ |
| | **Control** | $n = 10$ | $n = 10$ | $n = 10$ | $n = 10$ |

## STATISTICS IN ACTION *(continued)*

### ROACH
### TABLE 10.12  Average Trail Deviations for 120 Cockroaches

| Adult Males | | Adult Females | | Gravid Females | | Nymphs | |
|---|---|---|---|---|---|---|---|
| Extract | Control | Extract | Control | Extract | Control | Extract | Control |
| 3.1 | 42.0 | 7.2 | 70.2 | 78.7 | 54.6 | 7.7 | 132.9 |
| 6.2 | 22.7 | 17.3 | 49.0 | 70.3 | 54.3 | 27.7 | 19.7 |
| 34.0 | 93.1 | 9.1 | 40.5 | 79.9 | 63.5 | 18.4 | 32.1 |
| 2.1 | 17.5 | 13.2 | 13.3 | 51.0 | 52.6 | 47.6 | 66.4 |
| 2.4 | 78.1 | 101.2 | 31.8 | 13.6 | 95.1 | 22.4 | 126.0 |
| 4.4 | 74.1 | 4.6 | 116.0 | 20.4 | 117.9 | 8.3 | 131.1 |
| 2.4 | 50.3 | 18.1 | 164.0 | 51.2 | 53.3 | 13.5 | 50.7 |
| 7.6 | 8.9 | 73.0 | 30.2 | 27.5 | 84.0 | 45.6 | 93.8 |
| 5.5 | 11.3 | 4.8 | 44.3 | 63.1 | 103.5 | 8.4 | 59.4 |
| 6.9 | 82.0 | 20.5 | 72.6 | 4.8 | 53.0 | 3.3 | 25.6 |
| 25.4 | | 51.6 | | 23.4 | | 51.2 | |
| 2.2 | | 5.8 | | 48.2 | | 10.4 | |
| 2.5 | | 27.8 | | 13.3 | | 32.0 | |
| 4.9 | | 2.8 | | 57.4 | | 6.9 | |
| 18.5 | | 4.4 | | 65.4 | | 32.6 | |
| 4.6 | | 3.2 | | 10.5 | | 23.8 | |
| 7.7 | | 3.6 | | 59.9 | | 5.1 | |
| 3.2 | | 1.7 | | 38.4 | | 3.8 | |
| 2.4 | | 29.8 | | 27.0 | | 3.1 | |
| 1.5 | | 21.7 | | 76.6 | | 2.8 | |

*Source:* Dini Miller, Department of Entomolgy, Virginia Polytechnic Institute and State University.

## QUICK REVIEW

## Key Terms

Analysis of variance (ANOVA) 454
Balanced design 487
Blocks 469
Bonferroni multiple comparisons procedure 464
Comparisonwise error rate 464
Complete factorial experiment 485
Completely randomized design 448
Dependent variable 442
Designed experiment 444
Experimental unit 443
Experimentwise error rate 464

Factor interaction 486
Factor levels 443
Factor main effect 486
Factors 442
$F$ statistic 450
$F$-test 451
Mean square for error 449
Mean square for treatments 449
Multiple comparisons of means 464
Observational experiment 444
Qualitative factors 442
Quantitative factors 442
Randomized block design 469

Replicates of the experiment 487
Response variable 442
Robust method 456
Scheffé multiple comparisons procedure 464
Single-factor experiment 484
Sum of squares for blocks 470
Sum of squares for error 449
Sum of squares for treatment 449
Treatments 443
Tukey multiple comparisons procedure 464
Two-factor experiment 487
Two-way classification 485

## Key Formulas

*Note: Computing formulas for sums of squares (SS) and mean squares (MS) in ANOVA are provided in Appendix C.*

Completely randomized design:

$$F = \frac{MST}{MSE}$$    Testing treatments    451

Randomized block design:

$$F = \frac{MST}{MSE}$$    Testing treatments    472

$$F = \frac{MSB}{MSE}$$    Testing blocks    477

Factorial design with 2 factors:

$$F = \frac{MS(A)}{MSE}$$    Testing main effect $A$    491

$$F = \frac{MS(B)}{MSE}$$    Testing main effect $B$    492

$$F = \frac{MS(AB)}{MSE}$$    Testing $A \times B$ interaction    492

$c = p(p-1)/2$    Number of pairwise comparisons for $p$ treatment means    463

## Language Lab

| Symbol | Description |
| --- | --- |
| ANOVA | Analysis of variance |
| SST | Sum of Squares for Treatments (i.e., the variation among treatment means) |
| SSE | Sum of Squares for Error (i.e., the variability around the treatment means due to sampling error) |
| MST | Mean Square for Treatments |
| MSE | Mean Square for Error (an estimate of $\sigma^2$) |
| SSB | Sum of Squares for Blocks |
| MSB | Mean Square for Blocks |
| $a \times b$ factorial | Two-factor factorial experiment with one factor at $a$ levels and the other at $b$ levels (thus, there are $a \times b$ treatments in the experiment) |
| SS($A$) | Sum of Squares for Factor $A$ |
| MS($A$) | Mean Square for Factor $A$ |
| SS($B$) | Sum of Squares for Factor $B$ |
| MS($B$) | Mean Square for Factor $B$ |
| SS($AB$) | Sum of Square for $A \times B$ interaction |
| MS($AB$) | Mean Square for $A \times B$ interaction |

 SUPPLEMENTARY EXERCISES 10.69–10.90

### Learning the Mechanics

**10.69** Explain the difference between an experiment that utilizes a completely randomized design and one that utilizes a randomized block design.

**10.70** What are the treatments in a two-factor experiment, with factor $A$ at three levels and factor $B$ at two levels?

**10.71** Why does the experimentwise error rate of a multiple comparisons procedure differ from the significance level for each comparison (assuming the experiment has more than two treatments)?

**10.72** A completely randomized design is utilized to compare four treatment means. The data are shown in the table (p. 503).

**a.** Given that SST = 36.95 and SS(Total) = 62.55, complete an ANOVA table for this experiment.

**b.** Is there evidence that the treatment means differ? Use $\alpha = .10$.

⊘ LM10_72

| Treatment 1 | Treatment 2 | Treatment 3 | Treatment 4 |
|:---:|:---:|:---:|:---:|
| 8 | 6 | 9 | 12 |
| 10 | 9 | 10 | 13 |
| 9 | 8 | 8 | 10 |
| 10 | 8 | 11 | 11 |
| 11 | 7 | 12 | 11 |

**10.73** An experiment utilizing a randomized block design was conducted to compare the mean responses for four treatments, A, B, C, and D. The treatments were randomly assigned to the four experimental units in each of five blocks. The data are shown in the following table.

⊘ EX10_73

| | Block | | | | |
|:---:|:---:|:---:|:---:|:---:|:---:|
| Treatment | 1 | 2 | 3 | 4 | 5 |
| A | 8.6 | 7.5 | 8.7 | 9.8 | 7.4 |
| B | 7.3 | 6.3 | 7.3 | 8.4 | 6.3 |
| C | 9.1 | 8.3 | 9.0 | 9.9 | 8.2 |
| D | 9.3 | 8.2 | 9.2 | 10.0 | 8.4 |

**a.** Given that SS(Total) = 22.31 and SS(Block) = 10.688 and SSE = .288, complete an ANOVA table for the experiment.

**b.** Do the data provide sufficient evidence to indicate a difference among treatment means? Test using $\alpha = .05$

**c.** Does the result of the test in part **b** warrant further comparison of the treatment means? If so, how many pairwise comparisons need to be made?

**d.** Is there evidence that the block means differ? Use $\alpha = .05$.

**10.74** The table shows a partially completed ANOVA table for a two-factor factorial experiment.

| Source | df | SS | MS | F |
|:---|:---:|:---:|:---:|:---:|
| A | 3 | 2.6 | | |
| B | 5 | 9.2 | | |
| A × B | | | 3.1 | |
| Error | | 18.7 | | |
| Total | 47 | | | |

**a.** Complete the ANOVA table.

**b.** How many levels were used for each factor? How many treatments were used? How many replications were performed?

**c.** Find the value of the Sum of Squares for Treatments. Test to determine whether the data provide evidence that the treatment means differ. Use $\alpha = .05$.

**d.** Is further testing of the nature of factor effects warranted? If so, test to determine whether the factors interact. Use $\alpha = .05$. Interpret the result.

## Applying the Concepts—Basic

**10.75** The *Journal of Testing and Evaluation* (July 1992) published an investigation of the mean compression strength of corrugated fiberboard shipping containers. Comparisons were made for boxes of five different sizes: A, B, C, D, and E. Twenty identical boxes of each size were tested and the peak compression strength (pounds) was recorded for each box. The figure shows the sample means for the five box types as well as the variation around each sample mean.

**a.** Explain why the data are collected as a completely randomized design.

**b.** Refer to box types B and D. Based on the graph, does it appear that the mean compressive strengths of these two box types are significantly different? Explain.

**c.** Based on the graph, does it appear that the mean compressive strengths of all five box types are significantly different? Explain.

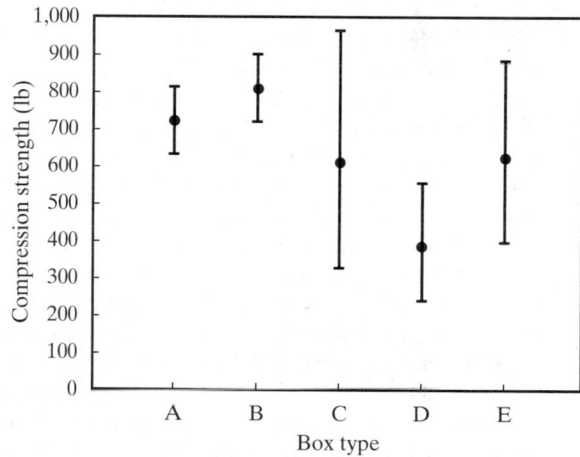

*Source:* Singh, S. P., *et al.* "Compression of single-wall corrugated shipping containers using fixed and floating test platens." *Journal of Testing and Evaluation*, Vol. 20, No. 4, July 1992, p. 319 (Figure 3). Copyright American Society for Testing and Materials.

**10.76** *Physical Therapy* (Aug. 1986) reported on a study of the medial rotation of the shoulder. Ten college students, who exhibited no evidence of disease or limitation of movement in their shoulders, served as the subjects for the study. For each subject, the medial rotation was measured (in degrees) at each of three positions: (1) beginning position, (2) point at which rotation changed directions, and (3) ending position. The goal of the analysis is to compare the mean medial rotation measurements of the three positions.

**a.** Explain why a randomized block design is appropriate in this experiment.

**b.** Identify the treatments in this experiment.

**c.** Identify the blocks in this experiment.

**d.** Identify the response variable.

**10.77** Refer to the *Journal of Communication Disorders* (Mar. 1995) study of pronoun misusage by specifically language-impaired (SLI) children, Exercise 2.51 (p. 53). Recall that percentage of pronoun errors was recorded for three groups of children: ten 5-year-old SLI children, ten younger (3-year-old) normally developing (YND) children, and ten older (5-year-old) normally developing (OND) children. A completely randomized design ANOVA with multiple comparisons of means was conducted on the data, producing the results shown here. Interpret these results.

| Source | df | SS | MS | F | p-value |
|---|---|---|---|---|---|
| Groups | 2 | 11,292.45 | 5,646.22 | 8.295 | .0016 |
| Error | 27 | 18,377.65 | 680.65 | | |
| Total | 29 | 29,670.10 | | | |

| Mean percentage of errors: | | 0.00 | 30.17 | 46.88 |
|---|---|---|---|---|
| Group: | | OND | SLI | YND |

*Source:* Moore, M. E. "Error analysis of pronouns by normal and language-impaired children." *Journal of Communication Disorders*, Vol. 28, No. 1, Mar. 1995, p. 67 (Table 6).

**10.78** In a study published in the *American Journal of Psychology* (Winter 1991), 14-year-old students from New York City were divided into four groups: (1) learning-disabled, low socioeconomic status; (2) nondisabled, low socioeconomic status; (3) learning-disabled, high socioeconomic status; and (4) nondisabled, high socioeconomic status. Each student was asked to read and retell a "story problem"; however, the way in which the problem was presented was varied. Some students read a "no-priority" problem (i.e., a problem with no clear goals and/or objectives), while others read a "priority" problem (i.e., a problem with a clear statement of the character's priority). The experiment was designed as a $4 \times 2$ factorial, with Group at 4 levels and Problem type (priority or no-priority) at 2 levels. One of the dependent variables measured was proportion of ideas recalled correctly.

  **a.** The test for Group by Problem type interaction was nonsignificant at $\alpha = .01$. Interpret this result.

  **b.** The test for Problem type main effect was nonsignificant at $\alpha = .01$. Interpret this result.

  **c.** The test for Group main effect was statistically significant at $\alpha = .01$. Interpret this result.

## Applying the Concepts—Intermediate

**10.79** The Minnesota Multiphasic Personality Inventory (MMPI) is a questionnaire used to gauge personality type. *Psychological Assessment* (Mar. 1995) published a study that investigated the effectiveness of the MMPI scales in detecting deliberately distorted responses. A completely randomized design with four treatments was employed. The treatments consisted of independent random samples of females in the following four groups: nonforensic psychiatric patients ($n_1 = 65$), forensic psychiatric patients ($n_2 = 28$), college students who were requested to respond honestly ($n_3 = 140$), and college students who were instructed to provide "fake bad" responses ($n_4 = 45$). All 278 participants were given the MMPI and the scores on several scales designed to assess response distortion were recorded for each. Each scale was treated as a response variable, and an analysis of variance was conducted. The ANOVA $F$-values are reported in the following table.

| Response (Scale) Variable | ANOVA F-Value |
|---|---|
| Infrequency | 155.8 |
| Obvious | 49.7 |
| Subtle | 10.3 |
| Obvious-subtle | 45.4 |
| Dissimulation | 39.1 |

  **a.** For each response variable, determine whether the mean scores of the four groups completing the MMPI differ significantly. Use $\alpha = .05$ for each test.

  **b.** If the MMPI is effective in detecting distorted responses, then the mean score for the "fake bad" treatment group will be largest. Based on the information provided, can the researchers make an inference about the effectiveness of the MMPI? Explain.

**Response Variable (Exercise 10.79)**

| Infrequency (F) | Mean | 7.1 | 11.3 | 14.6 | 33.6 |
|---|---|---|---|---|---|
| | Group | CSH | NFP | FP | CSFB |
| Obvious (O) | Mean | 240.9 | 270.7 | 287.5 | 341.9 |
| | Group | CSH | FP | NFP | CSFB |
| Subtle (S) | Mean | 231.4 | 244.1 | 244.5 | 259.7 |
| | Group | CSFB | CSH | FP | NFP |
| Obvious-subtle (O-S) | Mean | 51.3 | 88.2 | 93.4 | 198.8 |
| | Group | CSH | NFP | FP | CSFB |
| Dissimulation (D) | Mean | 12.0 | 12.1 | 20.8 | 21.0 |
| | Group | FP | NFP | CSH | CSFB |

*Source:* Bagby, R. M., Burs, T., and Nicholson, R. A. "Relative effectiveness of the standard validity scales in detecting fake-bad and fake-good responding: Replication and extension." *Psychological Assessment*, Vol. 7, No. 1, Mar. 1995, p. 86 (Table 1).

**c.** The Bonferroni method was used to rank the means of the four groups using an experimentwise error rate of $\alpha = .05$. The results for each response variable are shown in the table at the bottom of p. 504. Interpret the results for each response variable.

**10.80** An article in the *Journal of Family Violence* (Mar. 1992) investigated the differences in the opinions of adolescents and their parents when determining who should make decisions about a variety of issues. Thirty-five adolescents and their parents were independently given a list of 44 issues common to adolescent children (e.g., curfew, homework, chores, etc.). Each was asked to state how many of the 44 issues the adolescent should make decisions about on his or her own. The results are presented below.

|          | Mother | Father | Adolescent |
|----------|--------|--------|------------|
| Mean     | 7.57   | 7.51   | 15.94      |
| Std. dev.| 6.07   | 5.93   | 9.11       |

**a.** State the hypotheses necessary to determine whether parents and adolescents disagree on who should make decisions about issues common to adolescent children.

**b.** The article reported an $F$ statistic of $F = 15.92$. Conduct the test using $\alpha = .05$. [*Hint:* A total of 105 people were sampled.]

**c.** The Bonferroni technique was used to compare the pairs of means at an overall significance level of $\alpha = .06$. The resulting confidence intervals are shown here. Interpret the intervals.

| Comparison | Confidence Interval |
|------------|---------------------|
| (Mother − Father) | $.06 \pm 4.06$ |
| (Mother − Adolescent) | $-8.37 \pm 4.06$ |
| (Father − Adolescent) | $-8.43 \pm 4.06$ |

**10.81** Ducks inhabiting the Great Salt Lake marshes feed on a variety of animals, including water boatmen, brine shrimp, beetles, and snails. The changes in the availability of these animal species for ducks during the summer was investigated (*Wetlands*, March 1995). The goal was to compare the mean amount (measured as biomass) of a particular duck food species across four different summer time periods: (1) July 9–23, (2) July 24–Aug. 8, (3) Aug. 9–23, and (4) Aug. 24–31. Ten stations in the marshes were randomly selected, and the biomass density in a water specimen collected

from each was measured. Biomass measurements (milligrams per square meter) were collected during each of the four summer time periods at each station, with stations treated as a blocking factor. Thus, the data were analyzed as a randomized block design.

**a.** Fill in the missing degrees of freedom in the randomized block ANOVA table shown here.

| Source | df | F | *p*-value |
|--------|-----|-------|-----------|
| Time Period | – | 11.25 | .0001 |
| Station | – | – | – |
| Error | – | | |
| Total | 39 | | |

**b.** The $F$ value (and corresponding $p$-value) shown in the ANOVA table, part **a**, were computed from an analysis of biomass of water boatmen nymphs (a common duck food). Interpret these results.

**c.** A multiple comparisons of time period means was conducted using an experimentwise error rate of .05. The results are summarized at the bottom of the page. Identify the time period(s) with the largest and smallest mean biomass.

**10.82** A research psychologist wishes to investigate the difference in maze test scores for a strain of laboratory mice trained under different laboratory conditions. The experiment is conducted using 18 randomly selected mice of this strain, with six receiving no training at all (control group), six trained under condition 1, and six trained under condition 2. Then each of the mice is given a test score between 0 and 100, depending on its performance in a test maze. The experiment produced the results shown in the following table.

⚙ MICE

| Control | Condition 1 | Condition 2 |
|---------|-------------|-------------|
| 58 | 73 | 53 |
| 32 | 70 | 74 |
| 59 | 68 | 72 |
| 64 | 71 | 62 |
| 55 | 60 | 58 |
| 49 | 62 | 61 |

**a.** Identify the following elements of this experiment: experimental units, response, factor(s) and factor type(s), and treatments.

**Multiple Comparisons for Exercise 10.81**

| Mean biomass(mg/m²): | 19 | 54.5 | 90 | 148 |
|----------------------|----|------|-----|-----|
| Time period: | 8/24–8/31 | 8/9–8/23 | 7/24–8/8 | 7/9–7/23 |

**b.** Is this a designed or observational experiment?

**c.** Is there sufficient evidence to indicate a difference among mean maze test scores for mice trained under the three different laboratory conditions? Test at $\alpha = .10$.

**d.** Use statistical software with a multiple comparisons routine to compare the control mean to each of the condition means. How many comparisons are involved? Use an overall significance level of $\alpha = .10$.

**10.83** An evaluation of diffusion bonding of zircaloy components is performed. The main objective is to determine which of three elements—nickel, iron, or copper—is the best bonding agent. A series of zircaloy components are bonded with each of the possible bonding agents. Since there is a great deal of variation in components machined from different ingots, a randomized block design is used, blocking on the ingots. A pair of components from each ingot are bonded together using each of the three agents, and the pressure (in units of 1,000 pounds per square inch) required to separate the bonded components is measured. The data in the table are obtained.

⊘ INGOT

| | **Bonding Agent** | | |
|---|---|---|---|
| **Ingot** | **Nickel** | **Iron** | **Copper** |
| 1 | 67.0 | 71.9 | 72.2 |
| 2 | 67.5 | 68.8 | 66.4 |
| 3 | 76.0 | 82.6 | 74.5 |
| 4 | 72.7 | 78.1 | 67.3 |
| 5 | 73.1 | 74.2 | 73.2 |
| 6 | 65.8 | 70.8 | 68.7 |
| 7 | 75.6 | 84.9 | 69.0 |

**a.** Identify the following elements of the experiment: experimental units, blocks, response, factor(s) and factor type(s), and treatments.

**b.** Is the experiment designed or observational? Explain.

**c.** Construct an ANOVA summary table for the experiment.

**d.** Is there sufficient evidence that the mean pressure required to separate the components differs for the three bonding agents? Use $\alpha = .05$.

**e.** If warranted, perform multiple comparisons of bonding agents means using $\alpha = .05$.

**f.** What assumptions must hold to ensure the validity of the inferences in parts **d** and **e**?

**10.84** Sixteen workers were randomly selected to participate in an experiment to determine the effects of work scheduling and method of payment on attitude toward the job. Two types of scheduling were employed, the standard 8-to-5 workday and a modification in which the worker was permitted to start the day at either 7 or 8 a.m. and to vary the starting time as desired; in addition, the worker was allowed to choose, on a daily basis, either a $\frac{1}{2}$-hour or 1-hour lunch period. The two methods of payment were a standard hourly rate and a reduced hourly rate with an added piece rate based on the worker's production. Four workers were randomly assigned to each of the four Scheduling–Payment combinations, and each completed an attitude test after 1 month on the job. The test scores are shown in the table.

⊘ WORKSCHD

| Scheduling | Hourly Rate | Hourly and Piece Rate |
|---|---|---|
| 8–5 | 54, 68, 55, 63 | 89, 75, 71, 83 |
| Worker-modified | 79, 65, 62, 74 | 83, 94, 91, 86 |

**a.** What type of experiment was performed? Identify the response, factor(s), factor type(s), treatments, and experimental units.

**b.** Perform an ANOVA to compare the treatment means. Use $\alpha = .05$.

**c.** If the test in part **b** warrants further analysis, conduct the appropriate tests of interaction and main effects. Interpret your results.

**d.** What assumptions are necessary to ensure the validity of the inferences? State the assumptions in terms of this experiment.

**10.85** Complete factorial designs are commonly employed in marketing research to evaluate the effectiveness of sales strategies. At one supermarket, two of the factors were Price level (regular, reduced price, cost to supermarket) and Display level (normal display space, normal display space plus end-of-aisle display, twice the normal display space). Each of the $3 \times 3 = 9$ treatments was applied three times to a particular product for a full week. The dependent variable of interest was unit sales for the week. To minimize treatment carryover effects, each treatment was preceded and followed by a week in which the product was priced at its regular price and was displayed in its normal manner. The accompanying table reports the data collected.

**a.** Form an ANOVA table for the study.

**b.** Do the data indicate that the mean sales differ among the nine treatments? Test using $\alpha = .10$.

⊘ SUPMARK

| Display | Regular | Reduced | Cost to Supermarket |
|---|---|---|---|
| Normal | 989 1,025 1,030 | 1,211 1,215 1,182 | 1,577 1,559 1,598 |
| Normal Plus | 1,191 1,233 1,221 | 1,860 1,910 1,926 | 2,492 2,527 2,511 |
| Twice Normal | 1,226 1,202 1,180 | 1,516 1,501 1,498 | 1,801 1,833 1,852 |

**c.** Is the test of interaction between the factors Price and Display warranted as a result of the test in part **b**? If so, conduct the test using $\alpha = .10$.

**d.** Are the tests of the main effects for Price and Display warranted as a result of the previous tests? If so, conduct them using $\alpha = .10$.

**e.** Which pairs of treatment means should be compared as a result of the tests in parts **b–d**? Perform the comparisons.

**10.86** Three anticoagulant drugs are studied to compare their effectiveness in dissolving blood clots. Each of five subjects receives the drugs at equally spaced time intervals and in random order. Time periods between drug applications permit a drug to be passed out of a subject's body before the subject receives the next drug. After each drug is in the bloodstream, the length of time (in seconds) required for a cut of specified size to stop bleeding is recorded. The results are shown in the following table.

**CLOTS**

| Person | Drug | | |
|---|---|---|---|
| | **A** | **B** | **C** |
| 1 | 127.5 | 129.0 | 135.5 |
| 2 | 130.6 | 129.1 | 138.0 |
| 3 | 118.3 | 111.7 | 110.1 |
| 4 | 155.5 | 144.3 | 162.3 |
| 5 | 180.7 | 174.4 | 181.8 |

**a.** What type of experimental design was used in this study? Identify the response, factor(s), factor type(s), treatments, and experimental units.

**b.** Is there evidence of a difference in mean clotting time among the three drugs? Test using $\alpha = .10$.

**c.** What is the observed significance level of the test you conducted in part **a**? Interpret it.

**d.** Was blocking effective in reducing the variation among the data? That is, do the data support the contention that the mean clotting time varies from person to person?

**e.** If warranted, use a multiple comparisons technique to determine whether one of the drugs is most effective. Use an overall significance level of $\alpha = .10$.

**10.87** An experiment is conducted to determine whether there is a difference among the mean increases in growth produced by five inoculins of growth hormones for plants. The experimental material consists of 20 cuttings of a shrub (all of equal weight), with four cuttings randomly assigned to each of the five different inoculins. The results of the experiment are given in the table; all measurements represent an increase in weight (in grams).

**a.** Construct an ANOVA table for this experiment.

**b.** Is there evidence of a difference among the mean increases in weight for the five inoculins of growth hormone? Test using $\alpha = .05$.

**INOCULIN**

| A | B | C | D | E |
|---|---|---|---|---|
| 15 | 21 | 22 | 10 | 6 |
| 18 | 13 | 19 | 14 | 11 |
| 9 | 20 | 24 | 21 | 15 |
| 16 | 17 | 21 | 13 | 8 |

**c.** Find the observed significance level for the test in part **a**, and interpret its value.

**d.** What assumptions are necessary to ensure the validity of the inferences in parts **a** and **b**? Do they appear to be satisfied?

**10.88** Many temperate-zone animal species exhibit physiological and morphological changes when the hours of daylight begin to decrease during autumn months. A study was conducted to investigate the "short-day" traits of collared lemmings (*The Journal of Experimental Zoology*, Sept. 1993). A total of 124 lemmings were bred in a colony maintained with a photoperiod of 22 hours of light per day. At weaning (19 days of age), the lemmings were weighed and randomly assigned to live under one of two photoperiods: 16 hours or less of light per day and more than 16 hours of light per day. (Each group was assigned the same number of males and females.) After 10 weeks, the lemmings were weighed again. The response variable of interest was the gain in body weight (measured in grams) over the 10-week experimental period. The researchers analyzed the data using an ANOVA for a $2 \times 2$ factorial design, where the two factors are Photoperiod (at two levels) and Gender (at two levels).

**a.** Construct an ANOVA table for the experiment, listing the sources of variation and associated degrees of freedom.

**b.** The *F*-test for interaction was not significant. Interpret this result practically.

**c.** The *p*-values for testing for Photoperiod and Gender main effects were both smaller than .001. Interpret these results practically.

**10.89** The cattle inhabiting the Biological Reserve of Doñana (Spain), live under free-range conditions, with virtually no human interference. The cattle population is organized into four herds (LGN, MTZ, PLC, and QMD). The *Journal of Zoology* (July 1995) investigated the ranging behavior of the four herds across the four seasons. Thus, a $4 \times 4$ factorial experiment was

| Source | df | F | *p*-Value |
|---|---|---|---|
| Herd (H) | 3 | 17.2 | $p < .001$ |
| Season (S) | 3 | 3.0 | $p < .05$ |
| H × S | 9 | 1.2 | $p > .05$ |
| Error | 32 | | |
| Total | 47 | | |

employed, with Herd and Season representing the two factors. Three animals from each herd during each season were sampled and the home range of each individual was measured (in square kilometers). The data were subjected to an ANOVA, with the results shown in the table on p. 507.

**a.** Conduct the appropriate ANOVA *F*-tests and interpret the results.

**b.** The researcher ranked the four herd means independently of season. Do you agree with this strategy? Explain.

**c.** Refer to part **b**. The Bonferroni rankings of the four herd means (at $\alpha = .05$) are shown below. Interpret the results.

| Mean (km)$^2$: | .75 | 1.0 | 2.7 | 3.8 |
|---|---|---|---|---|
| Herd: | PLC | LGN | QMD | MTZ |

### Applying the Concepts—Advanced

**10.90** Traditionally, people protect themselves from mosquito bites by applying insect repellent to their skin and clothing. Recent research suggests that peremethrin, an insecticide with low toxicity to humans, can provide protection from mosquitoes. A study in the *Journal of the American Mosquito Control Association* (Mar. 1995) investigated whether a tent sprayed with a commercially available 1% peremethrin formulation would protect people, both inside and outside the tent, against biting mosquitoes. Two canvas tents—one treated with peremethrin, the other untreated—were positioned 25 meters apart on flat dry ground in an area infested with mosquitoes. Eight people participated in the experiment, with four randomly assigned to each tent. Of the four stationed at each tent, two were randomly assigned to stay inside the tent (at opposite corners) and two to stay outside the tent (at opposite corners). During a specified 20-minute period during the night, each person kept count of the number of mosquito bites received. The goal of the study was to determine the effect of both Tent type (treated or untreated) and Location (inside or outside the tent) on the mean mosquito bite count.

**a.** What type of design was employed in the study?

**b.** Identify the factors and treatments.

**c.** Identify the response variable.

**d.** The study found statistical evidence of interaction between Tent type and Location. Give a practical interpretation of this result.

---

## STUDENT PROJECTS

Owing to ever-increasing food costs, consumers are becoming more discerning in their choice of supermarkets. It is usually more convenient to shop at just one market, as opposed to buying different items at different markets. Thus, it would be useful to compare the mean food expenditure for a market basket of food items from store to store. Since there is a great deal of variability in the prices of products sold at any supermarket, we will consider an experiment that blocks on products.

Choose three (or more) supermarkets in your area that you want to compare. Then choose approximately 10 (or more) food products you typically purchase. For each food item, record the price each store charges in the following manner:

| Food Item 1 | Food Item 2 | $\cdots$ | Food Item 10 |
|---|---|---|---|
| Price store 1 | Price store 1 | $\cdots$ | Price store 1 |
| Price store 2 | Price store 2 | $\cdots$ | Price store 2 |
| Price store 3 | Price store 3 | $\cdots$ | Price store 3 |

Use the data you obtain to test

$H_0$: Mean expenditures at the stores are the same

$H_a$: Mean expenditures for at least two of the stores are different

Also, test to determine whether blocking on food items is advisable in this kind of experiment. Fully interpret the results of your analysis.

---

## REFERENCES

Mendenhall, W. *Introduction to Linear Models and the Design and Analysis of Experiments*. Belmont, Calif.: Wadsworth, 1968.

Miller, R. G. *Simultaneous Statistical Inference*. New York: Springer-Verlag, 1981.

Scheffé, H. *The Analysis of Variance*. New York: Wiley, 1959.

Snedecor, G. W., and Cochran, W. G. *Statistical Methods*, 7th ed. Ames: Iowa State University Press, 1980.

Steele, R. G. D., and Torrie, J. H. *Principles and Procedures of Statistics: A Biometrical Approach*, 2d ed. New York: McGraw-Hill, 1980.

Tukey, J. W. "Comparing individual means in the analysis of variance." *Biometrics*, Vol. 5, 1949, pp. 99–114.

Winer, B. J. *Statistical Principles in Experimental Design*, 2d ed. New York: McGraw-Hill, 1971.

# CHAPTER 11

# Simple Linear Regression

## Contents

## Statistics in Action

Can "Dowsers" Really Detect Water?

## 🔖 Where We've Been

We've learned how to estimate and test hypotheses about population parameters based on a random sample of observations from the population. We've also seen how to extend these methods to allow for a comparison of parameters from two or more populations.

## ☞ Where We're Going

Suppose we want to predict the rainfall at a given location on a given day. We could select a single random sample of $n$ daily rainfalls, use the methods of Chapter 7 to estimate the mean daily rainfall $\mu$, and then use this quantity to predict the day's rainfall. A better method uses information that is available to any forecaster, e.g., barometric pressure and cloud cover. If we measure barometric pressure and cloud cover at the same time as daily rainfall, we establish the relationship between these variables—one that lets us use these variables for prediction. This chapter covers the simplest situation—relating two variables. The more complex problem of relating more than two variables is the topic of Chapter 12.

In Chapters 7–10 we described methods for making inferences about population means. The mean of a population has been treated as a *constant*, and we have shown how to use sample data to estimate or to test hypotheses about this constant mean. In many applications, the mean of a population is not viewed as a constant, but rather as a variable. For example, the mean sale price of residences in a large city last year can be treated as a constant and might be equal to $150,000. But we might also treat the mean sale price as a variable that depends on the square feet of living space in the residence. For example, the relationship might be

$$\text{Mean sale price} = \$30,000 + \$60(\text{Square feet})$$

This formula implies that the mean sale price of 1,000-square-foot homes is $90,000, the mean sale price of 2,000-square-foot homes is $150,000, and the mean sale price of 3,000-square-foot homes is $210,000.

What do we gain by treating the mean as a variable rather than a constant? In many practical applications we will be dealing with highly variable data, data for which the standard deviation is so large that a constant mean is almost "lost" in a sea of variability. For example, if the mean residential sale price is $150,000 but the standard deviation is $75,000, then the actual sale prices will vary considerably, and the mean price is not a very meaningful or useful characterization of the price distribution. On the other hand, if the mean sale price is treated as a variable that depends on the square feet of living space, the standard deviation of sale prices for any given size of home might be only $10,000. In this case, the mean price will provide a much better characterization of sale prices when it is treated as a variable rather than a constant.

In this chapter we discuss situations in which the mean of the population is treated as a variable, dependent on the value of another variable. The dependence of residential sale price on the square feet of living space is one illustration. Other examples include the dependence of mean reaction time on the amount of a drug in the bloodstream, the dependence of mean starting salary of a college graduate on the student's GPA, and the dependence of mean number of years to which a criminal is sentenced on the number of previous convictions.

In this chapter we discuss the simplest of all models relating a population mean to another variable, the *straight-line model*. We show how to use the sample data to estimate the straight-line relationship between the mean value of one variable, $y$, as it relates to a second variable, $x$. The methodology of estimating and using a straight-line relationship is referred to as *simple linear regression analysis*.

## 11.1   PROBABILISTIC MODELS

An important consideration when taking a drug is how it may affect one's perception or general awareness. Suppose you want to model the length of time it takes to respond to a stimulus (a measure of awareness) as a function of the percentage of a certain drug in the bloodstream. The first question to be answered is this: "Do you think an exact relationship exists between these two variables?" That is, do you think it is possible to state the exact length of time it takes an individual (subject) to respond if the amount of the drug in the bloodstream is known? We think you will agree with us that this is *not* possible for several reasons. The reaction time depends on many variables other than the percentage of the drug in the bloodstream—for example, the time of day, the amount of sleep the subject had the night before, the subject's visual acuity, the subject's general reaction time without the drug, and the subject's age would all probably affect reaction time. Even if many variables are included in a model (the topic of Chapter 12), it is still unlikely that we would be able to predict *exactly* the subject's reaction time. There will almost certainly be some variation in response times due strictly to *random phenomena* that cannot be modeled or explained.

If we were to construct a model that hypothesized an exact relationship between variables, it would be called a **deterministic model**. For example, if we believe that $y$, the reaction time (in seconds), will be exactly one and one-half times $x$, the amount of drug in the blood, we write

$$y = 1.5x$$

This represents a **deterministic relationship** between the variables $y$ and $x$. It implies that $y$ can always be determined exactly when the value of $x$ is known. *There is no allowance for error in this prediction.*

If, on the other hand, we believe there will be unexplained variation in reaction times—perhaps caused by important but unincluded variables or by random phenomena—we discard the deterministic model and use a model that accounts for this **random error**. This **probabilistic model** includes both a deterministic component and a random error component. For example, if we hypothesize that the response time $y$ is related to the percentage of drug $x$ by

$$y = 1.5x + \text{Random error}$$

we are hypothesizing a **probabilistic relationship** between $y$ and $x$. Note that the deterministic component of this probabilistic model is $1.5x$.

Figure 11.1a shows the possible responses for five different values of $x$, the percentage of drug in the blood, when the model is deterministic. All the responses must fall exactly on the line because a deterministic model leaves no room for error.

Figure 11.1b shows a possible set of responses for the same values of $x$ when we are using a probabilistic model. Note that the deterministic part of the model (the straight line itself) is the same. Now, however, the inclusion of a random error component allows the response times to vary from this line. Since we know that the response time does vary randomly for a given value of $x$, the probabilistic model provides a more realistic model for $y$ than does the deterministic model.

**FIGURE 11.1**

*Possible Reaction Times, y, for Five Different Drug Percentages, x*

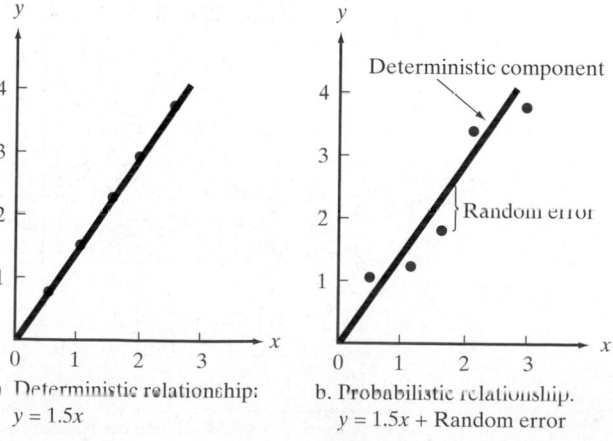

a. Deterministic relationship: $y = 1.5x$

b. Probabilistic relationship: $y = 1.5x + \text{Random error}$

## General Form of Probabilistic Models

$y$ = Deterministic component + Random error

where $y$ is the variable of interest. We always assume that the mean value of the random error equals 0. This is equivalent to assuming that the mean value of $y$, $E(y)$, equals the deterministic component of the model; that is,

$E(y)$ = Deterministic component

In this chapter we present the simplest of probabilistic models—the **straight-line model**—which derives its name from the fact that the deterministic portion of the model graphs as a straight line. Fitting this model to a set of data is an example of **regression analysis**, or **regression modeling**. The elements of the straight-line model are summarized in the next box.

## A First-Order (Straight-Line) Probabilistic Model

$y = \beta_0 + \beta_1 x + \varepsilon$

where $y$ = **Dependent** *or* **response variable** (variable to be modeled)

$x$ = **Independent** *or* **predictor variable** (variable used as a predictor of $y$)*

$E(y) = \beta_0 + \beta_1 x$ = Deterministic component

$\varepsilon$ (epsilon) = Random error component

$\beta_0$ (beta zero) = **y-intercept of the line**, that is, the point at which the line intersects or cuts through the $y$-axis (see Figure 11.2)

$\beta_1$ (beta one) = **Slope of the line**, that is, the amount of increase (or decrease) in the deterministic component of $y$ for every 1-unit increase in $x$. [As you can see in Figure 11.2, $E(y)$ increases by the amount $\beta_1$ as $x$ increases from 2 to 3.]

**FIGURE 11.2**

*The Straight-Line Model*

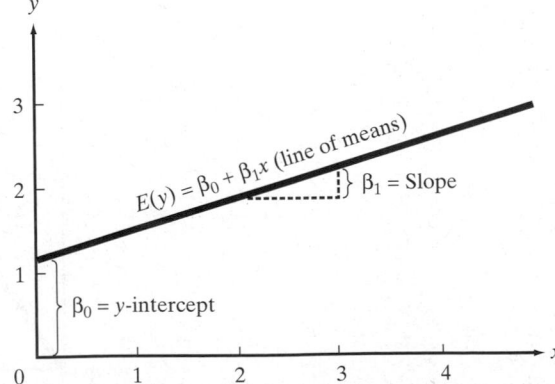

---

*The word *independent* should not be interpreted in a probabilistic sense, as defined in Chapter 3. The phrase *independent variable* is used in regression analysis to refer to a predictor variable for the response $y$.

In the probabilistic model, the deterministic component is referred to as the **line of means**, because the mean of $y$, $E(y)$, is equal to the straight-line component of the model. That is,

$$E(y) = \beta_0 + \beta_1 x$$

Note that the Greek symbols $\beta_0$ and $\beta_1$, respectively, represent the $y$-intercept and slope of the model. They are population parameters that will be known only if we have access to the entire population of $(x, y)$ measurements. Together with a specific value of the independent variable $x$, they determine the mean value of $y$, which is just a specific point on the line of means (Figure 11.2).

The values of $\beta_0$ and $\beta_1$ will be unknown in almost all practical applications of regression analysis. The process of developing a model, estimating the unknown parameters, and using the model can be viewed as the five-step procedure shown in the next box.

Step 1   Hypothesize the deterministic component of the model that relates the mean, $E(y)$, to the independent variable $x$ (Section 11.1)

Step 2   Use the sample data to estimate unknown parameters in the model (Section 11.2).

Step 3   Specify the probability distribution of the random error term and estimate the standard deviation of this distribution (Sections 11.3 and 11.4).

Step 4   Statistically evaluate the usefulness of the model (Sections 11.5, 11.6, and 11.7).

Step 5   When satisfied that the model is useful, use it for prediction, estimation, and other purposes (Section 11.8).

 EXERCISES 11.1–11.9

## Learning the Mechanics

**11.1**   In each case, graph the line that passes through the given points.
  **a.** $(1, 1)$ and $(5, 5)$
  **b.** $(0, 3)$ and $(3, 0)$
  **c.** $(-1, 1)$ and $(4, 2)$
  **d.** $(-6, -3)$ and $(2, 6)$

**11.2**   Give the slope and $y$-intercept for each of the lines graphed in Exercise 11.1.

**11.3**   The equation for a straight line (deterministic) is

$$y = \beta_0 + \beta_1 x$$

If the line passes through the point $(-2, 4)$, then $x = -2$, $y = 4$ must satisfy the equation; that is,

$$4 = \beta_0 + \beta_1(-2)$$

Similarly, if the line passes through the point $(4, 6)$, then $x = 4$, $y = 6$ must satisfy the equation; that is,

$$6 = \beta_0 + \beta_1(4)$$

Use these two equations to solve for $\beta_0$ and $\beta_1$; then find the equation of the line that passes through the points $(-2, 4)$ and $(4, 6)$.

**11.4**   Refer to Exercise 11.3. Find the equations of the lines that pass through the points listed in Exercise 11.1.

**11.5**   Plot the following lines:
  **a.** $y = 4 + x$
  **b.** $y = 5 - 2x$
  **c.** $y = -4 + 3x$
  **d.** $y = -2x$
  **e.** $y = x$
  **f.** $y = .50 + 1.5x$

**11.6**   Give the slope and $y$-intercept for each of the lines defined in Exercise 11.5.

**11.7**   Why do we generally prefer a probabilistic model to a deterministic model? Give examples for which the two types of models might be appropriate.

**11.8**   What is the line of means?

**11.9**   If a straight-line probabilistic relationship relates the mean $E(y)$ to an independent variable $x$, does it imply that every value of the variable $y$ will always fall exactly on the line of means? Why or why not?

## 11.2  FITTING THE MODEL: THE LEAST SQUARES APPROACH

After the straight-line model has been hypothesized to relate the mean $E(y)$ to the independent variable $x$, the next step is to collect data and to estimate the (unknown) population parameters, the $y$-intercept $\beta_0$ and the slope $\beta_1$.

To begin with a simple example, suppose an experiment involving five subjects is conducted to determine the relationship between the percentage of a certain drug in the bloodstream and the length of time it takes to react to a stimulus. The results are shown in Table 11.1. (The number of measurements and the measurements themselves are unrealistically simple in order to avoid arithmetic confusion in this introductory example.) This set of data will be used to demonstrate the five-step procedure of regression modeling given in Section 11.1. In this section we hypothesize the deterministic component of the model and estimate its unknown parameters (steps 1 and 2). The model assumptions and the random error component (step 3) are the subjects of Sections 11.3 and 11.4, whereas Sections 11.5–11.7 assess the utility of the model (step 4). Finally, we use the model for prediction and estimation (step 5) in Section 11.8.

⊘ STIMULUS

**TABLE 11.1    Reaction Time Versus Drug Percentage**

| Subject | Amount of Drug $x$ (%) | Reaction Time $y$ (seconds) |
|---------|------------------------|------------------------------|
| 1 | 1 | 1 |
| 2 | 2 | 1 |
| 3 | 3 | 2 |
| 4 | 4 | 2 |
| 5 | 5 | 4 |

**Step 1**    *Hypothesize the deterministic component of the probabilistic model.* As stated before, we will consider only straight-line models in this chapter. Thus the complete model to relate mean response time $E(y)$ to drug percentage $x$ is given by

$$E(y) = \beta_0 + \beta_1 x$$

**Step 2**    *Use sample data to estimate unknown parameters in the model.* This step is the subject of this section—namely, how can we best use the information in the sample of five observations in Table 11.1 to estimate the unknown $y$-intercept $\beta_0$ and slope $\beta_1$?

To determine whether a linear relationship between $y$ and $x$ is plausible, it is helpful to plot the sample data in a **scattergram**. Recall (Section 2.9) that a scattergram locates each of the five data points on a graph, as shown in Figure 11.3. Note that the scattergram suggests a general tendency for $y$ to increase as $x$ increases. If you place a ruler on the scattergram, you will see that a line may be

**FIGURE 11.3**

*Scattergram for Data in Table 11.1*

## FIGURE 11.4

*Visual Straight Line Fitted to the Data in Figure 11.3*

drawn through three of the five points, as shown in Figure 11.4. To obtain the equation of this visually fitted line, note that the line intersects the $y$-axis at $y = -1$, so the $y$-intercept is $-1$. Also, $y$ increases exactly 1 unit for every 1-unit increase in $x$, indicating that the slope is $+1$. Therefore, the equation is

$$\tilde{y} = -1 + 1(x) = -1 + x$$

where $\tilde{y}$ is used to denote the predicted $y$ from the visual model.

One way to decide quantitatively how well a straight line fits a set of data is to note the extent to which the data points deviate from the line. For example, to evaluate the model in Figure 11.4, we calculate the magnitude of the *deviations*, i.e., the differences between the observed and the predicted values of $y$. These deviations, or **errors of prediction**, are the vertical distances between observed and predicted values (see Figure 11.4). The observed and predicted values of $y$, their differences, and their squared differences are shown in Table 11.2. Note that the *sum of errors* equals 0 and the *sum of squares of the errors* (SSE), which gives greater emphasis to large deviations of the points from the line, is equal to 2.

**TABLE 11.2  Comparing Observed and Predicted Values for the Visual Model**

| $x$ | $y$ | $\tilde{y} = -1 + x$ | $(y - \tilde{y})$ | $(y - \tilde{y})^2$ |
|---|---|---|---|---|
| 1 | 1 | 0 | $(1 - 0) = \phantom{-}1$ | 1 |
| 2 | 1 | 1 | $(1 - 1) = \phantom{-}0$ | 0 |
| 3 | 2 | 2 | $(2 - 2) = \phantom{-}0$ | 0 |
| 4 | 2 | 3 | $(2 - 3) = -1$ | 1 |
| 5 | 4 | 4 | $(4 - 4) = \phantom{-}0$ | 0 |
| | | | Sum of errors = 0 | Sum of squared errors (SSE) = 2 |

You can see by shifting the ruler around the graph that it is possible to find many lines for which the sum of errors is equal to 0, but it can be shown that there is one (and only one) line for which the SSE is a *minimum*. This line is called the **least squares line**, the **regression line**, or the **least squares prediction equation**. The methodology used to obtain this line is called the **method of least squares**.

To find the least squares prediction equation for a set of data, assume that we have a sample of $n$ data points consisting of pairs of values of $x$ and $y$, say $(x_1, y_1)$, $(x_2, y_2), \ldots, (x_n, y_n)$. For example, the $n = 5$ data points shown in Table 11.2 are $(1, 1), (2, 1), (3, 2), (4, 2)$, and $(5, 4)$. The fitted line, which we will calculate based on the five data points, is written as

$$\hat{y} = \hat{\beta}_0 + \hat{\beta}_1 x$$

The "hats" indicate that the symbols below them are estimates: $\hat{y}$ ($y$-hat) is an estimator of the mean value of $y$, $E(y)$, and a predictor of some future value of $y$; and $\hat{\beta}_0$ and $\hat{\beta}_1$ are estimators of $\beta_0$ and $\beta_1$, respectively.

For a given data point, say the point $(x_i, y_i)$, the observed value of $y$ is $y_i$ and the predicted value of $y$ would be obtained by substituting $x_i$ into the prediction equation:

$$\hat{y}_i = \hat{\beta}_0 + \hat{\beta}_1 x_i$$

And the deviation of the $i$th value of $y$ from its predicted value is

$$(y_i - \hat{y}_i) = [y_i - (\hat{\beta}_0 + \hat{\beta}_1 x_i)]$$

Then the sum of squares of the deviations of the $y$-values about their predicted values for all the $n$ data points is

$$\text{SSE} = \sum [y_i - (\hat{\beta}_0 + \hat{\beta}_1 x_i)]^2$$

The quantities $\hat{\beta}_0$ and $\hat{\beta}_1$ that make the SSE a minimum are called the **least squares estimates** of the population parameters $\beta_0$ and $\beta_1$, and the prediction equation $\hat{y} = \hat{\beta}_0 + \hat{\beta}_1 x$ is called the *least squares line*.

---

**DEFINITION 11.1**

The **least squares line** is one that has the following two properties:

1. the sum of the errors (SE) equals 0
2. the sum of squared errors (SSE) is smaller than that for any other straight-line model

---

The values of $\hat{\beta}_0$ and $\hat{\beta}_1$ that minimize the SSE are (proof omitted) given by the formulas in the next box.*

---

**Formulas for the Least Squares Estimates**

*Slope:* $\quad \hat{\beta}_1 = \dfrac{\text{SS}_{xy}}{\text{SS}_{xx}}$

*y-intercept:* $\quad \hat{\beta}_0 = \bar{y} - \hat{\beta}_1 \bar{x}$

where $\quad \text{SS}_{xy} = \sum (x_i - \bar{x})(y_i - \bar{y}) = \sum x_i y_i - \dfrac{\left(\sum x_i\right)\left(\sum y_i\right)}{n}$

$$\text{SS}_{xx} = \sum (x_i - \bar{x})^2 = \sum x_i^2 - \dfrac{\left(\sum x_i\right)^2}{n}$$

$\quad n = $ Sample size

---

Preliminary computations for finding the least squares line for the drug reaction example are presented in Table 11.3. We can now calculate

---

*Students who are familiar with calculus should note that the values of $\beta_0$ and $\beta_1$ that minimize SSE $= \Sigma(y_i - \hat{y}_i)^2$ are obtained by setting the two partial derivatives $\partial\text{SSE}/\partial\beta_0$ and $\partial\text{SSE}/\partial\beta_1$ equal to 0. The solutions to these two equations yield the formulas shown in the box. Furthermore, we denote the *sample* solutions to the equations by $\hat{\beta}_0$ and $\hat{\beta}_1$, where the "hat" denotes that these are sample estimates of the true population intercept $\beta_0$ and slope $\beta_1$.

**TABLE 11.3    Preliminary Computations for the Drug Reaction Example**

| $x_i$ | $y_i$ | $x_i^2$ | $x_i y_i$ |
|---|---|---|---|
| 1 | 1 | 1 | 1 |
| 2 | 1 | 4 | 2 |
| 3 | 2 | 9 | 6 |
| 4 | 2 | 16 | 8 |
| 5 | 4 | 25 | 20 |
| Totals   $\sum x_i = 15$ | $\sum y_i = 10$ | $\sum x_i^2 = 55$ | $\sum x_i y_i = 37$ |

$$SS_{xy} = \sum x_i y_i - \frac{\left(\sum x_i\right)\left(\sum y_i\right)}{5} = 37 - \frac{(15)(10)}{5} = 37 - 30 = 7$$

$$SS_{xx} = \sum x_i^2 - \frac{\left(\sum x_i\right)^2}{5} = 55 - \frac{(15)^2}{5} = 55 - 45 = 10$$

Then the slope of the least squares line is

$$\hat{\beta}_1 = \frac{SS_{xy}}{SS_{xx}} = \frac{7}{10} = .7$$

and the *y*-intercept is

$$\hat{\beta}_0 = \bar{y} - \hat{\beta}_1 \bar{x} = \frac{\sum y_i}{5} - \hat{\beta}_1 \frac{\sum x_i}{5}$$

$$= \frac{10}{5} - (.7)\left(\frac{15}{5}\right) = 2 - (.7)(3) = 2 - 2.1 = -.1$$

The least squares line is thus

$$\hat{y} = \hat{\beta}_0 + \hat{\beta}_1 x = -.1 + .7x$$

The graph of this line is shown in Figure 11.5.

The predicted value of *y* for a given value of *x* can be obtained by substituting into the formula for the least squares line. Thus, when $x = 2$ we predict *y* to be

$$\hat{y} = -.1 + .7x = -.1 + .7(2) = 1.3$$

We show how to find a prediction interval for *y* in Section 11.8.

The observed and predicted values of *y*, the deviations of the *y* values about their predicted values, and the squares of these deviations are shown in Table 11.4. Note that the sum of squares of the deviations, SSE, is 1.10, and (as we would expect) this is less than the SSE $= 2.0$ obtained in Table 11.2 for the visually fitted line.

**FIGURE 11.5**

*The Line $\hat{y} = -.1 + .7x$
Fitted to the Data*

**TABLE 11.4** **Comparing Observed and Predicted Values for the Least Squares Prediction Equation**

| $x$ | $y$ | $\hat{y} = -.1 + .7x$ | $(y - \hat{y})$ | $(y - \hat{y})^2$ |
|---|---|---|---|---|
| 1 | 1 | .6 | $(1 - .6) = .4$ | .16 |
| 2 | 1 | 1.3 | $(1 - 1.3) = -.3$ | .09 |
| 3 | 2 | 2.0 | $(2 - 2.0) = 0$ | .00 |
| 4 | 2 | 2.7 | $(2 - 2.7) = -.7$ | .49 |
| 5 | 4 | 3.4 | $(4 - 3.4) = .6$ | .36 |
| | | | Sum of errors $= 0$ | SSE $= 1.10$ |

The calculations required to obtain $\hat{\beta}_0$, $\hat{\beta}_1$, and SSE in simple linear regression, although straightforward, can become rather tedious. Even with the use of a pocket calculator, the process is laborious and susceptible to error, especially when the sample size is large. Fortunately, the use of statistical computer software can significantly reduce the labor involved in regression calculations. The SAS, SPSS, MINITAB, and STATISTIX outputs for the simple linear regression of the data in Table 11.1 are displayed in Figure 11.6a–d. The values of $\hat{\beta}_0$ and $\hat{\beta}_1$ are highlighted on the printouts. These values, $\hat{\beta}_0 = -.1$ and $\hat{\beta}_1 = .7$, agree exactly with our hand-calculated values. The value of SSE $= 1.10$ is also highlighted on the printouts.

Whether you use a hand calculator or a computer, it is important that you be able to interpret the intercept and slope in terms of the data being utilized to fit the model. In the drug reaction example, the estimated $y$-intercept, $\hat{\beta}_0 = -.1$, appears to imply that the estimated mean reaction time is equal to $-.1$ second when the amount of drug, $x$, is equal to 0%. Since negative reaction times are not possible, this seems to make the model nonsensical. However, *the model parameters should be interpreted only within the sampled range of the independent variable*—in this case, for amounts of drug in the bloodstream between 1% and 5%. Thus, the $y$-intercept—which is, by definition, at $x = 0$ (0% drug)—is not within the range of the sampled values of $x$ and is not subject to meaningful interpretation.

**FIGURE 11.6a**

*SAS Printout for Time–Drug Regression*

The REG Procedure
Dependent Variable: TIME_Y

Analysis of Variance

| Source | DF | Sum of Squares | Mean Square | F Value | Pr > F |
|---|---|---|---|---|---|
| Model | 1 | 4.90000 | 4.90000 | 13.36 | 0.0354 |
| Error | 3 | 1.10000 | 0.36667 | | |
| Corrected Total | 4 | 6.00000 | | | |

| | | | | |
|---|---|---|---|---|
| Root MSE | 0.60553 | R-Square | 0.8167 | |
| Dependent Mean | 2.00000 | Adj R-Sq | 0.7556 | |
| Coeff Var | 30.27650 | | | |

Parameter Estimates

| Variable | DF | Parameter Estimate | Standard Error | t Value | Pr > \|t\| |
|---|---|---|---|---|---|
| Intercept | 1 | -0.10000 | 0.63509 | -0.16 | 0.8849 |
| DRUG_X | 1 | 0.70000 | 0.19149 | 3.66 | 0.0354 |

**FIGURE 11.6b**

*SPSS Printout for Time–Drug Regression*

**Model Summary**

| Model | R | R Square | Adjusted R Square | Std. Error of the Estimate |
|---|---|---|---|---|
| 1 | .904[a] | .817 | .756 | .61 |

a. Predictors: (Constant), DRUG_X

**ANOVA[b]**

| Model | | Sum of Squares | df | Mean Square | F | Sig. |
|---|---|---|---|---|---|---|
| 1 | Regression | 4.900 | 1 | 4.900 | 13.364 | .035[a] |
| | Residual | 1.100 | 3 | .367 | | |
| | Total | 6.000 | 4 | | | |

a. Predictors: (Constant), DRUG_X
b. Dependent Variable: Time_Y

**Coefficients[a]**

| Model | | Unstandardized Coefficients | | Standardized Coefficients | | |
|---|---|---|---|---|---|---|
| | | B | Std. Error | Beta | t | Sig. |
| 1 | (Constant) | -1.00E-01 | .635 | | -.157 | .885 |
| | DRUG_X | .700 | .191 | .904 | 3.656 | .035 |

a. Dependent Variable: Time_Y

**FIGURE 11.6c**

*MINITAB Printout for Time–Drug Regression*

```
The regression equation is
TIME_Y = - 0.100 + 0.700 DRUG_X


Predictor      Coef      SE Coef        T        P
Constant     -0.1000      0.6351     -0.16    0.885
DRUG_X        0.7000      0.1915      3.66    0.035


S = 0.6055     R-Sq = 81.7%    R-Sq(adj) = 75.6%

Analysis of Variance


Source          DF         SS          MS        F        P
Regression       1      4.9000      4.9000    13.36    0.035
Residual Error   3      1.1000      0.3667
Total            4      6.0000
```

**FIGURE 11.6d**

*STATISTIX Printout for Time–Drug Regression*

```
UNWEIGHTED LEAST SQUARES LINEAR REGRESSION OF TIME_Y

PREDICTOR
VARIABLES     COEFFICIENT     STD ERROR     STUDENT'S T        P
---------     -----------     ---------     -----------     ------

CONSTANT       -0.10000        0.63509        -0.16         0.8849
DRUG_X          0.70000        0.19149         3.66         0.0354


R-SQUARED             0.8167    RESID. MEAN SQUARE (MSE)    0.36667
ADJUSTED R-SQUARED    0.7556    STANDARD DEVIATION          0.60553


SOURCE        DF       SS           MS         F        P
---------     ---    ---------    ---------    -----    ------

REGRESSION     1     4.90000      4.90000     13.36     0.0354
RESIDUAL       3     1.10000      0.36667
TOTAL          4     6.00000


CASES INCLUDED 5    MISSING CASES 0
```

# Using the TI-83 Graphing Calculator

## Straight-Line (Linear) Regression on the TI-83

### A. Finding the Least Squares Regression Equation

Step 1   *Enter the data*
Press **STAT** **1** for **STAT Edit**
Enter your *x*-data in **L1** and your *y*-data in **L2**

Step 2   *Find the equation*
Press **STAT** and highlight **CALC**
Press **4** for **LinReg(ax + b)**
Press **ENTER**

The screen will show the values for *a* and *b* in the equation
$y = ax + b$.

Example  The figures below show a table of data entered on the TI-83 and the regression equation obtained using the steps given above.

### B. Finding *r* and $r^2$

(*Note:* We discuss these statistics in Sections 11.6 and 11.7.)

If *r* and $r^2$ do not already appear on the LinReg screen from part A,

Step 1   *Turn the diagnostics feature on*
Press **2nd 0** for **CATALOG**
Press the $x^{-1}$ key for **D**
Press down the arrow key until **DiagnosticsOn** is highlighted
Press **ENTER** twice

Step 2   *Find the regression equation as shown in part A above*
The values for *r* and $r^2$ will appear on the screen as well.

Example:  The figures below show a table of data entered on the TI-83 and the regression equation, *r*, and $r^2$ obtained using the steps given above.

---

**USING THE TI-83 GRAPHING CALCULATOR** *(continued)*

**C.  Graphing the Least Squares Line With the Scatterplot**

Step 1    *Enter the data as shown in part A above*

Step 2    *Set up the data plot*
Press    **2nd**   **Y=** for **STATPLOT**
Press    **1** for **PLOT1**

Use the arrow and **ENTER** keys to set up the screen as shown below.

Step 3    *Find the regression equation as shown in part A above*

Step 4    *Enter the regression equation*
Press **Y=**
Press **VARS 5** for **Statistics**...
Highlight **EQ** and press **ENTER**

You should see the regression equation in the **Y=** window.

Step 5    *View the scatterplot and regression line*
Press ZOOM 9 for ZoomStat

You should see the data graphed along with the regression line.

Example   The figures below show a table of data entered on the TI-83, and the graph of the scatterplot and least squares line obtained using the steps given above.

---

The slope of the least squares line, $\hat{\beta} = .7$, implies that for every unit increase of $x$, the mean value of $y$ is estimated to increase by .7 unit. In terms of this example, for every 1% increase in the amount of drug in the bloodstream, the mean reaction time is estimated to increase by .7 second *over the sampled range of drug amounts from 1%* *to 5%*. Thus, the model does not imply that increasing the drug amount from 5% to 10% will result in an increase in mean reaction time of 3.5 seconds, because the range of $x$ in the sample does not extend to 10% ($x = 10$). In fact, 10% might be such a high concentration that the drug would kill the subject! Be careful to interpret the estimated parameters only within the sampled range of $x$.

Even when the interpretations of the estimated parameters are meaningful, we need to remember that they are only estimates based on the sample. As such, their values will typically change in repeated sampling. How much confidence do we have that the estimated slope, $\hat{\beta}_1$, accurately approximates the true slope, $\beta_1$? This requires statistical inference, in the form of confidence intervals and tests of hypotheses, which we address in Section 11.5.

To summarize, we defined the best-fitting straight line to be the one that minimizes the sum of squared errors around the line, and we called it the least squares line. We should interpret the least squares line only within the sampled range of the independent variable. In subsequent sections we show how to make statistical inferences about the model.

 EXERCISES 11.10–11.24

### Learning the Mechanics

**11.10** The following table is similar to Table 11.3. It is used for making the preliminary computations for finding the least squares line for the given pairs of $x$ and $y$ values.
   **a.** Complete the table.
   **b.** Find $SS_{xy}$.
   **c.** Find $SS_{xx}$.
   **d.** Find $\hat{\beta}_1$.
   **e.** Find $\bar{x}$ and $\bar{y}$.
   **f.** Find $\hat{\beta}_0$.
   **g.** Find the least squares line.

| $x_i$ | $y_i$ | $x_i^2$ | $x_iy_i$ |
|---|---|---|---|
| 7 | 2 | — | — |
| 4 | 4 | — | — |
| 6 | 2 | — | — |
| 2 | 5 | — | — |
| 1 | 7 | — | — |
| 1 | 6 | — | — |
| 3 | 5 | — | — |
| Totals $\sum x_i =$ | $\sum y_i =$ | $\sum x_i^2 =$ | $\sum x_iy_i =$ |

**11.11** Refer to Exercise 11.10. After the least squares line has been obtained, the table below (which is similar to Table 11.4) can be used for (1) comparing the observed and the predicted values of $y$, and (2) computing SSE.

| $x$ | $y$ | $\hat{y}$ | $(y - \hat{y})$ | $(y - \hat{y})^2$ |
|---|---|---|---|---|
| 7 | 2 | — | — | — |
| 4 | 4 | — | — | — |
| 6 | 2 | — | — | — |
| 2 | 5 | — | — | — |
| 1 | 7 | — | — | — |
| 1 | 6 | — | — | — |
| 3 | 5 | — | — | — |
| | | | $\sum (y - \hat{y}) =$ | SSE $= \sum (y - \hat{y})^2 =$ |

**a.** Complete the table.
**b.** Plot the least squares line on a scattergram of the data. Plot the following line on the same graph:

$$\hat{y} = 14 - 2.5x$$

**c.** Show that SSE is larger for the line in part **b** than it is for the least squares line.

**11.12** Construct a scattergram for the data in the following table.

| $x$ | .5 | 1 | 1.5 |
|---|---|---|---|
| $y$ | 2 | 1 | 3 |

**a.** Plot the following two lines on your scattergram:

$$y = 3 - x \quad \text{and} \quad y = 1 + x$$

**b.** Which of these lines would you choose to characterize the relationship between $x$ and $y$? Explain.
**c.** Show that the sum of errors for both of these lines equals 0.
**d.** Which of these lines has the smaller SSE?
**e.** Find the least squares line for the data and compare it to the two lines described in part **a**.

**11.13** Consider the following pairs of measurements:

| $x$ | 5 | 3 | −1 | 2 | 7 | 6 | 4 |
|---|---|---|---|---|---|---|---|
| $y$ | 4 | 3 | 0 | 1 | 8 | 5 | 3 |

**a.** Construct a scattergram for these data.
**b.** What does the scattergram suggest about the relationship between $x$ and $y$?
**c.** Given that $SS_{xx} = 43.4286$, $SS_{xy} = 39.8571$, $\bar{y} = 3.4286$, and $\bar{x} = 3.7143$, calculate the least squares estimates of $\beta_0$ and $\beta_1$.

**d.** Plot the least squares line on your scattergram. Does the line appear to fit the data well? Explain.

**e.** Interpret the $y$-intercept and slope of the least squares line. Over what range of $x$ are these interpretations meaningful?

## Applying the Concepts—Basic

**11.14** Real estate investors, home buyers, and home owners often use the appraised value of a property as a basis for predicting sale price. Data on sale prices and total appraised values of 92 residential properties sold in 1999 in an upscale Tampa, Florida, neighborhood named Tampa Palms are saved in the TAMPALMS file. The first five and last five observations of the data set are listed in the accompanying table.

⊘TAMPALMS

| Property | Appraised Value | Sale Price |
|---|---|---|
| 1 | $ 170,432 | $ 180,000 |
| 2 | 212,827 | 245,100 |
| 3 | 68,130 | 85,400 |
| 4 | 65,505 | 87,900 |
| 5 | 68,655 | 84,200 |
| ⋮ | ⋮ | ⋮ |
| 88 | 195,862 | 244,000 |
| 89 | 176,850 | 219,000 |
| 90 | 95,718 | 132,000 |
| 91 | 137,108 | 156,900 |
| 92 | 183,704 | 263,000 |

*Source:* Hillsborough County (Florida) Property Appraiser's Office.

**MINITAB Scatterplot for Exercise 11.14**

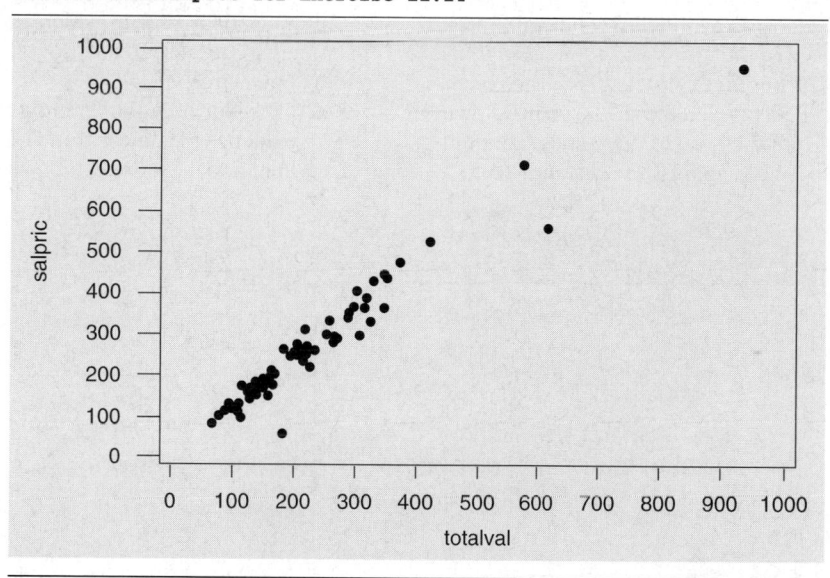

**MINITAB Simple Linear Regression Output for Exercise 11.14**

```
The regression equation is
salepric = 20.9 + 1.07 totalval

Predictor      Coef      SE Coef        T          P
Constant      20.942       6.446      3.25      0.002
totalval     1.06873     0.02709     39.45      0.000

S = 32.79      R-Sq = 94.5%      R-Sq(adj) = 94.5%

Analysis of Variance

Source         DF        SS          MS         F         P
Regression      1    1673142     1673142   1556.48     0.000
Residual Error 90      96746        1075
Total          91    1769888
```

a. Propose a straight-line model to relate the appraised property value $x$ to the sale price $y$ for residential properties in this neighborhood.

b. A MINITAB scatterplot of the data is shown on p. 523. [*Note*: Both sale price and total appraised value are shown in thousands of dollars.] Does it appear that a straight-line model will be an appropriate fit to the data?

c. A MINITAB simple linear regression printout is also shown on p. 523. Find the equation of the best-fitting line through the data on the printout.

d. Interpret the $y$-intercept of the least squares line. Does it have a practical meaning for this application? Explain.

e. Interpret the slope of the least squares line. Over what range of $x$ is the interpretation meaningful?

f. Use the least squares model to estimate the mean sale price of a property appraised at $300,000.

**11.15** In Denver, Colorado, environmentalists have discovered a link between high arsenic levels in soil and a crabgrass killer used in the 1950s and 1960s (*Environmental Science & Technology*, Sept. 1, 2000). The recent discovery was based, in part, on the scattergrams shown below. The graphs plot the level of the metals cadmium and arsenic, respectively, against the distance from a former smelter plant for samples of soil taken from Denver residential properties.

a. Normally, the metal level in soil decreases as distance from the source (e.g., a smelter plant) increases. Propose a straight-line model relating metal level $y$ to distance from the plant $x$. Based on the theory, would you expect the slope of the line to be positive or negative?

b. Examine the scatterplot for cadmium. Does the plot support the theory, part **a**?

c. Examine the scatterplot for arsenic. Does the plot support the theory, part **a**? (*Note:* This finding led investigators to discover the link between high arsenic levels and the use of the crabgrass killer.)

**11.16** In *Chance* (Winter, 2000), statistician Howard Wainer and two students compared men's and women's winning times in the Boston Marathon. One of the graphs used to illustrate gender differences is reproduced on p. 525. The scattergram plots the winning times (in minutes) against year in which the race was run. Men's times are represented by solid dots and women's times by open circles.

a. Consider only the winning times for men. Is there evidence of a linear trend? If so, propose a straight-line model for predicting winning time ($y$) based on

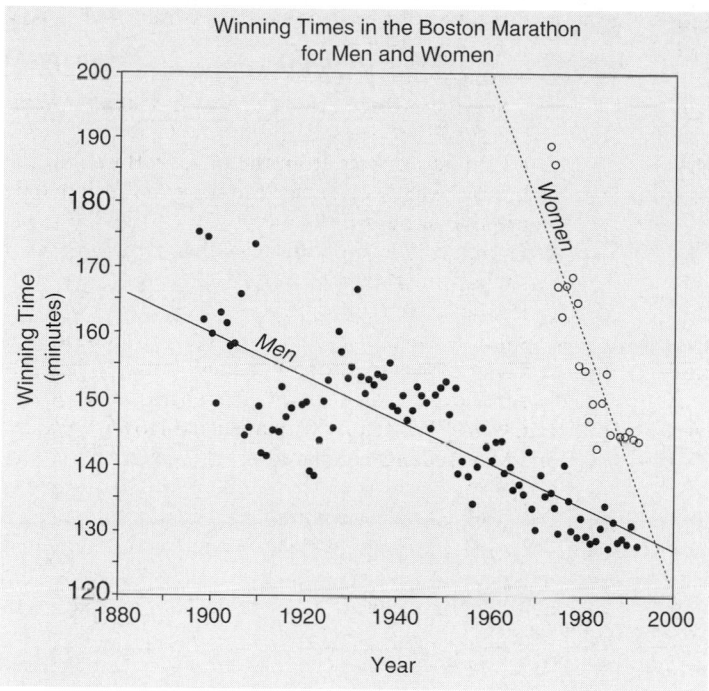

year ($x$). Would you expect the slope of this line to be positive or negative?

**b.** Repeat part **b** for women's times.

**c.** Which slope, men's or women's, will be greater in absolute value?

**d.** Would you recommend using the straight-line models to predict the winning time in the 2020 Boston Marathon? Why or why not?

**11.17** Refer to the *Brain and Behavior Evolution* (Apr. 2000) study of the feeding behavior of blackbream fish, Exercise 2.110 (p. 89). Recall that the zoologists recorded the number of aggressive strikes of two blackbream fish feeding at the bottom of an aquarium in the 10-minute period following the addition of food. The table listing the weekly number of strikes and age of the fish (in days) is reproduced here.

**⊘ BLACKBREAM**

| Week | Number of Strikes | Age of Fish (days) |
|------|-------------------|--------------------|
| 1 | 85 | 120 |
| 2 | 63 | 136 |
| 3 | 34 | 150 |
| 4 | 39 | 155 |
| 5 | 58 | 162 |
| 6 | 35 | 169 |
| 7 | 57 | 178 |
| 8 | 12 | 184 |
| 9 | 15 | 190 |

*Source:* Shand, J., *et al.* "Variability in the location of the retinal ganglion cell area centralis is correlated with ontogenetic changes in feeding behavior in the Blackbream, Acanthopagrus 'butcher'." *Brain and Behavior*, Vol. 55, No. 4, Apr. 2000 (Figure H).

**a.** Write the equation of a straight-line model relating number of strikes ($y$) to age of fish ($x$).

**b.** Fit the model to the data using the method of least squares and give the least squares prediction equation.

**c.** Give a practical interpretation of the value of $\hat{\beta}_0$, if possible.

**d.** Give a practical interpretation of the value of $\hat{\beta}_1$, if possible.

**11.18** Is the number of games won by a major league baseball team in a season related to the team's batting average? The table on p. 526, reproduced from Exercise 2.111 (p. 90), shows the number of games won and the batting averages for the 14 teams in the American League for the 2000 Major League Baseball season.

**a.** If you were to model the relationship between the mean (or expected) number of games won by a major league team and the team's batting average $x$, using a straight line, would you expect the slope of the line to be positive or negative? Explain.

**b.** In Exercise 2.111 you constructed a scattergram of the data. Does the pattern revealed by the scattergram agree with your answer to part **a**?

**c.** An SPSS printout of the simple linear regression is provided on p. 526. Find the estimates of the $\beta$'s on the printout and write the equation of the least squares line.

**d.** Graph the least squares line on your scattergram. Does your least squares line seem to fit the points on your scattergram?

**e.** Does the mean (or expected) number of games won appear to be strongly related to a team's batting average? Explain.

**f.** Interpret the values of $\hat{\beta}_0$ and $\hat{\beta}_1$ in the words of the problem.

🔵 ALWINS

| Team | Games Won | Batting Ave. |
|------|-----------|--------------|
| New York | 87 | .277 |
| Toronto | 83 | .275 |
| Baltimore | 74 | .272 |
| Boston | 85 | .267 |
| Tampa Bay | 69 | .257 |
| Cleveland | 90 | .288 |
| Detroit | 79 | .275 |
| Chicago | 95 | .286 |
| Kansas City | 77 | .288 |
| Minnesota | 69 | .270 |
| Anaheim | 82 | .280 |
| Texas | 71 | .283 |
| Seattle | 91 | .269 |
| Oakland | 91 | .270 |

*Source:* Major League Baseball, 2001.

## Applying the Concepts—Intermediate

**11.19** The quality of the orange juice produced by a manufacturer (e.g., Minute Maid, Tropicana) is constantly monitored. There are numerous sensory and chemical components that combine to make the best tasting orange juice. For example, one manufacturer has developed a quantitative index of the "sweetness" of orange juice. (The higher the index, the sweeter the juice.) Is there a relationship between the sweetness index and a chemical measure such as the amount of water soluble pectin (parts per million) in the orange juice? Data collected on these two variables for 24 production runs at a juice manufacturing plant are shown in the next table. Suppose a manufacturer wants to use simple linear regression to predict the sweetness ($y$) from the amount of pectin ($x$).
**a.** Find the least squares line for the data.
**b.** Interpret $\hat{\beta}_0$ and $\hat{\beta}_1$ in the words of the problem.
**c.** Predict the sweetness index if amount of pectin in the orange juice is 300 ppm. [*Note:* A measure of reliability of such a prediction is discussed in Section 11.8.]

🔵 OJUICE

| Run | Sweetness Index | Pectin (ppm) |
|-----|-----------------|--------------|
| 1 | 5.2 | 220 |
| 2 | 5.5 | 227 |
| 3 | 6.0 | 259 |
| 4 | 5.9 | 210 |
| 5 | 5.8 | 224 |
| 6 | 6.0 | 215 |
| 7 | 5.8 | 231 |
| 8 | 5.6 | 268 |
| 9 | 5.6 | 239 |
| 10 | 5.9 | 212 |
| 11 | 5.4 | 410 |
| 12 | 5.6 | 256 |
| 13 | 5.8 | 306 |
| 14 | 5.5 | 259 |
| 15 | 5.3 | 284 |
| 16 | 5.3 | 383 |
| 17 | 5.7 | 271 |
| 18 | 5.5 | 264 |
| 19 | 5.7 | 227 |
| 20 | 5.3 | 263 |
| 21 | 5.9 | 232 |
| 22 | 5.8 | 220 |
| 23 | 5.8 | 246 |
| 24 | 5.9 | 241 |

*Note:* The data in the table are authentic. For confidentiality reasons, the manufacturer cannot be disclosed.

**11.20** Chemists at Kyushu University (Japan) examined the linear relationship between the maximum absorption rate $y$ (in nanomoles) and the Hammett substituent constant $x$ for metacyclophane compounds (*Journal of Organic Chemistry*, July 1995). The data for variants of two compounds are given in the table on p. 527. The variants of compound 1 are labeled 1a, 1b, 1d, 1e, 1f, 1g, and 1h; the variants of compound 2 are 2a, 2b, 2c, and 2d.
**a.** Plot the data in a scattergram. Use two different plotting symbols for the two compounds. What do you observe?
**b.** Using only the data for compound 1, find $\hat{\beta}_0$ and $\hat{\beta}_1$. Interpret the results.

SPSS Output for Exercise 11.18

**Coefficients**[a]

| Model | | Unstandardized Coefficients | | Standardized Coefficients | t | Sig. |
|-------|--|--------|-----------|------|------|------|
| | | B | Std. Error | Beta | | |
| 1 | (Constant) | -.219 | 74.082 | | -.003 | .998 |
| | BATAVG | 297.140 | 268.768 | .304 | 1.106 | .291 |

a. Dependent Variable: WINS

**c.** Using only the data for compound 2, find $\hat{\beta}_0$ and $\hat{\beta}_1$. Interpret the results.

⊘ ORGCHEM

| Compound | Maximum Absorption $y$ | Hammett Constant $x$ |
|---|---|---|
| 1a | 298 | 0.00 |
| 1b | 346 | .75 |
| 1d | 303 | .06 |
| 1e | 314 | −.26 |
| 1f | 302 | .18 |
| 1g | 332 | .42 |
| 1h | 302 | −.19 |
| 2a | 343 | .52 |
| 2b | 367 | 1.01 |
| 2c | 325 | .37 |
| 2d | 331 | .53 |

*Source:* Adapted from Tsuge, A., *et al.* "Preparation and spectral properties of disubstituted [2-2] metacyclophanes," *Journal of Organic Chemistry*, Vol 60, No. 15, July 1995, pp. 4390–4391 (Table 1 and Figure 1).

**11.21** A study was conducted to investigate how people with a hearing impairment communicate with their conversational partners (*Journal of the Academy of Rehabilitative Audiology*, Vol. 27, 1994). Each of thirteen hearing-impaired subjects, all fitted with a cochlear implant, participated in a structured communication interaction with a familiar conversational partner (a family member) and with an unfamiliar conversational partner (who was instructed not to take the initiative to repair breakdowns in communication). The total number of words used by the subject in each of the two conversations is given in the next table.

  **a.** Plot the data in a scattergram. Is there visual evidence of a linear relationship between $x$ and $y$? If so, is it positive or negative?

**b.** Propose a straight-line model relating $y$ to $x$.

**c.** A STATISTIX printout of the simple linear regression analysis is provided at the bottom of the page. Find the estimate of $\beta_0$ and $\beta_1$ on the printout.

**d.** Interpret the values of $\hat{\beta}_0$ and $\hat{\beta}_1$.

⊘ HEARAID

| Subject | Words with Familiar Partner $x$ | Words with Unfamiliar Partner $y$ |
|---|---|---|
| 1 | 65 | 47 |
| 2 | 160 | 78 |
| 3 | 55 | 90 |
| 4 | 83 | 75 |
| 5 | 0 | 6 |
| 6 | 140 | 101 |
| 7 | 49 | 40 |
| 8 | 164 | 215 |
| 9 | 62 | 29 |
| 10 | 56 | 75 |
| 11 | 207 | 121 |
| 12 | 207 | 139 |
| 13 | 93 | 83 |

*Source:* Tye-Murray, N., *et al.* "Communication breakdowns: Partner contingencies and partner reactions." *Journal of the Academy of Rehabilitative Audiology*, Vol. 27, 1994, pp. 116–117 (Tables 6 & 7).

**11.22** Two species of predatory birds, collard flycatchers and tits, compete for nest holes during breeding season on the island of Gotland, Sweden. Frequently, dead flycatchers are found in nest boxes occupied by tits. A field study examined whether the risk of mortality to flycatchers is related to the degree of competition between the two bird species for nest sites (*The Condor*, May 1995). The table (p. 528) gives data on the number $y$ of flycatchers killed at each of 14 discrete locations

**STATISTIX Output for Exercise 11.21**

```
UNWEIGHTED LEAST SQUARES LINEAR REGRESSION OF UNFWORDS

PREDICTOR
VARIABLES     COEFFICIENT     STD ERROR     STUDENT'S T        P
---------     -----------     ---------     -------------     ------

CONSTANT        20.1275        19.2194          1.05          0.3174
FAMWORDS        0.62442        0.15899          3.93          0.0024


R-SQUARED              0.5837     RESID. MEAN SQUARE (MSE)     1305.22
ADJUSTED R-SQUARED     0.5459     STANDARD DEVIATION           36.1279


SOURCE        DF       SS          MS         F        P
----------    ---    ---------   ---------   -----    ------

REGRESSION     1     20131.8     20131.8     15.42    0.0024
RESIDUAL      11     14357.5     1305.22
TOTAL         12     34489.2


CASES INCLUDED 13    MISSING CASES 0
```

(plots) on the island as well as the nest box tit occupancy $x$ (that is, the percentage of nest boxes occupied by tits) at each plot. SAS was used to conduct a simple linear regression analysis for the model, $E(y) = \beta_0 + \beta_1 x$. The printout is shown below.

⊘ CONDOR2

| Plot | Number of Flycatchers Killed $y$ | Nest Box Tit Occupancy $x$ (%) |
|------|------|------|
| 1 | 0 | 24 |
| 2 | 0 | 33 |
| 3 | 0 | 34 |
| 4 | 0 | 43 |
| 5 | 0 | 50 |
| 6 | 1 | 35 |
| 7 | 1 | 35 |
| 8 | 1 | 38 |
| 9 | 1 | 40 |
| 10 | 2 | 31 |
| 11 | 2 | 43 |
| 12 | 3 | 55 |
| 13 | 4 | 57 |
| 14 | 5 | 64 |

*Source:* Merila, J., and Wiggins, D. A. "Interspecific competition for nest holes causes adult mortality in the collard flycatcher." *The Condor*, Vol. 97, No. 2, May 1995, p. 449 (Figure 2), Cooper Ornithological Society.

**a.** Plot the data in a scattergram. Does the frequency of flycatcher casualties per plot appear to increase linearly with increasing proportion of nest boxes occupied by tits?

**b.** Find the estimates of $\beta_0$ and $\beta_1$ in the SAS printout. Interpret their values.

**SAS Output for Exercise 11.22**

The REG Procedure
Dependent Variable: NOKILLED

Analysis of Variance

| Source | DF | Sum of Squares | Mean Square | F Value | Pr > F |
|--------|----|----|----|----|----|
| Model | 1 | 19.11669 | 19.11669 | 16.03 | 0.0018 |
| Error | 12 | 14.31188 | 1.19266 | | |
| Corrected Total | 13 | 33.42857 | | | |

| | | | | |
|---|---|---|---|---|
| Root MSE | 1.09209 | R-Square | 0.5719 | |
| Dependent Mean | 1.42857 | Adj R-Sq | 0.5362 | |
| Coeff Var | 76.44618 | | | |

Parameter Estimates

| Variable | DF | Parameter Estimate | Standard Error | t Value | Pr > \|t\| |
|----------|----|----|----|----|----|
| Intercept | 1 | -3.04686 | 1.15533 | -2.64 | 0.0217 |
| TITPCT | 1 | 0.10766 | 0.02689 | 4.00 | 0.0018 |

**11.23** Refer to the *Journal of Experimental Psychology-Applied* (June 2000) study in which the "name game" was used to help groups of students learn the names of other students in the group, Exercise 10.26 (p. 462). Recall that the "name game" requires the first student in the group to state his/her full name, the second student to say his/her name and the name of the first student, the third student to say his/her name and the names of the first two students, etc. After making their introductions, the students listened to a seminar speaker for 30 minutes. At the end of the seminar, all students were asked to remember the full name of each of the other students in their group and the researchers measured the proportion of names recalled for each. One goal of the study was to investigate the linear trend between $y$ = recall proportion and $x$ = position (order) of the student during the game. The data (simulated based on summary statistics provided in the research article) for 144 students in the first eight positions are saved in the NAMEGAME2 file. The first five and last five observations in the data set are listed below. [*Note*: Since the student in position 1 actually must recall the names of all the other students, he or she is assigned the position number 9 in the data set.] Use the method of least squares to estimate the line, $E(y) = \beta_0 + \beta_1 x$. Interpret the $\beta$ estimates in the words of the problem.

⊘ NAMEGAME2

| Position | Recall |
|----------|--------|
| 2 | 0.04 |
| 2 | 0.37 |
| 2 | 1.00 |
| 2 | 0.99 |
| 2 | 0.79 |
| ⋮ | ⋮ |
| 9 | 0.72 |
| 9 | 0.88 |
| 9 | 0.46 |
| 9 | 0.54 |
| 9 | 0.99 |

*Source:* Morris, P.E., and Fritz, C.O. "The name game: Using retrieval practice to improve the learning of names." *Journal of Experimental Psychology-Applied*, Vol. 6, No. 2, June 2000 (data simulated from Figure 2).

## Applying the Concepts—Advanced

**11.24** The long jump is a track and field event in which a competitor attempts to jump a maximum distance into a sand pit after a running start. At the edge of the sand pit is a takeoff board. Jumpers usually try to plant their toes at the front edge of this board to maximize jumping distance. The absolute distance between the front edge of the takeoff board and the spot where the toe actually lands on the board prior to jumping is called "takeoff error." Is takeoff error in the long jump linearly related to best jumping distance? To answer this question, kinesiology researchers videotaped the performances of 18 novice long jumpers at a high school track meet (*Journal of Applied Biomechanics*, May 1995). The average takeoff error, *x*, and best jumping distance (out of three jumps), *y*, for each jumper are recorded in the table. If a jumper can reduce his/her average takeoff error by .1 meter, how much would you estimate his/her best jumping distance to change? Based on your answer, comment on the usefulness of the model for predicting best jumping distance.

⦿ LONGJUMP

| Jumper | Best Jumping Distance $y$ (meters) | Average Takeoff Error $x$ (meters) |
|--------|------------------------------------|------------------------------------|
| 1 | 5.30 | .09 |
| 2 | 5.55 | .17 |
| 3 | 5.47 | .19 |
| 4 | 5.45 | .24 |
| 5 | 5.07 | .16 |
| 6 | 5.32 | .22 |
| 7 | 6.15 | .09 |
| 8 | 4.70 | .12 |
| 9 | 5.22 | .09 |
| 10 | 5.77 | .09 |
| 11 | 5.12 | .13 |
| 12 | 5.77 | .16 |
| 13 | 6.22 | .03 |
| 14 | 5.82 | .50 |
| 15 | 5.15 | .13 |
| 16 | 4.92 | .04 |
| 17 | 5.20 | .07 |
| 18 | 5.42 | .04 |

*Source:* Berg, W. P., and Greer, N. L. "A kinematic profile of the approach run of novice long jumpers." *Journal of Applied Biomechanics*, Vol. 11, No. 2, May 1995, p. 147 (Table 1).

## 11.3 MODEL ASSUMPTIONS

In Section 11.2 we assumed that the probabilistic model relating drug reaction time *y* to the percentage of drug *x* in the bloodstream is

$$y = \beta_0 + \beta_1 x + \varepsilon$$

We also recall that the least squares estimate of the deterministic component of the model, $\beta_0 + \beta_1 x$, is

$$\hat{y} = \hat{\beta}_0 + \hat{\beta}_1 x = -.1 + .7x$$

Now we turn our attention to the random component $\varepsilon$ of the probabilistic model and its relation to the errors in estimating $\beta_0$ and $\beta_1$. We will use a probability distribution to characterize the behavior of $\varepsilon$. We will see how the probability distribution of $\varepsilon$ determines how well the model describes the relationship between the dependent variable *y* and the independent variable *x*.

Step 3 in a regression analysis requires us to specify the probability distribution of the random error $\varepsilon$. We will make four basic assumptions about the general form of this probability distribution:

**Assumption 1** The mean of the probability distribution of $\varepsilon$ is 0. That is, the average of the values of $\varepsilon$ over an infinitely long series of experiments is 0 for each setting of the independent variable *x*. This assumption implies that the mean value of *y*, $E(y)$, for a given value of *x* is $E(y) = \beta_0 + \beta_1 x$.

**Assumption 2** The variance of the probability distribution of $\varepsilon$ is constant for all settings of the independent variable *x*. For our straight-line model, this assumption means that the variance of $\varepsilon$ is equal to a constant, say $\sigma^2$, for all values of *x*.

**FIGURE 11.7**

*The Probability Distribution of ε*

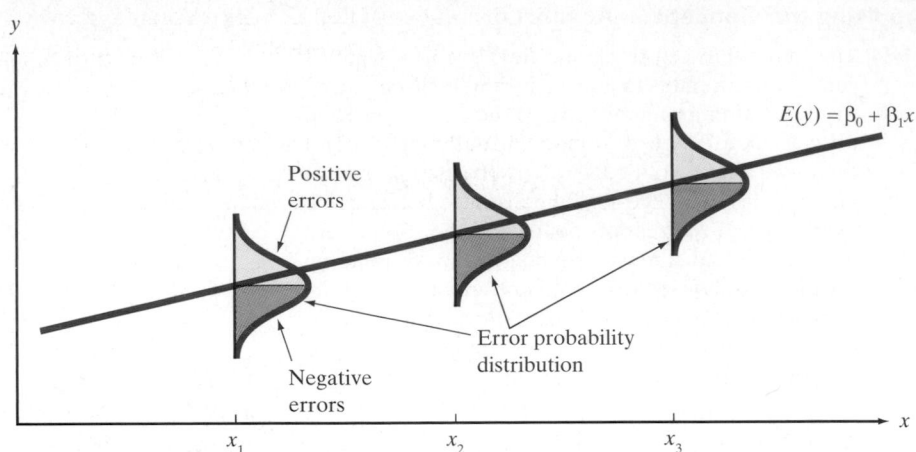

**Assumption 3**    The probability distribution of ε is normal.

**Assumption 4**    The values of ε associated with any two observed values of $y$ are independent. That is, the value of ε associated with one value of $y$ has no effect on the values of ε associated with other $y$ values.

The implications of the first three assumptions can be seen in Figure 11.7, which shows distributions of errors for three values of $x$, namely, $x_1$, $x_2$, and $x_3$. Note that the relative frequency distributions of the errors are normal with a mean of 0 and a constant variance $\sigma^2$. (All the distributions shown have the same amount of spread or variability.) The straight line shown in Figure 11.7 is the line of means. It indicates the mean value of $y$ for a given value of $x$. We denote this mean value as $E(y)$. Then, the line of means is given by the equation

$$E(y) = \beta_0 + \beta_1 x$$

These assumptions make it possible for us to develop measures of reliability for the least squares estimators and to develop hypothesis tests for examining the usefulness of the least squares line. We have various techniques for checking the validity of these assumptions, and we have remedies to apply when they appear to be invalid. Several of these remedies are discussed in Chapter 12. Fortunately, the assumptions need not hold exactly in order for least squares estimators to be useful. The assumptions will be satisfied adequately for many applications encountered in practice.

## 11.4    AN ESTIMATOR OF $\sigma^2$

It seems reasonable to assume that the greater the variability of the random error ε (which is measured by its variance $\sigma^2$), the greater will be the errors in the estimation of the model parameters $\beta_0$ and $\beta_1$ and in the error of prediction when $\hat{y}$ is used to predict $y$ for some value of $x$. Consequently, you should not be surprised, as we proceed through this chapter, to find that $\sigma^2$ appears in the formulas for all confidence intervals and test statistics that we will be using.

In most practical situations, $\sigma^2$ is unknown and we must use our data to estimate its value. The best estimate of $\sigma^2$, denoted by $s^2$, is obtained by dividing the sum of squares of the deviations of the $y$ values from the prediction line,

$$\text{SSE} = \sum (y_i - \hat{y}_i)^2$$

by the number of degrees of freedom associated with this quantity. We use 2 df to estimate the two parameters $\beta_0$ and $\beta_1$ in the straight-line model, leaving $(n - 2)$ df for the error variance estimation.

---

### Estimation of $\sigma^2$ for a (First-Order) Straight-Line Model

$$s^2 = \frac{\text{SSE}}{\text{Degrees of freedom for error}} = \frac{\text{SSE}}{n - 2}$$

where $\quad \text{SSE} = \sum (y_i - \hat{y}_i)^2 = \text{SS}_{yy} - \hat{\beta}_1 \text{SS}_{xy}$

$$\text{SS}_{yy} = \sum (y_i - \bar{y})^2 = \sum y_i^2 - \frac{\left(\sum y_i\right)^2}{n}$$

To estimate the standard deviation $\sigma$ of $\varepsilon$, we calculate

$$s = \sqrt{s^2} = \sqrt{\frac{\text{SSE}}{n - 2}}$$

We will refer to $s$ as the **estimated standard error of the regression model**.

*Warning:* When performing these calculations, you may be tempted to round the calculated values of $\text{SS}_{yy}$, $\hat{\beta}_1$, and $\text{SS}_{xy}$. Be certain to carry at least six significant figures for each of these quantities to avoid substantial errors in calculation of the SSE.

---

In the drug reaction example, we previously calculated SSE $= 1.10$ for the least squares line $\hat{y} = -.1 + .7x$. Recalling that there were $n = 5$ data points, we have $n - 2 = 5 - 2 = 3$ df for estimating $\sigma^2$. Thus,

$$s^2 = \frac{\text{SSE}}{n - 2} = \frac{1.10}{3} = .367$$

is the estimated variance, and

$$s = \sqrt{.367} = .61$$

is the standard error of the regression model.

The values of $s^2$ and $s$ can also be obtained from a simple linear regression printout. The SAS printout for the drug reaction example is reproduced in Figure 11.8. The value of $s^2$ is highlighted on the printout (in the **Mean Square** column in the row labeled **Error**). The value, $s^2 = .36667$, rounded to three decimal places, agrees with the one calculated by hand. The value of $s$ is also highlighted in Figure 11.8 (next to the heading **Root MSE**). This value, $s = .60553$, agrees (except for rounding) with our hand-calculated value.

You may be able to grasp $s$ intuitively by recalling the interpretation of a standard deviation given in Chapter 2 and remembering that the least squares line estimates the mean value of $y$ for a given value of $x$. Since $s$ measures the spread of the distribution of $y$ values about the least squares line and these errors of prediction are assumed to be normally distributed, we should not be surprised to find that most (about 95%) of the observations lie within $2s$, or $2(.61) = 1.22$, of the least squares line. For this simple example (only five data points), all five data points fall within $2s$ of the least squares line. In Section 11.8, we use $s$ to evaluate the error of prediction when the least squares line is used to predict a value of $y$ to be observed for a given value of $x$.

**FIGURE 11.8**

*SAS Printout for Time–Drug Regression*

The REG Procedure
Dependent Variable: TIME_Y

Analysis of Variance

| Source | DF | Sum of Squares | Mean Square | F Value | Pr > F |
|--------|----|----|----|----|----|
| Model | 1 | 4.90000 | 4.90000 | 13.36 | 0.0354 |
| Error | 3 | 1.10000 | 0.36667 | | |
| Corrected Total | 4 | 6.00000 | | | |

| | | | |
|---|---|---|---|
| Root MSE | 0.60553 | R-Square | 0.8167 |
| Dependent Mean | 2.00000 | Adj R-Sq | 0.7556 |
| Coeff Var | 30.27650 | | |

Parameter Estimates

| Variable | DF | Parameter Estimate | Standard Error | t Value | Pr > \|t\| |
|----------|----|----|----|----|----|
| Intercept | 1 | -0.10000 | 0.63509 | -0.16 | 0.8849 |
| DRUG_X | 1 | 0.70000 | 0.19149 | 3.66 | 0.0354 |

---

## Interpretation of $s$, the Estimated Standard Deviation of $\varepsilon$

We expect most ($\approx 95\%$) of the observed $y$ values to lie within $2s$ of their respective least squares predicted values, $\hat{y}$.

---

### EXERCISES 11.25–11.33

### Learning the Mechanics

**11.25** Calculate SSE and $s^2$ for each of the following cases:
   **a.** $n = 20, SS_{yy} = 95, SS_{xy} = 50, \hat{\beta}_1 = .75$

   **b.** $n = 40, \sum y^2 = 860, \sum y = 50,$
   $SS_{xy} = 2,700, \hat{\beta}_1 = .2$

   **c.** $n = 10, \sum (y_i - \bar{y})^2 = 58,$
   $SS_{xy} = 91, SS_{xx} = 170$

**11.26** Suppose you fit a least squares line to 12 data points and the calculated value of SSE is .429.
   **a.** Find $s^2$, the estimator of $\sigma^2$ (the variance of the random error term $\varepsilon$).
   **b.** What is the largest deviation that you might expect between any one of the 12 points and the least squares line?

**11.27** Visually compare the scattergrams shown at right. If a least squares line were determined for each data set, which do you think would have the smallest variance, $s^2$? Explain.

**11.28** Refer to Exercises 11.10 and 11.13 (p. 522). Calculate SSE, $s^2$, and $s$ for the least squares lines obtained in those exercises. Interpret the standard error of the regression model, $s$, for each.

### Applying the Concepts—Basic

**11.29** Refer to the simple linear regression relating number of agressive strikes $y$ to age $x$ for blackbream fish, Exercise 11.17 (p. 525).

**a.**

**b.**

**c.**
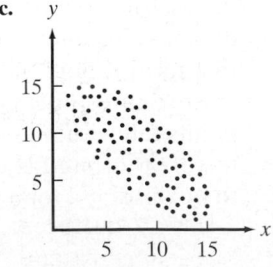

**a.** Find SSE, $s^2$, and $s$.

**b.** Interpret the value of $s$.

**11.30** *Statistical Bulletin* (Oct.–Dec. 1999) reported the average hospital charge and the average length of hospital stay for patients undergoing radical prostatectomies in a sample of 12 states. The data are listed in the accompanying table.

● HOSPITAL

| State | Average Hospital Charge ($) | Average Length of Stay (days) |
|---|---|---|
| Massachusetts | 11,680 | 3.64 |
| New Jersey | 11,630 | 4.20 |
| Pennsylvania | 9,850 | 3.84 |
| Minnesota | 9,950 | 3.11 |
| Indiana | 8,490 | 3.86 |
| Michigan | 9,020 | 3.54 |
| Florida | 13,820 | 4.08 |
| Georgia | 8,440 | 3.57 |
| Tennessee | 8,790 | 3.80 |
| Texas | 10,400 | 3.52 |
| Arizona | 12,860 | 3.77 |
| California | 16,740 | 3.78 |

*Source: Statistical Bulletin*, Vol. 80, No. 4, Oct.–Dec. 1999, p. 13.

**a.** Plot the data on a scattergram.

**b.** Use the method of least squares to model the relationship between average hospital charge ($y$) and length of hospital stay ($x$).

**c.** Find the estimated standard error of the regression model and interpret its value in the context of the problem.

**d.** For a hospital stay of length $x = 4$ days, find $\hat{y} \pm 2s$.

**e.** What fraction of the states in the sample have average hospital charges within $\pm 2s$ of the least squares line?

**11.31** Refer to the simple linear regression relating total number of words in a conversation with an unfamiliar partner $y$ to total number of words with a familiar partner $x$ for hearing-impaired subjects, Exercise 11.21 (p. 527).

**a.** Find SSE, $s^2$, and $s$ on the STATISTIX printout (p. 527).

**b.** Interpret the value of $s$.

**11.32** Refer to the simple linear regression relating number of flycatchers killed $y$ to nest box tit occupancy $x$ for nest sites, Exercise 11.22 (p. 527).

**a.** Find SSE, $s^2$, and $s$ on the SAS printout (p. 528).

**b.** Interpret the value of $s$.

### Applying the Concepts—Advanced

**11.33** To improve the quality of the output of any production process, it is necessary first to understand the capabilities of the process (*Out of the Crisis*, Deming, 1982). In a particular manufacturing process, the useful life of a cutting tool is related to the speed at which the tool is operated. The data in the accompanying table were derived from life tests for the two different brands of cutting tools currently used in the production process.

**a.** Construct a scattergram for each brand of cutting tool.

**b.** For each brand, use the method of least squares to model the relationship between useful life and cutting speed.

**c.** Find SSE, $s^2$, and $s$ for each least squares line.

**d.** For a cutting speed of 70 meters per minute, find $\hat{y} \pm 2s$ for each least squares line.

**e.** For which brand would you feel more confident in using the least squares line to predict useful life for a given cutting speed? Explain.

● CUTTOOL

| Cutting Speed (meters per minute) | Useful Life (hours) Brand A | Brand B |
|---|---|---|
| 30 | 4.5 | 6.0 |
| 30 | 3.5 | 6.5 |
| 30 | 5.2 | 5.0 |
| 40 | 5.2 | 6.0 |
| 40 | 4.0 | 4.5 |
| 40 | 2.5 | 5.0 |
| 50 | 4.4 | 4.5 |
| 50 | 2.8 | 4.0 |
| 50 | 1.0 | 3.7 |
| 60 | 4.0 | 3.8 |
| 60 | 2.0 | 3.0 |
| 60 | 1.1 | 2.4 |
| 70 | 1.1 | 1.5 |
| 70 | .5 | 2.0 |
| 70 | 3.0 | 1.0 |

## 11.5 ASSESSING THE UTILITY OF THE MODEL: MAKING INFERENCES ABOUT THE SLOPE $\beta_1$

Now that we have specified the probability distribution of $\varepsilon$ and found an estimate of the variance $\sigma^2$, we are ready to make statistical inferences about the model's usefulness for predicting the response $y$. This is step 4 in our regression modeling procedure.

**FIGURE 11.9**

*Graphing the Model*
*with $\beta_1 = 0$: $y = \beta_0 + \varepsilon$*

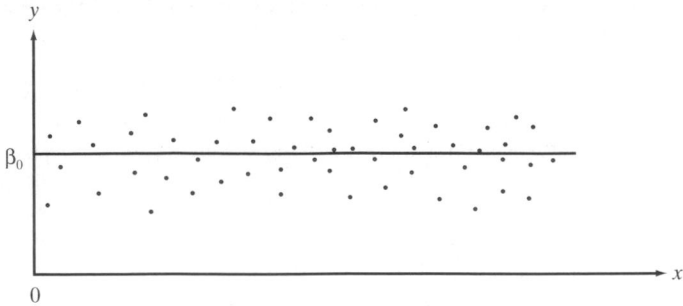

Refer again to the data of Table 11.1 and suppose the reaction times are *completely unrelated* to the percentage of drug in the bloodstream. What could be said about the values of $\beta_0$ and $\beta_1$ in the hypothesized probabilistic model

$$y = \beta_0 + \beta_1 x + \varepsilon$$

if $x$ contributes no information for the prediction of $y$? The implication is that the mean of $y$—that is, the deterministic part of the model $E(y) = \beta_0 + \beta_1 x$—does not change as $x$ changes. In the straight-line model, this means that the true slope, $\beta_1$, is equal to 0 (see Figure 11.9). Therefore, to test the null hypothesis that the linear model contributes no information for the prediction of $y$ against the alternative hypothesis that the linear model is useful for predicting $y$, we test

$$H_0: \beta_1 = 0$$
$$H_a: \beta_1 \neq 0$$

If the data support the alternative hypothesis, we will conclude that $x$ does contribute information for the prediction of $y$ using the straight-line model [although the true relationship between $E(y)$ and $x$ could be more complex than a straight line]. Thus, in effect, this is a test of the usefulness of the hypothesized model.

The appropriate test statistic is found by considering the sampling distribution of $\hat{\beta}_1$, the least squares estimator of the slope $\beta_1$, as shown in the following box.

---

### Sampling Distribution of $\hat{\beta}_1$

If we make the four assumptions about $\varepsilon$ (see Section 11.3), the sampling distribution of the least squares estimator $\hat{\beta}_1$ of the slope will be normal with mean $\beta_1$ (the true slope) and standard deviation

$$\sigma_{\hat{\beta}_1} = \frac{\sigma}{\sqrt{SS_{xx}}} \quad \text{(see Figure 11.10)}$$

We estimate $\sigma_{\hat{\beta}_1}$ by $s_{\hat{\beta}_1} = \dfrac{s}{\sqrt{SS_{xx}}}$ and refer to this quantity as the estimated

standard error of the least squares slope $\hat{\beta}_1$.

---

**FIGURE 11.10**

*Sampling Distribution of $\hat{\beta}_1$*

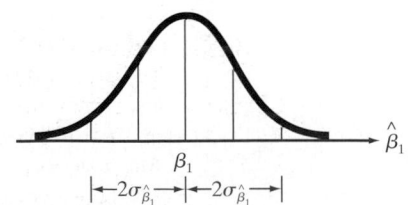

Since $\sigma$ is usually unknown, the appropriate test statistic is a $t$ statistic, formed as follows:

$$t = \frac{\hat{\beta}_1 - \text{Hypothesized value of } \beta_1}{s_{\hat{\beta}_1}} \qquad \text{where} \qquad s_{\hat{\beta}_1} = \frac{s}{\sqrt{SS_{xx}}}$$

Thus,

$$t = \frac{\hat{\beta}_1 - 0}{s/\sqrt{SS_{xx}}}$$

Note that we have substituted the estimator $s$ for $\sigma$ and then formed the estimated standard error $s_{\hat{\beta}_1}$ by dividing $s$ by $\sqrt{SS_{xx}}$. The number of degrees of freedom associated with this $t$ statistic is the same as the number of degrees of freedom associated with $s$. Recall that this number is $(n - 2)$ df when the hypothesized model is a straight line (see Section 11.4). The setup of our test of the usefulness of the straight-line model is summarized in the next box.

---

### A Test of Model Utility: Simple Linear Regression

| **ONE-TAILED TEST** | **TWO-TAILED TEST** |
|---|---|
| $H_0: \beta_1 = 0$ | $H_0: \beta_1 = 0$ |
| $H_a: \beta_1 < 0 \quad (\text{or } H_a: \beta_1 > 10)$ | $H_a: \beta_1 \neq 0$ |

*Test statistic:* $\quad t = \dfrac{\hat{\beta}_1}{s_{\hat{\beta}_1}} = \dfrac{\hat{\beta}_1}{s/\sqrt{SS_{xx}}}$

| *Rejection region:* $\quad t < -t_\alpha$ | *Rejection region:* $\quad |t| > t_{\alpha/2}$ |
|---|---|
| $(\text{or } t > t_\alpha \text{ when } H_a: \beta_1 > 0)$ | |

where $t_\alpha$ and $t_{\alpha/2}$ are based on $(n - 2)$ degrees of freedom
*Assumptions:* The four assumptions about $\varepsilon$ listed in Section 11.3.

---

For the drug reaction example, we will choose $\alpha = .05$ and, since $n = 5$, $t$ will be based on $n - 2 = 3$ df and the rejection region will be

$$|t| > t_{.025} = 3.182$$

We previously calculated $\hat{\beta}_1 = .7$, $s = .61$, and $SS_{xx} = 10$. Thus,

$$t = \frac{\hat{\beta}_1}{s/\sqrt{SS_{xx}}} = \frac{.7}{.61\sqrt{10}} = \frac{.7}{.19} = 3.7$$

Since this calculated $t$ value falls in the upper-tail rejection region (see Figure 11.11), we reject the null hypothesis and conclude that the slope $\beta_1$ is not 0. The sample evidence indicates that amount of drug $x$ in the blood stream contributes information for the prediction of reaction time $y$ when a linear model is used.

We can reach the same conclusion by using the observed significance level ($p$-value) of the test from a computer printout. The MINITAB printout for the drug reaction example is reproduced in Figure 11.12. The test statistic and two-tailed $p$-value are highlighted on the printout. Since the $p$-value $= .035$ is smaller than $\alpha = .05$, we will reject $H_0$.

What conclusion can be drawn if the calculated $t$ value does not fall in the rejection region or if the observed significance level of the test exceeds $\alpha$? We know

**FIGURE 11.11**

*Rejection Region and Calculated t Value for Testing $H_0$: $\beta_1 = 0$ Versus $H_a$: $\beta_1 \neq 0$*

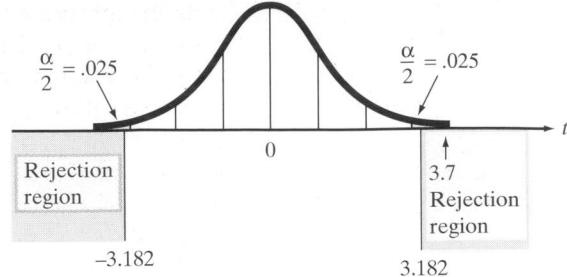

**FIGURE 11.12**

*MINITAB Printout for Time–Drug Regression*

```
The regression equation is
TIME_Y = - 0.100 + 0.700 DRUG_X


Predictor         Coef     SE Coef         T       P
Constant       -0.1000      0.6351     -0.16   0.885
DRUG_X          0.7000      0.1915      3.66   0.035


S = 0.6055       R-Sq = 81.7%     R-Sq(adj) = 75.6%


Analysis of Variance


Source            DF          SS         MS        F       P
Regression         1      4.9000     4.9000    13.36   0.035
Residual Error     3      1.1000     0.3667
Total              4      6.0000
```

from previous discussions of the philosophy of hypothesis testing that such a *t* value does *not* lead us to accept the null hypothesis. That is, we do not conclude that $\beta_1 = 0$. Additional data might indicate that $\beta_1$ differs from 0, or a more complex relationship may exist between *x* and *y*, requiring the fitting of a model other than the straight-line model. We discuss several such models in Chapter 12.

Another way to make inferences about the slope $\beta_1$ is to estimate it using a confidence interval. This interval is formed as shown in the next box.

---

### A 100(1 − $\alpha$)% Confidence Interval for the Simple Linear Regression Slope $\beta_1$

$$\hat{\beta}_1 \pm t_{\alpha/2}s_{\hat{\beta}_1}$$

where the estimated standard error of $\hat{\beta}_1$ is calculated by

$$s_{\hat{\beta}_1} = \frac{s}{\sqrt{SS_{xx}}}$$

and $t_{\alpha/2}$ is based on $(n - 2)$ degrees of freedom.

*Assumptions:* The four assumptions about $\varepsilon$ listed in Section 11.3.

---

For the drug reaction example, $t_{\alpha/2}$ is based on $(n - 2) = 3$ degrees of freedom. Therefore, a 95% confidence interval for the slope $\beta_1$, the expected change in reaction time for a 1% increase in the amount of drug in the bloodstream, is

$$\hat{\beta}_1 \pm t_{.025}s_{\hat{\beta}_1} = .7 \pm 3.182\left(\frac{s}{\sqrt{SS_{xx}}}\right) = .7 \pm 3.182\left(\frac{.61}{\sqrt{10}}\right) = .7 \pm .61$$

Thus, the interval estimate of the slope parameter $\beta_1$ is .09 to 1.31. In terms of this example, the implication is that we can be 95% confident that the *true* mean increase in reaction time per additional 1% of the drug is between .09 and 1.31 seconds. This inference is meaningful only over the sampled range of $x$—that is, from 1% to 5% of the drug in the bloodstream. Since all the values in this interval are positive, it appears that $\beta_1$ is positive and that the mean of $y$, $E(y)$, increases as $x$ increases. However, the rather large width of the confidence interval reflects the small number of data points (and, consequently, a lack of information) in the experiment. We would expect a narrower interval if the sample size were increased.

 EXERCISES 11.34–11.47

## Learning the Mechanics

**11.34** Construct both a 95% and a 90% confidence interval for $\beta_1$ for each of the following cases:
  **a.** $\hat{\beta}_1 = 31, s = 3, SS_{xx} = 35, n = 12$
  **b.** $\hat{\beta}_1 = 64, SSE = 1,960, SS_{xx} = 30, n = 18$
  **c.** $\hat{\beta}_1 = -8.4, SSE = 146, SS_{xx} = 64, n = 24$

**11.35** Consider the following pairs of observations:

| x | 1 | 5 | 3 | 2 | 6 | 6 | 0 |
|---|---|---|---|---|---|---|---|
| y | 1 | 3 | 3 | 1 | 4 | 5 | 1 |

  **a.** Construct a scattergram for the data.
  **b.** Use the method of least squares to fit a straight line to the seven data points in the table.
  **c.** Plot the least squares line on your scattergram of part **a**.
  **d.** Specify the null and alternative hypotheses you would use to test whether the data provide sufficient evidence to indicate that $x$ contributes information for the (linear) prediction of $y$.
  **e.** What is the test statistic that should be used in conducting the hypothesis test of part **d**? Specify the degrees of freedom associated with the test statistic.
  **f.** Conduct the hypothesis test of part **d** using $\alpha = .05$.

**11.36** Refer to Exercise 11.35. Construct an 80% and a 98% confidence interval for $\beta_1$.

**11.37** Do the accompanying data provide sufficient evidence to conclude that a straight line is useful for characterizing the relationship between $x$ and $y$?

| y | 4 | 2 | 5 | 3 | 2 | 4 |
|---|---|---|---|---|---|---|
| x | 1 | 4 | 5 | 3 | 2 | 4 |

## Applying the Concepts—Basic

**11.38** Refer to the data on sale prices and total appraised values of 92 residential properties in an upscale Tampa, Florida, neighborhood, Exercise 11.14 (p. 523). An SAS simple linear regression printout for the analysis is given here.

SAS Simple Linear Regression Output for Exercise 11.38

The REG Procedure
Dependent Variable: SALEPRIC

Analysis of Variance

| Source | DF | Sum of Squares | Mean Square | F Value | Pr > F |
|---|---|---|---|---|---|
| Model | 1 | 1673142 | 1673142 | 1556.48 | <.0001 |
| Error | 90 | 96746 | 1074.95355 | | |
| Corrected Total | 91 | 1769888 | | | |

| | | | | | |
|---|---|---|---|---|---|
| Root MSE | | 32.78648 | R-Square | 0.9453 | |
| Dependent Mean | | 236.55761 | Adj R-Sq | 0.9447 | |
| Coeff Var | | 13.85983 | | | |

Parameter Estimates

| Variable | DF | Parameter Estimate | Standard Error | t Value | Pr > |t| | 95% Confidence Limits | |
|---|---|---|---|---|---|---|---|
| Intercept | 1 | 20.94193 | 6.44617 | 3.25 | 0.0016 | 8.13550 | 33.74837 |
| TOTALVAL | 1 | 1.06873 | 0.02709 | 39.45 | <.0001 | 1.01491 | 1.12254 |

a. Use the printout to determine whether there is a positive linear relationship between appraised property value $x$ and sale price $y$ for residential properties sold in this neighborhood. That is, determine if there is sufficient evidence (at $\alpha = .01$) to indicate that $\beta_1$, the slope of the straight-line model, is positive.

b. Find a 95% confidence interval for the slope, $\beta_1$, on the printout. Interpret the result practically.

c. What can be done to obtain a narrower confidence interval in part **b**?

**11.39** Refer to Exercise 11.19 (p. 526) and the simple linear regression relating the sweetness index ($y$) of an orange juice sample to the amount of water soluble pectin ($x$) in the juice. Find a 90% confidence interval for the true slope of the line. Interpret the result.

**11.40** Refer to Exercises 11.22 and 11.32 (pp. 527, 533) and the simple linear regression relating number of flycatchers killed $y$ to nest box tit occupancy $x$.

a. Refer to the SAS printout (p. 528) and test whether $y$ is positively linearly related to $x$. Use $\alpha = .01$.

b. Construct a 99% confidence interval for $\beta_1$. Practically interpret the result.

**11.41** The abundance of tropical seagrasses in Australia was the subject of research published in *Aquatic Botany* (Mar. 1995). Simple linear regression was used to relate the standing crop ($y$) of a seagrass species (measured in grams per meter squared) to the percentage ($x$) of land covered by the plants. Data collected for $n = 12$ sites at Rowes Bay, Australia, yielded the following results:

$$\hat{y} = .031 + .089x \qquad t = 2.34 \qquad p\text{-value} = .042$$

a. Give the practical interpretation, if possible, of the estimated $y$-intercept of the line.

b. Give the practical interpretation, if possible, of the estimated slope of the line.

c. At what $\alpha$-level is there sufficient evidence of a linear relationship between standing crop and percentage cover of a seagrass species? Explain.

## Applying the Concepts—Intermediate

**11.42** Refer to the *British Journal of Sports Medicine* (Apr. 2000) study of the effect of massage on boxing performance, Exercise 10.48 (p. 482). Two other variables measured on the boxers were blood lactate concentration (mM) and the boxer's perceived recovery (28-point scale). Based on information provided in the article, the following data were obtained for 16 five-round boxing performances, where a massage was given to the boxer between rounds. Conduct a test to determine whether blood lactate level ($y$) is linearly related to perceived recovery ($x$). Use $\alpha = .10$.

🔵 BOXING2

| Blood Lactate Level | Perceived Recovery |
|---|---|
| 3.8 | 7 |
| 4.2 | 7 |
| 4.8 | 11 |
| 4.1 | 12 |
| 5.0 | 12 |
| 5.3 | 12 |
| 4.2 | 13 |
| 2.4 | 17 |
| 3.7 | 17 |
| 5.3 | 17 |
| 5.8 | 18 |
| 6.0 | 18 |
| 5.9 | 21 |
| 6.3 | 21 |
| 5.5 | 20 |
| 6.5 | 24 |

*Source:* Hemmings, B., Smith, M., Graydon, J., and Dyson, R. "Effects of massage on physiological restoration, perceived recovery, and repeated sports performance." *British Journal of Sports Medicine*, Vol. 34, No. 2, Apr. 2000 (data adapted from Figure 3).

**11.43** How do eye and head movements relate to body movements when reacting to a visual stimulus? Scientists at the California Institute of Technology designed an experiment to answer this question and reported their results in *Nature* (Aug. 1998). Adult male rhesus monkeys were exposed to a visual stimulus (i.e., a panel of light-emitting diodes) and their eye, head, and body movements were electronically recorded. In one variation of the experiment, two variables were measured: active head movement ($x$, percent per degree) and body plus head rotation ($y$, percent per degree). The data for $n = 39$ trials were subjected to a simple linear regression analysis, with the following results: $\hat{\beta}_1 = .88, s_{\hat{\beta}_1} = .14$

a. Conduct a test to determine whether the two variables, active head movement $x$ and body plus head rotation $y$ are positively linearly related. Use $\alpha = .05$.

b. Construct and interpret a 90% confidence interval for $\beta_1$.

c. The scientists want to know if the true slope of the line differs significantly from 1. Based on your answer to part **b**, make the appropriate inference.

**11.44** The U.S. Department of Agriculture has developed and adopted the Universal Soil Loss Equation (USLE) for predicting water erosion of soils. In geographic areas where runoff from melting snow is common, calculating the USLE requires an accurate estimate of snowmelt runoff erosion. An article in the *Journal of Soil and Water Conservation* (Mar.–Apr. 1995) used simple linear regression to develop a snowmelt erosion index. Data for 54 climatological stations in Canada were used to model the McCool winter-adjusted rainfall erosivity index, $y$, as a straight-line function of the once-in-5-year snowmelt runoff amount, $x$ (measured in millimeters).

a. The data points are plotted in the scattergram shown on p. 539. Is there visual evidence of a linear trend?

**b.** The data for seven stations were removed from the analysis due to lack of snowfall during the study period. Why is this strategy advisable?

**c.** The simple linear regression on the remaining $n = 47$ data points yielded the following results: $\hat{y} = -6.72 + 1.39x$, $s_{\hat{\beta}_1} = .06$. Use this information to construct a 90% confidence interval for $\beta_1$.

**d.** Interpret the interval, part **c**.

⊘NAMEGAME2

**11.45** Refer to the *Journal of Experimental Psychology-Applied* (June 2000) name retrieval study, Exercise 11.23 (p. 528). Recall that the goal of the study was to investigate the linear trend between proportion of names recalled ($y$) and position (order) of the student ($x$) during the "name game." Is there sufficient evidence (at $\alpha = .01$) of a linear trend? Answer the question by analyzing the data for 144 students saved in the NAMEGAME2 file.

**Applying the Concepts—Advanced**

**11.46** One of the most difficult tasks of developing and managing a global portfolio is assessing the risks of potential foreign investments. Duke University researcher C. R. Henry collaborated with two First Chicago Investment Management Company directors to examine the use of country credit ratings as a means of evaluating foreign investments (*Journal of Portfolio Management*, Winter 1995). To be effective, such a measure should help explain and predict the volatility of the foreign market in question. The researchers analyzed data on annualized risk ($y$) and average credit rating ($x$) for the 40 countries shown in the table on p. 540. A MINITAB printout for a simple linear regression analysis conducted on the data is shown below.

**a.** Locate the least squares estimates of $\beta_0$ and $\beta_1$ on the printout.

**MINITAB Output for Exercise 11.46**

```
The regression equation is
RISK = 57.8 - 0.400 RATING


Predictor        Coef      SE Coef        T       P
Constant       57.755        6.128     9.43   0.000
RATING        -0.39961      0.09152    -4.37   0.000


S = 12.68        R-Sq = 33.4%     R-Sq(adj) = 31.7%


Analysis of Variance

Source           DF          SS          MS        F       P
Regression        1      3064.4      3064.4    19.07   0.000
Residual Error   38      6107.5       160.7
Total            39      9171.9
```

⊙ GLOBRISK

| Country | Annualized Risk (%) | Average Credit Rating |
|---|---|---|
| Argentina | 87.0 | 31.8 |
| Australia | 26.9 | 78.2 |
| Austria | 26.3 | 83.8 |
| Belgium | 22.0 | 78.4 |
| Brazil | 64.8 | 36.2 |
| Canada | 19.2 | 87.1 |
| Chile | 31.6 | 38.6 |
| Colombia | 31.5 | 44.4 |
| Denmark | 20.6 | 72.6 |
| Finland | 26.1 | 76.0 |
| France | 23.8 | 85.3 |
| Germany | 23.0 | 93.4 |
| Greece | 39.6 | 51.9 |
| Hong Kong | 34.3 | 69.6 |
| India | 30.0 | 46.6 |
| Ireland | 23.4 | 66.4 |
| Italy | 28.0 | 75.5 |
| Japan | 25.7 | 94.5 |
| Jordan | 17.6 | 33.6 |
| Korea | 30.7 | 62.2 |
| Malaysia | 26.7 | 64.4 |
| Mexico | 46.3 | 43.3 |
| Netherlands | 18.5 | 87.6 |
| New Zealand | 26.3 | 68.9 |
| Nigeria | 41.4 | 30.6 |
| Norway | 28.3 | 83.0 |
| Pakistan | 24.4 | 26.4 |
| Philippines | 38.4 | 29.6 |
| Portugal | 47.5 | 56.7 |
| Singapore | 26.4 | 77.6 |
| Spain | 24.8 | 70.8 |
| Sweden | 24.5 | 79.5 |
| Switzerland | 19.6 | 94.7 |
| Taiwan | 53.7 | 72.9 |
| Thailand | 27.0 | 55.8 |
| Turkey | 74.1 | 32.6 |
| United Kingdom | 21.8 | 87.6 |
| United States | 15.4 | 93.4 |
| Venezuela | 46.0 | 45.0 |
| Zimbabwe | 35.6 | 24.5 |

*Source:* Erb, C. B., Harvey, C. R., and Viskanta, T. E. "Country risk and global equity selection." *Journal of Portfolio Management*, Vol. 21, No. 2, Winter 1995, p. 76.

**b.** Plot the data in a scattergram, then sketch the least squares line on the graph.

**c.** Do the data provide sufficient evidence to conclude that credit rating ($x$) contributes information for the prediction of annualized risk ($y$)?

**d.** Use the plot, part **b**, to locate any unusual data points (outliers).

**e.** Eliminate the outlier(s), part **d**, from the data set and rerun the simple linear regression analysis. Note any dramatic changes in the results.

**11.47** A recent civil suit revolved around a 5-building brick apartment complex located in the Bronx, New York, which began to suffer *spalling* damage (i.e., a separation of some portion of the face of a brick from its body). The owner of the complex alleged that the bricks were defectively manufactured. The brick manufacturer countered that poor design and shoddy management led to the damage. To settle the suit, an estimate of the rate of damage per 1,000 bricks, called the spall rate, was required (*Chance*, Summer 1994). The owner estimated the spall rate using several *scaffold-drop* surveys. (With this method, an engineer lowers a scaffold down at selected places on building walls and counts the number of visible spalls for every 1,000 bricks in the observation area.) The brick manufacturer conducted its own survey by dividing the walls of the complex into 83 wall segments and taking a photograph of each wall segment. (The number of spalled bricks that could be made out from each photo was recorded and the sum over all 83 wall segments used as an estimate of total spall damage.) In this court case, the jury was faced with the following dilemma: The scaffold-drop survey provided the most accurate estimate of spall rates in a given wall segment. Unfortunately, the drop areas were not selected at random from the entire complex; rather, drops were made at areas with high spall concentrations, leading to an overestimate of the total damage. On the other hand, the photo survey was complete in that all 83 wall segments in the complex were checked for spall damage. But the spall rate estimated by the photos, at least in areas of high spall concentration, was biased low (spalling damage cannot always be seen from a photo), leading to an underestimate of the total damage.

The data in the table are the spall rates obtained using the two methods at 11 drop locations. Use the data, as did expert statisticians who testified in the case, to help the jury estimate the true spall rate at a given wall segment. Then explain how this information, coupled with the data (not given here) on all 83 wall segments, can provide a reasonable estimate of the total spall damage (i.e., total number of damaged bricks).

⊙ BRICKS

| Drop Location | Drop Spall Rate (per 1,000 bricks) | Photo Spall Rate (per 1,000 bricks) |
|---|---|---|
| 1 | 0 | 0 |
| 2 | 5.1 | 0 |
| 3 | 6.6 | 0 |
| 4 | 1.1 | .8 |
| 5 | 1.8 | 1.0 |
| 6 | 3.9 | 1.0 |
| 7 | 11.5 | 1.9 |
| 8 | 22.1 | 7.7 |
| 9 | 39.3 | 14.9 |
| 10 | 39.9 | 13.9 |
| 11 | 43.0 | 11.8 |

*Source:* Fairley, W. B., *et al.* "Bricks, buildings, and the Bronx: Estimating masonry deterioration." *Chance*, Vol. 7. No. 3, Summer 1994, p. 36 (Figure 3). [*Note:* The data points are estimated from the points shown on a scatterplot.]

## 11.6   THE COEFFICIENT OF CORRELATION

Recall (from optional Section 2.9) that a **bivariate relationship** describes a relationship   or correlation—between two variables, $x$ and $y$. Scattergrams are used to graphically describe a bivariate relationship. In this section we will discuss the concept of **correlation** and how it can be used to measure the linear relationship between two variables, $x$ and $y$. A numerical descriptive measure of correlation is provided by the *Pearson product moment coefficient of correlation, r*.

---

**DEFINITION 11.2**

The **Pearson product moment coefficient of correlation,** $r$, is a measure of the strength of the *linear* relationship between two variables $x$ and $y$. It is computed (for a sample of $n$ measurements on $x$ and $y$) as follows:

$$r = \frac{SS_{xy}}{\sqrt{SS_{xx}SS_{yy}}}$$

---

Note that the computational formula for the correlation coefficient $r$ given in Definition 11.2 involves the same quantities that were used in computing the least squares prediction equation. In fact, since the numerators of the expressions for $\hat{\beta}_1$ and $r$ are identical, you can see that $r = 0$ when $\hat{\beta}_1 = 0$ (the case where $x$ contributes no information for the prediction of $y$) and that $r$ is positive when the slope is positive and negative when the slope is negative. Unlike $\hat{\beta}_1$, the correlation coefficient $r$ is *scaleless* and assumes a value between $-1$ and $+1$, regardless of the units of $x$ and $y$.

A value of $r$ near or equal to 0 implies little or no linear relationship between $y$ and $x$. In contrast, the closer $r$ comes to 1 or $-1$, the stronger the linear relationship between $y$ and $x$. And if $r = 1$ or $r = -1$, all the sample points fall exactly on the least squares line. Positive values of $r$ imply a positive linear relationship between $y$ and $x$; that is, $y$ increases as $x$ increases. Negative values of $r$ imply a negative linear relationship between $y$ and $x$; that is, $y$ decreases as $x$ increases. Each of these situations is portrayed in Figure 11.13.

We demonstrate how to calculate the coefficient of correlation $r$ using the data in Table 11.1 for the drug reaction example. The quantities needed to calculate $r$ are $SS_{xy}$, $SS_{xx}$, and $SS_{yy}$. The first two quantities have been calculated previously and are repeated here for convenience:

$$SS_{xy} = 7, \quad SS_{xx} = 10, \quad SS_{yy} = \sum y^2 - \frac{\left(\sum y\right)^2}{n}$$

$$= 26 - \frac{(10)^2}{5} = 26 - 20 = 6$$

We now find the coefficient of correlation:

$$r = \frac{SS_{xy}}{\sqrt{SS_{xx}SS_{yy}}} = \frac{7}{\sqrt{(10)(6)}} = \frac{7}{\sqrt{60}} = .904$$

The fact that $r$ is positive and near 1 in value indicates that the reaction time tends to increase as the amount of drug in the bloodstream increases—*for this sample of five subjects*. This is the same conclusion we reached when we found the calculated value of the least squares slope to be positive.

**FIGURE 11.13**

*Values of r and Their Implications*

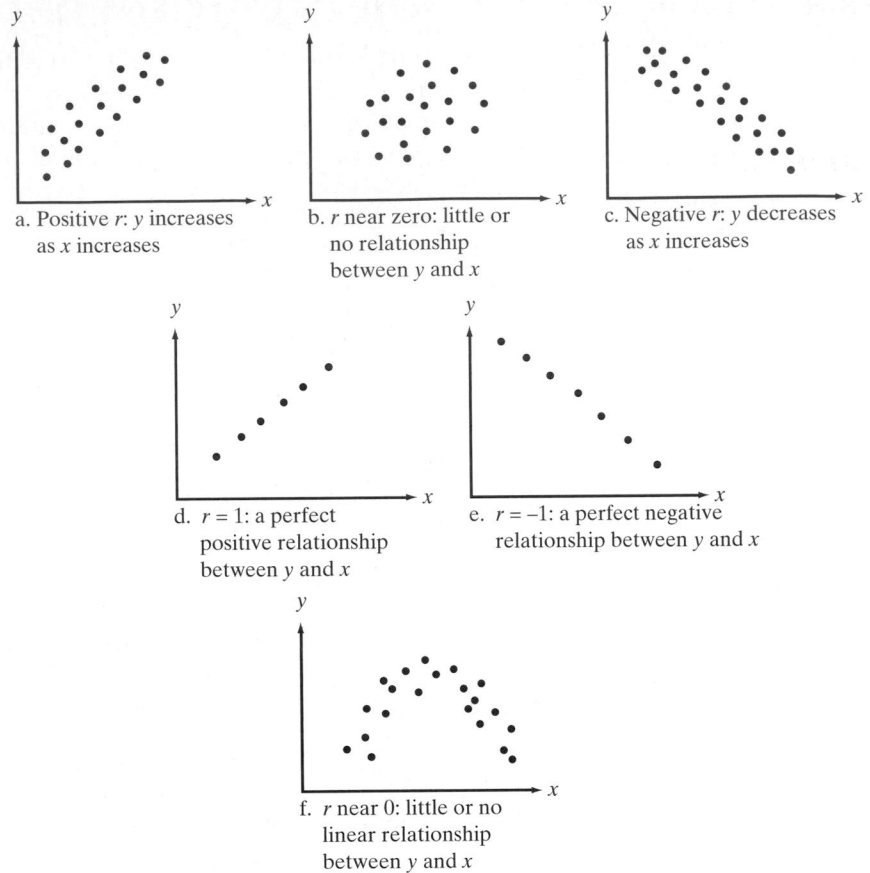

a. Positive *r*: *y* increases as *x* increases

b. *r* near zero: little or no relationship between *y* and *x*

c. Negative *r*: *y* decreases as *x* increases

d. *r* = 1: a perfect positive relationship between *y* and *x*

e. *r* = −1: a perfect negative relationship between *y* and *x*

f. *r* near 0: little or no linear relationship between *y* and *x*

**EXAMPLE 11.1**

Legalized gambling is available on several riverboat casinos operated by a city in Mississippi. The mayor of the city wants to know the correlation between the number of casino employees and yearly crime rate. The records for the past 10 years are examined, and the results listed in Table 11.5 are obtained. Calculate the coefficient of correlation *r* for the data.

⦿ CASINO

**TABLE 11.5    Data on Casino Employees and Crime Rate, Example 11.1**

| Year | Number of Casino Employees *x* (thousands) | Crime Rate *y* (number of crimes per 1,000 population) |
|------|---------------------------------------------|---------------------------------------------------------|
| 1993 | 15 | 1.35 |
| 1994 | 18 | 1.63 |
| 1995 | 24 | 2.33 |
| 1996 | 22 | 2.41 |
| 1997 | 25 | 2.63 |
| 1998 | 29 | 2.93 |
| 1999 | 30 | 3.41 |
| 2000 | 32 | 3.26 |
| 2001 | 35 | 3.63 |
| 2002 | 38 | 4.15 |

**FIGURE 11.14**

*MINITAB Correlation Printout for Example 11.1*

Pearson correlation of EMPLOY and CRIME = 0.987
P-Value = 0.000

**FIGURE 11.15**

*MINITAB Scattergram for Example 11.1*

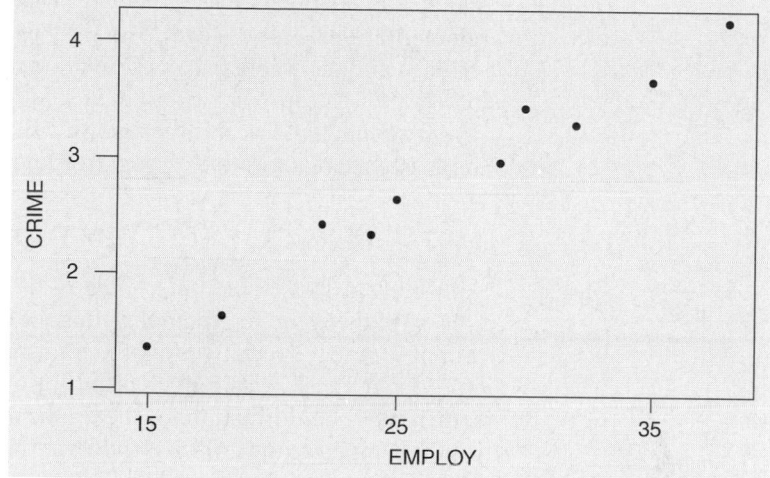

**Solution**

Rather than use the computing formula given in Definition 11.2, we resort to a statistical software package. The data of Table 11.5 were entered into a computer and MINITAB was used to compute $r$. The MINITAB printout is shown in Figure 11.14.

The coefficient of correlation, highlighted on the printout, is $r = .987$. Thus, the size of the casino workforce and crime rate in this city are very highly correlated—at least over the past 10 years. The implication is that a strong positive linear relationship exists between these variables (see Figure 11.15). We must be careful, however, not to jump to any unwarranted conclusions. For instance, the mayor may be tempted to conclude that hiring more casino workers next year will increase the crime rate—that is, that there is a *causal relationship* between the two variables. However, high correlation does not imply causality. The fact is, many things have probably contributed both to the increase in the casino workforce and to the increase in crime rate. The city's tourist trade has undoubtedly grown since legalizing riverboat casinos and it is likely that the casinos have expanded both in services offered and in number. *We cannot infer a causal relationship on the basis of high sample correlation. When a high correlation is observed in the sample data, the only safe conclusion is that a linear trend may exist between x and y.* Another variable, such as the increase in tourism, may be the underlying cause of the high correlation between $x$ and $y$. ∎

Keep in mind that the correlation coefficient $r$ measures the linear correlation between $x$ values and $y$ values in the sample, and a similar linear coefficient of correlation exists for the population from which the data points were selected. The **population correlation coefficient** is denoted by the symbol $\rho$ (rho). As you might expect, $\rho$ is estimated by the corresponding sample statistic, $r$. Or, instead of estimating $\rho$, we might want to test the null hypothesis $H_0: \rho = 0$ against $H_a: \rho \neq 0$—that is, we can test the hypothesis that $x$ contributes no information for the prediction of $y$ by using the straight-line model against the alternative that the two variables are at least linearly related.

However, we already performed this *identical* test in Section 11.5 when we tested $H_0: \beta_1 = 0$ against $H_a: \beta_1 \neq 0$. That is, the null hypothesis $H_0: \rho = 0$ is equivalent to the hypothesis $H_0: \beta_1 = 0$.* When we tested the null hypothesis $H_0: \beta_1 = 0$ in connection with the drug reaction example, the data led to a rejection of the null hypothesis at the $\alpha = .05$ level. This rejection implies that the null hypothesis of a 0 linear correlation between the two variables (drug and reaction time) can also be rejected at the $\alpha = .05$ level. The only real difference between the least squares slope $\hat{\beta}_1$ and the coefficient of correlation $r$ is the measurement scale. Therefore, the information they provide about the usefulness of the least squares model is to some extent redundant. For this reason, we will use the slope to make inferences about the existence of a positive or negative linear relationship between two variables.

## 11.7    THE COEFFICIENT OF DETERMINATION

Another way to measure the usefulness of the model is to measure the contribution of $x$ in predicting $y$. To accomplish this, we calculate how much the errors of prediction of $y$ were reduced by using the information provided by $x$. To illustrate, consider the sample shown in the scattergram of Figure 11.16a. If we assume that $x$ contributes no information for the prediction of $y$, the best prediction for a value of $y$ is the sample mean $\bar{y}$, which is shown as the horizontal line in Figure 11.16b. The vertical line segments in Figure 11.16b are the deviations of the points about the mean $\bar{y}$. Note that the sum of squares of deviations for the prediction equation $\hat{y} = \bar{y}$ is

$$SS_{yy} = \sum (y_i - \bar{y})^2$$

**FIGURE 11.16**

*A Comparison of the Sum of Squares of Deviations for Two Models*

a. Scattergram of data

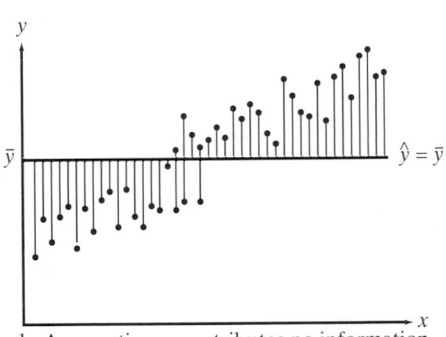

b. Assumption: $x$ contributes no information for predicting $y$, $\hat{y} = \bar{y}$

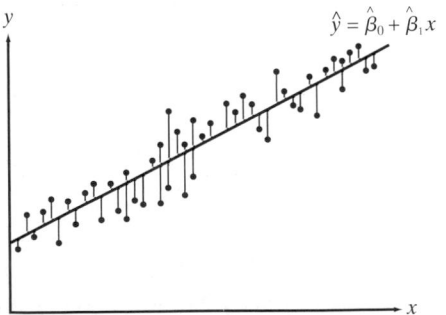

c. Assumption: $x$ contributes information for predicting $y$, $\hat{y} = \hat{\beta}_0 + \hat{\beta}_1 x$

---

*The correlation test statistic that is equivalent to $t = \hat{\beta}_1/s_{\hat{\beta}_1}$ is $t = \dfrac{r}{\sqrt{(1 - r^2)/(n - 2)}}$

Now suppose you fit a least squares line to the same set of data and locate the deviations of the points about the line as shown in Figure 11.16c. Compare the deviations about the prediction lines in Figures 11.16b and 11.16c. You can see that

1. If $x$ contributes little or no information for the prediction of $y$, the sums of squares of deviations for the two lines,

$$\text{SS}_{yy} = \sum (y_i - \bar{y})^2 \quad \text{and} \quad \text{SSE} = \sum (y_i - \hat{y}_i)^2$$

will be nearly equal.
2. If $x$ does contribute information for the prediction of $y$, the SSE will be smaller than $\text{SS}_{yy}$. In fact, if all the points fall on the least squares line, then SSE = 0.

Then, the reduction in the sum of squares of deviations that can be attributed to $x$, expressed as a proportion of $\text{SS}_{yy}$, is

$$\frac{\text{SS}_{yy} - \text{SSE}}{\text{SS}_{yy}}$$

Note that $\text{SS}_{yy}$ is the "total sample variation" of the observations around the mean $\bar{y}$ and that SSE is the remaining "unexplained sample variability" after fitting the line $\hat{y}$. Thus, the difference $(\text{SS}_{yy} - \text{SSE})$ is the "explained sample variability" attributable to the linear relationship with $x$. Then a verbal description of the proportion is

$$\frac{\text{SS}_{yy} - \text{SSE}}{\text{SS}_{yy}} = \frac{\text{Explained sample variability}}{\text{Total sample variability}}$$
$$= \text{Proportion of total sample variability explained by the linear relationship}$$

In simple linear regression, it can be shown that this proportion—called the *coefficient of determination*—is equal to the square of the simple linear coefficient of correlation $r$ (the Pearson product moment coefficient of correlation).

---

**DEFINITION 11.3**

The **coefficient of determination** is

$$r^2 = \frac{\text{SS}_{yy} - \text{SSE}}{\text{SS}_{yy}} = 1 - \frac{\text{SSE}}{\text{SS}_{yy}}$$

It represents the proportion of the total sample variability around $\bar{y}$ that is explained by the linear relationship between $y$ and $x$. (In simple linear regression, it may also be computed as the square of the coefficient of correlation $r$.)

---

Note that $r^2$ is always between 0 and 1, because $r$ is between $-1$ and $+1$. Thus, an $r^2$ of .60 means that the sum of squares of deviations of the $y$ values about their predicted values has been reduced 60% by the use of the least squares equation $\hat{y}$, instead of $\bar{y}$, to predict $y$.

**E X A M P L E   1 1 . 2**

Calculate the coefficient of determination for the drug reaction example. The data are repeated in Table 11.6 for convenience. Interpret the result.

**STIMULUS**
**TABLE 11.6**

| Amount of Drug $x$ (%) | Reaction Time $y$ (seconds) |
|:---:|:---:|
| 1 | 1 |
| 2 | 1 |
| 3 | 2 |
| 4 | 2 |
| 5 | 4 |

**Solution**

From previous calculations,

$$SS_{yy} = 6 \quad \text{and} \quad SSE = \sum (y - \hat{y})^2 = 1.10$$

Then, from Definition 11.3, the coefficient of determination is given by

$$r^2 = \frac{SS_{yy} - SSE}{SS_{yy}} = \frac{6.0 - 1.1}{6.0} = \frac{4.9}{6.0} = .817$$

Another way to compute $r^2$ is to recall (Section 11.6) that $r = .904$. Then we have $r^2 = (.904)^2 = .817$. A third way to obtain $r^2$ is from a computer printout. This value is highlighted on the SPSS printout in Figure 11.17. Our interpretation is as follows: We know that using the amount of drug in the blood, $x$, to predict $y$ with the least squares line

$$\hat{y} = -.1 + .7x$$

accounts for nearly 82% of the total sum of squares of deviations of the five sample $y$ values about their mean. Or, stated another way, 82% of the sample variation in reaction time ($y$) can be "explained" by using amount ($x$) of drug in a straight-line model.

**FIGURE 11.17**

*Portion of SPSS Printout for Time–Drug Regression*

**Model Summary**

| Model | R | R Square | Adjusted R Square | Std. Error of the Estimate |
|:---:|:---:|:---:|:---:|:---:|
| 1 | .904[a] | .817 | .756 | .61 |

a. Predictors: (Constant), DRUG_X

---

**Practical Interpretation of the Coefficient of Determination, $r^2$**

$100(r^2)\%$ of the sample variation in $y$ (measured by the total sum of squares of deviations of the sample $y$ values about their mean $\bar{y}$) can be explained by (or attributed to) using $x$ to predict $y$ in the straight-line model.

EXERCISES 11.48–11.61

## Learning the Mechanics

**11.48** Describe the slope of the least squares line if
a. $r = .7$
b. $r = -.7$
c. $r = 0$
d. $r^2 = .64$

**11.49** Explain what each of the following sample correlation coefficients tells you about the relationship between the $x$ and $y$ values in the sample:
a. $r = 1$
b. $r = -1$
c. $r = 0$
d. $r = .90$
e. $r = .10$
f. $r = -.88$

**11.50** Calculate $r^2$ for the least squares line in each of the following exercises. Interpret their values.
a. Exercise 11.10 (p. 522)
b. Exercise 11.13 (p. 522)

**11.51** Construct a scattergram for each data set. Then calculate $r$ and $r^2$ for each data set. Interpret their values.

a.

| $x$ | −2 | −1 | 0 | 1 | 2 |
|---|---|---|---|---|---|
| $y$ | −2 | 1 | 2 | 5 | 6 |

b.

| $x$ | −2 | −1 | 0 | 1 | 2 |
|---|---|---|---|---|---|
| $y$ | 6 | 5 | 3 | 2 | 0 |

c.

| $x$ | 1 | 2 | 2 | 3 | 3 | 3 | 4 |
|---|---|---|---|---|---|---|---|
| $y$ | 2 | 1 | 3 | 1 | 2 | 3 | 2 |

d.

| $x$ | 0 | 1 | 3 | 5 | 6 |
|---|---|---|---|---|---|
| $y$ | 0 | 1 | 2 | 1 | 0 |

## Applying the Concepts—Basic

⊘ TAMPALMS

**11.52** Refer to the data on sale prices and total appraised values of 92 residential properties recently sold in an upscale Tampa, Florida, neighborhood, Exercise 11.14 (p. 523). The MINITAB simple linear regression printout relating sale price ($y$) to appraised property value ($x$) is reproduced below, followed by a MINITAB correlation printout.
a. Find the coefficient of correlation between appraised property value and sale price on the printout. Interpret this value.
b. Find the coefficient of determination between appraised property value and sale price on the printout. Interpret this value.

**11.53** Is there a relationship between leisure activities and high school performance? This question was investigated in the *Journal of Leisure Research* (Vol. 24, 1992). A list of 43 leisure activities was presented to 159 high school students and the students were asked to identify the activities they participated in each week. The article reports that the correlation between high school GPA and the number of leisure activities is $r = .13$, which has a two-tailed $p$-value of .0512.
a. What are the appropriate null and alternative hypotheses to test whether the number of leisure activities and GPA are linearly related?
b. Interpret the $p$-value in terms of this test.

**MINITAB Simple Linear Regression Output for Exercise 11.52**

```
The regression equation is
SALEPRIC = 20.9 + 1.07 TOTALVAL

Predictor        Coef       SE Coef        T        P
Constant        20.942        6.446       3.25    0.002
TOTALVAL       1.06873      0.02709      39.45    0.000

S = 32.79       R-Sq = 94.5%      R-Sq(adj) = 94.5%

Analysis of Variance

Source            DF          SS          MS        F        P
Regression         1     1673142     1673142   1556.48    0.000
Residual Error    90       96746        1075
Total             91     1769888

Pearson correlation of SALEPRIC and TOTALVAL = 0.972
P-Value = 0.000
```

**11.54** Many high school students experience "math anxiety." Does such an attitude carry over to learning computer skills? A researcher at Duquesne University investigated this question and published her results in *Educational Technology* (May–June 1995). A sample of high school students—902 boys and 828 girls—from public schools in Pittsburgh, Pennsylvania, participated in the study. Using 5-point Likert scales, where 1 = "strongly disagree" and 5 = "strongly agree," the researcher measured the students' interest and confidence in both mathematics and computers.

**a.** For boys, math confidence and computer interest were correlated at $r = .14$. Fully interpret this result.

**b.** For girls, math confidence and computer interest were correlated at $r = .33$. Fully interpret this result.

## Applying the Concepts—Intermediate

**11.55** *Perception & Psychophysics* (July 1998) reported on a study of how people view the 3-dimensional objects projected onto a rotating 2-dimensional image. Each in a sample of 25 university students viewed various depth-rotated objects (e.g., hairbrush, duck, shoe) until they recognized the object. The recognition exposure time—that is, the minimum time (in milliseconds) required for the subject to recognize the object—was recorded for each. In addition, each subject rated the "goodness of view" of the object on a numerical scale, where lower scale values correspond to better views.

The table gives the correlation coefficient, $r$, between recognition exposure time and goodness of view for several different rotated objects.

| Object | $r$ | $t$ |
|---|---|---|
| Piano | .447 | 2.40 |
| Bench | −.057 | .27 |
| Motorbike | .619 | 3.78 |
| Armchair | .294 | 1.47 |
| Teapot | .949 | 14.50 |

**a.** Interpret the value of $r$ for each object.

**b.** Calculate and interpret the value of $r^2$ for each object.

**c.** The table also includes the $t$-value for testing the null hypothesis of no correlation (i.e., for testing $H_0: \beta_1 = 0$). Interpret these results.

**11.56** The fertility rate of a country is defined as the number of children a woman citizen bears, on average, in her lifetime. *Scientific American* (Dec. 1993) reported on the declining fertility rate in developing countries. The researchers found that family planning can have a great effect on fertility rate. The table on p. 549 gives the fertility rate, $y$, and contraceptive prevalence, $x$, (measured as the percentage of married women who use contraception) for each of 27 developing countries. An SAS printout of the simple linear regression analysis is provided below.

**a.** According to the researchers, "the data reveal that differences in contraceptive prevalence explain about 90% of the variation in fertility rates." Do you concur?

**SAS Output for Exercise 11.56**

The REG Procedure
Dependent Variable: FERTRATE

Analysis of Variance

| Source | DF | Sum of Squares | Mean Square | F Value | Pr > F |
|---|---|---|---|---|---|
| Model | 1 | 35.96633 | 35.96633 | 74.31 | <.0001 |
| Error | 25 | 12.10033 | 0.48401 | | |
| Corrected Total | 26 | 48.06667 | | | |

| | | | | |
|---|---|---|---|---|
| Root MSE | 0.69571 | R-Square | 0.7483 | |
| Dependent Mean | 4.51111 | Adj R-Sq | 0.7382 | |
| Coeff Var | 15.42216 | | | |

Parameter Estimates

| Variable | DF | Parameter Estimate | Standard Error | t Value | Pr > |t| |
|---|---|---|---|---|---|
| Intercept | 1 | 6.73193 | 0.29034 | 23.19 | <.0001 |
| CONTPREV | 1 | -0.05461 | 0.00634 | -8.62 | <.0001 |

## FERTRATE

| Country | Contraceptive Prevalence $x$ | Fertility Rate $y$ |
|---|---|---|
| Mauritius | 76 | 2.2 |
| Thailand | 69 | 2.3 |
| Colombia | 66 | 2.9 |
| Costa Rica | 71 | 3.5 |
| Sri Lanka | 63 | 2.7 |
| Turkey | 62 | 3.4 |
| Peru | 60 | 3.5 |
| Mexico | 55 | 4.0 |
| Jamaica | 55 | 2.9 |
| Indonesia | 50 | 3.1 |
| Tunisia | 51 | 4.3 |
| El Salvador | 48 | 4.5 |
| Morocco | 42 | 4.0 |
| Zimbabwe | 46 | 5.4 |
| Egypt | 40 | 4.5 |
| Bangladesh | 40 | 5.5 |
| Botswana | 35 | 4.8 |
| Jordan | 35 | 5.5 |
| Kenya | 28 | 6.5 |
| Guatemala | 24 | 5.5 |
| Cameroon | 16 | 5.8 |
| Ghana | 14 | 6.0 |
| Pakistan | 13 | 5.0 |
| Senegal | 13 | 6.5 |
| Sudan | 10 | 4.8 |
| Yemen | 9 | 7.0 |
| Nigeria | 7 | 5.7 |

*Source:* Robey, B., *et al.* "The fertility decline in developing countries." *Scientific American,* Dec. 1993, p. 62. [*Note:* The data values are estimated from a scatterplot.]

**b.** The researchers also concluded that "if contraceptive use increases by 18 percent, women bear, on average, one fewer child." Is this statement supported by the data? Explain.

**11.57** In cotherapy two or more therapists lead a group. An article in the *American Journal of Dance Therapy* (Spring/Summer 1995) examined the use of cotherapy in dance/movement therapy. Two of several variables measured on each of a sample of 136 professional dance/movement therapists were years of formal training $x$ and reported success rate $y$ (measured as a percentage) of coleading dance/movement therapy groups.

**a.** Propose a linear model relating $y$ to $x$.

**b.** The researcher hypothesized that dance/movement therapists with more years in formal dance training will report higher perceived success rates in cother-

apy relationships. State the hypothesis in terms of the parameter of the model, part **a**.

**c.** The correlation coefficient for the sample data was reported as $r = -.26$. Interpret this result.

**d.** Does the value of $r$ in part **c** support the hypothesis in part **b**? Test using $\alpha = .05$. [*Hint:* See the last paragraph and accompanying footnote of Section 11.6]

## NAMEGAME2

**11.58** Refer to the *Journal of Experimental Psychology-Applied* (June 2000) name retrieval study, Exercises 11.23 (p. 528) and 11.45 (p. 539). Find and interpret the values of $r$ and $r^2$ for the simple linear regression relating the proportion of names recalled ($y$) and position (order) of the student ($x$) during the "name game."

## BOXING2

**11.59** Refer to the *British Journal of Sports Medicine* (April 2000) study of the effect of massage on boxing performance, Exercise 11.42 (p. 538). Find and interpret the values of $r$ and $r^2$ for the simple linear regression relating the blood lactate concentration and the boxer's perceived recovery.

## Applying the Concepts—Advanced

**11.60** Botanists at the University of Toronto conducted a series of experiments to investigate the feeding habits of baby snow geese (*Journal of Applied Ecology*, Vol. 32, 1995). Goslings were deprived of food until their guts were empty, then were allowed to feed for 6 hours on a diet of plants or Purina Duck Chow. For each feeding trial, the change in the weight of the gosling after 2.5 hours was recorded as a percentage of initial weight. Two other variables recorded were digestion efficiency (measured as a percentage) and amount of acid-detergent fibre in the digestive tract (also measured as a percentage). The data for 42 feeding trials are listed in the table on p. 550.

**a.** The botanists were interested in the correlation between weight change ($y$) and digestion efficiency ($x$). Plot the data for these two variables in a scattergram. Do you observe a trend?

**b.** Find the coefficient of correlation relating weight change $y$ to digestion efficiency $x$. Interpret this value.

**c.** Conduct a test to determine whether weight change $y$ is correlated with a digestion efficiency $x$. Use $\alpha = .01$.

**d.** Repeat parts **b** and **c**, but exclude the data for trials that used duck chow from the analysis. What do you conclude?

**e.** The botanists were also interested in the correlation between digestion efficiency $y$ and acid-detergent fibre $x$. Repeat parts **a–d** for these two variables.

⊘SNOWGEESE

| Feeding Trial | Diet | Weight Change (%) | Digestion Efficiency (%) | Acid-Detergent Fibre (%) |
|---|---|---|---|---|
| 1 | Plants | −6 | 0 | 28.5 |
| 2 | Plants | −5 | 2.5 | 27.5 |
| 3 | Plants | −4.5 | 5 | 27.5 |
| 4 | Plants | 0 | 0 | 32.5 |
| 5 | Plants | 2 | 0 | 32 |
| 6 | Plants | 3.5 | 1 | 30 |
| 7 | Plants | −2 | 2.5 | 34 |
| 8 | Plants | −2.5 | 10 | 36.5 |
| 9 | Plants | −3.5 | 20 | 28.5 |
| 10 | Plants | −2.5 | 12.5 | 29 |
| 11 | Plants | −3 | 28 | 28 |
| 12 | Plants | −8.5 | 30 | 28 |
| 13 | Plants | −3.5 | 18 | 30 |
| 14 | Plants | −3 | 15 | 31 |
| 15 | Plants | −2.5 | 17.5 | 30 |
| 16 | Plants | −.5 | 18 | 22 |
| 17 | Plants | 0 | 23 | 22.5 |
| 18 | Plants | 1 | 20 | 24 |
| 19 | Plants | 2 | 15 | 23 |
| 20 | Plants | 6 | 31 | 21 |
| 21 | Plants | 2 | 15 | 24 |
| 22 | Plants | 2 | 21 | 23 |
| 23 | Plants | 2.5 | 30 | 22.5 |
| 24 | Plants | 2.5 | 33 | 23 |
| 25 | Plants | 0 | 27.5 | 30.5 |
| 26 | Plants | .5 | 29 | 31 |
| 27 | Plants | −1 | 32.5 | 30 |
| 28 | Plants | −3 | 42 | 24 |
| 29 | Plants | −2.5 | 39 | 25 |
| 30 | Plants | −2 | 35.5 | 25 |
| 31 | Plants | .5 | 39 | 20 |
| 32 | Plants | 5.5 | 39 | 18.5 |
| 33 | Plants | 7.5 | 50 | 15 |
| 34 | Duck Chow | 0 | 62.5 | 8 |
| 35 | Duck Chow | 0 | 63 | 8 |
| 36 | Duck Chow | 2 | 69 | 7 |
| 37 | Duck Chow | 8 | 42.5 | 7.5 |
| 38 | Duck Chow | 9 | 59 | 8.5 |
| 39 | Duck Chow | 12 | 52.5 | 8 |
| 40 | Duck Chow | 8.5 | 75 | 6 |
| 41 | Duck Chow | 10.5 | 72.5 | 6.5 |
| 42 | Duck Chow | 14 | 69 | 7 |

*Source:* Gadallah, F. L., and Jefferies, R. L. "Forage quality in brood rearing areas of the lesser snow goose and the growth of captive goslings." *Journal of Applied Biology*, Vol. 32, No. 2, 1995, pp. 281–282 (adapted from Figures 2 and 3).

**11.61** A study published in *Psychosomatic Medicine* (Mar./Apr. 2001) explored the relationship between reported severity of pain and actual pain tolerance in 337 patients who suffer from chronic pain. Each patient reported his/her severity of chronic pain on a 7-point scale (1 = no pain, 7 = extreme pain). To obtain a pain tolerance level, a tourniquet was applied to the arm of each patient and twisted. The maximum pain level tolerated was measured on a quantitative scale.

**a.** According to the researchers, "correlational analysis revealed a small but significant inverse relationship between [actual] pain tolerance and the reported severity of chronic pain." Based on this statement, is the value of $r$ for the 337 patients positive or negative?

**b.** Suppose that the result reported in part **a** is significant at $\alpha = .05$. Find the approximate value of $r$ for the sample of 337 patients. [*Hint:* Use the formula $t = r\sqrt{(n-2)}/\sqrt{(1-r^2)}$.]

## 11.8  USING THE MODEL FOR ESTIMATION AND PREDICTION

If we are satisfied that a useful model has been found to describe the relationship between reaction time and amount of drug in the bloodstream, we are ready for step 5 in our regression modeling procedure: using the model for estimation and prediction.

> *The most common uses of a probabilistic model for making inferences can be divided into two categories. The first is the use of the model for estimating the mean value of $y$, $E(y)$, for a specific value of $x$.*

For our drug reaction example, we may want to estimate the mean response time for all people whose blood contains 4% of the drug.

> *The second use of the model entails predicting a new individual $y$ value for a given $x$.*

That is, we may want to predict the reaction time for a specific person who possesses 4% of the drug in the bloodstream.

In the first case, we are attempting to estimate the mean value of $y$ for a very large number of experiments at the given $x$ value. In the second case, we are trying to predict the outcome of a single experiment at the given $x$ value. Which of these model uses—estimating the mean value of $y$ or predicting an individual new value of $y$ (for the same value of $x$)—can be accomplished with the greater accuracy?

Before answering this question, we first consider the problem of choosing an estimator (or predictor) of the mean (or a new individual) $y$ value. We will use the least squares prediction equation

$$\hat{y} = \hat{\beta}_0 + \hat{\beta}_1 x$$

both to estimate the mean value of $y$ and to predict a specific new value of $y$ for a given value of $x$. For our example, we found

$$\hat{y} = -.1 + .7x$$

so that the estimated mean reaction time for all people when $x = 4$ (drug is 4% of blood content) is

$$\hat{y} = -.1 + .7(4) = 2.7 \text{ seconds}$$

The same value is used to predict a new $y$ value when $x = 4$. That is, both the estimated mean and the predicted value of $y$ are $\hat{y} = 2.7$ when $x = 4$, as shown in Figure 11.18.

The difference between these two model uses lies in the relative accuracy of the estimate and the prediction. These accuracies are best measured by using the sampling errors of the least squares line when it is used as an estimator and as a predictor, respectively. These errors are reflected in the standard deviations given in the next box.

### FIGURE 11.18

*Estimated Mean Value and Predicted Individual Value of Reaction Time $y$ for $x = 4$*

**Sampling Errors for the Estimator of the Mean of $y$ and the Predictor of an Individual New Value of $y$**

1. The standard deviation of the sampling distribution of the estimator $\hat{y}$ of the mean value of $y$ at a specific value of $x$, say $x_p$, is

$$\sigma_{\hat{y}} = \sigma \sqrt{\frac{1}{n} + \frac{(x_p - \bar{x})^2}{SS_{xx}}}$$

where $\sigma$ is the standard deviation of the random error $\varepsilon$. We refer to $\sigma_{\hat{y}}$ as the standard error of $\hat{y}$.

2. The standard deviation of the prediction error for the predictor $\hat{y}$ of an individual new $y$ value at a specific value of $x$ is

$$\sigma_{(y - \hat{y})} = \sigma \sqrt{1 + \frac{1}{n} + \frac{(x_p - \bar{x})^2}{SS_{xx}}}$$

where $\sigma$ is the standard deviation of the random error $\varepsilon$. We refer to $\sigma_{(y - \hat{y})}$ as the standard error of prediction.

The true value of $\sigma$ is rarely known, so we estimate $\sigma$ by $s$ and calculate the estimation and prediction intervals as shown in the next two boxes.

**A 100$(1 - \alpha)$% Confidence Interval for the Mean Value of $y$ at $x = x_p$**

$$\hat{y} + t_{\alpha/2}(\text{Estimated standard error of } \hat{y})$$

or

$$\hat{y} \pm t_{\alpha/2}s \sqrt{\frac{1}{n} + \frac{(x_p - \bar{x})^2}{SS_{xx}}}$$

where $t_{\alpha/2}$ is based on $(n - 2)$ degrees of freedom.

**A 100$(1 - \alpha)$% Prediction Interval\* for an Individual New Value of $y$ at $x = x_p$**

$$\hat{y} \pm t_{\alpha/2}(\text{Estimated standard error of prediction})$$

or

$$\hat{y} \pm t_{\alpha/2}s \sqrt{1 + \frac{1}{n} + \frac{(x_p - \bar{x})^2}{SS_{xx}}}$$

where $t_{\alpha/2}$ is based on $(n - 2)$ degrees of freedom.

---

\*The term *prediction interval* is used when the interval formed is intended to enclose the value of a random variable. The term *confidence interval* is reserved for estimation of population parameters (such as the mean).

**E X A M P L E   1 1 . 3**

Find a 95% confidence interval for the mean reaction time when the concentration of the drug in the bloodstream is 4%.

**Solution**

For a 4% concentration, $x = 4$ and the confidence interval for the mean value of $y$ is

$$\hat{y} \pm t_{\alpha/2}s \sqrt{\frac{1}{n} + \frac{(x_p - \bar{x})^2}{SS_{xx}}} = \hat{y} \pm t_{.025}s \sqrt{\frac{1}{5} + \frac{(4 - \bar{x})^2}{SS_{xx}}}$$

where $t_{.025}$ is based on $n - 2 = 5 - 2 = 3$ degrees of freedom. Recall that $\hat{y} = 2.7$, $s = .61$, $\bar{x} = 3$, and $SS_{xx} = 10$. From Table VI in Appendix A, $t_{.025} = 3.182$. Thus, we have

$$2.7 \pm (3.182)(.61)\sqrt{\frac{1}{5} + \frac{(4 - 3)^2}{10}} = 2.7 \pm (3.182)(.61)(.55)$$
$$= 2.7 \pm (3.182)(.34)$$
$$= 2.7 \pm 1.1$$

Therefore, when the percentage of drug in the bloodstream is 4%, the standard error of $\hat{y}$ is .34 and the corresponding 95% confidence interval for the mean reaction time for all possible subjects is 1.6 to 3.8 seconds. Note that we used a small amount of data (small sample size) for purposes of illustration in fitting the least squares line. The interval would probably be narrower if more information had been obtained from a larger sample.

**E X A M P L E   1 1 . 4**

Predict the reaction time for the next performance of the experiment for a subject with a drug concentration of 4%. Use a 95% prediction interval.

**Solution**

To predict the response time for an individual new subject for whom $x = 4$, we calculate the 95% prediction interval as

$$\hat{y} \pm t_{\alpha/2}s \sqrt{1 + \frac{1}{n} + \frac{(x_p - \bar{x})^2}{SS_{xx}}} = 2.7 \pm (3.182)(.61)\sqrt{1 + \frac{1}{5} + \frac{(4 - 3)^2}{10}}$$
$$= 2.7 \pm (3.182)(.61)(1.14)$$
$$= 2.7 \pm (3.182)(.70)$$
$$= 2.7 \pm 2.2$$

Therefore, the standard error of prediction when the drug concentration is 4% is .70, and we predict with 95% confidence that the reaction time for this new individual will fall in the interval from .5 to 4.9 seconds. Like the confidence interval for the mean value of $y$, the prediction interval for $y$ is quite large. This is because we have chosen a simple example (only five data points) to fit the least squares line. The width of the prediction interval could be reduced by using a larger number of data points.

Both the confidence interval for $E(y)$ and the prediction interval for $y$ can be obtained using a statistical software package. Figures 11.19 and 11.20 are SAS printouts showing confidence intervals and prediction intervals, respectively, for the data in the drug example.

**FIGURE 11.19**

*SAS Printout Showing 95%*
*Confidence Intervals for E(y)*

Dependent Variable: TIME_Y

Output Statistics

| Obs | DRUG_X | Dep Var TIME_Y | Predicted Value | Std Error Mean Predict | 95% CL Mean | | Residual |
|-----|--------|----------------|-----------------|------------------------|---------|--------|----------|
| 1 | 1 | 1.0000 | 0.6000 | 0.4690 | -0.8927 | 2.0927 | 0.4000 |
| 2 | 2 | 1.0000 | 1.3000 | 0.3317 | 0.2445 | 2.3555 | -0.3000 |
| 3 | 3 | 2.0000 | 2.0000 | 0.2708 | 1.1382 | 2.8618 | 0 |
| 4 | 4 | 2.0000 | 2.7000 | 0.3317 | 1.6445 | 3.7555 | -0.7000 |
| 5 | 5 | 4.0000 | 3.4000 | 0.4690 | 1.9073 | 4.8927 | 0.6000 |

**FIGURE 11.20**

*SAS Printout Showing 95%*
*Prediction Intervals for y*

Dependent Variable: TIME_Y

Output Statistics

| Obs | DRUG_X | Dep Var TIME_Y | Predicted Value | Std Error Mean Predict | 95% CL Predict | | Residual |
|-----|--------|----------------|-----------------|------------------------|---------|--------|----------|
| 1 | 1 | 1.0000 | 0.6000 | 0.4690 | -1.8376 | 3.0376 | 0.4000 |
| 2 | 2 | 1.0000 | 1.3000 | 0.3317 | -0.8972 | 3.4972 | -0.3000 |
| 3 | 3 | 2.0000 | 2.0000 | 0.2708 | -0.1110 | 4.1110 | 0 |
| 4 | 4 | 2.0000 | 2.7000 | 0.3317 | 0.5028 | 4.8972 | -0.7000 |
| 5 | 5 | 4.0000 | 3.4000 | 0.4690 | 0.9624 | 5.8376 | 0.6000 |

The 95% confidence interval for $E(y)$ when $x = 4$ is highlighted in Figure 11.19. The interval shown on the printout, $(1.6445, 3.7555)$, agrees (except for rounding) with the interval calculated in Example 11.3. The 95% prediction interval for $y$ when $x = 4$ is highlighted in Figure 11.20. Again, except for rounding, the SAS interval $(.5028, 4.8972)$ agrees with the one computed in Example 11.4.

A comparison of the confidence interval for the mean value of $y$ and the prediction interval for a new value of $y$ for 4% drug concentration $(x = 4)$ is shown in Figure 11.21. Note that the prediction interval for an individual new value of $y$ is *always* wider than the corresponding confidence interval for the mean value of $y$. You can see this by examining the formulas for the two intervals and by studying Figure 11.21.

**FIGURE 11.21**

*A 95% Confidence Interval and*
*Prediction Interval for Response*
*Time when x = 4*

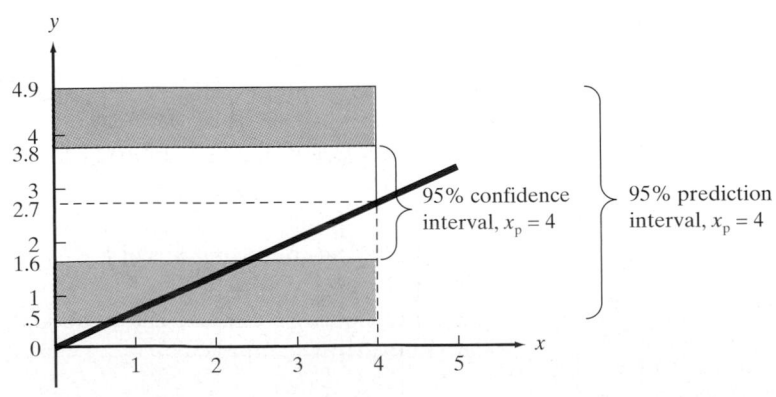

**FIGURE 11.22**

*Error of Estimating the Mean Value of y for a Given Value of x*

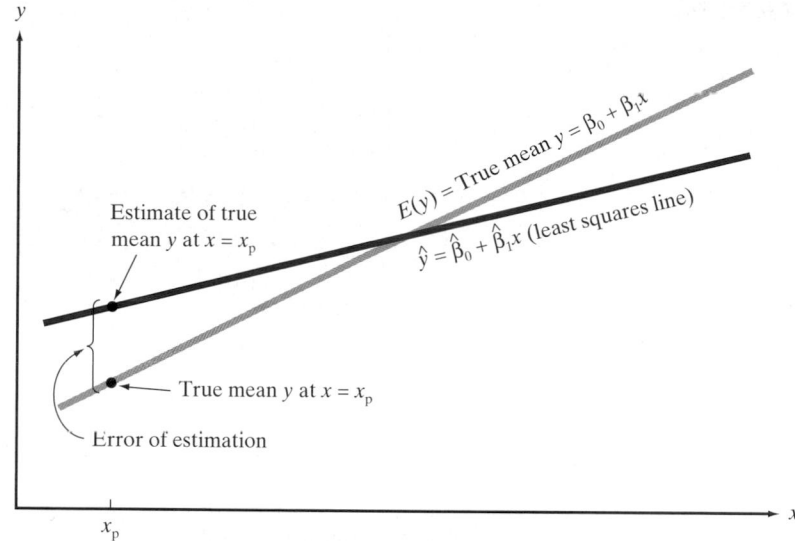

The error in estimating the mean value of $y$, $E(y)$, for a given value of $x$, say $x_p$, is the distance between the least squares line and the true line of means, $E(y) = \beta_0 + \beta_1 x$. This error, $[\hat{y} - E(y)]$, is shown in Figure 11.22. In contrast, *the error* $(y_p - \hat{y})$ *in predicting some future value of y is the sum of two errors*—the error of estimating the mean of $y$, $E(y)$, shown in Figure 11.22, plus the random error that is a component of the value of $y$ to be predicted (see Figure 11.23). Consequently, the error of predicting a particular value of $y$ will be larger than the error of estimating the mean value of $y$ for a particular value of $x$. Note from their formulas that both the error of estimation and the error of prediction take their smallest values when $x_p = \bar{x}$. The farther $x_p$ lies from $\bar{x}$, the larger will be the errors of estimation and prediction. You can see why this is true by noting the deviations for different values of $x_p$ between the line of means $E(y) = \beta_0 + \beta_1 x$ and the predicted line of means $\hat{y} = \hat{\beta}_0 + \hat{\beta}_1 x$ shown in Figure 11.23. The deviation is larger at the extremes of the interval where the largest and smallest values of $x$ in the data set occur.

**FIGURE 11.23**

*Error of Predicting a Future Value of y for a Given Value of x*

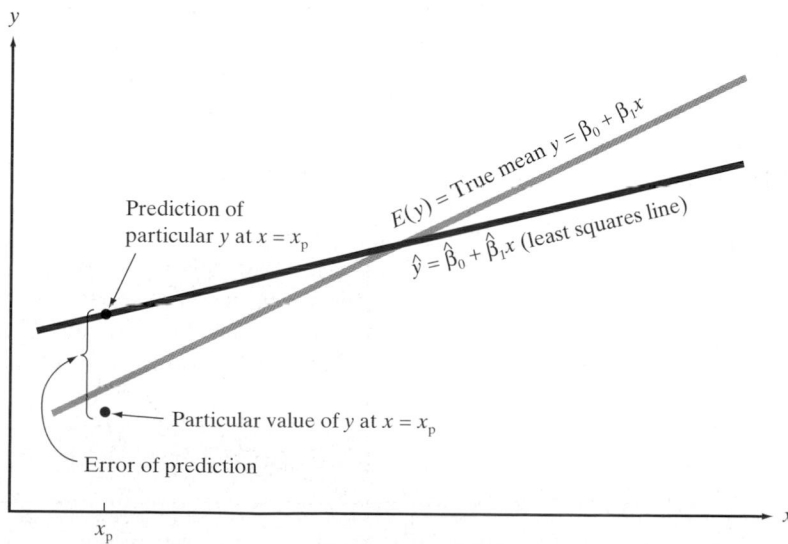

**FIGURE 11.24**

*Confidence Intervals for Mean Values and Prediction Intervals for New Values*

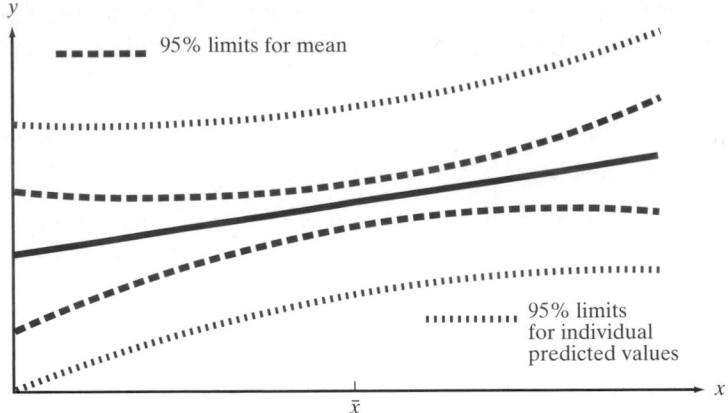

Both the confidence intervals for mean values and the prediction intervals for new values are depicted over the entire range of the regression line in Figure 11.24. You can see that the confidence interval is always narrower than the prediction interval, and that they are both narrowest at the mean $\bar{x}$, increasing steadily as the distance $|x - \bar{x}|$ increases. In fact, when $x$ is selected far enough away from $\bar{x}$ so that it falls outside the range of the sample data, it is dangerous to make any inferences about $E(y)$ or $y$.

---

### Caution

Using the least squares prediction equation to estimate the mean value of $y$ or to predict a particular value of $y$ for values of $x$ that fall outside the range of the values of $x$ contained in your sample data may lead to errors of estimation or prediction that are much larger than expected. Although the least squares model may provide a very good fit to the data over the range of $x$ values contained in the sample, it could give a poor representation of the true model for values of $x$ outside this region.

---

The confidence interval width grows smaller as $n$ is increased; thus, in theory, you can obtain as precise an estimate of the mean value of $y$ as desired (at any given $x$) by selecting a large enough sample. The prediction interval for a new value of $y$ also grows smaller as $n$ increases, but there is a lower limit on its width. If you examine the formula for the prediction interval, you will see that the interval can get no smaller than $\hat{y} \pm z_{\alpha/2}\sigma$.* Thus, the only way to obtain more accurate predictions for new values of $y$ is to reduce the standard deviation of the regression model, $\sigma$. This can be accomplished only by improving the model, either by using a curvilinear (rather than linear) relationship with $x$ or by adding new independent variables to the model, or both. Methods of improving the model are discussed in Chapter 12.

---

*The result follows from the facts that, for large $n$, $t_{\alpha/2} \approx z_{\alpha/2}$, $s \approx \sigma$, and the last two terms under the radical in the standard error of the predictor are approximately 0.

✓ EXERCISES 11.62–11.73

## Learning the Mechanics

**11.62** Consider the following pairs of measurements:

| x | 1 | 2 | 3 | 4 | 5 | 6 | 7 |
|---|---|---|---|---|---|---|---|
| y | 3 | 5 | 4 | 6 | 7 | 7 | 10 |

    **a.** Construct a scattergram for these data.
    **b.** Find the least squares line, and plot it on your scattergram.
    **c.** Find $s^2$.
    **d.** Find a 90% confidence interval for the mean value of $y$ when $x = 4$. Plot the upper and lower bounds of the confidence interval on your scattergram.
    **e.** Find a 90% prediction interval for a new value of $y$ when $x = 4$. Plot the upper and lower bounds of the prediction interval on your scattergram.
    **f.** Compare the widths of the intervals you constructed in parts **d** and **e**. Which is wider and why?

**11.63** Consider the following pairs of measurements:

| x | y | x | y |
|---|----|----|----|
| 1 | −1 | −2 | −6 |
| −1 | −5 | 3 | 4 |
| 2 | 1 | −1 | −4 |
| 0 | −3 | 5 | 4 |
| 4 | 7 | 1 | 0 |

    Given that $SS_{xx} = 47.6$, $SS_{yy} = 168.1$, $SS_{xy} = 85.6$, and $\hat{y} = -2.458 + 1.7983x$:
    **a.** Construct a scattergram for the data.
    **b.** Plot the least squares line on your scattergram.
    **c.** Use a 95% confidence interval to estimate the mean value of $y$ when $x = 5$. Plot the upper and lower bounds of the interval on your scattergram.
    **d.** Repeat part **c** for $x = 1.2$ and $x = -2$.
    **e.** Compare the widths of the three confidence intervals you constructed in parts **c** and **d** and explain why they differ.

**11.64** Refer to Exercise 11.63.
    **a.** Using no information about $x$, estimate and calculate a 95% confidence interval for the mean value of $y$. [*Hint:* Use the one-sample $t$ methodology of Section 7.3.]
    **b.** Plot the estimated mean value and the confidence interval as horizontal lines on your scattergram.
    **c.** Compare the confidence intervals you calculated in parts **c** and **d** of Exercise 11.63 with the one you calculated in part **a** of this exercise. Does $x$ appear to contribute information about the mean value of $y$?

    **d.** Check the answer you gave in part **c** with a statistical test of the null hypothesis $H_0: \beta_1 = 0$ against $H_a: \beta_1 \neq 0$. Use $\alpha = .05$.

**11.65** Consider the pairs of measurements shown in the following table.

| x | 4 | 6 | 0 | 5 | 2 | 3 | 2 | 6 | 2 | 1 |
|---|---|---|----|---|---|---|---|---|---|---|
| y | 3 | 5 | −1 | 4 | 3 | 2 | 0 | 4 | 1 | 1 |

    For these data, $SS_{xx} = 38.900$, $SS_{yy} = 33.600$, $SS_{xy} = 32.8$, and $\hat{y} = -4.14 + .843x$.
    **a.** Construct a scattergram for these data.
    **b.** Plot the least squares line on your scattergram.
    **c.** Use a 95% confidence interval to estimate the mean value of $y$ when $x_p = 6$. Plot the upper and lower bounds of the interval on your scattergram.
    **d.** Repeat part **c** for $x_p = 3.2$ and $x_p = 0$.
    **e.** Compare the widths of the three confidence intervals you constructed in parts **c** and **d** and explain why they differ.

**11.66** In fitting a least squares line to $n = 10$ data points, the following quantities were computed:

$$SS_{xx} = 32, \bar{x} = 3, SS_{yy} = 26, \bar{y} = 4, SS_{xy} = 28$$

    **a.** Find the least squares line.
    **b.** Graph the least squares line.
    **c.** Calculate SSE.
    **d.** Calculate $s^2$.
    **e.** Find a 95% confidence interval for the mean value of $y$ when $x_p = 2.5$.
    **f.** Find a 95% prediction interval for $y$ when $x_p = 4$.

## Applying the Concepts—Basic

**11.67** Refer to the data on sale prices and total appraised values of 92 residential properties in an upscale Tampa, Florida, neighborhood, Exercises 11.14 and 11.38 (pp. 523, 537).
    **a.** In Exercise 11.38, you determined that appraised property value $x$ and sale price $y$ are positively related for homes sold in the Tampa Palms subdivision. Does this result guarantee that appraised value will yield accurate predictions of sale price? Explain.
    **b.** MINITAB was used to predict the sale price of a residential property in the subdivision with a total appraised value of $300,000. Locate a 95% prediction interval for the sale price of this property on the printout (p. 558) and interpret the result.
    **c.** Locate a 95% confidence interval for $E(y)$ on the printout and interpret the result.

```
MINITAB Output for Exercise 11.67

Predicted Values for New Observations

New Obs      Fit      SE Fit        95.0% CI              95.0% PI
1         341.56       4.33    ( 332.95, 350.17) ( 275.86, 407.26)

Values of Predictors for New Observations

New Obs    TOTALVAL
1              300
```

**11.68** Refer to the simple linear regression of sweetness index *y* and amount of pectin *x* for *n* = 24 orange juice samples, Exercise 11.19 (p. 526). The SPSS printout of the analysis is shown on p. 559. A 90% confidence interval for the mean sweetness index, $E(y)$, for each value of *x* is shown on the SPSS spreadsheet. Select an observation and interpret this interval.

**11.69** Refer to the simple linear regression of number of flycatchers killed *y* and nest box tit occupancy *x* for *n* = 14 nest sites, Exercises 11.22 and 11.32 (pp. 527, 533).

   **a.** A 95% prediction interval for *y* when *x* = 64 is shown at the bottom of the SAS printout shown below. Interpret this interval.

   **b.** How would the width of a 95% confidence interval for $E(y)$ when *x* = 64 compare to the interval, part **a**?

   **c.** Would you recommend using the model to predict the number of flycatchers killed at a site with a nest box tit occupancy of 15%? Explain.

## Applying the Concepts—Intermediate

### NAMEGAME2

**11.70** Refer to the *Journal of Experimental Psychology-Applied* (June 2000) name retrieval study, Exercises 11.23, 11.45, and 11.58 (pp. 528, 539, 549).

   **a.** Find a 99% confidence interval for the mean recall proportion for students in the fifth position during the "name game." Interpret the result.

   **b.** Find a 99% prediction interval for the recall proportion of a particular student in the fifth position during the "name game." Interpret the result.

   **c.** Compare the two intervals, parts **a** and **b**. Which interval is wider? Will this always be the case? Explain.

**11.71** The Sasakawa Sports Foundation conducted a national survey to assess the physical activity patterns of Japanese adults. The table below lists the frequency (average number of days in the past year) and duration of time (average number of minutes per single activity) Japanese adults spent participating in a sample of 11 sports activities.

### JAPANSPORTS

| Activity | Frequency *x* (days/year) | Duration *y* (minutes) |
|---|---|---|
| Jogging | 135 | 43 |
| Cycling | 68 | 99 |
| Aerobics | 44 | 61 |
| Swimming | 39 | 60 |
| Volleyball | 30 | 80 |
| Tennis | 21 | 100 |
| Softball | 16 | 91 |
| Baseball | 19 | 127 |
| Skating | 7 | 115 |
| Skiing | 10 | 249 |
| Golf | 5 | 262 |

*Source:* J. Bennett, ed. *Statistics in Sport.* London: Arnold, 1998 (adapted from Figure 11.6).

```
SAS Output for Exercise 11.69
```

| | | | Output Statistics | | | |
|---|---|---|---|---|---|---|
| Obs | titpct | Dep Var nokilled | Predicted Value | Std Error Mean Predict | 95% CL Predict | |
| 1 | 24 | 0 | -0.4631 | 0.5554 | -3.1326 | 2.2064 |
| 2 | 33 | 0 | 0.5058 | 0.3719 | -2.0078 | 3.0194 |
| 3 | 34 | 0 | 0.6135 | 0.3559 | -1.8891 | 3.1161 |
| 4 | 43 | 0 | 1.5824 | 0.2944 | -0.8820 | 4.0468 |
| 5 | 50 | 0 | 2.3360 | 0.3695 | -0.1760 | 4.8479 |
| 6 | 35 | 1.0000 | 0.7211 | 0.3412 | -1.7718 | 3.2140 |
| 7 | 35 | 1.0000 | 0.7211 | 0.3412 | -1.7718 | 3.2140 |
| 8 | 38 | 1.0000 | 1.0441 | 0.3073 | -1.4278 | 3.5159 |
| 9 | 40 | 1.0000 | 1.2594 | 0.2949 | -1.2053 | 3.7241 |
| 10 | 31 | 2.0000 | 0.2905 | 0.4074 | -2.2492 | 2.8301 |
| 11 | 43 | 2.0000 | 1.5824 | 0.2944 | -0.8820 | 4.0468 |
| 12 | 55 | 3.0000 | 2.8742 | 0.4643 | 0.2887 | 5.4598 |
| 13 | 57 | 4.0000 | 3.0896 | 0.5073 | 0.4659 | 5.7132 |
| 14 | 64 | 5.0000 | 3.8431 | 0.6700 | 1.0516 | 6.6347 |

SPSS Output for Exercise 11.68

**Model Summary[b]**

| Model | R | R Square | Adjusted R Square | Std. Error of the Estimate |
|---|---|---|---|---|
| 1 | .478[a] | .229 | .194 | .215 |

a. Predictors: (Constant), PECTIN

b. Dependent Variable: SWEET

**ANOVA[b]**

| Model | | Sum of Squares | df | Mean Square | F | Sig. |
|---|---|---|---|---|---|---|
| 1 | Regression | .301 | 1 | .301 | 6.520 | .018[a] |
| | Residual | 1.017 | 22 | 4.622E-02 | | |
| | Total | 1.318 | 23 | | | |

a. Predictors: (Constant), PECTIN

b. Dependent Variable: SWEET

**Coefficients[a]**

| Model | | Unstandardized Coefficients | | Standardized Coefficients | t | Sig. |
|---|---|---|---|---|---|---|
| | | B | Std. Error | Beta | | |
| 1 | (Constant) | 6.252 | .237 | | 26.422 | .000 |
| | PECTIN | -2.31E-03 | .001 | -.478 | -2.554 | .018 |

a. Dependent Variable: SWEET

| | run | sweet | pectin | lower90m | upper90m |
|---|---|---|---|---|---|
| 1 | 1 | 5.2 | 220 | 5.64898 | 5.83848 |
| 2 | 2 | 5.5 | 227 | 5.63898 | 5.81613 |
| 3 | 3 | 6.0 | 259 | 5.57819 | 5.72904 |
| 4 | 4 | 5.9 | 210 | 5.66194 | 5.87173 |
| 5 | 5 | 5.8 | 224 | 5.64337 | 5.82560 |
| 6 | 6 | 6.0 | 215 | 5.65564 | 5.85493 |
| 7 | 7 | 5.8 | 231 | 5.63284 | 5.80379 |
| 8 | 8 | 5.6 | 268 | 5.55553 | 5.71011 |
| 9 | 9 | 5.6 | 239 | 5.61947 | 5.78019 |
| 10 | 10 | 5.9 | 212 | 5.65946 | 5.86497 |
| 11 | 11 | 5.4 | 410 | 5.05526 | 5.55416 |
| 12 | 12 | 5.6 | 256 | 5.58517 | 5.73592 |
| 13 | 13 | 5.8 | 306 | 5.43785 | 5.65219 |
| 14 | 14 | 5.5 | 259 | 5.57819 | 5.72904 |
| 15 | 15 | 5.3 | 284 | 5.50957 | 5.68213 |
| 16 | 16 | 5.3 | 383 | 5.15725 | 5.57694 |
| 17 | 17 | 5.7 | 271 | 5.54743 | 5.70434 |
| 18 | 18 | 5.5 | 264 | 5.56591 | 5.71821 |
| 19 | 19 | 5.7 | 227 | 5.63898 | 5.81613 |
| 20 | 20 | 5.3 | 263 | 5.56843 | 5.72031 |
| 21 | 21 | 5.9 | 232 | 5.63125 | 5.80075 |
| 22 | 22 | 5.8 | 220 | 5.64898 | 5.83848 |
| 23 | 23 | 5.8 | 246 | 5.60640 | 5.76091 |
| 24 | 24 | 5.9 | 241 | 5.61587 | 5.77454 |

a. Write the equation of a straight-line model relating duration ($y$) to frequency ($x$).

b. Find the least squares prediction equation.

c. Is there evidence of a linear relationship between $y$ and $x$? Test using $\alpha = .05$.

d. Use the least squares line to predict the duration of time Japanese adults participate in a sport that they play 25 times a year. Form a 95% confidence interval around the prediction and interpret the result.

### SNOWGEESE

**11.72** Refer to the *Journal of Applied Ecology* feeding study of the relationship between weight change $y$ of baby snow geese and digestion efficiency $x$, Exercise 11.60 (p. 549).

a. Fit the simple linear regression model to the data.

b. Do you recommend using the model to predict weight change $y$? Explain.

c. Use the model to form a 95% confidence interval for the mean weight change of all baby snow geese with a digestion efficiency of $x = 15\%$. Interpret the interval.

### Applying the Concepts—Advanced

**11.73** Refer to Exercise 11.33 (p. 533). The data are reproduced in the table.

a. Use a 90% confidence interval to estimate the mean useful life of a brand A cutting tool when the cutting speed is 45 meters per minute. Repeat for brand B. Compare the widths of the two intervals and comment on the reasons for any difference.

b. Use a 90% prediction interval to predict the useful life of a brand A cutting tool when the cutting speed is 45 meters per minute. Repeat for brand B. Compare the widths of the two intervals to each other and to the two intervals you calculated in part **a**. Comment on the reasons for any differences.

c. Note that the estimation and prediction you performed in parts **a** and **b** were for a value of $x$ that was not included in the original sample. That is, the value $x = 45$ was not part of the sample. However, the value is within the range of $x$ values in the sample, so that the regression model spans the $x$ value for which the estimation and prediction were made. In such situations, estimation and prediction represent **interpolations**.

Suppose you were asked to predict the useful life of a brand A cutting tool for a cutting speed of $x = 100$ meters per minute. Since the given value of $x$ is outside the range of the sample $x$ values, the prediction is an example of **extrapolation**. Predict the useful life of a brand A cutting tool that is operated at 100 meters per minute, and construct a 95% confidence interval for the actual useful life of the tool. What additional assumption do you have to make in order to ensure the validity of an extrapolation?

### CUTTOOL

| Cutting Speed (meters per minute) | Useful Life (Hours) | |
|---|---|---|
| | Brand A | Brand B |
| 30 | 4.5 | 6.0 |
| 30 | 3.5 | 6.5 |
| 30 | 5.2 | 5.0 |
| 40 | 5.2 | 6.0 |
| 40 | 4.0 | 4.5 |
| 40 | 2.5 | 5.0 |
| 50 | 4.4 | 4.5 |
| 50 | 2.8 | 4.0 |
| 50 | 1.0 | 3.7 |
| 60 | 4.0 | 3.8 |
| 60 | 2.0 | 3.0 |
| 60 | 1.1 | 2.4 |
| 70 | 1.1 | 1.5 |
| 70 | .5 | 2.0 |
| 70 | 3.0 | 1.0 |

## 11.9   A COMPLETE EXAMPLE

In the preceding sections we presented the basic elements necessary to fit and use a straight-line regression model. In this section we will assemble these elements by applying them in an example with the aid of computer software.

Suppose a fire insurance company wants to relate the amount of fire damage in major residential fires to the distance between the burning house and the nearest fire station. The study is to be conducted in a large suburb of a major city; a sample of 15 recent fires in this suburb is selected. The amount of damage, $y$, and the distance between the fire and the nearest fire station, $x$, are recorded for each fire. The results are given in Table 11.7.

⬤ FIREDAM
**TABLE 11.7** Fire Damage Data

| Distance from Fire Station $x$ (miles) | Fire Damage $y$ (thousands of dollars) |
|:---:|:---:|
| 3.4 | 26.2 |
| 1.8 | 17.8 |
| 4.6 | 31.3 |
| 2.3 | 23.1 |
| 3.1 | 27.5 |
| 5.5 | 36.0 |
| .7 | 14.1 |
| 3.0 | 22.3 |
| 2.6 | 19.6 |
| 4.3 | 31.3 |
| 2.1 | 24.0 |
| 1.1 | 17.3 |
| 6.1 | 43.2 |
| 4.8 | 36.4 |
| 3.8 | 26.1 |

**Step 1**  First, we hypothesize a model to relate fire damage, $y$, to the distance from the nearest fire station, $x$. We hypothesize a straight-line probabilistic model:

$$y = \beta_0 + \beta_1 x + \varepsilon$$

**Step 2**  Next, we enter the data of Table 11.7 into a computer and use statistical software to estimate the unknown parameters in the deterministic component of the hypothesized model. The STATISTIX printout for the simple linear regression analysis is shown in Figure 11.25. The least squares estimate of the slope $\beta_1$ and intercept $\beta_0$, highlighted on the printout, are

$$\hat{\beta}_1 = 4.91933$$

$$\hat{\beta}_0 = 10.2779$$

and the least squares equation is (rounded)

$$\hat{y} = 10.28 + 4.92x$$

This prediction equation is graphed in Figure 11.26 along with a plot of the data points.

The least squares estimate of the slope, $\hat{\beta}_1 = 4.92$, implies that the estimated mean damage increases by \$4,920 for each additional mile from the fire station. This interpretation is valid over the range of $x$, or from .7 to 6.1 miles from the station. The estimated $y$-intercept, $\hat{\beta}_0 = 10.28$, has the interpretation that a fire 0 miles from the fire station has an estimated mean damage of \$10,280. Although this would seem to apply to the fire station itself, remember that the $y$-intercept is meaningfully interpretable only if $x = 0$ is within the sampled range of the independent variable. Since $x = 0$ is outside the range, $\hat{\beta}_0$ has no practical interpretation.

**FIGURE 11.25**

*STATISTIX Printout for Fire Damage Regression*

```
UNWEIGHTED LEAST SQUARES LINEAR REGRESSION OF DAMAGE

PREDICTOR
VARIABLES     COEFFICIENT    STD ERROR     STUDENT'S T        P
---------     -----------    ---------    -------------    ------
CONSTANT        10.2779        1.42028         7.24        0.0000
DISTANCE        4.91933        0.39275        12.53        0.0000


R-SQUARED              0.9235    RESID. MEAN SQUARE (MSE)    5.36546
ADJUSTED R-SQUARED     0.9176    STANDARD DEVIATION          2.31635

SOURCE          DF        SS           MS          F         P
----------     ---    ----------   ----------    -----    ------
REGRESSION       1     841.766      841.766      156.89    0.0000
RESIDUAL        13     69.7510      5.36546
TOTAL           14     911.517


PREDICTED/FITTED VALUES OF DAMAGE

LOWER PREDICTED BOUND    22.324    LOWER FITTED BOUND    26.190
PREDICTED VALUE          27.496    FITTED VALUE          27.496
UPPER PREDICTED BOUND    32.667    UPPER FITTED BOUND    28.801
SE (PREDICTED VALUE)     2.3939    SE (FITTED VALUE)     0.6043


UNUSUALNESS (LEVERAGE)   0.0681
PERCENT COVERAGE          95.0
CORRESPONDING T           2.16

PREDICTOR VALUES: DISTANCE = 3.5000
```

**FIGURE 11.26**

*STATISTIX Plot of Least Squares Line for Fire Damage Regression*

Scatter Plot of DAMAGE vs DISTANCE

**Step 3**  Now we specify the probability distribution of the random error component $\varepsilon$. The assumptions about the distribution are identical to those listed in Section 11.3. Although we know that these assumptions are not completely satisfied (they rarely are for practical problems), we are willing to assume they are approximately satisfied for this example. The estimate of the standard deviation $\sigma$ of $\varepsilon$, highlighted on the printout, is

$$s = 2.31635$$

This implies that most of the observed fire damage ($y$) values will fall within approximately $2s = 4.64$ thousand dollars of their respective predicted values when using the least squares line.

**Step 4** We can now check the usefulness of the hypothesized model—that is, whether $x$ really contributes information for the prediction of $y$ using the straight-line model. First, test the null hypothesis that the slope $\beta_1$ is 0, that is, that there is no linear relationship between fire damage and the distance from the nearest fire station, against the alternative hypothesis that fire damage increases as the distance increases. We test

$$H_0: \beta_1 = 0$$
$$H_a: \beta_1 > 0$$

The two-tailed observed significance level for testing $H_a: \beta_1 \neq 0$, highlighted on the printout, is .0000. Thus, the $p$-value for our one-tailed test is half of this value. This small $p$-value leaves little doubt that mean fire damage and distance between the fire and station are at least linearly related, with mean fire damage increasing as the distance increases.

We gain additional information about the relationship by forming a confidence interval for the slope $\beta_1$. A 95% confidence interval is

$$\hat{\beta}_1 \pm t_{.025}s_{\hat{\beta}_1}$$

where $\hat{\beta}_1 = 4.919$ and its standard error, $s_{\hat{\beta}_1} = .393$, are both obtained from the printout. The value $t_{.025}$, based on $n - 2 = 13$ df, is 2.160. Therefore, the 95% confidence interval is

$$\hat{\beta}_1 \pm t_{.025}s_{\hat{\beta}_1} = 4.919 \pm (2.160)(.393) = 4.919 \pm .849 = (4.070, 5.768)$$

We estimate that the interval from $4,070 to $5,768 encloses the mean increase ($\beta_1$) in fire damage per additional mile distance from the fire station.

Another measure of the utility of the model is the coefficient of determination, $r^2$. The value (highlighted on the printout) is $r^2 = .9235$, which implies that about 92% of the sample variation in fire damage ($y$) is explained by the distance ($x$) between the fire and the fire station.

The coefficient of correlation, $r$, that measures the strength of the linear relationship between $y$ and $x$ is not shown on the STATISTIX printout and must be calculated. Using the facts that $r = \sqrt{r^2}$ in simple linear regression and that $r$ and $\hat{\beta}_1$ have the same sign, we find

$$r = +\sqrt{r^2} = \sqrt{.9235} = .96$$

The high correlation confirms our conclusion that $\beta_1$ is greater than 0; it appears that fire damage and distance from the fire station are positively correlated. All signs point to a strong linear relationship between $y$ and $x$.

**Step 5** We are now prepared to use the least squares model. Suppose the insurance company wants to predict the fire damage if a major residential fire were to occur 3.5 miles from the nearest fire station. The predicted value (highlighted at the bottom of the printout) is $\hat{y} = 27.496$, while the 95% prediction interval (also highlighted) is $(22.324, 32.667)$. Therefore, with 95% confidence we predict fire damage in a major residential fire 3.5 miles from the nearest station to be between $22,324 and $32,667.

*One caution before closing*: We would not use this prediction model to make predictions for homes less than .7 mile or more than 6.1 miles from the nearest fire station. A look at the data in Table 11.7 reveals that all the *x* values fall between .7 and 6.1. It is dangerous to use the model to make predictions outside the region in which the sample data fall. A straight line might not provide a good model for the relationship between the mean value of *y* and the value of *x* when stretched over a wider range of *x* values.

## STATISTICS IN ACTION
### Can "Dowsers" Really Detect Water?

The act of searching for and finding underground supplies of water using nothing more than a divining rod is commonly known as "dowsing." Although widely regarded among scientists as no more than a superstitious relic from medieval times, dowsing remains popular in folklore and, to this day, there are individuals who claim to have this mysterious skill.

Many dowsers in Germany claim that they respond to "earthrays" that emanate from the water source. These earthrays, say the dowsers, are a subtle form of radiation potentially hazardous to human health. As a result of these claims, the German government in the mid-1980s conducted a 2-year experiment to investigate the possibility that dowsing is a genuine skill. If such a skill could be demonstrated, reasoned government officials, then dangerous levels of radiation in Germany could be detected, avoided, and disposed of.

A group of university physicists in Munich, Germany, were provided a grant of 400,000 marks ($\approx$\$250,000) to conduct the study. Approximately 500 candidate dowsers were recruited to participate in preliminary tests of their skill. To avoid fraudulent claims, the 43 individuals who seemed to be the most successful in the preliminary tests were selected for the final, carefully controlled, experiment.

The researchers set up a 10-meter long line on the ground floor of a vacant barn, along which a small wagon could be moved. Attached to the wagon was a short length of pipe, perpendicular to the test line, that was connected by hoses to a pump with running water. The location of the pipe along the line for each trial of the experiment was assigned using a computer-generated random number. On the upper floor of the barn, directly above the experimental line, a 10-meter test line was painted. In each trial, a dowser was admitted to this upper level and required, with his or her rod, stick, or other tool of choice, to ascertain where the pipe with running water on the ground floor was located.

Each dowser participated in at least one test series, that is, a sequence of from 5 to 15 trials (typically 10), with the pipe randomly repositioned after each trial. (Some dowsers undertook only one test series, selected others underwent more than 10 test series.) Over the 2-year experimental period, the 43 dowsers participated in a total of 843 tests. The experiment was "double-blind" in that neither the observer (researcher) on the top floor nor the dowser knew the pipe's location, even after a guess was made. [*Note:* Before the experiment began, a professional magician inspected the entire arrangement for potential deception or cheating by the dowsers.]

For each trial, two variables were recorded: the actual pipe location (in decimeters from the beginning of the line) and the dowser's guess (also measured in decimeters). Based on an examination of these data, the German physicists concluded in their final report that although most dowsers did not do particularly well in the experiments, "some few dowsers, in particular tests, showed an extraordinarily high rate of success, which can scarcely if at all be explained as due to chance . . . a real core of dowser-phenomena can be regarded as empirically proven . . ." (Wagner, Betz, and König, 1990).

This conclusion was critically assessed by J. T. Enright, a professor of behavioral physiology at the University of California–San Diego (*Skeptical Inquirer*, Jan./Feb. 1999). Using scatterplots and the notion of correlation, Enright concluded exactly the opposite of the German physicists. According to Enright, "the Munich experiments constitute as decisive and complete a failure as can be imagined of dowsers to do what they claim they can."

### Focus

a. Using scatterplots, Enright provided several hypothetical examples of outcomes that might be expected from the dowsing experiments, assuming various arbitrary categories of dowser skill. Let *x* = dowser's guess and *y* = pipe location. Construct a scatterplot of hypothetical data that would reflect perfect prediction by the dowsers.

b. Repeat part **a** for dowsers that have good (but not perfect) skill.

c. Repeat part **a** for dowsers that have no skill (i.e., that are making random guesses).

d. Enright presented a scatterplot of the data for all 843 tests performed in the Munich barn. A reproduction of this plot is displayed in Figure 11.27. Based on this graph, what inference would you make about the overall ability of the dowsers? Explain. [*Note:* In this plot, dowser's guess is shown on the vertical axis to make the graph comparable to the scatterplots you drew in parts **a–c**.]

**STATISTICS IN ACTION** *(continued)*

e. Recall that the certain German physicists found that "some few dowsers ... showed an extraordinarily high rate of success." Consequently, they might argue that the scatterplot in Figure 11.27 obscures these outstanding performances since it lumps all 43 dowsers (the majority of which are unskilled) together. In the researchers' final report, they identified three dowsers (those numbered 99, 18, and 108) as having particularly impressive results. All three of these "best" dowsers performed the experiment multiple times. The best test series (sequence of trials) for each of these three dowsers was identified; these data are listed in Table 11.8 and stored in the DOWSING file. Conduct a complete simple linear regression analysis of the data in order to make an inference about the overall performance of each of the three best dowsers.

f. The data in Table 11.8 represent the outcome of the dowsing experiment in its most favorable light. Can these results be reproduced by the best dowsers in comparable tests? Remember, each of these three dowsers did participate in other test series. Enright plotted the data for these other series in which the three best dowsers performed; this scatterplot is reproduced in Figure 11.28. Comment on the performance of the three best dowsers during these "rest of the best" trials.

g. According to Enright, "there is another way of evaluating the results from those dowsers who produced the best test series .... Suppose that they had always left their dowsing equipment at home in the closet, and had simply, in each and every test, just guessed that the pipe was located exactly at the middle of the test line." Replace the dowsers' guesses in Table 11.8 with the midpoint of the test line, 50 decimeters, and repeat the analysis of part **e**. What do you conclude from the analysis?

h. Give a critical assessment of the Munich dowsing experiments. With whom do you tend to agree, the German physicists or J. T. Enright?

 DOWSING

**TABLE 11.8   Dowsing Trial Results: Best Series for the Three Best Dowsers**

| Trial | Dowser Number | Pipe Location | Dowser's Guess |
|-------|---------------|---------------|----------------|
| 1  | 99  | 4  | 4   |
| 2  | 99  | 5  | 87  |
| 3  | 99  | 30 | 95  |
| 4  | 99  | 35 | 74  |
| 5  | 99  | 36 | 78  |
| 6  | 99  | 58 | 65  |
| 7  | 99  | 40 | 39  |
| 8  | 99  | 70 | 75  |
| 9  | 99  | 74 | 32  |
| 10 | 99  | 98 | 100 |
| 11 | 18  | 7  | 10  |
| 12 | 18  | 38 | 40  |
| 13 | 18  | 40 | 30  |
| 14 | 18  | 49 | 47  |
| 15 | 18  | 75 | 9   |
| 16 | 18  | 82 | 95  |
| 17 | 108 | 5  | 52  |
| 18 | 108 | 18 | 16  |
| 19 | 108 | 33 | 37  |
| 20 | 108 | 45 | 40  |
| 21 | 108 | 38 | 66  |
| 22 | 108 | 50 | 58  |
| 23 | 108 | 52 | 74  |
| 24 | 108 | 63 | 65  |
| 25 | 108 | 72 | 60  |
| 26 | 108 | 95 | 49  |

*Source:* Enright, J. T. "Testing dowsing: The failure of the Munich experiments." *Skeptical Inquirer*, Jan./Feb. 1999, p. 45 (Figure 6a).

*(continues on next page)*

**STATISTICS IN ACTION (continued)**

**FIGURE 11.27**

*Results from All 843 Dowsing Trials*

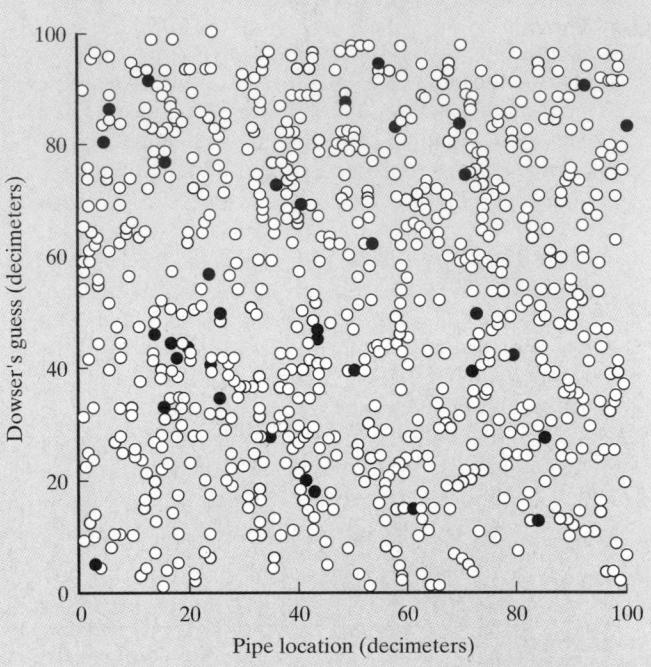

**FIGURE 11.28**

*Results for Other Test Series that Three Best Dowsers Participated In*

## QUICK REVIEW

### Key Terms

Bivariate relationship   541
Coefficient of correlation   541
Coefficient of determination   545
Confidence interval for mean
  of $y$   552
Dependent variable   512
Deterministic model   511
Deterministic relationship   511
Errors of prediction   515
Estimated standard error of the
  regression model   531
Independent variable   512

Least squares estimates   516
Least squares line (or prediction
  equation)   515
Line of means   513
Method of least squares   515
Pearson product moment
  coefficient of correlation   541
Population correlation
  coefficient   543
Prediction interval for $y$   552
Predictor variable   512
Probabilistic model   511

Probabilistic relationship   511
Random error   511
Regression analysis   512
Regression line   515
Response variable   512
Scattergram   514
Slope   512
Straight-line (first-order)
  model   512
$y$-intercept   512

### Key Formulas

$\hat{\beta}_1 = \dfrac{SS_{xy}}{SS_{xx}}, \quad \hat{\beta}_0 = \bar{y} - \hat{\beta}_1\bar{x}$

where $SS_{xy} = \sum xy - \dfrac{(\sum x)(\sum y)}{n}$

$SS_{xx} = \sum x^2 - \dfrac{(\sum x)^2}{n}$

Least squares estimates of $\beta$'s   516

$\hat{y} = \hat{\beta}_0 + \hat{\beta}_1 x$

Least squares line   515

$SSE = \sum (y_i - \hat{y}_i)^2 = SS_{yy} - \hat{\beta}_1 SS_{xy}$

where $SS_{yy} = \sum y^2 - \dfrac{(\sum y)^2}{n}$

Sum of squared errors   531

$s^2 = \dfrac{SSE}{n-2}$

Estimated variance of $\sigma^2$ of $\varepsilon$   531

$s_{\hat{\beta}_1} = \dfrac{s}{\sqrt{SS_{xx}}}$

Estimated standard error of $\hat{\beta}_1$   534

$t = \dfrac{\hat{\beta}_1}{s_{\hat{\beta}_1}}$

Test statistic for $H_0: \beta_1 = 0$   535

$\hat{\beta}_1 \pm (t_{\alpha/2})s_{\hat{\beta}_1}$

$(1 - \alpha)100\%$ confidence interval for $\beta_1$   536

$r = \dfrac{SS_{xy}}{\sqrt{SS_{xx}SS_{yy}}} = \pm\sqrt{r^2} \text{ (same sign as } \hat{\beta}_1)$

Coefficient of correlation   541

$r^2 = \dfrac{SS_{yy} - SSE}{SS_{yy}}$

Coefficient of determination   545

$\hat{y} \pm (t_{\alpha/2})s\sqrt{\dfrac{1}{n} + \dfrac{(x_p - \bar{x})^2}{SS_{xx}}}$

$(1 - \alpha)100\%$ confidence interval for $E(y)$ when $x = x_p$   552

$\hat{y} \pm (t_{\alpha/2})s\sqrt{1 + \dfrac{1}{n} + \dfrac{(x_p - \bar{x})^2}{SS_{xx}}}$

$(1 - \alpha)100\%$ prediction interval for $y$ when $x = x_p$   552

## Language Lab

| Symbol | Pronunciation | Description |
|---|---|---|
| $y$ | | Dependent variable (variable to be predicted or modeled) |
| $x$ | | Independent (predictor) variable |
| $E(y)$ | | Expected (mean) value of $y$ |
| $\beta_0$ | beta-zero | $y$-intercept of true line |
| $\beta_1$ | beta-one | Slope of true line |
| $\hat{\beta}_0$ | beta-zero hat | Least squares estimate of $y$-intercept |
| $\hat{\beta}_1$ | beta-one hat | Least squares estimate of slope |
| $\varepsilon$ | epsilon | Random error |
| $\hat{y}$ | y-hat | Predicted value of $y$ |
| $(y - \hat{y})$ | | Error of prediction |
| SE | | Sum of errors (will equal zero with least squares line) |
| SSE | | Sum of squared errors (will be smallest for least squares line) |
| $SS_{xx}$ | | Sum of squares of $x$-values |
| $SS_{yy}$ | | Sum of squares of $y$-values |
| $SS_{xy}$ | | Sum of squares of cross-products, $x \cdot y$ |
| $r$ | | Coefficient of correlation |
| $r^2$ | R-squared | Coefficient of determination |
| $x_p$ | | Value of $x$ used to predict $y$ |

 SUPPLEMENTARY EXERCISES 11.74–11.88

### Learning the Mechanics

**11.74** In fitting a least squares line to $n = 15$ data points, the following quantities were computed: $SS_{xx} = 55$, $SS_{yy} = 198$, $SS_{xy} = -88$, $\bar{x} = 1.3$, and $\bar{y} = 35$.
  **a.** Find the least squares line.
  **b.** Graph the least squares line.
  **c.** Calculate SSE.
  **d.** Calculate $s^2$.
  **e.** Find a 90% confidence interval for $\beta_1$. Interpret this estimate.
  **f.** Find a 90% confidence interval for the mean value of $y$ when $x = 15$.
  **g.** Find a 90% prediction interval for $y$ when $x = 15$.

**11.75** Consider the following sample data:

| $y$ | 5 | 1 | 3 |
|---|---|---|---|
| $x$ | 5 | 1 | 3 |

  **a.** Construct a scattergram for the data.
  **b.** It is possible to find many lines for which $\Sigma(y - \hat{y}) = 0$. For this reason, the criterion $\Sigma(y - \hat{y}) = 0$ is not used for identifying the "best-fitting" straight line. Find two lines that have $\Sigma(y - \hat{y}) = 0$.
  **c.** Find the least squares line.

  **d.** Compare the value of SSE for the least squares line to that of the two lines you found in part **b.** What principle of least squares is demonstrated by this comparison?

**11.76** Consider the following 10 data points:

LM11_76

| $x$ | 3 | 5 | 6 | 4 | 3 | 7 | 6 | 5 | 4 | 7 |
|---|---|---|---|---|---|---|---|---|---|---|
| $y$ | 4 | 3 | 2 | 1 | 2 | 3 | 3 | 5 | 4 | 2 |

  **a.** Plot the data on a scattergram.
  **b.** Calculate the values of $r$ and $r^2$.
  **c.** Is there sufficient evidence to indicate that $x$ and $y$ are linearly correlated? Test at the $\alpha = .10$ level of significance.

### Applying the Concepts—Basic

**11.77** A study was conducted to determine whether a student's final grade in an introductory sociology course is linearly related to his or her performance on the verbal ability test administered before college entrance. The verbal test scores and final grades for a random sample of 10 students are shown in the table on p. 569. A STATISTIX printout of the simple linear regression is also provided.

STATISTIX Output for Exercise 11.77

```
UNWEIGHTED LEAST SQUARES LINEAR REGRESSION OF FINGRADE

PREDICTOR
VARIABLES     COEFFICIENT    STD ERROR     STUDENT'S T       P
---------     -----------    ---------     -----------     ------
CONSTANT        40.7842       8.50686          4.79         0.0014
VERBAL           0.76556      0.17498          4.38         0.0024


R-SQUARED              0.7052     RESID. MEAN SQUARE (MSE)    75.7532
ADJUSTED R-SQUARED     0.6684     STANDARD DEVIATION           8.70363


SOURCE         DF       SS           MS          F       P
----------     ---    ---------   ---------   -----   ------
REGRESSION      1     1449.97      1449.97     19.14   0.0024
RESIDUAL        8      606.026      75.7532
TOTAL           9     2056.00


PREDICTED/FITTED VALUES OF FINGRADE

LOWER PREDICTED BOUND    57.950      LOWER FITTED BOUND    72.513
PREDICTED VALUE          79.062      FITTED VALUE          79.062
UPPER PREDICTED BOUND   100.17       UPPER FITTED BOUND    85.611
SE (PREDICTED VALUE)      9.1552     SE (FITTED VALUE)      2.8399


UNUSUALNESS (LEVERAGE)    0.1065
PERCENT COVERAGE         95.0
CORRESPONDING T           2.31

PREDICTOR VALUES: VERBAL = 50.000
```

🌐 SOCIOL

| Student | Verbal Ability Test Score $x$ | Final Sociology Grade $y$ |
|---------|------------------------------|----------------------------|
| 1  | 39 | 65 |
| 2  | 43 | 78 |
| 3  | 21 | 52 |
| 4  | 64 | 82 |
| 5  | 57 | 92 |
| 6  | 47 | 89 |
| 7  | 28 | 73 |
| 8  | 75 | 98 |
| 9  | 34 | 56 |
| 10 | 52 | 75 |

a. Find the least squares line relating $y$ to $x$.
b. Plot the data points and graph the least squares line.
c. Do the data provide sufficient evidence to indicate that a positive correlation exists between verbal score and final grade? Use $\alpha = .01$.
d. Find a 95% confidence interval for the slope $\beta_1$.
e. Predict a student's final grade in the introductory course when his or her verbal test score is 50. Use a 95% prediction interval.

f. Find a 95% confidence interval for the mean final grade for all students scoring 50 on the college entrance verbal exam.

**11.78** A breeder of thoroughbred horses wishes to model the relationship between the gestation period and the life span of a horse. The breeder believes that the two variables may follow a linear trend. The information in the table was supplied to the breeder from various thoroughbred stables across the state. (Note that the horse has the greatest variation of gestation period of any species owing to seasonal and nutritional factors.) A MINITAB printout of the simple linear regression analysis is shown on p. 570.

🌐 HORSES

| Horse | Gestation Period $x$ (days) | Life Span $y$ (years) |
|-------|------------------------------|------------------------|
| 1 | 416 | 24 |
| 2 | 279 | 25.5 |
| 3 | 298 | 20 |
| 4 | 307 | 21.5 |
| 5 | 356 | 22 |
| 6 | 403 | 23.5 |
| 7 | 265 | 21 |

MINITAB Output for Exercise 11.78

```
The regression equation is
LIFESPAN = 18.9 + 0.0109 GESTATE

Predictor        Coef     SE Coef       T       P
Constant       18.890       4.499    4.20   0.009
GESTATE       0.01087     0.01337    0.81   0.453

S = 1.971       R-Sq = 11.7%    R-Sq(adj) = 0.0%

Analysis of Variance

Source          DF          SS         MS       F       P
Regression       1       2.571      2.571    0.66   0.453
Residual Error   5      19.429      3.886
Total            6      22.000
```

**a.** Do the data provide sufficient evidence to support the breeder's hypothesis? Test using $\alpha = .05$.

**b.** Find a 90% confidence interval for $\beta_1$. Interpret this interval. Does the interval support part **a**?

**11.79** Studies of Asian (particularly Japanese) and U.S. managers in the 1970s and 1980s found sharp differences of opinion and attitude toward quality management. Do these differences continue to exist? To find out, two California State University researchers (B. F. Yavas and T. M. Burrows) surveyed 100 U.S. and 96 Asian managers in the electronics manufacturing industry (*Quality Management Journal*, Fall 1994). The table gives the percentage of U.S. and Asian managers who agree with each of the 13 randomly selected statements regarding quality. (For example, one statement is "Quality is a problem in my company." Another is "Improving quality is expensive.")

◉ QLAGREE

| Statement | Percent of Managers who Agree | |
|---|---|---|
| | **United States** | **Japan** |
| 1 | 36 | 38 |
| 2 | 31 | 42 |
| 3 | 28 | 43 |
| 4 | 27 | 48 |
| 5 | 78 | 58 |
| 6 | 74 | 49 |
| 7 | 43 | 46 |
| 8 | 50 | 56 |
| 9 | 31 | 65 |
| 10 | 66 | 58 |
| 11 | 18 | 21 |
| 12 | 61 | 69 |
| 13 | 53 | 45 |

*Source:* Yavas, B. F., and Burrows, T. M. "A comparative study of attitudes of U.S. and Asian managers toward product quality." *Quality Management Journal*, Fall 1994, p. 49 (Table 5).

**a.** Find the coefficient of correlation $r$ for these data.

**b.** Interpret $r$ in the context of the problem. Do U.S. and Asian managers tend to have similar attitudes toward quality and quality management?

**11.80** Refer to the *American Scientist* (July–Aug. 1998) study of the relationship between self-avoiding and unrooted walks, Exercise 2.113 (p. 90). Recall that in a self-avoiding walk you never retrace or cross your own path; while an unrooted walk is a path in which the starting and ending points are impossible to distinguish. The possible number of walks of each type of various lengths are reproduced in the table. Consider the straight-line model $y = \beta_0 + \beta_1 x + \varepsilon$, where $x$ is walk length (number of steps).

◉ WALKS

| Walk Length (number of steps) | Unrooted Walks | Self-Avoiding Walks |
|---|---|---|
| 1 | 1 | 4 |
| 2 | 2 | 12 |
| 3 | 4 | 36 |
| 4 | 9 | 100 |
| 5 | 22 | 284 |
| 6 | 56 | 780 |
| 7 | 147 | 2,172 |
| 8 | 388 | 5,916 |

*Source:* Hayes, B. "How to avoid yourself." *American Scientist*, Vol. 86, No. 4, July–Aug. 1988, p. 317 (Figure 5).

**a.** Use the method of least squares to fit the model to the data if $y$ is the possible number of unrooted walks.

**b.** Interpret $\hat{\beta}_0$ and $\hat{\beta}_1$ in the estimated model, part **a**.

**c.** Repeat parts **a** and **b** if $y$ is the possible number of self-avoiding walks.

**d.** Find a 99% confidence interval for the number of unrooted walks possible when walk length is four steps.

**e.** Would you recommend using simple linear regression to predict the number of walks possible when walk length is 15 steps? Explain.

## Applying the Concepts—Intermediate

**11.81** Refer to Exercise 2.127 (p. 98). The data on the value of 50 Beanie Babies collector's items, published in *Beanie World Magazine*, are reproduced here. Can age of a Beanie Baby be used to accurately predict its market value? Answer this question by conducting a complete simple linear regression analysis on the data.

⊘ BEANIE

| | Name | Age (months) as of Sept. 1998 | Retired (R)/ Current (C) | Value ($) |
|---|---|---|---|---|
| 1. | Ally the Alligator | 52 | R | 55.00 |
| 2. | Batty the Bat | 12 | C | 12.00 |
| 3. | Bongo the Brown Monkey | 28 | R | 40.00 |
| 4. | Blackie the Bear | 52 | C | 10.00 |
| 5. | Bucky the Beaver | 40 | R | 45.00 |
| 6. | Bumble the Bee | 28 | R | 600.00 |
| 7. | Crunch the Shark | 21 | C | 10.00 |
| 8. | Congo the Gorilla | 28 | C | 10.00 |
| 9. | Derby the Coarse Mane Horse | 28 | R | 30.00 |
| 10. | Digger the Red Crab | 40 | R | 150.00 |
| 11. | Echo the Dolphin | 17 | R | 20.00 |
| 12. | Fetch the Golden Retriever | 5 | C | 15.00 |
| 13. | Early the Robin | 5 | C | 20.00 |
| 14. | Flip the White Cat | 28 | R | 40.00 |
| 15. | Garcia the Teddy | 28 | R | 200.00 |
| 16. | Happy the Hippo | 52 | R | 20.00 |
| 17. | Grunt the Razorback | 28 | R | 175.00 |
| 18. | Gigi the Poodle | 5 | C | 15.00 |
| 19. | Goldie the Goldfish | 52 | R | 45.00 |
| 20. | Iggy the Iguana | 10 | C | 10.00 |
| 21. | Inch the Inchworm | 28 | R | 20.00 |
| 22. | Jake the Mallard Duck | 5 | C | 20.00 |
| 23. | Kiwi the Toucan | 40 | R | 165.00 |
| 24. | Kuku the Cockatoo | 5 | C | 20.00 |
| 25. | Mystic the Unicorn | 11 | R | 45.00 |
| 26. | Mel the Koala Bear | 21 | C | 10.00 |
| 27. | Nanook the Husky | 17 | C | 15.00 |
| 28. | Nuts the Squirrel | 21 | C | 10.00 |
| 29. | Peace the Tie Dyed Teddy | 17 | C | 25.00 |
| 30. | Patty the Platypus | 64 | R | 800.00 |
| 31. | Quacker the Duck | 40 | R | 15.00 |
| 32. | Puffer the Penguin | 10 | C | 15.00 |
| 33. | Princess the Bear | 12 | C | 65.00 |
| 34. | Scottie the Scottie | 28 | R | 28.00 |
| 35. | Rover the Dog | 28 | R | 15.00 |
| 36. | Rex the Tyrannosaurus | 40 | R | 825.00 |
| 37. | Sly the Fox | 28 | C | 10.00 |
| 38. | Slither the Snake | 52 | R | 1,900.00 |
| 39. | Skip the Siamese Cat | 21 | C | 10.00 |
| 40. | Splash the Orca Whale | 52 | R | 150.00 |
| 41. | Spooky the Ghost | 28 | R | 40.00 |
| 42. | Snowball the Snowman | 12 | R | 40.00 |
| 43. | Stinger the Scorpion | 5 | C | 15.00 |
| 44. | Spot the Dog | 52 | R | 65.00 |
| 45. | Tank the Armadillo | 28 | R | 85.00 |
| 46. | Stripes the Tiger (Gold/Black) | 40 | R | 400.00 |
| 47. | Teddy the 1997 Holiday Bear | 12 | R | 50.00 |
| 48. | Tuffy the Terrier | 17 | C | 10.00 |
| 49. | Tracker the Basset Hound | 5 | C | 15.00 |
| 50. | Zip the Black Cat | 28 | R | 40.00 |

*Source: Beanie World Magazine, Sept. 1998.*

**11.82** Common maize rust is a serious disease of sweet corn. Researchers in New York state have developed an action threshold for initiation of fungicide applications based on a regression equation relating maize rust incidence to severity of the disease (*Phytopathology*, Vol. 80, 1990). In one particular field, data were collected on more than 100 plants of the sweet corn hybrid Jubilee. For each plant, incidence was measured as the percentage of leaves infected ($x$) and severity was calculated as the log (base 10) of the average number of infections per leaf ($y$). A simple linear regression analysis of the data produced the following results:

$$\hat{y} = -.939 + .020x$$
$$r^2 = .816$$
$$s = .288$$

  **a.** Interpret the value of $\hat{\beta}_1$.
  **b.** Interpret the value of $r^2$.
  **c.** Interpret the value of $s$.
  **d.** Calculate the value of $r$ and interpret it.
  **e.** Use the result, part **d**, to test the utility of the model. Use $\alpha = .05$. (Assume $n = 100$.)
  **f.** Predict the severity of the disease when the incidence of maize rust for a plant is 80%. [*Note:* Take the antilog (base 10) of $\hat{y}$ to obtain the predicted average number of infections per leaf.

**11.83** At temperatures approaching absolute zero ($-273°$C), helium exhibits traits that seem to defy many laws of Newtonian physics. An experiment has been conducted with helium in solid form at various temperatures near absolute zero. The solid helium is placed in a dilution refrigerator along with a solid impure substance, and the fraction (in weight) of the impurity passing through the solid helium is recorded. (This phenomenon of solids passing directly through solids is known as *quantum tunneling*.) The data are given in the table.

🔘HELIUM

| Temperature $x$ (°C) | Proportion of Impurity |
|---|---|
| −262.0 | .315 |
| −265.0 | .202 |
| −256.0 | .204 |
| −267.0 | .620 |
| −270.0 | .715 |
| −272.0 | .935 |
| −272.4 | .957 |
| −272.7 | .906 |
| −272.8 | .985 |
| −272.9 | .987 |

  **a.** Find the least squares estimates of the intercept and slope. Interpret them.

  **b.** Use a 95% confidence interval to estimate the slope $\beta_1$. Interpret the interval in terms of this application. Does the interval support the hypothesis that temperature contributes information about the proportion of impurity passing through helium?
  **c.** Interpret the coefficient of determination for this model.
  **d.** Find a 95% prediction interval for the percentage of impurity passing through solid helium at $-273°$C. Interpret the result.
  **e.** Note that the value of $x$ in part **d** is outside the experimental region. Why might this lead to an unreliable prediction?

**11.84** Neurologists have found that the hippocampus, a structure found in the brain, plays an important role in short-term memory. Refer to the *American Journal of Psychiatry* (July 1995) study of the relationship between hippocampal volume and short-term verbal memory of 21 Vietnam vets with combat related post-traumatic stress disorder (PTSD), Exercise 2.114 (p. 90). Recall that magnetic resonance imaging was used to measure the volume, $x$, of the right hippocampus (in cubic millimeters) of each subject while verbal memory retention, $y$, of each subject was measured by the percent retention subscale of the Wechsler Memory Scale. The scattergram for the data is reproduced here.

  **a.** Propose a straight-line model relating verbal memory $y$ with right hippocampal volume $x$ to support the theory that smaller hippocampal volume would be associated with deficits in short-term verbal memory in patients with PTSD.
  **b.** Based on the theory, would you expect the slope of the line to be positive or negative? Explain.
  **c.** The coefficient of correlation between right hippocampal volume and verbal retention was $r = .64$. Interpret this value.

**d.** A statistical test of $H_0$: $\beta_1 = 0$ versus $H_a$: $\beta_1 > 0$ resulted in a *p*-value smaller than .05. Interpret this result.

**11.85** Researchers at the University of North Carolina-Greensboro investigated a model for the rate of seed germination (*Journal of Experimental Botany*, Jan. 1993). In one experiment, alfalfa seeds were placed in a specially constructed germination chamber. Eleven hours later, the seeds were examined and the change in free energy (a measure of germination rate) was recorded. The results for seeds germinated at seven different temperatures are given in the table. The data were used to fit a simple linear regression model, with $y =$ change in free energy and $x =$ temperature.

**🌀 SEEDGERM**

| Change in Free Energy (kJ/mol) y | Temperature (K) x |
|---|---|
| 7 | 295 |
| 6.2 | 297.5 |
| 9 | 291 |
| 9.5 | 289 |
| 8.5 | 301 |
| 7.8 | 293 |
| 11.2 | 286.5 |

*Source:* Hageseth, G. T., and Cody, A. L. "Energy-level model for isothermal seed germination." *Journal of Experimental Botany*, Vol. 44, No. 258, Jan. 1993, p. 123 (Figure 9).

**a.** Plot the points in a scattergram.
**b.** Find the least squares prediction equation.
**c.** Plot the least squares line, part **b** on the scattergram of part **a**.
**d.** Conduct a test of model adequacy. Use $\alpha = .01$.
**e.** Use the plot, part **c**, to locate any unusual data points (outliers).
**f.** Eliminate the outlier, part **e**, from the data set and repeat parts **a–d**.

**11.86** One of the most common types of "information retrieval" processes is document-database searching. An experiment was conducted to investigate the variables that influence search performance in the Medline database and retrieval system (*Journal of Information Science*, Vol. 21, 1995). Simple linear regression was used to model the fraction *y* of the set of potentially informative documents that are retrieved using Medline as a function of the number *x* of terms in the search query, based on a sample of $n = 124$ queries. The results are summarized below:

$$\hat{y} = .202 + .135x;$$
$$t(\text{for testing } H_0\text{: } \beta_1 = 0) = 4.98$$
$$\text{Two-tailed } p\text{-value} = .001$$

**a.** Is there sufficient evidence to indicate that *x* and *y* are linearly related? Test using $\alpha = .01$.
**b.** If appropriate, use the model to predict the fraction of documents retrieved for a search query with $x = 3$ terms.

**11.87** Is there a link between the loneliness of parents and that of their offspring? This question was examined in the *Journal of Marriage and Family* (Aug. 1986). The participants in the study were 130 female college undergraduates and their parents. Each triad of daughter, mother, and father completed the UCLA Loneliness Scale, a 20-item questionnaire designed to assess loneliness and several variables theoretically related to loneliness, such as social accessibility to others, difficulty in making friends, and depression.

**a.** The correlation between daughter's loneliness score *y* and mother's loneliness score *x* was determined to be $r = .26$. Interpret this value.
**b.** The correlation between daughter's loneliness score *y* and father's loneliness score *x* was determined to be $r = .19$. Interpret this value.
**c.** The correlation between daughter's loneliness score *y* and mother's self-esteem score *x* was determined to be $r = .14$. Interpret this value.
**d.** The correlation between daughter's loneliness score *y* and father's assertiveness score *x* was determined to be $r = .01$. Interpret this value.
**e.** Calculate the coefficient of determination $r^2$, for parts **a–d**. Interpret the results.

**Applying the Concepts—Advanced**

**11.88** Sometimes it is known from theoretical considerations that the straight-line relationship between two variables, *x* and *y*, passes through the origin of the *xy*-plane. Consider the relationship between the total weight of a shipment of 50-pound bags of flour, *y*, and the number of bags in the shipment, *x*. Since a shipment containing $x = 0$ bags (i.e., no shipment at all) has a total weight of $y = 0$, a straight-line model of the relationship between *x* and *y* should pass through the point $x = 0$, $y = 0$. In such a case you could assume $\beta_0 = 0$ and characterize the relationship between *x* and *y* with the following model:

$$y = \beta_1 x + \varepsilon$$

The least squares estimate of $\beta_1$ for this model is

$$\hat{\beta}_1 = \frac{\Sigma x_i y_i}{\Sigma x_i^2}$$

From the records of past flour shipments, 15 shipments were randomly chosen and the data shown in the table on p. 574 were recorded.

**a.** Find the least squares line for the given data under the assumption that $\beta_0 = 0$. Plot the least squares line on a scattergram of the data.

⊘ FLOUR

| Weight of Shipment | Number of 50-Pound Bags in Shipment |
|---|---|
| 5,050 | 100 |
| 10,249 | 205 |
| 20,000 | 450 |
| 7,420 | 150 |
| 24,685 | 500 |
| 10,206 | 200 |
| 7,325 | 150 |
| 4,958 | 100 |
| 7,162 | 150 |
| 24,000 | 500 |
| 4,900 | 100 |
| 14,501 | 300 |
| 28,000 | 600 |
| 17,002 | 400 |
| 16,100 | 400 |

**b.** Find the least squares line for the given data using the model

$$y = \beta_0 + \beta_1 x + \varepsilon$$

(i.e., do not restrict $\beta_0$ to equal 0). Plot this line on the same scatterplot you constructed in part **a**.

**c.** Refer to part **b**. Why might $\hat{\beta}_0$ be different from 0 even though the true value of $\beta_0$ is known to be 0?

**d.** The estimated standard error of $\hat{\beta}_0$ is equal to

$$s\sqrt{\frac{1}{n} + \frac{\bar{x}^2}{SS_{xx}}}$$

Use the $t$ statistic

$$t = \frac{\hat{\beta}_0 - 0}{s\sqrt{(1/n) + (\bar{x}^2/SS_{xx})}}$$

to test the null hypothesis $H_0$: $\beta_0 = 0$ against the alternative $H_a$: $\beta_0 \neq 0$. Use $\alpha = .10$. Should you include $\beta_0$ in your model?

---

## ⫸ STUDENT PROJECTS

Many dependent variables in all areas of research serve as the subjects of regression modeling efforts. We list five such variables here:

1. Crime rate in various communities
2. Daily maximum temperature in your town
3. Grade point average of students who have completed one academic year at your college
4. Gross Domestic Product of the United States
5. Points scored by your favorite football team in a single game

Choose one of these dependent variables or choose some other dependent variable for which you want to construct a prediction model. There may be a large number of independent variables that should be included in a prediction equation for the dependent variable you choose. List three potentially important independent variables, $x_1$, $x_2$, and $x_3$, that you think might be (individually) strongly related to your dependent variable. Next, obtain 10 data values, each of which

consists of a measure of your dependent variable $y$ and the corresponding values of $x_1$, $x_2$, and $x_3$.

a. Use the least squares formulas given in this chapter to fit three straight-line models—one for each independent variable—for predicting $y$.
b. Interpret the sign of the estimated slope coefficient $\hat{\beta}_1$ in each case, and test the utility of each model by testing $H_0$: $\beta_1 = 0$ against $H_a$: $\beta_1 \neq 0$. What assumptions must be satisfied to ensure the validity of these tests?
c. Calculate the coefficient of determination $r^2$ for each model. Which of the independent variables predicts $y$ best for the 10 sampled sets of data? Is this variable necessarily best in general (that is, for the entire population)? Explain.

Be sure to keep the data and the results of your calculations, since you will need them for the Student Projects section in Chapter 12.

---

## ⫸ REFERENCES

Chatterjee, S., and Price, B. *Regression Analysis by Example*, 2nd ed. New York: Wiley, 1991.

Draper, N., and Smith, H. *Applied Regression Analysis*, 2nd ed. New York: Wiley, 1981.

Mendenhall, W. *Introduction to Linear Models and the Design and Analysis of Experiments*. Belmont, Ca.: Wadsworth, 1968.

Mendenhall, W., and Sincich, T. *A Second Course in Statistics: Regression Analysis*, 5th ed. Upper Saddle River, N. J.: Prentice Hall, 1996.

Mosteller, F., and Tukey, J. W. *Data Analysis and Regression: A Second Course in Statistics*. Reading, Mass.: Addison-Wesley, 1977.

Neter, J., Kutner, M., Cjnachtsheim, C., and Wasserman, W. *Applied Linear Statistical Models*, 4th ed. Homewood, Ill.: Richard Irwin, 1996.

Rousseeuw, P. J., and Leroy, A. M. *Robust Regression and Outlier Detection*. New York: Wiley, 1987.

# CHAPTER 12

# Multiple Regression and Model Building

## Contents

## Statistics in Action

### ✊ *Where We've Been*

In Chapter 11 we demonstrated how to model the relationship between a dependent variable $y$ and an independent variable $x$ using a straight line. We fit the straight line to the data points, used $r$ and $r^2$ to measure the strength of the relationship between $y$ and $x$, and used the resulting prediction equation to estimate the mean value of $y$ or to predict some future value of $y$ for a given value of $x$.

### ☛ *Where We're Going*

This chapter extends the basic concept of Chapter 11, converting it into a powerful estimation and prediction device by modeling the mean value of $y$ as a function of two or more independent variables. The techniques developed will enable you to build a model for a response, $y$, as a function of both quantitative and qualitative variables. As in the case of a simple linear regression, a multiple regression analysis involves fitting the model to a data set, testing the utility of the model, and using it for estimation and prediction. Also, we demonstrate how to select an appropriate model and use the data to modify and improve it.

## 12.1 MULTIPLE REGRESSION MODELS

Most practical applications of regression analysis utilize models that are more complex than the simple straight-line model. For example, a realistic probabilistic model for reaction time would include more than just the amount of a particular drug in the bloodstream. Factors such as age, a measure of visual perception, and sex of the subject are a few of the many variables that might be related to reaction time. Thus, we would want to incorporate these and other potentially important independent variables into the model in order to make accurate predictions.

Probabilistic models that include more than one independent variable are called **multiple regression models**. The general form of these models is

$$y = \beta_0 + \beta_1 x_1 + \beta_2 x_2 + \cdots + \beta_k x_k + \varepsilon$$

The dependent variable $y$ is now written as a function of $k$ independent variables, $x_1, x_2, \ldots, x_k$. The random error term is added to make the model probabilistic rather than deterministic. The value of the coefficient $\beta_i$ determines the contribution of the independent variable $x_i$, and $\beta_0$ is the $y$-intercept. The coefficients $\beta_0, \beta_1, \ldots, \beta_k$ are usually unknown because they represent population parameters.

At first glance it might appear that the regression model shown above would not allow for anything other than straight-line relationships between $y$ and the independent variables, but this is not true. Actually, $x_1, x_2, \ldots, x_k$ can be functions of variables as long as the functions do not contain unknown parameters. For example, the reaction time, $y$, of a subject to a visual stimulus could be a function of the independent variables

$$x_1 = \text{Age of the subject}$$
$$x_2 = (\text{Age})^2 = x_1^2$$
$$x_3 = 1 \text{ if male subject}, 0 \text{ if female subject}$$

The $x_2$ term is called a **higher-order term**, since it is the value of a quantitative variable $(x_1)$ squared (i.e., raised to the second power). The $x_3$ term is a **coded variable** representing a qualitative variable (gender). The multiple regression model is quite versatile and can be made to model many different types of response variables.

---

### The General Multiple Regression Model

$$y = \beta_0 + \beta_1 x_1 + \beta_2 x_2 + \cdots + \beta_k x_k + \varepsilon$$

where

$y$ is the dependent variable

$x_1, x_2, \ldots, x_k$ are the independent variables

$E(y) = \beta_0 + \beta_1 x_1 + \beta_2 x_2 + \cdots + \beta_k x_k$ is the deterministic portion of the model

$\beta_i$ determines the contribution of the independent variable $x_i$

*Note:* The symbols $x_1, x_2, \ldots, x_k$ may represent higher-order terms for quantitative predictors or terms that represent qualitative predictors.

---

As shown in the next box, the steps used to develop the multiple regression model are similar to those used for the simple regression model.

> ## Analyzing a Multiple Regression Model
>
> Step 1   Hypothesize the deterministic component of the model. This component relates the mean, $E(y)$, to the independent variables $x_1, x_2, \ldots, x_k$. This involves the choice of the independent variables to be included in the model (Sections 12.2, 12.7–12.12).
>
> Step 2   Use the sample data to estimate the unknown model parameters $\beta_0, \beta_1, \beta_2, \ldots, \beta_k$ in the model (Section 12.2).
>
> Step 3   Specify the probability distribution of the random error term, $\varepsilon$, and estimate the standard deviation of this distribution, $\sigma$ (Section 12.3).
>
> Step 4   Check that the assumptions on $\varepsilon$ are satisfied, and make model modifications if necessary (Section 12.13).
>
> Step 5   Statistically evaluate the usefulness of the model (Sections 12.4 and 12.5).
>
> Step 6   When satisfied that the model is useful, use it for prediction, estimation, and other purposes (Section 12.6).

Throughout this chapter, we introduce several different types of models that form the foundation of **model building** (or useful model construction). In the next several sections, we consider the most basic multiple regression model, called the *first-order model*.

## 12.2   THE FIRST-ORDER MODEL: ESTIMATING AND INTERPRETING THE $\beta$ PARAMETERS

A model that includes only terms for *quantitative* independent variables, called a **first-order model**, is described in the box. Note that the first-order model does not include any higher-order terms (such as $x_1^2$).

> ## A First-Order Model in Five Quantitative Independent Variables*
>
> $$E(y) = \beta_0 + \beta_1 x_1 + \beta_2 x_2 + \beta_3 x_3 + \beta_4 x_4 + \beta_5 x_5$$
>
> where $x_1, x_2, \ldots, x_5$ are all quantitative variables that *are not* functions of other independent variables.
> *Note:* $\beta_i$ represents the slope of the line relating $y$ to $x_i$ when all the other $x$'s are held fixed.

The method of fitting first-order models—and multiple regression models in general—is identical to that of fitting the simple straight-line model: the method of least squares. That is, we choose the estimated model

$$\hat{y} = \hat{\beta}_0 + \hat{\beta}_1 x_1 + \cdots + \hat{\beta}_k x_k$$

---

*The terminology "first-order" is derived from the fact that each $x$ in the model is raised to the first power.

that minimizes

$$\text{SSE} = \sum (y - \hat{y})^2$$

As in the case of the simple linear model, the sample estimates $\hat{\beta}_0, \hat{\beta}_1, \ldots, \hat{\beta}_k$ are obtained as a solution to a set of simultaneous linear equations.*

The primary difference between fitting the simple and multiple regression models is computational difficulty. The $(k + 1)$ simultaneous linear equations that must be solved to find the $(k + 1)$ estimated coefficients $\hat{\beta}_0, \hat{\beta}_1, \ldots, \hat{\beta}_k$ are difficult (sometimes nearly impossible) to solve with a calculator. Consequently, we resort to the use of computers. Instead of presenting the tedious hand calculations required to fit the models, we present output from several statistical software packages, including SAS, SPSS, STATISTIX and MINITAB. Since these printouts are similar to most other packages, you should have little trouble interpreting regression output from other packages as you encounter them in the future.

---

**EXAMPLE 12.1**

Suppose a property appraiser wants to model the relationship between the sale price of a residential property in a mid-size city and the following three independent variables: (1) appraised land value of the property, (2) appraised value of improvements (i.e., home value) on the property, and (3) area of living space on the property (i.e., home size). Consider the first-order model

$$y = \beta_0 + \beta_1 x_1 + \beta_2 x_2 + \beta_3 x_3 + \varepsilon$$

where

$$
\begin{aligned}
y &= \text{Sale price (dollars)} \\
x_1 &= \text{Appraised land value (dollars)} \\
x_2 &= \text{Appraised improvements (dollars)} \\
x_3 &= \text{Area (square feet)}
\end{aligned}
$$

To fit the model, the appraiser selected a random sample of $n = 20$ properties from the thousands of properties that were sold in a particular year. The resulting data are given in Table 12.1.

a. Use scattergrams to plot the sample data. Interpret the plots.
b. Use the method of least squares to estimate the unknown parameters $\beta_0, \beta_1, \beta_2$, and $\beta_3$ in the model.
c. Find the value of SSE that is minimized by the least squares method.

### Solution

a. MINITAB scatterplots for examining the bivariate relationships between $y$ and $x_1$, $y$ and $x_2$, and $y$ and $x_3$ are shown in Figure 12.1. Of the three variables, appraised improvements $(x_2)$ appears to have the strongest linear relationship with sale price $(y)$.

b. The model hypothesized above is fit to the data of Table 12.1 using SAS. A portion of the SAS printout is reproduced in Figure 12.2. The least squares estimates of the $\beta$ parameters (highlighted on the printout) are $\hat{\beta}_0 = 1,470.27592$, $\hat{\beta}_1 = .81449$, $\hat{\beta}_2 = .82044$, and $\hat{\beta}_3 = 13.52865$. Therefore, the equation that minimizes SSE for this data set (i.e., the **least squares prediction equation**) is

$$\hat{y} = 1,470.28 + .8145x_1 + .8204x_2 + 13.53x_3$$

---

*Students who are familiar with calculus should note that $\hat{\beta}_0, \hat{\beta}_1, \ldots, \hat{\beta}_k$ are the solutions to the set of equations $\partial \text{SSE}/\partial \hat{\beta}_0 = 0, \partial \text{SSE}/\partial \hat{\beta}_1 = 0, \ldots, \partial \text{SSE}/\partial \hat{\beta}_k = 0$. The solution is usually given in matrix form, but we do not present the details here. See the references for details.

**TABLE 12.1   Real Estate Appraisal Data for 20 Properties**

| Property # (Obs.) | Sale Price $y$ | Land Value $x_1$ | Improvements Value $x_2$ | Area $x_3$ |
|---|---|---|---|---|
| 1  | 68,900  | 5,960  | 44,967 | 1,873 |
| 2  | 48,500  | 9,000  | 27,860 | 928 |
| 3  | 55,500  | 9,500  | 31,439 | 1,126 |
| 4  | 62,000  | 10,000 | 39,592 | 1,265 |
| 5  | 116,500 | 18,000 | 72,827 | 2,214 |
| 6  | 45,000  | 8,500  | 27,317 | 912 |
| 7  | 38,000  | 8,000  | 29,856 | 899 |
| 8  | 83,000  | 23,000 | 47,752 | 1,803 |
| 9  | 59,000  | 8,100  | 39,117 | 1,204 |
| 10 | 47,500  | 9,000  | 29,349 | 1,725 |
| 11 | 40,500  | 7,300  | 40,166 | 1,080 |
| 12 | 40,000  | 8,000  | 31,679 | 1,529 |
| 13 | 97,000  | 20,000 | 58,510 | 2,455 |
| 14 | 45,500  | 8,000  | 23,454 | 1,151 |
| 15 | 40,900  | 8,000  | 20,897 | 1,173 |
| 16 | 80,000  | 10,500 | 56,248 | 1,960 |
| 17 | 56,000  | 4,000  | 20,859 | 1,344 |
| 18 | 37,000  | 4,500  | 22,610 | 988 |
| 19 | 50,000  | 3,400  | 35,948 | 1,076 |
| 20 | 22,400  | 1,500  | 5,779  | 962 |

*Source:* Alachua County (Florida) Property Appraisers Office.

c. The minimum value of the sum of squared errors, highlighted in Figure 12.2, is SSE = 1,003,491,259.

————————————■

After obtaining the least squares prediction equation, the analyst will usually want to make meaningful interpretations of the β estimates. Recall that in the straight-line model (Chapter 11)

$$y = \beta_0 + \beta_1 x + \varepsilon$$

$\beta_0$ represents the $y$-intercept of the line and $\beta_1$ represents the slope of the line. From our discussion in Chapter 11, $\beta_1$ has a practical interpretation—it represents the mean change in $y$ for every 1-unit increase in $x$. When the independent variables are quantitative, the β parameters in the first-order model specified in Example 12.1 have similar interpretations. The difference is that when we interpret the β that multiplies one of the variables (e.g., $x_1$), we must be certain to hold the values of the remaining independent variables (e.g., $x_2, x_3$) fixed.

To see this, suppose that the mean $E(y)$ of a response $y$ is related to two quantitative independent variables, $x_1$ and $x_2$, by the first-order model

$$E(y) = 1 + 2x_1 + x_2$$

In other words, $\beta_0 = 1, \beta_1 = 2$, and $\beta_2 = 1$.

Now, when $x_2 = 0$, the relationship between $E(y)$ and $x_1$ is given by

$$E(y) = 1 + 2x_1 + (0) = 1 + 2x_1$$

**FIGURE 12.1**

*MINITAB Scatterplots for the
Data of Table 12.1*

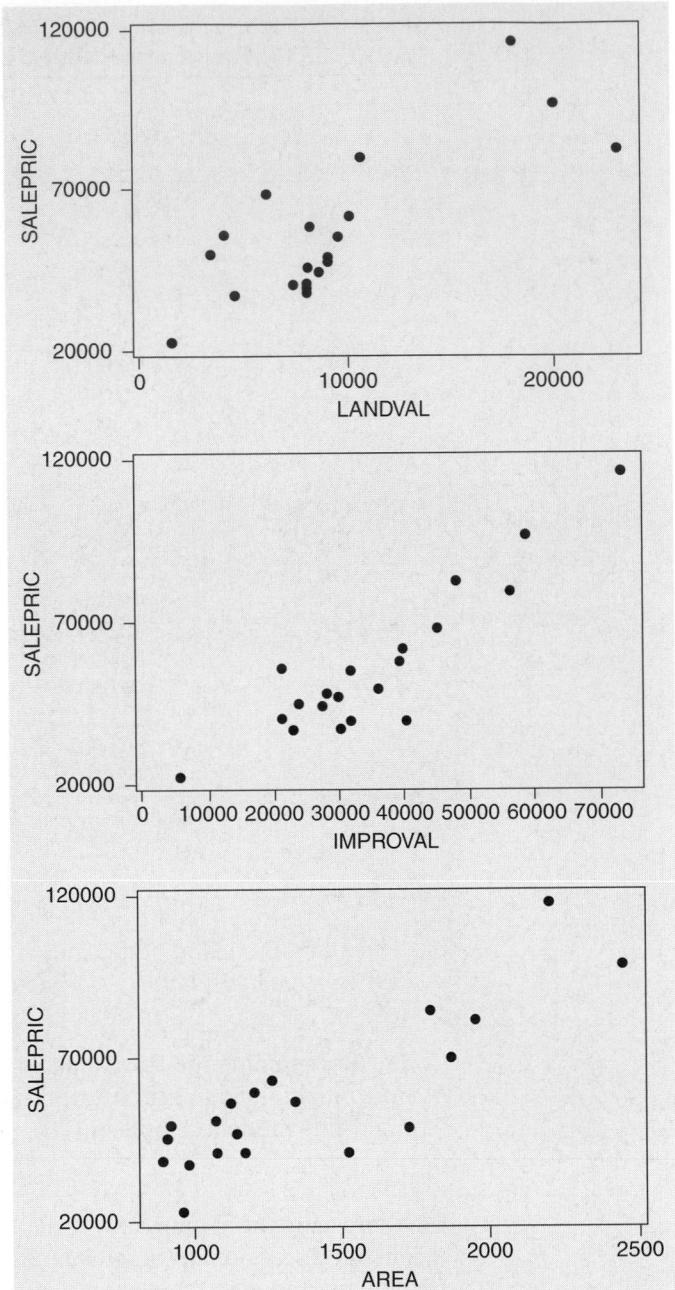

A graph of this relationship (a straight line) is shown in Figure 12.3. Similar graphs of the relationship between $E(y)$ and $x_1$ for $x_2 = 1$,

$$E(y) = 1 + 2x_1 + (1) = 2 + 2x_1$$

and for $x_2 = 2$,

$$E(y) = 1 + 2x_1 + (2) = 3 + 2x_1$$

also are shown in Figure 12.3. Note that the slopes of the three lines are all equal to $\beta_1 = 2$, the coefficient that multiplies $x_1$.

**FIGURE 12.2**

*SAS Regression Printout for the Sale Price Model, Example 12.1*

The REG Procedure
Dependent Variable: SALEPRIC

Analysis of Variance

| Source | DF | Sum of Squares | Mean Square | F Value | Pr > F |
|---|---|---|---|---|---|
| Model | 3 | 8779676741 | 2926558914 | 46.66 | <.0001 |
| Error | 16 | 1003491259 | 62718204 | | |
| Corrected Total | 19 | 9783168000 | | | |

| | | | | |
|---|---|---|---|---|
| Root MSE | 7919.48254 | R-Square | 0.8974 |
| Dependent Mean | 56660 | Adj R-Sq | 0.8782 |
| Coeff Var | 13.97720 | | |

Parameter Estimates

| Variable | DF | Parameter Estimate | Standard Error | t Value | Pr > \|t\| |
|---|---|---|---|---|---|
| Intercept | 1 | 1470.27592 | 5746.32458 | 0.26 | 0.8013 |
| LANDVAL | 1 | 0.81440 | 0.51222 | 1.59 | 0.1314 |
| IMPROVAL | 1 | 0.82044 | 0.21118 | 3.88 | 0.0013 |
| AREA | 1 | 13.52865 | 6.58568 | 2.05 | 0.0567 |

**FIGURE 12.3**

*Graphs of*
$E(y) = 1 + 2x_1 + x_2$
*for $x_2 = 0, 1, 2$*

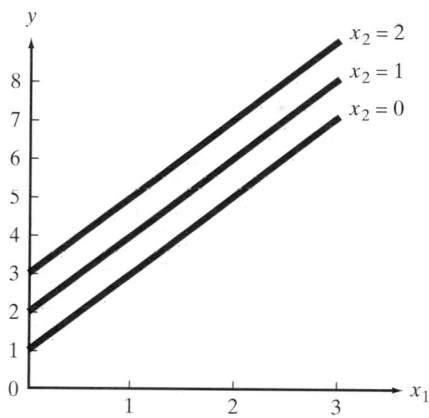

Figure 12.3 exhibits a characteristic of all first-order models: If you graph $E(y)$ versus any one variable—say, $x_1$—for fixed values of the other variables, the result will always be a *straight line* with slope equal to $\beta_1$. If you repeat the process for other values of the fixed independent variables, you will obtain a set of *parallel* straight lines. This indicates that the effect of the independent variable $x_i$ on $E(y)$ is independent of all the other independent variables in the model, and this effect is measured by the slope $\beta_i$ (see the box on p. 577).

A three-dimensional graph of the model $E(y) = 1 + 2x_1 + x_2$ is shown in Figure 12.4. Note that the model graphs as a plane. If you slice the plane at a particular value of $x_2$ (say, $x_2 = 0$), you obtain a straight line relating $E(y)$ to $x_1$ (e.g., $E(y) = 1 + 2x_1$). Similarly, if you slice the plane at a particular value of $x_1$, you obtain a straight line relating $E(y)$ to $x_2$. Since it is more difficult to visualize

**FIGURE 12.4**

*The Plane*
$E(y) = 1 + 2x_1 + x_2$

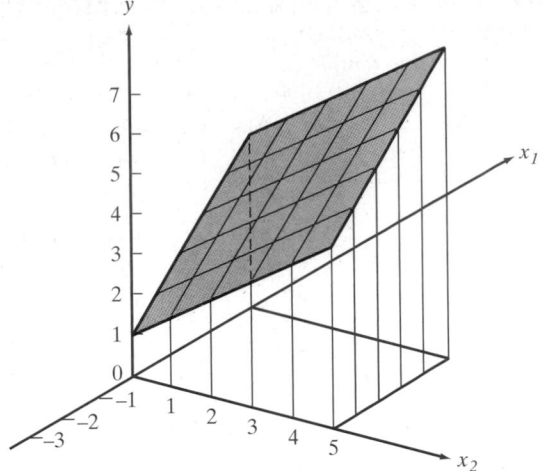

three-dimensional and, in general, $k$-dimensional surfaces, we will graph all the models presented in this chapter in two dimensions. The key to obtaining these graphs is to hold fixed all but one of the independent variables in the model.

**E X A M P L E   12.2** Refer to the first-order model for sale price $y$ considered in Example 12.1. Interpret the estimates of the $\beta$ parameters in the model.

**Solution**

The least squares prediction equation, as given in Example 12.1, is $\hat{y} = 1{,}470.28 + .8145x_1 + .8204x_2 + 13.53x_3$. We know that with first-order models $\beta_1$ represents the slope of the $y - x_1$ line for fixed $x_2$ and $x_3$. That is, $\beta_1$ measures the change in $E(y)$ for every 1-unit increase in $x_1$ when all other independent variables in the model are held fixed. Similar statements can be made about $\beta_2$ and $\beta_3$; e.g., $\beta_2$ measures the change in $E(y)$ for every 1-unit increase in $x_2$ when all other $x$'s in the model are held fixed. Consequently, we obtain the following interpretations:

$\hat{\beta}_1 = .8145$: We estimate the mean sale price of a property, $E(y)$, to increase .8145 dollar for every \$1 increase in appraised land value $(x_1)$ when both appraised improvements $(x_2)$ and area $(x_3)$ are held fixed.

$\hat{\beta}_2 = .8204$: We estimate the mean sale price of a property, $E(y)$, to increase .8204 dollar for every \$1 increase in appraised improvements $(x_2)$ when both appraised land value $(x_1)$ and area $(x_3)$ are held fixed.

$\hat{\beta}_3 = 13.53$: We estimate the mean sale price of a property, $E(y)$, to increase \$13.53 for each additional square foot of living area $(x_3)$ when both appraised land value $(x_1)$ and appraised improvements $(x_2)$ are held fixed.

The value $\hat{\beta}_0 = 1{,}470.28$ does not have a meaningful interpretation in this example. To see this, note that $\hat{y} = \hat{\beta}_0$ when $x_1 = x_2 = x_3 = 0$. Thus, $\hat{\beta}_0 = 1{,}470.28$ represents the estimated mean sale price when the values of all the independent variables are set equal to 0. Since a residential property with these characteristics—appraised land value of \$0, appraised improvements of \$0, and 0 square feet of living area—is not practical, the value of $\hat{\beta}_0$ has no meaningful interpretation. In general, $\hat{\beta}_0$ will not have a practical interpretation unless it makes sense to set the values of the $x$'s simultaneously equal to 0.

> **Caution**
>
> The interpretation of the $\beta$ parameters in a multiple regression model will depend on the terms specified in the model. The interpretations in Example 12.2 are for a first-order linear model only. In practice, you should be sure that a first-order model is the correct model for $E(y)$ before making these $\beta$ interpretations. [We discuss alternative models for $E(y)$ in Sections 12.7–12.10.]

## 12.3  MODEL ASSUMPTIONS

We noted in Section 12.1 that the general multiple regression model is of the form

$$y = \beta_0 + \beta_1 x_1 + \beta_2 x_2 + \cdots + \beta_k x_k + \varepsilon$$

where $y$ is the response variable that we wish to predict; $\beta_0, \beta_1, \ldots, \beta_k$ are parameters with unknown values; $x_1, x_2, \ldots, x_k$ are information-contributing variables that are measured without error; and $\varepsilon$ is a random error component. Since $\beta_0, \beta_1, \ldots, \beta_k$ and $x_1, x_2, \ldots, x_k$ are nonrandom, the quantity

$$\beta_0 + \beta_1 x_1 + \beta_2 x_2 + \cdots + \beta_k x_k$$

represents the deterministic portion of the model. Therefore, $y$ is composed of two components—one fixed and one random—and, consequently, $y$ is a random variable.

$$y = \overbrace{\beta_0 + \beta_1 x_1 + \cdots + \beta_k x_k}^{\substack{\text{Deterministic} \\ \text{portion of model}}} + \overbrace{\varepsilon}^{\substack{\text{Random} \\ \text{error}}}$$

We will assume (as in Chapter 11) that the random error can be positive or negative and that for any setting of the $x$ values, $x_1, x_2, \ldots, x_k$, the random error $\varepsilon$ has a normal probability distribution with mean equal to 0 and variance equal to $\sigma^2$. Further, we assume that the random errors associated with any (and every) pair of $y$ values are probabilistically independent. That is, the error, $\varepsilon$, associated with any one $y$ value is independent of the error associated with any other $y$ value. These assumptions are summarized in the next box.

> **Assumptions for Random Error $\varepsilon$**
>
> 1. For any given set of values of $x_1, x_2, \ldots, x_k$, the random error $\varepsilon$ has a normal probability distribution with mean equal to 0 and variance equal to $\sigma^2$.
> 2. The random errors are independent (in a probabilistic sense).

Note that $\sigma^2$ represents the variance of the random error $\varepsilon$. As such, $\sigma^2$ is an important measure of the usefulness of the model for the estimation of the mean and the prediction of actual values of $y$. If $\sigma^2 = 0$, all the random errors will equal 0 and the predicted values, $\hat{y}$, will be identical to $E(y)$; that is $E(y)$ will be estimated without error. In contrast, a large value of $\sigma^2$ implies large (absolute) values of $\varepsilon$ and larger deviations between the predicted values, $\hat{y}$, and the mean value, $E(y)$. Consequently, the larger the value of $\sigma^2$, the greater will be the error in estimating the model parameters $\beta_0, \beta_1, \ldots, \beta_k$ and the error in predicting a value of $y$ for a specific set of values of $x_1, x_2, \ldots, x_k$. Thus, $\sigma^2$ plays a major role in making inferences about $\beta_0, \beta_1, \ldots, \beta_k$, in estimating $E(y)$, and in predicting $y$ for specific values of $x_1, x_2, \ldots, x_k$.

Since the variance, $\sigma^2$, of the random error, $\varepsilon$, will rarely be known, we must use the results of the regression analysis to estimate its value. Recall that $\sigma^2$ is the variance of the probability distribution of the random error $\varepsilon$, for a given set of values for $x_1, x_2, \ldots, x_k$; hence it is the mean value of the squares of the deviations of the $y$ values (for given values of $x_1, x_2, \ldots, x_k$) about the mean value $E(y)$.* Since the predicted value, $\hat{y}$, estimates $E(y)$ for each of the data points, it seems natural to use

$$\text{SSE} = \sum (y_i - \hat{y}_i)^2$$

to construct an estimator of $\sigma^2$.

For example, in the first-order model of Example 12.2, we found that SSE = 1,003,491,259. We now want to use this quantity to estimate the variance of $\varepsilon$. Recall that the estimator for the straight-line model is $s^2 = \text{SSE}/(n-2)$ and note that the denominator is ($n$ − Number of estimated $\beta$ parameters), which is ($n - 2$) in the straight-line model. Since we must estimate four parameters, $\beta_0, \beta_1, \beta_2$, and $\beta_3$ for the first-order model, the estimator of $\sigma^2$ is

$$s^2 = \frac{\text{SSE}}{n - 4}$$

The numerical estimate for this example is

$$s^2 = \frac{\text{SSE}}{20 - 4} = \frac{1,003,491,259}{16} = 62,718,204$$

In many computer printouts and textbooks, $s^2$ is called the **mean square for error (MSE)**. This estimate of $\sigma^2$ is highlighted in the SAS printout in Figure 12.2 (p. 581).

The units of the estimated variance are squared units of the dependent variable $y$. Since the dependent variable $y$ in this example is sale price in dollars, the units of $s^2$ are (dollars)$^2$. This makes meaningful interpretation of $s^2$ difficult, so we use the standard deviation $s$ to provide a more meaningful measure of variability. In this example,

$$s = \sqrt{62,718,204} = 7,919.5$$

which is highlighted on the SAS printout in Figure 12.2 (next to **Root MSE**). One useful interpretation of the estimated standard deviation $s$ is that the interval $\pm 2s$ will provide a rough approximation to the accuracy with which the model will predict future values of $y$ for given values of $x$. Thus, in Example 12.2, we expect the model to provide predictions of sale price to within about $\pm 2s = \pm 2(7,919.5) = \pm 15,839$ dollars.[†]

For the general multiple regression model

$$y = \beta_0 + \beta_1 x_1 + \beta_2 x_2 + \cdots + \beta_k x_k + \varepsilon$$

we must estimate the ($k + 1$) parameters $\beta_0, \beta_1, \beta_2, \ldots, \beta_k$. Thus, the estimator of $\sigma^2$ is SSE divided by the quantity ($n$ − Number of estimated $\beta$ parameters).

---

[*]Since $y = E(y) + \varepsilon$, $\varepsilon$ is equal to the deviation $y - E(y)$. Also, by definition, the variance of a random variable is the expected value of the square of the deviation of the random variable from its mean. According to our model, $E(\varepsilon) = 0$. Therefore, $\sigma^2 = E(\varepsilon^2)$.

[†]The $\pm 2s$ approximation will improve as the sample size is increased. We will provide more precise methodology for the construction of prediction intervals in Section 12.6.

We will use the estimator of $\sigma^2$ both to check the utility of the model (Sections 12.4 and 12.5) and to provide a measure of reliability of predictions and estimates when the model is used for those purposes (Section 12.6). Thus, you can see that the estimation of $\sigma^2$ plays an important part in the development of a regression model.

---

### Estimator of $\sigma^2$ for a Multiple Regression Model with $k$ Independent Variables

$$s^2 = \frac{\text{SSE}}{n - \text{Number of estimated } \beta \text{ parameters}} = \frac{\text{SSE}}{n - (k + 1)}$$

---

## 12.4 INFERENCES ABOUT THE $\beta$ PARAMETERS

Inferences about the individual $\beta$ parameters in a model are obtained using either a confidence interval or a test of hypothesis, as outlined in the following two boxes.*

---

### Test of an Individual Parameter Coefficient in the Multiple Regression Model

**ONE-TAILED TEST**

$H_0: \beta_i = 0$

$H_a: \beta_i < 0 \, [\text{or } H_a: \beta_i > 0]$

**TWO-TAILED TEST**

$H_0: \beta_i = 0$

$H_a: \beta_i \neq 0$

*Test statistic:* $t = \dfrac{\hat{\beta}_i}{s_{\hat{\beta}_i}}$

*Rejection region:* $t < -t_\alpha$

$[\text{or } t > t_\alpha \text{ when } H_a: \beta_i > 0]$

*Rejection region:* $|t| > t_{\alpha/2}$

where $t_\alpha$ and $t_{\alpha/2}$ are based on $n - (k + 1)$ degrees of freedom and

$n$ = Number of observations
$k + 1$ = Number of $\beta$ parameters in the model

*Assumptions:* See Section 12.3 for assumptions about the probability distribution for the random error component $\varepsilon$. (p. 583)

---

### A $100(1 - \alpha)\%$ Confidence Interval for a $\beta$ Parameter

$$\hat{\beta}_i \pm t_{\alpha/2} s_{\hat{\beta}_i}$$

where $t_{\alpha/2}$ is based on $n - (k + 1)$ degrees of freedom and

$n$ = Number of observations
$k + 1$ = Number of $\beta$ parameters in the model

---

*The formulas for computing $\hat{\beta}_i$ and its standard error are so complex, the only reasonable way to present them is by using matrix algebra. We do not assume a prerequisite of matrix algebra for this text and, in any case, we think the formulas can be omitted in an introductory course without serious loss. They are programmed into almost all statistical software packages with multiple regression routines and are presented in some of the texts listed in the references.

We illustrate these methods with another example.

**EXAMPLE 12.3**

A collector of antique grandfather clocks knows that the price received for the clocks increases linearly with the age of the clocks. Moreover, the collector hypothesizes that the auction price of the clocks will increase linearly as the number of bidders increases. Thus, the following first-order model is hypothesized:

$$y = \beta_0 + \beta_1 x_1 + \beta_2 x_2 + \varepsilon$$

where

$y =$ Auction price
$x_1 =$ Age of clock (years)
$x_2 =$ Number of bidders

A sample of 32 auction prices of grandfather clocks, along with their age and the number of bidders, is given in Table 12.2. The model $y = \beta_0 + \beta_1 x_1 + \beta_2 x_2 + \varepsilon$ is fit to the data, and a portion of the MINITAB printout is shown in Figure 12.5.

⊘GFCLOCKS
**TABLE 12.2   Auction Price Data**

| Age $x_1$ | Number of Bidders $x_2$ | Auction Price $y$ | Age $x_1$ | Number of Bidders $x_2$ | Auction Price $y$ |
|---|---|---|---|---|---|
| 127 | 13 | $1,235 | 170 | 14 | $2,131 |
| 115 | 12 | 1,080 | 182 | 8 | 1,550 |
| 127 | 7 | 845 | 162 | 11 | 1,884 |
| 150 | 9 | 1,522 | 184 | 10 | 2,041 |
| 156 | 6 | 1,047 | 143 | 6 | 845 |
| 182 | 11 | 1,979 | 159 | 9 | 1,483 |
| 156 | 12 | 1,822 | 108 | 14 | 1,055 |
| 132 | 10 | 1,253 | 175 | 8 | 1,545 |
| 137 | 9 | 1,297 | 108 | 6 | 729 |
| 113 | 9 | 946 | 179 | 9 | 1,792 |
| 137 | 15 | 1,713 | 111 | 15 | 1,175 |
| 117 | 11 | 1,024 | 187 | 8 | 1,593 |
| 137 | 8 | 1,147 | 111 | 7 | 785 |
| 153 | 6 | 1,092 | 115 | 7 | 744 |
| 117 | 13 | 1,152 | 194 | 5 | 1,356 |
| 126 | 10 | 1,336 | 168 | 7 | 1,262 |

**FIGURE 12.5**

*MINITAB Regression Printout for Grandfather Clock Model, Example 12.3*

```
The regression equation is
PRICE = - 1339 + 12.7 AGE + 86.0 BIDDERS


Predictor        Coef      SE Coef         T        P
Constant      -1339.0        173.8     -7.70    0.000
AGE           12.7406       0.9047     14.08    0.000
BIDDERS        85.953        8.729      9.85    0.000


S = 133.5       R-Sq = 89.2%     R-Sq(adj) = 88.5%


Analysis of Variance

Source           DF          SS          MS         F       P
Regression        2     4283063     2141531    120.19   0.000
Residual Error   29      516727       17818
Total            31     4799790
```

a. Test the hypothesis that the mean auction price of a clock increases as the number of bidders increases when age is held constant, that is, $\beta_2 > 0$. Use $\alpha = .05$.
b. Form a 90% confidence interval for $\beta_1$ and interpret the result.

### Solution

a. The hypotheses of interest concern the parameter $\beta_2$. Specifically,

$$H_0: \beta_2 = 0$$
$$H_a: \beta_2 > 0$$

The test statistic is a $t$ statistic formed by dividing the sample estimate $\hat{\beta}_2$ of the parameter $\beta_2$ by the estimated standard error of $\hat{\beta}_2$ (denoted $s_{\hat{\beta}_2}$). These estimates, $\hat{\beta}_2 = 85.953$ and $s_{\hat{\beta}_2} = 8.729$, as well as the calculated $t$ value, are highlighted on the MINITAB printout, Figure 12.5.

$$\text{Test statistic: } t = \frac{\hat{\beta}_2}{s_{\hat{\beta}_2}} = \frac{85.953}{8.729} = 9.85$$

The rejection region for the test is found in exactly the same way as the rejection regions for the $t$ tests in previous chapters. That is, we consult Table VI in Appendix A to obtain an upper-tail value of $t$. This is a value $t_\alpha$ such that $P(t > t_\alpha) = \alpha$. We can then use this value to construct rejection regions for either one-tailed or two-tailed tests.

For $\alpha = .05$ and $n - (k + 1) = 32 - (2 + 1) = 29$ df, the critical $t$ value obtained from Table VI is $t_{.05} = 1.699$. Therefore,

*Rejection region:* $t > 1.699$   (see Figure 12.6)

Since the test statistic value, $t = 9.85$, falls in the rejection region, we have sufficient evidence to reject $H_0$. Thus, the collector can conclude that the mean auction price of a clock increases as the number of bidders increases, when age is held constant. Note that the observed significance level of the test is also highlighted on the printout. Since $p$-value $= 0$, any nonzero $\alpha$ will lead us to reject $H_0$.

### FIGURE 12.6

*Rejection Region for*
$H_0: \beta_2 = 0$ *vs* $H_a: \beta_2 > 0$

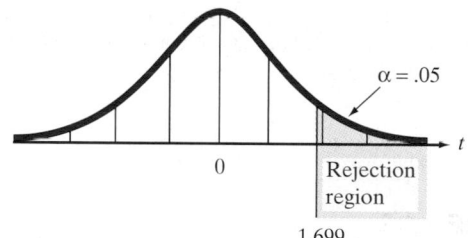

b. A 90% confidence interval for $\beta_1$ is (from the box):

$$\hat{\beta}_1 \pm t_{\alpha/2} s_{\hat{\beta}_1} = \hat{\beta}_1 \pm t_{.05} s_{\hat{\beta}_1}$$

Substituting $\hat{\beta}_1 = 12.74$, $s_{\hat{\beta}_i} = .905$ (both obtained from the MINITAB printout, Figure 12.5) and $t_{.05} = 1.699$ (from part **a**) into the equation, we obtain

$$12.74 \pm 1.699(.905) = 12.74 \pm 1.54$$

or (11.20,14.28). Thus, we are 90% confident that $\beta_1$ falls between 11.20 and 14.28. Since $\beta_1$ is the slope of the line relating auction price ($y$) to age of the clock ($x_1$), we conclude that price increases between \$11.20 and \$14.28 for every 1-year increase in age, holding number of bidders ($x_2$) constant.

We conclude this section with a caution about *t*-tests on individual $\beta$ parameters in a model.

---

### Caution

It is dangerous to conduct *t*-tests on the individual $\beta$ parameters in a *first-order linear model* for the purpose of determining which independent variables are useful for predicting *y* and which are not. If you fail to reject $H_0: \beta_i = 0$, several conclusions are possible:

1. There is no relationship between *y* and $x_i$.
2. A straight-line relationship between *y* and *x* exists (holding the other *x*'s in the model fixed), but a Type II error occurred.
3. A relationship between *y* and $x_i$ (holding the other *x*'s in the model fixed) exists, but is more complex than a straight-line relationship (e.g., a curvilinear relationship may be appropriate). The most you can say about a $\beta$ parameter test is that there is either sufficient (if you reject $H_0: \beta_i = 0$) or insufficient (if you do not reject $H_0: \beta_i = 0$) evidence of a *linear* (*straight-line*) relationship between *y* and $x_i$.

---

 EXERCISES 12.1–12.15

### Learning the Mechanics

**12.1** Write a first-order model relating $E(y)$ to:
 **a.** two quantitative independent variables
 **b.** four quantitative independent variables
 **c.** five quantitative independent variables

**12.2** MINITAB was used to fit the model $y = \beta_0 + \beta_1 x_1 + \beta_2 x_2 + \varepsilon$ to $n = 20$ data points and the printout shown here was obtained.

 **a.** What are the sample estimates of $\beta_0, \beta_1$, and $\beta_2$?
 **b.** What is the least squares prediction equation?
 **c.** Find SSE, MSE, and *s*. Interpret the standard deviation in the context of the problem.
 **d.** Test $H_0: \beta_1 = 0$ against $H_a: \beta_1 \neq 0$. Use $\alpha = .05$.
 **e.** Use a 95% confidence interval to estimate $\beta_2$.

**12.3** Suppose you fit the multiple regression model

$$y = \beta_0 + \beta_1 x_1 + \beta_2 x_2 + \beta_3 x_3 + \varepsilon$$

**MINITAB Output for Exercise 12.2**

```
The regression equation is
Y = 506.34 - 941.90X1 - 429.06X2


Predictor        Coef     SE Coef        T        P
Constant      506.346     45.1694    11.21    0.000
X1           -941.900    275.0856    -3.42    0.003
X2           -429.060    379.8256    -1.13    0.274


S = 94.251      R-Sq = 45.9%     R-Sq(adj) = 39.6%


Analysis of Variance

Source          DF        SS        MS       F       P
Regression       2    128329     64165    7.22   0.005
Residual Error  17    151015      8883
Total           19    279344
```

to $n = 30$ data points and obtain the following result:

$$\hat{y} = 3.4 - 4.6x_1 + 2.7x_2 + 93x_3$$

The estimated standard errors of $\hat{\beta}_2$ and $\hat{\beta}_3$ are 1.86 and .29, respectively.

**a.** Test the null hypothesis $H_0: \beta_2 = 0$ against the alternative hypothesis $H_a: \beta_2 \neq 0$. Use $\alpha = .05$.

**b.** Test the null hypothesis $H_0: \beta_3 = 0$ against the alternative hypothesis $H_a: \beta_3 \neq 0$. Use $\alpha = .05$.

**c.** The null hypothesis $H_0: \beta_2 = 0$ is not rejected. In contrast, the null hypothesis $H_0: \beta_3 = 0$ is rejected. Explain how this can happen even though $\hat{\beta}_2 > \hat{\beta}_3$.

**12.4** Suppose you fit the first-order multiple regression model

$$y = \beta_0 + \beta_1 x_1 + \beta_2 x_2 + \varepsilon$$

to $n = 25$ data points and obtain the prediction equation

$$\hat{y} = 6.4 + 3.1x_1 + .92x_2$$

The estimated standard deviations of the sampling distributions of $\hat{\beta}_1$ and $\hat{\beta}_2$ are 2.3 and .27, respectively.

**a.** Test $H_0: \beta_1 = 0$ against $H_a: \beta_1 > 0$. Use $\alpha = .05$.

**b.** Test $H_0: \beta_2 = 0$ against $H_a: \beta_2 \neq 0$. Use $\alpha = .05$.

**c.** Find a 90% confidence interval for $\beta_1$. Interpret the interval.

**d.** Find a 99% confidence interval for $\beta_2$. Interpret the interval.

**12.5** How is the number of degrees of freedom available for estimating $\sigma^2$ (the variance of $\varepsilon$) related to the number of independent variables in a regression model?

**12.6** Consider the first-order model equation in three quantitative independent variables

$$E(y) = 1 + 2x_1 + x_2 - 3x_3$$

**a.** Graph the relationship between $y$ and $x_1$ for $x_2 = 1$ and $x_3 = 3$.

**b.** Repeat part **a** for $x_2 = -1$ and $x_3 = 1$.

**c.** How do the graphed lines in parts **a** and **b** relate to each other? What is the slope of each line?

**d.** If a linear model is first-order in three independent variables, what type of geometric relationship will you obtain when $E(y)$ is graphed as a function of one of the independent variables for various combinations of values of the other independent variables?

## Applying the Concepts—Basic

**12.7** In *Chance* (Fall 2000), statistician Scott Berry built a multiple regression model for predicting total number of runs scored by a Major League Baseball team during a season. Using data on all teams from 1990–1998 (a sample of $n = 234$), the results in the next table were obtained.

**a.** Write the least squares prediction equation for $y =$ total number of runs scored by a team in a season.

**b.** Conduct a test of $H_0: \beta_7 = 0$ against $H_a: \beta_7 < 0$ at $\alpha = .05$. Interpret the results.

**c.** Form a 95% confidence interval for $\beta_5$. Interpret the results.

| Independent Variable | β Estimate | Standard Error |
|---|---|---|
| Intercept | 3.70 | 15.00 |
| Walks ($x_1$) | .34 | .02 |
| Singles ($x_2$) | .49 | .03 |
| Doubles ($x_3$) | .72 | .05 |
| Triples ($x_4$) | 1.14 | .19 |
| Home Runs ($x_5$) | 1.51 | .05 |
| Stolen Bases ($x_6$) | .26 | .05 |
| Caught Stealing ($x_7$) | −.14 | .14 |
| Strikeouts ($x_8$) | −.10 | .01 |
| Outs ($x_9$) | −.10 | .01 |

*Source:* Berry, S.M. "A statistician reads the sports pages: Modeling offensive ability in baseball." *Chance*, Vol. 13, No. 4, Fall 2000 (Table 2).

**d.** Predict the number of runs scored by your favorite Major League Baseball team last year. How close is the predicted value to the actual number of runs scored by your team? (Note: You can find data on your favorite team on the Internet at www.major-leaguebaseball.com.)

**12.8** Detailed interviews were conducted with over 1,000 street vendors in the city of Puebla, Mexico, in order to study the factors influencing vendors' incomes (*World Development*, Feb. 1998). Vendors were defined as individuals working in the street, and included vendors with carts and stands on wheels and excluded beggars, drug dealers, and prostitutes. The researchers collected data on gender, age, hours worked per day, annual earnings, and education level. A subset of these data appears in the table.

**a.** Write a first-order model for mean annual earnings, $E(y)$, as a function of age ($x_1$) and hours worked ($x_2$).

STREETVEN

| Vendor Number | Annual Earnings y | Age $x_1$ | Hours Worked per Day $x_2$ |
|---|---|---|---|
| 21 | $2841 | 29 | 12 |
| 53 | 1876 | 21 | 8 |
| 60 | 2934 | 62 | 10 |
| 184 | 1552 | 18 | 10 |
| 263 | 3065 | 40 | 11 |
| 281 | 3670 | 50 | 11 |
| 354 | 2005 | 65 | 5 |
| 401 | 3215 | 44 | 8 |
| 515 | 1930 | 17 | 8 |
| 633 | 2010 | 70 | 6 |
| 677 | 3111 | 20 | 9 |
| 710 | 2882 | 29 | 9 |
| 800 | 1683 | 15 | 5 |
| 914 | 1817 | 14 | 7 |
| 997 | 4066 | 33 | 12 |

*Source:* Adapted from Smith, P. A., and Metzger, M. R. "The return to education: Street vendors in Mexico." *World Development*, Vol. 26, No. 2, Feb. 1998, pp. 289–296.

**SAS Output for Exercise 12.8**

The REG Procedure
Dependent Variable: EARNINGS

Analysis of Variance

| Source | DF | Sum of Squares | Mean Square | F Value | Pr > F |
|---|---|---|---|---|---|
| Model | 2 | 5018232 | 2509116 | 8.36 | 0.0053 |
| Error | 12 | 3600196 | 300016 | | |
| Corrected Total | 14 | 8618428 | | | |

| | | | |
|---|---|---|---|
| Root MSE | 547.73748 | R-Square | 0.5823 |
| Dependent Mean | 2577.13333 | Adj R-Sq | 0.5126 |
| Coeff Var | 21.25375 | | |

Parameter Estimates

| Variable | DF | Parameter Estimate | Standard Error | t Value | Pr > \|t\| | 95% Confidence Limits | |
|---|---|---|---|---|---|---|---|
| Intercept | 1 | -20.35201 | 652.74532 | -0.03 | 0.9756 | -1442.56189 | 1401.85787 |
| AGE | 1 | 13.35045 | 7.67168 | 1.74 | 0.1074 | -3.36470 | 30.06559 |
| HOURS | 1 | 243.71446 | 63.51174 | 3.84 | 0.0024 | 105.33428 | 382.09465 |

**b.** The model was fit to the data using SAS. Find the least squares prediction equation on the printout shown above.

**c.** Interpret the estimated $\beta$ coefficients in your model.

**d.** Is age $(x_1)$ a statistically useful predictor of annual earnings? Test using $\alpha = .01$.

**e.** Find a 95% confidence interval for $\beta_2$. Interpret the interval in the words of the problem.

**12.9** A Minneapolis, Minnesota, real estate appraiser used regression analysis to explore the relationship between the sale prices of apartment buildings sold in Minneapolis and various characteristics of the properties. Twenty-five apartment buildings were randomly sampled from all apartment buildings that were sold during a recent year. The table on p. 591 lists the data collected by the appraiser. The real estate appraiser hypothesized that the sale price (that is, market value) of an apartment building is related to the other variables in the table according to the model $y = \beta_0 + \beta_1 x_1 + \beta_2 x_2 + \beta_3 x_3 + \beta_4 x_4 + \beta_5 x_5 + \varepsilon$.

**a.** Fit the real estate appraiser's model to the data in the table. Report the least squares prediction equation.

**b.** Find the standard deviation of the regression model and interpret its value in the context of this problem.

**c.** Do the data provide sufficient evidence to conclude that value increases with the number of units in an apartment building? Report the observed significance level and reach a conclusion using $\alpha = .05$.

**d.** Interpret the value of $\hat{\beta}_1$ in terms of these data. Remember that your interpretation must recognize the presence of the other variables in the model.

**e.** Construct a scattergram of sale price versus age. What does your scattergram suggest about the relationship between these variables?

**f.** Test $H_0: \beta_2 = 0$ against $H_a: \beta_2 < 0$ using $\alpha = .01$. Interpret the result in the context of the problem. Does the result agree with your observation in part **e**? Why is it reasonable to conduct a one-tailed rather than a two-tailed test of this null hypothesis?

**g.** What is the observed significance level of the hypothesis test of part **f**?

**12.10** Empirical research was conducted to investigate the variables that impact the size distribution of manufacturing firms in international markets (*World Development*, Vol. 20, 1992). Data collected on $n = 54$ countries were used to model the country's size distribution $y$, measured as the share of manufacturing firms in the country with 100 or more workers. The model studied was $E(y) = \beta_0 + \beta_1 x_1 + \beta_2 x_2 + \beta_3 x_3 + \beta_4 x_4 + \beta_5 x_5$, where

$x_1 =$ Natural logarithm of Gross National Product (LGNP)

$x_2 =$ Geographic area per capita (in thousands of square meters) (AREAC)

$x_3 =$ Share of heavy industry in manufacturing value added (SVA)

$x_4 =$ Ratio of credit claims on the private sector to Gross Domestic Product (CREDIT)

$x_5 =$ Ratio of stock equity shares to Gross Domestic Product (STOCK)

**a.** The researchers hypothesized that the higher the credit ratio of a country, the smaller the size distribution of manufacturing firms. Explain how to test this hypothesis.

 MNSALES

| Code No. | Sale Price y ($) | No. of Apartments $x_1$ | Age of Structure $x_2$ (years) | Lot Size $x_3$ (sq. ft) | No. of On-Site Parking Spaces $x_4$ | Gross Building Area $x_5$ (sq. ft) |
|---|---|---|---|---|---|---|
| 0229 | 90,300 | 4 | 82 | 4,635 | 0 | 4,266 |
| 0094 | 384,000 | 20 | 13 | 17,798 | 0 | 14,391 |
| 0043 | 157,500 | 5 | 66 | 5,913 | 0 | 6,615 |
| 0079 | 676,200 | 26 | 64 | 7,750 | 6 | 34,144 |
| 0134 | 165,000 | 5 | 55 | 5,150 | 0 | 6,120 |
| 0179 | 300,000 | 10 | 65 | 12,506 | 0 | 14,552 |
| 0087 | 108,750 | 4 | 82 | 7,160 | 0 | 3,040 |
| 0120 | 276,538 | 11 | 23 | 5,120 | 0 | 7,881 |
| 0246 | 420,000 | 20 | 18 | 11,745 | 20 | 12,600 |
| 0025 | 950,000 | 62 | 71 | 21,000 | 3 | 39,448 |
| 0015 | 560,000 | 26 | 74 | 11,221 | 0 | 30,000 |
| 0131 | 268,000 | 13 | 56 | 7,818 | 13 | 8,088 |
| 0172 | 290,000 | 9 | 76 | 4,900 | 0 | 11,315 |
| 0095 | 173,200 | 6 | 21 | 5,424 | 6 | 4,461 |
| 0121 | 323,650 | 11 | 24 | 11,834 | 8 | 9,000 |
| 0077 | 162,500 | 5 | 19 | 5,246 | 5 | 3,828 |
| 0060 | 353,500 | 20 | 62 | 11,223 | 2 | 13,680 |
| 0174 | 134,400 | 4 | 70 | 5,834 | 0 | 4,680 |
| 0084 | 187,000 | 8 | 19 | 9,075 | 0 | 7,392 |
| 0031 | 155,700 | 4 | 57 | 5,280 | 0 | 6,030 |
| 0019 | 93,600 | 4 | 82 | 6,864 | 0 | 3,840 |
| 0074 | 110,000 | 4 | 50 | 4,510 | 0 | 3,092 |
| 0057 | 573,200 | 14 | 10 | 11,192 | 0 | 23,704 |
| 0104 | 79,300 | 4 | 82 | 7,425 | 0 | 3,876 |
| 0024 | 272,000 | 5 | 82 | 7,500 | 0 | 9,542 |

*Source:* Robinson Appraisal Co., Inc., Mankato, Minnesota.

**b.** The researchers hypothesized that the higher the stock ratio of a country, the larger the size distribution of manufacturing firms. Explain how to test this hypothesis.

### Applying the Concepts—Intermediate

**12.11** Refer to the *Journal of Applied Ecology* (Vol. 32, 1995) study of the feeding habits of baby snow geese, Exercise 11.60 (p. 549). The data on gosling weight change,

```
STATISTIX Output for Exercise 12.11

UNWEIGHTED LEAST SQUARES LINEAR REGRESSION OF WTCHANGE

PREDICTOR
VARIABLES    COEFFICIENT    STD ERROR    STUDENT'S T        P       VIF
---------    -----------    ---------    -------------    ------    -----

CONSTANT       12.1804       4.40236         2.77        0.0086
DIGEST         -0.02654      0.05349        -0.50        0.6226     4.4
ADFIBRE        -0.45783      0.12828        -3.57        0.0010     4.4


R-SQUARED             0.5288    RESID. MEAN SQUARE (MSE)    12.3868
ADJUSTED R-SQUARED    0.5046    STANDARD DEVIATION           3.51948


SOURCE        DF        SS           MS         F         P
----------    ---    ----------    ----------    -----    ------

REGRESSION     2      542.035       271.017     21.88    0.0000
RESIDUAL      39      483.084        12.3868
TOTAL         41     1025.12


CASES INCLUDED 42    MISSING CASES 0
```

digestion efficiency, acid-detergent fibre (all measured as percentages) and diet (plants or duck chow) for 42 feeding trials are reproduced in the accompanying table. The botanists were interested in predicting weight change ($y$) as a function of the other variables. The first-order model $E(y) = \beta_0 + \beta_1 x_1 + \beta_2 x_2$, where $x_1$ is digestion efficiency and $x_2$ is acid-detergent fibre, was fit to the data. The STATISTIX printout is given on p. 591.

**a.** Find the least squares prediction equation for weight change, $y$.
**b.** Interpret the $\beta$-estimates in the equation, part **a**.
**c.** Conduct a test to determine if digestion efficiency, $x_1$, is a useful linear predictor of weight change. Use $\alpha = .01$.
**d.** Form a 99% confidence interval for $\beta_2$. Interpret the result.

🦢 SNOWGEESE

| Feeding Trial | Diet | Weight Change (%) | Digestion Efficiency (%) | Acid-Detergent Fibre (%) |
|---|---|---|---|---|
| 1 | Plants | −6 | 0 | 28.5 |
| 2 | Plants | −5 | 2.5 | 27.5 |
| 3 | Plants | −4.5 | 5 | 27.5 |
| 4 | Plants | 0 | 0 | 32.5 |
| 5 | Plants | 2 | 0 | 32 |
| 6 | Plants | 3.5 | 1 | 30 |
| 7 | Plants | −2 | 2.5 | 34 |
| 8 | Plants | −2.5 | 10 | 36.5 |
| 9 | Plants | −3.5 | 20 | 28.5 |
| 10 | Plants | −2.5 | 12.5 | 29 |
| 11 | Plants | −3 | 28 | 28 |
| 12 | Plants | −8.5 | 30 | 28 |
| 13 | Plants | −3.5 | 18 | 30 |
| 14 | Plants | −3 | 15 | 31 |
| 15 | Plants | −2.5 | 17.5 | 30 |
| 16 | Plants | −.5 | 18 | 22 |
| 17 | Plants | 0 | 23 | 22.5 |
| 18 | Plants | 1 | 20 | 24 |
| 19 | Plants | 2 | 15 | 23 |
| 20 | Plants | 6 | 31 | 21 |
| 21 | Plants | 2 | 15 | 24 |
| 22 | Plants | 2 | 21 | 23 |
| 23 | Plants | 2.5 | 30 | 22.5 |
| 24 | Plants | 2.5 | 33 | 23 |
| 25 | Plants | 0 | 27.5 | 30.5 |
| 26 | Plants | .5 | 29 | 31 |
| 27 | Plants | −1 | 32.5 | 30 |
| 28 | Plants | −3 | 42 | 24 |
| 29 | Plants | −2.5 | 39 | 25 |
| 30 | Plants | −2 | 35.5 | 25 |
| 31 | Plants | .5 | 39 | 20 |
| 32 | Plants | 5.5 | 39 | 18.5 |
| 33 | Plants | 7.5 | 50 | 15 |
| 34 | Duck Chow | 0 | 62.5 | 8 |
| 35 | Duck Chow | 0 | 63 | 8 |
| 36 | Duck Chow | 2 | 69 | 7 |
| 37 | Duck Chow | 8 | 42.5 | 7.5 |
| 38 | Duck Chow | 9 | 59 | 8.5 |
| 39 | Duck Chow | 12 | 52.5 | 8 |
| 40 | Duck Chow | 8.5 | 75 | 6 |
| 41 | Duck Chow | 10.5 | 72.5 | 6.5 |
| 42 | Duck Chow | 14 | 69 | 7 |

*Source:* Gadallah, F.L., and Jefferies, R. L. "Forage quality in brood rearing areas of the lesser snow goose and the growth of captive goslings." *Journal of Applied Biology*, Vol. 32, No. 2, 1995, pp. 281–282 (adapted from Figures 2 and 3).

**12.12** A quasar is a distant celestial object (at least four billion light-years away) that provides a powerful source of radio energy. The *Astronomical Journal* (July 1995) reported on a study of 90 quasars detected by a deep space survey. The survey enabled astronomers to measure several different quantitative characteristics of each quasar, including redshift range, line flux (erg/cm$^2 \cdot$ s), line luminosity (erg/s), AB$_{1450}$ magnitude, absolute magnitude, and rest frame equivalent width. The data for a sample of 25 large (redshift) quasars is listed in the table below.

**a.** Hypothesize a first-order model for equivalent width, $y$, as a function of the first four variables in the table.

**b.** The first-order model is fit to the data using SPSS. The printout is provided. Give the least squares prediction equation.

**c.** Interpret the $\beta$ estimates in the model.

**SPSS Output for Exercise 12.12**

**Coefficients[a]**

| Model | | Unstandardized Coefficients B | Std. Error | Standardized Coefficients Beta | t | Sig. |
|---|---|---|---|---|---|---|
| 1 | (Constant) | 21087.951 | 18553.161 | | 1.137 | .269 |
| | REDSHIFT | 108.451 | 88.740 | 1.102 | 1.222 | .236 |
| | LINEFLUX | 557.910 | 316.000 | 2.700 | 1.766 | .093 |
| | LINELUM | -340.166 | 320.763 | -1.412 | -1.060 | .302 |
| | AB1450 | 85.681 | 6.273 | 1.230 | 13.658 | .000 |

[a.] Dependent Variable: RFEWIDTH

🌐 QUASAR

| Quasar | Redshift $(x_1)$ | Line Flux $(x_2)$ | Line Luminosity $(x_3)$ | AB$_{1450}$ $(x_4)$ | Absolute Magnitude $(x_5)$ | Rest Frame Equivalent Width $(y)$ |
|---|---|---|---|---|---|---|
| 1 | 2.81 | −13.48 | 45.29 | 19.50 | −26.27 | 117 |
| 2 | 3.07 | −13.73 | 45.13 | 19.65 | −26.26 | 82 |
| 3 | 3.45 | −13.87 | 45.11 | 18.93 | −27.17 | 33 |
| 4 | 3.19 | −13.27 | 45.63 | 18.59 | −27.39 | 92 |
| 5 | 3.07 | −13.56 | 45.30 | 19.59 | −26.32 | 114 |
| 6 | 4.15 | −13.95 | 45.20 | 19.42 | −26.97 | 50 |
| 7 | 3.26 | −13.83 | 45.08 | 19.18 | −26.83 | 43 |
| 8 | 2.81 | −13.50 | 45.27 | 20.41 | −25.36 | 259 |
| 9 | 3.83 | −13.66 | 45.41 | 18.93 | −27.34 | 58 |
| 10 | 3.32 | −13.71 | 45.23 | 20.00 | −26.04 | 126 |
| 11 | 2.81 | −13.50 | 45.27 | 18.45 | −27.32 | 42 |
| 12 | 4.40 | −13.96 | 45.25 | 20.55 | −25.94 | 146 |
| 13 | 3.45 | −13.91 | 45.07 | 20.45 | −25.65 | 124 |
| 14 | 3.70 | −13.85 | 45.19 | 19.70 | −26.51 | 75 |
| 15 | 3.07 | −13.67 | 45.19 | 19.54 | −26.37 | 85 |
| 16 | 4.34 | −13.93 | 45.27 | 20.17 | −26.29 | 109 |
| 17 | 3.00 | −13.75 | 45.08 | 19.30 | −26.58 | 55 |
| 18 | 3.88 | −14.17 | 44.92 | 20.68 | −25.61 | 91 |
| 19 | 3.07 | −13.92 | 44.94 | 20.51 | −25.41 | 116 |
| 20 | 4.08 | −14.28 | 44.86 | 20.70 | −25.67 | 75 |
| 21 | 3.62 | −13.82 | 45.20 | 19.45 | −26.73 | 63 |
| 22 | 3.07 | −14.08 | 44.78 | 19.90 | −26.02 | 46 |
| 23 | 2.94 | −13.82 | 44.99 | 19.49 | −26.35 | 55 |
| 24 | 3.20 | −14.15 | 44.75 | 20.89 | −25.09 | 99 |
| 25 | 3.24 | −13.74 | 45.17 | 19.17 | −26.83 | 53 |

*Source:* Schmidt, M., Schneider, D. P., and Gunn, J. E. "Spectroscopic CCD surveys for quasars at large redshift." *The Astronomical Journal,* Vol. 110, No. 1, July 1995, p. 70 (Table 1).

**d.** Test to determine whether redshift $(x_1)$ is a useful linear predictor of equivalent width $(y)$, using $\alpha = .05$.

**12.13** In the oil industry, water that mixes with crude oil during production and transportation must be removed. Chemists have found that the oil can be extracted from the water/oil mix electrically. Researchers at the University of Bergen (Norway) conducted a series of experiments to study the factors that influence the voltage $(y)$ required to separate the water from the oil (*Journal of Colloid and Interface Science*, Aug.1995). The seven independent variables investigated in the study are listed in the table. (Each variable was measured at two levels—a "low" level and a "high" level.) Sixteen water/oil mixtures were prepared using different combinations of the independent variables; then each emulsion was exposed to a high electric field. In addition, three mixtures were tested when all independent variables were set to 0. The data for all 19 experiments are also given in the table below.

**a.** Propose a first-order model for $y$ as a function of all seven independent variables.

**b.** Use a statistical software package to fit the model to the data in the table.

**c.** Fully interpret the $\beta$ estimates.

**12.14** *Sociology of Sport Journal* (Spring 1992) published research on Proposition 48, the NCAA regulation that requires a minimum of 700 on the Scholastic Assessment Test (SAT) for eligibility for college athletics. Critics have claimed that the SAT is biased against black student athletes and is not a valid predictor of academic success. To investigate the validity of the SAT as a predictor of academic success, high school GPA and study time were also considered as potential predictors of college GPA. The following variables are defined:

$y$ = College GPA (the best measure of academic success)
$x_1$ = High school GPA
$x_2$ = Hours of studying in a season
$x_3$ = SAT score

The study conducted separate regression analyses for black and white student athletes. Some of the results are shown in the table on p. 595.

**a.** Write the model relating the mean value of the college GPA as a linear function of the three identified explanatory variables. Explain the meaning of each of the $\beta$ parameters in your model.

**b.** Note that the estimate of the $\beta$ parameter for SAT score for both models is .001. Interpret this estimate.

**c.** The reported *p*-value for testing $H_0$: $\beta_1 = 0$ is .734 for the black athletes' model and .000 for the white athletes' model. Interpret these values.

⊘WATEROIL

| Experiment Number | Voltage<br>$y$<br>(kw/cm) | Disperse Phase Volume<br>$x_1$<br>(%) | Salinity<br>$x_2$<br>(%) | Temperature<br>$x_3$<br>(°C) | Time Delay<br>$x_4$<br>(hours) | Surfactant Concentration<br>$x_5$<br>(%) | Span: Triton<br>$x_6$ | Solid Particles<br>$x_7$<br>(%) |
|---|---|---|---|---|---|---|---|---|
| 1 | .64 | 40 | 1 | 4 | .25 | 2 | .25 | .5 |
| 2 | .80 | 80 | 1 | 4 | .25 | 4 | .25 | 2 |
| 3 | 3.20 | 40 | 4 | 4 | .25 | 4 | .75 | .5 |
| 4 | .48 | 80 | 4 | 4 | .25 | 2 | .75 | 2 |
| 5 | 1.72 | 40 | 1 | 23 | .25 | 4 | .75 | 2 |
| 6 | .32 | 80 | 1 | 23 | .25 | 2 | .75 | .5 |
| 7 | .64 | 40 | 4 | 23 | .25 | 2 | .25 | 2 |
| 8 | .68 | 80 | 4 | 23 | .25 | 4 | .25 | .5 |
| 9 | .12 | 40 | 1 | 4 | 24 | 2 | .75 | 2 |
| 10 | .88 | 80 | 1 | 4 | 24 | 4 | .75 | .5 |
| 11 | 2.32 | 40 | 4 | 4 | 24 | 4 | .25 | 2 |
| 12 | .40 | 80 | 4 | 4 | 24 | 2 | .25 | .5 |
| 13 | 1.04 | 40 | 1 | 23 | 24 | 4 | .25 | .5 |
| 14 | .12 | 80 | 1 | 23 | 24 | 2 | .25 | 2 |
| 15 | 1.28 | 40 | 4 | 23 | 24 | 2 | .75 | .5 |
| 16 | .72 | 80 | 4 | 23 | 24 | 4 | .75 | 2 |
| 17 | 1.08 | 0 | 0 | 0 | 0 | 0 | 0 | 0 |
| 18 | 1.08 | 0 | 0 | 0 | 0 | 0 | 0 | 0 |
| 19 | 1.04 | 0 | 0 | 0 | 0 | 0 | 0 | 0 |

*Source:* Førdedal, H., *et al.* "A multivariate analysis of W/O emulsions in high external electric fields as studied by means of dielectric time domain spectroscopy." *Journal of Colloid and Interface Science*, Vol. 173, No. 2, Aug. 1995, p. 398 (Table 2).

**Results for Exercise 12.14**

|  | Parameter Estimate | Standard Error |
|---|---|---|
| **Blacks** | | |
| Intercept | 2.245 | .09 |
| High School GPA | −.040 | .01 |
| Study Hours | .14 | .10 |
| SAT Score | .001 | .0025 |
| **Whites** | | |
| Intercept | 1.970 | .16 |
| High School GPA | −.089 | .01 |
| Study Hours | .20 | .23 |
| SAT Score | .001 | .0002 |

## Applying the Concepts—Advanced

**12.15** The vineyards in the Bordeaux region of France are known for producing excellent red wines. However, the uncertainty of the weather during the growing season, the phenomenon that wine tastes better with age, and the fact that some Bordeaux vineyards produce better wines than others, encourages speculation concerning the value of a case of wine produced by a certain vineyard during a certain year (or vintage). As a result, many wine experts attempt to predict the auction price of a case of Bordeaux wine. The publishers of a newsletter titled *Liquid Assets: The International Guide to Fine Wine* discussed a multiple regression approach to predicting the London auction price of red Bordeaux wine in *Chance* (Fall 1995). The natural logarithm of the price $y$ (in dollars) of a case containing a dozen bottles of red wine was modeled as a function of weather during growing season and age of vintage using data collected for the vintages of 1952–1980. Three models were fit to the data. The results of the regressions are summarized in the table below.

**a.** For each model, conduct a $t$-test for each of the $\beta$ parameters in the model. Interpret the results.

**b.** When the natural log of $y$ is used as a dependent variable, the antilogarithm of a $\beta$ coefficient minus 1, that is $e^{\beta_i} - 1$, represents the percentage change in $y$ for every 1-unit increase in the associated $x$ value.* Use this information to interpret the $\beta$ estimates of each model.

| Independent Variables | Beta Estimates (Standard Errors) | | |
|---|---|---|---|
|  | **Model 1** | **Model 2** | **Model 3** |
| $x_1$ = Vintage year | .0354 (.0137) | .0238 (.00717) | .0240 (.00747) |
| $x_2$ = Average growing season temperature (°C) | (not included) | .616 (.0952) | .608 (.116) |
| $x_3$ = Sept./Aug. rainfall (cm) | (not included) | −.00386 (.00081) | −.00380 (.00095) |
| $x_4$ = Rainfall in months preceding vintage (cm) | (not included) | .0001173 (.000482) | .00115 (.000505) |
| $x_5$ = Average Sept. temperature (°C) | (not included) | (not included) | .00765 (.565) |

*Source:* Ashenfelter, O., Ashmore, D., and LaLonde, R. "Bordeaux wine vintage quality and weather." *Chance*, Vol. 8, No. 4, Fall 1995, p. 116 (Table 2).

## 12.5 CHECKING THE OVERALL UTILITY OF A MODEL

Conducting $t$-tests on each $\beta$ parameter in a model is *not* the best way to determine whether the overall model is contributing information for the prediction of $y$. If we were to conduct a series of $t$-tests to determine whether the independent variables are contributing to the predictive relationship, we would be very likely to make one or more errors in deciding which terms to retain in the model and which to exclude.

For example, suppose you fit a first-order model in 10 quantitative $x$ variables and decide to conduct $t$-tests on all 10 of the individual $\beta$'s in the model, each at $\alpha = .05$. Even if all the $\beta$ parameters (except $\beta_0$) are equal to 0, approximately 40% of the time you will incorrectly reject the null hypothesis at least once and conclude that

---

*The result is derived by expressing the percentage change in price $y$ as $(y_1 - y_0)/y_0$, where $y_1$ = the value of $y$ when, say, $x = 1$, and $y_0$ = the value of $y$ when $x = 0$. Now let $y^* = \ln(y)$ and assume the model is $y^* = \beta_0 + \beta_1 x$. Then

$$y = e^{y^*} = e^{\beta_0}e^{\beta_1 x} = \begin{cases} e^{\beta_0} & \text{when } x = 0 \\ e^{\beta_0}e^{\beta_1} & \text{when } x = 1 \end{cases}$$

Substituting, we have

$$\frac{y_1 - y_0}{y_0} = \frac{e^{\beta_0}e^{\beta_1} - e^{\beta_0}}{e^{\beta_0}} = e^{\beta_1} - 1$$

some $\beta$ parameter differs from 0.* Thus, in multiple regression models for which a large number of independent variables are being considered, conducting a series of *t*-tests may include a large number of insignificant variables and exclude some useful ones. If we want to test the utility of a multiple regression model, we will need a *global test* (one that encompasses all the $\beta$ parameters). We would also like to find some statistical quantity that measures how well the model fits the data.

We commence with the easier problem—finding a measure of how well a linear model fits a set of data. For this we use the multiple regression equivalent of $r^2$, the coefficient of determination for the straight-line model (Chapter 11), as shown in the box.

---

**DEFINITION 12.1**

The **multiple coefficient of determination, $R^2$**, is defined as

$$R^2 = 1 - \frac{\text{SSE}}{\text{SS}_{yy}} = \frac{\text{SS}_{yy} - \text{SSE}}{\text{SS}_{yy}} = \frac{\text{Explained variablilty}}{\text{Total variability}}$$

---

Just as for the simple linear model, $R^2$ represents the fraction of the sample variation of the *y* values (measured by $\text{SS}_{yy}$) that is explained by the least squares prediction equation. Thus, $R^2 = 0$ implies a complete lack of fit of the model to the data and $R^2 = 1$ implies a perfect fit with the model passing through every data point. In general, the larger the value of $R^2$, the better the model fits the data.

To illustrate, the value $R^2 = .8974$ for the sale price model of Example 12.1 is highlighted in Figure 12.7. This high value of $R^2$ implies that using the independent variables land value, appraised improvements, and home size in a first-order model explains 89.7% of the total *sample variation* (measured by $\text{SS}_{yy}$) of sale price *y*. Thus, $R^2$ is a sample statistic that tells how well the model fits the data and thereby represents a measure of the usefulness of the entire model.

A large value of $R^2$ computed from the *sample* data does not necessarily mean that the model provides a good fit to all of the data points in the *population*. For example, a first-order linear model that contains three parameters will provide a perfect fit to a sample of three data points and $R^2$ will equal 1. Likewise, you will always obtain a perfect fit ($R^2 = 1$) to a set of *n* data points if the model contains exactly *n* parameters. Consequently, if you want to use the value of $R^2$ as a measure of how useful the model will be for predicting *y*, it should be based on a sample that contains substantially more data points than the number of parameters in the model.

---

**Caution**

In a multiple regression analysis, use the value of $R^2$ as a measure of how useful a linear model will be for predicting *y* only if the sample contains substantially more data points than the number of $\beta$ parameters in the model.

---

*The proof of this result proceeds as follows:

$P(\text{Reject } H_0 \text{ at least once} | \beta_1 = \beta_2 = \cdots = \beta_{10} = 0)$

$= 1 - P(\text{Reject } H_0 \text{ no times} | \beta_1 = \beta_2 = \cdots = \beta_{10} = 0)$

$\leq 1 - [P(\text{Accept } H_0 : \beta_1 = 0 | \beta_1 = 0) \cdot P(\text{Accept } H_0 : \beta_2 = 0 | \beta_2 = 0)| \cdots \cdot P(\text{Accept } H_0 : \beta_{10} = 0 | \beta_{10} = 0)]$

$= 1 - [(1 - \alpha)^{10}] = 1 - (.95)^{10} = .401$ .

**FIGURE 12.7**

*SAS Regression Printout for the Sale Price Model*

The REG Procedure
Dependent Variable: SALEPRIC

Analysis of Variance

| Source | DF | Sum of Squares | Mean Square | F Value | Pr > F |
|---|---|---|---|---|---|
| Model | 3 | 8779676741 | 2926558914 | 46.66 | <.0001 |
| Error | 16 | 1003491259 | 62718204 | | |
| Corrected Total | 19 | 9783168000 | | | |

| | | | | |
|---|---|---|---|---|
| Root MSE | 7919.48254 | R-Square | 0.8974 |
| Dependent Mean | 56660 | Adj R-Sq | 0.8782 |
| Coeff Var | 13.97720 | | |

Parameter Estimates

| Variable | DF | Parameter Estimate | Standard Error | t Value | Pr > \|t\| |
|---|---|---|---|---|---|
| Intercept | 1 | 1470.27592 | 5746.32458 | 0.26 | 0.8013 |
| LANDVAL | 1 | 0.81449 | 0.51222 | 1.59 | 0.1314 |
| IMPROVAL | 1 | 0.82044 | 0.21118 | 3.88 | 0.0013 |
| AREA | 1 | 13.52865 | 6.58568 | 2.05 | 0.0567 |

As an alternative to using $R^2$ as a measure of model adequacy, the *adjusted multiple coefficient of determination*, denoted $R_a^2$, is often reported. The formula for $R_a^2$ is shown in the box.

> ### DEFINITION 12.2
> The **adjusted multiple coefficient of determination** is given by
>
> $$R_a^2 = 1 - \left[\frac{(n-1)}{n-(k+1)}\right]\left(\frac{\text{SSE}}{\text{SS}_{yy}}\right)$$
>
> $$= 1 - \left[\frac{(n-1)}{n-(k+1)}\right](1 - R^2)$$
>
> *Note:* $R_a^2 \leq R^2$

$R^2$ and $R_a^2$ have similar interpretations. However, unlike $R^2$, $R_a^2$ takes into account ("adjusts" for) both the sample size $n$ and the number of $\beta$ parameters in the model. $R_a^2$ will always be smaller than $R^2$, and more importantly, cannot be "forced" to 1 by simply adding more and more independent variables to the model. Consequently, analysts prefer the more conservative $R_a^2$ when choosing a measure of model adequacy. The value of $R_a^2$ is also highlighted in Figure 12.7. Note that $R_a^2 = .8782$, a value only slightly smaller than $R^2$.

Despite their utility, $R^2$ and $R_a^2$ are only sample statistics. Therefore, it is dangerous to judge the global usefulness of the model based solely on these values. A better method is to conduct a test of hypothesis involving *all* the $\beta$ parameters (except $\beta_0$) in a model. In particular, for the sale price model (Example 12.1), we would test

$$H_0: \beta_1 = \beta_2 = \beta_3 = 0$$
$$H_a: \text{At least one of the coefficients is nonzero}$$

The test statistic used to test this hypothesis is an $F$ statistic, and several equivalent versions of the formula can be used (although we will usually rely on the computer to calculate the $F$ statistic):

$$\textit{Test statistic:} \ \ F = \frac{(SS_{yy} - SSE)/k}{SSE/[n - (k + 1)]} = \frac{R^2/k}{(1 - R^2)/[n - (k + 1)]}$$

Both these formulas indicate that the $F$ statistic is the ratio of the *explained* variability divided by the model degrees of freedom to the *unexplained* variability divided by the error degrees of freedom. Thus, the larger the proportion of the total variability accounted for by the model, the larger the $F$ statistic.

To determine when the ratio becomes large enough that we can confidently reject the null hypothesis and conclude that the model is more useful than no model at all for predicting $y$, we compare the calculated $F$ statistic to a tabulated $F$ value with $k$ df in the numerator and $[n - (k + 1)]$ df in the denominator. Recall that tabulations of the $F$-distribution for various values of $\alpha$ are given in Tables VIII, IX, X, and XI of Appendix A.

*Rejection region:* $F > F_\alpha$, where $F$ is based on $k$ numerator and $n - (k + 1)$ denominator degrees of freedom.

For the sale price example $[n = 20, k = 3, n - (k + 1) = 16, \text{ and } \alpha = .05]$, we will reject $H_0: \beta_1 = \beta_2 = \beta_3 = 0$ if

$$F > F_{.05} = 3.24$$

The computed $F$ value, highlighted in Figure 12.7, is $F = 46.66$. Since this value greatly exceeds the tabulated value of 3.24, we conclude that at least one of the model coefficients $\beta_1$, $\beta_2$, and $\beta_3$ is nonzero. Therefore, this **global $F$-test** indicates that the first-order model $y = \beta_0 + \beta_1 x_1 + \beta_2 x_2 + \beta_3 x_3 + \varepsilon$ is useful for predicting sale price.

Like SAS, most other regression packages give the $F$ value in a portion of the printout called the "Analysis of Variance." This is an appropriate descriptive term, since the $F$ statistic relates the explained and unexplained portions of the total variance of $y$. For example, the elements of the SAS printout in Figure 12.7 that lead to the calculation of the $F$ value are:

$$\begin{aligned} F \text{ value} &= \frac{\text{Sum of squares (Model)}/\text{df(Model)}}{\text{Sum of squares (Error)}/\text{df(Error)}} = \frac{\text{Mean square (Model)}}{\text{Mean square (Error)}} \\[2mm] &= \frac{8,779,676,741/3}{1,003,491,259/16} = \frac{2,926,558,914}{62,718,204} = 46.66 \end{aligned}$$

Note, too, that the observed significance level for the $F$ statistic (highlighted in Figure 12.7) is .0001, which means that we would reject the null hypothesis $H_0: \beta_1 = \beta_2 = \beta_3 = 0$ at any $\alpha$ value greater than .0001.

The analysis of variance $F$-test for testing the usefulness of the model is summarized in the next box.

## Testing Global Usefulness of the Model: The Analysis of Variance *F*-Test

$H_0: \beta_1 = \beta_2 = \cdots = \beta_k = 0$   (All model terms are unimportant for predicting $y$)
$H_a$: At least one $\beta_i \neq 0$       (At least one model term is useful for predicting $y$)

$$\text{Test statistic: } F = \frac{(\text{SS}_{yy} - \text{SSE})/k}{\text{SSE}/[n - (k + 1)]} = \frac{R^2/k}{(1 - R^2)/[n - (k + 1)]}$$

$$= \frac{\text{Mean square (Model)}}{\text{Mean square (Error)}}$$

where $n$ is the sample size and $k$ is the number of terms in the model.
*Rejection region:* $F > F_\alpha$, with $k$ numerator degrees of freedom and $[n - (k + 1)]$ denominator degrees of freedom.
*Assumptions:* The standard regression assumptions about the random error component (Section 12.3).

## Caution

A rejection of the null hypothesis $H_0: \beta_1 = \beta_2 = \cdots = \beta_k$ in the global *F*-test leads to the conclusion [with $100(1 - \alpha)\%$ confidence] that the model is statistically useful. However, statistically "useful" does not necessarily mean "best." Another model may prove even more useful in terms of providing more reliable estimates and predictions. This global *F*-test is usually regarded as a test that the model *must* pass to merit further consideration.

**EXAMPLE 12.4**

Refer to Example 12.3, in which an antique collector modeled the auction price $y$ of grandfather clocks as a function of the age of the clock, $x_1$, and the number of bidders, $x_2$. The hypothesized first-order model is

$$y = \beta_0 + \beta_1 x_1 + \beta_2 x_2 + \varepsilon$$

A sample of 32 observations is obtained, with the results summarized in the MINITAB printout repeated in Figure 12.8.

a. Find and interpret the adjusted coefficient of determination $R_a^2$ for this example.
b. Conduct the global *F*-test of model usefulness at the $\alpha = .05$ level of significance.

**FIGURE 12.8**

*MINITAB Regression Printout for Grandfather Clock Model*

```
The regression equation is
PRICE = - 1339 + 12.7 AGE + 86.0 BIDDERS


Predictor        Coef     SE Coef         T       P
Constant      -1339.0       173.8     -7.70   0.000
AGE           12.7406      0.9047     14.08   0.000
BIDDERS        85.953       8.729      9.85   0.000


S = 133.5       R-Sq = 89.2%     R-Sq(adj) = 88.5%


Analysis of Variance


Source           DF          SS         MS        F       P
Regression        2     4283063    2141531   120.19   0.000
Residual Error    29     516727      17818
Total             31    4799790
```

### Solution

a. The $R_a^2$ value (highlighted in Figure 12.8) is .885. This implies that the least squares model has explained about 88.5% of the total sample variation in $y$ values (auction prices), after adjusting for sample size and number of independent variables in the model.

b. The elements of the global test of the model follow:

$H_0: \beta_1 = \beta_2 = 0 \qquad [\textit{Note: } k = 2]$
$H_a:$ At least one of the two model coefficients is nonzero

*Test statistic:* $F = 120.19$ (see Figure 12.8)

$p$-value: .000

*Conclusion:* Since $\alpha = .05$ exceeds the observed significance level, $p = .000$, the data provide strong evidence that at least one of the model coefficients is nonzero. The overall model appears to be statistically useful for predicting auction prices. ∎

Can we be sure that the best prediction model has been found if the global $F$-test indicates that a model is useful? Unfortunately, we cannot. The addition of other independent variables may improve the usefulness of the model. (See the box, p. 599.) We consider more complex multiple regression models in Sections 12.7–12.10.

To summarize the discussion in this section, both $R^2$ and $R_a^2$ are indicators of how well the prediction equation fits the data. Intuitive evaluations of the contribution of the model based on $R^2$ must be examined with care. Unlike $R_a^2$, the value of $R^2$ increases as more and more variables are added to the model. Consequently, you could force $R^2$ to take a value very close to 1 even though the model contributes no information for the prediction of $y$. In fact, $R^2$ equals 1 when the number of terms in the model (including $\beta_0$) equals the number of data points. Therefore, you should not rely solely on the value of $R^2$ (or even $R_a^2$) to tell you whether the model is useful for predicting $y$. Use the $F$-test for testing the global utility of the model.

After we have determined that the overall model is useful for predicting $y$ using the $F$-test, we may elect to conduct one or more $t$-tests on the individual $\beta$ parameters (see Section 12.4). However, the test (or tests) to be conducted should be decided *a priori*, that is, prior to fitting the model. Also, we should limit the number of $t$-tests conducted to avoid the potential problem of making too many Type I errors. Generally, the regression analyst will conduct $t$-tests only on the "most important" $\beta$'s. We provide insight in identifying the most important $\beta$'s in a linear model in Sections 12.7–12.10.

---

### Recommendation for Checking the Utility of a Multiple Regression Model

1. First, conduct a test of overall model adequacy using the $F$-test, that is, test $H_0: \beta_1 = \beta_2 = \cdots = \beta_k = 0$
   If the model is deemed adequate (that is, if you reject $H_0$), then proceed to step 2. Otherwise, you should hypothesize and fit another model. The new model may include more independent variables or higher-order terms.

2. Conduct $t$-tests on those $\beta$ parameters in which you are particularly interested (that is, the "most important" $\beta$'s). These usually involve only the $\beta$'s associated with higher-order terms ($x_2, x_1x_2$, etc.). However, it is a safe practice to limit the number of $\beta$'s that are tested. Conducting a series of $t$-tests leads to a high overall Type I error rate $\alpha$.

## EXERCISES 12.16–12.30

### Learning the Mechanics

**12.16** Suppose you fit the first-order model

$$y = \beta_0 + \beta_1 x_1 + \beta_2 x_2 + \beta_3 x_3 + \beta_4 x_4 + \beta_5 x_5 + \varepsilon$$

to $n = 30$ data points and obtain

$$\text{SSE} = .33 \quad R^2 = .92$$

**a.** Do the values of SSE and $R^2$ suggest that the model provides a good fit to the data? Explain.

**b.** Is the model of any use in predicting $y$? Test the null hypothesis $H_0: \beta_1 = \beta_2 = \cdots = \beta_5 = 0$ against the alternative hypothesis $H_a$: At least one of the parameters $\beta_1, \beta_2, \cdots, \beta_5$ is nonzero. Use $\alpha = .05$.

**12.17** Refer to Exercise 12.2. Recall that the first-order model, $y = \beta_0 + \beta_1 x_1 + \beta_2 x_2 + \varepsilon$, was fit to $n = 20$ data using MINITAB.

**a.** Find $R^2$ on the MINITAB printout (p. 588) and interpret its value.

**b.** Find $R_a^2$ on the MINITAB printout and interpret its value.

**c.** Test the null hypothesis that $\beta_1 = \beta_2 = 0$ against the alternative hypothesis that at least one of $\beta_1$ and $\beta_2$ is nonzero. Calculate the test statistic using the two formulas given in this section, and compare your results to each other and to that given on the printout. Use $\alpha = .05$ and interpret the result of your test.

**d.** Find the observed significance level for this test on the printout and interpret it.

**12.18** If the analysis of variance $F$-test leads to the conclusion that at least one of the model parameters is nonzero, can you conclude that the model is the best predictor for the dependent variable $y$? Can you conclude that all of the terms in the model are important for predicting $y$? What is the appropriate conclusion?

**12.19** Suppose you fit the first-order model

$$y = \beta_0 + \beta_1 x_1 + \beta_2 x_2 + \varepsilon$$

to $n = 20$ data points and obtain

$$\Sigma(y_i - \hat{y}_i)^2 = 12.35 \qquad \Sigma(y_i - \bar{y})^2 = 24.44$$

**a.** Construct an analysis of variance table for this regression analysis, using the printout from Exercise 12.2 as a model. Be sure to include the sources of variability, the degrees of freedom, the sums of squares, the mean squares, and the $F$ statistic. Calculate $R^2$ and $R_a^2$ for the regression analysis.

**b.** Test the null hypothesis that $\beta_1 = \beta_2 = 0$ against the alternative hypothesis that at least one of the parameters differs from 0. Calculate the test statistic in two different ways and compare the results. Use $\alpha = .05$ to reach a conclusion about whether the model contributes information for the prediction of $y$.

### Applying the Concepts—Basic

**12.20** Refer to the *World Development* (Feb. 1998) study of street vendors in the city of Puebla, Mexico, Exercise 12.8 (p. 589). Recall that the vendors' mean annual earnings, $E(y)$, was modeled as a first-order function of age $(x_1)$ and hours worked $(x_2)$. Refer to the SAS printout on p. 590 and answer the following:

**a.** Interpret the value of $R^2$.

**b.** Interpret the value of $R_a^2$. Explain the relationship between $R^2$ and $R_a^2$.

**c.** Conduct a test of the global utility of the model (at $\alpha = .01$). Interpret the result.

**12.21** Refer to the *Professional Geographer* (Feb. 2000) study of urban and rural counties in the western United States, Exercise 3.17 (p. 118). University of Nevada (Reno) researchers asked a sample of 256 county commissioners to rate their "home" county on a scale of 1 (most rural) to 10 (most urban). The urban/rural rating $(y)$ was used as the dependent variable in a first-order multiple regression model with six independent variables: total county population $(x_1)$, population density $(x_2)$, population concentration $(x_3)$, population growth $(x_4)$, proportion of county land in farms $(x_5)$, and 5-year change in agricultural land base $(x_6)$. Some of the regression results are shown in the table.

| Independent Variable | $\beta$ Estimate | $p$-value |
|---|---|---|
| $x_1$: Total population | 0.110 | 0.045 |
| $x_2$: Population density | 0.065 | 0.230 |
| $x_3$: Population concentration | 0.540 | 0.000 |
| $x_4$: Population growth | −0.009 | 0.860 |
| $x_5$: Farm land | −0.150 | 0.003 |
| $x_6$: Agricultural change | −0.027 | 0.58 |

Overall model: $R^2 = .44$    $R_a^2 = .43$    $F = 32.47$    $p$-value $< .001$

*Source:* Berry, K. A., *et al.* "Interpreting what is rural and urban for western U.S. counties." *Professional Geographer*, Vol. 52, No. 1, Feb. 2000 (Table 2).

a. Write the least squares prediction equation for $y$.
b. Give the null hypothesis for testing overall model adequacy.
c. Conduct the test, part **b**, at $\alpha = .01$ and give the appropriate conclusion.
d. Interpret the values of $R^2$ and $R_a^2$.
e. Give the null hypothesis for testing the contribution of population growth $(x_4)$ to the model.
f. Conduct the test, part **e**, at $\alpha = .01$ and give the appropriate conclusion.

**12.22** Refer to the *Journal of Applied Ecology* study of the feeding habits of baby snow geese, Exercise 12.11 (p. 591). Recall that weight change $(y)$ was related to digestion-efficiency $(x_1)$ and acid-detergent fibre $(x_2)$ using a first-order model.
a. Locate $R^2$ and $R_a^2$ on the STATISTIX printout (p. 591). Interpret these values. Which statistic is the preferred measure of model fit? Explain.
b. Locate the global $F$ value for testing the overall model on the STATISTIX printout. Use the statistic to test the null hypothesis $H_0: \beta_1 = \beta_2 = 0$.

**12.23** Refer to the *Astronomical Journal* study of quasars, Exercise 12.12 (p. 593). A portion of the SPSS printout for the multiple regression model relating equivalent width $(y)$ to four independent variables is reproduced below.
a. Locate $R^2$ and $R_a^2$ on the SPSS printout. Interpret these values. Which statistic is the preferred measure of model fit? Explain.
b. Locate the global $F$-value for testing the overall model on the SPSS printout. Use the statistic to test the null hypothesis $H_0: \beta_1 = \beta_2 = \cdots = \beta_5 = 0$.

## Applying the Concepts—Intermediate

**12.24** An article published in the *American Educational Research Journal* (Fall 1998) used multiple regression to model the students' perceptions of their ability in science classes. The sample consisted of 165 Grade 5–Grade 8 students in six performance-based science classrooms, all of which use hands-on activities as the main teaching tool. The dependent variable of interest, the student's ability perception $(y)$, was measured on a 4-point scale (where 1 = little or no ability and 4 = high ability). The independent variables included in the model are listed in the table at the top of p. 603, as well as the results of the multiple regression.
a. Which independent variables appear to contribute to the prediction of a student's perception of his/her ability in science? Explain.
b. Construct a 95% confidence interval for the $\beta$ coefficient of the variable active-leading behavior. Interpret the interval.
c. The following statistics for evaluating the overall prediction power of the model were reported: $R^2 = .48$, $F = 12.84$, $p < .001$. Interpret the results.

**12.25** The *Journal of Quantitative Criminology* (Vol. 8, 1992) published a paper on the determinants of area property crime levels in the United Kingdom. Several multiple regression models for property crime prevalence, $y$, measured as the percentage of residents in a geographical area who were victims of at least one property crime, were examined. The results for one of the models, based on a sample of $n = 313$ responses collected for the British Crime Survey, are shown in the second table on p. 603. [*Note:* All variables except Density are expressed as a percentage of the base area.]
a. Test the hypothesis that the density $(x_1)$ of a region is positively linearly related to crime prevalence $(y)$, holding the other independent variables constant.
b. Do you advise conducting $t$-tests on each of the 18 independent variables in the model to determine which variables are important predictors of crime prevalence? Explain.
c. The model yielded $R^2 = .411$. Use this information to conduct a test of the global utility of the model. Use $\alpha = .05$.

**12.26** External auditors are hired to review and analyze the financial and other records of an organization and to attest to the integrity of the organization's financial statements. In recent years, the fees charged by auditors

**SPSS Output for Exercise 12.23**

**Model Summary**

| Model | R | R Square | Adjusted R Square | Std. Error of the Estimate |
|---|---|---|---|---|
| 1 | .955[a] | .912 | .894 | 15.42 |

a. Predictors: (Constant), AB1450, REDSHIFT, LINELUM, LINEFLUX

**ANOVA[b]**

| Model | | Sum of Squares | df | Mean Square | F | Sig. |
|---|---|---|---|---|---|---|
| 1 | Regression | 49162.671 | 4 | 12290.668 | 51.720 | .000[a] |
| | Residual | 4752.769 | 20 | 237.638 | | |
| | Total | 53915.440 | 24 | | | |

a. Predictors: (Constant), AB1450, REDSHIFT, LINELUM, LINEFLUX
b. Dependent Variable: RFEWIDTH

**Results for Exercise 12.24**

| Independent Variable | $\beta$ Estimate | Standard Error | $t$ | $p$-Value |
|---|---|---|---|---|
| Prior science attitude (4-point scale) | .60 | .07 | 8.13 | $p < .001$ |
| Science ability (standardized test score) | .003 | .005 | .57 | NS |
| Gender (1 if boy, 0 if girl) | .22 | .07 | 2.98 | $p < .01$ |
| Class 1 (1 if classroom 1, 0 if not) | .15 | .153 | 0.98 | NS |
| Class 3 (1 if classroom 3, 0 if not) | −.16 | .13 | −1.17 | NS |
| Class 4 (1 if classroom 4, 0 if not) | .00 | .13 | 0.00 | NS |
| Class 5 (1 if classroom 5, 0 if not) | .08 | .13 | 0.64 | NS |
| Class 6 (1 if classroom 6, 0 if not) | −.15 | .14 | −1.12 | NS |
| Active-leading behavior (0 to 1 scale) | .88 | .34 | 2.62 | $p < .01$ |
| Passive-assisting behavior (0 to 1 scale) | .43 | .51 | 0.85 | NS |
| Active-manipulating behavior (0 to 1 scale) | .77 | .61 | 1.25 | NS |

*Note:* Base level for classroom dummy variables is classroom 2.

NS = not significant

*Source:* Jovanovic, J., and King, S. S. "Boys and girls in the performance-based science classroom: Who's doing the performing?" *American Educational Research Journal*, Vol. 35, No. 3, Fall 1998, p. 489 (Table 8).

**Results for Exercise 12.25**

| Variable | $\hat{\beta}$ | $t$ | $p$-value |
|---|---|---|---|
| $x_1$ = Density (population per hectare) | .331 | 3.88 | $p < .01$ |
| $x_2$ = Unemployed male population | −.121 | −1.17 | $p > .10$ |
| $x_3$ = Professional population | −.187 | −1.90 | $.01 < p < .10$ |
| $x_4$ = Population aged less than 5 | −.151 | −1.51 | $p > .10$ |
| $x_5$ = Population aged between 5 and 15 | .353 | 3.42 | $p < .01$ |
| $x_6$ = Female population | .095 | 1.31 | $p > .10$ |
| $x_7$ = 10-year change in population | .130 | 1.40 | $p > .10$ |
| $x_8$ = Minority population | −.122 | −1.51 | $p > .10$ |
| $x_9$ = Young adult population | .163 | 5.62 | $p < .01$ |
| $x_{10}$ = 1 if North region, 0 if not | .369 | 1.72 | $.01 < p < .10$ |
| $x_{11}$ = 1 if Yorkshire region, 0 if not | −.210 | −1.39 | $p > .10$ |
| $x_{12}$ = 1 if East Midlands region, 0 if not | −.192 | −0.78 | $p > .10$ |
| $x_{13}$ = 1 if East Anglia region, 0 if not | −.548 | −2.22 | $.01 < p < .10$ |
| $x_{14}$ = 1 if South East region, 0 if not | .152 | 1.37 | $p > .10$ |
| $x_{15}$ = 1 if South West region, 0 if not | −.151 | −0.88 | $p > .10$ |
| $x_{16}$ = 1 if West Midlands region, 0 if not | −.308 | −1.93 | $.01 < p < .10$ |
| $x_{17}$ = 1 if North West region, 0 if not | .311 | 2.13 | $.01 < p < .10$ |
| $x_{18}$ = 1 if Wales region, 0 if not | −.019 | −0.08 | $p > .10$ |

*Source:* Osborn, D. R., Tickett, Al., and Elder, R. "Area characteristics and regional variates as determinants of area property crime." *Journal of Quantitative Criminology*, Vol. 8, No. 3, 1992, Plenum Publishing Corp.

have come under increasing scrutiny. S. Butterworth and K. A. Houghton, two University of Melbourne (Australia) researchers, investigated the effects of several variables on the fee charged by auditors. These variables are described here.

$y$ = Logarithm of audit fee charged to auditee (FEE)

$x_1 = \begin{cases} 1 \text{ if auditee changed auditors after one year} \\ \quad \text{(CHANGE)} \\ 0 \text{ if not} \end{cases}$

$x_2$ = Logarithm of auditee's total assets (SIZE)

$x_3$ = Number of subsidiaries of auditee (COMPLEX)

$x_4 = \begin{cases} 1 \text{ if auditee receives an audit qualification (RISK)} \\ 0 \text{ if not} \end{cases}$

$x_5 = \begin{cases} 1 \text{ if auditee in mining industry (INDUSTRY)} \\ 0 \text{ if not} \end{cases}$

$x_6 = \begin{cases} 1 \text{ if auditor is a member of a "Big 8" firm (BIG8)} \\ 0 \text{ if not} \end{cases}$

$x_7$ = Logarithm of dollar-value of non-audit services provided by auditor (NAS)

The multiple regression model $E(y) = \beta_0 + \beta_1 x_1 + \beta_2 x_2 + \beta_3 x_3 + \cdots + \beta_7 x_7$ was fit to data collected for $n = 268$ companies. The results are summarized in the table on p. 604.

| Independent Variable | Expected Sign of $\beta$ | $\beta$ Estimate | $t$ Value | Level of Significance ($p$-value) |
|---|---|---|---|---|
| Constant | − | −4.30 | −3.45 | .001 (two-tailed) |
| CHANGE | + | −.002 | −0.049 | .961 (one-tailed) |
| SIZE | + | .336 | 9.94 | .000 (one-tailed) |
| COMPLEX | + | .384 | 7.63 | .000 (one-tailed) |
| RISK | + | .067 | 1.76 | .079 (one-tailed) |
| INDUSTRY | − | −.143 | −4.05 | .000 (one-tailed) |
| BIG8 | + | .081 | 2.18 | .030 (one-tailed) |
| NAS | +/− | .134 | 4.54 | .000 (two-tailed) |

$R^2 = .712$  $F = 111.1$

*Source:* Butterworth, S., and Houghton, K. A. "Auditor switching: The pricing of audit services." *Journal of Business Finance and Accounting*, Vol. 22, No. 3, April 1995, p. 334 (Table 4).

a. Write the least squares prediction equation.

b. Assess the overall fit of the model.

c. Interpret the estimate of $\beta_3$.

d. The researchers hypothesized the direction of the effect of each independent variable on audit fees. These hypotheses are given in the "Expected Sign of $\beta$" column in the table. (For example, if the expected sign is negative, the alternative hypothesis is $H_a: \beta_i < 0$.) Interpret the results of the hypothesis test for $\beta_4$. Use $\alpha = .05$.

e. The main objective of the analysis was to determine whether new auditors charge less than incumbent auditors in a given year. If this hypothesis is true, then the true value of $\beta_1$ is negative. Is there evidence to support this hypothesis? Explain.

12.27 An important goal in occupational safety is "active caring." Employees demonstrate active caring (AC) about the safety of their co-workers when they identify environmental hazards and unsafe work practices and then implement appropriate corrective actions for these unsafe conditions or behaviors. Three factors hypothesized to increase the propensity for an employee to actively care for safety are (1) high self-esteem, (2) optimism, and (3) group cohesiveness. *Applied & Preventive Psychology* (Winter 1995) attempted to establish empirical support for the AC hypothesis by fitting the model $E(y) = \beta_0 + \beta_1 x_1 + \beta_2 x_2 + \beta_3 x_3$, where

$y$ = AC score (measuring active caring on a 15-point scale)
$x_1$ = Self-esteem score
$x_2$ = Optimism score
$x_3$ = Group cohesion score

The regression analysis, based on data collected for $n = 31$ hourly workers at a large fiber-manufacturing plant, yielded a multiple coefficient of determination of $R^2 = .362$.

a. Interpret the value of $R^2$.

b. Use the $R^2$ value to test the global utility of the model. Use $\alpha = .05$.

12.28 Because the coefficient of determination $R^2$ always increases when a new independent variable is added to the model, it is tempting to include many variables in a model to force $R^2$ to be near 1. However, doing so reduces the degrees of freedom available for estimating $\sigma^2$, which adversely affects our ability to make reliable inferences. Suppose you want to use 18 psychological and sociological factors to predict a student's Scholastic Assessment Test (SAT) score. You fit the model

$$y = \beta_0 + \beta_1 x_1 + \beta_2 x_2 + \cdots + \beta_{17} x_{17} + \beta_{18} x_{18} + \varepsilon$$

where $y$ = SAT score and $x_1, x_2, \ldots, x_{18}$ are the psychological and sociological factors. Only 20 years of data ($n = 20$) are used to fit the model, and you obtain $R^2 = .95$. Test to see whether this impressive-looking $R^2$ is large enough for you to infer that the model is useful, that is, that at least one term in the model is important for predicting SAT scores. Use $\alpha = .05$.

12.29 Much research—and much litigation—has been conducted on the disparity between the salary levels of men and women. Research reported in *Work and Occupations* (Nov. 1992) analyzes the salaries for a sample of 191 Illinois managers using a regression analysis with the following independent variables:

$$x_1 = \text{Gender of manager} = \begin{cases} 1 & \text{if male} \\ 0 & \text{if not} \end{cases}$$

$$x_2 = \text{Race of manager} = \begin{cases} 1 & \text{if white} \\ 0 & \text{if not} \end{cases}$$

$x_3$ = Education level (in years)
$x_4$ = Tenure with firm (in years)
$x_5$ = Number of hours worked per week

The regression results are shown in the table as they were reported in the article.

a. Write the hypothesized model that was used, and interpret each of the $\beta$ parameters in the model.

b. Write the least squares equation that estimates the model in part a, and interpret each of the $\beta$ estimates.

| Variable | $\hat{\beta}$ | p-Value |
|----------|---------------|---------|
| $x_1$ | 12.774 | < .05 |
| $x_2$ | .713 | > .10 |
| $x_3$ | 1.519 | < .05 |
| $x_4$ | .320 | < .05 |
| $x_5$ | .205 | < .05 |
| Constant | 15.491 | — |

$$R^2 = .240 \qquad n = 191$$

**c.** Interpret the value of $R^2$. Test to determine whether the model is useful for predicting annual salary. Test using $\alpha = .05$.

**d.** Test to determine whether the gender variable indicates that male managers are paid more than female managers, even after adjusting for and holding constant the other four factors in the model. Test using $\alpha = .05$. [*Note:* The p-values given in the table are two-tailed.]

**c.** Why would one want to adjust for these other factors before conducting a test for salary discrimination?

**Applying the Concepts—Advanced**

**12.30** Refer to the *Liquid Assets* study of auction prices of red Bordeaux wine, Exercise 12.15 (p. 595). The three models for auction price ($y$) have $R^2$ and $s$ values as shown in the accompanying table. Based on this information, which of the three models would you use to predict red Bordeaux wine prices? Explain.

| | $R^2$ | $s$ |
|---------|-------|------|
| Model 1 | .212 | .575 |
| Model 2 | .828 | .287 |
| Model 3 | .828 | .293 |

## 12.6 USING THE MODEL FOR ESTIMATION AND PREDICTION

In Section 11.8 we discussed the use of the least squares line for estimating the mean value of $y$, $E(y)$, for some particular value of $x$, say $x = x_p$. We also showed how to use the same fitted model to predict, when $x = x_p$, some new value of $y$ to be observed in the future. Recall that the least squares line yielded the same value for both the estimate of $E(y)$ and the prediction of some future value of $y$. That is, both are the result of substituting $x_p$ into the prediction equation $\hat{y} = \hat{\beta}_0 + \hat{\beta}_1 x$ and calculating $\hat{y}_p$. There the equivalence ends. The confidence interval for the mean $E(y)$ is narrower than the prediction interval for $y$ because of the additional uncertainty attributable to the random error $\varepsilon$ when predicting some future value of $y$.

These same concepts carry over to the multiple regression model. Consider, again, the first-order model relating sale price of a residential property to land value ($x_1$), improvements ($x_2$), and home size ($x_3$). Suppose we want to estimate the mean sale price for a given property with $x_1 = \$15,000$, $x_2 = \$50,000$, and $x_3 = 1,800$ square feet. Assuming that the first-order model represents the true relationship between sale price and the three independent variables, we want to estimate

$$E(y) = \beta_0 + \beta_1 x_1 + \beta_2 x_2 + \beta_3 x_3$$
$$= \beta_0 + \beta_1(15,000) + \beta_2(50,000) + \beta_3(1,800)$$

Substituting into the least squares prediction equation, we find the estimate of $E(y)$ to be

$$\hat{y} = \hat{\beta}_0 + \hat{\beta}_1(15,000) + \hat{\beta}_2(50,000) + \hat{\beta}_3(1,800)$$
$$= 1,470.27 + .814(50,000) + .820(50,000) + 13.529(1,800) = 79,061.4$$

To form a confidence interval for the mean, we need to know the standard deviation of the sampling distribution for the estimator $\hat{y}$. For multiple regression models, the form of this standard deviation is rather complex. However, the regression routines of statistical computer software packages allow us to obtain the confidence intervals for mean values of $y$ for any given combination of values of the independent variables. A portion of the SAS output for the sale price example is shown in Figure 12.9a.

**FIGURE 12.9a**

*Portion of SAS Printout Giving Estimated Mean Sale Price and Corresponding 95% Confidence Interval*

| | | | | | Output Statistics | | | |
|---|---|---|---|---|---|---|---|---|
| Obs | LANDVAL | IMPROVAL | AREA | Dep Var SALEPRIC | Predicted Value | Std Error Mean Predict | 95% CL Mean | |
| 21 | 15000 | 50000 | 1800 | . | 79061 | 2680 | 73381 | 84742 |

**FIGURE 12.9b**

*Portion of SAS Printout Giving Predicted Sale Price and Corresponding 95% Prediction Interval*

| | | | | | Output Statistics | | | |
|---|---|---|---|---|---|---|---|---|
| Obs | LANDVAL | IMPROVAL | AREA | Dep Var SALEPRIC | Predicted Value | Std Error Mean Predict | 95% CL Predict | |
| 21 | 15000 | 50000 | 1800 | . | 79061 | 2680 | 61338 | 96785 |

The estimated mean value (i.e., predicted value) and corresponding 95% confidence interval for the selected $x$ values are highlighted in the SAS printout. We observe that $\hat{y} = 79{,}061$, which agrees (except for rounding) with our calculation. The corresponding 95% confidence interval for the true mean of $y$, is (73,381, 84,742). Thus, with 95% confidence, we conclude that the mean sale price for all properties with $x_1 = \$15{,}000$, $x_2 = \$50{,}000$, and $x_3 = 1{,}800$ square feet will fall between $73,381 and $84,742.

If we were interested in predicting the sale price for a particular (single) property with $x_1 = \$15{,}000$, $x_2 = \$50{,}000$, and $x_3 = 1{,}800$ square feet, we would use $\hat{y} = \$79{,}061$ as the predicted value. However, the prediction interval for a new value of $y$ is wider than the confidence interval for the mean value. This is reflected by the SAS printout shown in Figure 12.9b, which gives the predicted value of $y$ and corresponding 95% prediction interval. Note that the prediction interval is (61,338, 96,785). Thus, with 95% confidence, we conclude that the sale price for an individual property with the characteristics $x_1 = \$15{,}000$, $x_2 = \$50{,}000$, and $x_3 = 1{,}800$ square feet will fall between $61,338 and $96,785.

 **EXERCISES 12.31–12.36**

**Applying the Concepts–Basic**

**12.31** Refer to the *World Development* (Feb. 1998) study of street vendors' earnings ($y$), Exercises 12.8 and 12.20 (pp. 589, 601). The first STATISTIX printout on p. 607 shows both a 95% prediction interval for $y$ and a 95% confidence interval for $E(y)$ for a 45-year-old vendor who works 10 hours a day (i.e., for $x_1 = 45$ and $x_2 = 10$).

   **a.** Interpret the 95% prediction interval for $y$ in the words of the problem.

   **b.** Interpret the 95% confidence interval for $E(y)$ in the words of the problem.

   **c.** Note that the interval of part **a** is wider than the interval of part **b**. Will this always be true? Explain.

**12.32** Refer to the *Journal of Applied Ecology* study of the feeding habits of baby snow geese, Exercises 12.11 and 12.22 (pp. 591, 602). Recall that a first-order model was used to relate gosling weight change ($y$) to digestion efficiency ($x_1$) and acid-detergent fibre ($x_2$). The second STATISTIX printout on p. 607 shows both a confidence interval for $E(y)$ and a prediction interval for $y$ when $x_1 = 20\%$ and $x_2 = 5\%$.

   **a.** Interpret the confidence interval for $E(y)$.

   **b.** Interpret the prediction interval for $y$.

**STATISTIX Output for Exercise 12.31**

```
PREDICTED/FITTED VALUES OF EARNINGS

LOWER PREDICTED BOUND      1759.7      LOWER FITTED BOUND      2620.3
PREDICTED VALUE            3017.6      FITTED VALUE            3017.6
UPPER PREDICTED BOUND      4275.4      UPPER FITTED BOUND      3414.9
SE (PREDICTED VALUE)        577.29     SE (FITTED VALUE)        182.35


UNUSUALNESS (LEVERAGE)      0.1108
PERCENT COVERAGE            95.0
CORRESPONDING T             2.18

PREDICTOR VALUES: AGE = 45.000, HOURS = 10.000
```

**STATISTIX Output for Exercise 12.32**

```
PREDICTED/FITTED VALUES OF WTCHANGE

LOWER PREDICTED BOUND      0.2838      LOWER FITTED BOUND      3.7294
PREDICTED VALUE            9.3605      FITTED VALUE            9.3605
UPPER PREDICTED BOUND      18.437      UPPER FITTED BOUND      14.992
SE (PREDICTED VALUE)        4.4874     SE (FITTED VALUE)        2.7839


UNUSUALNESS (LEVERAGE)      0.6257
PERCENT COVERAGE            95.0
CORRESPONDING T             2.02

PREDICTOR VALUES: DIGEST = 20.000, ADFIBRE = 5.0000
```

**12.33** Refer to *The Astronomical Journal* study of quasars, Exercises 12.12 and 12.23 (pp. 593, 602). Recall that a first-order model was used to relate a quasar's equivalent width ($y$) to redshift ($x_1$), line flux ($x_2$), line luminosity ($x_3$), and $AB_{1450}$ ($x_4$). A portion of the SPSS spreadsheet showing 95% prediction intervals for $y$ for the first five observations in the data set is reproduced below. Interpret the interval corresponding to the fifth observation.

**12.34** An article published in *Geography* (July 1980) used multiple regression to predict annual rainfall levels in California. Data on the average annual precipitation ($y$), altitude ($x_1$), latitude ($x_2$), and distance from the Pacific coast ($x_3$) for 30 meteorological stations scattered throughout California are listed in the table (p. 608). Initially, the first-order model $y = \beta_0 + \beta_1 x_1 + \beta_2 x_2 + \beta_3 x_3 + \varepsilon$ was fit to the data. The SAS printout of the analysis is provided on p. 609.

**SPSS Output for Exercise 12.33**

|   | redshift | lineflux | linelum | ab1450 | rfewidth | lcl_95 | ucl_95 |
|---|----------|----------|---------|--------|----------|--------|--------|
| 1 | 2.81 | -13.48 | 45.29 | 19.50 | 117 | 101.29 | 172.22 |
| 2 | 3.07 | -13.73 | 45.13 | 19.65 | 82 | 59.62 | 125.89 |
| 3 | 3.45 | -13.87 | 45.11 | 18.93 | 33 | -35.09 | 37.04 |
| 4 | 3.19 | -13.27 | 45.63 | 18.59 | 92 | 63.76 | 139.25 |
| 5 | 3.07 | -13.56 | 45.30 | 19.59 | 114 | 90.69 | 158.57 |

## CALIRAIN

| Station | Avg. Annual Precipitation y (inches) | Altitude $x_1$ (feet) | Latitude $x_2$ (degrees) | Distance from Coast $x_3$ (miles) |
|---|---|---|---|---|
| 1.  Eureka | 39.57 | 43 | 40.8 | 1 |
| 2.  Red Bluff | 23.27 | 341 | 40.2 | 97 |
| 3.  Thermal | 18.20 | 4152 | 33.8 | 70 |
| 4.  Fort Bragg | 37.48 | 74 | 39.4 | 1 |
| 5.  Soda Springs | 49.26 | 6752 | 39.3 | 150 |
| 6.  San Francisco | 21.82 | 52 | 37.8 | 5 |
| 7.  Sacramento | 18.07 | 25 | 38.5 | 80 |
| 8.  San Jose | 14.17 | 95 | 37.4 | 28 |
| 9.  Giant Forest | 42.63 | 6360 | 36.6 | 145 |
| 10.  Salinas | 13.85 | 74 | 36.7 | 12 |
| 11.  Fresno | 9.44 | 331 | 36.7 | 114 |
| 12.  Pt. Piedras | 19.33 | 57 | 35.7 | 1 |
| 13.  Pasa Robles | 15.67 | 740 | 35.7 | 31 |
| 14.  Bakersfield | 6.00 | 489 | 35.4 | 75 |
| 15.  Bishop | 5.73 | 4108 | 37.3 | 198 |
| 16.  Mineral | 47.82 | 4850 | 40.4 | 142 |
| 17.  Santa Barbara | 17.95 | 120 | 34.4 | 1 |
| 18.  Susanville | 18.20 | 4152 | 40.3 | 198 |
| 19.  Tule Lake | 10.03 | 4036 | 41.9 | 140 |
| 20.  Needles | 4.63 | 913 | 34.8 | 192 |
| 21.  Burbank | 14.74 | 699 | 34.2 | 47 |
| 22.  Los Angeles | 15.02 | 312 | 34.1 | 16 |
| 23.  Long Beach | 12.36 | 50 | 33.8 | 12 |
| 24.  Los Banos | 8.26 | 125 | 37.8 | 74 |
| 25.  Blythe | 4.05 | 268 | 33.6 | 155 |
| 26.  San Diego | 9.94 | 19 | 32.7 | 5 |
| 27.  Daggett | 4.25 | 2105 | 34.1 | 85 |
| 28.  Death Valley | 1.66 | −178 | 36.5 | 194 |
| 29.  Crescent City | 74.87 | 35 | 41.7 | 1 |
| 30.  Colusa | 15.95 | 60 | 39.2 | 91 |

*Source:* Taylor, P.J. "A pedagogic application of multiple regression analysis." *Geography*, July 1980, Vol. 65, pp. 203–212.

a. Is there evidence that the first-order model is useful for predicting annual precipitation y? Test using $\alpha = .05$.

b. A 95% prediction interval for y for the Giant Forest meteorological station (station #9) is shown at the bottom of the printout. Locate and interpret the interval.

## Applying the Concepts—Intermediate

### WATEROIL

12.35  Refer to Exercise 12.13 (p. 594). The researchers concluded that "in order to break a water-oil mixture with the lowest possible voltage, the volume fraction of the disperse phase ($x_1$) should be high, while the salinity ($x_2$) and the amount of surfactant ($x_5$) should be low." Use this information and the first-order model of Exercise 12.13 to find a 95% prediction interval for this "low" voltage y. Interpret the interval.

12.36  In a production facility, an accurate estimate of hours needed to complete a task is crucial to management in making such decisions as the proper number of workers to hire, an accurate deadline to quote a client, or cost-analysis decisions regarding budgets. A manufacturer of boiler drums wants to use regression to predict the number of hours needed to erect the drums in future projects. To accomplish this, data for 35 boilers were collected. In addition to hours (y), the variables measured were boiler capacity ($x_1$ = lb/hr), boiler design pressure ($x_2$ = pounds per square inch or psi), boiler type ($x_3$ = 1 if industry field erected, 0 if utility field erected), and drum type ($x_4$ = 1 if steam, 0 if mud). The data are provided in the table on pp. 609–610. A MINITAB printout for the model $E(y) = \beta_0 + \beta_1 x_1 + \beta_2 x_2 + \beta_3 x_3 + \beta_4 x_4$ is given on p. 610.

a. Conduct a test for the global utility of the model. Use $\alpha = .01$.

SAS Output for Exercise 12.34

The REG Procedure
Dependent Variable: PRECIP

Analysis of Variance

| Source | DF | Sum of Squares | Mean Square | F Value | Pr > F |
|---|---|---|---|---|---|
| Model | 3 | 4809.35596 | 1603.11865 | 13.02 | <.0001 |
| Error | 26 | 3202.29762 | 123.16529 | | |
| Corrected Total | 29 | 8011.65359 | | | |

| | | | | |
|---|---|---|---|---|
| Root MSE | 11.09799 | R-Square | 0.6003 | |
| Dependent Mean | 19.80733 | Adj R-Sq | 0.5542 | |
| Coeff Var | 56.02968 | | | |

Parameter Estimates

| Variable | DF | Parameter Estimate | Standard Error | t Value | Pr > |t| |
|---|---|---|---|---|---|
| Intercept | 1 | -102.35743 | 29.20548 | -3.50 | 0.0017 |
| ALTITUDE | 1 | 0.00409 | 0.00122 | 3.36 | 0.0024 |
| LONGTUDE | 1 | 3.45108 | 0.79486 | 4.34 | 0.0002 |
| COAST | 1 | -0.14286 | 0.03634 | -3.93 | 0.0006 |

Predictions

| Obs | STATION | PRECIP | ALTITUDE | LONGTUDE | COAST | Predicted PRECIP | Lower prediction limit of PRECIP | Upper prediction limit of PRECIP |
|---|---|---|---|---|---|---|---|---|
| 9 | GiantFor | 42.63 | 6360 | 36.6 | 145 | 29.2534 | 17.7512 | 40.7556 |

**b.** Both a 95% confidence interval for $E(y)$ and a 95% prediction interval for $y$ when $x_1 = 150{,}000$, $x_2 = 500$, $x_3 = 1$, and $x_4 = 0$ are shown at the bottom of the MINITAB printout. Interpret both of these intervals.

**c.** Which of the intervals, part **b**, would you use if you want to estimate the average hours required to erect all industrial, mud boilers with a capacity of 150,000 lb/hr and a design pressure of 500 psi?

## BOILERS

| Hours $y$ | Boiler Capacity $x_1$ | Design Pressure $x_2$ | Boiler Type $x_3$ | Drum Type $x_4$ |
|---|---|---|---|---|
| 3,137 | 120,000 | 375 | 1 | 1 |
| 3,590 | 65,000 | 750 | 1 | 1 |
| 4,526 | 150,000 | 500 | 1 | 1 |
| 10,825 | 1,073,877 | 2,170 | 0 | 1 |
| 4,023 | 150,000 | 325 | 1 | 1 |
| 7,606 | 610,000 | 1,500 | 0 | 1 |
| 3,748 | 88,200 | 399 | 1 | 1 |
| 2,972 | 88,200 | 399 | 1 | 1 |
| 3,163 | 88,200 | 399 | 1 | 1 |
| 4,065 | 90,000 | 1,140 | 1 | 1 |
| 2,048 | 30,000 | 325 | 1 | 1 |
| 6,500 | 441,000 | 410 | 1 | 1 |

*(continues on next page)*

**⊘ BOILERS** *(continued)*

| Hours y | Boiler Capacity $x_1$ | Design Pressure $x_2$ | Boiler Type $x_3$ | Drum Type $x_4$ |
|---|---|---|---|---|
| 5,651 | 441,000 | 410 | 1 | 1 |
| 6,565 | 441,000 | 410 | 1 | 1 |
| 6,387 | 441,000 | 410 | 1 | 1 |
| 6,454 | 627,000 | 1,525 | 0 | 1 |
| 6,928 | 610,000 | 1,500 | 0 | 1 |
| 4,268 | 150,000 | 500 | 1 | 1 |
| 14,791 | 1,089,490 | 2,170 | 0 | 1 |
| 2,680 | 125,000 | 750 | 1 | 1 |
| 2,974 | 120,000 | 375 | 1 | 0 |
| 1,965 | 65,000 | 750 | 1 | 0 |
| 2,566 | 150,000 | 500 | 1 | 0 |
| 1,515 | 150,000 | 250 | 1 | 0 |
| 2,000 | 150,000 | 500 | 1 | 0 |
| 2,735 | 150,000 | 325 | 1 | 0 |
| 3,698 | 610,000 | 1,500 | 0 | 0 |
| 2,635 | 90,000 | 1,140 | 1 | 0 |
| 1,206 | 30,000 | 325 | 1 | 0 |
| 3,775 | 441,000 | 410 | 1 | 0 |
| 3,120 | 441,000 | 410 | 1 | 0 |
| 4,206 | 441,000 | 410 | 1 | 0 |
| 4,006 | 441,000 | 410 | 1 | 0 |
| 3,728 | 627,000 | 1,525 | 0 | 0 |
| 3,211 | 610,000 | 1,500 | 0 | 0 |
| 1,200 | 30,000 | 325 | 1 | 0 |

*Source:* Dr. Kelly Uscategui, University of Connecticut.

```
MINITAB Output for Exercise 12.36

The regression equation is
MANHRS = - 3783 + 0.00875 CAPACITY + 1.93 PRESSURE + 3444 BOILER + 2093 DRUM

Predictor        Coef      SE Coef        T        P
Constant        -3783         1205    -3.14    0.004
CAPACITY    0.0087490    0.0009035     9.68    0.000
PRESSURE       1.9265       0.6489     2.97    0.006
BOILER         3444.3        911.7     3.78    0.001
DRUM           2093.4        305.6     6.85    0.000

S = 894.6      R-Sq = 90.3%    R-Sq(adj) = 89.0%

Analysis of Variance

Source            DF            SS           MS        F        P
Regression         4     230854854     57713714    72.11    0.000
Residual Error    31      24809761       800315
Total             35     255664615

Predicted Values for New Observations

New Obs    Fit    SE Fit       95.0% CI             95.0% PI
1         1936       239    ( 1449,    2424)    (   48,    3825)

Values of Predictors for New Observations

New Obs  CAPACITY  PRESSURE   BOILER       DRUM
1          150000       500     1.00   0.000000
```

## 12.7  MODEL BUILDING: INTERACTION MODELS

In Section 12.2, we demonstrated the relationship between $E(y)$ and the independent variables in a first-order model. When $E(y)$ is graphed against any one variable (say, $x_1$) for fixed values of the other variables, the result is a set of *parallel* straight lines (see Figure 12.3, p. 581). When this situation occurs (as it always does for a first-order model), we say that the relationship between $E(y)$ and any one independent variable *does not depend* on the values of the other independent variables in the model.

However, if the relationship between $E(y)$ and $x_1$ does, in fact, depend on the values of the remaining $x$'s held fixed, then the first-order model is not appropriate for predicting $y$. In this case, we need another model that will take into account this dependence. Such a model includes the *cross products* of two or more $x$'s.

For example, suppose that the mean value $E(y)$ of a response $y$ is related to two quantitative independent variables, $x_1$ and $x_2$, by the model

$$E(y) = 1 + 2x_1 - x_2 + x_1x_2$$

A graph of the relationship between $E(y)$ and $x_1$ for $x_2 = 0, 1$, and 2 is displayed in Figure 12.10.

**FIGURE 12.10**

*Graphs of*
$1 + 2x_1 - x_2 + x_1x_2$
*for $x_2 = 0, 1, 2$*

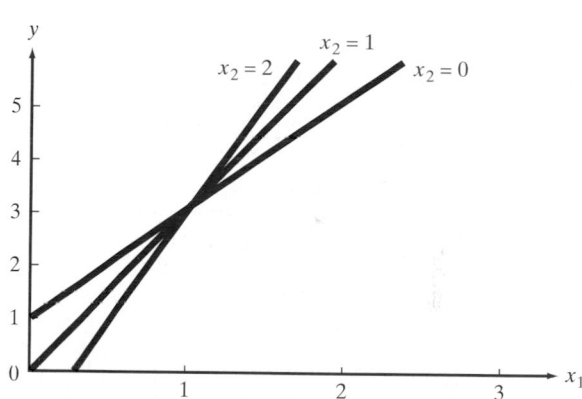

Note that the graph shows three nonparallel straight lines. You can verify that the slopes of the lines differ by substituting each of the values $x_2 = 0, 1$, and 2 into the equation. For $x_2 = 0$:

$$E(y) = 1 + 2x_1 - (0) + x_1(0) = 1 + 2x_1 \qquad (\text{slope} = 2)$$

For $x_2 = 1$:

$$E(y) = 1 + 2x_1 - (1) + x_1(1) = 3x_1 \qquad (\text{slope} = 3)$$

For $x_2 = 2$:

$$E(y) = 1 + 2x_1 - (2) + x_1(2) = -1 + 4x_1 \qquad (\text{slope} = 4)$$

Note that the slope of each line is represented by $\beta_1 + \beta_3x_2 = 2 + x_2$. Thus, the effect on $E(y)$ of a change in $x_1$ (i.e., the slope) now *depends* on the value of $x_2$. When this situation occurs, we say that $x_1$ and $x_2$ **interact**. The cross-product term, $x_1x_2$, is called an **interaction term**, and the model $E(y) = \beta_0 + \beta_1x_1 + \beta_2x_2 + \beta_3x_1x_2$ is called an **interaction model** with two quantitative variables.

> ### An Interaction Model Relating $E(y)$ to Two Quantitative Independent Variables
>
> $$E(y) = \beta_0 + \beta_1 x_1 + \beta_2 x_2 + \beta_3 x_1 x_2$$
>
> where
>
> $(\beta_1 + \beta_3 x_2)$ represents the change in $E(y)$ for every 1-unit increase in $x_1$, holding $x_2$ fixed
>
> $(\beta_2 + \beta_3 x_1)$ represents the change in $E(y)$ for every 1-unit increase in $x_2$, holding $x_1$ fixed

**FIGURE 12.11**

*Computer-Generated Graph for an Interaction Model (Second-Order)*

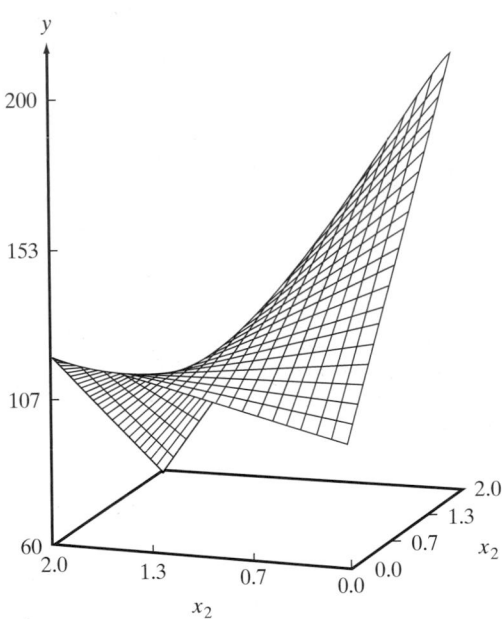

A three-dimensional graph (generated by computer) of an interaction model in two quantitative $x$'s is shown in Figure 12.11. Unlike the flat planar surface displayed in Figure 12.4 (p. 582), the interaction model traces a ruled surface (twisted plane) in three-dimensional space. If we slice the twisted plane at a fixed value of $x_2$, we obtain a straight line relating $E(y)$ to $x_1$; however, the slope of the line will change as we change the value of $x_2$.

■ EXAMPLE 12.5   Refer to Examples 12.3 and 12.4. Suppose the collector of grandfather clocks, having observed many auctions, believes that the *rate of increase* of the auction price with age will be driven upward by a large number of bidders. Thus, instead of a relationship like that shown in Figure 12.12a, in which the rate of increase in price with age is the same for any number of bidders, the collector believes the relationship is like that shown in Figure 12.12b. Note that as the number of bidders increases from 5 to 15, the slope of the price versus age line increases.

Consequently, the interaction model is proposed:

$$y = \beta_0 + \beta_1 x_1 + \beta_2 x_2 + \beta_3 x_1 x_2 + \varepsilon$$

## FIGURE 12.12

*Examples of No-Interaction and Interaction Models*

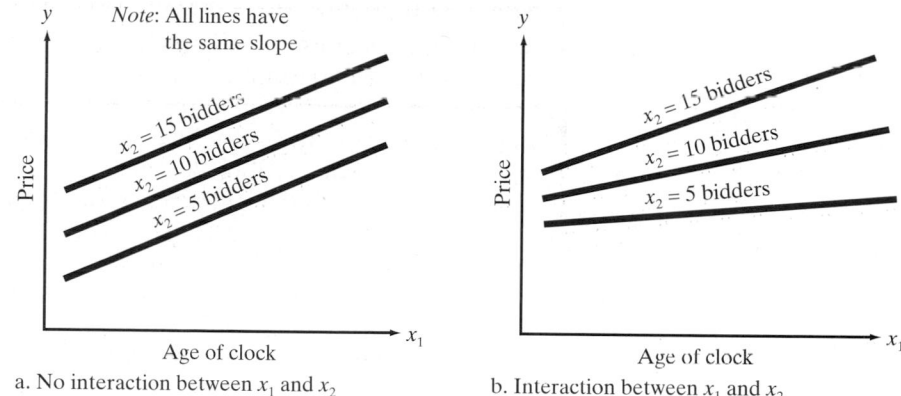

a. No interaction between $x_1$ and $x_2$

b. Interaction between $x_1$ and $x_2$

The 32 data points listed in Table 12.2 were used to fit the model with interaction. A portion of the MINITAB printout is shown in Figure 12.13.

a. Test the overall utility of the model using the global $F$-test at $\alpha = .05$.

b. Test the hypothesis (at $\alpha = .05$) that the price-age slope increases as the number of bidders increases—that is, that age and number of bidders, $x_2$, interact positively.

c. Estimate the change in auction price of a 150-year-old grandfather clock, $y$, for each additional bidder.

## FIGURE 12.13

*MINITAB Regression Printout for Grandfather Clock Model with Interaction*

```
The regression equation is
PRICE = 320 + 0.88 AGE - 93.3 BIDDERS + 1.30 AGEBID

Predictor        Coef      SE Coef         T        P
Constant        320.5        295.1      1.09    0.287
AGE             0.878        2.032      0.43    0.669
BIDDERS        -93.26        29.89     -3.12    0.004
AGEBID         1.2978       0.2123      6.11    0.000

S = 88.91      R-Sq = 95.4%      R-Sq(adj) = 94.9%

Analysis of Variance

Source            DF           SS          MS        F        P
Regression         3      4578427     1526142   193.04    0.000
Residual Error    28       221362        7906
Total             31      4799790
```

## Solution

a. The global $F$-test is used to test the null hypothesis

$$H_0: \beta_1 = \beta_2 = \beta_3 = 0$$

The test statistic and $p$-value of the test (highlighted on the MINITAB printout) are $F = 193.04$ and $p = 0$, respectively. Since $\alpha = .05$ exceeds the $p$-value, there is sufficient evidence to conclude that the model fit is a statistically useful predictor of auction price, $y$.

b. The hypotheses of interest to the collector concern the interaction parameter $\beta_3$. Specifically,

$$H_0: \beta_3 = 0$$
$$H_a: \beta_3 > 0$$

Since we are testing an individual $\beta$ parameter, a $t$-test is required. The test statistic and two-tailed $p$-value (highlighted on the printout) are $t = 6.11$ and $p = 0$, respectively. The upper-tailed $p$-value, obtained by dividing the two-tailed $p$-value in half, is $0/2 = 0$. Since $\alpha = .05$ exceeds the $p$-value, the collector can reject $H_0$ and conclude that the rate of change of the mean price of the clocks with age increases as the number of bidders increases; that is, $x_1$ and $x_2$ interact positively. Thus, it appears that the interaction term should be included in the model.

c. To estimate the change in auction price, $y$, for every 1-unit increase in number of bidders, $x_2$, we need to estimate the slope of the line relating $y$ to $x_2$ when the age of the clock, $x_1$, is 150 years old. An analyst who is not careful may estimate this slope as $\hat{\beta}_2 = -93.26$. Although the coefficient of $x_2$ is negative, this does *not* imply that auction price decreases as the number of bidders increases. Since interaction is present, the rate of change (slope) of mean auction price with the number of bidders *depends* on $x_1$, the age of the clock. Thus, the estimated rate of change of $y$ for a unit increase in $x_2$ (one new bidder) for a 150-year-old clock is

$$\text{Estimated } x_2 \text{ slope} = \hat{\beta}_2 + \hat{\beta}_3 x_1 = -93.26 + 1.30(150) = 101.74$$

In other words, we estimate that the auction price of a 150-year-old clock will *increase* by about \$101.74 for every additional bidder. Although the rate of increase will vary as $x_1$ is changed, it will remain positive for the range of values of $x_1$ included in the sample. Extreme care is needed in interpreting the signs and sizes of coefficients in a multiple regression model.

■

Example 12.5 illustrates an important point about conducting $t$-tests on the $\beta$ parameters in the interaction model. The "most important" $\beta$ parameter in this model is the interaction $\beta$, $\beta_3$. [Note that this $\beta$ is also the one associated with the highest-order term in the model, $x_1 x_2$.*] Consequently, we will want to test $H_0 : \beta_3 = 0$ after we have determined that the overall model is useful for predicting $y$. Once interaction is detected (as in Example 12.5), however, tests on the first-order terms $x_1$ and $x_2$ should *not* be conducted since they are meaningless tests; the presence of interaction implies that both $x$'s are important.

---

### Caution

Once interaction has been deemed important in the model $E(y) = \beta_0 + \beta_1 x_1 + \beta_2 x_2 + \beta_3 x_1 x_2$, do not conduct $t$-tests on the $\beta$ coefficients of the first-order terms $x_1$ and $x_2$. These terms should be kept in the model regardless of the magnitude of their associated $p$-values shown on the printout.

---

 EXERCISES 12.37–12.47

**Learning the Mechanics**

**12.37** Write an interaction model relating the mean value of $y, E(y)$ to:
  **a.** two quantitative independent variables

  **b.** three quantitative independent variables [*Hint:* Include all possible two-way cross-product terms.]

**12.38** If two variables, $x_1$ and $x_2$, do not interact, how would you describe their effect on the mean response, $E(y)$?

---

*The order of a term is equal to the sum of the exponents of the quantitative variables included in the term. Thus, when $x_1$ and $x_2$ are both quantitative variables, the cross product, $x_1 x_2$, is a second-order term.

**12.39** Suppose the true relationship between $E(y)$ and the quantitative independent variables $x_1$ and $x_2$ is

$$E(y) = 3 + x_1 + 2x_2 - x_1x_2$$

a. Describe the corresponding three-dimensional response surface.

b. Plot the linear relationship between $y$ and $x_2$ for $x_2 = 0, 1, 2$, where $0 \leq x_2 \leq 5$.

c. Explain why the lines you plotted in part **b** are not parallel.

d. Use the lines you plotted in part **b** to explain how changes in the settings of $x_1$ and $x_2$ affect $E(y)$.

e. Use your graph from part **b** to determine how much $E(y)$ changes when $x_1$ is changed from 2 to 0 and $x_2$ is simultaneously changed from 4 to 5.

**12.40** Suppose you fit the interaction model

$$y = \beta_0 + \beta_1x_1 + \beta_2x_2 + \beta_3x_1x_2 + \varepsilon$$

to $n = 32$ data points and obtain the following results:

$$\text{SS}_{yy} = 479 \qquad \text{SSE} = 21 \qquad \hat{\beta}_3 = 10 \qquad s_{\hat{\beta}_3} = 4$$

a. Find $R^2$ and interpret its value.

b. Is the model adequate for predicting $y$? Test at $\alpha = .05$.

c. Use a graph to explain the contribution of the $x_1x_2$ term to the model.

d. Is there evidence that $x_1$ and $x_2$ interact? Test at $\alpha = .05$.

**12.41** MINITAB was used to fit the model

$$y = \beta_0 + \beta_1x_1 + \beta_2x_2 + \beta_3x_1x_2 + \varepsilon$$

to $n = 15$ data points. The printout is shown here.

a. What is the prediction equation for the response surface?

b. Describe the geometric form of the response surface of part **a**.

c. Plot the prediction equation for the case when $x_2 = 1$. Do this twice more on the same graph for the cases when $x_2 = 3$ and $x_2 = 5$.

d. Explain what it means to say that $x_1$ and $x_2$ interact. Explain why your graph of part **c** suggests that $x_1$ and $x_2$ interact.

e. Specify the null and alternative hypotheses you would use to test whether $x_1$ and $x_2$ interact.

f. Conduct the hypothesis test of part **e** using $\alpha = .01$.

## Applying the Concepts—Basic

**12.42** Refer to the *World Development* (Feb. 1998) study of street vendors in the city of Puebla, Mexico, Exercise 12.8 (p. 589). Recall that the vendors' mean annual earnings, $E(y)$, was modeled as a first-order function of age ($x_1$) and hours worked ($x_2$). Now, consider the interaction model $E(y) = \beta_0 + \beta_1x_1 + \beta_2x_2 + \beta_3x_1x_2$. The SAS printout for the model is displayed on p. 616.

a. Give the least squares prediction equation.

b. What is the estimated slope relating annual earnings ($y$) to age ($x_1$) when number of hours worked ($x_2$) is 10? Interpret the result.

c. What is the estimated slope relating annual earnings ($y$) to hours worked ($x_2$) when age ($x_1$) is 40? Interpret the result.

d. Give the null hypothesis for testing whether age ($x_1$) and hours worked ($x_2$) interact.

e. Find the $p$-value of the test, part **d**.

f. Refer to part **e**. Give the appropriate conclusion in the words of the problem.

**12.43** To what degree do the attitudes of your peers influence your behavior? A study presented in *Social Psychology Quarterly* (Vol. 50, 1987) included a sample of $n = 143$ adult drinkers in an urban setting characterized by high physical availability of alcoholic beverages. The goal of the study was to build a model relating

**MINITAB Output for Exercise 12.41**

```
The regression equation is
Y = - 2.55 + 3.82 X1 + 2.63 X2 - 1.29 X1X2

Predictor       Coef       SE Coef       T       P
Constant       -2.550        1.142    -2.23   0.043
X1              3.815        0.529     7.22   0.000
X2              2.630        0.334     7.64   0.000
X1X2           -1.285        0.159    -8.06   0.000

S = 0.713       R-Sq = 85.6%       R-Sq(adj) = 81.6%

Analysis of Variance

Source           DF         SS        MS       F       P
Regression        3     33.149    11.050   21.75   0.000
Residual Error   11      5.587     0.508
Total            14     38.736
```

SAS Output for Exercise 12.42

The REG Procedure
Dependent Variable: EARNINGS

Analysis of Variance

| Source | DF | Sum of Squares | Mean Square | F Value | Pr > F |
|---|---|---|---|---|---|
| Model | 3 | 5287427 | 1762476 | 5.82 | 0.0124 |
| Error | 11 | 3331000 | 302818 | | |
| Corrected Total | 14 | 8618428 | | | |

| | | | |
|---|---|---|---|
| Root MSE | 550.28921 | R-Square | 0.6135 |
| Dependent Mean | 2577.13333 | Adj R-Sq | 0.5081 |
| Coeff Var | 21.35276 | | |

Parameter Estimates

| Variable | DF | Parameter Estimate | Standard Error | t Value | Pr > |t| |
|---|---|---|---|---|---|
| Intercept | 1 | 1041.89440 | 1303.59326 | 0.80 | 0.4411 |
| AGE | 1 | -13.23762 | 29.23395 | -0.45 | 0.6595 |
| HOURS | 1 | 103.30564 | 162.01356 | 0.64 | 0.5368 |
| AGEHRS | 1 | 3.62096 | 3.84044 | 0.94 | 0.3660 |

frequency of drinking alcoholic beverages, $y$, to attitude toward drinking $(x_1)$ and social support $(x_2)$. Consider the interaction model

$$E(y) = \beta_0 + \beta_1 x_1 + \beta_2 x_2 + \beta_3 x_1 x_2$$

**a.** Interpret the phrase "$x_1$ and $x_2$ interact" in terms of the problem.

**b.** Write the null and alternative hypotheses for determining whether attitude $(x_1)$ and social support $(x_2)$ interact.

**c.** The reported $p$-value for the test, part **b**, was $p < .001$. Interpret this result.

## Applying the Concepts—Intermediate

**12.44** Licensed therapists are mandated, by law, to report child abuse by their clients. This requires the therapist to breach confidentiality and possibly lose the client's trust. A national survey of licensed psychotherapists was conducted to investigate clients' reactions to legally mandated child-abuse reports (*American Journal of Orthopsychiatry*, Jan. 1997). The sample consisted of 303 therapists who had filed a child-abuse report against one of their clients. The researchers were interested in finding the best predictors of a client's reaction $(y)$ to the report, where $y$ is measured on a 30-point scale. (The higher the value, the more favorable the client's response to the report.) The independent variables found to have the most predictive power are listed here.

$x_1$: Therapist's age (years)
$x_2$: Therapist's gender (1 if male, 0 if female)

$x_3$: Degree of therapist's role strain (25-point scale)
$x_4$: Strength of client-therapist relationship (40-point scale)
$x_5$: Type of case (1 if family, 0 if not)
$x_1 x_2$: Age × Gender interaction

**a.** Hypothesize a first-order model relating $y$ to each of the five independent variables.

**b.** Give the null hypothesis for testing the contribution of $x_4$, strength of client-therapist relationship, to the model.

**c.** The test statistic for the test, part **b**, was $t = 4.408$ with an associated $p$-value of .001. Interpret this result.

**d.** The estimated $\beta$ coefficient for the $x_1 x_2$ interaction term was positive and highly significant ($p < .001$). According to the researchers, "this interaction suggests that ... as the age of the therapist increased, ... male therapists were less likely to get negative client reactions than were female therapists." Do you agree?

**e.** For this model, $R^2 = .2946$. Interpret this value.

**12.45** *Trichuristrichiura*, a parasitic worm, affects millions of school-age children each year, especially children from developing countries. A study was conducted to determine the effects of treatment of the parasite on school achievement in 407 school-age Jamaican children infected with the disease (*Journal of Nutrition*, July 1995). About half the children in the sample received the treatment, while the others received a placebo. Multiple regression was used to model spelling test score $y$, measured as number correct, as a function of the following independent variables:

Treatment (T): $x_1 = \begin{cases} 1 & \text{if treatment} \\ 0 & \text{if placebo} \end{cases}$

Disease intensity (I): $x_2 = \begin{cases} 1 & \text{if more than 7,000 eggs} \\ & \text{per gram of stool} \\ 0 & \text{if not} \end{cases}$

**a.** Propose a model for $E(y)$ that includes interaction between treatment and disease intensity.

**b.** The estimates of the $\beta$'s in the model, part **a**, and the respective $p$-values for $t$-tests on the $\beta$'s are given in the table. Is there sufficient evidence to indicate that the effect of the treatment on spelling score depends on disease intensity? Test using $\alpha = .05$.

| Variable | $\beta$ Estimate | $p$-Value |
|---|---|---|
| Treatment $(x_1)$ | $-.1$ | .62 |
| Intensity $(x_2)$ | $-.3$ | .57 |
| T $\times$ I $(x_1 x_2)$ | 1.6 | .02 |

**c.** Based on the result, part **b**, explain why the analyst should avoid conducting $t$-tests for the treatment $(x_1)$ and intensity $(x_2)$ $\beta$'s or interpreting these $\beta$'s individually.

**12.46** Does extensive media coverage of a military crisis influence public opinion on how to respond to the crisis? Political scientists at UCLA researched this question and reported their results in *Communication Research* (June 1993). The military crisis of interest was the 1990 Persian Gulf War, precipitated by Iraqi leader Saddam Hussein's invasion of Kuwait. The researchers used multiple regression analysis to model the level $y$ of support Americans had for a military (rather than a diplomatic) response to the crisis. Values of $y$ ranged from 0 (preference for a diplomatic response) to 4 (preference for a military response). The following independent variables were used in the model:

$x_1 = $ Level of TV news exposure in a selected week (number of days)

$x_2 = $ Knowledge of seven political figures (1 point for each correct answer)

$x_3 = $ Gender (1 if male, 0 if female)

$x_4 = $ Race (1 if nonwhite, 0 if white)

$x_5 = $ Partisanship (0–6 scale, where 0 = strong Democrat and 6 = strong Republican)

$x_6 = $ Defense spending attitude (1–7 scale, where 1 = greatly decrease spending and 7 = greatly increase spending)

$x_7 = $ Education level (1–7 scale, where 1 = less than eight grades and 7 = college)

Data from a survey of 1,763 U.S. citizens were used to fit the model

$$E(y) = \beta_0 + \beta_1 x_1 + \beta_2 x_2 + \beta_3 x_3 + \beta_4 x_4$$
$$+ \beta_5 x_5 + \beta_6 x_6 + \beta_7 x_7 + \beta_8 x_2 x_3 + \beta_9 x_2 x_4$$

The regression results are shown in the table at the bottom of the page.

**a.** Interpret the $\beta$ estimate for the variable $x_1$, TV news exposure.

**b.** Conduct a test to determine whether an increase in TV news exposure is associated with an increase in support for a military resolution of the crisis. Use $\alpha = .05$.

**c.** Is there sufficient evidence to indicate that the relationship between support for a military resolution $(y)$ and gender $(x_1)$ depends on political knowledge $(x_2)$? Test using $\alpha = .05$.

**d.** Is there sufficient evidence to indicate that the relationship between support for a military resolution $(y)$ and race $(x_4)$ depends on political knowledge $(x_2)$? Test using $\alpha = .05$.

**e.** The coefficient of determination for the model was $R^2 = .194$. Interpret this value.

**f.** Use the value of $R^2$, part **e**, to conduct a global test for model utility. Use $\alpha = .05$.

⊙ WATEROIL

**12.47** Refer to the *Journal of Colloid and Interface Science* study of water/oil mixtures, Exercises 12.13 and 12.35 (pp. 594, 608). Recall that three of the seven variables used to predict voltage $(y)$ were volume $(x_1)$, salinity

**Results for Exercise 12.46**

| Variable | $\beta$ Estimate | Standard Error | Two-Tailed $p$-Value |
|---|---|---|---|
| TV news exposure $(x_1)$ | .02 | .01 | .03 |
| Political knowledge $(x_2)$ | .07 | .03 | .03 |
| Gender $(x_3)$ | .67 | .11 | $< .001$ |
| Race $(x_4)$ | $-.76$ | .13 | $< .001$ |
| Partisanship $(x_5)$ | .07 | .01 | $< .001$ |
| Defense spending $(x_6)$ | .20 | .02 | $< .001$ |
| Education $(x_7)$ | .07 | .02 | $< .001$ |
| Knowledge $\times$ Gender $(x_2 x_3)$ | $-.09$ | .04 | .02 |
| Knowledge $\times$ Race $(x_2 x_4)$ | .10 | .06 | .08 |

*Source:* Iyengar, S., and Simon, A. "News coverage of the Gulf Crisis and public opinion." *Communication Research*, Vol. 20, No. 3, June 1993, p. 380 (Table 2).

$(x_2)$, and surfactant concentration $(x_5)$. The model the researchers fit is

$$E(y) = \beta_0 + \beta_1 x_1 + \beta_2 x_2 + \beta_3 x_5 + \beta_4 x_1 x_2 + \beta_5 x_1 x_5.$$

**a.** Note that the model includes interaction between disperse phase volume $(x_1)$ and salinity $(x_2)$ as well as interaction between disperse phase volume $(x_1)$ and surfactant concentration $(x_5)$. Discuss how these interaction terms affect the hypothetical relationship between $y$ and $x_1$. Draw a sketch to support your answer.

**b.** Fit the interaction model to the data on page 594. Does this model appear to fit the data better than the first-order model in Exercise 12.13? Explain.

**c.** Interpret the $\beta$ estimates of the interaction model.

## 12.8  MODEL BUILDING: QUADRATIC AND OTHER HIGHER-ORDER MODELS

All of the models discussed in the previous sections proposed straight-line relationships between $E(y)$ and each of the independent variables in the model. In this section, we consider models that allow for curvature in the relationships. Each of these models is a **second-order model** because it will include an $x^2$ term.

First, we consider a model that includes only one independent variable $x$. The form of this model, called the **quadratic model**, is

$$y = \beta_0 + \beta_1 x + \beta_2 x^2 + \varepsilon$$

The term involving $x^2$, called a **quadratic term** (or **second-order term**), enables us to hypothesize curvature in the graph of the response model relating $y$ to $x$. Graphs of the quadratic model for two different values of $\beta_2$ are shown in Figure 12.14. When the curve opens upward, the sign of $\beta_2$ is positive (see Figure 12.14a); when the curve opens downward, the sign of $\beta_2$ is negative (see Figure 12.14b).

---

**A Quadratic (Second-Order) Model in a Single Quantitative Independent Variable**

$$E(y) = \beta_0 + \beta_1 x + \beta_2 x^2$$

where

$\beta_0$ is the $y$-intercept of the curve

$\beta_1$ is a shift parameter

$\beta_2$ is the rate of curvature

---

**FIGURE 12.14**

*Graphs for Two Quadratic Models*

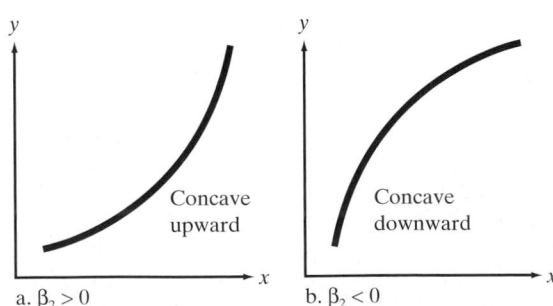

Concave upward

a. $\beta_2 > 0$

Concave downward

b. $\beta_2 < 0$

---

**EXAMPLE 12.6**   In all-electric homes, the amount of electricity expended is of interest to consumers, builders, and groups involved with energy conservation. Suppose we wish to investigate the monthly electrical usage, $y$, in all-electric homes and its relationship to the size, $x$, of the home. Moreover, suppose we think that

⊙ELECTRIC

**TABLE 12.3** **Home Size–Electrical Usage Data**

| Size of Home x (sq. ft) | Monthly Usage y (kilowatt-hours) |
|---|---|
| 1,290 | 1,182 |
| 1,350 | 1,172 |
| 1,470 | 1,264 |
| 1,600 | 1,493 |
| 1,710 | 1,571 |
| 1,840 | 1,711 |
| 1,980 | 1,804 |
| 2,230 | 1,840 |
| 2,400 | 1,956 |
| 2,930 | 1,954 |

monthly electrical usage in all-electric homes is related to the size of the home by the quadratic model

$$y = \beta_0 + \beta_1 x + \beta_2 x^2 + \varepsilon$$

To fit the model, the values of $y$ and $x$ are collected for 10 homes during a particular month. The data are shown in Table 12.3.

a. Construct a scatterplot for the data. Is there evidence to support the use of a quadratic model?
b. Use the method of least squares to estimate the unknown parameters $\beta_0, \beta_1$, and $\beta_2$ in the quadratic model.
c. Graph the prediction equation and assess how well the model fits the data, both visually and numerically.
d. Interpret the $\beta$ estimates.
e. Is the overall model useful (at $\alpha = .01$) for predicting electrical usage $y$?
f. Is there sufficient evidence of concave downward curvature in the electrical usage–home size relationship? Test using $\alpha = .01$.

**Solution**

a. A scattergram for the data of Table 12.3, produced using STATISTIX, is shown in Figure 12.15. The figure illustrates that the electrical usage appears to increase in a curvilinear manner with the size of the home. This provides some support for the inclusion of the quadratic term $x^2$ in the model.
b. We also used STATISTIX to fit the model to the data in Table 12.3. Part of the STATISTIX regression output is displayed in Figure 12.16. The least squares estimates of the $\beta$ parameters (highlighted) are $\hat{\beta}_0 = -1,216.14$, $\hat{\beta}_1 = 2.39893$, and $\hat{\beta}_2 = -.00045$. Therefore, the equation that minimizes the SSE for the data is

$$\hat{y} = -1,216.14 + 2.39893x - .00045x^2$$

c. Figure 12.17 is a graph of the least squares prediction equation. Note that the graph provides a good fit to the data of Table 12.3. A numerical measure of fit is obtained with the adjusted coefficient of determination, $R_a^2$. From the STATISTIX printout, $R_a^2 = .9767$. This implies that almost 98% of the sample variation in electrical usage ($y$) can be explained by the quadratic model (after adjusting for sample size and degrees of freedom).

**FIGURE 12.15**

*STATISTIX Scattergram of Home Size–Electrical Usage Data*

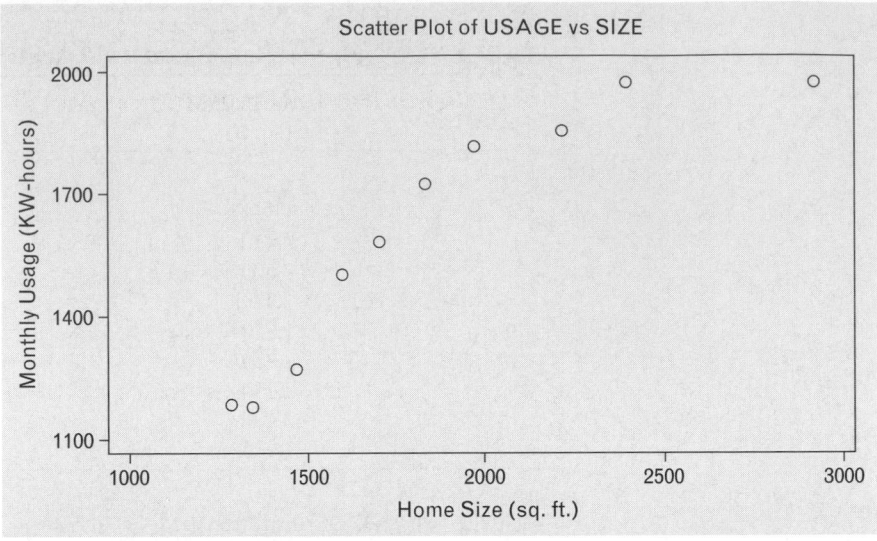

**FIGURE 12.16**

*STATISTIX Regression Printout for Home Size–Electrical Usage Model*

```
UNWEIGHTED LEAST SQUARES LINEAR REGRESSION OF USAGE

PREDICTOR
VARIABLES      COEFFICIENT      STD ERROR      STUDENT'S T       P
---------      -----------      ---------      -----------     ------

CONSTANT        -1216.14         242.806          -5.01        0.0016
SIZE             2.39893         0.24584           9.76        0.0000
SIZESQ          -0.000450        .00005908        -7.62        0.0001

R-SQUARED              0.9819      RESID. MEAN SQUARE (MSE)    2190.36
ADJUSTED R-SQUARED     0.9767      STANDARD DEVIATION          46.8013

SOURCE          DF        SS            MS         F         P
----------      ---    ----------    ----------   -----    ------

REGRESSION       2       831070        415535     189.71   0.0000
RESIDUAL         7       15332.6       2190.36
TOTAL            9       846402

CASES INCLUDED 10    MISSING CASES 0
```

**FIGURE 12.17**

*Least Squares Model for the Home Size–Electrical Usage Data*

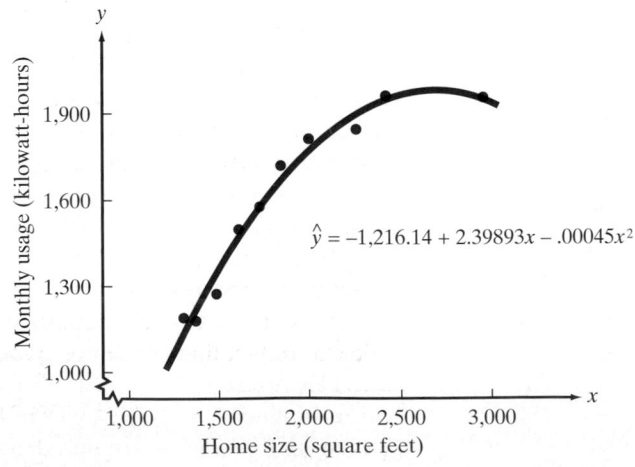

$$\hat{y} = -1{,}216.14 + 2.39893x - .00045x^2$$

# Using the TI-83 Graphing Calculator

## Quadratic Regression on the TI-83

Step 1    *Enter the data in L1 and L2*

Step 2    *Find the quadratic regression equation*
Press **STAT** and highlight **CALC**
Press **5** for **QuadReg**
Press **ENTER**
The screen will show the values for $a$, $b$, and $c$ in the equation
$y = ax^2 + bx + c$
If the diagnostics are on, the screen will also give the value for $r^2$

Step 3    *Graph the regression curve with the scatterplot* (see Section 11.2)

Example    The figures below show a table of data entered on the TI-83, the quadratic regression equation and the graph obtained using the steps given above.

d. The interpretation of the estimated coefficients in a quadratic model must be undertaken cautiously. First, the estimated $y$-intercept, $\hat{\beta}_0$, can be meaningfully interpreted only if the range of the independent variable includes zero—that is, if $x = 0$ is included in the sampled range of $x$. Although $\hat{\beta}_0 = -1,216.14$ seems to imply that the estimated electrical usage is negative when $x = 0$, this zero point is not in the range of the sample (the lowest value of $x$ is 1,290 square feet), and the value is nonsensical (a home with 0 square feet); thus the interpretation of $\hat{\beta}_0$ is not meaningful.

The estimated coefficient of $x$ is $\hat{\beta}_1 = 2.39893$, but it no longer represents a slope in the presence of the quadratic term $x^2$.* The estimated coefficient of the first-order term $x$ will not, in general, have a meaningful interpretation in the quadratic model.

The sign of the coefficient, $\hat{\beta}_2 = -.00045$, of the quadratic term, $x^2$, is the indicator of whether the curve is concave downward (mound-shaped) or concave

---

*For students with knowledge of calculus, note that the slope of the quadratic model is the first derivative $\partial y / \partial x = \beta_1 + 2\beta_2 x$. Thus, the slope varies as a function of $x$, rather than the constant slope associated with the straight-line model.

**FIGURE 12.18**

*Potential Misuse of Quadratic Model*

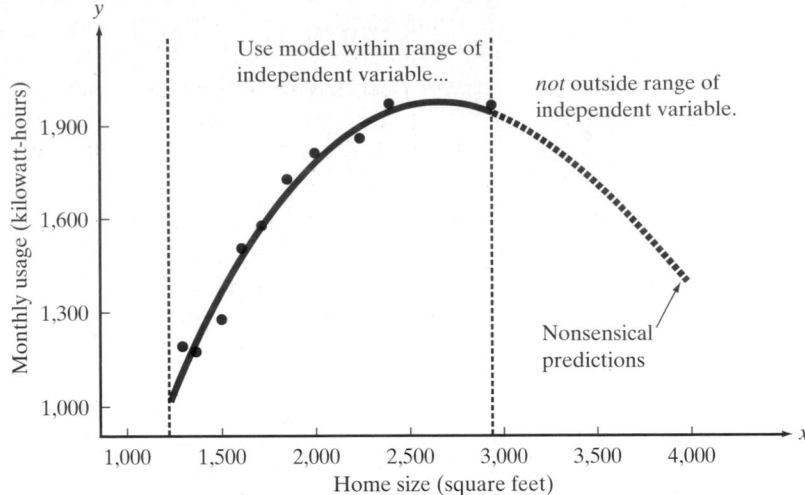

upward (bowl-shaped). A negative $\hat{\beta}_2$ implies downward concavity, as in this example (Figure 12.17), and a positive $\hat{\beta}_2$ implies upward concavity. Rather than interpreting the numerical value of $\hat{\beta}_2$ itself, we utilize a graphical representation of the model, as in Figure 12.17, to describe the model.

Note that Figure 12.17 implies that the estimated electrical usage is leveling off as the home sizes increase beyond 2,500 square feet. In fact, the concavity of the model would lead to decreasing usage estimates if we were to display the model out to 4,000 square feet and beyond (see Figure 12.18). However, model interpretations are not meaningful outside the range of the independent variable, which has a maximum value of 2,930 square feet in this example. Thus, although the model appears to support the hypothesis that the *rate of increase* per square foot *decreases* for the home sizes near the high end of the sampled values, the conclusion that usage will actually begin to decrease for very large homes would be a *misuse* of the model, since no homes of 3,000 square feet or more were included in the sample.

e. To test whether the quadratic model is statistically useful, we conduct the global $F$-test:

$$H_0: \beta_1 = \beta_2 = 0$$
$$H_a: \text{At least one of the above coefficients is nonzero}$$

From the STATISTIX printout, Figure 12.16, the test statistic is $F = 189.71$ with an associated $p$-value near 0. For any reasonable $\alpha$, we reject $H_0$ and conclude that the overall model is a useful predictor of electrical usage, $y$.

f. Figure 12.17 shows concave downward curvature in the relationship between size of a home and electrical usage in the sample of 10 data points. To determine if this type of curvature exists in the population, we want to test

$$H_0: \beta_2 = 0 \text{ (no curvature in the response curve)}$$
$$H_a: \beta_2 < 0 \text{ (downward concavity exists in the response curve)}$$

The test statistic for testing $\beta_2$, highlighted on the printout, is $t = -7.62$ and the associated two-tailed $p$-value is .0001. Since this is a one-tailed test, the appropriate $p$-value is $.0001/2 = .00005$. Now $\alpha = .01$ exceeds this $p$-value. Thus, there is very strong evidence of downward curvature in the population, that is, electrical usage increases more slowly per square foot for large homes than for small homes.

Note that the STATISTIX printout in Figure 12.16 also provides the *t*-test statistic and corresponding two-tailed *p*-values for the tests of $H_0$: $\beta_0 = 0$ and $H_0$: $\beta_1 = 0$. Since the interpretation of these parameters is not meaningful for this model, the tests are not of interest. ∎

When two or more quantitative independent variables are included in a second-order model, we can incorporate squared terms for each *x* in the model, as well as the interaction between the two independent variables. A model that includes all possible second-order terms in two independent variables—called a **complete second-order model**—is given in the box.

---

### Complete Second-Order Model with Two Quantitative Independent Variables

$$E(y) = \beta_0 + \beta_1 x_1 + \beta_2 x_2 + \beta_3 x_1 x_2 + \beta_4 x_1^2 + \beta_5 x_2^2$$

**COMMENTS ON THE PARAMETERS**

$\beta_0$: *y*-intercept, the value of $E(y)$ when $x_1 = x_2 = 0$

$\beta_1, \beta_2$: Changing $\beta_1$ and $\beta_2$ causes the surface to shift along the $x_1$- and $x_2$-axes

$\beta_3$: Controls the rotation of the surface

$\beta_4, \beta_5$: Signs and values of these parameters control the type of surface and the rates of curvature.

---

Three types of surfaces are produced by a second-order model:* a **paraboloid** that opens upward (Figure 12.19a), a paraboloid that opens downward (Figure 12.19b), and a **saddle-shaped surface** (Figure 12.19c).

A complete second-order model is the three-dimensional equivalent of a quadratic model in a single quantitative variable. Instead of tracing parabolas, it traces paraboloids and saddle-shaped surfaces. Since only a portion of the complete surface is used to fit the data, this model provides a very large variety of gently curving surfaces that can be used to fit data. It is a good choice for a model if you expect curvature in the response surface relating $E(y)$ to $x_1$ and $x_2$.

**FIGURE 12.19**

*Graphs for Three Second-Order Surfaces*

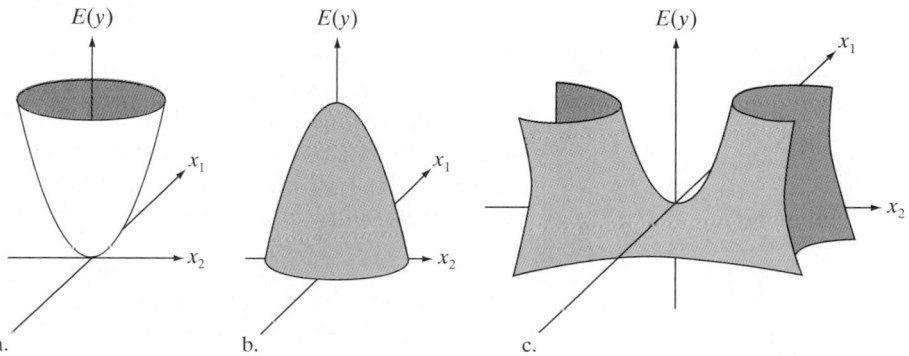

a.   b.   c.

---

*The saddle-shaped surface (Figure 12.19c) is produced when $\beta_3^2 > 4\beta_4\beta_5$; the paraboloid opens upward (Figure 12.19a) when $\beta_4 + \beta_5 > 0$ and opens downward (Figure 12.19b) when $\beta_4 + \beta_5 < 0$.

**EXAMPLE 12.7**

A social scientist would like to relate the number of hours worked per week (outside the home) by a married woman to the number of years of formal education she has completed and the number of children in her family.

a. Identify the dependent variable and the independent variables.
b. Write the first-order model for this example.
c. Modify the model in part b to include an interaction term.
d. Write a complete second-order model for $E(y)$

**Solution**

a. The dependent variable is

$$y = \text{Number of hours worked per week by a married woman}$$

The two independent variables, both quantitative in nature, are

$$x_1 = \text{Number of years of formal education completed by the woman}$$
$$x_2 = \text{Number of children in the family}$$

b. The first-order model is

$$E(y) = \beta_0 + \beta_1 x_1 + \beta_2 x_2$$

This model would probably not be appropriate in this situation because $x_1$ and $x_2$ may interact and/or curvature terms corresponding to $x_1^2$ and $x_2^2$ may be needed to obtain a good model for $E(y)$.

c. Adding the interaction term, we obtain

$$E(y) = \beta_0 + \beta_1 x_1 + \beta_2 x_2 + \beta_3 x_1 x_2$$

This model should be better than the model in part b, since we have now allowed for interaction between $x_1$ and $x_2$.

d. The complete second-order model is

$$E(y) = \beta_0 + \beta_1 x_1 + \beta_2 x_2 + \beta_3 x_1 x_2 + \beta_4 x_1^2 + \beta_5 x_2^2$$

Since it would not be surprising to find curvature in the response surface, the complete second-order model would be preferred to the models in parts b and c. How can we tell whether the complete second-order model really does provide better predictions of hours worked than the models in parts b and c? The answers to these and similar questions are examined in Section 12.11. ∎

Most relationships between $E(y)$ and two or more quantitative independent variables are second-order and require the use of either the interactive or the complete second-order model to obtain a good fit to a data set. As in the case of a single quantitative independent variable, however, the curvature in the response surface may be very slight over the range of values of the variables in the data set. When this happens, a first-order model may provide a good fit to the data.

EXERCISES 12.48–12.61

## Learning the Mechanics

**12.48** Write a second-order model relating the mean of $y$, $E(y)$, to:
  **a.** one quantitative independent variable
  **b.** two quantitative independent variables
  **c.** three quantitative independent variables [*Hint:* Include all possible two-way cross-product terms and squared terms.]

**12.49** Suppose you fit the second-order model

$$y = \beta_0 + \beta_1 x + \beta_2 x^2 + \varepsilon$$

to $n = 25$ data points. Your estimate of $\beta_2$ is $\hat{\beta}_2 = .47$, and the estimated standard error of the estimate is .15.
  **a.** Test $H_0: \beta_2 = 0$ against $H_a: \beta_2 \neq 0$. Use $\alpha = .05$.
  **b.** Suppose you want to determine only whether the quadratic curve opens upward; that is, as $x$ increases, the slope of the curve increases. Give the test statistic and the rejection region for the test for $\alpha = .05$. Do the data support the theory that the slope of the curve increases as $x$ increases? Explain.

**12.50** Suppose you fit the quadratic model

$$E(y) = \beta_0 + \beta_1 x + \beta_2 x^2$$

to a set of $n = 20$ data points and found $R^2 = .91$, $SS_{yy} = 29.94$, and SSE = 2.63.
  **a.** Is there sufficient evidence to indicate that the model contributes information for predicting $y$? Test using $\alpha = .05$.

  **b.** What null and alternative hypotheses would you test to determine whether upward curvature exists?
  **c.** What null and alternative hypotheses would you test to determine whether downward curvature exists?

**12.51** Consider the following quadratic models:

$$
\begin{aligned}
&(1)\ \ y = 1 - 2x + x^2\\
&(2)\ \ y = 1 + 2x + x^2\\
&(3)\ \ y = 1 + x^2\\
&(4)\ \ y = 1 - x^2\\
&(5)\ \ y = 1 + 3x^2
\end{aligned}
$$

  **a.** Graph each of the quadratic models, side by side, on the same sheet of graph paper.
  **b.** What effect does the first-order term $(2x)$ have on the graph of the curve?
  **c.** What effect does the second-order term $(x^2)$ have on the graph of the curve?

**12.52** MINITAB was used to fit the complete second-order model

$$E(y) = \beta_0 + \beta_1 x_1 + \beta_2 x_2 + \beta_3 x_1 x_2 + \beta_4 x_1^2 + \beta_5 x_2^2$$

to $n = 39$ data points. The printout is shown below.
  **a.** Is there sufficient evidence to indicate that at least one of the parameters, $\beta_1, \beta_2, \beta_3, \beta_4,$ and $\beta_5$ is nonzero? Test using $\alpha = .05$.
  **b.** Test $H_0: \beta_4 = 0$ against $H_a: \beta_4 \neq 0$. Use $\alpha = .01$.
  **c.** Test $H_0: \beta_5 = 0$ against $H_a: \beta_5 \neq 0$. Use $\alpha = .01$.
  **d.** Use graphs to explain the consequences of the tests in parts **b** and **c**.

**MINITAB Output for Exercise 12.52**

```
The regression equation is
Y = -24.56 + 1.12 X1 + 27.99 X2 - 0.54 X1X2 - 0.004 X1SQ + 0.002 X2SQ

Predictor        Coef      SE Coef        T       P
Constant       -24.563       6.531     -3.76   0.001
X1             1.19848       0.1103     10.86   0.000
X2              27.988      79.489      0.35   0.727
X1X2           -0.5397       1.0338     -0.52   0.605
X1SQ           -0.0043       0.0004    -10.74   0.000
X2SQ            0.0020       0.0033      0.60   0.550

S = 2.762     R-Sq = 79.7%     R-Sq(adj) = 76.6%

Analysis of Variance

Source           DF          SS         MS       F       P
Regression        5      989.30     197.86   25.93   0.000
Residual Error   33      251.81       7.63
Total            38     1241.11
```

## Applying the Concepts—Basic

**12.53** Underinflated or overinflated tires can increase tire wear. A new tire was tested for wear at different pressures with the results shown in the table.

⊘TIRES

| Pressure $x$ (pounds per square inch) | Mileage $y$ (thousands) |
|:---:|:---:|
| 30 | 29 |
| 31 | 32 |
| 32 | 36 |
| 33 | 38 |
| 34 | 37 |
| 35 | 33 |
| 36 | 26 |

  **a.** Plot the data on a scattergram.

  **b.** If you were given only the information for $x = 30$, 31, 32, 33, what kind of model would you suggest? For $x = 33, 34, 35, 36$? For all the data?

**12.54** A quadratic model was applied to motor vehicle toxic emissions data collected between 1984 and 1999 in Mexico City (*Environmental Science & Engineering*, Sept. 1, 2000). The following equation was used to predict the percentage ($y$) of motor vehicles without catalytic converters in the Mexico City fleet for a given year ($x$): $\hat{y} = 325{,}790 - 321.67x + 0.794x^2$.

  **a.** Explain why the value $\hat{\beta}_0 = 325{,}790$ has no practical interpretation.

  **b.** Explain why the value $\hat{\beta}_1 = -321.67$ should not be interpreted as a slope.

  **c.** Examine the value of $\hat{\beta}_2$ to determine the nature of the curvature (upward or downward) in the sample data.

  **d.** The researchers used the model to estimate "that just after the year 2021 the fleet of cars with catalytic converters will completely disappear." Comment on the danger of using the model to predict $y$ in the year 2021.

**12.55** The *Journal of Hazardous Materials* (July 1995) presented a literature review of models designed to predict oil spill evaporation. One model discussed in the article used boiling point ($x_1$) and API specific gravity ($x_2$) to predict the molecular weight ($y$) of the oil that is spilled. A complete second-order model for $y$ was proposed.

  **a.** Write the equation of the model.

  **b.** Identify the terms in the model that allow for curvilinear relationships.

**12.56** *Fisheries Science* (Feb. 1995) reported on a study of the variables that affect endogenous nitrogen excretion (ENE) in carp raised in Japan. Carp were divided into groups of 2 to 15 fish each according to body weight and each group placed in a separate tank. The carp were then fed a protein-free diet three times daily for a period of 20 days. One day after terminating the feeding experiment, the amount of ENE in each tank was measured. The table gives the mean body weight (in grams) and ENE amount (in milligrams per 100 grams of body weight per day) for each carp group.

⊘CARP

| Tank | Body Weight $x$ | ENE $y$ |
|:---:|:---:|:---:|
| 1 | 11.7 | 15.3 |
| 2 | 25.3 | 9.3 |
| 3 | 90.2 | 6.5 |
| 4 | 213.0 | 6.0 |
| 5 | 10.2 | 15.7 |
| 6 | 17.6 | 10.0 |
| 7 | 32.6 | 8.6 |
| 8 | 81.3 | 6.4 |
| 9 | 141.5 | 5.6 |
| 10 | 285.7 | 6.0 |

*Source:* Watanabe, T., and Ohta, M. "Endogenous nitrogen excretion and non-fecal energy losses in carp and rainbow trout." *Fisheries Science*, Vol. 61, No. 1, Feb. 1995, p. 56 (Table 5).

  **a.** Plot the data in a scattergram. Do you detect a pattern?

  **b.** The quadratic model $E(y) = \beta_0 + \beta_1 x + \beta_2 x^2$ was fit to the data using MINITAB. The MINITAB printout is displayed on p. 627. Use the printout information to test $H_0: \beta_2 = 0$ against $H_a: \beta_2 \neq 0$ using $\alpha = .10$. Give the conclusion in the words of the problem.

## Applying the Concepts—Intermediate

**12.57** Does exercise improve the human immune system? An experiment was conducted by a physiologist at the University of Florida to determine whether such a relationship exists. Thirty subjects volunteered to participate in the study. The amount of immunoglobulin known as IgG (an indicator of long-term immunity) and the maximal oxygen uptake (a measure of aerobic fitness level) were recorded for each subject. The resulting data are given in the table on p. 627.

  **a.** Construct a scattergram for the IgG–maximal oxygen uptake data.

  **b.** Hypothesize a probabilistic model relating IgG to maximal oxygen uptake.

  **c.** The STATISTIX printout for fitting the model to the data is provided on p. 627. Is there sufficient evidence to indicate that the model provides information for the prediction of IgG, $y$? Test using $\alpha = .05$.

  **d.** Give the observed significance level for the test of part **c**, and interpret it.

  **e.** Does the second-order term contribute information for the prediction of $y$? Test using $\alpha = .05$.

  **f.** Give the observed significance level for the test of part **e**, and interpret it.

**MINITAB Output for Exercise 12.56**

```
The regression equation is
ENE = 13.7 - 0.102 BODYWT +0.000273 BODYWTSQ

Predictor          Coef      SE Coef        T        P
Constant         13.713        1.306    10.50    0.000
BODYWT         -0.10184      0.02881    -3.53    0.010
BODYWTSQ      0.0002735    0.0001016     2.69    0.031


S = 2.194      R-Sq = 73.7%     R-Sq(adj) = 66.2%

Analysis of Variance

Source            DF           SS          MS        F       P
Regression         2       94.659      47.329     9.83   0.009
Residual Error     7       33.705       4.815
Total              9      128.364
```

**AEROBIC**

| Subject | IgG y | Maximum Oxygen Uptake x | Subject | IgG y | Maximum Oxygen Uptake x |
|---------|-------|-------------------------|---------|-------|-------------------------|
| 1 | 881 | 34.6 | 16 | 1,660 | 52.5 |
| 2 | 1,290 | 45.0 | 17 | 2,121 | 69.9 |
| 3 | 2,147 | 62.3 | 18 | 1,382 | 38.8 |
| 4 | 1,909 | 58.9 | 19 | 1,714 | 50.6 |
| 5 | 1,282 | 42.5 | 20 | 1,959 | 69.4 |
| 6 | 1,530 | 44.3 | 21 | 1,158 | 37.4 |
| 7 | 2,067 | 67.9 | 22 | 965 | 35.1 |
| 8 | 1,982 | 58.5 | 23 | 1,456 | 43.0 |
| 9 | 1,019 | 35.6 | 24 | 1,273 | 44.1 |
| 10 | 1,651 | 49.6 | 25 | 1,418 | 49.8 |
| 11 | 752 | 33.0 | 26 | 1,743 | 54.4 |
| 12 | 1,687 | 52.0 | 27 | 1,997 | 68.5 |
| 13 | 1,782 | 61.4 | 28 | 2,177 | 69.5 |
| 14 | 1,529 | 50.2 | 29 | 1,965 | 63.0 |
| 15 | 969 | 34.1 | 30 | 1,264 | 43.2 |

**STATISTIX Output for Exercise 12.57**

```
UNWEIGHTED LEAST SQUARES LINEAR REGRESSION OF IGG


PREDICTOR
VARIABLES    COEFFICIENT    STD ERROR     STUDENT'S T        P
---------    -----------    ---------    -------------     ------

CONSTANT       -1464.40      411.401         -3.56        0.0014
MAXOXY          88.3071      16.4735          5.36        0.0000
MAXOXYSQ       -0.53625      0.15819         -3.39        0.0022


R-SQUARED              0.9377    RESID. MEAN SQUARE (MSE)     11326.6
ADJUSTED R-SQUARED     0.9331    STANDARD DEVIATION           106.427


SOURCE        DF       SS          MS         F        P
----------    ---    ----------    ----------    -----    ------

REGRESSION     2      4602211     2301105    203.16    0.0000
RESIDUAL      27       305818       11326.6
TOTAL         29      4908029


CASES INCLUDED 30    MISSING CASES 0
```

**12.58** *the movie times* periodically lists the all-time gross revenues for major motion pictures produced in North America. The table below gives both the U.S. and worldwide gross revenues for the top 25 movies of all time.

  **a.** Write a first-order model for worldwide gross revenues, $y$, as a function of movie budget, $x$.

  **b.** Write a second-order model for worldwide gross revenues, $y$, as a function of movie budget, $x$.

  **c.** Fit the model of part **c** to the data and investigate its usefulness. Is there evidence of a curvilinear relationship between worldwide gross revenues and movie budget? Test using $\alpha = .05$.

  **d.** Based on your analysis in part **c**, which of the two models better explains the variation in worldwide gross revenues?

  **e.** Repeat parts **a–d**, but use U.S. gross revenues as the dependent variable, $y$.

**12.59** Spinocerebellar ataxia type 1 (SCA1) is an inherited neurodegenerative disorder characterized by dysfunction of the brain. From a DNA analysis of SCA1 chromosomes, researchers discovered the presence of repeat gene sequences (*Cell Biology*, Feb. 1995). In general, the more repeat sequences observed, the earlier the onset of the disease (in years of age). The scatterplot on p. 629 shows this relationship for data collected on 113 individuals diagnosed with SCA1.

  **a.** Suppose you want to model the age $y$ of onset of the disease as a function of number $x$ of repeat gene sequences in SCA1 chromosomes. Propose a quadratic model for $y$.

  **b.** Will the sign of $\beta_2$ in the model, part **a**, be positive or negative? Base your decision on the results shown in the scatterplot.

  **c.** The researchers reported a correlation of $r = -.815$ between age and number of repeats. Since $r^2 = (-.815)^2 = .664$, they concluded that about "66% of the variability in the age of onset can be accounted for by the number of repeats." Does this statement apply to the quadratic model $E(y) = \beta_0 + \beta_1 x + \beta_2 x^2$? If not, give the equation of the model for which it does apply.

⊘ MOVIEGROSS

| Rank | Movie | Year Released | Gross In U.S. Dollars Since Opening (millions) | MPAA Rating USA | Est. Budget (millions) | Est. Worldwide Gross (millions) |
|---|---|---|---|---|---|---|
| 1 | Titanic | 1997 | $600.8 | PG-13 | $200 | $1,835 |
| 2 | Star Wars | 1977 | $460.9 | PG | $11 | $784 |
| 3 | Star Wars: The Phantom Menace | 1999 | $431.1 | PG | $110 | $922 |
| 4 | E.T. | 1982 | $399.8 | PG | NA | $705 |
| 5 | Jurassic Park | 1993 | $356.8 | PG-13 | $63 | $920 |
| 6 | Forrest Gump | 1994 | $329.7 | PG-13 | $55 | $680 |
| 7 | The Lion King | 1994 | $312.9 | G | NA | $767 |
| 8 | Return of the Jedi | 1983 | $309.1 | PG | $33 | $573 |
| 9 | Independence Day | 1996 | $306.1 | PG-13 | $75 | $811 |
| 10 | The Sixth Sense | 1999 | $293.5 | PG-13 | $40 | $660 |
| 11 | The Empire Strikes Back | 1980 | $290.2 | PG | $18 | $513 |
| 12 | Home Alone | 1990 | $285.0 | PG | NA | $533 |
| 13 | How The Grinch Stole Christmas | 2000 | $260.0 | PG | $123 | $340 |
| 14 | Jaws | 1975 | $260.0 | PG | $12 | $471 |
| 15 | Batman | 1989 | $251.2 | PG-13 | NA | $413 |
| 16 | Men in Black | 1997 | $250.1 | PG-13 | $90 | $586 |
| 17 | Toy Story 2 | 1999 | $245.8 | G | $90 | $485 |
| 18 | Raiders of the Lost Ark | 1981 | $242.4 | PG | $20 | $384 |
| 19 | Twister | 1996 | $241.7 | PG-13 | $92 | $495 |
| 20 | Shrek | 2001 | $240.6 | PG | $60 | $289 |
| 21 | Beverly Hills Cop | 1984 | $234.8 | R | NA | $316 |
| 22 | Cast Away | 2000 | $233.6 | PG-13 | $90 | $408 |
| 23 | The Lost World | 1997 | $229.1 | PG-13 | $73 | $614 |
| 24 | Ghostbusters | 1984 | $220.9 | PG | $30 | $274 |
| 25 | Mrs. Doubtfire | 1993 | $219.2 | PG-13 | NA | $423 |

*Source: the movie times, July 11, 2001 (http://www.the-movie-times.com)*

Scatterplot for Exercise 12.59

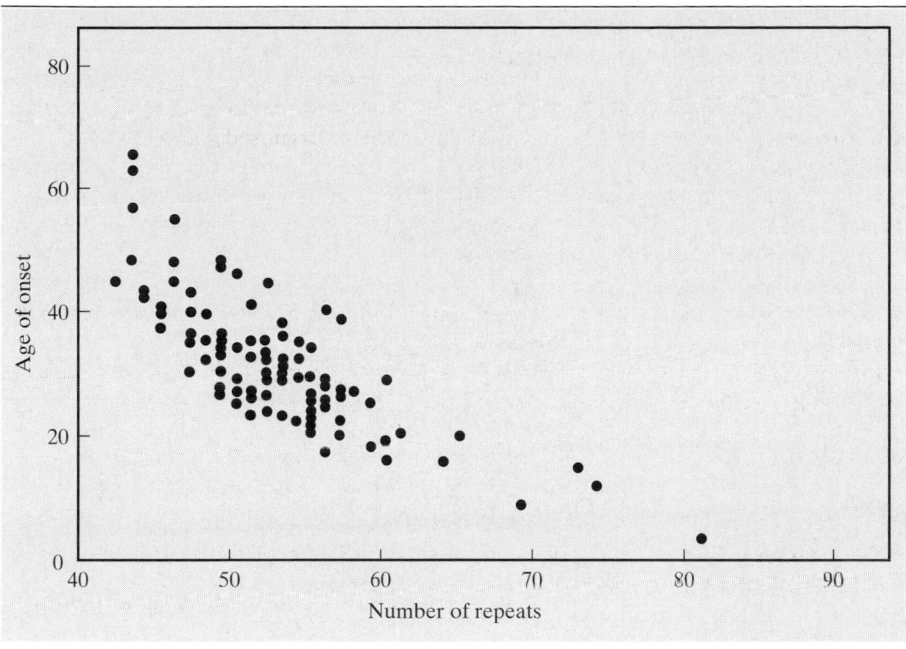

**12.60** Refer to the *American Scientist* (July–Aug. 1998) study of the relationship between self-avoiding and unrooted walks, Exercise 2.113 (p. 90). Recall that in a self-avoiding walk you never retrace or cross your own path, while an unrooted walk is a path in which the starting and ending points are impossible to distinguish. The possible number of walks of each type of various lengths are reproduced in the table. In Exercise 11.80 (p. 570) you analyzed the straight-line model relating number of unrooted walks ($y$) to walk length $(x)$. Now consider the quadratic model $E(y) = \beta_0 + \beta_1 x + \beta_2 x^2$. A SAS printout for the regression analysis is provided on p. 630. Is there sufficient evidence of an upward concave curvilinear relationship between $y$ and $x$? Test at $\alpha = .10$.

⬤ WALK

| Walk Length (Number of steps) | Unrooted Walks | Self-Avoiding Walks |
|---|---|---|
| 1 | 1 | 4 |
| 2 | 2 | 12 |
| 3 | 4 | 36 |
| 4 | 9 | 100 |
| 5 | 22 | 284 |
| 6 | 56 | 780 |
| 7 | 147 | 2,172 |
| 8 | 388 | 5,916 |

*Source:* Hayes, B. "How to avoid yourself." *American Scientist*, Vol. 86, No. 4, July–Aug. 1998, p. 317 (Figure 5).

## Applying the Concepts—Advanced

**12.61** The optomotor responses of tree frogs were studied in the *Journal of Experimental Zoology* (Sept. 1993). Microspectrophotometry was used to measure the threshold quantal flux (the light intensity at which the optomotor response was first observed) of tree frogs tested at different spectral wavelengths. The data revealed the relationship between the log of quantal flux ($y$) and wavelength ($x$) shown in the accompanying graph.

**a.** Explain why a first-order model would not be appropriate for modeling $E(y)$.

**b.** Explain why a second-order model would not be appropriate for modeling $E(y)$.

**c.** Demonstrate that the third-order model $E(y) = \beta_0 + \beta_1 x + \beta_2 x^2 + \beta_3 x^3$ may be the most appropriate model for $E(y)$.

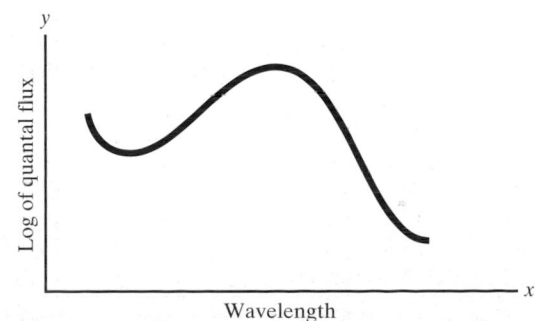

SAS Output for Exercise 12.60

The REG Procedure
Dependent Variable: UNROOT

Analysis of Variance

| Source | DF | Sum of Squares | Mean Square | F Value | Pr > F |
|---|---|---|---|---|---|
| Model | 2 | 115583 | 57792 | 26.66 | 0.0022 |
| Error | 5 | 10837 | 2167.33452 | | |
| Corrected Total | 7 | 126420 | | | |

| | | | | |
|---|---|---|---|---|
| Root MSE | 46.55464 | R-Square | 0.9143 | |
| Dependent Mean | 78.62500 | Adj R-Sq | 0.8800 | |
| Coeff Var | 59.21099 | | | |

Parameter Estimates

| Variable | DF | Parameter Estimate | Standard Error | t Value | Pr > \|t\| |
|---|---|---|---|---|---|
| Intercept | 1 | 112.12500 | 64.95046 | 1.73 | 0.1449 |
| LENGTH | 1 | -93.01786 | 33.11447 | -2.81 | 0.0376 |
| LENGTHSQ | 1 | 15.10119 | 3.59177 | 4.20 | 0.0085 |

## 12.9 MODEL BUILDING: QUALITATIVE (DUMMY) VARIABLE MODELS

Multiple regression models can also be written to include **qualitative** (or **categorical**) independent variables. Qualitative variables, unlike quantitative variables, cannot be measured on a numerical scale. Therefore, we must code the values of the qualitative variable (called **levels**) as numbers before we can fit the model. These coded qualitative variables are called **dummy** (or **indicator**) **variables** since the numbers assigned to the various levels are arbitrarily selected.

To illustrate, suppose a female executive at a certain company claims that male executives earn higher salaries, on average, than female executives with the same education, experience, and responsibilities. To support her claim, she wants to model the salary $y$ of an executive using a qualitative independent variable representing the gender of an executive (male or female).

A convenient method of coding the values of a qualitative variable at two levels involves assigning a value of 1 to one of the levels and a value of 0 to the other. For example, the dummy variable used to describe gender could be coded as follows:

$$x = \begin{cases} 1 & \text{if male} \\ 0 & \text{if female} \end{cases}$$

The choice of which level is assigned to 1 and which is assigned to 0 is arbitrary. The model then takes the following form:

$$E(y) = \beta_0 + \beta_1 x$$

The advantage of using a 0–1 coding scheme is that the $\beta$ coefficients are easily interpreted. The model above allows us to compare the mean executive salary $E(y)$ for males with the corresponding mean for females.

*Males* $(x = 1)$: $\qquad E(y) = \beta_0 + \beta_1(1) = \beta_0 + \beta_1$
*Females* $(x = 0)$: $\qquad E(y) = \beta_0 + \beta_1(0) = \beta_0$

**FIGURE 12.20**

*Bar Chart Comparing $E(y)$
for Males and Females*

These two means are illustrated in the bar graph in Figure 12.20.

First note that $\beta_0$ represents the mean salary for females (say, $\mu_F$). When a 0–1 coding convention is used, $\beta_0$ will always represent the mean response associated with the level of the qualitative variable assigned the value 0 (called the **base level**). The difference between the mean salary for males and the mean salary for females, $\mu_M - \mu_F$, is represented by $\beta_1$—that is,

$$\mu_M - \mu_F = (\beta_0 + \beta_1) - (\beta_0) = \beta_1$$

This difference is shown in Figure 12.20.* With a 0–1 coding convention, $\beta_1$ will always represent the difference between the mean response for the level assigned the value 1 and the mean for the base level. Thus, for the executive salary model, we have

$$\beta_0 = \mu_F$$
$$\beta_1 = \mu_M - \mu_F$$

Now carefully examine the model with a single qualitative independent variable at two levels, because we will use exactly the same pattern for any number of levels. Moreover, the interpretation of the parameters will always be the same.

One level (say, level A) is selected as the *base level*. Then for the 0–1 coding[†] for the dummy variables,

$$\mu_A = \beta_0$$

The coding for all dummy variables is as follows: To represent the mean value of $y$ for a particular level, let that dummy variable equal 1; otherwise, the dummy variable is set equal to 0. Using this system of coding,

$$\mu_B = \beta_0 + \beta_1$$
$$\mu_C = \beta_0 + \beta_2$$

and so on. Because $\mu_A = \beta_0$, any other model parameter will represent the difference between means for that level and the base level:

$$\beta_1 = \mu_B - \mu_A$$
$$\beta_2 = \mu_C - \mu_A$$

and so on.

---

*Note that $\beta_1$ could be negative. If $\beta_1$ were negative, the height of the bar corresponding to males would be *reduced* (rather than increased) from the height of the bar for females by the amount $\beta_1$. Figure 12.20 is constructed assuming that $\beta_1$ is a positive quantity.

[†]You do not have to use a 0–1 system of coding for the dummy variables. Any two-value system will work, but the interpretation given to the model parameters will depend on the code. Using the 0–1 system makes the model parameters easy to interpret.

## Procedure for Writing a Model with One Qualitative Independent Variable with $k$ Levels

Always use a number of dummy variables that is one less than the number of levels of the qualitative variable. Thus, for a qualitative variable with $k$ levels, use $k - 1$ dummy variables:

$$y = \beta_0 + \beta_1 x_1 + \beta_2 x_2 + \cdots + \beta_{k-1} x_{k-1} + \varepsilon$$

where $x_i$ is the dummy variable for level $i + 1$ and

$$x_i = \begin{cases} 1 & \text{if } y \text{ is observed at level } i + 1 \\ 0 & \text{otherwise} \end{cases}$$

Then, for this system of coding

$$\mu_A = \beta_0 \qquad \text{and} \qquad \beta_1 = \mu_B - \mu_A$$

$$\mu_B = \beta_0 + \beta_1 \qquad\qquad\qquad \beta_2 = \mu_C - \mu_A$$

$$\mu_C = \beta_0 + \beta_2 \qquad\qquad\qquad \beta_3 = \mu_D - \mu_A$$

$$\mu_D = \beta_0 + \beta_3 \qquad\qquad\qquad \vdots$$

$$\vdots$$

**E X A M P L E   1 2 . 8**

Refer to Example 10.3 (p. 452). Recall that the USGA wants to compare the mean driving distances of four different golf ball brands (A, B, C, and D). Iron Byron, the USGA's robotic golfer, is used to hit a sample of 10 balls of each brand. The distance data are reproduced in Table 12.4. Use a regression model to determine if the mean driving distances for the four brands are significantly different at $\alpha = .05$.

⊘ GOLFCRD

**TABLE 12.4   Driving Distances (in feet) for Four Golf Ball Brands**

| Brand A | Brand B | Brand C | Brand D |
|---------|---------|---------|---------|
| 251.2 | 263.2 | 269.7 | 251.6 |
| 245.1 | 262.9 | 263.2 | 248.6 |
| 248.0 | 265.0 | 277.5 | 249.4 |
| 251.1 | 254.5 | 267.4 | 242.0 |
| 260.5 | 264.3 | 270.5 | 246.5 |
| 250.0 | 257.0 | 265.5 | 251.3 |
| 253.9 | 262.8 | 270.7 | 261.8 |
| 244.6 | 264.4 | 272.9 | 249.0 |
| 254.6 | 260.6 | 275.6 | 247.1 |
| 248.8 | 255.9 | 266.5 | 245.9 |

### Solution

Note that golf ball brand (A, B, C, and D) is a qualitative variable (measured on a nominal scale). According to the box, for a four-level qualitative variable we require three dummy variables in the regression model. The model relating $E(y)$, where $y$ is the distance the ball is driven by Iron Byron, to this single qualitative variable, golf ball brand, is

$$E(y) = \beta_0 + \beta_1 x_1 + \beta_2 x_2 + \beta_3 x_3$$

where

$$x_1 = \begin{cases} 1 & \text{if Brand B} \\ 0 & \text{if not} \end{cases} \qquad x_2 = \begin{cases} 1 & \text{if Brand C} \\ 0 & \text{if not} \end{cases} \qquad x_3 = \begin{cases} 1 & \text{if Brand D} \\ 0 & \text{if not} \end{cases}$$

for this model

$$\beta_1 = \mu_B - \mu_A$$
$$\beta_2 = \mu_C - \mu_A$$
$$\beta_3 = \mu_D - \mu_A$$

where $\mu_A, \mu_B, \mu_C$, and $\mu_D$ are the mean distances for Brands A, B, C, and D, respectively. Testing the null hypothesis that the means for the four brands are equal, that is, $\mu_A = \mu_B = \mu_C = \mu_D$, is equivalent to testing

$$H_0: \beta_1 = \beta_2 = \beta_3 = 0$$

You can see this by observing that if $\beta_1 = \mu_B - \mu_A = 0$, then $\mu_A = \mu_B$. Similarly, $\beta_2 = \mu_C - \mu_A = 0$ implies that $\mu_A = \mu_C$, and $\beta_3 = \mu_D - \mu_A = 0$ implies that $\mu_A = \mu_D$. The alternative hypothesis is

$$H_a: \text{At least one of the parameters, } \beta_1, \beta_2 \text{ and } \beta_3, \text{ differs from 0}$$

which implies that at least two of the four means ($\mu_A, \mu_B, \mu_C$, and $\mu_D$) differ.

To test this hypothesis, we conduct the global $F$-test for the model. The SAS printout for fitting the model.

$$E(y) = \beta_0 + \beta_1 x_1 + \beta_2 x_2 + \beta_3 x_3$$

is shown in Figure 12.21. The value of the $F$ statistic for testing the adequacy of the model, $F = 43.99$, and the observed significance level of the test, $p < .0001$, are

**FIGURE 12.21**

*SAS Regression Printout for Golf Ball Brand Model*

The REG Procedure
Dependent Variable: DISTANCE

Analysis of Variance

| Source | DF | Sum of Squares | Mean Square | F Value | Pr > F |
|---|---|---|---|---|---|
| Model | 3 | 2794.38875 | 931.46292 | 43.99 | <.0001 |
| Error | 36 | 762.30100 | 21.17503 | | |
| Corrected Total | 39 | 3556.68975 | | | |

| | | | |
|---|---|---|---|
| Root MSE | 4.60163 | R-Square | 0.7857 |
| Dependent Mean | 257.77750 | Adj R-Sq | 0.7678 |
| Coeff Var | 1.78512 | | |

Parameter Estimates

| Variable | DF | Parameter Estimate | Standard Error | t Value | Pr > |t| |
|---|---|---|---|---|---|
| Intercept | 1 | 250.78000 | 1.45516 | 172.34 | <.0001 |
| X1 | 1 | 10.28000 | 2.05791 | 5.00 | <.0001 |
| X2 | 1 | 19.17000 | 2.05791 | 9.32 | <.0001 |
| X3 | 1 | -1.46000 | 2.05791 | -0.71 | 0.4826 |

both highlighted. Since $\alpha = .05$ exceeds the *p*-value, we reject $H_0$ and conclude that at least one of the parameters differs from 0. Or, equivalently, we conclude that the data provide sufficient evidence to indicate that the mean driving distance does vary from one golf ball brand to another. ∎

*Note:* The *F*-test in Example 12.8 can also be obtained using the ANOVA (analysis of variance) procedure for a completely randomized design, described in Chapter 10. The *F* value and *p*-value shown on the ANOVA printout, Figure 10.6 (p. 453) are both identical to the values in Figure 12.21.

 **EXERCISES 12.62–12.74**

## Learning the Mechanics

**12.62** Write a regression model relating the mean value of *y* to a qualitative independent variable that can assume two levels. Interpret all the terms in the model.

**12.63** Write a regression model relating $E(y)$ to a qualitative independent variable that can assume three levels. Interpret all the terms in the model.

**12.64** The following model was used to relate $E(y)$ to a single qualitative variable with four levels: $E(y) = \beta_0 + \beta_1 x_1 + \beta_2 x_2 + \beta_3 x_3$, where

$$x_1 = \begin{cases} 1 & \text{if level 2} \\ 0 & \text{if not} \end{cases}$$

$$x_2 = \begin{cases} 1 & \text{if level 3} \\ 0 & \text{if not} \end{cases}$$

$$x_3 = \begin{cases} 1 & \text{if level 4} \\ 0 & \text{if not} \end{cases}$$

This model was fit to $n = 30$ data points and the following result was obtained:

$$\hat{y} = 10.2 - 4x_1 + 12x_2 + 2x_3$$

**a.** Use the least squares prediction equation to find the estimate of $E(y)$ for each level of the qualitative independent variable.

**b.** Specify the null and alternative hypotheses you would use to test whether $E(y)$ is the same for all four levels of the independent variable.

**12.65** MINITAB (see printout, p. 635) was used to fit the following model to $n = 15$ data points:

$$y = \beta_0 + \beta_1 x_1 + \beta_2 x_2 + \varepsilon$$

where

$$x_1 = \begin{cases} 1 & \text{if level 2} \\ 0 & \text{if not} \end{cases}$$

$$x_2 = \begin{cases} 1 & \text{if level 3} \\ 0 & \text{if not} \end{cases}$$

**a.** Report the least squares prediction equation.

**b.** Interpret the values of $\beta_1$ and $\beta_2$.

**c.** Interpret the following hypotheses in terms of $\mu_1, \mu_2,$ and $\mu_3$:

$H_0: \beta_1 = \beta_2 = 0$
$H_a:$ At least one of the parameters $\beta_1$ and $\beta_2$ differs from 0

**d.** Conduct the hypothesis test of part **c**.

## Applying the Concepts—Basic

**12.66** Refer to the *Chance* (Winter 2001) study of students who paid a private tutor (or coach) to help them improve their Scholastic Assessment Test (SAT) scores, Exercise 2.81 (p. 70). Multiple regression was used to estimate the effect of coaching on SAT-Mathematics scores. Data on 3,492 students (573 of whom were coached) were used to fit the model $E(y) = \beta_0 + \beta_1 x_1 + \beta_2 x_2$, where $y$ = SAT-Math score, $x_1$ = score on PSAT, and $x_2 = \{1$ if student was coached, 0 if not}.

**a.** The fitted model had an adjusted $R^2$ value of .76. Interpret this result.

**b.** The estimate of $\beta_2$ in the model was 19, with a standard error of 3. Use this information to form a 95% confidence interval for $\beta_2$. Interpret the interval.

**c.** Based on the interval, part **b**, what can you say about the effect of coaching on SAT-Math scores?

**12.67** *New Scientist* (Apr. 3, 1993) published an article on strategies for foiling assassination attempts on politicians. The strategies are based on the findings of researchers at Middlesex University (United Kingdom), who used a multiple regression model for predicting the level *y* of assassination risk. One of the variables used in the model was political status of a country (communist, democratic, or dictatorship).

**a.** Propose a model for $E(y)$ as a function of political status.

**b.** Interpret the $\beta$'s in the model, part **a**.

MINITAB Output for Exercise 12.65

```
The regression equation is
Y = 80.0 + 16.8 X1 + 40.4 X2


Predictor        Coef     SE Coef        T        P
Constant       80.000       4.082    19.60    0.000
X1             16.800       5.774     2.91    0.013
X2             40.400       5.774     7.00    0.000


S = 9.129        R-Sq = 80.5%     R-Sq(adj) = 77.2%


Analysis of Variance


Source          DF          SS        MS        F        P
Regression       2      4118.9    2059.5    24.72    0.000
Residual Error  12      1000.0      83.3
Total           14      5118.9
```

**12.68** Which insect repellents protect best against mosquitoes? *Consumer Reports* (June 2000) tested 14 products that all claim to be an effective mosquito repellent. Each product was classified as either lotion/cream or aerosol/spray. The cost of the product (in dollars) was divided by the amount of the repellent needed to cover exposed areas of the skin (about 1/3 ounce) to obtain a cost-per-use value. Effectiveness was measured as the maximum number of hours of protection (in half-hour increments) provided when human testers exposed their arms to 200 mosquitoes. The data from the report are listed in the table below.

**a.** Suppose you want to use repellent type to model the cost per use ($y$). Create the appropriate number of dummy variables for repellent type and write the model.

**b.** Fit the model, part **a**, to the data.

**c.** Give the null hypothesis for testing whether repellent type is a useful predictor of cost per use ($y$).

**d.** Conduct the test, part **c**, and give the appropriate conclusion. Use $\alpha = .10$.

**e.** Repeat parts **a**–**d** if the dependent variable is maximum number of hours of protection ($y$).

**12.69** A fisheries researcher believes that the growth rate of fish is related to the age of the fish, its feeding location, and other factors. Suppose the researcher wishes to collect fish of a certain species from four different locations.

**a.** Write a model relating the mean growth rate $E(y)$ to location, a qualitative variable.

**b.** Use your model to find the mean growth rate of the fish at location 1.

**c.** Use your model to find the mean growth rate of the fish at location 4.

**d.** In terms of your model, what is the difference in mean growth rates between locations 2 and 3?

⊘ REPELLENT

| Insect Repellent | Type | Cost/Use | Maximum Protection |
|---|---|---|---|
| Amway HourGuard 12 | Lotion/Cream | $2.08 | 13.5 hours |
| Avon Skin-So-Soft | Aerosol/Spray | 0.67 | 0.5 |
| Avon BugGuard Plus | Lotion/Cream | 1.00 | 2.0 |
| Ben's Backyard Formula | Lotion/Cream | 0.75 | 7.0 |
| Bite Blocker | Lotion/Cream | 0.46 | 3.0 |
| BugOut | Aerosol/Spray | 0.11 | 6.0 |
| Cutter Skinsations | Aerosol/Spray | 0.22 | 3.0 |
| Cutter Unscented | Aerosol/Spray | 0.19 | 5.5 |
| Muskoll Ultra6Hours | Aerosol/Spray | 0.24 | 6.5 |
| Natrapel | Aerosol/Spray | 0.27 | 1.0 |
| Off! Deep Woods | Aerosol/Spray | 1.77 | 14.0 |
| Off! Skintastic | Lotion/Cream | 0.67 | 3.0 |
| Sawyer Deet Formula | Lotion/Cream | 0.36 | 7.0 |
| Repel Permanone | Aerosol/Spray | 2.75 | 24.0 |

*Source:* "Buzz off." *Consumer Reports*, June 2000.

## Applying the Concepts—Intermediate

**12.70** Where do you look when you are listening to someone speak? Researchers have discovered that listeners tend to gaze at the eyes or mouth of the speaker. In a study published in *Perception & Psychophysics* (Aug. 1998), subjects watched a videotape of a speaker giving a series of short monologues at a social gathering (e.g., a party). The level of background noise (multilingual voices and music) was varied during the listening sessions. Each subject wore a pair of clear plastic goggles on which an infrared corneal detection system was mounted, enabling the researchers to monitor the subject's eye movements. One response variable of interest was the proportion $y$ of times the subject's eyes fixated on the speaker's mouth.

  **a.** The researchers wanted to estimate $E(y)$ for four different noise levels: none, low, medium, and high. Hypothesize a model that will allow the researchers to obtain these estimates.

  **b.** Interpret the $\beta$'s in the model, part **a**.

  **c.** Explain how to test the hypothesis of no differences in the mean proportions of mouth fixations for the four background noise levels.

**12.71** A field experiment was conducted to assess the effect of organic enrichment on the mean density of mosquito larvae (*Journal of the American Mosquito Control Association*, June 1995). Larval specimens were collected from a pond three days after the pond was flooded with canal water. A second sample of specimens was collected three weeks after flooding and enriching the pond with rabbit pellets. All specimens were returned to the laboratory and the number $y$ of mosquito larvae counted in each specimen.

  **a.** Write a model that will allow you to compare the mean number of mosquito larvae found in the enriched pond to the corresponding mean for the natural pond.

  **b.** Interpret the $\beta$ coefficients in the model, part **a**.

  **c.** Set up the null and alternative hypotheses for testing whether the mean larval density for the enriched pond exceeds the mean for the natural pond.

  **d.** The $p$-value associated with the global $F$-test for the model, part **a**, was determined to be .004. Interpret this result.

**12.72** Five varieties of peas (A, B, C, D, and E) are currently being tested by a large agribusiness cooperative to determine which is best suited for production. A field was divided into 20 plots, with each variety of peas planted in four plots. The yields (in bushels of peas) produced from each plot are shown in the table.

⊘ PEAS

| A | B | C | D | E |
|------|------|------|------|------|
| 26.2 | 29.2 | 29.1 | 21.3 | 20.1 |
| 24.3 | 28.1 | 30.8 | 22.4 | 19.3 |
| 21.8 | 27.3 | 33.9 | 24.3 | 19.9 |
| 28.1 | 31.2 | 32.8 | 21.8 | 22.1 |

The SAS printout below resulted from fitting the model

$$y = \beta_0 + \beta_1 x_1 + \beta_2 x_2 + \beta_3 x_3 + \beta_4 x_4 + \varepsilon$$

**SAS Output for Exercise 12.72**

The REG Procedure
Dependent Variable: YIELD

Analysis of Variance

| Source | DF | Sum of Squares | Mean Square | F Value | Pr > F |
|--------|-----|-----------|-----------|-------|--------|
| Model | 4 | 342.04000 | 85.51000 | 23.97 | <.0001 |
| Error | 15 | 53.52000 | 3.56800 | | |
| Corrected Total | 19 | 395.56000 | | | |

| | | | | |
|---|---|---|---|---|
| Root MSE | 1.88892 | R-Square | 0.8647 |
| Dependent Mean | 25.70000 | Adj R-Sq | 0.8286 |
| Coeff Var | 7.34986 | | |

Parameter Estimates

| Variable | DF | Parameter Estimate | Standard Error | t Value | Pr > \|t\| |
|----------|-----|-----------|----------|-------|---------|
| Intercept | 1 | 20.35000 | 0.94446 | 21.55 | <.0001 |
| X1 | 1 | 4.75000 | 1.33566 | 3.56 | 0.0029 |
| X2 | 1 | 8.60000 | 1.33566 | 6.44 | <.0001 |
| X3 | 1 | 11.30000 | 1.33566 | 8.46 | <.0001 |
| X4 | 1 | 2.10000 | 1.33566 | 1.57 | 0.1367 |

to these data, using the coding $x_1 = 1$ for variety A, $x_2 = 1$ for variety B, $x_3 = 1$ for variety C, and $x_4 = 1$ for variety D.

**a.** Interpret all the parameters in the model.

**b.** Report the least squares model from the SAS printout.

**c.** What null and alternative hypotheses are tested by the global $F$-test for this model? Interpret the hypotheses both in terms of the $\beta$ parameters and the mean yields for the five varieties of peas.

**d.** Test the hypotheses of part **c** using $\alpha = .05$.

**e.** Place a 95% confidence interval on the differences in the mean yields of varieties D and E.

**f.** Perform an analysis of variance for a completely randomized design on the data. Show that the ANOVA results are equivalent to the regression results.

**12.73** Do grizzly bears segregate on the basis of sex? One hypothesis is that female grizzlies avoid male-occupied habitats because of competition for food and cannibalism. A competing theory is that females do not avoid males, but simply have different habitats available to them. These hypotheses were investigated in the *Journal of Wildlife Management* (July 1995). Grizzly bears were trapped, radio-collared, and released in the Highwood trapping zone (HTZ) in Alberta, Canada. The percentage of time, $y$, each bear used the HTZ as a habitat over a 4-year period was recorded. The researchers modeled $E(y)$ as a function of reproductive class at five levels: estrous adult females, adult females with offspring, independent subadult females, adult males, and independent subadult males. One goal was to compare the mean percent use of HTZ for the five classes of grizzly bears.

**a.** Write a model for $E(y)$ that will enable the researchers to carry out the comparison.

| Grizzly Class | $n$ | Mean Percent Use |
|---|---|---|
| Estrous adult females | 5 | 38 |
| Adult females with offspring | 7 | 16 |
| Independent subadult females | 2 | 89 |
| Adult males | 7 | 43 |
| Independent subadult males | 8 | 58 |

*Source:* Wielgus, R. B., and Bunnell, F. L. "Tests of hypotheses for sexual segregation in grizzly bears." *Journal of Wildlife Management*, Vol. 59, No. 3, July 1995, p. 555 (Table 1).

**b.** The sample sizes and sample means for the five classes are shown in the table above. Use this information to find estimates of the $\beta$'s in the model, part **a**.

**c.** Give the null hypothesis for a test to determine whether the mean percent use of HTZ differs among the grizzly bear classes.

**d.** The $p$-value for the test, part **c**, was reported as .15. Interpret this result.

### Applying the Concepts—Advanced

**12.74** Refer to *The Archives of Disease in Childhood* (Apr. 2000) study of whether height influences a child's progression through elementary school, Exercise 10.23 (p. 460). Recall that Australian school children were divided into equal thirds (tertiles) based on age (youngest third, middle third, and oldest third). The average heights of the three groups (where all height measurements were standardized using $z$-scores), by gender, are shown in the table below.

**a.** Propose a regression model that will enable you to compare the average heights of the three age groups for boys.

**b.** Find the estimates of the $\beta$'s in the model, part **a**.

**c.** Repeat parts **a** and **b** for girls.

| | Sample Size | Youngest Tertile Mean Height | Middle Tertile Mean Height | Oldest Tertile Mean Height |
|---|---|---|---|---|
| **Boys** | 1439 | 0.33 | 0.33 | 0.16 |
| **Girls** | 1409 | 0.27 | 0.18 | 0.21 |

*Source:* Wake, M., Coghlan, D., and Hesketh, K. "Does height influence progression through primary school grades?" *The Archives of Disease in Childhood*, Vol. 82, Apr. 2000 (Table 2).

## 12.10   MODEL BUILDING: MODELS WITH BOTH QUANTITATIVE AND QUALITATIVE VARIABLES

Suppose you want to relate the mean monthly sales $E(y)$ of a company to monthly advertising expenditure $x$ for three different advertising media (say newspaper, radio, and television) and you wish to use first-order (straight-line) models to model the responses for all three media. Graphs of these three relationships might appear as shown in Figure 12.22.

Since the lines in Figure 12.22 are hypothetical, a number of practical questions arise. Is one advertising medium as effective as any other? That is, do the three mean sales lines differ for the three advertising media? Do the increases in mean sales per dollar input in advertising differ for the three advertising

**FIGURE 12.22**

*Graphs of the Relationship Between Mean Sales $E(y)$ and Advertising Expenditure x*

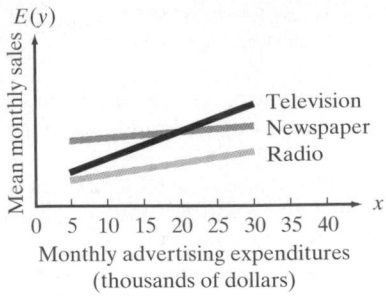

media? That is, do the slopes of the three lines differ? Note that the two practical questions have been rephrased into questions about the parameters that define the three lines of Figure 12.22. To answer them, we must write a single regression model that will characterize the three lines of Figure 12.22 and that, by testing hypotheses about the lines, will answer the questions.

The response described previously, monthly sales, is a function of *two* independent variables, one quantitative (advertising expenditure $x_1$) and one qualitative (type of medium). We will proceed, in stages, to build a model relating $E(y)$ to these variables and will show graphically the interpretation we would give to the model at each stage. This will help you to see the contributions of the various terms in the model.

1. The straight-line relationship between mean sales $E(y)$ and advertising expenditure is the same for all three media—that is, a single line will describe the relationship between $E(y)$ and advertising expenditure $x_1$ for all the media (see Figure 12.23).

$$E(y) = \beta_0 + \beta_1 x_1 \quad \text{where } x_1 = \text{Advertising expenditure}$$

**FIGURE 12.23**

*The Relationship Between $E(y)$ and $x_1$ is the Same for All Media*

2. The straight lines relating mean sales $E(y)$ to advertising expenditure $x_1$ differ from one medium to another, but the rate of increase in mean sales per increase in dollar advertising expenditure $x_1$ is the same for all media—that is, the lines are parallel but possess different $y$-intercepts (see Figure 12.24).

$$E(y) = \beta_0 + \beta_1 x_1 + \beta_2 x_2 + \beta_3 x_3$$

**FIGURE 12.24**

*Parallel Response Lines for the Three Media*

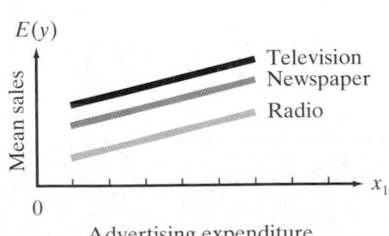

where

$$x_1 = \text{Advertising expenditure}$$

$$x_2 = \begin{cases} 1 & \text{if radio medium} \\ 0 & \text{if not} \end{cases}$$

$$x_3 = \begin{cases} 1 & \text{if television medium} \\ 0 & \text{if not} \end{cases}$$

Notice that this model is essentially a combination of a first-order model with a single quantitative variable and the model with a single qualitative variable:

First-order model with a single
quantitative variable:     $E(y) = \beta_0 + \beta_1 x_1$

Model with single qualitative
variable at three levels:     $E(y) = \beta_0 + \beta_2 x_2 + \beta_3 x_3$

where $x_1, x_2$, and $x_3$ are as just defined. The model described here implies no interaction between the two independent variables, which are advertising expenditure $x_1$ and the qualitative variable (type of advertising medium). The change in $E(y)$ for a 1-unit increase in $x_1$ is identical (the slopes of the lines are equal) for all three advertising media. The terms corresponding to each of the independent variables are called **main effect terms** because they imply no interaction.

3. The straight lines relating mean sales $E(y)$ to advertising expenditure $x_1$ differ for the three advertising media—that is, both the line intercepts and the slopes differ (see Figure 12.25).

**FIGURE 12.25**

*Different Response Lines
for the Three Media*

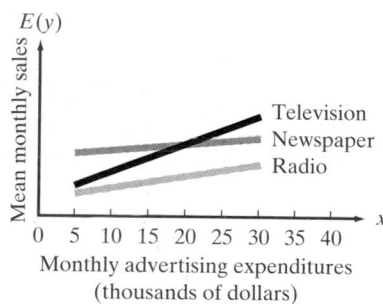

As you will see, this interaction model is obtained by adding terms involving the cross-product terms, one each from each of the two independent variables:

$$E(y) = \beta_0 + \underbrace{\beta_1 x_1}_{\substack{\text{Main effect,}\\\text{advertising}\\\text{expenditure}}} + \underbrace{\beta_2 x_2 + \beta_3 x_3}_{\substack{\text{Main effect,}\\\text{type of}\\\text{medium}}} + \underbrace{\beta_4 x_1 x_2 + \beta_5 x_1 x_3}_{\text{Interaction}}$$

Note that each of the preceding models is obtained by adding terms to model 1, the single first-order model used to model the responses for all three media. Model 2 is obtained by adding the main effect terms for type of medium, the qualitative variable. Model 3 is obtained by adding the interaction terms to model 2.

**EXAMPLE 12.9**

Substitute the appropriate values of the dummy variables in model 3 to obtain the equations of the three response lines in Figure 12.25.

**Solution**

The complete model that characterizes the three lines in Figure 12.25 is

$$E(y) = \beta_0 + \beta_1 x_1 + \beta_2 x_2 + \beta_3 x_3 + \beta_4 x_1 x_2 + \beta_5 x_1 x_3$$

where

$$x_1 = \text{Advertising expenditure}$$

$$x_2 = \begin{cases} 1 & \text{if radio medium} \\ 0 & \text{if not} \end{cases}$$

$$x_3 = \begin{cases} 1 & \text{if television medium} \\ 0 & \text{if not} \end{cases}$$

Examining the coding, you can see that $x_2 = x_3 = 0$ when the advertising medium is newspaper. Substituting these values into the expression for $E(y)$, we obtain the newspaper medium line:

$$E(y) = \beta_0 + \beta_1 x_1 + \beta_2(0) + \beta_3(0) + \beta_4 x_1(0) + \beta_5 x_1(0) = \beta_0 + \beta_1 x_1$$

Similarly, we substitute the appropriate values of $x_2$ and $x_3$ into the expression for $E(y)$ to obtain the radio medium line ($x_2 = 1, x_3 = 0$):

$$E(y) = \beta_0 + \beta_1 x_1 + \beta_2(1) + \beta_3(0) + \beta_4 x_1(1) + \beta_5 x_1(0)$$

$$= \underbrace{(\beta_0 + \beta_2)}_{y\text{-intercept}} + \underbrace{(\beta_1 + \beta_4)}_{\text{Slope}} x_1$$

and the television medium line ($x_2 = 0, x_3 = 1$):

$$E(y) = \beta_0 + \beta_1 x_1 + \beta_2(0) + \beta_3(1) + \beta_4 x_1(0) + \beta_5 x_1(1)$$

$$= \underbrace{(\beta_0 + \beta_3)}_{y\text{-intercept}} + \underbrace{(\beta_1 + \beta_5)}_{\text{Slope}} x_1$$

If you were to fit the general equation (model 3), obtain estimates of $\beta_0, \beta_1, \beta_2, \ldots, \beta_5$, and substitute them into the equations for the three media lines derived in Example 12.9, you would obtain exactly the same prediction equations as you would obtain if you were to fit three separate straight lines, one to each of the three sets of media data. You may ask why we would not fit the three lines separately. Why bother fitting a model that combines all three lines (model 3) into the same equation? The answer is that you need to use this procedure if you wish to use statistical tests to compare the three media lines. We need to be able to express a practical question about the lines in terms of a hypothesis that a set of parameters in the model equals 0. (We demonstrate this procedure in the next section.) You could not do this if you were to perform three separate regression analyses and fit a line to each set of media data.

**EXAMPLE 12.10**   An industrial psychologist conducted an experiment to investigate the relationship between worker productivity and a measure of salary incentive for two manufacturing plants; one plant had union representation and the other plant had nonunion

● CASTING
**TABLE 12.5   Productivity Data (Number of Castings) for Example 12.10**

| Type of Plant | Incentive | | | | | | | | |
|---|---|---|---|---|---|---|---|---|---|
| | 20¢/casting | | | 30¢/casting | | | 40¢/casting | | |
| Union | 1,435 | 1,512 | 1,491 | 1,583 | 1,529 | 1,610 | 1,601 | 1,574 | 1,636 |
| Nonunion | 1,575 | 1,512 | 1,488 | 1,635 | 1,589 | 1,661 | 1,645 | 1,616 | 1,689 |

representation. The productivity $y$ per worker was measured by recording the number of machined castings that a worker could produce in a 4-week period of 40 hours per week. The incentive was the amount $x_1$ of bonus (in cents per casting) paid for all castings produced in excess of 1,000 per worker for the 4-week period. Nine workers were selected from each plant, and three from each group of nine were assigned to receive a 20¢ bonus per casting, three a 30¢ bonus, and three a 40¢ bonus. The productivity data for the 18 workers, three for each plant type and incentive combination, are shown in Table 12.5.

a. Write a model for mean productivity, $E(y)$, assuming that the relationship between $E(y)$ and incentive, $x_1$, is first-order.
b. Fit the model and graph the prediction equations for the union and nonunion plants.
c. Do the data provide sufficient evidence to indicate that the rate of increase of worker productivity is different for union and nonunion plants? Test at $\alpha = .10$.

### Solution

a. If we assume that a first-order model* is adequate to detect a change in mean productivity as a function of incentive $x_1$, then the model that produces two straight lines, one for each plant, is

$$E(y) = \beta_0 + \beta_1 x_1 + \beta_2 x_2 + \beta_3 x_1 x_2$$

where

$$x_1 = \text{Incentive} \qquad x_2 = \begin{cases} 1 & \text{if nonunion plant} \\ 0 & \text{if union plant} \end{cases}$$

b. The SPSS printout for the regression analysis is shown in Figure 12.26. Reading the parameter estimates highlighted at the bottom of the printout, you can see that

$$\hat{y} = 1,365.833 + 6.217 x_1 + 47.778 x_2 + .033 x_1 x_2$$

The prediction equation for the union plant can be obtained (see the coding) by substituting $x_2 = 0$ into the general prediction equation. Then

$$\hat{y} = \hat{\beta}_0 + \hat{\beta}_1 x_1 + \hat{\beta}_2(0) + \hat{\beta}_3 x_1(0) = \hat{\beta}_0 + \hat{\beta}_1 x_1$$
$$= 1,365.833 + 6.217 x_1$$

Similarly, the prediction equation for the nonunion plant can be obtained by substituting $x_2 = 1$ into the general prediction equation. Then

$$\hat{y} = \hat{\beta}_0 + \hat{\beta}_1 x_1 + \hat{\beta}_2 x_2 + \hat{\beta}_3 x_1 x_2$$
$$= \hat{\beta}_0 + \hat{\beta}_1 x_1 + \hat{\beta}_2(1) + \hat{\beta}_3 x_1(1)$$

---

*Although the model contains a term involving $x_1 x_2$, it is first-order (graphs as a straight line) in the quantitative variable $x_1$. The variable $x_2$ is a dummy variable that introduces or deletes terms in the model. The order of a model is determined only by the quantitative variables that appear in the model.

**FIGURE 12.26**

*SPSS Regression Printout for Complete Model, Example 12.10*

**Model Summary**

| Model | R | R Square | Adjusted R Square | Std. Error of the Estimate |
|---|---|---|---|---|
| 1 | .843[a] | .711 | .649 | 40.84 |

a. Predictors: (Constant), INC_PDUM, INCENTIV, PDUMMY

**ANOVA[b]**

| Model | | Sum of Squares | df | Mean Square | F | Sig. |
|---|---|---|---|---|---|---|
| 1 | Regression | 57332.389 | 3 | 19110.796 | 11.459 | .000[a] |
| | Residual | 23349.222 | 14 | 1667.802 | | |
| | Total | 80681.611 | 17 | | | |

a. Predictors: (Constant), INC_PDUM, INCENTIV, PDUMMY
b. Dependent Variable: CASTINGS

**Coefficients[a]**

| Model | | Unstandardized Coefficients | | Standardized Coefficients | t | Sig. |
|---|---|---|---|---|---|---|
| | | B | Std. Error | Beta | | |
| 1 | (Constant) | 1365.833 | 51.836 | | 26.349 | .000 |
| | INCENTIV | 6.217 | 1.667 | .758 | 3.729 | .002 |
| | PDUMMY | 47.778 | 73.308 | .357 | .652 | .525 |
| | INC_PDUM | 3.333E-02 | 2.358 | .008 | .014 | .989 |

a. Dependent Variable: CASTINGS

$$\overbrace{= (\hat{\beta}_0 + \hat{\beta}_2)}^{y\text{-intercept}} + \overbrace{(\hat{\beta}_1 + \hat{\beta}_3)x_1}^{\text{Slope}}$$
$$= (1,365.833 + 47.778) + (6.217 + .033)x_1$$
$$= 1,413.611 + 6.250x_1$$

A graph of these prediction equations is shown in Figure 12.27. Note that the slopes of the two lines are nearly identical (6.217 for union and 6.250 for nonunion).

**FIGURE 12.27**

*Graphs of the Prediction Equations for the Two Productivity Lines, Example 12.10*

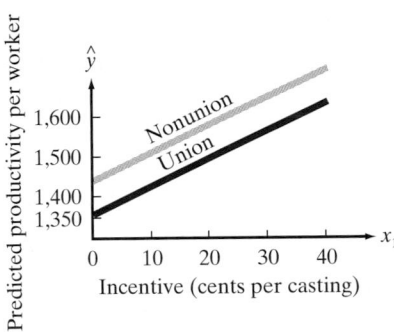

c. If the rate of increase of productivity with incentive (i.e, the slope) for nonunion plants is different than the corresponding slope for union plants, then the interaction $\beta$ (i.e., $\beta_3$) will differ from 0. Consequently, we want to test

$$H_0: \beta_3 = 0$$
$$H_a: \beta_3 \neq 0$$

This test is conducted using the *t*-test of Section 12.4. From the SPSS printout, the test statistic and corresponding *p*-value (highlighted) are:

$$t = .014 \qquad p\text{-value} = .989$$

Since $\alpha = .10$ is less than the $p$-value, we fail to reject $H_0$. There is insufficient evidence to conclude that the union and nonunion shapes differ. Thus, the test supports our observation of two nearly identical slopes in part b.   ∎

Models with both quantitative and qualitative $x$'s may also include higher-order (e.g., second-order) terms. In the problem of relating mean monthly sales $E(y)$ of a company to monthly advertising expenditure $x_1$ and type of medium, suppose we think that the relationship between $E(y)$ and $x_1$ is curvilinear. We will construct the model, stage by stage, to enable you to compare the procedure with the stage-by-stage construction of the first-order model in the beginning of this section. The graphical interpretations will help you understand the contributions of the model terms.

1.  The mean sales curves are identical for all three advertising media, that is, a single second-order curve will suffice to describe the relationship between $E(y)$ and $x_1$ for all the media (see Figure 12.28):

$$E(y) = \beta_0 + \beta_1 x_1 + \beta_2 x_1^2$$

where $x_1$ = Advertising expenditure

**FIGURE 12.28**

*The Relationship Between $E(y)$ and $x_1$ is the Same for All Media*

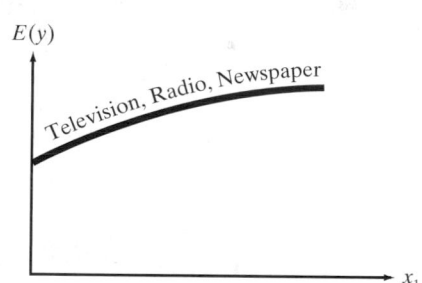

2.  The response curves possess the same shapes but different $y$-intercepts (see Figure 12.29):

$$E(y) = \beta_0 + \beta_1 x_1 + \beta_2 x_1^2 + \beta_3 x_2 + \beta_4 x_3$$

where

$$x_1 = \text{Advertising expenditure}$$

$$x_2 = \begin{cases} 1 & \text{if radio medium} \\ 0 & \text{if not} \end{cases}$$

$$x_3 = \begin{cases} 1 & \text{if television medium} \\ 0 & \text{if not} \end{cases}$$

**FIGURE 12.29**

*The Response Curves Have the Same Shapes But Different y-Intercepts*

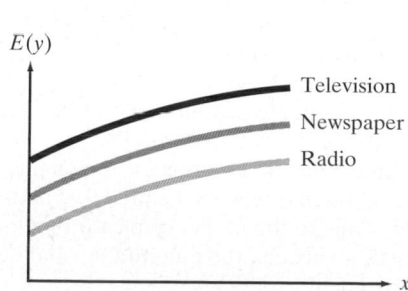

**FIGURE 12.30**

*The Response Curves for
the Three Media Differ*

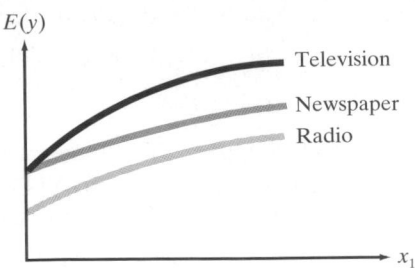

3. The response curves for the three advertising media are different (i.e., Advertising expenditure and Type of medium interact), as shown in Figure 12.30:

$$E(y) = \beta_0 + \beta_1 x_1 + \beta_2 x_1^2 + \beta_3 x_2 + \beta_4 x_3 + \beta_5 x_1 x_2 + \beta_6 x_1 x_3 + \beta_7 x_1^2 x_2 + \beta_8 x_1^2 x_3$$

Now that you know how to write a model with two independent variables—one qualitative and one quantitative—we ask a question. Why do it? Why not write a separate second-order model for each type of medium where $E(y)$ is a function of only advertising expenditure? As stated earlier, one reason we wrote the single model representing all three response curves is so that we can test to determine whether the curves are different. We illustrate this procedure in Section 12.11. A second reason for writing a single model is that we obtain a pooled estimate of $\sigma^2$, the variance of the random error component $\varepsilon$. If the variance of $\varepsilon$ is truly the same for each type of medium, the pooled estimate is superior to three separate estimates calculated by fitting a separate model for each type of medium.

 EXERCISES 12.75–12.86

### Learning the Mechanics

**12.75** Consider a multiple regression model for a response $y$, with one quantitative independent variable $x_1$, and one qualitative variable at three levels.

**a.** Write a first-order model that relates the mean response $E(y)$ to the quantitative independent variable.

**b.** Add the main effect terms for the qualitative independent variable to the model of part **a**. Specify the coding scheme you use.

**c.** Add terms to the model of part **b** to allow for interaction between the quantitative and qualitative independent variables.

**d.** Under what circumstances will the response lines of the model in part **c** be parallel?

**e.** Under what circumstances will the model in part **c** have only one response line?

**12.76** Refer to Exercise 12.75.

**a.** Write a complete second-order model that relates $E(y)$ to the quantitative variable.

**b.** Add the main effect terms for the qualitative variable (at three levels) to the model of part **a**.

**c.** Add terms to the model of part **b** to allow for interaction between the quantitative and qualitative independent variables.

**d.** Under what circumstances will the response curves of the model have the same shape but different $y$-intercepts?

**e.** Under what circumstances will the response curves of the model be parallel lines?

**f.** Under what circumstances will the response curves of the model be identical?

**12.77** Consider the model

$$y = \beta_0 + \beta_1 x_1 + \beta_2 x_2 + \beta_3 x_3 + \varepsilon$$

where $x_1$ is a quantitative variable and $x_2$ and $x_3$ are dummy variables describing a qualitative variable at three levels using the coding scheme

$$x_2 = \begin{cases} 1 & \text{if level 2} \\ 0 & \text{otherwise} \end{cases} \qquad x_3 = \begin{cases} 1 & \text{if level 3} \\ 0 & \text{otherwise} \end{cases}$$

The resulting least squares prediction equation is

$$\hat{y} = 44.8 + 2.2 x_1 + 9.4 x_2 + 15.6 x_3$$

**a.** What is the response line (equation) for $E(y)$ when $x_2 = x_3 = 0$? When $x_2 = 1$ and $x_3 = 0$? When $x_2 = 0$ and $x_3 = 1$?

**b.** What is the least squares prediction equation associated with level 1? Level 2? Level 3? Plot these on the same graph.

**12.78** Consider the model

$$y = \beta_0 + \beta_1 x_1 + \beta_2 x_1^2 + \beta_3 x_2 + \beta_4 x_3 + \beta_5 x_1 x_2 +$$
$$\beta_6 x_1 x_3 + \beta_7 x_1^2 x_2 + \beta_8 x_1^2 x_3 + \varepsilon$$

where $x_1$ is a quantitative variable and

$$x_2 = \begin{cases} 1 & \text{if level 2} \\ 0 & \text{otherwise} \end{cases} \qquad x_3 = \begin{cases} 1 & \text{if level 3} \\ 0 & \text{otherwise} \end{cases}$$

The resulting least squares prediction equation is

$$\hat{y} = 48.8 - 3.4x_1 + .07x_1^2$$
$$- 2.4x_2 - 7.5x_3 + 3.7x_1 x_2$$
$$+ 2.7x_1 x_3 - .02x_1^2 x_2$$
$$- .04x_1^2 x_3$$

**a.** What is the equation of the response curve for $E(y)$ when $x_2 = 0$ and $x_3 = 0$? When $x_2 = 1$ and $x_3 = 0$? When $x_2 = 0$ and $x_3 = 1$?

**b.** On the same graph, plot the least squares prediction equation associated with level 1, with level 2, and with level 3.

**12.79** Write a model that relates $E(y)$ to two independent variables—one quantitative and one qualitative at four levels. Construct a model that allows the associated response curves to be second-order but does not allow for interaction between the two independent variables.

## Applying the Concepts—Basic

**12.80** Refer to the *Chance* (Winter 2000) study of men's and women's winning times in the Boston Marathon, Exercise 11.16 (p. 524). Suppose the researchers want to build a model for predicting winning time ($y$) of the marathon as a function of year ($x_1$) in which race is run and gender of winning runner ($x_2$).

**a.** Set up the appropriate dummy variables (if necessary) for $x_1$ and $x_2$.

**b.** Write the equation of a model that proposes parallel straight-line relationships between winning time ($y$) and year ($x_1$), one line for each gender.

**c.** Write the equation of a model that proposes non-parallel straight-line relationships between winning time ($y$) and year ($x_1$), one line for each gender.

**d.** Which of the models do you think will provide the best predictions of winning time ($y$)? Base your answer on the graph displayed in Exercise 11.16.

**12.81** The influence of cigarette smoking on resting energy expenditure (REE) in normal-weight and obese smokers was recently investigated (*Health Psychology*, Mar. 1995). The researchers hypothesized that the relationship between a smoker's REE and length of time since smoking differs for normal-weight and obese smokers. Consequently, the interaction model was examined:

$$E(y) = \beta_0 + \beta_1 x_1 + \beta_2 x_2 + \beta_3 x_1 x_2$$

where

$y$ = REE, measured in kilocalories per day
$x_1$ = Time, in minutes after smoking, of metabolic energy reading (levels = 10, 20, and 30 minutes)

$$x_2 = \begin{cases} 1 & \text{if normal weight} \\ 0 & \text{if obese} \end{cases}$$

**a.** Give the equation of the hypothesized line relating mean REE to time after smoking for obese smokers. What is the slope of the line?

**b.** Repeat part **a** for normal-weight smokers.

**c.** A test for interaction resulted in an observed significance level of .044. Interpret this value.

## SNOWGEESE

**12.82** Refer to the *Journal of Applied Ecology* study of feeding habits of baby snow geese, Exercise 12.11 (p. 592).

**a.** Write a first-order model relating gosling weight change ($y$) to digestion efficiency ($x_1$) and diet (plants or duck chow) that allows for different slopes for each diet.

**b.** A MINITAB printout for the model, part **a**, is shown here. Find the least squares prediction equation.

MINITAB Output for Exercise 12.82

```
The regression equation is
WTCHANGE = 8.14 - 0.016 DIGEST - 10.4 PLANTS + 0.095 DIGEST*PLANTS

Predictor            Coef       SE Coef           T         P
Constant            8.144         8.446        0.96     0.341
DIGEST            -0.0165        0.1329       -0.12     0.902
PLANTS           -10.354         8.538       -1.21     0.233
DIGEST*PLANTS      0.0948        0.1418        0.67     0.508

S = 3.882        R-Sq = 44.1%        R-Sq(adj) = 39.7%

Analysis of Variance

Source           DF          SS          MS          F         P
Regression        3      452.54      150.85      10.01     0.000
Residual Error   38      572.58       15.07
Total            41     1025.12
```

**c.** Find the estimated slope of the line for goslings fed a diet of plants. Interpret its value.

**d.** Find the estimated slope of the line for goslings fed a diet of duck chow. Interpret its value.

**e.** Conduct a test to determine whether the slopes associated with the two diets are significantly different. Use $\alpha = .05$.

## Applying the Concepts—Intermediate

**12.83** A study of the atmospheric pollution on the slopes of the Blue Ridge Mountains (Tennessee) was conducted. The file LEADMOSS contains the levels of lead found in 70 fern moss specimens (in micrograms of lead per gram of moss tissue) collected from the mountain slopes, as well as the elevation of the moss specimen (in feet) and the direction (1 if east, 0 if west) of the slope face. The first five and last five observations of the data set are listed in the table.

**LEADMOSS**

| Specimen | Lead Level | Elevation | Slope Face |
|---|---|---|---|
| 1 | 3.475 | 2000 | 0 |
| 2 | 3.359 | 2000 | 0 |
| 3 | 3.877 | 2000 | 0 |
| 4 | 4.000 | 2500 | 0 |
| 5 | 3.618 | 2500 | 0 |
| ⋮ | ⋮ | ⋮ | ⋮ |
| 66 | 5.413 | 2500 | 1 |
| 67 | 7.181 | 2500 | 1 |
| 68 | 6.589 | 2500 | 1 |
| 69 | 6.182 | 2000 | 1 |
| 70 | 3.706 | 2000 | 1 |

*Source:* Schilling, J. "Bioindication of atmospheric heavy metal deposition in the Blue Ridge using the moss, *Thuidium delicatulum*." Master of Science Thesis, Spring 2000.

**a.** Write the equation of a first-order model relating mean lead level, $E(y)$, to elevation $(x_1)$ and slope face $(x_2)$. Include interaction between elevation and slope face in the model.

**b.** Graph the relationship between mean lead level and elevation for the different slope faces that is hypothesized by the model, part **a**.

**c.** In terms of the $\beta$'s of the model, part **a**, give the change in lead level for every one foot increase in elevation for moss specimens on the east slope.

**d.** Fit the model, part **a**, to the data using an available statistical software package. Is the overall model statistically useful for predicting lead level? Test using $\alpha = .10$.

**12.84** Excessive exposure to solar radiation is known to increase the risk of developing skin cancer, yet many people do not practice "sun safety." A group of University of Arizona researchers examined the feasibility of educating preschool (4- to 5-year-old) children about sun safety (*American Journal of Public Health*, July 1995). A sample of 122 preschool children was di-

vided into two groups, the control group and the intervention group. Children in the intervention group received a *Be Sun Safe* curriculum in preschool, while the control group did not. All children were tested for their knowledge, comprehension, and application of sun safety at two points in time: prior to the sun safety curriculum (pretest, $x_1$) and seven weeks following the curriculum (posttest, $y$).

**a.** Write a first-order model for mean posttest score, $E(y)$, as a function of pretest score, $x_1$ and group. Assume that no interaction exists between pretest score and group.

**b.** For the model, part **a**, show that the slope of the line relating posttest score to pretest score is the same for both groups of children.

**c.** Repeat part **a**, but assume that pretest score and group interact.

**d.** For the model, part **c**, show that the slope of the line relating posttest score to pretest score differs for the two groups of children.

**12.85** Many women suffer from anemia. A female physician, who is also an avid jogger, wanted to know if women who exercise regularly have a different mean red blood cell count than women who do not. She also wanted to know if the amount of a particular iron supplement a woman takes has any effect and whether the effect is the same for both groups. Write a model that will reflect the relationship between red blood cell count and the two independent variables described previously, assuming that

**a.** the effect of the iron supplement on mean blood cell count is the same regardless of whether a woman exercises regularly.

**b.** the effect of the iron supplement on mean blood cell count depends on whether a woman exercises regularly.

## Applying the Concepts—Advanced

**12.86** Multiple regression analysis was used to model the abundance $y$ of an individual bird species in transects in the United Kingdom (*Journal of Applied Ecology*, Vol. 32, 1995). Three of the independent variables used in the model, all field boundary attributes, are

(1) Transect location (small pasture field, small arable field, or large arable field)

(2) Land use (pasture or arable) adjacent to the transect

(3) Total number of trees in transect

**a.** Identify each of the independent variables as quantitative or qualitative variables.

**b.** Write a first-order model for $E(y)$ as a function of total number of trees.

**c.** Add main effect terms for transect location to the model, part **b**. Graph the hypothesized relationships of the new model.

**d.** Add main effect terms for land use to the model, part **c**. In terms of the $\beta$'s of the new model, what is the slope of the relationship between $E(y)$ and number of trees for any combination of transect location and land use?

e. Add terms for interaction between transect location and land use to the model, part **d**. Do these interaction terms affect the slope of the relationship between $E(y)$ and number of trees? Explain.

f. Add terms for interaction between number of trees and all coded dummy variables to the model, part **e**. In terms of the $\beta$'s of the new model, give the slope of the relationship between $E(y)$ and number of trees for each combination of transect location and land use.

## 12.11 MODEL BUILDING: COMPARING NESTED MODELS

To be successful model builders, we require a statistical method that will allow us to determine (with a high degree of confidence) which one among a set of candidate models best fits the data. In this section, we present such a technique for *nested models*.

**DEFINITION 12.3**

Two models are **nested** if one model contains all the terms of the second model and at least one additional term. The more complex of the two models is called the **complete** model, and the simpler of the two is called the **reduced** model.

To illustrate the concept of nested models, consider the straight-line interaction model for the mean auction price $E(y)$ of a grandfather clock as a function of two quantitative variables: age of the clock $(x_1)$ and number of bidders $(x_2)$. The interaction model, fit in Example 12.5, is

$$E(y) = \beta_0 + \beta_1 x_1 + \beta_2 x_2 + \beta_3 x_1 x_2$$

If we assume that the relationship between auction price $(y)$, age $(x_1)$, and bidders $(x_2)$ is curvilinear, then the complete second-order model is more appropriate:

$$E(y) = \overbrace{\beta_0 + \beta_1 x_1 + \beta_2 x_2 + \beta_3 x_1 x_2}^{\text{Terms in interaction model}} + \overbrace{\beta_4 x_1^2 + \beta_5 x_2^2}^{\text{Quadratic terms}}$$

Note that the curvilinear model contains quadratic terms for $x_1$ and $x_2$, as well as the terms in the interaction model. Therefore, the models are nested models. In this case, the interaction model is nested within the more complex curvilinear model. Thus, the curvilinear model is the *complete* model and the interaction model is the *reduced* model.

Suppose we want to know whether the more complex curvilinear model contributes more information for the prediction of $y$ than the straight-line interaction model. This is equivalent to determining whether the quadratic terms $\beta_4$ and $\beta_5$ should be retained in the model. To test whether these terms should be retained, we set up the null and alternative hypotheses as follows:

$H_0$: $\beta_4 = \beta_5 = 0$ (i.e., quadratic terms are not important for predicting $y$)

$H_a$: At least one of the parameters $\beta_4$ and $\beta_5$ is nonzero (i.e., at least one of the quadratic terms is useful for predicting $y$)

Note that the terms being tested are those additional terms in the complete (curvilinear) model that are not in the reduced (straight-line interaction) model.

In Section 12.4, we presented the *t*-test for a single $\beta$ coefficient and in Section 12.5 we gave the global *F*-test for *all* the $\beta$ parameters (except $\beta_0$) in the model. We now need a test for a *subset* of the $\beta$ parameters in the complete model. The test procedure is intuitive. First, we use the method of least squares to fit the reduced model and calculate the corresponding sum of squares for error, $SSE_R$ (the sum of squares of the deviations between observed and predicted *y* values). Next, we fit the complete model and calculate its sum of squares for error, $SSE_C$. Then, we compare $SSE_R$ to $SSE_C$ by calculating the difference, $SSE_R - SSE_C$. If the additional terms in the complete model are significant, then $SSE_C$ should be much smaller than $SSE_R$, and the difference $SSE_R - SSE_C$ will be large.

Since SSE will always decrease when new terms are added to the model, the question is whether the difference $SSE_R - SSE_C$ is large enough to conclude that it is due to more than just an increase in the number of model terms and to chance. The formal statistical test utilizes an *F* statistic, as shown in the box.

---

### *F*-Test for Comparing Nested Models

*Reduced model:*  $E(y) = \beta_0 + \beta_1 x_1 + \cdots + \beta_g x_g$

*Complete model:*  $E(y) = \beta_0 + \beta_1 x_1 + \cdots + \beta_g x_g + \beta_{g+1} x_{g+1} + \cdots + \beta_k x_k$

$H_0: \beta_{g+1} = \beta_{g+2} = \cdots = \beta_k = 0$

$H_a:$ At least one of the $\beta$ parameters specified in $H_0$ is nonzero.

*Test statistic:* 
$$F = \frac{(SSE_R - SSE_C)/(k - g)}{SSE_C/[n - (k + 1)]}$$
$$= \frac{(SSE_R - SSE_C)/\#\beta\text{'s tested in } H_0}{MSE_C}$$

where

$SSE_R$ = Sum of squared errors for the reduced model
$SSE_C$ = Sum of squared errors for the complete model
$MSE_C$ = Mean square error ($s^2$) for the complete model
$k - g$ = Number of $\beta$ parameters specified in $H_0$ (i.e., number of $\beta$'s tested)
$k + 1$ = Number of $\beta$ parameters in the complete model (including $\beta_0$)
$\quad n$ = Total sample size

*Rejection region:* $F > F_\alpha$

where *F* is based on $\nu_1 = k - g$ numerator degrees of freedom and $\nu_2 = n - (k + 1)$ denominator degrees of freedom.

---

When the assumptions listed in Section 12.3 about the random error term are satisfied, this *F* statistic has an *F*-distribution with $\nu_1$ and $\nu_2$ df. Note that $\nu_1$ is the number of $\beta$ parameters being tested and $\nu_2$ is the number of degrees of freedom associated with $s^2$ in the complete model.

■ **EXAMPLE   12.11**    A botanist conducted an experiment to study the growth of carnations as a function of the temperature $x_1$ (°F) in a greenhouse and the amount of fertilizer $x_2$ [kilograms (kg) per plot] applied to the soil. A total of 27 plots of equal size are treated with fertilizer in amounts varying between 50 and 60 kg per plot, and these plots are mechanically kept at constant temperatures between 80 and 100°F. Small carnation plants [approximately 15 centimeters (cm) in height] are planted in each

● CARNATIONS

**TABLE 12.6**   **Temperature ($x_1$), Amount of Fertilizer ($x_2$), and Height of Carnations (y)**

| $x_1$ | $x_2$ | y | $x_1$ | $x_2$ | y | $x_1$ | $x_2$ | y |
|---|---|---|---|---|---|---|---|---|
| 80 | 50 | 50.8 | 90 | 50 | 63.4 | 100 | 50 | 46.6 |
| 80 | 50 | 50.7 | 90 | 50 | 61.6 | 100 | 50 | 49.1 |
| 80 | 50 | 49.4 | 90 | 50 | 63.4 | 100 | 50 | 46.4 |
| 80 | 55 | 93.7 | 90 | 55 | 93.8 | 100 | 55 | 69.8 |
| 80 | 55 | 90.9 | 90 | 55 | 92.1 | 100 | 55 | 72.5 |
| 80 | 55 | 90.9 | 90 | 55 | 97.4 | 100 | 55 | 73.2 |
| 80 | 60 | 74.5 | 90 | 60 | 70.9 | 100 | 60 | 38.7 |
| 80 | 60 | 73.0 | 90 | 60 | 68.8 | 100 | 60 | 42.5 |
| 80 | 60 | 71.2 | 90 | 60 | 71.3 | 100 | 60 | 41.4 |

plot, and their height y (cm) is measured after a 6-week growing period. The resulting data are shown in Table 12.6.

a. Fit a complete second-order model to the data.
b. Sketch the fitted model in three dimensions.
c. Do the data provide sufficient evidence to indicate that the second-order terms, $\beta_3$, $\beta_4$, and $\beta_5$, contribute information for the prediction of y?

**Solution**

a. The complete second-order model is

$$E(y) = \beta_0 + \beta_1 x_1 + \beta_2 x_2 + \beta_3 x_1 x_2 + \beta_4 x_1^2 + \beta_5 x_2^2$$

The data in Table 12.6 were used to fit this model, and a portion of the SAS output is shown in Figure 12.31.

The least squares prediction equation (rounded) is

$$\hat{y} = -5{,}127.90 + 31.10 x_1 + 139.75 x_2 - .146 x_1 x_2 - .133 x_1^2 - 1.14 x_2^2$$

b. A three-dimensional graph of this prediction model, called a **response surface**, is shown in Figure 12.32. Note that the height seems to be greatest for temperatures of about 85–90°F and for applications of about 55–57 kg of fertilizer per plot.* Further experimentation in these ranges might lead to a more precise determination of the optimal temperature–fertilizer combination.

c. To determine whether the data provide sufficient information to indicate that the second-order terms contribute information for the prediction of y, we wish to test

$$H_0: \beta_3 = \beta_4 = \beta_5 = 0$$

against the alternative hypothesis

$$H_a: \text{At least one of the parameters, } \beta_3, \beta_4, \beta_5, \text{ differs from } 0$$

The first step in conducting the test is to drop the second-order terms out of the complete (second-order) model and fit the reduced model

$$E(y) = \beta_0 + \beta_1 x_1 + \beta_2 x_2$$

---

*Students with knowledge of calculus should note that we can solve for the exact temperature and amount of fertilizer that maximize height in the least squares model by solving $\partial \hat{y}/\partial x_1 = 0$ and $\partial \hat{y}/\partial x_2 = 0$ for $x_1$ and $x_2$. Sample estimates of these estimated optimal values are $x_1 = 86.25°F$ and $x_2 = 55.58$ kg per plot.

**FIGURE 12.31**

*SAS Regression Printout for Complete Second-Order Model, Example 12.11*

The REG Procedure
Dependent Variable: HEIGHT

Analysis of Variance

| Source | DF | Sum of Squares | Mean Square | F Value | Pr > F |
|---|---|---|---|---|---|
| Model | 5 | 8402.26454 | 1680.45291 | 596.32 | <.0001 |
| Error | 21 | 59.17843 | 2.81802 | | |
| Corrected Total | 26 | 8461.44296 | | | |

| | | | | |
|---|---|---|---|---|
| Root MSE | 1.67870 | R-Square | 0.9930 | |
| Dependent Mean | 66.96296 | Adj R-Sq | 0.9913 | |
| Coeff Var | 2.50690 | | | |

Parameter Estimates

| Variable | DF | Parameter Estimate | Standard Error | t Value | Pr > |t| |
|---|---|---|---|---|---|
| Intercept | 1 | -5127.89907 | 110.29601 | -46.49 | <.0001 |
| TEMP | 1 | 31.09639 | 1.34441 | 23.13 | <.0001 |
| FERT | 1 | 139.74722 | 3.14005 | 44.50 | <.0001 |
| TEMPFERT | 1 | -0.14550 | 0.00969 | -15.01 | <.0001 |
| TEMPSQ | 1 | -0.13339 | 0.00685 | -19.46 | <.0001 |
| FERTSQ | 1 | -1.14422 | 0.02741 | -41.74 | <.0001 |

Test HIGHORD Results for Dependent Variable HEIGHT

| Source | DF | Mean Square | F Value | Pr > F |
|---|---|---|---|---|
| Numerator | 3 | 2204.11003 | 782.15 | <.0001 |
| Denominator | 21 | 2.81802 | | |

**FIGURE 12.32**

*Plot of Second-Order Least Squares Model for Example 12.11*

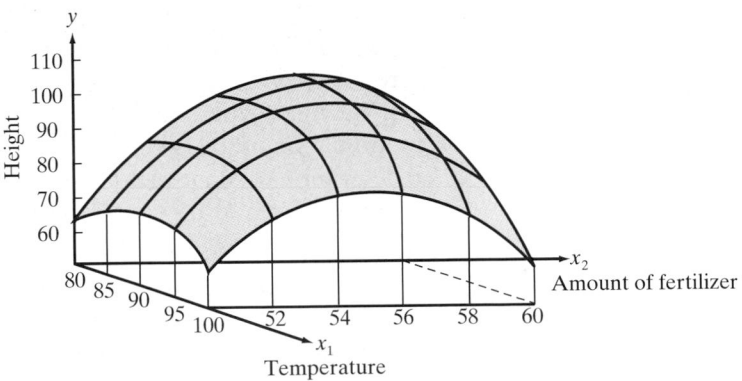

to the data. The SAS printout for this model is shown in Figure 12.33.

The sums of squares for error, highlighted in Figures 12.31 and 12.33 for the complete and reduced models, respectively, are

$$SSE_C = 59.17843$$
$$SSE_R = 6,671.50852$$

and that $s^2$ for the complete model (highlighted on Figure 12.31) is

$$s^2 = MSE_C = 2.81802$$

**FIGURE 12.33**

*SAS Regression Printout for Reduced (First-Order) Model, Example 12.11*

The REG Procedure
Dependent Variable: HEIGHT

Analysis of Variance

| Source | DF | Sum of Squares | Mean Square | F Value | Pr > F |
|--------|----|----------------|-------------|---------|--------|
| Model | 2 | 1789.93444 | 894.96722 | 3.22 | 0.0577 |
| Error | 24 | 6671.50852 | 277.97952 | | |
| Corrected Total | 26 | 8461.44296 | | | |

| | | | | |
|---|---|---|---|---|
| Root MSE | 16.67272 | R-Square | 0.2115 |
| Dependent Mean | 66.96296 | Adj R-Sq | 0.1458 |
| Coeff Var | 24.89842 | | |

Parameter Estimates

| Variable | DF | Parameter Estimate | Standard Error | t Value | Pr > |t| |
|----------|----|--------------------|-----------------|---------|---------|
| Intercept | 1 | 106.08519 | 55.94500 | 1.90 | 0.0700 |
| TEMP | 1 | -0.91011 | 0.09290 | -2.33 | 0.0285 |
| FERT | 1 | 0.78778 | 0.78596 | 1.00 | 0.3262 |

Recall that $n = 27$, $k = 5$, and $g = 2$. Therefore, the calculated value of the $F$ statistic, based on $\nu_1 = (k - g) = 3$ numerator df and $\nu_2 = [n - (k + 1)] = 21$ denominator df, is

$$F = \frac{(\text{SSE}_R - \text{SSE}_C)/(k - g)}{\text{SSE}_C/[n - (k + 1)]} = \frac{(\text{SSE}_R - \text{SSE}_C)/(k - g)}{\text{MSE}_C}$$

where $\nu_1 = (k - g)$ is equal to the number of parameters involved in $H_0$. Therefore,

*Test statistic:*   $$F = \frac{(6{,}671.50852 - 59.17843)/3}{2.81802} = 782.15$$

The final step in the test is to compare this computed value of $F$ with the tabulated value based on $\nu_1 = 3$ and $\nu_2 = 21$ df. If we choose $\alpha = .05$, then $F_{.05} = 3.07$ and the rejection region is

*Rejection region:*   $F > 3.07$.

Since the computed value of $F$ falls in the rejection region (i.e., it exceeds $F_{.05} = 3.07$), we reject $H_0$ and conclude that at least one of the second-order terms contributes information for the prediction of $y$. The second-order model appears to provide better predictions of $y$ than a first-order model.

*Note:* Using special commands, you can get SAS to perform the desired nested model $F$-test. The test statistic and $p$-value for the test above are highlighted at the bottom of the SAS printout, Figure 12.31. ■

The nested model $F$-test can be used to determine whether *any* subset of terms should be included in a complete model by testing the null hypothesis that a particular set of $\beta$ parameters simultaneously equals 0. For example, we may want to test to determine whether a set of interaction terms for quantitative variables or a set of main effect terms for a qualitative variable should be included in a model. If we reject $H_0$, the complete model is the better of the two nested models.

Suppose the *F*-test in Example 12.11 yielded a test statistic that did not fall in the rejection region. Although we must be cautious about accepting $H_0$, most practitioners of regression analysis adopt the principle of *parsimony*. That is, in situations where two competing models are found to have essentially the same predictive power (as in this case), the model with the fewer number of $\beta$'s (i.e., the more parsimonious model) is selected. Based on this principle, we would drop the three second-order terms and select the straight-line (reduced) model over the second-order (complete) model.

> ### DEFINITION 12.4
> A **parsimonious model** is a general linear model with a small number of $\beta$ parameters. In situations where two competing models have essentially the same predictive power (as determined by an *F*-test), choose the more parsimonious of the two.

When the candidate models in model building are nested models, the *F*-test developed in this section is the appropriate procedure to apply to compare the models. However, if the models are not nested, this *F*-test is not applicable. In this situation, the analyst must base the choice of the best model on statistics such as $R_a^2$ and $s$. It is important to remember that decisions based on these and other numerical descriptive measures of model adequacy cannot be supported with a measure of reliability and are often very subjective in nature.

 EXERCISES 12.87–12.97

### Learning the Mechanics

**12.87** Determine which pairs of the following models are "nested" models. For each pair of nested models, identify the complete and reduced model.
  **a.** $E(y) = \beta_0 + \beta_1 x_1 + \beta_2 x_2$
  **b.** $E(y) = \beta_0 + \beta_1 x_1$
  **c.** $E(y) = \beta_0 + \beta_1 x_1 + \beta_2 x_1^2$
  **d.** $E(y) = \beta_0 + \beta_1 x_1 + \beta_2 x_2 + \beta_3 x_1 x_2$
  **e.** $E(y) = \beta_0 + \beta_1 x_1 + \beta_2 x_2 + \beta_3 x_1 x_2 + \beta_4 x_1^2 + \beta_5 x_2^2$

**12.88** Suppose you fit the regression model

$$y = \beta_0 + \beta_1 x_1 + \beta_2 x_2 + \beta_3 x_1 x_2 + \beta_4 x_1^2 + \beta_5 x_2^2 + \varepsilon$$

to $n = 30$ data points and you wish to test

$$H_0: \beta_3 = \beta_4 = \beta_5 = 0$$

  **a.** State the alternative hypothesis $H_a$.
  **b.** Give the reduced model appropriate for conducting the test.
  **c.** What are the numerator and denominator degrees of freedom associated with the *F* statistic?
  **d.** Suppose the SSE's for the complete and reduced models are $SSE_R = 1{,}250.2$ and $SSE_C = 1{,}125.2$. Conduct the hypothesis test and interpret the results of your test. Test using $\alpha = .05$.

**12.89** MINITAB was used to fit the complete model

$$y = \beta_0 + \beta_1 x_1 + \beta_2 x_2 + \beta_3 x_3 + \beta_4 x_4 + \varepsilon$$

to $n = 20$ data points. The independent variables $x_3$ and $x_4$ were dropped from the preceding model, and MINITAB was also used to fit the resulting reduced model. (See the printouts on p. 653.)
  **a.** Report the least squares prediction equations for the complete and reduced models.
  **b.** Find $SSE_R$ and $SSE_C$. Interpret each of the quantities.
  **c.** How many $\beta$ parameters are in the complete model? The reduced model?
  **d.** Specify the null and alternative hypotheses you would use to investigate whether the complete model contributes more information for the prediction of *y* than the reduced model.
  **e.** Conduct the hypothesis test of part **d**. Use $\alpha = .05$.
  **f.** What is the approximate *p*-value of the test of part **e**?

**12.90** Explain why the *F*-test used to compare complete and reduced models is a one-tailed, upper-tailed test.

### Applying the Concepts—Basic

**12.91** Refer to the *American Educational Research Journal* (Fall 1998) study of students' perceptions of their

**MINITAB Output for Complete Model, Exercise 12.89**

```
The regression equation is
Y = 14.6 - 0.61 X1 + 0.44 X2 - 0.08 X3 - 0.06 X4

Predictor        Coef      SE Coef           T          P
Constant       14.575        4.887        2.98      0.010
X1             -0.611        0.177       -3.44      0.005
X2              0.439        0.220        2.00      0.062
X3             -0.080        0.108       -0.74      0.886
X4             -0.064        0.125       -0.51      0.921


S = 3.190      R-Sq = 84.5%      R-Sq(adj) = 80.3%

Analysis of Variance

Source          DF          SS          MS         F          P
Regression       4      821.09      207.77     20.41      0.002
Residual Error  15      152.66       10.18
Total           19      983.75
```

**MINITAB Output for Reduced Model, Exercise 12.89**

```
The regression equation is
Y = 14.0 - 0.64 X1 + 0.40 X2

Predictor        Coef      SE Coef           T          P
Constant       13.968        4.626        3.02      0.006
X1             -0.642        0.168       -3.84      0.001
X2              0.396        0.206        1.92      0.072


S = 3.072      R-Sq = 83.7%      R-Sq(adj) = 81.8%

Analysis of Variance

Source          DF          SS          MS         F          P
Regression       2      823.31      411.66     43.61      0.000
Residual Error  17      160.44        9.44
Total           19      983.75
```

science ability in hands-on classrooms, Exercise 12.24 (p. 602). Recall that the first-order, main effects model that was used to predict ability perception ($y$) included the following independent variables:

*Control Variables*

$x_1$ = Prior science attitude score
$x_2$ = Science ability test score
$x_3$ = 1 if boy, 0 if girl
$x_4$ = 1 if classroom 1 student, 0 if not
$x_5$ = 1 if classroom 3 student, 0 if not
$x_6$ = 1 if classroom 4 student, 0 if not
$x_7$ = 1 if classroom 5 student, 0 if not
$x_8$ = 1 if classroom 6 student, 0 if not

*Performance Behaviors*

$x_9$ = Active-leading behavior score
$x_{10}$ = Passive-assisting behavior score
$x_{11}$ = Active-manipulating behavior score

**a.** Hypothesize the equation of the first-order, main effects model for $E(y)$.

**b.** The researchers also considered a model that included all possible interactions between the control variables and the performance behavior variables. Write the equation for this model for $E(y)$.

**c.** The researchers determined that the interaction terms in the model, part **b**, were not significant, and therefore used the model, part **a**, to make inferences. Explain the best way to conduct this test for interaction. Give the null hypothesis of the test.

**12.92** An article in the *Community Mental Health Journal* (Aug. 2000) used multiple regression analysis to model the level of community adjustment of clients of the Department of Mental Health and Addiction Services in Connecticut. The dependent variable, community adjustment ($y$), was measured quantitatively based on staff ratings of the clients. (Lower scores indicate better adjustment.) The complete model was a first-order model with 21 independent variables. The independent variables were categorized as Demographic (four variables), Diagnostic (seven variables), Treatment (four variables), and Community (six variables).

a. Write the equation of $E(y)$ for the complete model.

b. Give the null hypothesis for testing whether the seven Diagnostic variables contribute information for the prediction of $y$.

c. Give the equation of the reduced model appropriate for the test, part **b**.

d. The test, part **b**, resulted in a test statistic of $F = 59.3$ and $p$-value $< .0001$. Interpret this result in the words of the problem.

### ⊘ BEANIE

**12.93** Refer to Exercise 11.81 (p. 571) and the data on the values of 50 beanie babies collector's items, published in *Beanie World Magazine*. Suppose we want to predict the market value of a beanie baby using age (in months since Sept. 1998) and whether the beanie baby has been retired or is current (i.e., still in production).

a. Write a complete second-order model for market value as a function of age and current/retired status.

b. Specify the null hypothesis for testing whether the quadratic terms in the model, part **a**, are important for predicting market value.

c. Specify the null hypothesis for testing whether the interaction terms in the model, part **a**, are important for predicting market value.

d. Three models were fit to the data using SPSS. The relevant SPSS printouts are shown below. Use this information to conduct the tests specified in parts **b** and **c**. Interpret the results.

### Applying the Concepts—Intermediate

**12.94** Refer to the *American Journal of Public Health* study of preschool children's awareness of sun safety, Exercise 12.84 (p. 646). Consider the first-order, interaction model

$$E(y) = \beta_0 + \beta_1 x_1 + \beta_2 x_2 + \beta_3 x_1 x_2$$

where

$y$ = sun safety posttest score

$x_1$ = sun safety pretest score

$$x_2 = \begin{cases} 1 & \text{if in the } Be \ Sun \ Safe \text{ intervention group} \\ 0 & \text{if in the control group} \end{cases}$$

a. Assuming interaction exists, give the reduced model for testing whether the mean posttest scores differ for the intervention and control groups.

b. When sun safety knowledge was used as the dependent variable, the test of part **a** resulted in a $p$-value of .03. Interpret this result.

**SPSS Output for Exercise 12.93**

**ANOVA**[b]

| Model | | Sum of Squares | df | Mean Square | F | Sig. |
|---|---|---|---|---|---|---|
| 1 | Regression | 1186549 | 5 | 237309.713 | 2.885 | .024[a] |
| | Residual | 3618994 | 44 | 82249.862 | | |
| | Total | 4805543 | 49 | | | |

a. Predictors: (Constant), AGESQ_ST, RCSTATUS, AGE, AGESQ, AGE_STAT
b. Dependent Variable: VALUE

**ANOVA**[b]

| Model | | Sum of Squares | df | Mean Square | F | Sig. |
|---|---|---|---|---|---|---|
| 2 | Regression | 1116017 | 3 | 372005.590 | 4.638 | .006[a] |
| | Residual | 3689526 | 46 | 80207.081 | | |
| | Total | 4805543 | 49 | | | |

a. Predictors: (Constant), AGE_STAT, AGE, RCSTATUS
b. Dependent Variable: VALUE

**ANOVA**[b]

| Model | | Sum of Squares | df | Mean Square | F | Sig. |
|---|---|---|---|---|---|---|
| 3 | Regression | 1082210 | 3 | 360736.761 | 4.457 | .008[a] |
| | Residual | 3723332 | 46 | 80942.005 | | |
| | Total | 4805543 | 49 | | | |

a. Predictors: (Constant), RCSTATUS, AGESQ, AGE
b. Dependent Variable: VALUE

**c.** When sun safety comprehension was used as the dependent variable, the test of part **a** resulted in a $p$-value of .033. Interpret this result.

**d.** When sun safety application was used as the dependent variable, the test of part **a** resulted in a $p$-value of .322. Interpret this result.

**12.95** Refer to the *Chance* (Winter 2001) study of students who paid a private tutor (or coach) to help them improve their Scholastic Assessment Test (SAT) scores, Exercise 12.66 (p. 634). Recall that the baseline model, $E(y) = \beta_0 + \beta_1 x_1 + \beta_2 x_2$, where $y$ = SAT-Math score, $x_1$ = score on PSAT, and $x_2$ = {1 if student was coached, 0 if not}, had the following results: $R_a^2 = .76$, $\hat{\beta}_2 = 19$, and $s_{\hat{\beta}_2} = 3$. As an alternative model, the researcher added several "control" variables, including dummy variables for student ethnicity ($x_3$, $x_4$, and $x_5$), a socioeconomic status index variable ($x_6$), two variables that measured high school performance ($x_7$ and $x_8$), the number of math courses taken in high school ($x_9$), and the overall GPA for the math courses ($x_{10}$).

**a.** Write the hypothesized equation for $E(y)$ for the alternative model.

**b.** Give the null hypothesis for a nested model $F$-test comparing the initial and alternative models.

**c.** The nested model $F$-test, part **b**, was statistically significant at $\alpha = .05$. Practically interpret this result.

**d.** The alternative model, part **a**, resulted in $R_a^2 = .79$, $\hat{\beta}_2 = 14$, and $s_{\hat{\beta}_2} = 3$. Interpret the value of $R_a^2$.

**e.** Refer to part **d**. Find and interpret a 95% confidence interval for $\beta_2$.

**f.** The researcher concluded that "the estimated effect of SAT coaching decreases from the baseline model when control variables are added to the model." Do you agree? Justify your answer.

**g.** As a modification to the model of part **a**, the researcher added all possible interactions between the coaching variable ($x_2$) and the other independent variables in the model. Write the equation for $E(y)$ for this modified model.

**h.** Give the null hypothesis for comparing the models, parts **a** and **g**. How would you perform this test?

**12.96** Since glass is not subject to radiation damage, encapsulation of waste in glass is considered to be one of the most promising solutions to the problem of low-level nuclear waste in the environment. However, chemical reactions may weaken the glass. This concern led to a study undertaken jointly by the Department of Materials Science and Engineering at the University of Florida and the U.S. Department of Energy to assess the utility of glass as a waste encapsulant.* Corrosive chemical solutions (called corrosion baths) were prepared and applied directly to glass samples containing one of three types of waste (TDS-3A, FE, and AL); the

chemical reactions were observed over time. A few of the key variables measured were

$y$ = Amount of silicon (in parts per million) found in solution at end of experiment. (This is both a measure of the degree of breakdown in the glass and a proxy for the amount of radioactive species released into the environment.)

$x_1$ = Temperature (°C) of the corrosion bath

$x_2$ = 1 if waste type TDS-3A, 0 if not

$x_3$ = 1 if waste type FE, 0 if not

(Waste type AL is the base level.) Suppose we want to model amount $y$ of silicon as a function of temperature ($x_1$) and type of waste ($x_2$, $x_3$).

**a.** Write a model that proposes parallel straight-line relationships between amount of silicon and temperature, one line for each of the three waste types.

**b.** Add terms for the interaction between temperature and waste type to the model of part **a**.

**c.** Refer to the model of part **b**. For each waste type, give the slope of the line relating amount of silicon to temperature.

**d.** Explain how you could test for the presence of temperature–waste type interaction.

**12.97** The *Journal of Human Stress* (Summer 1987) reported on a study of "psychological response of firefighters to chemical fire." It is thought that the following complete second-order model will be adequate to describe the relationship between emotional distress and years of experience for two groups of firefighters—those exposed to a chemical fire and those unexposed.

$$E(y) = \beta_0 + \beta_1 x_1 + \beta_2 x_1^2 + \beta_3 x_2 + \beta_4 x_1 x_2 + \beta_5 x_1^2 x_2$$

where

$y$ = Emotional distress

$x_1$ = Experience (years)

$x_2$ = 1 if exposed to chemical fire, 0 if not

**a.** What hypothesis would you test to determine whether the *rate* of increase of emotional distress with experience is different for the two groups of firefighters?

**b.** What hypothesis would you test to determine whether there are differences in mean emotional distress levels that are attributable to exposure group?

**c.** Data for a sample of 200 firefighters were used to fit the complete model as well as the reduced model, $E(y) = \beta_0 + \beta_1 x_1 + \beta_2 x_1^2$. The resulting SAS printouts are shown on p. 656. Is there sufficient evidence to support the claim that the mean emotional distress levels differ for the two groups of firefighters? Use $\alpha = .05$.

---

*The background information for this exercise was provided by Dr. David Clark, Department of Materials Science and Engineering, University of Florida.

SAS Output for Complete Model, Exercise 12.97

The REG Procedure
Dependent Variable: DISTRESS

Analysis of Variance

| Source | DF | Sum of Squares | Mean Square | F Value | Pr > F |
|--------|-----|------|------|------|------|
| Model | 5 | 2351.70 | 470.34 | 116.42 | <.0001 |
| Error | 194 | 783.90 | 4.04 | | |
| Corrected Total | 199 | 3135.60 | | | |

| | | | | |
|---|---|---|---|---|
| Root MSE | 2.0102 | R-Square | 0.7500 |
| Dependent Mean | 24.221 | Adj R-Sq | 0.7436 |
| Coeff Var | 8.229 | | |

SAS Output for Reduced Model, Exercise 12.97

The REG Procedure
Dependent Variable: DISTRESS

Analysis of Variance

| Source | DF | Sum of Squares | Mean Square | F Value | Pr > F |
|--------|-----|------|------|------|------|
| Model | 2 | 2340.37 | 1170.18 | 289.87 | <.0001 |
| Error | 197 | 795.23 | 4.04 | | |
| Corrected Total | 199 | 3135.60 | | | |

| | | | | |
|---|---|---|---|---|
| Root MSE | 2.0092 | R-Square | 0.7464 |
| Dependent Mean | 24.221 | Adj R-Sq | 0.7438 |
| Coeff Var | 8.295 | | |

## 12.12  MODEL BUILDING: STEPWISE REGRESSION

Consider the problem of predicting the salary *y* of an executive. Perhaps the biggest problem in building a model to describe executive salaries is choosing the important independent variables to be included in the model. The list of potentially important independent variables is extremely long (e.g., age, experience, tenure, education level, etc.), and we need some objective method of screening out those that are not important.

The problem of deciding which of a large set of independent variables to include in a model is a common one. Trying to determine which variables influence the profit of a firm, affect blood pressure of humans, or are related to a student's performance in college are only a few examples.

A systematic approach to building a model with a large number of independent variables is difficult because the interpretation of multivariable interactions and higher-order terms is tedious. We therefore turn to a screening procedure, available in most statistical software packages, known as *stepwise regression*.

The most commonly used **stepwise regression** procedure works as follows. The user first identifies the response, $y$, and the set of potentially important independent variables, $x_1, x_2, \ldots, x_k$, where $k$ is generally large. [*Note:* This set of variables could include both first-order and higher-order terms. However, we may often include only the main effects of both quantitative variables (first-order terms) and qualitative variables (dummy variables), since the inclusion of second-order terms greatly increases the number of independent variables.] The response and independent variables are then entered into the computer software, and the stepwise procedure begins.

**Step 1** The software program fits all possible one-variable models of the form

$$E(y) = \beta_0 + \beta_1 x_i$$

to the data, where $x_i$ is the $i$th independent variable, $i = 1, 2, \ldots, k$. For each model, the test of the null hypothesis

$$H_0: \beta_1 = 0$$

against the alternative hypothesis

$$H_a: \beta_1 \neq 0$$

is conducted using the $t$-test (or the equivalent $F$-test) for a single $\beta$ parameter. The independent variable that produces the largest (absolute) $t$ value is declared the best one-variable predictor of $y$.* Call this independent variable $x_1$.

**Step 2** The stepwise program now begins to search through the remaining $(k - 1)$ independent variables for the best two-variable model of the form

$$E(y) = \beta_0 + \beta_1 x_1 + \beta_2 x_i$$

This is done by fitting all two-variable models containing $x_1$ and each of the other $(k - 1)$ options for the second variable $x_i$. The $t$ values for the test $H_0: \beta_2 = 0$ are computed for each of the $(k - 1)$ models (corresponding to the remaining independent variables, $x_i, i = 2, 3, \ldots, k$), and the variable having the largest $t$ is retained. Call this variable $x_2$. At this point, some software packages diverge in methodology. The better packages now go back and check the $t$ value of $\hat{\beta}_1$ after $\hat{\beta}_2 x_2$ has been added to the model. If the $t$ value has become nonsignificant at some specified $\alpha$ level (say $\alpha = .05$), the variable $x_1$ is removed and a search is made for the independent variable with a $\beta$ parameter that will yield the most significant $t$ value in the presence of $\hat{\beta}_2 x_2$. Other packages do not recheck the significance of $\hat{\beta}_1$ but proceed directly to step 3.

The reason the $t$ value for $x_1$ may change from step 1 to step 2 is that the meaning of the coefficient $\hat{\beta}_1$ changes. In step 2, we are approximating a complex response surface in two variables with a plane. The best-fitting plane may yield a different value for $\hat{\beta}_1$ than that obtained in step 1. Thus, both the value of $\hat{\beta}_1$ and its significance usually changes from step 1 to step 2. For this reason, the software packages that recheck the $t$ values at each step are preferred.

---

*Note that the variable with the largest $t$ value is also the one with the largest (absolute) Pearson product moment correlation, $r$ (Section 11.6), with $y$.

**Step 3** The stepwise procedure now checks for a third independent variable to include in the model with $x_1$ and $x_2$. That is, we seek the best model of the form

$$E(y) = \beta_0 + \beta_1 x_1 + \beta_2 x_2 + \beta_3 x_i$$

To do this, we fit all the $(k - 2)$ models using $x_1, x_2$, and each of the $(k - 2)$ remaining variables, $x_i$, as a possible $x_3$. The criterion is again to include the independent variable with the largest $t$ value. Call this best third variable $x_3$.

The better programs now recheck the $t$ values corresponding to the $x_1$ and $x_2$ coefficients, replacing the variables that yield nonsignificant $t$ values. This procedure is continued until no further independent variables can be found that yield significant $t$ values (at the specified $\alpha$ level) in the presence of the variables already in the model.

The result of the stepwise procedure is a model containing only those terms with $t$ values that are significant at the specified $\alpha$ level. Thus, in most practical situations only several of the large number of independent variables remain. However, it is very important *not* to jump to the conclusion that all the independent variables important for predicting $y$ have been identified or that the unimportant independent variables have been eliminated. Remember, the stepwise procedure is using only *sample estimates* of the true model coefficients ($\beta$'s) to select the important variables. An extremely large number of single $\beta$ parameter $t$-tests have been conducted, and the probability is very high that one or more errors have been made in including or excluding variables. That is, we have very probably included some unimportant independent variables in the model (Type I errors) and eliminated some important ones (Type II errors).

There is a second reason why we might not have arrived at a good model. When we choose the variables to be included in the stepwise regression, we may often omit high-order terms (to keep the number of variables manageable). Consequently, we may have initially omitted several important terms from the model. Thus, we should recognize stepwise regression for what it is: an objective *variable screening* procedure.

Successful model builders will now consider second-order terms (for quantitative variables) and other interactions among variables screened by the stepwise procedure. It would be best to develop this response surface model with a second set of data independent of that used for the screening, so the results of the stepwise procedure can be partially verified with new data. This is not always possible, however, because in many modeling situations only a small amount of data is available.

Do not be deceived by the impressive-looking $t$ values that result from the stepwise procedure—it has retained only the independent variables with the largest $t$ values. Also, be certain to consider second-order terms in systematically developing the prediction model. Finally, if you have used a first-order model for your stepwise procedure, remember that it may be greatly improved by the addition of higher-order terms.

> **Caution**
> ___
>
> Be wary of using the results of stepwise regression to make inferences about the relationship between $E(y)$ and the independent variables in the resulting first-order model. First, an extremely large number of $t$-tests have been conducted, leading to a high probability of making one or more Type I or Type II errors. Second, the stepwise model does not include any higher-order or interaction terms. Stepwise regression should be used only when necessary, that is, when you want to determine which of a large number of potentially important independent variables should be used in the model-building process.

**E X A M P L E  1 2 . 1 2**

An international management consulting company develops multiple regression models for executive salaries of its client firms. The consulting company has found that models that use the natural logarithm of salary as the dependent variable have better predictive power than those using salary as the dependent variable.* A preliminary step in the construction of these models is the determination of the most important independent variables. For one firm, 10 potential independent variables (seven quantitative and three qualitative) were measured in a sample of 100 executives as shown in Table 12.7. Since it would be very difficult to construct a complete second-order model with all of the 10 independent variables, use stepwise regression to decide which of the 10 variables should be included in the building of the final model for the log of executive salaries.

**Solution**

We will use stepwise regression with the main effects of the 10 independent variables to identify the most important variables. The dependent variable $y$ is the natural logarithm of the executive salaries. The STATISTIX stepwise regression printout is shown in Figure 12.34. STATISTIX automatically enters the constant term $(\beta_0)$ into the model in the first step. The remaining steps follow the procedure outlined earlier in this section.

**TABLE 12.7**  **Independent Variables in the Executive Salary Example**

| Independent Variable | Description |
|---|---|
| $x_1$ | Experience (years)—quantitative |
| $x_2$ | Education (years)—quantitative |
| $x_3$ | Bonus eligibility (1 if yes, 0 if no)—qualitative |
| $x_4$ | Number of employees supervised—quantitative |
| $x_5$ | Corporate assets (millions of dollars)—quantitative |
| $x_6$ | Board member (1 if yes, 0 if no)—qualitative |
| $x_7$ | Age (years)—quantitative |
| $x_8$ | Company profits (past 12 months, millions of dollars)—quantitative |
| $x_9$ | Has international responsibility (1 if yes, 0 if no)—qualitative |
| $x_{10}$ | Company's total sales (past 12 months, millions of dollars)—quantitative |

___

*This is probably because salaries tend to be incremented in *percentages* rather than dollar values. When a response variable undergoes percentage changes as the independent variables are varied, the logarithm of the response variable will be more suitable as a dependent variable.

**FIGURE 12.34**

*A Portion of the STATISTIX
Stepwise Regression Printout for
Example 12.12*

```
STEPWISE REGRESSION OF LNSALARY
UNFORCED VARIABLES: X1 X2 X3 X4 X5 X6 X7 X8 X9 X10
  P to ENTER 0.0500
  P to EXIT  0.0500
```

| STEP | VARIABLE | COEFFICIENT | T | P | R SQ | MSE |
|------|----------|-------------|---|---|------|-----|
| 1 | CONSTANT | 10.3987 | 4.57 | | 0.0000 | 0.35281 |
| 2 | CONSTANT | 10.2008 | 5.76 | | 0.4207 | 0.16114 |
|   | X4 | 0.00057 | 8.44 | 0.0001 | | |
| 3 | CONSTANT | 9.8770 | 5.09 | | 0.7830 | 0.06098 |
|   | X4 | 0.00058 | 13.97 | 0.0001 | | |
|   | X5 | 0.00184 | 12.73 | 0.0001 | | |
| 4 | CONSTANT | 9.6645 | 5.07 | | 0.8967 | 0.02934 |
|   | X4 | 0.00055 | 18.96 | 0.0001 | | |
|   | X5 | 0.00191 | 19.04 | 0.0001 | | |
|   | X1 | 0.01871 | 10.28 | 0.0001 | | |
| 5 | CONSTANT | 9.4008 | 5.15 | | 0.9482 | 0.01488 |
|   | X4 | 0.00055 | 26.65 | 0.0001 | | |
|   | X5 | 0.00191 | 26.70 | 0.0001 | | |
|   | X1 | 0.02075 | 15.80 | 0.0001 | | |
|   | X3 | 0.30012 | 9.71 | 0.0001 | | |
| 6 | CONSTANT | 8.8539 | 4.89 | | 0.9604 | 0.01149 |
|   | X4 | 0.00056 | 30.66 | 0.0001 | | |
|   | X5 | 0.00194 | 30.72 | 0.0001 | | |
|   | X1 | 0.02142 | 18.46 | 0.0001 | | |
|   | X3 | 0.31928 | 11.66 | 0.0001 | | |
|   | X2 | 0.03316 | 5.39 | 0.0001 | | |
| 7 | CONSTANT | 8.8751 | 4.96 | | 0.9610 | 0.01143 |
|   | X4 | 0.00056 | 30.42 | 0.0001 | | |
|   | X5 | 0.00194 | 30.81 | 0.0001 | | |
|   | X1 | 0.02133 | 18.40 | 0.0001 | | |
|   | X3 | 0.31094 | 11.04 | 0.0001 | | |
|   | X2 | 0.03272 | 5.32 | 0.0001 | | |
|   | X6 | 0.03866 | 1.21 | 0.2295 | | |
| 8 | CONSTANT | 8.8539 | 4.89 | | 0.9604 | 0.01149 |
|   | X4 | 0.00056 | 30.66 | 0.0001 | | |
|   | X5 | 0.00194 | 30.72 | 0.0001 | | |
|   | X1 | 0.02142 | 18.46 | 0.0001 | | |
|   | X3 | 0.31928 | 11.66 | 0.0001 | | |
|   | X2 | 0.03316 | 5.39 | 0.0001 | | |

```
RESULTING STEPWISE MODEL
```

| VARIABLE | COEFFICIENT | STD ERROR | STUDENT'S T | P |
|----------|-------------|-----------|-------------|---|
| CONSTANT | 8.85388 | 1.8106094 | 4.89 | 0.0000 |
| X4 | 0.00056 | 0.0000183 | 30.66 | 0.0000 |
| X5 | 0.00194 | 0.0000632 | 30.72 | 0.0000 |
| X1 | 0.02142 | 0.0011603 | 18.46 | 0.0000 |
| X3 | 0.31928 | 0.0273825 | 11.66 | 0.0000 |
| X2 | 0.03316 | 0.0061521 | 5.39 | 0.0000 |

```
CASES INCLUDED   100      R SQUARED  0.9604    MSE  0.01148
MISSING CASES      0      ADJ R SQ   0.9041    SD   0.10714
```

Note that the first variable included in the model is $x_4$, Number of employees supervised by the executive. At the next step (step 3), $x_5$, Corporate assets, enters the model. By the seventh step, $x_6$, a dummy variable for the qualitative variable Board member or not, is brought into the model. However, because the significance

(.2295) of the $t$ statistic for $x_6$ is above the preassigned $\alpha = .05$, $x_6$ is then removed from the model. Thus, at step 8 the procedure indicates that the five-variable model including $x_1$, $x_2$, $x_3$, $x_4$, and $x_5$ is best. That is, none of the other independent variables can meet the $\alpha = .05$ criterion for admission to the model.

Thus, in our final modeling effort we should concentrate on these five independent variables. Models with second-order terms and interactions should be proposed and evaluated to determine the best model for predicting executive salaries. ∎

We conclude this section with some advice on the use of stepwise regression.

## Recommendation

Do *not* use the stepwise regression model as the final model for predicting $y$. Recall that the stepwise procedure tends to perform a large number of $t$-tests, inflating the overall probability of a Type I error, and does not automatically include higher-order terms (e.g., interactions and squared terms) in the final model. Use stepwise regression as a variable screening tool when there exists a large number of potentially important independent variables. Then begin building models for $y$ using the variables identified by stepwise regression.

 EXERCISES 12.98–12.101

### Learning the Mechanics

**12.98** There are six independent variables, $x_1, x_2, x_3, x_4, x_5$, and $x_6$, that might be useful in predicting a response $y$. A total of $n = 50$ observations are available, and it is decided to employ stepwise regression to help in selecting the independent variables that appear to be useful. The computer fits all possible one-variable models of the form

$$E(y) = \beta_0 + \beta_1 x_i$$

where $x_i$ is the $i$th independent variable, $i = 1, 2, \ldots, 6$. The information in the table is provided from the computer printout.

| Independent Variable | $\hat{\beta}_i$ | $s_{\hat{\beta}_i}$ |
|---|---|---|
| $x_1$ | 1.6 | .42 |
| $x_2$ | −.9 | .01 |
| $x_3$ | 3.4 | 1.14 |
| $x_4$ | 2.5 | 2.06 |
| $x_5$ | −4.4 | .73 |
| $x_6$ | .3 | .35 |

**a.** Which independent variable is declared the best one-variable predictor of $y$? Explain.

**b.** Would this variable be included in the model at this stage? Explain.

**c.** Describe the next phase that a stepwise procedure would execute.

### Applying the Concepts—Basic

**12.99** *Benefits Quarterly* (First Quarter, 1995) published a study of entry level job preferences. A number of independent variables were used to model the job preferences (measured on a 10-point scale) of 164 business school graduates. Suppose stepwise regression is used to build a model for job preference score ($y$) as a function of the following independent variables:

$$x_1 = \begin{cases} 1 & \text{if flextime position} \\ 0 & \text{if not} \end{cases}$$

$$x_2 = \begin{cases} 1 & \text{if day care support required} \\ 0 & \text{if not} \end{cases}$$

$$x_3 = \begin{cases} 1 & \text{if spousal transfer support required} \\ 0 & \text{if not} \end{cases}$$

$x_4 =$ Number of sick days allowed

$$x_5 = \begin{cases} 1 & \text{if applicant married} \\ 0 & \text{if not} \end{cases}$$

$x_6 =$ Number of children of applicant

$$x_7 = \begin{cases} 1 & \text{if male applicant} \\ 0 & \text{if female applicant} \end{cases}$$

a. How many models are fit to the data in step 1? Give the general form of these models.
b. How many models are fit to the data in step 2? Give the general form of these models.
c. How many models are fit to the data in step 3? Give the general form of these models.
d. Explain how the procedure determines when to stop adding independent variables to the model.
e. Describe two major drawbacks to using the final stepwise model as the "best" model for job preference score ($y$).

## Applying the Concepts—Intermediate

12.100 A marine biologist was hired by the EPA to determine whether the hot-water runoff from a particular power plant located near a large gulf is having an adverse effect on the marine life in the area. The biologist's goal is to acquire a prediction equation for the number of marine animals located at certain designated areas, or stations, in the gulf. Based on past experience, the EPA considered the following environmental factors as predictors for the number of animals at a particular station:

**SAS Output for Exercise 12.100**

The REG Procedure
Dependent Variable: NUMBER

Stepwise Selection: Step 1

Variable ST_DEPTH Entered: R-Square = 0.1223 and C(p) = 51.57

Analysis of Variance

| Source | DF | Sum of Squares | Mean Square | F Value | Pr > F |
|---|---|---|---|---|---|
| Model | 1 | 57.44 | 57.44 | 99.47 | <.0001 |
| Error | 714 | 412.33 | 0.58 | | |
| Corrected Total | 715 | 469.77 | | | |

| Variable | Parameter Estimate | Standard Error | Type II SS | F Value | Pr > F |
|---|---|---|---|---|---|
| Intercept | 8.38559 | 5.60286 | 1.299 | 2.24 | 0.1356 |
| ST_DEPTH | -0.43678 | 0.05763 | 57.44 | 99.03 | <.0001 |

Stepwise Selection: Step 2

Variable TGRSWT Entered: R-Square = 0.1821 and C(p) = 1.52

Analysis of Variance

| Source | DF | Sum of Squares | Mean Square | F Value | Pr > F |
|---|---|---|---|---|---|
| Model | 2 | 85.55 | 42.78 | 79.38 | <.0001 |
| Error | 713 | 384.22 | 0.54 | | |
| Corrected Total | 715 | 469.77 | | | |

| Variable | Parameter Estimate | Standard Error | Type II SS | F Value | Pr > F |
|---|---|---|---|---|---|
| Intercept | 8.07682 | 7.41024 | 1.188 | 2.20 | 0.1411 |
| ST_DEPTH | -0.35355 | 0.05966 | 35.11 | 65.02 | <.0001 |
| TGRSWT | 0.00271 | 0.00051 | 28.17 | 52.16 | <.0001 |

**SAS Output for Exercise 12.100** *(continued)*

Stepwise Selection: Step 3

Variable TI Entered: R-Square = 0.1870 and C(p) = 3.51

Analysis of Variance

| Source | DF | Sum of Squares | Mean Square | F Value | Pr > F |
|---|---|---|---|---|---|
| Model | 3 | 87.85 | 29.28 | 54.59 | <.0001 |
| Error | 712 | 381.92 | 0.54 | | |
| Corrected Total | 715 | 469.77 | | | |

| Variable | Parameter Estimate | Standard Error | Type II SS | F Value | Pr > F |
|---|---|---|---|---|---|
| Intercept | 7.38864 | 5.03901 | 1.161 | 2.15 | 0.1660 |
| ST_DEPTH | -0.31451 | 0.04764 | 80.70 | 43.58 | <.0001 |
| TGRSWT | 0.00262 | 0.00038 | 88.39 | 47.73 | <.0001 |
| TI | 0.65774 | 0.31793 | 7.93 | 4.28 | 0.0389 |

Stepwise Selection: Step 4

Variable DO Entered: R-Square = 0.1889 and C(p) = 1.03

Analysis of Variance

| Source | DF | Sum of Squares | Mean Square | F Value | Pr > F |
|---|---|---|---|---|---|
| Model | 4 | 88.76 | 22.19 | 41.40 | <.0001 |
| Error | 711 | 381.01 | 0.54 | | |
| Corrected Total | 715 | 469.77 | | | |

| Variable | Parameter Estimate | Standard Error | Type II SS | F Value | Pr > F |
|---|---|---|---|---|---|
| Intercept | 7.33576 | 5.35015 | 1.015 | 1.88 | 0.1782 |
| ST_DEPTH | -0.30417 | 0.04828 | 21.43 | 39.69 | <.0001 |
| TGRSWT | 0.00267 | 0.00038 | 26.58 | 49.23 | <.0001 |
| TI | 0.67347 | 0.31783 | 2.42 | 4.49 | 0.0345 |
| DO | 0.01769 | 0.01361 | 0.91 | 1.69 | 0.1946 |

$x_1$ = Temperature of water (TEMP)
$x_2$ = Salinity of water (SAL)
$x_3$ = Dissolved oxygen content of water (DO)
$x_4$ = Turbidity index, a measure of the turbidity of the water (TI)
$x_5$ = Depth of the water at the station (ST_DEPTH)
$x_6$ = Total weight of sea grasses in sampled area (TGRSWT)

As a preliminary step in the construction of this model, the biologist used a stepwise regression procedure to identify the most important of these six variables. A total of 716 samples were taken at different stations in the gulf, producing the SAS printout on pp. 662–664. (The response measured was $y$, the logarithm of the number of marine animals found in the sampled area.)

**a.** According to the SAS printout, which of the six independent variables should be used in the model? [*Note:* SAS uses the $F = t^2$ statistic for testing terms.]

**b.** Are we able to assume that the marine biologist has identified all the important independent variables for the prediction of $y$? Why?

**c.** Using the variables identified in part **a**, write the first-order model with interaction that may be used to predict $y$.

SAS Output for Exercise 12.100 *(continued)*

Stepwise Selection: Step 5

Variable DO Removed: R-Square = 0.1870 and C(p) = 3.51

Analysis of Variance

| Source | DF | Sum of Squares | Mean Square | F Value | Pr > F |
|---|---|---|---|---|---|
| Model | 3 | 87.85 | 29.28 | 54.59 | <.0001 |
| Error | 712 | 381.92 | 0.54 | | |
| Corrected Total | 715 | 469.77 | | | |

| Variable | Parameter Estimate | Standard Error | Type II SS | F Value | Pr > F |
|---|---|---|---|---|---|
| Intercept | 7.38864 | 5.03901 | 1.161 | 2.15 | 0.1660 |
| ST_DEPTH | -0.31451 | 0.04764 | 80.70 | 43.58 | <.0001 |
| TGRSWT | 0.00262 | 0.00038 | 88.39 | 47.73 | <.0001 |
| TI | 0.65774 | 0.31793 | 7.93 | 4.28 | 0.0389 |

All variables left in the model are significant at the 0.0500 level.

No other variable met the 0.0500 significance level for entry into the model.

**d.** How would the marine biologist determine whether the model specified in part **c** is better than the first-order model?

**e.** Note the small value of $R_2$. What action might the biologist take to improve the model?

## Applying the Concepts—Advanced

**12.101** Road construction contracts in the state of Florida are awarded on the basis of competitive, sealed bids; the contractor who submits the lowest bid price wins the contract. During the 1980's, the Office of the Florida Attorney General (FLAG) suspected numerous contractors of practicing bid collusion, i.e., setting the winning bid price above the fair, or competitive, price in order to increase profit margin. By comparing the bid prices (and other important bid variables) of the fixed (or rigged) contracts to the competitively bid contracts, FLAG was able to establish invaluable benchmarks for detecting future bid-rigging. FLAG collected data for 279 road construction contracts. For each contract, the following variables were measured:

⚫FLAG

1. Price of contract ($) bid by lowest bidder
2. Department of Transportation (DOT) engineer's estimate of fair contract price ($)
3. Ratio of low (winning) bid price to DOT engineer's estimate of fair price
4. Status of contract (1 if fixed, 0 if competitive)
5. District (1, 2, 3, 4, or 5) in which construction project is located
6. Number of bidders on contract
7. Estimated number of days to complete work
8. Length of road project (miles)
9. Percentage of costs allocated to liquid asphalt
10. Percentage of costs allocated to base material
11. Percentage of costs allocated to excavation
12. Percentage of costs allocated to mobilization
13. Percentage of costs allocated to structures
14. Percentage of costs allocated to traffic control
15. Subcontractor utilization (1 if yes, 0 if no)

**a.** Consider building a model for the low-bid price ($y$). Apply stepwise regression to the data to find the independent variables most suitable for modeling $y$.

**b.** Interpret the $\beta$ estimates in the resulting stepwise regression model.

**c.** What are the dangers associated with drawing inferences from the stepwise model?

## 12.13   RESIDUAL ANALYSIS: CHECKING THE REGRESSION ASSUMPTIONS

When we apply regression analysis to a set of data, we never know for certain whether the assumptions of Section 12.3 are satisfied. How far can we deviate from the assumptions and still expect regression analysis to yield results that will have the reliability stated in this chapter? How can we detect departures (if they exist) from the assumptions and what can we do about them? We provide some answers to these questions in this section.

Recall from Section 12.3 that for any given set of values of $x_1, x_2, \ldots, x_k$ we assume that the random error term $\varepsilon$ has a normal probability distribution with mean equal to 0 and variance equal to $\sigma^2$. Also, we assume that the random errors are probabilistically independent. It is unlikely that these assumptions are ever satisfied exactly in a practical application of regression analysis. Fortunately, experience has shown that least squares regression analysis produces reliable statistical tests, confidence intervals, and prediction intervals as long as the departures from the assumptions are not too great. In this section we present some methods for determining whether the data indicate significant departures from the assumptions.

Because the assumptions all concern the random error component, $\varepsilon$, of the model, the first step is to estimate the random error. Since the actual random error associated with a particular value of $y$ is the difference between the actual $y$ value and its unknown mean, we estimate the error by the difference between the actual $y$ value and the *estimated* mean. This estimated error is called the *regression residual*, or simply the **residual**, and is denoted by $\hat{\varepsilon}$. The actual error $\varepsilon$ and residual $\hat{\varepsilon}$ are shown in Figure 12.35.

---

**DEFINITION 12.5**

A **regression residual**, $\hat{\varepsilon}$, is defined as the difference between an observed $y$ value and its corresponding predicted value:

$$\hat{\varepsilon} = (y - \hat{y}) = y - (\hat{\beta}_0 + \hat{\beta}_1 x_1 + \hat{\beta}_2 x_2 + \cdots + \hat{\beta}_k x_k)$$

---

**FIGURE 12.35**

*Actual Random Error $\varepsilon$ and Regression Residual $\hat{\varepsilon}$*

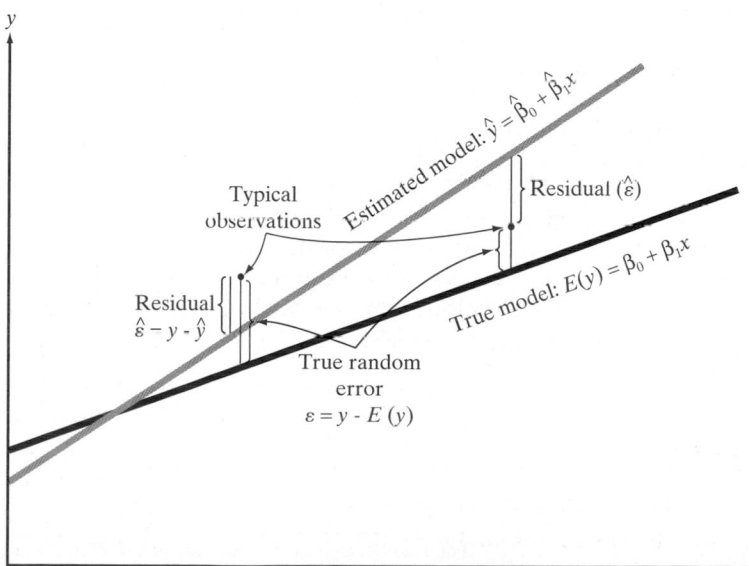

Since the true mean of $y$ (that is, the true regression model) is not known, the actual random error cannot be calculated. However, because the residual is based on the estimated mean (the least squares regression model), it can be calculated and used to estimate the random error and to check the regression assumptions. Such checks are generally referred to as **residual analyses**. Two useful properties of residuals are given in the next box.

The following examples show how a graphical analysis of regression residuals can be used to verify the assumptions associated with the model and to support improvements to the model when the assumptions do not appear to be satisfied. Although the residuals can be calculated and plotted by hand, we rely on the statistical software for these tasks in the examples and exercises.

---

### Properties of Regression Residuals

1. The mean of the residuals is equal to 0. This property follows from the fact that the sum of the differences between the observed $y$ values and their least squares predicted $\hat{y}$ values is equal to 0.

$$\sum (\text{Residuals}) = \sum (y - \hat{y}) = 0$$

2. The standard deviation of the residuals is equal to the standard deviation of the fitted regression model, $s$. This property follows from the fact that the sum of the squared residuals is equal to SSE, which when divided by the error degrees of freedom is equal to the variance of the fitted regression model, $s^2$. The square root of the variance is both the standard deviation of the residuals and the standard deviation of the regression model.

$$\sum (\text{Residuals})^2 = \sum (y - \hat{y})^2 = \text{SSE}$$

$$s = \sqrt{\frac{\sum (\text{Residuals})^2}{n - (k + 1)}} = \sqrt{\frac{\text{SSE}}{n - (k + 1)}} = \sqrt{\text{MSE}}$$

---

First, we demonstrate how a residual plot can detect a model in which the hypothesized relationship between $E(y)$ and an independent variable $x$ is misspecified. The assumption of mean error of 0 is violated in these types of models.*

**EXAMPLE 12.13**

Refer to the problem of modeling the relationship between home size $(x)$ and electrical usage $(y)$ in Example 12.6 (p. 618). The data for $n = 10$ homes are repeated in Table 12.8. SAS printouts for a straight-line model and a quadratic model fitted to the data are shown in Figures 12.36a and 12.36b, respectively. The residuals from these models are highlighted in the printouts. The residuals are then plotted on the vertical axis against the variable $x$, size of home, on the horizontal axis in Figures 12.37a and 12.37b, respectively.

a. Verify that each residual is equal to the difference between the observed $y$ value and the estimated mean value, $\hat{y}$.
b. Analyze the residual plots.

---

*For a misspecified model, the hypothesized mean of $y$, denoted by $E_h(y)$, will not equal the true mean of $y$, $E(y)$. Since $y = E_h(y) + \varepsilon$, then $\varepsilon = y - E_h(y)$ and $E(\varepsilon) = E\{y - E_h(y)\} = E(y) - E_h(y) \neq 0$.

⊘ ELECTRIC

**TABLE 12.8    Home Size–Electrical Usage Data**

| Size of Home<br>x (sq. ft) | Monthly Usage<br>y (kilowatt-hours) |
|---|---|
| 1,290 | 1,182 |
| 1,350 | 1,172 |
| 1,470 | 1,264 |
| 1,600 | 1,493 |
| 1,710 | 1,571 |
| 1,840 | 1,711 |
| 1,980 | 1,804 |
| 2,230 | 1,840 |
| 2,400 | 1,956 |
| 2,930 | 1,954 |

**FIGURE 12.36a**

*SAS Regression Printout for Straight-Line Electrical Usage Model*

The REG Procedure
Dependent Variable: USAGE

Analysis of Variance

| Source | DF | Sum of<br>Squares | Mean<br>Square | F Value | Pr > F |
|---|---|---|---|---|---|
| Model | 1 | 703957 | 703957 | 39.54 | 0.0002 |
| Error | 8 | 142445 | 17806 | | |
| Corrected Total | 9 | 846402 | | | |

| | | | | |
|---|---|---|---|---|
| Root MSE | 133.43768 | R-Square | 0.8317 |
| Dependent Mean | 1594.70000 | Adj R-Sq | 0.8107 |
| Coeff Var | 8.36757 | | |

Parameter Estimates

| Variable | DF | Parameter<br>Estimate | Standard<br>Error | t Value | Pr > \|t\| |
|---|---|---|---|---|---|
| Intercept | 1 | 578.92775 | 166.96806 | 3.47 | 0.0085 |
| SIZE | 1 | 0.54030 | 0.08593 | 6.29 | 0.0002 |

Predictions

| Obs | SIZE | USAGE | SIZESQ | Predicted<br>USAGE | Residual<br>of USAGE |
|---|---|---|---|---|---|
| 1 | 1290 | 1182 | 1664100 | 1275.92 | -93.920 |
| 2 | 1350 | 1172 | 1822500 | 1308.34 | -136.339 |
| 3 | 1470 | 1264 | 2160900 | 1373.18 | -109.175 |
| 4 | 1600 | 1493 | 2560000 | 1443.41 | 49.585 |
| 5 | 1710 | 1571 | 2924100 | 1502.85 | 68.152 |
| 6 | 1840 | 1711 | 3385600 | 1573.09 | 137.912 |
| 7 | 1980 | 1804 | 3920400 | 1648.73 | 155.270 |
| 8 | 2230 | 1840 | 4972900 | 1783.81 | 56.193 |
| 9 | 2400 | 1956 | 5760000 | 1875.66 | 80.342 |
| 10 | 2930 | 1954 | 8584900 | 2162.02 | -208.020 |

**FIGURE 12.36b**

*SAS Regression Printout for Quadratic Electrical Usage Model*

The REG Procedure
Dependent Variable: USAGE

Analysis of Variance

| Source | DF | Sum of Squares | Mean Square | F Value | Pr > F |
|--------|-----|----------------|-------------|---------|--------|
| Model | 2 | 831070 | 415535 | 189.71 | <.0001 |
| Error | 7 | 15333 | 2190.36480 | | |
| Corrected Total | 9 | 846402 | | | |

| | | | |
|---|---|---|---|
| Root MSE | 46.80133 | R-Square | 0.9819 |
| Dependent Mean | 1594.70000 | Adj R-Sq | 0.9767 |
| Coeff Var | 2.93480 | | |

Parameter Estimates

| Variable | DF | Parameter Estimate | Standard Error | t Value | Pr > \|t\| |
|----------|-----|-------------------|----------------|---------|-----------|
| Intercept | 1 | -1216.14389 | 242.80637 | -5.01 | 0.0016 |
| SIZE | 1 | 2.39893 | 0.24584 | 9.76 | <.0001 |
| SIZESQ | 1 | -0.00045004 | 0.00005908 | -7.62 | 0.0001 |

Predictions

| Obs | SIZE | USAGE | SIZESQ | Predicted USAGE | Residual of USAGE |
|-----|------|-------|---------|-----------------|-------------------|
| 1 | 1290 | 1182 | 1664100 | 1129.56 | 52.4359 |
| 2 | 1350 | 1172 | 1822500 | 1202.21 | -30.2136 |
| 3 | 1470 | 1264 | 2160900 | 1337.79 | -73.7916 |
| 4 | 1600 | 1493 | 2560000 | 1470.04 | 22.9586 |
| 5 | 1710 | 1571 | 2924100 | 1570.06 | 0.9359 |
| 6 | 1840 | 1711 | 3385600 | 1674.23 | 36.7685 |
| 7 | 1980 | 1804 | 3920400 | 1769.40 | 34.5998 |
| 8 | 2230 | 1840 | 4972900 | 1895.47 | -55.4654 |
| 9 | 2400 | 1956 | 5760000 | 1949.06 | 6.9431 |
| 10 | 2930 | 1954 | 8584900 | 1949.17 | 4.8287 |

**FIGURE 12.37a**

*SAS Residual Plot for Straight-Line Electrical Usage Model*

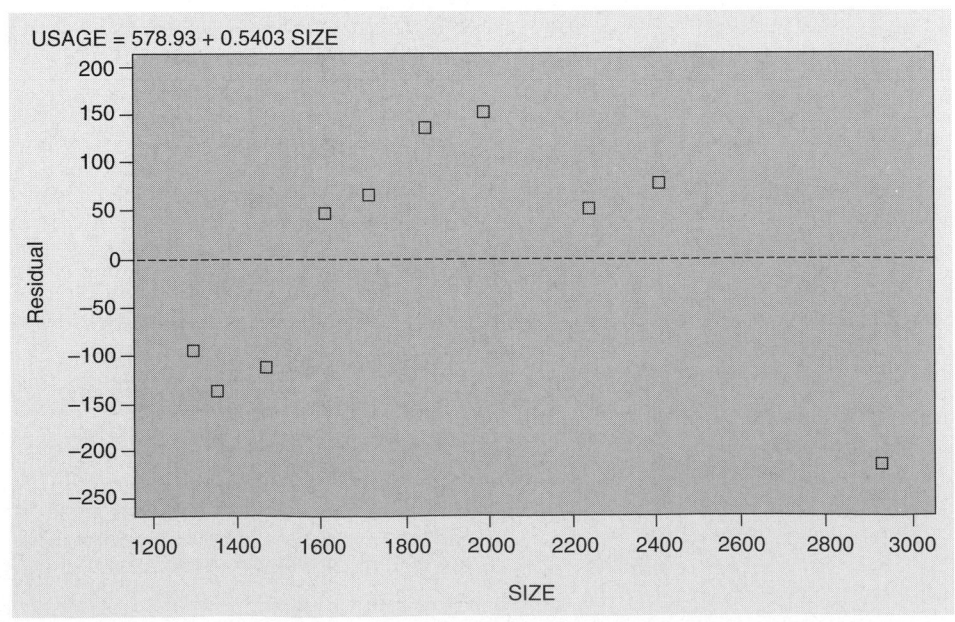

## FIGURE 12.37b

*SAS Residual Plot for Quadratic Electrical Usage Model*

USAGE = –1216.1 + 2.3989 SIZE –0.0005 SIZESQ

### Solution

a. Consider the first usage value in the data set, $y = 1,182$. For the straight-line model the predicted value for this observation is $\hat{y} = 1,275.92$ (highlighted on Figure 12.36a). Consequently, the residual is

$$\hat{\varepsilon} = (y - \hat{y}) = 1,182 - 1,275.9 = -93.9$$

Similarly, the residual for the first $y$ value using the quadratic model (Figure 12.36b) is

$$\hat{\varepsilon} = 1,182 - 1,129.6 = 52.4$$

Both residuals agree (after rounding) with the first values given in the column labeled **Residual of USAGE** in Figures 12.36a and 12.36b, respectively. Although the residuals both correspond to the same observed $y$ value, 1,182, they differ because the predicted mean value changes depending on whether the straight-line model or quadratic model is used. Similar calculations produce the remaining residuals.

b. The plot of the residuals for the straight-line model (Figure 12.37a) reveals a nonrandom pattern. The residuals exhibit a curved shape, with the residuals for the small values of $x$ below the horizontal 0 (mean of the residuals) line, the residuals corresponding to the middle values of $x$ above the 0 line, and the residual for the largest value of $x$ again below the 0 line. The indication is that the mean value of the random error, *within* each of these ranges of $x$ (small, medium, large) may not be equal to 0. Such a pattern usually indicates that curvature needs to be added to the model.

When the second-order term is added to the model, the nonrandom pattern disappears. In Figure 12.37b, the residuals appear to be randomly distributed around the 0 line, as expected. The implication is that the quadratic model provides a considerably better model for predicting electrical usage, verifying our conclusions from Example 12.6. ■

Residual analyses are also useful for detecting one or more observations that deviate significantly from the regression model. We expect approximately 95% of the residuals to fall within 2 standard deviations of the 0 line, and all or almost all of them to lie within 3 standard deviations of their mean of 0. Residuals that are extremely far from the 0 line, and disconnected from the bulk of the other residuals, are called *outliers*, and should receive special attention from the regression analyst.

---

**DEFINITION 12.6**

A residual that is larger than $3s$ (in absolute value) is considered to be an **outlier**.

---

**EXAMPLE 12.14**

Refer to Example 12.5 (p. 612) in which we modeled the auction price $y$ of a grand-father clock as a function of age $x_1$ and number of bidders $x_2$. The data for this example are repeated in Table 12.9, with one important difference: The auction price of the clock at the top of the second column has been changed from $2,131 to $1,131 (highlighted in Table 12.9). The interaction model

$$E(y) = \beta_0 + \beta_1 x_1 + \beta_2 x_2 + \beta_3 x_1 x_2$$

is again fit to these (modified) data, with the MINITAB printout shown in Figure 12.38. The residuals are shown highlighted in the printout and then plotted against the number of bidders, $x_2$, in Figure 12.39. Analyze the residual plot.

**Solution**

The residual plot dramatically reveals the one altered measurement. Note that one of the two residuals at $x_2 = 14$ bidders falls more than 3 standard deviations below 0. Note that no other residual falls more than 2 standard deviations from 0.

What do we do with outliers once we identify them? First, we try to determine the cause. Were the data entered into the computer incorrectly? Was the observation recorded incorrectly when the data were collected? If so, we correct the

⊙ GFCLOCKALT

**TABLE 12.9   Altered Auction Price Data**

| Age $x_1$ | Number of Bidders $x_2$ | Auction Price $y$ | Age $x_1$ | Number of Bidders $x_2$ | Auction Price $y$ |
|---|---|---|---|---|---|
| 127 | 13 | $1,235 | 170 | 14 | $1,131 |
| 115 | 12 | 1,080 | 182 | 8 | 1,550 |
| 127 | 7 | 845 | 162 | 11 | 1,884 |
| 150 | 9 | 1,522 | 184 | 10 | 2,041 |
| 156 | 6 | 1,047 | 143 | 6 | 845 |
| 182 | 11 | 1,979 | 159 | 9 | 1,483 |
| 156 | 12 | 1,822 | 108 | 14 | 1,055 |
| 132 | 10 | 1,253 | 175 | 8 | 1,545 |
| 137 | 9 | 1,297 | 108 | 6 | 729 |
| 113 | 9 | 946 | 179 | 9 | 1,792 |
| 137 | 15 | 1,713 | 111 | 15 | 1,175 |
| 117 | 11 | 1,024 | 187 | 8 | 1,593 |
| 137 | 8 | 1,147 | 111 | 7 | 785 |
| 153 | 6 | 1,092 | 115 | 7 | 744 |
| 117 | 13 | 1,152 | 194 | 5 | 1,356 |
| 126 | 10 | 1,336 | 168 | 7 | 1,262 |

**FIGURE 12.38**

*MINITAB Regression Printout for Grandfather Clock Example with Altered Data*

```
The regression equation is
PRICE = - 513 + 8.17 AGE + 19.9 BIDDERS + 0.320 AGEBID

Predictor        Coef      SE Coef          T         P
Constant       -512.8        665.9      -0.77     0.448
AGE             8.165        4.585       1.78     0.086
BIDDERS         19.89        67.44       0.29     0.770
AGEBID         0.3196       0.4790       0.67     0.510

S = 200.6      R-Sq = 72.9%     R-Sq(adj) = 70.0%

Analysis of Variance

Source           DF          SS         MS         F         P
Regression        3     3033587    1011196     25.13     0.000
Residual Error   28     1126703      40239
Total            31     4160290

Obs     AGE      PRICE      Fit     SE Fit    Residual   St Resid
  1     127     1235.0   1310.4      59.3       -75.4      -0.39
  2     115     1080.0   1105.9      62.1       -25.9      -0.14
  3     127      845.0    947.5      61.1      -102.5      -0.54
  4     150     1522.0   1322.5      37.1       199.5       1.01
  5     156     1047.0   1179.5      60.3      -132.5      -0.69
  6     182     1979.0   1831.9      82.9       147.1       0.81
  7     156     1822.0   1598.0      61.9       224.0       1.17
  8     132     1253.0   1185.8      39.7        67.2       0.34
  9     137     1297.0   1178.9      39.0       118.1       0.60
 10     113      946.0    913.9      58.6        32.1       0.17
 11     137     1713.0   1561.0      78.4       152.0       0.82
 12     117     1024.0   1072.6      53.1       -48.6      -0.25
 13     137     1147.0   1115.2      44.3        31.8       0.16
 14     153     1092.0   1149.2      59.0       -57.2      -0.30
 15     117     1152.0   1187.2      69.7       -35.2      -0.19
 16     126     1336.0   1117.6      43.4       218.4       1.12
 17     170     1131.0   1914.4     116.7      -783.4      -4.80R
 18     182     1550.0   1597.7      62.8       -47.7      -0.25
 19     162     1884.0   1598.3      57.0       258.7       1.49
 20     184     2041.0   1776.6      70.7       264.4       1.41
 21     143      845.0   1048.4      58.9      -203.4      -1.06
 22     159     1483.0   1421.8      40.6        61.2       0.31
 23     108     1055.0   1130.7      97.9       -75.7      -0.43
 24     175     1545.0   1522.7      55.4        22.3       0.12
 25     108      729.0    695.5      99.6        33.5       0.19
 26     179     1792.0   1642.7      57.6       149.3       0.78
 27     111     1175.0   1224.0     107.2       -49.0      -0.29
 28     187     1593.0   1651.3      68.6       -58.3      -0.31
 29     111      785.0    781.1      80.9         3.9       0.02
 30     115      744.0    822.7      75.5       -78.7      -0.42
 31     194     1356.0   1480.7     133.6      -124.7      -0.83 X
 32     168     1262.0   1374.0      57.7      -112.0      -0.58

R denotes an observation with a large standardized residual
X denotes an observation whose X value gives it large influence.
```

observation and rerun the analysis. Another possibility is that the observation is not representative of the conditions we are trying to model. For example, in this case the low price may be attributable to extreme damage to the clock, or to a clock of inferior quality compared to the others. In these cases we probably would exclude the observation from the analysis. In many cases you may not be able to determine the cause of the outlier. Even so, you may want to rerun the regression analysis excluding the outlier in order to assess the effect of that observation on the results of the analysis.

**FIGURE 12.39**

*MINITAB Residual Plot
Against Number of Bidders*

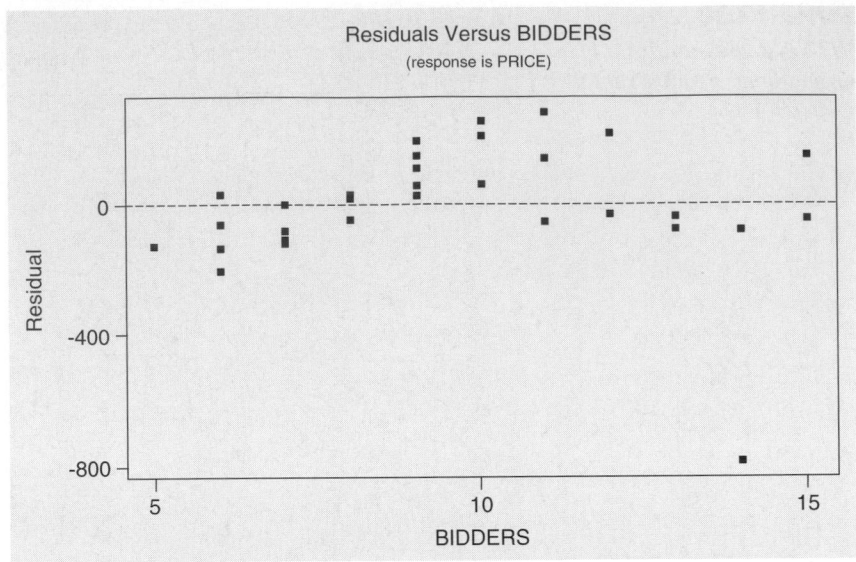

**FIGURE 12.40**

*MINITAB Regression Printout
for Grandfather Clock Example
(Outlier Deleted)*

```
The regression equation is
PRICE = 474 - 0.46 AGE - 114 BIDDERS + 1.48 AGEBID

Predictor        Coef      SE Coef          T        P
Constant        474.0        298.2       1.59    0.124
AGE            -0.465        2.107      -0.22    0.827
BIDDERS       -114.12        31.23      -3.65    0.001
AGEBID         1.4781       0.2295       6.44    0.000

S = 85.83       R-Sq = 95.2%      R-Sq(adj) = 94.7%

Analysis of Variance

Source          DF          SS          MS        F        P
Regression       3     3933417     1311139   177.99    0.000
Residual Error  27      198897        7367
Total           30     4132314
```

Figure 12.40 shows the printout when the outlier observation is excluded from the grandfather clock analysis, and Figure 12.41 shows the new plot of the residuals against the number of bidders. Now only one of the residuals lies beyond 2 standard deviations from 0, and none of them lies beyond 3 standard deviations. Also, the model statistics indicate a much better model without the outlier. Most notably, the standard deviation ($s$) has decreased from 200.6 to 85.83, indicating a model that will provide more precise estimates and predictions (narrower confidence and prediction intervals) for clocks that are similar to those in the reduced sample. But remember that if the outlier is removed from the analysis when in fact it belongs to the same population as the rest of the sample, the resulting model may provide misleading estimates and predictions. ⬛

Outlier analysis is another example of testing the assumption that the expected (mean) value of the random error $\varepsilon$ is 0, since this assumption is in doubt for the error terms corresponding to the outliers. The next example in this section checks the assumption of the normality of the random error component.

**FIGURE 12.41**

*MINITAB Residual Plot Against Number of Bidders (Outlier Deleted)*

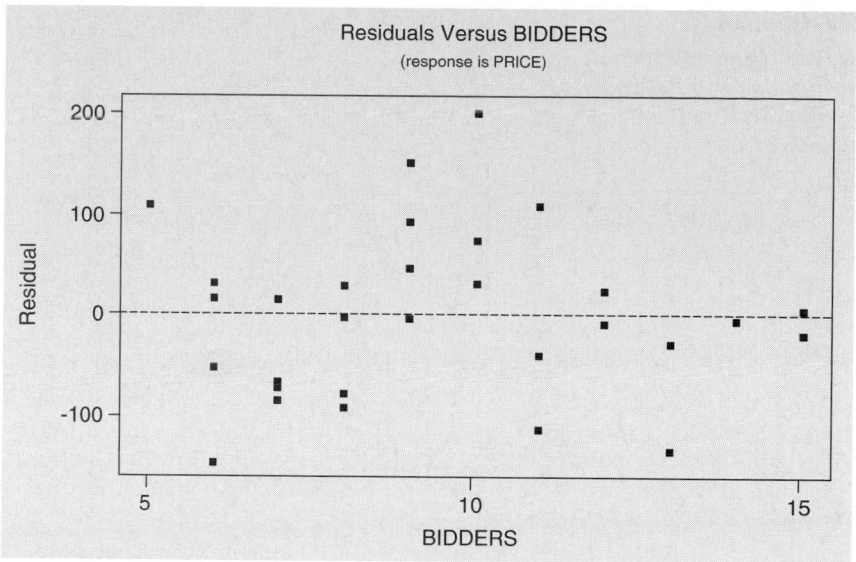

**EXAMPLE 12.15**

Refer to Example 12.14. Analyze the distribution of the residuals in the grandfather clock example, both before and after the outlier residual is removed. Determine whether the assumption of a normally distributed error term is reasonable.

**Solution**

A histogram and normal probability plot for the two sets of residuals are constructed using MINITAB and are shown in Figures 12.42 and 12.43. Note that the outlier appears to skew the histogram in Figure 12.42, whereas the histogram in Figure 12.43 appears to be more mound-shaped. Similarly, the pattern of the residuals in the normal probability plot in Figure 12.43 (outlier deleted) is more nearly a straight line than the pattern in Figure 12.42 (outlier included). Although the graphs do not provide formal statistical tests of normality, they do provide a descriptive display. In this example the normality assumption appears to be more plausible after the outlier is removed. Consult the references for methods to conduct statistical tests of normality using the residuals.

Of all the assumptions in Section 12.3, the assumption that the random error is normally distributed is the least restrictive when we apply regression analysis in practice. That is, moderate departures from a normal distribution have very little effect on the validity of the statistical tests, confidence intervals, and prediction intervals presented in this chapter. In this case, we say that regression analysis is **robust** with respect to nonnormal errors. However, great departures from normality cast doubt on any inferences derived from the regression analysis.

Residual plots can also be used to detect violations of the assumption of constant error variance. For example, a plot of the residuals versus the predicted value $\hat{y}$ may display a pattern as shown in Figure 12.44. In this figure, the range in values of the residuals increases as $\hat{y}$ increases, thus indicating that the variance of the random error, $\varepsilon$, becomes larger as the estimate of $E(y)$ increases in value. Since $E(y)$ depends on the $x$ values in the model, this implies that the variance of $\varepsilon$ is not constant for all settings of the $x$'s.

**FIGURE 12.42**

*MINITAB Histogram and Normal Probability Plot of Regression Residuals for Grandfather Clock Model (Outlier Included)*

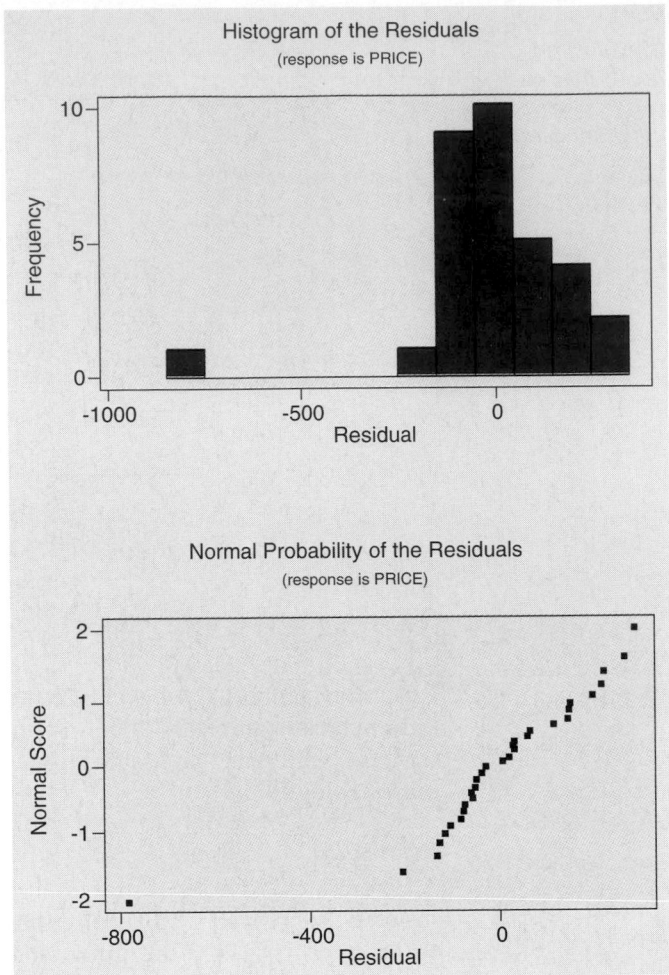

In the final example of this section, we demonstrate how to use this plot to detect a nonconstant variance and suggest a useful remedy.

**E X A M P L E   1 2 . 1 6**   The data in Table 12.10 are the salaries, $y$, and years of experience, $x$, for a sample of 50 social workers. The first-order model $E(y) = \beta_0 + \beta_1 x$ was fitted to the data using MINITAB. The MINITAB printout is shown in Figure 12.45, followed by a plot of the residuals versus $\hat{y}$ in Figure 12.46. Interpret the results. Make model modifications, if necessary.

**Solution**

The MINITAB printout, Figure 12.45, suggests that the first-order model provides an adequate fit to the data. The $R^2$ value indicates that the model explains about 79% of the sample variation in salaries. The $t$ value for testing $\beta_1$, 13.31, is highly significant ($p$-value $\approx 0$) and indicates that the model contributes information for the prediction of $y$. However, an examination of the residuals plotted against $\hat{y}$ (Figure 12.46) reveals a potential problem. Note the "cone" shape of the residual variability; the size of the residuals increases as the estimated mean salary increases, implying that the constant variance assumption is violated.

**FIGURE 12.43**

*MINITAB Histogram and Normal Probability Plot of Regression Residuals for Grandfather Clock Model (Outlier Deleted)*

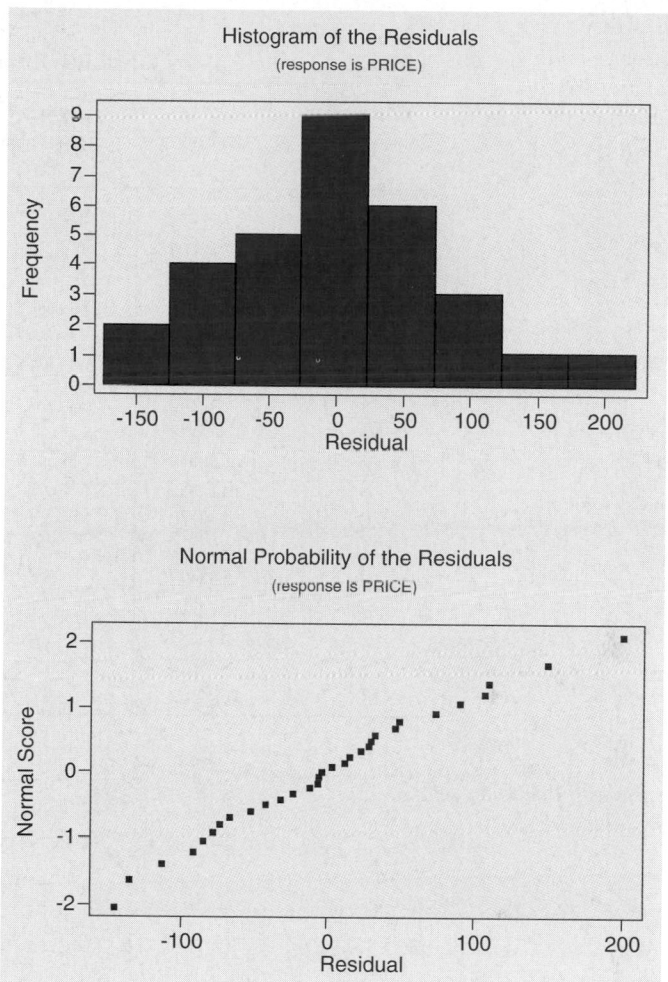

**FIGURE 12.44**

*Residual Plot Showing Changes in the Variance of ε*

$(y-\hat{y})$

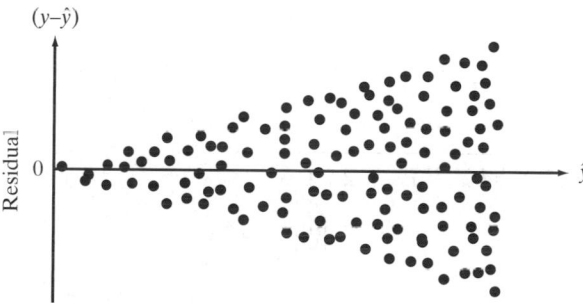

One way to stabilize the variance of $\varepsilon$ is to refit the model using a transformation on the dependent variable $y$. With economic data (e.g., salaries) a useful **variance-stabilizing transformation** is the natural logarithm of $y$, denoted $\ln(y)$.* We fit the model

$$\ln(y) = \beta_0 + \beta_1 x + \varepsilon$$

---

*Other variance-stabilizing transformations that are used successfully in practice are $\sqrt{y}$ and $\sin^{-1}\sqrt{y}$. Consult the references for more details on these transformations.

SOCWORK
**TABLE 12.10   Salary Data for Example 12.16**

| Years of Experience x | Salary y | Years of Experience x | Salary y | Years of Experience x | Salary y |
|---|---|---|---|---|---|
| 7 | $26,075 | 21 | $43,628 | 28 | $99,139 |
| 28 | 79,370 | 4 | 16,105 | 23 | 52,624 |
| 23 | 65,726 | 24 | 65,644 | 17 | 50,594 |
| 18 | 41,983 | 20 | 63,022 | 25 | 53,272 |
| 19 | 62,308 | 20 | 47,780 | 26 | 65,343 |
| 15 | 41,154 | 15 | 38,853 | 19 | 46,216 |
| 24 | 53,610 | 25 | 66,537 | 16 | 54,288 |
| 13 | 33,697 | 25 | 67,447 | 3 | 20,844 |
| 2 | 22,444 | 28 | 64,785 | 12 | 32,586 |
| 8 | 32,562 | 26 | 61,581 | 23 | 71,235 |
| 20 | 43,076 | 27 | 70,678 | 20 | 36,530 |
| 21 | 56,000 | 20 | 51,301 | 19 | 52,745 |
| 18 | 58,667 | 18 | 39,346 | 27 | 67,282 |
| 7 | 22,210 | 1 | 24,833 | 25 | 80,931 |
| 2 | 20,521 | 26 | 65,929 | 12 | 32,303 |
| 18 | 49,727 | 20 | 41,721 | 11 | 38,371 |
| 11 | 33,233 | 26 | 82,641 | | |

**FIGURE 12.45**

*MINITAB Regression Printout for First-Order Salary Model*

```
The regression equation is
SALARY = 11369 + 2141 EXP


Predictor      Coef      SE Coef         T        P
Constant      11369         3160      3.60    0.001
EXP          2141.4        160.8     13.31    0.000


S = 8642       R-Sq = 78.7%      R-Sq(adj) = 78.2%


Analysis of Variance


Source         DF           SS           MS        F        P
Regression      1   13239655469  13239655469   177.26    0.000
Residual Error 48    3585206077     74691793
Total          49   16824861546


Unusual Observations

Obs      EXP     SALARY        Fit      SE Fit    Residual   St Resid
 31      1.0      24833      13510        3013       11323      1.40 X
 35     28.0      99139      71327        2005       27812      3.31R
 45     20.0      36530      54196        1259      -17666     -2.07R


R denotes an observation with a large standardized residual
X denotes an observation whose X value gives it large influence.
```

to the data of Table 12.10. Figure 12.47 shows the regression analysis printout for the $n = 50$ measurements, while Figure 12.48 shows a plot of the residuals from the log model.

You can see that the logarithmic transformation has stabilized the error variances. Note that the cone shape is gone; there is no apparent tendency of the residual variance to increase as mean salary increases. We therefore are confident that inferences using the logarithmic model are more reliable than those using the untransformed model.                                                                                              ∎

**FIGURE 12.46**

*MINITAB Residual Plot for First-Order Salary Model*

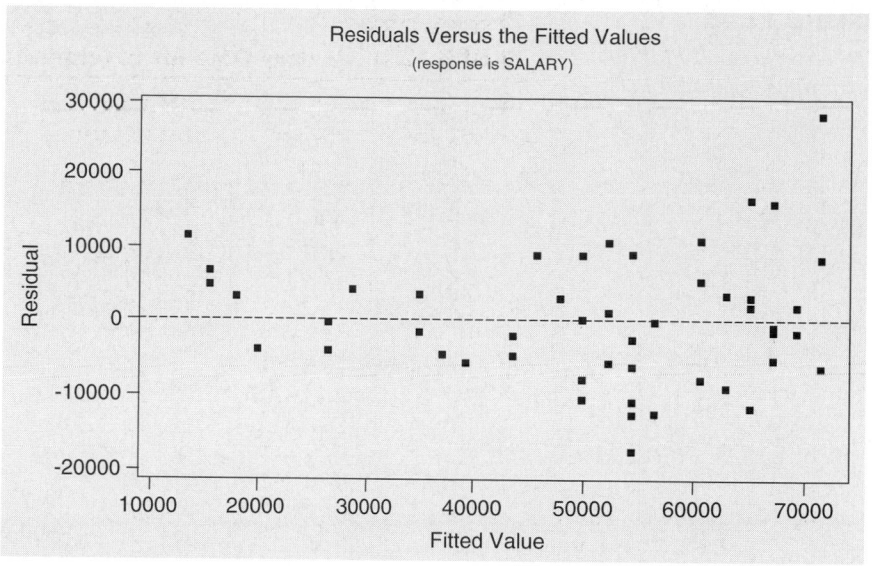

**FIGURE 12.47**

*MINITAB Regression Printout for Transformed Salary Model*

```
The regression equation is
LNSALARY = 9.84 + 0.0500 EXP

Predictor        Coef      SE Coef          T          P
Constant      9.84132      0.05636     174.63      0.000
EXP          0.049979     0.002868      17.43      0.000

S = 0.1541      R-Sq = 86.4%      R-Sq(adj) = 86.1%

Analysis of Variance

Source           DF           SS           MS        F        P
Regression        1       7.2121       7.2121   303.65    0.000
Residual Error   48       1.1401       0.0238
Total            49       8.3522

Unusual Observations

Obs      EXP    LNSALARY       Fit     SE Fit     Residual    St Resid
 19      4.0      9.6869    10.0412     0.0460      -0.3543       -2.41R
 31      1.0     10.1199     9.8913     0.0537       0.2286        1.58 X
 45     20.0     10.5059    10.8409     0.0225      -0.3350       -2.20R

R denotes an observation with a large standardized residual
X denotes an observation whose X value gives it large influence.
```

Residual analysis is a useful tool for the regression analyst, not only to check the assumptions, but also to provide information about how the model can be improved. A summary of the residual analysis presented in this section to check the assumption that the random error $\varepsilon$ is normally distributed with mean 0 and constant variance is presented in the next box.

**FIGURE 12.48**

*MINITAB Residual Plot for Transformed Salary Model*

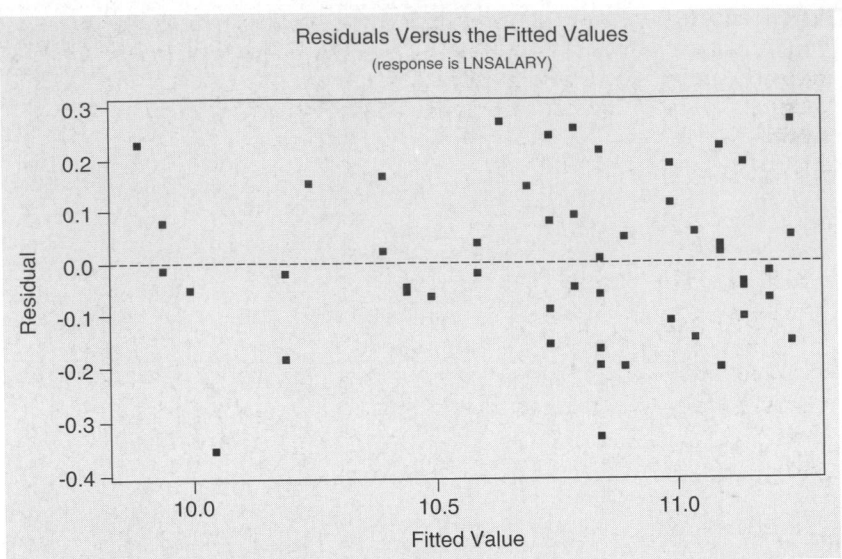

## Steps in a Residual Analysis

1. Check for a misspecified model by plotting the residuals against each of the quantitative independent variables. Analyze each plot, looking for a curvilinear trend. This shape signals the need for a quadratic term in the model. Try a second-order term in the variable against which the residuals are plotted.

2. Examine the residual plots for outliers. Draw lines on the residual plots at 2- and 3-standard-deviation distances below and above the 0 line. Examine residuals outside the 3-standard-deviation lines as potential outliers, and check to see that approximately 5% of the residuals exceed the 2-standard-deviation lines. Determine whether each outlier can be explained as an error in data collection or transcription, or corresponds to a member of a population different from that of the remainder of the sample, or simply represents an unusual observation. If the observation is determined to be an error, fix it or remove it. Even if you can't determine the cause, you may want to rerun the regression analysis without the observation to determine its effect on the analysis.

3. Check for nonnormal errors by plotting a frequency distribution of the residuals, using a stem-and-leaf display or a histogram. Check to see if obvious departures from normality exist. Extreme skewness of the frequency distribution may be due to outliers or could indicate the need for a transformation of the dependent variable. (Normalizing transformations are beyond the scope of this book, but you can find it in the references.)

4. Check for unequal error variances by plotting the residuals against the predicted values, $\hat{y}$. If you detect a cone-shaped pattern or some other pattern that indicates that the variance of $\varepsilon$ is not constant, refit the model using an appropriate variance-stabilizing transformation on $y$, such as $\ln(y)$. (Consult the references for other useful variance-stabilizing transformations.)

**Using the TI-83 Graphing Calculator**

## Plotting Residuals on the TI-83

When you compute a regression equation on the TI-83 the residuals are automatically computed and saved to a list called **RESID**. **RESID** can be found under the **LIST menu** (**2nd STAT**).

To make a scatterplot of the residuals,

Step 1    *Enter the data in L1 and L2*

Step 2    *Compute the regression equation* (see Section 11.2)

Step 3    *Set up the data plot*
          Press **2nd Y=** for **STATPLOT**
          Press **1** for **Plot1**
          Use the arrow and **ENTER** keys to set up the screen as shown below.

Note:   To enter the **RESID** as your **Ylist**:
          1. Use the arrow keys to move the cursor after **Ylist:**
          2. Press **2nd STAT** for **LIST**
          3. Highlight the listname **RESID** and press **ENTER**

Step 4    *View the scatterplot of the residuals*
          Press **ZOOM 9** for **ZoomStat**

Example   The figures below show a table of data entered on the TI-83 and the scatterplot of the residuals obtained using the steps given above.

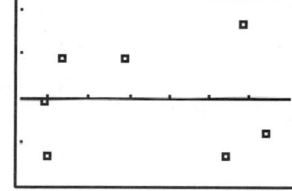

## 12.14 SOME PITFALLS: ESTIMABILITY, MULTICOLLINEARITY, AND EXTRAPOLATION

You should be aware of several potential problems when constructing a prediction model for some response $y$. A few of the most important are discussed in this final section.

### PROBLEM 1 PARAMETER ESTIMABILITY

Suppose you want to fit a model relating annual crop yield $y$ to the total expenditure for fertilizer, $x$. We propose the first-order model

$$E(y) = \beta_0 + \beta_1 x$$

Now suppose we have three years of data and \$1,000 is spent on fertilizer each year. The data are shown in Figure 12.49. You can see the problem: The parameters of the model cannot be estimated when all the data are concentrated at a single $x$ value. Recall that it takes two points ($x$ values) to fit a straight line. Thus, the parameters are not estimable when only one $x$ is observed.

**FIGURE 12.49**

*Yield and Fertilizer
Expenditure Data:
Three Years*

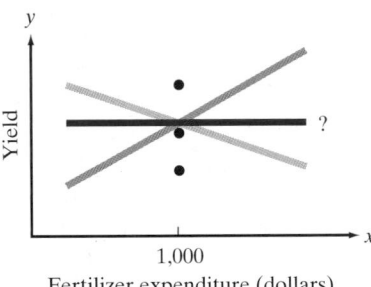

A similar problem would occur if we attempted to fit the quadratic model

$$E(y) = \beta_0 + \beta_1 x + \beta_2 x^2$$

to a set of data for which only one or two different $x$ values were observed (see Figure 12.50). At least three different $x$ values must be observed before a quadratic model can be fit to a set of data (that is, before all three parameters are estimable).

*In general, the number of levels of observed x values must be one more than the order of the polynomial in x that you want to fit.*

For controlled experiments, the researcher can select experimental designs that will permit estimation of the model parameters. Even when the values of the independent variables cannot be controlled by the researcher, the independent variables are almost always observed at a sufficient number of levels to permit estimation of the model parameters. When the statistical software you use suddenly refuses to fit a model, however, the problem is probably inestimable parameters.

**FIGURE 12.50**

*Only Two x Values
Observed: Quadratic Model
Is Not Estimable*

## PROBLEM 2   MULTICOLLINEARITY

Often, two or more of the independent variables used in a regression model contribute redundant information. That is, the independent variables are correlated with each other. For example, suppose we want to construct a model to predict the gas mileage rating of a truck as a function of its load, $x_1$ (in tons), and the horsepower, $x_2$ (in foot-pounds per second), of its engine. We would expect heavy loads to require greater horsepower and to result in lower mileage ratings. Thus, although both $x_1$ and $x_2$ contribute information for the prediction of mileage rating, $y$, some of the information is overlapping because $x_1$ and $x_2$ are correlated.

When the independent variables are correlated, we say *multicollinearity* exists. In practice, it is not uncommon to observe correlations among the independent variables. However, a few problems arise when serious multicollinearity is present in the regression variables.

> **DEFINITION 12.7**
> **Multicollinearity** exists when two or more of the independent variables used in regression are correlated.

First, high correlations among the independent variables increase the likelihood of rounding errors in the calculations of the $\beta$ estimates, standard errors, and so forth. Second, and more important, the regression results may be confusing and misleading. Consider the model for gasoline mileage rating ($y$) of a truck,

$$E(y) = \beta_0 + \beta_1 x_1 + \beta_2 x_2$$

where $x_1 =$ load and $x_2 =$ horsepower. Fitting the model to a sample data set, we might find that the $t$-tests for testing $\beta_1$ and $\beta_2$ are both nonsignificant at the $\alpha = .05$ level, while the $F$-test for $H_0: \beta_1 = \beta_2 = 0$ is highly significant ($p = .001$). The tests may seem to be contradictory, but really they are not. The $t$-tests indicate that the contribution of one variable, say $x_1 =$ load, is not significant after the effect of $x_2 =$ horsepower has been accounted for (because $x_2$ is also in the model). The significant $F$-test, on the other hand, tells us that at least one of the two variables is making a contribution to the prediction of $y$ (i.e., either $\beta_1$ or $\beta_2$ or both differ from 0). In fact, both are probably contributing, but the contribution of one overlaps with that of the other.

Multicollinearity can also have an effect on the signs of the parameter estimates. More specifically, a value of $\hat{\beta}_i$ may have the opposite sign from what is expected. In the truck gasoline mileage example, we expect heavy loads to result in lower mileage ratings and we expect higher horsepowers to result in lower mileage ratings; consequently, we expect the signs of both the parameter estimates to be negative. Yet, we may actually see a positive value of $\hat{\beta}_1$ and be tempted to claim that heavy loads result in *higher* mileage ratings. This is the danger of interpreting a $\beta$ coefficient when the independent variables are correlated. Because the variables contribute redundant information, the effect of $x_1 =$ load on $y =$ mileage rating is measured only partially by $\beta_1$.

How can you avoid the problems of multicollinearity in regression analysis? One way is to conduct a designed experiment (Chapter 10) so that the levels of the $x$ variables are uncorrelated. Unfortunately, time and cost constraints may prevent you from collecting data in this manner. Consequently, most data is collected observationally. Since observational data frequently consist of correlated independent variables, you will need to recognize when multicollinearity is present and, if necessary, make modifications in the regression analysis.

Several methods are available for detecting multicollinearity in regression. A simple technique is to calculate the coefficient of correlation, $r$, between each pair of independent variables in the model and use the procedure outlined in Section 11.6 to test for significantly correlated variables. If one or more of the $r$ values is statistically different from 0, the variables in question are correlated and a multicollinearity problem may exist.* Other indications of the presence of multicollinearity include those mentioned above—namely, nonsignificant $t$-tests for the individual parameter estimates when the $F$-test for overall model adequacy is significant, and estimates with opposite signs from what is expected.†

---

### Detecting Multicollinearity in the Regression Model

1. Significant correlations between pairs of independent variables
2. Nonsignificant $t$-tests for all (or nearly all) of the individual $\beta$ parameters when the $F$-test for overall model adequacy is significant
3. Signs opposite from what is expected in the estimated $\beta$ parameters

---

**EXAMPLE 12.17**

The Federal Trade Commission (FTC) annually ranks varieties of domestic cigarettes according to their tar, nicotine, and carbon monoxide contents. The U.S. surgeon general considers each of these three substances hazardous to a smoker's health. Past studies have shown that increases in the tar and nicotine contents of a cigarette are accompanied by an increase in the carbon monoxide emitted from the cigarette smoke. Table 12.11 presents data on tar, nicotine, and carbon monoxide contents (in milligrams) and weight (in grams) for a sample of 25 (filter) brands tested in a recent year. Suppose we want to model carbon monoxide content, $y$, as a function of tar content, $x_1$, nicotine content, $x_2$, and weight, $x_3$, using the model

$$E(y) = \beta_0 + \beta_1 x_1 + \beta_2 x_2 + \beta_3 x_3$$

The model is fit to the 25 data points in Table 12.11, and a portion of the STATISTIX printout is shown in Figure 12.51. Examine the printout. Do you detect any signs of multicollinearity?

### Solution

First, note that the $F$-test for overall model utility is highly significant. The test statistic ($F = 78.98$) and observed significance level ($p$-value $= .0000$) are highlighted on the STATISTIX printout, Figure 12.51. Therefore, we can conclude at, say, $\alpha = .01$, that at least one of the parameters, $\beta_1$, $\beta_2$, or $\beta_3$, in the model is nonzero. The $t$-tests for two of three individual $\beta$'s, however, are nonsignificant.

---

*Remember that $r$ measures only the pairwise correlation between $x$ values. Three variables, $x_1$, $x_2$, and $x_3$, may be highly correlated as a group but may not exhibit large pairwise correlations. Thus, multicollinearity may be present even when all pairwise correlations are not significantly different from 0.

†More formal methods for detecting multicollinearity, such as variance-inflation factors (VIFs), are available. Independent variables with a VIF of 10 or above are usually considered to be highly correlated with one or more of the other independent variables in the model. Calculation of VIFs are beyond the scope of this introductory text. Consult the chapter references for a discussion of VIFs and other formal methods of detecting multicollinearity.

⊘FTC
**TABLE 12.11    FTC Cigarette Data for Example 12.17**

| Tar ($x_1$) | Nicotine ($x_2$) | Weight ($x_3$) | Carbon Monoxide ($y$) |
|---|---|---|---|
| 14.1 | .86 | .9853 | 13.6 |
| 16.0 | 1.06 | 1.0938 | 16.6 |
| 29.8 | 2.03 | 1.1650 | 23.5 |
| 8.0 | .67 | .9280 | 10.2 |
| 4.1 | .40 | .9462 | 5.4 |
| 15.0 | 1.04 | .8885 | 15.0 |
| 8.8 | .76 | 1.0267 | 9.0 |
| 12.4 | .95 | .9225 | 12.3 |
| 16.6 | 1.12 | .9372 | 16.3 |
| 14.9 | 1.02 | .8858 | 15.4 |
| 13.7 | 1.01 | .9643 | 13.0 |
| 15.1 | .90 | .9316 | 14.4 |
| 7.8 | .57 | .9705 | 10.0 |
| 11.4 | .78 | 1.1240 | 10.2 |
| 9.0 | .74 | .8517 | 9.5 |
| 1.0 | .13 | .7851 | 1.5 |
| 17.0 | 1.26 | .9186 | 18.5 |
| 12.8 | 1.08 | 1.0395 | 12.6 |
| 15.8 | .96 | .9573 | 17.5 |
| 4.5 | .42 | .9106 | 4.9 |
| 14.5 | 1.01 | 1.0070 | 15.9 |
| 7.3 | .61 | .9806 | 8.5 |
| 8.6 | .69 | .9693 | 10.6 |
| 15.2 | 1.02 | .9496 | 13.9 |
| 12.0 | .82 | 1.1184 | 14.9 |

*Source:* Federal Trade Commission

(The *p*-values for these tests are highlighted on the printout.) Unless tar ($x_1$) is the only one of the three variables useful for predicting carbon monoxide content, these results are the first indication of a potential multicollinearity problem.

The negative values for $\hat{\beta}_2$ and $\hat{\beta}_3$ (highlighted on the printout) are a second clue to the presence of multicollinearity. From past studies, the FTC expects carbon monoxide content ($y$) to increase when either nicotine content ($x_2$) or weight ($x_3$) increases—that is, the FTC expects *positive* relationships between $y$ and $x_2$ and between $y$ and $x_3$, not negative ones.

All signs indicate that a serious multicollinearity problem exists.* To confirm our suspicions, we had STATISTIX produce the coefficient of correlation, $r$, for each of the three pairs of independent variables in the model. The resulting output is shown (highlighted) at the bottom of Figure 12.51. You can see that tar ($x_1$) and nicotine ($x_2$) are highly correlated ($r = .9766$) while weight ($x_3$) is moderately correlated with the other two $x$'s ($r \approx .5$). All three correlations have *p*-values less than .05; consequently, all three are significantly different from 0 at $\alpha = .05$. ∎

Once you have detected that multicollinearity exists, there are several alternative measures available for solving the problem. The appropriate measure to take depends on the severity of the multicollinearity and the ultimate goal of the regression analysis.

---

*Note also that the variance-inflation factors (VIFs) for both tar and nicotine, given on the STATISTIX printout, Figure 12.51, exceed 10.

**FIGURE 12.51**

*STATISTIX Printout for Model of CO Content, Example 12.17*

```
UNWEIGHTED LEAST SQUARES LINEAR REGRESSION OF CO

PREDICTOR
VARIABLES    COEFFICIENT     STD ERROR     STUDENT'S T      P        VIF
---------    -----------     ---------     -----------    ------    -----
CONSTANT       3.20219        3.46175          0.93       0.3655
TAR            0.96257        0.24224          3.97       0.0007    21.6
NICOTINE      -2.63166        3.90056         -0.67       0.5072    21.9
WEIGHT        -0.13048        3.88534         -0.03       0.9735     1.3

R-SQUARED              0.9186     RESID. MEAN SQUARE (MSE)     2.09012
ADJUSTED R-SQUARED     0.9070     STANDARD DEVIATION           1.44573

SOURCE         DF        SS          MS          F        P
----------     ---    ----------   ----------   -----    ------
REGRESSION      3      495.258      165.086      78.98    0.0000
RESIDUAL       21      43.8926      2.09012
TOTAL          24      539.150

CASES INCLUDED 25   MISSING CASES 0
```
```
CORRELATIONS (PEARSON)

              TAR       NICOTINE
NICOTINE    0.9766
  P-VALUE   0.0000

WEIGHT      0.4908      0.5002
            0.0127      0.0109
```

---

## Solutions to Some Problems Created by Multicollinearity in Regression*

---

1. Drop one or more of the correlated independent variables from the model. One way to decide which variables to keep in the model is to employ stepwise regression (Section 12.12).

2. If you decide to keep all the independent variables in the model:

   a) Avoid making inferences about the individual $\beta$ parameters based on the *t*-tests.

   b) Restrict inferences about $E(y)$ and future $y$ values to values of the *x*'s that fall within the range of the sample data.

---

Some researchers, when confronted with highly correlated independent variables, choose to include only one of the correlated variables in the final model. If you are interested only in using the model for estimation and prediction (step 6), you may decide not to drop any of the independent variables from the model. In the presence of multicollinearity, we have seen that it is dangerous to interpret the

---

*Several other solutions are available. For example, in the case where higher-order regression models are fit, the analyst may want to code the independent variables so that higher-order terms (e.g., $x^2$) for a particular *x* variable are not highly correlated with *x*. One transformation that works is $z = (x - \bar{x})/s$. Other, more sophisticated procedures for addressing multicollinearity (such as *ridge regression*) are beyond the scope of this text. Consult the references at the end of this chapter.

individual $\beta$ parameters. However, confidence intervals for $E(y)$ and prediction intervals for $y$ generally remain unaffected *as long as the values of the x's used to predict y follow the same pattern of multicollinearity exhibited in the sample data.* That is, you must take strict care to ensure that the values of the $x$ variables fall within the range of the sample data.

## PROBLEM 3 PREDICTION OUTSIDE THE EXPERIMENTAL REGION

Many research economists have developed highly technical models to relate the state of the economy to various economic indices and other independent variables. Many of these models are multiple regression models, where, for example, the dependent variable $y$ might be next year's growth in GDP and the independent variables might include this year's rate of inflation, this year's Consumer Price Index (CPI), etc. In other words, the model might be constructed to predict next year's economy using this year's knowledge.

Unfortunately, models such as these were almost all unsuccessful in predicting the recession in the early 1970s and the late 1990s. What went wrong? One of the problems was that many of the regression models were used to **extrapolate**, i.e., predict $y$ for values of the independent variables that were outside the region in which the model was developed. For example, the inflation rate in the late 1960s, when many of the models were developed, ranged from 6% to 8%. When the double-digit inflation of the early 1970s became a reality, some researchers attempted to use the same models to predict future growth in GDP. As you can see in Figure 12.52, the model may be very accurate for predicting $y$ when $x$ is in the range of experimentation, but the use of the model outside that range is a dangerous practice.

**FIGURE 12.52**

*Using a Regression Model Outside the Experimental Region*

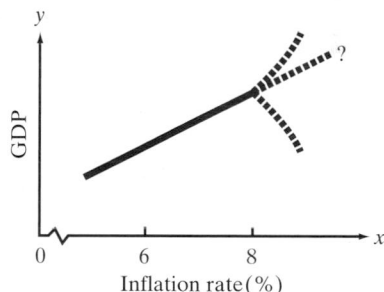

## PROBLEM 4 CORRELATED ERRORS

Another problem associated with using a regression model to predict a variable $y$ based on independent variables $x_1, x_2, \ldots, x_k$ arises from the fact that the data are frequently *time series*. That is, the values of both the dependent and independent variables are observed sequentially over a period of time. The observations tend to be correlated over time, which in turn often causes the prediction errors of the regression model to be correlated. Thus, the assumption of independent errors is violated, and the model tests and prediction intervals are no longer valid. One solution to this problem is to construct a **time series model**; the interested reader should consult the references for details of time series analysis.

EXERCISES 12.102–12.110

## Learning the Mechanics

**12.102** Identify the problem(s) in each of the residual plots shown at the bottom of the page.

**12.103** Consider fitting the multiple regression model

$$E(y) = \beta_0 + \beta_1 x_1 + \beta_2 x_2 + \beta_3 x_3 + \beta_4 x_4 + \beta_5 x_5$$

A matrix of correlations for all pairs of independent variables is given below. Do you detect a multi-collinearity problem? Explain.

|       | $x_1$ | $x_2$ | $x_3$ | $x_4$ | $x_5$ |
|-------|-------|-------|-------|-------|-------|
| $x_1$ | —     | .17   | .02   | −.23  | .19   |
| $x_2$ |       | —     | .45   | .93   | .02   |
| $x_3$ |       |       | —     | .22   | −.01  |
| $x_4$ |       |       |       | —     | .86   |
| $x_5$ |       |       |       |       | —     |

## Applying the Concepts—Basic

**12.104** Refer to the *Professional Geographer* (Feb. 2000) study of urban and rural counties in the western United States, Exercise 12.21 (p. 601). Recall that six independent variables —total county population ($x_1$), population density ($x_2$), population concentration ($x_3$), population growth ($x_4$), proportion of county land in farms ($x_5$), and 5-year change in agricultural land base ($x_6$)—were used to model the urban/rural

rating ($y$) of a county. Prior to running the multiple regression analysis, the researchers were concerned about possible multicollinearity in the data. Below is a correlation matrix, that is, a table of correlations between all pairs of the independent variables.

| Independent Variable |                         | $x_1$ | $x_2$ | $x_3$ | $x_4$ | $x_5$ |
|----------------------|-------------------------|-------|-------|-------|-------|-------|
| $x_1$                | Total population        |       |       |       |       |       |
| $x_2$                | Population density      | .20   |       |       |       |       |
| $x_3$                | Population concentration| .45   | .43   |       |       |       |
| $x_4$                | Population growth       | −.05  | −.14  | −.01  |       |       |
| $x_5$                | Farm land               | −.16  | −.15  | −.07  | −.20  |       |
| $x_6$                | Agricultural change     | −.12  | −.12  | −.22  | −.06  | −.06  |

*Source:* Berry, K. A., *et al.* "Interpreting what is rural and urban for western U.S. counties." *Professional Geographer*, Vol. 52, No. 1, Feb. 2000 (Table 2).

**a.** Based on the correlation matrix, is there any evidence of extreme multicollinearity?

**b.** Refer to the multiple regression results in the table given in Exercise 12.21 (p. 601). Based on the reported tests, is there any evidence of extreme multicollinearity?

**12.105** Refer to the *World Development* study of the variables impacting the size distribution of manufacturing firms in international markets, Exercise 12.10 (p. 590). Five independent variables, LGNP, AREAC, SVA, CREDIT, and STOCK were used to model the share, $y$, of firms with 100 or more workers. The

Residual plots for Exercise 12.102

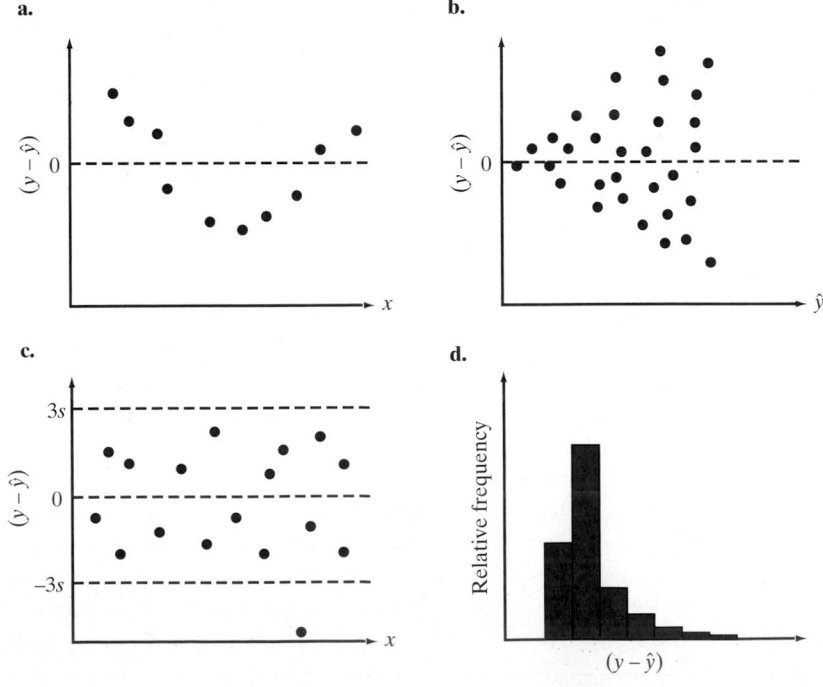

## CFSMOKE

| Weight Percentile y | No. of Cigarettes Smoked per Day x | Weight Percentile y | No. of Cigarettes Smoked per Day x |
|---|---|---|---|
| 6 | 0 | 43 | 0 |
| 6 | 15 | 49 | 0 |
| 2 | 40 | 50 | 0 |
| 8 | 23 | 49 | 22 |
| 11 | 20 | 46 | 30 |
| 17 | 7 | 54 | 0 |
| 24 | 3 | 58 | 0 |
| 25 | 0 | 62 | 0 |
| 17 | 25 | 66 | 0 |
| 25 | 20 | 66 | 23 |
| 25 | 15 | 83 | 0 |
| 31 | 23 | 87 | 44 |
| 35 | 10 | | |

*Source:* Rubin, B. K. "Exposure of children with cystic fibrosis to environmental tobacco smoke." *The New England Journal of Medicine*, Sept. 20, 1990. Vol. 323, No. 12, p. 85 (data extracted from Figure 3).

researchers detected a high correlation between pairs of the following independent variables: LGNP and SVA, LGNP and STOCK, and CREDIT and STOCK. Describe the problems that may arise if these high correlations are ignored in the multiple regression analysis of the model.

**12.106** Passive exposure to environmental tobacco smoke has been associated with growth suppression and an increased frequency of respiratory tract infections in normal children. Is this association more pronounced in children with cystic fibrosis? To answer this question, 43 children (18 girls and 25 boys) attending a 2-week summer camp for cystic fibrosis patients were studied (*The New England Journal of Medicine*, Sept. 20, 1990). Researchers investigated the correlation between a child's weight percentile ($y$) and the number of cigarettes smoked per day in the child's home ($x$). The table above lists the data for the 25 boys. A MINITAB regression printout (with residuals) for the straight-line model relating $y$ to $x$ is provided on p. 688. Examine the residuals. Do you detect any outliers?

### Applying the Concepts—Intermediate

**12.107** A physiologist wishes to investigate the relationship between the physical characteristics of preadolescent boys and their maximum oxygen uptake (measured in milliliters of oxygen per kilogram of body weight). The data shown below were collected on a random sample of ten preadolescent boys.

**a.** Fit the regression model

$$y = \beta_0 + \beta_1 x_1 + \beta_2 x_2 + \beta_3 x_3 + \beta_4 x_4 + \varepsilon$$

to the data and give the least squares prediction equation.

**b.** It seems reasonable to assume that the greater a child's weight, the greater should be the maximum oxygen uptake. Is $\hat{\beta}_3$, the estimated coefficient of weight, $x_3$, positive as expected? Give an explanation for this result.

**c.** It would seem that the chest depth of a child should be positively correlated to lung volume and hence to maximum oxygen uptake. Is $\hat{\beta}_4$ significantly different from 0, as expected? If not, explain why.

## BOYS10

| Maximum Oxygen Uptake y | Age $x_1$ (years) | Height $x_2$ (centimeters) | Weight $x_3$ (kilograms) | Chest Depth $x_4$ (centimeters) |
|---|---|---|---|---|
| 1.54 | 8.4 | 132.0 | 29.1 | 14.4 |
| 1.74 | 8.7 | 135.5 | 29.7 | 14.5 |
| 1.32 | 8.9 | 127.7 | 28.4 | 14.0 |
| 1.50 | 9.9 | 131.1 | 28.8 | 14.2 |
| 1.46 | 9.0 | 130.0 | 25.9 | 13.6 |
| 1.35 | 7.7 | 127.6 | 27.6 | 13.9 |
| 1.53 | 7.3 | 129.9 | 29.0 | 14.0 |
| 1.71 | 9.9 | 138.1 | 33.6 | 14.6 |
| 1.27 | 9.3 | 126.6 | 27.7 | 13.9 |
| 1.50 | 8.1 | 131.8 | 30.8 | 14.5 |

MINITAB Output for Exercise 12.106

```
The regression equation is
WTPCTILE = 41.2 - 0.262 SMOKED

Predictor         Coef       SE Coef           T         P
Constant        41.153         6.843        6.01     0.000
SMOKED         -0.2619         0.3702       -0.71     0.486

S = 24.68       R-Sq = 2.1%       R-Sq(adj) = 0.0%

Analysis of Variance

Source            DF          SS          MS         F         P
Regression         1       304.9       304.9      0.50     0.486
Residual Error    23     14011.1       609.2
Total             24     14316.0

Obs     SMOKED   WTPCTILE        Fit     SE Fit     Residual    St Resid
  1        0.0       6.00      41.15       6.84       -35.15       -1.48
  2       15.0       6.00      37.22       5.00       -31.22       -1.29
  3       40.0       2.00      30.68      11.22       -28.68       -1.30
  4       23.0       8.00      35.13       6.22       -27.13       -1.14
  5       20.0      11.00      35.91       5.61       -24.91       -1.04
  6        7.0      17.00      39.32       5.38       -22.32       -0.93
  7        3.0      24.00      40.37       6.13       -16.37       -0.68
  8        0.0      25.00      41.15       6.84       -16.15       -0.68
  9       25.0      17.00      34.60       6.69       -17.60       -0.74
 10       20.0      25.00      35.91       5.61       -10.91       -0.45
 11       15.0      25.00      37.22       5.00       -12.22       -0.51
 12       23.0      31.00      35.13       6.22        -4.13       -0.17
 13       10.0      35.00      38.53       5.04        -3.53       -0.15
 14        0.0      43.00      41.15       6.84         1.85        0.08
 15        0.0      49.00      41.15       6.84         7.85        0.33
 16        0.0      50.00      41.15       6.84         8.85        0.37
 17       22.0      49.00      35.39       6.00        13.61        0.57
 18       30.0      46.00      33.29       8.06        12.71        0.54
 19        0.0      54.00      41.15       6.84        12.85        0.54
 20        0.0      58.00      41.15       6.84        16.85        0.71
 21        0.0      62.00      41.15       6.84        20.85        0.88
 22        0.0      66.00      41.15       6.84        24.85        1.05
 23       23.0      66.00      35.13       6.22        30.87        1.29
 24        0.0      83.00      41.15       6.84        41.85        1.76
 25       44.0      87.00      29.63      12.56        57.37        2.70RX

R denotes an observation with a large standardized residual
X denotes an observation whose X value gives it large influence.
```

**d.** Calculate the Pearson product moment correlation coefficients between all pairs of the independent variables $x_1$–$x_4$. Do these correlations provide an explanation for the confusing signs and small $t$ values associated with the estimated regression coefficients of the model?

**12.108** Chemical engineers at Tokyo Metropolitan University analyzed urban air specimens for the presence of low-molecular-weight dicarboxylic acid (*Environmental Science & Engineering*, Oct. 1993). The dicarboxylic acid (as a percentage of total carbon) and oxidant concentrations for 19 air specimens collected from urban Tokyo are listed in the table on p. 689. SAS printouts for the straight-line model relating dicarboxylic acid percentage ($y$) to oxidant concentration ($x$) are also provided (pp. 689, 690). Conduct a complete residual analysis.

## URBANAIR

| Dicarboxylic Acid (%) | Oxidant (ppm) |
|:---:|:---:|
| .85 | 78 |
| 1.45 | 80 |
| 1.80 | 74 |
| 1.80 | 78 |
| 1.60 | 60 |
| 1.20 | 62 |
| 1.30 | 57 |
| .20 | 49 |
| .22 | 34 |
| .40 | 36 |
| .50 | 32 |
| .38 | 28 |
| .30 | 25 |
| .70 | 45 |
| .80 | 40 |
| .90 | 45 |
| 1.22 | 41 |
| 1.00 | 34 |
| 1.00 | 25 |

*Source:* Kawamura, K., and Ikushima, K. "Seasonal changes in the distribution of dicarboxylic acids in the urban atmosphere." *Environmental Science & Technology*, Vol. 27. No. 10, Oct. 1993, p. 2232 (data extracted from Figure 4).

**SAS Output for Exercise 12.108**

The REG Procedure
Dependent Variable: DICARBOX

### Analysis of Variance

| Source | DF | Sum of Squares | Mean Square | F Value | Pr > F |
|:---|:---:|:---:|:---:|:---:|:---:|
| Model | 1 | 2.41362 | 2.41362 | 17.08 | 0.0007 |
| Error | 17 | 2.40234 | 0.14131 | | |
| Corrected Total | 18 | 4.81597 | | | |

| | | | | |
|:---|:---:|:---|:---:|
| Root MSE | 0.37592 | R-Square | 0.5012 |
| Dependent Mean | 0.92737 | Adj R-Sq | 0.4718 |
| Coeff Var | 40.53600 | | |

### Parameter Estimates

| Variable | DF | Parameter Estimate | Standard Error | t Value | Pr > \|t\| |
|:---|:---:|:---:|:---:|:---:|:---:|
| Intercept | 1 | -0.02374 | 0.24577 | -0.10 | 0.9242 |
| OXIDANT | 1 | 0.01958 | 0.00474 | 4.13 | 0.0007 |

### Predictions

| Obs | DICARBOX | OXIDANT | Predicted DICARBOX | Residual of DICARBOX |
|:---:|:---:|:---:|:---:|:---:|
| 1 | 0.85 | 78 | 1.50339 | -0.65339 |
| 2 | 1.45 | 80 | 1.54255 | -0.09255 |
| 3 | 1.80 | 74 | 1.42508 | 0.37492 |
| 4 | 1.80 | 78 | 1.50339 | 0.29661 |
| 5 | 1.60 | 60 | 1.15098 | 0.44902 |
| 6 | 1.20 | 62 | 1.19013 | 0.00987 |
| 7 | 1.30 | 57 | 1.09224 | 0.20776 |
| 8 | 0.20 | 49 | 0.93561 | -0.73561 |
| 9 | 0.22 | 34 | 0.64193 | -0.42193 |
| 10 | 0.40 | 36 | 0.68109 | -0.28109 |
| 11 | 0.50 | 32 | 0.60278 | -0.10278 |
| 12 | 0.38 | 28 | 0.52446 | -0.14446 |
| 13 | 0.30 | 25 | 0.46573 | -0.16573 |
| 14 | 0.70 | 45 | 0.85730 | -0.15730 |
| 15 | 0.80 | 40 | 0.75941 | 0.04059 |
| 16 | 0.90 | 45 | 0.85730 | 0.04270 |
| 17 | 1.22 | 41 | 0.77898 | 0.44102 |
| 18 | 1.00 | 34 | 0.64193 | 0.35807 |
| 19 | 1.00 | 25 | 0.46573 | 0.53427 |

## FLAG

**12.109** Refer to the Florida Attorney General (FLAG) Office's investigation of bid-rigging in the road construction industry, Exercise 12.101 (p. 664). Recall that FLAG wants to model the price (**y**) of the contract bid by lowest bidder in hopes of preventing price-fixing in the future.

**a.** Consider the independent variables selected by the stepwise regression run in Exercise 12.101. Do you detect any multicollinearity in these variables? If so, do you recommend that all of these variables be used to predict low-bid price, $y$?

**b.** Using the variables selected in part **a**, fit a full interaction model for $E(y)$ to the data. [*Note:* Do not include squared terms in the model,

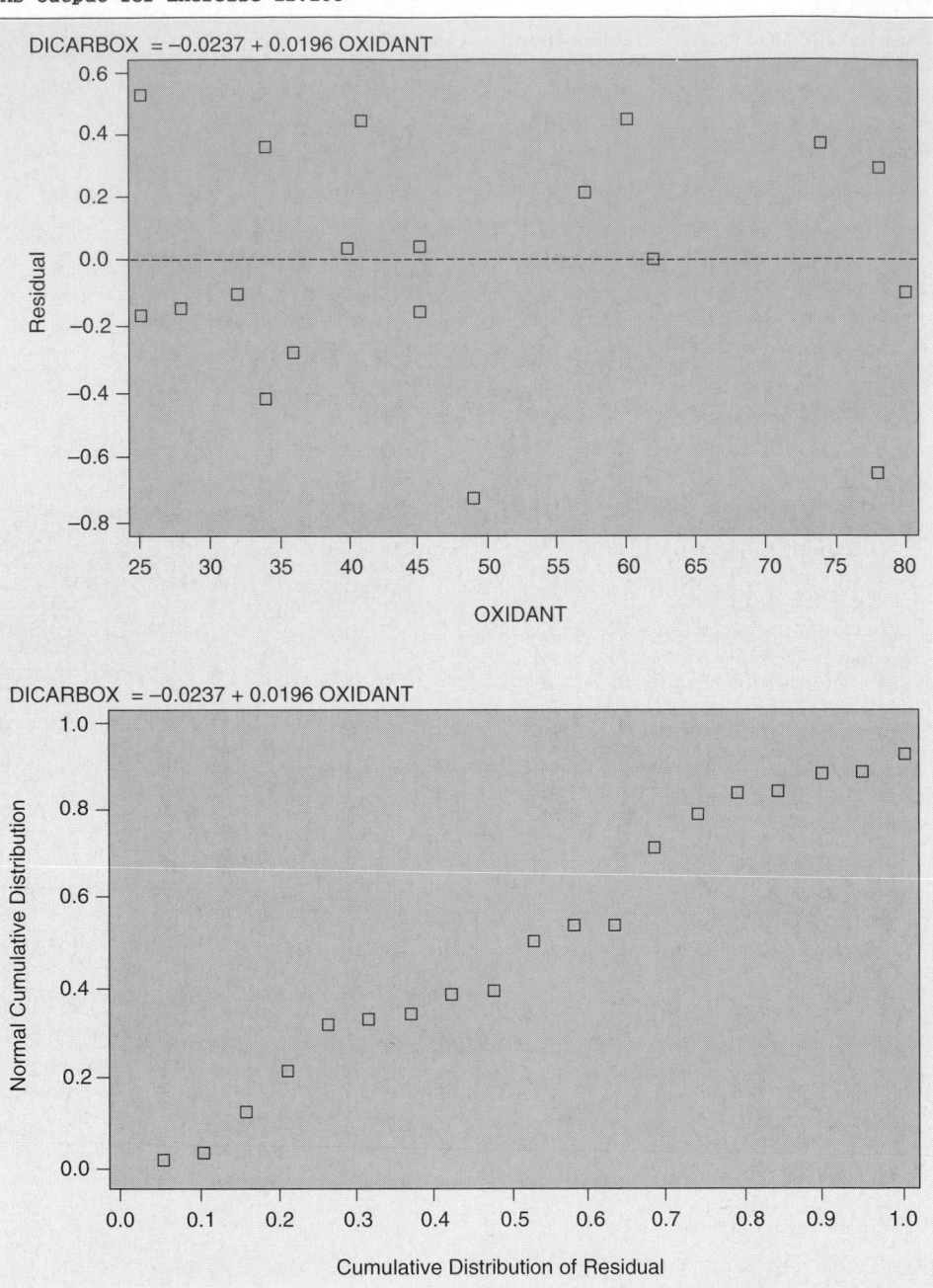

SAS Output for Exercise 12.108

but be sure to include all possible two-variable interactions.]

c. Conduct a complete residual analysis on the model fitted in part **b**. Do you detect any outliers? Are the standard regression assumptions reasonably satisfied?

**12.110** *Teaching Sociology* (July 1995) developed a model for the professional socialization of graduate students working toward a Ph.D. in sociology. One of the dependent variables modeled was professional confidence, $y$, measured on a 5-point scale. The model included over 20 independent variables and was fit to data collected for a sample of 309 sociology graduate students. One concern is whether multicollinearity exists in the data. A matrix of Pearson product moment correlations for ten of the independent variables is shown on p. 691. [*Note:* Each entry in the table is the correlation coefficient $r$ between the variable in the corresponding row and corresponding column.]

a. Examine the correlation matrix and find the independent variables that are moderately or highly correlated.

b. What modeling problems may occur if the variables, part **a**, are left in the model? Explain.

**Matrix of correlations for Exercise 12.110**

| Independent Variable | | (1) | (2) | (3) | (4) | (5) | (6) | (7) | (8) | (9) | (10) |
|---|---|---|---|---|---|---|---|---|---|---|---|
| **(1)** | Father's occupation | 1.000 | .363 | .099 | −.110 | −.047 | −.053 | −.111 | .178 | .078 | .049 |
| **(2)** | Mother's education | .363 | 1.000 | .228 | −.139 | −.216 | .084 | −.118 | .192 | .125 | .068 |
| **(3)** | Race | .099 | .228 | 1.000 | .036 | −.515 | .014 | −.120 | .112 | .117 | .337 |
| **(4)** | Sex | −.110 | −.139 | .036 | 1.000 | .165 | −.256 | .173 | −.106 | −.117 | .073 |
| **(5)** | Foreign status | −.047 | −.216 | −.515 | .165 | 1.000 | −.041 | .159 | −.130 | −.165 | −.171 |
| **(6)** | Undergraduate GPA | −.053 | .084 | .014 | −.256 | −.041 | 1.000 | .032 | .028 | −.034 | .092 |
| **(7)** | Year GRE taken | −.111 | −.118 | −.120 | .173 | .159 | .032 | 1.000 | −.086 | −.602 | .016 |
| **(8)** | Verbal GRE score | .178 | .192 | .112 | −.106 | −.130 | .028 | −.086 | 1.000 | .132 | .087 |
| **(9)** | Years in graduate program | .078 | .125 | .117 | −.117 | −.165 | −.034 | −.602 | .132 | 1.000 | −.071 |
| **(10)** | First-year graduate GPA | .049 | .068 | .337 | .073 | −.171 | .092 | .016 | .087 | −.071 | 1.000 |

*Source:* Keith, B., and Moore, H. A. "Training sociologists: An assessment of professional socialization and the emergence of career aspirations." *Teaching Sociology*, Vol. 23, No. 3, July 1995, p. 205 (Table 1).

---

## STATISTICS IN ACTION 12.1
### *"Wringing" the Bell Curve*

In Chapter 5's Statistics in Action (p. 255), we introduced *The Bell Curve* (Free Press, 1994) by Richard Herrnstein and Charles Murray, a controversial book about race, genes, IQ, and economic mobility. The book heavily employs statistics and statistical methodology in an attempt to support the authors' positions on the relationships among these variables and their social consequences. The main theme of *The Bell Curve* can be summarized as follows:

1. Measured intelligence (IQ) is largely genetically inherited.
2. IQ is correlated positively with a variety of socioeconomic status success measures, such as prestigious job, high annual income, and high educational attainment.
3. From 1 and 2, it follows that socioeconomic successes are largely genetically caused and therefore resistant to educational and environmental interventions (such as affirmative action).

With the help of a major marketing campaign, the book became a best-seller shortly after its publication in October 1994. The underlying theme of the book—that intelligence is hereditary and tied to race and class—apparently appealed to many readers. However, reviews of *The Bell Curve* in popular magazines and newspapers were mostly negative. Social critics have described the authors as "un-American" and "pseudo-scientific racists," and their book as "alien and repellant." (On the other hand, there were defenders who labeled the book as "powerfully written" and "overwhelmingly convincing.") This Statistics in Action is based on two reviews of *The Bell Curve* that critique the statistical methodology employed by the authors and the inferences derived

from the statistics. Both reviews, one published in *Chance* (Summer 1995) and the other in *The Journal of the American Statistical Association* (Dec. 1995), were written by Carnegie Mellon University professors Bernie Devlin, Stephen Fienberg, Daniel Resnick, and Kathryn Roeder. (Devlin, Fienberg, and Roeder are all statisticians; Resnick, a historian.)

Here, our focus is on the statistical method used repeatedly by Herrnstein and Murray (H&M) to support their conclusions in *The Bell Curve*: regression analysis. The following are just a few of the problems with H&M's use of regression that are identified by the Carnegie Mellon professors:

**Problem 1** H&M consistently use a trio of independent variables—IQ, socioeconomic status, and age—in a series of first-order models designed to predict dependent social outcome variables such as income and unemployment. (Only on a single occasion are interaction terms incorporated.) Consider, for example, the model:

$$E(y) = \beta_0 + \beta_1 x_1 + \beta_2 x_2 + \beta_3 x_3,$$

where $y$ = income, $x_1$ = IQ, $x_2$ = socioeconomic status, and $x_3$ = age. H&M utilize $t$-tests on the individual $\beta$ parameters to assess the importance of the independent variables. As with most of the models considered in *The Bell Curve*, the estimate of $\beta_1$ in the income model is positive and statistically significant at $\alpha = .05$, and the associated $t$ value is larger (in absolute value) than the $t$ values associated with the other independent variables. Consequently, *H&M claim that IQ is a better predictor of income than the other two independent variables*. No attempt was made to determine whether the model was properly specified or whether the model provides an adequate fit to the data.

## STATISTICS IN ACTION 12.1 *(continued)*

**Problem 2**  In an appendix, the authors describe multiple regression as a "mathematical procedure that yields coefficients for each of [the independent variables], indicating how much of a change in [the dependent variable] can be anticipated for a given change in any particular [independent] variable, with all the others held constant." Armed with this information and the fact that the estimate of $\beta_1$ in the model above is positive, *H&M infer that a high IQ necessarily implies (or causes) a high income, and a low IQ inevitably leads to a low income.* (Cause-and-effect inferences like this are made repeatedly throughout the book.)

**Problem 3**  The title of the book refers to the normal distribution and its well-known "bell-shaped" curve. There is a misconception among the general public that scores on intelligence tests (IQ) are normally distributed. In fact, most IQ scores have distributions that are decidedly skewed. Traditionally, psychologists and psychometricians have transformed these scores so that the resulting numbers have a precise normal distribution. H&M make a special point to do this. Consequently, *the measure of IQ used in all the regression models is normalized (i.e., transformed so that the*

*resulting distribution is normal), despite the fact that regression methodology does not require predictor (independent) variables to be normally distributed.*

**Problem 4**  A variable that is not used as a predictor of social outcome in any of the models in *The Bell Curve* is level of education. H&M purposely omit education from the models, arguing that IQ causes education, not the other way around. Other researchers who have examined H&M's data report that *when education is included as an independent variable in the model, the effect of IQ on the dependent variable (say, income) is diminished.*

### Focus

a. Comment on each of the problems identified by the Carnegie Mellon University professors in their review of *The Bell Curve*. Why do each of these problems cast a shadow on the inferences made by the authors?

b. Using the variables specified in the model above, describe how you would conduct the multiple regression analysis. (Propose a more complex model and describe the appropriate model tests, including a residual analysis.)

## STATISTICS IN ACTION 12.2
### Building a Model for Condo Sale Prices

This application involves an investigation of the factors that affect the sale price of oceanside condominium units. It represents an extension of an analysis of the same data by Herman Kelting (1979). Although condo sale prices have increased dramatically over the past 20 years, the relationship between these factors and sale price remain about the same. Consequently, the data provide valuable insight into today's condominium sales market.

The sales data were obtained for a newly built oceanside condominium complex consisting of two adjacent and connected eight-floor buildings. The complex contains 200 units of equal size (approximately 500 square feet each). The locations of the buildings relative to the ocean, the swimming pool, the parking lot, etc., are shown in Figure 12.53.

There are several features of the complex that you should note:

1. The units facing south, called *oceanview*, face the beach and ocean. In addition, units in building 1 have a good view of the pool. Units to the rear of the building, called *bayview*, face the parking lot and an area of land that, ultimately, borders a bay. The view from the upper floors of these units is primarily of wooded, sandy terrain. The bay is very distant and barely visible.

2. The only elevator in the complex is located at the east end of building 1, next to the office and the

game room. People moving to or from the higher floor units in building 2 would probably use the elevator, then move through the passages to their units. Thus, units on the higher floors and at a greater distance from the elevator would be less convenient for their occupants; they would expend greater effort in moving baggage, groceries, etc., and would be farther away from the game room, the office, and the swimming pool. These units also possess an advantage: Because traffic through the hallways in the area would be minimal, these units would be the most private.

3. Lower-floor oceanside units are most suited to active people; they open onto the beach, ocean, and pool. They are within easy reach of the game room and they are easily reached from the parking area.

4. Checking Figure 12.53, you will see that the views in some of the units at the center of the complex, units ending in numbers 11 and 14, are partially blocked.

5. The condominium complex was completed at the time of the 1975 recession; sales were slow and the developer was forced to sell most of the units at auction approximately 18 months after opening. Many unsold units were furnished by the developer and rented prior to the auction.

*(continues on next page)*

**STATISTICS IN ACTION 12.2** *(continued)*

This condominium complex is particularly suited to our study. Because the single elevator is located at one end of the complex, it is the source of a remarkably high level of both inconvenience and privacy for the people occupying units on the top floors in building 2. Consequently, the data provide a good opportunity to investigate the relationship that might exist between sale price, height of the unit (floor number), distance of the unit from the elevator, and presence or absence of an ocean view. In addition, the presence or absence of furniture in each of the units permits an investigation of the effect of the availability of furniture on sale price. Finally, the auction data are completely buyer-specified and hence consumer-oriented, in contrast to most other real estate sales data that are, to a high degree, seller-oriented and broker-specified.

In addition to the sale price, the following data were recorded for each of the 106 units sold at the auction:

🔵 CONDO

1. *Floor height* ($x_1$): The floor location of the unit; the variable levels are 1, 2, ..., 8.
2. *Distance from elevator* ($x_2$): This distance, measured along the length of the complex, is expressed in number of condominium units. An additional two units of distance was added to the units in building 2 to account for the walking distance in the connecting area between the two buildings. Thus, the distance of unit 105 from the elevator would be 3, and the distance between unit 113 and the elevator would be 9. The variable levels are 1, 2, ..., 15.
3. *View of ocean* ($x_3$): The presence or absence of an ocean view is recorded for each unit and specified with a dummy variable, where $x_3 = 1$ if the unit offered an ocean view and $x_3 = 0$ if not. Note that units not offering an ocean view would face the parking lot.
4. *End unit* ($x_4$): We expect the partial reduction of view of end units on the ocean side (numbers ending in 11) to reduce their sale price. The ocean view of these end units is partially blocked by building 2. This qualitative variable is specified with a dummy variable, where $x_4 = 1$ if the unit has a unit number ending in 11 and $x_4 = 0$ if not.
5. *Furniture* ($x_5$): The presence or absence of furniture is recorded for each unit. This qualitative variable is represented with a single dummy variable, where $x_5 = 1$ if the unit is furnished and $x_5 = 0$ if not.

These data are stored in the CONDO file.

**FIGURE 12.53**
*Layout of Condominium Complex*

## STATISTICS IN ACTION 12.2 *(continued)*

Three models relating mean sale price to these five factors are postulated. The models, numbered 1–3, are given below:

**Model 1**

$$E(y) = \beta_0 + \beta_1 x_1 + \beta_2 x_2 + \beta_3 x_3 + \beta_4 x_4 + \beta_5 x_5$$

**Model 2**

Second-order model in $x_1$ and $x_2$

$$E(y) = \beta_0 + \beta_1 x_1 + \beta_2 x_2 + \beta_3 x_1 x_2 + \beta_4 x_1^2 + \beta_5 x_2^2$$

View of ocean    End unit    Furniture

$$+ \quad \beta_6 x_3 \quad + \quad \beta_7 x_4 \quad + \quad \beta_8 x_5$$

**Model 3**

Second-order model in $x_1$ and $x_2$

$$E(y) = \beta_0 + \beta_1 x_1 + \beta_2 x_2 + \beta_3 x_1 x_2 + \beta_4 x_1^2 + \beta_5 x_2^2$$

View of ocean    End unit    Furniture

$$+ \quad \beta_6 x_3 \quad + \quad \beta_7 x_4 \quad + \quad \beta_8 x_5$$

Interaction of the second-order model with view of ocean

$$+ \beta_9 x_1 x_3 + \beta_{10} x_2 x_3 + \beta_{11} x_1 x_2 x_3$$
$$+ \beta_{12} x_1^2 x_3 + \beta_{13} x_2^2 x_3$$

### Focus

a. Verbally describe the theoretical relationships hypothesized by each of the three models.

b. For each model, sketch the theoretical relationship between sale price $y$ and floor height $x_1$ for each level of view of ocean.

c. Fit Model 1 to the data and interpret the $p$-value for the variable floor height $x_1$. Why is it dangerous to conclude, at this stage, that floor height is *not* an important predictor of sale price?

d. Fit Models 2 and 3, then conduct nested $F$-tests to determine which of the three models is "best" for predicting sale price.

e. Compare the value of $s$ for the preferred model to the $s$ values for the other models. Do the results support your choice of models?

f. Construct a plot of the residuals for the preferred model. Do you detect any outliers? If these outliers are not found to have strong influence on the analysis, would you recommend they be deleted?

g. Assume the outliers identified in part f were not found to be very influential. Refit the model with these outliers deleted. Interpret the new model's standard deviation, $s$.

h. Conduct a residual analysis on the refitted model. Interpret the results.

i. Write the prediction equation for the preferred model, with outliers deleted.

j. Refer to the prediction equation, part i. Suppose we wish to examine the relationship between $\hat{y}$, floor height $(x_1)$, and distance from elevator $(x_2)$ for bayview $(x_3 = 0)$, non-end units $(x_4 = 0)$ with no furniture $(x_5 = 0)$. Substitute $x_3 = x_4 = x_5 = 0$ into the equation to obtain the predicted relationship. Now obtain the prediction curve relating $\hat{y}$ to distance from elevator $x_2$ for each floor by first substituting $x_1 = 1$, then $x_1 = 2, \ldots, x_1 = 8$ into the prediction equation.

k. Refer to part j. Graph the curves for the bayview $(x_3 = 0)$ units. There should be eight curves, one for each level of floor height, on the graph. Describe the relationships predicted by the model for the bayview units.

l. Repeat part k for the oceanview $(x_3 = 1)$ units.

m. Use the patterns depicted in the graphs, parts k and l, to explain why the $t$-test for floor height $x_1$ is nonsignificant in Model 1.

---

## QUICK REVIEW

## Key Terms

## Key Formulas

| | |
|---|---|
| $E(y) = \beta_0 + \beta_1 x_1 + \beta_2 x_2$ | First-order model with two quantitative independent variables   577 |
| $s^2 = \text{MSE} = \dfrac{\text{SSE}}{n - (k + 1)}$ | Estimator of $\sigma^2$ for a model with $k$ independent variables   585 |
| $t = \dfrac{\hat{\beta}_i}{s_{\hat{\beta}_i}}$ | Test statistic for testing $H_0: \beta_i = 0$   585 |
| $\hat{\beta}_i \pm (t_{\alpha/2}) s_{\hat{\beta}_i}$, where $t$ depends on $n - (k + 1)$ df | $100(1 - \alpha)\%$ confidence interval for $\beta_i = 0$   585 |
| $R^2 = \dfrac{\text{SS}_{yy} - \text{SSE}}{\text{SS}_{yy}}$ | Multiple coefficient of determination   596 |
| $R_a^2 = 1 - \left[ \dfrac{(n - 1)}{n - (k + 1)} \right](1 - R^2)$ | Adjusted multiple coefficient of determination   597 |
| $y - \hat{y}$ | Regression residual   665 |
| $F = \dfrac{\text{MS(Model)}}{\text{MSE}} = \dfrac{R^2/k}{(1 - R^2)/[n - (k + 1)]}$ | Test statistic for testing $H_0: \beta_i = \beta_2 = \cdots = \beta_k = 0$   599 |
| $E(y) = \beta_0 + \beta_1 x_1 + \beta_2 x_2 + \beta_3 x_1 x_2$ | Interaction model with two quantitative independent variables   612 |
| $E(y) = \beta_0 + \beta_1 x + \beta_2 x^2$ | Quadratic model   618 |
| $E(y) = \beta_0 + \beta_1 x_1 + \beta_2 x_2 + \beta_3 x_1 x_2 + \beta_4 x_1^2 + \beta_5 x_2^2$ | Complete second-order model with two quantitative independent variables   623 |
| $E(y) = \beta_0 + \beta_1 x_1 + \beta_2 x_2 + \cdots + \beta_{k-1} x_{k-1}$, where $x_i = \begin{cases} 1 & \text{if level } i + 1 \\ 0 & \text{if not} \end{cases}$ | Model with one qualitative variable at $k$ levels   632 |
| $F = \dfrac{(\text{SSE}_R - \text{SSE}_C)/\text{number of } \beta\text{'s tested}}{\text{MSE}_C}$ | Test statistic for comparing reduced and complete models   648 |

## Language Lab

| Symbol | Pronunciation | Description |
|---|---|---|
| $x_1^2$ | x-1 squared | Quadratic term that allows for curvature in the relationship between $y$ and $x$ |
| $x_1 x_2$ | x-1 x-2 | Interaction term |
| MSE | M-S-E | Mean square for error (estimates $\sigma^2$) |
| $\beta_i$ | beta-$i$ | Coefficient of $x_i$ in the model |
| $\hat{\beta}_i$ | beta-$i$-hat | Least squares estimate of $\beta_i$ |
| $s_{\hat{\beta}_i}$ | $s$ of beta-$i$-hat | Estimated standard error of $\hat{\beta}_i$ |
| $R^2$ | R-squared | Multiple coefficient of determination |
| $R_a^2$ | R-squared adjusted | Adjusted multiple coefficient of determination |
| $F$ | | Test statistic for testing global usefulness of model |
| $\hat{\varepsilon}$ | epsilon-hat | Estimated random error, or residual |
| $\text{SSE}_R$ | | Sum of squared errors for reduced model |
| $\text{SSE}_C$ | | Sum of squared errors for complete model |
| $\text{MSE}_C$ | | Mean square error for complete model |
| $\ln(y)$ | Natural log of $y$ | Natural logarithm of dependent variable |

## SUPPLEMENTARY EXERCISES 12.111–12.140

### Learning the Mechanics

**12.111** When a multiple regression model is used for estimating the mean of the dependent variable and for predicting a new value of $y$, which will be narrower— the confidence interval for the mean or the prediction interval for the new $y$ value? Why?

**12.112** Suppose you fit the model

$$y = \beta_0 + \beta_1 x_1 + \beta_2 x_1^2 + \beta_3 x_2 + \beta_4 x_1 x_2 + \varepsilon$$

to $n = 25$ data points and find that

$$\hat{\beta}_0 = 1.26 \quad \hat{\beta}_1 = -2.43 \quad \hat{\beta}_2 = .05 \quad \hat{\beta}_3 = .62 \quad \hat{\beta}_4 = 1.81$$

$$s_{\hat{\beta}_1} = 1.21 \quad s_{\hat{\beta}_2} = .16 \quad s_{\hat{\beta}_3} = .26 \quad s_{\hat{\beta}_4} = 1.49$$

$$\text{SSE} = .41 \quad R^2 = .83$$

**a.** Is there sufficient evidence to conclude that at least one of the parameters $\beta_1, \beta_2, \beta_3,$ or $\beta_4$ is nonzero? Test using $\alpha = .05$.
**b.** Test $H_0: \beta_1 = 0$ against $H_a: \beta_1 < 0$. Use $\alpha = .05$.
**c.** Test $H_0: \beta_2 = 0$ against $H_a: \beta_2 > 0$. Use $\alpha = .05$.
**d.** Test $H_0: \beta_3 = 0$ against $H_a: \beta_3 \neq 0$. Use $\alpha = .05$.

**12.113** Suppose you used MINITAB to fit the model

$$y = \beta_0 + \beta_1 x_1 + \beta_2 x_2 + \varepsilon$$

to $n = 15$ data points and you obtained the printout shown below.

**a.** What is the least squares prediction equation?
**b.** Find $R^2$ and interpret its value.
**c.** Is there sufficient evidence to indicate that the model is useful for predicting $y$? Conduct an $F$-test using $\alpha = .05$.
**d.** Test the null hypothesis $H_0: \beta_1 = 0$ against the alternative hypothesis $H_a: \beta_1 \neq 0$. Test using $\alpha = .05$. Draw the appropriate conclusions.

**e.** Find the standard deviation of the regression model and interpret it.

**12.114** Suppose you have developed a regression model to explain the relationship between $y$ and $x_1, x_2,$ and $x_3$. The ranges of the variables you observed were as follows: $10 \le y \le 100, 5 \le x_1 \le 55, .5 \le x_2 \le 1,$ and $1,000 \le x_3 \le 2,000$. Will the error of prediction be smaller when you use the least squares equation to predict $y$ when $x_1 = 30, x_2 = .6,$ and $x_3 = 1,300,$ or when $x_1 = 60, x_2 = .4,$ and $x_3 = 900$? Why?

**12.115** Why is model building the key to the success or failure of a regression analysis?

**12.116** The first-order model $E(y) = \beta_0 + \beta_1 x_1$ was fit to $n = 19$ data points. A residual plot for the model is shown here. Is the need for a quadratic term in the model evident from the residual plot? Explain.

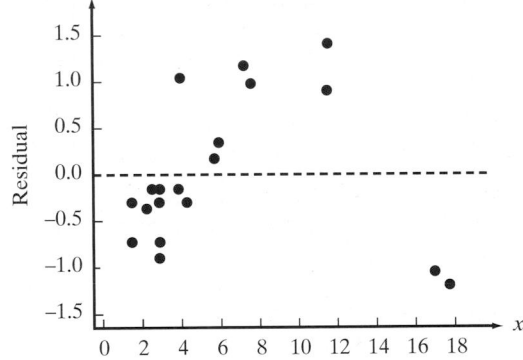

**12.117** Write a model relating $E(y)$ to one qualitative independent variable that is at four levels. Define all the terms in your model.

**MINITAB Output for Exercise 12.113**

```
The regression equation is
Y = 90.1 - 1.84 X1 + .285 X2


Predictor        Coef      SE Coef         T         P
Constant        90.10        23.10      3.90     0.002
X1              -1.836       0.367     -5.01     0.001
X2               0.285       0.231      1.24     0.465


S = 10.68      R-Sq = 91.6%      R-Sq(adj) = 90.2%


Analysis of Variance

Source           DF         SS        MS       F        P
Regression        2      14801      7400   64.91    0.001
Residual Error   12       1364       114
Total            14      16165
```

**12.118** Suppose you fit the regression model

$$E(y) = \beta_0 + \beta_1 x_1 + \beta_2 x_2 + \beta_3 x_2^2 + \beta_4 x_1 x_2 + \beta_5 x_1 x_2^2$$

to $n - 35$ data points and wish to test the null hypothesis $H_0: \beta_4 = \beta_5 = 0$.

**a.** State the alternative hypothesis.

**b.** Explain in detail how to compute the $F$ statistic needed to test the null hypothesis.

**c.** What are the numerator and denominator degrees of freedom associated with the $F$ statistic in part **b**?

**d.** Give the rejection region for the test if $\alpha = .05$.

**12.119** Explain why stepwise regression is used. What is its value in the model-building process?

**12.120** It is desired to relate $E(y)$ to a quantitative variable $x_1$ and a qualitative variable at three levels.

**a.** Write a first-order model.

**b.** Write a model that will graph as three different second-order curves—one for each level of the qualitative variable.

**12.121** To model the relationship between $y$, a dependent variable, and $x$, an independent variable, a researcher has taken one measurement on $y$ at each of three different $x$ values. Drawing on his mathematical expertise, the researcher realizes that he can fit the second-order model

$$E(y) = \beta_0 + \beta_1 x + \beta_2 x^2$$

and it will pass exactly through all three points, yielding SSE = 0. The researcher, delighted with the "excellent" fit of the model, eagerly sets out to use it to make inferences. What problems will he encounter in attempting to make inferences?

**12.122 a.** Write a first-order model relating $E(y)$ to two quantitative independent variables, $x_1$ and $x_2$.

**b.** Write a complete second-order model.

## Applying the Concepts—Basic

**12.123** A disabled person's acceptance of a disability is critical to the rehabilitation process. The *Journal of Rehabilitation* (Sept. 1989) published a study that investigated the relationship between assertive behavior level and acceptance of disability in 160 disabled adults. The dependent variable, assertiveness ($y$), was measured using the Adult Self Expression Scale (ASES). Scores on the ASES range from 0 (no assertiveness) to 192 (extreme assertiveness). The model analyzed was $E(y) = \beta_0 + \beta_1 x_1 + \beta_2 x_2 + \beta_3 x_3$, where

$x_1 =$ Acceptance of disability (AD) score
$x_2 =$ Age (years)
$x_3 =$ Length of disability (years)

The regression results are shown in the table.

| Independent Variable | $t$ | Two-Tailed $p$-Value |
|---|---|---|
| AD score ($x_1$) | 5.96 | .0001 |
| Age ($x_2$) | 0.01 | .9620 |
| Length ($x_3$) | 1.91 | .0576 |

**a.** Is there sufficient evidence to indicate that AD score is positively linearly related to assertiveness level, once age and length of disability are accounted for? Test using $\alpha = .05$.

**b.** Test the hypothesis $H_0: \beta_2 = 0$ against $H_a: \beta_2 \neq 0$. Use $\alpha = .05$. Give the conclusion in the words of the problem.

**c.** Test the hypothesis $H_0: \beta_3 = 0$ against $H_a: \beta_3 > 0$. Use $\alpha = .05$. Give the conclusion in the words of the problem.

**12.124** An Educational Testing Service (ETS) research scientist used multiple regression analysis to model $y$, the final grade point average (GPA) of business and management doctoral students (*Journal of Educational Statistics*, Spring 1993). A list of the potential independent variables measured for each doctoral student in the study is given below.

**1.** Quantitative Graduate Management Aptitude Test (GMAT) score

**2.** Verbal GMAT score

**3.** Undergraduate GPA

**4.** First-year graduate GPA

**5.** Student cohort (i.e., year in which student entered doctoral program: 1988, 1990, or 1992)

**a.** Identify the variables as quantitative or qualitative.

**b.** For each quantitative variable, give your opinion on whether the variable is positively or negatively related to final GPA.

**c.** For each of the qualitative variables, set up the appropriate dummy variable.

**d.** Write a first-order, main-effects model relating final GPA, $y$, to the five independent variables.

**e.** Interpret the $\beta$'s in the model, part **d**.

**f.** Write a first-order model for final GPA, $y$, that allows for a different slope for each student cohort.

**g.** For each quantitative independent variable in the model, part **f**, give the slope of the line (in terms of the $\beta$'s) for the 1988 cohort.

**12.125** The *Journal of Consulting and Clinical Psychology* (June 1995) reported on a study of emergency service (EMS) rescue workers who responded to the I-880 freeway collapse during the 1989 San Francisco earthquake. The goal of the study was to identify the predictors of symptomatic distress in the EMS workers. One of the distress variables studied was the Global Symptom Index (GSI). Several models for GSI, $y$, were considered based on the following independent variables:

$x_1 =$ Critical Incident Exposure scale (CIE)
$x_2 =$ Hogan Personality Inventory-Adjustment scale (HPI-A)
$x_3 =$ Years of experience (EXP)
$x_4 =$ Locus of Control scale (LOC)
$x_5 =$ Social Support scale (SS)
$x_6 =$ Dissociative Experiences scale (DES)
$x_7 =$ Peritraumatic Dissociation Experiences Questionnaire, self-report (PDEQ-SR)

**a.** Write a first-order model for $E(y)$ as a function of the first five independent variables, $x_1$–$x_5$.

**b.** The model of part **a**, fitted to data collected for $n = 147$ EMS workers, yielded the following results: $R^2 = .469$, $F = 34.47$, $p$-value $< .001$. Interpret these results.

**c.** Write a first-order model for $E(y)$ as a function of all seven independent variables, $x_1$–$x_7$.

**d.** The model, part **c**, yielded $R^2 = .603$. Interpret this result.

**e.** The $t$-tests for testing the DES and PDEQ-SR variables both yielded a $p$-value of .001. Interpret these results.

**12.126** The *Journal of Applied Phycology* (Dec. 1994) published research on the seasonal growth activity of algae in an indoor environment. The daily growth rate $y$ was regressed against temperature $x$ using the quadratic model $E(y) = \beta_0 + \beta_1 x + \beta_2 x^2$. A particular algal strain was grown in a glass dish indoors at temperatures ranging from 10° to 32° Celsius. The data for $n = 33$ such experiments were used to fit the quadratic model, with the following results:

$$\hat{y} = -2.51 + .55x - .01x^2, R^2 = .67$$

**a.** Sketch the least squares prediction equation. Interpret the graph.

**b.** Interpret the value of $R^2$.

**c.** Is the model useful for predicting algae growth rate, $y$? Test using $\alpha = .05$.

**12.127** A medical researcher wants to model the extent of lung damage in emphysema patients as a function of two variables: the number of years the patient has smoked ($x_1$) and the sex of the patient ($x_2$). Fifty emphysema patients are used in the study, and the response $y$ for each patient is a subjective score ranging from 0 to 50. (High scores indicate extensive lung damage.)

**a.** Identify the independent variables as qualitative or quantitative.

**b.** Write the main effects (first-order) model for predicting the mean lung damage score, $E(y)$, from the two variables identified in part **a**.

**c.** Using a female patient as the base level, the researcher obtained the prediction equation

$$\hat{y} = -.85 + .65x_1 + .09x_2$$

What is the estimated difference between the mean lung damage scores for a male and female who have both smoked for 20 years?

**d.** Find the estimated mean lung damage score for a female patient who has smoked for 15 years.

**e.** It would not be surprising if the functional relation between lung damage and years of smoking were second-order. Further, it is conceivable that years of smoking might affect men differently than women. Write a model that allows for these conditions.

**12.128** The amount of fructose in an athlete's bloodstream is critical to performance. A researcher ran an experiment to determine whether diet had any effect on the level of fructose in the blood after 10 minutes of running on a treadmill. Twenty-one subjects were selected, and each subject's fructose level was measured after 10 minutes on a treadmill. Each subject was randomly assigned to one of three diets. One diet was high in protein, one high in carbohydrates, and one high in fruits. After one month on the diet, each subject was again asked to run on the treadmill for 10 minutes, and the fructose level in the bloodstream was measured. The dependent variable was the difference in the level of fructose in the blood between the second and the first runs on the treadmill.

**a.** Identify the independent variables in the experiment.

**b.** Write an appropriate regression model relating mean difference in fructose in the blood, $E(y)$, to the independent variables. Identify and code all dummy variables.

**12.129** Location is one of the most important decisions for hotel chains and lodging firms. Researchers S. E. Kimes (Cornell University) and J. A. Fitzsimmons (University of Texas) studied the site selection process of La Quinta Motor Inns, a moderately priced hotel chain (*Interfaces*, Mar.–Apr. 1990). Using data collected on 57 mature inns owned by La Quinta, the researchers built a regression model designed to predict the profitability for sites under construction. The least squares model is:

$$\hat{y} = 39.05 - 5.41x_1 + 5.86x_2 + 3.09x_3 + 1.75x_4$$

where

$y$ = operating margin (measured as a percentage)

$x_1$ = state population (in thousands) divided by the total number of inns in the state

$x_2$ = room rate ($) for the inn

$x_3$ = square root of the median income of the area (in $ thousands)

$x_4$ = number of college students within four miles of the inn

All variables were "standardized" to have a mean of 0 and a standard deviation of 1.

**a.** Interpret the $\beta$ estimates of the model. Comment on the effect of each independent variable on operating margin, $y$. [*Note:* A profitable inn is defined as one with an operating margin of over 50%.]

**b.** The model yielded $R^2 = .51$. Give a descriptive measure of model adequacy.

**c.** Make an inference about model adequacy by conducting the appropriate test. Use $\alpha = .05$.

### Applying the Concepts—Intermediate

**12.130** Since the Great Depression of the 1930s, the link between the suicide rate and the state of the economy has been the subject of much research. Research exploring this link using regression analysis was reported in an article in the *Journal of Socio-Economics* (Spring, 1992). The researchers collected data from a 45-year period on the following variables:

$y$ = Suicide rate
$x_1$ = Unemployment rate
$x_2$ = Percentage of females in the labor force
$x_3$ = Divorce rate
$x_4$ = Logarithm of Gross National Product (GNP)
$x_5$ = Annual percent change in GNP

One of the models explored by the researchers was a multiple regression model relating $y$ to linear terms in $x_1$ through $x_5$. The least squares model shown below resulted (the observed significance levels of the $\beta$ estimates are shown in parentheses beneath the estimates):

$$\hat{y} = .002 + .0204x_1 - .0231x_2 + .0765x_3 + .2760x_4 + .0018x_5$$
$$(.002) \quad (.02) \quad (> .10) \quad (> .10) \quad (> .10)$$
$$R^2 = .45$$

**a.** Interpret the value of $R^2$. Is there sufficient evidence to indicate that the model is useful for predicting the suicide rate? Use $\alpha = .05$.

**b.** Interpret each of the coefficients in the model, and each of the corresponding significance levels.

**c.** Is there sufficient evidence to indicate that the unemployment rate is a useful predictor of the suicide rate? Use $\alpha = .05$.

**d.** Discuss each of the following terms with respect to potential problems with the above model: curvature (second-order terms), interaction, and multicollinearity.

**12.131** Perfectionists are persons who set themselves standards and goals that cannot be reasonably met or accomplished. One theory suggests that those individuals who are depressed have a tendency toward perfectionism. To study this phenomenon, 76 members of an introductory psychology class completed four questionnaires: (1) the ASO scale, designed to measure self-acceptance, (2) the Burns scale, designed to measure perfectionism, (3) the Zung scale, designed to measure depression, and (4) the Rotter scale, designed to measure perceptions between actions and reinforcement (*The Journal of Adlerian Theory, Research, and Practice*, Mar. 1986).

**a.** Write a first-order model relating depression (Zung scale) to self-acceptance (ASO scale), perfectionism (Burns scale), and reinforcement (Rotter scale).

**b.** The model, part **a**, was fitted to the $n = 76$ points and resulted in a coefficient of determination of $R^2 = .70$. Interpret this value.

**c.** Is there sufficient evidence to indicate that the model is useful for predicting depression (Zung scale) score? Test using $\alpha = .05$.

**d.** A $t$-test for the perfectionism (Burns scale) variable resulted in a (two-tailed) $p$-value of .87. Interpret this value.

**12.132** To meet the increasing demand for new software products, many systems development experts have adopted a prototyping methodology. The effects of prototyping on the system development life cycle (SDLC) was investigated in the *Journal of Computer Information Systems* (Spring 1993). A survey of 500 randomly selected corporate level MIS managers was conducted. Three potential independent variables were: (1) *importance* of prototyping to each phase of the SDLC; (2) degree of *support* prototyping provides for the SDLC; and (3) degree to which prototyping *replaces* each phase of the SDLC. The table gives the pairwise correlations of the three variables in the survey data for one particular phase of the SDLC. Use this information to assess the degree of multicollinearity in the survey data. Would you recommend using all three independent variables in a regression analysis? Explain.

| Variable Pairs | Correlation Coefficient, $r$ |
|---|---|
| Importance–Replace | .2682 |
| Importance–Support | .6991 |
| Replace–Support | −.0531 |

*Source:* Hardgrave, B. C., Doke, E. R., and Swanson, N. E. "Prototyping effects of the system development life cycle: An empirical study." *Journal of Computer Information Systems*, Vol. 33, No. 3, Spring 1993, p. 16 (Table 1).

**12.133** Newspaper cartoons, although designed to be funny, often invoke hostility, pain, and/or aggression in readers, especially when those cartoons depict violence. A study was undertaken to determine how violence in cartoons is related to aggression or pain (*Motivation and Emotion*, Vol. 10, 1986). A group of volunteers (psychology students) rated each of 32 violent newspaper cartoons (16 "Herman" and 16 "Far Side" cartoons) on three dimensions:

$y$ = Funniness (0 = not funny, . . . , 9 = very funny)
$x_1$ = Pain (0 = none, . . . , 9 = a very great deal)
$x_2$ = Aggression/hostility (0 = none, . . . , 9 = a very great deal)

The ratings of the students on each dimension were averaged and the resulting $n = 32$ observations were subjected to a multiple regression analysis. Based on the underlying theory (called the *inverted-U theory*) that the funniness of a joke will increase at low levels of aggression or pain, level off, and then decrease at high levels of aggression or pain, the following quadratic models were proposed:

Model 1:
$$E(y) = \beta_0 + \beta_1 x_1 + \beta_2 x_1^2, R^2 = .099, F = 1.60$$

Model 2:
$$E(y) = \beta_0 + \beta_1 x_2 + \beta_2 x_2^2, R^2 = .100, F = 1.61$$

**a.** According to the theory, what is the expected sign of $\beta_2$ in either model?

**b.** Is there sufficient evidence to indicate that the quadratic model relating pain to funniness rating is useful? Test at $\alpha = .05$.

**c.** Is there sufficient evidence to indicate that the quadratic model relating aggression/hostility to funniness rating is useful? Test at $\alpha = .05$.

**12.134** An operations manager is interested in modeling $E(y)$, the expected length of time per month (in hours) that a machine will be shut down for repairs, as a function of the type of machine (001 or 002) and the age of the machine (in years). The manager has proposed the following model:

$$E(y) = \beta_0 + \beta_1 x_1 + \beta_2 x_1^2 + \beta_3 x_2$$

where

$x_1$ = Age of machine

$x_2$ = 1 if machine type 001, 0 if machine type 002

**a.** Use the data obtained on $n = 20$ machine breakdowns, shown here, to estimate the parameters of this model.

● SHUTDOWN

| Downtime (hours per month) | Machine Age $x_1$ (years) | Machine Type | $x_2$ |
|---|---|---|---|
| 10 | 1.0 | 001 | 1 |
| 20 | 2.0 | 001 | 1 |
| 30 | 2.7 | 001 | 1 |
| 40 | 4.1 | 001 | 1 |
| 9 | 1.2 | 001 | 1 |
| 25 | 2.5 | 001 | 1 |
| 19 | 1.9 | 001 | 1 |
| 41 | 5.0 | 001 | 1 |
| 22 | 2.1 | 001 | 1 |
| 12 | 1.1 | 001 | 1 |
| 10 | 2.0 | 002 | 0 |
| 20 | 4.0 | 002 | 0 |
| 30 | 5.0 | 002 | 0 |
| 44 | 8.0 | 002 | 0 |
| 9 | 2.4 | 002 | 0 |
| 25 | 5.1 | 002 | 0 |
| 20 | 3.5 | 002 | 0 |
| 42 | 7.0 | 002 | 0 |
| 20 | 4.0 | 002 | 0 |
| 13 | 2.1 | 002 | 0 |

**b.** Do these data provide sufficient evidence to conclude that the second-order term ($x_1^2$) in the model proposed by the operations manager is necessary? Test using $\alpha = .05$.

**c.** Test the null hypothesis that $\beta_1 = \beta_2 = 0$ using $\alpha = .10$. Interpret the results of the test in the context of the problem.

**12.135** Phosphorus used in soil fertilizers can contaminate freshwater sources during rainfall runoff. Consequently, it is important for water quality engineers to estimate the amount of dissolved phosphorus in the water. *Geoderma* (June 1995) presented an investigation of the relationship between soil loss and percentage of dissolved phosphorus in water samples collected at 20 fertilized watersheds in Oklahoma. The data are given in the table.

● PHOSPHOR

| Watershed | Soil Loss $x$ (kilometers per half-acre) | Dissolved Phosphorus Percentage $y$ |
|---|---|---|
| 1 | 18 | 42.3 |
| 2 | 17 | 50.2 |
| 3 | 35 | 52.7 |
| 4 | 16 | 77.1 |
| 5 | 14 | 36.8 |
| 6 | 54 | 17.5 |
| 7 | 153 | 66.4 |
| 8 | 81 | 67.5 |
| 9 | 183 | 28.9 |
| 10 | 284 | 15.1 |
| 11 | 767 | 20.1 |
| 12 | 148 | 38.3 |
| 13 | 649 | 5.6 |
| 14 | 479 | 8.6 |
| 15 | 1,371 | 5.5 |
| 16 | 9,150 | 4.6 |
| 17 | 15,022 | 2.2 |
| 18 | 69 | 77.9 |
| 19 | 4,392 | 7.8 |
| 20 | 312 | 42.9 |

*Source:* Sharpley, A. N., Robinson, J. S., and Smith, S. J. "Bioavailable phosphorus dynamics in agricultural soils and effects on water quality." *Geoderma*, Vol. 67, No. 1–2, June 1995, p. 11 (Table 4).

**a.** Plot the data in a scattergram. Do you detect a linear or curvilinear trend?

**b.** Fit the quadratic model $E(y) = \beta_0 + \beta_1 x + \beta_2 x^2$ to the data.

**c.** Conduct a test to determine if a curvilinear relationship exists between dissolved phosphorus percentage ($y$) and soil loss ($x$). Test using $\alpha = .05$.

**12.136** A research physician wishes to find a model that will predict the mean time to relief after the administration of a certain drug. Two important independent variables considered to be good predictors of relief time are age of the patient and method of administration. Physicians administer a standard dose of a drug to a patient in one of three different ways: orally in liquid form (method 1), orally in pill form (method 2), and intravenously (method 3). Consider the interaction model

$$E(y) = \beta_0 + \beta_1 x_1 + \beta_2 x_2 + \beta_3 x_3 + \beta_4 x_1 x_2 + \beta_5 x_1 x_3$$

where

   $y$ = Time to relief (in minutes)
   $x_1$ = Age of patient (in years)
   $x_2$ = 1 if drug is administered orally in pill form,
      0 if not
   $x_3$ = 1 if drug is administered intravenously, 0 if not

**a.** The data in the table were obtained on 12 patients. Fit this model to the data using a statistical software package.

⦿ RELIEF

| Method 1 | | Method 2 | | Method 3 | |
|---|---|---|---|---|---|
| Age | Time to Relief | Age | Time to Relief | Age | Time to Relief |
| 51 | 22 | 46 | 28 | 37 | 19 |
| 36 | 25 | 40 | 24 | 60 | 17 |
| 31 | 20 | 26 | 23 | 25 | 21 |
| 20 | 25 | 32 | 25 | 38 | 20 |

**b.** Test whether the model is useful in predicting mean time to relief. Use $\alpha = .05$.

**c.** What hypothesis would you use to test whether the age–drug method interaction terms contribute to the prediction of mean time to relief?

**d.** Give the reduced model appropriate for conducting the test, part **c**.

**e.** Fit the reduced model to the data and use the results to conduct the test, part **c**.

**12.137** To determine whether extra personnel are needed for the day, the owners of a water adventure park would like to find a model that would allow them to predict the day's attendance each morning before opening based on the day of the week and weather conditions. The model is of the form

$$E(y) = \beta_0 + \beta_1 x_1 + \beta_2 x_2 + \beta_3 x_3$$

where

   $y$ = Daily admission
   $x_1 = \begin{cases} 1 & \text{if weekend} \\ 0 & \text{otherwise} \end{cases}$ (dummy variable)
   $x_2 = \begin{cases} 1 & \text{if sunny} \\ 0 & \text{if overcast} \end{cases}$ (dummy variable)
   $x_3$ = Predicted daily high temperature (°F)

These data were recorded for a random sample of 30 days, and a regression model was fitted to the data. The least squares analysis produced the following results:

$$\hat{y} = -105 + 25x_1 + 100x_2 + 10x_3$$

with

$$s_{\hat{\beta}_1} = 10 \qquad s_{\hat{\beta}_2} = 30 \qquad s_{\hat{\beta}_3} = 4 \qquad R^2 = .65$$

**a.** Interpret the estimated model coefficients.

**b.** Is there sufficient evidence to conclude that this model is useful for the prediction of daily attendance? Use $\alpha = .05$.

**c.** Is there sufficient evidence to conclude that the mean attendance increases on weekends? Use $\alpha = .10$.

**d.** Use the model to predict the attendance on a sunny weekday with a predicted high temperature of 95°F.

**e.** Suppose the 90% prediction interval for part **d** is (645, 1,245). Interpret this interval.

**12.138** Refer to Exercise 12.137. The owners of the water adventure park are advised that the prediction model could probably be improved if interaction terms were added. In particular, it is thought that the *rate* at which mean attendance increases as predicted high temperature increases will be greater on weekends than on weekdays. The following model is therefore proposed:

$$E(y) = \beta_0 + \beta_1 x_1 + \beta_2 x_2 + \beta_3 x_3 + \beta_4 x_1 x_3$$

The same 30 days of data used in Exercise 12.137 are again used to obtain the least squares model

$$\hat{y} = 250 - 700x_1 + 100x_2 + 5x_3 + 15x_1 x_3$$

with

$$s_{\hat{\beta}_4} = 3.0 \qquad R^2 = .96$$

**a.** Graph the predicted day's attendance, $y$, against the day's predicted high temperature, $x_3$, for a sunny weekday and for a sunny weekend day. Plot both on the same paper for $x_3$ between 70 and 100°F. Note the increase in slope for the weekend day. Interpret this.

**b.** Do the data indicate that the interaction term is a useful addition to the model? Use $\alpha = .05$.

**c.** Use this model to predict the attendance for a sunny weekday with a predicted high temperature of 95°F.

**d.** Suppose the 90% prediction interval for part **c** is (800, 850). Compare this result with the prediction interval for the model without interaction in Exercise 12.137, part **e**. Do the relative widths of the confidence intervals support or refute your conclusion about the utility of the interaction term (part **b**)?

**e.** The owners, noting that the coefficient $\hat{\beta}_1 = -700$, conclude the model is ridiculous because it seems to imply that the mean attendance will be 700 less on weekends than on weekdays. Explain why this is *not* the case.

**12.139** The *American Sociological Review* (Summer 1993) published a study of the variables that affect academic achievement in high school. Two variables hypothesized to have an impact on achievement were track (college program of courses or not) and sector (Catholic school or public school). One of the models considered in the study was:

$$E(y) = \beta_0 + \beta_1 x_1 + \beta_2 x_2 + \beta_3 x_1 x_2$$

where

$$y = \text{Verbal achievement score}$$
$$x_1 = 1 \text{ if college track, 0 if not}$$
$$x_2 = 1 \text{ if Catholic school, 0 if not}$$

**a.** How would you determine whether the impact of track $(x_1)$ on verbal achievement score $(y)$ depends on sector $(x_2)$? Set up the null and alternative hypotheses for the test.

**b.** In terms of the $\beta$'s, give the mean verbal achievement score, $E(y)$, for public school students not on a college program.

**c.** Repeat part **b** for public school students on a college program.

**d.** Subtract the results, parts **b** and **c**, to obtain the difference between the means of college-track and regular-track students in public schools.

**e.** In terms of the $\beta$'s, give the mean verbal achievement score, $E(y)$, for Catholic school students not on a college program.

**f.** Repeat part **e** for Catholic school students on a college program.

**g.** Subtract the results, parts **e** and **f**, to obtain the difference between the means of college-track and regular-track students in Catholic schools.

**12.140** In Exercise 12.9 (p. 590), a real estate appraiser used regression analysis to explore the relationship between the sale prices of apartment buildings and various characteristics of the apartments. The data from Exercise 12.9 are reproduced in the table, which also contains data on the physical condition of each apartment building (E: excellent; G: good; F: fair).

**a.** Write a model that describes the relationship between sale price and number of apartment units as three parallel lines, one for each level of physical condition. Be sure to specify the dummy variable coding scheme you use.

### MNSALES

| Code No. | Sale Price $y$ ($) | No. of Apartments $x_1$ | Age of Structure $x_2$ (years) | Lot Size $x_3$ (sq. ft.) | No. of On-Site Parking Spaces $x_4$ | Gross Building Area $x_5$ (sq. ft.) | Condition of Apartment Building |
|---|---|---|---|---|---|---|---|
| 0229 | 90,300 | 4 | 82 | 4,635 | 0 | 4,266 | F |
| 0094 | 384,000 | 20 | 13 | 17,798 | 0 | 14,391 | G |
| 0043 | 157,500 | 5 | 66 | 5,913 | 0 | 6,615 | G |
| 0079 | 676,200 | 26 | 64 | 7,750 | 6 | 34,144 | E |
| 0134 | 165,000 | 5 | 55 | 5,150 | 0 | 6,120 | G |
| 0179 | 300,000 | 10 | 65 | 12,506 | 0 | 14,552 | G |
| 0087 | 108,750 | 4 | 82 | 7,160 | 0 | 3,040 | G |
| 0120 | 276,538 | 11 | 23 | 5,120 | 0 | 7,881 | G |
| 0246 | 420,000 | 20 | 18 | 11,745 | 20 | 12,600 | G |
| 0025 | 950,000 | 62 | 71 | 21,000 | 3 | 39,448 | G |
| 0015 | 560,000 | 26 | 74 | 11,221 | 0 | 30,000 | G |
| 0131 | 268,000 | 13 | 56 | 7,818 | 13 | 8,088 | F |
| 0172 | 290,000 | 9 | 76 | 4,900 | 0 | 11,315 | E |
| 0095 | 173,200 | 6 | 21 | 5,424 | 6 | 4,461 | G |
| 0121 | 323,650 | 11 | 24 | 11,834 | 8 | 9,000 | G |
| 0077 | 162,500 | 5 | 19 | 5,246 | 5 | 3,828 | G |
| 0060 | 353,500 | 20 | 62 | 11,223 | 2 | 13,680 | F |
| 0174 | 134,400 | 4 | 70 | 5,834 | 0 | 4,680 | E |
| 0084 | 187,000 | 8 | 19 | 9,075 | 0 | 7,392 | G |
| 0031 | 155,700 | 4 | 57 | 5,280 | 0 | 6,030 | E |
| 0019 | 93,600 | 4 | 82 | 6,864 | 0 | 3,840 | F |
| 0074 | 110,000 | 4 | 50 | 4,510 | 0 | 3,092 | G |
| 0057 | 573,200 | 14 | 10 | 11,192 | 0 | 23,704 | E |
| 0104 | 79,300 | 4 | 82 | 7,425 | 0 | 3,876 | F |
| 0024 | 272,000 | 5 | 82 | 7,500 | 0 | 9,542 | E |

*Source:* Robinson Appraisal Co., Inc., Mankato, Minnesota.

**b.** Plot $y$ against $x_1$ (number of apartment units) for all buildings in excellent condition. On the same graph, plot $y$ against $x_1$ for all buildings in good condition. Do this again for all buildings in fair condition. Does it appear that the model you specified in part **a** is appropriate? Explain.

**c.** Fit the model from part **a** to the data. Report the least squares prediction equation for each of the three building condition levels.

**d.** Plot the three prediction equations of part **c** on a scattergram of the data.

**e.** Do the data provide sufficient evidence to conclude that the relationship between sale price and number of units differs depending on the physical condition of the apartments? Test using $\alpha = .05$.

**f.** Check the data set for multicollinearity. How does this impact your choice of independent variables to use in a model for sale price?

**g.** Consider the first-order model $E(y) = \beta_0 + \beta_1 x_1 + \cdots + \beta_5 x_5$, analyzed in Exercise 12.9 (p. 590). Conduct a complete residual analysis for the model to check the assumptions on $\varepsilon$.

## STUDENT PROJECTS

*Note: The use of a computer is required for this project.*

This is a continuation of the Student Projects section in Chapter 11, in which you selected three independent variables as predictors of a dependent variable of your choice and obtained at least 10 data values. Now fit the multiple regression model using an available software package:

$$y = \beta_0 + \beta_1 x_1 + \beta_2 x_2 + \beta_3 x_3 + \varepsilon$$

where

$y$ = Dependent variable you chose
$x_1$ = First independent variable you chose
$x_2$ = Second independent variable you chose
$x_3$ = Third independent variable you chose

**a.** Compare the coefficients $\hat{\beta}_1$, $\hat{\beta}_2$, and $\hat{\beta}_3$ to their corresponding slope coefficients in the Chapter 11 Student Projects, where you fit three separate straight-line models. How do you account for the differences?

**b.** Calculate the coefficient of determination $R^2$, and conduct the $F$-test of the null hypothesis $H_0: \beta_1 = \beta_2 = \beta_3 = 0$. What is your conclusion?

**c.** Check the data for multicollinearity. If multicollinearity exists, how should you proceed?

**d.** Now increase your list of three variables to include approximately 10 that you think would be useful in predicting the dependent variable. With the aid of statistical software, employ a stepwise regression program to choose the important variables among those you have listed. To test your intuition, list the variables in the order you think they will be selected before you conduct the analysis. How does your list compare with the stepwise regression results?

**e.** After the group of 10 variables has been narrowed to a smaller group of variables by the stepwise analysis, try to improve the model by including interactions and quadratic terms. Be sure to consider the meaning of each interaction or quadratic term before adding it to the model—a quick sketch can be very helpful. See if you can systematically construct a useful model for prediction. If you have a large data set, you might want to hold out the last observations to test the predictive ability of your model after it is constructed. (As noted in Section 12.12, using the same data to construct *and* to evaluate predictive ability can lead to invalid statistical tests and a false sense of security.)

## REFERENCES

Barnett, V., and Lewis, T. *Outliers in Statistical Data.* New York: Wiley, 1978.

Belsley, D. A., Kuh, E., and Welsch, R. E. *Regression Diagnostics: Identifying Influential Data and Sources of Collinearity.* New York: Wiley, 1980.

Chatterjee, S., and Price, B. *Regression Analysis by Example*, 2nd ed. New York: Wiley, 1991.

Draper, N., and Smith, H. *Applied Regression Analysis*, 2nd ed. New York: Wiley, 1981.

Graybill, F. *Theory and Application of the Linear Model.* North Scituate, Mass.: Duxbury, 1976.

Mendenhall, W. *Introduction to Linear Models and the Design and Analysis of Experiments.* Belmont, Ca.: Wadsworth, 1968.

Mendenhall, W., and Sincich, T. *A Second Course in Statistics: Regression Analysis*, 5th ed. Upper Saddle River, N. J.: Prentice Hall, 1996.

Mosteller, F., and Tukey, J. W. *Data Analysis and Regression: A Second Course in Statistics.* Reading, Mass.: Addison-Wesley, 1977.

Neter, J., Kutner, M., Cjnachtsheim, C., and Wasserman, W. *Applied Linear Statistical Models*, 4th ed. Homewood, Ill.: Richard Irwin, 1996.

Rousseeuw, P. J., and Leroy, A. M. *Robust Regression and Outlier Detection.* New York: Wiley, 1987.

Weisberg, S. *Applied Linear Regression*, 2nd ed. New York: Wiley, 1985.

# Categorical Data Analysis

## Contents

## Statistics in Action

The Level of Agreement Among Movie Reviewers: Thumbs Up or Thumbs Down?

### 👍 Where We've Been

In the preceding chapters we've examined several statistical methods for analyzing many types of data. Except for those that pertain to binomial proportions, most of the techniques presented in Chapters 7–10 are appropriate for populations of data on quantitative random variables. Regression models were presented in Chapters 11 and 12 for the purpose of predicting a quantitative dependent variable from the values of both quantitative and qualitative independent variables.

### ☞ Where We're Going

The methods we discuss in this chapter are appropriate for *qualitative* data, exemplified by the binomial proportions of Chapters 7–9. Recall that the binomial experiment (Chapter 4) consists of $n$ trials, each of which results in *two* outcomes. Thus, we have two classifications into which all the data fall. In this chapter, we consider the analysis of data that fall into *two or more* categories.

## 13.1   CATEGORICAL DATA AND THE MULTINOMIAL EXPERIMENT

Recall from Section 1.4 (p. 9) that observations on a qualitative variable can only be categorized. For example, consider the highest level of education attained by a professional hockey player. Level of education is a qualitative variable with several categories, including some high school, high school diploma, some college, college undergraduate degree, and graduate degree. If we were to record education level for all professional hockey players, the result of the categorization would be a count of the numbers of players falling in the respective categories.

When the qualitative variable of interest results in one of two responses (e.g., yes or no, success or failure, favor or do not favor), the data—called *counts*—can be analyzed using the binomial probability distribution discussed in Section 4.4. However, qualitative variables, such as level of education, that allow for more than two categories for a response are much more common, and these must be analyzed using a different method. Qualitative data with more than two levels often result from a **multinomial experiment**. The characteristics for a multinomial experiment with $k$ outcomes are described in the box. You can see that the binomial experiment of Chapter 4 is a multinomial experiment with $k = 2$.

---

### Properties of the Multinomial Experiment

1. The experiment consists of $n$ identical trials.
2. There are $k$ possible outcomes to each trial. These outcomes are sometimes called **classes, categories**, or **cells**.
3. The probabilities of the $k$ outcomes, denoted by $p_1, p_2, \ldots, p_k$, remain the same from trial to trial, where $p_1 + p_2 + \cdots + p_k = 1$.
4. The trials are independent.
5. The random variables of interest are the **cell counts**, $n_1, n_2, \ldots, n_k$, of the number of observations that fall in each of the $k$ categories.

---

**EXAMPLE 13.1**

Consider the problem of determining the highest level of education attained by each of a sample of $n = 40$ National Hockey League (NHL) players. Suppose we categorize level of education into one of five categories—some high school, high school diploma, some college, college undergraduate degree, and graduate degree—and count the number of the 40 players that fall into each category. Is this a multinomial experiment to a reasonable degree of approximation?

### Solution

Checking the five properties of a multinomial experiment shown in the box, we have

1. The experiment consists of $n = 40$ identical trials, where each trial is to determine the education level of an NHL player.
2. There are $k = 5$ possible outcomes to each trial corresponding to the five education-level responses.
3. The probabilities of the $k = 5$ outcomes, $p_1, p_2, p_3, p_4$, and $p_5$, remain the same from trial to trial (to a reasonable degree of approximation), where $p_i$ represents the true probability that an NHL player attains level-of-education category $i$.

4. The trials are independent, that is, the education level attained by one NHL player does not affect the level attained by any other player.
5. We are interested in the count of the number of hockey players who fall into each of the five education-level categories. These five cell counts are denoted $n_1, n_2, n_3, n_4$, and $n_5$.

Thus, the properties of a multinomial experiment are satisfied. ■

In this chapter, we are concerned with the analysis of categorical data—specifically, the data that represent the counts for each category of a multinomial experiment. In Section 13.2, we learn how to make inferences about category probabilities for data classified according to a single qualitative (or categorical) variable. Then, in Section 13.3, we consider inferences about categorical probabilities for data classified according to two qualitative variables. The statistic used for these inferences is one that possesses, approximately, the familiar chi-square distribution.

## 13.2 TESTING CATEGORICAL PROBABILITIES: ONE-WAY TABLE

In this section, we consider a multinomial experiment with $k$ outcomes that corresponds to the categories of a *single* qualitative variable. The results of such an experiment are summarized in a **one-way table**. The term "one-way" is used because only one variable is classified. Typically, we want to make inferences about the true percentages that occur in the $k$ categories based on the sample information in the one-way table.

To illustrate, suppose three political candidates are running for the same elective position. Prior to the election, we conduct a survey to determine the voting preferences of a random sample of 150 eligible voters. The qualitative variable of interest is *preferred candidate*, which has three possible outcomes: candidate 1, candidate 2, and candidate 3. Suppose the number of voters preferring each candidate is tabulated and the resulting count data appear as in Table 13.1.

**TABLE 13.1    Results of Voter-Preference Survey**

| Candidate | | |
|---|---|---|
| **1** | **2** | **3** |
| 61 | 53 | 36 |

Note that our voter-preference survey satisfies the properties of a multinomial experiment for the qualitative variable, preferred candidate. The experiment consists of randomly sampling $n = 150$ voters from a large population of voters containing an unknown proportion $p_1$ who favor candidate 1, a proportion $p_2$ who favor candidate 2, and a proportion $p_3$ who favor candidate 3. Each voter sampled represents a single trial that can result in one of three outcomes: The voter will favor candidate 1, 2, or 3 with probabilities $p_1$, $p_2$, and $p_3$, respectively. (Assume that all voters will have a preference.) The voting preference of any single voter in the sample does not affect the preference of another; consequently, the trials are independent. And, finally, you can see that the recorded data are the numbers of voters in each of the three voter-preference categories. Thus, the voter-preference survey satisfies the five properties of a multinomial experiment.

In the voter-preference survey, and in most practical applications of the multinomial experiment, the $k$ outcome probabilities $p_1, p_2, \ldots, p_k$ are unknown and we want to use the survey data to make inferences about their values. The unknown probabilities in the voter-preference survey are

$$p_1 = \text{Proportion of all voters who favor candidate 1}$$
$$p_2 = \text{Proportion of all voters who favor candidate 2}$$
$$p_3 = \text{Proportion of all voters who favor candidate 3}$$

To decide whether the voters have a preference for any of the candidates, we will want to test the null hypothesis that the candidates are equally preferred (that is, $p_1 = p_2 = p_3 = \frac{1}{3}$) against the alternative hypothesis that one candidate is preferred (that is, at least one of the probabilities $p_1, p_2$, and $p_3$ exceeds $\frac{1}{3}$). Thus, we want to test

$H_0$: $p_1 = p_2 = p_3 = \frac{1}{3}$   (no preference)

$H_a$: At least one of the proportions exceeds $\frac{1}{3}$   (a preference exists)

If the null hypothesis is true and $p_1 = p_2 = p_3 = \frac{1}{3}$, the expected value (mean value) of the number of voters who prefer candidate 1 is given by

$$E(n_1) = np_1 = (n)\frac{1}{3} = (150)\frac{1}{3} = 50$$

Similarly, $E_{(n_2)} = E_{(n_3)} = 50$ if the null hypothesis is true and no preference exists.

The following test statistic—the **chi-square test**—measures the degree of disagreement between the data and the null hypothesis:

$$\chi^2 = \frac{[n_1 - E(n_1)]^2}{E(n_1)} + \frac{[n_2 - E(n_2)]^2}{E(n_2)} + \frac{[n_3 - E(n_3)]^2}{E(n_3)}$$
$$= \frac{(n_1 - 50)^2}{50} + \frac{(n_2 - 50)^2}{50} + \frac{(n_3 - 50)^2}{50}$$

Note that the farther the observed numbers $n_1, n_2$, and $n_3$ are from their expected value (50), the larger $\chi^2$ will become. That is, large values of $\chi^2$ imply that the null hypothesis is false.

We have to know the distribution of $\chi^2$ in repeated sampling before we can decide whether the data indicate that a preference exists. When $H_0$ is true, $\chi^2$ can be shown to have (approximately) the familiar chi-square distribution of Sections 8.7 and 11.4. For this one-way classification, the $\chi^2$ distribution has $(k-1)$ degrees of freedom.* The rejection region for the voter-preference survey for $\alpha = .05$ and $k - 1 = 3 - 1 = 2$ df is

*Rejection region:*   $\chi^2 > \chi^2_{.05}$

This value of $\chi^2_{.05}$ (found in Table VII) is 5.99147. (See Figure 13.1.) The computed value of the test statistic is

$$\chi^2 = \frac{(n_1 - 50)^2}{50} + \frac{(n_2 - 50)^2}{50} + \frac{(n_3 - 50)^2}{50}$$
$$= \frac{(61 - 50)^2}{50} + \frac{(53 - 50)^2}{50} + \frac{(36 - 50)^2}{50} = 6.52$$

---

*The derivation of the degrees of freedom for $\chi^2$ involves the number of linear restrictions imposed on the count data. In the present case, the only constraint is that $\sum n_i = n$, where $n$ (the sample size) is fixed in advance. Therefore, df $= k - 1$. For other cases, we will give the degrees of freedom for each usage of $\chi^2$ and refer the interested reader to the references for more detail.

**FIGURE 13.1**

*Rejection Region for Voter-Preference Survey*

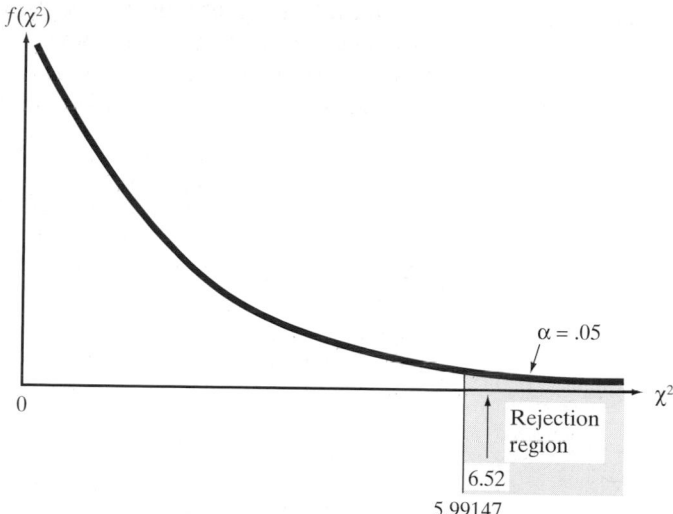

Since the computed $\chi^2 = 6.52$ exceeds the critical value of 5.99147, we conclude at the $\alpha = .05$ level of significance that there does exist a voter preference for one or more of the candidates.

Now that we have evidence to indicate that the proportions $p_1, p_2,$ and $p_3$ are unequal, we can make inferences concerning their individual values using the methods of Section 7.3. [*Note:* We cannot use the methods of Section 9.3 to compare two proportions because the cell counts are dependent random variables.] The general form for a test of a hypothesis concerning multinomial probabilities is shown in the next box.

---

## A Test of a Hypothesis About Multinomial Probabilities: One-way Table

$H_0$: $p_1 = p_{1,0}, p_2 = p_{2,0}, \ldots, p_k = p_{k,0}$

where $p_{1,0}, p_{2,0}, \ldots, p_{k,0}$ represent the hypothesized values of the multinomial probabilities

$H_a$: At least one of the multinomial probabilities does not equal its hypothesized value

*Test statistic:* $\chi^2 = \sum \dfrac{[n_i - E(n_i)]^2}{E(n_i)}$

where $E(n_i) = np_{i,0}$ is the **expected cell count**, that is, the expected number of outcomes of type $i$ assuming that $H_0$ is true. The total sample size is $n$.

*Rejection region:* $\chi^2 > \chi_\alpha^2$, where $\chi_\alpha^2$ has $(k - 1)$ df

*Assumptions:* 1. A multinomial experiment has been conducted. This is generally satisfied by taking a random sample from the population of interest.

2. The sample size $n$ will be large enough so that for every cell, the expected cell count $E(n_i)$ will be equal to 5 or more.*

---

*The assumption that all expected cell counts are at least 5 is necessary in order to ensure that the $\chi^2$ approximation is appropriate. Exact methods for conducting the test of a hypothesis exist and may be used for small expected cell counts, but these methods are beyond the scope of this text.

**EXAMPLE  13.2**   Suppose an educational television station has broadcast a series of programs on the physiological and psychological effects of smoking marijuana. Now that the series is finished, the station wants to see whether the citizens within the viewing area have changed their minds about how possession of marijuana should be considered legally. Before the series was shown, it was determined that 7% of the citizens favored legalization, 18% favored decriminalization, 65% favored the existing law (an offender could be fined or imprisoned), and 10% had no opinion.

A summary of the opinions (after the series was shown) of a random sample of 500 people in the viewing area is given in Table 13.2. Test at the $\alpha = .01$ level to see whether these data indicate that the distribution of opinions differs significantly from the proportions that existed before the educational series was aired.

**TABLE 13.2    Distribution of Opinions About Marijuana Possession**

| Legalization | Decriminalization | Existing Laws | No Opinion |
|---|---|---|---|
| 39 | 99 | 336 | 26 |

**Solution**

Define the proportions after the airing to be

$p_1$ = Proportion of citizens favoring legalization
$p_2$ = Proportion of citizens favoring decriminalization
$p_3$ = Proportion of citizens favoring existing laws
$p_4$ = Proportion of citizens with no opinion

Then the null hypothesis representing no change in the distribution of percentages is

$H_0$: $p_1 = .07, p_2 = .18, p_3 = .65, p_4 = .10$

and the alternative is

$H_a$: At least one of the proportions differs from its null hypothesized value

Test statistic: $\chi^2 = \sum \dfrac{[n_i - E(n_i)]^2}{E(n_i)}$

where

$E(n_1) = np_{1,0} = 500(.07) = 35$
$E(n_2) = np_{2,0} = 500(.18) = 90$
$E(n_3) = np_{3,0} = 500(.65) = 325$
$E(n_4) = np_{4,0} = 500(.10) = 50$

Since all these values are larger than 5, the $\chi^2$ approximation is appropriate. Also, if the citizens in the sample were randomly selected, the properties of the multinomial probability distribution are satisfied.

*Rejection region:* For $\alpha = .01$ and df $= k - 1 = 3$, reject $H_0$ if $\chi^2 > \chi^2_{.01}$, where (from Table VII in Appendix A) $\chi^2_{.01} = 11.3449$.

We now calculate the test statistic:

$$\chi^2 = \frac{(39 - 35)^2}{35} + \frac{(99 - 90)^2}{90} + \frac{(336 - 325)^2}{325} + \frac{(26 - 50)^2}{50} = 13.249$$

Since this value exceeds the table value of $\chi^2$ (11.3449), the data provide sufficient evidence ($\alpha = .01$) that the opinions on legalization of marijuana have changed since the series was aired.

The $\chi^2$-test can also be conducted using an available statistical software package. Figure 13.2 is an SPSS printout of the analysis of the data in Table 13.2. The test statistic and $p$-value of the test are highlighted on the printout. Since $\alpha = .01$ exceeds $p = .004$, there is sufficient evidence to reject $H_0$. ∎

**FIGURE 13.2**

*SPSS Analysis of Data in Table 13.2*

**OPINION**

| | Observed N | Expected N | Residual |
|---|---|---|---|
| LEGAL | 39 | 35.0 | 4.0 |
| DECRIM | 99 | 90.0 | 9.0 |
| EXISTLAW | 336 | 325.0 | 11.0 |
| NONE | 26 | 50.0 | -24.0 |
| Total | 500 | | |

**Test Statistics**

| | OPINION |
|---|---|
| Chi-Square[a] | 13.249 |
| df | 3 |
| Asymp. Sig. | .004 |

a. 0 cells (.0%) have expected frequencies less than 5. The minimum expected cell fequency is 35.0.

If we focus on one particular outcome of a multinomial experiment, we can use the methods developed in Section 7.4 for a binomial proportion to establish a confidence interval for any one of the multinomial probabilities.* For example, if we want a 95% confidence interval for the proportion of citizens in the viewing area who have no opinion about the issue, we calculate

$$\hat{p}_4 \pm 1.96\sigma_{\hat{p}_4}$$

where

$$\hat{p}_4 = \frac{n_4}{n} = \frac{26}{500} = .052 \quad \text{and} \quad \sigma_{\hat{p}_4} \approx \sqrt{\frac{\hat{p}_4(1 - \hat{p}_4)}{n}}$$

Thus, we get

$$.052 \pm 1.96\sqrt{\frac{(.052)(.948)}{500}} = .052 \pm .019$$

or (.033, .071). Thus, we estimate that between 3.3% and 7.1% of the citizens now have no opinion on the issue of marijuana legalization. The series of programs may have helped citizens who formerly had no opinion on the issue to form an opinion, since it appears that the proportion of "no opinions" is now less than 10%.

---

*Note that focusing on one outcome has the effect of lumping the other $(k - 1)$ outcomes into a single group. Thus, we obtain, in effect, two outcomes—or a binomial experiment.

EXERCISES 13.1–13.16

## Learning the Mechanics

**13.1** Use Table VII of Appendix A to find each of the following $\chi^2$ values:
 a. $\chi^2_{.05}$ for df = 10
 b. $\chi^2_{.990}$ for df = 50
 c. $\chi^2_{.10}$ for df = 16
 d. $\chi^2_{.005}$ for df = 50

**13.2** Use Table VII of Appendix A to find the following probabilities:
 a. $P(\chi^2 \leq 1.063623)$ for df = 4
 b. $P(\chi^2 > 30.5779)$ for df = 15
 c. $P(\chi^2 \geq 82.3581)$ for df = 100
 d. $P(\chi^2 < 18.4926)$ for df = 30

**13.3** Find the rejection region for a one-dimensional $\chi^2$-test of a null hypothesis concerning $p_1, p_2, \ldots, p_k$ if
 a. $k = 3; \alpha = .05$
 b. $k = 5; \alpha = .10$
 c. $k = 4; \alpha = .01$

**13.4** What are the characteristics of a multinomial experiment? Compare the characteristics to those of a binomial experiment.

**13.5** What conditions must $n$ satisfy to make the $\chi^2$-test valid?

**13.6** A multinomial experiment with $k = 3$ cells and $n = 320$ produced the data shown in the following table. Do these data provide sufficient evidence to contradict the null hypothesis that $p_1 = .25$, $p_2 = .25$, and $p_3 = .50$? Test using $\alpha = .05$.

| | **Cell** | | |
|---|---|---|---|
| | **1** | **2** | **3** |
| $n_i$ | 78 | 60 | 182 |

**13.7** A multinomial experiment with $k = 4$ cells and $n = 205$ produced the data shown in the table.

| | **Cell** | | | |
|---|---|---|---|---|
| | **1** | **2** | **3** | **4** |
| $n_i$ | 43 | 56 | 59 | 47 |

 a. Do these data provide sufficient evidence to conclude that the multinomial probabilities differ? Test using $\alpha = .05$.
 b. What are the Type I and Type II errors associated with the test of part **a**?

**13.8** Refer to Exercise 13.7. Construct a 95% confidence interval for the multinomial probability associated with cell 3.

**13.9** A multinomial experiment with $k = 4$ cells and $n = 400$ produced the data shown in the following table. Do these data provide sufficient evidence to contradict the null hypothesis that $p_1 = .2$, $p_2 = .4$, $p_3 = .1$, and $p_4 = .3$? Test using $\alpha = .05$.

| | **Cell** | | | |
|---|---|---|---|---|
| | **1** | **2** | **3** | **4** |
| $n_i$ | 70 | 196 | 46 | 88 |

## Applying the Concepts—Basic

**13.10** There has been a recent trend for professional sports franchises in Major League Baseball (MLB), the National Football League (NFL), the National Basketball Association (NBA), and the National Hockey League (NHL) to build new stadiums and ballparks in urban, downtown venues. An article in *Professional Geographer* (Feb. 2000) investigated whether there has been a significant suburban-to-urban shift in the location of major sport facilities. In 1985, 40% of all major sport facilities were located downtown, 30% in central city, and 30% in suburban areas. In contrast, of the 113 major sports franchises that existed in 1997, 58 were built downtown, 26 in central city, and 29 in a suburban area.
 a. Describe the qualitative variable of interest in the study. Give the levels (categories) associated with the variable.
 b. Give the null hypothesis for a test to determine whether the proportions of major sports facilities in downtown, central city, and suburban areas in 1997 are the same as in 1985.
 c. If the null hypothesis, part **b**, is true, how many of the 113 sports facilities in 1997 would you expect to be located in downtown, central city, and suburban areas, respectively?
 d. Find the value of the chi-square statistic for testing the null hypothesis, part **b**.
 e. Find the (approximate) p-value of the test and give the appropriate conclusion in the words of the problem. Assume $\alpha = .05$.

**13.11** Refer to the *Chance* (Fall 2000) study of ancient Greek pottery, Exercise 2.8 (p. 26). Recall that 837 pottery pieces were uncovered at the excavation site. The table describing the types of pottery found is reproduced here.

| Pot Category | Number Found |
|---|---|
| Burnished | 133 |
| Monochrome | 460 |
| Painted | 183 |
| Other | 61 |
| **Total** | 837 |

*Source:* Berg, I., and Bliedon, S. "The pots of Phyiakopi: Applying statistical techniques to archaeology." *Chance,* Vol. 13, No. 4, Fall 2000.

```
STATISTIX Output for Exercise 13.11

MULTINOMIAL TEST

HYPOTHESIZED PROPORTIONS VARIABLE: HYPPROP
OBSERVED FREQUENCIES VARIABLE:      FOUND

              HYPOTHESIZED   OBSERVED     EXPECTED     CHI-SQUARE
CATEGORY       PROPORTION    FREQUENCY    FREQUENCY    CONTRIBUTION
   1            0.25000         133        209.25         27.79
   2            0.25000         460        209.25        300.48
   3            0.25000         183        209.25          3.29
   4            0.25000          61        209.25        105.03


OVERALL CHI-SQUARE    436.59
P-VALUE                0.0000
DEGREES OF FREEDOM        3
```

a. Describe the qualitative variable of interest in the study. Give the levels (categories) associated with the variable.

b. Assume the four types of pottery occur with equal probability at the excavation site. What are the values of $p_1$, $p_2$, $p_3$, and $p_4$, the probabilities associated with the four pottery types?

c. Give the null and alternative hypotheses for testing whether one type of pottery is more likely to occur at the site than any of the other types.

d. Use the STATISTIX printout provided above to find the test statistic for testing the hypotheses, part **c**.

e. Find and interpret the *p*-value of the test. State the conclusion in the words of the problem if you use $\alpha = .10$.

**13.12** *Inc. Technology* (Mar. 18, 1997) reported the results of an Equifax/Harris Consumer Privacy Survey in which 328 Internet users indicated their level of agreement with the following statement: "The government needs to be able to scan Internet messages and user communications to prevent fraud and other crimes." The number of users in each response category is summarized below.

| Agree Strongly | Agree Somewhat | Disagree Somewhat | Disagree Strongly |
|---|---|---|---|
| 59 | 108 | 82 | 79 |

a. Specify the null and alternative hypotheses you would use to determine if the opinions of Internet users are evenly divided among the four categories.

b. Conduct the test of part **a** using $\alpha = .05$.

c. In the context of this exercise, what is a Type I error? A Type II error?

d. What assumptions must hold in order to ensure the validity of the test you conducted in part **b**?

## Applying the Concepts—Intermediate

**13.13** Each year, approximately 1.3 million people in the United States suffer adverse drug effects (ADEs), that is, unintended injuries caused by prescribed medication. A study in the *Journal of the American Medical Association* (July 5, 1995) identified the cause of 247 ADEs that occurred at two Boston hospitals. The researchers found that dosing errors (that is, wrong dosage prescribed and/or dispensed) were the most common. The table summarizes the proximate cause of 95 ADEs that resulted from a dosing error. Conduct a test (at $\alpha = .10$) to determine whether the true percentages of ADEs in the five "cause" categories are different. Use the accompanying SPSS printout to arrive at your decision.

| Wrong Dosage Cause | Number of ADEs |
|---|---|
| (1) Lack of knowledge of drug | 29 |
| (2) Rule violation | 17 |
| (3) Faulty dose checking | 13 |
| (4) Slips | 9 |
| (5) Other | 27 |

**SPSS Output for Exercise 13.13**

**CAUSE**

|  | Observed N | Expected N | Residual |
|---|---|---|---|
| LACKKNOW | 29 | 19.0 | 10.0 |
| RULEVIOLATE | 17 | 19.0 | -2.0 |
| FAULTYDOSE | 13 | 19.0 | -6.0 |
| SLIPS | 9 | 19.0 | -10.0 |
| OTHER | 27 | 19.0 | 8.0 |
| Total | 95 |  |  |

**Test Statistics**

|  | CAUSE |
|---|---|
| Chi-Square[a] | 16.000 |
| df | 4 |
| Asymp. Sig. | .003 |

a. 0 cells (.0%) have expected frequencies less than 5. The minimum expected cell frequency is 19.0.

**13.14** In education, the term *instructional technology* refers to products such as computers, spreadsheets, CD-ROMs, videos, and presentation software. How frequently do professors use instructional technology in the classroom? To answer this question, researchers at Western Michigan University surveyed 306 of their fellow faculty (*Educational Technology*, Mar.–Apr. 1995). Responses to the frequency-of-technology use in teaching were recorded as "weekly to every class," "once a semester to monthly," or "never." The faculty responses (number in each response category) for the three technologies are summarized in the table.

| Technology | Weekly | Once a Semester/ Monthly | Never |
|---|---|---|---|
| Computer spreadsheets | 58 | 67 | 181 |
| Word processing | 168 | 61 | 77 |
| Statistical software | 37 | 82 | 187 |

a. Determine whether the percentages in the three frequency-of-use response categories differ for computer spreadsheets. Use $\alpha = .01$.
b. Repeat part **a** for word processing.
c. Repeat part **a** for statistical software.
d. Construct a 99% confidence interval for the true percentage of faculty who never use computer spreadsheets in the classroom. Interpret the interval.

**13.15** Interferons are proteins produced naturally by the human body that help fight infections and regulate the immune system. A drug developed from interferons, called Avonex, is now available for treating patients with multiple sclerosis (MS). In a clinical study, 85 MS patients received weekly injections of Avonex over a 2-year period. The number of exacerbations (i.e., flareups of symptoms) was recorded for each patient and is summarized in the next table. For MS patients who take a placebo (no drug) over a similar 2-year period, it is known from previous studies that 26% will experience no exacerbations, 30% one exacerbation, 11% two exacerbations, 14% three exacerbations, and 19% four or more exacerbations.

| Number of Exacerbations | Number of Patients |
|---|---|
| 0 | 32 |
| 1 | 26 |
| 2 | 15 |
| 3 | 6 |
| 4 or more | 6 |

*Source:* Biogen, Inc., 1997.

a. Conduct a test to determine whether the exacerbation distribution of MS patients who take Avonex differs from the percentages reported for placebo patients. Test using $\alpha = .05$.
b. Find a 95% confidence interval for the true percentage of Avonex MS patients who are exacerbation-free during a 2-year period.
c. Refer to part **b**. Is there evidence that Avonex patients are more likely to have no exacerbations than placebo patients? Explain.

**13.16** *Nature* (Sept. 1993) reported on a study of animal and plant species "hotspots" in Great Britain. A hotspot is defined as a 10-km$^2$ area that is species-rich, that is, is heavily populated by the species of interest. Analogously, a coldspot is a 10-km$^2$ area that is species-poor. The table gives the number of butterfly hotspots and the number of butterfly coldspots in a sample of 2,588 10-km$^2$ areas. In theory, 5% of the areas should be butterfly hotspots and 5% should be butterfly coldspots, while the remaining areas (90%) are neutral. Test the theory using $\alpha = .01$.

| | |
|---|---|
| Butterfly hotspots | 123 |
| Butterfly coldspots | 147 |
| Neutral areas | 2,318 |
| **Total** | 2,588 |

*Source:* Prendergast, J. R., *et al.* "Rare species, the coincidence of diversity hotspots and conservation strategies." *Nature*, Vol. 365, No. 6444, Sept. 23, 1993, p. 335 (Table 1).

## 13.3 TESTING CATEGORICAL PROBABILITIES: TWO-WAY (CONTINGENCY) TABLE

In Section 13.1, we introduced the multinomial probability distribution and considered data classified according to a single criterion. We now consider multinomial experiments in which the data are classified according to two criteria, that is, *classification with respect to two qualitative factors*.

For example, consider a study in the *Journal of Marketing* (Fall 1992) on the impact of using celebrities in television advertisements. The researchers investigated the relationship between gender of a viewer and the viewer's brand awareness. Three hundred TV viewers were asked to identify products advertised by male celebrity spokespersons. The data are summarized in the **two-way table** shown in Table 13.3. This table is called a **contingency table**; it presents multinomial count data classified on two scales, or **dimensions, of classification**—namely, gender of viewer and brand awareness.

The symbols representing the cell counts for the multinomial experiment in Table 13.3 are shown in Table 13.4a; and the corresponding cell, row, and column probabilities are shown in Table 13.4b. Thus, $n_{11}$ represents the number of viewers who are male and could identify the brand and $p_{11}$ represents the corresponding cell probability. Note the symbols for the row and column totals and also the symbols for the probability totals. The latter are called **marginal probabilities** for each row and column. The marginal probability $p_{r1}$ is the probability that a TV viewer identifies the product; the marginal probability $p_{c1}$ is the probability that the TV viewer is male. Thus,

$$p_{r1} = p_{11} + p_{12} \text{ and } p_{c1} = p_{11} + p_{21}$$

Thus, we can see that this really is a multinomial experiment with a total of 300 trials, $(2)(2) = 4$ cells or possible outcomes, and probabilities for each cell as shown in Table 13.4b. If the 300 TV viewers are randomly chosen, the trials are considered independent and the probabilities are viewed as remaining constant from trial to trial.

Suppose we want to know whether the two classifications, gender and brand awareness, are dependent. That is, if we know the gender of the TV viewer, does

**TABLE 13.3    Contingency Table for Marketing Example**

|  |  | Gender | | |
| --- | --- | --- | --- | --- |
|  |  | Male | Female | Totals |
| **Brand Awareness** | **Could Identify Product** | 95 | 41 | 136 |
|  | **Could Not Identify Product** | 55 | 109 | 164 |
|  | Totals | 150 | 150 | 300 |

**TABLE 13.4a    Observed Counts for Contingency Table 13.3**

|  |  | Gender | | |
| --- | --- | --- | --- | --- |
|  |  | Male | Female | Totals |
| **Brand Awareness** | **Could Identify Product** | $n_{11}$ | $n_{12}$ | $r_1$ |
|  | **Could Not Identify Product** | $n_{21}$ | $n_{22}$ | $r_2$ |
|  | Totals | $c_1$ | $c_2$ | $n$ |

**TABLE 13.4b    Probabilities for Contingency Table 13.3**

|  |  | Gender | | |
| --- | --- | --- | --- | --- |
|  |  | Male | Female | Totals |
| **Brand Awareness** | **Could Identify Product** | $p_{11}$ | $p_{12}$ | $p_{r1}$ |
|  | **Could Not Identify Product** | $p_{21}$ | $p_{22}$ | $p_{r2}$ |
|  | Totals | $p_{c1}$ | $p_{c2}$ | 1 |

that information give us a clue about the viewer's brand awareness? In a probabilistic sense we know (Chapter 3) that independence of events A and B implies $P(AB) = P(A)P(B)$. Similarly, in the contingency table analysis, if the **two classifications are independent**, the probability that an item is classified in any particular cell of the table is the product of the corresponding marginal probabilities. Thus, under the hypothesis of independence, in Table 13.4b, we must have

$$p_{11} = p_{r1}p_{c1} \quad p_{12} = p_{r1}p_{c2}$$
$$p_{21} = p_{r2}p_{c1} \quad p_{22} = p_{r2}p_{c2}$$

To test the hypothesis of independence, we use the same reasoning employed in the one-dimensional tests of Section 13.2. First, we calculate the *expected*, or *mean, count in each cell* assuming that the null hypothesis of independence is true. We do this by noting that the expected count in a cell of the table is just the total number of multinomial trials, $n$, times the cell probability. Recall that $n_{ij}$ represents the **observed count** in the cell located in the $i$th row and $j$th column. Then the expected cell count for the upper-left-hand cell (first row, first column) is

$$E(n_{11}) = np_{11}$$

or, when the null hypothesis (the classifications are independent) is true,

$$E(n_{11}) = np_{r1}p_{c1}$$

Since these true probabilities are not known, we estimate $p_{r1}$ and $p_{c1}$ by the same proportions $\hat{p}_{r1} = r_1/n$ and $\hat{p}_{c1} = c_1/n$. Thus, the estimate of the expected value $E(n_{11})$ is

$$\hat{E}(n_{11}) = n\left(\frac{r_1}{n}\right)\left(\frac{c_1}{n}\right) = \frac{r_1 c_1}{n}$$

Similarly, for each $i, j$,

$$\hat{E}(n_{ij}) = \frac{(\text{Row total})(\text{Column total})}{\text{Total sample size}}$$

Thus,

$$\hat{E}(n_{12}) = \frac{r_1 c_2}{n}$$

$$\hat{E}(n_{21}) = \frac{r_2 c_1}{n}$$

$$\hat{E}(n_{22}) = \frac{r_2 c_2}{n}$$

Using the data in Table 13.3, we find

$$\hat{E}(n_{11}) = \frac{r_1 c_1}{n} = \frac{(136)(150)}{300} = 68$$

$$\hat{E}(n_{12}) = \frac{r_1 c_2}{n} = \frac{(136)(150)}{300} = 68$$

$$\hat{E}(n_{12}) = \frac{r_2 c_1}{n} = \frac{(164)(150)}{300} = 82$$

$$\hat{E}(n_{22}) = \frac{r_2 c_2}{n} = \frac{(164)(150)}{300} = 82$$

The observed data and the estimated expected values (in parentheses) are shown in Table 13.5.

**TABLE 13.5   Observed and Estimated Expected (in Parentheses) Counts**

|  |  | Gender | | Totals |
|---|---|---|---|---|
|  |  | **Male** | **Female** | |
| **Brand Awareness** | **Could Identify Product** | 95 (68) | 41 (68) | 136 |
| | **Could Not Identify Product** | 55 (82) | 109 (82) | 164 |
| | **Totals** | 150 | 150 | 300 |

We now use the $\chi^2$ statistic to compare the observed and expected (estimated) counts in each cell of the contingency table:

$$\chi^2 = \frac{[n_{11} - \hat{E}(n_{11})]^2}{\hat{E}(n_{11})} + \frac{[n_{12} - \hat{E}(n_{12})]^2}{\hat{E}(n_{12})} + \frac{[n_{21} - \hat{E}(n_{21})]^2}{\hat{E}(n_{21})} + \frac{[n_{22} - \hat{E}(n_{22})]^2}{\hat{E}(n_{22})}$$

$$= \sum \frac{[n_{ij} - \hat{E}(n_{ij})]^2}{\hat{E}(n_{ij})}$$

*Note:* The use of $\sum$ in the context of a contingency table analysis refers to a sum over all cells in the table.

Substituting the data of Table 13.5 into this expression, we get

$$\chi^2 = \frac{(95 - 68)^2}{68} + \frac{(41 - 68)^2}{68} + \frac{(55 - 82)^2}{82} + \frac{(109 - 82)^2}{82} = 39.22$$

Large values of $\chi^2$ imply that the observed counts do not closely agree and hence that the hypothesis of independence is false. To determine how large $\chi^2$ must be before it is too large to be attributed to chance, we make use of the fact that the sampling distribution of $\chi^2$ is approximately a $\chi^2$ probability distribution when the classifications are independent.

When testing the null hypothesis of independence in a two-way contingency table, the appropriate degrees of freedom will be $(r - 1)(c - 1)$, where $r$ is the number of rows and $c$ is the number of columns in the table.

For the brand awareness example, the degrees of freedom for $\chi^2$ is $(r - 1)(c - 1) = (2 - 1)(2 - 1) = 1$. Then, for $\alpha = .05$, we reject the hypothesis of independence when

$$\chi^2 > \chi^2_{.05} = 3.84146$$

Since the computed $\chi^2 = 39.22$ exceeds the value 3.84146, we conclude that viewer gender and brand awareness are dependent events.

The pattern of **dependence** can be seen more clearly by expressing the data as percentages. We first select one of the two classifications to be used as the base variable. In the above example, suppose we select gender of the TV viewer as the classificatory variable to be the base. Next, we represent the responses for each level of the second categorical variable (brand awareness in our example) as a percentage of the subtotal for the base variable. For example, from Table 13.5 we convert the response for males who identify the brand (95) to a percentage of the total number of male viewers (150). That is,

$$\left(\frac{95}{150}\right)100\% = 63.3\%$$

**TABLE 13.6** **Percentage of TV Viewers Who Identify Brand, by Gender**

| | | Gender | | |
|---|---|---|---|---|
| | | **Male** | **Female** | **Totals** |
| **Brand Awareness** | **Could Identify Product** | 63.3 | 27.3 | 45.3 |
| | **Could Not Identify Product** | 36.7 | 72.7 | 54.7 |
| | **Totals** | 100 | 100 | 100 |

The conversions of all Table 13.5 entries are similarly computed, and the values are shown in Table 13.6. The value shown at the right of each row is the row's total expressed as a percentage of the total number of responses in the entire table. Thus, the percentage of TV viewers who identify the product is: $\left(\frac{136}{300}\right)100\% = 45.3\%$ (rounded to the nearest percent).

If the gender and brand awareness variables are independent, then the percentages in the cells of the table are expected to be approximately equal to the corresponding row percentages. Thus, we would expect the percentage who identify the brand for each gender to be approximately 45% if the two variables are independent. The extent to which each gender's percentage departs from this value determines the dependence of the two classifications, with greater variability of the row percentages meaning a greater degree of dependence. A plot of the percentages helps summarize the observed pattern. In the SPSS bar graph in Figure 13.3, we show the gender of the viewer (the base variable) on the horizontal axis, and the percentage of TV viewers who identify the brand on the vertical axis. The "expected" percentage under the assumption of independence is shown as a dotted horizontal line.

Figure 13.3 clearly indicates the reason that the test resulted in the conclusion that the two classifications in the contingency table are dependent. The percentage of male TV viewers who identify the brand promoted by a male celebrity is more than twice as high as the percentage of female TV viewers who identify the

**FIGURE 13.3**

*SPSS Bar Graph Showing Percentage of Viewers Who Identified the TV Product*

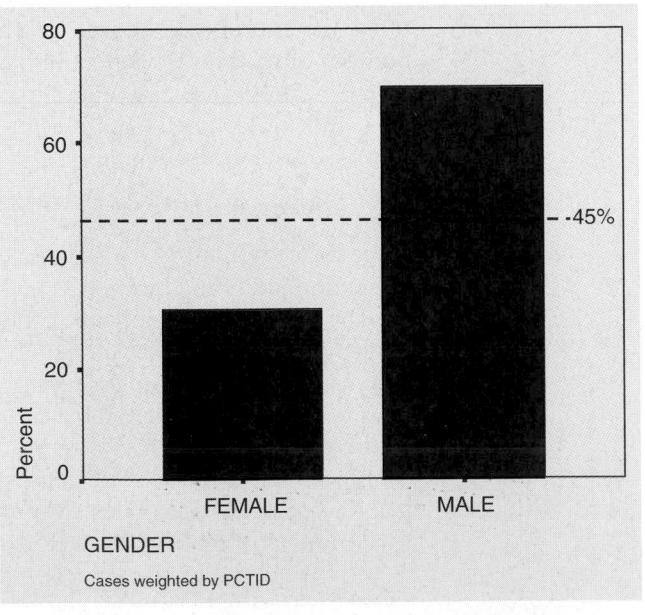

**TABLE 13.7   General $r \times c$ Contingency Table**

|  |  | Column | | | | Row Totals |
|---|---|---|---|---|---|---|
|  |  | **1** | **2** | $\cdots$ | **$c$** |  |
| **Row** | **1** | $n_{11}$ | $n_{12}$ | $\cdots$ | $n_{1c}$ | $r_1$ |
|  | **2** | $n_{21}$ | $n_{22}$ | $\cdots$ | $n_{2c}$ | $r_2$ |
|  | $\vdots$ | $\vdots$ | $\vdots$ | | $\vdots$ | $\vdots$ |
|  | **r** | $n_{r1}$ | $n_{r2}$ | $\cdots$ | $n_{rc}$ | $r_r$ |
| **Column Totals** |  | $c_1$ | $c_2$ | $\cdots$ | $c_c$ | $n$ |

brand. Statistical measures of the degree of dependence and procedures for making comparisons of pairs of levels for classifications are available. They are beyond the scope of this text, but can be found in the references. We will, however, utilize descriptive summaries such as Figure 13.3 to examine the degree of dependence exhibited by the sample data.

The general form of a two-way contingency table containing $r$ rows and $c$ columns (called an $r \times c$ contingency table) is shown in Table 13.7. Note that the observed count in the *(ij)* cell is denoted by $n_{ij}$, the $i$th row total is $r_i$, the $j$th column total is $c_j$, and the total sample size is $n$. Using this notation, we give the general form of the contingency table test for independent classifications in the next box.

---

### General Form of a Two-way (Contingency) Table Analysis: A Test for Independence

$H_0$: The two classifications are independent
$H_a$: The two classifications are dependent

*Test statistic:*   $\chi^2 = \sum \dfrac{[n_{ij} - \hat{E}(n_{ij})]^2}{\hat{E}(n_{ij})}$

where $\hat{E}(n_{ij}) = \dfrac{r_i c_j}{n}$

*Rejection region:*   $\chi^2 > \chi^2_\alpha$, where $\chi^2_\alpha$ has $(r-1)(c-1)$ df

*Assumptions:*  1. The $n$ observed counts are a random sample from the population of interest. We may then consider this to be a multinomial experiment with $r \times c$ possible outcomes.

2. The sample size, $n$, will be large enough so that, for every cell, the expected count, $E(n_{ij})$, will be equal to 5 or more.

---

**EXAMPLE 13.3**

A social scientist wants to determine whether the marital status (divorced or not divorced) of U.S. men is independent of their religious affiliation (or lack thereof). A sample of 500 U.S. men is surveyed and the results are tabulated as shown in Table 13.8.

a. Test to see whether there is sufficient evidence to indicate that the marital status of men who have been or are currently married is dependent on religious affiliation. Test $\alpha = .01$.

b. Graph the data and describe the patterns revealed. Is the result of the test supported by the graph?

**TABLE 13.8    Survey Results (Observed Counts), Example 13.2**

|  |  | **Religious Affiliation** | | | | | |
| --- | --- | --- | --- | --- | --- | --- | --- |
|  |  | **A** | **B** | **C** | **D** | **None** | **Totals** |
| **Marital Status** | **Divorced** | 39 | 19 | 12 | 28 | 18 | 116 |
|  | **Never divorced** | 172 | 61 | 44 | 70 | 37 | 384 |
|  | **Totals** | 211 | 80 | 56 | 98 | 55 | 500 |

### Solution

a. The first step is to calculate estimated expected cell frequencies under the assumption that the classifications are independent. Rather than compute these values by hand, we resort to a computer. The SAS printout of the analysis of Table 13.8 is displayed in Figure 13.4. Each cell in Figure 13.4 contains the observed (top) and expected (bottom) frequency in that cell. Note that $\hat{E}(n_{11})$, the estimated expected count for the Divorced, A cell, is 48.952. Similarly, the estimated expected count for the Divorced, B cell, is $\hat{E}(n_{12}) = 18.56$. Since all the estimated expected cell frequencies are greater than 5, the $\chi^2$ approximation for the test statistic is appropriate. Assuming the men chosen were randomly selected from all married or previously married American men, the characteristics of the multinomial probability distribution are satisfied.

**FIGURE 13.4**

*SAS Contingency Table Printout*

```
                        The FREQ Procedure

                   Table of MARITAL by RELIGION

MARITAL      RELIGION

Frequency
Expected
Col Pct    A        B        C        D        NONE      Total

DIVORCED        39       19       12       28       18       116
             48.952    18.56    12.992   22.736   12.76
             18.48     23.75    21.43    28.57    32.73

NEVER          172       61       44       70       37       384
             162.05    60.44    43.008   75.264   42.24
             81.52     76.25    78.57    71.43    67.27

Total          211       80       56       98       55       500


              Statistics for Table of MARITAL by RELIGION

   Statistic                        DF      Value     Prob

   Chi-Square                        4      7.1355    0.1289
   Likelihood Ratio Chi-Square       4      6.9854    0.1367
   Mantel-Haenszel Chi-Square        1      6.4943    0.0108
   Phi Coefficient                          0.1195
   Contingency Coefficient                  0.1186
   Cramer's V                               0.1195

                     Sample Size = 500
```

**FIGURE 13.5**

*SAS Bar Graph Showing Percentage of Divorced Males by Religion*

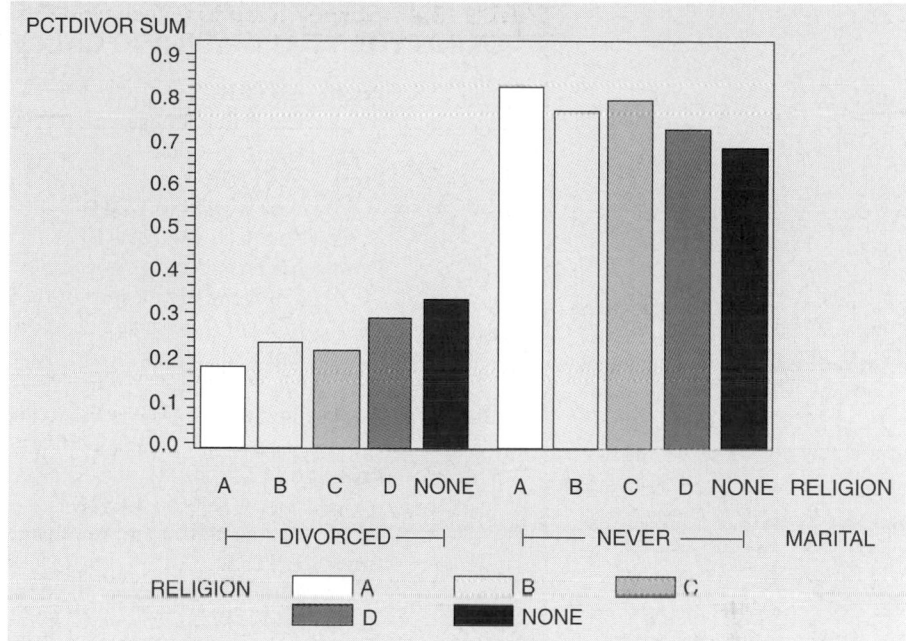

The null and alternative hypotheses we want to test are

$H_0$: The marital status of U.S. men and their religious affiliation are independent
$H_a$: The marital status of U.S. men and their religious affiliation are dependent

The test statistic, $\chi^2 = 7.135$, is highlighted at the bottom of the printout, as is the observed significance level (*p*-value) of the test. Since $\alpha = .01$ is less than $p = .129$, we fail to reject $H_0$; that is, we cannot conclude that the marital status of U.S. men depends on their religious affiliation. (Note that we could not reject $H_0$ even with $\alpha = .10$.)

b. The marital status frequencies are expressed as percentages of the number of men in each religious affiliation category and highlighted in the SAS printout, Figure 13.4. The expected percentage of divorced men under the assumption of independence is $\left(\frac{116}{500}\right)100\% = 23\%$. A SAS graph of the percentages is shown in Figure 13.5. Note that the percentages of divorced men (see the bars in the "DIVORCED" block of the SAS graph) deviate only slightly from that expected under the assumption of independence, supporting the result of the test in part a. That is, neither the descriptive bar graph nor the statistical test provides evidence that the male divorce rate depends on (varies with) religious affiliation. ■

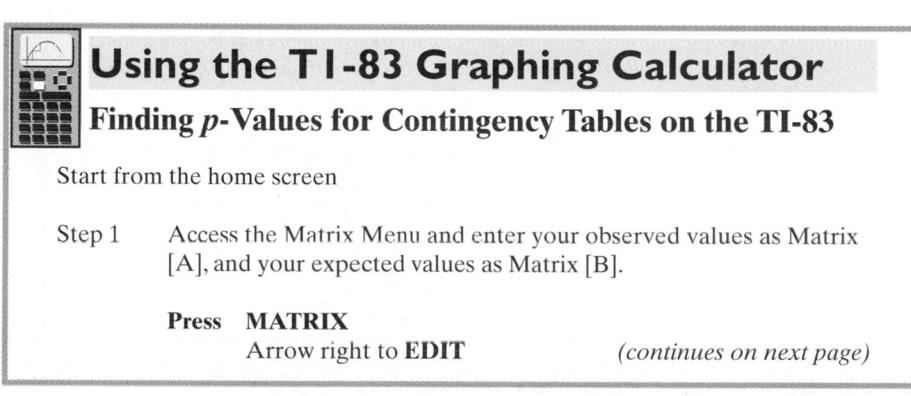

# Using the TI-83 Graphing Calculator

## Finding *p*-Values for Contingency Tables on the TI-83

Start from the home screen

Step 1    Access the Matrix Menu and enter your observed values as Matrix [A], and your expected values as Matrix [B].

**Press   MATRIX**
Arrow right to **EDIT**                          *(continues on next page)*

## USING THE TI-83 GRAPHING CALCULATOR *(continued)*

**Press** **ENTER**
Enter the dimensions of your observed matrix
Enter all values
**Press** **MATRIX**
Arrow right to **EDIT**
Arrow down to **2:[B]**
**Press** **ENTER**
Enter the dimensions of the expected values matrix
(*it will be the same size as your observed matrix above*)
Enter all values

Step 2   Access the Statistical Tests Menu and perform the $\chi^2$ test.

**Press** **STAT**
Arrow right to **TESTS**
Arrow up to **C: $\chi^2$ − Tests...**
**Press** **ENTER**
Enter the names for the observed and expected matrices
(*Press Matrix arrow to desired Matrix, Press ENTER*)
Arrow down to **Calculate**
**Press** **ENTER**

Step 3   *View Display*
Reject $H_0$ if the *p*-value $< \alpha$.

Example   Consider Example 13.2 from the Text.

Our observed matrix will be $[A] = \begin{vmatrix} 39 & 19 & 12 & 28 & 18 \\ 172 & 61 & 44 & 70 & 37 \end{vmatrix}$

Our expected matrix will be $[B] =$

$$\begin{vmatrix} 48.952 & 18.56 & 12.992 & 22.736 & 12.76 \\ 162.05 & 61.44 & 43.008 & 75.264 & 42.24 \end{vmatrix}$$

The hypotheses for this problem are as follows:

$H_0$: The matrix entries represent independent events.
$H_a$: The matrix entries represent events that *are not* independent.

Step 1   Access the Matrix Menu and enter all values

**Press** **MATRIX**
Arrow right to **EDIT**
**Press** **ENTER**
Enter the dimensions of the observed matrix (see screens below).

```
MATRIX[A]  2 ×5
[ 39      19      12     _
[ 172     61      44     _



```

```
MATRIX[A]  2 ×5
_ 12      28      18     ]
_ 44      70      37     ]




2,5=37
```

## USING THE TI-83 GRAPHING CALCULATOR *(continued)*

Enter all values (see screens above).

**Press  MATRIX**
Arrow right to **EDIT**
**Press  ENTER**
Enter the dimensions of the expected matrix (same as above).
Enter all expected values* (see screen below).

Each expected value entry is the product of (that row total)* (that column total) /grand total.

Step 2  Perform the $\chi^2 -$ Test.

**Press  STAT**
Arrow right to **TESTS**
Arrow up to **C: $\chi^2 -$ Test . . .** (see screen)

**Press  ENTER(see screen)**
**Set**  Observed**: [A]**
Expected**: [B]**
Arrow down to **Calculate**
**Press  ENTER**(see screen)

Step 3  *View Display*
As you can see from the screen, the *p*-value is 0.129. Since the *p*-value is *not less* than we *do not* reject $H_0$. There is no statistical evidence at the $\alpha = 0.01$ level to conclude that the male divorce rate depends on religious affiliation.

Step 4  Clear screen for the next problem. Return to the home screen.
Press  **CLEAR**

**EXERCISES 13.17–13.31**

### Learning the Mechanics

**13.17** Find the rejection region for a test of independence of two classifications where the contingency table contains $r$ rows and $c$ columns and
- **a.** $r = 5$, $c = 5$, $\alpha = .05$
- **b.** $r = 3$, $c = 6$, $\alpha = .10$
- **c.** $r = 2$, $c = 3$, $\alpha = .01$

**13.18** Consider the accompanying $2 \times 3$ (i.e., $r = 2$ and $c = 3$) contingency table.

|  |  | Column | | |
|---|---|---|---|---|
|  |  | **1** | **2** | **3** |
| **Row** | **1** | 9 | 34 | 53 |
|  | **2** | 16 | 30 | 25 |

- **a.** Specify the null and alternative hypotheses that should be used in testing the independence of the row and column classifications.
- **b.** Specify the test statistic and the rejection region that should be used in conducting the hypothesis test of part **a**. Use $\alpha = .01$.
- **c.** Assuming the row classification and the column classification are independent, find estimates for the expected cell counts.
- **d.** Conduct the hypothesis test of part **a**. Interpret your result.

**13.19** Refer to Exercise 13.18.
- **a.** Convert the frequency responses to percentages by calculating the percentage of each column total falling in each row. Also convert the row totals to percentages of the total number of responses. Display the percentages in a table.
- **b.** Create a bar graph with row 1 percentage on the vertical axis and column number on the horizontal axis. Show the row 1 total percentage as a horizontal line on the graph.
- **c.** What pattern do you expect to see if the rows and columns are independent? Does the plot support the result of the test of independence in Exercise 13.18?

**13.20** Test the null hypothesis of independence of the two classifications, A and B, of the $3 \times 3$ contingency table shown here. Test using $\alpha = .05$.

|  |  | B | | |
|---|---|---|---|---|
|  |  | **B₁** | **B₂** | **B₃** |
|  | **A₁** | 40 | 72 | 42 |
| **A** | **A₂** | 63 | 53 | 70 |
|  | **A₃** | 31 | 38 | 30 |

**13.21** Refer to Exercise 13.20. Convert the responses to percentages by calculating the percentage of each B class total falling into each A classification. Also, calculate the percentage of the total number of responses that constitute each of the A classification totals.
- **a.** Create a bar graph with row $A_1$ percentage on the vertical axis and B classification on the horizontal axis. Does the graph support the result of the test of hypothesis in Exercise 13.20? Explain.
- **b.** Repeat part **a** for the row $A_2$ percentages.
- **c.** Repeat part **a** for the row $A_3$ percentages.

### Applying the Concepts—Basic

**13.22** The *Journal of the American Medical Association* (Apr. 18, 2001) published the results of a study of alcohol consumption in patients suffering from acute myocardial infarction (AMI). The patients were classified according to average number of alcoholic drinks per week and whether or not they had congestive heart failure. A summary of the results for 1,913 AMI patients is shown in the table.

⊘ AMI

|  |  | Alcohol Consumption | | |
|---|---|---|---|---|
|  |  | **Abstainers** | **Less than 7 drinks/week** | **7 or more drinks/week** |
| **Congestive** | **Yes** | 146 | 106 | 29 |
| **Heart Failure** | **No** | 750 | 590 | 292 |
|  | **Totals** | 896 | 696 | 321 |

*Source:* Mukamal, K. J., *et al.* "Prior alcohol consumption and mortality following acute myocardial infarction." *Journal of the American Medical Association*, Vol. 285, No. 15, Apr. 18, 2001 (Table 1).

- **a.** Find the sample proportion of abstainers with congestive heart failure.
- **b.** Find the sample proportion of moderate drinkers (patients who have less than 7 drinks per week) with congestive heart failure.
- **c.** Find the sample proportion of heavy drinkers (patients who have 7 or more drinks per week) with congestive heart failure.
- **d.** Compare the sample proportions, parts **a–c**. Does it appear that the proportion of AMI patients with congestive heart failure depends on alcohol consumption?
- **e.** Give the null hypothesis for testing whether the proportion of AMI patients with congestive heart failure depends on alcohol consumption.
- **f.** Use the MINITAB printout on p. 723 to conduct the test, part **e**. Test at $\alpha = .05$.

**13.23** One of the most common musculoskeletal disorders is lumbar disk disease (LDD). Medical researchers

```
MINITAB Output for Exercise 13.22

Expected counts are printed below observed counts

       ABSTAIN    LESS7   7ORMORE    Total
   1       146      106        29      281
        131.61   102.24     47.15

   2       750      590       292     1632
        764.39   593.76    273.85

Total      896      696       321     1913

Chi-Sq =   1.573 +   0.139 +   6.988 +
           0.271 +   0.024 +   1.203 = 10.197
DF = 2, P-Value = 0.006
```

reported finding a common genetic risk factor for LDD (*Journal of the American Medical Association*, Apr. 11, 2001). The study included 171 Finnish patients diagnosed with LDD (the patient group) and 321 Finnish individuals without LDD (the control group). Of the 171 LDD patients, 21 were discovered to have the genetic trait. Of the 321 people in the control group, 15 had the genetic trait.

**a.** Consider the two categorical variables, group and presence/absence of genetic trait. Form a $2 \times 2$ contingency table for these variables.

**b.** Conduct a test to determine whether the genetic trait occurs at a higher rate in LDD patients than in the controls without LDD. Use $\alpha = .01$.

**c.** Construct a bar graph that will visually support your conclusion in part **b**.

**13.24** The *American Journal of Public Health* (July 1995) reported on a population-based study of trauma in Hispanic children. One of the objectives of the study was to compare the use of protective devices in motor vehicles used to transport Hispanic and non-Hispanic white children. On the basis of data collected from the San Diego County Regionalized Trauma System, 792 children treated for injuries sustained in vehicular accidents were classified according to ethnic status (Hispanic or non-Hispanic white) and seatbelt usage (worn

or not worn) during the accident. The data are summarized in the table.

⬤ TRAUMA

|  | Hispanic | Non-Hispanic White | Totals |
|---|---|---|---|
| **Seatbelts worn** | 31 | 148 | 179 |
| **Seatbelts not worn** | 283 | 330 | 613 |
| **Totals** | 314 | 478 | 792 |

*Source:* Matteneci, R.M., *et al.* "Trauma among Hispanic children: A population-based study in a regionalized system of trauma care." *American Journal of Public Health*, Vol. 85, No. 7, July 1995, p. 1007 (Table 2).

**a.** Calculate the sample proportion of injured Hispanic children who were not wearing seatbelts during the accident.

**b.** Calculate the sample proportion of injured non-Hispanic white children who were not wearing seatbelts during the accident.

**c.** Compare the two sample proportions, parts **a** and **b**. Do you think the true population proportions differ?

**d.** Conduct a test to determine whether seatbelt usage in motor vehicle accidents depends on ethnic status in the San Diego County Regionalized Trauma System. Use $\alpha = .01$.

**e.** Construct a 99% confidence interval for the difference between the proportions, parts **a** and **b**. Interpret the interval.

### Applying the Concepts—Intermediate

**13.25** University of Maryland professor Ted R. Gurr examined the political strategies used by ethnic groups worldwide in their fight for minority rights (*Political Science & Politics*, June 2000). Each in a sample of 275 ethnic groups was classified according to world region and highest level of political action reported. The data are summarized in the accompanying contingency table. Conduct a test at $\alpha = .10$ to determine whether political strategy of ethnic groups depends on world region. Support your answer with a graph.

⬤ ETHNIC

| | | **Political Strategy** | | |
|---|---|---|---|---|
| | | **No Political Action** | **Mobilization, Mass Action** | **Terrorism, Rebellion, Civil War** |
| **World Region** | **Latin American** | 24 | 31 | 7 |
| | **Post-Communist** | 32 | 23 | 4 |
| | **South, Southeast, East Asia** | 11 | 22 | 26 |
| | **Africa/Middle East** | 39 | 36 | 20 |

*Source:* Gurr, T. R. "Nonviolence in ethnopolitics: Strategies for the attainment of group rights and autonomy." *Political Science & Politics*, Vol. 33, No. 2, June 2000 (Table 1).

**13.26** An article in *Sociological Methods & Research* (May 2001) analyzed the data presented in the table. A sample of 262 Kansas pig farmers were classified according to their education level (college or not) and size of their pig farm (number of pigs). Conduct a test to determine whether a pig farmer's education level has an impact on the size of the pig farm. Use $\alpha = .05$ and support your answer with a graph.

🔵 PIGFARM

| | | Education Level | | |
| --- | --- | --- | --- | --- |
| | | **No College** | **College** | **Totals** |
| **Farm Size** | **<1,000 pigs** | 42 | 53 | 95 |
| | **1,000–2,000 pigs** | 27 | 42 | 69 |
| | **2,001–5,000 pigs** | 22 | 20 | 42 |
| | **>5,000 pigs** | 27 | 29 | 56 |
| | **Totals** | 118 | 144 | 262 |

*Source:* Agresti, A., and Liu, I. "Strategies for modeling a categorical variable allowing multiple category choices." *Sociological Methods & Research*, Vol. 29, No. 4, May 2001 (Table I).

**13.27** A field study was conducted to identify the natural predators of the gypsy moth (*Environmental Entomology*, June 1995). For one part of the study, 24 black-capped chickadees (common wintering birds) were captured in mist nets and individually caged. Each bird was offered a mass of gypsy moth eggs attached to a piece of bark. Half the birds were offered no other food (no choice) and half were offered a variety of other naturally occurring foods such as spruce and pine seeds (choice). The numbers of birds that did and did not feed on the gypsy moth egg mass are given in the table. Analyze the data in the table to determine if a relationship exists between food choice and whether or not the chickadees fed on gypsy moth eggs. Use $\alpha = .10$.

🔵 MOTH

| | Fed on Egg Mass | |
| --- | --- | --- |
| | **Yes** | **No** |
| **Choice of foods** | 2 | 10 |
| **No choice** | 8 | 4 |

**13.28** In order to evaluate their situational awareness, fighter aircraft pilots participate in battle simulations. At a random point in the trial, the simulator is frozen and data on situation awareness are immediately collected. The simulation is then continued until, ultimately, performance (e.g., number of kills) is measured. A study reported in *Human Factors* (Mar. 1995) investigated whether temporarily stopping the simulation results in any change in pilot performance. Trials were designed so that some simulations were stopped to collect situation awareness data while others were not stopped. Each trial was then classified according to the number of kills made by the pilot. The data for 180 trials are summarized in the contingency table, and an SPSS printout of the analysis is provided on p. 725. Fully interpret the results.

🔵 SIMKILLS

| | Number of Kills | | | | | |
| --- | --- | --- | --- | --- | --- | --- |
| | **0** | **1** | **2** | **3** | **4** | **Totals** |
| **Stops** | 32 | 33 | 19 | 5 | 2 | 91 |
| **No Stops** | 24 | 36 | 18 | 8 | 3 | 89 |
| **Totals** | 56 | 69 | 37 | 13 | 5 | 180 |

**13.29** A standardized procedure for determining a person's susceptibility to hypnosis is the Stanford Hypnotic Susceptibility Scale, Form C (SHSS:C). Recently, a new method called the Computer-Assisted Hypnosis Scale (CAHS), which uses a computer as a facilitator of hypnosis, has been developed. Each scale classifies a person's hypnotic susceptibility as low, medium, high, and very high. Researchers at the University of Tennessee compared the two scales by administering both tests to each of 130 undergraduate volunteers (*Psychological Assessment*, Mar. 1995). The hypnotic classifications are summarized in the table below. A chi-square analysis of the data in the contingency table is shown in the SAS printout on p. 726. Note that the expected cell counts are provided in the computer printout.

**a.** Check to see if the assumption of expected cell counts of 5 or more is satisfied. Should you proceed with the analysis? Explain.

🔵 HYPNOSIS

| | | CAHS Level | | | | |
| --- | --- | --- | --- | --- | --- | --- |
| | | **Low** | **Medium** | **High** | **Very High** | **Totals** |
| **SHSS: C Level** | **Low** | 32 | 14 | 2 | 0 | 48 |
| | **Medium** | 11 | 14 | 6 | 0 | 31 |
| | **High** | 6 | 14 | 19 | 3 | 42 |
| | **Very High** | 0 | 2 | 4 | 3 | 9 |
| | **Totals** | 49 | 44 | 31 | 6 | 130 |

*Source:* Grant, C. D., and Nash, M. R. "The Computer-Assisted Hypnosis Scale: Standardization and norming of a computer-administered measure of hypnotic ability." *Psychological Assessment*, Vol. 7, No. 1, Mar. 1995, p. 53 (Table 4).

SPSS Output for Exercise 13.28

**STOPS *KILLS Crosstabulation**

| | | | KILLS 0 | 1 | 2 | 3 | 4 | Total |
|---|---|---|---|---|---|---|---|---|
| STOPS | NO | Count | 24 | 36 | 18 | 8 | 3 | 89 |
| | | Expected Count | 27.7 | 34.1 | 18.3 | 6.4 | 2.5 | 89.0 |
| | YES | Count | 32 | 33 | 19 | 5 | 2 | 91 |
| | | Expected Count | 28.3 | 34.9 | 18.7 | 6.6 | 2.5 | 91.0 |
| Total | | Count | 56 | 69 | 37 | 13 | 5 | 180 |
| | | Expected Count | 56.0 | 69.0 | 37.0 | 13.0 | 5.0 | 180.0 |

**Chi-Square Tests**

| | Value | df | Asymp. Sig. (2-sided) |
|---|---|---|---|
| Pearson Chi-Square | 2.171[a] | 4 | .704 |
| Likelihood Ratio | 2.182 | 4 | .702 |
| Linear-by-Linear Association | 1.401 | 1 | .237 |
| N of Valid Cases | 180 | | |

a. 2 cells (20.0%) have expected count less than 5. The minimum expected count is 2.47.

b. One way to satisfy the assumption, part **a**, is to combine the data for two or more categories in the contingency table (e.g., high and very high). Form a new contingency table by combining the data for the high and very high categories in both the rows and columns.

c. Calculate the expected cell counts in the new contingency table, part **c**. Is the assumption now satisfied?

d. Perform the chi-square test on the new contingency table. Use $\alpha = .05$. Interpret the results.

13.30 The National Gang Crime Research Center (NGCRC) has developed a 6-level gang classification system for both adults and juveniles. The six categories are shown in the table below. The classification system was developed as a potential predictor of a gang member's propensity for violence when in prison, jail, or a correctional facility. To test this theory, the NGCRC collected data on approximately 10,000 confined offenders and assigned each a score using the gang classification system (*Journal of Gang Research*, Winter 1997). One of several other variables measured by the NGCRC was whether or not the offender has ever carried a homemade weapon (e.g.,

knife) while in custody. The data on gang score and homemade weapon are summarized in the table. Conduct a test to determine if carrying a homemade weapon in custody depends on gang classification score. (Use $\alpha = .01$.) Support your conclusion with a graph.

**Applying the Concepts—Advanced**

13.31 New, effective AIDS vaccines are now being developed using the process of "sieving," that is, sifting out infections with some strains of HIV. Harvard School of Public Health statistician Peter Gilbert demonstrated

🖲 HIVVAC1

| | | MN Strain Positive | Negative | Totals |
|---|---|---|---|---|
| **Patient Group** | **Unvaccinated** | 22 | 9 | 31 |
| | **Vaccinated** | 2 | 5 | 7 |
| | **Totals** | 24 | 14 | 38 |

*Source:* Gilbert, P. "Developing an AIDS vaccine by sieving." *Chance*, Vol. 13, No. 4, Fall 2000.

🖲 GANGS

| Gang Classification Score | Homemade Weapon Carried Yes | No |
|---|---|---|
| 0 (Never joined a gang, no close friends in a gang) | 255 | 2,551 |
| 1 (Never joined a gang, 1–4 close friends in a gang) | 110 | 560 |
| 2 (Never joined a gang, 5 or more friends in a gang) | 151 | 636 |
| 3 (Inactive gang member) | 271 | 959 |
| 4 (Active gang member, no position of rank) | 175 | 513 |
| 5 (Active gang member, holds position of rank) | 476 | 831 |

*Source:* Knox, G. W., *et al.* "A gang classification system for corrections." *Journal of Gang Research*, Vol. 4, No. 2, Winter 1997, p. 54 (Table 4).

SAS Output for Exercise 13.29

The FREQ Procedure

Table of SHSSC by CAHS

SHSSC     CAHS

| Frequency<br>Expected<br>Col Pct | High | Low | Medium | VeryHi | Total |
|---|---|---|---|---|---|
| High | 19<br>10.015 | 6<br>15.831 | 14<br>14.215 | 3<br>1.9385 | 42 |
| Low | 2<br>11.446 | 32<br>18.092 | 14<br>16.246 | 0<br>2.2154 | 48 |
| Medium | 6<br>7.3923 | 11<br>11.685 | 14<br>10.492 | 0<br>1.4308 | 31 |
| VeryHi | 4<br>2.1462 | 0<br>3.3923 | 2<br>3.0462 | 3<br>0.4154 | 9 |
| Total | 31 | 49 | 44 | 6 | 130 |

Statistics for Table of SHSSC by CAHS

| Statistic | DF | Value | Prob |
|---|---|---|---|
| Chi-Square | 9 | 60.1026 | <.0001 |
| Likelihood Ratio Chi-Square | 9 | 59.6382 | <.0001 |
| Mantel-Haenszel Chi-Square | 1 | 2.3505 | 0.1252 |
| Phi Coefficient | | 0.6799 | |
| Contingency Coefficient | | 0.5623 | |
| Cramer's V | | 0.3926 | |

WARNING: 44% of the cells have expected counts less than 5. Chi-Square may not be a valid test.

Sample Size = 130

how to test the efficacy of an HIV vaccine in *Chance* (Fall 2000). As an example, Gilbert reported the results of VaxGen's preliminary HIV vaccine trial using the $2 \times 2$ table on p. 725. The vaccine was designed to eliminate a particular strain of the virus, called the "MN strain." The trial consisted of 7 AIDS patients vaccinated with the new drug and 31 AIDS patients who were treated with a placebo (no vaccination). The table shows the number of patients who tested positive and negative for the MN strain in the trial follow-up period.

a. Conduct a test to determine whether the vaccine is effective in treating the MN strain of HIV. Use $\alpha = .05$.

b. Are the assumptions for the test, part **a**, satisfied? What are the consequences if the assumptions are violated?

c. In the case of a $2 \times 2$ contingency table, R.A. Fisher (1935) developed a procedure for computing the exact $p$-value for the test (called *Fishers exact test*).

The method utilizes the hypergeometric probability distribution of Chapter 4 (p. 208). Consider the hypergeometric probability

$$\frac{\binom{7}{2}\binom{31}{22}}{\binom{38}{24}}$$

This represents the probability that 2 out of 7 vaccinated AIDS patients test positive and 22 out of 31 unvaccinated patients test positive, that is, the probability of the table result given that the null hypothesis of independence is true. Compute this probability (called the *probability of the contingency table*).

d. Refer to part **c**. Two contingency tables (with the same marginal totals as the original table) that are more contradictory to the null hypothesis of independence than the observed table are shown at right. First, explain why these tables provide more evidence to reject $H_0$ than the original table. Then, compute the probability of each table using the hypergeometric formula.

e. The *p*-value of Fisher's exact test is the probability of observing a result at least as contradictory to the null hypothesis as the observed contingency table, given the same marginal totals. Sum the probabilities of parts **c** and **d** to obtain the *p*-value of Fisher's exact test. (To verify your calculations, check the *p*-value at the bottom of the accompanying SAS printout labeled **Left-sided Pr < = F**.) Interpret this value in the context of the vaccine trial.

🌐 HIVVAC2

|  |  | MN Strain | | |
| --- | --- | --- | --- | --- |
|  |  | **Positive** | **Negative** | **Totals** |
| **Patient Group** | **Unvaccinated** | 23 | 8 | 31 |
|  | **Vaccinated** | 1 | 6 | 7 |
|  | **Totals** | 24 | 14 | 38 |

🌐 HIVVAC3

|  |  | MN Strain | | |
| --- | --- | --- | --- | --- |
|  |  | **Positive** | **Negative** | **Totals** |
| **Patient Group** | **Unvaccinated** | 24 | 7 | 31 |
|  | **Vaccinated** | 0 | 7 | 7 |
|  | **Totals** | 24 | 14 | 38 |

**SAS Output for Exercise 13.31**

---

The FREQ Procedure

Table of group by MNSTRAIN

GROUP        MNSTRAIN

| Frequency Expected | NEG | POS | Total |
| --- | --- | --- | --- |
| UNVACC | 9 11.421 | 22 19.579 | 31 |
| VACC | 5 2.5789 | 2 4.4211 | 7 |
| Total | 14 | 24 | 38 |

Statistics for Table of group by MNSTRAIN

| Statistic | DF | Value | Prob |
| --- | --- | --- | --- |
| Chi-Square | 1 | 4.4112 | 0.0357 |
| Likelihood Ratio Chi-Square | 1 | 4.2893 | 0.0384 |
| Continuity Adj. Chi-Square | 1 | 2.7773 | 0.0956 |
| Mantel-Haenszel Chi-Square | 1 | 4.2952 | 0.0382 |
| Phi Coefficient |  | -0.3407 |  |
| Contingency Coefficient |  | 0.3225 |  |
| Cramer's V |  | -0.3407 |  |

WARNING: 50% of the cells have expected counts less than 5. Chi-Square may not be a valid test.

Fisher's Exact Test

| | |
| --- | --- |
| Cell (1, 1) Frequency (F) | 9 |
| Left-sided Pr <= F | 0.0498 |
| Right-sided Pr >= F | 0.9940 |
| Table Probability (P) | 0.0438 |
| Two-sided Pr <= P | 0.0772 |

Sample Size = 38

---

## 13.4 A WORD OF CAUTION ABOUT CHI-SQUARE TESTS

Because the $\chi^2$ statistic for testing hypotheses about multinomial probabilities is one of the most widely applied statistical tools, it is also one of the most abused statistical procedures. Consequently, the user should always be certain that the experiment satisfies the assumptions given with each procedure. Furthermore, the user should be certain that the sample is drawn from the correct population—that is, from the population about which the inference is to be made.

The use of the $\chi^2$ probability distribution as an approximation to the sampling distribution for $\chi^2$ should be avoided when the expected counts are very small. The approximation can become very poor when these expected counts are small, and thus the true $\alpha$ level may be quite different from the tabled value. As a rule of thumb, an expected cell count of at least 5 means that the $\chi^2$ probability distribution can be used to determine an approximate critical value.

If the $\chi^2$ value does not exceed the established critical value of $\chi^2$, *do not accept the hypothesis of independence*. You would be risking a Type II error (accepting $H_0$ if it is false), and the probability $\beta$ of committing such an error is unknown. The usual alternative hypothesis is that the classifications are dependent. Because the number of ways in which two classifications can be dependent is virtually infinite, it is difficult to calculate one or even several values of $\beta$ to represent such a broad alternative hypothesis. Therefore, we avoid concluding that two classifications are independent, even when $\chi^2$ is small.

Finally, if a contingency table $\chi^2$ value does exceed the critical value, we must be careful to avoid inferring that a *causal* relationship exists between the classifications. Our alternative hypothesis states that the two classifications are statistically dependent—and a statistical dependence does not imply causality. Therefore, *the existence of a causal relationship cannot be established by a contingency table analysis.*

### STATISTICS IN ACTION
#### The Level of Agreement Among Movie Reviewers: Thumbs Up or Thumbs Down?

For over 20 years, movie critics Gene Siskel of the *Chicago Tribune* and Roger Ebert of the *Chicago Sun-Times* evaluated the latest movie releases on national television, first on PBS with *Sneak Previews*, then in syndication with *Siskel and Ebert at the Movies*. During their entertaining 30-minute shows, Siskel (now deceased) and Ebert would face each other across a theater aisle and discuss the merits or faults of a movie. After engaging in a lively debate about the film, the reviewers would give the movie either a "thumbs up" or a "thumbs down." To film advertisers and producers, "two thumbs up" from Siskel and Ebert would almost assure a money-making movie.

The popularity of Siskel and Ebert's television shows have encouraged other well-known critics to imitate the face-to-face format. For example, Michael Medved and Jeffrey Lyons star on PBS's *Sneak Previews*, and Gene Shalit (NBC's *Today Show*) and Rex Reed (*New York Observer*) face off in a syndicated TV show. (Currently, Roger Ebert teams with Richard Roeper, Siskel's replacement.) The confrontational nature of these programs leave the viewer with the impression that the critics disagree about most movies. But how strongly do these popular movie reviewers really disagree?

To answer this question, University of Florida statisticians Alan Agresti and Larry Winner collected data on the reviews of movie critics (*Chance*, Spring 1997). They built a database from the weekly reviews that were published in *Variety* magazine during 1996. Each critic's review was categorized as Pro ("thumbs up"), Con ("thumbs down") or Mixed (neutral opinion). Consequently, for the movies evaluated by Siskel and Ebert, each film has a Siskel rating (Pro, Con, or Mixed) and an Ebert rating (Pro, Con, or Mixed).

⊘ SISEBERT contains the data for 160 movie reviews by Siskel and Ebert

⊘ MEDLYONS contains the data for 123 movie reviews by Medved and Lyons

### Focus

a. Conduct an analysis to determine whether the movie reviews of Gene Siskel and Roger Ebert are independent. Support your conclusion with a graph.

b. Based on your analysis, part a, do Siskel and Ebert tend to agree or disagree? Explain.

c. Repeat parts a and b for the movie reviews of Michael Medved and Jeffrey Lyons.

## QUICK REVIEW

### Key Terms

Categories  704
Cell 704
Cell counts  704
Chi-square test  706
Classes  704
Contingency table  713

Dependence  715
Dimensions of classification  713
Expected cell count  707
Independence of two
  classifications  714
Marginal probabilities  713

Multinomial experiment  704
Observed cell count  714
One-way table  705
Two-way table  713

### Key Formulas

$$\chi^2 = \sum \frac{[n_i - E(n_i)]^2}{E(n_i)} \qquad \text{One-way table} \quad 707$$

where      $n_i$ = count for cell $i$

$$E(n_i) = np_{i,0}$$

$p_{i,0}$ = hypothesized value of $p_i$ in $H_0$

$$\chi^2 = \sum \frac{[n_{ij} - \hat{E}(n_{ij})]^2}{\hat{E}(n_{ij})} \qquad \text{Two-way table} \quad 717$$

where      $n_{ij}$ = count for cell in row $i$, column $j$

$$\hat{E}(n_{ij}) = r_i c_j / n$$

$r_i$ = total for row $i$

$c_j$ = total for column $j$

$n$ = total sample size

### Language Lab

| Symbol | Pronunciation | Description |
|---|---|---|
| $p_{i,0}$ | *p-i*-zero | Value of multinomial probability $p_i$ hypothesized in $H_0$ |
| $\chi^2$ | Chi-square | Test statistic used in analysis of count data |
| $n_i$ | *n-i* | Number of observed outcomes in cell $i$ of one-way table |
| $E(n_i)$ | Expected value of $n_i$ | Expected number of outcomes in cell $i$ of one-way table when $H_0$ is true |
| $p_{ij}$ | *p-i-j* | Probability of an outcome in row $i$ and column $j$ of a two-way contingency table |
| $n_{ij}$ | *n-i-j* | Number of observed outcomes in row $i$ and column $j$ of a two-way contingency table |
| $\hat{E}(n_{ij})$ | Estimated expected value of $n_{ij}$ | Estimated expected number of outcomes in row $i$ and column $j$ of a two-way contingency table |
| $r_i$ | *r-i* | Total number of outcomes in row $i$ of a contingency table |
| $c_j$ | *c-j* | Total number of outcomes in column $j$ of a contingency table |

## SUPPLEMENTARY EXERCISES 13.32–13.50

### Learning the Mechanics

**13.32** A random sample of 250 observations was classified according to the row and column categories shown in the table.

  **a.** Do the data provide sufficient evidence to conclude that the rows and columns are dependent? Test using $\alpha = .05$.

| | | Column | | |
|---|---|---|---|---|
| | | **1** | **2** | **3** |
| | **1** | 20 | 20 | 10 |
| **Row** | **2** | 10 | 20 | 70 |
| | **3** | 20 | 50 | 30 |

**b.** Would the analysis change if the row totals were fixed before the data were collected?

**c.** Do the assumptions required for the analysis to be valid differ according to whether the row (or column) totals are fixed? Explain.

**d.** Convert the table entries to percentages by using each column total as a base and calculating each row response as a percentage of the corresponding column total. In addition, calculate the row totals and convert them to percentages of all 250 observations.

**e.** Create a bar graph with row 1 percentages on the vertical axis and the column number on the horizontal axis. Draw a horizontal line corresponding to the row 1 total percentage. Does the graph support the result of the test conducted in part **a**?

**f.** Repeat part **e** for the row 2 percentages.

**g.** Repeat part **e** for the row 3 percentages.

**13.33** A random sample of 150 observations was classified into the categories shown in the table.

| | **Category** | | | | |
|---|---|---|---|---|---|
| | **1** | **2** | **3** | **4** | **5** |
| $n_i$ | 28 | 35 | 33 | 25 | 29 |

**a.** Do the data provide sufficient evidence that the categories are not equally likely? Use $\alpha = .10$.

**b.** Form a 90% confidence interval for $p_2$, the probability that an observation will fall in category 2.

## Applying the Concepts—Basic

**13.34** Westinghouse Electric Company has experimented with different means of evaluating the performance of solder joint inspectors. One approach involves comparing an individual inspector's classifications with those of the group of experts that comprise Westinghouse's Work Standards Committee. In one experiment, 153 solder connections were evaluated by the committee and 111 were classified as acceptable. An inspector evaluated the same 153 connections and classified 124 as acceptable. Of the items rejected by the inspector, the committee agreed with 19.

**a.** Construct a contingency table that summarizes the classifications of the committee and the inspector.

**b.** Based on a visual examination of the table you constructed in part **a**, does it appear that there is a relationship between the inspector's classifications and the committee's? Explain. (A bar graph of the percentage rejected by committee and inspector will aid your examination.)

**c.** Conduct a chi-square test of independence for these data. Use $\alpha = .05$. Carefully interpret the results of your test in the context of the problem.

**13.35** According to research reported in the *Journal of the National Cancer Institute* (Apr. 1991), eating foods high in fiber may help protect against breast cancer. The researchers randomly divided 120 laboratory rats into four groups of 30 each. All rats were injected with a drug that causes breast cancer, then each rat was fed a diet of fat and fiber for 15 weeks. However, the levels of fat and fiber varied from group to group. At the end of the feeding period, the number of rats with cancer tumors was determined for each group. The data are summarized in the table at the bottom of the page.

**a.** Does the sampling appear to satisfy the assumptions for a multinomial experiment (see Section 13.1)? Explain.

**b.** Calculate the expected cell counts for the contingency table.

**c.** Calculate the $\chi^2$ statistic.

**d.** Is there evidence to indicate that diet and presence/absence of cancer are independent? Test using $\alpha = 05$.

**e.** Compare the percentage of rats on a high-fat/no-fiber diet with cancer to the percentage of rats on a high-fat/fiber diet with cancer using a 95% confidence interval. Interpret the result.

**13.36** A study in the *Annals of Tourism Research* (Vol. 19, 1992) investigates the relationship of retirement status (pre- and postretirement) to various items related to the travel industry. One part of the study investigated the differences in the length of stay of a trip for pre- and postretirees. A sample of 703 travelers were asked how long they stayed on a typical trip. The results are shown in the table on p. 731. Use the information in the table to determine whether the retirement status of a traveler and the duration of a typical trip are dependent. Test using $\alpha = .05$.

**TUMORS**

| | | **Diet** | | | | |
|---|---|---|---|---|---|---|
| | | **High-Fat/No-Fiber** | **High-Fat/Fiber** | **Low-Fat/No-Fiber** | **Low-Fat/Fiber** | **Totals** |
| **Cancer Tumors** | **Yes** | 27 | 20 | 19 | 14 | 80 |
| | **No** | 3 | 10 | 11 | 16 | 40 |
| | **Totals** | 30 | 30 | 30 | 30 | 120 |

*Source: Tampa Tribune,* Apr. 3, 1991.

Data for Exercise 13.39

| FATHSON | | Son | | | |
|---|---|---|---|---|---|
| | | Professional or Business | Skilled | Unskilled | Farmer |
| **Father** | **Professional or Business** | 55 | 38 | 7 | 0 |
| | **Skilled** | 79 | 71 | 25 | 0 |
| | **Unskilled** | 22 | 75 | 38 | 10 |
| | **Farmer** | 15 | 23 | 10 | 32 |

Data for Exercise 13.36

TRAVEL

| Number of Nights | Preretirement | Postretirement |
|---|---|---|
| 4–7 | 247 | 172 |
| 8–13 | 82 | 67 |
| 14–21 | 35 | 52 |
| 22 or more | 16 | 32 |
| **Total** | 380 | 323 |

**13.37** The *Journal of Intellectual Disability Research* (Feb. 1995) published a longitudinal study of hearing impairment in a group of elderly patients with intellectual disability. The hearing function of each patient was screened each year over a 10-year period. At the study's conclusion, the hearing loss of each patient was categorized as severe, moderate, mild, or none. The classifications of the 28 surviving patients are summarized in the table below.

HEARIMP

| Hearing Loss | Number of Patients |
|---|---|
| None | 7 |
| Mild | 7 |
| Moderate | 9 |
| Severe | 5 |
| **Total** | 28 |

**a.** Conduct a test to determine whether the true proportions of intellectually disabled elderly patients in each of the hearing-loss categories differ. Use $\alpha = .05$.

**b.** Use a 90% confidence interval to estimate the proportion of disabled elderly patients with severe hearing loss.

**13.38** According to one geneticist's theory, a crossing of red and white snapdragons should produce offspring that are 25% red, 50% pink, and 25% white. An experiment conducted to test the theory produced 30 red, 78 pink, and 36 white offspring in 144 crossings. Do the data provide sufficient evidence to contradict the geneticist's theory?

**13.39** A sociologist was interested in knowing whether sons have a tendency to choose the same occupation as their fathers. To investigate this question, 500 males were polled and questioned concerning their occupations and those of their fathers. A summary of the numbers of father–son pairs falling in each occupational category is shown in the table above. Do the data provide sufficient evidence at $\alpha = .05$ to indicate a dependence between a son's choice of occupation and his father's occupation? Use the SAS printout on p. 734 to conduct the analysis.

## Applying the Concepts—Intermediate

**13.40** An article in the *Annals of the Association of American Geographers* (June 1992) investigated many factors that are considered when purchasing earthquake insurance. One factor investigated was the proximity to a major earthquake fault. Surveys were mailed to residents in four California counties. The data collected are shown in the table below.

QUAKE

| | Counties | | | |
|---|---|---|---|---|
| | Contra Costa | Santa Clara | Los Angeles | San Bernardino |
| **Sample Size** | 521 | 556 | 337 | 372 |
| **Number with Earthquake Insurance** | 117 | 222 | 133 | 109 |

SAS Output for Exercise 13.39

The FREQ Procedure

Table of FATHER by SON

| FATHER | SON | | | | |
|---|---|---|---|---|---|
| Frequency<br>Expected | FARMER | PROF/BUS | SKILLED | UNSKILL | Total |
| FARMER | 32<br>6.72 | 15<br>27.36 | 23<br>33.12 | 10<br>12.8 | 80 |
| PROF/BUS | 0<br>8.4 | 55<br>34.2 | 38<br>41.4 | 7<br>16 | 100 |
| SKILLED | 0<br>14.7 | 79<br>59.85 | 71<br>72.45 | 25<br>28 | 175 |
| UNSKILL | 10<br>12.18 | 22<br>49.59 | 75<br>60.03 | 38<br>23.2 | 145 |
| Total | 42 | 171 | 207 | 80 | 500 |

Statistics for Table of FATHER by SON

| Statistic | DF | Value | Prob |
|---|---|---|---|
| Chi-Square | 9 | 180.8739 | <.0001 |
| Likelihood Ratio Chi-Square | 9 | 160.8323 | <.0001 |
| Mantel-Haenszel Chi-Square | 1 | 52.0404 | <.0001 |
| Phi Coefficient | | 0.6015 | |
| Contingency Coefficient | | 0.5154 | |
| Cramer's V | | 0.3473 | |

Sample Size = 500

a. State the hypothesis necessary to test the theory that the proportion of earthquake-insured residents differs for the four sampled counties.

b. Do the data provide evidence of a difference in the proportion of earthquake-insured residents for the four sampled counties? Test using $\alpha = .05$.

**13.41** In chess, the first few moves often play a determining role in the final outcome. Five different opening strategies are highly favored by chess experts. To determine whether one or more of these strategies is most preferred by grand masters in international competition, a random sample of 100 grand masters is taken, and each is asked which of the strategies he or she would prefer to employ. A summary of their responses is shown below. Do these data present sufficient evidence to indicate a preference for one or more of the strategies? Use $\alpha = 05$.

| Strategy | A | B | C | D | E |
|---|---|---|---|---|---|
| Frequency | 17 | 27 | 22 | 15 | 19 |

**13.42** Marketing researchers often use the purchase of a sample of households, called a *scanner panel*, to understand the purchasing patterns and preferences of consumers. When shopping, these households present a magnetic identification card that permits their purchase data to be identified and aggregated. Researchers recently studied the extent to which panel households' purchase behavior is representative of the population of households shopping at the same stores (*Journal of Marketing Research*, Nov. 1996). The table on p. 735 reports the peanut butter purchase data collected by A.C. Nielson Company for the panel of 2,500 households in Sioux Falls, South Dakota, over a 102-week period. The market share percentages in the right-hand column are derived from all peanut butter purchases at the same 15 stores at which the panel shopped during the same 102-week period.

a. Do the data provide sufficient evidence to conclude that the purchases of the household panel are representative of the population of households? Test using $\alpha = .05$.

**b.** What assumptions must hold to ensure the validity of the testing procedure you used in part **a**?

**c.** Find the approximate *p*-value for the test of part **a** and interpret it in the context of the problem.

🔵 SCANNER

| Brand | Size | Number of Purchases by Household Panel | Market Shares |
|---|---|---|---|
| Jif | 18 oz. | 3,165 | 20.10% |
| Jif | 28 | 1,892 | 10.10 |
| Jif | 40 | 726 | 5.42 |
| Peter Pan | 10 | 4,079 | 16.01 |
| Skippy | 18 | 6,206 | 28.65 |
| Skippy | 28 | 1,627 | 12.38 |
| Skippy | 40 | 1,420 | 7.32 |
| **Total** | | 19,115 | |

*Source:* Gupta, S., Chintagunta, P., Kaul, A., and Wittink, D. "Do household scanner data provide representative inferences from brand choices: A comparison with store data." *Journal of Marketing Research*, Vol. 33, Nov. 1996, p. 393 (Table 6).

**13.43** An experimenter wishes to determine whether there is a relationship between hair color and eye color. One hundred people are randomly sampled and the eyes and hair of each person are judged to be light or dark. A summary of the number of people in each of the four categories is shown in the table. Do the data provide sufficient evidence at $\alpha = .10$ to indicate a relationship between eye and hair color? Use the MINITAB printout provided here to perform the analysis.

🔵 HAIREYE

| | Light Hair | Dark Hair | Totals |
|---|---|---|---|
| **Light Eyes** | 31 | 21 | 52 |
| **Dark Eyes** | 14 | 34 | 48 |
| **Totals** | 45 | 55 | 100 |

**MINITAB Output for Exercise 13.43**

```
Expected counts are printed below observed counts

        LGHTHAIR DARKHAIR    Total
   1        31       21         52
         23.40    28.60

   2        14       34         48
         21.60    26.40

Total      45       55        100

Chi-Sq =   2.468 +  2.020 +
           2.674 +  2.188 =  9.350
DF = 1,  P-Value = 0.002
```

**13.44** Researchers conducted a study of 58 children admitted to the child psychiatry ward of the University of Iowa for aggressive conduct (*Journal of Child Psychology and Psychiatry*, Vol. 25, 1984). The purpose of the study was to determine the influence of social class, sex, and age on the clinical characteristics of the children. A small portion of their data, which lists the numbers of children exhibiting antisocial behavior for two categories of social class, is shown in the table. Classes I to III represent children from middle-class families. Those in Classes IV and V were from poor families.

| | Classes I–III | Classes IV–V |
|---|---|---|
| **Numbers exhibiting antisocial behavior** | 24 | 17 |
| **Sample Size** | 28 | 30 |

*Source:* Behar, D., and Stewart, M. A. "Aggressive conduct disorder: The influence of social class, sex, and age on the clinical picture." *Journal of Child Psychology and Psychiatry*, Vol. 25, 1984, p. 25.

**a.** Do the data provide sufficient evidence to indicate a dependence between antisocial behavior and social class? Test using $\alpha = .05$.

**b.** The authors give the approximate *p*-value for the test as .015. Calculate the approximate *p*-value for the test and compare it with the authors' value. Interpret the *p*-value.

**13.45** The *Journal of Adlerian Theory, Research and Practice* (Mar. 1986) investigated the role lifestyles play in forming a marriage. Adlerian psychologists group people into one of the following five lifestyle types:

1. *Pleasing priority:* a strong desire to make others happy and win their approval
2. *Achieving priority:* industrial, responsible, orderly, with an active approach to life (particularly in the workplace)
3. *Outdoing priority:* a strong desire to be on top, in a position of superiority over others; likes to be considered the best at whatever he/she does
4. *Suppressing (control) priority:* maintains control over his/her emotions; discloses very little of himself/herself
5. *Avoiding priority:* is easily hurt; avoids any potentially painful (physical or emotional) situation

Data on both wife's and husband's lifestyles were obtained for 202 married couples living in family housing at the University of Georgia. The results are summarized in the table on p. 736.

**a.** The key question of the study is: "Do the personality profiles contribute to marriage pairings?" In other words, is there a relationship between the personality priorities of the husband and wife pairs? Analyze the data for the researchers.

**b.** If no relationship between the personality priorities of the husband and wife pairs is found, part **a**, then analyze the data for husbands and wives separately.

⊘LIFESTYLE

| | | Wife's Lifestyle | | | | | |
| --- | --- | --- | --- | --- | --- | --- | --- |
| | | Pleasing | Outdoing | Avoiding | Control | Achieving | Totals |
| | Pleasing | 9 | 6 | 6 | 3 | 9 | 33 |
| | Outdoing | 7 | 11 | 12 | 11 | 5 | 46 |
| **Husband's Lifestyle** | Avoiding | 8 | 8 | 6 | 11 | 5 | 38 |
| | Control | 8 | 10 | 7 | 15 | 11 | 51 |
| | Achieving | 4 | 6 | 7 | 10 | 7 | 34 |
| | **Totals** | 36 | 41 | 38 | 50 | 37 | 202 |

*Source:* Evans, T. D., and Bozarth, J. "Pairing of personality profiles in marriage." *Journal of Adlerian Theory, Research and Practice*, Vol. 42, No. 1, March 1986, pp. 59–64.

Are there differences in the proportions of husbands who fall in the five personality priorities? Are there differences in the proportions of wives who fall in the five personality profiles?

**13.46** The *American Journal on Mental Retardation* (Jan. 1992) published a study of the social interactions of two groups of children. Independent random samples of 15 children with and 15 children without developmental delays (i.e., mild mental retardation) were the subjects of the experiment. After observing the children during "freeplay," the number of children who exhibited disruptive behavior (e.g., ignoring or rejecting other children, taking toys from another child) was recorded for each group. The data are summarized in the two-way table below. Analyze the data and interpret the results.

**13.47** Two methods of treating moderately advanced cancer of the larynx are surgical removal and radiation therapy. A study of patients treated by these methods at the University of Florida's Shands Hospital was reported in the *International Journal of Radiation Oncology and Biological Physics* (Vol. 10, 1984). Eighteen patients with laryngeal cancer were treated by radiation alone and 23 were treated by surgery alone. Of those treated by radiation, 15 cancers were ultimately controlled at the primary site (the larynx). Ultimate control of the cancer was achieved for 21 of the 23 patients treated by surgery alone. The objective of the study was to determine whether the rate of controlling laryngeal cancer is independent of type of treatment (surgery or radiation therapy).

**a.** Form a contingency table for the data.

**b.** Assuming the null hypothesis of independence is true, how many patients are expected to fall in each cell of the table?

**c.** The researchers were unable to use the chi-square test to analyze the data. Show why.

**d.** A test of the null hypothesis can be conducted by using a small-sample method known as *Fisher's exact test*. (See Exercise 13.31, p. 725.) This method calculates the exact probability (*p*-value) of observing sample results at least as contradictory to the hypothesis of independence as those observed for the researchers' data. The researchers report the *p*-value for this test as .187. Interpret this result.

**e.** Calculate the $\chi^2$ statistic for the data using the formula provided in this chapter, then find its associated *p*-value. Compare the result to the *p*-value given in part **d**. Do the tests lead to the same conclusion?

**13.48** *Human Factors* (Dec. 1988) published a study of color brightness as a body orientation clue. Ninety college students, reclining on their backs in the dark, were disoriented when positioned on a rotating platform under a slowly rotating disk that blocked their field of vision. The subjects were asked to say "Stop" when they felt as if they were right-side up. The position of the brightness pattern on the disk in relation to each student's body orientation was then recorded. Subjects selected only three disk brightness patterns as subjective vertical clues: (1) brighter side up, (2) darker side up, and

⊘DISRUPT

| | Disruptive Behavior | Nondisruptive Behavior | Totals |
| --- | --- | --- | --- |
| **With Developmental Delays** | 12 | 3 | 15 |
| **Without Developmental Delays** | 5 | 10 | 15 |
| **Total** | 17 | 13 | 30 |

*Source:* Kopp, C. B., Baker, B., and Brown, K. W. "Social skills and their correlates: Preschoolers with developmental delays." *American Journal on Mental Retardation*, Vol. 96, No. 4, Jan. 1992.

(3) brighter and darker sides aligned on either side of the subjects' heads. The frequency counts for the experiment are given in the accompanying table. Conduct a test to compare the proportions of subjects that fall in the three disk-orientation categories. Assume you want to determine whether the three proportions differ. Use $\alpha = .05$.

**Disk Orientation**

| Brighter Side Up | Darker Side Up | Bright and Dark Sides Aligned |
|---|---|---|
| 58 | 15 | 17 |

## Applying the Concepts—Advanced

**13.49** A statistical analysis is to be done on a set of data consisting of 1,000 monthly salaries. The analysis requires the assumption that the sample was drawn from a normal distribution. A preliminary test, called the $\chi^2$ *goodness-of-fit test*, can be used to help determine whether it is reasonable to assume that the sample is from a normal distribution. Suppose the mean and standard deviation of the 1,000 salaries are hypothesized to be $1,200 and $200, respectively. Using the standard normal table, we can approximate the probability of a salary being in the intervals listed in the next table. The third column represents the expected number of the 1,000 salaries to be found in each interval if the sample was drawn from a normal distribution with $\mu = \$1,200$ and $\sigma = \$200$. Suppose the last column contains the actual observed frequencies

in the sample. Large differences between the observed and expected frequencies cast doubt on the normality assumption.

| Interval | Probability | Expected Frequency | Observed Frequency |
|---|---|---|---|
| Less than $800 | .023 | 23 | 26 |
| $800 < $1,000 | .136 | 136 | 146 |
| $1,000 < $1,200 | .341 | 341 | 361 |
| $1,200 < $1,400 | .341 | 341 | 311 |
| $1,400 < $1,600 | .136 | 136 | 143 |
| $1,600 or above | .023 | 23 | 13 |

**a.** Compute the $\chi^2$ statistic based on the observed and expected frequencies.
**b.** Find the tabulated $\chi^2$ value when $\alpha = .05$ and there are 5 degrees of freedom. (There are $k - 1 = 5$ df associated with this $\chi^2$ statistic.)
**c.** Based on the $\chi^2$ statistic and the tabulated $\chi^2$ value, is there evidence that the salary distribution is nonnormal?
**d.** Find the approximate observed significance level for the test in part **c**.

**13.50** Suppose a random variable is hypothesized to be normally distributed with mean 0 and standard deviation 1. A random sample of 200 observations on the variable yields frequencies in the intervals listed in the table below. Do the data provide sufficient evidence to contradict the hypothesis that $x$ is normally distributed with $\mu = 0$ and $\sigma = 1$? Use the technique developed in Exercise 13.49.

| Interval | $x < -2$ | $-2 \leq x < -1$ | $-1 \leq x < 0$ | $0 \leq x < 1$ | $1 \leq x < 2$ | $x \geq 2$ |
|---|---|---|---|---|---|---|
| Frequency | 7 | 20 | 61 | 77 | 26 | 9 |

# STUDENT PROJECTS

Many researchers rely on surveys to estimate the proportions of experimental units in populations that possess certain specified characteristics. A political scientist may want to estimate the proportion of an electorate in favor of a certain legislative bill. A social scientist may be interested in the proportions of people in a geographical region who fall in certain socioeconomic classifications. A psychologist might want to compare the proportions of patients who have different psychological disorders.

Choose a specific topic, similar to those described above, that interests you. Clearly define the population of interest, identify data categories of specific interest, and identify the proportions associated with them. Now *guesstimate* the pro-

portions of the population that you think fall in each of the categories. For example, you might guess that all the proportions are equal, or that the first proportion is twice as large as the second but equal to the third, etc.

You are now ready to collect the data by obtaining a random sample from your population of interest. Select a sample size so that all expected cell counts are at least 5 (preferably larger), then collect the data.

Use the count data you have obtained to test the null hypothesis that the true proportions in the population are equal to your presampling guesstimates. Would failure to reject this null hypothesis imply that your guesstimates are correct?

## REFERENCES

Agresti, A. *Categorical Data Analysis*. New York: Wiley, 1990.

Cochran, W. G. "The $\chi^2$ test of goodness of fit." *Annals of Mathematical Statistics*, 1952, 23.

Conover, W. J. *Practical Nonparametric Statistics*, 2nd ed. New York: Wiley, 1980.

Fisher, R. A. "The logic of inductive inference (with discussion)." *Journal of the Royal Statistical Society*, Vol. 98, 1935, pp. 39–82.

Hollander, M., and Wolfe, D. A. *Nonparametric Statistical Methods*. New York: Wiley, 1973.

Savage, I. R. "Bibliography of nonparametric statistics and related topics." *Journal of the American Statistical Association*, 1953, 48.

# Nonparametric Statistics

## Contents

## Statistics in Action

Deadly Exposure: Agent Orange and Vietnam Vets

### ✊ Where We've Been

In Chapters 7–10, we explored techniques for making inferences about the mean of a single population and for comparing the means of two or more populations. Most of the techniques discussed in those chapters are based on the assumption that the sampled populations have probability distributions that are approximately normal with equal variances. But how can you analyze data that evolve from populations that do not satisfy these assumptions? How can you make comparisons between populations when you cannot assign specific numerical values to your observations?

### ☞ Where We're Going

In this chapter, we present statistical techniques for comparing two or more populations that are based on an ordering of the sample measurements according to their relative magnitudes. These techniques, which require fewer or less stringent assumptions concerning the nature of the probability distributions of the populations, are called *nonparametric statistical methods*.

## 14.1 INTRODUCTION: DISTRIBUTION-FREE TESTS

The confidence interval and testing procedures developed in Chapters 7–10 all involve making inferences about population parameters. Consequently, they are often referred to as **parametric statistical tests**. Many of these parametric methods (e.g., the small sample *t*-test of Chapter 8 or the ANOVA *F*-test of Chapter 10) rely on the assumption that the data are sampled from a normally distributed population. When the data are normal, these tests are *most powerful*. That is, the use of these parametric tests maximizes power—the probability of the researcher correctly rejecting the null hypothesis.

Consider a population of data that is decidedly nonnormal. For example, the distribution might be very flat, peaked, or strongly skewed to the right or left (see Figure 14.1). Applying the small sample *t*-test to such a data set may result in serious consequences. Since the normality assumption is clearly violated, the results of the *t*-test are unreliable: (1) the probability of a Type I error (i.e., rejecting $H_0$ when it is true) may be larger than the value of $\alpha$ selected; and (2) the power of the test, $1 - \beta$, is not maximized.

**FIGURE 14.1**

*Some nonnormal distributions for which the t statistic is invalid*

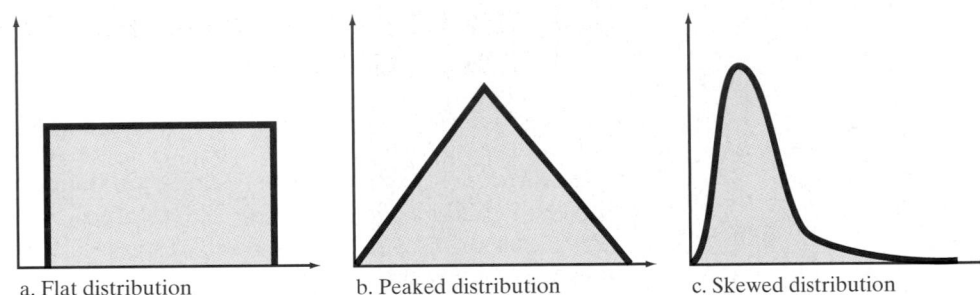

a. Flat distribution          b. Peaked distribution          c. Skewed distribution

A host of *nonparametric* techniques are available for analyzing data that do not follow a normal distribution. Nonparametric tests do not depend on the distribution of the sampled population; thus, they are called *distribution-free tests*. Also, nonparametric methods focus on the location of the probability distribution of the population, rather than on specific parameters of the population, such as the mean (hence, the name "nonparametrics").

> **DEFINITION 14.1**
> **Distribution-free tests** are statistical tests that do not rely on any underlying assumptions about the probability distribution of the sampled population.
>
> **DEFINITION 14.2**
> The branch of inferential statistics devoted to distribution-free tests is called **nonparametrics**.

Nonparametric tests are also appropriate when the data are nonnumerical in nature but can be ranked.* For example, when taste-testing foods or in other types of consumer product evaluations, we can say we like product A better than product B, and B better than C, but we cannot obtain exact quantitative values for the respective measurements. Nonparametric tests based on the ranks of measurements are called *rank tests*.

---

*Qualitative data that can be ranked in order of magnitude are called *ordinal* data.

---

**DEFINITION 14.3**

Nonparametric statistics (or tests) based on the ranks of measurements are called **rank statistics** (or **rank tests**).

---

In this chapter, we present several useful nonparametric methods. Keep in mind that these nonparametric tests are more powerful than their corresponding parametric counterparts in those situations where either the data are nonnormal or the data are ranked.

In Section 14.2, we develop a test to make inferences about the central tendency of a single population. In Sections 14.3 and 14.5, we present rank statistics for comparing two or more probability distributions using independent samples. In Sections 14.4 and 14.6, the matched-pairs and randomized block designs are used to make nonparametric comparisons of populations. Finally, in Section 14.7, we present a nonparametric measure of correlation between two variables.

## 14.2  SINGLE POPULATION INFERENCES: THE SIGN TEST

In Chapter 8 we utilized the $z$ and $t$ statistics for testing hypotheses about a population mean. The $z$ statistic is appropriate for large random samples selected from "general" populations—that is, with few limitations on the probability distribution of the underlying population. The $t$ statistic was developed for small-sample tests in which the sample is selected at random from a *normal* distribution. The question is: How can we conduct a test of hypothesis when we have a small sample from a *nonnormal* distribution?

The **sign test** is a relatively simple nonparametric procedure for testing hypotheses about the central tendency of a nonnormal probability distribution. Note that we used the phrase *central tendency* rather than *population mean*. This is because the sign test, like many nonparametric procedures, provides inferences about the population *median* rather than the population mean $\mu$. Denoting the population median by the Greek letter $\eta$, we know (Chapter 2) that $\eta$ is the 50th percentile of the distribution (Figure 14.2) and as such is less affected by the skewness of the distribution and the presence of outliers (extreme observations). Since the nonparametric test must be suitable for all distributions, not just the normal, it is reasonable for nonparametric tests to focus on the more robust (less sensitive to extreme values) measure of central tendency, the median.

**FIGURE 14.2**

*Location of the Population Median, $\eta$*

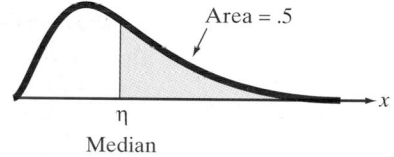

For example, increasing numbers of both private and public agencies are requiring their employees to submit to tests for substance abuse. One laboratory that conducts such testing has developed a system with a normalized measurement scale, in which values less than 1.00 indicate "normal" ranges and values equal to or greater than 1.00 are indicative of potential substance abuse. The lab reports a normal result as long as the median level for an individual is less than 1.00. Eight independent measurements of each individual's sample are made. One individual's results are shown in Table 14.1.

⊘SUBABUSE
**TABLE 14.1** **Substance Abuse Test Results**

| .78 | .51 | 3.79 | .23 | .77 | .98 | .96 | .89 |
|---|---|---|---|---|---|---|---|

If the objective is to determine whether the *population* median (that is, the true median level if an infinitely large number of measurements were made on the same individual sample) is less than 1.00, we establish that as our alternative hypothesis and test

$$H_0: \eta = 1.00$$
$$H_a: \eta < 1.00$$

The one-tailed sign test is conducted by counting the number of sample measurements that "favor" the alternative hypothesis—in this case, the number that are less than 1.00. If the null hypothesis is true, we expect approximately half of the measurements to fall on each side of the hypothesized median and if the alternative is true, we expect significantly more than half to favor the alternative—that is, to be less than 1.00. Thus,

*Test statistic:* $S$ = Number of measurements less than 1.00, the null hypothesized median

If we wish to conduct the test at the $\alpha = .05$ level of significance, the rejection region can be expressed in terms of the observed significance level, or *p*-value, of the test:

*Rejection region: p*-value $\leq .05$

In this example, $S = 7$ of the 8 measurements are less than 1.00. To determine the observed significance level associated with this outcome, we note that the number of measurements less than 1.00 is a binomial random variable (check the binomial characteristics presented in Chapter 4), and *if $H_0$ is true*, the binomial probability $p$ that a measurement lies below (or above) the median 1.00 is equal to .5 (Figure 14.2). What is the probability that a result is *as contrary to or more contrary to $H_0$* than the one observed? That is, what is the probability that 7 *or more* of 8 binomial measurements will result in Success (be less than 1.00) if the probability of Success is .5? Binomial Table II in Appendix A (using $n = 8$ and $p = .5$) indicates that

$$P(x \geq 7) = 1 - P(x \leq 6) = 1 - .965 = .035$$

Thus, the probability that at least 7 of 8 measurements would be less than 1.00 *if the true median were 1.00* is only .035. The *p*-value of the test is therefore .035.

This *p*-value can also be obtained by using a statistical software package. The MINITAB printout of the analysis is shown in Figure 14.3, with the *p*-value highlighted on the printout. Since $p = .035$ is less than $\alpha = .05$, we conclude that this sample provides sufficient evidence to reject the null hypothesis. The implication of this rejection is that the laboratory can conclude at the $\alpha = .05$ level of significance that the true median level for the tested individual is less than 1.00. However, we note that one of the measurements greatly exceeds the others, with a value of 3.79, and deserves special attention. Note that this large measurement is an outlier that would make the use of a *t*-test and its

**FIGURE 14.3**

*MINITAB Printout of Sign Test*

```
Sign test of median = 1.000 versus  <  1.000


              N  Below  Equal  Above         P      Median
READING       8      7      0      1    0.0352      0.8350
```

concomitant assumption of normality dubious. The only assumption necessary to ensure the validity of the sign test is that the probability distribution of measurements is continuous.

The use of the sign test for testing hypotheses about population medians is summarized in the next box.

---

### Sign Test for a Population Median $\eta$

**ONE-TAILED TEST**

$H_0$: $\eta = \eta_0$

$H_a$: $\eta > \eta_0$ [or $H_a$: $\eta < \eta_0$]

*Test statistic:*

$S$ = Number of sample measurements greater than $\eta_0$ [or $S$ = number of measurements less than $\eta_0$]

*Observed significance level:*

$p$-value = $P(x \geq S)$

**TWO-TAILED TEST**

$H_0$: $\eta = \eta_0$

$H_a$: $\eta \neq \eta_0$

*Test statistic:*

$S$ = Larger of $S_1$ and $S_2$, where $S_1$ is the number of measurements less than $\eta_0$ and $S_2$ is the number of measurements greater than $\eta_0$

*Observed significance level:*

$p$-value = $2P(x \geq S)$

where $x$ has a binomial distribution with parameters $n$ and $p = .5$. (Use Table II, Appendix A.)

*Rejection region:* Reject $H_0$ if $p$-value $\leq \alpha$

*Assumption:* The sample is selected randomly from a continuous probability distribution. [*Note:* No assumptions need to be made about the shape of the probability distribution.]

---

Recall that the normal probability distribution provides a good approximation for the binomial distribution when the sample size is large. For tests about the median of a distribution, the null hypothesis implies that $p = .5$, and the normal distribution provides a good approximation if $n \geq 10$. (Samples with $n \geq 10$ satisfy the condition that $np \pm 2\sqrt{npq}$ is contained in the interval 0 to $n$.) Thus, we can use the standard normal $z$-distribution to conduct the sign test for large samples. The large-sample sign test is summarized in the next box.

**EXAMPLE 14.1**

A manufacturer of compact disk (CD) players has established that the median time to failure for its players is 5,250 hours of utilization. A sample of 20 CDs from a competitor is obtained, and they are continuously tested until each fails. The 20 failure times range from 5 hours (a "defective" player) to 6,575 hours, and 14 of the 20 exceed 5,250 hours. Is there evidence that the median failure time of the competitor differs from 5,250 hours? Use $\alpha = .10$.

### Solution

The null and alternative hypotheses of interest are

$$H_0: \eta = 5,250 \text{ hours}$$

$$H_a: \eta \neq 5,250 \text{ hours}$$

---

### Large-Sample Sign Test for a Population Median $\eta$

ONE-TAILED TEST

$H_0: \eta = \eta_0$

$H_a: \eta > \eta_0$
[or $H_a: \eta < \eta_0$]

TWO-TAILED TEST

$H_0: \eta = \eta_0$

$H_a: \eta \neq \eta_0$

*Test statistic:*   $z = \dfrac{(S - .5) - .5n}{.5\sqrt{n}}$

[*Note:* $S$ is calculated as shown in the previous box. We subtract .5 from $S$ as the "correction for continuity." The null hypothesized mean value is $np = .5n$, and the standard deviation is

$$\sqrt{npq} = \sqrt{n(.5)(.5)} = .5\sqrt{n}$$

See Chapter 5 for details on the normal approximation to the binomial distribution.]

*Rejection region:*   $z > z_\alpha$

*Rejection region:*   $z > z_{\alpha/2}$

where tabulated $z$ values can be found in Table IV, Appendix A.

---

Since $n \geq 10$, we use the standard normal $z$ statistic:

$$\text{Test statistic:}\quad z = \frac{(S - .5) - .5n}{.5\sqrt{n}}$$

where $S$ is the maximum of $S_1$, the number of measurements greater than 5,250, and $S_2$, the number of measurements less than 5,250.

$$\text{Rejection region: } z > 1.645, \text{ where } z_{\alpha/2} = z_{.05} = 1.645$$

*Assumptions:* The distribution of the failure times is continuous (time is a continuous variable), but nothing is assumed about the shape of its probability distribution.

Since the number of measurements exceeding 5,250 is $S_2 = 14$, then the number of measurements less than 5,250 is $S_1 = 6$. Consequently, $S = 14$, the greater of $S_1$ and $S_2$. The calculated $z$ statistic is therefore

$$z = \frac{(S - .5) - .5n}{.5\sqrt{n}} = \frac{13.5 - 10}{.5\sqrt{20}} = \frac{3.5}{2.236} = 1.565$$

The value of $z$ is not in the rejection region, so we cannot reject the null hypothesis at the $\alpha = .10$ level of significance. Thus, the CD manufacturer should not conclude, on the basis of this sample, that its competitor's CDs have a median failure time that differs from 5,250 hours.   ∎

The one-sample nonparametric sign test for a median provides an alternative to the *t*-test for small samples from nonnormal distributions. However, if the distribution is approximately normal, the *t*-test provides a more powerful test about the central tendency of the distribution.

EXERCISES 14.1–14.11

## Learning the Mechanics

**14.1** Under what circumstances is the sign test preferred to the *t*-test for making inferences about the central tendency of a population?

**14.2** What is the probability that a randomly selected observation exceeds the
  **a.** Mean of a normal distribution?
  **b.** Median of a normal distribution?
  **c.** Mean of a nonnormal distribution?
  **d.** Median of a nonnormal distribution?

**14.3** Use Table II of Appendix A to calculate the following binomial probabilities:
  **a.** $P(x \geq 6)$ when $n = 7$ and $p = .5$
  **b.** $P(x \geq 5)$ when $n = 9$ and $p = .5$
  **c.** $P(x \geq 8)$ when $n = 8$ and $p = .5$
  **d.** $P(x \geq 10)$ when $n = 15$ and $p = .5$. Also use the normal approximation to calculate this probability, then compare the approximation with the exact value.
  **e.** $P(x \geq 15)$ when $n = 25$ and $p = .5$. Also use the normal approximation to calculate this probability, then compare the approximation with the exact value.

**14.4** Consider the following sample of 10 measurements:

⊚ LM14_4

| 8.4 | 16.9 | 15.8 | 12.5 | 10.3 | 4.9 | 12.9 | 9.8 | 23.7 | 7.3 |
|-----|------|------|------|------|-----|------|-----|------|-----|

Use these data to conduct each of the following sign tests using the binomial tables (Table II, Appendix A) and $\alpha = .05$.
  **a.** $H_0: \eta = 9$ versus $H_a: \eta > 9$
  **b.** $H_0: \eta = 9$ versus $H_a: \eta \neq 9$
  **c.** $H_0: \eta = 20$ versus $H_a: \eta < 20$
  **d.** $H_0: \eta = 20$ versus $H_a: \eta \neq 20$
  **e.** Repeat each of the preceding tests using the normal approximation to the binomial probabilities. Compare the results.
  **f.** What assumptions are necessary to ensure the validity of each of the preceding tests?

**14.5** Suppose you wish to conduct a test of the research hypothesis that the median of a population is greater than 80. You randomly sample 25 measurements from the population and determine that 16 of them exceed 80. Set up and conduct the appropriate test of hypothesis at the .10 level of significance. Be sure to specify all necessary assumptions.

## Applying the Concepts—Basic

**14.6** In *The American Statistician* (May 2001), the nonparametric sign test was used to analyze data on the quality of white shrimp. One measure of shrimp quality is cohesiveness. Since freshly caught shrimp are usually stored on ice, there is concern that cohesiveness will deteriorate after storage. For a sample of 20 newly caught white shrimp, cohesiveness was measured both before storage and after storage on ice for two weeks. The difference in the cohesiveness measurements (before minus after) was obtained for each shrimp. If storage has no effect on cohesiveness, the population median of the differences will be 0. If cohesiveness deteriorates after storage, the population median of the differences will be positive.
  **a.** Set up the null and alternative hypotheses to test whether cohesiveness will deteriorate after storage.
  **b.** In the sample of 20 shrimp, there were 13 positive differences. Use this value to find the *p*-value of the test.
  **c.** Make the appropriate conclusion (in the words of the problem) if $\alpha = .05$.

**14.7** Refer to the *Environmental Science & Technology* (Sept. 1, 2000) study of ammonia levels near the exit ramp of a San Francisco highway tunnel, Exercise 2.45 (p. 52). The daily ammonia concentrations (parts per million) on eight randomly selected days during afternoon drive-time are reproduced in the table. Suppose you want to determine if the median daily ammonia concentration for all afternoon drive-time days exceeds 1.5 ppm.

⊚ AMMONIA

| 1.53 | 1.50 | 1.37 | 1.51 | 1.55 | 1.42 | 1.41 | 1.48 |
|------|------|------|------|------|------|------|------|

  **a.** Set up the null and alternative hypotheses for the test.
  **b.** Find the value of the test statistic.
  **c.** Find the *p*-value of the test.
  **d.** Give the appropriate conclusion (in the words of the problem) if $\alpha = .05$.

**14.8** The table on p. 746 lists the lymphocyte (LYMPHO) count results from hematology tests administered to a sample of 50 West Indian or African workers. Test (at $\alpha = .05$) the hypothesis that the median lymphocyte count of all West Indian/African workers exceeds 20. Use the MINITAB printout below to make your conclusion.

**MINITAB Output for Exercise 14.8**

Sign test of median = 20.00 versus > 20.00

|        | N  | Below | Equal | Above | P      | Median |
|--------|----|-------|-------|-------|--------|--------|
| LYMPHO | 50 | 19    | 6     | 25    | 0.2257 | 20.50  |

⬤ LYMPHO

| Case Number | LYMPHO | Case Number | LYMPHO | Case Number | LYMPHO |
|---|---|---|---|---|---|
| 1 | 14 | 18 | 28 | 35 | 11 |
| 2 | 15 | 19 | 17 | 36 | 25 |
| 3 | 19 | 20 | 14 | 37 | 30 |
| 4 | 23 | 21 | 8 | 38 | 32 |
| 5 | 17 | 22 | 25 | 39 | 17 |
| 6 | 20 | 23 | 37 | 40 | 22 |
| 7 | 21 | 24 | 20 | 41 | 20 |
| 8 | 16 | 25 | 15 | 42 | 20 |
| 9 | 27 | 26 | 9 | 43 | 20 |
| 10 | 34 | 27 | 16 | 44 | 26 |
| 11 | 26 | 28 | 18 | 45 | 40 |
| 12 | 28 | 29 | 17 | 46 | 22 |
| 13 | 24 | 30 | 23 | 47 | 61 |
| 14 | 26 | 31 | 43 | 48 | 12 |
| 15 | 23 | 32 | 17 | 49 | 20 |
| 16 | 9 | 33 | 23 | 50 | 35 |
| 17 | 18 | 34 | 31 | | |

*Source:* Royston, J. P. "Some techniques for assessing multivariate normality based on the Shapiro-Wilk W." *Applied Statistics*, Vol. 32, No. 2, pp. 121–133.

## Applying the Concepts—Intermediate

**14.9** The biting rate of a particular species of fly was investigated in a study reported in the *Journal of the American Mosquito Control Association* (Mar. 1995). Biting rate was defined as the number of flies biting a volunteer during 15 minutes of exposure. This species of fly is known to have a median biting rate of 5 bites per 15 minutes on Stanbury Island, Utah. However, it is theorized that the median biting rate is higher in bright, sunny weather. To test this theory, 122 volunteers were exposed to the flies during a sunny day on Stanbury Island. Of these volunteers, 95 experienced biting rates greater than 5.
a. Set up the null and alternative hypotheses for the test.
b. Calculate the approximate *p*-value of the test. [*Hint:* Use the normal approximation for a binomial probability.]
c. Make the appropriate conclusion at $\alpha = .01$.

**14.10** In a study of the excessive transitory migration of guppy populations, 40 adult female guppies were placed in the left compartment of an experimental aquarium tank which was divided in half by a glass plate. After the plate was removed, the numbers of fish passing through the slit from the left compartment to the right one, and vice versa, were monitored every minute for 30 minutes

(*Zoological Science*, Vol. 6, 1989). If an equilibrium is reached, the researchers would expect the median number of fish remaining in the left compartment to be 20. The data for the 30 observations (that is, numbers of fish in the left compartment at the end of each one-minute interval) are shown in the table below. Use the large-sample sign test to determine whether the median is less than 20. Test using $\alpha = .05$.

**14.11** According to the National Restaurant Association, hamburgers are the number-one selling fast-food item in the United States. An economist studying the fast-food buying habits of Americans paid graduate students to stand outside two suburban McDonald's restaurants near Boston and ask departing customers whether they spent more than $2.25 on hamburger products for their lunch. Twenty answered yes; 50 said no; and 10 refused to answer the question (*Newark Star-Ledger*, Mar. 17, 1997).
a. Is there sufficient evidence to conclude that the median amount spent for hamburgers at lunch at McDonald's is less than $2.25?
b. Does your conclusion apply to all Americans who eat lunch at McDonald's? Justify your answer.
c. What assumptions must hold to ensure the validity of your test in part **a**?

⬤ GUPPY

| | | | | | | | | | |
|---|---|---|---|---|---|---|---|---|---|
| 16 | 11 | 12 | 15 | 14 | 16 | 18 | 15 | 13 | 15 |
| 14 | 14 | 16 | 13 | 17 | 17 | 14 | 22 | 18 | 19 |
| 17 | 17 | 20 | 23 | 18 | 19 | 21 | 17 | 21 | 17 |

*Source:* Terami, H., and Watanabe, M. "Excessive transitory migration of guppy populations. III. Analysis of perception of swimming space and a mirror effect." *Zoological Science*, Vol. 6, 1989, p. 977 (Figure 2).

## 14.3  COMPARING TWO POPULATIONS: THE WILCOXON RANK SUM TEST FOR INDEPENDENT SAMPLES

Suppose two independent random samples are to be used to compare two populations and the *t*-test of Chapter 9 is inappropriate for making the comparison. We may be unwilling to make assumptions about the form of the underlying population probability distributions or we may be unable to obtain exact values of the sample measurements. If the data can be ranked in order of magnitude for either of these situations, the **Wilcoxon rank sum test** (developed by Frank Wilcoxon) can be used to test the hypothesis that the probability distributions associated with the two populations are equivalent.

Suppose, for example, an experimental psychologist wants to compare reaction times for adult males under the influence of drug A to those under the influence of drug B. Experience has shown that the populations of reaction time measurements often possess probability distributions that are skewed to the right, as shown in Figure 14.4. Consequently, a *t*-test should not be used to compare the mean reaction times for the two drugs because the normality assumption that is required for the *t*-test may not be valid.

**FIGURE 14.4**

*Typical Probability Distribution of Reaction Times*

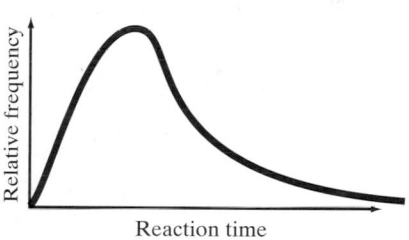

Suppose the psychologist randomly assigns seven subjects to each of two groups, one group to receive drug A and the other to receive drug B. The reaction time for each subject is measured at the completion of the experiment. These data (with the exception of the measurement for one subject in group A who was eliminated from the experiment for personal reasons) are shown in Table 14.2.

The population of reaction times for either of the drugs, say drug A, is that which could conceptually be obtained by giving drug A to all adult males. To compare the probability distributions for populations A and B, *we first rank the sample observations as though they were all drawn from the same population*. That is, we pool the measurements from both samples and then rank all the measurements from the smallest (a rank of 1) to the largest (a rank of 13). The results of this ranking process are also shown in Table 14.2.

⦿ REACTION

**TABLE 14.2  Reaction Times of Subjects Under the Influence of Drug A or B**

| Drug A | | Drug B | |
|---|---|---|---|
| **Reaction Time (seconds)** | **Rank** | **Reaction Time (seconds)** | **Rank** |
| 1.96 | 4 | 2.11 | 6 |
| 2.24 | 7 | 2.43 | 9 |
| 1.71 | 2 | 2.07 | 5 |
| 2.41 | 8 | 2.71 | 11 |
| 1.62 | 1 | 2.50 | 10 |
| 1.93 | 3 | 2.84 | 12 |
| | | 2.88 | 13 |

If the two populations were identical, we would expect the ranks to be *randomly mixed* between the two samples. If, on the other hand, one population tends to have longer reaction times than the other, we would expect the larger ranks to be mostly in one sample and the smaller ranks mostly in the other. Thus, the test statistic for the Wilcoxon test is based on the totals of the ranks for each of the two samples—that is, on the **rank sums**. When the sample sizes are equal, for example, the greater the difference in the rank sums, the greater will be the weight of evidence to indicate a difference between the probability distributions of the populations. In the reaction times example, we denote the rank sum for drug A by $T_1$ and that for drug B by $T_2$. Then

$$T_1 = 4 + 7 + 2 + 8 + 1 + 3 = 25$$
$$T_2 = 6 + 9 + 5 + 11 + 10 + 12 + 13 = 66$$

The sum of $T_1$ and $T_2$ will always equal $n(n + 1)/2$, where $n = n_1 + n_2$. So, for this example, $n_1 = 6, n_2 = 7$, and

$$T_1 + T_2 = \frac{13(13 + 1)}{2} = 91$$

Since $T_1 + T_2$ is fixed, a small value for $T_1$ implies a large value for $T_2$ (and vice versa) and a large difference between $T_1$ and $T_2$. Therefore, the smaller the value of one of the rank sums, the greater the evidence to indicate that the samples were selected from different populations.

The test statistic for this test is the rank sum for the smaller sample; or, in the case where $n_1 = n_2$, either rank sum can be used. Values that locate the rejection region for this rank sum are given in Table XII of Appendix A. A partial reproduction of this table is shown in Table 14.3. The columns of the table represent $n_1$, the first sample size, and the rows represent $n_2$, the second sample size. *The $T_L$ and $T_U$ entries in the table are the boundaries of the lower and upper regions, respectively, for the rank sum associated with the sample that has fewer measurements.* If the sample sizes $n_1$ and $n_2$ are the same, either rank sum may be used as the test statistic. To illustrate, suppose $n_1 = 8$ and $n_2 = 10$. For a two-tailed test with $\alpha = .05$, we consult part a of the table and find that the null hypothesis will be rejected if the rank sum of sample 1 (the sample with fewer measurements), $T$, is less than or equal to $T_L = 54$ *or* greater than or equal to $T_U = 98$. The Wilcoxon rank sum test is summarized in the next box.

**TABLE 14.3    Reproduction of Part of Table XII in Appendix A: Critical Values for the Wilcoxon Rank Sum Test**

$\alpha = .025$ one-tailed; $\alpha = .05$ two-tailed

| $n_2$ \ $n_1$ | 3 | | 4 | | 5 | | 6 | | 7 | | 8 | | 9 | | 10 | |
|---|---|---|---|---|---|---|---|---|---|---|---|---|---|---|---|---|
| | $T_L$ | $T_U$ | $T_L$ | $T_U$ | $T_L$ | $T_U$ | $T_L$ | $T_U$ | $T_L$ | $T_U$ | $T_L$ | $T_U$ | $T_L$ | $T_U$ | $T_L$ | $T_U$ |
| 3 | 5 | 16 | 6 | 18 | 6 | 21 | 7 | 23 | 7 | 26 | 8 | 28 | 8 | 31 | 9 | 33 |
| 4 | 6 | 18 | 11 | 25 | 12 | 28 | 12 | 32 | 13 | 35 | 14 | 38 | 15 | 41 | 16 | 44 |
| 5 | 6 | 21 | 12 | 28 | 18 | 37 | 19 | 41 | 20 | 45 | 21 | 49 | 22 | 53 | 24 | 56 |
| 6 | 7 | 23 | 12 | 32 | 19 | 41 | 26 | 52 | 28 | 56 | 29 | 61 | 31 | 65 | 32 | 70 |
| 7 | 7 | 26 | 13 | 35 | 20 | 45 | 28 | 56 | 37 | 68 | 39 | 73 | 41 | 78 | 43 | 83 |
| 8 | 8 | 28 | 14 | 38 | 21 | 49 | 29 | 61 | 39 | 73 | 49 | 87 | 51 | 93 | 54 | 98 |
| 9 | 8 | 31 | 15 | 41 | 22 | 53 | 31 | 65 | 41 | 78 | 51 | 93 | 63 | 108 | 66 | 114 |
| 10 | 9 | 33 | 16 | 44 | 24 | 56 | 32 | 70 | 43 | 83 | 54 | 98 | 66 | 114 | 79 | 131 |

## Wilcoxon Rank Sum Test: Independent Samples*

Let $D_1$ and $D_2$ represent the probability distributions for populations 1 and 2, respectively.

| ONE-TAILED TEST | TWO-TAILED TEST |
|---|---|
| $H_0$: $D_1$ and $D_2$ are identical | $H_0$: $D_1$ and $D_2$ are identical |
| $H_a$: $D_1$ is shifted to the right of $D_2$ | $H_a$: $D_1$ is shifted either to the left or to the right of $D_2$ |
| [or $H_a$: $D_1$ is shifted to the left of $D_2$] | |

*Test statistic:*

$T_1$, if $n_1 < n_2$; $T_2$, if $n_2 < n_1$ (Either rank sum can be used if $n_1 = n_2$.)

*Test statistic:*

$T_1$, if $n_1 < n_2$; $T_2$, if $n_2 < n_1$ (Either rank sum can be used if $n_1 = n_2$.) We will denote this rank sum as $T$.

*Rejection region:*

$T_1$: $T_1 \geq T_U$ [or $T_1 \leq T_L$]

$T_2$: $T_2 \leq T_L$ [or $T_2 \geq T_U$]

*Rejection region:*

$T \leq T_L$ or $T \geq T_U$

where $T_L$ and $T_U$ are obtained from Table XII of Appendix A.

*Assumptions:* 1. The two samples are random and independent.

2. The two probability distributions from which the samples are drawn are continuous.

*Ties:* Assign tied measurements the average of the ranks they would receive if they were unequal but occurred in successive order. For example, if the third-ranked and fourth-ranked measurements are tied, assign each a rank of $(3 + 4)/2 = 3.5$.

Note that the assumptions necessary for the validity of the Wilcoxon rank sum test do not specify the shape or type of probability distribution. However, the distributions are assumed to be continuous so that the probability of tied measurements is 0 (see Chapter 5), and each measurement can be assigned a unique rank. In practice, however, rounding of continuous measurements will sometimes produce ties. As long as the number of ties is small relative to the sample sizes, the Wilcoxon test procedure will still have approximate significance level $\alpha$. The test is not recommended to compare discrete distributions for which many ties are expected.

**EXAMPLE 14.2**

Do the data given in Table 14.2 provide sufficient evidence to indicate a shift in the probability distributions for drugs A and B, that is, that the probability distribution corresponding to drug A lies either to the right or left of the probability distribution corresponding to drug B? Test at the .05 level of significance.

**Solution**

$H_0$: The two populations of reaction times corresponding to drug A and drug B have the same probability distribution

---

*Another statistic used for comparing two populations based on independent random samples is the *Mann–Whitney U statistic*. The $U$ statistic is a simple function of the rank sums. It can be shown that the Wilcoxon rank sum test and the Mann–Whitney $U$-test are equivalent.

$H_a$:  The probability distribution for drug A is shifted to the right or left
of the probability distribution corresponding to drug B.*

*Test statistic:*  Since drug A has fewer subjects than drug B, the test statistic is $T_1$,
the rank sum of drug A's reaction times.

*Rejection region:*  Since the test is two-sided, we consult part a of Table XII for the
rejection region corresponding to $\alpha = .05$. We will reject $H_0$ for
$T_1 \leq T_L$ or $T_1 \geq T_U$. Thus, we will reject $H_0$ if $T_1 \leq 28$ or $T_1 \geq 56$.

Since $T_1$, the rank sum of drug A's reaction times in Table 14.2, is 25, it is in the
rejection region (see Figure 14.5).[†] Therefore, there is sufficient evidence to reject
$H_0$. This same conclusion can be reached using a statistical software package. The
SAS printout of the analysis is shown in Figure 14.6. Both the test statistic
($T_1 = 25$) and one-tailed *p*-value ($p = .007$) are highlighted on the printout. The
one-tailed *p*-value is less than $\alpha = .05$, leading us to reject $H_0$.

Our conclusion is that the probability distributions for drugs A and B are not
identical. In fact, it appears that drug B tends to be associated with reaction times
that are larger than those associated with drug A (because $T_1$ falls in the lower tail
of the rejection region). ∎

**FIGURE 14.5**

*Alternative Hypothesis and
Rejection Region for
Example 14.2*

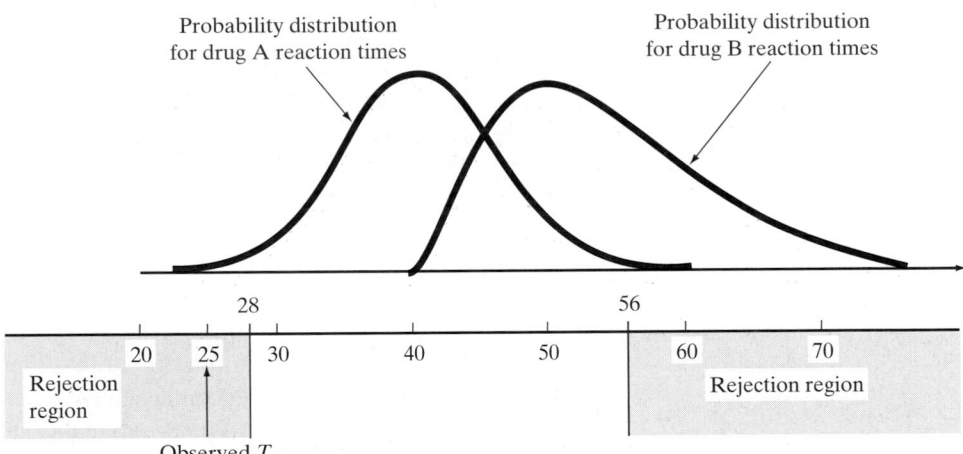

Table XII in Appendix A gives values of $T_L$ and $T_U$ for values of $n_1$ and $n_2$ less
than or equal to 10. When both sample sizes, $n_1$ and $n_2$, are 10 or larger, the sampling distribution of $T_1$ can be approximated by a normal distribution with mean
and variance

$$E(T_1) = \frac{n_1(n_1 + n_2 + 1)}{2} \quad \text{and} \quad \sigma_{T_1}^2 = \frac{n_1 n_2(n_1 + n_2 + 1)}{12}$$

Therefore, for $n_1 \geq 10$ and $n_2 \geq 10$ we can conduct the Wilcoxon rank sum test
using the familiar *z*-test of Chapters 9 and 10. The test is summarized in the next
box.

---

[*]The alternative hypotheses in this chapter will be stated in terms of a difference in the *location* of the distributions.
However, since the shapes of the distributions may also differ under $H_a$, some of the figures (e.g., Figure 14.5) depicting
the alternative hypothesis will show probability distributions with different shapes.
[†]Figure 14.5 depicts only one side of the two-sided alternative hypothesis. The other would show the distribution for
drug A shifted to the right of the distribution for drug B.

**FIGURE 14.6**

*SAS Printout for Example 14.2*

The NPAR1WAY Procedure

Wilcoxon Scores (Rank Sums) for Variable REACTIME
Classified by Variable DRUG

| DRUG | N | Sum of Scores | Expected Under H0 | Std Dev Under H0 | Mean Score |
|------|---|---------------|-------------------|------------------|------------|
| A | 6 | 25.0 | 42.0 | 7.0 | 4.166667 |
| B | 7 | 66.0 | 49.0 | 7.0 | 9.428571 |

Wilcoxon Two-Sample Test

| | |
|---|---|
| Statistic | 25.0000 |

Normal Approximation
| | |
|---|---|
| Z | -2.4286 |
| One-Sided Pr < Z | 0.0076 |
| Two-Sided Pr > \|Z\| | 0.0152 |

t Approximation
| | |
|---|---|
| One-Sided Pr < Z | 0.0159 |
| Two-Sided Pr > \|Z\| | 0.0318 |

Exact Test
| | |
|---|---|
| One-Sided Pr <=  S | 0.0070 |
| Two-Sided Pr >= \|S - Mean\| | 0.0140 |

Kruskal-Wallis Test

| | |
|---|---|
| Chi-Square | 5.8980 |
| DF | 1 |
| Pr > Chi-Square | 0.0152 |

---

## The Wilcoxon Rank Sum Test for Large Samples ($n_1 \geq 10$ *and* $n_2 \geq 10$)

Let $D_1$ and $D_2$ represent the probability distributions for populations 1 and 2, respectively.

**ONE-TAILED TEST**

$H_0$: $D_1$ and $D_2$ are identical

$H_a$: $D_1$ is shifted to the right of $D_2$
  (or $H_a$: $D_1$ is shifted to the left of $D_2$)

**TWO-TAILED TEST**

$H_0$: $D_1$ and $D_2$ are identical

$H_a$: $D_1$ is shifted to the right or to the left of $D_2$

$$\text{Test statistic:} \quad z = \frac{T_1 - \dfrac{n_1(n_1 + n_2 + 1)}{2}}{\sqrt{\dfrac{n_1 n_2(n_1 + n_2 + 1)}{12}}}$$

*Rejection region:*

$z > z_\alpha$ (or $z < -z_\alpha$)

*Rejection region:*

$|z| > z_{\alpha/2}$

## EXERCISES 14.12–14.23

### Learning the Mechanics

**14.12** Specify the test statistic and the rejection region for the Wilcoxon rank sum test for independent samples in each of the following situations:

**a.** $H_0$: Two probability distributions, 1 and 2, are identical
$H_a$: Probability distribution for population 1 is shifted to the right or left of the probability distribution for population 2
$n_1 = 7, n_2 = 8, \alpha = .10$

**b.** $H_0$: Two probability distributions, 1 and 2, are identical
$H_a$: Probability distribution for population 1 is shifted to the right of the probability distribution for population 2
$n_1 = 6, n_2 = 6, \alpha = .05$

**c.** $H_0$: Two probability distributions, 1 and 2, are identical
$H_a$: Probability distribution for population 1 is shifted to the left of the probability distribution for population 2
$n_1 = 7, n_2 = 10, \alpha = .025$

**d.** $H_0$: Two probability distributions, 1 and 2, are identical
$H_a$: Probability distribution for population 1 is shifted to the right or left of the probability distribution for population 2
$n_1 = 20, n_2 = 20, \alpha = .05$

**14.13** Suppose you want to compare two treatments, A and B. In particular, you wish to determine whether the distribution for population B is shifted to the right of the distribution for population A. You plan to use the Wilcoxon rank sum test.

**a.** Specify the null and alternative hypotheses you would test.

**b.** Suppose you obtained the following independent random samples of observations on experimental units subjected to the two treatments:

⊘ LM14_13

| Sample A | 37, 40, 33, 29, 42, 33, 35, 28, 34 |
|----------|-------------------------------------|
| Sample B | 65, 35, 47, 52 |

Conduct a test of the hypotheses described in part **a**. Test using $\alpha = .05$.

**14.14** Random samples of sizes $n_1 = 16$ and $n_2 = 12$ were drawn from populations 1 and 2, respectively. The measurements obtained are listed in the table.

⊘ LM14_14

| Sample 1 | | | | Sample 2 | | |
|------|------|------|------|------|------|------|
| 9.0 | 15.6 | 25.6 | 31.1 | 10.1 | 11.1 | 13.5 |
| 21.1 | 26.9 | 24.6 | 20.0 | 12.0 | 18.2 | 10.3 |
| 24.8 | 16.5 | 26.0 | 25.1 | 9.2 | 7.0 | 14.2 |
| 17.2 | 30.1 | 18.7 | 26.1 | 15.8 | 13.6 | 13.2 |

**a.** Conduct a hypothesis test to determine whether the probability distribution for population 2 is shifted to the left of the probability distribution for population 1. Use $\alpha = .05$.

**b.** What is the approximate *p*-value of the test of part **a**?

**14.15** Independent random samples are selected from two populations. The data are shown in the table.

| Sample 1 | | Sample 2 | | |
|------|------|------|------|------|
| 15 | 16 | 5 | 9 | 5 |
| 10 | 13 | 12 | 8 | 10 |
| 12 | 8 | 9 | 4 | |

**a.** Use the Wilcoxon rank sum test to determine whether the data provide sufficient evidence to indicate a shift in the locations of the probability distributions of the sampled populations. Test using $\alpha = .05$.

**b.** Do the data provide sufficient evidence to indicate that the probability distribution for population 1 is shifted to the right of the probability distribution for population 2? Use the Wilcoxon rank sum test with $\alpha = .05$.

**14.16** Suppose you wish to compare two treatments, A and B, based on independent random samples of 15 observations selected from each of the two populations. If $T_1 = 173$, do the data provide sufficient evidence to indicate that distribution A is shifted to the left of distribution B? Test using $\alpha = .05$.

### Applying the Concepts—Basic

**14.17** The "fear of negative evaluation" (FNE) scores for 11 female students known to suffer from the eating disorder bulimia and 14 female students with normal eating habits, first presented in Exercise 2.27 (p. 39), are reproduced on p. 753. (Recall that the higher the score, the greater the fear of negative evaluation.) Suppose you want to determine if the distribution of the FNE scores for bulimic female students is shifted above the corresponding distribution for female students with normal eating habits.

**a.** Specify $H_0$ and $H_a$ for the test.

**b.** Rank all 25 FNE scores in the data set from smallest to largest.

**c.** Sum the ranks of the 11 FNE scores for bulimic students.

**d.** Sum the ranks of the 14 FNE scores for students with normal eating habits.

**e.** Give the rejection region for a nonparametric test of the data if $\alpha = .10$.

**f.** Conduct the test and give the conclusion in the words of the problem.

### ⊘ BULIMIA

| Bulimic students | 21 | 13 | 10 | 20 | 25 | 19 | 16 | 21 | 24 | 13 | 14 | | | |
|---|---|---|---|---|---|---|---|---|---|---|---|---|---|---|
| Normal students | 13 | 6 | 16 | 13 | 8 | 19 | 23 | 18 | 11 | 19 | 7 | 10 | 15 | 20 |

*Source:* Randles, R. H. "On neutral responses (zeros) in the sign test and ties in the Wilcoxon–Mann–Whitney test." *The American Statistician*, Vol. 55, No. 2, May 2001 (Figure 3).

**14.18** Refer to the *Organizational Science* (May–June 2000) study of new artists on *Billboard* magazine's chart of the top selling music CDs, Exercise 9.19 (p. 395). Recall that in 1991, *Billboard* began using the SoundScan system to chart CD and album sales. The number of new artists entering the weekly *Billboard* chart for 12 weeks prior to the introduction of SoundScan and for 12 weeks after the introduction of SoundScan are reproduced in the table.

### ⊘ BILLBOARD

| Pre-SoundScan | 1 | 1 | 0 | 2 | 3 | 4 | 3 | 4 | 5 | 5 | 2 | 3 |
|---|---|---|---|---|---|---|---|---|---|---|---|---|---|
| Post-SoundScan | 0 | 2 | 3 | 1 | 3 | 1 | 0 | 1 | 4 | 2 | 2 | 1 |

  **a.** In Exercise 9.19 you analyzed the data using the Student's *t*-distribution. Why might a nonparametric procedure be a better test to apply?
  **b.** Use nonparametrics to test for a shift in the distributions of number of new artists entering the *Billboard* chart prior to and after SoundScan. Test at $\alpha = .10$.

**14.19** In a comparison of visual acuity of deaf and hearing children, eye movement rates are taken on 10 deaf and 10 hearing children. The data are shown in the table. A clinical psychologist believes that deaf children have greater visual acuity than hearing children. (The larger a child's eye movement rate, the more visual acuity the child possesses.)

### ⊘ EYEMOVE

| Deaf Children | | Hearing Children | |
|---|---|---|---|
| 2.75 | 1.95 | 1.15 | 1.23 |
| 3.14 | 2.17 | 1.65 | 2.03 |
| 3.23 | 2.45 | 1.43 | 1.64 |
| 2.30 | 1.83 | 1.83 | 1.96 |
| 2.64 | 2.23 | 1.75 | 1.37 |

  **a.** Use the Wilcoxon rank sum procedure to test the psychologist's claim at $\alpha = .05$.
  **b.** Conduct the test by using the large-sample approximation for the Wilcoxon rank sum test. Compare the results with those found in part **a**.

## Applying the Concepts—Intermediate

**14.20** The data in the table, extracted from *Technometrics* (Feb. 1986), represent daily accumulated stream flow and precipitation (in inches) for two U.S. Geological Survey stations in Colorado. Conduct a test to determine whether the distributions of daily accumulated stream flow and precipitation for the two stations differ in location. Use $\alpha = .10$. Why is a nonparametric test appropriate for these data?

### ⊘ COLORAIN

| Station 1 | | | Station 2 | | |
|---|---|---|---|---|---|
| 127.96 | 108.91 | 100.85 | 114.79 | 85.54 | 280.55 |
| 210.07 | 178.21 | 85.89 | 109.11 | 117.64 | 145.11 |
| 203.24 | 285.37 | | 330.33 | 302.74 | 95.36 |

*Source:* Gastwirth, J. L., and Mahmoud, H. "An efficient robust nonparametric test for scale change for data from a gamma distribution." *Technometrics*, Vol. 28, No. 1, Feb. 1986, p. 83 (Table 2).

**14.21** University of Queensland researchers sampled private sector and public sector organizations in Australia to study the planning undertaken by their information systems departments (*Management Science*, July 1996). As part of the process they asked each sample organization how much it had spent on information systems and technology in the previous fiscal year as a percentage of the organization's total revenues. The results are reported in the table.

### ⊘ INFOSYS

| Private Sector | Public Sector |
|---|---|
| 2.58% | 5.40% |
| 5.05 | 2.55 |
| .05 | 9.00 |
| 2.10 | 10.55 |
| 4.30 | 1.02 |
| 2.25 | 5.11 |
| 2.50 | 24.42 |
| 1.94 | 1.67 |
| 2.33 | 3.33 |

Adapted from Hann, J., and Weber, R. "Information systems planning: A model and empirical tests." *Management Science*, Vol. 42, No. 2 (July, 1996), pp. 1043–1064.

  **a.** Do the two sampled populations have identical probability distributions or is the distribution for public sector organizations in Australia located to the right of Australia's private sector firms? Test using $\alpha = .05$.
  **b.** Is the *p*-value for the test less than or greater than .05? Justify your answer.
  **c.** What assumption must be met to ensure the validity of the test you conducted in part **a**?

**14.22** The term *brood-parasitic intruder* is used to describe birds that search for and lay eggs in the nests built by another bird species. For example, the brown-headed cowbird is known to be a brood parasite of the smaller-sized willow flycatcher. Ornithologists theorize that those flycatchers that recognize but do not vocally react to cowbird calls are more apt to defend their nests and less likely to be found and parasitized. In a study published in *The Condor* (May 1995), each of 13 active flycatcher nests was categorized as parasitized (if at least one cowbird egg was present) or nonparasitized. Cowbird songs were taped and played back while the flycatcher pairs were sitting in the nest prior to incubation. The vocalization rate (number of calls per minute) of each flycatcher pair was recorded. The data for the two groups of flycatchers are given in the table. Do the data suggest (at $\alpha = .05$) that the vocalization rates of parasitized flycatchers are higher than those of nonparasitized flycatchers? Use the accompanying SAS printout to form your conclusion.

**⊘ COWBIRD**

| Parasitized | Not Parasitized |
|---|---|
| 2.00 | 1.00 |
| 1.25 | 1.00 |
| 8.50 | 0 |
| 1.10 | 3.25 |
| 1.25 | 1.00 |
| 3.75 | .25 |
| 5.50 | |

*Source:* Uyehara, J. C., and Narins, P. M. "Nest defense by Willow Flycatchers to brood-parasitic intruders." *The Condor*, Vol. 97, No. 2, May 1995, p. 364 (Figure 1).

**14.23** Refer to the impact study of the interactive Teachers Involve Parents in Schoolwork (TIPS) program, Exercise 9.17 (p. 394). Recall that 128 middle school students were assigned to complete TIPS homework assignments, while 98 students were assigned traditional, noninteractive, homework assignments (ATIPS). At the

SAS Output for Exercise 14.22

The NPAR1WAY Procedure

Wilcoxon Scores (Rank Sums) for Variable VOCRATE
Classified by Variable PARGROUP

| PARGROUP | N | Sum of Scores | Expected Under HO | Std Dev Under HO | Mean Score |
|---|---|---|---|---|---|
| PAR | 7 | 66.0 | 49.0 | 6.951757 | 9.428571 |
| NOT | 6 | 25.0 | 42.0 | 6.951757 | 4.166667 |

Average scores were used for ties.

Wilcoxon Two-Sample Test

| | |
|---|---|
| Statistic (S) | 25.0000 |

Normal Approximation
| | |
|---|---|
| Z | -2.4454 |
| One-Sided Pr < Z | 0.0072 |
| Two-Sided Pr > \|Z\| | 0.0145 |

t Approximation
| | |
|---|---|
| One-Sided Pr < Z | 0.0154 |
| Two-Sided Pr > \|Z\| | 0.0309 |

Exact Test
| | |
|---|---|
| One-Sided Pr <=  S | 0.0064 |
| Two-Sided Pr >= \|S - Mean\| | 0.0128 |

Kruskal-Wallis Test

| | |
|---|---|
| Chi-Square | 5.9801 |
| DF | 1 |
| Pr > Chi-Square | 0.0145 |

end of the study, all students reported on the level of family involvement in their homework on a 5-point scale (0 = Never, 1 = Rarely, 2 = Sometimes, 3 = Frequently, 4 = Always). The data for the science, math, and language arts homework are saved in the HWSTUDY file. (The first five and last five observations in the data set are reproduced in the table.)

**a.** Why might a nonparametric test be the most appropriate test to apply in order to compare the levels of family involvement in homework assignments of TIPS and ATIPS students?

**b.** Conduct a nonparametric analysis to compare the science homework assignments of TIPS and ATIPS students. Use $\alpha = .05$.

**c.** Repeat part **a** for mathematics homework assignments.

**d.** Repeat part **a** for language arts homework assignments.

⬤ HWSTUDY

| Homework Condition | Science | Math | Language |
|---|---|---|---|
| ATIPS | 1 | 0 | 0 |
| ATIPS | 0 | 1 | 1 |
| ATIPS | 0 | 1 | 0 |
| ATIPS | 1 | 2 | 0 |
| ATIPS | 1 | 1 | 2 |
| ⋮ | ⋮ | ⋮ | ⋮ |
| TIPS | 2 | 3 | 2 |
| TIPS | 1 | 4 | 2 |
| TIPS | 2 | 4 | 2 |
| TIPS | 4 | 0 | 3 |
| TIPS | 2 | 0 | 1 |

*Source:* Van Voorhis, F. L. "Teachers' use of interactive homework and its effects on family involvement and science achievement of middle grade students." Paper presented at the annual meeting of the American Educational Research Association, Seattle, April 2001.

## 14.4 COMPARING TWO POPULATIONS: THE WILCOXON SIGNED RANK TEST FOR THE PAIRED DIFFERENCE EXPERIMENT

Nonparametric techniques may also be employed to compare two probability distributions when a paired difference design is used. For example, consumer preferences for two competing products are often compared by having each of a sample of consumers rate both products. Thus, the ratings have been paired on each consumer. Here is an example of this type of experiment.

For some paper products, softness is an important consideration in determining consumer acceptance. One method of determining softness is to have judges give a sample of the products a softness rating. Suppose each of 10 judges is given a sample of two products that a company wants to compare. Each judge rates the softness of each product on a scale from 1 to 10, with higher ratings implying a softer product. The results of the experiment are shown in Table 14.4.

⬤ SOFTPAPER

**TABLE 14.4** **Softness Ratings of Paper**

| Judge | Product A | Product B | Difference (A − B) | Absolute Value of Difference | Rank of Absolute Value |
|---|---|---|---|---|---|
| 1 | 6 | 4 | 2 | 2 | 5 |
| 2 | 8 | 5 | 3 | 3 | 7.5 |
| 3 | 4 | 5 | −1 | 1 | 2 |
| 4 | 9 | 8 | 1 | 1 | 2 |
| 5 | 4 | 1 | 3 | 3 | 7.5 |
| 6 | 7 | 9 | −2 | 2 | 5 |
| 7 | 6 | 2 | 4 | 4 | 9 |
| 8 | 5 | 3 | 2 | 2 | 5 |
| 9 | 6 | 7 | −1 | 1 | 2 |
| 10 | 8 | 2 | 6 | 6 | 10 |

$T_+ =$ Sum of positive ranks $= 46$
$T_- =$ Sum of negative ranks $= 9$

Since this is a paired difference experiment, we analyze the differences between the measurements (see Section 9.2). However, a nonparametric approach developed by Wilcoxon requires that we calculate the ranks of the absolute values of the differences between the measurements, that is, the ranks of the differences after removing any minus signs. *Note that tied absolute differences are assigned the average of the ranks they would receive if they were unequal but successive measurements.* After the absolute differences are ranked, the sum of the ranks of the positive differences of the original measurements, $T_+$, and the sum of the ranks of the negative differences of the original measurements, $T_-$, are computed.

We are now prepared to test the nonparametric hypotheses:

$H_0$: The probability distributions of the ratings for products A and B are identical.

$H_a$: The probability distributions of the ratings differ (in location) for the two products. (Note that this is a two-sided alternative and that it implies a two-tailed test.)

*Test statistic:*   $T$ = Smaller of the positive and negative rank sums $T_+$ and $T_-$

The smaller the value of $T$, the greater the evidence to indicate that the two probability distributions differ in location. The rejection region for $T$ can be determined by consulting Table XIII in Appendix A (part of the table is shown in Table 14.5). This table gives a value $T_0$ for both one-tailed and two-tailed tests for each value of $n$, the number of matched pairs. For a two-tailed test with $\alpha = .05$, we will reject $H_0$ if $T \leq T_0$. You can see in Table 14.5 that the value of $T_0$ that locates the boundary of the rejection region for the judges' ratings for $\alpha = .05$ and $n = 10$ pairs of observations is 8. Thus, the rejection region for the test (see Figure 14.7) is

*Rejection region:*   $T \leq 8$   for $\alpha = .05$

**TABLE 14.5**   **Reproduction of Part of Table XIII of Appendix A: Critical Values for the Wilcoxon Paired Difference Signed Rank Test**

| One-Tailed | Two-Tailed | $n = 5$ | $n = 6$ | $n = 7$ | $n = 8$ | $n = 9$ | $n = 10$ |
|---|---|---|---|---|---|---|---|
| $\alpha = .05$ | $\alpha = .10$ | 1 | 2 | 4 | 6 | 8 | 11 |
| $\alpha = .025$ | $\alpha = .05$ | | 1 | 2 | 4 | 6 | 8 |
| $\alpha = .01$ | $\alpha = .02$ | | | 0 | 2 | 3 | 5 |
| $\alpha = .005$ | $\alpha = .01$ | | | | 0 | 2 | 3 |
| | | $n = 11$ | $n = 12$ | $n = 13$ | $n = 14$ | $n = 15$ | $n = 16$ |
| $\alpha = .05$ | $\alpha = .10$ | 14 | 17 | 21 | 26 | 30 | 36 |
| $\alpha = .025$ | $\alpha = .05$ | 11 | 14 | 17 | 21 | 25 | 30 |
| $\alpha = .01$ | $\alpha = .02$ | 7 | 10 | 13 | 16 | 20 | 24 |
| $\alpha = .005$ | $\alpha = .01$ | 5 | 7 | 10 | 13 | 16 | 19 |
| | | $n = 17$ | $n = 18$ | $n = 19$ | $n = 20$ | $n = 21$ | $n = 22$ |
| $\alpha = .05$ | $\alpha = .10$ | 41 | 47 | 54 | 60 | 68 | 75 |
| $\alpha = .025$ | $\alpha = .05$ | 35 | 40 | 46 | 52 | 59 | 66 |
| $\alpha = .01$ | $\alpha = .02$ | 28 | 33 | 38 | 43 | 49 | 56 |
| $\alpha = .005$ | $\alpha = .01$ | 23 | 28 | 32 | 37 | 43 | 49 |
| | | $n = 23$ | $n = 24$ | $n = 25$ | $n = 26$ | $n = 27$ | $n = 28$ |
| $\alpha = .05$ | $\alpha = .10$ | 83 | 92 | 101 | 110 | 120 | 130 |
| $\alpha = .025$ | $\alpha = .05$ | 73 | 81 | 90 | 98 | 107 | 117 |
| $\alpha = .01$ | $\alpha = .02$ | 62 | 69 | 77 | 85 | 93 | 102 |
| $\alpha = .005$ | $\alpha = .01$ | 55 | 61 | 68 | 76 | 84 | 92 |

**FIGURE 14.7**

*Rejection Region for Paired Difference Experiment*

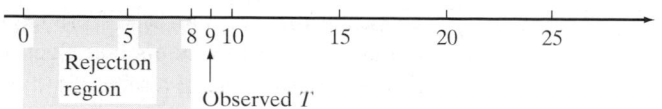

Since the smaller rank sum for the paper data, $T_- = 9$, does not fall within the rejection region, the experiment has not provided sufficient evidence to indicate that the two paper products differ with respect to their softness ratings at the $\alpha = .05$ level.

Note that if a significance level of $\alpha = .10$ had been used, the rejection region would have been $T \le 11$ and we would have rejected $H_0$. In other words, the samples do provide evidence that the probability distributions of the softness ratings differ at the $\alpha = .10$ significance level.

The **Wilcoxon signed rank test** is summarized in the following box. Note that the difference measurements are assumed to have a continuous probability distribution so that the absolute differences will have unique ranks. Although tied (absolute) differences can be assigned ranks by averaging, the number of ties should be small relative to the number of observations to ensure the validity of the test.

---

### Wilcoxon Signed Rank Test for a Paired Difference Experiment

Let $D_1$ and $D_2$ represent the probability distributions for populations 1 and 2, respectively.

**ONE-TAILED TEST**

$H_0$: $D_1$ and $D_2$ are identical

$H_a$: $D_1$ is shifted to the right of $D_2$
   [or $H_a$: $D_1$ is shifted to the left of $D_2$]

**TWO-TAILED TEST**

$H_0$: $D_1$ and $D_2$ are identical

$H_a$: $D_1$ is shifted either to the left or to the right of $D_2$

Calculate the difference within each of the $n$ matched pairs of observations. Then rank the absolute value of the $n$ differences from the smallest (rank 1) to the highest (rank $n$) and calculate the rank sum $T_-$ of the negative differences and the rank sum $T_+$ of the positive differences. [*Note:* Differences equal to 0 are eliminated, and the number $n$ of differences is reduced accordingly.]

*Test statistic:*

$T_-$, the rank sum of the negative differences [or $T_+$, the rank sum of the positive differences]

*Test statistic:*

$T$, the smaller of $T_+$ or $T_-$

*Rejection region:*

$T_- \le T_0$  [or $T_+ \le T_0$]

*Rejection region:*

$T \le T_0$

where $T_0$ is given in Table XIII in Appendix A.

*Assumptions:*  1. The sample of differences is randomly selected from the population of differences.

              2. The probability distribution from which the sample of paired differences is drawn is continuous.

*Ties:* Assign tied absolute differences the average of the ranks they would receive if they were unequal but occurred in successive order. For example, if the third-ranked and fourth-ranked differences are tied, assign both a rank of $(3 + 4)/2 = 3.5$.

**EXAMPLE 14.3**

Suppose the police commissioner in a small community must chose between two plans for patrolling the town's streets. Plan A, the less expensive plan, uses voluntary citizen groups to patrol certain high-risk neighborhoods. In contrast, plan B would utilize police patrols. As an aid in reaching a decision, both plans are examined by 10 trained criminologists, each of whom is asked to rate the plans on a scale from 1 to 10. (High ratings imply a more effective crime prevention plan.) The city will adopt plan B (and hire extra police) only if the data provide sufficient evidence that criminologists tend to rate plan B more effective than plan A.

The results of the survey are shown in Table 14.6. Do the data provide evidence at the $\alpha = .05$ level that the distribution of ratings for plan B lies above that for plan A?

◎ CRIMEPLAN

**TABLE 14.6  Effectiveness Ratings by 10 Qualified Crime Prevention Experts**

| | Plan | | Difference | |
|---|:---:|:---:|:---:|:---:|
| **Crime Prevention Expert** | **A** | **B** | **(A − B)** | **Rank of Absolute Difference** |
| 1 | 7 | 9 | −2 | 4.5 |
| 2 | 4 | 5 | −1 | 2 |
| 3 | 8 | 8 | 0 | (Eliminated) |
| 4 | 9 | 8 | 1 | 2 |
| 5 | 3 | 6 | −3 | 6 |
| 6 | 6 | 10 | −4 | 7.5 |
| 7 | 8 | 9 | −1 | 2 |
| 8 | 10 | 8 | 2 | 4.5 |
| 9 | 9 | 4 | 5 | 9 |
| 10 | 5 | 9 | −4 | 7.5 |
| | | | | Positive rank sum $= T_+ = 15.5$ |

**Solution**

The null and alternative hypotheses are

$H_0$: The two probability distributions of effectiveness ratings are identical

$H_a$: The effectiveness ratings of the more expensive plan (B) tend to exceed those of plan A

Observe that the alternative hypothesis is one-sided (that is, we only wish to detect a shift in the distribution of the B ratings to the right of the distribution of A ratings) and therefore it implies a one-tailed test of the null hypothesis (see Figure 14.8).

**FIGURE 14.8**

*The Alternative Hypothesis for Example 14.3*

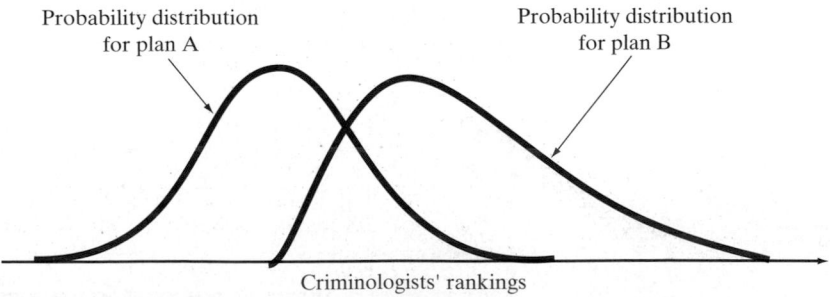

Probability distribution for plan A

Probability distribution for plan B

Criminologists' rankings

If the alternative hypothesis is true, the B ratings will tend to be larger than the paired A ratings, more negative differences in pairs will occur, $T_-$ will be large, and $T_+$ will be small. Because Table XIII is constructed to give lower-tail values of $T_0$, we will use $T_+$ as the test statistic and reject $H_0$ for $T_+ \leq T_0$.

The differences in ratings for the pairs $(A - B)$ are shown in Table 14.6. Note that one of the differences equals 0. Consequently, we eliminate this pair from the ranking and reduce the number of pairs to $n = 9$. Looking in Table XIII, we have $T_0 = 8$ for a one-tailed test with $\alpha = .05$ and $n = 9$. Therefore, the test statistic and rejection region for the test are

$$\text{Test statistic:} \quad T_+, \text{ the positive rank sum}$$

$$\text{Rejection region:} \quad T_+ \leq 8$$

Summing the ranks of the positive differences from Table 14.6, we find $T_+ = 15.5$. Since this value exceeds the critical value, $T_0 = 8$, we conclude that this sample provides insufficient evidence at the $\alpha = .05$ level to support the alternative hypothesis. The commissioner *cannot* conclude that the plan utilizing police patrols tends to be rated higher than the plan using citizen volunteers. That is, on the basis of this study, extra police will not be hired.

A STATISTIX printout of the analysis, shown in Figure 14.9, confirms this conclusion. Both the test statistic and *p*-value are highlighted on the printout. Since the *p*-value, .2168, exceeds $\alpha = .05$, we fail to reject $H_0$.

**FIGURE 14.9**

*STATISTIX Printout for Example 14.3*

```
WILCOXON SIGNED RANK TEST FOR A - B

SUM OF NEGATIVE RANKS                                    -29.500
SUM OF POSITIVE RANKS                                     15.500

EXACT PROBABILITY OF A RESULT AS OR MORE
EXTREME THAN THE OBSERVED RANKS (1 TAILED P-VALUE) 0.2168

NORMAL APPROXIMATION WITH CONTINUITY CORRECTION           0.770
TWO TAILED P-VALUE FOR NORMAL APPROXIMATION               0.4413

TOTAL NUMBER OF VALUES THAT WERE TIED        7
NUMBER OF ZERO DIFFERENCES DROPPED           1
MAX. DIFF. ALLOWED BETWEEN TIES        0.00001

CASES INCLUDED 9   MISSING 1
```

As is the case for the rank sum test for independent samples, the sampling distribution of the signed rank statistic can be approximated by a normal distribution when the number $n$ of paired observations is large (say $n \geq 25$). The large-sample *z*-test is summarized in the next box.

---

### Wilcoxon Signed Ranks Test for Large Samples ($n \geq 25$)

Let $D_1$ and $D_2$ represent the probability distributions for populations 1 and 2, respectively. *(continues on next page)*

## Wilcoxon Signed Ranks Test for Large Samples ($n \geq 25$) (*continued*)

**ONE-TAILED TEST**

$H_0$: $D_1$ and $D_2$ are identical

$H_a$: $D_1$ is shifted to the right of $D_2$
   [or $H_a$: $D_1$ is shifted to the left of $D_2$]

**TWO-TAILED TEST**

$H_0$: $D_1$ and $D_2$ are identical

$H_a$: $D_1$ is shifted either to the left
   or to the right of $D_2$

*Test statistic:*  $z = \dfrac{T_+ - [n(n+1)/4]}{\sqrt{[n(n+1)(2n+1)]/24}}$

*Rejection region:*

$z > z_\alpha$  [or $z < -z_\alpha$]

*Rejection region:*

$|z| > z_{\alpha/2}$

*Assumptions:* The sample size $n$ is greater than or equal to 25. Differences equal to 0 are eliminated and the number $n$ of differences is reduced accordingly. Tied absolute differences receive ranks equal to the average of the ranks they would have received had they not been tied.

## EXERCISES 14.24–14.36

### Learning the Mechanics

**14.24** Specify the test statistic and the rejection region for the Wilcoxon signed rank test for the paired difference design in each of the following situations:

   **a.** $H_0$: Two probability distributions, A and B, are identical

      $H_a$: Probability distribution for population A is shifted to the right or left of the probability distribution for population B

      $n = 20, \alpha = .10$

   **b.** $H_0$: Two probability distributions, A and B, are identical

      $H_a$: Probability distribution for population A is shifted to the right of the probability distribution for population B

      $n = 39, \alpha = .05$

   **c.** $H_0$: Two probability distributions, A and B, are identical

      $H_a$: Probability distribution for population A is shifted to the left of the probability distribution for population B

      $n = 7, \alpha = .005$

**14.25** Suppose you want to test a hypothesis that two treatments, A and B, are equivalent against the alternative hypothesis that the responses for A tend to be larger than those for B. You plan to use a paired difference experiment and to analyze the resulting data using the Wilcoxon signed rank test.

   **a.** Specify the null and alternative hypotheses you would test.

   **b.** Suppose the paired difference experiment yielded the data in the table. Conduct the test of part **a**. Test using $\alpha = .025$.

**LM14_25**

| Pair | A | B | Pair | A | B |
|------|-----|-----|------|-----|-----|
| 1 | 54 | 45 | 6 | 77 | 75 |
| 2 | 60 | 45 | 7 | 74 | 63 |
| 3 | 98 | 87 | 8 | 29 | 30 |
| 4 | 43 | 31 | 9 | 63 | 59 |
| 5 | 82 | 71 | 10 | 80 | 82 |

**14.26** Explain the difference between the one- and two-tailed versions of the Wilcoxon signed rank test for the paired difference experiment.

**14.27** In order to conduct the Wilcoxon signed rank test, why do we need to assume the probability distribution of differences is continuous?

**14.28** Suppose you wish to test a hypothesis that two treatments, A and B, are equivalent against the alternative that the responses for A tend to be larger than those for B.

   **a.** If the number of pairs equals 25, give the rejection region for the large-sample Wilcoxon signed rank test for $\alpha = .05$.

   **b.** Suppose that $T_+ = 273$. State your test conclusions.

   **c.** Find the $p$-value for the test and interpret it.

**14.29** A paired difference experiment with $n = 30$ pairs yielded $T_+ = 354$.

  **a.** Specify the null and alternative hypotheses that should be used in conducting a hypothesis test to determine whether the probability distribution for population A is located to the right of that for population B.

  **b.** Conduct the test of part **a** using $\alpha = .05$.

  **c.** What is the approximate $p$-value of the test of part **b**?

  **d.** What assumptions are necessary to ensure the validity of the test you performed in part **b**?

**14.30** A random sample of nine pairs of measurements is shown in the table.

| Pair | Sample Data from Population 1 | Sample Data from Population 2 |
|------|------|------|
| 1 | 8 | 7 |
| 2 | 10 | 1 |
| 3 | 6 | 4 |
| 4 | 10 | 10 |
| 5 | 7 | 4 |
| 6 | 8 | 3 |
| 7 | 4 | 6 |
| 8 | 9 | 2 |
| 9 | 8 | 4 |

  **a.** Use the Wilcoxon signed rank test to determine whether the data provide sufficient evidence to indicate that the probability distribution for population 1 is shifted to the right of the probability distribution for population 2. Test using $\alpha = .05$.

  **b.** Use the Wilcoxon signed rank test to determine whether the data provide sufficient evidence to indicate that the probability distribution for population 1 is shifted either to the right or to the left of the probability distribution for population 2. Test using $\alpha = .05$.

## Applying the Concepts—Basic

**14.31** Refer to the *Journal of Genetic Psychology* (March 1998) study of attitudes of male college students toward their parents, Exercise 9.34 (p. 407). Recall that each of 13 students was asked to complete the following statement: My relationship with my father (mother) can best be described as: (1) Awful, (2) Poor, (3) Average, (4) Good, or (5) Great. The study data are reproduced in the next table. The researchers want to compare male students' attitudes toward their fathers with their attitudes toward their mothers.

  **a.** Why is a nonparametric test applicable for analyzing this data set?

  **b.** Specify $H_0$ and $H_a$ for the test.

  **c.** Compute and rank the differences between the father and mother attitudes.

  **d.** Sum the ranks of the positive and negative differences.

  **e.** Find the rejection region for the test at $\alpha = .05$.

**FMATTITUDES**

| Student | Attitude toward Father | Attitude toward Mother |
|------|------|------|
| 1 | 2 | 3 |
| 2 | 5 | 5 |
| 3 | 4 | 3 |
| 4 | 4 | 5 |
| 5 | 3 | 4 |
| 6 | 5 | 4 |
| 7 | 4 | 5 |
| 8 | 2 | 4 |
| 9 | 4 | 5 |
| 10 | 5 | 4 |
| 11 | 4 | 5 |
| 12 | 5 | 4 |
| 13 | 3 | 3 |

*Source:* Adapted from Vitulli, W. F., and Richardson, D. K. "College student's attitudes toward relationships with parents: A five-year comparative analysis." *Journal of Genetic Psychology,* Vol. 159, No. 1, (March 1998), pp. 45–52.

  **f.** Give the conclusion of the test in the context of the problem. Compare your answer to the test results of Exercise 9.34b.

**14.32** A study published in the *Journal of Business Communications* (Fall 1985) found that video teleconferencing may be a more effective method of dealing with complex group problem-solving tasks than face-to-face meetings. Ten groups of four people each were randomly assigned both to a specific communication setting (face-to-face or video teleconferencing) and to one of two specific complex problems. Upon completion of the problem-solving task, the same groups were placed in the alternative communication setting and asked to complete the second problem-solving task. The percentage of each problem task correctly completed was recorded for each group, with the results given in the accompanying table.

**FACEVT**

| Group | Face-to-Face | Video Teleconferencing |
|------|------|------|
| 1 | 65% | 75% |
| 2 | 82 | 80 |
| 3 | 54 | 60 |
| 4 | 69 | 65 |
| 5 | 40 | 55 |
| 6 | 85 | 90 |
| 7 | 98 | 98 |
| 8 | 35 | 40 |
| 9 | 85 | 89 |
| 10 | 70 | 80 |

  **a.** What type of experimental design was used in this study?

  **b.** Specify the null and alternative hypotheses that should be used in determining whether the data provide sufficient evidence to conclude that the problem-solving performance of video teleconferencing groups is superior to that of groups that interact face-to-face.

SPSS Output for Exercise 14.33

### Ranks

| | | N | Mean Rank | Sum of Ranks |
|---|---|---|---|---|
| SECBORN - FRSTBORN | Negative Ranks | 7[a] | 5.93 | 41.50 |
| | Positive Ranks | 4[b] | 6.13 | 24.50 |
| | Ties | 1[c] | | |
| | Total | 12 | | |

a. SECBORN < FRSTBORN
b. SECBORN > FRSTBORN
a. FRSTBORN = SECBORN

### Test Statistics[b]

| | SECBORN-FRSTBORN |
|---|---|
| Z | -.756[a] |
| Asymp. Sig. (2-tailed) | .449 |

a. Based on positive ranks.
b. Wilcoxon Signed Ranks Test

**c.** Conduct the hypothesis test of part **b**. Use $\alpha = .05$. Interpret the results of your test in the context of the problem.

**d.** What is the *p*-value of the test in part **c**?

**14.33** Twelve sets of identical twins are given psychological tests to determine whether the firstborn of the twins tends to be more aggressive than the secondborn. The test scores are shown in the table, where the higher score indicates greater aggressiveness. Do the data provide sufficient evidence (at $\alpha = .05$) to indicate that the firstborn of a pair of twins is more aggressive than the other? Use the accompanying SPSS printout above to make your conclusion.

In a survey of atlas users (*Journal of Geography*, May/June 1995), a large sample of high school teachers in British Columbia ranked 12 thematic atlas topics for usefulness. The consensus rankings of the teachers (based on the percentage of teachers who responded they "would definitely use" the topic) are given in the accompanying table. These teacher rankings were compared to the rankings a group of university geography alumni made three years earlier. Compare the distributions of theme rankings for the two groups with an appropriate nonparametric test. Use $\alpha = .05$. Interpret the results practically.

### AGGTWINS

| Set | Firstborn | Secondborn |
|---|---|---|
| 1 | 86 | 88 |
| 2 | 71 | 77 |
| 3 | 77 | 76 |
| 4 | 68 | 64 |
| 5 | 91 | 96 |
| 6 | 72 | 72 |
| 7 | 77 | 65 |
| 8 | 91 | 90 |
| 9 | 70 | 65 |
| 10 | 71 | 80 |
| 11 | 88 | 81 |
| 12 | 87 | 72 |

## Applying the Concepts—Intermediate

**14.34** The regional atlas is an important educational resource that is updated on a periodic basis. One of the most critical aspects of a new atlas design is its thematic content.

### ATLAS

| | Rankings | |
|---|---|---|
| **Theme** | **High School Teachers** | **Geography Alumni** |
| Tourism | 10 | 2 |
| Physical | 2 | 1 |
| Transportation | 7 | 3 |
| People | 1 | 6 |
| History | 2 | 5 |
| Climate | 6 | 4 |
| Forestry | 5 | 8 |
| Agriculture | 7 | 10 |
| Fishing | 9 | 7 |
| Energy | 2 | 8 |
| Mining | 10 | 11 |
| Manufacturing | 12 | 12 |

*Source:* Keller, C. P., *et al.* "Planning the next generation of regional atlases: Input from educators." *Journal of Geography*, Vol. 94, No. 3, May/June 1995, p. 413 (Table 1).

**14.35** Hypoglycemia is a condition in which blood sugar is below normal limits. To compare the effectiveness of two compounds, X and Y, for treating hypoglycemia, each compound is applied to half the diaphragms of each of seven white mice. Blood glucose uptake in milligrams per gram of tissue is measured for each half, producing the results listed in the table below. Do the data provide sufficient evidence to indicate that one of the compounds tends to produce higher blood sugar uptake readings than the other? Test using $\alpha = .10$.

⊘HYPOGLY

| Mouse | X | Y | Mouse | X | Y |
|-------|-----|-----|-------|-----|-----|
| 1 | 4.7 | 5.1 | 5 | 7.0 | 6.1 |
| 2 | 3.3 | 4.6 | 6 | 4.7 | 4.1 |
| 3 | 8.5 | 8.7 | 7 | 5.2 | 5.1 |
| 4 | 3.9 | 3.6 | | | |

**14.36** Eleven prisoners of war during the war in Croatia were evaluated for neurological impairment after their release from a Serbian detention camp (*Collegium Antropologicum*, June 1997). All 11 released POWs received blows to the head and neck and/or loss of consciousness during imprisonment. Neurological impairment was assessed by measuring the amplitude of the visual evoked potential (VEP) in both eyes at two points in time: 157 days and 379 days after release from prison. (The higher the VEP value, the greater the neurological impairment.) The data for the 11 POWs are shown in the table. Determine whether the VEP measurements of POWs 379 days after their release tend to be greater than the VEP measurements of POWs 157 days after their release. Test using $\alpha = .05$.

⊘POWVEP

| POW | 157 Days After Release | 379 Days After Release |
|-----|------------------------|------------------------|
| 1 | 2.46 | 3.73 |
| 2 | 4.11 | 5.46 |
| 3 | 3.93 | 7.04 |
| 4 | 4.51 | 4.73 |
| 5 | 4.96 | 4.71 |
| 6 | 4.42 | 6.19 |
| 7 | 1.02 | 1.42 |
| 8 | 4.30 | 8.70 |
| 9 | 7.56 | 7.37 |
| 10 | 7.07 | 8.46 |
| 11 | 8.00 | 7.16 |

*Source:* Vrca, A., *et al.* "The use of visual evoked potentials to follow-up prisoners of war after release from detention camps." *Collegium Antropologicum*, Vol. 21, No. 1, June 1997, p. 232. (Data simulated from information provided in Table 3.)

## 14.5 THE KRUSKAL–WALLIS *H*-TEST FOR A COMPLETELY RANDOMIZED DESIGN

In Chapter 10 we used an analysis of variance and the *F*-test to compare the means of *p* populations (treatments) based on random sampling from populations that were normally distributed with a common variance $\sigma^2$. We now present a nonparametric technique—the **Kruskal–Wallis *H*-test**—for comparing the populations that requires no assumptions concerning the population probability distributions.

Suppose a health administrator wants to compare the unoccupied bed space for three hospitals located in the same city. She randomly selects 10 different days from the records of each hospital and lists the number of unoccupied beds for each day (see Table 14.7). Because the number of unoccupied beds per day may occasionally be quite large, it is conceivable that the population distributions of data may be skewed to the right and that this type of data may not satisfy the assumptions necessary for a parametric comparison of the population means. We therefore use a nonparametric analysis and base our comparison on the rank sums for the three sets of sample data. Just as with two independent samples (Section 14.2), the ranks are computed for each observation according to the relative magnitude of the measurements *when the data for all the samples are combined* (see Table 14.7). Ties are treated as they were for the Wilcoxon rank sum and signed rank tests by assigning the average value of the ranks to each of the tied observations.

HOSPBEDS
**TABLE 14.7    Number of Available Beds**

| Hospital 1 | | Hospital 2 | | Hospital 3 | |
| --- | --- | --- | --- | --- | --- |
| **Beds** | **Rank** | **Beds** | **Rank** | **Beds** | **Rank** |
| 6 | 5 | 34 | 25 | 13 | 9.5 |
| 38 | 27 | 28 | 19 | 35 | 26 |
| 3 | 2 | 42 | 30 | 19 | 15 |
| 17 | 13 | 13 | 9.5 | 4 | 3 |
| 11 | 8 | 40 | 29 | 29 | 20 |
| 30 | 21 | 31 | 22 | 0 | 1 |
| 15 | 11 | 9 | 7 | 7 | 6 |
| 16 | 12 | 32 | 23 | 33 | 24 |
| 25 | 17 | 39 | 28 | 18 | 14 |
| 5 | 4 | 27 | 18 | 24 | 16 |
| $R_1 = 120$ | | $R_2 = 210.5$ | | $R_3 = 134.5$ | |

We test

$H_0$: The probability distributions of the number of unoccupied beds are the same for all three hospitals

$H_a$: At least two of the three hospitals have probability distributions of number of unoccupied beds that differ in location

If we denote the three sample rank sums by $R_1$, $R_2$ and $R_3$, the test statistic is given by

$$H = \frac{12}{n(n+1)} \sum \frac{R_j^2}{n_j} - 3(n+1)$$

where $n_j$ is the number of measurements in the $j$th sample and $n$ is the total sample size ($n = n_1 + n_2 + \cdots + n_p$). For the data in Table 14.7, we have $n_1 = n_2 = n_3 = 10$ and $n = 30$. The rank sums are $R_1 = 120$, $R_2 = 210.5$, and $R_3 = 134.5$. Thus,

$$H = \frac{12}{30(31)} \left[ \frac{(120)^2}{10} + \frac{(210.5)^2}{10} + \frac{(134.5)^2}{10} \right] - 3(31)$$
$$= 99.097 - 93 = 6.097$$

The $H$ statistic measures the extent to which the $p$ samples differ with respect to their relative ranks. This is more easily seen by writing $H$ in an alternative but equivalent form:

$$H = \frac{12}{n(n+1)} \sum n_j (\overline{R}_j - \overline{R})^2$$

where $\overline{R}_j$ is the mean rank corresponding to sample $j$ and $\overline{R}$ is the mean of all the ranks [that is, $\overline{R} = \frac{1}{2}(n+1)$]. Thus, the $H$ statistic is 0 if all samples have the same mean rank and becomes increasingly large as the distance between the sample mean ranks grows.

If the null hypothesis is true, the distribution of $H$ in repeated sampling is approximately a $\chi^2$ (chi-square) distribution. This approximation for the sampling distribution of $H$ is adequate as long as one of the $p$ sample sizes exceeds 5. (See the references for more detail.) The degrees of freedom corresponding to the approximate sampling distribution of $H$ will always be $(p-1)$—one less than the number

**FIGURE 14.10**

*Rejection Region for the Comparison of Three Probability Distributions*

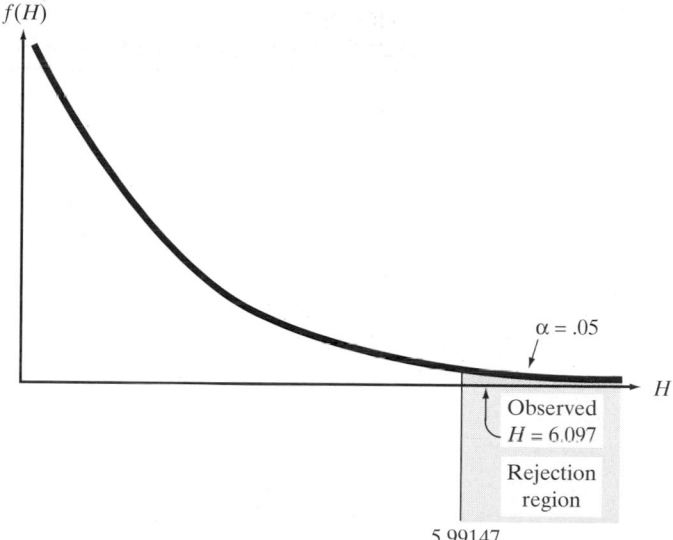

of probability distributions being compared. Because large values of $H$ support the alternative hypothesis that the populations have different probability distributions, the rejection region for the test is located in the upper tail of the $\chi^2$ distribution.

For the data of Table 14.7, the approximate distribution of the test statistic $H$ is $\chi^2$ with $(p - 1) = 2$ df. To determine how large $H$ must be before we will reject the null hypothesis, we consult Table VII in Appendix A. For $\alpha = .05$ and df = 2, $\chi^2_{.05} = 5.99147$. Therefore, we can reject the null hypothesis that the three probability distributions are the same if

$$H > 5.99147$$

The rejection region is pictured in Figure 14.10. Since the calculated $H = 6.097$ exceeds the critical value of 5.99147, we conclude that at least one of the three hospitals tends to have a larger number of unoccupied beds than the others.

The same conclusion can be reached from a computer printout of the analysis. The average ranks test statistic, and $p$-value of the nonparametric test are highlighted on the SPSS printout in Figure 14.11. Since $\alpha = .05$ exceeds $p$-value $= .047$, there is sufficient evidence to reject $H_0$.

**FIGURE 14.11**

*SPSS Kruskal–Wallis Printout*

**Ranks**

| | HOSPITAL | N | Mean Rank |
|---|---|---|---|
| BEDS | 1 | 10 | 12.00 |
| | 2 | 10 | 21.05 |
| | 3 | 10 | 13.45 |
| | Total | 30 | |

**Test Statistics[a,b]**

| | BEDS |
|---|---|
| Chi-Square | 6.099 |
| df | 2 |
| Asymp. Sig. | .047 |

a. Kruskal Wallis Test
b. Grouping Variable: HOSPITAL

The Kruskal–Wallis *H*-test for comparing more than two probability distributions is summarized in the next box. Note that we can use the Wilcoxon rank sum test of Section 14.2 to compare a pair of populations (selected a priori) if the Kruskal–Wallis *H*-test supports the alternative hypothesis that at least two of the probability distributions differ.*

---

### Kruskal–Wallis *H*-Test for Comparing *p* Probability Distributions

$H_0$: The *p* probability distributions are identical

$H_a$: At least two of the *p* probability distributions differ in location

*Test statistic:*   $H = \dfrac{12}{n(n + 1)} \sum \dfrac{R_j^2}{n_j} - 3(n + 1)$

where

$n_j$ = Number of measurements in sample *j*

$R_j$ = Rank sum for sample *j*, where the rank of each measurement is computed according to its relative magnitude in the totality of data for the *p* samples

$n$ = Total sample size = $n_1 + n_2 + \cdots + n_p$

*Rejection region:*   $H > \chi_\alpha^2$ with $(p - 1)$ degrees of freedom.

*Assumptions:*   1. The *p* samples are random and independent.

2. There are 5 or more measurements in each sample.

3. The *p* probability distributions from which the samples are drawn are continuous.

*Ties:*   Assign tied measurements the average of the ranks they would receive if they were unequal but occurred in successive order. For example, if the third-ranked and fourth ranked measurements are tied, assign both a rank of $(3 + 4)/2 = 3.5$. The number of ties should be small relative to the total number of observations.

---

## EXERCISES 14.37–14.46

### Learning the Mechanics

**14.37** Data were collected from three populations, A, B, and C, using a completely randomized design. The following describes the sample data:

$$n_A = n_B = n_C = 15$$
$$R_A = 235 \quad R_B = 439 \quad R_C = 361$$

**a.** Specify the null and alternative hypotheses that should be used in conducting a test of hypothesis to determine whether the probability distributions of populations A, B, and C differ in location.

**b.** Conduct the test of part **a**. Use $\alpha = .05$.

**c.** What is the approximate *p*-value of the test of part **b**?

**d.** Calculate the mean rank for each sample and compute *H* according to the formula that utilizes these means. Verify that this formula yields the same value of *H* that you obtained in part **b**.

**14.38** Suppose you want to use the Kruskal–Wallis *H*-test to compare the probability distributions of three populations. The following are independent random samples selected from the three populations:

|  |  |
|---|---|
| *I*: | 34, 56, 65, 59, 82, 70, 45 |
| *II*: | 24, 18, 27, 41, 34, 42, 33 |
| *III*: | 72, 101, 91, 76, 80, 75 |

**a.** What experimental design was used?

---

*A method similar to the multiple comparisons procedure of Chapter 10 can be used to rank the treatment medians. This nonparametric multiple comparisons of medians will control the experimentwise error rate selected by the analyst. Consult the references [Dantel (1990) and Dunn (1964)] for details.

**b.** Specify the null and alternative hypotheses you would test.

**c.** Specify the rejection region you would use for your hypothesis test at $\alpha = .01$.

**d.** Conduct the test at $\alpha = .01$.

**14.39** Under what circumstances does the $\chi^2$ distribution provide an appropriate characterization of the sampling distribution of the Kruskal–Wallis $H$ statistic?

## Applying the Concepts—Basic

**14.40** Real estate market cycles are commonly divided into four phases that are based on the rate of change of the demand and supply of properties: Phase I, Recovery; Phase II, Expansion; Phase III, Hypersupply; and Phase IV, Recession. Glenn Mueller of Johns Hopkins University studied the office market cycles of U.S. real es-

⊘ MKTCYCLE

| Phase I | Phase II | Phase III | Phase IV |
|---------|----------|-----------|----------|
| 2.7 | 10.5 | 6.1 | −1.0 |
| −1.0 | 11.5 | 1.2 | 6.2 |
| 1.1 | 9.4 | 11.4 | −10.8 |
| 3.4 | 12.2 | 4.4 | 2.0 |
| 4.2 | 8.6 | 6.2 | −1.1 |
| 3.5 | 10.9 | 7.6 | −2.3 |

*Source:* Adapted from Mueller, G. R. "Real estate rental growth rates at different points in the physical market cycle." *Journal of Real Estate Research*, Vol. 18, No. 1, (Jul/Aug 1999), pp. 131–150.

tate markets (*Journal of Real Estate Research* (Jul/Aug 1999). For each of the four market cycles, office rental growth rates (i.e., growth rates for asking rents) were measured for a sample of six different real estate markets. These data (in percentages) are presented in the table. A STATISTIX printout of a Kruskal–Wallis analysis of the data is also provided.

**a.** Specify the null hypothesis for the Kruskal–Wallis test.

**b.** Find the mean rank for each market cycle phase on the printout. Multiply this value by the sample size, 6, to obtain the rank sum for each phase. Verify these rank sums by ranking the data.

**c.** Find the Kruskal–Wallis test statistic on the printout. Substitute the rank sums from part **a** into the appropriate formula to verify this value.

**d.** Give the rejection region for the test at $\alpha = .05$.

**e.** Is there sufficient evidence to conclude that the distributions of office rental growth rates differ among the four market cycle phases?

**f.** Locate the $p$-value of the test on the printout and interpret this result.

**g.** What are the advantages and disadvantages of applying the Kruskal–Wallis $H$-test rather than the ANOVA $F$-test of Chapter 10?

**14.41** Sixth graders in an elementary school are taught reading by three different methods. Students were randomly assigned to three classes. One class used programmed instruction, a second used standard memorization

STATISTIX Output for Exercise 14.40

```
KRUSKAL-WALLIS ONE-WAY NONPARAMETRIC AOV FOR GROWRATE BY CYCLE

                  MEAN      SAMPLE
        CYCLE     RANK      SIZE
        -------   ------    ------
        I          8.8        6
        II        20.8        6
        III       14.8        6
        IV         5.7        6
        TOTAL     12.5       24

KRUSKAL-WALLIS STATISTIC                      16.2458
P-VALUE, USING CHI-SQUARED APPROXIMATION       0.0010

PARAMETRIC AOV APPLIED to RANKS

SOURCE      DF        SS          MS         F       P
-------     ----    --------    --------   ------   ------
BETWEEN      3       811.583     270.528    16.04   0.0000
WITHIN      20       337.417      16.8708
TOTAL       23      1149.00

TOTAL NUMBER OF VALUES THAT WERE TIED      4
MAX. DIFF. ALLOWED BETWEEN TIES       0.00001

CASES INCLUDED 24      MISSING CASES 0
```

techniques, and the third used an open classroom approach. The increases in reading levels attained by five students randomly selected from each of the three classes are shown in the table.

⊘ READ3

| Programmed | Standard | Open |
|---|---|---|
| .9 | 1.0 | 1.7 |
| 1.5 | .8 | .5 |
| .7 | .9 | 1.6 |
| 1.1 | 1.2 | 1.4 |
| .5 | 1.4 | 1.0 |

a. Rank the 15 data points in the table.
b. Obtain the rank sums for the three methods.
c. Use the rank sums to find the Kruskal–Wallis test statistic.
d. Do the data provide sufficient evidence to indicate that the probability distributions of increases in reading level differ for at least two of the methods? Use $\alpha = .05$.

**14.42** Refer to the *Journal of the American Mosquito Control Association* study of biting flies, Exercise 14.9 (p. 746). The effect of wind speeds in kilometers per hour (kph) on the biting rate of the fly on Stanbury Island, Utah, was investigated by exposing samples of volunteers to one of six wind speed conditions. The distributions of the biting rates for the six wind speeds were compared using the Kruskal–Wallis test. The rank sums of the biting rates for the six conditions are shown in the table.

| Wind Speed (kph) | Number of Volunteers ($n_j$) | Rank Sum of Biting Rates ($R_j$) |
|---|---|---|
| < 1 | 11 | 1,804 |
| 1 – 2.9 | 49 | 6,398 |
| 3 – 4.9 | 62 | 7,328 |
| 5 – 6.9 | 39 | 4,075 |
| 7 – 8.9 | 35 | 2,660 |
| 9 – 20 | 21 | 1,388 |
| **Totals** | **217** | **23,653** |

*Source:* Strickman, D., *et al.* "Meteorological effects on the biting activity of *Leptoconops americanus* (Diptera: Ceratopogonidae)." *Journal of the American Mosquito Control Association*, Vol. II, No. 1, Mar. 1995, p. 17 (Table 1).

a. The researchers reported the test statistic as $H = 35.2$. Verify this value.
b. Find the rejection region for the test using $\alpha = .01$.
c. Make the proper conclusions.
d. The researchers reported that the *p*-value of the test is less than .01. Does this value support your inference in part **c**? Explain.

## Applying the Concepts—Intermediate

**14.43** Two University of Georgia researchers studied the effect of social reinforcement on exercise duration in adolescents with moderate mental retardation (*Clinical Kinesiology*, Spring 1995). Eleven adolescents with IQs ranging from 32 to 61 were divided into two groups. All participated in a 6-week exercise program. Group A (4 subjects) received verbal and social reinforcement during the program, while Group B (7 subjects) received verbal and social reinforcement and kept a self-record of their individual performances. The researchers theorized that Group B subjects would exercise for longer periods than Group A. Upon completion of the exercise program, all 11 subjects participated in a run/walk "race" in which the goal was to complete as many laps as possible during a 15-minute period. The number of laps completed (to the nearest quarter lap) was used as a measure of exercise duration.

a. Specify the null and alternative hypotheses for a nonparametric analysis of the data.
b. The Kruskal–Wallis H-test was applied to the data. The researchers reported the test statistic as $H = 5.1429$ and the observed significance level of the test as *p*-value $= .0233$. Interpret these results.
c. Are the assumptions for the test, part **b**, satisfied? If not, propose an alternative nonparametric method for the analysis.

**14.44** An experiment was conducted to determine whether a test designed to identify a certain form of mental disturbance could be easily interpreted by a person with little psychological training. Thirty judges were selected to review the results of 100 tests, half of which were given to disturbed patients and half to nondisturbed people. Of the 30 judges chosen, 10 were staff members of a mental hospital, 10 were trainees at the hospital, and 10 were undergraduate psychology majors. The results in the table give the number of the 100 tests correctly classified by each judge. Do the data provide sufficient evidence (at $\alpha = .05$) of a difference in the probability distributions of the number of correct identifications among the three types of judges?

⊘ MENTALID

| Staff | | Trainees | | Undergraduates | |
|---|---|---|---|---|---|
| 78 | 76 | 80 | 69 | 65 | 74 |
| 79 | 86 | 75 | 81 | 70 | 80 |
| 85 | 88 | 72 | 76 | 74 | 73 |
| 93 | 84 | 68 | 72 | 78 | 75 |
| 90 | 81 | 75 | 76 | 68 | 73 |

**14.45** Refer to the *Journal of Experimental Psychology-Applied* (June 2000) study of different methods of learning names, Exercise 10.26 (p. 462). Recall that three groups of students used a different method to learn the names of the other students in their group. Group 1 used the "simple name game," Group 2 used the "elaborate name game," and Group 3 used "pairwise introductions." The accompanying table lists the percentage of names recalled (after one year) for each student respondent.

### NAMEGAME

**Simple Name Game:**

| 24 | 43 | 38 | 65 | 35 | 15 | 44 | 44 | 18 | 27 | 0 | 38 | 50 | 31 |
|----|----|----|----|----|----|----|----|----|----|----|----|----|----|
| 7 | 46 | 33 | 31 | 0 | 29 | 0 | 0 | 52 | 0 | 29 | 42 | 39 | 26 |
| 51 | 0 | 42 | 20 | 37 | 51 | 0 | 30 | 43 | 30 | 99 | 39 | 35 | 19 |
| 24 | 34 | 3 | 60 | 0 | 29 | 40 | 40 | | | | | | |

**Elaborate Name Game:**

| 39 | 71 | 9 | 86 | 26 | 45 | 0 | 38 | 5 | 53 | 29 | 0 | 62 | 0 |
|----|----|----|----|----|----|----|----|----|----|----|----|----|----|
| 1 | 35 | 10 | 6 | 33 | 48 | 9 | 26 | 83 | 33 | 12 | 5 | 0 | 0 |
| 25 | 36 | 39 | 1 | 37 | 2 | 13 | 26 | 7 | 35 | 3 | 8 | 55 | 50 |

**Pairwise Intro:**

| 5 | 21 | 22 | 3 | 32 | 29 | 32 | 0 | 4 | 41 | 0 | 27 | 5 | 9 |
|----|----|----|----|----|----|----|----|----|----|----|----|----|----|
| 66 | 54 | 1 | 15 | 0 | 26 | 1 | 30 | 2 | 13 | 0 | 2 | 17 | 14 |
| 5 | 29 | 0 | 45 | 35 | 7 | 11 | 4 | 9 | 23 | 4 | 0 | 8 | 2 |
| 18 | 0 | 5 | 21 | 14 | | | | | | | | | |

*Source:* Morris, P. E., and Fritz, C. O. "The name game: Using retrieval practice to improve the learning of names." *Journal of Experimental Psychology-Applied*, Vol. 6, No. 2, June 2000 (data simulated from Figure 1).

**a.** Consider an analysis of variance *F*-test to determine whether the mean percentages of names recalled differ for the three name retrieval methods. Demonstrate that the ANOVA assumptions are likely to be violated.

**b.** Use an alternative nonparametric test to compare the distributions of the percentages of names recalled for the three name retrieval methods. Use $\alpha = .05$.

**14.46** In disagreements between the Internal Revenue Service (IRS) and the taxpayer that end up in litigation, taxpayers are permitted by law to choose the court forum. Three trial courts are available: (1) U.S. Tax Court, (2) Federal District Court, and (3) U.S. Claims Court. Each court possesses different requirements and restrictions that make the choice an important one for the taxpayer. A study of taxpayers' choice of forum in litigating tax issues was published in the *Journal of Applied Business Research* (Fall 1996). In a random sample of 161 litigated tax disputes, the researchers measured the taxpayer's choice of forum (Tax, District, or Claims Court) and tax deficiency (i.e., the disputed amount, in dollars). One of the objectives of the study was to determine those factors taxpayers consider important in their choice of forum. If tax deficiency (called DEF by the researchers) is an important factor, then the mean DEF values for the three tax courts should be significantly different.

**a.** The researchers applied a nonparametric test rather than a parametric test to compare the DEF distributions of the three tax litigation forums. Give a plausible reason for their choice.

**b.** What nonparametric test is appropriate for this analysis? Explain.

**c.** The table below summarizes the data analyzed by the researchers. Use the information in the table to compute the appropriate test statistic.

**d.** The observed significance level of the test was reported as *p*-value = .0037. Fully interpret this result.

| Court Selected by Taxpayer | Sample Size | Sample Mean DEF | Rank Sum of DEF Values |
|----------------------------|-------------|-----------------|------------------------|
| Tax | 67 | $ 80,357 | 5,335 |
| District | 57 | 74,213 | 3,937 |
| Claims | 37 | 184,648 | 3,769 |

*Source:* Billings, B. A., Green, B. P., and Volz, W. H. "Selection of forum for litigated tax issues." *Journal of Applied Business Research*, Vol. 12, No. 4, Fall, 1996, p. 38 (Table 2).

## 14.6 THE FRIEDMAN $F_r$-TEST FOR A RANDOMIZED BLOCK DESIGN

In Section 10.4 we employed an analysis of variance to compare *p* population (treatment) means when the data were collected using a randomized block design. The *Friedman $F_r$-test* provides another method for testing to detect a shift in location of a set of *p* populations.* Like other nonparametric tests, it requires no assumptions concerning the nature of the populations other than the capacity of individual observations to be ranked.

---

*The Friedman $F_r$-test was developed by the Nobel prize–winning economist Milton Friedman.

In Section 14.2, we gave an example in which a completely randomized design was used to compare the reaction times of subjects under the influence of one of two drugs. When the effect of a drug is short-lived (there is no carryover effect) and when the drug effect varies greatly from person to person, it may be useful to employ a *randomized block design*. Using the subjects as blocks, we would hope to eliminate the variability among subjects and thereby increase the amount of information in the experiment. Suppose that three drugs, A, B, and C, are to be compared using a randomized block design. Each of the three drugs is administered to the *same subject* with suitable time lags between the three doses. The order in which the drugs are administered is randomly determined for each subject. Thus, one drug would be administered to a subject and its reaction time would be noted; then after a sufficient length of time, the second drug administered; etc.

⊘ REACTION2

**TABLE 14.8   Reaction Time for Three Drugs**

| Subject | Drug A | Rank | Drug B | Rank | Drug C | Rank |
|---------|--------|------|--------|------|--------|------|
| 1 | 1.21 | 1 | 1.48 | 2 | 1.56 | 3 |
| 2 | 1.63 | 1 | 1.85 | 2 | 2.01 | 3 |
| 3 | 1.42 | 1 | 2.06 | 3 | 1.70 | 2 |
| 4 | 2.43 | 2 | 1.98 | 1 | 2.64 | 3 |
| 5 | 1.16 | 1 | 1.27 | 2 | 1.48 | 3 |
| 6 | 1.94 | 1 | 2.44 | 2 | 2.81 | 3 |
| | | $R_1 = 7$ | | $R_2 = 12$ | | $R_3 = 17$ |

Suppose six subjects are chosen and that the reaction times for each drug are as shown in Table 14.8. To compare the three drugs, we rank the observations within each subject (block) and then compute the rank sums for each of the drugs (treatments). Tied observations within blocks are handled in the usual manner by assigning the average value of the ranks to each of the tied observations.

The null and alternative hypotheses are

$H_0$:  The populations of reaction times are identically distributed for all three drugs

$H_a$:  At least two of the drugs have probability distributions of reaction times that differ in location

The **Friedman $F_r$ statistic**, which is based on the rank sums for each treatment, is

$$F_r = \frac{12}{bp(p+1)} \sum R_j^2 - 3b(p+1)$$

where $b$ is the number of blocks, $p$ is the number of treatments, and $R_j$ is the $j$th rank sum. For the data in Table 14.8,

$$F_r = \frac{12}{(6)(3)(4)}[(7)^2 + (12)^2 + (17)^2] - 3(6)(4) = 80.33 - 72 = 8.33$$

The Friedman $F_r$ statistic measures the extent to which the $p$ samples differ with respect to their relative ranks within the blocks. This is more easily seen by writing $F_r$ in an alternative, but equivalent, form:

$$F_r = \frac{12}{bp(p+1)} \sum b(\bar{R}_j - \bar{R})^2$$

where $\bar{R}_j$ is the mean rank corresponding to treatment $j$ and $\bar{R}$ is the mean of all the ranks (i.e., $\bar{R} = \frac{1}{2}(p+1)$]. Thus, the $F_r$ statistic is 0 if all treatments have the same mean rank and becomes increasingly large as the distance between the sample mean ranks grows.

As for the Kruskal–Wallis $H$ statistic, the Friedman $F_r$ statistic has approximately a $\chi^2$ sampling distribution with $(p - 1)$ degrees of freedom. Empirical results show the approximation to be adequate if either $b$ (the number of blocks) or $p$ (the number of treatments) exceeds 5. The Friedman $F_r$-test for a randomized block design is summarized in the next box.

---

### Friedman $F_r$-Test for a Randomized Block Design

$H_0$: The probability distributions for the $p$ treatments are identical
$H_a$: At least two of the probability distributions differ in location

*Test statistic:* $\quad F_r = \dfrac{12}{bp(p + 1)} \sum R_j^2 - 3b(p + 1)$

where

$b$ = Number of blocks
$p$ = Number of treatments
$R_j$ = Rank sum of the $j$th treatment, where the rank of each measurement is computed relative to its position *within its own block*

*Rejection region:* $\quad F_r > \chi_\alpha^2$ with $(p - 1)$ degrees of freedom

*Assumptions:* 1. The treatments are randomly assigned to experimental units within the blocks.

2. The measurements can be ranked within blocks.

3. The $p$ probability distributions from which the samples within each block are drawn are continuous.

*Ties:* Assign tied measurements within a block the average of the ranks they would receive if they were unequal but occurred in successive order. For example, if the third-ranked and fourth-ranked measurements are tied, assign each a rank of $(3 + 4)/2 = 3.5$. The number of ties should be small relative to the total number of observations.

---

For the drug example, we will use $\alpha = .05$ to form the rejection region:

$$F_r > \chi_{.05}^2 = 5.99147 \quad \text{(see Figure 14.12)}$$

**FIGURE 14.12**

*Rejection Region for Reaction Time Example*

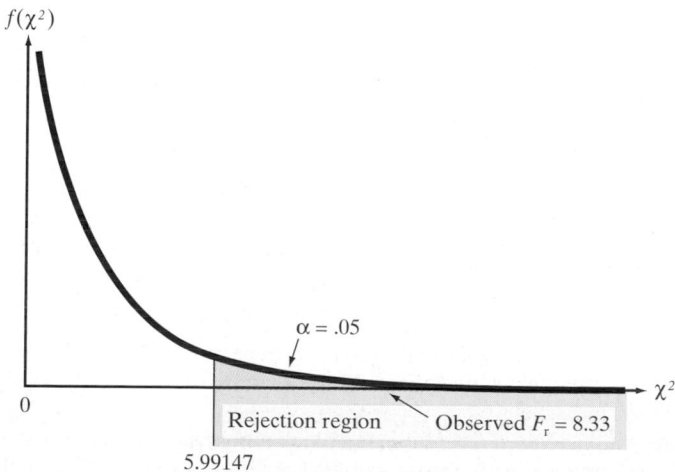

**FIGURE 14.13**

*STATISTIX Friedman Printout*

```
FRIEDMAN TWO WAY NONPARAMETRIC AOV

  FACTOR 1       MEAN     SAMPLE
  VARIABLE       RANK      SIZE

  ---------     ------    ------
  A              1.17        6
  B              2.00        6
  C              2.83        6


  FRIEDMAN STATISTIC                        8.3333
  P-VALUE, CHI-SQUARED APPROXIMATION        0.0155
  DEGREES OF FREEDOM                             2


  CASES INCLUDED 18    MISSING CASES 0
```

where $\chi^2_{.05}$ is based on $(p - 1) = 2$ degrees of freedom. Consequently, because the observed value, $F_r = 8.33$, exceeds 5.99147, we conclude that at least two of the three drugs have probability distributions of reaction times that differ in location.

A STATISTIX printout of the nonparametric analysis, shown in Figure 14.13, confirms our inference. Both the test statistic and *p*-value are highlighted on the printout. Since *p*-value =.0155 is less than our selected $\alpha = .05$, there is evidence to reject $H_0$.

Clearly, the assumptions for this test—that the measurements are ranked within blocks and that the number of blocks (subjects) is greater than 5—are satisfied. However, we must be sure that the treatments are randomly assigned to blocks. For the procedure to be valid, we assume that the three drugs are administered in a random order to each subject. If this were not true, the difference in the reaction times for the three drugs might be due to the order in which the drugs are given.

## EXERCISES 14.47–14.56

### Learning the Mechanics

**14.47** Data were collected using a randomized block design with four treatments (A, B, C, and D) and $b = 6$. The following rank sums were obtained:

$$R_A = 11 \quad R_B = 21 \quad R_C = 21 \quad R_D = 7$$

a. How many blocks were used in the experimental design?
b. Specify the null and alternative hypotheses that should be used in conducting a hypothesis test to determine whether the probability distributions for at least two of the treatments differ in location.
c. Conduct the test of part **b**. Use $\alpha = .10$.
d. What is the approximate *p*-value of the test of part **c**?
e. Calculate the mean rank for each of the four treatments and compute the value of the $F_r$-test statistic according to the formula that utilizes those means. Verify that the test statistic is the same as that you obtained in part **c**.

**14.48** Suppose you have used a randomized block design to help you compare the effectiveness of three different treatments, A, B, and C. You obtained the data given in the table and plan to conduct a Friedman $F_r$-test.

🔵 LM14_48

|  | **Treatment** | | |
| **Block** | **A** | **B** | **C** |
| 1 | 9 | 11 | 18 |
| 2 | 13 | 13 | 13 |
| 3 | 11 | 12 | 12 |
| 4 | 10 | 15 | 16 |
| 5 | 9 | 8 | 10 |
| 6 | 14 | 12 | 16 |
| 7 | 10 | 12 | 15 |

a. Specify the null and alternative hypotheses you will test.
b. Specify the rejection region for the test. Use $\alpha = .10$.
c. Conduct the test and interpret the results.

**14.49** An experiment was conducted using a randomized block design with four treatments and six blocks. The ranks of the measurements within each block are shown in the table. Use the Friedman $F_r$-test for a randomized block design to determine whether the data provide sufficient evidence to indicate that at least two of the treatment probability distributions differ in location. Test using $\alpha = .05$.

⊘ LM14_49

| Treatment | Block | | | | | |
|-----------|---|---|---|---|---|---|
| | **1** | **2** | **3** | **4** | **5** | **6** |
| 1 | 3 | 3 | 2 | 3 | 2 | 3 |
| 2 | 1 | 1 | 1 | 2 | 1 | 1 |
| 3 | 4 | 4 | 3 | 4 | 4 | 4 |
| 4 | 2 | 2 | 4 | 1 | 3 | 2 |

## Applying the Concepts—Basic

**14.50** A review of farmer involvement in agricultural research was presented in the *Journal of Agricultural, Biological, and Environmental Statistics* (Mar. 2001). In one study, each of six farmers ranked the level of farm production constraint imposed by five conditions: drought, pest damage, weed interference, farming costs, and labor shortage. The rankings, ranging from 1 (least severe) to 5 (most severe), and rank sums for the five conditions are listed in the table below.
   **a.** Use the rank sums shown in the table to compute the Friedman $F_r$ statistic.
   **b.** At $\alpha = .05$, find the rejection region for a test to compare the farmer opinion distributions for the five conditions.
   **c.** Make the proper conclusion, in the words of the problem.

**14.51** Refer to the *Journal of Geography*'s published rankings of regional atlas theme topics, Exercise 14.34 (p. 762). In addition to high school teachers and university geography alumni, university geography students and representatives of the general public also ranked the 12 thematic topics. The rankings of all four groups are shown in the table on p. 774. A MINITAB analysis is provided to compare the atlas theme ranking distributions of the four groups.

```
MINITAB Output for Exercise 14.51

Friedman test for RANK by GROUP blocked by THEME

S = 0.93 DF = 3 P = 0.819
S = 1.08 DF = 3 P = 0.782 (adjusted for ties)

                      Est      Sum of
GROUP       N      Median      Ranks
1          12      5.1250       27.0
2          12      6.1250       32.5
3          12      5.6250       29.0
4          12      6.1250       31.5

Grand median =   5.7500
```

   **a.** Locate the rank sums on the printout.
   **b.** Use the rank sums to find the Friedman $F_r$ statistic.
   **c.** Locate the test statistic and associated *p*-value on the printout.
   **d.** Conduct the test and give the conclusion in the words of the problem.

**14.52** Suppose we want to compare the mean flood crests on the Suwannee and Santa Fe rivers for four flood level years in Florida. The data (in feet), matched by year at six locations on the rivers, are shown in the second table on p. 774. Do the data indicate that the flood crests on the rivers were higher or lower in any one of the flood level years than any other? Test using the Friedman $F_r$-test (at $\alpha = .05$) shown on the SPSS printout (p. 774).

⊘ FARM6

| | | Condition | | | | |
|---|---|---|---|---|---|---|
| | | **Drought** | **Pest Damage** | **Weed Interference** | **Farming Costs** | **Labor Shortage** |
| **Farmer** | **1** | 5 | 4 | 3 | 2 | 1 |
| | **2** | 5 | 3 | 4 | 1 | 2 |
| | **3** | 3 | 5 | 4 | 2 | 1 |
| | **4** | 5 | 4 | 1 | 2 | 3 |
| | **5** | 4 | 5 | 3 | 2 | 1 |
| | **6** | 5 | 4 | 3 | 2 | 1 |
| | **Rank sum** | 27 | 25 | 18 | 11 | 9 |

*Source:* Riley, J., and Fielding, W. J. "An illustrated review of some farmer participatory research techniques." *Journal of Agricultural, Biological, and Environmental Statistics*, Vol. 6, No. 1, Mar. 2001 (Table 1).

⊘ ATLAS2

| | Rankings | | | |
|---|---|---|---|---|
| **Theme** | **High School Teachers** | **Geography Alumni** | **Geography Students** | **General Public** |
| Tourism | 10 | 2 | 5 | 1 |
| Physical | 2 | 1 | 1 | 5 |
| Transportation | 7 | 3 | 7 | 2 |
| People | 1 | 6 | 2 | 3 |
| History | 2 | 5 | 9 | 4 |
| Climate | 6 | 4 | 4 | 8 |
| Forestry | 5 | 8 | 2 | 7 |
| Agriculture | 7 | 10 | 6 | 9 |
| Fishing | 9 | 7 | 10 | 6 |
| Energy | 2 | 8 | 7 | 10 |
| Mining | 10 | 11 | 11 | 11 |
| Manufacturing | 12 | 12 | 12 | 12 |

*Source:* Keller, C. P., *et al.* "Planning the next generation of regional atlases: Input from educators." *Journal of Geography*, Vol. 94, No. 3, May/June 1995, p. 413 (Table 1).

⊘ FLOOD

| | | Flood Year | | | |
|---|---|---|---|---|---|
| **Station** | **Flood Stage** | **1** | **2** | **3** | **4** |
| Suwannee River | | | | | |
| White Springs | 77.0 | 79.9 | 74.4 | 88.5 | 85.2 |
| Ellaville | 54.0 | 59.8 | 52.8 | 65.0 | 68.1 |
| Branford | 29.0 | 29.3 | 29.2 | 35.6 | 38.9 |
| Wilcox | 14.0 | 11.6 | 13.0 | 18.6 | 22.3 |
| Santa Fe River | | | | | |
| Three Rivers | 19.0 | 24.1 | 24.3 | 27.2 | 30.6 |
| U.S. 129 Bridge | 21.0 | 23.8 | 23.8 | 30.7 | 34.2 |

**SPSS Output for Exercise 14.52**

**Ranks**

| | Mean Rank |
|---|---|
| CRESTYR1 | 1.58 |
| CRESTYR2 | 1.42 |
| CRESTYR3 | 3.17 |
| CRESTYR4 | 3.83 |

**Test Statistics**[a]

| N | 6 |
|---|---|
| Chi-Square | 15.508 |
| df | 3 |
| Asymp. Sig. | .001 |

[a] Friedman Test

## Applying the Concepts—Intermediate

**14.53** Refer to the *British Journal of Sports Medicine* (Apr. 2000) experiment to investigate the effect of massage on boxing performance, Exercise 10.48 (p. 482). Recall

that the punching power of each of eight amateur boxers was measured (in Newtons) after each of four rounds: (M1), round 1 following a pre-bout sports massage; (R1), round 1 following a pre-bout period of rest; (M5), round 5 following a sports massage between

⊘ BOXING

| | | Intervention | | | |
|---|---|---|---|---|---|
| | | **M1** | **R1** | **M5** | **R5** |
| **Boxer** | **1** | 1243 | 1244 | 1291 | 1262 |
| | **2** | 1147 | 1053 | 1169 | 1177 |
| | **3** | 1247 | 1375 | 1309 | 1321 |
| | **4** | 1274 | 1235 | 1290 | 1285 |
| | **5** | 1177 | 1139 | 1233 | 1238 |
| | **6** | 1336 | 1313 | 1366 | 1362 |
| | **7** | 1238 | 1279 | 1275 | 1261 |
| | **8** | 1261 | 1152 | 1289 | 1266 |

*Source:* Hemmings, B., Smith, M., Graydon, J., and Dyson, R. "Effects of massage on physiological restoration, perceived recovery, and repeated sports performance." *British Journal of Sports Medicine*, Vol. 34, No. 2, Apr. 2000 (adapted from Table 3).

rounds; and (R5), round 5 following a period of rest between rounds. The data are reproduced in the table. Use the appropriate nonparametric test to compare the punching power means of the four interventions. Compare the results to those of Exercise 10.48.

**14.54** Refer to the Kansas State study designed to investigate the effects of plants on human stress levels, Exercise 10.49 (p. 483). The data below are given as finger temperatures for each of ten students in a dimly lit room under three experimental conditions: presence of a live plant, presence of a plant photo, and absence of a plant (either live or photo). Analyze the data using a nonparametric procedure. Do the students' finger temperatures depend on the experimental condition?

### ⊘ PLANTS

| Student | Live Plant | Plant Photo | No Plant (control) |
|---|---|---|---|
| 1 | 91.4 | 93.5 | 96.6 |
| 2 | 94.9 | 96.6 | 90.5 |
| 3 | 97.0 | 95.8 | 95.4 |
| 4 | 93.7 | 96.2 | 96.7 |
| 5 | 96.0 | 96.6 | 93.5 |
| 6 | 96.7 | 95.5 | 94.8 |
| 7 | 95.2 | 94.6 | 95.7 |
| 8 | 96.0 | 97.2 | 96.2 |
| 9 | 95.6 | 94.8 | 96.0 |
| 10 | 95.6 | 92.6 | 96.6 |

*Source:* Elizabeth Schreiber, Department of Statistics, Kansas State University, Manhattan, Kansas

**14.55** Refer to the *Journal of the American Mosquito Control Association* (Mar. 1995) study of the effectiveness of five different types of insecticides in controlling a species of Caribbean mosquito, Exercise 10.51 (p. 483). The resistance ratios (i.e., the dosage of insecticide required to kill 50% of the larvae divided by the known dosage for a susceptible mosquito strain) of the insecticides at each of seven Caribbean locations are reproduced in the table at the bottom of the page. Compare the resistance ratio distributions of the five insecticides using a nonparametric procedure. Are any of the insecticides more effective than any of the others?

**14.56** Corrosion of different metals is a problem in many mechanical devices. Three sealers used to help retard the corrosion of metals were tested to see whether there were any differences among them. Samples of 10 different metal compositions were treated with each of the three sealers, and the amount of corrosion was measured after exposure to the same environmental conditions for 1 month. The data are given in the next table. Is there any evidence of a difference in the probability distributions of the amounts of corrosion among the three types of sealer? Use $\alpha = .05$.

### ⊘ CORRODE

| Metal | Sealer | | |
|---|---|---|---|
| | 1 | 2 | 3 |
| 1 | 4.6 | 4.2 | 4.9 |
| 2 | 7.2 | 6.4 | 7.0 |
| 3 | 3.4 | 3.5 | 3.4 |
| 4 | 6.2 | 5.3 | 5.9 |
| 5 | 8.4 | 6.8 | 7.8 |
| 6 | 5.6 | 4.8 | 5.7 |
| 7 | 3.7 | 3.7 | 4.1 |
| 8 | 6.1 | 6.2 | 6.4 |
| 9 | 4.9 | 4.1 | 4.2 |
| 10 | 5.2 | 5.0 | 5.1 |

### ⊘ MOSQUITO

| | | Insecticide | | | | |
|---|---|---|---|---|---|---|
| | | Temephos | Malathion | Fenitrothion | Fenthion | Chlorpyrifos |
| | Anguilla | 4.6 | 1.2 | 1.5 | 1.8 | 1.5 |
| | Antigua | 9.2 | 2.9 | 2.0 | 7.0 | 2.0 |
| | Dominica | 7.8 | 1.4 | 2.4 | 4.2 | 4.1 |
| Location | Guyana | 1.7 | 1.9 | 2.2 | 1.5 | 1.8 |
| | Jamaica | 3.4 | 3.7 | 2.0 | 1.5 | 7.1 |
| | St. Lucia | 6.7 | 2.7 | 2.7 | 4.8 | 8.7 |
| | Suriname | 1.4 | 1.9 | 2.0 | 2.1 | 1.7 |

*Source:* Rawlins, S. C., and Oh Hing Wan, J. "Resistance in some Caribbean populations of *Aedes aegypti* to several insecticides." *Journal of the American Mosquito Control Association*, Vol. 11, No. 1, Mar. 1995 (Table 1).

## 14.7 SPEARMAN'S RANK CORRELATION COEFFICIENT

Suppose 10 new paintings are shown to two art critics and each critic ranks the paintings from 1 (best) to 10 (worst). We want to determine whether the critics' ranks are related. Does a correspondence exist between their ratings? If a painting is ranked high by critic 1, is it likely to be ranked high by critic 2? Or do high rankings by one critic correspond to low rankings by the other? That is, are the rankings of the critics *correlated*?

If the rankings are as shown in the "Perfect Agreement" columns of Table 14.9, we immediately notice that the critics agree on the rank of every painting. High ranks correspond to high ranks and low ranks to low ranks. This is an example of *perfect positive correlation* between the ranks. In contrast, if the rankings appear as shown in the "Perfect Disagreement" columns of Table 14.9, high ranks for one critic correspond to low ranks for the other. This is an example of *perfect negative correlation*.

In practice, you will rarely see perfect positive or negative correlation between the ranks. In fact, it is quite possible for the critics' ranks to appear as shown in Table 14.10. You will note that these rankings indicate some agreement between the critics, but not perfect agreement, thus indicating a need for a measure of rank correlation.

**TABLE 14.9  Rankings of 10 Paintings by Two Critics**

| | **Perfect Agreement** | | **Perfect Disagreement** | |
|---|---|---|---|---|
| **Painting** | **Critic 1** | **Critic 2** | **Critic 1** | **Critic 2** |
| 1 | 4 | 4 | 9 | 2 |
| 2 | 1 | 1 | 3 | 8 |
| 3 | 7 | 7 | 5 | 6 |
| 4 | 5 | 5 | 1 | 10 |
| 5 | 2 | 2 | 2 | 9 |
| 6 | 6 | 6 | 10 | 1 |
| 7 | 8 | 8 | 6 | 5 |
| 8 | 3 | 3 | 4 | 7 |
| 9 | 10 | 10 | 8 | 3 |
| 10 | 9 | 9 | 7 | 4 |

**TABLE 14.10  Rankings of Paintings: Less Than Perfect Agreement**

| | **Critic** | | **Difference Between Rank 1 and Rank 2** | |
|---|---|---|---|---|
| **Painting** | **1** | **2** | **$d$** | **$d^2$** |
| 1 | 4 | 5 | $-1$ | 1 |
| 2 | 1 | 2 | $-1$ | 1 |
| 3 | 9 | 10 | $-1$ | 1 |
| 4 | 5 | 6 | $-1$ | 1 |
| 5 | 2 | 1 | 1 | 1 |
| 6 | 10 | 9 | 1 | 1 |
| 7 | 7 | 7 | 0 | 0 |
| 8 | 3 | 3 | 0 | 0 |
| 9 | 6 | 4 | 2 | 4 |
| 10 | 8 | 8 | 0 | 0 |
| | | | | $\Sigma d^2 = 10$ |

**Spearman's rank correlation coefficient**, $R_s$, provides a measure of correlation between ranks. The formula for this measure of correlation is given in the next box. We also give a formula that is identical to $r_s$ when there are no ties in rankings; this provides a good approximation to $r_s$ when the number of ties is small relative to the number of pairs.

Note that if the ranks for the two critics are identical, as in the second and third columns of Table 14.9, the differences between the ranks, $d$, will all be 0. Thus,

$$r_s = 1 - \frac{6 \sum d^2}{n(n^2 - 1)} = 1 - \frac{6(0)}{10(99)} = 1$$

That is, *perfect positive correlation* between the pairs of ranks is characterized by a Spearman correlation coefficient of $r_s = 1$. When the ranks indicate perfect disagreement, as in the fourth and fifth columns of Table 14.9, $\sum d_i^2 = 330$ and

$$r_s = 1 - \frac{6(330)}{10(99)} = -1.$$

Thus, *perfect negative correlation* is indicated by $r_s = -1$.

---

## Spearman's Rank Correlation Coefficient

$$r_s = \frac{\text{SS}_{uv}}{\sqrt{\text{SS}_{uu}\text{SS}_{vv}}}$$

where

$$\text{SS}_{uv} = \sum (u_i - \bar{u})(v_i - \bar{v}) = \sum u_i v_i - \frac{\left(\sum u_i\right)\left(\sum v_i\right)}{n}$$

$$\text{SS}_{uu} = \sum (u_i - \bar{u})^2 = \sum u_i^2 - \frac{\left(\sum u_i\right)^2}{n}$$

$$\text{SS}_{vv} = \sum (v_i - \bar{v})^2 = \sum v_i^2 - \frac{\left(\sum v_i\right)^2}{n}$$

$u_i$ = Rank of the $i$th observation in sample 1
$v_i$ = Rank of the $i$th observation in sample 2
$n$ = Number of pairs of observations (number of observations in each sample)

---

## Shortcut Formula for $r_s$*

$$r_s = 1 - \frac{6 \sum d_i^2}{n(n^2 - 1)}$$

where

$d_i = u_i - v_i$   (difference in the ranks of the $i$th observations and 2)

---

*The shortcut formula is not exact when there are tied measurements, but it is a good approximation when the total number of ties is not large relative to $n$.

For the data of Table 14.10,

$$r_s = 1 - \frac{6 \sum d^2}{n(n^2 - 1)} = 1 - \frac{6(10)}{10(99)} = 1 - \frac{6}{99} = .94$$

The fact that $r_s$ is *close* to 1 indicates that the critics tend to agree, but the agreement is not perfect.

*The value of $r_s$ always falls between −1 and +1, with +1 indicating perfect positive correlation and −1 indicating perfect negative correlation.* The closer $r_s$ falls to +1 or −1, the greater the correlation between the ranks. Conversely, the nearer $r_s$ is to 0, the less the correlation.

Note that the concept of correlation implies that two responses are obtained for each experimental unit. In the art critics example, each painting received two ranks (one for each critic) and the objective of the study was to determine the degree of positive correlation between the two rankings. Rank correlation methods can be used to measure the correlation between any pair of variables. If two variables are measured on each of *n* experimental units, we rank the measurements associated with each variable separately. Ties receive the average of the ranks of the tied observations. Then we calculate the value of $r_s$ for the two rankings. This value measures the rank correlation between the two variables. We illustrate the procedure in Example 14.4.

**EXAMPLE 14.4**

A study is conducted to investigate the relationship between cigarette smoking during pregnancy and the weights of newborn infants. A sample of 15 women smokers kept accurate records of the number of cigarettes smoked during their pregnancies, and the weights of their children were recorded at birth. The data are given in Table 14.11. Calculate and interpret Spearman's rank correlation coefficient for the data.

NEWBORN
**TABLE 14.11  Data and Calculations for Example 14.4**

| Woman | Cigarettes per Day | Rank | Baby's Weight (pounds) | Rank | $d$ | $d^2$ |
|-------|-------------------|------|------------------------|------|-----|-------|
| 1 | 12 | 1 | 7.7 | 5 | −4 | 16 |
| 2 | 15 | 2 | 8.1 | 9 | −7 | 49 |
| 3 | 35 | 13 | 6.9 | 4 | 9 | 81 |
| 4 | 21 | 7 | 8.2 | 10 | −3 | 9 |
| 5 | 20 | 5.5 | 8.6 | 13.5 | −8 | 64 |
| 6 | 17 | 3 | 8.3 | 11.5 | −8.5 | 72.25 |
| 7 | 19 | 4 | 9.4 | 15 | −11 | 121 |
| 8 | 46 | 15 | 7.8 | 6 | 9 | 81 |
| 9 | 20 | 5.5 | 8.3 | 11.5 | −6 | 36 |
| 10 | 25 | 8.5 | 5.2 | 1 | 7.5 | 56.25 |
| 11 | 39 | 14 | 6.4 | 3 | 11 | 121 |
| 12 | 25 | 8.5 | 7.9 | 7 | 1.5 | 2.25 |
| 13 | 30 | 12 | 8.0 | 8 | 4 | 16 |
| 14 | 27 | 10 | 6.1 | 2 | 8 | 64 |
| 15 | 29 | 11 | 8.6 | 13.5 | −2.5 | 6.25 |
| | | | | | | Total = 795 |

**Solution**

We first rank the number of cigarettes smoked per day, assigning a 1 to the smallest number (12) and a 15 to the largest (46). Note that the two ties receive the averages of their respective ranks. Similarly, we assign ranks to the 15 babies' weights. Since the number of ties is relatively small, we will use the shortcut formula to calculate $r_s$. The differences $d$ between the ranks of the babies' weights and the ranks of the number of cigarettes smoked per day are shown in Table 14.11. The squares of the differences, $d^2$, are also given. Thus,

$$r_s = 1 - \frac{6 \sum d_i^2}{n(n^2 - 1)} = 1 - \frac{6(795)}{15(15^2 - 1)} = 1 - 1.42 = -.42$$

The value of $r_s$ can also be obtained using a computer. A SAS printout of the analysis is shown in Figure 14.14. The value of $r_s$, highlighted on the printout, agrees (except for rounding) with our hand-calculated value of $-.42$.

**FIGURE 14.14**

*SAS Printout for Example 14.14*

The CORR Procedure

2 Variables:  CIGARETS WEIGHT

Spearman Correlation Coefficients, N = 15
Prob > |r| under H0: Rho=0

|  | CIGARETS | WEIGHT |
|---|---|---|
| CIGARETS | 1.00000 | -0.42473 |
|  |  | 0.1145 |
| WEIGHT | -0.42473 | 1.00000 |
|  | 0.1145 |  |

This negative correlation coefficient indicates that in this sample an increase in the number of cigarettes smoked per day is *associated with* (but is not necessarily the *cause of*) a decrease in the weight of the newborn infant. Can this conclusion be generalized from the sample to the population? That is, can we conclude that weights of newborns and the number of cigarettes smoked per day are negatively correlated for the populations of observations for *all* smoking mothers?

If we define $\rho$ as the **population rank correlation coefficient** [i.e., the rank correlation coefficient that could be calculated from all $(x, y)$ values in the population], this question can be answered by conducting the test

$H_0: \rho = 0$ (no population correlation between ranks)

$H_a: \rho < 0$ (negative population correlation between ranks)

*Test statistic:*  $r_s$ (the *sample* Spearman rank correlation coefficient)

To determine a rejection region, we consult Table XIV in Appendix A, which is partially reproduced in Table 14.12. Note that the left-hand column gives values of $n$, the number of pairs of observations. The entries in the table are values for an upper-tail rejection region, since only positive values are given. Thus, for $n = 15$ and $\alpha = .05$, the value .441 is the boundary of the upper-tailed rejection region, so that $P(r_s > .441) = .05$ if $H_0: \rho = 0$ is true. Similarly, for negative values of $r_s$, we have $P(r_s < -.441) = .05$ if $\rho = 0$. That is, we expect to see $r_s < -.441$ only 5% of the time if there is really no relationship between the ranks of the variables.

**TABLE 14.12 Reproduction of Part of Table XIV in Appendix A: Critical Values of Spearman's Rank Correlation Coefficient**

| $n$ | $\alpha = .05$ | $\alpha = .025$ | $\alpha = .01$ | $\alpha = .005$ |
|---|---|---|---|---|
| 5 | .900 | — | — | — |
| 6 | .829 | .886 | .943 | — |
| 7 | .714 | .786 | .893 | — |
| 8 | .643 | .738 | .833 | .881 |
| 9 | .600 | .683 | .783 | .833 |
| 10 | .564 | .648 | .745 | .794 |
| 11 | .523 | .623 | .736 | .818 |
| 12 | .497 | .591 | .703 | .780 |
| 13 | .475 | .566 | .673 | .745 |
| 14 | .457 | .545 | .646 | .716 |
| 15 | .441 | .525 | .623 | .689 |
| 16 | .425 | .507 | .601 | .666 |
| 17 | .412 | .490 | .582 | .645 |
| 18 | .399 | .476 | .564 | .625 |
| 19 | .388 | .462 | .549 | .608 |
| 20 | .377 | .450 | .534 | .591 |

The lower-tailed rejection region is therefore

$$\text{Rejection region } (\alpha = .05): \quad r_s < -.441$$

Since the calculated $r_s = -.42$ is not less than $-.441$, we cannot reject $H_0$ at the $\alpha = .05$ level of significance. That is, this sample of 15 smoking mothers provides insufficient evidence to conclude that a negative correlation exists between number of cigarettes smoked and the weight of newborns for the populations of measurements corresponding to all smoking mothers. (Note that the $p$-value of .1145 shown on the SAS printout, Figure 14.14, confirms this result.) This does not, of course, mean that no relationship exists. A study using a larger sample of smokers and taking other factors into account (father's weight, sex of newborn child, etc.) would be more likely to reveal whether smoking and the weight of a newborn child are related. ∎

A summary of Spearman's nonparametric test for correlation is given in the next box.

---

### Spearman's Nonparametric Test for Rank Correlation

**ONE-TAILED TEST**

$H_0: \rho = 0$

$H_a: \rho > 0$
[or $H_a: \rho < 0$]

*Test statistic:* $r_s$ the sample rank correlation (see the formulas for calculating $r_s$)

*Rejection region:* $r_s > r_{s,\alpha}$
[or $r_s < -r_{s,\alpha}$ when $H_a: \rho_s < 0$]

where $r_{s,\alpha}$ is the value from Table XIV corresponding to the upper-tail area $\alpha$ and $n$ pairs of observations

**TWO-TAILED TEST**

$H_0: \rho = 0$

$H_a: \rho \neq 0$

*Rejection region:* $|r_s| > r_{s,\alpha/2}$

where $r_{s,\alpha/2}$ is the value from Table XIV corresponding to the upper-tail area $\alpha/2$ and $n$ pairs of observations

> ### Spearman's Nonparametric Test for Rank Correlation (*continued*)
>
> *Assumptions:* 1. The sample of experimental units on which the two variables are measured is randomly selected.
> 2. The probability distributions of the two variables are continuous.
>
> *Ties:* Assign tied measurements the average of the ranks they would receive if they were unequal but occurred in successive order. For example, if the third-ranked and fourth-ranked measurements are tied, assign each a rank of $(3 + 4)/2 = 3.5$. The number of ties should be small relative to the total number of observations.

 EXERCISES 14.57–14.68

### Learning the Mechanics

**14.57** Use Table XIV of Appendix A to find each of the following probabilities:
  **a.** $P(r_s > .508)$ when $n = 22$
  **b.** $P(r_s > .448)$ when $n = 28$
  **c.** $P(r_s \leq .648)$ when $n = 10$
  **d.** $P(r_s < -.738 \text{ or } r_s > .738)$ when $n = 8$

**14.58** Specify the rejection region for Spearman's nonparametric test for rank correlation in each of the following situations:
  **a.** $H_0: \rho = 0, H_a: \rho \neq 0, n = 10, \alpha = .05$
  **b.** $H_0: \rho = 0, H_a: \rho > 0, n = 20, \alpha = .025$
  **c.** $H_0: \rho = 0, H_a: \rho < 0, n = 30, \alpha = .01$

**14.59** Compute Spearman's rank correlation coefficient for each of the following pairs of sample observations:

| **a.** *x* | 33 | 61 | 20 | 19 | 40 |
|---|---|---|---|---|---|
| *y* | 26 | 36 | 65 | 25 | 35 |

| **b.** *x* | 89 | 102 | 120 | 137 | 41 |
|---|---|---|---|---|---|
| *y* | 81 | 94 | 75 | 52 | 136 |

| **c.** *x* | 2 | 15 | 4 | 10 |
|---|---|---|---|---|
| *y* | 11 | 2 | 15 | 21 |

| **d.** *x* | 5 | 20 | 15 | 10 | 3 |
|---|---|---|---|---|---|
| *y* | 80 | 83 | 91 | 82 | 87 |

**14.60** The following sample data were collected on variables *x* and *y:*

LM14_60

| *x* | 0 | 3 | 0 | −4 | 3 | 0 | 4 |
|---|---|---|---|---|---|---|---|
| *y* | 0 | 2 | 2 | 0 | 3 | 1 | 2 |

  **a.** Specify the null and alternative hypotheses that should be used in conducting a hypothesis test to determine whether the variables *x* and *y* are correlated.
  **b.** Conduct the test of part **a** using $\alpha = .05$.
  **c.** What is the approximate *p*-value of the test of part **b**?
  **d.** What assumptions are necessary to ensure the validity of the test of part **b**?

### Applying the Concepts—Basic

**14.61** Refer to the *British Journal of Sports Medicine* (Apr. 2000) study of the effect of massaging boxers between rounds, Exercise 11.42 (p. 538). Two variables measured on the boxers were blood lactate level (*y*) and the boxer's perceived recovery (*x*). The data for 16 five-round boxing performances are reproduced here.

BOXING2

| **Blood Lactate Level** | **Perceived Recovery** |
|---|---|
| 3.8 | 7 |
| 4.2 | 7 |
| 4.8 | 11 |
| 4.1 | 12 |
| 5.0 | 12 |
| 5.3 | 12 |
| 4.2 | 13 |
| 2.4 | 17 |
| 3.7 | 17 |
| 5.3 | 17 |
| 5.8 | 18 |
| 6.0 | 18 |
| 5.9 | 21 |
| 6.3 | 21 |
| 5.5 | 20 |
| 6.5 | 24 |

*Source:* Hemmings, B., Smith, M., Graydon, J., and Dyson, R. "Effects of massage on physiological restoration, perceived recovery, and repeated sports performance." *British Journal of Sports Medicine*, Vol. 34, No. 2, Apr. 2000 (data adapted from Figure 3).

a. Rank the values of the 16 blood lactate levels.
b. Rank the values of the 16 perceived recovery values.
c. Use the ranks, parts **a** and **b**, to compute Spearman's rank correlation coefficient. Give a practical interpretation of the result.
d. Find the rejection region for a test to determine whether $y$ and $x$ are rank correlated. Use $\alpha = .10$.
e. What is the conclusion of the test, part **d**? State your answer in the words of the problem.

**14.62** The Milan Overall Dementia Assessment (MODA) is a neuropsychologically oriented test that provides a measure of an individual's cognitive deterioration. MODA scores range from 0 to 100, with higher scores indicating a greater degree of deterioration. A team of psychologists and neurologists administered the MODA to a sample of 30 patients with Alzheimer's disease (*Neuropsychologia*, June 1995). In addition, all patients were given a "face matching" test in which they were requested to match photographs of unknown faces. Scores on this test were recorded as the number of matching errors, with a maximum score of 27.
a. The researchers reported Spearman's rank correlation between the MODA and face-matching scores as $r_s = .48$. Interpret this result.
b. Test the hypothesis of a positive correlation between MODA score and face-matching score in Alzheimer's patients. Use $\alpha = .05$.

**14.63** Two expert wine tasters were asked to rank six brands of wine. Their rankings are shown below.

⊙ WINETASTE

| Brand | Expert 1 | Expert 2 |
|-------|----------|----------|
| A | 6 | 5 |
| B | 5 | 6 |
| C | 1 | 2 |
| D | 3 | 1 |
| E | 2 | 4 |
| F | 4 | 3 |

a. Calculate Spearman's rank correlation between the rankings of the two experts.
b. Do the data present sufficient evidence to indicate a positive correlation in the rankings of the two experts? Test Using $\alpha = .10$.

**14.64** The metabolic and cardiopulmonary responses during maximal effort exercise in persons with multiple sclerosis (MS) was studied (*Clinical Kinesiology*, Spring 1995). The following variables were measured for each of 10 MS patients:
1. Expanded Disability Status Scale (EDSS)—Ratings range from 1 to 4.5
2. Peak oxygen uptake (liters per minute) during an arm cranking exercise (ARM)
3. Peak oxygen uptake (liters per minute) during a leg cycling exercise (LEG)

4. Peak oxygen uptake (liters per minute) during a combined leg cycling and arm cranking exercise (LEG/ARM)

Spearman's rank correlation was calculated for each pair of variables. The accompanying table gives the correlation $r_s$ between the variables in the corresponding row and column.

| | EDSS | ARM | LEG/ARM | LEG |
|---|------|------|---------|------|
| **EDSS** | 1.000 | −.439 | −.655 | −.512 |
| **ARM** | | 1.000 | .634 | .722 |
| **LEG/ARM** | | | 1.000 | .890 |
| **LEG** | | | | 1.000 |

*Source:* Ponichtera-Mulcare, J. A., *et al.* "Maximal aerobic exercise of individuals with multiple sclerosis using three modes of energy." *Clinical Kinesiology*, Vol. 49, No. 1, Spring 1995, p. 10 (Table 2).

a. Explain why the table shows rank correlations of 1.000 in the diagonal.
b. Practically interpret each of the other rank correlations in the table.
c. Is there sufficient evidence (at $\alpha = .01$) to conclude that EDSS and peak oxygen uptake are negatively correlated? Perform the test for each of the three exercises.

**14.65** Refer to the *Brain and Behavior Evolution* (Apr. 2000) study of the feeding behavior of blackbream fish, Exercises 2.110 (p. 89) and 11.17 (p. 525). Recall that the zoologists recorded the number of aggressive strikes of two blackbream fish feeding at the bottom of an aquarium in the 10-minute period following the addition of food. The table listing the weekly number of strikes and age of the fish (in days) is reproduced here.

⊙ BLACKBREAM

| Week | Number of Strikes | Age of Fish (days) |
|------|-------------------|--------------------|
| 1 | 85 | 120 |
| 2 | 63 | 136 |
| 3 | 34 | 150 |
| 4 | 39 | 155 |
| 5 | 58 | 162 |
| 6 | 35 | 169 |
| 7 | 57 | 178 |
| 8 | 12 | 184 |
| 9 | 15 | 190 |

*Source:* Shand, J., *et al.* "Variability in the location of the retinal ganglion cell area centralis is correlated with ontogenetic changes in feeding behavior in the Blackbream, Acanthopagrus 'butcher'." *Brain and Behavior*, Vol. 55, No. 4, Apr. 2000 (Figure H).

a. Find Spearman's correlation coefficient relating number of strikes ($y$) to age of fish ($x$).
b. Conduct a nonparametric test to determine whether number of strikes ($y$) and age ($x$) are negatively correlated. Test using $\alpha = .01$.

## Applying the Concepts—Intermediate

**14.66** Refer to the *Journal of Experimental Psychology-Applied* (June 2000) study in which the "name game" was used to help groups of students learn the names of other students in the group, Exercise 11.23 (p. 528). Recall that one goal of the study was to investigate the relationship between proportion of names recalled ($y$) by a student and position (order) of the student during the game ($x$). The data for 144 students in the first eight positions are saved in the NAMEGAME2 file. (The first five and last five observations in the data set are listed here.)

🔘 NAMEGAME2

| Position | Recall |
|---|---|
| 2 | 0.04 |
| 2 | 0.37 |
| 2 | 1.00 |
| 2 | 0.99 |
| 2 | 0.79 |
| ⋮ | ⋮ |
| 9 | 0.72 |
| 9 | 0.88 |
| 9 | 0.46 |
| 9 | 0.54 |
| 9 | 0.99 |

*Source:* Morris, P.E., and Fritz, C.O. "The name game: Using retrieval practice to improve the learning of names." *Journal of Experimental Psychology-Applied*, Vol. 6, No. 2, June 2000 (data simulated from Figure 2).

**a.** To properly apply the parametric test for correlation based on the Pearson coefficient of correlation, $r$ (Section 11.6), both the $x$ and $y$ variables must be normally distributed. Demonstrate that this assumption is violated for this data. What are the consequences of this violation?

**b.** Find Spearman's rank correlation coefficient on the SAS printout provided below and interpret its value.

**c.** Find the observed significance level for testing for zero rank correlation on the SAS printout and interpret its value.

**SAS Output for Exercise 14.66**

```
            The CORR Procedure

    2  Variables:   POSITION RECALL

 Spearman Correlation Coefficients,  N = 144
       Prob > |r| under HO: Rho=0.

                  POSITION      RECALL

  POSITION        1.00000      0.20652
                               0.0130

  RECALL          0.20652      1.00000
                  0.0130
```

**d.** At $\alpha = .05$, is there sufficient evidence of rank correlation between proportion of names recalled ($y$) by a student and position (order) of the student during the game ($x$)?

**14.67** Psychologists theorize that certain types of recreation have a tranquilizing effect on highly excitable mentally retarded patients. In one experiment, nine mentally retarded patients were monitored for 1 week, and the total number of hours each spent in a specially designed recreation room was recorded. Also recorded was the number of times during the week that some type of tranquilizing medication had to be given to a patient in a highly excited state. Using the data in the table, determine whether there is sufficient evidence to indicate that recreation is negatively rank correlated with the number of times a tranquilizer has to be given. Test using $\alpha = .05$.

🔘 RECTIME

| Patient | Recreation Time (hours) | Tranquilizers |
|---|---|---|
| 1 | 16 | 3 |
| 2 | 22 | 1 |
| 3 | 10 | 4 |
| 4 | 8 | 9 |
| 5 | 14 | 5 |
| 6 | 34 | 2 |
| 7 | 26 | 0 |
| 8 | 13 | 10 |
| 9 | 5 | 9 |

**14.68** *The Wall Street Journal* (Feb. 7, 2001) reported on a Harris Interactive, Inc. survey of consumers to rate the reputations of America's most visible companies. The 1999 and 2000 ranks of 15 randomly selected companies are listed in the accompanying table. Conduct a test to determine if the 1999 and 2000 reputation ranks are positively correlated. Use $\alpha = .05$.

🔘 COREP

| Company | 1999 Rank | 2000 Rank |
|---|---|---|
| Johnson & Johnson | 1 | 1 |
| Anheuser-Busch | 14 | 6 |
| Disney | 10 | 8 |
| Microsoft | 15 | 9 |
| FedEx | 21 | 13 |
| Wal-Mart | 6 | 14 |
| Coca-Cola | 2 | 16 |
| McDonalds | 24 | 24 |
| Yahoo! | 19 | 27 |
| Sears | 29 | 35 |
| America Online | 26 | 39 |
| Kmart | 37 | 41 |
| Toyota | 28 | 19 |
| Home Depot | 8 | 4 |
| IBM | 17 | 7 |

*Source:* Harris Interactive, the Reputation Institute.

## STATISTICS IN ACTION
### Deadly Exposure: Agent Orange and Vietnam Vets

During the Vietnam War, American soldiers were exposed to Agent Orange, a defoliant used by the U.S. Armed Forces to destroy the dense plant and tree cover of the Asian jungle. As a result of this exposure, many Vietnam veterans have dangerously high levels of the dioxin 2, 3, 7, 8-tetrachlorodibenzo-*p* (TCDD) in their blood and adipose (fatty) tissue.

A study published in *Chemosphere* (Vol. 20, 1990) reported on the TCDD levels of 20 Massachusetts Vietnam vets who were possibly exposed to Agent Orange. The amounts of TCDD (measured in parts per trillion) in fat tissue drawn from each veteran are shown in Table 14.13 and are stored in the TCDDFAT file. The data provide us with an opportunity to apply the nonparametric tests of this chapter.

**TCDDFAT**
### TABLE 14.13   TCDD Levels in Fat Tissue of Vietnam Veterans

| | | | | | | |
|---|---|---|---|---|---|---|
| 4.9 | 6.9 | 10.0 | 4.4 | 4.6 | 1.1 | 2.3 |
| 5.9 | 7.0 | 5.5 | 7.0 | 1.4 | 11.0 | 2.5 |
| 4.4 | 4.2 | 41.0 | 2.9 | 7.7 | 2.5 | |

*Source:* Schecter, A., *et al.* "Partitioning of 2, 3, 7, 8-chlorinated dibenzo-*p*-dioxins and dibenzofurans between adipose tissue and plasma lipid of 20 Massachusetts Vietnam veterans." *Chemosphere*, Vol. 20, Nos. 7–9, 1990, pp. 954–955 (Table II).

### Focus

a. Suppose we want to use this small sample ($n = 20$) to test whether the median TCDD level in fat tissue of Vietnam vets exceeds 3 parts per trillion. Set up the null and alternative hypotheses of this test.

b. If the population of TCDD levels is normally distributed, how are the median $\eta$ and the mean $\mu$ related? Why is the parametric small-sample *t*-test of Section 8.4 appropriate in this case?

c. Check the validity of the normality assumption required to conduct the *t*-test of part **b**.

d. Test the hypothesis of part **a** using the nonparametric sign test discussed in this chapter. Use $\alpha = .05$. Interpret the result.

e. Repeat part **d**, using $\alpha = .01$.

f. Although the sign test of parts **d** and **e** is easy to apply, it is not the most powerful nonparametric test available. Recall (Section 8.6) that the "power" of a test is the probability that the test rejects the null hypothesis when it is false. Define "power" in practical terms of this case.

A nonparametric test with higher power than the sign test is the Wilcoxon signed rank test. Although we presented the signed rank procedure as a test to compare two populations in a matched-pairs design, it can also be used to test the location (median) of a single population. To conduct the signed rank test, we calculate the differences ($y_i - 3$) for the sample, where $y_i$ is the TCDD level for veteran *i* and 3 is the hypothesized value of the median $\eta$. The absolute values of the differences are ranked, and the rank sum of the negative differences represents the test statistic.

### Focus

g. Compute the differences ($y_i - 3$) for the sample data in Table 14.13 and rank the absolute values of these differences.

h. Use the results of part **g** to calculate the rank sum of the negative differences, $T_-$. This is the Wilcoxon signed rank test statistic.

i. Find and interpret the *p*-value of the test, part **h**.

j. Compare the results, parts **d**, **e**, and **i**. Comment on the power of the signed rank test as compared to the sign test.

The researchers also measured the TCDD levels in blood plasma drawn from the 20 Vietnam veterans. These values (also measured in parts per trillion) are shown in Table 14.14 and saved in the TCDDPLAS file.

**TCDDPLAS**
### TABLE 14.14   TCDD Levels in Plasma of Vietnam Veterans

| | | | | | | |
|---|---|---|---|---|---|---|
| 2.5 | 6.9 | 3.1 | 3.3 | 2.1 | 4.6 | 3.5 |
| 7.2 | 1.8 | 1.8 | 6.0 | 20.0 | 3.0 | 2.0 |
| 4.7 | 4.1 | 3.1 | 2.5 | 1.6 | 36.0 | |

*Source:* Schecter, A., *et al.* "Partitioning of 2, 3, 7, 8-chlorinated dibenzo-*p*-dioxins and dibenzofurans between adipose tissue and plasma lipid of 20 Massachusetts Vietnam veterans." *Chemosphere*, Vol. 20, Nos. 7–9, 1990, pp. 954–955 (Table I).

### Focus

k. Conduct an appropriate test to determine whether the true median TCDD level in plasma of Vietnam vets exceeds 3 parts per trillion. Justify your choice of a test and fully interpret the results.

# QUICK REVIEW

## Key Terms

Distribution-free tests 740
Friedman $F_r$ statistic 770
Kruskal–Wallis $H$-test 763
Nonparametrics 740
Parametric statistical tests 740

Population rank correlation
  coefficient 779
Rank statistics 741
Rank sum 748
Sign test 741

Spearman's rank correlation
  coefficient 777
Wilcoxon rank sum test 747
Wilcoxon signed rank test 757

## Key Formulas

| Test | Test Statistic | Large Sample Approximation |
|------|----------------|----------------------------|
| Sign | $S$ = number of sample measurements greater than (or less than) hypothesized mean, $\eta_0$ | $z = \dfrac{(S - .5) - .5n}{.5\sqrt{n}}$  744 |
| Wilcoxon rank sum | $T_1$ = rank sum of sample 1 or $T_2$ = rank sum of sample 2 | $z = \dfrac{T_1 - \dfrac{n_1(n_1 + n_2 + 1)}{2}}{\sqrt{\dfrac{n_1 n_2(n_1 + n_2 + 1)}{12}}}$  751 |
| Wilcoxon signed ranks | $T_-$ = negative rank sum or $T_+$ = positive rank sum | $z = \dfrac{T_+ - \dfrac{n(n + 1)}{4}}{\sqrt{\dfrac{n(n + 1)(2n + 1)}{24}}}$  760 |
| Kuskal–Wallis | $H = \dfrac{12}{n(n + 1)} \sum \dfrac{R_j^2}{n_j} - 3(n + 1)$   766 | |
| Friedman | $F_r = \dfrac{12}{bp(p + 1)} \sum R_j^2 - 3b(p + 1)$   771 | |
| Spearman rank correlation (shortcut formula) | $r_s = 1 - \dfrac{6 \sum d_i^2}{n(n^2 - 1)}$   777 | |

where $d_i$ = difference in ranks of $i$th observations for samples 1 and 2

## Language Lab

| Symbol | Description |
|--------|-------------|
| $\eta$(eta) | Population median |
| $S$ | Test statistic for sign test (see Key Formulas) |
| $T_1$ | Sum of ranks of observations in sample 1 |
| $T_2$ | Sum of ranks of observations in sample 2 |
| $T_L$ | Critical lower Wilcoxon rank sum value |
| $T_U$ | Critical upper Wilcoxon rank sum value |
| $T_+$ | Sum of ranks of positive differences of paired observations |
| $T_-$ | Sum of ranks of negative differences of paired observations |
| $T_0$ | Critical value of Wilcoxon signed rank test |
| $R_j$ | Rank sum of observations in sample $j$ |
| $H$ | Test statistic for Kruskal–Wallis test (see Key Formulas) |
| $F_r$ | Test statistic for Friedman test (see Key Formulas) |
| $r_s$ | Spearman's rank correlation coefficient (see Key Formulas) |
| $\rho$(rho) | Population correlation coefficient |

## SUPPLEMENTARY EXERCISES 14.69–14.92

### Learning the Mechanics

**14.69** When is it appropriate to use the *t*- and *F*-tests of Chapters 9 and 10 for comparing two or more population means?

**14.70** The data for three independent random samples are shown in the table. It is known that the sampled populations are not normally distributed. Use an appropriate test to determine whether the data provide sufficient evidence to indicate that at least two of the populations differ in location. Test using $\alpha = .05$.

⊘ LM14_70

| Sample 1 | | Sample 2 | | Sample 3 | |
|---|---|---|---|---|---|
| 18 | 15 | 12 | 34 | 87 | 50 |
| 32 | 63 | 33 | 18 | 53 | 64 |
| 43 | | 10 | | 65 | 77 |

**14.71** A random sample of nine pairs of observations are recorded on two variables, *x* and *y*. The data are shown in the table.

⊘ LM14_71

| Pair | *x* | *y* | Pair | *x* | *y* |
|---|---|---|---|---|---|
| 1 | 19 | 12 | 6 | 29 | 10 |
| 2 | 27 | 19 | 7 | 16 | 16 |
| 3 | 15 | 7 | 8 | 22 | 10 |
| 4 | 35 | 25 | 9 | 16 | 18 |
| 5 | 13 | 11 | | | |

   **a.** Do the data provide sufficient evidence to indicate that $\rho$, the rank correlation between *x* and *y*, differs from 0? Test using $\alpha = .05$.

   **b.** Do the data provide sufficient evidence to indicate that the probability distribution for *x* is shifted to the right of that for *y*? Test using $\alpha = .05$.

**14.72** Two independent random samples produced the measurements listed in the table. Do the data provide sufficient evidence to conclude that there is a difference between the locations of the probability distributions for the sampled populations? Test using $\alpha = .05$.

⊘ LM14_72

| Sample 1 | | Sample 2 | |
|---|---|---|---|
| 1.2 | 1.0 | 1.5 | 1.9 |
| 1.9 | 1.8 | 1.3 | 2.7 |
| .7 | 1.1 | 2.9 | 3.5 |
| 2.5 | | | |

**14.73** An experiment was conducted using a randomized block design with five treatments and four blocks. The data are shown in the table. Do the data provide sufficient evidence to conclude that at least two of the treatment probability distributions differ in location? Test using $\alpha = .05$.

⊘ LM14_73

| Treatment | Block | | | |
|---|---|---|---|---|
| | **1** | **2** | **3** | **4** |
| 1 | 75 | 77 | 70 | 80 |
| 2 | 65 | 69 | 63 | 69 |
| 3 | 74 | 78 | 69 | 80 |
| 4 | 80 | 80 | 75 | 86 |
| 5 | 69 | 72 | 63 | 77 |

### Applying the Concepts—Basic

**14.74** The American Cancer Society funds medical research focused on finding treatments for various forms of cancer. One new treatment is being tested to determine whether the average remission time for a particularly aggressive form of cancer is extended by the treatment. Assume that current treatments have been shown to provide a median remission time of 4.5 years. Seven patients with this cancer are given the new treatment, and their remission times (in years) are as follows:

⊘ REMISSION

| 5.3 | 7.3 | 3.6 | 5.2 | 6.1 | 4.8 | 8.4 |
|---|---|---|---|---|---|---|

   Use the sign test to determine if the median remission time is increased by the new treatment. Test at $\alpha = .05$.

**14.75** An educational psychologist claims that the order in which test questions are asked affects a student's ability to answer correctly. To investigate this assertion, a professor randomly divides a class of 13 students into two groups—7 in one group and 6 in the other. The professor prepares one set of test questions but arranges the questions in two different orders. On test A the questions are arranged in order of increasing difficulty (that is, from easiest to most difficult), while on test B the order is reversed. One group of students is given test A, the other test B, and the test score is recorded for each student. The results are shown in the table.

⊘ TESTORDER

| Test A | 90 | 71 | 83 | 82 | 75 | 91 | 65 |
|---|---|---|---|---|---|---|---|
| Test B | 66 | 78 | 50 | 68 | 80 | 60 | |

Use the Wilcoxon rank sum procedure to test for a difference (a shift in location) in the probability distributions of student scores on the two tests. Test using $\alpha = .05$.

**14.76** A union wants to determine the preferences of its members before negotiating with management. Ten union members are randomly selected, and an extensive questionnaire is completed by each member. The responses to the various aspects of the questionnaire will enable the union to rank in order of importance the items to be negotiated. The rankings are shown in the table.

⊘ UNION

| Person | More Pay | Job Stability | Fringe Benefits | Shorter Hours |
|--------|----------|---------------|-----------------|---------------|
| 1 | 2 | 1 | 3 | 4 |
| 2 | 1 | 2 | 3 | 4 |
| 3 | 4 | 3 | 2 | 1 |
| 4 | 1 | 4 | 2 | 3 |
| 5 | 1 | 2 | 3 | 4 |
| 6 | 1 | 3 | 4 | 2 |
| 7 | 2.5 | 1 | 2.5 | 4 |
| 8 | 3 | 1 | 4 | 2 |
| 9 | 1.5 | 1.5 | 3 | 4 |
| 10 | 2 | 3 | 1 | 4 |

Use the Friedman $F_r$ test to determine whether evidence exists that the probability distributions of ratings differ for at least two of the four negotiable items. Use $\alpha = .05$.

**14.77** For many years, the Girl Scouts of America have sold cookies using various sales techniques. One troop experimented with several techniques and reported the number of boxes of cookies sold per scout, as listed in the table.

⊘ COOKIES

| Door-to-Door | Telephone | Grocery Store Stand | Department Store Stand |
|--------------|-----------|---------------------|------------------------|
| 47 | 63 | 113 | 25 |
| 93 | 19 | 50 | 36 |
| 58 | 29 | 68 | 21 |
| 37 | 24 | 37 | 27 |
| 62 | 33 | 39 | 18 |
|    |    | 77 | 31 |

Use the Kruskal–Wallis $H$-test to determine if at least one sales method tends to produce a larger number of sales per scout than the others. Use $\alpha = .10$.

**14.78** A drug company has synthesized two new compounds to be used in sleeping pills. The data in the next table represent the additional hours of sleep gained by 10

patients through the use of the two drugs. Do the data present sufficient evidence to indicate that the probability distributions of additional hours of sleep differ for the two drugs? Test using the Wilcoxon signed rank test at $\alpha = .10$.

⊘ SLEEPILL

| Patient | Drug A | Drug B | Patient | Drug A | Drug B |
|---------|--------|--------|---------|--------|--------|
| 1 | .4 | .7 | 6 | 2.9 | 3.4 |
| 2 | −.7 | −1.6 | 7 | 4.0 | 3.7 |
| 3 | −.4 | −.2 | 8 | .1 | .8 |
| 4 | −1.4 | −1.4 | 9 | 3.1 | .0 |
| 5 | −1.6 | −.2 | 10 | 1.9 | 2.0 |

**14.79** A psychologist ranked a random sample of 10 children on two subjective scales according to the amount of paranoid behavior and the amount of aggressiveness they exhibit. The rankings according to these two criteria are given in the table. Compute Spearman's rank correlation coefficient. Is there evidence of a relationship between aggression and paranoia in children as judged by this psychologist? Test using $\alpha = .05$.

⊘ PARANOIA

| Child | Paranoia | Aggression |
|-------|----------|------------|
| 1 | 7 | 5 |
| 2 | 3 | 1 |
| 3 | 6 | 4 |
| 4 | 1 | 2 |
| 5 | 2 | 8 |
| 6 | 4 | 7 |
| 7 | 10 | 9 |
| 8 | 8 | 3 |
| 9 | 5 | 6 |
| 10 | 9 | 10 |

## Applying the Concepts—Intermediate

**14.80** A major razor blade manufacturer advertises that its twin-blade disposable razor will "get you more shaves" than any single-blade disposable razor on the market. A rival blade company that has been very successful in selling single-blade razors wishes to test this claim. Marketing managers in this company randomly sampled eight single-blade shavers and eight twin-blade shavers, and counted the number of shaves that each got before a change of blades was indicated. The results are shown in the table.

⊘ RAZORS

| Twin Blades | | Single Blade | |
|-------------|----|--------------|----|
| 8 | 15 | 10 | 13 |
| 17 | 10 | 6 | 14 |
| 9 | 6 | 3 | 5 |
| 11 | 12 | 7 | 7 |

a. Do the data support the twin-blade manufacturer's claim? Use $\alpha = .05$.

b. Do you think this experiment was designed in the best possible way? If not, what design might have been better?

c. What assumptions are necessary for the test performed in part **a** to be valid? Do the assumptions seem reasonable for this application?

**14.81** An experiment was conducted to compare two print types, A and B, to determine whether type A is easier to read. Ten subjects were randomly divided into two groups of five. Each subject was given the same material to read, one group receiving the material printed with type A, the other group receiving print type B. The times necessary for each subject to read the material (in seconds) are shown below:

⬤ PRINTYPE

| Type A: | 95 | 122 | 101 | 99 | 108 |
|---------|-----|-----|-----|-----|-----|
| Type B: | 110 | 102 | 115 | 112 | 120 |

Do the data provide sufficient evidence to indicate that print type A is easier to read? Test using $\alpha = .05$.

**14.82** Many states now require that mathematics teachers pass a basic math literacy test in order to qualify to teach math in the state. An important part of the process is the development of a fair test. Suppose that one state board of education has developed a test and that a sample of 20 math teachers are asked to take it. The test will be considered too hard if at least 50% of *all* math teachers in the state score less than 60 (on a scale of 0 to 100).

a. Set up the appropriate null and alternative hypotheses to test whether the test is too hard.

b. Establish the appropriate test statistic and rejection region for the test using $\alpha = .05$.

c. What assumptions are necessary to ensure the validity of the test?

d. Suppose that 14 of the 20 sampled teachers score less than 60. What is the appropriate conclusion?

e. Calculate the observed significance level of the test. Interpret it.

f. Considering your answers to parts **d** and **e**, what would your recommendation be concerning the adoption of the test statewide?

**14.83** A sociologist conducted an experiment to investigate the general public's perception of certain occupations. Each of a random sample of 15 people was asked to rank, in order of prestige, five leading professions: lawyer, politician, physician, corporate president, and college professor. Do the data shown in the table below provide sufficient evidence to indicate a difference in the amount of prestige the public attaches to these five professions? Test using $\alpha = .025$.

**14.84** The length of time required for a human subject to respond to a new painkiller was tested. Seven randomly selected subjects were assigned to receive both aspirin and the new drug. The two treatments were spaced in time and assigned in random order. The length of time (in minutes) required for a subject to indicate that he or she could feel pain relief was recorded for both the aspirin and the drug. The data are shown in the table on p. 789. Do the data provide sufficient evidence to indicate that the probability distribution of the times required to obtain relief with

⬤ PRESTIGE

| Person | Lawyer | Politician | Physician | Corporate President | College Professor |
|--------|--------|-----------|-----------|---------------------|-------------------|
| 1 | 3 | 5 | 1 | 4 | 2 |
| 2 | 4 | 5 | 1 | 2 | 3 |
| 3 | 1 | 4 | 2 | 5 | 3 |
| 4 | 3 | 5 | 2 | 4 | 1 |
| 5 | 4 | 5 | 1 | 3 | 2 |
| 6 | 5 | 4 | 3 | 2 | 1 |
| 7 | 1 | 5 | 3 | 4 | 2 |
| 8 | 4 | 5 | 1 | 3 | 2 |
| 9 | 3 | 5 | 1 | 4 | 2 |
| 10 | 4 | 5 | 2 | 3 | 1 |
| 11 | 5 | 4 | 2 | 3 | 1 |
| 12 | 3 | 5 | 1 | 4 | 2 |
| 13 | 4 | 5 | 2 | 3 | 1 |
| 14 | 3 | 4 | 1 | 5 | 2 |
| 15 | 4 | 5 | 1 | 2 | 3 |

aspirin is shifted to the right of the probability distribution of the times required to obtain relief with the new drug? Test using $\alpha = .05$.

**⊘ PAINKILL**

| Subject | Aspirin | New Drug |
|---------|---------|----------|
| 1 | 15 | 7 |
| 2 | 20 | 14 |
| 3 | 12 | 13 |
| 4 | 20 | 11 |
| 5 | 17 | 10 |
| 6 | 14 | 16 |
| 7 | 17 | 11 |

**14.85** Three lists of words, representing three levels of abstractness, are randomly assigned to 21 experimental subjects so that seven subjects receive each list. The subjects are asked to respond to each word on their list with as many associated words as possible within a given period of time. A subject's score is the total number of word associates, summing over all words in the list. Scores for each list are given in the accompanying table. Do the data provide sufficient evidence to indicate a difference (shift in location) between at least two of the probability distributions of the numbers of word associates that subjects can name for the three lists? Use $\alpha = .05$.

**⊘ WORDLIST**

| List 1 | List 2 | List 3 |
|--------|--------|--------|
| 48 | 41 | 18 |
| 43 | 36 | 42 |
| 39 | 29 | 28 |
| 57 | 40 | 38 |
| 21 | 35 | 15 |
| 47 | 45 | 33 |
| 58 | 32 | 31 |

**14.86** An experiment was designed to study whether eye pupil size is related to a person's attempt at deception. Eight students were asked to respond verbally to a series of questions. Before the questioning began, the pupil size of each student was noted and the students were instructed to answer some of the questions dishonestly. (The number of questions answered dishonestly was left to individual choice.) During questioning, the percentage increase in pupil size was recorded. Each student was then given a deception score based on the proportion of questions answered dishonestly. (High scores indicate a large number of deceptive responses.) The results are given in the next table. Can

you conclude that the percentage increase in eye pupil size is positively correlated with deception score? Use $\alpha = .05$.

**⊘ DECEPEYE**

| Student | Deception Score | Percentage Increase in Pupil Size |
|---------|-----------------|-----------------------------------|
| 1 | 87 | 10 |
| 2 | 63 | 6 |
| 3 | 95 | 11 |
| 4 | 50 | 7 |
| 5 | 43 | 0 |
| 6 | 89 | 15 |
| 7 | 33 | 4 |
| 8 | 55 | 5 |

**14.87** A state highway patrol was interested in knowing whether frequent patrolling of highways substantially reduces the number of speeders. Two similar interstate highways were selected for the study—one heavily patrolled and the other only occasionally patrolled. After 1 month, random samples of 100 cars were chosen on each highway, and the number of cars exceeding the speed limit was recorded. This process was repeated on 5 randomly selected days. The data are shown in the table.

**⊘ HWPATROL**

| Day | Highway 1 Heavily Patrolled | Highway 2 Occasionally Patrolled |
|-----|------------------------------|-----------------------------------|
| 1 | 35 | 60 |
| 2 | 40 | 36 |
| 3 | 25 | 48 |
| 4 | 38 | 54 |
| 5 | 47 | 63 |

**a.** Do the data provide evidence to indicate that the heavily patrolled highway tends to have fewer speeders per 100 cars than the occasionally patrolled highway? Test using $\alpha = .05$.

**b.** Use the paired $t$-test with $\alpha = .05$ to compare the population mean number of speeders per 100 cars for the two highways. What assumptions are necessary for this procedure to be valid?

**14.88** Many water treatment facilities supplement the natural fluoride concentration with hydrofluosilicic acid in order to reach a target concentration of fluoride in drinking water. Certain levels are thought to enhance dental health, but very high concentrations can be dangerous. Suppose that one such treatment plant targets .75 milligrams per liter (mg/L) for their water. The plant tests 25 samples each day to determine whether the median level differs from the target.

**a.** Set up the null and alternative hypotheses.
**b.** Set up the test statistic and rejection region using $\alpha = .10$.
**c.** Explain the implication of a Type I error in the context of this application. A Type II error.
**d.** Suppose that one day's samples result in 18 values that exceed .75 mg/L. Conduct the test and state the appropriate conclusion in the context of this application.
**e.** When it was suggested to the plant's supervisor that a $t$-test should be used to conduct the daily test, she replied that the probability distribution of the fluoride concentrations was "heavily skewed to the right." Show graphically what she meant by this, and explain why this is a reason to prefer the sign test to the $t$-test.

**14.89** A manufacturer wants to determine whether the number of defectives produced by its employees tends to increase as the day progresses. Unknown to the employees, a complete inspection is made of every item that was produced on one day, and the hourly fraction defective is recorded. The resulting data are given in the table. Do they provide evidence that the fraction defective increases as the day progresses? Test at the $\alpha = .05$ level.

⦿ DEFECTS

| Hour | Fraction Defective |
|---|---|
| 1 | .02 |
| 2 | .05 |
| 3 | .03 |
| 4 | .08 |
| 5 | .06 |
| 6 | .09 |
| 7 | .11 |
| 8 | .10 |

**14.90** Refer to Exercise 14.89. Assume that the eight measurements of the fraction defective constitute a random sample from the probability distribution of all fraction defective measurements. The manufacturer wants to determine whether the median hourly fraction defective exceeds .05.
**a.** Conduct the appropriate test to make this determination at the .10 level of significance.
**b.** Why might the manufacturer wish to conduct the test with this relatively high level of significance?
**c.** What assumptions are necessary to ensure the validity of this test? How do they differ from the assumptions that would be necessary to conduct a $t$-test about the mean fraction defective?

**14.91** An experiment was conducted to compare the length of time it takes a human to recover from each of three types of influenza—Victoria A, Texas, and Russian. Twenty-one human subjects were selected at random

from a group of volunteers and divided into three groups of seven each. Each group was randomly assigned a strain of the virus, and the influenza was induced in the subjects. All the subjects were then cared for under identical conditions, and the recovery time (in days) was recorded. The results are shown in the accompanying table.

⦿ FLU

| Victoria A | Texas | Russian |
|---|---|---|
| 12 | 9 | 7 |
| 6 | 10 | 3 |
| 13 | 5 | 7 |
| 10 | 4 | 5 |
| 8 | 9 | 6 |
| 11 | 8 | 4 |
| 7 | 11 | 8 |

**a.** Do the data provide sufficient evidence to indicate that the recovery times for one or more types of influenza tend to be longer than for the other types? Test using $\alpha = .05$.
**b.** Do the data provide sufficient evidence to indicate a difference in locations of the distributions of recovery times for the Victoria A and Russian types? Test using $\alpha = .05$.

**14.92** A serious drought-related problem for farmers is the spread of aflatoxin, a highly toxic substance generated by mold, which contaminates field corn. Three sprays, A, B, and C, have been developed to control aflatoxin in field corn. To determine whether differences exist among the sprays, 10 ears of corn are randomly chosen from a contaminated corn field and each is divided into three pieces of equal size. The sprays are then randomly assigned to the pieces for each ear of corn, thus setting up a randomized block design. The table gives the amount (in parts per billion) of aflatoxin present in the corn samples after spraying. Determine whether there is evidence that the distributions of the levels of aflatoxin in corn differ for at least two of the three sprays. Test at $\alpha = .05$.

⦿ AFLATOX

| | Spray | | |
|---|---|---|---|
| Ear | A | B | C |
| 1 | 21 | 23 | 15 |
| 2 | 29 | 30 | 21 |
| 3 | 16 | 19 | 18 |
| 4 | 20 | 19 | 18 |
| 5 | 13 | 10 | 14 |
| 6 | 5 | 12 | 6 |
| 7 | 18 | 18 | 12 |
| 8 | 26 | 32 | 21 |
| 9 | 17 | 20 | 9 |
| 10 | 4 | 10 | 2 |

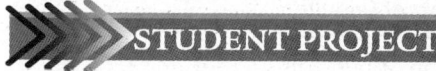

## STUDENT PROJECTS

In Chapters 10 and 14 we discussed two methods of analyzing a randomized block design. When the populations have normal probability distributions and their variances are equal, we can employ the analysis of variance described in Chapter 10. Otherwise, we can use the Friedman $F_r$-test.

In the Student Projects section of Chapter 10, we asked you to conduct a randomized block design to compare su-

permarket prices, and to use an analysis of variance to interpret the data. Now use the Friedman $F_r$-test to compare the supermarket prices.

How do the results of the two analyses compare? Explain the similarity (or lack of similarity) between the two results.

## REFERENCES

Agresti, A., and Agresti, B. F. *Statistical Methods for the Social Sciences*, 2nd ed. San Francisco: Dellen, 1986.

Conover, W. J. *Practical Nonparametric Statistics*, 2nd ed. New York: Wiley, 1980.

Daniel, W. W. *Applied Nonparametric Statistics*, 2nd ed. Boston: PWS-Kent, 1990.

Dunn, O. J. "Multiple comparisons using rank sums." *Technometrics*, Vol. 6, 1964.

Friedman, M. "The use of ranks to avoid the assumption of normality implicit in the analysis of variance." *Journal of the American Statistical Association*, Vol. 32, 1937.

Gibbons, J. D. *Nonparametric Statistical Inference*, 2nd ed. New York: McGraw-Hill, 1985.

Hollander, M., and Wolfe, D. A. *Nonparametric Statistical Methods*. New York: Wiley, 1973.

Kruskal, W. H., and Wallis, W. A. "Use of ranks in one-criterion variance analysis." *Journal of the American Statistical Association*, Vol. 47, 1952.

Lehmann, E. L. *Nonparametrics: Statistical Methods Based on Ranks*. San Francisco: Holden-Day, 1975.

Marascuilo, L. A., and McSweeney, M. *Nonparametric and Distribution-Free Methods for the Social Sciences*. Monterey, Ca.: Brooks/Cole, 1977.

Wilcoxon, F., and Wilcox, R. A. "Some rapid approximate statistical procedures." The American Cyanamid Co., 1964.

# APPENDIX A

## Tables

**TABLE I**   Random Numbers

| Row / Column | 1 | 2 | 3 | 4 | 5 | 6 | 7 | 8 | 9 | 10 | 11 | 12 | 13 | 14 |
|---|---|---|---|---|---|---|---|---|---|---|---|---|---|---|
| 1 | 10480 | 15011 | 01536 | 02011 | 81647 | 91646 | 69179 | 14194 | 62590 | 36207 | 20969 | 99570 | 91291 | 90700 |
| 2 | 22368 | 46573 | 25595 | 85393 | 30995 | 89198 | 27982 | 53402 | 93965 | 34095 | 52666 | 19174 | 39615 | 99505 |
| 3 | 24130 | 48360 | 22527 | 97265 | 76393 | 64809 | 15179 | 24830 | 49340 | 32081 | 30680 | 19655 | 63348 | 58629 |
| 4 | 42167 | 93093 | 06243 | 61680 | 07856 | 16376 | 39440 | 53537 | 71341 | 57004 | 00849 | 74917 | 97758 | 16379 |
| 5 | 37570 | 39975 | 81837 | 16656 | 06121 | 91782 | 60468 | 81305 | 49684 | 60672 | 14110 | 06927 | 01263 | 54613 |
| 6 | 77921 | 06907 | 11008 | 42751 | 27756 | 53498 | 18602 | 70659 | 90655 | 15053 | 21916 | 81825 | 44394 | 42880 |
| 7 | 99562 | 72905 | 56420 | 69994 | 98872 | 31016 | 71194 | 18738 | 44013 | 48840 | 63213 | 21069 | 10634 | 12952 |
| 8 | 96301 | 91977 | 05463 | 07972 | 18876 | 20922 | 94595 | 56869 | 69014 | 60045 | 18425 | 84903 | 42508 | 32307 |
| 9 | 89579 | 14342 | 63661 | 10281 | 17453 | 18103 | 57740 | 84378 | 25331 | 12566 | 58678 | 44947 | 05585 | 56941 |
| 10 | 85475 | 36857 | 53342 | 53988 | 53060 | 59533 | 38867 | 62300 | 08158 | 17983 | 16439 | 11458 | 18593 | 64952 |
| 11 | 28918 | 69578 | 88231 | 33276 | 70997 | 79936 | 56865 | 05859 | 90106 | 31595 | 01547 | 85590 | 91610 | 78188 |
| 12 | 63553 | 40961 | 48235 | 03427 | 49626 | 69445 | 18663 | 72695 | 52180 | 20847 | 12234 | 90511 | 33703 | 90322 |
| 13 | 09429 | 93969 | 52636 | 92737 | 88974 | 33488 | 36320 | 17617 | 30015 | 08272 | 84115 | 27156 | 30613 | 74952 |
| 14 | 10365 | 61129 | 87529 | 85689 | 48237 | 52267 | 67689 | 93394 | 01511 | 26358 | 85104 | 20285 | 29975 | 89863 |
| 15 | 07119 | 97336 | 71048 | 08178 | 77233 | 13916 | 47564 | 81056 | 97735 | 85977 | 29372 | 74461 | 28551 | 90707 |
| 16 | 51085 | 12765 | 51821 | 51259 | 77452 | 16308 | 60756 | 92144 | 49442 | 53900 | 70960 | 63990 | 75601 | 40719 |
| 17 | 02368 | 21382 | 52404 | 60268 | 89368 | 19885 | 55322 | 44819 | 01188 | 65255 | 64835 | 44919 | 05944 | 55157 |
| 18 | 01011 | 54092 | 33362 | 94904 | 31273 | 04146 | 18594 | 29852 | 71585 | 85030 | 51132 | 01915 | 92747 | 64951 |
| 19 | 52162 | 53916 | 46369 | 58586 | 23216 | 14513 | 83149 | 98736 | 23495 | 64350 | 94738 | 17752 | 35156 | 35749 |
| 20 | 07056 | 97628 | 33787 | 09998 | 42698 | 06691 | 76988 | 13602 | 51851 | 46104 | 88916 | 19509 | 25625 | 58104 |
| 21 | 48663 | 91245 | 85828 | 14346 | 09172 | 30168 | 90229 | 04734 | 59193 | 22178 | 30421 | 61666 | 99904 | 32812 |
| 22 | 54164 | 58492 | 22421 | 74103 | 47070 | 25306 | 76468 | 26384 | 58151 | 06646 | 21524 | 15227 | 95909 | 44592 |
| 23 | 32639 | 32363 | 05597 | 24200 | 13363 | 38005 | 94342 | 28728 | 35806 | 06912 | 17012 | 64161 | 18296 | 22851 |
| 24 | 29334 | 27001 | 87637 | 87308 | 58731 | 00256 | 45834 | 15398 | 46557 | 41135 | 10367 | 07684 | 36188 | 18510 |
| 25 | 02488 | 33062 | 28834 | 07351 | 19731 | 92420 | 60952 | 61280 | 50001 | 67658 | 32586 | 86679 | 50720 | 94953 |
| 26 | 81525 | 72295 | 04839 | 96423 | 24878 | 82651 | 66566 | 14778 | 76797 | 14780 | 13300 | 87074 | 79666 | 95725 |
| 27 | 29676 | 20591 | 68086 | 26432 | 46901 | 20849 | 89768 | 81536 | 86645 | 12659 | 92259 | 57102 | 80428 | 25280 |
| 28 | 00742 | 57392 | 39064 | 66432 | 84673 | 40027 | 32832 | 61362 | 98947 | 96067 | 64760 | 64584 | 96096 | 98253 |
| 29 | 05366 | 04213 | 25669 | 26422 | 44407 | 44048 | 37937 | 63904 | 45766 | 66134 | 75470 | 66520 | 34693 | 90449 |
| 30 | 91921 | 26418 | 64117 | 94305 | 26766 | 25940 | 39972 | 22209 | 71500 | 64568 | 91402 | 42416 | 07844 | 69618 |
| 31 | 00582 | 04711 | 87917 | 77341 | 42206 | 35126 | 74087 | 99547 | 81817 | 42607 | 43808 | 76655 | 62028 | 76630 |
| 32 | 00725 | 69884 | 62797 | 56170 | 86324 | 88072 | 76222 | 36086 | 84637 | 93161 | 76038 | 65855 | 77919 | 88006 |
| 33 | 69011 | 65795 | 95876 | 55293 | 18988 | 27354 | 26575 | 03625 | 40801 | 59920 | 29841 | 80150 | 12777 | 48501 |
| 34 | 25976 | 57948 | 29888 | 88604 | 67917 | 48708 | 18912 | 82271 | 65424 | 69774 | 33611 | 54262 | 85963 | 03547 |
| 35 | 09763 | 83473 | 73577 | 12908 | 30883 | 18317 | 28290 | 35797 | 05998 | 41688 | 34952 | 37888 | 38917 | 88050 |

*(continued)*

**TABLE I  Random Numbers**

| Row / Column | 1 | 2 | 3 | 4 | 5 | 6 | 7 | 8 | 9 | 10 | 11 | 12 | 13 | 14 |
|---|---|---|---|---|---|---|---|---|---|---|---|---|---|---|
| 36 | 91576 | 42595 | 27958 | 30134 | 04024 | 86385 | 29880 | 99730 | 55536 | 84855 | 29080 | 09250 | 79656 | 73211 |
| 37 | 17955 | 56349 | 90999 | 49127 | 20044 | 59931 | 06115 | 20542 | 18059 | 02008 | 73708 | 83517 | 36103 | 42791 |
| 38 | 46503 | 18584 | 18845 | 49618 | 02304 | 51038 | 20655 | 58727 | 28168 | 15475 | 56942 | 53389 | 20562 | 87338 |
| 39 | 92157 | 89634 | 94824 | 78171 | 84610 | 82834 | 09922 | 25417 | 44137 | 48413 | 25555 | 21246 | 35509 | 20468 |
| 40 | 14577 | 62765 | 35605 | 81263 | 39667 | 47358 | 56873 | 56307 | 61607 | 49518 | 89656 | 20103 | 77490 | 18062 |
| 41 | 98427 | 07523 | 33362 | 64270 | 01638 | 92477 | 66969 | 98420 | 04880 | 45585 | 46565 | 04102 | 46880 | 45709 |
| 42 | 34914 | 63976 | 88720 | 82765 | 34476 | 17032 | 87589 | 40836 | 32427 | 70002 | 70663 | 88863 | 77775 | 69348 |
| 43 | 70060 | 28277 | 39475 | 46473 | 23219 | 53416 | 94970 | 25832 | 69975 | 94884 | 19661 | 72828 | 00102 | 66794 |
| 44 | 53976 | 54914 | 06990 | 67245 | 68350 | 82948 | 11398 | 42878 | 80287 | 88267 | 47363 | 46634 | 06541 | 97809 |
| 45 | 76072 | 29515 | 40980 | 07391 | 58745 | 25774 | 22987 | 80059 | 39911 | 96189 | 41151 | 14222 | 60697 | 59583 |
| 46 | 90725 | 52210 | 83974 | 29992 | 65831 | 38857 | 50490 | 83765 | 55657 | 14361 | 31720 | 57375 | 56228 | 41546 |
| 47 | 64364 | 67412 | 33339 | 31926 | 14883 | 24413 | 59744 | 92351 | 97473 | 89286 | 35931 | 04110 | 23726 | 51900 |
| 48 | 08962 | 00358 | 31662 | 25388 | 61642 | 34072 | 81249 | 35648 | 56891 | 69352 | 48373 | 45578 | 78547 | 81788 |
| 49 | 95012 | 68379 | 93526 | 70765 | 10592 | 04542 | 76463 | 54328 | 02349 | 17247 | 28865 | 14777 | 62730 | 92277 |
| 50 | 15664 | 10493 | 20492 | 38391 | 91132 | 21999 | 59516 | 81652 | 27195 | 48223 | 46751 | 22923 | 32261 | 85653 |
| 51 | 16408 | 81899 | 04153 | 53381 | 79401 | 21438 | 83035 | 92350 | 36693 | 31238 | 59649 | 91754 | 72772 | 02338 |
| 52 | 18629 | 81953 | 05520 | 91962 | 04739 | 13092 | 97662 | 24822 | 94730 | 06496 | 35090 | 04822 | 86774 | 98289 |
| 53 | 73115 | 35101 | 47498 | 87637 | 99016 | 71060 | 88824 | 71013 | 18735 | 20286 | 23153 | 72924 | 35165 | 43040 |
| 54 | 57491 | 16703 | 23167 | 49323 | 45021 | 33132 | 12544 | 41035 | 80780 | 45393 | 44812 | 12512 | 98931 | 91202 |
| 55 | 30405 | 83946 | 23792 | 14422 | 15059 | 45799 | 22716 | 19792 | 09983 | 74353 | 68668 | 30429 | 70735 | 25499 |
| 56 | 16631 | 35006 | 85900 | 98275 | 32388 | 52390 | 16815 | 69290 | 82732 | 38480 | 73817 | 32523 | 41961 | 44437 |
| 57 | 96773 | 20206 | 42559 | 78985 | 05300 | 22164 | 24369 | 54224 | 35083 | 19687 | 11052 | 91491 | 60383 | 19746 |
| 58 | 38935 | 64202 | 14349 | 82674 | 66523 | 44133 | 00697 | 35552 | 35970 | 19124 | 63318 | 29686 | 03387 | 59846 |
| 59 | 31624 | 76384 | 17403 | 53363 | 44167 | 64486 | 64758 | 75366 | 76554 | 31601 | 12614 | 33072 | 60332 | 92325 |
| 60 | 78919 | 19474 | 23632 | 27889 | 47914 | 02584 | 37680 | 20801 | 72152 | 39339 | 34806 | 08930 | 85001 | 87820 |
| 61 | 03931 | 33309 | 57047 | 74211 | 63445 | 17361 | 62825 | 39908 | 05607 | 91284 | 68833 | 25570 | 38818 | 46920 |
| 62 | 74426 | 33278 | 43972 | 10110 | 89917 | 15665 | 52872 | 73823 | 73144 | 88662 | 88970 | 74492 | 51805 | 99378 |
| 63 | 09066 | 00903 | 20795 | 95452 | 92648 | 45454 | 09552 | 88815 | 16553 | 51125 | 79375 | 97596 | 16296 | 66092 |
| 64 | 42238 | 12426 | 87025 | 14267 | 20979 | 04508 | 64535 | 31355 | 86064 | 29472 | 47689 | 05974 | 52468 | 16834 |
| 65 | 16153 | 08002 | 26504 | 41744 | 81959 | 65642 | 74240 | 56302 | 00033 | 67107 | 77510 | 70625 | 28725 | 34191 |
| 66 | 21457 | 40742 | 29820 | 96783 | 29400 | 21840 | 15035 | 34537 | 33310 | 06116 | 95240 | 15957 | 16572 | 06004 |
| 67 | 21581 | 57802 | 02050 | 89728 | 17937 | 37621 | 47075 | 42080 | 97403 | 48626 | 68995 | 43805 | 33386 | 21597 |
| 68 | 55612 | 78095 | 83197 | 33732 | 05810 | 24813 | 86902 | 60397 | 16489 | 03264 | 88525 | 42786 | 05269 | 92532 |
| 69 | 44657 | 66999 | 99324 | 51281 | 84463 | 60563 | 79312 | 93454 | 68876 | 25471 | 93911 | 25650 | 12682 | 73572 |
| 70 | 91340 | 84979 | 46949 | 81973 | 37949 | 61023 | 43997 | 15263 | 80644 | 43942 | 89203 | 71795 | 99533 | 50501 |

*(continued)*

796

**TABLE I   Random Numbers**

| Row | 1 | 2 | 3 | 4 | 5 | 6 | 7 | 8 | 9 | 10 | 11 | 12 | 13 | 14 |
|---|---|---|---|---|---|---|---|---|---|---|---|---|---|---|
| 71 | 91227 | 21199 | 31935 | 27022 | 84067 | 05462 | 35216 | 14486 | 29891 | 68607 | 41867 | 14951 | 91696 | 85065 |
| 72 | 50001 | 38140 | 66321 | 19924 | 72163 | 09538 | 12151 | 06878 | 91903 | 18749 | 34405 | 56087 | 82790 | 70925 |
| 73 | 65390 | 05224 | 72958 | 28609 | 81406 | 39147 | 25549 | 48542 | 42627 | 45233 | 57202 | 94617 | 23772 | 07896 |
| 74 | 27504 | 96131 | 83944 | 41575 | 10573 | 08619 | 64482 | 73923 | 36152 | 05184 | 94142 | 25299 | 84387 | 34925 |
| 75 | 37169 | 94851 | 39117 | 89632 | 00959 | 16487 | 65536 | 49071 | 39782 | 17095 | 02330 | 74301 | 00275 | 48280 |
| 76 | 11508 | 70225 | 51111 | 38351 | 19444 | 66499 | 71945 | 05422 | 13442 | 78675 | 84081 | 66938 | 93654 | 59894 |
| 77 | 37449 | 30362 | 06694 | 54690 | 04052 | 53115 | 62757 | 95348 | 78662 | 11163 | 81651 | 50245 | 34971 | 52924 |
| 78 | 46515 | 70331 | 85922 | 38329 | 57015 | 15765 | 97161 | 17369 | 45349 | 61796 | 66345 | 81073 | 45106 | 79860 |
| 79 | 30986 | 81223 | 42416 | 58353 | 21532 | 30502 | 32305 | 86482 | 05174 | 07901 | 54339 | 58861 | 74818 | 46942 |
| 80 | 63798 | 64995 | 46583 | 09785 | 44160 | 78128 | 83991 | 42365 | 92520 | 83531 | 80377 | 35909 | 81250 | 54238 |
| 81 | 82486 | 84846 | 99254 | 67632 | 43218 | 50076 | 21361 | 64816 | 51202 | 88124 | 41870 | 52689 | 51275 | 83556 |
| 82 | 21885 | 32906 | 92431 | 09060 | 64297 | 51674 | 64126 | 62570 | 26123 | 05155 | 59194 | 52799 | 28225 | 85762 |
| 83 | 60336 | 98782 | 07408 | 53458 | 13564 | 59089 | 26445 | 29789 | 85205 | 41001 | 12535 | 12133 | 14645 | 23541 |
| 84 | 43937 | 46891 | 24010 | 25560 | 86355 | 33941 | 25786 | 54990 | 71899 | 15475 | 95434 | 98227 | 21824 | 19585 |
| 85 | 97655 | 63175 | 89303 | 16275 | 07100 | 92063 | 21942 | 18611 | 47348 | 20203 | 18534 | 03862 | 78095 | 50136 |
| 86 | 03299 | 01221 | 05418 | 38982 | 55758 | 92237 | 26759 | 86367 | 21216 | 98442 | 08303 | 56613 | 91511 | 75928 |
| 87 | 79625 | 06486 | 03574 | 17668 | 07785 | 76020 | 79924 | 25651 | 83325 | 88428 | 85076 | 72811 | 22717 | 50585 |
| 88 | 85635 | 68335 | 47539 | 03129 | 65651 | 11977 | 02510 | 26113 | 99447 | 68645 | 34327 | 15152 | 55230 | 93448 |
| 89 | 18039 | 14367 | 64337 | 06177 | 12143 | 46609 | 32989 | 74014 | 64708 | 00533 | 35398 | 58408 | 13261 | 47908 |
| 90 | 08362 | 15656 | 60627 | 36478 | 65648 | 16764 | 53412 | 09013 | 07832 | 41574 | 17639 | 82163 | 60859 | 75567 |
| 91 | 79556 | 29068 | 04142 | 16268 | 15387 | 12856 | 66227 | 38358 | 22478 | 73373 | 88732 | 09443 | 82558 | 05250 |
| 92 | 92608 | 82674 | 27072 | 32534 | 17075 | 27698 | 98204 | 63863 | 11951 | 34648 | 88022 | 56148 | 34925 | 57031 |
| 93 | 23982 | 25835 | 40055 | 67006 | 12293 | 02753 | 14827 | 23235 | 35071 | 99704 | 37543 | 11601 | 35503 | 85171 |
| 94 | 09915 | 96306 | 05908 | 97901 | 28395 | 14186 | 00821 | 80703 | 70426 | 75647 | 76310 | 88717 | 37890 | 40129 |
| 95 | 59037 | 33300 | 26695 | 62247 | 69927 | 76123 | 50842 | 43834 | 86654 | 70959 | 79725 | 93872 | 28117 | 19233 |
| 96 | 42488 | 78077 | 69882 | 61657 | 34136 | 79180 | 97526 | 43092 | 04098 | 73571 | 80799 | 76536 | 72255 | 64239 |
| 97 | 46764 | 86273 | 63003 | 93017 | 31204 | 36692 | 40202 | 35275 | 57306 | 55543 | 53203 | 18098 | 47625 | 88684 |
| 98 | 03237 | 45430 | 55417 | 63282 | 90816 | 17349 | 88298 | 90183 | 36600 | 78406 | 06216 | 95787 | 42579 | 90733 |
| 99 | 86591 | 81482 | 52667 | 61582 | 14972 | 90053 | 89534 | 76036 | 49199 | 43716 | 97548 | 04379 | 46370 | 28672 |
| 100 | 38534 | 01715 | 94964 | 87288 | 65680 | 43772 | 39560 | 12918 | 86537 | 62738 | 19636 | 51132 | 25739 | 56947 |

*Source:* Abridged from W. H. Beyer (ed.). *CRC Standard Mathematical Tables*, 24th edition. (Cleveland: The Chemical Rubber Company), 1976. Reproduced by permission of the publisher.

## TABLE II    Binomial Probabilities

*Tabulated values are* $\sum\limits_{x=0}^{k} p(x)$. *(Computations are rounded at the third decimal place.)*

**a.** $n = 5$

| k | .01 | .05 | .10 | .20 | .30 | .40 | .50 | .60 | .70 | .80 | .90 | .95 | .99 |
|---|-----|-----|-----|-----|-----|-----|-----|-----|-----|-----|-----|-----|-----|
| 0 | .951 | .774 | .590 | .328 | .168 | .078 | .031 | .010 | .002 | .000 | .000 | .000 | .000 |
| 1 | .999 | .977 | .919 | .737 | .528 | .337 | .188 | .087 | .031 | .007 | .000 | .000 | .000 |
| 2 | 1.000 | .999 | .991 | .942 | .837 | .683 | .500 | .317 | .163 | .058 | .009 | .001 | .000 |
| 3 | 1.000 | 1.000 | 1.000 | .993 | .969 | .913 | .812 | .663 | .472 | .263 | .081 | .023 | .001 |
| 4 | 1.000 | 1.000 | 1.000 | 1.000 | .998 | .990 | .969 | .922 | .832 | .672 | .410 | .226 | .049 |

**b.** $n = 6$

| k | .01 | .05 | .10 | .20 | .30 | .40 | .50 | .60 | .70 | .80 | .90 | .95 | .99 |
|---|-----|-----|-----|-----|-----|-----|-----|-----|-----|-----|-----|-----|-----|
| 0 | .941 | .735 | .531 | .262 | .118 | .047 | .016 | .004 | .001 | .000 | .000 | .000 | .000 |
| 1 | .999 | .967 | .886 | .655 | .420 | .233 | .109 | .041 | .011 | .002 | .000 | .000 | .000 |
| 2 | 1.000 | .998 | .984 | .901 | .744 | .544 | .344 | .179 | .070 | .017 | .001 | .000 | .000 |
| 3 | 1.000 | 1.000 | .999 | .983 | .930 | .821 | .656 | .456 | .256 | .099 | .016 | .002 | .000 |
| 4 | 1.000 | 1.000 | 1.000 | .998 | .989 | .959 | .891 | .767 | .580 | .345 | .114 | .033 | .001 |
| 5 | 1.000 | 1.000 | 1.000 | 1.000 | .999 | .996 | .984 | .953 | .882 | .738 | .469 | .265 | .059 |

**c.** $n = 7$

| k | .01 | .05 | .10 | .20 | .30 | .40 | .50 | .60 | .70 | .80 | .90 | .95 | .99 |
|---|-----|-----|-----|-----|-----|-----|-----|-----|-----|-----|-----|-----|-----|
| 0 | .932 | .698 | .478 | .210 | .082 | .028 | .008 | .002 | .000 | .000 | .000 | .000 | .000 |
| 1 | .998 | .956 | .850 | .577 | .329 | .159 | .063 | .019 | .004 | .000 | .000 | .000 | .000 |
| 2 | 1.000 | .996 | .974 | .852 | .647 | .420 | .227 | .096 | .029 | .005 | .000 | .000 | .000 |
| 3 | 1.000 | 1.000 | .997 | .967 | .874 | .710 | .500 | .290 | .126 | .033 | .003 | .000 | .000 |
| 4 | 1.000 | 1.000 | 1.000 | .995 | .971 | .904 | .773 | .580 | .353 | .148 | .026 | .004 | .000 |
| 5 | 1.000 | 1.000 | 1.000 | 1.000 | .996 | .981 | .937 | .841 | .671 | .423 | .150 | .044 | .002 |
| 6 | 1.000 | 1.000 | 1.000 | 1.000 | 1.000 | .998 | .992 | .972 | .918 | .790 | .522 | .302 | .068 |

*(continued)*

**TABLE II**   *(continued)*

**d.** $n = 8$

| k | .01 | .05 | .10 | .20 | .30 | .40 | .50 | .60 | .70 | .80 | .90 | .95 | .99 |
|---|-----|-----|-----|-----|-----|-----|-----|-----|-----|-----|-----|-----|-----|
| 0 | .923 | .663 | .430 | .168 | .058 | .017 | .004 | .001 | .000 | .000 | .000 | .000 | .000 |
| 1 | .997 | .943 | .813 | .503 | .255 | .106 | .035 | .009 | .001 | .000 | .000 | .000 | .000 |
| 2 | 1.000 | .994 | .962 | .797 | .552 | .315 | .145 | .050 | .011 | .001 | .000 | .000 | .000 |
| 3 | 1.000 | 1.000 | .995 | .944 | .806 | .594 | .363 | .174 | .058 | .010 | .000 | .000 | .000 |
| 4 | 1.000 | 1.000 | 1.000 | .990 | .942 | .826 | .637 | .406 | .194 | .056 | .005 | .000 | .000 |
| 5 | 1.000 | 1.000 | 1.000 | .999 | .989 | .950 | .855 | .685 | .448 | .203 | .038 | .006 | .000 |
| 6 | 1.000 | 1.000 | 1.000 | 1.000 | .999 | .991 | .965 | .894 | .745 | .497 | .187 | .057 | .003 |
| 7 | 1.000 | 1.000 | 1.000 | 1.000 | 1.000 | .999 | .996 | .983 | .942 | .832 | .570 | .337 | .077 |

**e.** $n = 9$

| k | .01 | .05 | .10 | .20 | .30 | .40 | .50 | .60 | .70 | .80 | .90 | .95 | .99 |
|---|-----|-----|-----|-----|-----|-----|-----|-----|-----|-----|-----|-----|-----|
| 0 | .914 | .630 | .387 | .134 | .040 | .010 | .002 | .000 | .000 | .000 | .000 | .000 | .000 |
| 1 | .997 | .929 | .775 | .436 | .196 | .071 | .020 | .004 | .000 | .000 | .000 | .000 | .000 |
| 2 | 1.000 | .992 | .947 | .738 | .463 | .232 | .090 | .025 | .004 | .000 | .000 | .000 | .000 |
| 3 | 1.000 | .999 | .992 | .914 | .730 | .483 | .254 | .099 | .025 | .003 | .000 | .000 | .000 |
| 4 | 1.000 | 1.000 | .999 | .980 | .901 | .733 | .500 | .267 | .099 | .020 | .001 | .000 | .000 |
| 5 | 1.000 | 1.000 | 1.000 | .997 | .975 | .901 | .746 | .517 | .270 | .086 | .008 | .001 | .000 |
| 6 | 1.000 | 1.000 | 1.000 | 1.000 | .996 | .975 | .910 | .768 | .537 | .262 | .053 | .008 | .000 |
| 7 | 1.000 | 1.000 | 1.000 | 1.000 | 1.000 | .996 | .980 | .929 | .804 | .564 | .225 | .071 | .003 |
| 8 | 1.000 | 1.000 | 1.000 | 1.000 | 1.000 | 1.000 | .998 | .990 | .960 | .866 | .613 | .370 | .086 |

**f.** $n = 10$

| k | .01 | .05 | .10 | .20 | .30 | .40 | .50 | .60 | .70 | .80 | .90 | .95 | .99 |
|---|-----|-----|-----|-----|-----|-----|-----|-----|-----|-----|-----|-----|-----|
| 0 | .904 | .599 | .349 | .107 | .028 | .006 | .001 | .000 | .000 | .000 | .000 | .000 | .000 |
| 1 | .996 | .914 | .376 | .376 | .149 | .046 | .011 | .002 | .000 | .000 | .000 | .000 | .000 |
| 2 | 1.000 | .988 | .930 | .678 | .383 | .167 | .055 | .012 | .002 | .000 | .000 | .000 | .000 |
| 3 | 1.000 | .999 | .987 | .879 | .650 | .382 | .172 | .055 | .011 | .001 | .000 | .000 | .000 |
| 4 | 1.000 | 1.000 | .998 | .967 | .850 | .633 | .377 | .166 | .047 | .006 | .000 | .000 | .000 |
| 5 | 1.000 | 1.000 | 1.000 | .999 | .953 | .834 | .623 | .367 | .150 | .033 | .002 | .000 | .000 |
| 6 | 1.000 | 1.000 | 1.000 | .999 | .989 | .945 | .828 | .618 | .350 | .121 | .013 | .001 | .000 |
| 7 | 1.000 | 1.000 | 1.000 | 1.000 | .998 | .988 | .945 | .833 | .617 | .322 | .070 | .012 | .000 |
| 8 | 1.000 | 1.000 | 1.000 | 1.000 | 1.000 | .998 | .989 | .954 | .851 | .624 | .264 | .086 | .004 |
| 9 | 1.000 | 1.000 | 1.000 | 1.000 | 1.000 | 1.000 | .999 | .994 | .972 | .893 | .651 | .401 | .096 |

*(continued)*

**TABLE II**    **(continued)**

**g.** $n = 15$

| k \ p | .01 | .05 | .10 | .20 | .30 | .40 | .50 | .60 | .70 | .80 | .90 | .95 | .99 |
|---|---|---|---|---|---|---|---|---|---|---|---|---|---|
| 0 | .860 | .463 | .206 | .035 | .005 | .000 | .000 | .000 | .000 | .000 | .000 | .000 | .000 |
| 1 | .990 | .829 | .549 | .167 | .035 | .005 | .000 | .000 | .000 | .000 | .000 | .000 | .000 |
| 2 | 1.000 | .964 | .816 | .398 | .127 | .027 | .004 | .000 | .000 | .000 | .000 | .000 | .000 |
| 3 | 1.000 | .995 | .944 | .648 | .297 | .091 | .018 | .002 | .000 | .000 | .000 | .000 | .000 |
| 4 | 1.000 | .999 | .987 | .838 | .515 | .217 | .059 | .009 | .001 | .000 | .000 | .000 | .000 |
| 5 | 1.000 | 1.000 | .998 | .939 | .722 | .403 | .151 | .034 | .004 | .000 | .000 | .000 | .000 |
| 6 | 1.000 | 1.000 | 1.000 | .982 | .869 | .610 | .304 | .095 | .015 | .001 | .000 | .000 | .000 |
| 7 | 1.000 | 1.000 | 1.000 | .996 | .950 | .787 | .500 | .213 | .050 | .004 | .000 | .000 | .000 |
| 8 | 1.000 | 1.000 | 1.000 | .999 | .985 | .905 | .696 | .390 | .131 | .018 | .000 | .000 | .000 |
| 9 | 1.000 | 1.000 | 1.000 | 1.000 | .996 | .966 | .849 | .597 | .278 | .061 | .002 | .000 | .000 |
| 10 | 1.000 | 1.000 | 1.000 | 1.000 | .999 | .991 | .941 | .783 | .485 | .164 | .013 | .001 | .000 |
| 11 | 1.000 | 1.000 | 1.000 | 1.000 | 1.000 | .998 | .982 | .909 | .703 | .352 | .056 | .005 | .000 |
| 12 | 1.000 | 1.000 | 1.000 | 1.000 | 1.000 | 1.000 | .996 | .973 | .873 | .602 | .184 | .036 | .000 |
| 13 | 1.000 | 1.000 | 1.000 | 1.000 | 1.000 | 1.000 | 1.000 | .995 | .965 | .833 | .451 | .171 | .010 |
| 14 | 1.000 | 1.000 | 1.000 | 1.000 | 1.000 | 1.000 | 1.000 | 1.000 | .995 | .965 | .794 | .537 | .140 |

**h.** $n = 20$

| k \ p | .01 | .05 | .10 | .20 | .30 | .40 | .50 | .60 | .70 | .80 | .90 | .95 | .99 |
|---|---|---|---|---|---|---|---|---|---|---|---|---|---|
| 0 | .818 | .358 | .122 | .012 | .001 | .000 | .000 | .000 | .000 | .000 | .000 | .000 | .000 |
| 1 | .983 | .736 | .392 | .069 | .008 | .001 | .000 | .000 | .000 | .000 | .000 | .000 | .000 |
| 2 | .999 | .925 | .677 | .206 | .035 | .004 | .000 | .000 | .000 | .000 | .000 | .000 | .000 |
| 3 | 1.000 | .984 | .867 | .411 | .107 | .016 | .001 | .000 | .000 | .000 | .000 | .000 | .000 |
| 4 | 1.000 | .997 | .957 | .630 | .238 | .051 | .006 | .000 | .000 | .000 | .000 | .000 | .000 |
| 5 | 1.000 | 1.000 | .989 | .804 | .416 | .126 | .021 | .002 | .000 | .000 | .000 | .000 | .000 |
| 6 | 1.000 | 1.000 | .998 | .913 | .608 | .250 | .058 | .006 | .000 | .000 | .000 | .000 | .000 |
| 7 | 1.000 | 1.000 | 1.000 | .968 | .772 | .416 | .132 | .021 | .001 | .000 | .000 | .000 | .000 |
| 8 | 1.000 | 1.000 | 1.000 | .990 | .887 | .596 | .252 | .057 | .005 | .000 | .000 | .000 | .000 |
| 9 | 1.000 | 1.000 | 1.000 | .997 | .952 | .755 | .412 | .128 | .017 | .001 | .000 | .000 | .000 |
| 10 | 1.000 | 1.000 | 1.000 | .999 | .983 | .872 | .588 | .245 | .048 | .003 | .000 | .000 | .000 |
| 11 | 1.000 | 1.000 | 1.000 | 1.000 | .995 | .943 | .748 | .404 | .113 | .010 | .000 | .000 | .000 |
| 12 | 1.000 | 1.000 | 1.000 | 1.000 | .999 | .979 | .868 | .584 | .228 | .032 | .000 | .000 | .000 |
| 13 | 1.000 | 1.000 | 1.000 | 1.000 | 1.000 | .994 | .942 | .750 | .392 | .087 | .002 | .000 | .000 |
| 14 | 1.000 | 1.000 | 1.000 | 1.000 | 1.000 | .998 | .979 | .874 | .584 | .196 | .011 | .000 | .000 |
| 15 | 1.000 | 1.000 | 1.000 | 1.000 | 1.000 | 1.000 | .994 | .949 | .762 | .370 | .043 | .003 | .000 |
| 16 | 1.000 | 1.000 | 1.000 | 1.000 | 1.000 | 1.000 | .999 | .984 | .893 | .589 | .133 | .016 | .000 |
| 17 | 1.000 | 1.000 | 1.000 | 1.000 | 1.000 | 1.000 | 1.000 | .996 | .965 | .794 | .323 | .075 | .001 |
| 18 | 1.000 | 1.000 | 1.000 | 1.000 | 1.000 | 1.000 | 1.000 | .999 | .992 | .931 | .608 | .264 | .017 |
| 19 | 1.000 | 1.000 | 1.000 | 1.000 | 1.000 | 1.000 | 1.000 | 1.000 | .999 | .988 | .878 | .642 | .182 |

*(continued)*

**TABLE II**   *(continued)*

i. *n* = 25

| k \ p | .01 | .05 | .10 | .20 | .30 | .40 | .50 | .60 | .70 | .80 | .90 | .95 | .99 |
|---|---|---|---|---|---|---|---|---|---|---|---|---|---|
| 0 | .778 | .277 | .072 | .004 | .000 | .000 | .000 | .000 | .000 | .000 | .000 | .000 | .000 |
| 1 | .974 | .642 | .271 | .027 | .002 | .000 | .000 | .000 | .000 | .000 | .000 | .000 | .000 |
| 2 | .998 | .873 | .537 | .098 | .009 | .000 | .000 | .000 | .000 | .000 | .000 | .000 | .000 |
| 3 | 1.000 | .966 | .764 | .234 | .033 | .002 | .000 | .000 | .000 | .000 | .000 | .000 | .000 |
| 4 | 1.000 | .993 | .902 | .421 | .090 | .009 | .000 | .000 | .000 | .000 | .000 | .000 | .000 |
| 5 | 1.000 | .999 | .967 | .617 | .193 | .029 | .002 | .000 | .000 | .000 | .000 | .000 | .000 |
| 6 | 1.000 | 1.000 | .991 | .780 | .341 | .074 | .007 | .000 | .000 | .000 | .000 | .000 | .000 |
| 7 | 1.000 | 1.000 | .998 | .891 | .512 | .154 | .022 | .001 | .000 | .000 | .000 | .000 | .000 |
| 8 | 1.000 | 1.000 | 1.000 | .953 | .677 | .274 | .054 | .004 | .000 | .000 | .000 | .000 | .000 |
| 9 | 1.000 | 1.000 | 1.000 | .983 | .811 | .425 | .115 | .013 | .000 | .000 | .000 | .000 | .000 |
| 10 | 1.000 | 1.000 | 1.000 | .994 | .902 | .586 | .212 | .034 | .002 | .000 | .000 | .000 | .000 |
| 11 | 1.000 | 1.000 | 1.000 | .998 | .956 | .732 | .345 | .078 | .006 | .000 | .000 | .000 | .000 |
| 12 | 1.000 | 1.000 | 1.000 | 1.000 | .983 | .846 | .500 | .154 | .017 | .000 | .000 | .000 | .000 |
| 13 | 1.000 | 1.000 | 1.000 | 1.000 | .994 | .922 | .655 | .268 | .044 | .002 | .000 | .000 | .000 |
| 14 | 1.000 | 1.000 | 1.000 | 1.000 | .998 | .966 | .788 | .414 | .098 | .006 | .000 | .000 | .000 |
| 15 | 1.000 | 1.000 | 1.000 | 1.000 | 1.000 | .987 | .885 | .575 | .189 | .017 | .000 | .000 | .000 |
| 16 | 1.000 | 1.000 | 1.000 | 1.000 | 1.000 | .996 | .946 | .726 | .323 | .047 | .000 | .000 | .000 |
| 17 | 1.000 | 1.000 | 1.000 | 1.000 | 1.000 | .999 | .978 | .846 | .488 | .109 | .002 | .000 | .000 |
| 18 | 1.000 | 1.000 | 1.000 | 1.000 | 1.000 | 1.000 | .993 | .926 | .659 | .220 | .009 | .000 | .000 |
| 19 | 1.000 | 1.000 | 1.000 | 1.000 | 1.000 | 1.000 | .998 | .971 | .807 | .383 | .033 | .001 | .000 |
| 20 | 1.000 | 1.000 | 1.000 | 1.000 | 1.000 | 1.000 | 1.000 | .991 | .910 | .579 | .098 | .007 | .000 |
| 21 | 1.000 | 1.000 | 1.000 | 1.000 | 1.000 | 1.000 | 1.000 | .998 | .967 | .766 | .236 | .034 | .000 |
| 22 | 1.000 | 1.000 | 1.000 | 1.000 | 1.000 | 1.000 | 1.000 | 1.000 | .991 | .902 | .463 | .127 | .002 |
| 23 | 1.000 | 1.000 | 1.000 | 1.000 | 1.000 | 1.000 | 1.000 | 1.000 | .998 | .973 | .729 | .358 | .026 |
| 24 | 1.000 | 1.000 | 1.000 | 1.000 | 1.000 | 1.000 | 1.000 | 1.000 | 1.000 | .996 | .928 | .723 | .222 |

## TABLE III    Poisson Probabilities

Tabulated values are $\sum_{x=0}^{k} p(x)$. (Computations are rounded at the third decimal place.)

| λ \ k | 0 | 1 | 2 | 3 | 4 | 5 | 6 | 7 | 8 | 9 |
|---|---|---|---|---|---|---|---|---|---|---|
| .02 | .980 | 1.000 | | | | | | | | |
| .04 | .961 | .999 | 1.000 | | | | | | | |
| .06 | .942 | .998 | 1.000 | | | | | | | |
| .08 | .923 | .997 | 1.000 | | | | | | | |
| .10 | .905 | .995 | 1.000 | | | | | | | |
| .15 | .861 | .990 | .999 | 1.000 | | | | | | |
| .20 | .819 | .982 | .999 | 1.000 | | | | | | |
| .25 | .779 | .974 | .998 | 1.000 | | | | | | |
| .30 | .741 | .963 | .996 | 1.000 | | | | | | |
| .35 | .705 | .951 | .994 | 1.000 | | | | | | |
| .40 | .670 | .938 | .992 | .999 | 1.000 | | | | | |
| .45 | .638 | .925 | .989 | .999 | 1.000 | | | | | |
| .50 | .607 | .910 | .986 | .998 | 1.000 | | | | | |
| .55 | .577 | .894 | .982 | .998 | 1.000 | | | | | |
| .60 | .549 | .878 | .977 | .997 | 1.000 | | | | | |
| .65 | .522 | .861 | .972 | .996 | .999 | 1.000 | | | | |
| .70 | .497 | .844 | .966 | .994 | .999 | 1.000 | | | | |
| .75 | .472 | .827 | .959 | .993 | .999 | 1.000 | | | | |
| .80 | .449 | .809 | .953 | .991 | .999 | 1.000 | | | | |
| .85 | .427 | .791 | .945 | .989 | .998 | 1.000 | | | | |
| .90 | .407 | .772 | .937 | .987 | .998 | 1.000 | | | | |
| .95 | .387 | .754 | .929 | .981 | .997 | 1.000 | | | | |
| 1.00 | .368 | .736 | .920 | .981 | .996 | .999 | 1.000 | | | |
| 1.1 | .333 | .699 | .900 | .974 | .995 | .999 | 1.000 | | | |
| 1.2 | .301 | .663 | .879 | .966 | .992 | .998 | 1.000 | | | |
| 1.3 | .273 | .627 | .857 | .957 | .989 | .998 | 1.000 | | | |
| 1.4 | .247 | .592 | .833 | .946 | .986 | .997 | .999 | 1.000 | | |
| 1.5 | .223 | .558 | .809 | .934 | .981 | .996 | .999 | 1.000 | | |

*(continued)*

**TABLE III** *(continued)*

| λ \ k | 0 | 1 | 2 | 3 | 4 | 5 | 6 | 7 | 8 | 9 |
|------|------|------|------|------|------|------|------|------|------|------|
| 1.6 | .202 | .525 | .783 | .921 | .976 | .994 | .999 | 1.000 | | |
| 1.7 | .183 | .493 | .757 | .907 | .970 | .992 | .998 | 1.000 | | |
| 1.8 | .165 | .463 | .731 | .891 | .964 | .990 | .997 | .999 | 1.000 | |
| 1.9 | .150 | .434 | .704 | .875 | .956 | .987 | .997 | .999 | 1.000 | |
| 2.0 | .135 | .406 | .677 | .857 | .947 | .983 | .995 | .999 | 1.000 | |
| 2.2 | .111 | .355 | .623 | .819 | .928 | .975 | .993 | .998 | 1.000 | |
| 2.4 | .091 | .308 | .570 | .779 | .904 | .964 | .988 | .997 | .999 | 1.000 |
| 2.6 | .074 | .267 | .518 | .736 | .877 | .951 | .983 | .995 | .999 | 1.000 |
| 2.8 | .061 | .231 | .469 | .692 | .848 | .935 | .976 | .992 | .998 | .999 |
| 3.0 | .050 | .199 | .423 | .647 | .815 | .916 | .966 | .988 | .996 | .999 |
| 3.2 | .041 | .171 | .380 | .603 | .781 | .895 | .955 | .983 | .994 | .998 |
| 3.4 | .033 | .147 | .340 | .558 | .744 | .871 | .942 | .977 | .992 | .997 |
| 3.6 | .027 | .126 | .303 | .515 | .706 | .844 | .927 | .969 | .988 | .996 |
| 3.8 | .022 | .107 | .269 | .473 | .668 | .816 | .909 | .960 | .984 | .994 |
| 4.0 | .018 | .092 | .238 | .433 | .629 | .785 | .889 | .949 | .979 | .992 |
| 4.2 | .015 | .078 | .210 | .395 | .590 | .753 | .867 | .936 | .972 | .989 |
| 4.4 | .012 | .066 | .185 | .359 | .551 | .720 | .844 | .921 | .964 | .985 |
| 4.6 | .010 | .056 | .163 | .326 | .513 | .686 | .818 | .905 | .955 | .980 |
| 4.8 | .008 | .048 | .143 | .294 | .476 | .651 | .791 | .887 | .944 | .975 |
| 5.0 | .007 | .040 | .125 | .265 | .440 | .616 | .762 | .867 | .932 | .968 |
| 5.2 | .006 | .034 | .109 | .238 | .406 | .581 | .732 | .845 | .918 | .960 |
| 5.4 | .005 | .029 | .095 | .213 | .373 | .546 | .702 | .822 | .903 | .951 |
| 5.6 | .004 | .024 | .082 | .191 | .342 | .512 | .670 | .797 | .886 | .941 |
| 5.8 | .003 | .021 | .072 | .170 | .313 | .478 | .638 | .771 | .867 | .929 |
| 6.0 | .002 | .017 | .062 | .151 | .285 | .446 | .606 | .744 | .847 | .916 |

| λ | 10 | 11 | 12 | 13 | 14 | 15 | 16 |
|------|------|------|------|------|------|------|------|
| 2.8 | 1.000 | | | | | | |
| 3.0 | 1.000 | | | | | | |
| 3.2 | 1.000 | | | | | | |
| 3.4 | .999 | 1.000 | | | | | |
| 3.6 | .999 | 1.000 | | | | | |
| 3.8 | .998 | .999 | 1.000 | | | | |
| 4.0 | .997 | .999 | 1.000 | | | | |
| 4.2 | .996 | .999 | 1.000 | | | | |
| 4.4 | .994 | .998 | .999 | 1.000 | | | |
| 4.6 | .992 | .997 | .999 | 1.000 | | | |
| 4.8 | .990 | .996 | .999 | 1.000 | | | |
| 5.0 | .986 | .995 | .998 | .999 | 1.000 | | |
| 5.2 | .982 | .993 | .997 | .999 | 1.000 | | |
| 5.4 | .977 | .990 | .996 | .999 | 1.000 | | |
| 5.6 | .972 | .988 | .995 | .998 | .999 | 1.000 | |
| 5.8 | .965 | .984 | .993 | .997 | .999 | 1.000 | |
| 6.0 | .957 | .980 | .991 | .996 | .999 | .999 | 1.000 |

*(continued)*

**TABLE III**    *(continued)*

| λ \ k | 0 | 1 | 2 | 3 | 4 | 5 | 6 | 7 | 8 | 9 |
|---|---|---|---|---|---|---|---|---|---|---|
| 6.2 | .002 | .015 | .054 | .134 | .259 | .414 | .574 | .716 | .826 | .902 |
| 6.4 | .002 | .012 | .046 | .119 | .235 | .384 | .542 | .687 | .803 | .886 |
| 6.6 | .001 | .010 | .040 | .105 | .213 | .355 | .511 | .658 | .780 | .869 |
| 6.8 | .001 | .009 | .034 | .093 | .192 | .327 | .480 | .628 | .755 | .850 |
| 7.0 | .001 | .007 | .030 | .082 | .173 | .301 | .450 | .599 | .729 | .830 |
| 7.2 | .001 | .006 | .025 | .072 | .156 | .276 | .420 | .569 | .703 | .810 |
| 7.4 | .001 | .005 | .022 | .063 | .140 | .253 | .392 | .539 | .676 | .788 |
| 7.6 | .001 | .004 | .019 | .055 | .125 | .231 | .365 | .510 | .648 | .765 |
| 7.8 | .000 | .004 | .016 | .048 | .112 | .210 | .338 | .481 | .620 | .741 |
| 8.0 | .000 | .003 | .014 | .042 | .100 | .191 | .313 | .453 | .593 | .717 |
| 8.5 | .000 | .002 | .009 | .030 | .074 | .150 | .256 | .386 | .523 | .653 |
| 9.0 | .000 | .001 | .006 | .021 | .055 | .116 | .207 | .324 | .456 | .587 |
| 9.5 | .000 | .001 | .004 | .015 | .040 | .089 | .165 | .269 | .392 | .522 |
| 10.0 | .000 | .000 | .003 | .010 | .029 | .067 | .130 | .220 | .333 | .458 |

| λ | 10 | 11 | 12 | 13 | 14 | 15 | 16 | 17 | 18 | 19 |
|---|---|---|---|---|---|---|---|---|---|---|
| 6.2 | .949 | .975 | .989 | .995 | .998 | .999 | 1.000 | | | |
| 6.4 | .939 | .969 | .986 | .994 | .997 | .999 | 1.000 | | | |
| 6.6 | .927 | .963 | .982 | .992 | .997 | .999 | .999 | 1.000 | | |
| 6.8 | .915 | .955 | .978 | .990 | .996 | .998 | .999 | 1.000 | | |
| 7.0 | .901 | .947 | .973 | .987 | .994 | .998 | .999 | 1.000 | | |
| 7.2 | .887 | .937 | .967 | .984 | .993 | .997 | .999 | .999 | 1.000 | |
| 7.4 | .871 | .926 | .961 | .980 | .991 | .996 | .998 | .999 | 1.000 | |
| 7.6 | .854 | .915 | .954 | .976 | .989 | .995 | .998 | .999 | 1.000 | |
| 7.8 | .835 | .902 | .945 | .971 | .986 | .993 | .997 | .999 | 1.000 | |
| 8.0 | .816 | .888 | .936 | .966 | .983 | .992 | .996 | .998 | .999 | 1.000 |
| 8.5 | .763 | .849 | .909 | .949 | .973 | .986 | .993 | .997 | .999 | .999 |
| 9.0 | .706 | .803 | .876 | .926 | .959 | .978 | .989 | .995 | .998 | .999 |
| 9.5 | .645 | .752 | .836 | .898 | .940 | .967 | .982 | .991 | .996 | .998 |
| 10.0 | .583 | .697 | .792 | .864 | .917 | .951 | .973 | .986 | .993 | .997 |

| λ | 20 | 21 | 22 |
|---|---|---|---|
| 8.5 | 1.000 | | |
| 9.0 | 1.000 | | |
| 9.5 | .999 | 1.000 | |
| 10.0 | .998 | .999 | 1.000 |

*(continued)*

**TABLE III** *(continued)*

| λ \ k | 0 | 1 | 2 | 3 | 4 | 5 | 6 | 7 | 8 | 9 |
|---|---|---|---|---|---|---|---|---|---|---|
| 10.5 | .000 | .000 | .002 | .007 | .021 | .050 | .102 | .179 | .279 | .397 |
| 11.0 | .000 | .000 | .001 | .005 | .015 | .038 | .079 | .143 | .232 | .341 |
| 11.5 | .000 | .000 | .001 | .003 | .011 | .028 | .060 | .114 | .191 | .289 |
| 12.0 | .000 | .000 | .001 | .002 | .008 | .020 | .046 | .090 | .155 | .242 |
| 12.5 | .000 | .000 | .000 | .002 | .005 | .015 | .035 | .070 | .125 | .201 |
| 13.0 | .000 | .000 | .000 | .001 | .004 | .011 | .026 | .054 | .100 | .166 |
| 13.5 | .000 | .000 | .000 | .001 | .003 | .008 | .019 | .041 | .079 | .135 |
| 14.0 | .000 | .000 | .000 | .000 | .002 | .006 | .014 | .032 | .062 | .109 |
| 14.5 | .000 | .000 | .000 | .000 | .001 | .004 | .010 | .024 | .048 | .088 |
| 15.0 | .000 | .000 | .000 | .000 | .001 | .003 | .008 | .018 | .037 | .070 |

| | 10 | 11 | 12 | 13 | 14 | 15 | 16 | 17 | 18 | 19 |
|---|---|---|---|---|---|---|---|---|---|---|
| 10.5 | .521 | .639 | .742 | .825 | .888 | .932 | .960 | .978 | .988 | .994 |
| 11.0 | .460 | .579 | .689 | .781 | .854 | .907 | .944 | .968 | .982 | .991 |
| 11.5 | .402 | .520 | .633 | .733 | .815 | .878 | .924 | .954 | .974 | .986 |
| 12.0 | .347 | .462 | .576 | .682 | .772 | .844 | .899 | .937 | .963 | .979 |
| 12.5 | .297 | .406 | .519 | .628 | .725 | .806 | .869 | .916 | .948 | .969 |
| 13.0 | .252 | .353 | .463 | .573 | .675 | .764 | .835 | .890 | .930 | .957 |
| 13.5 | .211 | .304 | .409 | .518 | .623 | .718 | .798 | .861 | .908 | .942 |
| 14.0 | .176 | .260 | .358 | .464 | .570 | .669 | .756 | .827 | .883 | .923 |
| 14.5 | .145 | .220 | .311 | .413 | .518 | .619 | .711 | .790 | .853 | .901 |
| 15.0 | .118 | .185 | .268 | .363 | .466 | .568 | .664 | .749 | .819 | .875 |

| | 20 | 21 | 22 | 23 | 24 | 25 | 26 | 27 | 28 | 29 |
|---|---|---|---|---|---|---|---|---|---|---|
| 10.5 | .997 | .999 | .999 | 1.000 | | | | | | |
| 11.0 | .995 | .998 | .999 | 1.000 | | | | | | |
| 11.5 | .992 | .996 | .998 | .999 | 1.000 | | | | | |
| 12.0 | .988 | .994 | .987 | .999 | .999 | 1.000 | | | | |
| 12.5 | .983 | .991 | .995 | .998 | .999 | .999 | 1.000 | | | |
| 13.0 | .975 | .986 | .992 | .996 | .998 | .999 | 1.000 | | | |
| 13.5 | .965 | .980 | .989 | .994 | .997 | .998 | .999 | 1.000 | | |
| 14.0 | .952 | .971 | .983 | .991 | .995 | .997 | .999 | .999 | 1.000 | |
| 14.5 | .936 | .960 | .976 | .986 | .992 | .996 | .998 | .999 | .999 | 1.000 |
| 15.0 | .917 | .947 | .967 | .981 | .989 | .994 | .997 | .998 | .999 | 1.000 |

*(continued)*

**TABLE III**   *(continued)*

| λ \ k | 0 | 1 | 2 | 3 | 4 | 5 | 6 | 7 | 8 | 9 |
|---|---|---|---|---|---|---|---|---|---|---|
| 16 | .000 | .001 | .004 | .010 | .022 | .043 | .077 | .127 | .193 | .275 |
| 17 | .000 | .001 | .002 | .005 | .013 | .026 | .049 | .085 | .135 | .201 |
| 18 | .000 | .000 | .001 | .003 | .007 | .015 | .030 | .055 | .092 | .143 |
| 19 | .000 | .000 | .001 | .002 | .004 | .009 | .018 | .035 | .061 | .098 |
| 20 | .000 | .000 | .000 | .001 | .002 | .005 | .011 | .021 | .039 | .066 |
| 21 | .000 | .000 | .000 | .000 | .001 | .003 | .006 | .013 | .025 | .043 |
| 22 | .000 | .000 | .000 | .000 | .001 | .002 | .004 | .008 | .015 | .028 |
| 23 | .000 | .000 | .000 | .000 | .000 | .001 | .002 | .004 | .009 | .017 |
| 24 | .000 | .000 | .000 | .000 | .000 | .000 | .001 | .003 | .005 | .011 |
| 25 | .000 | .000 | .000 | .000 | .000 | .000 | .001 | .001 | .003 | .006 |

| λ | 14 | 15 | 16 | 17 | 18 | 19 | 20 | 21 | 22 | 23 |
|---|---|---|---|---|---|---|---|---|---|---|
| 16 | .368 | .467 | .566 | .659 | .742 | .812 | .868 | .911 | .942 | .963 |
| 17 | .281 | .371 | .468 | .564 | .655 | .736 | .805 | .861 | .905 | .937 |
| 18 | .208 | .287 | .375 | .469 | .562 | .651 | .731 | .799 | .855 | .899 |
| 19 | .150 | .215 | .292 | .378 | .469 | .561 | .647 | .725 | .793 | .849 |
| 20 | .105 | .157 | .221 | .297 | .381 | .470 | .559 | .644 | .721 | .787 |
| 21 | .072 | .111 | .163 | .227 | .302 | .384 | .471 | .558 | .640 | .716 |
| 22 | .048 | .077 | .117 | .169 | .232 | .306 | .387 | .472 | .556 | .637 |
| 23 | .031 | .052 | .082 | .123 | .175 | .238 | .310 | .389 | .472 | .555 |
| 24 | .020 | .034 | .056 | .087 | .128 | .180 | .243 | .314 | .392 | .473 |
| 25 | .012 | .022 | .038 | .060 | .092 | .134 | .185 | .247 | .318 | .394 |

| λ | 24 | 25 | 26 | 27 | 28 | 29 | 30 | 31 | 32 | 33 |
|---|---|---|---|---|---|---|---|---|---|---|
| 16 | .978 | .987 | .993 | .996 | .998 | .999 | .999 | 1.000 |  |  |
| 17 | .959 | .975 | .985 | .991 | .995 | .997 | .999 | .999 | 1.000 |  |
| 18 | .932 | .955 | .972 | .983 | .990 | .994 | .997 | .998 | .999 | 1.000 |
| 19 | .893 | .927 | .951 | .969 | .980 | .988 | .993 | .996 | .998 | .999 |
| 20 | .843 | .888 | .922 | .948 | .966 | .978 | .987 | .992 | .995 | .997 |
| 21 | .782 | .838 | .883 | .917 | .944 | .963 | .976 | .985 | .991 | .994 |
| 22 | .712 | .777 | .832 | .877 | .913 | .940 | .959 | .973 | .983 | .989 |
| 23 | .635 | .708 | .772 | .827 | .873 | .908 | .936 | .956 | .971 | .981 |
| 24 | .554 | .632 | .704 | .768 | .823 | .868 | .904 | .932 | .953 | .969 |
| 25 | .473 | .553 | .629 | .700 | .763 | .818 | .863 | .900 | .929 | .950 |

| λ | 34 | 35 | 36 | 37 | 38 | 39 | 40 | 41 | 42 | 43 |
|---|---|---|---|---|---|---|---|---|---|---|
| 19 | .999 | 1.000 |  |  |  |  |  |  |  |  |
| 20 | .999 | .999 | 1.000 |  |  |  |  |  |  |  |
| 21 | .997 | .998 | .999 | .999 | 1.000 |  |  |  |  |  |
| 22 | .994 | .996 | .998 | .999 | .999 | 1.000 |  |  |  |  |
| 23 | .988 | .993 | .996 | .997 | .999 | .999 | 1.000 |  |  |  |
| 24 | .979 | .987 | .992 | .995 | .997 | .998 | .999 | .999 | 1.000 |  |
| 25 | .966 | .978 | .985 | .991 | .991 | .997 | .998 | .999 | .999 | 1.000 |

## TABLE IV   Normal Curve Areas

| z | .00 | .01 | .02 | .03 | .04 | .05 | .06 | .07 | .08 | .09 |
|---|-----|-----|-----|-----|-----|-----|-----|-----|-----|-----|
| .0 | .0000 | .0040 | .0080 | .0120 | .0160 | .0199 | .0239 | .0279 | .0319 | .0359 |
| .1 | .0398 | .0438 | .0478 | .0517 | .0557 | .0596 | .0636 | .0675 | .0714 | .0753 |
| .2 | .0793 | .0832 | .0871 | .0910 | .0948 | .0987 | .1026 | .1064 | .1103 | .1141 |
| .3 | .1179 | .1217 | .1255 | .1293 | .1331 | .1368 | .1406 | .1443 | .1480 | .1517 |
| .4 | .1554 | .1591 | .1628 | .1664 | .1700 | .1736 | .1772 | .1808 | .1844 | .1879 |
| .5 | .1915 | .1950 | .1985 | .2019 | .2054 | .2088 | .2123 | .2157 | .2190 | .2224 |
| .6 | .2257 | .2291 | .2324 | .2357 | .2389 | .2422 | .2454 | .2486 | .2517 | .2549 |
| .7 | .2580 | .2611 | .2642 | .2673 | .2704 | .2734 | .2764 | .2794 | .2823 | .2852 |
| .8 | .2881 | .2910 | .2939 | .2967 | .2995 | .3023 | .3051 | .3078 | .3106 | .3133 |
| .9 | .3159 | .3186 | .3212 | .3238 | .3264 | .3289 | .3315 | .3340 | .3365 | .3389 |
| 1.0 | .3413 | .3438 | .3461 | .3485 | .3508 | .3531 | .3554 | .3577 | .3599 | .3621 |
| 1.1 | .3643 | .3665 | .3686 | .3708 | .3729 | .3749 | .3770 | .3790 | .3810 | .3830 |
| 1.2 | .3849 | .3869 | .3888 | .3907 | .3925 | .3944 | .3962 | .3980 | .3997 | .4015 |
| 1.3 | .4032 | .4049 | .4066 | .4082 | .4099 | .4115 | .4131 | .4147 | .4162 | .4177 |
| 1.4 | .4192 | .4207 | .4222 | .4236 | .4251 | .4265 | .4279 | .4292 | .4306 | .4319 |
| 1.5 | .4332 | .4345 | .4357 | .4370 | .4382 | .4394 | .4406 | .4418 | .4429 | .4441 |
| 1.6 | .4452 | .4463 | .4474 | .4484 | .4495 | .4505 | .4515 | .4525 | .4535 | .4545 |
| 1.7 | .4554 | .4564 | .4573 | .4582 | .4591 | .4599 | .4608 | .4616 | .4625 | .4633 |
| 1.8 | .4641 | .4649 | .4656 | .4664 | .4671 | .4678 | .4686 | .4693 | .4699 | .4706 |
| 1.9 | .4713 | .4719 | .4726 | .4732 | .4738 | .4744 | .4750 | .4756 | .4761 | .4767 |
| 2.0 | .4772 | .4778 | .4783 | .4788 | .4793 | .4798 | .4803 | .4808 | .4812 | .4817 |
| 2.1 | .4821 | .4826 | .4830 | .4834 | .4838 | .4842 | .4846 | .4850 | .4854 | .4857 |
| 2.2 | .4861 | .4864 | .4868 | .4871 | .4875 | .4878 | .4881 | .4884 | .4887 | .4890 |
| 2.3 | .4893 | .4896 | .4898 | .4901 | .4904 | .4906 | .4909 | .4911 | .4913 | .4916 |
| 2.4 | .4918 | .4920 | .4922 | .4925 | .4927 | .4929 | .4931 | .4932 | .4934 | .4936 |
| 2.5 | .4938 | .4940 | .4941 | .4943 | .4945 | .4946 | .4948 | .4949 | .4951 | .4952 |
| 2.6 | .4953 | .4955 | .4956 | .4957 | .4959 | .4960 | .4961 | .4962 | .4963 | .4964 |
| 2.7 | .4965 | .4966 | .4967 | .4968 | .4969 | .4970 | .4971 | .4972 | .4973 | .4974 |
| 2.8 | .4974 | .4975 | .4976 | .4977 | .4977 | .4978 | .4979 | .4979 | .4980 | .4981 |
| 2.9 | .4981 | .4982 | .4982 | .4983 | .4984 | .4984 | .4985 | .4985 | .4986 | .4986 |
| 3.0 | .4987 | .4987 | .4987 | .4988 | .4988 | .4989 | .4989 | .4989 | .4990 | .4990 |

*Source:* Abridged from Table I of A. Hald, *Statistical Tables and Formulas* (New York: Wiley), 1952. Reproduced by permission of A. Hald.

## TABLE V  Exponentials

| $\lambda$ | $e^{-\lambda}$ | $\lambda$ | $e^{-\lambda}$ | $\lambda$ | $e^{-\lambda}$ | $\lambda$ | $e^{-\lambda}$ | $\lambda$ | $e^{-\lambda}$ |
|---|---|---|---|---|---|---|---|---|---|
| .00 | 1.000000 | 2.05 | .128735 | 4.05 | .017422 | 6.05 | .002358 | 8.05 | .000319 |
| .05 | .951229 | 2.10 | .122456 | 4.10 | .016573 | 6.10 | .002243 | 8.10 | .000304 |
| .10 | .904837 | 2.15 | .116484 | 4.15 | .015764 | 6.15 | .002133 | 8.15 | .000289 |
| .15 | .860708 | 2.20 | .110803 | 4.20 | .014996 | 6.20 | .002029 | 8.20 | .000275 |
| .20 | .818731 | 2.25 | .105399 | 4.25 | .014264 | 6.25 | .001930 | 8.25 | .000261 |
| .25 | .778801 | 2.30 | .100259 | 4.30 | .013569 | 6.30 | .001836 | 8.30 | .000249 |
| .30 | .740818 | 2.35 | .095369 | 4.35 | .012907 | 6.35 | .001747 | 8.35 | .000236 |
| .35 | .704688 | 2.40 | .090718 | 4.40 | .012277 | 6.40 | .001661 | 8.40 | .000225 |
| .40 | .670320 | 2.45 | .086294 | 4.45 | .011679 | 6.45 | .001581 | 8.45 | .000214 |
| .45 | .637628 | 2.50 | .082085 | 4.50 | .011109 | 6.50 | .001503 | 8.50 | .000204 |
| .50 | .606531 | 2.55 | .078082 | 4.55 | .010567 | 6.55 | .001430 | 8.55 | .000194 |
| .55 | .576950 | 2.60 | .074274 | 4.60 | .010052 | 6.60 | .001360 | 8.60 | .000184 |
| .60 | .548812 | 2.65 | .070651 | 4.65 | .009562 | 6.65 | .001294 | 8.65 | .000175 |
| .65 | .522046 | 2.70 | .067206 | 4.70 | .009095 | 6.70 | .001231 | 8.70 | .000167 |
| .70 | .496585 | 2.75 | .063928 | 4.75 | .008652 | 6.75 | .001171 | 8.75 | .000158 |
| .75 | .472367 | 2.80 | .060810 | 4.80 | .008230 | 6.80 | .001114 | 8.80 | .000151 |
| .80 | .449329 | 2.85 | .057844 | 4.85 | .007828 | 6.85 | .001059 | 8.85 | .000143 |
| .85 | .427415 | 2.90 | .055023 | 4.90 | .007447 | 6.90 | .001008 | 8.90 | .000136 |
| .90 | .406570 | 2.95 | .052340 | 4.95 | .007083 | 6.95 | .000959 | 8.95 | .000130 |
| .95 | .386741 | 3.00 | .049787 | 5.00 | .006738 | 7.00 | .000912 | 9.00 | .000123 |
| 1.00 | .367879 | 3.05 | .047359 | 5.05 | .006409 | 7.05 | .000867 | 9.05 | .000117 |
| 1.05 | .349938 | 3.10 | .045049 | 5.10 | .006097 | 7.10 | .000825 | 9.10 | .000112 |
| 1.10 | .332871 | 3.15 | .042852 | 5.15 | .005799 | 7.15 | .000785 | 9.15 | .000106 |
| 1.15 | .316637 | 3.20 | .040762 | 5.20 | .005517 | 7.20 | .000747 | 9.20 | .000101 |
| 1.20 | .301194 | 3.25 | .038774 | 5.25 | .005248 | 7.25 | .000710 | 9.25 | .000096 |
| 1.25 | .286505 | 3.30 | .036883 | 5.30 | .004992 | 7.30 | .000676 | 9.30 | .000091 |
| 1.30 | .272532 | 3.35 | .035084 | 5.35 | .004748 | 7.35 | .000643 | 9.35 | .000087 |
| 1.35 | .259240 | 3.40 | .033373 | 5.40 | .004517 | 7.40 | .000611 | 9.40 | .000083 |
| 1.40 | .246597 | 3.45 | .031746 | 5.45 | .004296 | 7.45 | .000581 | 9.45 | .000079 |
| 1.45 | .234570 | 3.50 | .030197 | 5.50 | .004087 | 7.50 | .000553 | 9.50 | .000075 |
| 1.50 | .223130 | 3.55 | .028725 | 5.55 | .003887 | 7.55 | .000526 | 9.55 | .000071 |
| 1.55 | .212248 | 3.60 | .027324 | 5.60 | .003698 | 7.60 | .000501 | 9.60 | .000068 |
| 1.60 | .201897 | 3.65 | .025991 | 5.65 | .003518 | 7.65 | .000476 | 9.65 | .000064 |
| 1.65 | .192050 | 3.70 | .024724 | 5.70 | .003346 | 7.70 | .000453 | 9.70 | .000061 |
| 1.70 | .182684 | 3.75 | .023518 | 5.75 | .003183 | 7.75 | .000431 | 9.75 | .000058 |
| 1.75 | .173774 | 3.80 | .022371 | 5.80 | .003028 | 7.80 | .000410 | 9.80 | .000056 |
| 1.80 | .165299 | 3.85 | .021280 | 5.85 | .002880 | 7.85 | .000390 | 9.85 | .000053 |
| 1.85 | .157237 | 3.90 | .020242 | 5.90 | .002739 | 7.90 | .000371 | 9.90 | .000050 |
| 1.90 | .149569 | 3.95 | .019255 | 5.95 | .002606 | 7.95 | .000353 | 9.95 | .000048 |
| 1.95 | .142274 | 4.00 | .018316 | 6.00 | .002479 | 8.00 | .000336 | 10.00 | .000045 |
| 2.00 | .135335 | | | | | | | | |

**TABLE VI   Critical Values of *t***

| $v$ | $t_{.100}$ | $t_{.050}$ | $t_{.025}$ | $t_{.010}$ | $t_{.005}$ | $t_{.001}$ | $t_{.0005}$ |
|---|---|---|---|---|---|---|---|
| 1 | 3.078 | 6.314 | 12.706 | 31.821 | 63.657 | 318.31 | 636.62 |
| 2 | 1.886 | 2.920 | 4.303 | 6.965 | 9.925 | 22.326 | 31.598 |
| 3 | 1.638 | 2.353 | 3.182 | 4.541 | 5.841 | 10.213 | 12.924 |
| 4 | 1.533 | 2.132 | 2.776 | 3.747 | 4.604 | 7.173 | 8.610 |
| 5 | 1.476 | 2.015 | 2.571 | 3.365 | 4.032 | 5.893 | 6.869 |
| 6 | 1.440 | 1.943 | 2.447 | 3.143 | 3.707 | 5.208 | 5.959 |
| 7 | 1.415 | 1.895 | 2.365 | 2.998 | 3.499 | 4.785 | 5.408 |
| 8 | 1.397 | 1.860 | 2.306 | 2.896 | 3.355 | 4.501 | 5.041 |
| 9 | 1.383 | 1.833 | 2.262 | 2.821 | 3.250 | 4.297 | 4.781 |
| 10 | 1.372 | 1.812 | 2.228 | 2.764 | 3.169 | 4.144 | 4.587 |
| 11 | 1.363 | 1.796 | 2.201 | 2.718 | 3.106 | 4.025 | 4.437 |
| 12 | 1.356 | 1.782 | 2.179 | 2.681 | 3.055 | 3.930 | 4.318 |
| 13 | 1.350 | 1.771 | 2.160 | 2.650 | 3.012 | 3.852 | 4.221 |
| 14 | 1.345 | 1.761 | 2.145 | 2.624 | 2.977 | 3.787 | 4.140 |
| 15 | 1.341 | 1.753 | 2.131 | 2.602 | 2.947 | 3.733 | 4.073 |
| 16 | 1.337 | 1.746 | 2.120 | 2.583 | 2.921 | 3.686 | 4.015 |
| 17 | 1.333 | 1.740 | 2.110 | 2.567 | 2.898 | 3.646 | 3.965 |
| 18 | 1.330 | 1.734 | 2.101 | 2.552 | 2.878 | 3.610 | 3.922 |
| 19 | 1.328 | 1.729 | 2.093 | 2.539 | 2.861 | 3.579 | 3.883 |
| 20 | 1.325 | 1.725 | 2.086 | 2.528 | 2.845 | 3.552 | 3.850 |
| 21 | 1.323 | 1.721 | 2.080 | 2.518 | 2.831 | 3.527 | 3.819 |
| 22 | 1.321 | 1.717 | 2.074 | 2.508 | 2.819 | 3.505 | 3.792 |
| 23 | 1.319 | 1.714 | 2.069 | 2.500 | 2.807 | 3.485 | 3.767 |
| 24 | 1.318 | 1.711 | 2.064 | 2.492 | 2.797 | 3.467 | 3.745 |
| 25 | 1.316 | 1.708 | 2.060 | 2.485 | 2.787 | 3.450 | 3.725 |
| 26 | 1.315 | 1.706 | 2.056 | 2.479 | 2.779 | 3.435 | 3.707 |
| 27 | 1.314 | 1.703 | 2.052 | 2.473 | 2.771 | 3.421 | 3.690 |
| 28 | 1.313 | 1.701 | 2.048 | 2.467 | 2.763 | 3.408 | 3.674 |
| 29 | 1.311 | 1.699 | 2.045 | 2.462 | 2.756 | 3.396 | 3.659 |
| 30 | 1.310 | 1.697 | 2.042 | 2.457 | 2.750 | 3.385 | 3.646 |
| 40 | 1.303 | 1.684 | 2.021 | 2.423 | 2.704 | 3.307 | 3.551 |
| 60 | 1.296 | 1.671 | 2.000 | 2.390 | 2.660 | 3.232 | 3.460 |
| 120 | 1.289 | 1.658 | 1.980 | 2.358 | 2.617 | 3.160 | 3.373 |
| ∞ | 1.282 | 1.645 | 1.960 | 2.326 | 2.576 | 3.090 | 3.291 |

**TABLE VII   Critical Values of $\chi^2$**

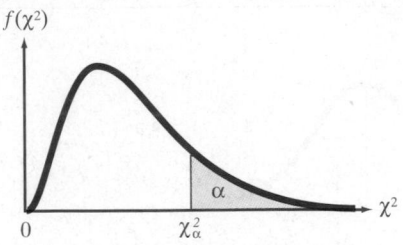

| Degrees of Freedom | $\chi^2_{.995}$ | $\chi^2_{.990}$ | $\chi^2_{.975}$ | $\chi^2_{.950}$ | $\chi^2_{.900}$ |
|---|---|---|---|---|---|
| 1 | .0000393 | .0001571 | .0009821 | .0039321 | .0157908 |
| 2 | .0100251 | .0201007 | .0506356 | .102587 | .210720 |
| 3 | .0717212 | .114832 | .215795 | .351846 | .584375 |
| 4 | .206990 | .297110 | .484419 | .710721 | 1.063623 |
| 5 | .411740 | .554300 | .831211 | 1.145476 | 1.61031 |
| 6 | .675727 | .872085 | 1.237347 | 1.63539 | 2.20413 |
| 7 | .989265 | 1.239043 | 1.68987 | 2.16735 | 2.83311 |
| 8 | 1.344419 | 1.646482 | 2.17973 | 2.73264 | 3.48954 |
| 9 | 1.734926 | 2.087912 | 2.70039 | 3.32511 | 4.16816 |
| 10 | 2.15585 | 2.55821 | 3.24697 | 3.94030 | 4.86518 |
| 11 | 2.60321 | 3.05347 | 3.81575 | 4.57481 | 5.57779 |
| 12 | 3.07382 | 3.57056 | 4.40379 | 5.22603 | 6.30380 |
| 13 | 3.56503 | 4.10691 | 5.00874 | 5.89186 | 7.04150 |
| 14 | 4.07468 | 4.66043 | 5.62872 | 6.57063 | 7.78953 |
| 15 | 4.60094 | 5.22935 | 6.26214 | 7.26094 | 8.54675 |
| 16 | 5.14224 | 5.81221 | 6.90766 | 7.96164 | 9.31223 |
| 17 | 5.69724 | 6.40776 | 7.56418 | 8.67176 | 10.0852 |
| 18 | 6.26481 | 7.01491 | 8.23075 | 9.39046 | 10.8649 |
| 19 | 6.84398 | 7.63273 | 8.90655 | 10.1170 | 11.6509 |
| 20 | 7.43386 | 8.26040 | 9.59083 | 10.8508 | 12.4426 |
| 21 | 8.03366 | 8.89720 | 10.28293 | 11.5913 | 13.2396 |
| 22 | 8.64272 | 9.54249 | 10.9823 | 12.3380 | 14.0415 |
| 23 | 9.26042 | 10.19567 | 11.6885 | 13.0905 | 14.8479 |
| 24 | 9.88623 | 10.8564 | 12.4011 | 13.8484 | 15.6587 |
| 25 | 10.5197 | 11.5240 | 13.1197 | 14.6114 | 16.4734 |
| 26 | 11.1603 | 12.1981 | 13.8439 | 15.3791 | 17.2919 |
| 27 | 11.8076 | 12.8786 | 14.5733 | 16.1513 | 18.1138 |
| 28 | 12.4613 | 13.5648 | 15.3079 | 16.9279 | 18.9392 |
| 29 | 13.1211 | 14.2565 | 16.0471 | 17.7083 | 19.7677 |
| 30 | 13.7867 | 14.9535 | 16.7908 | 18.4926 | 20.5992 |
| 40 | 20.7065 | 22.1643 | 24.4331 | 26.5093 | 29.0505 |
| 50 | 27.9907 | 29.7067 | 32.3574 | 34.7642 | 37.6886 |
| 60 | 35.5346 | 37.4848 | 40.4817 | 43.1879 | 46.4589 |
| 70 | 43.2752 | 45.4418 | 48.7576 | 51.7393 | 55.3290 |
| 80 | 51.1720 | 53.5400 | 57.1532 | 60.3915 | 64.2778 |
| 90 | 59.1963 | 61.7541 | 65.6466 | 69.1260 | 73.2912 |
| 100 | 67.3276 | 70.0648 | 74.2219 | 77.9295 | 82.3581 |

*Source:* From C. M. Thompson, "Tables of the Percentage Points of the $\chi^2$-Distribution," *Biometrika*, 1941, 32, 188–189. Reproduced by permission of the *Biometrika* Trustees.

*(continued)*

**TABLE VII**   *(continued)*

| Degrees of Freedom | $\chi^2_{.100}$ | $\chi^2_{.050}$ | $\chi^2_{.025}$ | $\chi^2_{.010}$ | $\chi^2_{.005}$ |
|---|---|---|---|---|---|
| *1* | 2.70554 | 3.84146 | 5.02389 | 6.63490 | 7.87944 |
| *2* | 4.60517 | 5.99147 | 7.37776 | 9.21034 | 10.5966 |
| *3* | 6.25139 | 7.81473 | 9.34840 | 11.3449 | 12.8381 |
| *4* | 7.77944 | 9.48773 | 11.1433 | 13.2767 | 14.8602 |
| *5* | 9.23635 | 11.0705 | 12.8325 | 15.0863 | 16.7496 |
| *6* | 10.6446 | 12.5916 | 14.4494 | 16.8119 | 18.5476 |
| *7* | 12.0170 | 14.0671 | 16.0128 | 18.4753 | 20.2777 |
| *8* | 13.3616 | 15.5073 | 17.5346 | 20.0902 | 21.9550 |
| *9* | 14.6837 | 16.9190 | 19.0228 | 21.6660 | 23.5893 |
| *10* | 15.9871 | 18.3070 | 20.4831 | 23.2093 | 25.1882 |
| *11* | 17.2750 | 19.6751 | 21.9200 | 24.7250 | 26.7569 |
| *12* | 18.5494 | 21.0261 | 23.3367 | 26.2170 | 28.2995 |
| *13* | 19.8119 | 22.3621 | 24.7356 | 27.6883 | 29.8194 |
| *14* | 21.0642 | 23.6848 | 26.1190 | 29.1413 | 31.3193 |
| *15* | 22.3072 | 24.9958 | 27.4884 | 30.5779 | 32.8013 |
| *16* | 23.5418 | 26.2962 | 28.8454 | 31.9999 | 34.2672 |
| *17* | 24.7690 | 27.5871 | 30.1910 | 33.4087 | 35.7185 |
| *18* | 25.9894 | 28.8693 | 31.5264 | 34.8053 | 37.1564 |
| *19* | 27.2036 | 30.1435 | 32.8523 | 36.1908 | 38.5822 |
| *20* | 28.4120 | 31.4104 | 34.1696 | 37.5662 | 39.9968 |
| *21* | 29.6151 | 32.6705 | 35.4789 | 38.9321 | 41.4010 |
| *22* | 30.8133 | 33.9244 | 36.7807 | 40.2894 | 42.7956 |
| *23* | 32.0069 | 35.1725 | 38.0757 | 41.6384 | 44.1813 |
| *24* | 33.1963 | 36.4151 | 39.3641 | 42.9798 | 45.5585 |
| *25* | 34.3816 | 37.6525 | 40.6465 | 44.3141 | 46.9278 |
| *26* | 35.5631 | 38.8852 | 41.9232 | 45.6417 | 48.2899 |
| *27* | 36.7412 | 40.1133 | 43.1944 | 46.9630 | 49.6449 |
| *28* | 37.9159 | 41.3372 | 44.4607 | 48.2782 | 50.9933 |
| *29* | 39.0875 | 42.5569 | 45.7222 | 49.5879 | 52.3356 |
| *30* | 40.2560 | 43.7729 | 46.9792 | 50.8922 | 53.6720 |
| *40* | 51.8050 | 55.7585 | 59.3417 | 63.6907 | 66.7659 |
| *50* | 63.1671 | 67.5048 | 71.4202 | 76.1539 | 79.4900 |
| *60* | 74.3970 | 79.0819 | 83.2976 | 88.3794 | 91.9517 |
| *70* | 85.5271 | 90.5312 | 95.0231 | 100.425 | 104.215 |
| *80* | 96.5782 | 101.879 | 106.629 | 112.329 | 116.321 |
| *90* | 107.565 | 113.145 | 118.136 | 124.116 | 128.299 |
| *100* | 118.498 | 124.342 | 129.561 | 135.807 | 140.169 |

**TABLE VIII    Percentage Points of the *F*-Distribution, $\alpha = .10$**

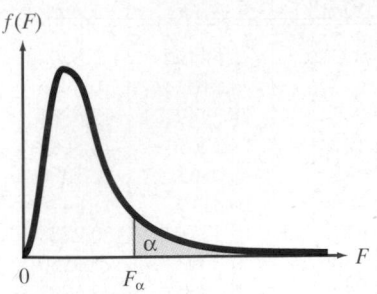

| $v_2$ \ $v_1$ | NUMERATOR DEGREES OF FREEDOM | | | | | | | | |
|---|---|---|---|---|---|---|---|---|---|
| | **1** | **2** | **3** | **4** | **5** | **6** | **7** | **8** | **9** |
| *1* | 39.86 | 49.50 | 53.59 | 55.83 | 57.24 | 58.20 | 58.91 | 59.44 | 59.86 |
| *2* | 8.53 | 9.00 | 9.16 | 9.24 | 9.29 | 9.33 | 9.35 | 9.37 | 9.38 |
| *3* | 5.54 | 5.46 | 5.39 | 5.34 | 5.31 | 5.28 | 5.27 | 5.25 | 5.24 |
| *4* | 4.54 | 4.32 | 4.19 | 4.11 | 4.05 | 4.01 | 3.98 | 3.95 | 3.94 |
| *5* | 4.06 | 3.78 | 3.62 | 3.52 | 3.45 | 3.40 | 3.37 | 3.34 | 3.32 |
| *6* | 3.78 | 3.46 | 3.29 | 3.18 | 3.11 | 3.05 | 3.01 | 2.98 | 2.96 |
| *7* | 3.59 | 3.26 | 3.07 | 2.96 | 2.88 | 2.83 | 2.78 | 2.75 | 2.72 |
| *8* | 3.46 | 3.11 | 2.92 | 2.81 | 2.73 | 2.67 | 2.62 | 2.59 | 2.56 |
| *9* | 3.36 | 3.01 | 2.81 | 2.69 | 2.61 | 2.55 | 2.51 | 2.47 | 2.44 |
| *10* | 3.29 | 2.92 | 2.73 | 2.61 | 2.52 | 2.46 | 2.41 | 2.38 | 2.35 |
| *11* | 3.23 | 2.86 | 2.66 | 2.54 | 2.45 | 2.39 | 2.34 | 2.30 | 2.27 |
| *12* | 3.18 | 2.81 | 2.61 | 2.48 | 2.39 | 2.33 | 2.28 | 2.24 | 2.21 |
| *13* | 3.14 | 2.76 | 2.56 | 2.43 | 2.35 | 2.28 | 2.23 | 2.20 | 2.16 |
| *14* | 3.10 | 2.73 | 2.52 | 2.39 | 2.31 | 2.24 | 2.19 | 2.15 | 2.12 |
| *15* | 3.07 | 2.70 | 2.49 | 2.36 | 2.27 | 2.21 | 2.16 | 2.12 | 2.09 |
| *16* | 3.05 | 2.67 | 2.46 | 2.33 | 2.24 | 2.18 | 2.13 | 2.09 | 2.06 |
| *17* | 3.03 | 2.64 | 2.44 | 2.31 | 2.22 | 2.15 | 2.10 | 2.06 | 2.03 |
| *18* | 3.01 | 2.62 | 2.42 | 2.29 | 2.20 | 2.13 | 2.08 | 2.04 | 2.00 |
| *19* | 2.99 | 2.61 | 2.40 | 2.27 | 2.18 | 2.11 | 2.06 | 2.02 | 1.98 |
| *20* | 2.97 | 2.59 | 2.38 | 2.25 | 2.16 | 2.09 | 2.04 | 2.00 | 1.96 |
| *21* | 2.96 | 2.57 | 2.36 | 2.23 | 2.14 | 2.08 | 2.02 | 1.98 | 1.95 |
| *22* | 2.95 | 2.56 | 2.35 | 2.22 | 2.13 | 2.06 | 2.01 | 1.97 | 1.93 |
| *23* | 2.94 | 2.55 | 2.34 | 2.21 | 2.11 | 2.05 | 1.99 | 1.95 | 1.92 |
| *24* | 2.93 | 2.54 | 2.33 | 2.19 | 2.10 | 2.04 | 1.98 | 1.94 | 1.91 |
| *25* | 2.92 | 2.53 | 2.32 | 2.18 | 2.09 | 2.02 | 1.97 | 1.93 | 1.89 |
| *26* | 2.91 | 2.52 | 2.31 | 2.17 | 2.08 | 2.01 | 1.96 | 1.92 | 1.88 |
| *27* | 2.90 | 2.51 | 2.30 | 2.17 | 2.07 | 2.00 | 1.95 | 1.91 | 1.87 |
| *28* | 2.89 | 2.50 | 2.29 | 2.16 | 2.06 | 2.00 | 1.94 | 1.90 | 1.87 |
| *29* | 2.89 | 2.50 | 2.28 | 2.15 | 2.06 | 1.99 | 1.93 | 1.89 | 1.86 |
| *30* | 2.88 | 2.49 | 2.28 | 2.14 | 2.05 | 1.98 | 1.93 | 1.88 | 1.85 |
| *40* | 2.84 | 2.44 | 2.23 | 2.09 | 2.00 | 1.93 | 1.87 | 1.83 | 1.79 |
| *60* | 2.79 | 2.39 | 2.18 | 2.04 | 1.95 | 1.87 | 1.82 | 1.77 | 1.74 |
| *120* | 2.75 | 2.35 | 2.13 | 1.99 | 1.90 | 1.82 | 1.77 | 1.72 | 1.68 |
| *∞* | 2.71 | 2.30 | 2.08 | 1.94 | 1.85 | 1.77 | 1.72 | 1.67 | 1.63 |

*Denominator Degrees of Freedom*

*Source:* From M. Merrington and C. M. Thompson, "Tables of Percentage Points of the Inverted Beta (*F*)-Distribution," *Biometrika*, 1943, 33, 73–88.

*(continued)*

**TABLE VIII** *(continued)*

| $v_2$ | \ $v_1$ NUMERATOR DEGREES OF FREEDOM | | | | | | | | | |
|---|---|---|---|---|---|---|---|---|---|---|
| | **10** | **12** | **15** | **20** | **24** | **30** | **40** | **60** | **120** | **∞** |
| *1* | 60.19 | 60.71 | 61.22 | 61.74 | 62.00 | 62.26 | 62.53 | 62.79 | 63.06 | 63.33 |
| *2* | 9.39 | 9.41 | 9.42 | 9.44 | 9.45 | 9.46 | 9.47 | 9.47 | 9.48 | 9.49 |
| *3* | 5.23 | 5.22 | 5.20 | 5.18 | 5.18 | 5.17 | 5.16 | 5.15 | 5.14 | 5.13 |
| *4* | 3.92 | 3.90 | 3.87 | 3.84 | 3.83 | 3.82 | 3.80 | 3.79 | 3.78 | 3.76 |
| *5* | 3.30 | 3.27 | 3.24 | 3.21 | 3.19 | 3.17 | 3.16 | 3.14 | 3.12 | 3.10 |
| *6* | 2.94 | 2.90 | 2.87 | 2.84 | 2.82 | 2.80 | 2.78 | 2.76 | 2.74 | 2.72 |
| *7* | 2.70 | 2.67 | 2.63 | 2.59 | 2.58 | 2.56 | 2.54 | 2.51 | 2.49 | 2.47 |
| *8* | 2.54 | 2.50 | 2.46 | 2.42 | 2.40 | 2.38 | 2.36 | 2.34 | 2.32 | 2.29 |
| *9* | 2.42 | 2.38 | 2.34 | 2.30 | 2.28 | 2.25 | 2.23 | 2.21 | 2.18 | 2.16 |
| *10* | 2.32 | 2.28 | 2.24 | 2.20 | 2.18 | 2.16 | 2.13 | 2.11 | 2.08 | 2.06 |
| *11* | 2.25 | 2.21 | 2.17 | 2.12 | 2.10 | 2.08 | 2.05 | 2.03 | 2.00 | 1.97 |
| *12* | 2.19 | 2.15 | 2.10 | 2.06 | 2.04 | 2.01 | 1.99 | 1.96 | 1.93 | 1.90 |
| *13* | 2.14 | 2.10 | 2.05 | 2.01 | 1.98 | 1.96 | 1.93 | 1.90 | 1.88 | 1.85 |
| *14* | 2.10 | 2.05 | 2.01 | 1.96 | 1.94 | 1.91 | 1.89 | 1.86 | 1.83 | 1.80 |
| *15* | 2.06 | 2.02 | 1.97 | 1.92 | 1.90 | 1.87 | 1.85 | 1.82 | 1.79 | 1.76 |
| *16* | 2.03 | 1.99 | 1.94 | 1.89 | 1.87 | 1.84 | 1.81 | 1.78 | 1.75 | 1.72 |
| *17* | 2.00 | 1.96 | 1.91 | 1.86 | 1.84 | 1.81 | 1.78 | 1.75 | 1.72 | 1.69 |
| *18* | 1.98 | 1.93 | 1.89 | 1.84 | 1.81 | 1.78 | 1.75 | 1.72 | 1.69 | 1.66 |
| *19* | 1.96 | 1.91 | 1.86 | 1.81 | 1.79 | 1.76 | 1.73 | 1.70 | 1.67 | 1.63 |
| *20* | 1.94 | 1.89 | 1.84 | 1.79 | 1.77 | 1.74 | 1.71 | 1.68 | 1.64 | 1.61 |
| *21* | 1.92 | 1.87 | 1.83 | 1.78 | 1.75 | 1.72 | 1.69 | 1.66 | 1.62 | 1.59 |
| *22* | 1.90 | 1.86 | 1.81 | 1.76 | 1.73 | 1.70 | 1.67 | 1.64 | 1.60 | 1.57 |
| *23* | 1.89 | 1.84 | 1.80 | 1.74 | 1.72 | 1.69 | 1.66 | 1.62 | 1.59 | 1.55 |
| *24* | 1.88 | 1.83 | 1.78 | 1.73 | 1.70 | 1.67 | 1.64 | 1.61 | 1.57 | 1.53 |
| *25* | 1.87 | 1.82 | 1.77 | 1.72 | 1.69 | 1.66 | 1.63 | 1.59 | 1.56 | 1.52 |
| *26* | 1.86 | 1.81 | 1.76 | 1.71 | 1.68 | 1.65 | 1.61 | 1.58 | 1.54 | 1.50 |
| *27* | 1.85 | 1.80 | 1.75 | 1.70 | 1.67 | 1.64 | 1.60 | 1.57 | 1.53 | 1.49 |
| *28* | 1.84 | 1.79 | 1.74 | 1.69 | 1.66 | 1.63 | 1.59 | 1.56 | 1.52 | 1.48 |
| *29* | 1.83 | 1.78 | 1.73 | 1.68 | 1.65 | 1.62 | 1.58 | 1.55 | 1.51 | 1.47 |
| *30* | 1.82 | 1.77 | 1.72 | 1.67 | 1.64 | 1.61 | 1.57 | 1.54 | 1.50 | 1.46 |
| *40* | 1.76 | 1.71 | 1.66 | 1.61 | 1.57 | 1.54 | 1.51 | 1.47 | 1.42 | 1.38 |
| *60* | 1.71 | 1.66 | 1.60 | 1.54 | 1.51 | 1.48 | 1.44 | 1.40 | 1.35 | 1.29 |
| *120* | 1.65 | 1.60 | 1.55 | 1.48 | 1.45 | 1.41 | 1.37 | 1.32 | 1.26 | 1.19 |
| *∞* | 1.60 | 1.55 | 1.49 | 1.42 | 1.38 | 1.34 | 1.30 | 1.24 | 1.17 | 1.00 |

*Denominator Degrees of Freedom*

## TABLE IX    Percentage Points of the *F*-Distribution, $\alpha = .05$

|  $v_1$ | NUMERATOR DEGREES OF FREEDOM | | | | | | | | |
|---|---|---|---|---|---|---|---|---|---|
| $v_2$ | 1 | 2 | 3 | 4 | 5 | 6 | 7 | 8 | 9 |
| 1 | 161.4 | 199.5 | 215.7 | 224.6 | 230.2 | 234.0 | 236.8 | 238.9 | 240.5 |
| 2 | 18.51 | 19.00 | 19.16 | 19.25 | 19.30 | 19.33 | 19.35 | 19.37 | 19.38 |
| 3 | 10.13 | 9.55 | 9.28 | 9.12 | 9.01 | 8.94 | 8.89 | 8.85 | 8.81 |
| 4 | 7.71 | 6.94 | 6.59 | 6.39 | 6.26 | 6.16 | 6.09 | 6.04 | 6.00 |
| 5 | 6.61 | 5.79 | 5.41 | 5.19 | 5.05 | 4.95 | 4.88 | 4.82 | 4.77 |
| 6 | 5.99 | 5.14 | 4.76 | 4.53 | 4.39 | 4.28 | 4.21 | 4.15 | 4.10 |
| 7 | 5.59 | 4.74 | 4.35 | 4.12 | 3.97 | 3.87 | 3.79 | 3.73 | 3.68 |
| 8 | 5.32 | 4.46 | 4.07 | 3.84 | 3.69 | 3.58 | 3.50 | 3.44 | 3.39 |
| 9 | 5.12 | 4.26 | 3.86 | 3.63 | 3.48 | 3.37 | 3.29 | 3.23 | 3.18 |
| 10 | 4.96 | 4.10 | 3.71 | 3.48 | 3.33 | 3.22 | 3.14 | 3.07 | 3.02 |
| 11 | 4.84 | 3.98 | 3.59 | 3.36 | 3.20 | 3.09 | 3.01 | 2.95 | 2.90 |
| 12 | 4.75 | 3.89 | 3.49 | 3.26 | 3.11 | 3.00 | 2.91 | 2.85 | 2.80 |
| 13 | 4.67 | 3.81 | 3.41 | 3.18 | 3.03 | 2.92 | 2.83 | 2.77 | 2.71 |
| 14 | 4.60 | 3.74 | 3.34 | 3.11 | 2.96 | 2.85 | 2.76 | 2.70 | 2.65 |
| 15 | 4.54 | 3.68 | 3.29 | 3.06 | 2.90 | 2.79 | 2.71 | 2.64 | 2.59 |
| 16 | 4.49 | 3.63 | 3.24 | 3.01 | 2.85 | 2.74 | 2.66 | 2.59 | 2.54 |
| 17 | 4.45 | 3.59 | 3.20 | 2.96 | 2.81 | 2.70 | 2.61 | 2.55 | 2.49 |
| 18 | 4.41 | 3.55 | 3.16 | 2.93 | 2.77 | 2.66 | 2.58 | 2.51 | 2.46 |
| 19 | 4.38 | 3.52 | 3.13 | 2.90 | 2.74 | 2.63 | 2.54 | 2.48 | 2.42 |
| 20 | 4.35 | 3.49 | 3.10 | 2.87 | 2.71 | 2.60 | 2.51 | 2.45 | 2.39 |
| 21 | 4.32 | 3.47 | 3.07 | 2.84 | 2.68 | 2.57 | 2.49 | 2.42 | 2.37 |
| 22 | 4.30 | 3.44 | 3.05 | 2.82 | 2.66 | 2.55 | 2.46 | 2.40 | 2.34 |
| 23 | 4.28 | 3.42 | 3.03 | 2.80 | 2.64 | 2.53 | 2.44 | 2.37 | 2.32 |
| 24 | 4.26 | 3.40 | 3.01 | 2.78 | 2.62 | 2.51 | 2.42 | 2.36 | 2.30 |
| 25 | 4.24 | 3.39 | 2.99 | 2.76 | 2.60 | 2.49 | 2.40 | 2.34 | 2.28 |
| 26 | 4.23 | 3.37 | 2.98 | 2.74 | 2.59 | 2.47 | 2.39 | 2.32 | 2.77 |
| 27 | 4.21 | 3.35 | 2.96 | 2.73 | 2.57 | 2.46 | 2.37 | 2.31 | 2.25 |
| 28 | 4.20 | 3.34 | 2.95 | 2.71 | 2.56 | 2.45 | 2.36 | 2.29 | 2.24 |
| 29 | 4.18 | 3.33 | 2.93 | 2.70 | 2.55 | 2.43 | 2.35 | 2.28 | 2.22 |
| 30 | 4.17 | 3.32 | 2.92 | 2.69 | 2.53 | 2.42 | 2.33 | 2.27 | 2.21 |
| 40 | 4.08 | 3.23 | 2.84 | 2.61 | 2.45 | 2.34 | 2.25 | 2.18 | 2.12 |
| 60 | 4.00 | 3.15 | 2.76 | 2.53 | 2.37 | 2.25 | 2.17 | 2.10 | 2.04 |
| 120 | 3.92 | 3.07 | 2.68 | 2.45 | 2.29 | 2.17 | 2.09 | 2.02 | 1.96 |
| ∞ | 3.84 | 3.00 | 2.60 | 2.37 | 2.21 | 2.10 | 2.01 | 1.94 | 1.88 |

*Denominator Degrees of Freedom*

*Source:* From M. Merrington and C. M. Thompson, "Tables of Percentage Points of the Inverted Beta (*F*)-Distribution." *Biometrika*, 1943, 33, 73–88.

*(continued)*

**TABLE IX**  *(continued)*

| $v_2$ | 10 | 12 | 15 | 20 | 24 | 30 | 40 | 60 | 120 | ∞ |
|---|---|---|---|---|---|---|---|---|---|---|
| 1 | 241.9 | 243.9 | 245.9 | 248.0 | 249.1 | 250.1 | 251.1 | 252.2 | 253.3 | 254.3 |
| 2 | 19.40 | 19.41 | 19.43 | 19.45 | 19.45 | 19.46 | 19.47 | 19.48 | 19.49 | 19.50 |
| 3 | 8.79 | 8.74 | 8.70 | 8.66 | 8.64 | 8.62 | 8.59 | 8.57 | 8.55 | 8.53 |
| 4 | 5.96 | 5.91 | 5.86 | 5.80 | 5.77 | 5.75 | 5.72 | 5.69 | 5.66 | 5.63 |
| 5 | 4.74 | 4.68 | 4.62 | 4.56 | 4.53 | 4.50 | 4.46 | 4.43 | 4.40 | 4.36 |
| 6 | 4.06 | 4.00 | 3.94 | 3.87 | 3.84 | 3.81 | 3.77 | 3.74 | 3.70 | 3.67 |
| 7 | 3.64 | 3.57 | 3.51 | 3.44 | 3.41 | 3.38 | 3.34 | 3.30 | 3.27 | 3.23 |
| 8 | 3.35 | 3.28 | 3.22 | 3.15 | 3.12 | 3.08 | 3.04 | 3.01 | 2.97 | 2.93 |
| 9 | 3.14 | 3.07 | 3.01 | 2.94 | 2.90 | 2.86 | 2.83 | 2.79 | 2.75 | 2.71 |
| 10 | 2.98 | 2.91 | 2.85 | 2.77 | 2.74 | 2.70 | 2.66 | 2.62 | 2.58 | 2.54 |
| 11 | 2.85 | 2.79 | 2.72 | 2.65 | 2.61 | 2.57 | 2.53 | 2.49 | 2.45 | 2.40 |
| 12 | 2.75 | 2.69 | 2.62 | 2.54 | 2.51 | 2.47 | 2.43 | 2.38 | 2.34 | 2.30 |
| 13 | 2.67 | 2.60 | 2.53 | 2.46 | 2.42 | 2.38 | 2.34 | 2.30 | 2.25 | 2.21 |
| 14 | 2.60 | 2.53 | 2.46 | 2.39 | 2.35 | 2.31 | 2.27 | 2.22 | 2.18 | 2.13 |
| 15 | 2.54 | 2.48 | 2.40 | 2.33 | 2.29 | 2.25 | 2.20 | 2.16 | 2.11 | 2.07 |
| 16 | 2.49 | 2.42 | 2.35 | 2.28 | 2.24 | 2.19 | 2.15 | 2.11 | 2.06 | 2.01 |
| 17 | 2.45 | 2.38 | 2.31 | 2.23 | 2.19 | 2.15 | 2.10 | 2.06 | 2.01 | 1.96 |
| 18 | 2.41 | 2.34 | 2.27 | 2.19 | 2.15 | 2.11 | 2.06 | 2.02 | 1.97 | 1.92 |
| 19 | 2.38 | 2.31 | 2.23 | 2.16 | 2.11 | 2.07 | 2.03 | 1.98 | 1.93 | 1.88 |
| 20 | 2.35 | 2.28 | 2.20 | 2.12 | 2.08 | 2.04 | 1.99 | 1.95 | 1.90 | 1.84 |
| 21 | 2.32 | 2.25 | 2.18 | 2.10 | 2.05 | 2.01 | 1.96 | 1.92 | 1.87 | 1.81 |
| 22 | 2.30 | 2.23 | 2.15 | 2.07 | 2.03 | 1.98 | 1.94 | 1.89 | 1.84 | 1.78 |
| 23 | 2.27 | 2.20 | 2.13 | 2.05 | 2.01 | 1.96 | 1.91 | 1.86 | 1.81 | 1.76 |
| 24 | 2.25 | 2.18 | 2.11 | 2.03 | 1.98 | 1.94 | 1.89 | 1.84 | 1.79 | 1.73 |
| 25 | 2.24 | 2.16 | 2.09 | 2.01 | 1.96 | 1.92 | 1.87 | 1.82 | 1.77 | 1.71 |
| 26 | 2.22 | 2.15 | 2.07 | 1.99 | 1.95 | 1.90 | 1.85 | 1.80 | 1.75 | 1.69 |
| 27 | 2.20 | 2.13 | 2.06 | 1.97 | 1.93 | 1.88 | 1.84 | 1.79 | 1.73 | 1.67 |
| 28 | 2.19 | 2.12 | 2.04 | 1.96 | 1.91 | 1.87 | 1.82 | 1.77 | 1.71 | 1.65 |
| 29 | 2.18 | 2.10 | 2.03 | 1.94 | 1.90 | 1.85 | 1.81 | 1.75 | 1.70 | 1.64 |
| 30 | 2.16 | 2.09 | 2.01 | 1.93 | 1.89 | 1.84 | 1.79 | 1.74 | 1.68 | 1.62 |
| 40 | 2.08 | 2.00 | 1.92 | 1.84 | 1.79 | 1.74 | 1.69 | 1.64 | 1.58 | 1.51 |
| 60 | 1.99 | 1.92 | 1.84 | 1.75 | 1.70 | 1.65 | 1.59 | 1.53 | 1.47 | 1.39 |
| 120 | 1.91 | 1.83 | 1.75 | 1.66 | 1.61 | 1.55 | 1.50 | 1.43 | 1.35 | 1.25 |
| ∞ | 1.83 | 1.75 | 1.67 | 1.57 | 1.52 | 1.46 | 1.39 | 1.32 | 1.22 | 1.00 |

NUMERATOR DEGREES OF FREEDOM ($v_1$); *Denominator Degrees of Freedom*

**TABLE X**    **Percentage Points of the *F*-Distribution, $\alpha = .025$**

| $v_2$ | NUMERATOR DEGREES OF FREEDOM $v_1$ | | | | | | | | |
|---|---|---|---|---|---|---|---|---|---|
| | **1** | **2** | **3** | **4** | **5** | **6** | **7** | **8** | **9** |
| *1* | 647.8 | 799.5 | 864.2 | 899.6 | 921.8 | 937.1 | 948.2 | 956.7 | 963.3 |
| *2* | 38.51 | 39.00 | 39.17 | 39.25 | 39.30 | 39.33 | 39.36 | 39.37 | 39.39 |
| *3* | 17.44 | 16.04 | 15.44 | 15.10 | 14.88 | 14.73 | 14.62 | 14.54 | 14.47 |
| *4* | 12.22 | 10.65 | 9.98 | 9.60 | 9.36 | 9.20 | 9.07 | 8.98 | 8.90 |
| *5* | 10.01 | 8.43 | 7.76 | 7.39 | 7.15 | 6.98 | 6.85 | 6.76 | 6.68 |
| *6* | 8.81 | 7.26 | 6.60 | 6.23 | 5.99 | 5.82 | 5.70 | 5.60 | 5.52 |
| *7* | 8.07 | 6.54 | 5.89 | 5.52 | 5.29 | 5.12 | 4.99 | 4.90 | 4.82 |
| *8* | 7.57 | 6.06 | 5.42 | 5.05 | 4.82 | 4.65 | 4.53 | 4.43 | 4.36 |
| *9* | 7.21 | 5.71 | 5.08 | 4.72 | 4.48 | 4.32 | 4.20 | 4.10 | 4.03 |
| *10* | 6.94 | 5.46 | 4.83 | 4.47 | 4.24 | 4.07 | 3.95 | 3.85 | 3.78 |
| *11* | 6.72 | 5.26 | 4.63 | 4.28 | 4.04 | 3.88 | 3.76 | 3.66 | 3.59 |
| *12* | 6.55 | 5.10 | 4.47 | 4.12 | 3.89 | 3.73 | 3.61 | 3.51 | 3.44 |
| *13* | 6.41 | 4.97 | 4.35 | 4.00 | 3.77 | 3.60 | 3.48 | 3.39 | 3.31 |
| *14* | 6.30 | 4.86 | 4.24 | 3.89 | 3.66 | 3.50 | 3.38 | 3.29 | 3.21 |
| *15* | 6.20 | 4.77 | 4.15 | 3.80 | 3.58 | 3.41 | 3.29 | 3.20 | 3.12 |
| *16* | 6.12 | 4.69 | 4.08 | 3.73 | 3.50 | 3.34 | 3.22 | 3.12 | 3.05 |
| *17* | 6.04 | 4.62 | 4.01 | 3.66 | 3.44 | 3.28 | 3.16 | 3.06 | 2.98 |
| *18* | 5.98 | 4.56 | 3.95 | 3.61 | 3.38 | 3.22 | 3.10 | 3.01 | 2.93 |
| *19* | 5.92 | 4.51 | 3.90 | 3.56 | 3.33 | 3.17 | 3.05 | 2.96 | 2.88 |
| *20* | 5.87 | 4.46 | 3.86 | 3.51 | 3.29 | 3.13 | 3.01 | 2.91 | 2.84 |
| *21* | 5.83 | 4.42 | 3.82 | 3.48 | 3.25 | 3.09 | 2.97 | 2.87 | 2.80 |
| *22* | 5.79 | 4.38 | 3.78 | 3.44 | 3.22 | 3.05 | 2.93 | 2.84 | 2.76 |
| *23* | 5.75 | 4.35 | 3.75 | 3.41 | 3.18 | 3.02 | 2.90 | 2.81 | 2.73 |
| *24* | 5.72 | 4.32 | 3.72 | 3.38 | 3.15 | 2.99 | 2.87 | 2.78 | 2.70 |
| *25* | 5.69 | 4.29 | 3.69 | 3.35 | 3.13 | 2.97 | 2.85 | 2.75 | 2.68 |
| *26* | 5.66 | 4.27 | 3.67 | 3.33 | 3.10 | 2.94 | 2.82 | 2.73 | 2.65 |
| *27* | 5.63 | 4.24 | 3.65 | 3.31 | 3.08 | 2.92 | 2.80 | 2.71 | 2.63 |
| *28* | 5.61 | 4.22 | 3.63 | 3.29 | 3.06 | 2.90 | 2.78 | 2.69 | 2.61 |
| *29* | 5.59 | 4.20 | 3.61 | 3.27 | 3.04 | 2.88 | 2.76 | 2.67 | 2.59 |
| *30* | 5.57 | 4.18 | 3.59 | 3.25 | 3.03 | 2.87 | 2.75 | 2.65 | 2.57 |
| *40* | 5.42 | 4.05 | 3.46 | 3.13 | 2.90 | 2.74 | 2.62 | 2.53 | 2.45 |
| *60* | 5.29 | 3.93 | 3.34 | 3.01 | 2.79 | 2.63 | 2.51 | 2.41 | 2.33 |
| *120* | 5.15 | 3.80 | 3.23 | 2.89 | 2.67 | 2.52 | 2.39 | 2.30 | 2.22 |
| $\infty$ | 5.02 | 3.69 | 3.12 | 2.79 | 2.57 | 2.41 | 2.29 | 2.19 | 2.11 |

*Denominator Degrees of Freedom*

*Source:* From M. Merrington and C. M. Thompson, "Tables of Percentage Points of the Inverted Beta (*F*)-Distribution," *Biometrika*, 1943, 33, 73–88.

*(continued)*

**TABLE X** *(continued)*

| $v_2$ \ $v_1$ | 10 | 12 | 15 | 20 | 24 | 30 | 40 | 60 | 120 | ∞ |
|---|---|---|---|---|---|---|---|---|---|---|
| | \multicolumn{10}{NUMERATOR DEGREES OF FREEDOM} | | | | | | | | | |
| 1 | 968.6 | 976.7 | 984.9 | 993.1 | 997.2 | 1,001 | 1,006 | 1,010 | 1,014 | 1,018 |
| 2 | 39.40 | 39.41 | 39.43 | 39.45 | 39.46 | 39.46 | 39.47 | 39.48 | 39.49 | 39.50 |
| 3 | 14.42 | 14.34 | 14.25 | 14.17 | 14.12 | 14.08 | 14.04 | 13.99 | 13.95 | 13.90 |
| 4 | 8.84 | 8.75 | 8.66 | 8.56 | 8.51 | 8.46 | 8.41 | 8.36 | 8.31 | 8.26 |
| 5 | 6.62 | 6.52 | 6.43 | 6.33 | 6.28 | 6.23 | 6.18 | 6.12 | 6.07 | 6.02 |
| 6 | 5.46 | 5.37 | 5.27 | 5.17 | 5.12 | 5.07 | 5.01 | 4.96 | 4.90 | 4.85 |
| 7 | 4.76 | 4.67 | 4.57 | 4.47 | 4.42 | 4.36 | 4.31 | 4.25 | 4.20 | 4.14 |
| 8 | 4.30 | 4.20 | 4.10 | 4.00 | 3.95 | 3.89 | 3.84 | 3.78 | 3.73 | 3.67 |
| 9 | 3.96 | 3.87 | 3.77 | 3.67 | 3.61 | 3.56 | 3.51 | 3.45 | 3.39 | 3.33 |
| 10 | 3.72 | 3.62 | 3.52 | 3.42 | 3.37 | 3.31 | 3.26 | 3.20 | 3.14 | 3.08 |
| 11 | 3.53 | 3.43 | 3.33 | 3.23 | 3.17 | 3.12 | 3.06 | 3.00 | 2.94 | 2.88 |
| 12 | 3.37 | 3.28 | 3.18 | 3.07 | 3.02 | 2.96 | 2.91 | 2.85 | 2.79 | 2.72 |
| 13 | 3.25 | 3.15 | 3.05 | 2.95 | 2.89 | 2.84 | 2.78 | 2.72 | 2.66 | 2.60 |
| 14 | 3.15 | 3.05 | 2.95 | 2.84 | 2.79 | 2.73 | 2.67 | 2.61 | 2.55 | 2.49 |
| 15 | 3.06 | 2.96 | 2.86 | 2.76 | 2.70 | 2.64 | 2.59 | 2.52 | 2.46 | 2.40 |
| 16 | 2.99 | 2.89 | 2.79 | 2.68 | 2.63 | 2.57 | 2.51 | 2.45 | 2.38 | 2.32 |
| 17 | 2.92 | 2.82 | 2.72 | 2.62 | 2.56 | 2.50 | 2.44 | 2.38 | 2.32 | 2.25 |
| 18 | 2.87 | 2.77 | 2.67 | 2.56 | 2.50 | 2.44 | 2.38 | 2.32 | 2.26 | 2.19 |
| 19 | 2.82 | 2.72 | 2.62 | 2.51 | 2.45 | 2.39 | 2.33 | 2.27 | 2.20 | 2.13 |
| 20 | 2.77 | 2.68 | 2.57 | 2.46 | 2.41 | 2.35 | 2.29 | 2.22 | 2.16 | 2.09 |
| 21 | 2.73 | 2.64 | 2.53 | 2.42 | 2.37 | 2.31 | 2.25 | 2.18 | 2.11 | 2.04 |
| 22 | 2.70 | 2.60 | 2.50 | 2.39 | 2.33 | 2.27 | 2.21 | 2.14 | 2.08 | 2.00 |
| 23 | 2.67 | 2.57 | 2.47 | 2.36 | 2.30 | 2.24 | 2.18 | 2.11 | 2.04 | 1.97 |
| 24 | 2.64 | 2.54 | 2.44 | 2.33 | 2.27 | 2.21 | 2.15 | 2.08 | 2.01 | 1.94 |
| 25 | 2.61 | 2.51 | 2.41 | 2.30 | 2.24 | 2.18 | 2.12 | 2.05 | 1.98 | 1.91 |
| 26 | 2.59 | 2.49 | 2.39 | 2.28 | 2.22 | 2.16 | 2.09 | 2.03 | 1.95 | 1.88 |
| 27 | 2.57 | 2.47 | 2.36 | 2.25 | 2.19 | 2.13 | 2.07 | 2.00 | 1.93 | 1.85 |
| 28 | 2.55 | 2.45 | 2.34 | 2.23 | 2.17 | 2.11 | 2.05 | 1.98 | 1.91 | 1.83 |
| 29 | 2.53 | 2.43 | 2.32 | 2.21 | 2.15 | 2.09 | 2.03 | 1.96 | 1.89 | 1.81 |
| 30 | 2.51 | 2.41 | 2.31 | 2.20 | 2.14 | 2.07 | 2.01 | 1.94 | 1.87 | 1.79 |
| 40 | 2.39 | 2.29 | 2.18 | 2.07 | 2.01 | 1.94 | 1.88 | 1.80 | 1.72 | 1.64 |
| 60 | 2.27 | 2.17 | 2.06 | 1.94 | 1.88 | 1.82 | 1.74 | 1.67 | 1.58 | 1.48 |
| 120 | 2.16 | 2.05 | 1.94 | 1.82 | 1.76 | 1.69 | 1.61 | 1.53 | 1.43 | 1.31 |
| ∞ | 2.05 | 1.94 | 1.83 | 1.71 | 1.64 | 1.57 | 1.48 | 1.39 | 1.27 | 1.00 |

*Denominator Degrees of Freedom*

**TABLE XI   Percentage Points of the *F*-Distribution, $\alpha = .01$**

| $v_2$ \ $v_1$ | NUMERATOR DEGREES OF FREEDOM | | | | | | | | |
|---|---|---|---|---|---|---|---|---|---|
| | **1** | **2** | **3** | **4** | **5** | **6** | **7** | **8** | **9** |
| *1* | 4,052 | 4,999.5 | 5,403 | 5,625 | 5,764 | 5,859 | 5,928 | 5,982 | 6,022 |
| *2* | 98.50 | 99.00 | 99.17 | 99.25 | 99.30 | 99.33 | 99.36 | 99.37 | 99.39 |
| *3* | 34.12 | 30.82 | 29.46 | 28.71 | 28.24 | 27.91 | 27.67 | 27.49 | 27.35 |
| *4* | 21.20 | 18.00 | 16.69 | 15.98 | 15.52 | 15.21 | 14.98 | 14.80 | 14.66 |
| *5* | 16.26 | 13.27 | 12.06 | 11.39 | 10.97 | 10.67 | 10.46 | 10.29 | 10.16 |
| *6* | 13.75 | 10.92 | 9.78 | 9.15 | 8.75 | 8.47 | 8.26 | 8.10 | 7.98 |
| *7* | 12.25 | 9.55 | 8.45 | 7.85 | 7.46 | 7.19 | 6.99 | 6.84 | 6.72 |
| *8* | 11.26 | 8.65 | 7.59 | 7.01 | 6.63 | 6.37 | 6.18 | 6.03 | 5.91 |
| *9* | 10.56 | 8.02 | 6.99 | 6.42 | 6.06 | 5.80 | 5.61 | 5.47 | 5.35 |
| *10* | 10.04 | 7.56 | 6.55 | 5.99 | 5.64 | 5.39 | 5.20 | 5.06 | 4.94 |
| *11* | 9.65 | 7.21 | 6.22 | 5.67 | 5.32 | 5.07 | 4.89 | 4.74 | 4.63 |
| *12* | 9.33 | 6.93 | 5.95 | 5.41 | 5.06 | 4.82 | 4.64 | 4.50 | 4.39 |
| *13* | 9.07 | 6.70 | 5.74 | 5.21 | 4.86 | 4.62 | 4.44 | 4.30 | 4.19 |
| *14* | 8.86 | 6.51 | 5.56 | 5.04 | 4.69 | 4.46 | 4.28 | 4.14 | 4.03 |
| *15* | 8.68 | 6.36 | 5.42 | 4.89 | 4.56 | 4.32 | 4.14 | 4.00 | 3.89 |
| *16* | 8.53 | 6.23 | 5.29 | 4.77 | 4.44 | 4.20 | 4.03 | 3.89 | 3.78 |
| *17* | 8.40 | 6.11 | 5.18 | 4.67 | 4.34 | 4.10 | 3.93 | 3.79 | 3.68 |
| *18* | 8.29 | 6.01 | 5.09 | 4.58 | 4.25 | 4.01 | 3.84 | 3.71 | 3.60 |
| *19* | 8.18 | 5.93 | 5.01 | 4.50 | 4.17 | 3.94 | 3.77 | 3.63 | 3.52 |
| *20* | 8.10 | 5.85 | 4.94 | 4.43 | 4.10 | 3.87 | 3.70 | 3.56 | 3.46 |
| *21* | 8.02 | 5.78 | 4.87 | 4.37 | 4.04 | 3.81 | 3.64 | 3.51 | 3.40 |
| *22* | 7.95 | 5.72 | 4.82 | 4.31 | 3.99 | 3.76 | 3.59 | 3.45 | 3.35 |
| *23* | 7.88 | 5.66 | 4.76 | 4.26 | 3.94 | 3.71 | 3.54 | 3.41 | 3.30 |
| *24* | 7.82 | 5.61 | 4.72 | 4.22 | 3.90 | 3.67 | 3.50 | 3.36 | 3.26 |
| *25* | 7.77 | 5.57 | 4.68 | 4.18 | 3.85 | 3.63 | 3.46 | 3.32 | 3.22 |
| *26* | 7.72 | 5.53 | 4.64 | 4.14 | 3.82 | 3.59 | 3.42 | 3.29 | 3.18 |
| *27* | 7.68 | 5.49 | 4.60 | 4.11 | 3.78 | 3.56 | 3.39 | 3.26 | 3.15 |
| *28* | 7.64 | 5.45 | 4.57 | 4.07 | 3.75 | 3.53 | 3.36 | 3.23 | 3.12 |
| *29* | 7.60 | 5.42 | 4.54 | 4.04 | 3.73 | 3.50 | 3.33 | 3.20 | 3.09 |
| *30* | 7.56 | 5.39 | 4.51 | 4.02 | 3.70 | 3.47 | 3.30 | 3.17 | 3.07 |
| *40* | 7.31 | 5.18 | 4.31 | 3.83 | 3.51 | 3.29 | 3.12 | 2.99 | 2.89 |
| *60* | 7.08 | 4.98 | 4.13 | 3.65 | 3.34 | 3.12 | 2.95 | 2.82 | 2.72 |
| *120* | 6.85 | 4.79 | 3.95 | 3.48 | 3.17 | 2.96 | 2.79 | 2.66 | 2.56 |
| *∞* | 6.63 | 4.61 | 3.78 | 3.32 | 3.02 | 2.80 | 2.64 | 2.51 | 2.41 |

*Source:* From M. Merrington and C. M. Thompson, "Tables of Percentage Points of the Inverted Beta (*F*)-Distribution," *Biometrika*, 1943, 33, 73–88.

*(continued)*

**TABLE XI** *(continued)*

| $v_2$ \ $v_1$ | NUMERATOR DEGREES OF FREEDOM | | | | | | | | | |
|---|---|---|---|---|---|---|---|---|---|---|
| | 10 | 12 | 15 | 20 | 24 | 30 | 40 | 60 | 120 | ∞ |
| 1 | 6,056 | 6,106 | 6,157 | 6,209 | 6,235 | 6,261 | 6,287 | 6,313 | 6,339 | 6,366 |
| 2 | 99.40 | 99.42 | 99.43 | 99.45 | 99.46 | 99.47 | 99.47 | 99.48 | 99.49 | 99.50 |
| 3 | 27.23 | 27.05 | 26.87 | 26.69 | 26.60 | 26.50 | 26.41 | 26.32 | 26.22 | 26.13 |
| 4 | 14.55 | 14.37 | 14.20 | 14.02 | 13.93 | 13.84 | 13.75 | 13.65 | 13.56 | 13.46 |
| 5 | 10.05 | 9.89 | 9.72 | 9.55 | 9.47 | 9.38 | 9.29 | 9.20 | 9.11 | 9.02 |
| 6 | 7.87 | 7.72 | 7.56 | 7.40 | 7.31 | 7.23 | 7.14 | 7.06 | 6.97 | 6.88 |
| 7 | 6.62 | 6.47 | 6.31 | 6.16 | 6.07 | 5.99 | 5.91 | 5.82 | 5.74 | 5.65 |
| 8 | 5.81 | 5.67 | 5.52 | 5.36 | 5.28 | 5.20 | 5.12 | 5.03 | 4.95 | 4.86 |
| 9 | 5.26 | 5.11 | 4.96 | 4.81 | 4.73 | 4.65 | 4.57 | 4.48 | 4.40 | 4.31 |
| 10 | 4.85 | 4.71 | 4.56 | 4.41 | 4.33 | 4.25 | 4.17 | 4.08 | 4.00 | 3.91 |
| 11 | 4.54 | 4.40 | 4.25 | 4.10 | 4.02 | 3.94 | 3.86 | 3.78 | 3.69 | 3.60 |
| 12 | 4.30 | 4.16 | 4.01 | 3.86 | 3.78 | 3.70 | 3.62 | 3.54 | 3.45 | 3.36 |
| 13 | 4.10 | 3.96 | 3.82 | 3.66 | 3.59 | 3.51 | 3.43 | 3.34 | 3.25 | 3.17 |
| 14 | 3.94 | 3.80 | 3.66 | 3.51 | 3.43 | 3.35 | 3.27 | 3.18 | 3.09 | 3.00 |
| 15 | 3.80 | 3.67 | 3.52 | 3.37 | 3.29 | 3.21 | 3.13 | 3.05 | 2.96 | 2.87 |
| 16 | 3.69 | 3.55 | 3.41 | 3.26 | 3.18 | 3.10 | 3.02 | 2.93 | 2.84 | 2.75 |
| 17 | 3.59 | 3.46 | 3.31 | 3.16 | 3.08 | 3.00 | 2.92 | 2.83 | 2.75 | 2.65 |
| 18 | 3.51 | 3.37 | 3.23 | 3.08 | 3.00 | 2.92 | 2.84 | 2.75 | 2.66 | 2.57 |
| 19 | 3.43 | 3.30 | 3.15 | 3.00 | 2.92 | 2.84 | 2.76 | 2.67 | 2.58 | 2.49 |
| 20 | 3.37 | 3.23 | 3.09 | 2.94 | 2.86 | 2.78 | 2.69 | 2.61 | 2.52 | 2.42 |
| 21 | 3.31 | 3.17 | 3.03 | 2.88 | 2.80 | 2.72 | 2.64 | 2.55 | 2.46 | 2.36 |
| 22 | 3.26 | 3.12 | 2.98 | 2.83 | 2.75 | 2.67 | 2.58 | 2.50 | 2.40 | 2.31 |
| 23 | 3.21 | 3.07 | 2.93 | 2.78 | 2.70 | 2.62 | 2.54 | 2.45 | 2.35 | 2.26 |
| 24 | 3.17 | 3.03 | 2.89 | 2.74 | 2.66 | 2.58 | 2.49 | 2.40 | 2.31 | 2.21 |
| 25 | 3.13 | 2.99 | 2.85 | 2.70 | 2.62 | 2.54 | 2.45 | 2.36 | 2.27 | 2.17 |
| 26 | 3.09 | 2.96 | 2.81 | 2.66 | 2.58 | 2.50 | 2.42 | 2.33 | 2.23 | 2.13 |
| 27 | 3.06 | 2.93 | 2.78 | 2.63 | 2.55 | 2.47 | 2.38 | 2.29 | 2.20 | 2.10 |
| 28 | 3.03 | 2.90 | 2.75 | 2.60 | 2.52 | 2.44 | 2.35 | 2.26 | 2.17 | 2.06 |
| 29 | 3.00 | 2.87 | 2.73 | 2.57 | 2.49 | 2.41 | 2.33 | 2.23 | 2.14 | 2.03 |
| 30 | 2.98 | 2.84 | 2.70 | 2.55 | 2.47 | 2.39 | 2.30 | 2.21 | 2.11 | 2.01 |
| 40 | 2.80 | 2.66 | 2.52 | 2.37 | 2.29 | 2.20 | 2.11 | 2.02 | 1.92 | 1.80 |
| 60 | 2.63 | 2.50 | 2.35 | 2.20 | 2.12 | 2.03 | 1.94 | 1.84 | 1.73 | 1.60 |
| 120 | 2.47 | 2.34 | 2.19 | 2.03 | 1.95 | 1.86 | 1.76 | 1.66 | 1.53 | 1.38 |
| ∞ | 2.32 | 2.18 | 2.04 | 1.88 | 1.79 | 1.70 | 1.59 | 1.47 | 1.32 | 1.00 |

*Denominator Degrees of Freedom*

## TABLE XII   Critical Values of $T_L$ and $T_U$ for the Wilcoxon Rank Sum Test: Independent Samples

*Test statistic is the rank sum associated with the smaller sample (if equal sample sizes, either rank sum can be used).*

**a.** $\alpha = .025$ one-tailed; $\alpha = .05$ two-tailed

| $n_2 \backslash n_1$ | 3 $T_L$ | $T_U$ | 4 $T_L$ | $T_U$ | 5 $T_L$ | $T_U$ | 6 $T_L$ | $T_U$ | 7 $T_L$ | $T_U$ | 8 $T_L$ | $T_U$ | 9 $T_L$ | $T_U$ | 10 $T_L$ | $T_U$ |
|---|---|---|---|---|---|---|---|---|---|---|---|---|---|---|---|---|
| 3 | 5 | 16 | 6 | 18 | 6 | 21 | 7 | 23 | 7 | 26 | 8 | 28 | 8 | 31 | 9 | 33 |
| 4 | 6 | 18 | 11 | 25 | 12 | 28 | 12 | 32 | 13 | 35 | 14 | 38 | 15 | 41 | 16 | 44 |
| 5 | 6 | 21 | 12 | 28 | 18 | 37 | 19 | 41 | 20 | 45 | 21 | 49 | 22 | 53 | 24 | 56 |
| 6 | 7 | 23 | 12 | 32 | 19 | 41 | 26 | 52 | 28 | 56 | 29 | 61 | 31 | 65 | 32 | 70 |
| 7 | 7 | 26 | 13 | 35 | 20 | 45 | 28 | 56 | 37 | 68 | 39 | 73 | 41 | 78 | 43 | 83 |
| 8 | 8 | 28 | 14 | 38 | 21 | 49 | 29 | 61 | 39 | 73 | 49 | 87 | 51 | 93 | 54 | 98 |
| 9 | 8 | 31 | 15 | 41 | 22 | 53 | 31 | 65 | 41 | 78 | 51 | 93 | 63 | 108 | 66 | 114 |
| 10 | 9 | 33 | 16 | 44 | 24 | 56 | 32 | 70 | 43 | 83 | 54 | 98 | 66 | 114 | 79 | 131 |

**b.** $\alpha = .05$ one-tailed; $\alpha = .10$ two-tailed

| $n_2 \backslash n_1$ | 3 $T_L$ | $T_U$ | 4 $T_L$ | $T_U$ | 5 $T_L$ | $T_U$ | 6 $T_L$ | $T_U$ | 7 $T_L$ | $T_U$ | 8 $T_L$ | $T_U$ | 9 $T_L$ | $T_U$ | 10 $T_L$ | $T_U$ |
|---|---|---|---|---|---|---|---|---|---|---|---|---|---|---|---|---|
| 3 | 6 | 15 | 7 | 17 | 7 | 20 | 8 | 22 | 9 | 24 | 9 | 27 | 10 | 29 | 11 | 31 |
| 4 | 7 | 17 | 12 | 24 | 13 | 27 | 14 | 30 | 15 | 33 | 16 | 36 | 17 | 39 | 18 | 42 |
| 5 | 7 | 20 | 13 | 27 | 19 | 36 | 20 | 40 | 22 | 43 | 24 | 46 | 25 | 50 | 26 | 54 |
| 6 | 8 | 22 | 14 | 30 | 20 | 40 | 28 | 50 | 30 | 54 | 32 | 58 | 33 | 63 | 35 | 67 |
| 7 | 9 | 24 | 15 | 33 | 22 | 43 | 30 | 54 | 39 | 66 | 41 | 71 | 43 | 76 | 46 | 80 |
| 8 | 9 | 27 | 16 | 36 | 24 | 46 | 32 | 58 | 41 | 71 | 52 | 84 | 54 | 90 | 57 | 95 |
| 9 | 10 | 29 | 17 | 39 | 25 | 50 | 33 | 63 | 43 | 76 | 54 | 90 | 66 | 105 | 69 | 111 |
| 10 | 11 | 31 | 18 | 42 | 26 | 54 | 35 | 67 | 46 | 80 | 57 | 95 | 69 | 111 | 83 | 127 |

*Source:* From F. Wilcoxon and R. A. Wilcox, "Some Rapid Approximate Statistical Procedures," 1964, 20–23.

**TABLE XIII** **Critical Values of $T_0$ in the Wilcoxon Paired Difference Signed Ranks Test**

| One-Tailed | Two-Tailed | $n = 5$ | $n = 6$ | $n = 7$ | $n = 8$ | $n = 9$ | $n = 10$ |
|---|---|---|---|---|---|---|---|
| $\alpha = .05$ | $\alpha = .10$ | 1 | 2 | 4 | 6 | 8 | 11 |
| $\alpha = .025$ | $\alpha = .05$ | | 1 | 2 | 4 | 6 | 8 |
| $\alpha = .01$ | $\alpha = .02$ | | | 0 | 2 | 3 | 5 |
| $\alpha = .005$ | $\alpha = .01$ | | | | 0 | 2 | 3 |
| | | $n = 11$ | $n = 12$ | $n = 13$ | $n = 14$ | $n = 15$ | $n = 16$ |
| $\alpha = .05$ | $\alpha = .10$ | 14 | 17 | 21 | 26 | 30 | 36 |
| $\alpha = .025$ | $\alpha = .05$ | 11 | 14 | 17 | 21 | 25 | 30 |
| $\alpha = .01$ | $\alpha = .02$ | 7 | 10 | 13 | 16 | 20 | 24 |
| $\alpha = .005$ | $\alpha = .01$ | 5 | 7 | 10 | 13 | 16 | 19 |
| | | $n = 17$ | $n = 18$ | $n = 19$ | $n = 20$ | $n = 21$ | $n = 22$ |
| $\alpha = .05$ | $\alpha = .10$ | 41 | 47 | 54 | 60 | 68 | 75 |
| $\alpha = .025$ | $\alpha = .05$ | 35 | 40 | 46 | 52 | 59 | 66 |
| $\alpha = .01$ | $\alpha = .02$ | 28 | 33 | 38 | 43 | 49 | 56 |
| $\alpha = .005$ | $\alpha = .01$ | 23 | 28 | 32 | 37 | 43 | 49 |
| | | $n = 23$ | $n = 24$ | $n = 25$ | $n = 26$ | $n = 27$ | $n = 28$ |
| $\alpha = .05$ | $\alpha = .10$ | 83 | 92 | 101 | 110 | 120 | 130 |
| $\alpha = .025$ | $\alpha = .05$ | 73 | 81 | 90 | 98 | 107 | 117 |
| $\alpha = .01$ | $\alpha = .02$ | 62 | 69 | 77 | 85 | 93 | 102 |
| $\alpha = .005$ | $\alpha = .01$ | 55 | 61 | 68 | 76 | 84 | 92 |
| | | $n = 29$ | $n = 30$ | $n = 31$ | $n = 32$ | $n = 33$ | $n = 34$ |
| $\alpha = .05$ | $\alpha = .10$ | 141 | 152 | 163 | 175 | 188 | 201 |
| $\alpha = .025$ | $\alpha = .05$ | 127 | 137 | 148 | 159 | 171 | 183 |
| $\alpha = .01$ | $\alpha = .02$ | 111 | 120 | 130 | 141 | 151 | 162 |
| $\alpha = .005$ | $\alpha = .01$ | 100 | 109 | 118 | 128 | 138 | 149 |
| | | $n = 35$ | $n = 36$ | $n = 37$ | $n = 38$ | $n = 39$ | |
| $\alpha = .05$ | $\alpha = .10$ | 214 | 228 | 242 | 256 | 271 | |
| $\alpha = .025$ | $\alpha = .05$ | 195 | 208 | 222 | 235 | 250 | |
| $\alpha = .01$ | $\alpha = .02$ | 174 | 186 | 198 | 211 | 224 | |
| $\alpha = .005$ | $\alpha = .01$ | 160 | 171 | 183 | 195 | 208 | |
| | | $n = 40$ | $n = 41$ | $n = 42$ | $n = 43$ | $n = 44$ | $n = 45$ |
| $\alpha = .05$ | $\alpha = .10$ | 287 | 303 | 319 | 336 | 353 | 371 |
| $\alpha = .025$ | $\alpha = .05$ | 264 | 279 | 295 | 311 | 327 | 344 |
| $\alpha = .01$ | $\alpha = .02$ | 238 | 252 | 267 | 281 | 297 | 313 |
| $\alpha = .005$ | $\alpha = .01$ | 221 | 234 | 248 | 262 | 277 | 292 |
| | | $n = 46$ | $n = 47$ | $n = 48$ | $n = 49$ | $n = 50$ | |
| $\alpha = .05$ | $\alpha = .10$ | 389 | 408 | 427 | 446 | 466 | |
| $\alpha = .025$ | $\alpha = .05$ | 361 | 379 | 397 | 415 | 434 | |
| $\alpha = .01$ | $\alpha = .02$ | 329 | 345 | 362 | 380 | 398 | |
| $\alpha = .005$ | $\alpha = .01$ | 307 | 323 | 339 | 356 | 373 | |

*Source:* From F. Wilcoxon and R. A. Wilcox, "Some Rapid Approximate Statistical Procedures," 1964, p. 28.

## TABLE XIV Critical Values of Spearman's Rank Correlation Coefficient

*The $\alpha$ values correspond to a one-tailed test of $H_0$: $\rho = 0$. The value should be doubled for two-tailed tests.*

| $n$ | $\alpha = .05$ | $\alpha = .025$ | $\alpha = .01$ | $\alpha = .005$ | $n$ | $\alpha = .05$ | $\alpha = .025$ | $\alpha = .01$ | $\alpha = .005$ |
|-----|------|------|------|------|-----|------|------|------|------|
| 5 | .900 | | — | — | 18 | .399 | .476 | .564 | .625 |
| 6 | .829 | .886 | .943 | — | 19 | .388 | .462 | .549 | .608 |
| 7 | .714 | .786 | .893 | — | 20 | .377 | .450 | .534 | .591 |
| 8 | .643 | .738 | .833 | .881 | 21 | .368 | .438 | .521 | .576 |
| 9 | .600 | .683 | .783 | .833 | 22 | .359 | .428 | .508 | .562 |
| 10 | .564 | .648 | .745 | .794 | 23 | .351 | .418 | .496 | .549 |
| 11 | .523 | .623 | .736 | .818 | 24 | .343 | .409 | .485 | .537 |
| 12 | .497 | .591 | .703 | .780 | 25 | .336 | .400 | .475 | .526 |
| 13 | .475 | .566 | .673 | .745 | 26 | .329 | .392 | .465 | .515 |
| 14 | .457 | .545 | .646 | .716 | 27 | .323 | .385 | .456 | .505 |
| 15 | .441 | .525 | .623 | .689 | 28 | .317 | .377 | .448 | .496 |
| 16 | .425 | .507 | .601 | .666 | 29 | .311 | .370 | .440 | .487 |
| 17 | .412 | .490 | .582 | .645 | 30 | .305 | .364 | .432 | .478 |

*Source:* From E. G. Olds, "Distribution of Sums of Squares of Rank Differences for Small Samples," *Annals of Mathematical Statistics*, 1938, 9.

# APPENDIX B

## Data Sets on CD-ROM That Accompanies Text

| File Name | Location | Number of Variables | Description of Variables |
|---|---|---|---|
| APHASIA | Table 2.1 | 1 | Type of aphasia |
| BLOODLOSS | Example 2.1 | 3 | Patient, drug type, & complication type |
| CEODEGREES | Ex. 2.9 | 1 | Highest degree obtained |
| HEARLOSS | Ex. 2.10 | 1 | Type of hearing loss (coded) |
| DOLPHIN | Ex. 2.13 | 2 | Whistle category & number |
| DIGITS | Ex. 2.14 | 2 | 1st digit & number of occurrences |
| OILSPILL | Ex. 2.15 | 2 | Spillage amount & cause of spill |
| EPAGAS | Table 2.3 | 1 | Gas mileage rating |
| LOANDEFAULT | Table 2.5 | 1 | Percent of loans in default |
| SPIDERS | Ex. 2.24 | 1 | Fluid loss |
| BRAINPMI | Ex. 2.25 | 1 | Postmortem interval |
| BULIMIA | Ex. 2.27 | 2 | Group & FNE score |
| NEUROSPECT | Ex. 2.28 | 2 | Density ratio & patient type |
| SAT | Ex. 2.30 | 3 | State, 1990 SAT & 2000 SAT |
| LISTEN | Ex. 2.31 | 3 | Group, response rate & percent correct |
| STLRUNS | Ex. 2.32 | 1 | Runs scored by St. Louis |
| CUBSRUNS | Ex. 2.32 | 1 | Runs scored by Chicago |
| WOMENPOWER | Ex. 2.44 | 4 | Name, age, company & title |
| AMMONIA | Ex. 2.45 | 1 | Ammonia concentration |
| CONDOR | Ex. 2.46 | 3 | Plot, number killed & breeders |
| PAF | Ex. 2.48 | 1 | PAF inhibition percentage |
| SLI | Ex. 2.51 | 5 | Subject, gender, group, DIQ & percent pronoun errors |
| LEFTEYE | Ex. 2.52 | 1 | Cylinder power measurement |
| RATMAZE | Table 2.9 | 1 | Maze run time |
| SHIPSANIT | Ex. 2.72 | 2 | Ship name & sanitation score |
| REACTION | Table 2.10 | 2 | Stimulus type & reaction time |
| LM2_98 | Ex. 2.98 | 2 | Sample & value |
| MBASAL | Ex. 2.100 | 1 | Top salary offer |
| DOWNTIME | Ex. 2.103 | 2 | Customer & downtime |
| MEDFACTORS | Table 2.11 | 2 | Number of factors & length of stay |
| BLACKBREAM | Ex. 2.110 | 3 | Week, number of strikes & age |
| ALWINS | Ex. 2.111 | 3 | Team, wins & batting average |
| EYECUE | SIA Chap. 2 | 4 | Gender, group, deviation & judgment |
| WALKS | Ex. 2.113 | 3 | Steps, unrooted walks & self-avoiding walks |
| CRASH | Ex. 2.115 | 2 | Driver star rating & head injury rating |
| BEANIE | Ex. 2.127 | 5 | Number, name, age retired status & value |
| TORNADO | Ex. 2.130 | 3 | State, number of tornados & peak month |
| TRAPS | Ex. 2.134 | 2 | Species & number killed |
| EVOS | Ex. 2.135 | 4 | Transect, number of sea birds, length & oil status |
| GALAXY2 | Ex. 2.137 | 1 | Galaxy velocity |
| SUICIDE | Ex. 2.139 | 7 | Victim, jail days, marital status, race, charge, suicide time & year |
| RANUNI | Ex. 5.9 | 1 | Random number value |
| LM5_46 | Ex. 5.46 | 1 | Quantitative value |
| DARTS | Ex. 5.49 | 2 | Distance from target & frequency |
| ALBATAVG | Ex. 5.50 | 2 | AL player & batting average |

| File Name | Location | Number of Variables | Description of Variables |
|---|---|---|---|
| NLBATAVG | Ex. 5.50 | 2 | NL player & batting average |
| RUBIDIUM | Ex. 5.107 | 1 | Rubidium amount |
| INSOMNIA | Table 6.3 | 1 | Time to fall asleep |
| HOSPLOS | Table 7.1 | 1 | Length of stay |
| AIRNOSHOWS | Example 7.1 | 1 | Number of unoccupied seats |
| CHARITY | Ex. 7.14 | 1 | Charitable amount |
| CORR34 | Ex. 7.17 | 1 | Study correlation |
| ATTIMES | Ex. 7.18 | 1 | Twins attention time |
| PRINTHEAD | Table 7.4 | 1 | Number of characters printed |
| LEADCOPP | Ex. 7.27 | 2 | Lead level & copper level |
| DECAY | Ex. 7.28 | 1 | Daily decay rate |
| PIGEONS | Ex. 7.29 | 1 | Pigeon crop weight |
| SOUND | Ex. 7.30 | 1 | Sound pressure level |
| TRAFFIC | Ex. 7.31 | 4 | Station, route type, 30th hour volume & 100th hour volume |
| SNACK | Ex. 7.40 | 1 | Taste rating (coded) |
| SCALLOPS | Table 7.6 | 1 | Scallop weight measurement |
| MATHCPU | Ex. 7.81 | 4 | Solution time |
| PCB | Ex. 8.26 | 1 | Number of joints inspected |
| BONES | Ex. 8.27 | 1 | Length-to-width ratio |
| EMISSIONS | Example 8.5 | 1 | Engine emission level |
| IRONTEMP | Ex. 8.54 | 1 | Crankshaft pouring temperature |
| FILLAMOUNTS | Example 8.10 | 1 | Can fill weight |
| CO2 | Ex. 8.95 | 1 | Carbon dioxide amount |
| DIETSTUDY | Table 9.1 | 2 | Diet & weight loss |
| READING | Table 9.2 | 2 | Method & reading score |
| LM9_7 | Ex. 9.7 | 2 | Sample & value |
| HWSTUDY | Ex. 9.17 | 4 | Home work condition, level of involvement in science, math & language |
| BILLBOARD | Ex. 9.19 | 3 | Pre/Post SoundScan, week & number of new artists |
| PAIREDSCORES | Table 9.3 | 3 | Pair, new method score & standard method score |
| PAIREDGRADS | Table 9.5 | 3 | Pair, male salary & female salary |
| LM9_29 | Ex. 9.29 | 3 | Pair, value 1 & value 2 |
| LM9_30 | Ex. 9.30 | 3 | Pair, value 1 & value 2 |
| LM9_31 | Ex. 9.31 | 3 | Pair, x-value & y-value |
| DEPOPROV | Ex. 9.33 | 3 | Patient, pre-treatment level & post-treatment level |
| FMATTITUDES | Ex. 9.34 | 3 | Student, father attitude & mother attitude |
| MSSTUDY | Ex. 9.37 | 6 | Pair, gender, MS age, MS height, MS weight, nonMS age, nonMS height & nonMS weight |
| TWINSIQ | Ex. 9.38 | 3 | Pair, twin A IQ & twin B IQ |
| RATPUPS | Ex. 9.40 | 3 | Litter, male swims & female swims |
| HOMOPHONE | Ex. 9.41 | 3 | Patient, time 1 confusion error & time 2 confusion error |
| INOSITOL | Ex. 9.42 | 3 | Patient, placebo panic attacks & inositol panic attacks |
| MICEWTS | Table 9.7 | 2 | Supplier & mouse weight |
| INSPECT | Ex. 9.82 | 2 | Inspection type & number of errors |
| ORDTIMES | Ex. 9.84 | 2 | Date & order-to-delivery times |
| MILK | SIA Chap. 9 | 6 | Year, market, winner, whole white bid, low-fat white bid & low-fat chocolate bid |
| LM9_89 | Ex. 9.89 | 3 | Pair, value 1 & value 2 |
| BACTERIA | Ex. 9.97 | 2 | Location & bacteria count |
| PULLFORCE | Ex. 9.100 | 3 | Pair, female force & male force |
| PUPILL | Ex. 9.102 | 3 | Consumer, reading 1 & reading 2 |
| WOOLYCAT | Ex. 9.104 | 3 | Day, air temperature & cocoon temperature |
| BIOFEED | Ex. 9.105 | 3 | Subject, before blood pressure & after blood pressure |
| GOLFCRD | Table 10.1 | 2 | Golf ball brand & driving distance |
| SORPRATE | Ex. 10.24 | 2 | Organic solvent & sorption rate |

| File Name | Location | Number of Variables | Description of Variables |
|---|---|---|---|
| HAIRPAIN | Ex. 10.25 | 2 | Hair color & pain threshold |
| NAMEGAME | Ex. 10.26 | 2 | Group & percent of names recalled |
| FACES | Ex. 10.27 | 2 | Dominance rating & facial expression rating |
| GOLFRBD | Table 10.5 | 3 | Golfer, brand & driving distance |
| LM10_42 | Ex. 10.42 | 3 | Treatment, block & value |
| LM10_43 | Ex. 10.43 | 3 | Treatment, block & value |
| DECOY | Ex. 10.46 | 3 | Pit blind, decoy type & percent of goose flock |
| WALKERS | Ex. 10.47 | 3 | Week, prompting group & number of walkers/week |
| BOXING | Ex. 10.48 | 3 | Boxer, intervention & punching power |
| PLANTS | Ex. 10.49 | 3 | Student, plant type & finger temperature |
| EYEDOGS | Ex. 10.50 | 3 | Dog, drug & eye pressure |
| MOSQUITO | Ex. 10.51 | 6 | Location & resistance ratios for five insecticides |
| GOLFFAC1 | Table 10.10 | 3 | Golf club, ball brand & driving distance |
| GOLFFAC2 | Table 10.11 | 3 | Golf club, ball brand & driving distance |
| LM10_56 | Ex. 10.56 | 3 | A level, B level & value |
| LM10_58 | Ex. 10.58 | 3 | A level, B level & value |
| VITAMINB | Ex. 10.65 | 3 | Rat size, diet & kidney weight |
| PRACEXAM | Ex. 10.67 | 3 | Preparation type, class standing & helpfulness rating |
| ROACH | Table 10.12 | 3 | Trail deviation, roach group & extract type |
| LM10_72 | Ex. 10.72 | 2 | Treatment & value |
| LM10_73 | Ex. 10.73 | 3 | Treatment, block & value |
| MICE | Ex. 10.82 | 2 | Group & maze test score |
| INGOT | Ex. 10.83 | 3 | Ingot, bonding agent & pressure |
| WORKSCHD | Ex. 10.84 | 3 | Schedule, payment & attitude score |
| SUPMARK | Ex. 10.85 | 3 | Display type, price type & sales |
| CLOTS | Ex. 10.86 | 3 | Person, drug & clotting time |
| INOCULIN | Ex. 10.87 | 2 | Inoculin & weight increase |
| STIMULUS | Table 11.1 | 2 | Drug amount & reaction time |
| TAMPALMS | Ex. 11.14 | 2 | Appraisal value & sale price |
| OJUICE | Ex. 11.19 | 3 | Run, sweetness index & pectin amount |
| ORGCHEM | Ex. 11.20 | 3 | Compound, maximum absorption & Hammett constant |
| HEARAID | Ex. 11.21 | 3 | Subject, familiar words & unfamiliar words |
| CONDOR2 | Ex. 11.22 | 3 | Plot, number killed & tit occupancy |
| NAMEGAME2 | Ex. 11.23 | 2 | Position & name recall percentage |
| LONGJUMP | Ex. 11.24 | 3 | Jumper, best distance & takeoff error |
| HOSPITAL | Ex. 11.30 | 3 | State, average change & average length of stay |
| CUTTOOL | Ex. 11.33 | 3 | Speed, brand A life & brand B life |
| BOXING2 | Ex. 11.42 | 2 | Blood lactate level & perceived recovery |
| GLOBRISK | Ex. 11.46 | 3 | Country, risk & credit rating |
| BRICKS | Ex. 11.47 | 3 | Location, drop spall rate & photo spall rate |
| CASINO | Table 11.5 | 3 | Year, number of employees & crime rate |
| FERTRATE | Ex. 11.56 | 2 | Contraceptive prevalence & fertility rate |
| SNOWGEESE | Ex. 11.60 | 5 | Trial, diet, weight change, digest efficiency & acid fibre |
| JAPANSPORTS | Ex. 11.71 | 3 | Activity, frequency & duration |
| FIREDAM | Table 11.7 | 2 | Distance from station & fire damage |
| DOWSING | Table 11.8 | 4 | Trial, dowser, pipe location & guess |
| LM11_76 | Ex. 11.76 | 2 | x-value & y-value |
| SOCIOL | Ex. 11.77 | 2 | Verbal score & final grade |
| HORSES | Ex. 11.78 | 2 | Gestation period & life span |
| QLAGREE | Ex. 11.79 | 3 | Statement, US percent & Japan percent |
| HELIUM | Ex. 11.83 | 2 | Temperature & impurity proportion |
| SEEDGERM | Ex. 11.85 | 2 | Energy change & temperature |
| FLOUR | Ex. 11.88 | 2 | Shipment weight & number of bags |
| REALESTATE | Table 12.1 | 5 | Property, sale price, land value, improvements & area |
| GFCLOCKS | Table 12.2 | 3 | Age, number of bidders & price |

| File Name | Location | Number of Variables | Description of Variables |
|---|---|---|---|
| STREETVEN | Ex. 12.8 | 4 | Vendor, earnings, age & hours worked |
| MNSALES | Ex. 12.9 | 8 | Code, price, apartments, age, lot size, parking spaces, area & condition |
| QUASAR | Ex. 12.12 | 7 | Quasar, redshift, lineflux, luminosity, AB1450, magnitude & RFE width |
| WATEROIL | Ex. 12.13 | 9 | Experiment, voltage, volume, salinity, temperature, delay, surfactant, span & solid particles |
| CALIRAIN | Ex. 12.34 | 6 | Number, station, precipitation, altitude, longitude & distance from coast |
| BOILERS | Ex. 12.36 | 5 | Man-hours, capacity, pressure, boiler type & drum type |
| ELECTRIC | Table 12.3 | 2 | Home size & electrical usage |
| CARP | Ex. 12.52 | 3 | Tank, body weight & ENE |
| TIRES | Ex. 12.53 | 2 | Pressure & mileage |
| AEROBIC | Ex. 12.57 | 3 | Subject, IGG value & maximal oxygen uptake |
| MOVIEGROSS | Ex. 12.58 | 7 | Rank, movie, year, US gross, rating, budget & world gross |
| REPELLENT | Ex. 12.68 | 3 | Repellent type, cost/use & protection hours |
| PEAS | Ex. 12.72 | 2 | Variety & yield |
| CASTING | Table 12.5 | 3 | Plant, incentive & productivity |
| LEADMOSS | Ex. 12.83 | 3 | Lead level, elevation & slope face |
| CARNATIONS | Table 12.6 | 3 | Temperature, fertilizer amount & height |
| FLAG | Ex. 12.101 | 15 | Price, DOT estimate, ratio, status, district, number of bidders, days to complete, road length, asphalt, material, excavation, mobil, structures, traffic & sub status |
| GFCLOCKALT | Table 12.9 | 3 | Age, number of bidders & price |
| SOCWORK | Table 12.10 | 2 | Years experience & salary |
| FTC | Table 12.11 | 4 | Tar content, nicotine content, weight & CO content |
| CFSMOKE | Ex. 12.106 | 2 | Weight percentile & cigarettes smoked/day |
| BOYS10 | Ex. 12.107 | 5 | Max. oxygen uptake, age, height, weight & chest depth |
| URBANAIR | Ex. 12.108 | 2 | Acid percent & oxidant |
| CONDO | SIA 12.2 | 6 | Price, floor height, distance from elevator, view, end unit & furnish |
| SHUTDOWN | Ex. 12.134 | 4 | Downtime, age, machine type & type (coded) |
| PHOSPHOR | Ex. 12.135 | 3 | Watershed, soil loss & dissolved phosphorus |
| RELIEF | Ex. 12.136 | 3 | Method, age & time to relief |
| AMI | Ex. 13.22 | 3 | Heart status, alcohol consumption & number |
| TRAUMA | Ex. 13.24 | 3 | Seatbelt status, race & number |
| ETHNIC | Ex. 13.25 | 3 | Region, political strategy & number |
| PIGFARM | Ex. 13.26 | 3 | Size, education level & number |
| MOTH | Ex. 13.27 | 3 | Food choice, egg fed & number |
| SIMKILLS | Ex. 13.28 | 3 | Stop status, number of kills & number |
| HYPNOSIS | Ex. 13.29 | 3 | SHSS, CAHS & number |
| GANGS | Ex. 13.30 | 3 | Gang score, weapon & number |
| HIVVAC1 | Ex. 13.31 | 3 | Group, MN strain & number |
| HIVVAC2 | Ex. 13.31 | 3 | Group, MN strain & number |
| HIVVAC3 | Ex. 13.31 | 3 | Group, MN strain & number |
| TUMORS | Ex. 13.35 | 3 | Cancer, diet & number |
| TRAVEL | Ex. 13.36 | 3 | Nights, retire status & number |
| LIFESTYLE | Ex. 13.45 | 3 | Husband lifestyle, wife lifestyle & number |
| SISEBERT | SIA Chap. 13 | 3 | Movie, Siskel rating & Ebert rating |
| MEDLYONS | SIA Chap. 13 | 3 | Movie, Medved rating & Lyons rating |
| HEARIMP | Ex. 13.37 | 2 | Hearing loss & number of patients |
| FATHSON | Ex. 13.39 | 3 | Father's job, son's job & number |
| QUAKE | Ex. 13.40 | 3 | Insurance status, county & number |
| SCANNER | Ex. 13.42 | 3 | Brand, number of purchases & market share |
| HAIREYE | Ex. 13.43 | 3 | Eye color, hair color & number |
| DISRUPT | Ex. 13.46 | 3 | Delay status, behavior & number |
| SUBABUSE | Table 14.1 | 1 | Substance abuse score |

| File Name | Location | Number of Variables | Description of Variables |
|---|---|---|---|
| LM14_4 | Ex. 14.4 | 1 | Value |
| LYMPHO | Ex. 14.8 | 2 | Case & lymphocyte cell count |
| GUPPY | Ex. 14.10 | 1 | Number of fish |
| REACTION | Table 14.2 | 2 | Drug & reaction time |
| LM14_13 | Ex. 14.13 | 2 | Sample & value |
| LM14_14 | Ex. 14.14 | 2 | Sample & value |
| LM14_15 | Ex. 14.15 | 2 | Sample & value |
| EYEMOVE | Ex. 14.19 | 2 | Deaf status & eye movement rate |
| COLORAIN | Ex. 14.20 | 2 | Station & precipitation |
| INFOSYS | Ex. 14.21 | 2 | Sector & IS expenditures |
| COWBIRD | Ex. 14.22 | 2 | Flycatcher group & vocalization rate |
| SOFTPAPER | Table 14.4 | 3 | Judge, A rating & B rating |
| CRIMEPLAN | Table 14.6 | 3 | Expert, A rating & B rating |
| LM14_25 | Ex. 14.25 | 3 | Pair, A value & B value |
| FACEVT | Ex. 14.32 | 3 | Group, face score & video score |
| AGGTWINS | Ex. 14.33 | 3 | Twin set, 1st born score & 2nd born score |
| ATLAS | Ex. 14.34 | 3 | Theme, teacher ranking & alumni ranking |
| HYPOGLY | Ex. 14.35 | 3 | Mouse, X blood sugar & Y blood sugar |
| POWVEP | Ex. 14.36 | 3 | POW, VEP 157 days & VEP 379 days |
| HOSPBEDS | Table 14.7 | 2 | Hospital & number of beds |
| MKTCYCLE | Ex. 14.40 | 2 | Phase & growth rate |
| READ3 | Ex. 14.41 | 2 | Method & reading level |
| MENTALID | Ex. 14.44 | 2 | Identification & judge type |
| REACTION2 | Table 14.8 | 4 | Subject & times for drugs A, B, C |
| LM14_48 | Ex. 14.48 | 3 | Block, treatment & value |
| LM14_49 | Ex. 14.49 | 3 | Treatment, block & value |
| FARM6 | Ex. 14.50 | 6 | Farmer & constraint rankings for 5 conditions |
| ATLAS2 | Ex. 14.51 | 5 | Theme, group & ranking |
| FLOOD | Ex. 14.52 | 3 | Station, year & flood crest |
| CORRODE | Ex. 14.56 | 3 | Metal, sealer & corrosion amount |
| NEWBORN | Table 14.11 | 2 | Number of cigarettes & baby weight |
| LM14_60 | Ex. 14.60 | 2 | x-value & y-value |
| WINETASTE | Ex. 14.63 | 3 | Brand, expert 1 ranking & expert 2 ranking |
| RECTIME | Ex. 14.67 | 2 | Recreation time & number of tranquilizers |
| COREP | Ex. 14.68 | 3 | Company, 1999 rank & 2000 rank |
| TCDDFAT | Table 14.13 | 1 | TCDD level in fat |
| TCDDPLAS | Table 14.14 | 1 | TCDD level in plasma |
| LM14_70 | Ex. 14.70 | 2 | Sample & value |
| LM14_71 | Ex. 14.71 | 2 | x-value & y-value |
| LM14_72 | Ex. 14.72 | 2 | Sample & value |
| LM14_73 | Ex. 14.73 | 3 | Treatment, block & value |
| TESTORDER | Ex. 14.75 | 2 | Test & score |
| UNION | Ex. 14.76 | 3 | Person, item & ranking |
| COOKIES | Ex. 14.77 | 2 | Technique & boxes sold |
| SLEEPILL | Ex. 14.78 | 3 | Patient, Drug A hours & Drug B hours |
| PARANOIA | Ex. 14.79 | 3 | Child, paranoia ranking & aggression ranking |
| RAZORS | Ex. 14.80 | 2 | Blade type & number of shaves |
| PRINTYPE | Ex. 14.81 | 2 | Print type & read time |
| PRESTIGE | Ex. 14.83 | 3 | Person, job & rank |
| PAINKILL | Ex. 14.84 | 3 | Subject, aspirin relief & new drug relief |
| WORDLIST | Ex. 14.85 | 2 | List & number of words |
| DECEPEYE | Ex. 14.86 | 2 | Deception score & pupil increase |
| HWPATROL | Ex. 14.87 | 3 | Day, heavy number of cars & occasional number of cars |
| DEFECTS | Ex. 14.89 | 2 | Hours & fraction defective |
| FLU | Ex. 14.91 | 2 | Flu strain & recovery time |
| AFLATOX | Ex. 14.92 | 3 | Ear, spray & aflatoxin amount |

# APPENDIX C

## Calculation Formulas for Analysis of Variance

## C.1    FORMULAS FOR THE CALCULATIONS IN THE COMPLETELY RANDOMIZED DESIGN

$$\text{CM} = \text{Correction for mean}$$

$$= \frac{(\text{Total of all observations})^2}{\text{Total number of observations}} = \frac{\left(\sum y_i\right)^2}{n}$$

$$\text{SS(Total)} = \text{Total sum of squares}$$

$$= (\text{Sum of squares of all observations}) - \text{CM} = \sum y_i^2 - \text{CM}$$

$$\text{SST} = \text{Sum of squares for treatments}$$

$$= \left(\begin{array}{c} \text{Sum of squares of treatment totals with} \\ \text{each square divided by the number of} \\ \text{observations for that treatment} \end{array}\right) - \text{CM}$$

$$= \frac{T_1^2}{n_1} + \frac{T_2^2}{n_2} + \cdots + \frac{T_p^2}{n_p} - \text{CM}$$

$$\text{SSE} = \text{Sum of squares for error} = \text{SS(Total)} - \text{SST}$$

$$\text{MST} = \text{Mean square for treatments} = \frac{\text{SST}}{p - 1}$$

$$\text{MSE} = \text{Mean square for error} = \frac{\text{SSE}}{n - p}$$

$$F = \text{Test statistic} = \frac{\text{MST}}{\text{MSE}}$$

where

$$n = \text{Total number of observations}$$
$$p = \text{Number of treatments}$$
$$T_i = \text{Total for treatment } i \ (i = 1, 2, \ldots, p)$$

## C.2 FORMULAS FOR THE CALCULATIONS IN THE RANDOMIZED BLOCK DESIGN

$$CM = \text{Correction for mean}$$

$$= \frac{(\text{Total of all observations})^2}{\text{Total number of observations}} = \frac{(\sum y_i)^2}{n}$$

$$SS(\text{Total}) = \text{Total sum of squares}$$

$$= (\text{Sum of squares of all observations}) - CM = \sum y_i^2 - CM$$

$$SST = \text{Sum of squares for treatments}$$

$$= \left( \begin{array}{c} \text{Sum of squares of treatment totals with} \\ \text{each square divided by } b, \text{ the number of} \\ \text{observations for that treatment} \end{array} \right) - CM$$

$$= \frac{T_1^2}{b} + \frac{T_2^2}{b} + \cdots + \frac{T_p^2}{b} - CM$$

$$SST = \text{Sum of squares for blocks}$$

$$= \left( \begin{array}{c} \text{Sum of squares of block totals with} \\ \text{each square divided by } p, \text{ the number} \\ \text{of observations in that block} \end{array} \right) - CM$$

$$= \frac{B_1^2}{p} + \frac{B_2^2}{p} + \cdots + \frac{B_b^2}{p} - CM$$

$$SSE = \text{Sum of squares for error} = SS(\text{Total}) - SST - SSB$$

$$MST = \text{Mean square for treatments} = \frac{SST}{p-1}$$

$$MSB = \text{Mean square for blocks} = \frac{SSB}{b-1}$$

$$MSE = \text{Mean square for error} = \frac{SSE}{n-p-b+1}$$

$$F = \text{Test statistic} = \frac{MST}{MSE}$$

where

$$n = \text{Total} \quad \text{number of observations}$$
$$b = \text{Number of blocks}$$
$$p = \text{Number of treatments}$$
$$T_i = \text{Total} \quad \text{for treatment } i \ (i = 1, 2, \ldots, p)$$
$$B_i = \text{Total} \quad \text{for block } i \ (i = 1, 2, \ldots, b)$$

## C.3 FORMULAS FOR THE CALCULATIONS FOR A TWO-FACTOR FACTORIAL EXPERIMENT

CM = Correction for mean

$$= \frac{(\text{Total of all } n \text{ measurements})^2}{n} = \frac{\left(\sum_{i=1}^{n} y_i\right)^2}{n}$$

SS(Total) = Total sum of squares

$$= (\text{Sum of squares of all } n \text{ measurements}) - \text{CM} = \sum_{i=1}^{n} y_i^2 - \text{CM}$$

SS($A$) = Sum of squares for main effects, factor $A$

$$= \left(\begin{array}{c} \text{Sum of squares of the totals } A_1, A_2, \ldots, A_a \\ \text{divided by the number of measurements} \\ \text{in a single total, namely } br \end{array}\right) - \text{CM}$$

$$= \frac{\sum_{i=1}^{a} A_i^2}{br} - \text{CM}$$

SS($B$) = Sum of squares for main effects, factor $B$

$$= \left(\begin{array}{c} \text{Sum of squares of the totals } B_1, B_2, \ldots, B_b \\ \text{divided by the number of measurements} \\ \text{in a single total, namely } ar \end{array}\right) - \text{CM}$$

$$= \frac{\sum_{i=1}^{b} B_i^2}{ar} - \text{CM}$$

SS($AB$) = Sum of squares for $AB$ interaction

$$= \left(\begin{array}{c} \text{Sum of squares of the cell totals} \\ AB_{11}, AB_{12}, \ldots, AB_{ab} \text{ divided by} \\ \text{the number of measurements in} \\ \text{a single total, namely } r \end{array}\right) - \text{SS}(A) - \text{SS}(B) - \text{CM}$$

$$= \frac{\sum_{j=1}^{b} \sum_{i=1}^{a} AB_{ij}^2}{r} - \text{SS}(A) - \text{SS}(B) - \text{CM}$$

where

$a$ = Number of levels of factor $A$

$b$ = Number of levels of factor $B$

$r$ = Number of replicates (observations per treatment)

$A_i$ = Total for level $i$ of factor $A$ ($i = 1, 2, \ldots, a$)

$B_i$ = Total for level $i$ of factor $B$ ($i = 1, 2, \ldots, b$)

$AB_{ij}$ = Total for treatment ($i\,j$), i.e., for $i$th level of factor $A$ and $i$th level of factor $B$

# ANSWERS TO SELECTED EXERCISES

## CHAPTER 1

**1.11** qualitative; qualitative **1.13a.** quantitative **b.** qualitative **c.** qualitative **d.** quantitative **1.15b.** qualitative **e.** survey data **1.17a.** qualitative; qualitative; qualitative; quantitative **b.** sample **1.19a.** quantitative **b.** quantitative **c.** qualitative **d.** quantitative **e.** qualitative **f.** quantitative **g.** qualitative **1.21a.** designed experiment **b.** amateur boxers **c.** heart rate (quantitative); blood lactate level (quantitative) **d.** no difference between the two groups of boxers **e.** no **1.23a.** designed experiment **b.** inferential statistics **c.** all possible burn patients **1.25b.** number of headers per game and IQ **c.** both quantitative **1.27b.** answer to question posed; qualitative **d.** survey data

## CHAPTER 2

**2.1** relative frequencies: A, .08; B, .18; C, .45; D, .15; F, .14 **2.3a.** relative frequencies: Black, .191; White, .623; Sumatran, .029; Javan, .005; Indian, .151 **c.** .814; .185 **2.5a.** qualitative; infant, child, medical, infant & medical, child & medical, infant & child, infant & child & medical **b.** infant, .061; child, .565; medical, .276; infant & medical, .001; child & medical, .030; infant & child, .062; infant & child & medical, .004 **d.** .312 **2.7a.** pie chart **b.** breast cancer (19%) **c.** 19% **2.9** 68% of sampled CEOs had advanced degrees **2.11a.** "favorable/recommended"; .635 **b.** yes **2.13** Over half of the whistle types were "Type a."

**2.15a.**

| Cause of Spillage | Frequency |
|---|---|
| Collision | 1 |
| Grounding | 1 |
| Fire/explosion | 1 |
| Hull failure | 1 |
| Unknown | |
| Total | 5 |

**b.** It does not appear that one cause is more likely to occur than any other. **2.17** frequencies: 50, 75, 125, 100, 25, 50, 50, 25 **2.19a.** frequency histogram **b.** 14 **c.** 49 **2.23a.** 28.5% **b.** 71% **c.** slightly skewed to the right **d.** skewed to the left **2.25** most of the PMI's range from 3 to 7.5

**2.27a.**

| Stem | Leaf |
|---|---|
| 0 | 6 7 8 |
| 1 | 0 0 1 3 3 3 3 4 5 6 6 8 9 9 9 |
| 2 | 0 0 1 1 3 4 5 |
| 3 | |

**b.** yes

**2.29a.**

| Stem | Leaf |
|------|------|
| 0 | 1  1  2  3  4  5  5  9 |
| 1 | 1  2  3 |
| 2 | 0  0 |
| 3 | 9 |
| 4 | 6 |
| 5 | 6 |
| 6 | 1  5 |
| 7 | |
| 8 | |
| 9 | |
| 10 | 0 |

**c.** eclipses    **2.31a.** response rate of the familiarity group is higher    **b.** familiarity group had the highest accuracy rate; control group had the lowest accuracy rate    **2.33a.** 12    **b.** 40    **c.** 7    **d.** 21    **e.** 144    **2.35a.** 11.2    **b.** 12    **c.** 30    **2.37** mode $= 15$; mean $= 14.545$; median $= 15$    **2.41a.** 8.5    **b.** 25    **c.** .78    **d.** 13.44    **2.43a.** mean $<$ median    **b.** mean $>$ median    **c.** mean $=$ median    **2.45a.** 1.47    **b.** 1.49    **2.49b.** probably none    **2.51a.** qualitative, qualitative, quantitative, quantitative    **c.** mean $= 93.6$; median $= 91.5$; mode $= 86$    **d.** mean $= 95.3$; median $= 92$; mode $= 92$    **e.** mean $= 101.9$; median $= 103$; mode $= 113$    **g.** SLI: mean $= 30.17$, median $= 32.46$; YND: mean $= 46.88$, median $= 43.72$; OND: mean $= 0$, median $= 0$    **h.** three centers    **2.53a.** median    **b.** mean    **2.55a.** $s^2 = 4.8889$;    $s = 2.211$    **b.** $s^2 = 3.3333$;    $s = 1.826$    **c.** $s^2 = .1868$;    $s = .432$    **2.57a.** 5, 3.7, 1.92    **b.** 99, 1,949.25, 44.15    **c.** 98, 1,307.84, 36.16    **2.59** data set 1: 1, 1, 2, 2, 3, 3, 4, 4, 5, 5; data set 2: 1, 1, 1, 1, 1, 5, 5, 5, 5, 5    **2.61a.** 3, 1.3, 1.1402    **b.** 3, 1.3, 1.1402    **c.** 3, 1.3, 1.1402    **2.63a.** .18    **b.** .0041    **c.** .064    **d.** morning    **2.65a.** $s^2 = 36.178$, $s = 6.1048$    **b.** $s = 8$    **c.** $s = 5$    **d.** decrease    **2.67a.** dollars; quantitative    **b.** at least 3/4; at least 8/9; nothing; nothing    **2.69a.** $\approx 68\%$    **b.** $\approx 95\%$    **c.** $\approx$ all    **2.71** range/6 $= 104.17$, range/4 $= 156.25$;  no    **2.73b.** $59.82 \pm 3(53.362)$    **2.75a.** at most 25%    **b.** $\approx 2.5\%$    **2.77a.** at least 8/9 of the velocities will fall within $936 \pm 3(10)$    **b.** no    **2.79a.** adults: $6.45 \pm 2.89$, $6.45 \pm 2(2.89)$; adolescents: $10.89 \pm 2.48$, $10.89 \pm 2(2.48)$    **b.** adults: 31, 44; adolescents: 13, 18    **2.81a.** $19 \pm 195$    **b.** $7 \pm 147$    **c.** SAT-Math    **2.83a.** 25%; 75%    **b.** 50%; 50%    **c.** 80%; 20%    **d.** 16%; 84%    **2.85a.** $z = 2$    **b.** $z = -3$    **c.** $z = -2$    **d.** $z = 1.67$    **2.87a.** skewed right    **b.** 90% of the scores are below 660    **c.** 6% exceed your score    **2.89a.** $z = -2.92$    **b.** $z = .75$    **2.91a.** $-0.2$    **b.** $-0.06$    **c.** $z = -4.65$    **2.93a.** $z = 2.0: 3.7$; $z = -1.0: 2.2$; $z = .5: 2.95$; $z = -2.5: 1.45$    **b.** 1.9    **c.** mound-shaped; symmetric    **2.97a.** 4    **b.** 3, 5.75    **c.** 2.75    **d.** skewed to the right    **e.** 50%; 75%    **f.** potential outlier: 14; two outliers: 18.5 and 19.5    **2.99a.** $-1.26$    **b.** no    **2.101a.** median for 1990 is smaller than median for 2000    **b.** IQR for 1990 is smaller than IQR for 2000    **c.** no    **2.103b.** customers 238, 268, 269, and 264    **c.** $z = 1.92, 2.06, 2.13, 3.14$    **2.105** group 1: $M = 48$, IQR $= 12$; group 2:    $M = 47.5$, IQR $= 12$; group 3: $M = 46$, IQR $= 4$    **2.107** too little; $z = -2.5$    **2.109** slight positive linear trend    **2.111** very slight increasing trend    **2.113a.** increasing    **b.** increasing    **2.115a.** no trend    **b.** no trend    **2.117a.** $-1, 1, 2$    **b.** $-2, 2, 4$    **c.** 1, 3, 4    **d.** .1, .3, .4    **2.119a.** 3.1234    **b.** 9.0233    **c.** 9.7857    **2.121a.** $\bar{x} = 5.67$,    $s^2 = 1.0667$,    $s = 1.03$    **b.** $\bar{x} = -\$1.50$,    $s^2 = 11.5$,    $s = \$3.39$    **c.** $\bar{x} = .4125\%$,    $s^2 = .0883$, $s = .30\%$    **d.** 3; $10; .7375%    **2.123** yes, positive    **2.125** More than half (60.2%) of the cars received 4 star ratings.    **2.127a.** current: 42.0%; retired: 58.0%    **b.** Most (40 of 50) Beanie babies have values less than $100.    **c.** yes, positive    **2.129c.** $26.4 \pm 21.2$    **d.** $24.5 \pm 22.4$    **2.131a.** yes, Texas    **b.** $\bar{x} = 16.2046$, $M = 10$, $s^2 = 476.0939$, $s = 21.8196$    **c.** skewed to the right    **d.** at least 0; at least 3/4; at least 8/9    **e.** $\bar{x} \pm s$: $47/50 = .94$; $\bar{x} \pm 2s$: $49/50 = .98$; $\bar{x} \pm 3s$: $49/50 = .98$    **2.133a.** frequency histogram    **b.** drop in frequency in years following disaster    **2.135a.** seabirds: quantitative; length: quantitative; oil: qualitative    **b.** transect    **c.** oiled: 38%; unoiled: 62%    **e.** distributions are similar    **f.** $3.27 \pm 13.4$    **g.** $3.495 \pm 11.936$    **h.** unoiled    **2.137b.** yes    **c.** A1775A: $\bar{x} = 19,462.2$, $s = 532.29$; A1775B: $\bar{x} = 22,838.5$, $s = 560.98$    **d.** cluster A1775A    **2.139a.** quantitative: days and year    **b.** lesser crimes    **c.** yes    **d.** mean $= 41.4$ days; median $= 15$ days    **e.** no; $z = 2.38$    **f.** no

# CHAPTER 3

**3.1a.** .5 **b.** .3 **c.** .6 **3.3** $P(A) = .55; P(B) = .50; P(C) = .70$ **3.5a.** $P(A) = 4/9$ **b.** $P(B) = 1/3$ **c.** 0 **3.7** $P(A) = 1/12; P(B) = 1/4; P(C) = 1/2; P(D) = 1/2$ **3.9** $P(A) = 1/10; P(B) = 6/10; P(C) = 3/10$ **3.11** .05 **3.13a.** .01 **b.** yes **3.15a.** 1, 2, 3, 4, 5, 6, 7, 8, 9 **c.** .147, .101, .104, .133, .097, .157, .120, .083, .058 **d.** .248 **e.** .418 **3.17a.** Total population, Agricultural change, Presence of industry, Growth, Population concentration **b.** .18, .05, .27, .05, .45 **c.** .63 **3.19a.** .385 **b.** .378 **c.** .184 **3.21** probabilities are the same **3.23a.** 1 to 2 **b.** .5 **c.** .4 **3.25b.** $P(A) = 7/8$; $P(B) = 1/2$; $P(A \cup B) = 7/8$; $P(A^c) = 1/8$; $P(A \cap B) = 1/2$ **c.** 7/8 **d.** no **3.27a.** 3/4 **b.** 13/20 **c.** 1 **d.** 2/5 **e.** 1/4 **f.** 7/20 **g.** 1 **h.** 1/4 **3.29a.** .65 **b.** .72 **c.** .25 **d.** .35 **e.** .72 **f.** .65 **g.** $A$ and $C$, $B$ and $C$ **3.31** $A = \{$eighth-grader scores above 653 on mathematics assessment test$\}$; $P(A^c) = .95$ **3.33a.** $\{11, 13, 15, 17, 29, 31, 33, 35\}$ **b.** $\{2, 4, 6, 8, 10, 11, 13, 15, 17, 20, 22, 24, 26, 28, 29, 31, 33, 35, 1, 3, 5, 7, 9, 19, 21, 23, 25, 27\}$ **c.** $P(A) = 9/19$; $P(B) = 9/19; P(A \cap B) = 4/19$; $P(A \cup B) = 14/19; P(C) = 9/19$ **d.** $\{11, 13, 15, 17\}$ **e.** 14/19, no **f.** 2/19 **g.** $\{1, 2, 3, \ldots, 29, 31, 33, 35\}$ **h.** 16/19 **3.35a.** .15, no **b.** .36, no **c.** 0, yes **d.** .82, no **e.** .46, no **3.37a.** .216 **b.** .151 **c.** .547 **d.** .475 **3.39a.** 1/2 **b.** 5/6 **3.41** $P(A|B) = .25, P(B|A) = .5$ **3.43b.** $P(A) = 3/4$; $P(B) = 1/2$; $P(A \cap B) = 1/2$ **c.** $P(A|B) = 1; P(B|A) = 2/3$ **3.45** .364 **3.47a.** .857 **b.** .5 **c.** .333 **3.49a.** .28 **b.** .143 **3.51a.** .333 **b.** .120 **c.** no **d.** probability of playing Center depends on race **3.53a.** .557 **b.** .075 **c.** .028 **d.** .504 **3.55a.** .553 **b.** $P(W|CA) = 1$; $P(W|CB) = .873$; $P(W|CC) = .559$; $P(W|BA) = .741$; $P(W|BB) = .547$; $P(W|BC) = .342$; $P(W|AA) = .536$; $P(W|AB) = .216$; $P(W|AC) = .077$ **c.** .856 **d.** .552 **3.57a.** $P(A) = 7/8; P(B) = 3/8; P(C) = 3/8$; $P(D) = 1/2$; $P(A \cap B) = 3/8$; $P(A \cap D) = 3/8$; $P(B \cap C) = 0$; $P(B \cap D) = 3/8$ **b.** $P(B|A) = 3/7; P(A|D) = 3/4; P(C|B) = 0$ **c.** No pairs are independent. **3.59** no **3.61a.** false **b.** true **c.** false **3.63** .095 **3.65a.** .1 **b.** .522 **c.** The person is guessing. **3.67a.** .02 **b.** .08 **3.69** **b.** .147 **c.** .85 **d.** .65 **e.** yes **f.** African Americans are stopped for speeding more often than expected. **3.71a.** .116 **b.** .728 **3.73a.** .25 **b.** .0156 **c.** .4219 **d.** .001, .000000001, .9970 **3.75a.** probability of any sequence is .00098 **b.** .00196 **c.** .99804 **3.77a.** 35,820,200 **b.** 1/35,820,200 **c.** highly unlikely **3.81a.** .000186 **c.** no **3.83a.** 4 **b.** 8 **c.** 32 **d.** $2^n$ **3.85a.** 35 **b.** 15 **c.** 435 **d.** 45 **e.** $\binom{q}{r} = \dfrac{q!}{r!(q - r!)}$ **3.87a.** 4,200 **b.** 2,520 **c.** 1,260 **d.** 3,150 **3.89a.** 56 **b.** 1,680 **c.** 6,720 **3.91a.** 24 **b.** 12 **3.93a.** 30 **b.** 1,050 **3.95a.** 6 **b.** 1/3 **3.97a.** 3,003 **b.** .0186 **c.** .5734 **3.99a.** 21/252 **b.** 21/252 **c.** 105/252 **3.101a.** 1/22,957,480 **b.** 1/593,775 **c.** 1/3,838,380 **d.** Delaware **e.** 7 **f.** 7/3,838,380 **3.103c.** $P(A) = 1/4$; $P(B) = 1/2$ **e.** $P(A^c) = 3/4$; $P(B^c) = 1/2$; $P(A \cap B) = 1/4$; $P(A \cup B) = 1/2$; $P(A|B) = 1/2$; $P(B|A) = 1$ **f.** no, no **3.105a.** no **b.** yes **c.** no **3.107a.** no **b.** .3, .1 **c.** .37 **3.109a.** false **b.** true **c.** true **d.** false **3.111a.** .25 **b.** .13 **c.** .75 **d.** .0325 **3.113** **c.** .6637, .1815, .1263, .0214, .0071 **d.** .2029 **3.115a.** .03125 **b.** .15625 **c.** Pheromone has no effect. **3.117a.** .64, .32, .04 **b.** .72, .22, .06 **c.** dependent **3.119a.** .4096 **b.** .0016 **c.** .9984 **3.121a.** .9639 **b.** .3078 **c.** .0361 **d.** 3 **3.123a.** .3886 **b.** .3049 **c.** .1720 **d.** no **3.125b.** all 1/20 **c.** 1/20 **d.** 1/2 **3.127a.** 142,506 **b.** .236 **c.** .671 **3.129** .4167, .2431, .1418, $(21/36)^{n-1}(15/36)$ **3.131** .993 **3.133** $4.4739 \times 10^{-28}$ **3.135a.** 2/9 **b.** 1/12 **c.** 7/18

# CHAPTER 4

**4.3a.** discrete **b.** continuous **c.** continuous **d.** discrete **e.** continuous **f.** continuous **4.5a.** continuous **b.** discrete **c.** discrete **d.** discrete **e.** discrete **f.** continuous **4.7a.** .25 **b.** .40 **c.** .75 **4.9a.** .7 **b.** .3 **c.** 1 **d.** .2 **e.** .8 **4.11b.** $P(0) = 1/8; P(1) = 3/8; P(2) = 3/8; P(3) = 1/8$ **d.** 1/2 **4.13a.** 1 **b.** .24 **c.** .39 **4.15a.** .9985, .0015, 0, 0 **c.** 1 **4.17a.** .15 **b.** .15 **4.19a.** .51 **b.** .0225 **4.21** 7/8 **4.23a.** 3.8 **b.** 10.56 **c.** 3.2496 **e.** no **f.** yes **4.25a.** $\mu_x = 1, \mu_y = 1$ **b.** distribution of $x$ **c.** $\mu_x = 1, \sigma = .6; \mu_y = 1, \sigma = .2$ **4.27a.** 1.8 **b.** .9899 **c.** .96 **4.29a.** $p(0) = .7$, $p(300,000) = .3$ **b.** \$90,000 **4.31** $\mu = -\$.50$ **4.33a.** mean = 2.9; median = 3

**b.** 3, 4   **d.** 3, 3   **4.35a.** 15   **b.** 10   **c.** 1   **d.** 1   **e.** 4   **4.37a.** .4096   **b.** .3456   **c.** .027   **d.** .0081   **e.** .3456   **f.** .027   **4.39a.** $\mu = 12.5, \sigma^2 = 6.25, \sigma = 2.5$   **b.** $\mu = 16, \sigma^2 = 12.8,$ $\sigma = 3.578$   **c.** $\mu = 60, \sigma^2 = 24, \sigma = 4.899$   **d.** $\mu = 63, \sigma^2 = 6.3, \sigma = 2.510$   **e.** $\mu = 48,$ $\sigma^2 = 9.6, \sigma = 3.098$   **f.** $\mu = 40, \sigma^2 = 38.4, \sigma = 6.197$   **4.41b.** .03125, .15625, .3125, .3125, .15625, .03125   **4.43b.** .05   **c.** 10   **4.45a.** .001   **b.** .322   **c.** .994   **4.47** **b.** 1/30   **c.** .1455   **d.** .1559   **4.49a.** .383   **b.** .983   **c.** 20   **4.51a.** .7908   **b.** .0555   **4.53b.** $\mu = 2.4, \sigma = 1.47$   **c.** $p = .9, q = .1, n = 24, \mu = 21.6, \sigma = 1.47$   **4.55** no   **4.57a.** discrete   **b.** Poisson   **d.** $\mu = 3, \sigma = 1.7321$   **e.** $\mu = 3, \sigma = 1.7321$   **4.59a.** .934   **b.** .191   **c.** .125   **d.** .223   **e.** .777   **f.** .001   **4.61** **b.** $\mu = 3, \sigma = 1.7321$   **c.** .966   **4.63a.** .368   **b.** .264   **c.** .920   **4.65a.** .0224   **b.** .0850   **c.** $\mu = 3.8, \ \sigma = 1.9494$   **4.67a.** 2   **b.** no, $p = .003$   **4.69a.** $p(2) = .039, p(6) = .160, p(10) = .047$   **c.** $\mu = 6.2, \sigma = 2.490$   **d.** very unlikely   **4.71** yes, .96   **4.75a.** .3   **b.** .119   **c.** .167   **d.** .167   **4.77b.** $\mu = 4, \sigma = .853$   **d.** .939   **4.79a.** hypergeometric   **b.** binomial   **4.81a.** .383   **b.** .0002   **4.83** .2693   **4.85** $P(x = 1) = .25$   **4.87a.** $\mu = 113.24, \sigma = 4.194$   **b.** $z = 6.38$   **4.89a.** continuous   **b.** discrete   **c.** continuous   **d.** discrete   **4.91a.** Poisson   **b.** binomial   **c.** binomial   **4.93a.** .243   **b.** .131   **c.** .36   **d.** .157   **e.** .128   **f.** .121   **4.95a.** .180   **b.** .015   **c.** .076   **4.97** binomial   **4.99a.** yes   **b.** .051   **c.** .757   **d.** .192   **e.** 3.678   **4.101a.** $\mu = 520,$ $\sigma = 13.491$   **b.** no, $z = -8.895$   **4.103** .642   **4.105a.** .4019   **b.** .1608   **c.** independent calls   **d.** \$50,000   **4.107a.** .006   **b.** The insecticide is less effective than claimed.   **4.109a.** .021   **b.** 292   **4.111** .0005

# CHAPTER 5

**5.1b.** $\mu = 20, \sigma = 5.774$   **c.** (8.452, 31.548)   **5.3b.** $\mu = 3, \sigma = .577$   **c.** .577   **d.** .61   **e.** .65   **f.** 0   **5.5a.** 0   **b.** 1   **c.** 1   **5.7a.** .667   **b.** .333   **c.** 82.5°F   **5.9** yes   **5.11a.** continuous   **c.** $\mu = 7, \sigma = .2887$   **d.** .5   **e.** 0   **f.** .75   **g.** .0002   **5.13** .4444   **5.15a.** .4772   **b.** .3413   **c.** .4987   **d.** .2190   **e.** .4772   **f.** .3413   **g.** .4545   **h.** .2190   **5.17a.** 0   **b.** .8413   **c.** .8413   **d.** .1587   **e.** .6826   **f.** .9544   **g.** .6934   **h.** .6378   **5.19a.** 1.645   **b.** 1.96   **c.** −1.96   **d.** 1.28   **e.** 1.28   **5.21a.** $z = 1$   **b.** $z = -1$   **c.** $z = 0$   **d.** $z = -2.5$   **e.** $z = 3$   **5.23a.** .0456, .0026   **b.** .6826, .9544   **c.** $x_0 = 261.6$   **5.25a.** 30   **b.** 14.32   **c.** 40.24   **d.** 16.84   **e.** 19.76   **f.** 36.72   **g.** 48.64   **5.27** 182   **5.29a.** .9406   **b.** .9406   **c.** .1140   **5.31a.** .1515   **b.** .1660   **5.33a.** .5124   **b.** .2912   **c.** .0027   **5.35a.** .0735   **b.** .3651   **c.** 7.29   **5.37** $d = 56.24$   **5.39a.** .3745   **b.** .7553   **5.41a.** $z_L = -.675, z_U = .675$   **b.** −2.68, 2.68   **c.** −4.69, 4.69   **d.** .0074, 0   **5.43a.** .68   **b.** .95   **c.** .997   **5.45** plot c   **5.47a.** not normal   **5.49a.** no   **5.51a.** skewed to the right   **b.** approximately normal   **5.53** The lowest score of 0 is less than one standard deviation below the mean.   **5.55a.** yes   **b.** $\mu = 10, \sigma^2 = 6$   **c.** .726   **d.** .7291   **5.57a.** .1788   **b.** .5236   **c.** .6950   **5.59a.** 10   **b.** 3.082   **c.** .16   **d.** .4364   **5.61a.** $\mu = 6,000, \sigma^2 = 2,400$   **b.** .0202   **c.** no   **5.63a.** no   **b.** yes   **c.** yes   **5.65a.** 0   **b.** .4641   **c.** no   **5.67** .2676; no   **5.69a.** 300   **b.** 800   **c.** 1   **5.73a.** .367879   **b.** .950213   **c.** .223130   **d.** .993262   **5.75** .950213   **5.77a.** .449329   **b.** .864665   **5.79** .434598   **5.81a.** .367879   **b.** .606531   **c.** .135335   **d.** .814046   **5.83a.** 17   **b.** .5862   **5.87a.** .9821   **b.** .0179   **c.** .9505   **d.** .3243   **e.** .9107   **f.** .0764   **5.89a.** .6915   **b.** .0228   **c.** .5328   **d.** .3085   **e.** 0   **f.** .9938   **5.91a.** .3821   **b.** .5398   **c.** 0   **d.** .1395   **5.93** .8315   **5.95a.** .0901   **b.** .0384   **c.** .9573   **5.97b.** .9599   **c.** .982   **5.99** 0   **5.101** .3125   **5.103** .0154; very unlikely   **5.105a.** .0764   **b.** (65.024, 88.576)   **5.107** approx. normal   **5.109a.** (i) .384   (ii) .49   (iii). 212   (iv) .84   **b.** (i) .3849   (ii) .4938   (iii) .2119   (iv) .8413   **c.** (i) .0029   (ii) .0038   (iii) .0001   (iv) .0013   **5.111** 52 minutes   **5.113** 0; unlikely   **5.115a.** .135335   **b.** .594

# CHAPTER 6

**6.1c.** 1/16   **6.3c.** .05   **d.** no   **6.9a.** 5   **b.** $E(\bar{x}) = 5$   **c.** $E(M) = 4.778$   **d.** $\bar{x}$   **6.13b.** 1.61   **c.** $E(s^2) = 1.61$   **e.** $E(s) = 1.004$   **6.15a.** $\mu = 100, \sigma = 5$   **b.** $\mu = 100,$ $\sigma = 2$   **c.** $\mu = 100, \sigma = 1$   **d.** $\mu = 100, \sigma = 1.414$   **e.** $\mu = 100, \sigma = .447$   **f.** $\mu = 100,$ $\sigma = .316$   **6.17a.** $\mu = 2.9, \ \sigma^2 = 3.29, \ \sigma = 1.814$   **c.** $\mu_{\bar{x}} = 2.9, \ \sigma_{\bar{x}} = 1.283$   **6.19a.** $\mu = 20, \sigma = 2$   **b.** approximately   normal,   yes   **c.** $z = -2.25$   **d.** $z = 1.50$

**6.21a.** .8944  **b.** .0228  **c.** .1292  **d.** .9699  **6.25a.** $\mu_{\bar{x}} = 89.34, \sigma_{\bar{x}} = 1.3083$  **c.** .8461
**d.** .0367  **6.27** **.0838**  **6.29a.** $\mu_{\bar{x}} = 5.1, \sigma_{\bar{x}} = .4981$  **b.** yes  **c.** .2119  **d.** .4071
**6.31a.** .0013  **b.** program did decrease mean number of sick days  **6.33a.** .0031  **b.** more
likely if $\mu = 156$; less likely if $\mu = 158$  **c.** less likely if $\sigma = 2$; more likely if $\sigma - 6$
**6.37a.** .5  **b.** .0606  **c.** .0985  **d.** .8436  **6.43a.** $\mu_{\bar{x}} = 1.3, \sigma_{\bar{x}} = .240$  **b.** yes  **c.** .1056
**d.** .0062  **6.45a.** .008  **b.** $n$ large  **6.47a.** discrete  **b.** approximately normal with
$\mu_{\bar{x}} = 1.5, \sigma_{\bar{x}} = .01764$  **c.** .0023  **d.** no  **6.49a.** approximately normal with $\mu_{\bar{x}} = 107$,
$\sigma_{\bar{x}} = 1.637$  **b.** .0336  **c.** no  **6.51** **.9332**

# CHAPTER 7

**7.1a.** 1.645  **b.** 2.58  **c.** 1.96  **d.** 1.28  **7.3a.** $10.9375 \pm 2.0094$  **7.5a.** $83.2 \pm 1.25$
**c.** $83.2 \pm 1.65$  **d.** increases  **e.** yes  **7.9** yes  **7.11b.** no assumptions needed  **c.** 95%
confident claim is false  **7.13a.** $19 \pm 7.826$  **b.** $7 \pm 5.90$  **c.** SAT-Math
**7.15a.** $1.13 \pm .672$  **b.** yes  **7.17** $(.3526, .4921)$  **7.19a.** $\mu_Y: 4.17 \pm .095$;  $\mu_{MA}:$
$4.04 \pm .057$; $\mu_O: 4.31 \pm .062$  **b.** more likely  **7.23a.** 2.228  **b.** 2.567  **c.** $-3.707$
**d.** $-1.771$  **7.25a.** $5 \pm 1.876$  **b.** $5 \pm 2.394$  **c.** $5 \pm 3.754$  **d.** (a) $5 \pm .780$, (b) $5 \pm .941$,
(c) $5 \pm 1.276$; width decreases  **7.27a.** $2.8856 \pm 4.034$  **b.** $.4083 \pm .2564$
**7.29** $(0.61, 2.13)$  **7.31c.** $(1,633.05, 2,778.85)$  **e.** $(1,532.39, 2,658.81)$  **7.33a.** $17.82 \pm 3.31$

**b.** $14.14 \pm 3.05$  **c.** possibly  **7.35** approximately normal with $\mu_{\hat{p}} = p$ and $\sigma_{\hat{p}} = \sqrt{\dfrac{pq}{n}}$.

**7.37a.** yes  **b.** $.64 + .067$  **7.39a.** yes  **b.** no  **c.** yes  **d.** no  **7.41** $.556 \pm .044$
**7.43** $.219 \pm .024$  **7.45a.** $.368 \pm .129$  **b.** $.605 \pm .130$  **7.47a.** $.636$  **b.** $.636 \pm .167$
**7.49a.** $.524 \pm .095$  **7.51** 519  **7.53a.** 482  **b.** 214  **7.55** 34  **7.57a.** $n = 16: W = .98$;
$n = 25: W = .784$; $n = 49: W = .56$; $n = 100: W = .392$; $n = 400: W = .196$  **7.59** 984
**7.61** 14,785  **7.63** 129  **7.65** 271  **7.67a.** 1,083  **b.** wider  **c.** 38.3%  **7.69a.** $t = 2.086$
**b.** $z = 1.96$  **c.** $z = 1.96$  **d.** $z = 1.96$  **e.** neither $t$ nor $z$  **7.71** (1) $p$; (2) $\mu$; (3) $\mu$; (4) $\mu$
**7.73a.** $.257 \pm .003$  **b.** $.241 \pm .003$  **7.75** $2 \pm .47$  **7.77a.** $.660 \pm .029$  **b.** $.301 \pm .035$
**7.79a.** $.044 \pm .162$  **b.** no evidence to indicate that this species has tendency to inbreed
**7.81** $(.39, 1.23)$  **7.83** $49.70 \pm .1498$  **7.85a.** $.378 \pm .156$  **b.** $.703 \pm .147$
**c.** $41.405 \pm 21.493$  **d.** $.378 \pm .156$  **7.87a.** $.094$  **b.** yes  **c.** $.094 \pm .037$  **7.89a.** $49.3 \pm$
$8.60$  **c.** Population is normal.  **d.** 35

# CHAPTER 8

**8.1** null hypothesis  **8.3** $\alpha$  **8.7** no  **8.9a.** $H_0: p = .45$  **b.** $H_0: \mu = 2.5$  **8.11a.** $H_0:$
$p = .75, H_a: p > .75$  **8.13a.** $H_0: \mu = 15, H_a: \mu < 15$  **b.** conclude that mean level of
mercury in 2000 is less than 15 ppm when mean is equal to 15 ppm  **c.** conclude that mean
level of mercury in 2000 is equal to 15 ppm when mean is less than 15 ppm  **8.15b.** .37, .24
**8.17g.** (a) .025, (b) .05, (c) $\approx .005$, (d) $\approx .10$, (e) .10, (f) $\approx .01$  **8.19a.** $z = 1.67$, reject $H_0$
**b.** $z = 1.67$, fail to reject $H_0$  **8.21** yes, $z = 12.36$  **8.23a.** $H_0: \mu = 10, H_a: \mu > 10$; one-
tailed  **b.** $z > 1.645$  **c.** $z = 1.29$, fail to reject $H_0$  **8.25a.** $H_0: \mu = 16, H_a: \mu < 16$
**b.** $z = -4.31$, reject $H_0$  **8.27a.** $z = 4.03$, reject $H_0$  **8.29a.** $z = -3.72$, reject $H_0$
**8.31a.** fail to reject $H_0$  **b.** reject $H_0$  **c.** reject $H_0$  **d.** fail to reject $H_0$  **e.** fail to reject $H_0$
**8.33** .0150  **8.35** $p$-value $= .9279$, fail to reject $H_0$  **8.37a.** fail to reject $H_0$  **b.** fail to reject $H_0$  **c.** reject $H_0$  **d.** fail to reject $H_0$  **8.39a.** .0985; fail to reject $H_0$ at $\alpha = .05$
**b.** smaller  **8.41** .0002; reject $H_0$ at $\alpha = .05$  **8.43b.** reject $H_0$ at $\alpha = .05$  **c.** reject $H_0$ at
$\alpha - .05$  **8.45** small $n$, normal population  **8.49a.** $t < -2.160$ or $t > 2.160$  **b.** $t > 2.500$
**c.** $t > 1.397$  **d.** $t < -2.718$  **e.** $t < -1.729$  or  $t > 1.729$  **f.** $t < -2.353$
**8.51a.** $t = -2.064$; fail to reject $H_0$  **b.** $t = -2.064$; fail to reject $H_0$  **c.** (a) .05 $<$
$p$-value $< .10$; (b) .10 $< p$-value $< .20$  **8.53** $t = -.43$; fail to reject $H_0$  **8.55a.** $H_a:$
$\mu \neq 1.0$  **b.** $t = 1.44$  **c.** $p$-value $= .169$, fail to reject $H_0$  **8.57** $t = 2.97$, fail to reject
$H_0: \mu = .100$  **8.59** yes, $t = 1.95$  **8.61a.** yes  **b.** no  **c.** yes  **d.** no  **e.** no
**8.63a.** $z = -2.00$  **c.** reject $H_0$  **d.** .0228  **8.65a.** $z = 1.13$, fail to reject $H_0$  **b.** .1292
**8.67** yes, $z = 1.74$  **8.69** no, $z = -2.37$  **8.71a.** $H_0: p = .5, H_a: p \neq .5$  **b.** $z = 5.99$
**c.** yes  **d.** reject $H_0$  **8.73** $z = 33.47$, reject $H_0: p = .5$  **8.75a.** yes, $z = 1.85$  **b.** yes,

$z = 3.09$ **8.77b.** 1,032.9 **d.** .7422 **e.** .2578 **8.79a.** approximately normal, $\mu_{\bar{x}} = 50$, $\sigma_{\bar{x}} = 2.5$ **b.** approximately normal, $\mu_{\bar{x}} = 45, \sigma_{\bar{x}} = 2.5$ **c.** .2358 **d.** .7642 **8.81a.** approximately normal, $\mu_{\bar{x}} = 10, \sigma_{\bar{x}} = 0.1$ **b.** approximately normal, $\mu_{\bar{x}} = 9.9, \sigma_{\bar{x}} = 0.1$ **c.** .8300 **d.** .8300 **8.83** power increases **8.85** .1075 **8.87a.** $\chi^2 < 6.26214$ or $\chi^2 > 27.4884$ **b.** $\chi^2 > 40.2894$ **c.** $\chi^2 > 21.0642$ **d.** $\chi^2 < 3.57056$ **e.** $\chi^2 < 1.63539$ or $\chi^2 > 12.5916$ **f.** $\chi^2 < 13.8484$ **8.89a.** $\chi^2 = 479.16$, reject $H_0$ **8.91a.** $H_0: \sigma^2 = 225$, $H_a: \sigma^2 > 225$ **b.** $\chi^2 = 187.896$ **c.** $p$-value = .9938, fail to reject $H_0$ **8.93** $\chi^2 = 2.16$, fail to reject $H_0$ **8.95a.** no, $\chi^2 = 2.09$ **b.** The population of $CO_2$ amounts are normally distributed **8.97** alternative **8.99** $H_0, H_a, \alpha$ **8.101** null **8.103a.** $z = -1.78$, fail to reject $H_0$ **b.** $z = -1.78$, fail to reject $H_0$ **c.** $.29 \pm .063$ **d.** $.29 \pm .083$ **e.** 549 **8.105a.** $\chi^2 = 63.48$, reject $H_0$ **b.** $\chi^2 = 63.48$, reject $H_0$ **8.107a.** $H_0$: Drug is unsafe, $H_a$: Drug is safe **c.** $\alpha$ **8.109a.** $t = -3.46$, reject $H_0$ **b.** normal population **8.111a.** yes, $z = 6.11$ **b.** $p$-value $\approx 0$ **c.** .4247 **8.113a.** $H_0: \mu = 250, H_a: \mu > 250$ **b.** $p$-value = .0455, reject $H_0$ at $\alpha = .05$ **c.** $p$-value = .0910, fail to reject $H_0$ at $\alpha = .05$ **8.115a.** $t = 3.512; .025 < p < .05$ **b.** .0362 **c.** normal population **d.** Type I **8.117a.** $z = .70$, fail to reject $H_0$ **b.** yes **8.119a.** no, $z = 1.41$ **b.** small $\alpha$ **8.121a.** no **b.** $\beta = .5910$, power = .4090 **c.** power increases **8.123** $z = 21.66$, reject $H_0$ **8.125a.** .1949; Type II error **b.** .05; Type I error **c.** .8051

# CHAPTER 9

**9.1a.** $35 \pm 24.5$ **b.** $p$-value = .0052, reject $H_0$ **c.** $p$-value = .0026 **d.** $p$-value = .4238, fail to reject $H_0$ **e.** independent random samples **9.3a.** 14; .4 **b.** 10; .3 **c.** 4; .5 **d.** yes **9.5a.** no **b.** no **c.** no **d.** yes **e.** no **9.7a.** .5989 **b.** $t = -2.39$, reject $H_0$ **c.** $-1.24 \pm .98$ **9.9a.** $p$-value = .1150, fail to reject $H_0$ at $\alpha = .10$ **b.** $p$-value = .0575, reject $H_0$ at $\alpha = .10$ **9.11a.** $z = -2.64$, reject $H_0$ **b.** $z = .27$, fail to reject $H_0$ **9.13a.** $(-.60, 7.96)$ **b.** independent random samples from normal populations with equal variances **9.15a.** no, $z = 1.19$ **b.** $.5 \pm .69$; yes **c.** independent random samples **d.** .2340 **9.17a.** $t = -7.24$, $p$-value = 0; reject $H_0$ **b.** $t = -.50$, $p$-value = .62; fail to reject $H_0$ **c.** $t = -1.25$, $p$-value = .21; fail to reject $H_0$ **d.** independent random samples **9.19a.** $(.078, 2.082)$ **b.** .077; yes **9.21a.** $-.52 \pm 3.546$ **b.** both populations normal and $\sigma_1^2 = \sigma_2^2$ **c.** $8.59 \pm 3.152$ **9.23** $t = -.46$, fail to reject $H_0$ **9.25a.** no standard deviations reported **b.** $s_1 = s_2 = 5$ **c.** $s_1 = s_2 = 6$ **9.27a.** $t > 1.833$ **b.** $t > 1.328$ **c.** $t > 2.776$ **d.** $t > 2.896$ **9.29a.** $\bar{x}_D = 2, s_D^2 = 2$ **b.** $\mu_D = \mu_1 - \mu_2$ **c.** $2 \pm 1.484$ **d.** $t = 3.46$, reject $H_0$ **9.31a.** $t = .81$, fail to reject $H_0$ **b.** $p$-value > .20 **9.33a.** $749.5 \pm 278.1$ **9.35a.** $H_0: \mu_D = 0, H_a: \mu_D \neq 0$ **b.** $z = 2.08$, $p$-value = .0376 **c.** reject $H_0$ **9.37b.** $t = .66$, fail to reject $H_0$ **c.** height: $p$-value = .210, fail to reject $H_0$; weight: $p$-value = .149, fail to reject $H_0$ **9.39a.** $H_0: \mu_D = 0, H_a: \mu_D > 0$ **b.** paired difference test **d.** leadership: fail to reject $H_0$, at $\alpha = .05$; popularity: reject $H_0$ at $\alpha = .05$; intellectual self-confidence: reject $H_0$ at $\alpha = .05$ **9.41** $t = -2.306$, $p$-value = .0163, reject $H_0$ at $\alpha = .05$ **9.45a.** $z < -2.33$ **b.** $z < -1.96$ **c.** $z < -1.645$ **d.** $z < -1.28$ **9.47a.** $.07 \pm .067$ **b.** $.06 \pm .086$ **c.** $-.15 \pm .131$ **9.49** $z = 1.16$, fail to reject $H_0$ **9.51a.** .143 **b.** .049 **c.** $z = 2.27$, fail to reject $H_0$ **d.** reject $H_0$ **9.53** $z = -1.201$, fail to reject $H_0$ **9.55** yes; $p$-value = .033, reject $H_0$ at $\alpha = .05$ **9.57** $z = 1.21$, fail to reject $H_0$ at $\alpha = .01$ **9.59a.** yes, $z = 13.02$ **9.61** $n_1 = n_2 = 24$ **9.63a.** $n_1 = n_2 = 29,954$ **b.** $n_1 = n_2 = 2,165$ **c.** $n_1 = n_2 = 1,113$ **9.65** $n_1 = n_2 = 49$ **9.67** $n_1 = n_2 = 130$ **9.69** $n_1 = n_2 = 1,729$ **9.71** $n_1 = n_2 = 136$ **9.73a.** .025 **b.** .90 **c.** .99 **d.** .05 **9.75a.** $F > 2.36$ **b.** $F > 3.04$ **c.** $F > 3.84$ **d.** $F > 5.12$ **9.77a.** $F = 3.43$, reject $H_0$ **b.** $F = 3.43$, reject $H_0$ **9.79** $F = 1.16$, fail to reject $H_0$ **9.81** $F = 1.196$, reject $H_0$ **9.83** $F = 1.045$, fail to reject $H_0$ **9.85a.** $t = .78$, fail to reject $H_0$ **b.** $2.5 \pm 8.99$ **c.** $n_1 = n_2 = 225$ **9.87a.** $3.9 \pm .31$ **b.** $z = 20.60$, reject $H_0$ **c.** $n_1 = n_2 = 346$ **9.89a.** $t = 5.73$, reject $H_0$ **b.** $3.8 \pm 1.84$ **9.91a.** reject $H_0$ **b.** fail to reject $H_0$ **c.** reject $H_0$ **d.** reject $H_0$ **e.** fail to reject $H_0$ **f.** fail to reject $H_0$ **9.93a.** $H_0: \mu_1 - \mu_2 = 0, H_a: \mu_1 - \mu_2 \neq 0$ **b.** $z = -7.69$, reject $H_0$ **9.95a.** time needed **b.** climbers **c.** paired experiment **9.97b.** $H_0: \mu_1 - \mu_2 = 0$, $H_a: \mu_1 - \mu_2 > 0$ **c.** $p$-value = .148, fail to reject $H_0$ at $\alpha = .05$ **d.** both populations normal and $\sigma_{\text{plant}}^2 = \sigma_{\text{upstream}}^2$ **9.99a.** $H_0: \mu_1 - \mu_2 = 0, H_a: \mu_1 - \mu_2 \neq 0$ **b.** $p$-value = .007, reject $H_0$ at $\alpha = .01$ **9.101a.** $H_0: \mu_1 - \mu_2 = 0, H_a: \mu_1 - \mu_2 \neq 0$ **b.** hours of work: fail to reject $H_0$; free time: reject $H_0$ at $\alpha = .05$; breaks: reject $H_0$ at $\alpha = .01$ **9.103a.** $t$-test **b.** yes **c.** no **d.** fail to reject $H_0$ **e.** reject $H_0$ at $\alpha = .05$ **9.105a.** $H_0: \mu_D = 0, H_a$: $\mu_D > 0$ **b.** $p$-value $= .015$, reject $H_0$ at $\alpha = .025$ **c.** $(1.39, 19.01)$ **9.107** $z = -2.55$, reject $H_0$ **9.109** $-.33 \pm .22$ **9.111a.** yes, $F = 10$ **b.** both populations normal **9.113** $z = 2.45$, reject $H_0: p_1 - p_2 = 0$ **9.115** yes, $t = -2.19$

# CHAPTER 10

**10.1** A, B, C, D **10.5a.** observational **b.** designed **c.** designed **d.** observational **e.** observational **f.** observational **10.7a.** cockatiel **b.** yes **c.** experimental group **d.** 1, 2, 3 **e.** 3 **f.** total consumption **10.9a.** ethical behavior **b.** job type (high tech, low tech); sales task (new, maintenance) **c.** high/new, high/maintenance, low/new, low/maintenance **d.** college students **10.11a.** 6.59 **b.** 16.69 **c.** 1.61 **d.** 3.87 **10.13** dot plot **b** **10.15a.** (a) MSE $= 2$; (b) MSE $= 14.4$ **b.** (a) $t = -6.12$; (b) $t = -2.28$ **c.** (a) $|t| > 2.228$; (b) $|t| > 2.228$ **d.** (a) reject $H_0$; (b) reject $H_0$ **e.** both populations normal with equal variances **10.17a.** $F = 1.5625$; do not reject $H_0$ **b.** $F = 6.25$; reject $H_0$ **c.** $F = 25$; reject $H_0$ **d.** increases **10.19a.** 4; 38 **b.** $F = 14.80$; reject $H_0$ **c.** sample means **10.21a.** completely randomized **b.** treatments: 3, 6, 9, 12 robots; dep. var.: energy expanded **c.** $H_0: \mu_3 = \mu_6 = \mu_9 = \mu_{12}, H_a$: At least two $\mu$'s differ **d.** reject $H_0$ **10.23a.** $H_0: \mu_{young} = \mu_{middle} = \mu_{old}$ **b.** reject $H_0$ **c.** $H_0: \mu_{young} = \mu_{middle} = \mu_{old}$; fail to reject $H_0$ **10.25a.** completely randomized **b.** $F = 6.79$; reject $H_0$ **c.** .0041 **10.27** $F = 3.96$; do not reject $H_0$ **10.29a.** 52.32 **b.** 77.24

**c.**

| Source | df | SS | MS | F |
|---|---|---|---|---|
| Treatment | 2 | 52.32 | 26.16 | 12.89 |
| Error | 38 | 77.24 | 2.03 | |
| Total | 40 | 129.56 | | |

**d.** yes, $F = 12.89$ **f.** Subjects in the first two groups were not randomly and independently assigned to their groups. **10.31** probability of at least one Type I error for multiple comparisons; probability of a Type I error for a single comparison **10.33a.** 6 **b.** $\mu_{12}$ is the largest mean; $\mu_3, \mu_6,$ and $\mu_9$ are not significantly different **10.35a.** yes **b.** yes **c.** no **d.** .05 **e.** girl means not significantly different **10.37a.** reject $H_0$ **b.** Control (41.8) Slide (43.1) Questionnaire (53.8) **10.39** no significant differences **10.41a.** 3 blocks; 5 treatments **b.** 15 **c.** $H_0: \mu_1 = \mu_2 = \mu_3 = \mu_4 = \mu_5, H_a$: At least two $\mu$'s differ **d.** $F = 9.109$ **e.** $F > 7.01$ **f.** reject $H_0$

**10.43a.**

| Source | df | SS | MS | F |
|---|---|---|---|---|
| Treatment | 2 | 12.032 | 6.016 | 50.958 |
| Block | 3 | 71.749 | 23.916 | 202.586 |
| Error | 6 | .708 | .118 | |
| Total | 12 | 323.101 | | |

**b.** yes, $p$-value $= .000$ **c.** yes, $p$-value $= .000$ **d.** $\mu_C < \mu_A < \mu_B$ **10.45a.** randomized block design **c.** reject $H_0$ at $\alpha > .009$ **d.** $\mu_{control} < \mu_{burning}$

**10.47b.**

| Source | df | SS | MS | F |
|---|---|---|---|---|
| Prompt | 4 | 1185.00 | 296.25 | 39.87 |
| Week | 5 | 386.40 | 77.28 | 10.40 |
| Error | 20 | 148.60 | 7.43 | |
| Total | 29 | 1720.00 | | |

**c.** yes, $p$-value $= .0000$ **d.** $\mu_{Control} < (\mu_{Inf-Low}, \mu_{Inf-Hi}) < (\mu_{Freq-Low}, \mu_{Freq-Hi})$ **10.49** do not reject $H_0$, $F = 0.019$ **10.51a.** treatments: 5 insecticides; blocks: 7 locations **b.** no; fail to reject $H_0$ at $\alpha = .10$ **10.53a.** 2 **b.** no **c.** yes; 3 and 5 **d.** 15 **e.** df(Error) $= 0$; replication

**10.55a.**

| Source | df | SS | MS | F |
|--------|-----|------|--------|------|
| A | 2 | .8 | .4000 | 2.00 |
| B | 3 | 5.3 | 1.7667 | 8.83 |
| AB | 6 | 9.6 | 1.6000 | 8.00 |
| Error | 12 | 2.4 | .2000 | |
| Total | 23 | 18.1 | | |

**b.** SSA, SSB, SS(AB); yes, $F = 7.14$ **c.** yes **d.** The effects of one factor on the dependent variable are not the same at different levels of the second factor. **e.** $F = 8.00$, reject $H_0$ **f.** no **10.57a.** $F(AB) = .75$; $F(A) = 3$; $F(B) = 1.5$ **b.** $F(AB) = 7.5$; $F(A) = 3$; $F(B) = 3$ **c.** $F(AB) = 3$; $F(A) = 12$; $F(B) = 3$ **d.** $F(AB) = 4.5$; $F(A) = 36$; $F(B) = 36$ **10.59a.** $6 \times 6$ factorial **b.** coagulant $(5, 10, 20, 50, 100, 200)$; pH $(4, 5, 6, 7, 8, 9)$; 36 treatments **10.61a.** $2 \times 2$ factorial; color (blue, red), question (difficult, simple) **b.** sufficient evidence of interaction at $\alpha = .05$ **10.63b.** sufficient evidence of interaction at $\alpha = .01$ **c.** no

**10.65a.**

| Source | d | SS | MS | F | p |
|--------|-----|--------|--------|--------|----------|
| Diet ($D$) | 1 | 0.0124 | 0.0124 | 0.22 | 0.645 |
| Size ($S$) | 1 | 8.0679 | 8.0679 | 141.18 | < 0.0001 |
| DS | 1 | 0.0364 | 0.0364 | 0.64 | 0.432 |
| Error | 24 | 1.3715 | 0.0571 | | |
| Total | 27 | 9.4883 | | | |

**b.** main effect Diet: fail to reject $H_0$; main effect Size: reject $H_0$; Diet-Size interaction: fail to reject $H_0$ **10.67** main effect Exam Prep: reject $H_0$ ($p$-value $= .0002$); main effect Class Standing: fail to reject $H_0$ ($p$-value $= .1184$); Prep-Standing interaction: fail to reject $H_0$ ($p$-value $= .1742$)

**10.73a.**

| Source | df | SS | MS | F |
|--------|-----|--------|-------|--------|
| Treatment | 3 | 11.332 | 3.777 | 157.38 |
| Block | 4 | 10.688 | 2.672 | 111.33 |
| Error | 12 | .288 | .024 | |
| Total | 19 | 22.308 | | |

**b.** yes, reject $H_0$ **c.** yes; 6 **d.** yes, reject $H_0$ **10.75b.** yes **c.** yes **10.77** $F = 8.295$, reject $H_0$; $\mu_{OND} < (\mu_{SLI}, \mu_{YND})$ **10.79a.** Infrequency: $F = 155.8$, reject $H_0$; Obvious: $F = 49.7$, reject $H_0$; Subtle: $F = 10.3$, reject $H_0$; Obvious-Subtle: $F = 45.4$, reject $H_0$; Dissimulation: $F = 39.1$, reject $H_0$ **b.** no **c.** Infrequency: $\mu_{CSH} < (\mu_{NFP}, \mu_{FP}) < \mu_{CSFB}$; Obvious: $\mu_{CSH} < (\mu_{NFP}, \mu_{FP}) < \mu_{CSFB}$; Subtle: no significant differences; Obvious-Subtle: $(\mu_{CSH}, \mu_{NFP}, \mu_{FP}) < \mu_{CSFB}$; Dissimulation: $(\mu_{CSH}, \mu_{NFP}, \mu_{FP}) < \mu_{CSFB}$

**10.81a.**

| Source | df | F | p-value |
|--------|-----|-------|---------|
| Time Period | 3 | 11.25 | .0001 |
| Station | 9 | | |
| Error | 27 | | |
| Total | 39 | | |

**b.** reject $H_0$, $F = 11.25$ **c.** largest: either 7/9–7/23 or 7/24–8/8; smallest: 8/24–8/31 **10.83a.** experimental units: zircaloy components; blocks: ingots; response: pressure required to separate the bonded components; factor: bonding agent or element; factor type: qualitative; treatments: nickel, iron, copper **b.** designed

**c.**

| Source | df | SS | MS | F |
|--------|----|----|----|----|
| Agent | 2 | 131.9 | 65.95 | 6.36 |
| Ingot | 6 | 268.3 | 44.71 | 4.31 |
| Error | 12 | 124.4 | 10.37 | |
| Total | 20 | 524.6 | | |

**d.** yes, $F = 6.36$  **e.** $\mu_{\text{iron}}$ is largest mean  **10.85b.** yes, $F = 1{,}336.85$  **c.** yes; reject $H_0$, $F = 258.07$  **d.** no  **e.** nine treatment means

**10.87a.**

| Source | df | SS | MS | F | p-value |
|--------|----|----|----|----|---------|
| Treatment | 4 | 292.8 | 73.2 | 5.29 | 0.007 |
| Error | 15 | 207.7 | 13.9 | | |
| Total | 19 | 500.6 | | | |

**b.** yes, $F = 5.29$  **c.** .007  **10.89a.** no evidence of interaction at $\alpha = .01$; evidence of $H$ main effect at $\alpha = .01$; no evidence of $S$ main effect at $\alpha = .01$  **b.** yes  **c.** $(\mu_{\text{PLC}}, \mu_{\text{LGN}}) < (\mu_{\text{QMD}}, \mu_{\text{MTZ}})$

# CHAPTER 11

**11.3** $\beta_1 = 1/3, \beta_0 = 14/3, y - 14/3 + (1/3)x$  **11.9** no  **11.13c.** $\hat{\beta}_1 = .198, \hat{\beta}_0 = .020$  **e.** $-1$ to 7  **11.15a.** $y = \beta_0 + \beta_1 x + \varepsilon$; negative  **b.** yes  **c.** no  **11.17a.** $y = \beta_0 + \beta_1 x + \varepsilon$  **b.** $\hat{y} = 175.7033 - .8195x$  **11.19a.** $\hat{y} = 6.25 - .0023x$  **c.** 5.56  **11.21a.** yes; positive  **b.** $y = \beta_0 + \beta_1 x + \varepsilon$  **c.** $\hat{\beta}_0 = 20.1275, \hat{\beta}_1 = .62442$  **11.23** $\hat{y} = .5704 + .0264x$  **11.25a.** 57.5; 3.19444  **b.** 257.5; 6.7763  **c.** 9.288; 1.1610  **11.27** graph  **b**  **11.29a.** SSE $= 1{,}667.65, s^2 = 238.24, s = 15.43$  **11.31a.** SSE $= 14{,}357.5, s^2 = 1305.22$, $s = 36.1279$  **11.33b.** Brand A: $\hat{y} = 6.62 - .0727x$; Brand B: $\hat{y} = 9.31 - .1077x$  **c.** Brand A: SSE $= 19.056, s^2 = 1.466, s = 1.211$; Brand B: SSE $= 4.833, s^2 = .372$, $s = .610$  **d.** Brand A: $(-.891, 3.593)$; Brand B: $(.551, 2.991)$  **e.** Brand B  **11.35b.** $\hat{y} = .544 + .617x$  **d.** $H_0: \beta_1 = 0, H_a: \beta_1 \neq 0$  **e.** $t = 5.50, \text{df} = 5$  **f.** reject $H_0$  **11.37** no, $t = .627$  **11.39** $(-.0039, -.0007)$  **11.41c.** $\alpha > .042$  **11.43a.** $t = 6.286$, reject $H_0$  **b.** $.88 \pm .24$  **c.** no evidence that the true slope differs from 1  **11.45** yes, $t = 2.858$  **11.47** $\hat{y} = 2.55 + 2.76x$  **11.49a.** perfect positive linear  **b.** perfect negative linear  **c.** strong positive linear  **d.** weak positive linear  **e.** strong negative linear  **11.51a.** $r = .9853, r^2 = .9709$  **b.** $r = -.9934, r^2 = .9868$  **c.** $r = 0, r^2 = 0$  **d.** $r = 0$, $r^2 = 0$  **11.53a.** $H_0: \rho = 0, H_a: \rho \neq 0$  **b.** do not reject $H_0$ at $\alpha = .05$  **11.55b.** piano: $r^2 = .1998$; bench: $r^2 = .0032$; motorbike: $r^2 = .3832$, armchair: $r^2 = .0864$; teapot: $r^2 = .9006$  **c.** Reject $H_0$ for all objects except bench and armchair.  **11.57a.** $y = \beta_0 + \beta_1 x + \varepsilon$  **b.** $\beta_1 > 0$  **d.** no, $t = -3.12$  **11.59** $r = .5702, r^2 = .3251$  **11.61a.** negative  **b.** $r < -.0895$  **11.63c.** $6.5335 \pm 1.9487$  **d.** $-6.0546 \pm 1.7225$  **11.65c.** $4.644 \pm 1.118$  **d.** $-.414 \pm 1.717$  **11.67a.** no  **b.** $(275.86, 407.26)$  **c.** $(332.95, 350.17)$  **11.69a.** $(1.0516, 6.6347)$  **b.** narrower  **c.** no  **11.71a.** $y = \beta_0 + \beta_1 x + \varepsilon$  **b.** $\hat{y} = 155.912 - 1.086x$  **c.** no, $t = -2.05$  **d.** $(-21.56, 279.09)$  **11.73a.** Brand A: $3.349 \pm .587$; Brand B: $4.464 \pm .296$  **b.** Brand A: $3.349 \pm 2.224$; Brand B: $4.464 \pm 1.120$  **c.** $-.65 \pm 3.606$  **11.75b.** $\hat{y} = x; \hat{y} = 3$  **c.** $\hat{y} = x$  **d.** least squares line has the smallest SSE  **11.77a.** $\hat{y} = 40.7842 + .76556x$  **c.** yes, $t = 4.38$  **d.** $.76556 \pm .4035$  **e.** $(72.513, 85.611)$  **f.** $(57.95, 100.17)$  **11.79a.** .570  **11.81** $\hat{y} = -92.458 + 8.347x; t = 3.248$, reject $H_0$; $r = .42$  **11.83a.** $\hat{\beta}_0 = -13.49, \hat{\beta}_1 = -.0528$  **b.** $-.0528 \pm .0178$; yes  **c.** $r^2 = .8538$  **d.** $(.5987, 1.2653)$  **11.85b.** $\hat{y} = 78.516 - .2389x$  **d.** $t = -2.31$, do not reject $H_0$  **e.** $(301, 8.5)$  **f.** $\hat{y} = 139.759 - .4497x; t = -15.35$, reject $H_0$  **11.87e.** (a): $r^2 = .0676$; (b): $r^2 = .0361$; (c): $r^2 = .0196$; (d): $r^2 = .0001$

# CHAPTER 12

**12.1a.** $E(y) = \beta_0 + \beta_1 x_1 + \beta_2 x_2$  **b.** $E(y) = \beta_0 + \beta_1 x_1 + \beta_2 x_2 + \beta_3 x_3 + \beta_4 x_4$
**c.** $E(y) = \beta_0 + \beta_1 x_1 + \beta_2 x_2 + \beta_3 x_3 + \beta_4 x_4 + \beta_5 x_5$  **12.3a.** $t = 1.45$, do not reject $H_0$
**b.** $t = 3.21$, reject $H_0$  **12.5** df $= n - (k + 1)$  **12.7a.** $\hat{y} = 3.70 + .34x_1 + .49x_2 +$
$.72x_3 + 1.14x_4 + 1.51x_5 + .26x_6 - .14x_7 - .10x_8 - .10x_9$  **b.** $t = -1.00$, do not reject
$H_0$  **c.** $(1.412, 1.608)$  **12.9a.** $\hat{y} = 93{,}074 + 4{,}152x_1 - 855x_2 + .924x_3 + 2{,}692x_4 +$
$15.5x_5$  **b.** $33{,}225.9$  **c.** yes, $t = 2.78, p = .0059$  **f.** $t = -2.86$, reject $H_0$  **g.** $.00495$
**12.11a.** $\hat{y} = 12.1804 - .0265x_1 - .45783x_2$  **c.** $t = -.50, p = .6226$, do not reject $H_0$
**d.** $-.4578 \pm .3469$  **12.13a.** $E(y) = \beta_0 + \beta_1 x_1 + \beta_2 x_2 + \beta_3 x_3 + \beta_4 x_4 + \beta_5 x_5 +$
$\beta_6 x_6 + \beta_7 x_7$  **b.** $\hat{y} = .998 - .022x_1 + .156x_2 - .017x_3 - .0095x_4 + .421x_5 +$
$.417x_6 - .155x_7$  **12.15a.** model 1:$t = 2.58$, reject $H_0$: $\beta_1 = 0$; model 2:$t = 3.32$, reject $H_0$:
$\beta_1 = 0, t = 6.47$, reject $H_0$: $\beta_2 = 0, t = -4.77$, reject $H_0$: $\beta_3 = 0, t = 0.24$, do not reject
$H_0$: $\beta_4 = 0$; model 3:$t = 3.21$, reject $H_0$: $\beta_1 = 0, t = 5.24$, reject $H_0$: $\beta_2 = 0, t = -4.00$, reject $H_0$: $\beta_3 = 0, t = 2.28$, reject $H_0$: $\beta_4 = 0, t = .014$, do not reject $H_0$: $\beta_5 = 0$  **12.17a.** $.459$
**b.** $.396$  **c.** $F = 7.22$, reject $H_0$  **d.** $.005$

**12.19a.**

| Source | df | SS | MS | F |
|--------|----|----|----|----|
| Model | 2 | 12.09 | 6.045 | 8.321 |
| Error | 17 | 12.35 | .72647 | |
| Total | 19 | 24.44 | | |

**b.** $F = 8.321$, reject $H_0$  **12.21a.** $\hat{y} = \hat{\beta}_0 + .110x_1 + .065x_2 + .540x_3 - .009x_4 -$
$.150x_5 - .027x_6$  **b.** $H_0: \beta_1 = \beta_2 = \beta_3 = \beta_4 = \beta_5 = \beta_6 = 0$  **c.** $F = 32.47$, reject $H_0$
**d.** $.44; .43$  **e.** $H_0: \beta_4 = 0$  **f.** $p$-value $= .860$, fail to reject $H_0$  **12.23a.** $.912; .894; R^2_{adj}$
**b.** $F = 51.72$, reject $H_0$  **12.25a.** $t = 3.88, p$-value $= .005$, reject $H_0$  **b.** no  **c.** $F = 11.4$,
reject $H_0$  **12.27b.** $F = 5.11$, reject $H_0$  **12.29a.** $E(y) = \beta_0 + \beta_1 x_1 + \beta_2 x_2 + \beta_3 x_3$
$+ \beta_4 x_4 + \beta_5 x_5$  **b.** $\hat{y} = 15.491 + 12.774x_1 + .713x_2 + 1.519x_3 + .32x_4 + .205x_5$
**c.** $R^2 = .240, F = 11.68$, reject $H_0$  **d.** $p$-value $= .025$, reject $H_0$  **12.31c.** yes
**12.33** $(90.69, 158.57)$  **12.37a.** $E(y) = \beta_0 + \beta_1 x_1 + \beta_2 x_2 + \beta_3 x_1 x_2$  **b.** $E(y) = \beta_0 +$
$\beta_1 x_1 + \beta_2 x_2 + \beta_3 x_3 + \beta_4 x_1 x_2 + \beta_5 x_1 x_3 + \beta_6 x_2 x_3$  **12.39c.** interaction is present
**12.41a.** $\hat{y} = -2.550 + 3.815x_1 + 2.630x_2 - 1.285x_1 x_2$  **e.** $H_0: \beta_3 = 0, H_a: \beta_3 \neq 0$
**f.** $p$-value $= .000$, reject $H_0$  **12.43b.** $H_0: \beta_3 = 0, H_a: \beta_3 \neq 0$  **c.** reject $H_0$  **12.45a.** $E(y)$
$= \beta_0 + \beta_1 x_1 + \beta_2 x_2 + \beta_3 x_1 x_2$  **b.** yes, $t = 1.6$, $p$-value $= .02$, reject $H_0: \beta_3 = 0$
**12.47b.** $\hat{y} = .906 - .023x_1 + .350x_2 + .275x_5 - .003x_1 x_2 + .005x_1 x_5, F = 5.505$,
$R^2 = .6792, s = .505$  **12.49a.** $t = 3.133$, reject $H_0$  **b.** $t = 3.133, t > 1.717$, reject $H_0$
**12.51b.** moves the graph to the right or left  **c.** controls whether the graph opens up or
down  **12.53b.** first-order model where $\beta_1 > 0$; first-order model where $\beta_1 < 0$; second-
order model  **12.55a.** $E(y) = \beta_0 + \beta_1 x_1 + \beta_2 x_2 + \beta_3 x_1 x_2 + \beta_4 (x_1)^2 + \beta_5 (x_2)^2$
**b.** $\beta_4 (x_1)^2, \beta_5 (x_2)^2$  **12.57b.** $E(y) = \beta_0 + \beta_1 x + \beta_2 x^2$  **c.** yes, $F = 203.16$  **d.** 0
**e.** yes, $t = -3.39$  **f.** $.0022$  **12.59a.** $E(y) = \beta_0 + \beta_1 x + \beta_2 x^2$  **b.** positive  **c.** no;
$E(y) = \beta_0 + \beta_1 x$  **12.63** $E(y) = \beta_0 + \beta_1 x_1 + \beta_2 x_2; x_1 = 1$ if level 2, 0 if not; $x_2 = 1$ if
level 3, 0 if not  **12.65a.** $\hat{y} = 80 + 16.8x_1 + 40.4x_2$  **c.** $H_0: \mu_1 = \mu_2 = \mu_3$  **d.** $F = 24.72$,
reject $H_0$  **12.67a.** $E(y) = \beta_0 + \beta_1 x_1 + \beta_2 x_2; x_1 = 1$ if political status is democratic, 0
otherwise; $x_2 = 1$ if political status is dictatorship, 0 otherwise  **b.** $\beta_0 =$ mean assassination
risk level for communist $= \mu_{communist}; \beta_1 = \mu_{democrat} - \mu_{communist}; \beta_2 = \mu_{dictator} - \mu_{communist}$
**12.69a.** $E(y) = \beta_0 + \beta_1 x_1 + \beta_2 x_2 + \beta_3 x_3; x_1 = 1$ if location 2, 0 otherwise; $x_2 = 1$ if location 3, 0 otherwise; $x_3 = 1$ if location 4, 0 otherwise  **b.** $\beta_0$  **c.** $\beta_0 + \beta_3$  **d.** $\beta_1 - \beta_2$
**12.71a.** $E(y) = \beta_0 + \beta_1 x$, where $x = 1$ if enriched pond, 0 if natural pond  **b.** $\beta_0 =$ mean
larval density of natural pond $= \mu_{natural}; \beta_1 = \mu_{enriched} - \mu_{natural}$  **c.** $H_0: \beta_1 = 0, H_a:$
$\beta_1 > 0$  **d.** reject $H_0$  **12.73a.** $E(y) = \beta_0 + \beta_1 x_1 + \beta_2 x_2 + \beta_3 x_3 + \beta_4 x_4$, where $x_1 - x_4$
are dummy variables for the five reproductive classes  **b.** $\hat{\beta}_0 = 38, \hat{\beta}_1 = -22, \hat{\beta}_2 = 51$,
$\hat{\beta}_3 = 5, \hat{\beta}_4 = 20$,  **c.** $H_0: \beta_1 = \beta_2 = \beta_3 = \beta_4 = 0$  **d.** do not reject $H_0$  **12.75a.** $E(y)$
$= \beta_0 + \beta_1 x_1$  **b.** $E(y) = \beta_0 + \beta_1 x_1 + \beta_2 x_2 + \beta_3 x_3, x_2 = 1$ if level 2, 0 otherwise;
$x_3 = 1$ if level 3, 0 otherwise  **c.** $E(y) = \beta_0 + \beta_1 x_1 + \beta_2 x_2 + \beta_3 x_3 + \beta_4 x_1 x_2 + \beta_5 x_1 x_3$
**d.** $\beta_4 = \beta_5 = 0$  **e.** $\beta_2 = \beta_3 = \beta_4 = \beta_5 = 0$  **12.77a.** $E(y) = \beta_0 + \beta_1 x_1; E(y) = \beta_0$

$+ \beta_1 x_1 + \beta_2; E(y) = \beta_0 + \beta_1 x_1 + \beta_3$ **b.** $\hat{y} = 44.803 + 2.173 x_1; \hat{y} = 54.216 + 2.173 x_1; \hat{y} = 60.435 + 2.173 x_1$ **12.79** $E(y) = \beta_0 + \beta_1 x_1 + \beta_2 (x_1)^2 + \beta_3 x_2 + \beta_4 x_3 + \beta_5 x_4$, where $x_2$–$x_4$ are dummy variables **12.81a.** $E(y) = \beta_0 + \beta_1 x_1; \beta_1$ **b.** $E(y) = (\beta_0 + \beta_2) + (\beta_1 + \beta_3) x_1; \beta_1 + \beta_3$ **c.** reject $H_0: \beta_3 = 0$ **12.83a.** $E(y) = \beta_0 + \beta_1 x_1 + \beta_2 x_2 + \beta_3 x_1 x_2$ **c.** $\beta_1 + \beta_3$ **d.** $\hat{y} = 2.38 + .0018 x_1 + 3.20 x_2 - .00133 x_1 x_2$; no, $F = .26$, $p$-value $= .8567$ **12.85a.** $E(y) = \beta_0 + \beta_1 x_1 + \beta_2 x_2$ **b.** $E(y) = \beta_0 + \beta_1 x_1 + \beta_2 x_2 + \beta_3 x_1 x_2$ **12.87** (a and b); (a and d); (a and e); (b and c); (b and d); (b and e); (c and e); (d and e) **12.89a.** complete: $\hat{y} = 14.6 - .61 x_1 + .44 x_2 - .08 x_3 - .06 x_4$; reduced: $\hat{y} = 14.0 - .64 x_1 + .40 x_2$ **b.** $SSE_R = 160.44, SSE_c = 152.66$ **c.** $5; 3$ **d.** $H_0: \beta_3 = \beta_4 = 0$ **e.** $F = .38$, do not reject $H_0$ **f.** $p$-value $> .10$ **12.91a.** $E(y) = \beta_0 + \beta_1 x_1 + \beta_2 x_2 + \beta_3 x_3 + \beta_4 x_4 + \beta_5 x_5 + \beta_6 x_6 + \beta_7 x_7 + \beta_8 x_8 + \beta_9 x_9 + \beta_{10} x_{10} + \beta_{11} x_{11}$ **b.** $E(y) = \beta_0 + \beta_1 x_1 + \beta_2 x_2 + \beta_3 x_3 + \beta_4 x_4 + \beta_5 x_5 + \beta_6 x_6 + \beta_7 x_7 + \beta_8 x_8 + \beta_9 x_9 + \beta_{10} x_{10} + \beta_{11} x_{11} + \beta_{12} x_1 x_9 + \beta_{13} x_1 x_{10} + \beta_{14} x_1 x_{11} + \beta_{15} x_2 x_9 + \beta_{16} x_2 x_{10} + \beta_{17} x_2 x_{11} + \beta_{18} x_3 x_9 + \beta_{19} x_3 x_{10} + \beta_{20} x_3 x_{11} + \beta_{21} x_4 x_9 + \beta_{22} x_4 x_{10} + \beta_{23} x_4 x_{11} + \beta_{24} x_5 x_9 + \beta_{25} x_5 x_{10} + \beta_{26} x_5 x_{11} + \beta_{27} x_6 x_9 + \beta_{28} x_6 x_{10} + \beta_{29} x_6 x_{11} + \beta_{30} x_7 x_9 + \beta_{31} x_7 x_{10} + \beta_{32} x_7 x_{11} + \beta_{33} x_8 x_9 + \beta_{34} x_8 x_{10} + \beta_{35} x_8 x_{11}$ **c.** $H_0: \beta_{12} = \beta_{13} = \cdots = \beta_{35} = 0$ **12.93a.** $E(y) = \beta_0 + \beta_1 x_1 + \beta_2 (x_1)^2 + \beta_3 x_2 + \beta_4 x_1 x_2 + \beta_5 (x_1)^2 x_2; x_1 = $ age; $x_2 = 1$ if current, 0 if retired **b.** $H_0: \beta_2 = \beta_5 = 0$ **c.** $H_0: \beta_4 = \beta_5 = 0$ **d.** curvature: $F = .43$, do not reject $H_0$; interaction: $F = .63$, do not reject $H_0$ **12.95a.** $E(y) = \beta_0 + \beta_1 x_1 + \beta_2 x_2 + \beta_3 x_3 + \beta_4 x_4 + \beta_5 x_5 + \beta_6 x_6 + \beta_7 x_7 + \beta_8 x_8 + \beta_9 x_9 + \beta_{10} x_{10}$ **b.** $H_0: \beta_3 = \beta_4 = \cdots = \beta_{10} = 0$ **c.** at least one of the additional variables is important **e.** $(8.12, 19.88)$ **f.** yes **g.** $E(y) = \beta_0 + \beta_1 x_1 + \beta_2 x_2 + \beta_3 x_3 + \beta_4 x_4 + \beta_5 x_5 + \beta_6 x_6 + \beta_7 x_7 + \beta_8 x_8 + \beta_9 x_9 + \beta_{10} x_{10} + \beta_{11} x_1 x_2 + \beta_{12} x_2 x_3 + \beta_{13} x_2 x_4 + \beta_{14} x_2 x_5 + \beta_{15} x_2 x_6 + \beta_{16} x_2 x_7 + \beta_{17} x_2 x_8 + \beta_{18} x_2 x_9 + \beta_{19} x_2 x_{10}$ **h.** $H_0: \beta_{11} = \beta_{12} = \cdots = \beta_{19} = 0$; nested model $F$-test **12.97a.** $H_0: \beta_4 = \beta_5 = 0$ **b.** $H_0: \beta_3 = \beta_4 = \beta_5 = 0$ **c.** no, $F = .93$ **12.99a.** 6 **b.** 5 **c.** 4 **12.101a.** DOT estimate and low-bid ratio **12.103** yes; $x_4$ is highly correlated with both $x_2$ and $x_5$ **12.107a.** $\hat{y} = -4.77 - .0352 x_1 + .0516 x_2 - .0234 x_3 + .0345 x_4$ **b.** no; weight is highly correlated with another independent variable **c.** no; multicollinearity **d.** $r_{12} = .327, r_{13} = .231, r_{14} = .166, r_{23} = .790, r_{24} = .791, r_{34} = .881$ **12.109a.** no **b.** $\hat{y} = 3.77 - .006 x_1 - 3.174 x_2 + 1.005 x_1 x_2$ **c.** 9 outliers; equal variance assumption violated **12.111** confidence interval for the mean **12.113a.** $\hat{y} = 90.1 - 1.836 x_1 + .285 x_2$ **b.** .916 **c.** yes, $F = 64.91$ **d.** $t = -5.01$, reject $H_0$ **e.** 10.677 **12.117** $E(y) = \beta_0 + \beta_1 x_1 + \beta_2 x_2 + \beta_3 x_3; x_1 = 1$ if level 2, 0 otherwise; $x_2 = 1$ if level 3, 0 otherwise; $x_3 = 1$ if level 4, 0 otherwise **12.121** df(Error) $= 0$ **12.123a.** yes, $t = 5.96$, $p$-value $= .00005$ **b.** $t = 0.01$, $p$-value $= .9620$, do not reject $H_0$ **c.** $t = 1.91$, $p$-value $= .0288$, reject $H_0$ **12.125a.** $E(y) = \beta_0 + \beta_1 x_1 + \beta_2 x_2 + \beta_3 x_3 + \beta_4 x_4 + \beta_5 x_5$ **b.** model is statistically useful **c.** $E(y) = \beta_0 + \beta_1 x_1 + \beta_2 x_2 + \beta_3 x_3 + \beta_4 x_4 + \beta_5 x_5 + \beta_6 x_6 + \beta_7 x_7$ **12.127a.** $x_1 = $ quantitative; $x_2 = $ qualitative **b.** $E(y) = \beta_0 + \beta_1 x_1 + \beta_2 x_2$ **c.** .09 **d.** 8.90 **e.** $E(y) = \beta_0 + \beta_1 x_1 + \beta_2 x_2 + \beta_3 (x_1)^2 + \beta_4 x_1 x_2 + \beta_5 (x_1)^2 x_2$ **12.129c.** $F = 13.53$, reject $H_0$ **12.131a.** $E(y) = \beta_0 + \beta_1 x_1 + \beta_2 x_2 + \beta_3 x_3$ **c.** yes, $F = 56$ **d.** do not reject $H_0$ **12.133a.** negative **b.** no, $F = 1.60$ **c.** no, $F = 1.61$ **12.135b.** $\hat{y} = 42.25 - .0114 x + .000000608 x^2$ **c.** no, $t = 1.66$ **12.137b.** yes, $F = 16.10$ **c.** yes, $t = 2.5$ **d.** 945 **12.139a.** $H_0: \beta_3 = 0$ **b.** $\beta_0$ **c.** $\beta_0 + \beta_1$ **d.** $\beta_1$ **e.** $\beta_0 + \beta_2$ **f.** $\beta_0 + \beta_1 + \beta_2 + \beta_3$ **g.** $\beta_1 + \beta_3$

# CHAPTER 13

**13.1a.** 18.3070 **b.** 29.7067 **c.** 23.5418 **d.** 79.4900 **13.3a.** $\chi^2 > 5.99147$ **b.** $\chi^2 > 7.77944$ **c.** $\chi^2 > 11.3449$ **13.5** $E(n_j) \geq 5$ **13.7a.** no, $\chi^2 = 3.293$ **13.9** yes, $\chi^2 = 18.78$ **13.11a.** Pottery type; burnished, monochrome, painted, other **b.** $p_1 = p_2 = p_3 = p_4 = .25$ **c.** $H_0: p_1 = p_2 = p_3 = p_4 = .25$ **d.** $\chi^2 = 436.59$ **e.** $p$-value $= .0000$ **13.13** $\chi^2 = 16$, reject $H_0$ **13.15a.** $\chi^2 = 17.16$, reject $H_0$ **b.** $.376 \pm .103$ **c.** yes **13.17a.** $\chi^2 > 26.2962$ **b.** $\chi^2 > 15.9871$ **c.** $\chi^2 > 9.21034$ **13.19a.** column 1: 36%, 64%; column 2: 53.1%, 46.9%; column 3: 67.9%, 32.1%; totals: 57.5%, 42.5% **13.21a.** $B_1$: 29.9%; $B_2$: 44.2%; $B_3$: 29.6% **b.** $B_1$: 47.0%; $B_2$: 32.5%; $B_3$: 49.3% **c.** $B_1$: 23.1%; $B_2$: 23.3%; $B_3$: 21.1%

**13.23a.**

|  |  | LLD | | |
|---|---|---|---|---|
|  |  | Yes | No | Total |
| Genetic Trait | Yes | 21 | 15 | 36 |
|  | No | 150 | 306 | 456 |
|  | Total | 171 | 321 | 492 |

**b.** $\chi^2 = 9.52$, reject $H_0$    **13.25** $\chi^2 = 35.41$, reject $H_0$    **13.27** $\chi^2 = 6.171$, reject $H_0$
**13.29a.** no

**b.**

|  |  | CAHS | | |
|---|---|---|---|---|
|  |  | Low | Medium | High |
| SHSS:C | Low | 32 | 14 | 2 |
|  | Medium | 11 | 14 | 6 |
|  | High | 6 | 16 | 29 |

**c.** yes    **d.** $\chi^2 = 46.70$, reject $H_0$    **13.31a.** $\chi^2 = 4.407$, reject $H_0$    **b.** no    **c.** .0438
**d.** .0057; .0003    **e.** $p$-value = .0498, reject $H_0$    **13.33a.** no, $\chi^2 = 2.133$    **b.** .233 ± .057
**13.35a.** possibly not    **c.** $\chi^2 = 12.9$    **d.** yes, reject $H_0$    **e.** .233 ± .200
**13.37a.** $\chi^2 = 1.143$, do not reject $H_0$    **b.** .179 ± .119    **13.39** yes, $\chi^2 = 180.874$    **13.41** no,
$\chi^2 = 4.4$    **13.43** yes, $\chi^2 = 9.35$    **13.45a.** no, $\chi^2 = 13.715$    **b.** wife: no, $\chi^2 = 3.198$; husband: no, $\chi^2 = 6.069$

**13.47a.**

|  | Surgery | Radiation | Total |
|---|---|---|---|
| Cancer controlled | 21 | 15 | 36 |
| Cancer not controlled | 2 | 3 | 5 |
| Total | 23 | 18 | 41 |

**c.** Two of the four expected cell counts are less than 5.    **d.** do not reject $H_0$    **e.** $\chi^2 = .592$
**13.49a.** $\chi^2 = 9.647$    **b.** 11.0705    **c.** no    **d.** $.05 < p$-value $< .10$

# CHAPTER 14

**14.1** population not normal    **14.3a.** .063    **b.** .500    **c.** .004    **d.** .151; .1515    **e.** .212; .2119
**14.5** $S = 16$, $p$-value = .115, do not reject $H_0$    **14.7a.** $H_0: \eta = 1.5, H_a: \eta > 1.5$    **b.** 3
**c.** .855    **d.** do not reject $H_0$    **14.9a.** $H_0: \eta = 5$    **b.** $z = 6.07$, $p$-value = 0    **c.** reject $H_0$
**14.11a.** yes, $z = 2.12$    **b.** no    **14.13b.** $T_B = 42.5$, reject $H_0$    **14.15a.** $T_1 = 62.5$, reject $H_0$
**b.** yes, $T_1 = 62.5$    **14.17a.** $H_0: D_{female} = D_{normal}, H_a: D_{female} > D_{normal}$    **c.** 174.5    **d.** 150.5
**e.** $z > 1.28$    **f.** $z = 1.72$, reject $H_0$    **14.19a.** $T_1 = 150.5$, reject $H_0$    **b.** $z = 3.44$, reject $H_0$
**14.21a.** yes, $T_2 = 105$    **b.** less than    **14.23a.** data not normal    **b.** $z = -6.47$, reject $H_0$
**c.** $z = -.39$, do not reject $H_0$    **d.** $z = -1.35$, do not reject $H_0$    **14.25a.** $H_a: D_A > D_B$
**b.** $T_- = 3.5$, reject $H_0$    **14.29a.** $H_a: D_A > D_B$    **b.** $z = 2.499$, reject $H_0$    **c.** .0062
**14.31a.** data not normal    **b.** $H_0: D_{father} = D_{mother}, H_a: D_{father} \neq D_{mother}$    **d.** $T_+ = 22$;
$T_- = 44$    **e.** $T_+ \leq 11$    **f.** do not reject $H_0$    **14.33** no, $p$-value = .449    **14.35** no, $T_- = 13$
**14.37a.** $H_0$: The three probability distributions are identical    **b.** $H = 8.190$, reject $H_0$
**c.** $.010 < p$-value $< .025$    **14.39** sample size for each distribution is more than 5
**14.41** no, $H = 1.565$    **14.43b.** reject $H_0$    **14.45a.** normality assumption violated
**b.** $H = 13.66$, reject $H_0$    **14.47a.** 6    **b.** $H_0$: The probability distributions for the four treatments are identical    **c.** $F_r = 15.2$, reject $H_0$    **d.** $p$-value $< .005$    **14.49** $F_r = 13$, reject $H_0$
**14.51** $p$-value = .819, do not reject $H_0$    **14.53** $F_r = 11.1$, reject $H_0$    **14.55** no; $F_r = 2.09$,
do not reject $H_0$    **14.57a.** .01    **b.** .01    **c.** .975    **d.** .05    **14.59a.** .4    **b.** −.9    **c.** −.2    **d.** .2
**14.61c.** .713    **d.** $|r_s| > .425$    **e.** reject $H_0$    **14.63** no, $r_s = .657$    **14.65a.** −.733    **b.** do not
reject $H_0$    **14.67** $r_s = -.812$, reject $H_0$    **14.71a.** no, $r_s = .40$    **b.** yes, $T_- = 1.5$    **14.73** yes,
$F_r = 14.9$    **14.75** no, $T_2 = 29$    **14.77a.** yes, $H = 12.79$    **b.** no, $T_1 = 28.5$    **14.79** no,
$r_s = .479$    **14.81** no, $T_1 = 21$    **14.83** yes, $F_r = 39.413$    **14.85** yes, $H = 7.154$    **14.87a.** yes,
$T_+ = 1$    **b.** $t = -2.96$, reject $H_0$    **14.89** yes, $r_s = .929$    **14.91a.** yes, $H = 6.66$    **b.** yes,
$T = 71.0$

# INDEX

## A

Additive rule of probability, 126–127
Alternative hypothesis, 328
Analysis of variance (ANOVA), 454–458
　for analyzing two-factor factorial experiment, 491–492
　robustness of, 456
　steps for completely randomized design, 458
　steps for randomized block design, 477–478
ANOVA. *See* Analysis of variance

## B

$\beta$ (probability of type II error), 359–365
Balanced design, 487
Bar graph, 22
Base level, 631
Bell-shaped distribution, 223
Biased estimate, 268
Biases, 12
Binomial random variables, 189–199
　approximating with normal distribution, 244–248
Bivariate relationships, 86–89, 541
Block(ing), 400, 469
Bonferroni multiple comparisons procedure, 464–467, 475
Bound B, 316
Box plots, 76, 78

## C

Categorical independent variables, 630
Cell counts, 706
　expected, 707
Census, 5
Central limit theorem, 273–279
Central tendency, 44
Chebyshev's rule, 61, 64
　for discrete random variable, 184
Chi-square ($\chi^2$)
　distribution, 367
　proper use of test, 730
Class, 20
Class frequency, 20
Class relative frequency, 20
Coded variable, 576

Coefficient(s)
　of correlation, 541–544
　of determination, 544–546
　　adjusted multiple, 597
　　multiple, 596
Combinations rule, 158–159
Combinatorial mathematics, 150–151
Comparisonwise error rate, 464
Complementary events, 123–125
Complete factorial experiments, 485
Complete model, 647
Completely randomized design, 448. *See also* Experiment(s)
　Kruskal-Wallis $H$-test for, 763–766
Conditional probability, 131–134
Confidence coefficient, 290
Confidence interval(s)
　defined, 289, 290
　large-sample
　　for a population mean, 288–293
　　for a population proportion, 308–313
　small-sample, 297–303
Confidence level, 290
Contingency table, 714–723
Continuous random variables. *See* Random variable(s)
Correction for continuity, 244
Correlated errors, 685
Correlation, 541
Counting rules, 154–162

## D

Data, 2
　collecting, 10–13
　qualitative, 10, 20–25
　quantitative, 9–10, 29–36
　sets, 19. *See also* Qualitative data; Quantitative data
Degrees of freedom (df), 298
Dependent pattern, 717
Dependent variable, 442
Descriptive statistics, 1, 2–4, 8
　distortions of truth and, 91–94
Designed experiment(s), 11, 444
　elements of, 442–446
Deterministic relationship, 511
Dimensions, 715
Discrete random variables. *See* Random variable(s)
Distribution, symmetric, 48
Distribution-free tests, 740–741

Dot plot, 30
Dummy variable models, 630–634

## E

Empirical rule, 62, 64
Empirical rule, for discrete random variable, 184
Error of estimation, 269
Errors of prediction, 515
Estimability, 680
Estimated standard error of the regression model, 531
Event(s)
　complementary, 123–125
　compound, 120
　defined, 114
　dependent, 142
　independent, 139–146
　intersections of, 120–123
　mutually exclusive, 127–128
　probability of, 115–116
　unions of, 120–123
Expected cell counts, 709
Experiment(s)
　balanced design, 487
　binomial, 189
　completely randomized design, 447–458
　　steps for ANOVA for, 458
　defined, 108–109
　designed, 11
　　elements of, 442–446
　factorial, 484–494
　multinomial, 706–707
　multiple comparisons of means, 463–467
　observational, 444
　paired difference, 397–405
　　Wilcoxon signed rank test for, 755–760
　randomized block design, 400–401, 469–478
　replicates of, 487
　sample points of, 109–114
　sample space of, 109–110
Experimental unit, 443
Experimentwise error rate, 464
Extrapolation, 685

## F

Factor(s), 442
Factor interaction component, 486